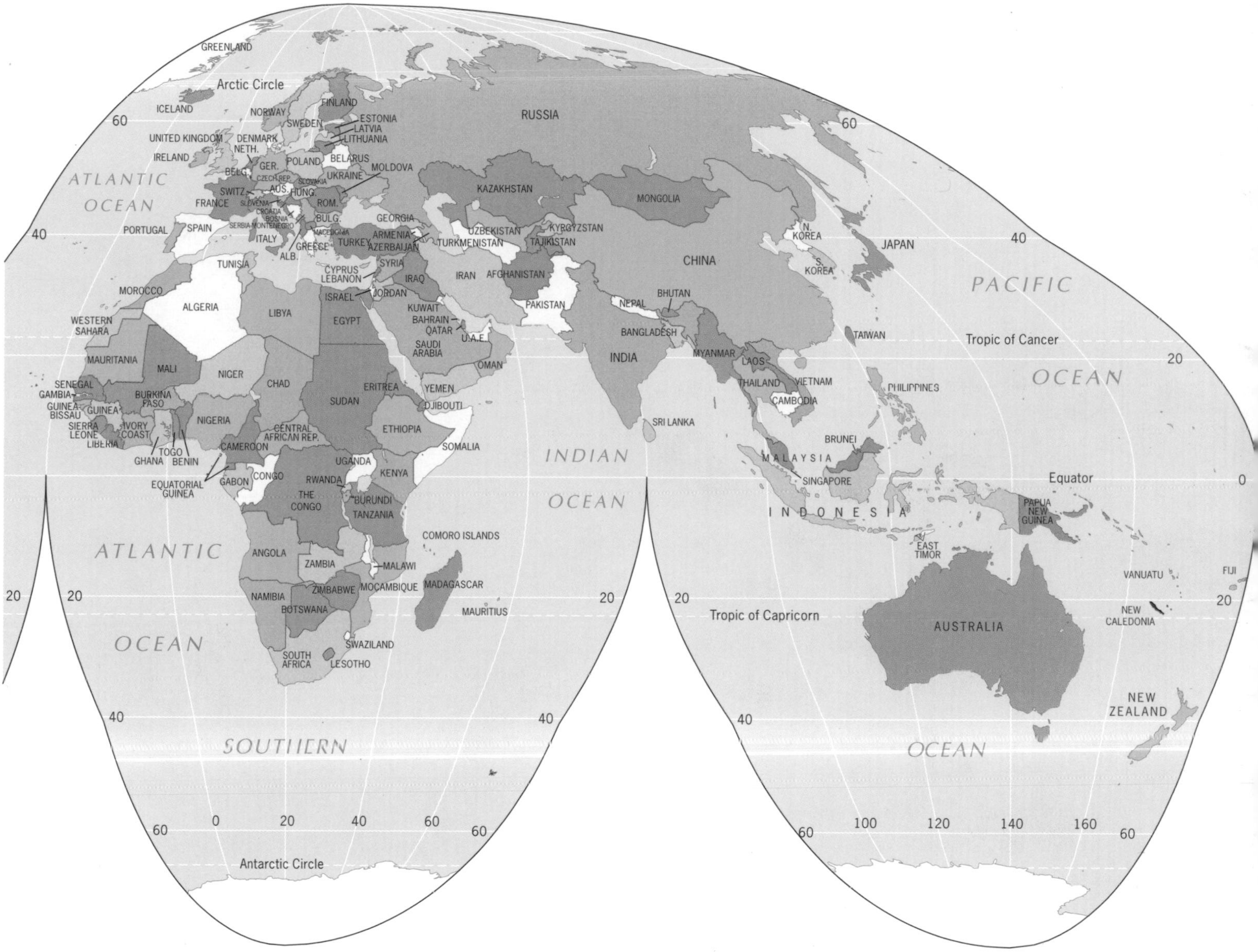
GREENLAND
Arctic Circle
ICELAND
NORWAY
SWEDEN
FINLAND
ESTONIA
LATVIA
LITHUANIA
RUSSIA
UNITED KINGDOM
DENMARK
NETH.
IRELAND
BELG.
GER.
POLAND
BELARUS
MOLDOVA
UKRAINE
CZECH REP.
SLOVAKIA
SWITZ.
AUS.
HUNG.
FRANCE
SLOVENIA
CROATIA
BOSNIA
ROM.
SERBIA-MONTENEGRO
BULG.
MACEDONIA
ALB.
ITALY
GREECE
ATLANTIC
OCEAN
PORTUGAL
SPAIN
GEORGIA
ARMENIA
AZERBAIJAN
TURKEY
KAZAKHSTAN
MONGOLIA
UZBEKISTAN
TURKMENISTAN
KYRGYZSTAN
TAJIKISTAN
N. KOREA
S. KOREA
JAPAN
CHINA
TUNISIA
MOROCCO
ALGERIA
LIBYA
EGYPT
CYPRUS
LEBANON
SYRIA
IRAQ
ISRAEL
JORDAN
IRAN
AFGHANISTAN
PAKISTAN
NEPAL
BHUTAN
KUWAIT
BAHRAIN
QATAR
U.A.E.
SAUDI ARABIA
OMAN
YEMEN
BANGLADESH
INDIA
MYANMAR
LAOS
TAIWAN
PACIFIC
OCEAN
Tropic of Cancer
WESTERN SAHARA
MAURITANIA
MALI
NIGER
CHAD
SUDAN
ERITREA
DJIBOUTI
ETHIOPIA
SOMALIA
SENEGAL
GAMBIA
GUINEA-BISSAU
GUINEA
BURKINA FASO
SIERRA LEONE
LIBERIA
IVORY COAST
GHANA
TOGO
BENIN
NIGERIA
CAMEROON
CENTRAL AFRICAN REP.
EQUATORIAL GUINEA
GABON
CONGO
THE CONGO
UGANDA
KENYA
RWANDA
BURUNDI
TANZANIA
THAILAND
VIETNAM
CAMBODIA
PHILIPPINES
SRI LANKA
BRUNEI
MALAYSIA
SINGAPORE
INDONESIA
PAPUA NEW GUINEA
EAST TIMOR
INDIAN
OCEAN
Equator
ANGOLA
ZAMBIA
MALAWI
COMORO ISLANDS
MADAGASCAR
NAMIBIA
ZIMBABWE
MOÇAMBIQUE
BOTSWANA
MAURITIUS
SWAZILAND
SOUTH AFRICA
LESOTHO
Tropic of Capricorn
AUSTRALIA
VANUATU
FIJI
NEW CALEDONIA
NEW ZEALAND
SOUTHERN
OCEAN
Antarctic Circle
0
20
40
60
100
120
140
160

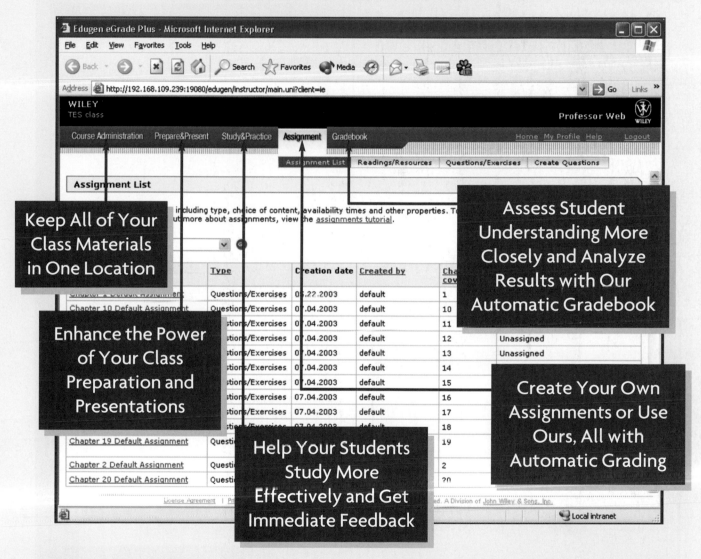

Students,
Wiley PLUS Allows You to:

Study More Effectively

Get Immediate Feedback When You Practice on Your Own

A **"Study & Practice"** area links directly to text content, allowing you to review the text while you study and complete homework assignments. Additional resources can include **virtual tours of regions covered, animations, video clips,** and **issues in geography highlights.**

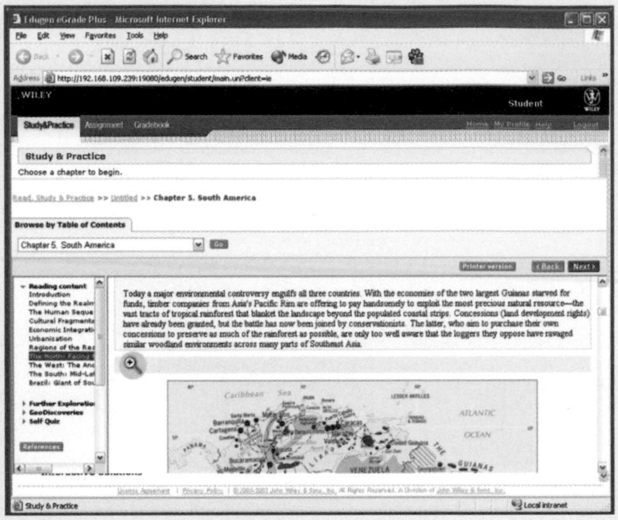

Complete Assignments / Get Help with Problem Solving

An **Assignment** area keeps all your assigned work in one location, making it easy for you to stay "on task." In addition, many homework problems contain a **link** to the relevant section of the **multimedia book,** providing you with a text explanation to help you conquer problem-solving obstacles as they arise. You will have access to a variety of **interactive problem-solving tools,** as well as other resources for building your confidence and understanding.

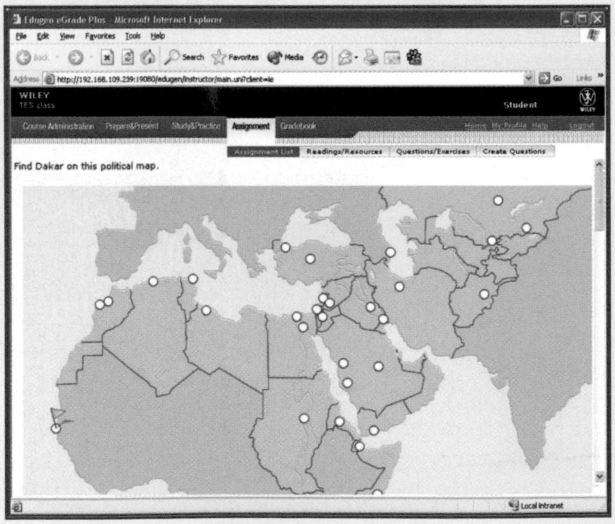

Keep Track of How You're Doing

A **Personal Gradebook** allows you to view your results from past assignments at any time.

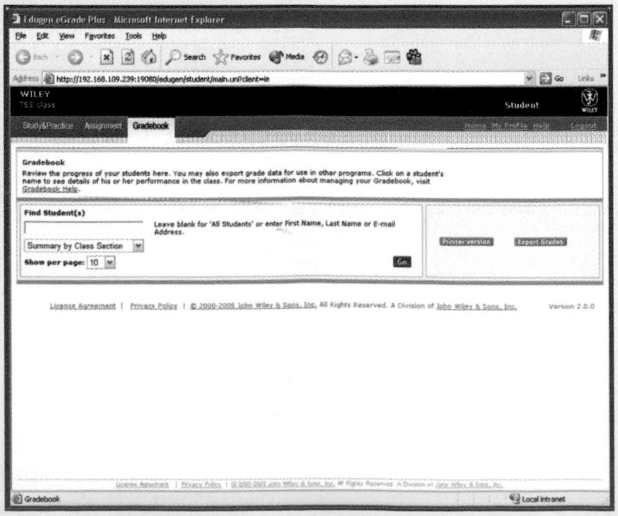

www.wiley.com/college/deblij

The Wiley Faculty Network

Where Faculty Connect

The Wiley Faculty Network is a faculty-to-faculty network promoting the effective use of technology to enrich the teaching experience. The Wiley Faculty Network facilitates the exchange of best practices, connects teachers with technology, and helps to enhance instructional efficiency and effectiveness. The network provides technology training and tutorials, including *Wiley PLUS* training, online seminars, peer-to-peer exchanges of experiences and ideas, personalized consulting, and sharing of resources.

Connect with a Colleague

Wiley Faculty Network mentors are faculty like you, from educational institutions around the country, who are passionate about enhancing instructional efficiency and effectiveness through best practices. You can engage a faculty mentor in an online conversation at **www.WhereFacultyConnect.com**

Participate in a Faculty-Led Online Seminar

The Wiley Faculty Network provides you with virtual seminars led by faculty using the latest teaching technologies. In these seminars, faculty share their knowledge and experiences on discipline-specific teaching and learning issues. All you need to participate in a virtual seminar is high-speed internet access and a phone line. To register for a seminar, go to **www.WhereFacultyConnect.com**

Connect with the Wiley Faculty Network

Web: **www.WhereFacultyConnect.com**

Phone: 1-866-4FACULTY

WILEY FACULTY NETWORK

GEOGRAPHY

Realms, Regions, and Concepts

Twelfth Edition

H. J. de Blij
Michigan State University

Peter O. Muller
University of Miami

Chapter-opening maps that appear in this text come from ***Goode's World Atlas***, 21st edition, © Rand McNally, R.L. 05-S-64, and are used with permission.

WILEY
John Wiley & Sons, Inc.

Executive Editor	Ryan Flahive
Executive Marketing Manager	Jeffrey Rucker
Marketing Manager	Emily Streutker
Editorial Assistant	Rachel Schneider
Production Editor	Barbara Russiello
Senior Illustration Editor	Sigmund Malinowski
Electronic Illustrations	Mapping Specialists, Ltd.
Senior Photo Editor	Jennifer MacMillan
Senior Media Editor	Tom Kulesa
Creative Designer	Harry Nolan
Interior Design	Lee Goldstein
Cover Design	Howard Grossman
Front Cover Photo	David R. Frazier/Danita Delimont.com
Back Cover Photo	Haley/Sipa Press

This book was set in Times Roman by GGS Book Services and printed and bound by Von Hoffmann Press. The cover was printed by Von Hoffmann Press.

This book is printed on acid free paper. ∞

To order books or for customer service please, call 1(800)-CALL-WILEY (225-5945).

ISBN-13 978-0-471-71786-7
ISBN-10 0-471-71786-X
Printed in the United States of America

10 9 8 7 6 5 4 3 2

Duke,

You'll Always Be a '61

Preface

FOR NEARLY FOUR DECADES, *Geography: Realms, Regions, and Concepts* has reported (and sometimes anticipated) trends in the discipline of Geography and developments in the world at large. In eleven preceding editions, *Regions*, as the book is generally called, has explained the modern world's great geographic realms and their physical and human contents, and has introduced geography itself, the discipline that links the study of human societies and natural environments through a fascinating, spatial approach. From old ideas to new, from environmental determinism to expansion diffusion, from decolonization to devolution, *Regions* has provided geographic perspective on our transforming world.

The book before you, therefore, is an information highway to geographic literacy. The first edition appeared in 1971, at a time when school geography in the United States (though not in Canada) was a subject in decline. It was a precursor of a dangerous isolationism in America, and geographers foresaw the looming cost of geographic illiteracy. Sure enough, the media during the 1980s began to report that polls, public surveys, tests, and other instruments were recording a lack of geographic knowledge at a time when our world was changing ever faster and becoming more competitive by the day. Various institutions, including the National Geographic Society, banks, airline companies, and a consortium of scholarly organizations mobilized to confront an educational dilemma that had resulted substantially from a neglect of the very topics this book is about.

Before we can usefully discuss such commonplace topics as our "shrinking world," our "global village," and our "distant linkages," we should know what the parts are, the components that do the shrinking and linking. This is not just an academic exercise. You will find that much of what you encounter in this book is of immediate, practical value to you—as a citizen, a consumer, a traveller, a voter, a jobseeker. North America is a geographic realm with intensifying global interests and involvements. Those interests and involvements require countless, often instantaneous decisions. Such decisions must be based on the best possible knowledge of the world beyond our continent. That knowledge can be gained by studying the layout of our world, its environments, societies, resources, policies, traditions, and other properties—in short, its regional geography.

REALMS AND CONCEPTS

This book is organized into thirteen chapters. The Introduction discusses the world as a whole, outlining the physical stage on which the human drama is being played out, providing environmental frameworks, demographic data, political background, and economic-geographical context. Each of the remaining twelve chapters focuses on one of the world's major geographic realms.

Geographic concepts and ideas are placed in their regional settings in all thirteen chapters. Most of these approximately 150 concepts are primarily geographical, but others are ideas about which, we believe, students of geography should have some knowledge. Although such concepts are listed on the opening page of every chapter, we have not, of course, enumerated every geographic notion used in that chapter. Many colleagues, we suspect, will want to make their own realm-concept associations, and as readers will readily perceive, the book's organization is quite flexible. It is possible, for example, to focus almost exclusively on substantive regional material, or, alternatively, to concentrate mainly on conceptual issues.

NEWS ABOUT THE TWELFTH EDITION

This Twelfth Edition of *Regions* continues a tradition of comprehensive revision that is a hallmark of this book's context and currency. As those who have long been familiar with it are aware, previous editions have invariably reflected major developments in the world as well as in the discipline, ranging from the collapse of the Soviet Union to the impact of globalization and from the rise of Asia's Pacific Rim to the war in Iraq. The previous edition brought significant structural changes and introduced the "Regional Issue" feature, presenting opposing viewpoints on matters of major concern and giving readers an opportunity to "vote" their opinion. To our pleasant surprise, thousands of "votes" were cast—a compliment to students who took the time to do so voluntarily. Also making its appearance in the previous edition was the concept of a "War on Terror," an unfortunate necessity in an era of growing intercultural conflict.

Another hallmark of this book is Table G-1, Area and Demographic Data for the World's States (see pp. 35-41). Colleagues as well as students comment frequently on Table G-1 as a valuable resource and information base, and several have suggested that we modify and expand it. To do so we have eliminated some categories (for example, the "doubling time" notion as a measure of population growth; it has lost its relevance in this era of declining rates of natural increase) and added others, including discrete male and female literacy rates, a national "corruption index," and the Big-Mac Index (yes, it refers to the price of hamburgers around the world). Another key change in this version of Table G-1 has to do with the last column, per-capita income by country. In the past, we reported this as per-capita gross national product (GNP), a country's total annual income divided by the population earning it, shown in U.S. dollars. But the new standard is to combine this number with the purchasing power of that income (one dollar is worth much less in America than in, say, Tanzania). This is the so-called GNI-PPP index, a more realistic reflection of the financial picture than GNP alone. In order to make all this possible, we have had to align the table sideways each of its page. We hope you will find it even more useful than before; please spend some time just reading the data, because you will encounter many surprises. The population of a growing number of countries is actually declining! Don't stop for a Big Mac in Kazakhstan unless you're on an expense account, nor in Iceland! Do you think that North America has the world's least corrupt countries? Well . . . not quite! For a brief discussion of the indices used in Table G-1, see p. 34.

A new feature in this Twelfth Edition occurs at the end of each chapter, under the headline What You Can Do. It is not difficult to formulate questions about the content of each chapter, as in "Describe the distribution of agriculture in North America." You do not need us to make these up; that is what study guides are for. But we do hear from readers about practical suggestions, ideas to get involved with, ways to contribute. Each chapter now closes with a recommendation—for example, to join an organization whose goals you endorse or to arrange meetings with fellow students from countries in which you have become interested—and a set of challenges of a practical nature, asking you to visualize how you would use what you have learned in a business, government, or educational setting. Geography's interconnections and your ability to communicate can make a powerful combination.

Significant text and cartographic revisions mark the individual chapters. In addition to those in Table G-1, the Introduction's Population Cartogram (Fig. G-10) reflects the changing global population balance (Russia and many European countries continue to shrink). The narrative on debt and globalization issues has been tightened.

The far-reaching changes in Europe (Chapter 1) since the preceding edition include the momentous expansion of the European Union (chronicled here with a revised map) in 2004 and the "Orange Revolution" in Ukraine, whose political-geographic division is represented on a new map. Another new map details developments in the northeastern Mediterranean, where the sensitive boundaries between Greece and Turkey, the continuing division of Cyprus, and the geography of the new EU ministate of Malta required cartographic elaboration. This edition was in press when the French and Dutch voted against the European Constitution, but the potential for such a setback is clear from this chapter's text, including disquiet over the prospective admission of Turkey. New material also covers the "Old" versus "New" Europe controversy, the continuing economic disparity between East and West Germany, and geographic developments in Ireland, Serbia-Montenegro, and Kosovo. The issue of Muslim immigration and accommodation receives detailed attention.

Russia (Chapter 2) incorporates a new map and expanded discussion of the vast country's climates and environmental challenges. Attention is also given to Russia's continuing post-Soviet reorganization, and its changing relationships with several of its neighbors (in addition to Ukraine) ranging from Estonia to Kazakhstan. The problems of minorities in Russia and on its borders are discussed in detail, including new text on the Southern Periphery and Chechnya and its terrorist threat. New material covers the energy situation, including the geography of pipeline routes. Given Russia's external initiatives, from Ukraine to Iran, some background is provided via President Putin's proclaimed intention to remake Russia into a world power, not merely a part of an expanding Europe.

North America (Chapter 3) benefits from an expanded discussion of the postindustrial revolution, notably the Internet Backbone Cities (including a new map), as well as broader coverage of the changing geography of U.S. agriculture. A new map displays the distribution of French speakers in Canada, and the continuing issue of Quebec's secession from the federation is updated. A new box focuses on popular (vernacular) regions of North America. Several changes affecting NAFTA are given attention in new text.

In Chapter 4 (Middle America), the extensively revised narrative required a new map of the regional geography of Mexico. We also expanded the discussion of Mexico's Amerindian minorities. The Caribbean region, notably the Greater Antilles, has enhanced coverage in both text and maps.

Chapter 5 (South America) includes a number of significant revisions. In the definitional segment of the chapter, this realm's halting progress toward economic integration and political cooperation receive emphasis. As the U.S.-inspired Free Trade Area of the Americas falters, South American leaders are taking alternative initiatives, one of these embodied by the new South American Community of Nations. Another issue roiling the realm is democracy, and this receives attention in both the definitional as well as the regional parts of the chapter (including the "Regional Issues" box). Further revisions appear in the section on Colombia, introducing the "failed state" concept and updating the box on the geography of cocaine. We have also given more space to the roles of indigenous peoples, notably in the rewritten coverage of Bolivia, where disputes over energy revenues are endangering the future of the state, as well as in Peru and Ecuador, where similar problems, though centered more on environmental issues, are arising. In Chile, new text discusses the conflict between the Mapuche and the central government over deforestation. A new map of Chile strengthens coverage of the Southern Cone. The discussion of Brazil is enhanced by a new map showing the regional disparity of wealth, another topic given greater emphasis, and readers' attention is directed to the massive, ecologically disruptive agricultural development of the *cerrado*.

Economic and developmental problems in Subsaharan Africa (Chapter 6) are given new focus, notably in agriculture and mineral extraction, both subject to infrastructure weaknesses. A new map of what is arguably this realm's most important country—South Africa—supports a stronger narrative; it replaces the pre- and post-liberation provincial maps representing a process that has now run its course. The status of destabilizing conflicts in Equatorial Africa, West Africa, and the African Horn is updated, in the last instances in context of the Islamic Front, a key geographic feature in the transition zone between Subsaharan and Arabized Africa.

Numerous revisions were required in Chapter 7 (North Africa/Southwest Asia), including new and revised cartography on Iraq, the West Bank (notably Israel's Security Barrier), and Afghanistan. New text focuses on regional reaction to events in Iraq, Afghanistan, and the Palestinian areas and analyzes the implications beyond the realm's borders, in turn affecting the War on Terror. The growing importance of the United Arab Emirates, notably Dubai, is reflected by additional text and photographic coverage. The geographic dimensions of the tragedy in Darfur expand the discussion of Sudan. Turkey's European-Union prospects are seen from this regional perspective. Iran's nuclear development alters the political geography of the "Empire States." The changing prospects for Turkestan, including restive Tajikistan, Kyrgyzstan, and Uzbekistan, are highlighted by new text.

The most important revision of South Asia (Chapter 8) involves a new map and text displaying the distribution of Muslim inhabitants of India, comparing the situation in 1991 with the latest census information (2001). The continuing consequences of the 1947 partition between India and Pakistan are thus given new perspective. Also in the definitional part of this chapter, the topic of population geography receives improved conceptual as well as substantive attention as changes in the nature and rates of population growth presage a new stage in the global demographic transition. In the regional segment, the volatile issue of Kashmir required revision. India's continuing flirtation with nationalist "*Hindutva*" is reconsidered. A brief section covers the impact of the late-2004 tsunami on Sri Lanka and the Maldives.

The chapter on East Asia (9) was exhaustively revised for the Eleventh Edition, and the revision for the current edition consists of a large number of details ranging from an expanded consideration of the effect of China's restrictive (and changing) population policies to the newest developments in Shanghai, which in June 2005 took over the position of leading port of the world from Rotterdam, Netherlands. Revisions also deal with the Three Gorges Dam/Sichuan Basin and China's efforts to diffuse its Pacific Rim success into the poorer interior, and the continued burgeoning of the Pearl River Delta economic engine. Against this background we raise the issue of China's increasingly problematic relations with Japan and the potential geopolitical implications. Also subject to revision was the text on the Koreas, not only in terms of the North's current militancy but also the South's continuing modernization (using the KTX bullet train as an example of the country's regional integration).

Southeast Asia (Chapter 10), especially Indonesia's Sumatera (Sumatra for those who still use colonial terminology) was severely hit by the 2004 tsunami, and new text describes this impact. This catastrophe occurred shortly after a momentous democratic presidential election in this realm's most populous country, which had geographic implications because of the devolutionary forces still at work in several parts of it. New text also focuses on the persistence of radical Islam and terrorism in the Thai/Malaysian borderland. The now resolved dispute between East Timor and Australia over energy resources under the sea required revised cartography as well as narrative.

The chapter on the Austral realm (Chapter 11) has further detail on the Timor Sea energy issues, but otherwise this chapter is marked by minor revisions from the Eleventh Edition relating, chiefly, to Australia's changing economic relations with Asian markets.

The major revision in the Pacific realm (Chapter 12) relates to the changing circumstances of the Solomon Islands.

The Twelfth Edition of *Regions* also includes new photography, field notes, captions, as well as updated bibliographies, pronunciation guide, and statistical tables (for example, of the populations of major cities in each realm). We hope that you will find this edition as satisfying and productive as you have our earlier ones.

Data Sources

For all matters geographical, of course, we consult *The Annals of the Association of American Geographers, The Professional Geographer, The Geographical Review, The Journal of Geography*, and many other academic journals published regularly in North America—plus an array of similar periodicals published in English-speaking countries from Scotland to New Zealand.

As with every new edition of this book, all quantitative information was updated to the year of publication and checked rigorously. In addition to the major revisions described above, hundreds of other modifications were made, many in response to readers' and reviewers' comments. Some readers found our habit of reporting urban population data within the text disruptive, so we continue to tabulate these at the beginning of the "Regions of the Realm" section of each chapter. The stream of new spellings of geographic names continues, and we pride ourselves in being a reliable source for current and correct usage.

The statistical data that constitute Table G-1 (pp. 35–41) are derived from numerous sources, and each data category is explained in the box on page 34. As users of such data are aware, considerable inconsistency marks the reportage by various agencies, and it is often necessary to make informed decisions on contradictory information. For example, some sources still do not reflect the rapidly declining rates of population increase or life expectancies in AIDS-stricken African countries. Others list demographic averages without accounting for differences between males and females in this regard. In formulating Table G-1 we have used among our sources the United Nations, the Population Reference Bureau, the World Bank, the Encyclopaedia Britannica *Books of the Year*, the *Economist* Intelligence Unit, the *Statesman's Year-Book*, and the *The New York Times Almanac*.

The urban population figures—which also entail major problems of reliability and comparability—are mainly drawn from the most recent database published by the United Nations' Population Division. For cities of less than 750,000, we developed our own estimates from a variety of other sources. At any rate, the urban population figures used here are estimates for 2006, and they represent metropolitan-area totals unless otherwise specified.

Photography

The map undoubtedly is geography's closest ally, but there are times when photography is not far behind. Whether from space, from an aircraft, from the tallest building in town, or on the ground, a photograph can, indeed, be worth a thousand words. When geographers perform field research in some area of the world, they are likely to maintain a written record that correlates with the photographic one.

This Twelfth Edition revision is not confined to the text and maps. As readers of *Regions* will note, the illustrations program was thoroughly overhauled with numerous new photographs, many by the authors, new accompanying "Field Notes" where appropriate, and new, detailed captions.

Pedagogy

We continue to devise ways to help students learn important geographic concepts and ideas, and to make sense of our complex and rapidly changing world. Our newest special feature (What You Can Do) is discussed above under News About the Twelfth Edition. Continuing special features include the following:

Atlas Maps. As in previous editions, a comprehensive map of the region opens each chapter. The maps are reproduced from the 21st revised edition (2005) of *Goode's World Atlas* (the maps for Chapters 7, 8, and 12 have been created in the *Atlas*'s style). In the Twelfth Edition, each of these maps is assigned the first figure number in each chapter, which better facilitates the integration of this cartographic material into the text.

Concepts, Ideas, and Terms. Each chapter begins with a boxed sequential listing of the key geographic concepts, ideas, and terms that appear in the pages that follow. These are noted by numbers in the margins that correspond to the introduction of each item in the text.

Two-Part Chapter Organization. To help the reader to logically organize the material within chapters, we have broken the regional chapters into two distinct parts: first, "Defining the Realm" includes the general physiographic, historical, and human-geographic background common to the realm, and the second section, "Regions of the Realm," presents each of the distinctive regions within the realm (denoted by the symbol ◗).

List of Regions. Also on the chapter-opening page, a list of the regions within the particular realm provides a preview and helps to organize the chapter. For ease of identification, the semi-circular symbol ◗ that denotes the regions list here also appears beside each region heading in the chapter.

Major Geographic Qualities. Near the beginning of each realm chapter, we list, in boxed format, the major geographic qualities that best summarize that portion of the Earth's surface.

Sidebar Boxes. Special topical issues are highlighted in boxed sections. These boxes allow us to include interesting and current topics without interrupting the flow of material within chapters.

Regional Issues (Boxes). In each regional chapter, a controversial issue has been singled out for debate by two opposing voices. As indicated above, these commentaries are drawn from the senior author's field notes and correspondence, and are intended to stimulate classroom discussion. Individual readers may "vote" their opinions by e-mail, as indicated at the bottom of each box.

Among the Realm's Great Cities (Boxes). This feature reflects the growing process and influence of urbanization worldwide. More than thirty profiles of the world's leading cities are presented, each accompanied by a specially drawn map.

Major Cities of the Realm Population Tables. Near the beginning of every "Regions" section of each chapter, we have included a table reporting the most up-to-date urban population data (based on 2006 estimates drawn from the sources listed above). Readers should find this format less disruptive than citation of the population when the city is mentioned in the text.

From the Field Notes. In the Eighth Edition we introduced a new feature that has proven effective in some of our other textbooks. Many of the photographs in this book were taken by the senior author while doing fieldwork. The more extensive captions and From the Field Notes provide valuable insights into how a geographer observes and interprets information in the field.

Appendices, References, and Glossary. At the end of the book, the reader will find five sections that enrich and/or supplement the main text: (1) Appendix A, a guide to Using the Maps; (2) Appendix B, an overview of Career Opportunities in Geography; (3) Appendix C, a Pronunciation Guide; (4) a detailed bibliography (References and Further Readings) that introduces the wide-ranging literature of the discipline and World Regional Geography; and (5) an extensive Glossary. The general index follows. A geographical index or gazetteer of the place names contained in our maps now appears in the book's website.

Ancillaries

A broad spectrum of print and electronic ancillaries are available to accompany the Twelfth Edition of *Regions*. (Additional information, including prices and ISBNs for ordering, can be obtained by contacting John Wiley & Sons.) These ancilliaries are described next.

THE TEACHING AND LEARNING PACKAGE

This Twelfth Edition of *Geography: Realms, Regions, and Concepts* is supported by a comprehensive supplements package that includes an extensive selection of print, visual, and electronic materials.

Resources That Help Teachers Teach

Videos and Podcasts. We have created a series of streaming video resources and podcasts to support *Geography: Realms, Regions, and Concepts 12e*. Our streaming videos provide short lecture-launching video clips that can be used to introduce new topics, enhance your presentations, and stimulate classroom discussion. To help you integrate these resources into your syllabus, we have developed a comprehensive library of teaching resources, study questions, and assignments.

Image Gallery. We provide online electronic files for the line illustrations and maps in the text, which the instructor can customize for presenting in class (for example, in handouts, overhead transparencies, or PowerPoints).

Wiley Faculty Network. This peer-to-peer network of faculty is ready to support your use of online course management tools and discipline-specific software/learning systems in the classroom. The WFN will help you apply innovative classroom techniques, implement software packages, tailor the technology experience to the needs of each individual class, and provide you with virtual training sessions led by faculty for faculty.

The *Regions* 12e Instructors' Site. This comprehensive website includes numerous resources to help you enhance your current presentations, create new presentations, and employ our pre-made PowerPoint presentations. These resources include:

- A complete collection of PowerPoint presentations (prepared by Eugene J. Palka). Three for each chapter.
- All of the line art from the text.
- Access to photographs.
- A comprehensive collection of animations.
- A comprehensive test bank.

Instructor's Manual and Test Bank. This manual includes a test bank, chapter overviews and outlines, and lecture suggestions. The *Test Bank* for the Twelfth Edition of *Regions* contains over 3,000 test items including multiple-choice, fill-in, matching, and essay questions. It is distributed via the secure Instructor's website as electronic files, which can be saved into all major word processing programs. The Instructors Manual is also available on the password protected Instructor's website.

Overhead Transparency Set. All of the book's maps and diagrams are available for transparency projection in beautifully rendered, 4-color format, and have been resized and edited for maximum effectiveness in large lecture halls.

Course Management. On-line course management assets are available to accompany the Twelfth Edition of *Regions*.

Resources That Help Students Learn

GeoDiscoveries Website ***www.wiley.com/college/deblij*** This easy to use and student-focused website helps reinforce and illustrate key concepts from the text. It also provides interactive media content that helps students prepare for tests and improve their grades. This website provides additional resources that complement the textbook and enhance your students' understanding of geography and improve their grade by using the following resources:

- **Videos** provide a first-hand look at life in other parts of the world.
- **Interactive Animations and Exercises** allow you to explore key concepts from the text.
- **Virtual Field Trips** invite you to discover different locations through photos, and help you better understand how geographers view the world.
- **Podcasts, Web Cams, and Live Radio** let you see and hear what is happening all over the world in real time.
- **Flashcards** offer an excellent way to drill and practice key concepts, ideas, and terms from the text.
- The **Learning Styles Survey** will help you identify your unique way of understanding and processing information.
- With the **Study Skills Chart**, you can create a customized learning plan that best suits your personal learning style.
- **Map Quizzes** help students master the place names that are crucial to their success in this course. Three game-formatted place name activities are provided for each chapter.
- **GeoDiscoveries Modules** allow students to explore key concepts in greater depth using videos, animations, and interactive exercises.
- **Chapter Review Quizzes** provide immediate feedback to true/false, multiple choice, and short answer questions.
- **Concepts, Ideas, and Terms Interactive Flashcards** help students review and quiz themselves on the concepts, ideas, and terms discussed in each chapter.
- **Interactive Drag-and-Drop Exercises** challenge students to correctly label important illustrations from the textbook.
- **Audio Pronunciation** is provided for over 2000 key words and place names from the text.
- **Annotated Web Links** put useful electronic resources into context.
- **Area and Demographic Data** are provided for every country and world realm.

Student Study Guide. Text co-author Peter O. Muller and his geographer daughter, Elizabeth Muller Hames, have written a popular Study Guide to accompany the book that is packed with useful study and review tools. For each chapter in the textbook, the Study Guide gives students and faculty access to chapter objectives, content questions-and-answers, outline maps of each realm, sample tests, and more.

***Goode's Atlas* from Rand McNally.** With the Twelfth Edition of *Regions*, we are delighted to be able to continue offering the *Goode's Atlas* (newly revised in 2005) at a deeply-discounted price when shrink-wrapped with the text. Economies of scale allow us to provide this at a net price that is close to our cost. Our partnership with Rand McNally and the widely-popular *Goode's Atlas* is an arrangement that is exclusive to John Wiley & Sons.

Social Geographies of the Modern World. This FREE 60-page supplement includes chapters that discuss:

- Global Disparities in Nutrition and Health;
- Geographies of Inequality: Race and Ethnicity; and
- Gender Inequities in Geographic Perspective.

To arrange for your students to receive this free supplement with their copy of *Regions* 12e, please contact your local Wiley Sales Representative.

Microsoft Encarta Interactive Atlas. This award-winning atlas will captivate the imaginations of your students and engage them in a spatial adventure, all the while exposing them to an abundance of resources appropriate for university-level geography. Our arrangement with Microsoft enables us to offer the Encarta Interactive Atlas at a cost that is less than one-third the suggested retail price when shrink-wrapped with our text.

ACKNOWLEDGMENTS

Over the 35 years since the publication of the First Edition of *Geography: Realms, Regions, and Concepts*, we have been fortunate to receive advice and assistance from literally hundreds of people. One of the rewards associated with the publication of a book of this kind is the steady stream of correspondence and other feedback it generates. Geographers, economists, political scientists, education specialists, and others have written us, often with fascinating enclosures. We make it a point to respond personally to every such letter, and our editors have communicated with many of our correspondents as well. Moreover, we have considered every suggestion made and many who wrote or transmitted their reactions through other channels will see their recommendations in print in this edition.

STUDENT RESPONSE

A major part of the correspondence we receive comes from student readers. We would like to take this opportunity to extend our deep appreciation to the several million students around the world who have studied from the first eleven editions of our text. In particular, we thank the students from more than 120 different colleges across the United States who took the time to send us their opinions.

Students told us they found the maps and graphics attractive and functional. We have not only enhanced the map program with exhaustive updating but have added a number of new maps to this Twelfth Edition as well as making significant changes in many others. Generally, students have told us that they found the pedagogical devices quite useful. We have kept the study aids the students cited as effective: a boxed list of each chapter's key concepts, ideas, and terms (numbered for quick reference in both the box and text margins); a box summarizing each realm's major geographic qualities; a pronunciation guide in Appendix C; and an extensive Glossary.

FACULTY FEEDBACK

In assembling the Twelfth Edition, we are indebted to the following people for advising us on a number of matters:

Thomas L. Bell, University of Tennessee
Kathleen Braden, Seattle Pacific University
Deborah Corcoran, Southwest Missouri State University
William V. Davidson, Louisiana State University
Chuck Fahrer, Georgia College and State University, Milledgeville
Gary A. Fuller, University of Hawai'i
Richard J. Grant, University of Miami
Margaret M. Gripshover, University of Tennessee
Richard Lisichenko, Fort Hays State University (Kansas)
Dalton E. Miller, Mississippi State University
Jan Nijman, University of Miami
Valiant C. Norman, Lexington Community College (Kentucky)
Pai Yung-Feng, New York City
Eugene J. Palka, U.S. Military Academy (New York)
Mika Roinila, State University of New York at New Paltz
Rinku Roy Chowdhury, University of Miami
Richard Slease, Oakland, North Carolina
Cathy Weidman, Austin, Texas

In addition, several faculty colleagues from around the world assisted us with earlier editions, and their contributions continue to grace the pages of this book. Among them are:

James P. Allen, California State University, Northridge
Stephen S. Birdsall, University of North Carolina
J. Douglas Eyre, University of North Carolina
Fang Yong-Ming, Shanghai, China
Edward J. Fernald, Florida State University
Ray Henkel, Arizona State University
Richard C. Jones, University of Texas at San Antonio
Gil Latz, Portland State University
Ian MacLachlan, University of Lethbridge (Alberta)
Melinda S. Meade, University of North Carolina
Henry N. Michael, Temple University (Pennsylvania)
Clifton W. Pannell, University of Georgia
J. R. Victor Prescott, University of Melbourne (Australia)

JOHN D. STEPHENS, University of Washington
CANUTE VANDER MEER, University of Vermont

Faculty members from a large number of North American colleges and universities continue to supply us with vital feedback and much-appreciated advice. Our publishers arranged several feedback sessions, and we are most grateful to the following professors for showing us where the text could be strengthened and made more precise:

HEIKE C. ALBERT, University of Wisconsin-Oshkosh
CHARLES AMISSAH, Hampton University (Virginia)
ROBERT ARTHUR, North Dakota State University
LARRY BECKER, Oregon State University
RICHARD BENFIELD, Central Connecticut State University
MICHAEL BIKERMAN, Community College of Allegheny County-South (Pennsylvania)
JOHN BOYER, Virginia Polytechnic Institute and State University
JOHN CHRISTOPHER BROWN, University of Kansas
JEFFREY BURY, San Francisco State University
JAYE ROGERS CALDWELL, Anderson University (Indiana)
SCOTT CAMPBELL, The University of Kansas
SCOTT CARLIN, Southampton College, Long Island University (New York)
NORMAN CARTER, California State University-Long Beach
IRENE CASAS, State University of New York-Buffalo
BEN CECIL, University of Regina (Saskatchewan)
JAMES CHAMERNICK, University of Wisconsin-Superior
GABE CHEREM, Eastern Michigan University
BRIAN CRAWFORD, West Liberty State University
BRUCE CREW, Ferris State University (Michigan)
FRANCIS HENRY DILLON III, George Mason University (Virginia)
DIMITAR DIMITROV, Virginia Commonwealth University
DAVID DOCKSTADER, Jefferson Community College (New York)
JOHN DONALDSON, Liberty University (Virginia)
MARK DRAYSE, California State University-Fullerton
DICKEN EVERSON, California State University-San Bernardino
BRIAN FARMER, Amarillo College (Texas)
THOMAS FLETCHER, Bishop's University (Quebec)
ALISTAIR FRASER, Ohio State University
ROBERT GOODRICH, University of Idaho
WILLIAM GORDON, University of Rhode Island
E. DAVID GREGORY, Athabasca University (Saskatchewan)
JOAN HACKELING, University of California-Los Angeles
JANET HALPIN, Chicago State University
JOHN HARRINGTON, Kansas State University
DAVID HOLT, Miami University Ohio)
JAMES HUGHES, Slippery Rock University (Pennsylvania)
GLEN HVENEGAARD, University of Alberta
STEVE GRAVES, California State University-Northridge
JOHN JEZIERSKI, Saginaw Valley State University (Michigan)
JAMES KEESE, California Polytechnic State University-San Luis Obispo
JOHN KEYANTASH, California State University-Dominguez Hills
ERIC KEYS, Arizona State University
JAMES KNOTWELL, Wayne State College (Nebraska)
OLAF KUHLKE, University of Minnesota-Duluth
GEORGES LABRECQUE, Royal Military College-Kingston (Ontario)
UNNA LASSITER, California State University-Long Beach
UTE LEHRER, Brock University (Ontario)
C.K. LEUNG, California State University-Fresno
IAN MACLACHLAN, University of Lethbridge (Alberta)
DEBORAH MATTHEWS, Boise State University (Idaho)
HELEN MURDOCH, Sault College of Applied Arts & Technology (Ontario)
PETER LI, Tennessee Tech University-Cookeville
DONALD LYONS, University of North Texas
EMMETT PANZELLA, Point Park College (Pennsylvania)
EVELYN RAVURI, Central Michigan University
FRED SUNDERMAN, Saginaw Valley State University (Michigan)
JEAN-CLAUDE THILL, State University of New York-Buffalo
MIRIAM WALL, Confederation College (Ontario)
MARK WELFORD, Georgia Southern University

We also received input from a much wider circle of academic geographers. The list that follows is merely representative of a group of colleagues across North America to whom we are grateful for taking the time to share their thoughts and opinions with us:

MEL AAMODT, California State University-Stanislaus
WILLIAM V. ACKERMAN, Ohio State University
W. FRANK AINSLEY, University of North Carolina, Wilmington
SIAW AKWAWUA, University of Northern Colorado
DONALD P. ALBERT, Sam Houston State University
TONI ALEXANDER, Louisiana State University
R. GABRYS ALEXSON, University of Wisconsin-Superior
KHALED MD. ALI, Geologist, Dhaka, Bangladesh
NIGEL ALLAN, University of California-Davis
JOHN L. ALLEN, University of Connecticut
KRISTIN J. ALVAREZ, Keene State College (New Hampshire)
DAVID L. ANDERSON, Louisiana State University, Shreveport
KEAN ANDERSON, Freed-Hardeman University
JEFF ARNOLD, Southwestern Illinois College
JERRY R. ASCHERMANN, Missouri Western State College
JOSEPH M. ASHLEY, Montana State University
PATRICK ASHWOOD, Hawkeye Community College
THEODORE P. AUFDEMBERGE, Concordia College (Michigan)
JAIME M. AVILA, Sacramento City College
EDWARD BABIN, University of South Carolina-Spartanburg
ROBERT BAERENT, Randolph-Macon College
MARVIN W. BAKER, University of Oklahoma
GOURI BANERJEE, Boston University (Massachusetts)
MICHELE BARNABY, Pittsburg State University (Kansas)
J. HENRY BARTON, Thiel College (Pennsylvania)
CATHY BARTSCH, Temple University (Pennsylvania)
STEVEN BASS, Paradise Valley Community College (Arizona)
THOMAS F. BAUCOM, Jacksonville State University (Alabama)
KLAUS J. BAYR, Keene State College
DENIS A. BEKAERT, Middle Tennessee State University
JAMES BELL, Linn Benton Community College (Oregon)
KEITH M BELL, Volunteer State Community College
THOMAS L. BELL, University of Tennessee

WILLIAM H. BERENTSEN, University of Connecticut
DONALD J. BERG, South Dakota State University
ROYAL BERGLEE, Morehead State University
RIVA BERLEANT-SCHILLER, University of Connecticut
LEE LUCAS BERMAN, Southern Connecticut State University
RANDY BERTOLAS, Wayne State College
THOMAS BITNER, University of Wisconsin
WARREN BLAND, California State University-Northridge
DAVIS BLEVINS, Huntington College (Alabama)
HUBERTUS BLOEMER, Ohio University
S. BO JUNG, Bellevue College (Nebraska)
R. DENISE BLANCHARD BOEHM, Texas State University
MARTHA BONTE, Clinton Community College (Idaho)
GEORGE R. BOTJER, University of Tampa (Florida)
KATHLEEN BRADEN, Seattle Pacific University
R. LYNN BRADLEY, Belleville Area College (Illinois)
KEN BREHOB, Elmhurst College, Illinois
JAMES A. BREY, University of Wisconsin-Fox Valley
ROBERT BRINSON, Santa Fe Community College (Florida)
REUBEN H. BROOKS, Tennessee State University
PHILLIP BROUGHTON, Arizona Western College
LARRY BROWN, Ohio State University
LAWRENCE A. BROWN, Troy State-Dothan (Alabama)
ROBERT N. BROWN, Delta State University (Mississippi)
STANLEY D. BRUNN, University of Kentucky
RANDALL L. BUCHMAN, Defiance College (Ohio)
MICHAELE ANN BUELL, NorthWest Arkansas Community College
DANIEL BUNYE, South Plains College
MICHAEL BUSBY, Murray State University
MARY CAMERON, Slippery Rock University
MICHAEL CAMILLE, Univ. of Louisiana at Monroe
MARY CARAVELIS, Barry University
DIANA CASEY, Muskegon Community College
DIANN CASTEEL, Tusculum College (Tennessee)
BILL CHAPPELL, Keystone College
MICHAEL SEAN CHENOWETH, University of Wisconsin, Milwaukee
STANLEY CLARK, California State University Bakersfield
JOHN E. COFFMAN, University of Houston (Texas)
DAWYNE COLE, Grand Rapids Baptist College (Michigan)
JERRY COLEMAN, Mississippi Gulf Coast Community College
JONATHAN C. COMER, Oklahoma State University
BARBARA CONNELLY, Westchester Community College (New York)
FRED CONNINGTON, Southern Wesleyan University
WILLIS M. CONOVER, University of Scranton (Pennsylvania)
OMAR CONRAD, Maple Woods Community College (Missouri)
ALLAN D. COOPER, Otterbein College
DEBORAH CORCORAN, Southwest Missouri State University
WILLIAM COUCH, University of Alabama, Huntsville
BARBARA CRAGG, Aquinas College (Michigan)
GEORGES G. CRAVINS, University of Wisconsin
ELLEN K. CROMLEY, University of Connecticut
JOHN A. CROSS, University of Wisconsin-Oshkosh
SHANNON CRUM, University of Texas at San Antonio
WILLIAM CURRAN, South Suburban College (Illinois)
KEVIN M. CURTIN, University of Texas at Dallas
JOSE A. DA CRUZ, Ozarks Technical Community College
ARMANDO DA SILVA, Towson State University (Maryland)
MARK DAMICO, Green Mountain College
DAVID D. DANIELS, Central Missouri State University
RUDOLPH L. DANIELS, Morningside College (Iowa)
SATISH K. DAVGUN, Bemidji State University (Minnesota)
WILLIAM V. DAVIDSON, Louisiana State University
CHARLES DAVIS, Mississippi Gulf Coast Community College
JAMES DAVIS, Illinois College
JAMES L. DAVIS, Western Kentucky University
PEGGY E. DAVIS, Pikeville College
ANN DEAKIN, State University of New York, College at Fredonia
KEITH DEBBAGE, University of North Carolina-Greensboro
MOLLY DEBYSINGH, California State University, Long Beach
DENNIS K. DEDRICK, Georgetown College (Kentucky)
JOEL DEICHMANN, Bentley College
STANFORD DEMARS, Rhode Island College
TOM DESULIS, Spoon River College
THOMAS DIMICELLI, William Paterson College (New Jersey)
MARY DOBBS, Highland Community College, Wamego
SCOTT DOBLER, Western Kentucky University
D.F. DOEPPERS, University of Wisconsin-Madison
JAMES DOERNER, University of Northern Colorado
ANN DOOLEN, Lincoln College (Illinois)
STEVEN DRIEVER, University of Missouri-Kansas City
WILLIAM DRUEN, Western Kentucky University
ALASDAIR DRYSDALE, University of New Hampshire
KEITH A. DUCOTE, Cabrillo Community College (California)
ELIZABETH DUDLEY-MURPHY, University of Utah
WALTER N. DUFFET, University of Arizona
MIKE DUNNING, University of Alaska Southeast
CHRISTINA DUNPHY, Champlain College (Vermont)
ANTHONY DZIK, Shawnee State University (Kansas)
DENNIS EDGELL, Firelands BGSU (Ohio)
RUTH M. EDIGER, Seattle Pacific University
JAMES H. EDMONSON, Union University (Tennessee)
M.H. EDNEY, State University of New York-Binghamton
HAROLD M. ELLIOTT, Weber State University (Utah)
JAMES ELSNES, Western State College
LINDA ESSENSON
ROBERT W. EVANS, Fresno City College
EVERSON DICKEN, California State University, San Bernardino
DINO FIABANE, Community College of Philadelphia (Pennsylvania)
G.A. FINCHUM, Milligan College (Tennessee)
IRA FOGEL, Foothill College (California)
RICHARD FOLEY, Cumberland College
ROBERT G. FOOTE, Wayne State College (Nebraska)
RONALD FORESTA, University of Tennessee
ELLEN J. FOSTER, Texas State University
G.S. FREEDOM, McNeese State University (Louisiana)
EDWARD T. FREELS, Carson-Newman College
PHILIP FRIEND, Inver Hills Community College
JAMES FRYMAN, University of Northern Iowa
GARY A. FULLER, University of Hawai'i
OWEN FURUSETH, University of North Carolina-Charlotte
RICHARD FUSCH, Ohio Wesleyan University
GARY GAILE, University of Colorado-Boulder
EVELYN GALLEGOS, Eastern Michigan University & Schoolcraft College
GAIL GARBRANDT, Mount Union College, University of Akron
RICHARD GARRETT, Marymount Manhattan College

JERRY GERLACH, Winona State University (Minnesota)
MARK GISMONDI, Northwest Nazarene University
LORNE E. GLAIM, Pacific Union College (California)
SHARLEEN GONZALEZ, Baker College (Michigan)
DANIEL B. GOOD, Georgia Southern University
GARY C. GOODWIN, Suffolk Community College (New York)
S. GOPAL, Boston University (Massachusetts)
MARVIN GORDON, Lake Forest Graduate School of Business
ROBERT GOULD, Morehead State University (Kentucky)
MARY GRAHAM, York College of Pennsylvania
GORDON GRANT, Texas A&M University
PAUL GRAY, Arkansas, Tech University
DONALD GREEN, Baylor University (Texas)
GARY M. GREEN, University of North Alabama
STANLEY C. GREEN, Laredo State University (Texas)
RAYMOND GREENE, Western Illinois University
MARK GREER, Laramie County Community College (Wyoming)
MARGARET M. GRIPSHOVER, University of Tennessee
WALT GUTTINGER, Flagler College
JOHN E. GYGAX, Wilkes Community College
LEE ANN HAGAN, College of Southern Idaho
RON HAGELMAN, University of New Orleans
W. GREGORY HAGER, Northwestern Connecticut Community College
RUTH F. HALE, University of Wisconsin-River Falls
JOHN W. HALL, Louisiana State University-Shreveport
PETER L. HALVORSON, University of Connecticut
DAVID HANSEN, Pennsylvania State University, Harrisburg & Schuylkill
MERVIN HANSON, Willmar Community College (Minnesota)
ROBERT C. HARDING, Lynchburg College
MICHELLE D. HARRIS, Bob Jones University
SCOTT HARRIS, CGCS Drury University
ROBERT J. HARTIG, Fort Valley State College (Georgia)
SUZANNA HARTLEY, Shelton State Community College
TRUMAN A. HARTSHORN, Georgia State University
CAROL HAZARD, Meredith College
HARLOW Z. HEAD, Barton College
DOUG HEFFINGTON, Middle Tennessee State University
JAMES G. HEIDT, University of Wisconsin Center-Sheboygan
CATHERINE HELGELAND, University of Wisconsin-Manitowoc
NORMA HENDRIX, East Arkansas Community College
JAMES E. HERRELL, Otero Junior College
JAMES HERTZLER, Goshen College (Indiana)
JOHN HICKEY, Inver Hills Community College (Minnesota)
THOMAS HIGGINS, San Jacinto College (Texas)
EUGENE HILL, Westminster College (Missouri)
LOUISE HILL, University of South Carolina-Spartanburg
MIRIAM HELEN HILL, Indiana University Southeast
SUZY HILL, University of South Carolina-Spartanburg
ROBERT HILT, Pittsburg State University (Kansas)
SOPHIA HINSHALWOOD, Montclair State University (New Jersey)
PRISCILLA HOLLAND, University of North Alabama
MARK R. HOOPER, Freed-Hardeman University
R. HOSTETLER, Fresno City College (California)
ERICK HOWENSTINE, Northeastern Illinois University
LLOYD E. HUDMAN, Brigham Young University (Utah)
JANIS W. HUMBLE, University of Kentucky
JUANA IBANEZ, University of New Orleans
TONY IJOMAH, Harrisburg Area Community College
WILLIAM IMPERATORE, Appalachian State University (North Carolina)
RICHARD JACKSON, Brigham Young University (Utah)
MARY JACOB, Mount Holyoke College (Massachusetts)
GREGORY JEANE, Samford University (Alabama)
SCOTT JEFFREY, Catonsville Community College (Maryland)
JERZY JEMIOLO, Ball State University (Indiana)
NILS I. JOHANSEN, University of Southern Indiana
DAVID JOHNSON, University of Southwestern Louisiana
INGRID JOHNSON, Towson University
RICHARD JOHNSON, Oklahoma City University
SHARON JOHNSON, Marymount College (New York)
JEFFREY JONES, University of Kentucky
KRIS JONES, Saddleback College
MARCUS E. JONES, Claflin College (South Carolina)
M. KAMIAR, Florida Community College, Jacksonville
MOHAMMAD S. KAMIAR, Florida Community College, Jacksonville
MATTI E. KAUPS, University of Minnesota-Duluth
JO ANNE W. KAY, Brigham Young University, Idaho
PHILIP L. KEATING, Indiana University
DAVID KEELING, Western Kentucky University
COLLEEN KEEN, Gustavus Adolphus College (Minnesota)
ARTIMUS KEIFFER, Wittenberg University
GORDON F. KELLS, Mott Community College
KAREN J. KELLY, Palm Beach Community College, Boca Raton
RYAN KELLY, Lexington Community College
VIRGINIA KERKHEIDE, Cuyahoga Community College and Cleveland State University
TOM KESSINGER, Xavier University
MASOUD KHEIRABADI, Maryhurst University
SUSANNE KIBLER-HACKER, Unity College (Maine)
CHANGJOO KIM, Ohio State University
JAMES W. KING, University of Utah
JOHN C. KINWORTHY, Concordia College (Nebraska)
ALBERT KITCHEN, Paine College
TED KLIMASEWSKI, Jacksonville State University (Alabama)
ROBERT D. KLINGENSMITH, Ohio State University-Newark
LAWRENCE M. KNOPP, JR., University of Minnesota-Duluth
LYNN KOEHNEMANN, Gulf Coast Community College
TERRILL J. KRAMER, University of Nevada
JOE KRAUSE, Community College of Indiana at Lafayette
BRENDA KREKELER, Northern Kentucky University, Miami University
ARTHUR J. KRIM, Cambridge, Massachusetts
MICHAEL A. KUKRAL, Rose-Hulman Institute of Technology
ELROY LANG, El Camino Community College (California)
RICHARD L. LANGILL, Saint Martin's College
CHRISTOPHER LANT, Southern Illinois University
A.J. LARSON, University of Illinois-Chicago
PAUL R. LARSON, Southern Utah University
LARRY LEAGUE, Dickinson State University (North Dakota)
DAVID R. LEE, Florida Atlantic University
WOOK LEE, Ohio State University
JOE LEEPER, Humboldt State University (California)
YECHIEL M. LEHAVY, Atlantic Community College (New Jersey)

Scott Leith, Central Connecticut State University
James Leonard, Marshall University
Elizabeth J. Leppman, St. Cloud State University
John C. Lewis, Northeast Louisiana University
Dan Lewman Jr., Southwest Mississippi Community College
Caedmon S. Liburd, University of Alaska-Anchorage
T. Ligibel, Eastern Michigan University
Z.L. Lipchinsky, Berea College (Kentucky)
Allan L. Lippert, Manatee Community College (Florida)
Richard Lisichenko, Fort Hays State University (Kansas)
John H. Litcher, Wake Forest University (North Carolina)
Lee Liu, Central Montana State University
Li Liu, Stephen F. Austin State University (Texas)
William R. Livingston, Baker College (Michigan)
Catherine M. Lockwood, Chadron State College
George E. Longenecker, Vermont Technical College
Cynthia Longstreet, Ohio State University
Jack Looney, University of Massachusetts, Boston
Tom Love, Linfield College (Oregon)
K.J. Lowrey, Miami University (Ohio)
James Lowry, Stephen F. Austin State University
Max Lu, Kansas State University
Robin R. Lyons, University of Hawai'i-Leeward Community College
Susan M. Macey, Texas State University
Michael Madsen, Brigham Young University, Idaho
Ronald Magden, Tacoma Community College
Christiane Mainzer, Oxnard College (California)
Laura Makey, California State University, San Bernardino
Harley I. Manner, University of Guam
Anthony Paul Mannion, Kansas State University
Gary Manson, Michigan State University
Charles Manyara, Radford University
James T. Markley, Lord Fairfax Community College (Virginia)
Sister May Lenore Martin, Saint Mary College (Kansas)
Kent Mathewson, Louisiana State University
Patrick May, Plymouth State College
Dick Mayer, Maui Community College (Hawai'i)
Sara Mayfield, San Jacinto College, Central (California)
Dean R. Mayhew, Maine Maritime Academy
J.P. McFadden, Orange Coast College (California)
Bernard McGonigle, Community College of Philadelphia (Pennsylvania)
Paul D. Meartz, Mayville State University (North Dakota)
Dianne Meredith, California State University-Sacramento
Gary C. Meyer, University of Wisconsin, Stevens Point
Judith L. Meyer, Southwest Missouri State University
Mark Micozzi, East Central University
John Milbauer, Northeastern State University
Dalton W. Miller Jr., W. Mississippi State University
Raoul Miller, University of Minnesota, Duluth
Roger Miller, Black Hills State University
James Mills, State University of New York, College at Oneonta
Ines Miyares, Hunter College, CUNY (New York)
Bob Monahan, Western Carolina University
Keith Montgomery, University of Wisconsin-Madison
Dan Morgan, University of South Carolina, Beaufort
Debbie Morimoto, Merced College
John Morton, Benedict College (South Carolina)
Anne Mosher, Syracuse University (New York)
Barry Mowell, Broward Community College (Florida)
Tom Mueller, California University of Pennsylvania
Donald Myers, Central Connecticut State University
Robert R. Myers, West Georgia College
Gary Nachtigall, Fresno Pacific University
Yaser M. Najjar, Framingham State College (Massachusetts)
Katherine Nashleanas, Southeast Community College
Jeffrey W. Neff, Western Carolina University
David Nemeth, University of Toledo (Ohio)
Robert Newcomer, East Central University
William Nieter, St. Johns University
William N. Noll, Highland Community College
Valiant C. Norman, Lexington Community College (Kentucky)
Raymond O'Brien, Bucks County Community College (Pennsylvania)
Patrick O'Sullivan, Florida State University
Diane O'Connell, Schoolcraft College
John Odland, Indiana University
Doug Oetter, Georgia College and State University
Anne O'Hara, Marygrove College
Patrick Olsen, University of Idaho
Joseph R. Oppong, University of North Texas
Lynn Orlando, Holy Family University
Mark A. Oumette, Hardin-Simmons University
Richard Outwater, California State University, Long Beach
Cissie Owen, Lamar University
Mary Ann Owoc, Mercyhurst College
Eugene J. Palka, U.S. Military Academy (New York)
Steve Palladino, Ventura College
Bimal K. Paul, Kansas State University
Selina Pearson, Northwest-Shoals Community College
Mauri Pelto, Nichols College
James Penn, Southeastern Louisiana University
Linda Pett-Conklin, University of St. Thomas
Diane Philen, Lower Brule Community College
Paul Phillips, Fort Hays State University (Kansas)
Michael Phoenix, ESRI, Redlands, California
Jerry Pitzl, Macalester College (Minnesota)
Brian Plaster, Texas State University
Armand Policicchio, Slippery Rock University
Rosann Poltrone, Arapahoe Community College
Billie E. Pool, Holmes Community College (Mississippi)
Gregory Pope, Montclair State University (New Jersey)
Jeff Popke, East Carolina University
Vinton M. Prince, Wilmington College (North Carolina)
George Puhrmann, Drury University
Donald N. Rallis, Mary Washington College
Rhonda Reagan, Blinn College (Texas)
Danny I. Reams, Southeast Community College (Nebraska)
Jim Reck, Golden West College (California)
Roger Reede, Southwest State University (Minnesota)
John Ressler, Central Washington University
John B. Richards, Southern Oregon State College
David C. Richardson, Evangel College (Missouri)
Gary Ringley, Virginia Highlands Community College
Susan Roberts, University of Kentucky

Curt Robinson, California State University, Sacramento
Wolf Roder, University of Cincinnati
James Rogers, University of Central Oklahoma
Paul A. Rollinson, AICP, Southwest Missouri State University
James C. Rose, Tompkins/Cortland Community College (New York)
Thomas E. Ross, Pembroke State University (North Carolina)
Thomas A. Rumney, State University of New York-Plattsburgh
George H. Russell, University of Connecticut
Bill Rutherford, Martin Methodist College
Rajagopal Ryali, Auburn University at Montgomery (Alabama)
Perry Ryan, Mott Community College
James Saku, Frostburg State University
David R. Sallee, North Lake College
Richard A. Sambrook, Eastern Kentucky University
Eduardo Sanchez, Grand Valley State University
John Santosuosso, Florida Southern College
Ginger Schmid, Texas State University
Brenda Thompson Schoolfield, Bob Jones University
Adena Schutzberg, Middlesex Community College (Massachusetts)
RogerM. Selya, University of Cincinnati
Renee Shaffer, University of South Carolina
Wendy Shaw, Southern Illinois University, Edwardsville
Sidney R. Sherter, Long Island University (New York)
Harold Shilk, Wharton County Jr. College
Nanda Shrestha, Florida A&M University
William R. Siddall, Kansas State University
David Silva, Bee County College (Texas)
Jose Antonio Simental, Marshall University
Morris Simon, Stillman College (Alabama)
Robert Mark Simpson, University of Tennessee at Martin
Kenn E. Sinclair, Holyoke Community College (Massachusetts)
Robert Sinclair, Wayne State University (Michigan)
Jim Skinner, Southwest Missouri State University
Bruce Smith, Bowling Green State University
Everett G. Smith Jr., University of Oregon
Peggy Smith, California State University, Fullerton
Richard V. Smith, Miami University (Ohio)
James Snaden, Charter Oak State College
David Sorenson, Augustana College
Sister Consuelo Sparks, Immaculata University
Carolyn D. Spatta, California State University-Hayward
M.R. Sponberg, Laredo Junior College (Texas)
Donald L. Stahl, Towson State University (Maryland)
David Stea, Texas State University
Elaine Steinberg, Central Florida Community College
D.J. Stephenson, Ohio University Eastern
Herschel Stern, Mira Costa College (California)
Reed F. Stewart, Bridgewater State College (Massachusetts)
Noel L. Stirrat, College of Lake County (Illinois)
Joseph P. Stoltman, Western Michigan University
Dean B. Stone, Scott Community College
William M. Stone, Saint Xavier University, Chicago
George Stoops, Mankato State University (Minnesota)
Debra Straussfogel, University of New Hampshire
Jamie Strickland, University of North Carolina, Charlotte
Wayne Strickland, Roanoke College
Philip Sturm, Ohio Valley College, Vienna
Philip Suckling, University of Northern Iowa
Allen Sullivan, Central Washington University
Selima Sultana, Auburn University
Ray Sumner, Long Beach City College
Christopher Sutton, Western Illinois University
T. L. Tarlos, Orange Coast College (California)
Wesley Teraoka, Leeward Community College
Michael Thede, North Iowa Area Community College
Derrick J. Thom, Utah State University
Curtis Thomson, University of Idaho
Ben Tillman, Texas Christian University
Cliff Todd, University of Nebraska, Omaha
S. Toops, Miami University (Ohio)
Richard J. Torz, St. Joseph's College (New York)
Harry Trendell, Kennesaw State University
Roger T. Trindell, Mansfield University of Pennsylvania
Dan Turbeville, East Oregon State College
Norman Tyler, Eastern Michigan University
George Van Otten, Northern Arizona University
Gregory Veeck, Western Michigan University
C.S. Verma, Weber State College (Utah)
Kelly Ann Victor, Eastern Michigan University
Sean Wagner, Tri-State University
Graham T. Walker, Metropolitan State College of Denver
Montgomery Walker, Yakima Valley Community College
Deborah Wallin, Skagit Valley College (Washington)
Mike Walters, Henderson Community College (Kentucky)
Linda Wang, University of South Carolina, Aiken
J.L. Watkins, Midwestern State University (Texas)
David Welk, Reedley College (California)
Kit W. Wesler, Murray State University
Peter W. Whaley, Murray State University
Macel Wheeler, Northern Kentucky University
P. Gary White, Western Carolina University (North Carolina)
W.R. White, Western Oregon University
Gary Whitton, Fairbanks, Alaska
Rebecca Wiechel, Wilmington College
Mark Wiljanen, Eastern Kentucky University
Gene C. Wilken, Colorado State University
Forrest Wilkerson, Texas State University
P. Williams, Baldwin-Wallace College
Stephen A. Williams, Methodist College
Deborah Wilson, Sandhills Community College
Morton D. Winsberg, Florida State University
Roger Winsor, Appalachian State University (North Carolina)
William A. Withington, University of Kentucky
A. Wolf, Appalachian State University, N.C.)
Joseph Wood, University of Southern Maine
Richard Wood, Seminole Junior College (Florida)
George I. Woodall, Winthrop College (North Carolina)
Stephen E. Wright, James Madison University (Virginia)
Leon Yacher, Southern Connecticut State University
Pai Yung-Feng, New York City
Kyong Yup Chu, Bergen Community College (New Jersey)
Firooz E. Zadeh, Colorado Mountain College
Donald J. Zeigler, Old Dominion University (Virginia)
Robert C. Ziegenfus, Kutztown University (Pennsylvania)

PERSONAL APPRECIATION

For assistance with the map of North American indigenous peoples, we are greatly indebted to Jack Weatherford, Professor of Anthropology at Macalester College (Minnesota); Henry T. Wright, Professor and Curator of Anthropology at the University of Michigan; and George E. Stuart, President of the Center for Maya Research (North Carolina), who also assisted with the Altun Ha site in Belize. The map of Russia's federal regions could not have been compiled without the invaluable help of David B. Miller, Senior Edit Cartographer at the National Geographic Society, and Leo Dillon of the Russia Desk of the U.S. Department of State. And special thanks, too, go to Charles Pirtle, Professor of Geography at Georgetown University's School of Foreign Service for his exhaustive review of the entire book.

We also record our appreciation to those geographers who ensured the quality of this book's ancillary products: Ira M. Sheskin (University of Miami) prepared the *Test Bank* and manipulated a large body of demographic data to derive the tabular display in Table G-1; Eugene J. Palka (U.S. Military Academy) prepared the *Instructor's Manual*; and Elizabeth Muller Hames (M.A. in Geography, University of Miami) co-authored the *Study Guide*, re-compiled the Geographical Index found on the website, and handled the data preparation for Table G-1. At the University of Miami's Department of Geography and Regional Studies, the authors are most grateful for the advice they received from faculty colleagues Douglas Fuller, Richard Grant, Jan Nijman, Rinku Roy Chowdhury, and Ira Sheskin as well as GIS Lab Manager Chris Hanson.

We are privileged to work with a team of professionals at John Wiley & Sons that is unsurpassed in the college textbook publishing industry. As authors we are acutely aware of these talents on a daily basis during the crucial production stage, especially the outstanding coordination and leadership skills of Production Editor Barbara Russiello, Senior Illustration Editor Sigmund Malinowski, and Senior Photo Editor Jennifer MacMillan. Others who played a leading role in this process were designers Karin Kincheloe, Lee Goldstein, and Creative Director Harry Nolan, copy editor Betty Pessagno, Joyce Franzen of GGS Book Services in Atlantic Highlands, New Jersey, and Don Larson and Emily Van Cleave of Mapping Specialists, Ltd., in Madison, Wisconsin. Executive Geosciences Editor Ryan Flahive was again the prime mover in introducing a number of innovations for the Twelfth Edition, and together with his assistant, Rachel Schneider, smoothly propelled and guided this latest revision. Our College Marketing Manager, Jeffrey Rucker, gave us considerable attention throughout the revision process, and his advice and enthusiasm were most welcome. We also thank Denise Powell, Christine Cordek, Tom Kulesa, Emily Streutker, and Lindsay Lovier for their contributions to various components of this revision project. Beyond this immediate circle, we acknowledge the support and encouragement we have received over the years from many others at Wiley including Vice President for Production Ann Berlin and Publisher Jay O'Callaghan.

Finally, and most of all, we thank our wives, Bonnie and Nancy, for once again seeing us through the challenging schedule of our ninth collaboration on this volume in the past 23 years.

H. J. de Blij
Boca Grande, Florida

Peter O. Muller
Coral Gables, Florida

July 22, 2005

Brief Table of Contents

Contents

CONCEPTS, IDEAS, AND TERMS

1. Geographic realm
2. Spatial perspective
3. Taxonomy
4. Transition zone
5. Geographic change
6. New World Order
7. Regional concept
8. Regional boundaries
9. Location
10. Absolute location
11. Relative location
12. Formal region
13. Spatial system
14. Hinterland
15. Functional region
16. Scale
17. Natural landscape
18. Physical geography
19. Continental drift
20. Tectonic plate
21. Subduction
22. Pacific Ring of Fire
23. Climate
24. Desertification
25. Glaciation
26. Ice age
27. Interglaciation
28. Hydrologic cycle
29. Climatic region
30. Physiography
31. Culture
32. Regional character
33. Cultural landscape
34. Central business district (CBD)
35. Ethnicity
36. Population distribution
37. Population density
38. Megalopolis
39. Cartogram
40. Urbanization
41. State
42. European state model
43. Development
44. Economic geography
45. Core area
46. Periphery
47. Regional disparity
48. Advantage
49. Neocolonialism
50. Globalization
51. Regional geography
52. Systematic geography

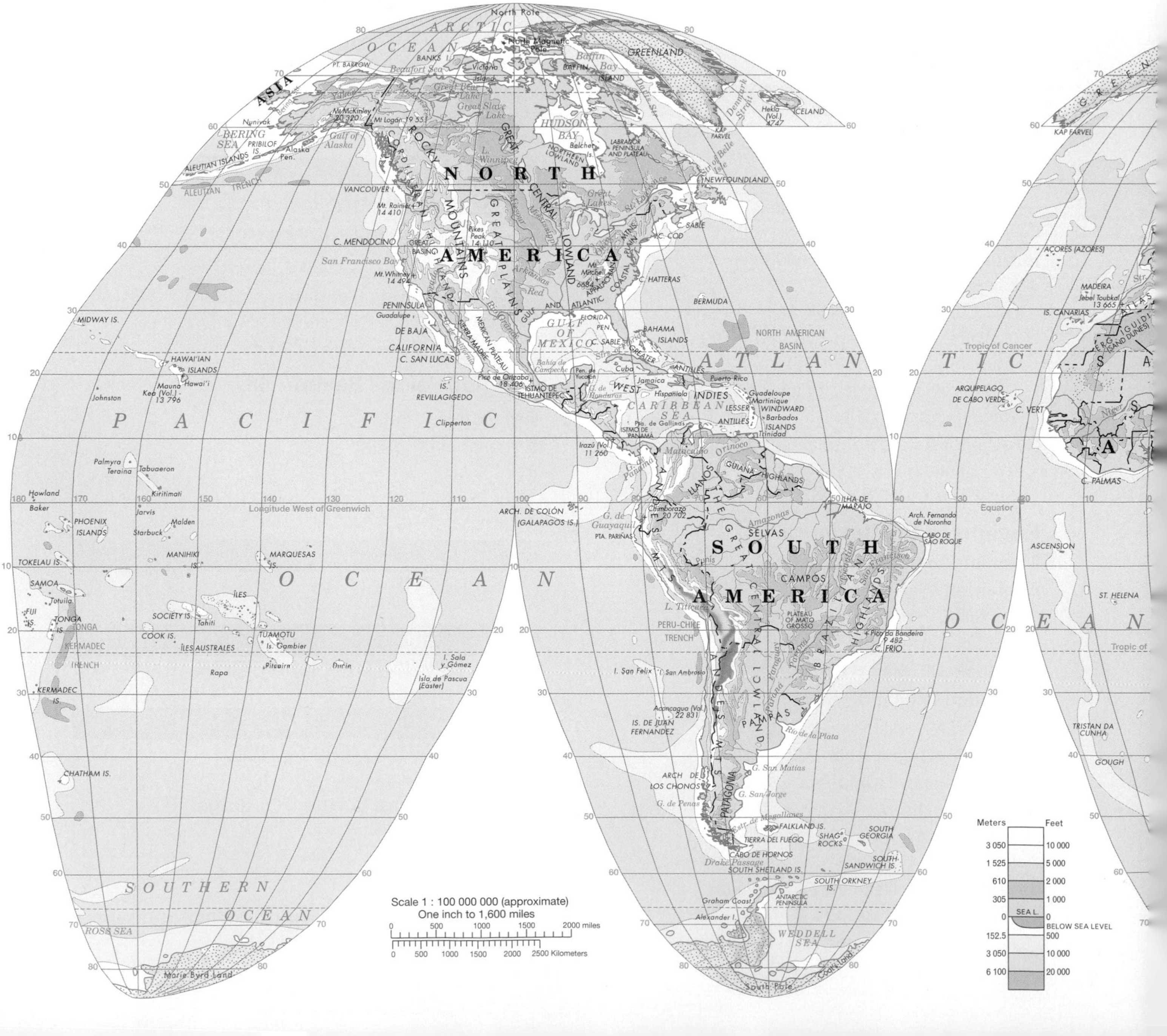

Introduction

World Regional Geography: Global Perspectives

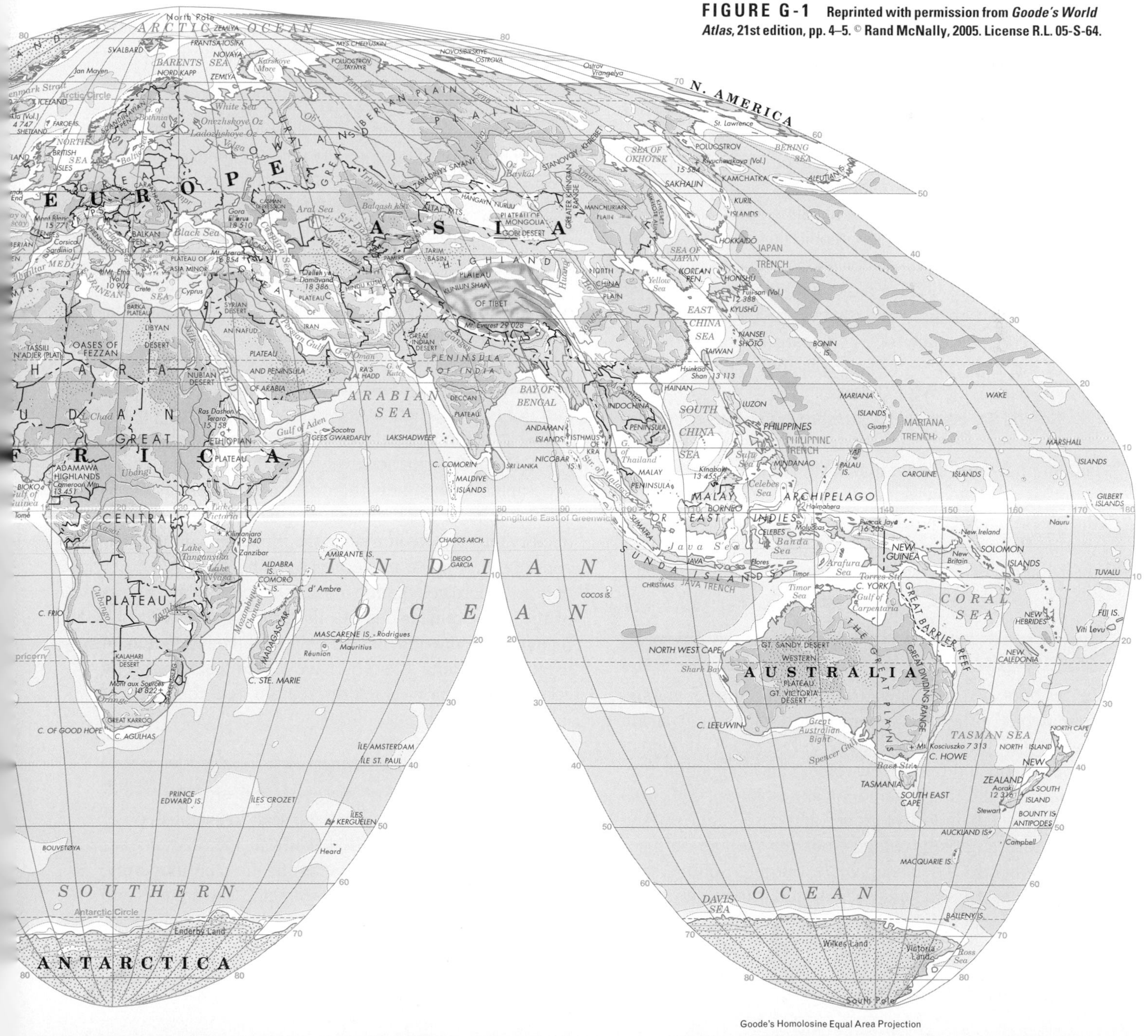

FIGURE G-1 Reprinted with permission from *Goode's World Atlas*, 21st edition, pp. 4–5. © Rand McNally, 2005. License R.L. 05-S-64.

YOU COULD NOT have chosen a better time to take a course in world regional geography than the present. Those of us in North America live in, or in the shadow of, the country that has become the world's sole superpower. Those who live elsewhere face forces unleashed by the United States—military, economic, cultural—and felt around the globe. As the November 2004 presidential election approached, foreign newspapers and other national media commented that the result would directly affect not just the United States but their countries as well. Some countries even conducted surveys and polls to determine which candidate was perceived as preferable. "When the United States votes for president," editorialized one European newspaper, "every citizen in the world has a stake in the outcome. You can't say that about any other nation."

The United States has emerged from the twentieth century as the planet's most powerful country, capable of influencing nations and peoples, lives and livelihoods from pole to pole. That power confers on Americans the responsibility to learn as much as they can about those nations and livelihoods, so that their decisions will be well-informed and carefully considered. But in this respect, the United States is no superpower. Geographic literacy is a measure of international comprehension and awareness, and Americans' geographic literacy ranks low among countries of consequence. For the world, that is not a good thing, because such geographic fogginess tends to afflict not only voters but also the representatives they elect, from the school board to the White House.

So you are about to strike a blow against geographic illiteracy, and at the end of this term you will find yourself among a still-small minority of Americans who have spent even one semester studying what you will study in the weeks ahead. When you get into discussions about world affairs, you will be able to bring to the table a dimension not usually weighed during such conversations.

GEOGRAPHIC PERSPECTIVES

In this book, we take a penetrating look at the geographic framework of the contemporary world, the grand design that is the product of thousands of years of human achievement and failure, movement and stagnation, revolution and stability, interaction and isolation. Ours is an interconnected world of travel and trade, tourism and television, a global village—but the village still has neighborhoods. Their names are Europe, South America, Southeast Asia, and others familiar to all of us. We
1 call such global neighborhoods **geographic realms**, and when we subject these realms to geographic scrutiny, we find that each has its own identity and distinctiveness.

Geographers study the location and distribution of features on the Earth's surface. These features may be the landmarks of human occupation, the properties of the natural environment, or both; one of the most interesting themes in geography is the relationship between natural environments and human societies, which is why the first map in this introductory chapter summarizes the prominent natural (physical) features of the continents we inhabit (Fig. G-1). Geographers investigate the reasons for these distributions. Their approach is guided by a **spatial** 2
perspective. Just as historians focus on chronology, geographers concentrate on space and place. The spatial structure of cities, the layout of farms and fields, the networks of transportation, the system of rivers, the pattern of climate—all these and more go into the study of a geographic realm. As you will discover, geography is full of spatial terms: area, distance, direction, clustering, proximity, accessibility, isolation, and many others that we will encounter in the pages that follow.

In this book we use the geographic perspective and geography's spatial terminology to investigate the world's great geographic realms. We will find that each of these realms possesses a special combination of cultural, organizational, and environmental properties. These characteristic qualities are imprinted on the landscape, giving each realm its own traditional attributes and social settings. As we come to understand the human and natural makeup of those geographic realms, we learn not only *where* they are located (always a crucial question in geography, and often the answer is not as simple as it may appear), but also *why they are located where they are*, how they are constituted, and what their future is likely to be in our changing world.

It is not enough, however, to study the world from a geographic viewpoint without learning something about the discipline of geography itself. Beginning in this introductory chapter, we introduce the fundamental ideas and concepts that make modern geography what it is. Not only will these geographic ideas enhance your awareness of the many dimensions of our complex, multicultural, interconnected world, but also many of them will remain useful to you long after you have closed this book. Welcome to geography . . . realms, regions, *and* concepts.

REALMS AND REGIONS

Geographers, like other scholars, seek to establish order from the countless data that confront them. Biologists have established a system of classification, or **taxonomy**, 3
to categorize the many millions of plants and animals into a hierarchical system of seven ranks. In descending order,

we humans belong to the animal *kingdom*, the *phylum* (division) named chordata, the *class* of mammals, the *order* of primates, the *family* of hominids, the *genus* designated *Homo*, and the *species* known as *Homo sapiens*. Geologists classify the Earth's rocks into three major (and many subsidiary) categories and then fit these categories into a complicated geologic time scale that spans hundreds of millions of years. Historians define eras, ages, and periods to conceptualize the sequence of the events they study.

Geography, too, employs systems of classification. When geographers deal with urban problems, for instance, they use a classification scheme based on the sizes and functions of the places involved. Some of the terms in this classification are part of our everyday language: megalopolis, metropolis, city, town, village, hamlet.

In regional geography, which is the focus of this book, our challenge is different. We, too, need a hierarchical framework for the areas of the world we study, from the largest to the smallest. But our classification scheme is horizontal, not vertical. It is *spatial*. Our equivalent of the biologists' overarching kingdoms (of plants and animals) is the Earth's natural partitioning into landmasses and oceans. The next level is the division of the inhabited landmasses into geographic realms based on human as well as natural properties. The great global realms, in turn, subdivide into regions.

FROM THE FIELD NOTES

"From the observation platform atop the Seoul Tower one would be able to see into North Korea except for the range of hills in the background: the capital lies in the shadow of the DMZ (demilitarized zone), relic of one of the hot conflicts of the Cold War. The vulnerable Seoul-Inchon metropolitan area ranks among the world's largest, its population approaching 20 million. I asked my colleague, a professor of geography at Seoul National University, why the central business district, the cluster of buildings in the center of this photo, does not seem to reflect the economic power of this city (note the sector of traditional buildings in the foreground, some with blue roofs, the culture's favorite color). 'Turn around and look across the [Han] River,' he said. 'That's the new Seoul, and there the skyline matches that of Singapore, Beijing, or Tokyo.' Height restrictions, disputes over land ownership, and congestion are among the factors that slowed growth in this part of Seoul and caused many companies to build in the Samsung area. Looking in that direction, you cannot miss the large U.S. military facility, right in the heart of the urban area, on some of the most valuable real estate and right next to an upscale shopping district. In the local press, the debate over the presence of American forces dominated the letters page day after day." © H. J. de Blij

The Criteria for Realms

In any classification system, criteria are the key. Not all animals are mammals; the criteria for inclusion in that biological class are more specific and restrictive. A dolphin may look and act like a fish, but both anatomically and functionally dolphins belong to the class of mammals.

Geographic realms are based on sets of spatial criteria. First, they are the largest units into which the inhabited world can be divided. The criteria on which such a broad regionalization is based include both physical (that is, natural) and human (or social) yardsticks. South America is a geographic realm, for example, because physically it is a continent and culturally it is dominated by a set of social norms. The realm called South Asia, on the other hand, lies on a Eurasian landmass shared by several other geographic realms; high mountains, wide deserts, and dense forests combine with a distinctive social fabric to create this well-defined realm centered on India.

Second, geographic realms are the result of the interaction of human societies and natural environments, a *functional* interaction revealed by farms, mines, fishing ports, transport routes, dams, bridges, villages, and countless other features that mark the landscape. According to this criterion, Antarctica is a continent but not a geographic realm.

Third, geographic realms must represent the most comprehensive and encompassing definition of the great clusters of humankind in the world today. China lies at the heart of such a cluster, as does India. Africa constitutes a geographic realm from the southern margin of the Sahara (an Arabic word for "desert") to the Cape of Good Hope and from its Atlantic to its Indian Ocean shores.

Figure G-2 displays the 12 world geographic realms based on the these criteria. As we will show in more detail later, waters, deserts, and mountains as well as cultural and

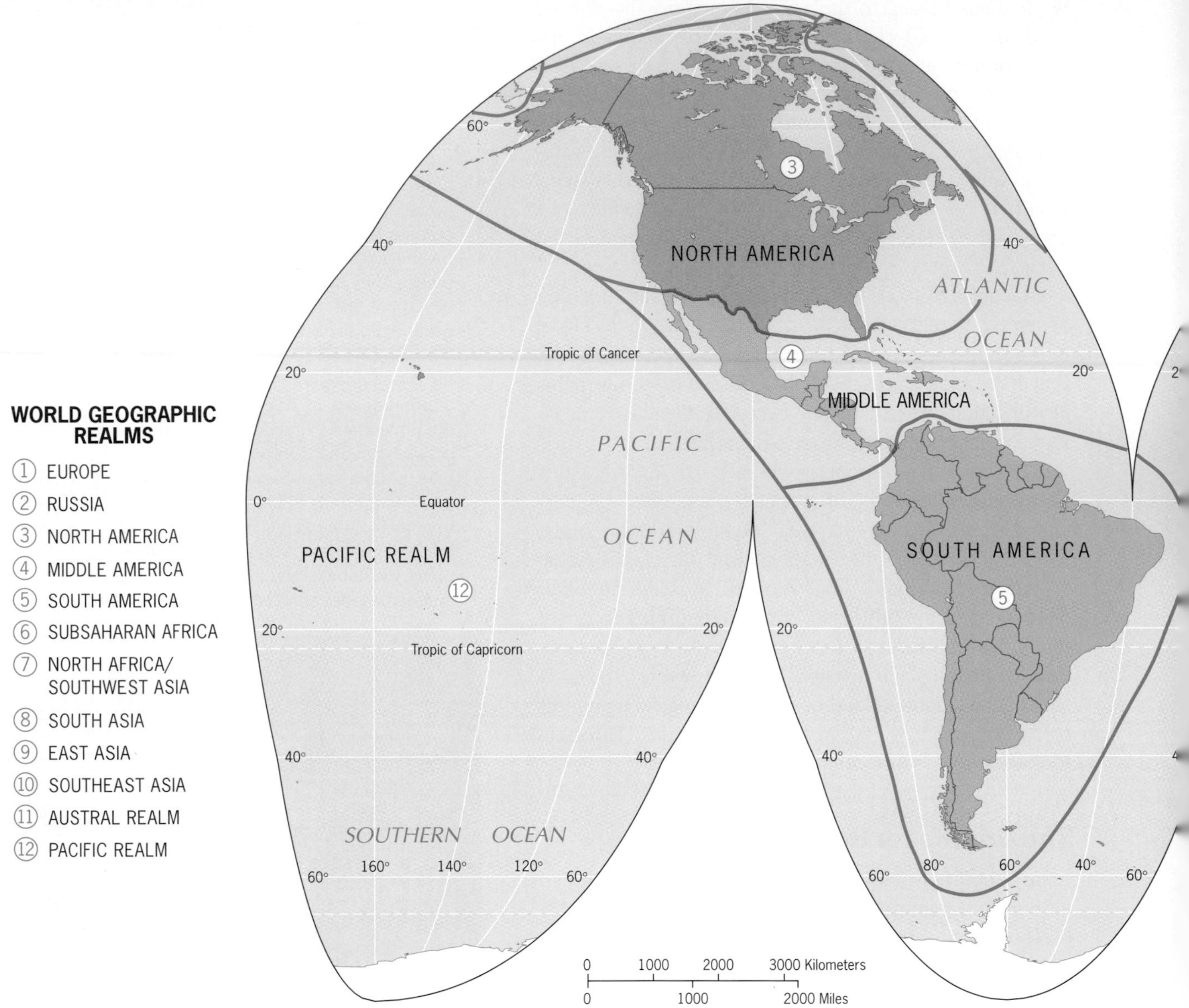

FIGURE G-2 © H. J. de Blij, P. O. Muller, and John Wiley & Sons, Inc.

political shifts mark the borders of these realms. We will discuss the position of these boundaries as we examine each realm. For the moment, keep in mind the following:

4 • Where geographic realms meet, **transition zones**, not sharp boundaries, mark their contacts.

We need only remind ourselves of the border zone between the geographic realm in which most of us live, North America, and the adjacent realm of Middle America. The line on Figure G-2 coincides with the boundary between Mexico and the United States, crosses the Gulf of Mexico, and then separates Florida from Cuba and the Bahamas. But Hispanic influences are strong in North America north of this boundary, and the U.S. economic influence is strong south of it. The line, therefore, represents an ever-changing zone of regional interaction. Again, there are many ties between South Florida and the Bahamas, but the Bahamas resemble a Caribbean more than a North American society.

In Africa, the transition zone from Subsaharan to North Africa is so wide and well defined that we have put it on the world map; elsewhere, transition zones tend to be narrower and less easily represented. In these early years of the twenty-first century, such countries as Belarus (between Europe and Russia) and Kazakhstan (between Russia and Muslim Southwest Asia) lie in inter-realm transition zones. Remember, over much (though not all) of their length, borders between realms are zones of regional change.

• **Geographic** realms **change** over time. 5

Had we drawn Figure G-2 before Columbus made his voyages (1492–1504), the map would have looked different: Amerindian territories and peoples would have deter-

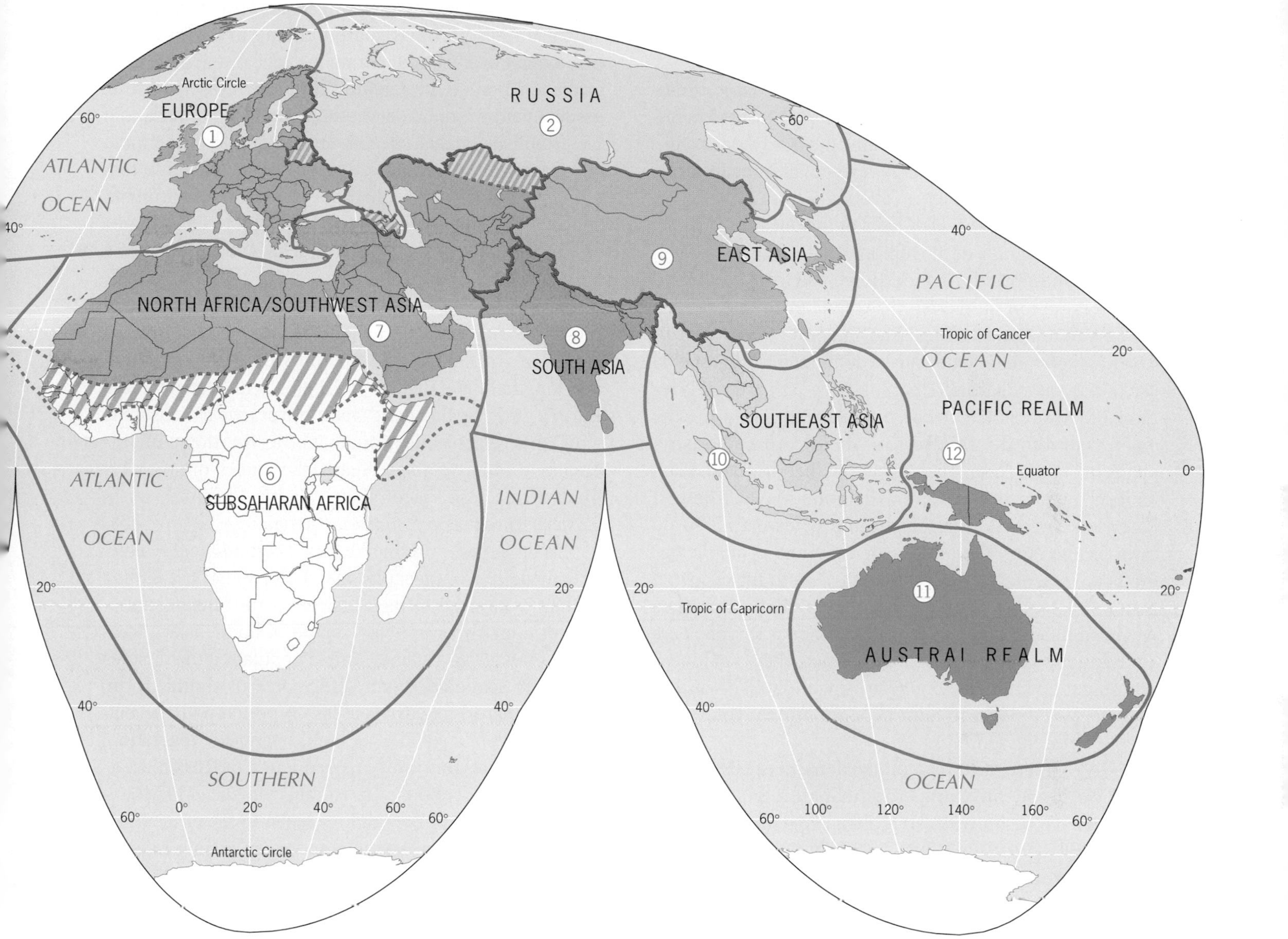

mined the boundaries in the Americas; Australia and New Guinea would have constituted one realm, and New Zealand would have been part of the Pacific Realm. The colonization, Europeanization, and Westernization of the world changed that map dramatically. During the four decades after World War II relatively little change took place, but since 1985 far-reaching realignments have again been occurring.

6 As we try to envisage what a **New World Order** will look like on the map, note that the 12 geographic realms can be divided into two groups: (1) those dominated by one major political entity, in terms of territory or population or both (North America/United States, Middle America/Mexico, South America/Brazil, South Asia/India, East Asia/China, Southeast Asia/Indonesia as well as Russia and Australia), and (2) those that contain many countries but no dominant state (Europe, North Africa/Southwest Asia, Subsaharan Africa, and the Pacific Realm). For several decades two major powers, the United States and the former Soviet Union, dominated the world and competed for global influence. Today, the United States is dominant in what has become a unipolar world. What lies ahead? Will China and/or some other power challenge U.S. supremacy? Is our map a prelude to a multipolar world? We will address such questions in the pages that follow.

The Criteria for Regions

The spatial division of the world into geographic realms establishes a broad global framework, but for our purposes a more refined level of spatial classification is needed. This brings us to an important organizing concept in geography: the **regional concept**. To continue the 7 analogy with biological taxonomy, we now go from phylum to order. To establish regions within geographic realms, we need more specific criteria.

Let us use the North American realm to demonstrate the regional idea. When we refer to a part of the United States or Canada (e.g., the South, the Midwest, or the Prairie Provinces), we employ a regional concept—not scientifically but as part of everyday communication. We reveal our *perception* of local or distant space as well as our mental image of the region we are describing.

But what exactly is the Midwest? How would you draw this region on the North American map? Regions are easy to imagine and describe, but they can be difficult to outline on a map. One way to define the Midwest is to use the borders of States: certain States are part of this region, others are not. You could also use agriculture as the chief criterion: the Midwest is where corn and/or soybeans occupy a certain percentage of the farmland. Each method results in a different delimitation; a Midwest based on States is different from a Midwest based on farm production. Therein lies an important principle: regions are scientific devices that allow us to make spatial generalizations, and they are based on artificial criteria we establish to help us construct them. If you were studying the geography behind politics, then a Midwest region defined by State boundaries would make sense. If you were studying agricultural distributions, you would need a different definition.

Given these different dimensions of the same region, we can identify properties that all regions have in common. To begin with, all regions have *area*. This observation would seem obvious, but there is more to this idea than meets the eye. Regions may be intellectual constructs, but they are not abstractions: they exist in the real world, and they occupy space on the Earth's surface.

It follows that regions have *boundaries*. Occasionally, nature itself draws sharp dividing lines, for example, along the crest of a mountain range or the margin of a forest. More often, **regional boundaries** are not self- 8
evident, and we must determine them using criteria that we establish for that purpose. For example, to define a citrus-growing agricultural region, we may decide that only areas where more than 50 percent of all farmland stands under citrus trees qualify to be part of that region.

All regions also have **location**. Often the name of a 9
region contains a locational clue, as in Amazon Basin or Indochina (a region of Southeast Asia lying between India and China). Geographers refer to the **absolute lo-** 10
cation of a place or region by providing the latitudinal and longitudinal extent of the region with respect to the Earth's grid coordinates. A far more practical measure is a region's **relative location**, that is, its location with ref- 11
erence to other regions. Again, the names of some regions reveal aspects of their relative locations, as in *Eastern* Europe and *Equatorial* Africa.

Many regions are marked by a certain *homogeneity* or sameness. Homogeneity may lie in a region's human (cultural) properties, its physical (natural) characteristics, or both. Siberia, a vast region of northeastern Russia, is marked by a sparse human population that resides in widely scattered, small settlements of similar form, frigid climates, extensive areas of permafrost (permanently frozen subsoil), and cold-adapted vegetation. This dominant uniformity makes it one of Russia's natural and cultural regions, extending from the Ural Mountains in the west to the Pacific Ocean in the east. When regions display a measurable and often visible internal homogeneity, they are called **formal regions**. But not all 12
formal regions are visibly uniform. For example, a region may be delimited by the area in which, say, 90 per-

Interconnections in the spatial system of the Corn Belt: Garden City, Iowa. Roads and grain elevators underscore the linkages between the town and its surroundings. The spatial homogeneity of the region is quite evident from the agricultural landscape beyond (evenly dispersed farmsteads, repetition of field size and land uses). Also visible is the signature geometry of the Township-and-Range system of land division which, beginning in 1785, superimposed a rectangular grid of 6-mile by 6-mile townships on all of the U.S. States opened for settlement after that date. Within that geographical matrix, townships were subdivided into 36 square-mile (640-acre) properties that were further split into 160-acre (and sometimes smaller) farms. As the photo shows, this land division scheme dominates every aspect of the cultural landscape: not only the rectangular fields and the north-south/east-west layout of roads and railways in the countryside, but also the configuration and spacing of streets, buildings, and properties in Garden City itself. ©WestLight/Corbis Images.

cent of the people or more speak a particular language. This cannot be seen in the landscape, but the region is a reality, and we can use this criterion to draw its boundaries accurately. It, too, is a formal region.

Other regions are marked *not* by their internal sameness but by their functional integration—that is, the way
13 they work. These regions are defined as **spatial systems** and are formed by the areal extent of the activities that define them. Take the case of a large city with its surrounding zone of suburbs, urban-fringe countryside, satellite towns, and farms. The city supplies goods and services to this encircling zone, and it buys farm products and other commodities from it. The city is the heart, the *core* of this region, and we call the surrounding zone of interaction
14 the city's **hinterland**. But the city's influence wanes on the outer periphery of that hinterland, and there lies the boundary of the functional region of which the city is the
15 focus. A **functional region**, therefore, is forged by a structured, urban-centered system of interaction. It has a core and a periphery. As we shall see, core-periphery contrasts in some parts of the world are becoming strong enough to endanger the stability of countries.

All human-geographic regions are *interconnected*, being linked to other regions. We know that the borders of geographic realms sometimes take on the character of transition zones, and so do neighboring regions. Trade, migration, education, television, computer linkages, and other interactions blur regional boundaries. These are just some of the links in the fast-growing interdependence among the world's peoples, and they reduce the differences that still divide us. Understanding these differences will lessen them further.

Realms and Regions on the Map

Maps are the language of geography. If it is true that a picture is worth a thousand words, then a map is often worth a million. But some maps convey much more local detail than others, and this is a function of scale. Obviously an atlas map of the world, at a small scale, cannot show you anything about the street layout of your hometown: for that, you need a large-scale map covering just the municipal area where you live. In this book, we must use mostly small-scale maps because regions are large and our page size is small, but from time to time we use a larger-scale map to highlight a particular regional issue or problem. And we also include maps of many of the world's great cities, employing a larger scale still.

The map is the geographer's strongest ally. It does for geography what taxonomic (and other) classification systems do for the other sciences. Maps display enormous amounts of information. They can suggest spatial relationships, they pose questions for researchers, and they often suggest where the answers may lie. Maps represent the surface of the Earth at various levels of generalization. A map's **scale** is the ratio of the distance be- 16
tween two locations on a map and the actual distance between those two locations on the Earth's surface. That ratio is expressed as a fraction.

In Figure G-3, note how that fraction grows larger as the level of detail increases—and the area displayed shrinks. Map A shows almost all of North America at the small scale of 1:103,000,000. The scale of map B is almost twice as large but shows only the northeast quadrant of the realm. At 1:24,000,000 (map C) only the Canadian province of Quebec can mostly be shown; at 1:1,000,000 (map D) you can see details of the urban area of Montreal at a scale about 100 times as large as map A. (Scale is a complicated concept, with more to it than areal size and coverage; some insights are provided in Appendix A.)

THE PHYSICAL SETTING

This book focuses on the geographic realms and regions produced by human activity over thousands of years. But we should never forget the natural environments in which all this activity took place because we can still recognize the role of these environments in how people make their living. Certain areas of the world, for example, presented opportunities for plant and animal domestication that other areas did not. The people who happened to live in those favored areas learned to grow wheat, rice, or root crops, and to domesticate oxen, goats, or llamas. We can still discern those early "patterns of opportunity" on the map in the twenty-first century. From such opportunities came adaptation and invention, and thus arose villages, towns, cities, and states. But people living in more difficult environments found it much harder to achieve this organization. Take tropical Africa. There, the human communities could not domesticate any of the many species of wildlife, from gazelles and zebras to giraffes and buffalo. Wild animals were a threat, not an opportunity. Eventually humans domesticated only one African animal, the guinea fowl. Early African peoples faced environmental disadvantages that persist today. The modern map carries many imprints of the past.

The landmasses of Planet Earth present a jumble of **natural landscapes** ranging from rugged mountain 17
chains to smooth coastal plains (Fig. G-1). The map of natural (physical) landscapes reminds us that the names of the continents and their mountain backbones are closely linked: North America and its Rocky Mountains, South America and the Andes, Eurasia with its Alps (in

EFFECT OF SCALE

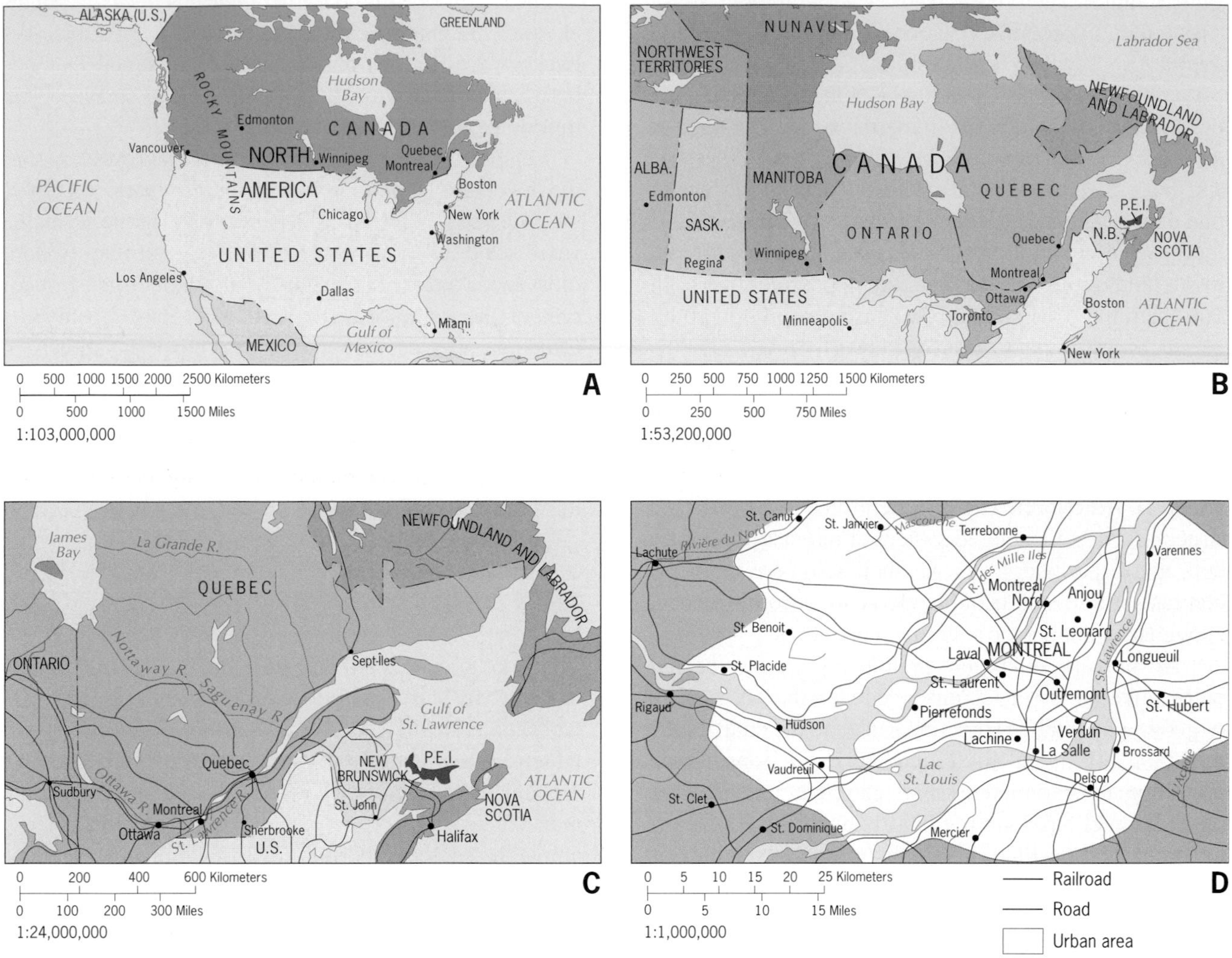

FIGURE G-3 © H. J. de Blij, P. O. Muller, and John Wiley & Sons, Inc.

the west) and Himalayas (toward the east), Australia and its Great Dividing Range. There is, however, a prominent exception: Africa, the Earth's second-largest landmass, has peripheral mountains in the north (the Atlas) and south (the Cape Ranges), but no linear chain from coast to coast. We will soon discover why.

In his book *Guns, Germs and Steel*, UCLA geographer Jared Diamond argues that the orientation of these continental "axes" has had "enormous, sometimes tragic, consequences [that] affected the rate of spread of crops and livestock, and possibly also of writing, wheels, and other inventions. That basic feature of geography thereby contributed heavily to the very different experiences of Native Americans, Africans, and Eurasians in the last 500 years." Inevitably it also affected the evolution of the regional map we are set to investigate. And it affected not only its evolution, but also its future. Mountain ranges formed barriers to movement but, as Diamond stresses, also channeled the spread of agricultural and technical innovations. Rugged mountains obstructed contact but also protected peoples and cultures against their enemies (today al-Qaeda, Taliban, and Chechnyan fugitives are using remote and rugged mountain refuges for such purposes). As we study each of the world's geographic realms, we will find that physical landscapes continue to play significant roles in this modern world. River basins in Asia still contain several of the planet's largest population concentrations: the advantages of fertile soils and ample water that first enabled clustered human settlement now sustain hundreds of millions in crowded South and East Asia. But river basins in Africa and South America support no such numbers because the combination of natural and historical factors is different. That is one reason the study of world regional geography is important: it puts the human map in environmental as well as regional perspective.

Continents and Plates

18 The study of natural landscapes is part of **physical geography**. Nearly a century ago a physical geographer, Alfred Wegener, used spatial analysis to explain the details he saw on maps like Figure G-1. He studied the outlines of the continents and, marshalling vast evidence, theorized that the Earth's landmasses are pieces of a giant supercontinent that existed hundreds of millions of years ago. He named that postulated supercontinent *Pangaea*, and Africa lay at the heart of it. When North and South America split away from Eurasia and Africa, the Atlantic Ocean opened—but you can still see how the opposite coastlines fit together.

19 Wegener's hypothesis of **continental drift** seemed to explain much of what Figure G-1 shows, including the global distribution of major mountain ranges. As South America "drifted" westward, away from Africa, the rocks of its leading edge crumpled like folds of an accordion, creating the Andes Mountains. On the opposite side of Africa, Australia moved eastward, so that its major mountain chain now lies near its eastern margin. Africa itself moved little, which is why no major linear mountain range formed there.

Wegener's hypothesis, based on spatial evidence, pointed the way for the research of others. Eventually geologists, building on the geographic evidence, formulated the notion of plate tectonics. The continental
landmasses, they reasoned, rest upon great slabs of 20
crust called **tectonic plates**. These plates are mobile, moving slowly a few centimeters (an inch or so) a year, propelled by heat-driven convection cells in the molten rock below.

Where these giant plates collide, the plate that consists of lighter (less heavy) rock rides up over the weightier plate, which is pushed downward in a process called
subduction. Such gigantic collisions create the moun- 21
tain chains that girdle the Earth, causing earthquakes and volcanic eruptions in the process. In Figure G-4, note that the Pacific Ocean is encircled by a nearly continuous
plate-collision zone called the **Pacific Ring of Fire**. Dis- 22
astrous earthquakes, volcanic eruptions, landslides, and other threats are facts of daily life here (Fig. G-5).

The danger associated with these plate-collision zones, in the Pacific as well as elsewhere, was underscored when, on December 26, 2004, a major earthquake beneath the Indian Ocean off western Sumatera (Sumatra) generated a sea wave called a *tsunami* that radiated outward to strike coastal Indonesia, Malaysia, Thailand, Myanmar, Sri Lanka, India, the Maldives, and even Somalia and Kenya in Africa. The site of this earthquake was the subduction zone where the Eurasian Plate collides with the Indian and Australian Plates (upper end of the pink plate fragment in Fig. G-4). It was the most powerful earthquake

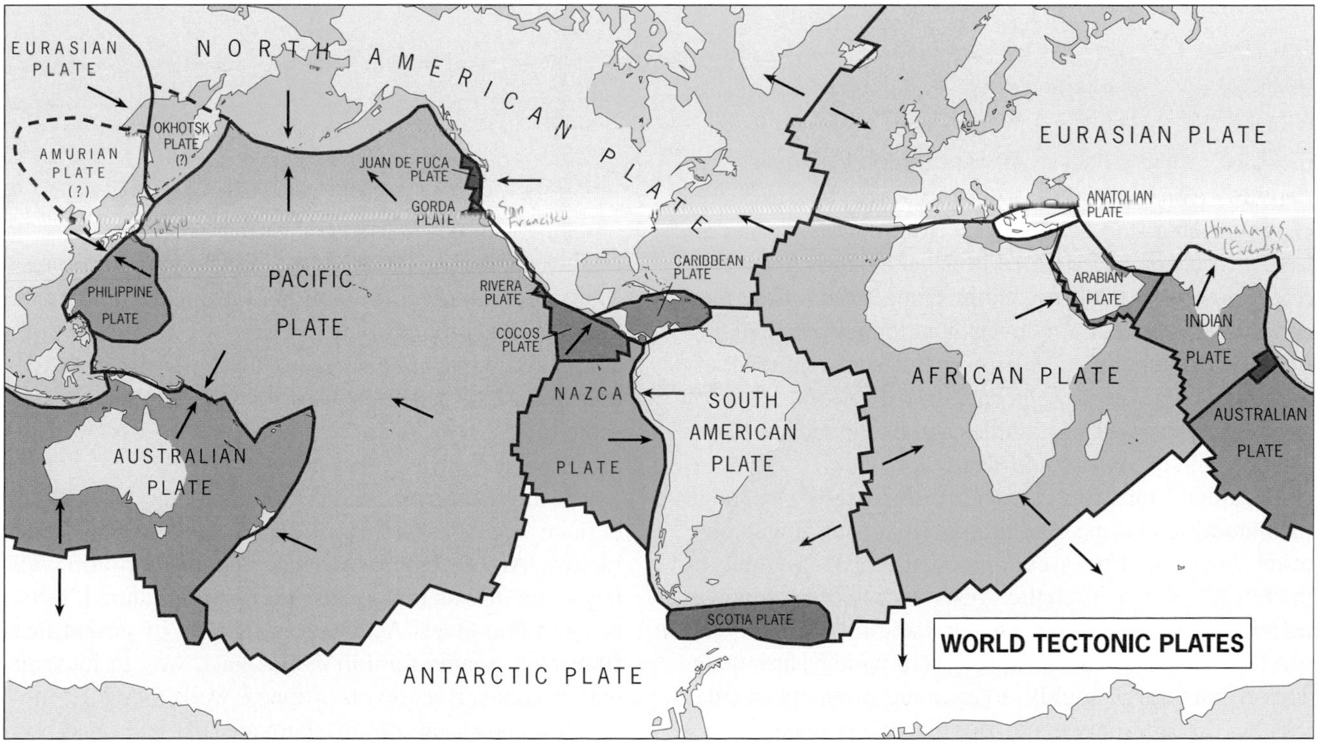

FIGURE G-4 © H. J. de Blij, P. O. Muller, and John Wiley & Sons, Inc.

Two dramatic images of the calamitous tsunami resulting from a submarine earthquake off the Indonesian island of Sumatera (Sumatra) on December 26, 2004. The doomed beachgoers wading in the suddenly shallow waters off the beach at Krabi in southern Thailand have just realized the magnitude of the first two of six waves rushing toward the coast; most have begun to run shoreward, but none would survive. . . . The tsunami devastated Banda Aceh, Sumatera's northernmost city and closest to the epicenter, sweeping away entire neighborhoods and leaving few buildings standing; similar damage occurred as far away as coastal Sri Lanka, one thousand miles (1600 km) to the west. An estimated 300,000 persons perished in this first great environmental disaster of the twenty-first century. (*left*) © AFP/Getty Images News and Sport Services; (*right*) © AP/Wide World Photos.

recorded on the planet in 40 years, and eyewitnesses reported that the resulting wall of water was as high as 50 feet (15 m) when it smashed into the coast in Sumatera and Sri Lanka. The exact death toll will never be known, but estimates exceed 300,000; the victims ranged from fishermen and farmers to hotel employees and tourists representing more than 50 countries. A massive international relief effort could only partially alleviate the misery of millions of survivors who lost their families and livelihoods; the devastation will take decades to clear.

Sudden, catastrophic events of this kind that change the natural landscape occur fairly frequently in geologic terms, but other processes are more gradual. The rocks of the Earth's crust vary in their resistance to breakdown by *weathering* and removal by *erosion*, in which rivers, glaciers, coastal waves, and desert winds participate. This results in the highly diverse and often spectacular scenery that uniquely marks our planet's surface.

Climate

It would be impossible to discuss the world's geographic realms and regions without reference to their **climates**. 23
Many of humanity's inventions were motivated by the need to overcome problems posed by climate: clothing against cold, irrigation against drought, air conditioning against heat. When we think of some distant region, the first question we are likely to ask is "What is its climate?"

A more appropriate question would be "What is its climate *today*?" Global climates change, sometimes rapidly. Archeological evidence tells us that farms, villages, and towns that sprang up in well-watered, fertile areas of Southwest Asia were overtaken by **desertifica-** 24
tion. Dependable rainfall cycles gave way to intermittent droughts, river levels dropped, wells failed. Eventually the expanding desert claimed what had once been

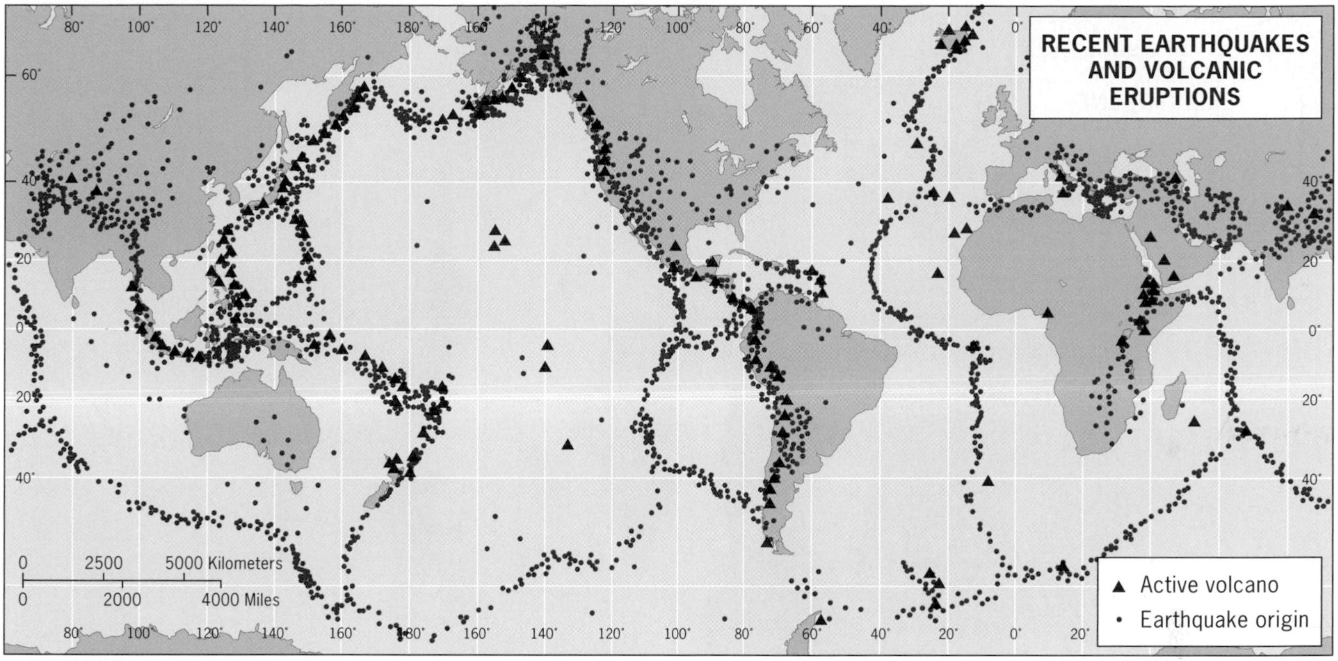

FIGURE G-5 Data courtesy U.S. National Oceanic and Atmospheric Administration (NOAA). © H. J. de Blij, P. O. Muller, and John Wiley & Sons, Inc.

thriving farmlands and bustling settlements. Some settlements were buried under advancing dunes.

Over the longer term, too, climate is subject to wide swings. Little more than 12,000 years ago, an eyeblink in geologic history, giant icesheets covered virtually all of Canada and much of the U.S. Midwest (Fig. G-6) as far south as the Ohio River. Most of Europe was covered from the Alps to Scandinavia. This was only the
25 latest phase of **glaciation**, a period of lowered temperatures, ice surges, and dropping sea levels that has prevailed over the past 12 million years. It all started with a cooling trend, perhaps as long as 20 million years
26 ago, the beginning of a global **ice age**. Eventually this gradual cooling gave way to massive advances of icesheets and valley glaciers, marking the beginning of what geologists call the Pleistocene epoch less than 3 million years ago. Ever since, frigid glaciations have
27 been followed by relatively warm spells called **interglaciations**. During the current interglaciation, named the *Holocene* and already over 10,000 years in duration, human civilizations have evolved from the smallest and simplest communities to the populous countries and megacities of today.

Climatic Regions

Climatic variability also occurs in the short term, and we should not be surprised that climates continue to change. Unless we prepare for it, these dynamics can produce social and political dislocation. Thus we need to understand the climatic environments of the world's geographic realms and regions. Our technological advances notwithstanding, climate still plays a key role in the lives and livelihoods of billions of people. A perturbation in the system—such as an El Niño event or a drought—can have widespread repercussions on economies and even politics.

It is no easier to generalize about climates than it is about natural landscapes. A regional climate is the average of countless weather observations among which measures of precipitation and temperature are key. Also taken into account are seasonal variations and extremes. In terms of precipitation, the Earth displays enormous spatial variation (Fig. G-7). Equatorial and tropical areas tend to be well-watered because the
hydrologic cycle—the system that carries evaporated 28
ocean water (leaving the salt behind) over the landmasses where it falls as rain or snow—is most efficient in the high heat and humidity of the low latitudes. But in Africa this equatorial bounty soon gives way to subtropical aridity. From over 80 inches (200 cm) of rain in The Congo near the equator, the annual total drops to 4 inches (10 cm) or less in the Sahara to the north and the Kalahari-Namib to the south. In higher latitudes, only certain favored areas (notably Western Europe) receive ample precipitation.

Average temperatures decline from equatorial toward polar latitudes, but the gradation is irregular, affected by elevation (mountains have a cooling effect) and

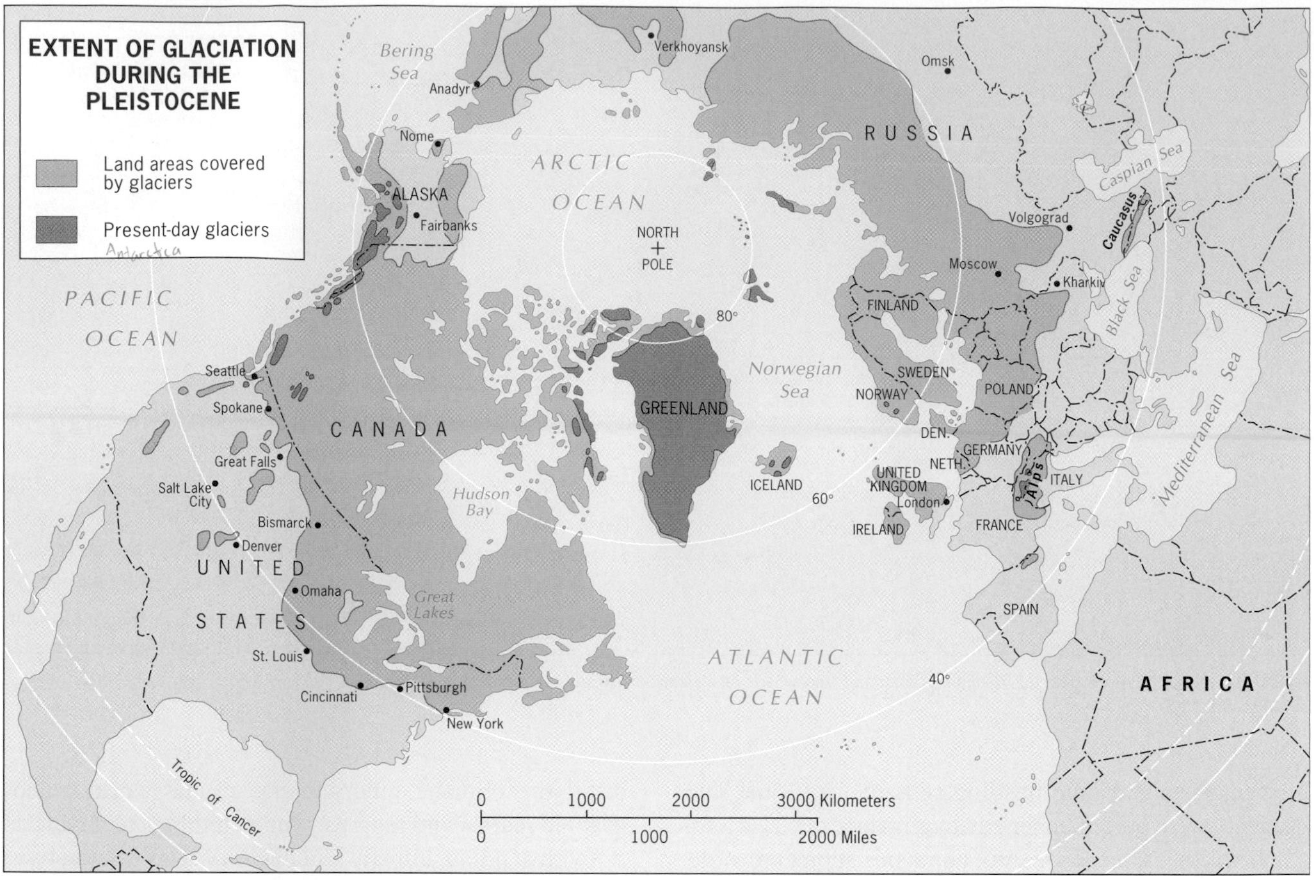

FIGURE G-6 © H. J. de Blij, P. O. Muller, and John Wiley & Sons, Inc.

FROM THE FIELD NOTES

"Walking the hilly countryside anywhere in Indonesia leaves you in no doubt about the properties of *Af* climate. The sweltering sun, the hot, humid air, the daily afternoon rains, and the lack of relief even when the sun goes down—the still atmosphere lies like a heavy blanket on the countryside to make this a challenging field trip. Deep, fertile soils form rapidly here on the volcanic rocks, and the entire landscape is green, most of it now draped by rice fields and terraced paddies. The farmer whose paddies I photographed here told me that his land produces three crops per year. Consider this: the island of Jawa, about the size of Louisiana, has a population of more than 125 million—its growth made possible by this combination of equatorial circumstances . . . Alaska, almost a dozen times as large as Jawa, has a population under three-quarters of a million. Here climates range from *Cfc* to *E*, soils are thin and take thousands of years to develop, and the air is arctic. We sailed slowly into Glacier Bay, in awe of the spectacular, unspoiled scenery, and turned into a bay filled with ice floes, some of them serving as rafts for sleeping seals. Calving (breaking up) into the bay was the Grand Pacific Glacier, with evidence of its recent recession all around. Less than 300 years ago, all of Glacier Bay was filled with ice; today you can sail miles to the Johns Hopkins (shown here) and Margerie tidewater glaciers' current outer edges. Global warming in action—and a reminder that any map of climate is a still picture of a changing world." © H. J. de Blij

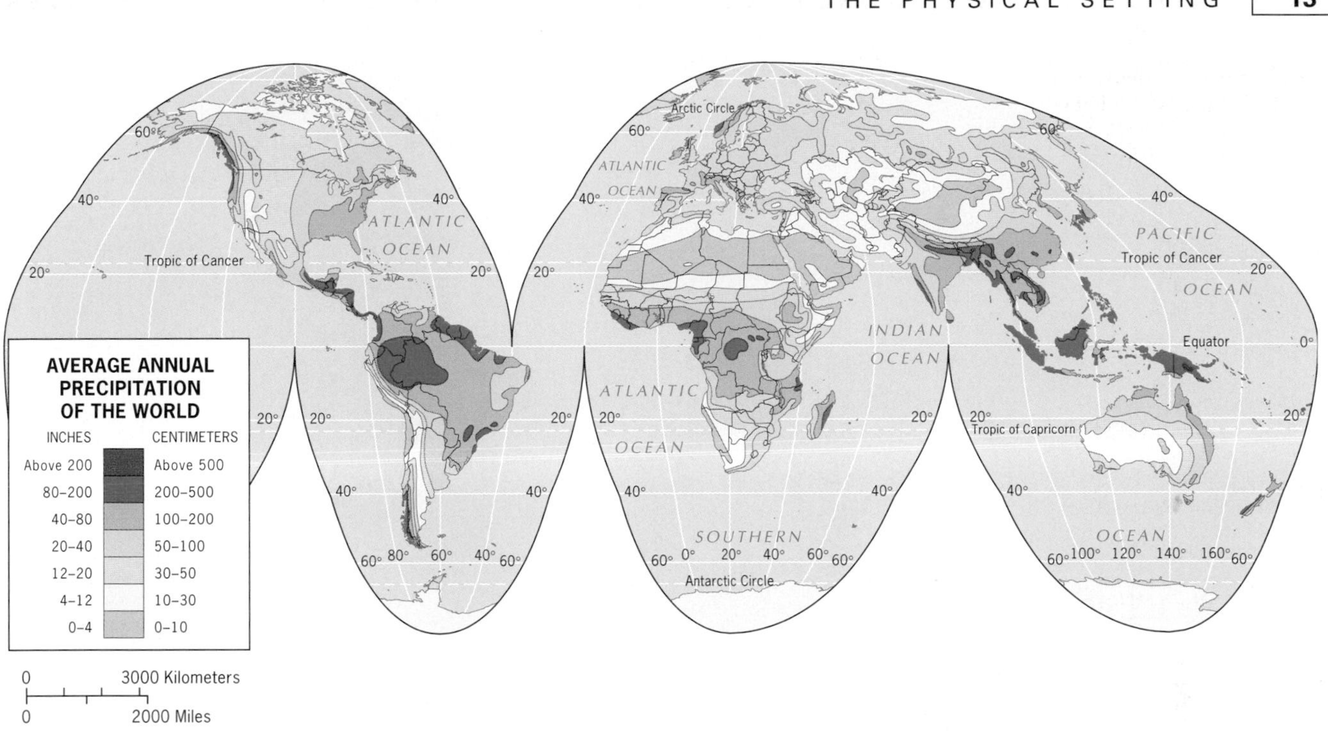

FIGURE G-7 © H. J. de Blij, P. O. Muller, and John Wiley & Sons, Inc.

location relative to coastlines (interiors of continents are hotter in summer, cooler in winter). The size, shape, and terrain of the landmasses complicate the map of average temperatures. Ocean currents also influence the pattern. Warm waters offshore, combined with onshore winds, give most of Europe a warmth that is not experienced at similar latitudes elsewhere.

How can we forge regional coherence out of such a jumble of data? The effort to create a relatively simple, small-scale world map of climates has continued for nearly a century. Figure G-8 is based on the system Wladimir Köppen devised and Rudolf Geiger then modified. This system, which has the advantage of simplicity, is represented by a set of letter symbols. The first (capital) letter is the critical one: the *A* climates are humid and tropical, the *B* climates are arid, the *C* climates are mild and humid, the *D* climates show increasing extremes of seasonal heat and cold, and the *E* climates reflect the frigid conditions at and near the poles.

Figure G-8 merits your attention because familiarity with it will help you understand much of what follows in this book. The map has practical utility, too. Although it
29 depicts **climatic regions**, daily weather in each color-coded region is relatively standard. If, for example, you are familiar with the weather in the large area mapped as *Cfa* in the southeastern United States, you will feel at home in Uruguay (South America), Kwazulu-Natal (South Africa), New South Wales (Australia), and Fujian Province (China). Let us look at the world's climatic regions in some detail.

Humid Equatorial (A) Climates

The humid equatorial, or tropical, climates are characterized by high temperatures all year and by heavy precipitation. In the *Af* subtype, the rainfall arrives in substantial amounts every month; but in the *Am* areas, the arrival of the annual wet *monsoon* (the Arabic word for "season" [see p. 384]) marks a sudden enormous increase in precipitation. The *Af* subtype is named after the vegetation that develops there—the tropical rainforest. The *Am* subtype, prevailing in part of peninsular India, in a coastal area of West Africa, and in sections of Southeast Asia, is appropriately referred to as the monsoon climate. A third tropical climate, the savanna (*Aw*), has a wider daily and annual temperature range and a more strongly seasonal distribution of rainfall. As Figure G-7 indicates, savanna rainfall totals tend to be lower than those in the rainforest zone, and savanna seasonality is often expressed in a "double maximum." Each year produces two periods of increased rainfall separated by pronounced dry spells. In many savanna zones, inhabitants refer to the "long rains" and the "short rains" to identify those seasons; a persistent problem is the unpredictability of the rain's arrival. Savanna soils are not among the most fertile, and when the rains fail hunger looms. Savanna regions are far more densely peopled than rainforest areas, and millions of residents of the savanna subsist

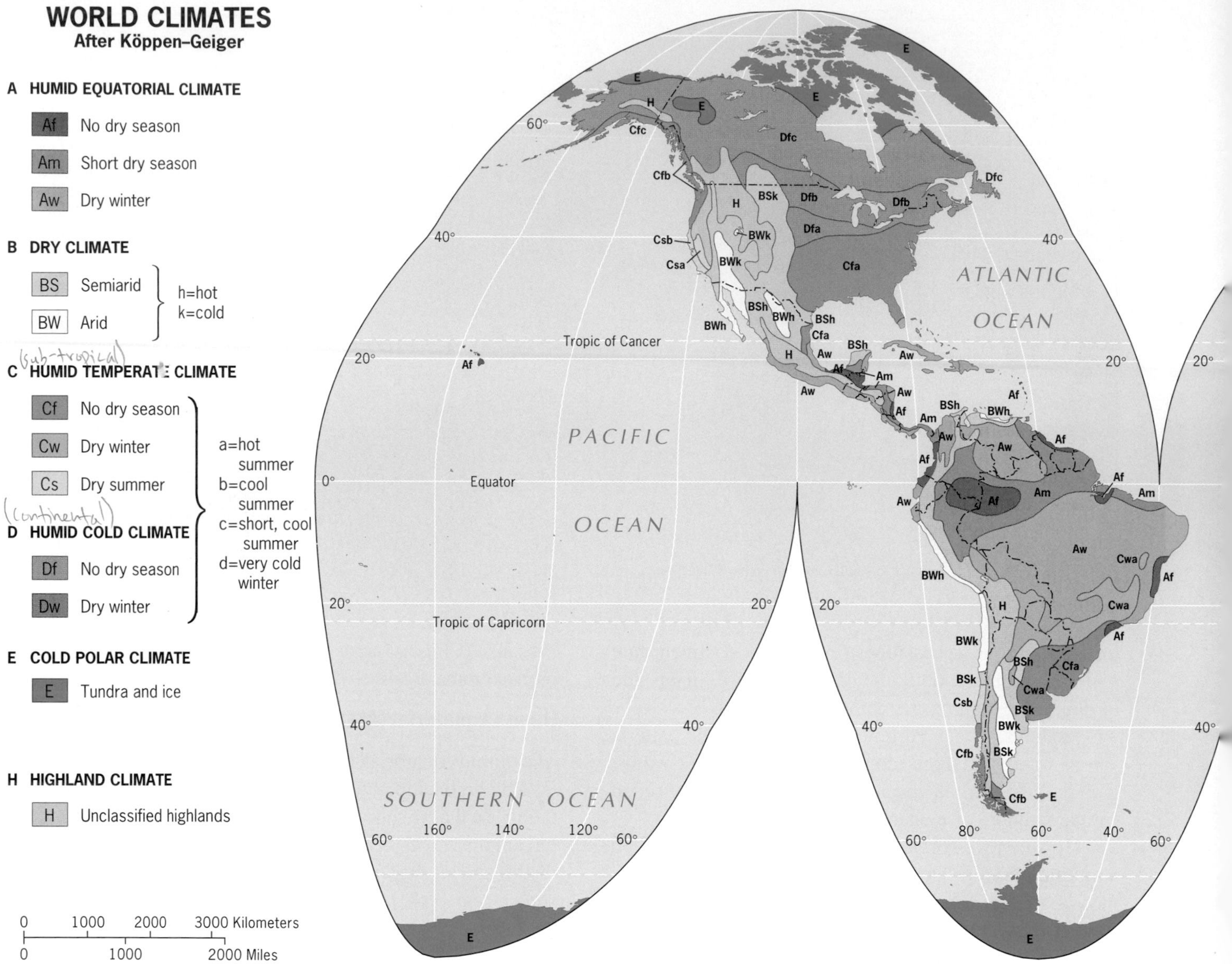

FIGURE G-8 © H. J. de Blij, P. O. Muller, and John Wiley & Sons, Inc.

on what they cultivate. Rainfall variability is their principal environmental problem.

Dry (B) Climates

Dry climates occur in both lower and higher latitudes. The difference between the *BW* (true desert) and the moister *BS* (semiarid steppe) varies but may be taken to lie at about 10 inches (25 cm) of annual precipitation. Parts of the central Sahara in North Africa receive less than 4 inches (10 cm) of rainfall. Most of the world's arid areas have an enormous daily temperature range, especially in subtropical deserts. In the Sahara, there are recorded instances of a maximum daytime shade temperature of over 120°F (49°C) followed by a nighttime low of 48°F (9°C). Soils in these arid areas tend to be thin and poorly developed; soil scientists have an appropriate name for them—aridisols.

Humid Temperate (C) Climates

As the map shows, almost all these mid-latitude climate areas lie just beyond the Tropics of Cancer and Capricorn (23½° North and South latitude, respectively). This is the prevailing climate in the southeastern United States from Kentucky to central Florida, on North America's west coast, in most of Europe and the Mediterranean, in southern Brazil and northern Argentina, in coastal South Africa, in eastern Australia, and in eastern China and southern Japan. None of these areas suffers

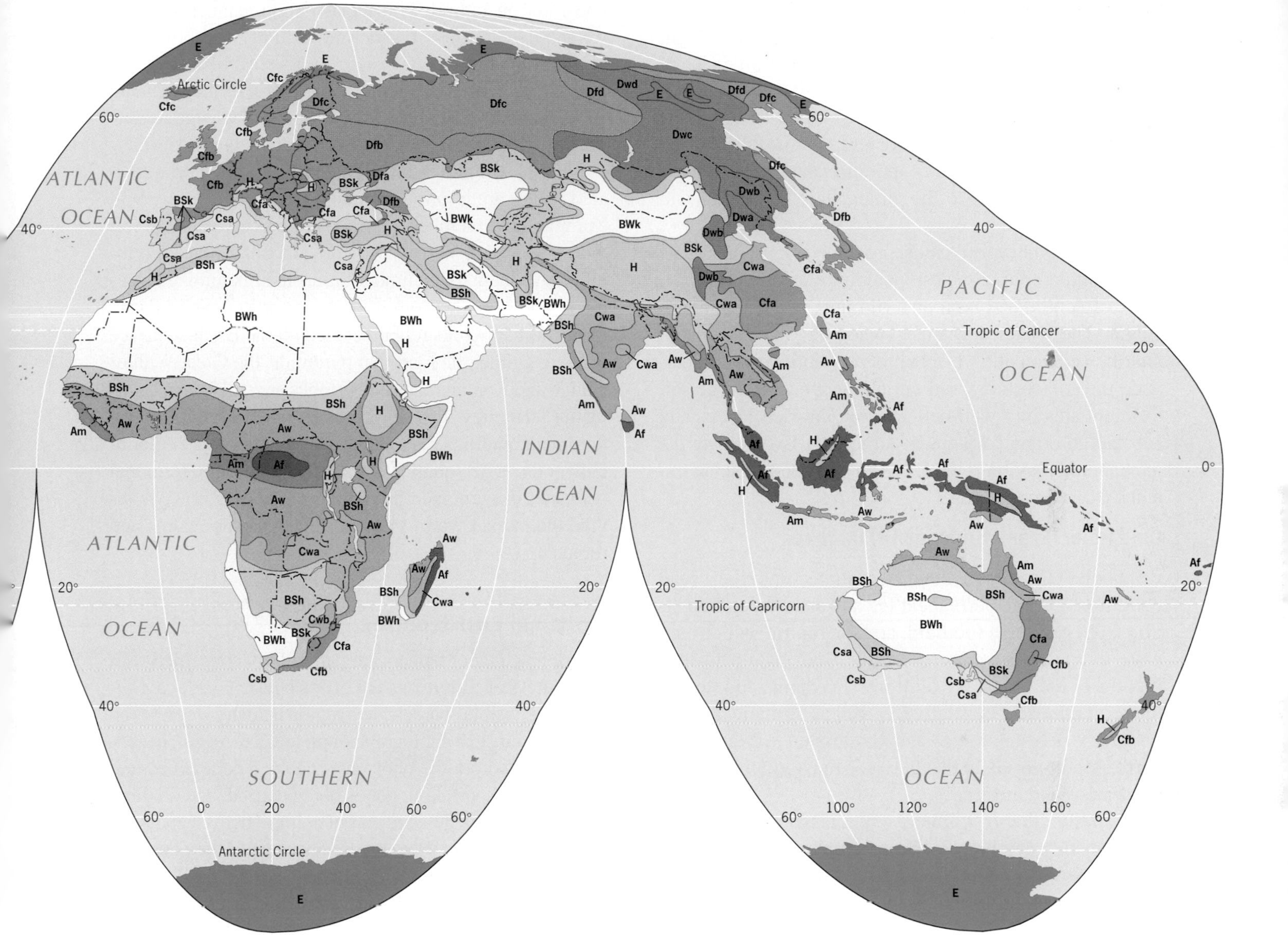

climatic extremes or severity, but the winters can be cold, especially away from water bodies that moderate temperatures. These areas lie midway between the winterless equatorial climates and the summerless polar zones. Fertile and productive soils have developed under this regime, as we will note in our discussion of the North American and European realms.

The humid temperate climates range from moist, as along the densely forested coasts of Oregon, Washington, and British Columbia, to relatively dry, as in the so-called Mediterranean (dry-summer) areas that include not only coastal southern Europe and northwestern Africa but also the southwestern tips of Australia and Africa, central Chile, and Southern California. In these Mediterranean environments, the scrubby, moisture-preserving vegetation creates a natural landscape different from that of richly green Western Europe.

Humid Cold (D) Climates

The humid cold (or "snow") climates may be called the continental climates, for they seem to develop in the interior of large landmasses, as in the heart of Eurasia or North America. No equivalent land areas at similar latitudes exist in the Southern Hemisphere; consequently, no *D* climates occur there.

Great annual temperature ranges mark these humid continental climates, and cold winters and relatively cool summers are the rule. In a *Dfa* climate, for instance, the warmest summer month (July) may average as high as 70°F (21°C), but the coldest month (January) might average only 12°F (−11°C). Total precipitation, much of it snow, is not high, ranging from over 30 inches (75 cm) to a steppe-like 10 inches (25 cm). Compensating for this paucity of precipitation are cool temperatures that inhibit the loss of moisture from evaporation and evapotranspiration (moisture loss to the atmosphere from soils and plants).

Some of the world's most productive soils lie in areas under humid cold climates, including the U.S. Midwest, parts of southern Russia and Ukraine, and northeastern China. The winter dormancy (when all water is frozen) and the accumulation of plant debris during the fall balance the soil-forming and enriching processes. The soil differentiates into well-defined, nutrient-rich layers, and substantial organic humus accumulates. Even where the annual precipitation is light, this environment sustains extensive coniferous forests.

Cold Polar (E) and Highland (H) Climates

Cold polar (*E*) climates are differentiated into true icecap conditions, where permanent ice and snow keep vegetation from gaining a foothold, and the tundra, which may have average temperatures above freezing up to four months of the year. Like rainforest, savanna, and steppe, the term *tundra* is vegetative as well as climatic, and the boundary between the *D* and *E* climates in Figure G-8 corresponds closely to that between the northern coniferous forests and the tundra.

Finally, the *H* climates—unclassified highlands mapped in gray (Fig. G-8)—resemble the *E* climates. High elevations and the complex topography of major mountain systems often produce near-Arctic climates above the tree line, even in the lowest latitudes such as the equatorial section of the high Andes of South America.

Let us not forget an important qualification concerning Figure G-8: this is a still-picture of a changing scene, a single frame from an ongoing film. Climate is still changing, and less than a century from now climatologists are likely to be modifying the climate maps to reflect new data. Who knows: we may have to redraw even those familiar coastlines. Environmental change is a never-ending challenge.

REGIONS AND CULTURES

Whenever we explore a geographic realm or region, we should assess the physical stage that forms its base—
30 its total physical geography or **physiography**. Still, the realms and regions we discuss in the chapters that follow are defined primarily by human-geographic criteria, and one criterion, culture, is especially significant. Therefore, we should carefully consider the culture concept in a regional context.

When anthropologists define **culture**, they tend to 31
concentrate on abstractions: learning, knowledge and its transmission, and behavior. Geographers are most interested in how culture and the patterns of behavior associated with it are imprinted on the landscape. Thus we will examine how members of a society perceive and exploit their available resources, how they maximize the opportunities and adapt to the limitations of their natural environment, and how they organize their portion of the Earth. Some human works remain etched on the Earth's surface for a long time: the Egyptian pyramids, the Great Wall of China, and Roman roads and bridges are still in place millennia after they were built. Over time, regions take on dominant qualities that collectively constitute a **regional** 32
character, a personality, a distinct atmosphere. Regional character is a crucial criterion in ascertaining how we divide the human world into major geographic realms and regions.

Cultural Landscapes

As we noted, geographers are particularly concerned with the impress of culture on the Earth's physical surface. Culture gives visible character to a region in many ways. Often a single scene in a photograph or picture can reveal, in general terms, where the photo was taken. The architecture, forms of transportation, goods being carried, and clothing of the people (all these are part of culture) help us guess the region in the photo. We can do this because the people of any culture are active agents of change: they transform the land they occupy by building structures, creating lines of transport and communication, parceling out fields, and tilling the soil (among countless other activities).

The composite of human imprints on the Earth's surface is called the **cultural landscape**, a term that 33
came into general use in geography during the 1920s. Carl Ortwin Sauer (a professor of geography at the University of California, Berkeley) developed a school of cultural geography that focused on the concept of cultural landscape. In a paper titled "Recent Developments in Cultural Geography" (1927), Sauer proposed his most straightforward definition of the cultural landscape: *the forms superimposed on the physical landscape by the activities of man*. Such forms result from cultural processes—causal forces that shape cultural patterns—that unfold over a long time and involve the cumulative influences of successive occupants.

When it comes to mapping a cultural landscape, its material, tangible properties are the easiest to observe and record. Take, for instance, the urban "townscape"—a

FROM THE FIELD NOTES

"The Atlantic-coast city of Bergen, Norway displayed the Norse cultural landscape more comprehensively, it seemed, than any other Norwegian city, even Oslo. The high-relief site of Bergen creates great vistas, but also long shadows; windows are large to let in maximum light. Red-tiled roofs are pitched steeply to enhance runoff and inhibit snow accumulation; streets are narrow and houses clustered, conserving warmth . . . The coastal village of Mengkabong on the Borneo coast of the South China Sea represents a cultural landscape seen all along the island's shores, a stilt village of the Bajau, a fishing people. Houses and canoes are built of wood as they have been for centuries. But we could see some evidence of modernization: windows filling wall openings, water piped in from a nearby well." © H. J. de Blij

prominent element of the overall cultural landscape—and compare a major U.S. city with, say, a major Japanese city. Photographs of these two metropolitan scenes would reveal the differences quickly, of course, but so would maps. The American city with its square-grid layout of the
34 **central business district (CBD)** and its widely dispersed, now heavily urbanized suburbs contrasts sharply with the clustered, space-conserving Japanese metropolis. In a rural example, the spatially lavish subdivision and ownership patterns of American farmland look unmistakably different from the traditional African countryside, with its irregular, often tiny patches of land surrounding a village. Still, the totality of a cultural landscape can never be captured by a photograph or map because the personality of a region involves more than its prevailing spatial organization. One must also include the region's visual appearance, its noises and odors, the shared experiences of its inhabitants, and even their pace of life.

Culture and Ethnicity

Language, religion, and other cultural traditions often are durable and persistent. Culture is not necessarily based on
35 **ethnicity**, however, and as we study the world's human-geographic regions, we should be aware that peoples of different ethnic stocks can achieve a common cultural landscape, while people of the same ethnic background can be divided along cultural lines.

The recent events in the former Eastern European country of Yugoslavia are a case in point. As the old Yugoslavia broke up in the early 1990s, its component parts were first affected and then engulfed by what was often described as "ethnic" conflict. When the crisis reached Bosnia-Herzegovina in the heart of Yugoslavia, three groups fought a bitter civil war. These groups were identified as Bosnian Muslims, Serbs, and Croats—but in fact all were ethnic Slavs (Yugoslavia means "Land of the South Slavs"). What distanced them from one another, and what kept the conflict going, was cultural tradition, not ethnicity. The Bosnian Muslims and the Serbs had developed different communities in different parts of the former Yugoslavia. The Muslims were descendants of Slavs who, a century ago or more, had been converted by the Turks (who once ruled here) from Christianity to Islam. Each of these groups feared domination by the others, and so South Slav turned against South Slav in what was, in truth, a culture-based conflict.

Even though the post–Cold War world is still young, it has already witnessed numerous intraregional conflicts; in later chapters, we will explain some of these conflicts in geographic context. Not all of them have ethnicity at their roots. Culture is a great unifier; it can also be a powerful divider.

REALMS OF POPULATION

Earlier we noted that population numbers by themselves do not define geographic realms or regions. Population distributions, and the functioning society that gives them common ground, are more significant criteria. That is why we can identify one geographic realm (the Austral) with just 25 million people and another (East Asia) with more than 1.5 billion inhabitants. Neither population numbers nor territorial size alone can delimit a geographic realm. Nevertheless, the map of world population distribution (Fig. G-9) suggests the relative location of several of the world's geographic realms, based on the strong clustering of population in certain areas. Before we examine these clusters in some detail, remember that the Earth's human population now totals well over 6 billion—6 thousand million people confined to the landmasses that constitute less than 30 percent of our planet's surface, much of which is arid desert, rugged mountain terrain, or frigid tundra. (Remember that Figure G-9 is an early twenty-first-century still picture of an ever-changing scene: the rapid growth of humankind continues.) After thousands of years of slow growth, world population during the nineteenth and twentieth centuries expanded at an increasing rate. That rate

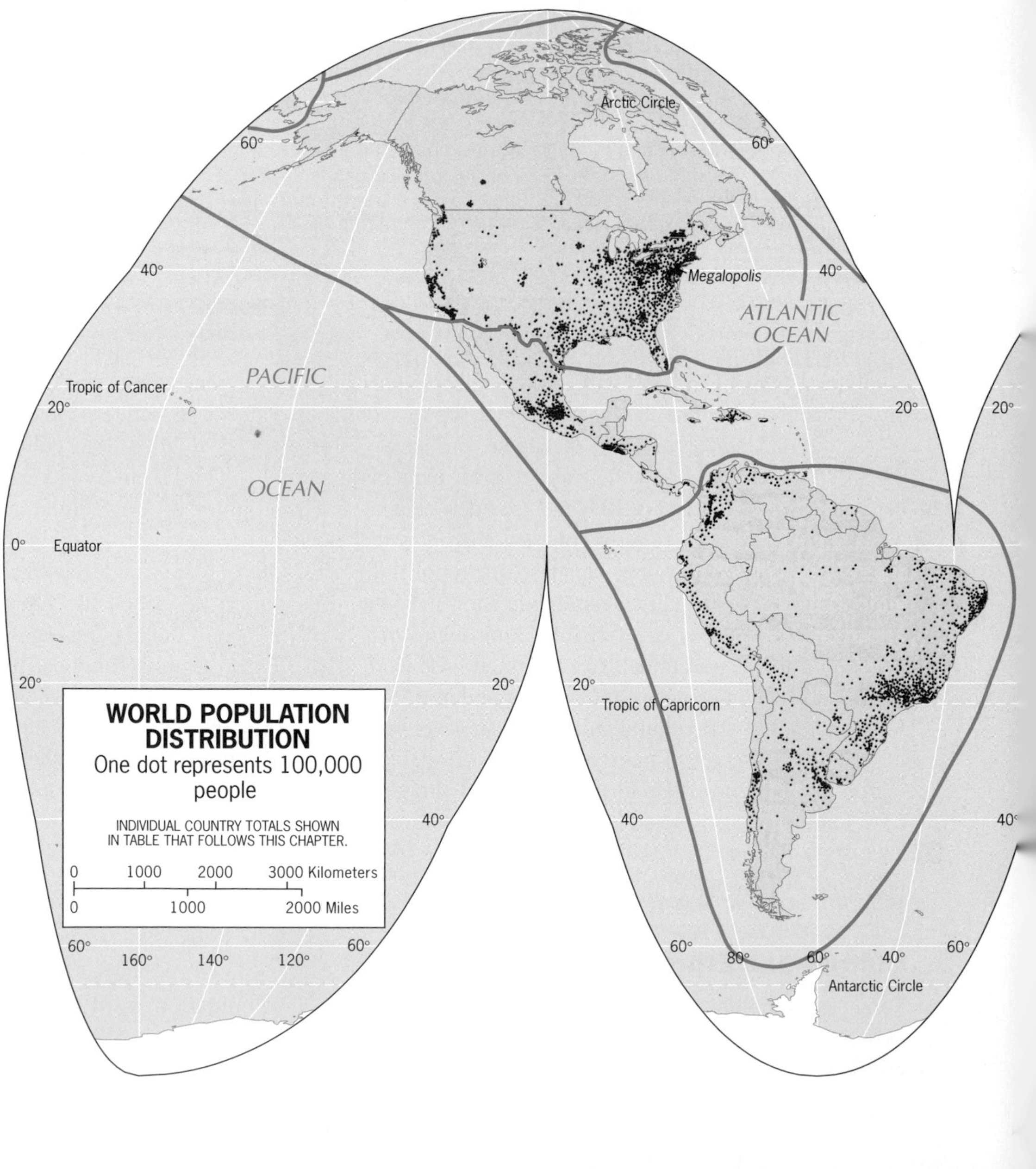

FIGURE G-9 © H. J. de Blij, P. O. Muller, and John Wiley & Sons, Inc.

has recently been slowing down somewhat, but consider this: it took about 17 centuries after the birth of Christ for the world to add 250 million people to its numbers; now we are adding 250 million about every *four years*.

Demographic (population-related) issues will arise repeatedly in later chapters as we survey the world's most crowded realms. For the moment, we will confine ourselves to the overall global situation. For that purpose
36 let us compare the map of world **population distribution** to the demographic data shown in Table G-1 (pp. 35–41). This table provides information about the total populations in each geographic realm as well as their individual countries. Also, if you compare Figure G-9 to the maps of terrain (Fig. G-1) and climate (Fig. G-8), you will note that some of the largest population concentrations lie in the fertile basins of major rivers (China's Huang He and Chang Jiang, India's Ganges). We live in a modern world, but old ways of life still prescribe where hundreds of millions on this Earth live.

Major Population Clusters

The world's greatest population cluster, *East Asia*, lies centered on China and includes the Pacific-facing Asian coastal zone from the Korean Peninsula to Vietnam. The map indicates that the number of people per unit area—the **population density**—tends to decline from the coastal 37
zone toward the interior. Note, however, the ribbon-like extensions marked ***A*** and ***B*** (Fig. G-9). As the map of world natural landscapes (Fig. G-1) confirms, these are populations concentrated in the valleys of China's major rivers, the Huang He (Yellow) and the Chang Jiang

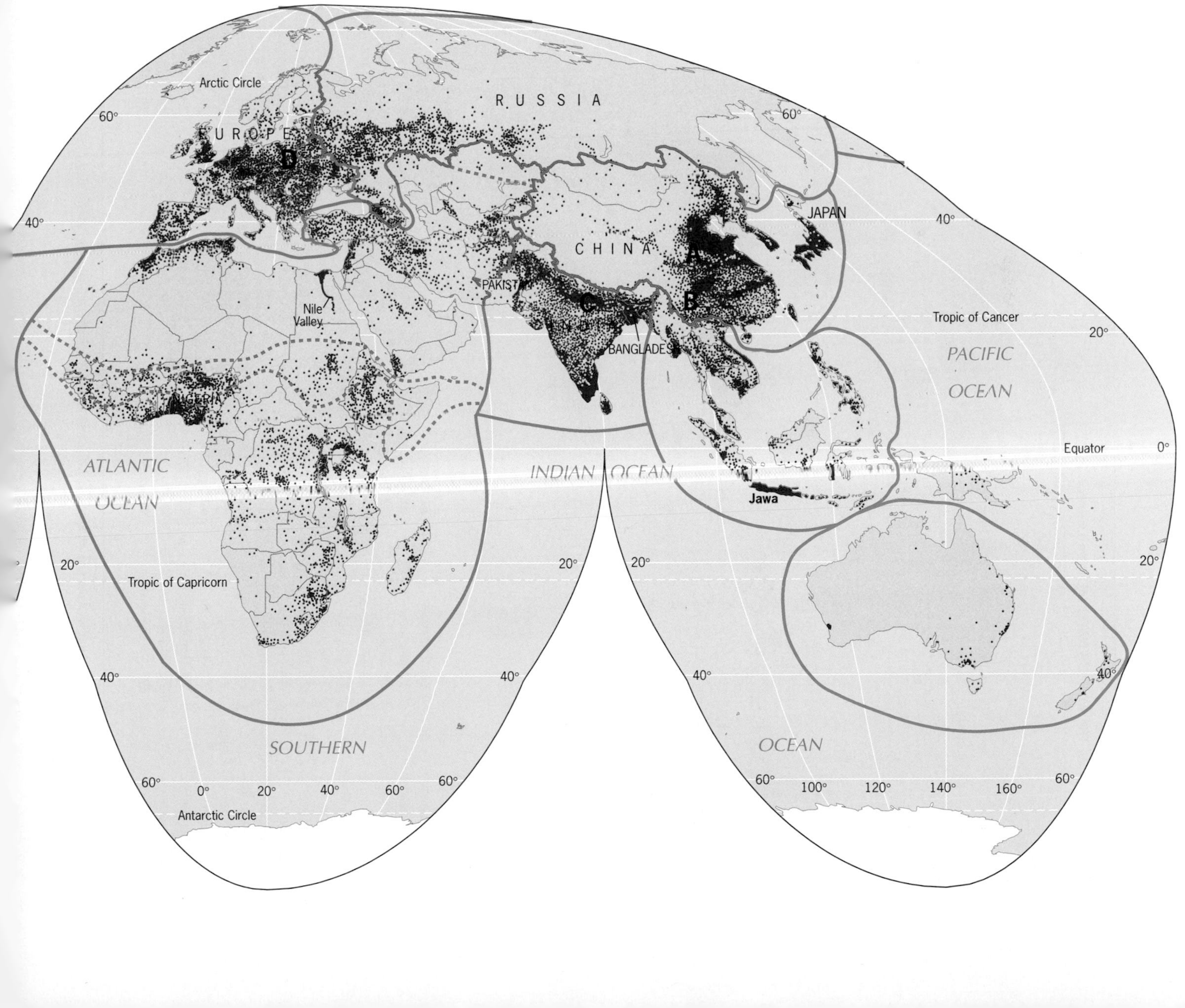

CARTOGRAM OF THE WORLD'S NATIONAL POPULATIONS, 2006

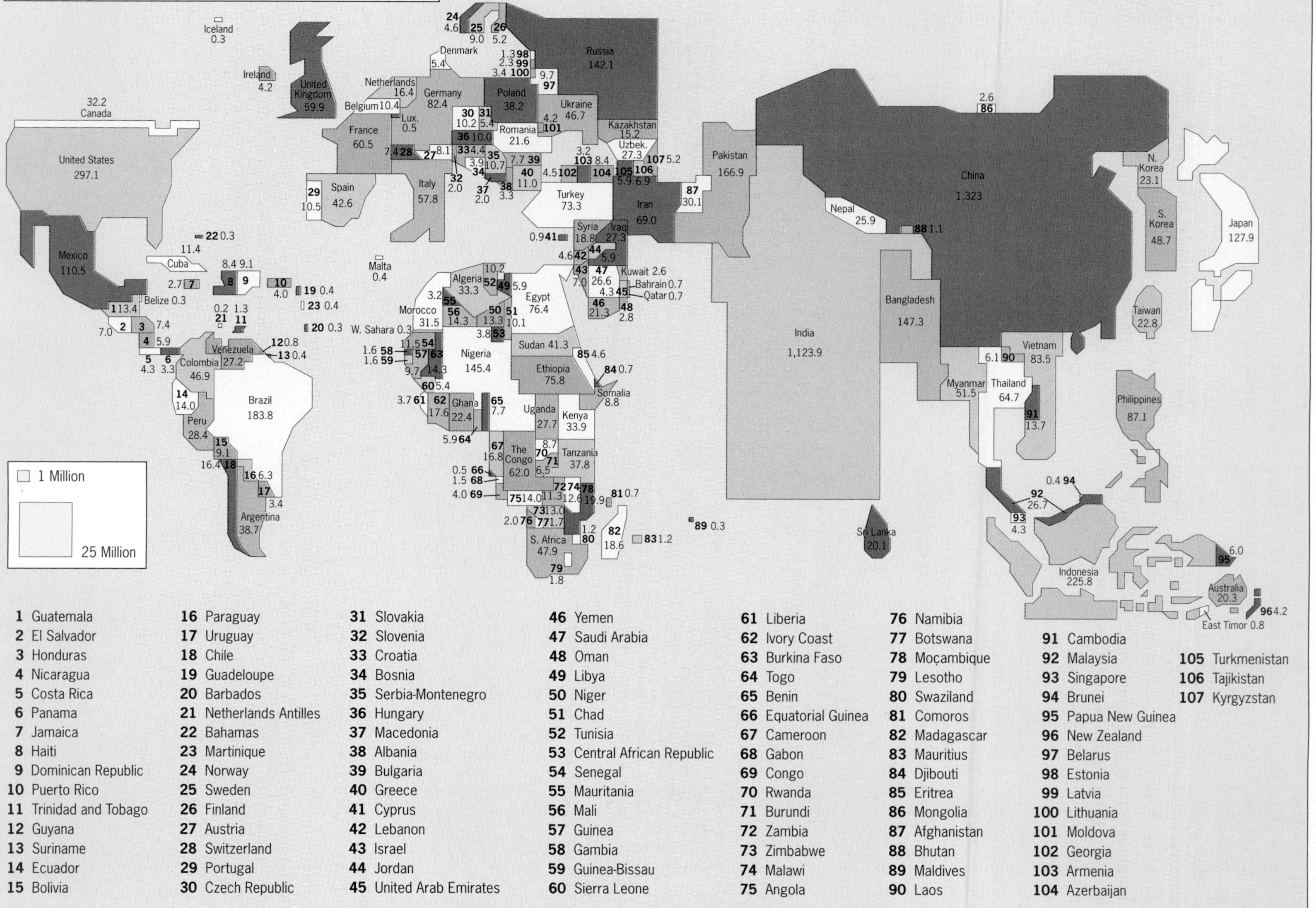

1 Guatemala
2 El Salvador
3 Honduras
4 Nicaragua
5 Costa Rica
6 Panama
7 Jamaica
8 Haiti
9 Dominican Republic
10 Puerto Rico
11 Trinidad and Tobago
12 Guyana
13 Suriname
14 Ecuador
15 Bolivia
16 Paraguay
17 Uruguay
18 Chile
19 Guadeloupe
20 Barbados
21 Netherlands Antilles
22 Bahamas
23 Martinique
24 Norway
25 Sweden
26 Finland
27 Austria
28 Switzerland
29 Portugal
30 Czech Republic
31 Slovakia
32 Slovenia
33 Croatia
34 Bosnia
35 Serbia-Montenegro
36 Hungary
37 Macedonia
38 Albania
39 Bulgaria
40 Greece
41 Cyprus
42 Lebanon
43 Israel
44 Jordan
45 United Arab Emirates
46 Yemen
47 Saudi Arabia
48 Oman
49 Libya
50 Niger
51 Chad
52 Tunisia
53 Central African Republic
54 Senegal
55 Mauritania
56 Mali
57 Guinea
58 Gambia
59 Guinea-Bissau
60 Sierra Leone
61 Liberia
62 Ivory Coast
63 Burkina Faso
64 Togo
65 Benin
66 Equatorial Guinea
67 Cameroon
68 Gabon
69 Congo
70 Rwanda
71 Burundi
72 Zambia
73 Zimbabwe
74 Malawi
75 Angola
76 Namibia
77 Botswana
78 Moçambique
79 Lesotho
80 Swaziland
81 Comoros
82 Madagascar
83 Mauritius
84 Djibouti
85 Eritrea
86 Mongolia
87 Afghanistan
88 Bhutan
89 Maldives
90 Laos
91 Cambodia
92 Malaysia
93 Singapore
94 Brunei
95 Papua New Guinea
96 New Zealand
97 Belarus
98 Estonia
99 Latvia
100 Lithuania
101 Moldova
102 Georgia
103 Armenia
104 Azerbaijan
105 Turkmenistan
106 Tajikistan
107 Kyrgyzstan

FIGURE G-10 Adapted and updated with permission from Jan O.M. Broek & John W. Webb, *A Geography of Mankind*, pp. 34–35. © McGraw-Hill Book Co., 1968.

(Yangzi), respectively. Here in East Asia, most people are farmers, not city dwellers. There are great cities in China (such as Beijing and Shanghai), but their total population still is far outnumbered by the farmers—those who live and work on the land and whose crops of rice and wheat feed not only themselves but also the people in those cities.

The *South Asia* population cluster lies centered on India and includes the populous neighboring countries of Bangladesh and Pakistan. This huge agglomeration of humanity focuses on the broad plain of the lower Ganges River (***C*** in Fig. G-9). This cluster is nearly as large as that of East Asia and at present growth rates will overtake East Asia in less than 20 years. As in East Asia, most people are farmers, but in South Asia pressure on the land is even greater, and farming is less efficient. As we note in Chapter 8, the population issue looms large in this realm.

The third-ranking population cluster, *Europe*, also lies on the world's biggest landmass but at the opposite end from China. The European cluster, including western Russia, counts over 700 million inhabitants, which puts it in a class with the two larger Eurasian concentrations—but there the similarity ends. In Europe, the key to the linear, east-west orientation of the axis of population (***D*** in Fig. G-9) is not a fertile river basin but a zone of raw materials for industry. Europe is among the world's most highly urbanized and industrialized realms, its human agglomeration sustained by industries and offices rather than paddies and pastures. The three world population concentrations just discussed (East Asia, South Asia, and Europe) account for more than 3.7 billion of the world's 6.6 billion people. No other cluster comes close to these numbers. The next-ranking cluster, *Eastern North America*, is only about one-quarter the size of the smallest of the Eurasian concentrations. As in Europe, the population in this area is concentrated in major metropolitan complexes; the rural areas are now relatively sparsely settled. The heart of the North American cluster lies in the urban complex that lines the U.S. northeastern seaboard from Boston to Washington, D.C., and includes New York, Philadelphia, and Baltimore—the great multimetropolitan agglomeration that urban geographers refer to as
38 **Megalopolis**.

The quantitative dominance of Eurasia's population clusters is revealed dramatically by a special map trans-
39 formation called a **cartogram** (Fig. G-10). Here the map of the world is redrawn, so that the area of each country reflects *not* territory but population numbers. China and India stand out because, unlike the European population cluster, their populations are not fragmented among many national political entities.

Neither Figure G-9 nor Figure G-10 can tell us how the populations they exhibit are changing. As we will see, varying rates of population growth mark the world's realms and regions; some countries' populations are growing rapidly, while populations in other countries are actually shrinking. Another key factor is **urbanization**, the 40
percentage of the population living in cities and towns. Not only is there much geographic variation in the level of urbanization among realms and regions, but some realms are urbanizing much more rapidly than others. Urbanization is a defining property of geographic regions, and we will explain why they differ so markedly in this respect.

REALMS, REGIONS, AND STATES

Our analysis of the world's regional geography requires data, and it is crucial to know the origin of these data. Unfortunately, we do not have a uniformly sized grid to superimpose over the globe: we must depend on the world's 180-plus countries to report vital information. Irregular as the boundary framework shown on the world map (see Front Endpaper or Fig. G-11) may be, it is all we have. Fortunately, all large and populous countries are subdivided into provinces, States, or other internal entities, and their governments provide information on each of these subdivisions when they conduct their census.

Although we often refer to the political entities on the world map as *countries*, the appropriate geographic term for them is *states*. To avoid confusion with subdivisions of the same name, we capitalize internal States (as in the preceding paragraph). Thus we refer to the State of California and the State of Victoria (Australia), but to the state of Japan and the state of Argentina.

The **state** as a political, social, and economic entity 41
has been developing for thousands of years, ever since agricultural surpluses made possible the growth of large and powerful towns that could command hinterlands and control peoples far beyond their walls. But the modern state is a relatively recent phenomenon. The boundary framework we see on the world political map today substantially came about during the nineteenth century, and the independence of dozens of former colonies (which made them states as well) occurred during the twentieth century. Today, the **European state model**—a clearly and 42
legally defined territory inhabited by a population governed from a capital city by a representative government—prevails in the aftermath of the collapse of colonial and communist empires.

As Figures G-2 and G-11 suggest, geographic realms are mostly assemblages of states, and the borders between realms frequently coincide with the boundaries between countries—for example, between North America and Middle America along the U.S.–Mexico boundary. But a realm boundary can also cut *across* a state, as does the one between Subsaharan Africa and the Muslim-dominated realm of North Africa/Southwest Asia. Here the boundary takes on the properties of a wide transition zone, but it still divides states such as Chad and Sudan. The transformation of the margins of the former Soviet Union is creating similar cross-country transitions. Newly independent states such as Belarus (between Europe and Russia) and Kazakhstan (between Russia and Muslim Southwest Asia) lie in zones of regional change.

Most often, however, geographic realms consist of groups of states whose boundaries also mark the limits of the realms. Look at Southeast Asia, for instance. Its northern border coincides with the political boundary that separates China (a realm practically unto itself) from Vietnam, Laos, and Myanmar (Burma). The boundary between Myanmar and Bangladesh (which is part of the South Asian realm) defines its western border. Here, the state boundary framework helps delimit geographic realms.

The global boundary framework is even more useful in delimiting regions *within* geographic realms. We shall discuss regional divisions every time we introduce a geographic realm, but an example is appropriate here. In the Middle American realm, we recognize four regions. Two of these lie on the mainland: Mexico, the giant of the realm, and Central America, which consists of the seven comparatively small countries located between Mexico and the Panama–Colombia border (which is the boundary with the South American realm). Central America often is misdefined in news reports; the correct regional definition is based on the politico-geographical framework.

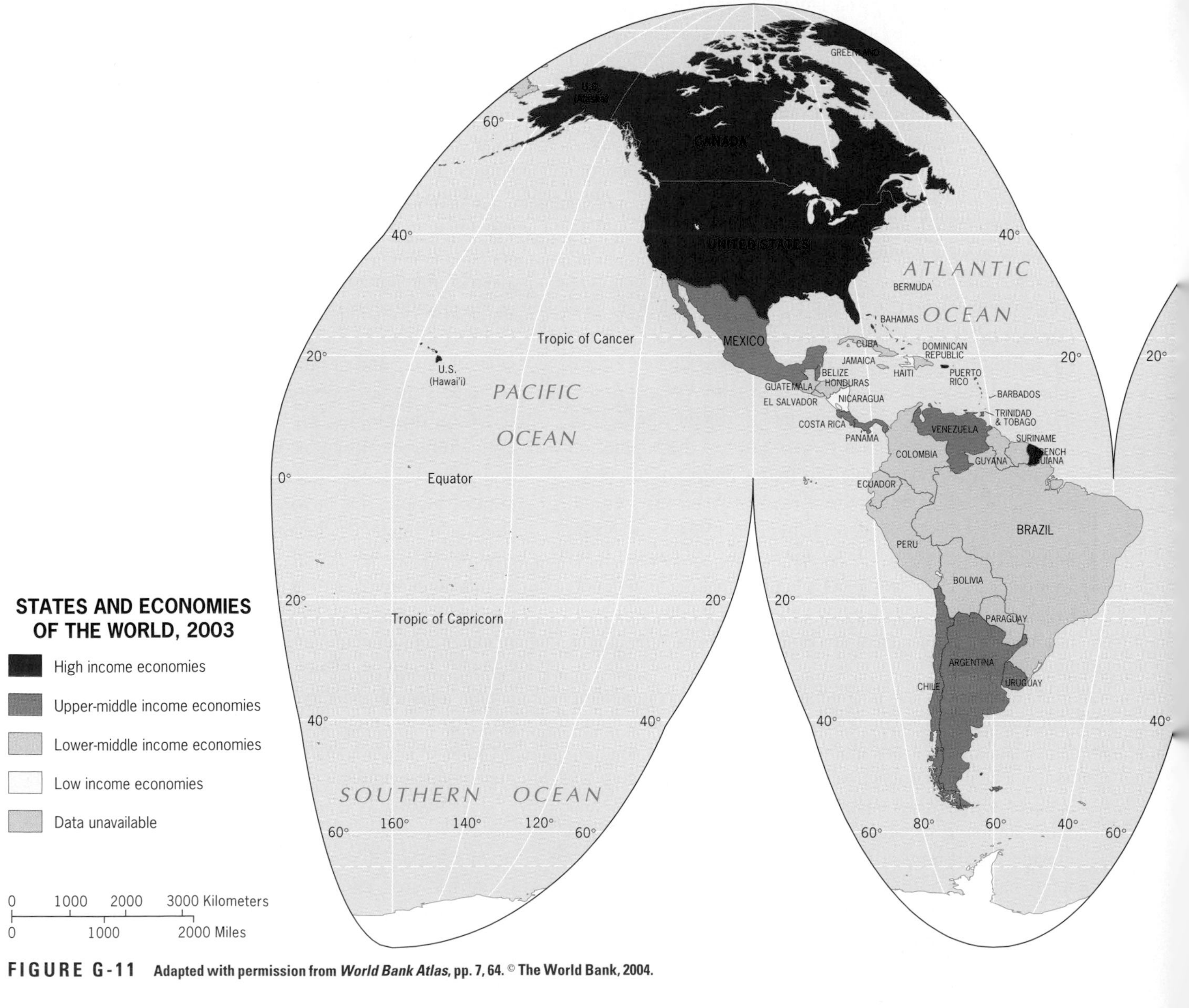

FIGURE G-11 Adapted with permission from *World Bank Atlas*, pp. 7, 64. © The World Bank, 2004.

To our earlier criteria of physiography, population distribution, and cultural geography, therefore, we now add political geography as a determinant of world-scale geographic regions. In doing so, we should be aware that the global boundary framework continues to change and that boundaries are created (as in 1993 between the Czech Republic and Slovakia) as well as eliminated (e.g., between former West and East Germany in 1990). But the overall system, much of it resulting from colonial and imperial expansionism, has endured, despite the predictions of some geographers that the "boundaries of imperialism" would be replaced by newly negotiated ones in the postcolonial period.

Toward the end of this book, we will discuss a recent, ominous development in boundary-making: the extension of boundaries onto and into the oceans and seas. This process has been consuming the last of the Earth's open frontiers, with uncertain consequences.

PATTERNS OF DEVELOPMENT

Finally, we turn to economic geography for criteria that allow us to place countries, regions, and realms in regional groupings based on their level of **development**. 43
Economic geography focuses on spatial aspects of how 44
people make a living; thus it deals with patterns of production, distribution, and consumption of goods and services. As with all else in this world, these patterns exhibit considerable variation. Individual states report

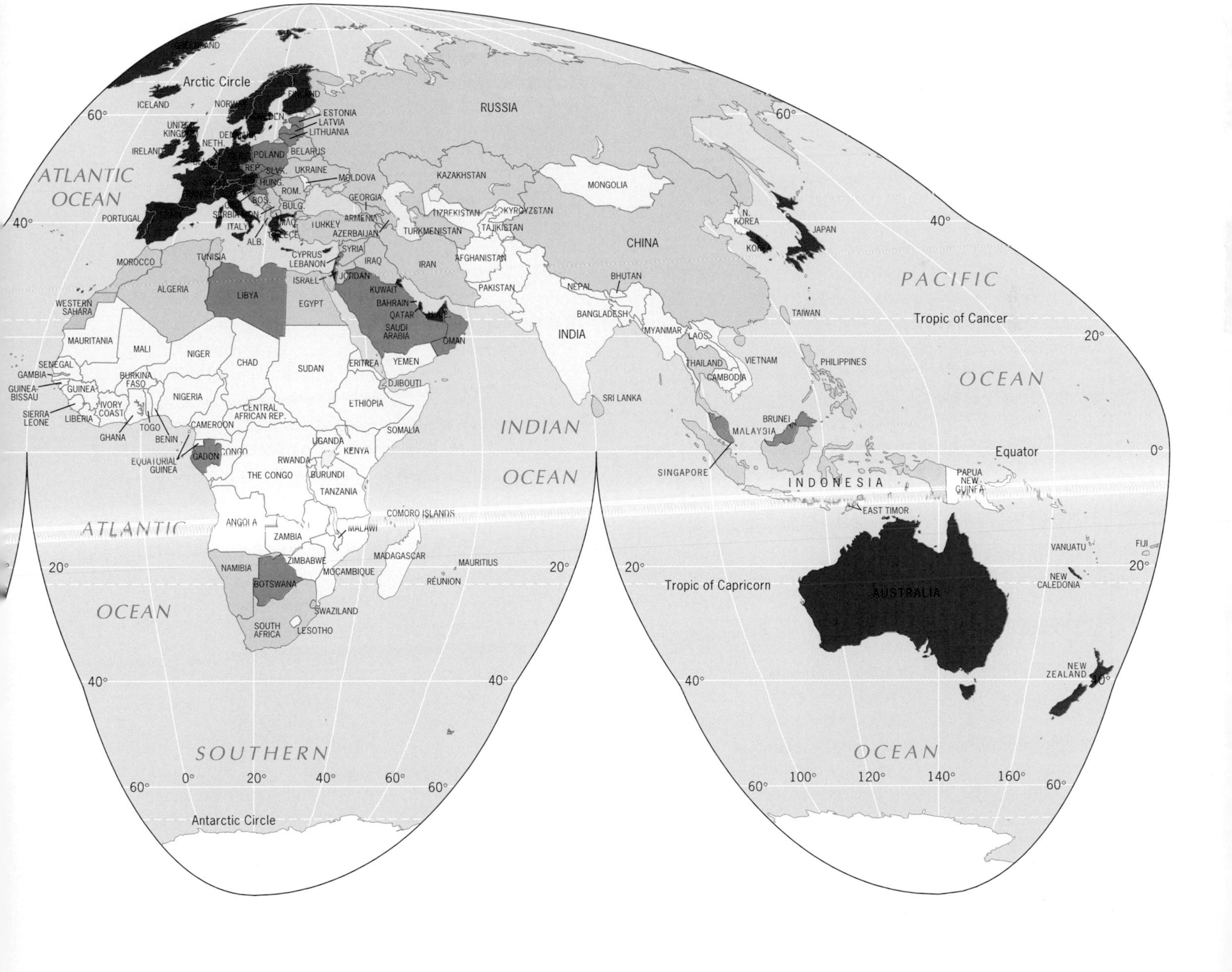

their imports and exports, farm and factory production, and many other economic data to the United Nations and other international agencies. From such information we can determine the comparative economic well-being of the world's countries.

The World Bank, one of the agencies that monitor economic conditions, classifies countries into four groups: (1) high-income, (2) upper-middle-income, (3) lower-middle-income, and (4) low-income countries. These groupings display regional clustering when mapped (Fig. G-11). The higher-income economies are concentrated in Europe, North America, and along the western Pacific Rim, most notably in Japan and Australia. The low-income countries dominate in Africa and parts of Asia.

As Figure G-11 shows, certain geographic-realm boundaries can be discerned on this economic map, including those between North America and Middle America, as well as Australia and Southeast Asia. Within the great geographic realms, you can also see some distinct regional boundaries: between Western and Eastern Europe,

Core-Periphery Regional Relationships

Ours is a world of uneven development. The advantages lie with the developed countries (DCs), and economic indicators show that the gap between the DCs and many underdeveloped countries (UDCs) is growing. How has this situation become so extreme? And what perpetuates it?

Regional contrasts in development have been attributed to many factors. Climate has been blamed (as a conditioner of human capacity), as has cultural heritage (e.g., resistance to innovation), colonial exploitation, and, more recently, neocolonialism. Obviously, the distribution of accessible natural resources and the territorial location of countries play a major part. Certain areas have always had more opportunities for interaction and exchange than others. Isolation from the mainstreams of change has always been a disadvantage. These factors, to a greater or lesser degree depending on location, have been important.

One hallmark of national spatial organization is the evolution of *core areas*, foci of human activity that function as the leading regions of control and change. Today we tend to associate this regional concept with a country's heartland—its largest population cluster, most productive and influential region, the area possessing the greatest centrality and accessibility that usually contains a powerful capital city.

This is nothing new. Archeologists have unearthed evidence of the earliest states, proving that city-dominated cores formed their foci. These were small urban centers, but later they were recast as city-states, culture hearths, seats of empires, and sources of modern technological revolutions. But such cores would not have developed without contributions from their surrounding areas. One of the earliest developments in ancient cities was the creation of organized armed forces to help rulers secure taxes and tribute from the people living in the countryside. Thus core and *periphery* (margin) became functionally tied together: the core had requirements, and the periphery supplied them.

Have modern times ended this core-periphery relationship? The answer not only is no—the situation has worsened. You can see core (have) and periphery (have-not) relationships in every country in the world. In many African countries (Kenya, Senegal, and Zimbabwe, to name examples in three different regions), the former colonial capital now lies at the focus of the national core area. Its tall buildings and government centers symbolize the concentration of privilege, power, and control. Here are the trappings of development, the paved roads and streetlights, hospitals and schools, corporate headquarters and businesses, industries and markets. Farms producing high-priced goods, especially perishables, for the city dwellers who can afford to buy them surround the capital. But farther out into the countryside, life changes. In the periphery, the core area is distant, even ominous—always a concern. How much will markets in the core pay for crops? How much will goods made there, goods that the countryside needs, cost? How high will taxes be? This core-periphery system keeps regional development contrasts high. The core may appear to be a thriving metropolis and look like a city in a developed country. But in the periphery underdevelopment reigns.

Looking at the world from this perspective, we see that core-periphery systems encompass not just individual countries but the entire globe. Today the whole of Western Europe, North America, and Japan function as core areas. Therefore, what we said about that powerful city-region in an otherwise underdeveloped country also applies, at the macroscale, to these global core areas. Power, money, influence, and decision-making organizations are concentrated in Europe, North America, and Japan.

Imagine that you, as a geographer, are asked to locate a factory (say, a textile plant) to sell products in core-area markets. Because salaries and wages in the core area are high, you look for an alternative site in the periphery, where wages are much lower. Raw materials also can be bought locally. So you advise the corporation to establish its factory in a certain country in the periphery, thereby initiating a relationship between core-based business and periphery-based producers. Now, however, the corporate management is concerned about political stability in that periphery country. Soon you are talking to govern-

between the oil-rich Arabian Peninsula and the countries of the Middle East to its northwest, and between South America's Southern Cone (Chile–Argentina–Uruguay) and its less affluent northern neighbors.

Figure G-11 reveals the level of economic development among the world's countries—based, we should remind ourselves, on the political framework as the reporting grid. Until recently, economists and economic geographers used this basis to divide the world into developed countries (DCs) and underdeveloped countries (UDCs). But that distinction has lost much of its relevance, for reasons Figure G-11 cannot show. Today, the world has **core areas** where the richer countries are clustered. 45
But many countries, whatever their level of income, display cores of development (where they often resemble the richest societies) as well as **peripheries** of poverty and un- 46
derdevelopment (see box titled "Core-Periphery Regional Relationships"). Even comparatively wealthy countries still have areas of severe underdevelopment. Other countries, mapped as low-income economies, have given rise to

ment representatives about guarantees, and the economic connection also becomes a political one.

Is all this necessarily disadvantageous to the countries in the periphery, the mainly underdeveloped countries, the countries of the disadvantaged world? After all, the core-area corporations and businesses make investments, stimulate economies, and put people to work in the countries of the periphery. However, a dependency relationship develops that soon diminishes any such advantage. The profits earned return mostly to the core area and benefit the periphery only minimally. When you see that ultramodern high-rise hotel towering above the sandy beaches of a Caribbean island and watch its staff at work, you observe both sides of it. An investment was made, and jobs were created. But the hotel's profits go back to the multinational corporation's bank account—in New York, Tokyo, or London.

Another problem concerns the sociopolitical effects. That textile factory we mentioned needs a manager and other administrative personnel. These people tend to be (or become) part of the elite, the "upper class" in that periphery country, a kind of cadre of the world core area in the disadvantaged realms. Because they represent core-area interests, these people have divided loyalties. What is best for core-area enterprises is not always best for their home country. Such examples of core-periphery interactions abound, and the entire world is now more interconnected than many of us realize. Superimposed on the traditional cultural landscapes of the disadvantaged countries is a growing global network of interaction that reaches into even the smallest village store in the remotest part of a UDC. The numerous geographic realms and regions we will come to recognize in this book are, above all, enmeshed in a worldwide economic-geographic system—a system tightly oriented to the world's core areas.

Thus the countries of the periphery find themselves locked within a global economic system over which they have no control. Even countries that possess commodities in high demand in the core (such as the oil of the OPEC states) have difficulty converting their temporary advantage into longer-term parity with core-area powers. Countries whose income depends heavily on the export of raw materials such as strategic minerals (or single crops like bananas or sugar) are at the mercy of those who do the buying. But there are other reasons for the depressing condition of many periphery countries. Like the contemporary city, the world core area constantly siphons off the skilled and professional people from the periphery. How many doctors, engineers, and teachers from India are working in England and the United States? And how badly India now needs such trained people! Every loss is magnified—but oppressive political circumstances in periphery countries contribute to this "brain drain."

The continuing underdevelopment of periphery countries also involves traditional cultural attitudes and values. The introduction of core-area tentacles into outlying regions where traditional culture is strong may cause a reaction, perhaps a resurgence of fundamentalism. This further restricts economic change. The invasions of modern ways can be unsettling. Do not forget that the long traditions of societies in the periphery have generated large bureaucracies that are threatened by the change development entails. Occasionally, the tangible presence of an external (core) installation such as an office building, airline agency, or retail establishment is violently attacked. Such incidents symbolize ideological opposition to the intrusion they represent.

The countries of the periphery, therefore, face severe problems. They are pawns in a global economic game whose rules they cannot change. Their internal problems are intensified by the aggressive involvement of core-area interests. The way they use resources (such as whether to produce food for local consumption or to produce export crops) is strongly affected by foreign interference. They suffer far more from environmental degradation, overpopulation, and mismanagement than core-area countries do. Their inherited disadvantages have grown, not lessened, over time. The widening gaps that result between enriching cores and persistently impoverished peripheries threaten the future of the world.

and Nicaragua, income per person has shrunk, and there globalization is seen as a culprit, not a cure.

Globalization in the economic sphere is proceeding under the auspices of the World Trade Organization (WTO), of which the United States is the leading architect. To join, countries must agree to open their economies to foreign trade and investment and to adhere to a set of economic rules discussed annually at ministerial meetings. By 2004, the WTO had approximately 150 member-states, all expecting benefits from their participation. But the driving forces of globalization, including European countries and Japan, did not always do their part when it came to giving the poorer countries the chances the WTO seemed to promise. The case of the Philippines is often cited: joining the WTO, its government assumed, would open world markets to Filipino farm products, which can be brought to those markets cheaply because farm workers' wages in the Philippines are very low. Economic planners forecast a huge increase in farm jobs, a slow rise in wages, and a major expansion in produce exports. Instead they found themselves competing against American and European farmers who receive subsidies toward the production as well as the export of their products—and losing out. Meanwhile, low-priced, subsidized American corn appeared on Filipino markets. As a result, the Philippine economy lost several hundred thousand farm jobs, wages went down, and WTO membership had the effect of severely damaging its agricultural sector. Not surprisingly, the notion of globalization is not popular among rural Filipinos.

This story has parallels among banana and sugar growers in the Caribbean, cotton and sugar producers in Africa, and fruit and vegetable farmers in Mexico. Economic globalization has not been the two-way street most WTO members, especially the poorer countries, had hoped. Nor is rich-country protectionism confined to agriculture. When cheap foreign steel began to threaten what remains of the U.S. steel industry in 2001, the American government erected tariffs to guard domestic producers against unwanted competition.

When meetings of economic ministers take place in major cities, they attract thousands of vocal and sometimes violent opponents who demonstrate and occasionally riot to express their opposition. This happens because globalization is seen as more than an economic strategy to perpetuate the advantages of rich countries: it is also viewed as a cultural threat. Globalization constitutes Americanization, these protesters say, eroding local traditions, endangering moral standards, and menacing the social fabric. Such views are not confined to the disadvantaged, poorer countries still losing out in our globalizing world: in France the leader of a so-called Peasant Federation became a national hero after his sympathizers destroyed a McDonald's restaurant in what he described as an act of cultural self-defense. McDonald's restaurants, he argued, promote the consumption of junk food over better, more traditional fare. Elsewhere the target is violent American movies or the lyrics of certain American popular music.

Nevertheless, the process of globalization—in cultural as well as economic spheres—continues to change our world, for better and for worse. (In the United States, opponents of globalization bemoan the loss of American jobs to foreign countries where labor is cheaper.) In theory, globalization breaks down barriers to international trade, stimulates commerce, brings jobs to remote places, and promotes social, cultural, political, and other kinds of exchanges. High-tech workers in India are employed by computer firms based in California. Japanese cars are assembled in Thailand. American shoes are made in China. McDonald's restaurants spread standards of service and hygiene as well as familiar (and standard) menus from Tokyo to Tel Aviv. If wages and standards of employment are lower in newly involved countries than in the global core, the gap is shrinking, employment conditions are more open to scrutiny, and global openness is a byproduct of globalization.

We should be aware that the current globalization process is a revolution—but also that it is not the first of its kind. The first "globalization revolution" occurred during the nineteenth and early twentieth centuries, when Europe's colonial expansion spread ideas, inventions, products, and habits around the world. Colonialism transformed the world as the European powers built cities, transport networks, dams, irrigation systems, power plants, and other facilities, often with devastating impact on local traditions, cultures, and economies. From goods to games (soap to soccer) people in much of the world started doing similar things. The largest of all colonial empires, that of Britain, made English a worldwide language, a key element in the current, second globalization process.

The current globalization is even more revolutionary than the colonial phase because it is driven by more modern, higher-speed communications. When the British colonists planned the construction of their ornate Victorian government and public buildings in (then) Bombay, the architectural drawings had to be prepared in London and sent by boat to India. When the Chinese government in the 1980s decided to create a Manhattan-like commercial district on the riverfront in Shanghai, the plans were drawn in the United States, Japan, and Western Europe and transmitted to Shanghai via the Internet. Today, one container ship carrying products from China to the American market hauls more cargo than a hundred colonial-era boats.

And, as the chapters that follow will show frequently, the world's national political boundaries are becoming increasingly porous. Economic alliances enable manufacturers to send raw materials and finished products across borders that once inhibited such exchanges. Groups of countries forge economic unions whose acronyms (NAFTA, Mercosur) stand for freer trade. The ultimate goal of the World Trade Organization is to lower the remaining trade barriers the world over, boosting not just regional commerce but also global trade.

As with all revolutions, the overall consequences of the ongoing globalization process are uncertain. Critics underscore that one of its outcomes is a growing gap between rich and poor, a polarization of wealth that will destabilize the world. Core-periphery contrasts are intensified, not lessened, by globalization as the poor in peripheral societies are exploited by core-based corporations. Proponents argue that, as with the Industrial Revolution, it will take time for the benefits to spread—but that globalization's ultimate effects will be advantageous to all.

Indeed, the world is functionally shrinking, and we will find evidence for this throughout the book. But the "global village" still retains its distinctive neighborhoods, and two revolutionary globalizations have failed to erase their particular properties. In the chapters that follow we use the vehicle of geography to visit and investigate them.

FROM THE FIELD NOTES

"It was an equatorial day here today, in hot and humid Singapore. Walked the three miles to the Sultan Mosque this morning and observed the activity arising from the Friday prayers, then sat in the shade of the large ficus tree on the corner of Arab Street and watched the busy pedestrian and vehicular traffic. An Indian man saw me taking notes and sat down beside me. For some time he said nothing, then asked whether I was writing a novel. 'If only I could!' I answered, and explained that I was noting locations and impressions for photos and text in a geography book. He said that he was Hindu but liked living in this area because some of what he called "Old Singapore" still survived. 'No problems between Hindus and Muslims here," he said. 'The police maintain strict security here around the mosques, especially after September 2001. The government tolerates all religions, but it doesn't tolerate religious conflict.' He offered to show me the Hindu shrine where he and his family joined others from Singapore's Hindu community. As we walked down Arab Street, he pointed skyward. 'That's the future,' he said. 'The symbol of globalization. Do you mention it in your book? It's in the papers every day, supposedly a good thing for Singapore. Well, maybe. But a lot of history has been lost to make space for what you see here.' I took the photo, and promised to mention his point." © H. J. de Blij

THE REGIONAL FRAMEWORK

At the beginning of this Introduction, we outlined a map of the great geographic realms of the world (Fig. G-2). We then addressed the task of dividing these realms into regions, and we used criteria ranging from physical geography to economic geography. The result is Figure G-12. Before we begin our survey, here is a summary of the 12 geographic realms and their regional components.

Europe (1)

Territorially small and politically fragmented, Europe remains disproportionately influential in global affairs. A core geographic realm, Europe has five regions: Western Europe, the British Isles, Northern (Nordic) Europe, Mediterranean Europe, and Eastern Europe.

Russia (2)

Territorially enormous and politically unified, Russia was the dominant force in the former Soviet Union that disbanded in 1991. Undergoing a difficult transition from

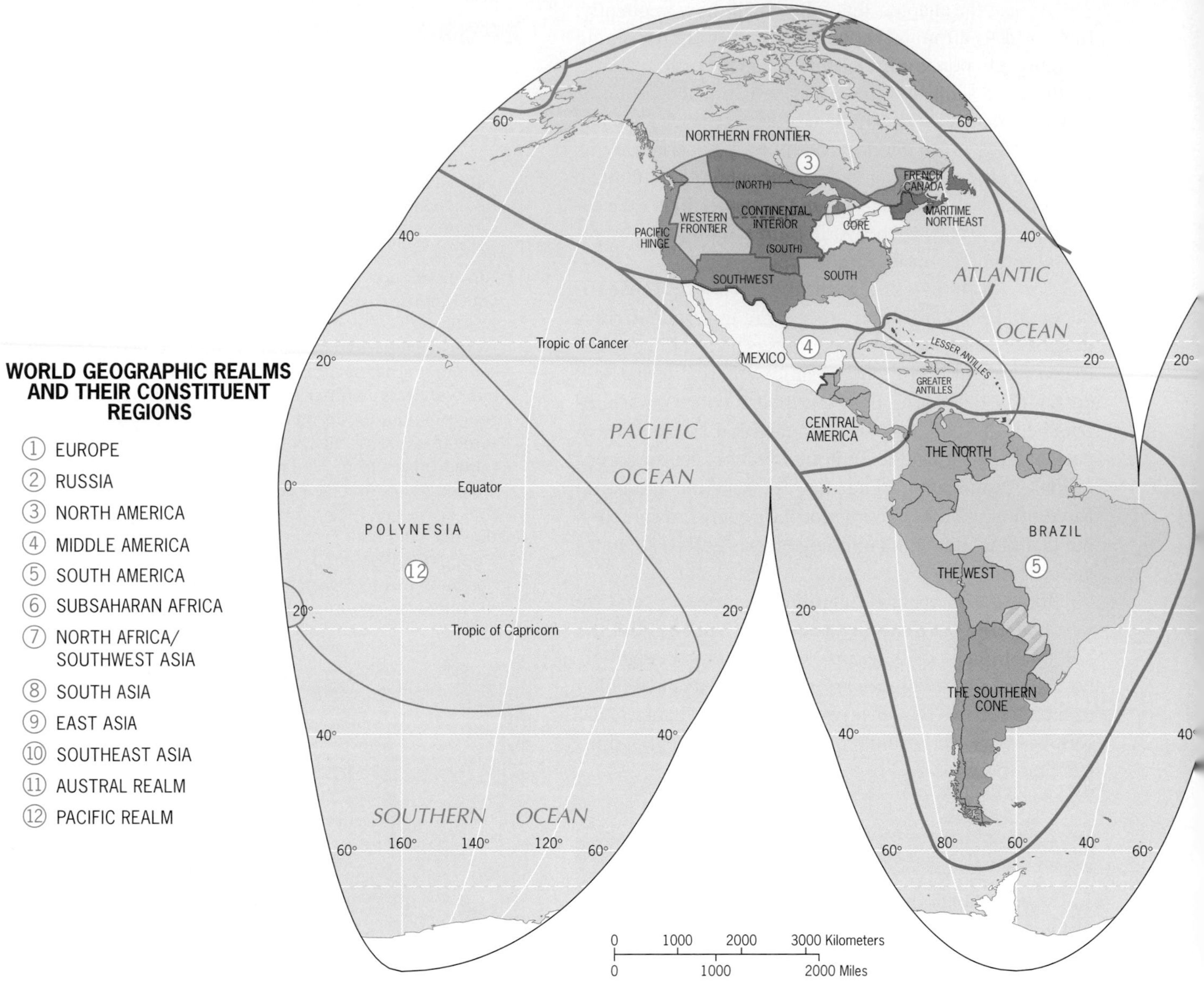

FIGURE G-12 © H. J. de Blij, P. O. Muller, and John Wiley & Sons, Inc.

dictatorship to democracy and from communism to capitalism, Russia is geographically complex and changing. We define four regions: the Russian Core, the Eastern Frontier, Siberia, and the Far East.

North America (3)

Another realm in the global core, North America consists of the United States and Canada. We identify nine regions: the North American Core, the Maritime Northeast, French Canada, the Continental Interior, the South, the Southwest, the Western Frontier, the Northern Frontier, and the Pacific Hinge. Five of these regions extend across the U.S.–Canada border.

Middle America (4)

Nowhere in the world is the contrast between core and periphery as sharply demarcated as it is between North and Middle America. This small, fragmented realm clearly divides into four regions: Mexico, Central America, and the Greater Antilles and Lesser Antilles of the Caribbean.

South America (5)

The continent of South America also defines a geographic realm in which Iberian (Spanish and Portuguese) influences dominate the cultural geography but Amerindian imprints survive. We recognize four regions: the North, composed of Caribbean-facing states; the Andean West, with

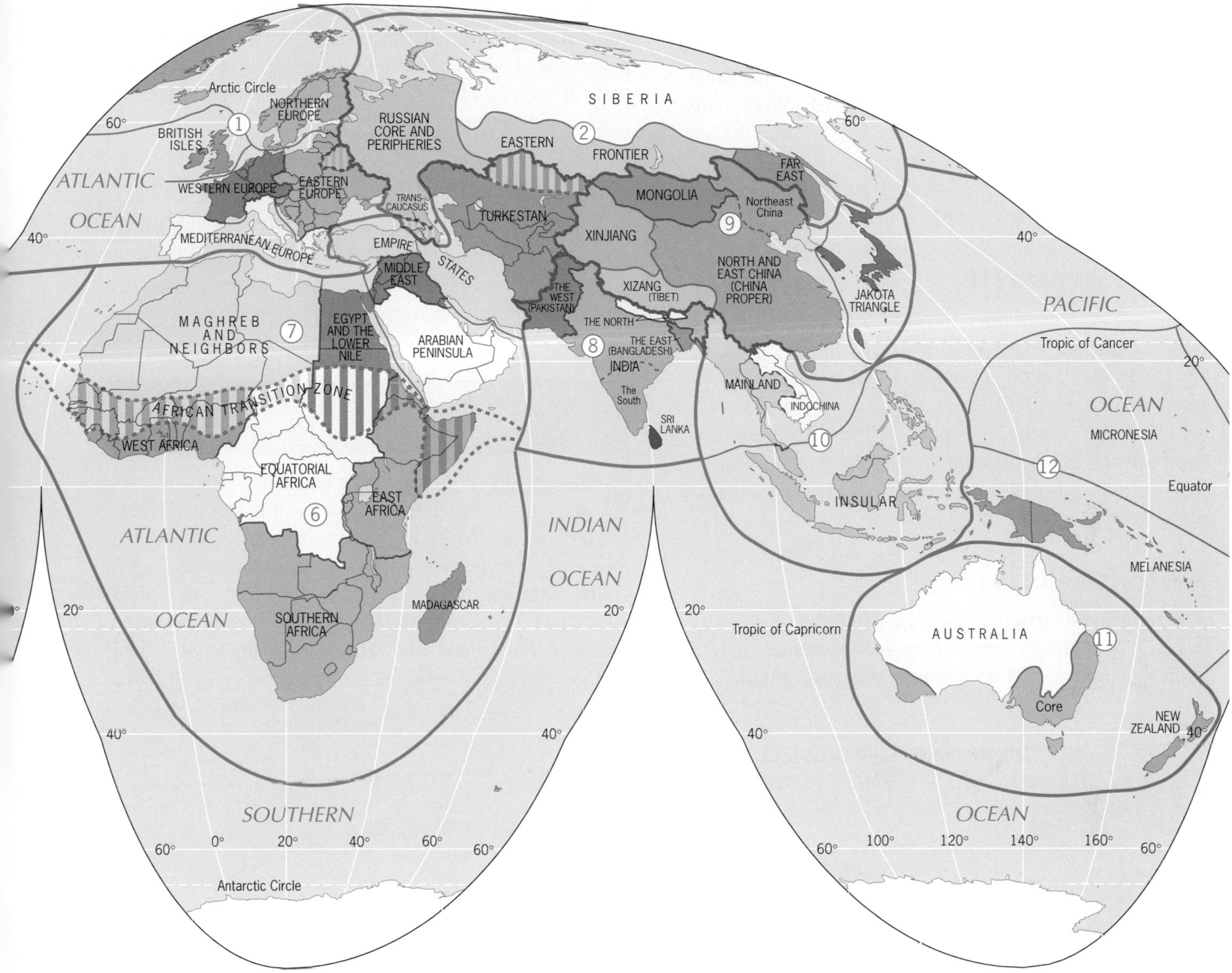

its strong Amerindian influences; the Southern Cone; and Brazil, the realm's giant.

Subsaharan Africa (6)

Between the African Transition Zone in the north and the southernmost Cape of South Africa lies Subsaharan Africa. The realm consists of five regions: Southern Africa, East Africa, Equatorial Africa, West Africa, and the African Transition Zone.

North Africa/Southwest Asia (7)

This vast geographic realm has several names, extending as it does from North Africa into Southwest and, indeed, Central Asia. Some geographers call it *Naswasia* or *Afrasia*. There are six regions: Egypt and the Lower Nile Basin plus the Maghreb in North Africa; the Middle East, the Arabian Peninsula, the Empire States, and Turkestan in Southwest Asia.

South Asia (8)

Physiographically one of the most clearly defined geographic realms, South Asia has a complex cultural geography. It consists of five regions: India at the center; Pakistan to the west; Bangladesh to the east; the mountainous North; and the peninsular South, which includes island Sri Lanka.

East Asia (9)

The vast East Asian geographic realm extends from the deserts of Central Asia to the tropical coasts of the South China Sea and from Japan to Xizang (Tibet). We identify five regions: China Proper, including North Korea; Xizang (Tibet) in the southwest; desert Xinjiang in the west; Mongolia in the north; and the Jakota Triangle (Japan, South Korea, and Taiwan) in the east.

Southeast Asia (10)

Southeast Asia is a varied mosaic of natural landscapes, cultures, and economies. Influenced by India, China, Europe, and the United States, it includes dozens of religions and hundreds of languages plus economies representing both core and periphery. Physically, Southeast Asia consists of a peninsular mainland and an arc consisting of thousands of islands. The two regions (Mainland and Insular) are based on this distinction.

Austral Realm (11)

Australia and its neighbor New Zealand form the Austral geographic realm by virtue of continental dimensions, insular separation, and dominantly Western cultural heritage. The four regions are defined by physical as well as cultural geography: a highly urbanized, two-part core and a vast, desert-dominated interior in Australia; and two main islands in New Zealand that exhibit considerable geographic contrast.

Pacific Realm (12)

The vast Pacific Ocean, larger than all the landmasses combined, contains tens of thousands of islands large and small. Dominant cultural criteria warrant three regions: Melanesia, Micronesia, and Polynesia.

THE PERSPECTIVE OF GEOGRAPHY

As this introductory chapter demonstrates, our world regional survey is no mere description of places and areas. We have combined the study of realms and regions with a look at geography's ideas and concepts—the notions, generalizations, and basic theories that make the discipline what it is. We continue this method in the chapters ahead so that we will become better acquainted with the world and with geography. By now you are aware that geography is a

THE RELATIONSHIP BETWEEN REGIONAL AND SYSTEMATIC GEOGRAPHY

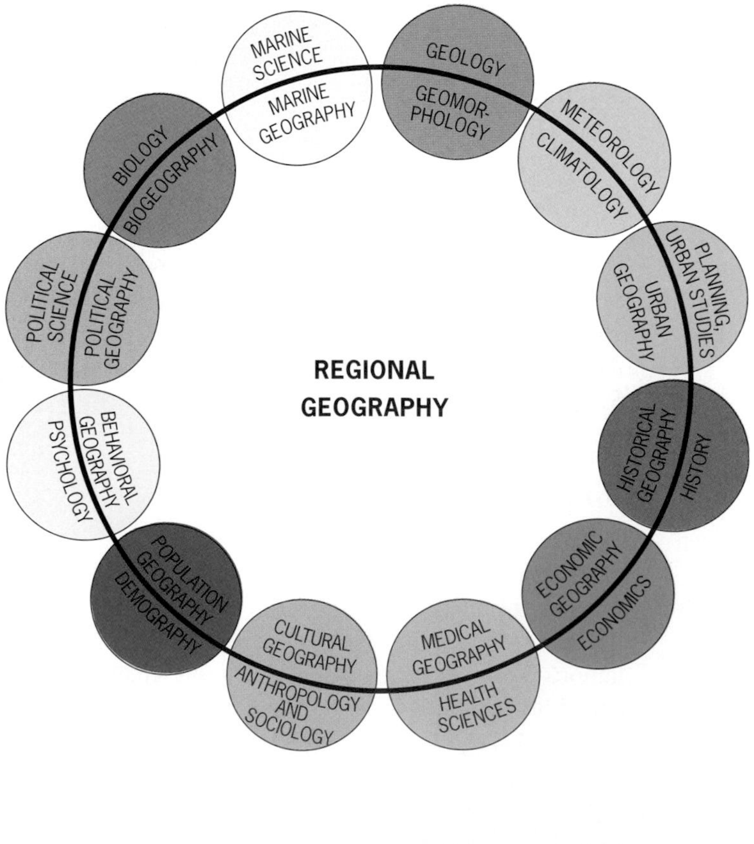

FIGURE G-13 © H. J. de Blij, P. O. Muller, and John Wiley & Sons, Inc.

What Do Geographers Do?

A systematic spatial perspective and an interest in regional study are the unifying themes and enthusiasms of geography. Geography's practitioners include physical geographers, whose principal interests are the study of geomorphology (land surfaces), research on climate and weather, vegetation and soils, and the management of water and other natural resources. There also are geographers whose research and teaching concentrate on the ecological interrelationships between the physical and human worlds. They study the impact of humankind on our globe's natural environments and the influences of the environment (including such artificial contents as air and water pollution) on human individuals and societies.

Other geographers are regional specialists, who often concentrate their work for governments, planning agencies, and multinational corporations on a particular region of the world. Still other geographers—who now constitute the largest group of practitioners—are devoted to topical or systematic subfields such as urban geography, economic geography, and cultural geography (see Fig. G-13). They perform numerous tasks associated with the identification and resolution (through policy-making and planning) of spatial problems in their specialized areas. And, as in the past, there remain many geographers who combine their fascination for spatial questions with technical know-how.

Computerized cartography, geographic information systems, remote sensing, and even environmental engineering are among the specializations listed by the 10,000-plus professional geographers of North America. In Appendix B, you will find considerable information on the discipline, how one trains to become a geographer, and the many exciting career options that are open to the young professional.

wide-ranging, multifaceted discipline. It is often described as a social science, but that is only half the story: in fact, geography straddles the divide between the social and the physical (natural) sciences. Many of the ideas and concepts you will encounter have to do with the many interactions between human societies and natural environments.

51 **Regional geography** allows us to view the world in an all-encompassing way. As we have seen, regional geography borrows information from many sources to create an overall image of our divided world. Those
52 sources are not random. They form topical or **systematic geography**. Research in the systematic fields of geography makes our world-scale generalizations possible. As Figure G-13 shows, these systematic fields relate closely to those of other disciplines. Cultural geography, for example, is allied with anthropology; it is the spatial perspective that distinguishes cultural geography. Economic geography focuses on the spatial dimensions of economic activity; political geography concentrates on the spatial imprints of political behavior. Other systematic fields include historical, medical, behavioral, environmental, agricultural, and coastal geography. We will also draw on information from biogeography, marine geography, population geography, and climatology (as we did earlier in this chapter).

These systematic fields of geography are so named because their approach is global, not regional. Take the geographic study of cities, urban geography. Urbanization is a worldwide process, and urban geographers can identify certain human activities that all cities in the world exhibit in one form or another. But cities also display regional properties. The model Japanese city is quite distinct from, say, the African city. Regional geography, therefore, borrows from the systematic field of urban geography, but it injects this regional perspective.

In the following chapters we call upon these systematic fields to give us a better understanding of the world's realms and regions. As a result, you will gain insights into the discipline of geography as well as the regions we investigate. This will prove that geography is a relevant and practical discipline when it comes to comprehending, and coping with, our fast-changing world (see box above titled "What Do Geographers Do?" and Appendix B for a discussion of career opportunities).

WHAT YOU CAN DO

RECOMMENDATION: If you are not already doing so, please allow us to urge you to begin keeping a diary. Several decades from now, the most interesting and perhaps even useful parts of your diary may not be time and place, but your feelings and opinions about events and experiences you are having and will have over the years to come. What is important enough to warrant recording your own views? The intervention in Iraq? The earthquake-generated tsunami disaster in Indonesia and elsewhere? The impact of globalization? The effect on you of an unusually stimulating course? The comments of a particularly opinionated visiting speaker? Many years from now you may be surprised about your concerns and perceptions in this first decade of our new century.

GEOGRAPHIC CONNECTIONS

1 If you were asked by someone without geographic background to define the geographic characteristics of your home region, how would you respond? Where do you perceive its boundaries to lie, on what criteria do you base these, and are they sharp or transitional? (You will be tempted to draw a sketch map to support this part of your answer.) Where is your home with respect to the urban framework in your region, and do you consider yourself a resident of a city, suburbia, exurbia, or a rural hinterland? What elements of the formal and the functional can you discern in your regional experience? How would you summarize your region's natural environment(s) and, in your lifetime, have you noticed changes that might be considered trends and if so, how have these affected you?

2 Imagine that you are working in your university's study-abroad office. Students will be planning to study for a semester in numerous countries, and they know that you have some geographic experience. How would you answer their questions about the climate and weather they are likely to experience in those countries, using Figure G-8? What should a student from Vancouver in western Canada's British Columbia expect in Auckland, New Zealand? Would a student from Charleston, South Carolina feel at home (weatherwise) in Shanghai, China? What can you tell a student from San Francisco about the weather in Cape Town, South Africa? How would you prepare a student from Phoenix, Arizona for the weather in Paris, France?

3 A well-known economist once wrote the following: "Virtually all of the rich countries of the world are outside the tropics, and virtually all the poor countries are in them . . . climate, then, accounts for quite a significant proportion of the cross-national and cross-regional disparities of world income." Comment on this assertion, using Figures G-8 and G-11 as well as the text and box on pp. 24–25. How might a geographer's perspective on such an issue differ from an economist's?

TABLE G-1 on pages 35–41 is a valuable resource, and should be consulted throughout your reading. Like all else in this book, Table G-1 is subject to continuous revision and modification. In this 12th edition, we have deleted some indices, elaborated others, and introduced new ones. For example, in a world with ever-slower population growth, the so-called Doubling Time index—the number of years it will take for a population to double in size based on its current rate of natural increase—has lost most of its utility. On the other hand, when it comes to **Life Expectancy** and **Literacy**, general averages conceal significant differences between male and female rates, which in turn reflect conditions in individual countries, so we report these by gender. Also in this edition, we introduce the **Corruption Index**, not available for all countries but an important reflection of a global problem. The **Big Mac Price** index, a measure introduced by the journal *The Economist*, tells you much more than what a hamburger with all the trimmings would cost in real dollars in various countries of the world—it also reflects whether those countries' currencies are overvalued or undervalued. And the final column, in which we formerly used to reported the per-capita GNP (Gross National Product), now reveals the **GNI**, that is, the Gross National Income per person and what this would buy in each country. In the language of economic geographers, this is called the GNI-PPP, the per-capita Gross National Income in terms of its Purchasing Power Parity.

Indexes that may not be immediately obvious to you include **Arithmetic Population Density**, the number of people per square mile in each country; **Physiologic Population Density**, the number of people per square mile of agriculturally productive land; **Birth** and **Death Rates** per thousand in the population, resulting in the national population's rate of **Natural Increase**; a population's **Infant Mortality**, the number of deaths per thousand in the first year of life, thus reflecting largely the number of deaths at birth; **Child Mortality**, the deaths per thousand of children in their first five years; the **Corruption Index**, based on Transparency International data in which 10.0 is perfect and 0.1 is the worst; the **Big Mac Price** index, which tells you why China in 2005 was the best place to buy a hamburger in U.S. dollars; and the **Per Capita GNI ($US)**, the GNI-PPP index referred to above, which tells you how spendable income varies around the globe. For additional details on sources and data, please consult the Data Sources section of the Preface.

Table G-1
AREA AND DEMOGRAPHIC DATA FOR THE WORLD'S STATES

	Land Area (sq mi)	Population 2006 (Millions)	Population 2025 (Millions)	Population Density		Birth Rate	Death Rate	Natural Increase %	Infant Mortality per 1,000	Child Mortality per 1,000	Life Expectancy		Percent Urban Pop.	Percent Male Literacy	Percent Female Literacy	2004 Corruption Index	Big Mac Price ($US)	Per Capita GNI ($US)
				Arithmetic	Physiologic						Male (years)	Female (years)						
WORLD	**51,789,601**	**6,550.4**	**7,934.0**	**126.5**		**21**	**9**	**1.2**	**56**		**65**	**69**	**48**					**$7,590**
Europe	**2,284,860**	**584.7**	**585.4**	**255.9**		**10**	**12**	**−0.2**	**7**		**70**	**78**	**73**					
Albania	11,100	3.3	3.7	295.2	1,405.9	17	5	1.2	11	40	72	76	42	95.5	88.0	2.5		$4,700
Austria	32,378	8.1	8.4	250.2	1,471.6	9	9	0.0	5	5	76	82	54	100.0	100.0	8.4	$3.28	$29,610
Belarus	80,154	9.7	9.4	120.8	402.7	9	15	−0.6	8	27	63	75	72	99.7	99.2	3.3	$1.37	$6,010
Belgium	11,787	10.4	10.8	884.1	3,536.4	11	10	0.1	4	7	75	82	97	100.0	100.0	7.5	$3.28	$28,930
Bosnia	19,741	3.9	3.9	198.3	1,983.5	10	8	0.2	9	19	71	76	43	96.5	76.6	3.1		$6,200
Bulgaria	42,822	7.7	6.5	180.3	462.4	9	14	−0.5	12	19	69	75	70	99.1	98.0	4.1	$1.85	$7,610
Croatia	21,830	4.4	4.3	200.8	836.5	9	11	−0.2	7	9	71	78	56	99.4	97.3	3.5	$2.42	$10,710
Cyprus	3,571	0.9	1.1	254.6	2,314.2	12	7	0.5	6	9	75	80	65	98.7	95.0	5.4		$19,530
Czech Republic	30,448	10.2	10.1	333.7	834.1	9	11	−0.2	4	7	72	79	77	100.0	100.0	4.2	$2.13	$15,650
Denmark	16,637	5.4	5.4	325.2	580.8	12	11	0.1	4	6	75	79	72	100.0	100.0	9.5	$4.46	$33,750
Estonia	17,413	1.3	1.2	74.2	274.9	10	13	−0.3	6	23	65	77	69	99.9	99.6	6.0	$2.27	$12,480
Finland	130,560	5.2	5.3	40.0	571.3	11	9	0.2	3	4	75	82	62	100.0	100.0	9.7	$3.28	$27,100
France	212,934	60.5	63.4	284.0	860.7	13	9	0.4	4	5	76	83	74	98.9	98.7	7.1	$3.28	$27,460
Germany	137,830	82.4	82.0	598.1	1,759.1	9	10	−0.1	4	5	75	81	88	100.0	100.0	8.2	$3.28	$27,460
Greece	50,950	11.0	10.4	215.9	981.4	9	9	0.0	6	8	76	81	60	98.6	96.0	4.3	$3.28	$19,920
Hungary	35,919	10.0	8.9	278.9	536.4	9	13	−0.4	7	11	68	77	65	99.5	99.3	4.8	$2.52	$13,780
Iceland	39,768	0.3	0.3	7.7		14	6	0.8	2	5	79	83	94	100.0	100.0	9.5	$6.01	$30,810
Ireland	27,135	4.2	4.5	153.8	769.1	16	7	0.9	5	7	75	80	60	100.0	100.0	7.5	$3.28	$30,450
Italy	116,320	57.8	57.6	496.9	1,774.7	10	10	0.0	5	6	77	83	90	98.9	98.1	4.8	$3.28	$26,760
Latvia	24,942	2.3	2.2	91.3	314.8	9	14	−0.5	9	22	65	77	68	99.8	99.6	4.0	$2.00	$10,130
Liechtenstein	62	0.1	0.1	489.7	1,958.8	12	6	0.6	4	7	79	82	21	100.0	100.0			
Lithuania	25,174	3.4	3.5	134.3	298.3	9	12	−0.3	7	24	66	78	67	99.7	99.4	4.6	$2.26	$11,090
Luxembourg	999	0.5	0.6	503.5	2,014.0	12	9	0.3	5	7	75	81	91	100.0	100.0	8.4	$3.28	$54,430
Macedonia	9,927	2.0	2.2	203.5	847.9	14	9	0.5	12	23	71	75	59	94.2	83.8	2.7	$1.84	$6,720
Malta	124	0.4	0.4	3,238.7	10,447.5	10	8	0.2	7	10	76	80	91	91.4	92.8	6.8		$17,870
Moldova	13,012	4.2	3.9	321.5	595.4	10	12	−0.2	18	35	65	72	45	99.6	98.3	2.3	$1.93	$1,750
Netherlands	15,768	16.4	17.4	1,040.0	3,851.7	12	9	0.3	5	6	76	81	62	100.0	100.0	8.7	$3.28	$28,600
Norway	125,050	4.6	5.1	37.0	1,233.5	12	9	0.3	3	4	77	82	78	100.0	100.0	8.9	$5.18	$43,350

Table G-1 *(Continued)*

	Land Area (sq mi)	Population 2006 (Millions)	Population 2025 (Millions)	Population Density		Birth Rate	Death Rate	Natural Increase %	Infant Mortality per 1,000	Child Mortality per 1,000	Life Expectancy		Percent Urban Pop.	Percent Male Literacy	Percent Female Literacy	2004 Corruption Index	Big Mac Price ($US)	Per Capita GNI ($US)
				Arithmetic	Physiologic						Male (years)	Female (years)						
Poland	124,087	38.2	36.6	307.8	669.2	9	9	0.0	8	11	70	79	62	99.8	99.8	3.5		$11,450
Portugal	35,514	10.5	10.4	296.2	1,410.7	11	10	0.1	5	8	74	81	53	94.8	90.0	6.3	$3.28	$17,980
Romania	92,042	21.6	18.1	234.8	572.7	10	12	−0.2	17	26	68	75	53	99.1	97.3	2.9		$7,140
Serbia-Montenegro	39,448	10.7	10.7	271.8	876.7	12	11	0.1	13	11	70	75	52	100.0	100.0	2.7		$2,400
Slovakia	18,923	5.4	5.2	285.4	951.1	10	10	0.0	8	6	70	78	56	100.0	100.0	4.0	$1.98	$13,420
Slovenia	7,819	2.0	2.0	255.3	2,411.2	9	10	−0.1	4	5	72	80	51	98.6	96.8	6.0	$2.42	$19,420
Spain	195,363	42.6	43.5	218.0	721.4	10	9	0.1	4	4	76	83	76	100.0	100.0	7.1	$3.28	$22,020
Sweden	173,730	9.0	9.9	51.9	811.2	11	10	0.1	3	5	78	82	84	100.0	100.0	9.2	$3.94	$28,840
Switzerland	15,942	7.4	7.4	465.1	4,710.1	10	9	0.1	4	23	77	83	68	99.5	97.4	9.1	$4.90	$39,880
Ukraine	233,089	46.7	45.1	200.5	381.1	9	16	−0.7	10	7	62	74	68	100.0	100.0	2.2	$1.36	$5,410
United Kingdom	94,548	59.9	64.0	634.0	1,761.0	12	10	0.2	5	21	76	80	89	97.6	89.2	8.6	$3.37	$28,350
Russia	**6,592,819**	**142.1**	**136.9**	**21.6**	**119.7**	**10**	**17**	**−0.7**	**13**	**22**	**58**	**72**	**73**	**99.8**	**99.2**	**2.8**	**$1.45**	**$8,920**
Armenia	11,506	3.2	3.0	279.2	1,469.6	10	8	0.2	36	30	70	76	64	99.4	98.1	3.1		$3,770
Azerbaijan	33,436	8.4	9.7	252.2	1,117.4	14	6	0.8	13	46	69	75	51	98.9	95.9	1.9		$3,380
Georgia	3,571	4.5	4.0	1,260.2	188.6	11	11	0.0	24	23	68	75	52	99.7	99.4	2.0		$2,540
North America	**7,567,466**	**329.4**	**385.4**	**43.5**	**902.4**	**14**	**8**	**0.6**	**7**	**7**	**75**	**80**	**79**					
Canada	3,849,670	32.2	36.0	8.4	168.7	11	7	0.4	5	8	77	82	79	95.7	95.3	8.5	$2.33	$29,740
United States	3,717,796	297.1	349.4	79.9	447.6	14	8	0.6	7		75	80	79			7.5	$2.90	$37,610
Middle America	**1,047,354**	**184.1**	**234.8**	**175.8**	**976.5**	**26**	**6**	**2.0**		**21**			**68**					
Antigua and Barbuda	170	0.1	0.1	609.6	2,881.5	24	6	1.8	17	21	68	73	37					$9,590
Bahamas	5,359	0.3	0.3	57.4	8,004.1	18	5	1.3	16	12	70	75	89	95.4	96.8			$16,140
Barbados	166	0.3	0.3	1,832.6	4,111.3	15	8	0.7	13	43	70	75	50	98.0	96.8	7.3		$15,060
Belize	8,865	0.3	0.4	35.4	1,601.0	28	5	2.3	20	14	67	74	49			3.8		$5,840
Costa Rica	19,730	4.3	5.6	218.9	3,418.1	18	4	1.4	10	8	76	81	59	95.5	95.7	4.9	$2.61	$9,040
Cuba	42,803	11.4	11.8	266.1	1,201.9	11	7	0.4	7	20	74	78	75	96.5	96.4	3.7		
Dominica	290	0.1	0.1	351.8	3,501.3	17	7	1.0	16	53	71	77	71					$5,090
Dominican Republic	18,815	9.1	11.1	485.7	2,201.1	25	6	1.9	31	36	67	70	64	84.0	83.7	2.9	$1.32	$6,210
El Salvador	8,124	7.0	8.5	858.0	3,311.4	26	6	2.0	25	29	67	73	58	81.6	76.1	4.2		$4,890
Grenada	131	0.1	0.1	781.8	7,107.2	19	7	1.2	17				39					$6,710
Guadeloupe	660	0.4	0.5	618.2	4,755.7	17	7	1.0	8	55	74	81	100	89.7	90.5			

	Land Area (sq mi)	Population 2006 (Millions)	Population 2025 (Millions)	Population Density		Birth Rate	Death Rate	Natural Increase %	Infant Mortality per 1,000	Child Mortality per 1,000	Life Expectancy		Percent Urban Pop.	Percent Male Literacy	Percent Female Literacy	2004 Corruption Index	Big Mac Price ($US)	Per Capita GNI ($US)
				Arithmetic	Physiologic						Male (years)	Female (years)						
Guatemala	42,042	13.4	19.8	318.6	2,491.7	34	7	2.7	39	132	63	69	39	76.2	61.1	2.2	$2.01	$4,060
Haiti	10,714	8.4	11.7	785.0	3,417.1	33	14	1.9	80	45	50	53	36	51.0	46.5	1.5		$1,630
Honduras	43,278	7.4	10.7	170.9	1,068.3	33	5	2.8	34	11	67	74	47	72.5	72.0	2.3	$1.98	$2,580
Jamaica	4,243	2.7	3.3	628.8	4,601.2	20	7	1.3	24		73	77	52	82.5	90.7	3.3	$1.88	$3,790
Martinique	425	0.4	0.4	952.5	7,327.0	14	8	0.6	8	35	76	82	95	96.0	97.1			
Mexico	756,052	110.5	131.7	146.1	1,461.4	25	5	2.0	25		73	78	75	93.1	89.1	3.6	$2.08	$8,950
Netherlands Antilles	309	0.2	0.2	657.6	6,472.6	15	7	0.8	6	57	73	79	69	96.6	96.6		$2.29	
Nicaragua	50,193	5.9	8.3	117.7	1,681.1	32	5	2.7	31	20	66	71	58	64.2	64.4	2.7	$2.19	$2,400
Panama	29,158	3.3	4.2	113.7	1,804.2	23	5	1.8	21	8	72	77	62	92.6	91.3	3.7		$6,310
Puerto Rico	3,456	4.0	4.1	1,144.3	22,886.5	14	7	0.7	10	29	73	82	71	93.7	94.0			
Saint Lucia	239	0.2	0.2	855.3	8,553.3	17	6	1.1	14	21	70	74	30					$5,220
St. Vincent and the Grenadines	151	0.1	0.1	676.9	4,512.7	18	7	1.1	19	17	71	74	44					$6,590
Trinidad and Tobago	1,981	1.3	1.3	664.1	4,606.4	13	7	0.6	19		68	73	74	99.0	97.5	4.2		$9,450
South America	**6,893,881**	**365.0**	**449.6**	**52.9**	**1,050.7**	**21**	**6**	**1.6**		**24**			**79**					
Argentina	1,073,514	38.7	45.9	36.1	396.1	19	8	1.1	16	96	71	78	89	96.9	96.9	2.5	$1.48	$10,920
Bolivia	424,162	9.1	12.2	21.5	1,101.4	28	9	1.9	54	44	61	64	63	92.1	79.4	2.2		$2,450
Brazil	3,300,154	183.8	211.2	55.7	1,856.3	20	7	1.3	33	13	67	75	81	85.5	85.4	3.9	$1.70	$7,480
Chile	292,135	16.4	19.5	56.1	1,216.1	17	5	1.2	8	30	73	79	87	95.9	95.5	7.4	$2.18	$9,810
Colombia	439,734	46.9	58.1	106.5	2,614.8	23	6	1.7	26	39	69	75	71	91.8	91.8	3.8	$2.35	$6,520
Ecuador	109,483	14.0	17.3	127.6	2,217.5	25	4	2.1	30		68	74	61	93.6	90.2	2.4		$3,440
French Guiana	34,749	0.2	0.3	6.1	607.4	31	4	2.7	12	82	73	78	75	83.6	82.3			
Guyana	83,000	0.8	0.7	9.9	541.3	23	9	1.4	53	33	60	67	36	99.0	98.1			$3,950
Paraguay	157,046	6.3	9.2	40.1	691.1	30	5	2.5	37	56	69	73	54	94.4	92.2	1.9		$4,740
Peru	496,224	28.4	35.7	57.3	1,917.4	23	6	1.7	33	36	66	71	72	94.7	85.4	3.5	$2.57	$5,090
Suriname	63,039	0.4	0.4	6.5	691.9	23	7	1.6	27	21	67	72	69	95.9	92.6	4.3		$1,990
Uruguay	68,498	3.4	3.8	50.3	716.1	16	9	0.7	14	25	71	79	93	97.4	98.2	6.2	$1.00	$7,980
Venezuela	352,143	27.2	35.3	77.3	1,937.5	24	5	1.9	20		70	76	87	93.3	92.7	2.3	$1.48	$4,740
Subsaharan Africa	**8,357,680**	**714.3**	**1,057.8**	**85.5**	**1,273.6**	**41**	**16**	**2.5**	**95**	**292**	**47**	**50**	**33**					
Angola	481,351	14.0	23.8	29.0	1,391.4	49	24	2.5	145	167	39	42	33	55.6	28.5	2.0		$1,890
Benin	43,483	7.7	11.8	177.1	1,251.1	41	14	2.7	89	49	50	52	40	47.8	23.6	3.2		$1,110

Table G-1 (Continued)

	Land Area (sq mi)	Population 2006 (Millions)	Population 2025 (Millions)	Population Density		Birth Rate	Death Rate	Natural Increase %	Infant Mortality per 1,000	Child Mortality per 1,000	Life Expectancy		Percent Urban Pop.	Percent Male Literacy	Percent Female Literacy	2004 Corruption Index	Big Mac Price ($US)	Per Capita GNI ($US)
				Arithmetic	Physiologic						Male (years)	Female (years)						
Botswana	224,606	1.7	1.1	7.6	363.1	27	26	0.1	62	169	35	36	54	74.4	79.8	6.0		$7,960
Burkina Faso	105,792	14.3	22.5	135.3	987.2	45	19	2.6	83	176	44	46	15	31.2	13.1			$1,180
Burundi	10,745	6.5	10.1	602.7	1,618.0	40	18	2.2	74	145	42	44	8	56.3	40.5			$620
Cameroon	183,568	16.8	22.4	91.6	3,202.4	37	15	2.2	77	73	47	49	48	81.8	69.2	2.1		$1,980
Cape Verde Islands	1,556	0.5	0.7	335.6	3,001.6	29	7	2.2	31	173	66	73	46	84.3	65.3			$5,440
Central African Republic	240,533	3.8	4.8	15.9	531.4	37	19	1.8	96	198	41	44	39	59.6	34.5			$1,080
Chad	495,753	10.1	16.7	20.4	651.8	49	16	3.3	103	93	47	51	24	66.9	40.8	1.7		$1,100
Comoros Islands	861	0.7	1.1	870.9	3,040.1	47	12	3.5	84	108	54	59	33	63.5	49.1			$1,760
Congo	132,046	4.0	6.8	30.5	2,511.6	44	15	2.9	84	207	47	50	24	87.5	74.4	2.3		$710
Congo, The	905,531	62.0	104.9	68.4	2,216.6	46	15	3.1	100	172	46	51	52	86.6	67.7	2.0		$640
Equatorial Guinea	10,830	0.5	0.8	48.6	957.2	43	17	2.6	105	175	47	50	30	92.5	74.5			$930
Eritrea	45,045	4.6	7.0	102.8	1,028.3	39	13	2.6	76	35	52	55	19	43.9	33.4	2.6		$1,110
Ethiopia	426,371	75.8	117.6	177.7	1,612.1	41	18	2.3	105	145	45	47	15	83.7	70.0	2.3		$710
Gabon	103,347	1.5	1.9	14.1	1,412.7	33	12	2.1	57	87	56	58	73	79.8	62.2	3.3		$5,700
Gambia	4,363	1.6	2.7	363.3	2,270.8	41	13	2.8	78	107	52	56	26	43.8	29.6	2.8		$1,820
Ghana	92,100	22.4	30.6	243.2	2,100.3	33	10	2.3	64	201	57	59	44	79.5	61.2	3.6		$2,190
Guinea	94,927	9.7	16.2	102.2	4,701.8	43	16	2.7	98	220	48	50	33	55.1	27.0			$2,100
Guinea-Bissau	13,946	1.6	2.8	114.1	1,267.9	50	20	3.0	125	150	43	47	32	53.0	21.4			$660
Ivory Coast	124,502	17.6	22.1	141.2	2,017.5	39	19	2.0	102	87	42	43	53	54.6	38.5	2.0		$1,390
Kenya	224,081	33.9	39.9	151.3	2,170.2	38	15	2.3	78	137	48	53	36	89.0	76.0	2.1		$1,020
Lesotho	11,718	1.8	2.1	157.0	1,811.4	33	22	1.1	90	235	37	38	17	73.6	93.6			$3,120
Liberia	43,000	3.7	6.1	86.0	9,511.9	50	22	2.8	150	158	41	43	45	69.9	36.8			$130
Madagascar	226,656	18.6	33.0	82.1	1,941.5	43	12	3.1	84	215	53	58	26	87.7	72.9	3.1		$800
Malawi	45,745	12.6	23.8	276.0	911.3	51	21	3.0	121	239	42	45	14	74.5	46.7	2.8		$600
Mali	478,838	14.3	25.7	29.9	1,371.1	50	17	3.3	123	183	48	49	30	47.9	33.2	3.2		$960
Mauritania	395,954	3.2	5.0	8.0	701.2	42	15	2.7	102	23	53	55	40	50.6	29.5			$2,010
Mauritius	788	1.2	1.4	1,550.4	3,101.1	16	7	0.9	13	209	68	75	42	87.7	81.0	4.1		$11,260
Moçambique	309,494	19.9	25.4	64.2	1,691.1	40	23	1.7	127	75	38	42	29	59.9	28.4	2.8		$1,070
Namibia	318,259	2.0	2.1	6.2	575.2	31	15	1.6	38	285	48	46	33	82.9	81.2	4.1		$6,620

	Land Area (sq mi)	Population 2006 (Millions)	Population 2025 (Millions)	Population Density		Birth Rate	Death Rate	Natural Increase %	Infant Mortality per 1,000	Child Mortality per 1,000	Life Expectancy		Percent Urban Pop.	Percent Male Literacy	Percent Female Literacy	2004 Corruption Index	Big Mac Price ($US)	Per Capita GNI ($US)
				Arithmetic	Physiologic						Male (years)	Female (years)						
Niger	489,189	13.3	25.5	27.2	876.8	55	20	3.5	123	187	45	46	21	23.5	8.3	2.2		$820
Nigeria	356,668	145.4	206.4	407.6	1,991.4	42	13	2.9	100		52	52	36	72.3	56.2	1.6		$900
Réunion	969	0.8	1.0	850.5	4,212.9	20	5	1.5	27	170	71	80	89	84.8	89.2			
Rwanda	10,170	8.7	11.7	857.6	2,418.6	40	21	1.9	107	78	39	41	17	73.7	60.6			$1,290
São Tomé and Principe	371	0.2	0.3	569.7	31,202.7	34	6	2.8	34	124	66	72	38	70.2	39.1			$320
Senegal	75,954	11.5	17.1	151.1	1,240.1	37	11	2.6	64	18	55	57	43	47.2	27.6	3.0		$1,660
Seychelles	174	0.1	0.1	586.3	28,375.2	18	8	1.0	18	316	67	76	50	82.9	85.7	4.4		$15,960
Sierra Leone	27,699	5.4	7.6	195.7	3,112.1	50	29	2.1	180	82	34	36	37	50.7	22.6	2.3		$530
Somalia	246,201	8.8	14.9	35.7	1,719.1	47	18	2.9	124	115	45	48	33	85.8	84.5			
South Africa	471,444	47.9	44.6	101.7	924.4	24	13	1.1	48	94	49	57	53	68.3	46.0	4.6	$1.86	$10,270
Swaziland	6,703	1.2	1.1	186.3	1,515.1	36	16	2.0	78	143	45	42	25	80.9	78.7			$4,850
Tanzania	364,900	37.8	52.1	103.5	3,891.2	40	17	2.3	105	125	44	46	22	84.1	66.6	2.8		$610
Togo	21,927	5.9	7.6	269.4	707.4	38	11	2.7	72	137	53	56	33	72.2	42.6			$1,500
Uganda	93,066	27.7	47.5	297.5	1,414.6	47	17	3.0	88	202	43	46	12	77.7	57.1	2.6		$1,440
Zambia	290,583	11.3	14.4	38.9	485.9	42	24	1.8	95	80	35	35	35	85.2	71.2	2.6		$850
Zimbabwe	105,873	13.0	12.8	122.9	1,177.9	32	20	1.2	65		43	40	32	95.5	89.9	2.3		$2,180
North Africa/ Southwest Asia	**7,468,022**	**572.8**	**760.6**	**76.7**	**997.4**	**27.1**	**8**	**2.0**	**53**	**257**	**65**	**68**	**52**					
Afghanistan	251,772	30.1	50.3	119.4	971.2	48	21	2.7	165	39	42	43	22	51.0	20.8			
Algeria	919,591	33.3	40.5	36.3	1,218.4	20	4	1.6	54	22	73	74	49	75.1	51.3	2.7		$5,940
Bahrain	266	0.7	1.0	2,721.8		20	3	1.7	7	156	73	75	87	91.0	82.7	5.8		$16,170
Djibouti	8,958	0.7	1.0	81.9		41	17	2.4	106	73	45	48	82	65.0	38.4			$2,200
Egypt	386,660	76.4	103.2	197.5	9,811.3	26	6	2.0	38	116	66	70	43	66.6	43.7	3.2	$1.62	$3,940
Iran	630,575	69.0	84.7	109.5	912.2	18	6	1.2	32	122	68	70	67			2.9		$7,190
Iraq	169,236	27.3	41.7	161.4	949.5	36	9	2.7	102	6	58	61	68	70.7	45.0	2.1		
Israel	8,131	7.0	9.3	863.3	5,114.1	22	6	1.6	5	24	77	81	92	97.9	94.3	6.4		$19,200
Jordan	34,444	5.9	8.1	170.5	4,101.4	29	5	2.4	22	44	71	72	79	94.9	84.4	5.3	$3.65	$4,290
Kazakhstan	1,049,151	15.2	15.8	14.5	186.6	17	11	0.6	52	13	58	70	57	99.1	96.1	2.2		$6,170
Kuwait	6,880	2.6	4.6	375.1		18	2	1.6	10	68	77	79	100	84.3	79.9	4.6	$7.33	$17,870
Kyrgyzstan	76,641	5.2	6.7	68.3	1,021.1	21	8	1.3	42	37	65	72	35	98.6	95.5	2.2		$1,660
Lebanon	4,015	4.6	5.7	1,156.9	5,316.8	23	7	1.6	27	25	72	75	87	92.3	80.4	2.7	$2.84	$4,840
Libya	679,359	5.9	8.3	8.6	818.4	28	4	2.4	28	72	74	78	86	90.9	67.6	2.5		
Morocco	172,413	31.5	39.2	182.8	870.1	21	6	1.5	40	18	68	72	57	61.9	36.0	3.2		$3,950

Table G-1 (Continued)

	Land Area (sq mi)	Population 2006 (Millions)	Population 2025 (Millions)	Population Density		Birth Rate	Death Rate	Natural Increase %	Infant Mortality per 1,000	Child Mortality per 1,000	Life Expectancy		Percent Urban Pop.	Percent Male Literacy	Percent Female Literacy	2004 Corruption Index	Big Mac Price ($US)	Per Capita GNI ($US)
				Arithmetic	Physiologic						Male (years)	Female (years)						
Oman	82,031	2.8	4.0	34.4		26	4	2.2	16				76	80.4	61.7	6.1		$13,000
Palestinian Territories	2,417	4.1	7.4	1,684.2		39	4	3.5	26	20	71	74	57			2.5		
Qatar	4,247	0.7	1.0	170.1	8,506.9	20	4	1.6	12	28	70	75	92	80.5	83.2	5.2	$0.68	
Saudi Arabia	829,996	26.6	40.1	32.0	1,601.0	32	3	2.9	25	211	71	73	86	84.1	67.2	3.4	$0.64	$12,850
Sudan	967,494	41.3	61.3	42.7	781.2	38	10	2.8	69	33	68	72	31	36.0	14.0	2.2		$1,880
Syria	71,498	18.8	27.6	263.5	914.4	28	5	2.3	18	76	69	71	50	88.3	60.4	3.4		$3,430
Tajikistan	55,251	6.9	8.6	124.0	1,981.2	25	6	1.9	50	33	66	71	27	99.6	98.9	2.0		$1,040
Tunisia	63,170	10.2	11.6	161.8	909.4	17	6	1.1	22	45	71	75	63	81.4	60.1	5.0		$6,840
Turkey	299,158	73.3	88.9	245.1	733.1	21	7	1.4	39	78	66	71	59	93.6	76.7	3.2	$2.58	$6,690
Turkmenistan	188,456	5.9	7.6	31.2	1,091.8	25	9	1.6	74	10	63	70	47	98.8	96.6	2.0		$5,840
United Arab Emirates	32,278	4.3	5.4	133.8		16	2	1.4	8	58	73	77	78	75.5	79.5	6.1	$0.67	$21,040
Uzbekistan	172,741	27.3	36.9	157.8	1,901.8	24	8	1.6	62		68	73	37	98.5	96.0	2.3		$1,720
Western Sahara	97,344	0.3	0.5	3.2		29	8	2.1	59	100								
Yemen	203,849	21.3	39.6	104.7	3,630.3	43	10	3.3	75		58	62	26	67.4	25.0	2.4		$820
South Asia	**1,732,734**	**1,485.5**	**1,857.9**	**857.3**	**1,905.4**	**26.6**	**8**	**1.8**	**66**	**109**	**61**	**63**	**28**					**$2,672**
Bangladesh	55,598	147.3	204.5	2,649.3	3,913.7	30	9	2.1	66	121	60	60	23	51.7	29.5	1.5		$1,870
Bhutan	18,147	1.1	1.5	57.9	2,516.1	34	9	2.5	61	108	66	66	21	61.1	33.6			$660
India	1,269,340	1,123.9	1,363.0	885.4	1,707.1	25	8	1.7	64	74	61	63	28	68.6	42.1	2.8		$2,880
Maldives	116	0.3	0.4	2,659.1	13,295.6	18	4	1.4	18	104	73	74	27	96.3	96.4			$3,900
Nepal	56,826	25.9	37.8	455.8	2,814.1	34	10	2.4	64	136	59	58	14	59.1	21.8	2.8		$1,420
Pakistan	307,375	166.9	228.8	543.1	1,871.6	34	10	2.4	85	19	60	62	34	57.6	27.8	2.1	$1.90	$2,060
Sri Lanka	25,332	20.1	21.9	794.0	5,451.6	19	6	1.3	10		70	74	30	94.5	88.9	3.5	$1.41	$3,730
East Asia	**4,546,050**	**1,546.3**	**1,709.1**	**340.2**	**3,811.1**	**12**	**7**	**0.5**	**30**	**47**	**70**	**75**	**46**					
China	3,696,521	1,323.0	1,484.9	357.9	3,612.2	12	6	0.6	32	6	70	73	41	92.3	77.4	3.4	$1.26	$4,990
Japan	145,869	127.9	121.1	876.5	8,104.4	9	8	0.1	3	30	78	85	78	100.0	100.0	6.9	$2.33	$34,510
Korea, North	46,541	23.1	24.7	495.8	2,916.4	17	11	0.6	45	6	61	66	60	99.0	99.0			$1,000
Korea, South	38,324	48.7	50.6	1,270.3	6,848.3	10	5	0.5	8	5	73	80	80	99.2	96.4	4.5	$2.72	$17,930
Mongolia	604,826	2.6	3.4	4.2	401.1	18	6	1.2	30	8	63	68	57	99.2	99.3	3.0		$1,800
Taiwan	13,969	22.8	24.4	1,630.8	6,991.7	10	6	0.4	6		73	79	78	97.6	90.2	5.6	$2.24	

	Land Area (sq mi)	Population 2006 (Millions)	Population 2025 (Millions)	Population Density		Birth Rate	Death Rate	Natural Increase %	Infant Mortality per 1,000	Child Mortality per 1,000	Life Expectancy		Percent Urban Pop.	Percent Male Literacy	Percent Female Literacy	2004 Corruption Index	Big Mac Price ($US)	Per Capita GNI ($US)
				Arithmetic	Physiologic						Male (years)	Female (years)						
Southeast Asia	**1,735,449**	**564.3**	**697.7**	**325.1**	**2,402.2**	**22**	**7**	**1.5**	**41**	**10**	**66**	**70**	**38**					**$3,690**
Brunei	2,228	0.4	0.5	186.4		22	3	1.9	7	167	74	79	74	94.7	88.2			$17,900
Cambodia	69,900	13.7	19.8	195.7	1,811.8	32	10	2.2	95		55	59	16	79.7	53.4			$2,060
East Timor	5,741	0.8	1.2	143.0		26	13	1.3	129	60	48	49	8					$430
Indonesia	735,355	225.8	275.5	307.0	3,272.4	22	6	1.6	46	122	66	70	42	91.9	82.1	2.0	$1.77	$3,210
Laos	91,429	6.1	8.6	66.4	2,218.3	36	13	2.3	104	11	52	55	19	73.6	50.5			$1,730
Malaysia	127,317	26.7	36.0	210.0	6,737.4	26	4	2.2	11	114	71	76	62	91.5	83.6	5.0	$1.33	$8,940
Myanmar/Burma	261,228	51.5	59.8	197.2	1,095.5	25	11	1.4	87	46	54	60	28	89.0	80.6	1.7		$1,490
Philippines	115,830	87.1	113.4	751.8	3,943.6	26	6	2.0	29	4	67	72	48	95.5	95.2	2.6	$1.23	$4,640
Singapore	239	4.3	4.8	17,784.7		10	4	0.6	2	38	77	81	100	96.4	88.5	9.3	$1.92	$24,180
Thailand	198,116	64.7	72.2	326.6	964.1	14	7	0.7	20	43	68	75	31	97.2	94.0	3.6	$1.45	$7,450
Vietnam	128,066	83.5	102.9	651.8	4,001.4	18	6	1.2	21		70	73	25	95.7	91.0	2.6		$2,490
Austral Realm	**3,093,340**	**24.5**	**28.9**	**7.9**		**13.2**	**7**	**0.6**	**5**	**6**	**77**	**83**	**89**					**$27,073**
Australia	2,988,888	20.3	24.2	6.8	113.4	13	7	0.6	5	7	77	83	91	100.0	100.0	8.8	$2.27	$28,290
New Zealand	104,452	4.2	4.7	39.8	27.3	14	7	0.7	6		76	81	78	100.0	100.0	9.6	$2.65	$21,120
Pacific Realm	**376,000**	**9.2**	**12.1**	**24.4**		**28.9**	**8**	**2.1**	**47**	**24**	**57**	**59**	**22**					**$2,291**
Federated States of Micronesia	270	0.1	0.1	386.1		28	7	2.1	40	24	67	67	22	67.0	87.2			$2,090
Fiji	7,054	0.8	1.0	117.8	5,888.1	25	6	1.9	22		65	69	39	95.0	90.9		$2.35	$5,410
French Polynesia	1,544	0.3	0.3	200.2	18,180.5	20	5	1.5	6		69	74	53	94.9	95.0			
Guam	212	0.2	0.2	973.8	5,728.4	20	4	1.6	6	92	76	80	93	99.0	99.0			
Marshall Islands	69	0.1	0.1	1,558.5		42	5	3.7	37		67	70	68	92.4	90.0			$2,710
New Caledonia	7,174	0.2	0.3	28.8	3,108.4	22	5	1.7	5	112	70	76	71	57.4	58.3			
Papua New Guinea	178,703	6.0	8.2	33.3		32	10	2.2	60	27	57	59	15					$2,240
Samoa	1,097	0.2	0.2	190.8		29	6	2.3	18	28	72	74	22	100.0	100.0			$5,700
Solomon Islands	11,158	0.5	0.7	47.3		36	9	2.7	66	50	61	62	16	62.4	44.9			$1,630
Vanuatu	4,707	0.2	0.4	44.4		28	6	2.2	27		66	69	21	57.3	47.8			$2,880

Continued on pages 230-231

Scale 1: 16 000 000; one inch to 250 miles. Conic Projection

Elevations and depressions are given in feet

0 50 100 200 300 400 500 Miles

0 100 200 400 600 800 Kilometers

Chapter 1

Europe

CONCEPTS, IDEAS, AND TERMS

1 **Land hemisphere**
2 **Infrastructure**
3 **Local functional specialization**
4 ***The Isolated State***
5 **Model**
6 **Industrial Revolution**
7 **Nation-state**
8 **Nation**
9 **Centrifugal forces**
10 **Centripetal forces**
11 **Indo-European languages**
12 **Complementarity**
13 **Transferability**
14 **Intervening opportunity**
15 **Primate city**
16 **Metropolis**
17 **Supranationalism**
18 **Devolution**
19 **Four Motors of Europe**
20 **Regional state**
21 **Site**
22 **Situation**
23 **Conurbation**
24 **Landlocked location**
25 **Break-of-bulk**
26 ***Entrepôt***
27 **Shatter belt**
28 **Balkanization**
29 **Exclave**
30 **Irredentism**

REGIONS

- **WESTERN EUROPE**
- **THE BRITISH ISLES**
- **NORTHERN (NORDIC) EUROPE**
- **MEDITERRANEAN EUROPE**
- **EASTERN EUROPE**

FIGURE 1-1 Reprinted with permission from *Goode's World Atlas*, 21st edition, pp. 156–157. © Rand McNally, 2005. License R.L. 05-S-64.

don't have to memorize, but study

*I*T IS APPROPRIATE to begin our investigation of the world's geographic realms in Europe because over the past five centuries Europe and Europeans have influenced and changed the rest of the world more than any other realm or people has done. European empires spanned the globe and transformed societies far and near. European colonialism propelled the first wave of globalization. Millions of Europeans migrated from their homelands to the Old World as well as the New, changing (and sometimes nearly obliterating) traditional communities and creating new societies from Australia to North America. Colonial power and economic incentive combined to impel the movement of millions of imperial subjects from their ancestral homes to distant lands: Africans to the Americas, Indians to Africa, Chinese to Southeast Asia, Malays to South Africa's Cape, Native Americans from east to west. In agriculture, industry, politics, and other spheres, Europe generated revolutions—and then exported those revolutions across the world, thereby consolidating the European advantage.

But throughout much of that 500-year period of European hegemony, Europe also was a cauldron of conflict. Religious, territorial, and political disputes precipitated bitter wars that even spilled over into the colonies. During the twentieth century, Europe twice plunged the world into war. The terrible, unprecedented toll of World War I (1914–1918) was not enough to stave off World War II (1939–1945), which ended with the first-ever use of nuclear weapons in Japan. In the aftermath of that war, Europe's weakened powers lost most of their colonial possessions and a new rivalry emerged: an ideological Cold War between the communist Soviet Union and the capitalist United States. This Cold War lowered an Iron Curtain across the heart of Europe, leaving most of the east under Soviet control and most of the west in the American camp. Western Europe proved resilient, overcoming the destruction of war and the loss of colonial power to regain economic strength. Meanwhile the Soviet communist experiment failed at home and abroad, and in 1990 the last

MAJOR GEOGRAPHIC QUALITIES OF
Europe

1. The European realm lies on the western extremity of the Eurasian landmass, a locale of maximum efficiency for contact with the rest of the world.

2. Europe's lingering and resurgent world influence results largely from advantages accrued over centuries of global political and economic domination.

3. The European natural environment displays a wide range of topographic, climatic, and soil conditions and is endowed with many industrial resources.

4. Europe is marked by strong internal regional differentiation (cultural as well as physical), exhibits a high degree of functional specialization, and provides multiple exchange opportunities.

5. European economies are dominated by manufacturing, and the level of productivity has been high; levels of development generally decline from west to east.

6. Europe's nation-states emerged from durable power cores that formed the headquarters of world colonial empires. A number of those states are now plagued by internal separatist movements. *Basques*

7. Europe's rapidly aging population is generally well off, highly urbanized, well educated, and enjoys long life expectancies.

8. A growing number of European countries are experiencing population declines; in many of these countries, the natural decrease is partially offset by immigration.

9. Europe has made significant progress toward international economic integration and, to a lesser extent, political coordination.

vestiges of the Iron Curtain were lifted. Since then, a massive effort has been underway to reintegrate and reunify Europe from the Atlantic coast to the Russian border, the key geographic story of this chapter.

DEFINING THE REALM

As Figure 1-1 shows, Europe is a realm of peninsulas and islands on the western margin of the world's largest landmass, Eurasia. It is a realm of 585 million people and 39 countries, but it is territorially quite small. Yet despite its modest proportions it has had—and continues to have—a major impact on world affairs. For many centuries Europe has been a hearth of achievement, innovation, and invention.

The European realm is bounded on the west, north, and south by Atlantic, Arctic, and Mediterranean waters, respectively. But where is Europe's eastern limit? Some scholars place it at the Ural Mountains, deep inside Russia, thereby recognizing a "European" Russia and, presumably, an "Asian" one as well. Our regional definition places Europe's eastern boundary between Russia and its numerous European neighbors to the west. This definition

EUROPE'S CLIMATES

B DRY CLIMATE

BS	Semiarid
BW	Arid

h=hot
k=cold

C HUMID TEMPERATE CLIMATE

Cfa
Cfb } No dry season
Cfc

Csa } Dry summer
Csb } (Mediterranean)

a=hot summer c=short, cool summer
b=cool summer d=very cold winter

D HUMID COLD CLIMATE

Dfa
Dfb } No dry season
Dfc

E COLD POLAR CLIMATE

E Tundra and ice

H HIGHLAND CLIMATE

H Undifferentiated highlands

0 200 400 600 800 Kilometers
0 100 200 300 400 500 Miles

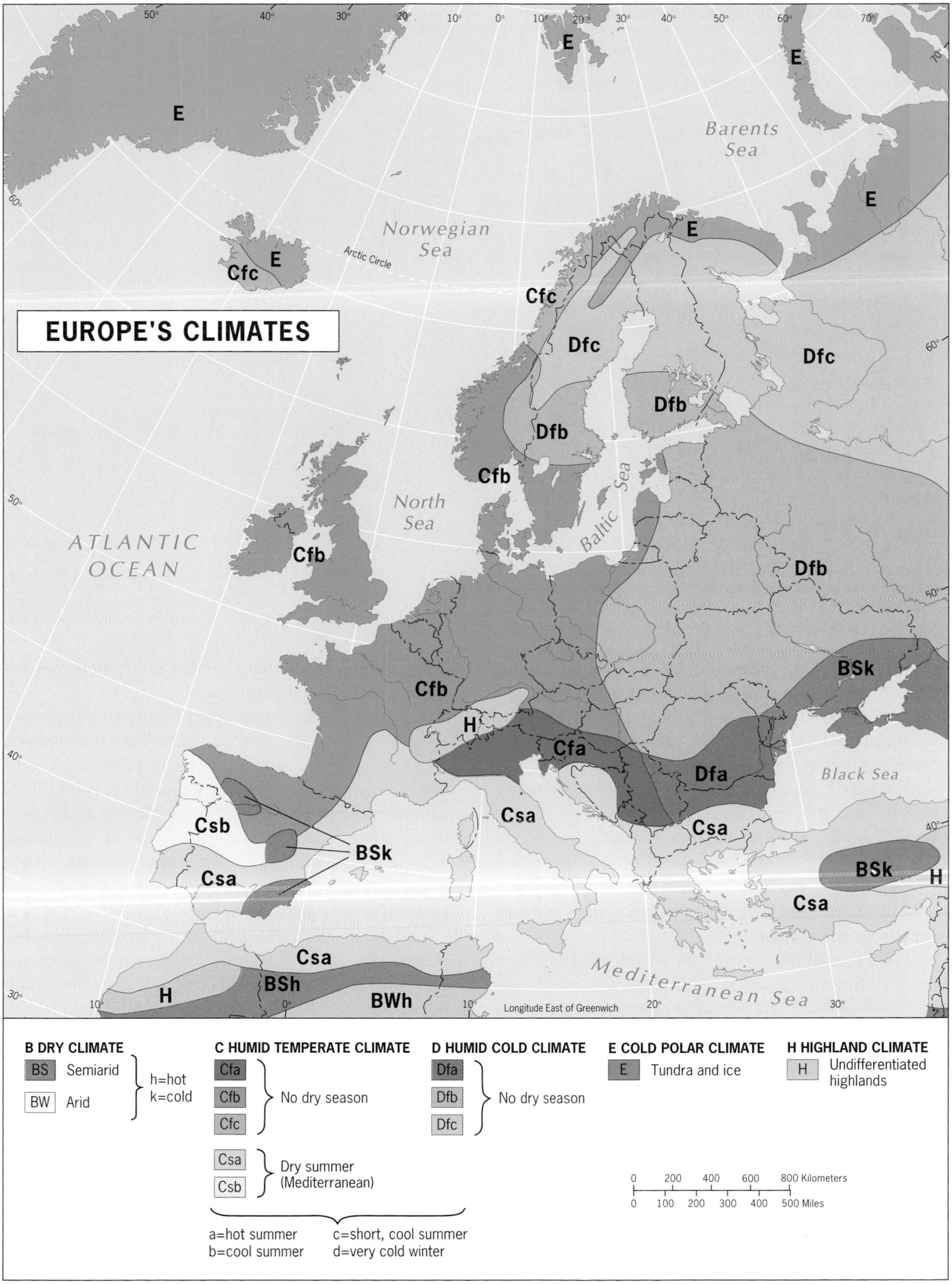

FIGURE 1-2 © H. J. de Blij, P. O. Muller, and John Wiley & Sons, Inc.

RELATIVE LOCATION: EUROPE IN THE LAND HEMISPHERE

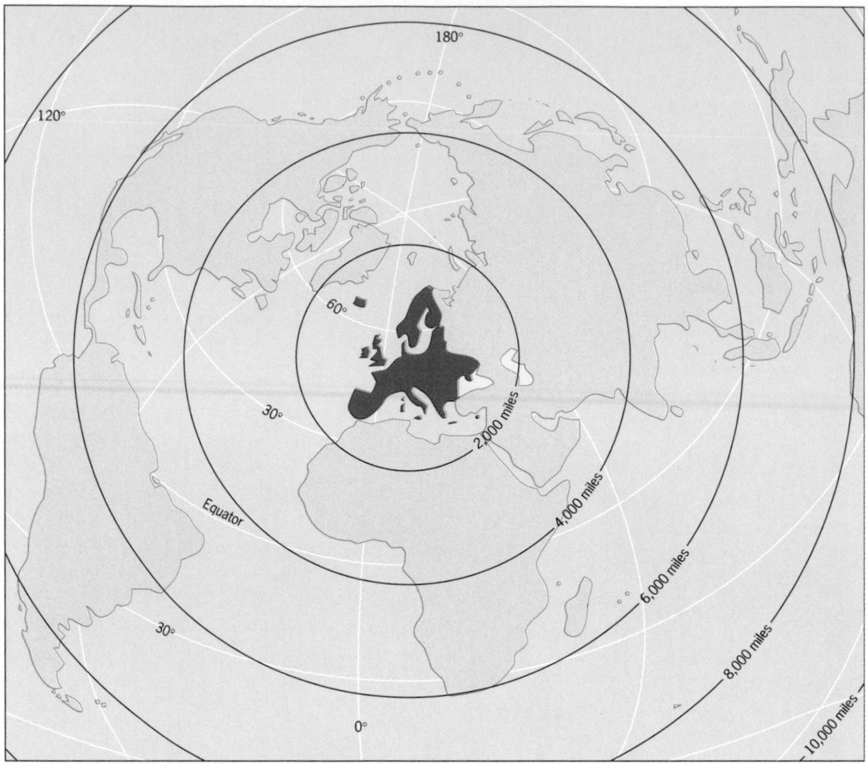

Azimuthal equidistant projection centered on Hamburg, Germany

FIGURE 1-3 © H. J. de Blij, P. O. Muller, and John Wiley & Sons, Inc.

is based on several geographic factors including European-Russian contrasts in territorial dimensions, population size, cultural properties, and historic aspects, all discernible on the maps in Chapters 1 and 2.

Europe's peoples have benefited from a large and varied store of raw materials. Whenever the opportunity or need arose, the realm proved to contain what was required. Early on, these requirements included cultivable soils, rich fishing waters, and wild animals that could be domesticated; in addition, extensive forests provided wood for houses and boats. Later, mineral fuels and ores propelled industrialization.

From the balmy shores of the Mediterranean Sea to the icy peaks of the Alps, and from the moist woodlands and moors of the Atlantic fringe to the semiarid prairies north of the Black Sea, Europe presents an almost infinite range of natural environments (Fig. 1-2). Compare Western Europe and eastern North America on Figure G-8 (pp. 14–15) and you will see the moderating influence of the warm ocean current known as the North Atlantic Drift and its onshore windflow.

The European realm is home to peoples of numerous cultural-linguistic stocks, including not only Latins, Germanics, and Slavs but also minorities such as Finns, Magyars (Hungarians), Basques, and Celts. This diversity of ancestries continues to be an asset as well as a liability. It has generated not only interaction and exchange, but also conflict and war.

Europe also has outstanding locational advantages. Its *relative location*, at the heart of the **land hemisphere**, 1
creates maximum efficiency for contact with much of the rest of the world (Fig. 1-3). A "peninsula of peninsulas," Europe is nowhere far from the ocean and its avenues of seaborne trade and conquest. Hundreds of miles of navigable rivers, augmented by an unmatched system of canals, open the interior of Europe to its neighboring seas and to the waterways of the world.

Also consider the scale of the maps of Europe in this chapter. Europe is a realm of moderate distances and close proximities. Short distances and large cultural differences make for intense interaction, the constant circulation of goods and ideas. That has been the hallmark of Europe's geography for over a millennium.

LANDSCAPES AND OPPORTUNITIES

Europe's area may be small, but its landscapes are varied and complex. Regionally, we identify four broad units: the Central Uplands, the southern Alpine Mountains, the Western Uplands, and the North European Lowland (Fig. 1-4).

FIGURE 1-4 © H. J. de Blij, P. O. Muller, and John Wiley & Sons, Inc.

The *Central Uplands* form the heart of Europe. It is a region of hills and low plateaus loaded with raw materials whose farm villages grew into towns and cities when the Industrial Revolution transformed this realm.

The *Alpine Mountains*, a highland region named after the Alps, extend from the Pyrenees on the French-Spanish border to the Balkan Mountains near the Black Sea, and include Italy's Appennines and Eastern Europe's Carpathians.

The *Western Uplands*, geologically older, lower, and more stable than the Alpine Mountains, extend from Scandinavia through western Britain and Ireland to the heart of the Iberian Peninsula in Spain.

The *North European Lowland* extends in a lengthy arc from southeast Britain and central France across Germany and Denmark into Poland and Ukraine, from where it continues well into Russia. Also known as the Great European Plain, this has been an avenue for human migration time after time, so that complex cultural and economic mosaics developed here together with a jigsaw-like political map. As Figure 1-4 shows, many of Europe's major rivers and connecting waterways serve this populous region, where many of Europe's leading cities (London, Paris, Amsterdam, Copenhagen, Berlin, Warsaw) are located.

HISTORICAL GEOGRAPHY

Modern Europe was peopled in the wake of the Pleistocene's most recent glacial retreat and global warming—a gradual warming that caused tundra to give way to deciduous forest and ice-filled valleys to turn into grassy vales. On Mediterranean shores, Europe witnessed the rise of its first great civilizations, on the islands and peninsulas of Greece and later in what is today Italy.

Ancient Greece lay exposed to influences radiating from the advanced civilizations of Mesopotamia and the Nile Valley, and in their fragmented habitat the Greeks laid the foundations of European civilization. Their achievements in political science, philosophy, the arts, and other spheres have endured for 25 centuries. But the ancient Greeks never managed to unify their domain, and their persistent conflicts proved fatal when the Romans challenged them from the west. By 147 B.C. the last of the sovereign Greek intercity leagues (alliances) had fallen to the Roman conquerors.

The center of civilization and power now shifted to Rome in present-day Italy. Borrowing from Greek culture, the Romans created an empire that stretched from Britain to the Persian Gulf and from the Black Sea to Egypt; they made the Mediterranean Sea a Roman lake carrying armies to distant shores and goods to imperial Rome. With an urban population that probably exceeded 1 million, Rome was the first metropolitan-scale urban center in Europe.

The Romans founded numerous other cities throughout their empire and linked them to the capital through a vast system of highway and water routes, facilitating political control and enabling economic growth in their
2 provinces. It was an unparalleled **infrastructure**, much of which long outlasted the empire itself.

Roman rule brought disparate, isolated peoples into the imperial political and economic sphere. By guiding (and often forcing) these groups to produce particular goods or materials, they launched Europe down a road for
3 which it would become famous: **local functional specialization**. The workers on Elba, a Mediterranean island, mined iron ore. Those near Cartagena in Spain produced silver and lead. Certain farmers were taught irrigation to produce specialty crops. Others raised livestock for meat or wool. The *production of particular goods by particular people in particular places* became and remained a hallmark of the realm.

The Romans also spread their language across the empire, setting the stage for the emergence of the *Romance* languages; they disseminated Christianity; and they established durable systems of education, administration, and commerce. But when their empire collapsed in the fifth century, disorder ensued, and massive migrations soon brought Germanic and Slavic peoples to their present positions on the European stage. Capitalizing on Europe's weakness, the Arab-Berber Moors from North Africa, energized by Islam, conquered most of Iberia and penetrated France. Later the Ottoman Turks invaded Eastern Europe and reached the outskirts of Vienna.

Europe's revival—its *Renaissance*—did not begin until the fifteenth century. After a thousand years of feudal turmoil marking the "Dark" and "Middle" Ages, powerful monarchies began to lay the foundations of modern states. The discovery of continents and riches across the oceans opened a new era of *mercantilism*, the competitive accumulation of wealth chiefly in the form of gold and silver. Best placed for this competition were the kingdoms of Western Europe. Europe was on its way to colonial expansion and world domination.

THE REVOLUTIONS OF MODERNIZING EUROPE

Even as Europe's rising powers reached for world domination overseas, they fought with each other in Europe itself. Powerful monarchies and land-owning ("landed") aristocracies had their status and privilege challenged by ever-wealthier merchants and businesspeople. Demands for political recognition grew; cities mushroomed with the development of industries; the markets for farm products burgeoned; and Europe's population, more or less stable at about 100 million since the sixteenth century, began to increase.

The Agrarian Revolution

We know Europe as the focus of the Industrial Revolution, but before this momentous development occurred another revolution was already in progress: the *agrarian revolution*. Port cities and capital cities thrived and expanded, and their growing markets created economic opportunities for farmers. This led to revolutionary changes in land ownership and agricultural methods. Improved farm practices, better equipment, superior storage facilities, and more efficient transport to the urban markets marked a revolution in the countryside. The colonial merchants brought back new crops (the American potato soon became a European staple), and market prices rose, drawing more and more farmers into the economy.

The transformation of Europe's farmlands reshaped its economic geography, producing new patterns of land use and market links. The economic geographer Johann Heinrich von Thünen (1783–1850), himself an estate farmer who had studied these changes for several decades, published his observations in 1826 in a pioneering work
4 entitled ***The Isolated State***, chronicling the geography of Europe's agricultural transformation.

Von Thünen used information from his own farm-
5 stead to build what today we call a **model** (an idealized representation of reality that demonstrates its most important properties) of the location of productive activities in Europe's farmlands. Since a model is an abstraction that must always involve assumptions, von Thünen postulated a self-contained area (hence the "isolation") with a single market center, flat and uninterrupted land without impediments to cultivation or transportation. In such a situation, transport costs would be directly proportional to distance.

Von Thünen's model revealed four zones or rings of land use encircling the market center (Fig. 1-5). Innermost and directly adjacent to the market would lie a zone of intensive farming and dairying, yielding the most perishable and highest-priced products. Immediately beyond lay a zone of forest used for timber and firewood (still a priority in von Thünen's time). Next there would be a ring of field crops, for example, grains or potatoes. A fourth zone would contain pastures and livestock. Beyond lay wilderness, from where the costs of transport to market would become prohibitive.

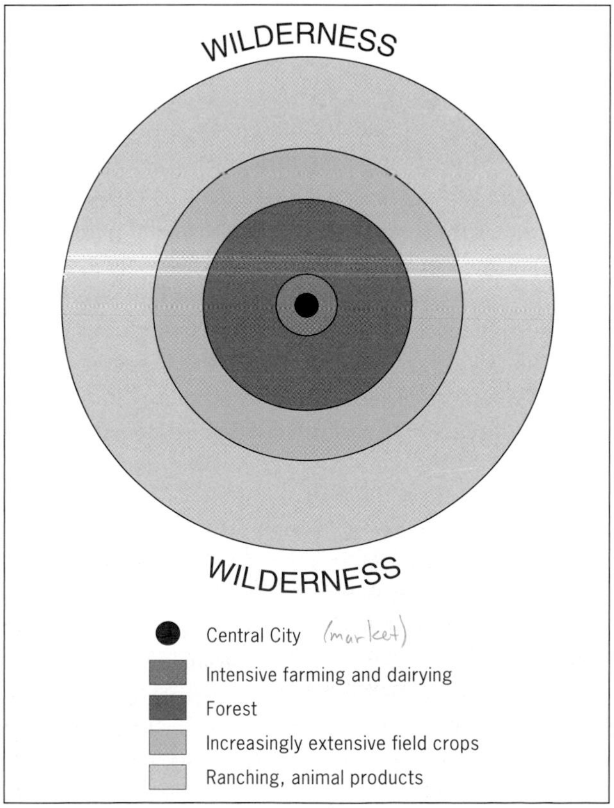

FIGURE 1-5 © H. J. de Blij, P. O. Muller, and John Wiley & Sons, Inc.

In many ways, von Thünen's model was the first analysis in a field that would eventually become known as *location theory*. Von Thünen knew, of course, that the real Europe did not present the conditions postulated in his model. But it did demonstrate the economic-geographic forces that shaped the new Europe, which is why it is still being discussed today. More than a century after the publication of *The Isolated State*, geographers Samuel van Valkenburg and Colbert Held produced a map of twentieth-century European agricultural intensity, revealing a striking, ring-like concentricity focused on the vast urbanized area lining the North Sea—now the dominant market for a realmwide, macroscale "Thünian" agricultural system (Fig. 1-6).

The Industrial Revolution

The term **Industrial Revolution** suggests that an agrarian 6
Europe was suddenly swept up in wholesale industrialization that changed the realm in a few decades. In reality, seventeenth- and eighteenth-century Europe had been industrializing in many spheres, long before the chain of events known as the Industrial Revolution began. From the textiles of England and Flanders to the iron farm implements of Saxony (in present-day Germany), from Scandinavian furniture to French linens, Europe had already entered a new era of local functional specialization. It would therefore be more appropriate to call what happened next the period of Europe's *industrial intensification*.

In the 1780s, the Scotsman James Watt and others devised a steam-driven engine, which was soon adopted for numerous industrial uses. At about the same time, coal (converted into carbon-rich coke) was recognized as a vastly superior substitute for charcoal in smelting iron. These momentous innovations had a rapid effect. The power loom revolutionized the weaving industry. Iron smelters, long dependent on Europe's dwindling forests for fuel, could now be concentrated near coalfields. Engines could move locomotives as well as power looms. Ocean shipping entered a new age.

Britain had an enormous advantage, for the Industrial Revolution occurred when British influence reigned worldwide and the significant innovations were achieved in Britain itself. The British controlled the flow of raw materials, they held a monopoly over products

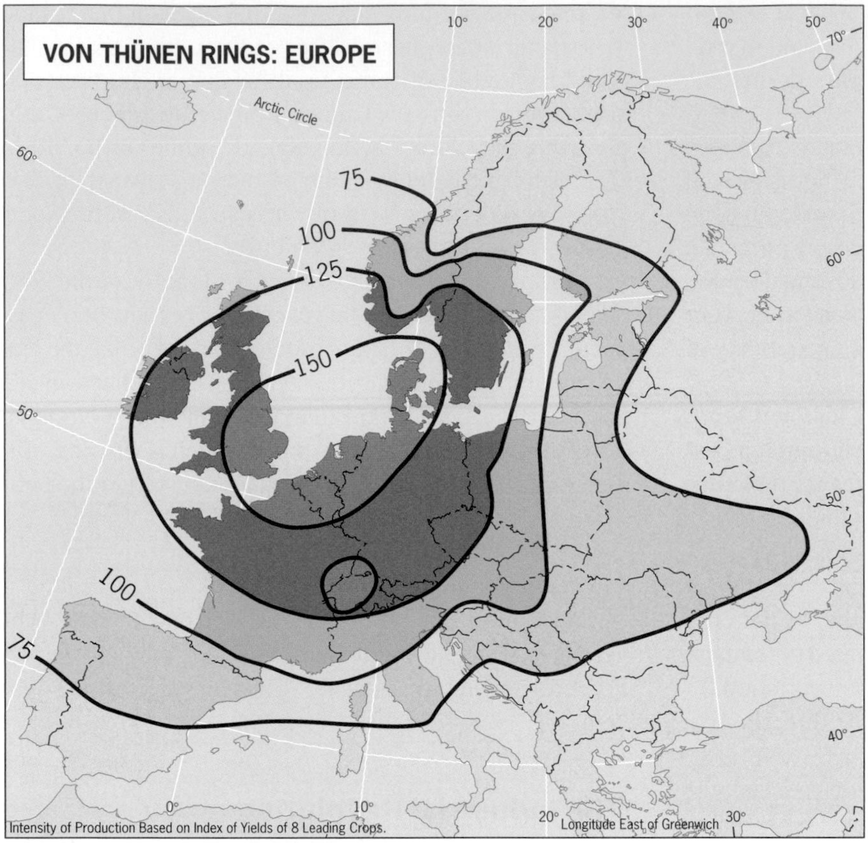

FIGURE 1-6 Adapted with permission from Samuel van Valkenburg & Colbert C. Held, *Europe*, 2 rev. ed. © John Wiley & Sons, Inc., 1952.

that were in global demand, and they alone possessed the skills necessary to make the machines that manufactured the products. Soon the fruits of the Industrial Revolution were being exported, and the modern industrial spatial organization of Europe began to take shape. In Britain, manufacturing regions, densely populated and heavily urbanized, developed near coalfields in the English Midlands, at Newcastle to the northeast, in southern Wales, and along Scotland's Clyde River around Glasgow.

In mainland Europe, a belt of major coalfields extends from west to east, roughly along the southern margins of the North European Lowland, due eastward from southern England across northern France and Belgium, Germany (the Ruhr), western Bohemia in the Czech Republic, Silesia in southern Poland, and the Donets Basin (Donbas) in eastern Ukraine. Iron ore is found in a broadly similar belt, and the industrial map of Europe reflects the resulting concentrations of economic activity (Fig. 1-7). Another set of manufacturing zones emerged in and near the growing urban centers of Europe, as the same map demonstrates.

London—already Europe's leading urban focus and Britain's richest domestic market—was typical of these developments. Many local industries were established here, taking advantage of the large supply of labor, the ready availability of capital, and the proximity of so great a number of potential buyers. Although the Industrial Revolution thrust other places into prominence, London did not lose its primacy: industries in and around the British capital multiplied.

The industrial transformation of Europe, like the agrarian revolution, became the focus of geographic research. One of the leaders in this area was the economic geographer Alfred Weber (1868–1958), who published a spatial analysis of the process titled *Concerning the Location of Industries* (1909). Unlike von Thünen, Weber focused on activities that take place at particular points rather than over large areas. His model, therefore, represented the factors of industrial location, the clustering or dispersal of places of intense manufacturing activity.

One of Weber's most interesting conclusions has to do with what he called *agglomerative* (concentrating) and *deglomerative* (dispersive) forces. It is often advantageous for certain industries to cluster together, sharing equipment, transport facilities, labor skills, and other assets of urban areas. This is what made London (as well as Paris and other cities that were not situated on rich deposits of industrial raw materials) attractive to many manufacturing plants that could benefit from agglomeration and from the large markets that these cities anchored. As Weber found, however, excessive agglomeration may lead to disadvantages such as competition for increasingly

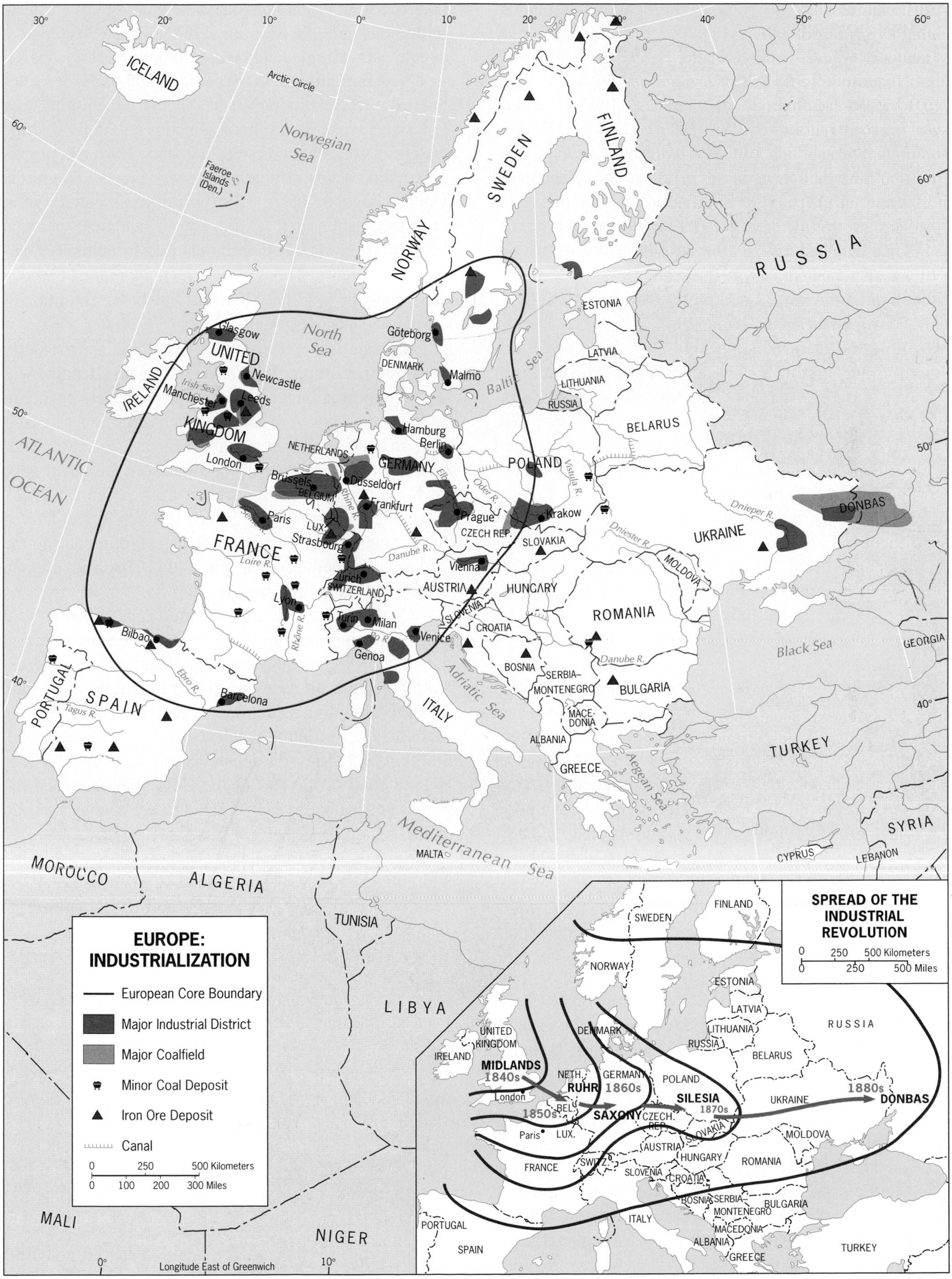

FIGURE 1-7 © H. J. de Blij, P. O. Muller, and John Wiley & Sons, Inc.

expensive space, congestion, overburdening of infrastructure, and environmental pollution. Manufacturers may then move away, and deglomerative forces will increase.

The Industrial Revolution spread eastward from Britain onto the European mainland throughout the middle and late nineteenth century (see inset, Fig. 1-7). Population skyrocketed, emigration mushroomed, and industrializing cities burst at the seams. European states already had acquired colonial empires before this revolution started; now colonialism gave Europe an unprecedented advantage in its dominance over the rest of the world.

Political Revolutions

Revolution in a third sphere—the political—had been going on in Europe even longer than the agrarian or industrial revolutions. Europe's *political revolution* took many different forms and affected diverse peoples and countries, but in general it headed toward parliamentary representation and democracy.

Historical geographers point to the Peace (Treaty) of Westphalia in 1648 as a key step in the evolution of Europe's state system, ending decades of war and recognizing territories, boundaries, and the sovereignty of countries. This treaty's stabilizing effect lasted until 1806, by which time revolutionary changes were again sweeping across Europe.

Most dramatic was the French Revolution (1789–1795), but political transformation had come much earlier to the Netherlands, Britain, and Scandinavian countries. Other parts of Europe remained under the control of authoritarian (dictatorial) regimes headed by monarchs or despots. Europe's patchy political revolution lasted into the twentieth century, and by then *nationalism* (national spirit, pride, patriotism) had become a powerful force in European politics.

When you look at the political map of Europe, the question that arises is how did so small a geographic realm come to be divided into so many political entities? Europe's map is a legacy of its feudal and royal periods, when powerful kings, barons, dukes, and other rulers, rich enough to fund armies and powerful enough to exact taxes and tribute from their domains, created bounded territories in which they reigned supreme. Royal marriages, alliances, and conquests actually simplified Europe's political map. In the early nineteenth century there still were 39 German states; Germany as we know it today did not emerge until the 1870s.

Europe's political revolution produced a form of
7 political-territorial organization known as the **nation-state**, a state embodied by its culturally distinctive population. But what is a nation-state and what is not? The term
nation has multiple meanings. In one sense it refers to a 8
people with a single language, a common history, a similar ethnic background. In the sense of *nationality* it relates to legal membership in the state, that is, citizenship. Very few states today are so homogeneous culturally that the culture is conterminous with the state. Europe's prominent nation-states of a century ago—France, Spain, the United Kingdom, Italy—have become multicultural societies, their nations defined more by an intangible "national spirit" and emotional commitment than by cultural or ethnic homogeneity. Today, Poland, Hungary, and Sweden are among the few states that still satisfy the definition of the nation-state in Europe.

Mercantilism and colonialism empowered the states of Western Europe; the United Kingdom (Britain) was the superpower of its day. But all countries, even Europe's nation-states in their heyday, are subject to divisive stresses. Political geographers use the term **cen-** 9
trifugal forces to identify and measure the strength of such division, which may result from religious, racial, linguistic, political, or regional factors. In the United States, racial issues form a centrifugal force; during the Vietnam (Indochina) War (1964–1975) politics created strong and dangerous disunity.

Centrifugal forces are measured against **centripetal** 10
forces, the binding, unifying glue of the state. General satisfaction with the system of government and administration, legal institutions, and other functions of the state (notably including its treatment of minorities) can ensure stability and continuity when centrifugal forces threaten. In the recent case of Yugoslavia, the centrifugal forces unleashed after the end of the Cold War exceeded the weak centripetal forces in that relatively young state, and it disintegrated.

Europe's political revolution continues. Today a growing group of European states is trying to create a realmwide union that might some day become a European superstate.

CONTEMPORARY EUROPE

Europe has been a regional laboratory of political revolution and evolution, and some of its nation-states were among the first of their kind to emerge on the world stage. Enriched and empowered by colonialism, European states competed and fought with each other, but Europe's nations survived and prospered. Strong cultural identities and historic durability gave European peoples a confidence that continues to mark the realm today, long after Europe's global empires collapsed.

FROM THE FIELD NOTES

"I went to London to observe the British elections of May 5, 2005 and to learn more about the issues that drove the campaign and its outcome. Here's something to consider: British national elections involve a six-week intensive campaign costing an estimated $80 million (the most recent U.S. presidential election with its year-long campaign is estimated to have cost about $4 billion). Polling stations ranged from the quaint to the modern (the one shown here on Ebury Street in Belgravia surely ranks among the former); polls were open until 10 p.m., giving everyone a chance to vote outside of working hours. Neither the EU nor Iraq dominated the debate; they were factors, to be sure—and the credibility issue (on Iraq) undoubtedly damaged Prime Minister Blair's prospects—but the key issues were domestic, ranging from immigration to health care and from education to public safety. As it happened, the election occurred two days before the 60-year commemoration of the end of World War II, and noteworthy was the outpouring of anti-German sentiment in the media, to the point that the German ambassador to the UK was quoted on the front page of one of London's leading newspapers with the headline 'GET OVER IT!' Neither the election nor the war's remembrance showed the EU at its finest hour in this corner of Europe." © H. J. de Blij

The European realm exhibits only limited geographic homogeneity, which is a challenge that confronts those leading states that want to create a more unified Europe. As Figure 1-8 shows, most Europeans speak **Indo-European** 11
languages, but in fact Europe remains a veritable Tower of Babel. Not only are many of those Indo-European languages not mutually understandable, but peoples such as the Hungarians and the Finns have other linguistic sources. English has become Europe's *lingua franca*, but generally with declining effectiveness from west to east.

Christian religious traditions, another factor that might serve as a unifier, have instead been the source of endless conflict. Shared Christian values, for example, have done little to bring peace to Northern Ireland, where longstanding sectarian strife continues. Religion and politics remain closely connected, and some national and regional political parties still have religious names, such as Germany's Christian Democratic Union and Bavaria's Christian Social Union.

The name "European" is sometimes taken to refer not just to someone residing in Europe, but as a racial reference, describing a common ancestry. But here again, Europe's purported homogeneity is more apparent than real. In terms of physical characteristics, Europeans, from Swede to Spaniard and from Scot to Sicilian, are as varied as any of the world's other geographic realms.

Spatial Interaction

If not culture, what does unify Europe? The answer lies in this realm's outstanding opportunities for productive contact and profitable interaction. The ancient Romans knew it, but they centered their system on the imperial capital. Modern Europeans have seized the same opportunities to create a regionwide structure of *spatial interaction* that links regions, countries, and places in countless ways. The American geographer Edward Ullman conceptualized this process around three operating principles: (1) complementarity, (2) transferability, and (3) intervening opportunity.

Complementarity occurs when one area has a sur- 12
plus of a commodity required by another area. The mere existence of a particular resource or product is no guarantee of trade: it must be needed elsewhere. One of Europe's countless examples (at various levels of scale) involves Italy, leader among Mediterranean countries in economic development but lacking adequate coal supplies. Italy imports coal from Western Europe, and in turn Italy exports to Western Europe its citrus fruits, olives, and grapes—which are in high demand on Western European markets. This is a case of *double complementarity*,

don't need to know this

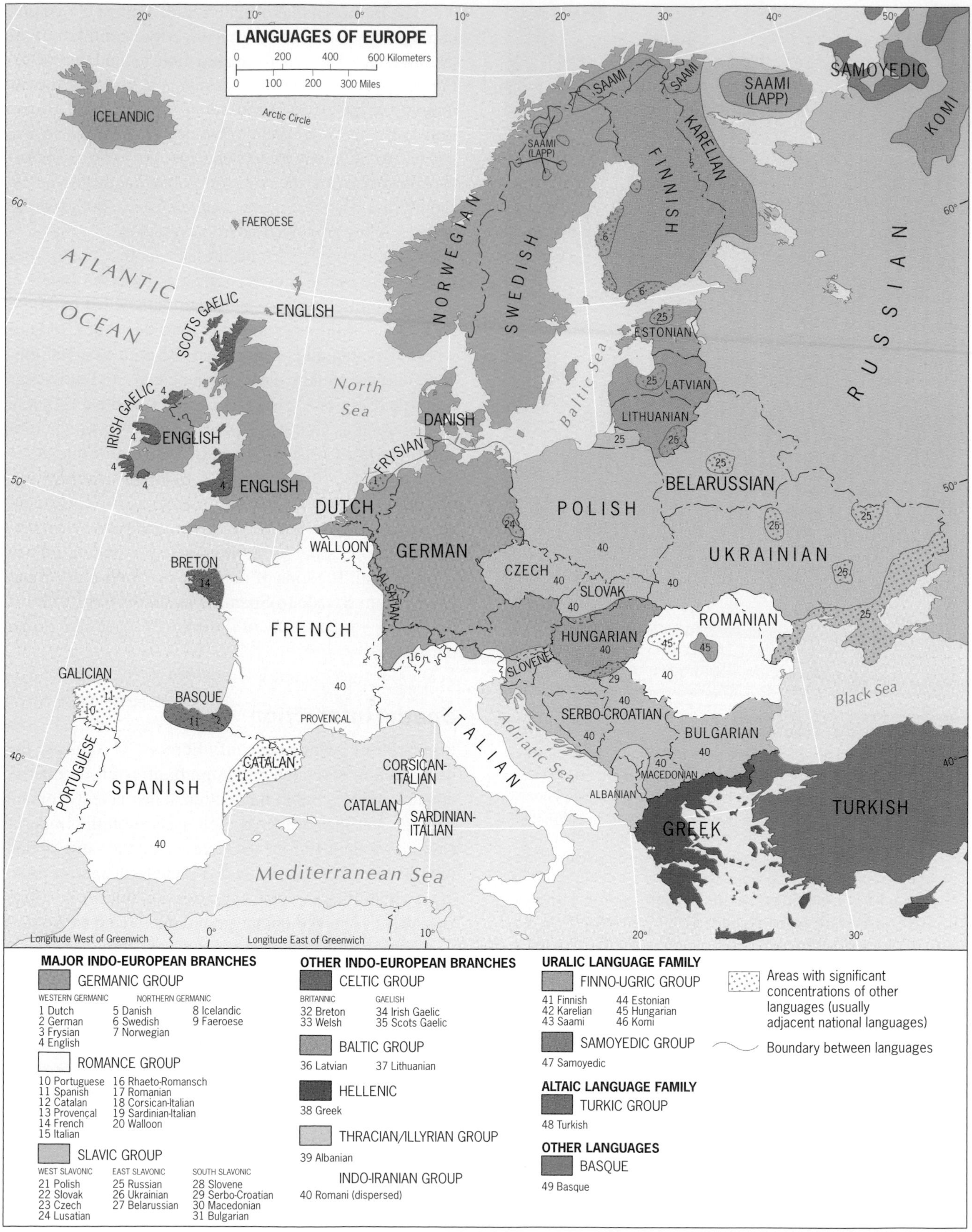

FIGURE 1-8 Based on a map adapted from several sources in Alexander B. Murphy, "European Languages," in Tim Unwin, ed., *A European Geography* (Harlow, UK: Addison Wesley Longman, 1998), p. 38.

FROM THE FIELD NOTES

"It seemed the quickest way to go from central London to central Paris: the Eurostar "Chunnel" Express from Waterloo Station to the *Gare du Nord* in less than three hours. But it was also a study in contrasts. On the British side of the English Channel, the train chugged along, sometimes at less than 40 mph (65 kph). Once in the tunnel beneath the Channel, the track was smoother and the speed picked up. On the French side, we moved at 150 mph (250 kph), not fast enough, however, to make up for lost time. The conductor's announcement pointedly referred to delays on the *Angleterre* segment of the journey being the cause of the late arrival at the Paris terminal. This money-losing link between the United Kingdom and the continent is nevertheless symbolic of unifying Europe, but it reveals some other contrasts: the dreadful breakfast out of London and the delightful dinner out of Paris, the chaotic security and immigration scene at Waterloo and the smooth process at *Gare du Nord* (wasn't this supposed to be a domestic intra-EU excursion?), and both ways the complete lack of passenger information—no map, no brochure on the train and its manufacture, no history of the link. Who's in charge here?" © H. J. de Blij

and even the physical barrier of the Alps has not restricted this two-way trade by rail and road.

13 **Transferability** refers to the ease with which a commodity is transported between producer and consumer. Sheer distance, in terms of cost and time, may make it economically impractical to transfer a product. This is not a problem in modestly sized Europe, where distances are short and transport systems efficient.

14 **Intervening opportunity**, the third of Ullman's spatial interaction principles, holds that potential trade between two places, even if they are in a position of complementarity and do not have problems of transferability, will develop only in the absence of a closer, intervening source of supply. Using our current example, if a major coal reserve were discovered in southern Switzerland, Italy would avail itself of that (intervening) opportunity, reducing or eliminating its imports from Western Europe.

Europe's internal spatial interaction is facilitated by what in many respects is the world's most effective network of railroads, highways, waterways, and air routes. This network is continuously improving as tunnels, bridges, high-speed rail lines (see photo at left), and augmented airports are built. The continent's burgeoning cities and their environs, however, are increasingly troubled by severe congestion.

An Urbanized Realm

Overall, 73 percent of Europe's population resides in cities and towns, but this average is exceeded by far in the west and not attained in the east.

Large cities are the crucibles of their nations' cultures. In his 1939 study of the pivotal role of great cities in the development of national cultures, American geographer Mark Jefferson postulated the law of the **pri-** 15
mate city, which stated that "a country's leading city is always disproportionately large and exceptionally expressive of national capacity and feeling." Though imprecise, this "law" can readily be demonstrated using European examples. Certainly Paris personifies France in countless ways, and nothing in England rivals London. In both of these primate cities, the culture and history of a nation and empire are indelibly etched in the urban landscape. Similarly, Vienna is a microcosm of Austria, Warsaw is the heart of Poland, Stockholm typifies Sweden, and Athens is Greece. Today, each of these (together with the other primate cities of Europe) sits atop a hierarchy of urban centers that has captured the lion's share of its national population growth since World War II.

Primate cities tend to be old, and in general the European cityscape looks quite different from the American. Seemingly haphazard inner-city street systems impede traffic; central cities may be picturesque, but they are also cramped. The urban layout of the London region (Fig. 1-9) reveals much about the internal spatial structure of the European **metropolis** (the central city and its suburban ring). 16
The metropolitan area remains focused on the large city at its center, especially the downtown *central business district* (*CBD*), which is the oldest part of the urban agglomeration and contains the region's largest concentration of business, government, and shopping facilities as well as its wealthiest and most prestigious residences. Wide residential sectors radiate outward from the CBD across the rest of the central city, each one home to a particular income group. Beyond the central city lies a sizeable suburban ring, but residential densities are much higher here than in the United States because the European tradition is one of setting aside recreational spaces (in "greenbelts") and living in apartments rather than in detached single-family

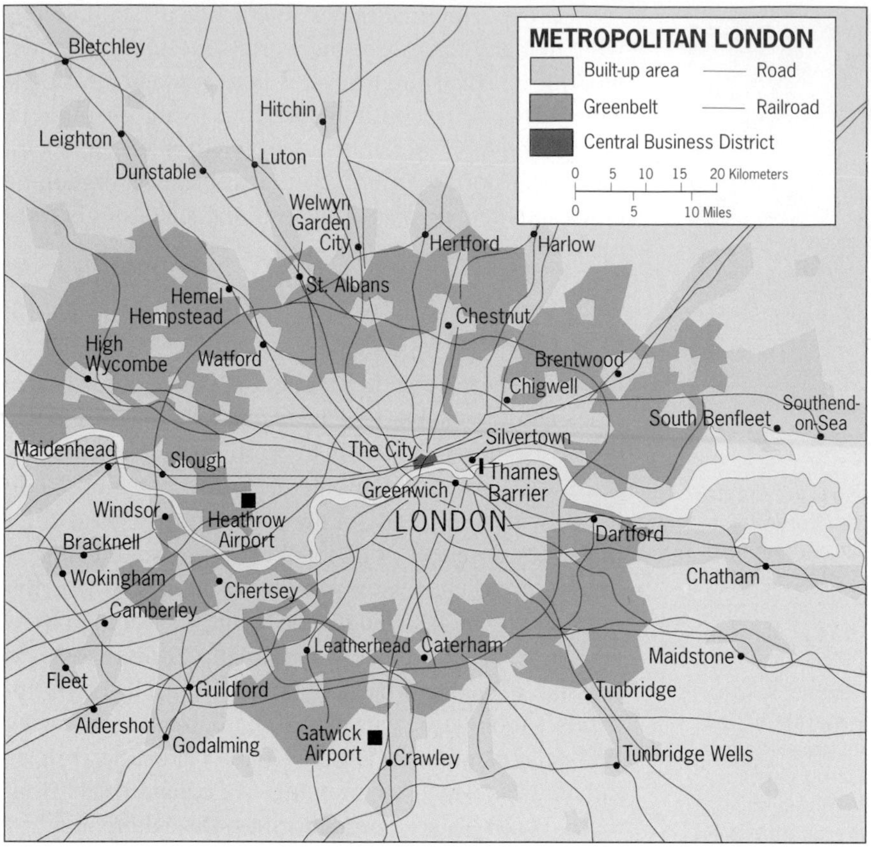

FIGURE 1-9 © H. J. de Blij, P. O. Muller, and John Wiley & Sons, Inc.

houses. There is also a greater reliance on public transportation, which further concentrates the suburban development pattern. That has allowed many nonresidential activities to suburbanize as well, and today ultramodern outlying business centers increasingly compete with the CBD in many parts of urban Europe (see photo p. 57).

A Changing Population

When a population urbanizes, average family size declines, and so does the overall rate of natural increase. There was a time when Europe's population was (in the terminology of population geographers) exploding, sending millions to the New World and the colonies and still growing at home. But today Europe's indigenous population, unlike most of the rest of the world's, is actually shrinking. To keep a population from declining, the (statistically) average woman must bear 2.1 children. For Europe as a whole, that figure was 1.4 in 2004. Several large countries recorded 1.3, including Germany and Italy. And no fewer than ten Eastern European countries recorded 1.2—the lowest ever seen in any human population.

Such *negative population growth* poses serious challenges for any nation. When the population pyramid becomes top-heavy, the number of workers whose taxes pay for the social services of the aged goes down, leading to reduced pensions and dwindling funds for health care. Governments that impose tax increases endanger the business climate; their options are limited. Europe, and especially Western Europe, is experiencing a *population implosion* that will be a formidable challenge in decades to come.

Meanwhile, *immigration* is partially offsetting the losses European countries face. Millions of Turkish Kurds (mainly to Germany), Algerians (France), Moroccans (Spain), West Africans (Britain), and Indonesians (Netherlands) are changing the social fabric of what once were unicultural nation-states. One key dimension of this change is the spread of Islam in Europe (Fig. 1-10). Muslim populations in Eastern Europe (such as Albania's, Kosovo's, and Bosnia's) are local, Slavic communities converted during the period of Ottoman Turkish rule. The Muslim sectors of Western European countries, on the other hand, represent more recent immigrations.

The vast majority of these immigrants are intensely devout, politically aware, and culturally insular. They continue to arrive in a Europe where native populations are stagnant or declining, where religious institutions are weakening, where secularism is rapidly rising, where

Booming *suburban downtowns* (see Chapter 3) are transforming the U.S. metropolitan landscape, and such "edge cities" are now multiplying in Western Europe as well. The biggest and most important of these complexes is *La Défense*, located just west of Paris, which has grown so robustly that it is now continental Europe's single largest business district, whose annual transactions exceed the GNI of the Netherlands! Besides its economic prowess, the cubic Grande Arche that forms the centerpiece of *La Défense* is an architectural jewel that is the western anchor of Paris's "Historic Axis" which includes the *Arc de Triomphe* (top center of photo), *Champs Elysées, Place de la Concorde*, and the *Louvre*. © Yann Arthus-Bertrand/Photo Researchers.

political positions often appear to be anti-Islamic, and where cultural norms are incompatible with Muslim traditions. Many young men, unskilled and uncompetitive in a European Union (EU) where unemployment is already high, get involved in petty crime or the drug trade, are harassed by the police, and turn to their faith for solace and reassurance. Unlike most other immigrant groups, Muslim communities also tend to resist assimilation, making Islam the essence of their identity. In Britain alone there are more than 1500 mosques, in themselves a transformation of local cultural landscapes.

In truth, European governments have not done enough to foster the very integration they see Muslims rejecting. The French assumed that their North African immigrants would aspire to assimilation as Muslim children went to public schools from their urban public-housing projects; instead they got into disputes over the dress codes of Muslim girls. The Germans for decades would not award German citizenship to children of immigrant parents born on German soil. The British found their modest efforts at integration stymied by the Muslim practice of importing brides from Islamic countries, which had the effect of perpetuating segregation. In many European cities, the housing projects that form the living space of Muslim residents (by now many of them born and retired there) are dreadful, barren, impoverished environments unseen by Europeans whose images of Paris, Manchester, or Frankfurt are far different.

The social and political implications of Europe's demographic transformation are numerous and far-reaching. Long known for tolerance and openness, European societies are attempting to restrict immigration in various ways; political parties with anti-immigrant platforms are gaining ground. Multiculturalism poses a growing challenge in a changing Europe.

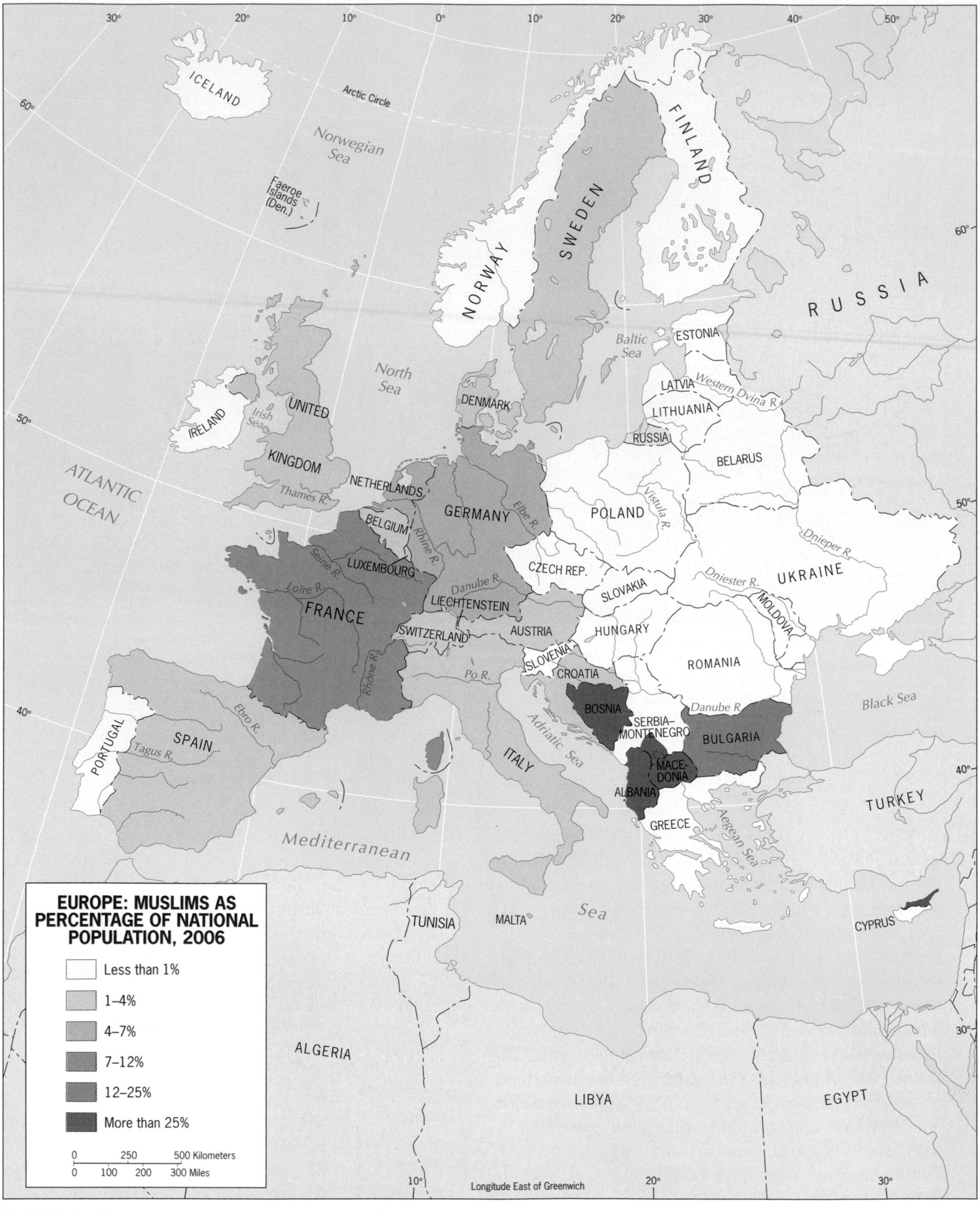

FIGURE 1-10 © H. J. de Blij, P. O. Muller, and John Wiley & Sons, Inc.

The growing Islamic presence in Europe is transforming urban cultural landscapes. The minarets of mosques now rise between the spires of churches in many major cities. A mosque often functions as the centerpiece for a multipurpose Muslim cultural center that serves as a social gathering place for a large neighborhood, as is the case here in eastern London along Whitechapel Road. Note the crowds on and across the street from the mosque; this was a day of Friday prayer in June 2004 led by the imam of the Grand Mosque of Mecca, Saudi Arabia, attracting thousands of worshippers. . . . Relations between Muslim residents and domestic authorities are sometimes strained by cultural practices such as the wearing of women's headscarves. In France, Muslim girls attending public schools were forbidden from wearing such headscarves, deemed to represent "overt" religious symbolism (at the same time Christian children were instructed not to wear "large" crosses as jewelry). The photo on the right shows Muslim women in Berlin walking past the election posters of two German leaders confronting such issues: one of Germany's states in 2004 was the first to ban Muslim public school teachers from wearing the headscarf. Europe's Muslim immigration poses a multifaceted challenge to its hosts. (*left*) © O. Anderson/AFP/Getty Images News and Sport Services, (*right*) © Eric Peterberg/AFP/Getty Images News and Sport Services.

EUROPE'S MODERN TRANSFORMATION

At the end of World War II, much of Europe lay shattered, its cities and towns devastated, its infrastructure wrecked, its economies devastated. The Soviet Union had taken control over the bulk of Eastern Europe, and communist parties seemed poised to dominate the political life of major Western European countries.

In 1947, U.S. Secretary of State George C. Marshall proposed a European Recovery Program designed to counter all this dislocation and to create stable political conditions in which democracy would survive. Over the next four years, the United States provided about $13 billion in assistance to Europe (almost $100 billion in today's money). Because the Soviet Union refused U.S. aid and forced its Eastern European satellites to do the same, the Marshall Plan applied solely to 16 European countries, including defeated (West) Germany, and Turkey.

European Unification

The Marshall Plan did far more than stimulate European economies. It showed European leaders that their countries needed a joint economic-administrative structure not only to coordinate the financial assistance, but also to ease the flow of resources and products across Europe's mosaic of boundaries, to lower restrictive trade tariffs, and to seek ways to effect political cooperation.

For all these needs Europe's governments had some guidelines. While in exile in Britain, leaders of three small countries—Belgium, the Netherlands, and Luxembourg—had been discussing an association of this kind even before the end of the war. There, in 1944, they formulated and signed the Benelux Agreement, intended to achieve total economic integration. When the Marshall Plan was launched, the Benelux precedent helped speed the creation of the Organization for European Economic Cooperation (OEEC), which was established to coordinate the investment of America's aid (see box titled "Supranationalism in Europe").

Soon the economic steps led to greater political cooperation as well. In 1949, the participating governments created the Council of Europe, the beginnings of what was to become a European Parliament meeting in Strasbourg, France. Europe was embarked on still another political revolution, the formation of a multinational union involving a growing number of European

Supranationalism in Europe

1944	Benelux Agreement signed.
1947	Marshall Plan created (effective 1948–1952).
1948	Organization for European Economic Cooperation (OEEC) established.
1949	Council of Europe created.
1951	European Coal and Steel Community (ECSC) Agreement signed (effective 1952).
1957	Treaty of Rome signed, establishing European Economic Community (EEC) (effective 1958), also known as the Common Market and "The Six."
	European Atomic Energy Community (EURATOM) Treaty signed (effective 1958).
1959	European Free Trade Association (EFTA) Treaty signed (effective 1960).
1961	United Kingdom, Ireland, Denmark, and Norway apply for EEC membership.
1963	France vetoes United Kingdom EEC membership; Ireland, Denmark, and Norway withdraw applications.
1965	EEC-ECSC-EURATOM Merger Treaty signed (effective 1967).
1967	European Community (EC) inaugurated.
1968	All customs duties removed for intra-EC trade; common external tariff established.
1973	United Kingdom, Denmark, and Ireland admitted as members of EC, creating "The Nine." Norway rejects membership in the EC by referendum.
1979	First general elections for a European Parliament held; new 410-member legislature meets in Strasbourg.
	European Monetary System established.
1981	Greece admitted as member of EC, creating "The Ten."
1985	Greenland, acting independently of Denmark, withdraws from EC.
1986	Spain and Portugal admitted as members of EC, creating "The Twelve."
	Single European Act ratified, targeting a functioning European Union in the 1990s.
1987	Turkey and Morocco make first application to join EC. Morocco is rejected; Turkey is told that discussions will continue.
1990	Charter of Paris signed by 34 members of the Conference on Security and Cooperation in Europe (CSCE).
	Former East Germany, as part of newly reunified Germany, incorporated into EC.
1991	Maastricht meeting charts European Union (EU) course for the 1990s.
1993	Single European Market goes into effect.
	Modified European Union Treaty ratified, transforming EC into EU.
1995	Austria, Finland, and Sweden admitted into EU, creating "The Fifteen."
1999	European Monetary Union (EMU) goes into effect.
	Helsinki summit discusses fast-track negotiations with six prospective members and applications from six others; prospects for Turkey considered in longer term.
2001	Denmark's voters reject EMU participation by 53 to 47 percent.
2002	The euro is introduced as historic national currencies disappear in 12 countries.
2003	First draft of a European Constitution is published to mixed reviews from member-states.
2004	Historic expansion of EU from 15 to 25 countries with the admission of Cyprus, the Czech Republic, Estonia, Hungary, Latvia, Lithuania, Malta, Poland, Slovakia, and Slovenia.

17 states. Geographers define **supranationalism** as the voluntary association in economic, political, or cultural spheres of three or more independent states willing to yield some measure of sovereignty for their mutual benefit. In later chapters we will encounter other supranational organizations, including the North American Free Trade Agreement (NAFTA), but none had reached the plateau achieved by the European Union (EU).

In Europe, the key initiatives arose from the Marshall Plan and lay in the economic arena; political integration came more haltingly. Under the Treaty of Rome, six countries joined to become the European Economic Community (EEC) in 1957, also called the "Common Market." In 1973 the United Kingdom, Ireland, and Denmark joined, and the renamed European Community (EC) had nine members. As Figure 1-11 shows, membership reached 15 countries in 1995, after the organization had been renamed yet one more time to become the *European Union* (*EU*).

The European Union is not just a paper organization for bankers and manufacturers. It has a major impact on the daily lives of its member countries' citizens in countless ways. (You will see one of these ways when you arrive at an EU airport and find that EU-passport holders move through inspection at a fast pace, while non-EU citizens wait in long immigration lines.) Taxes tend to be high in Europe, and those collected in the richer member-states

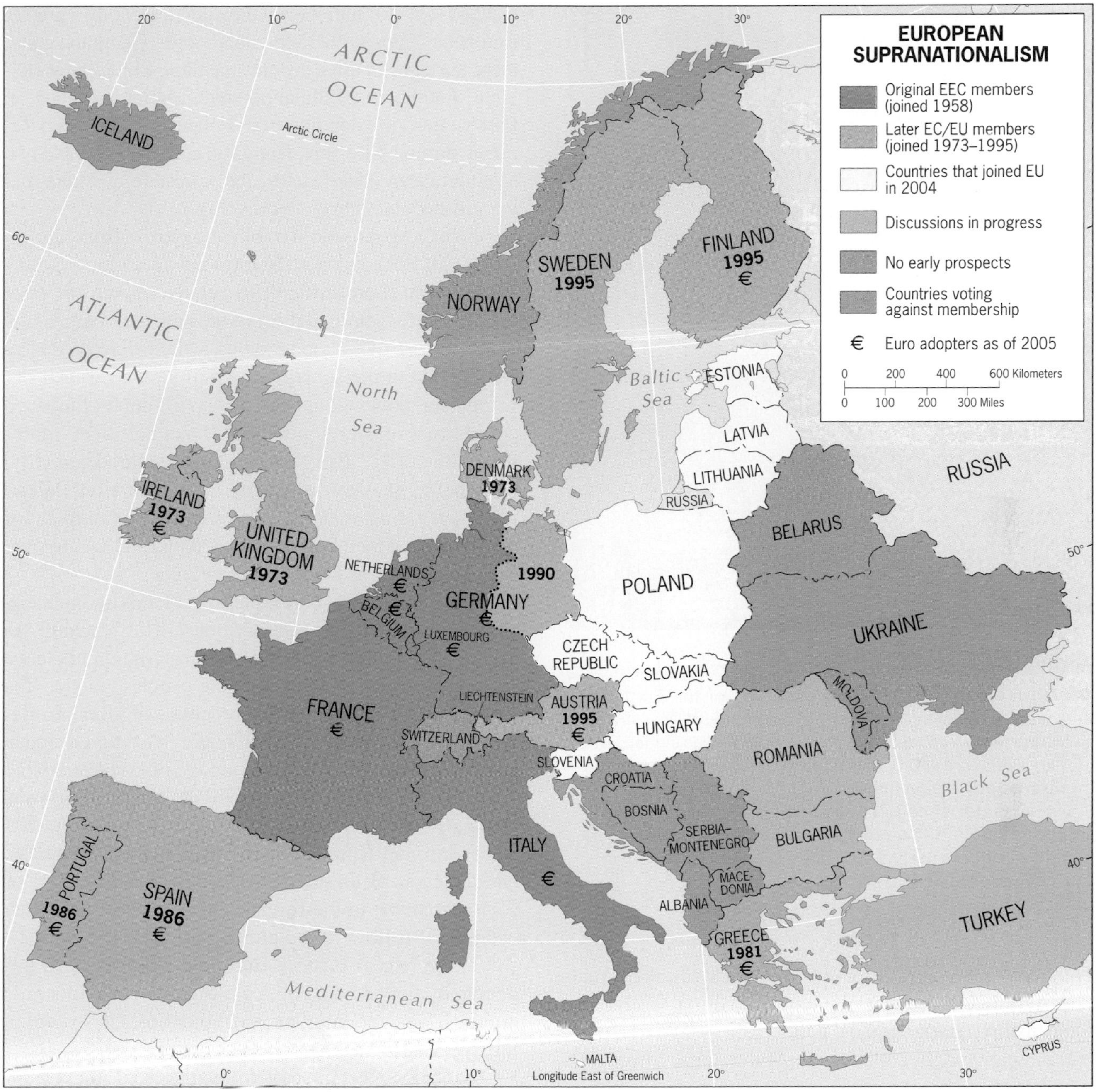

FIGURE 1-11 © H. J. de Blij, P. O. Muller, and John Wiley & Sons, Inc.

are used to subsidize growth and development in the less prosperous ones. This is one of the burdens of membership that is not universally popular in the EU, to say the least. But it has strengthened the economies of Portugal, Greece, and other national and regional economies to the betterment of the entire organization. Some countries also object to the terms and rules of the Common Agricultural Policy (CAP) which, according to critics, supports farmers far too much and, according to others, far too little. (France in particular obstructs efforts to move the CAP closer to consensus, subsidizing its agricultural industry relentlessly while arguing that this protects its rural cultural heritage as well as its farmers.)

New Money

Ever since the first steps toward European unification were taken, EU planners dreamed of a time when the EU would have a single currency not only to symbolize its strengthening unity but also to establish a joint counterweight to the mighty American dollar. A European Monetary Union (EMU) became a powerful objective of EU

FROM THE FIELD NOTES

"No place in Europe displays the flag of the European Union as liberally as does the Dutch city of Maastricht, where the European Union Treaty of 1991 was signed. The flag, seen here at City Hall, shows twelve yellow stars (representing the signatories to the Treaty) against a blue background. Although three additional states joined the EU in 1995 [and 10 more in 2004], the flag will remain as is, and will not, as the U.S. flag does, change to reflect changing times." © H. J. de Blij

planners, but many observers thought it unlikely that member-states would in the end be willing to give up their historic marks, francs, guilders, liras, and escudos. Yet it happened. In 2002, twelve of the (then) fifteen EU countries withdrew their currencies and began using the euro, with only the United Kingdom (Britain), Denmark, and Sweden staying out (Fig. 1-11). Add nonmember Norway to this group, and not a single Scandinavian country has converted to the euro.

Momentous Expansion

Expansion has always been an EU objective, and the subject has always aroused passionate debate (see the Issue Box titled "Europe: How Desirable Is Economic and Political Union?"). Will the incorporation of weaker economies undermine the strength of the whole? Despite such misgivings, negotiations to expand the EU have long been in progress, and in 2004 a momentous milestone was reached: ten new members were added, creating a greater European Union with 25 member-states. Geographically, these ten came in three groups: the three Baltic states (Estonia, Latvia, and Lithuania); five contiguous states in Eastern Europe extending from Poland and the Czech Republic through Slovakia, Hungary, and Slovenia; and two Mediterranean island-states, the ministate of Malta and the still-divided state of Cyprus (Fig. 1-11).

Numerous structural implications arise from this, affecting all EU countries. A common agricultural policy is now even more difficult to achieve, given the poor condition of farming in most of the new members. Some of the former EU's poorer states, which were on the receiving end of the subsidy program that aided their development, now will have to pay up to support the much poorer new eastern members. Disputes over representation at EU's Brussels headquarters arose quickly. Even before the new members' 2004 accession, Poland was demanding that the representative system favor medium-sized members (such as Poland and Spain) over larger ones such as Germany and France.

The geographic consequences of this momentous EU expansion for Europe as a whole are far-reaching. As Figure 1-11 shows, the EU's new eastern borders leave several groups of countries and peoples outside the Union: (1) the former Soviet republics of Ukraine, Belarus, and Moldova; (2) two major Eastern European states, Romania and Bulgaria; and (3) the states arising from the breakup of Yugoslavia along with impoverished Albania. Already, negotiations between the EU leadership and Romania and Bulgaria are in progress, and the entry of these countries is expected in 2007 or soon thereafter. In the former Yugoslavia, Croatia may eventually follow its neighbor Slovenia into the EU. But in the rest of Eastern Europe the process will undoubtedly slow down. Political instability, undemocratic regimes, weak economies, and other obstacles stand in the way.

We should take note of still another potential candidate for EU membership: Turkey. EU leaders would like to include a mainly Muslim country in what Islamic states sometimes call the "Christian Club," but Turkey needs to make progress on social standards, human rights, and economic policies before its accession can be contemplated. Still, the fact that the EU has now reached deeply into Eastern Europe, encompasses 25 members, has a common currency and a parliament, is considering a constitution, and is even contemplating expansion beyond the realm's borders constitutes a tremendous and historic achievement in this fragmented, fractious part of the world. The EU now has a combined population of almost 460 million constituting one of the world's richest markets; its member-states account for more than 40 percent of the

Europe: How Desirable Is Economic and Political Union?

Regional ISSUE

IN SUPPORT OF UNION

"As a public servant in the French government I have been personally affected by the events of the past half-century. I am in favor of European unification because Europe is a realm of comparatively small countries, all of which will benefit from economic and political union. Europe's history is bedeviled by its fragmentation—its borders and barriers and favored cores and disadvantaged peripheries, its different economies and diverse currencies, its numerous languages and varied cultures. Now we have the opportunity and determination to overcome these divisions and to create an entity whose sum is greater than its parts, a supranational Europe that will benefit all of its members.

"In the process we will not only establish a common currency, eliminate trade tariffs and customs barriers among our members, and enable citizens of one member-state to live and work in any of the others. We will also make European laws that will supersede the old "national" laws and create regulations that will be adhered to by all members on matters ranging from subsidies to help small farms survive to rules regarding environmental protection. In order to assist the poorer members of this European Union, the richer members will send billions of euros to the needy ones, paid for from the taxes levied on their citizens. This has already lifted Spain and Ireland from poverty to prosperity. But to ensure that "national" economies do not get out of line, there will be rules to govern their fiscal policies, including a debt limit of 3 percent of GDP.

"The EU will require sacrifices from people and governments, but the rewards will make these worth it. Our European Parliament, with representatives from all member countries, is the harbinger of a truly united Europe. The European Commission deals with the practical problems as we move toward unification, including the momentous EU enlargement of 2004. Less than a half-century after the first significant steps were taken, the EU has 25 member-states including former wartime enemies, reformed communist economies, and newly independent entities. Rather than struggling alone, these member-states will form part of an increasingly prosperous and economically powerful whole.

"I hope that administrative coordination and economic union will be followed by political unification. The ultimate goal of this great movement should be the creation of a federal United States of Europe, a worthy competitor and countervailing force in a world dominated by another federation, the United States of America."

IN OPPOSITION TO THE EUROPEAN UNION

"I am a taxi driver in Vienna, Austria, and here's something to consider: the bureaucrats, not the workers, are the great proponents of this European Union plan. Let me ask you this: how many member countries' voters have had the chance to tell their governments whether they want to join the EU and submit to all those confounded regulations? It all happened before we realized the implications. At least the Norwegians had common sense: they stayed out of it, because they didn't want some international body to tell them what to do with their oil and gas and their fishing industry. But I can tell you this: if the workers of Europe had had the chance to vote on each step in this bureaucratic plot, we'd have ended it long ago.

"So what are we getting for it? Higher prices for everything, higher taxes to pay for those poor people in southern Italy and needy Portugal, costly environmental regulations, and—here's the worst of it—a flood of cheap labor and an uncontrolled influx of immigrants. Notice that this new 2004 expansion wasn't voted on by the Germans (they're too small to matter much, but Germany is the largest EU member), and do you know why? Because the German government knew well and good that the people wouldn't support this ridiculous scheme. This whole European Union program is concocted by the elite, and we have to accept it whether we like it or not.

"Does make me wonder, though. Do these EU-planners have any notion of the real Europe they're dealing with? They talk about how those "young democracies" joining in 2004 will have their democratic systems and their market reforms locked into place by being part of the Union, but I have friends who try to do business in Poland and Slovakia, and they tell me that there may be democracy, but all they experience is corruption, from top to bottom. I'll bet that those subsidies we'll be paying for to help these "democracies" get on their feet will wind up in Swiss bank accounts. By the way, the Swiss, you'll notice, aren't EU members either. Wonder why?

"A United States of Europe? Not as far as I am concerned. I am Austrian, and I am and will always be a nationalist as far as my country is concerned. I speak German, but already English is becoming the major EU language, though the Brits were smart enough not to join the euro. I'm afraid that these European Commission bureaucrats have their heads in the clouds, which is why they can't see the real Europe they're putting in a straitjacket."

Vote your opinion at www.wiley.com/college/deblij

world's exports. Although, as we will note later, Europe is currently experiencing budgetary troubles, it remains an economic power of global significance.

It is remarkable that all this has been accomplished in little more than half a century. Some of the EU's leaders want more than economic union: they envisage a United States of Europe, a political as well as an economic competitor for the United States. To others, such a "federalist" notion is an abomination not even to be mentioned (the British in general are especially wary of such an idea) and certainly not to be made part of any European constitution. Whatever happens, Europe is going through still another of its revolutionary changes, and when you study its evolving map you are looking at history in the making.

Centrifugal Forces

For all its dramatic progress toward unification, Europe remains a realm of geographic contradictions. Europeans are well aware of their history of conflict, division, and repeated self-destruction. Will supranationalism finally overcome the centrifugal forces that have so long and so frequently afflicted this part of the world?

Even as Europe's states have been working to join forces in the EU, many of those same states are con-
18 fronting severe centrifugal stresses. The term **devolution** has come into use to describe the powerful centrifugal forces whereby regions or peoples within a state, through negotiation or active rebellion, demand and gain political strength and sometimes autonomy at the expense of the center. Most states exhibit some level of internal regionalism, but the process of devolution is set into motion when a key centripetal binding force—the nationally accepted idea of what a country stands for—erodes to the point that a regional drive for autonomy, or for outright secession, is launched.

As Figure 1-12 shows, numerous European countries are affected by devolution. States large and small, young and old, EU members and non-EU members must deal with the problem. Even the long-stable United Kingdom is affected. England, the historic core area of the British Isles, dominates the UK in terms of population as well as political and economic power. The country's three other entities—Scotland, Wales, and Northern Ireland—were acquired over several centuries and attached to England (hence the "United" Kingdom). But neither time nor representative democratic government was enough to eliminate all latent regionalism in these three components of the UK. During the 1960s and 1970s the British government confronted a virtual civil war in Northern Ireland and rising tides of nationalism in Scotland and Wales. In 1997, the government in London gave the Scots and Welsh the opportunity to vote for greater autonomy in new regional parliaments that would have limited but significant powers over local affairs. The Scots voted overwhelmingly in favor, and the Welsh by a slim majority—and thus a major devolutionary step was taken in one of Europe's oldest, most durable, and most unified states.

Even while this devolutionary process was ongoing, the British government still joined the European Union on behalf not only of the English, but also of the Scots, Welsh, and Northern Irelanders seeking a new relationship with London. As the map shows, the UK is not alone in such contradiction. Spain faces severe devolutionary forces in its Basque area and lesser ones in Catalonia and Galicia; France contends with a secessionist movement on its island of Corsica; Belgium is riven by Flemish-Walloon separatism; Italy confronts devolutionary pressures in South Tyrol and Lombardy. In recent decades Eastern Europe has been a cauldron of devolution as Yugoslavia and Czechoslovakia collapsed, Moldova fragmented, and Ukraine suffered from the stresses of its historical (Russian-penetrated) geography.

Political devolution is not the only centrifugal force to buffet European states. As the European Union materialized, its freedoms (in the form of money flows, labor movements, and transferabilities) led to the emergence of powerful urban regions as hubs of economic power and influence, in some ways beyond the control of their national governments. Examples include the Rhône-Alpes region in France, centered on Lyon; Lombardy in Italy, focused on Milan; Catalonia in Spain, anchored by Barcelona; and Baden-Württemberg in Germany, head-
quartered by Stuttgart. This group, known as the **Four** 19
Motors of Europe, bypasses not only their national governments in dealing with each other but even extends their business channels to span the world. The Japanese economist Kenichi Ohmae calls such economic power-
houses **regional states**, entities that defy old borders and 20
are shaped by the globalizing economy of which they have become a part.

Like devolutionary forces, these emerging regional states are changing the map of Europe. In some ways, they are beginning to supplant the old framework that arose from the nation-states of an earlier period. Elsewhere, economic development on both sides of international boundaries is having a similar effect, creating *Euroregions* that foster more localized, cross-border cooperation on a continent once rigidly partitioned.

All these developments underscore how far-reaching Europe's current transformation is. The Marshall Plan jump-started it; good economic times sustained it; and the end of the Cold War stimulated it. But Europe's gains remain clouded by violent discord in parts of the United Kingdom, Spain, France, and the Balkans, by devolutionary forces elsewhere, and by uncertainties over EU expansion and further political integration.

From Coast to Coast?

You cannot look at maps of the eastward expansion of the European Union without a thought crossing your mind: will the European Union eventually include even Russia, and will the EU, as a result, extend from the Atlantic coast to Pacific shores?

This optimistic projection is popular with those who see a European superpower emerge as a counterweight to

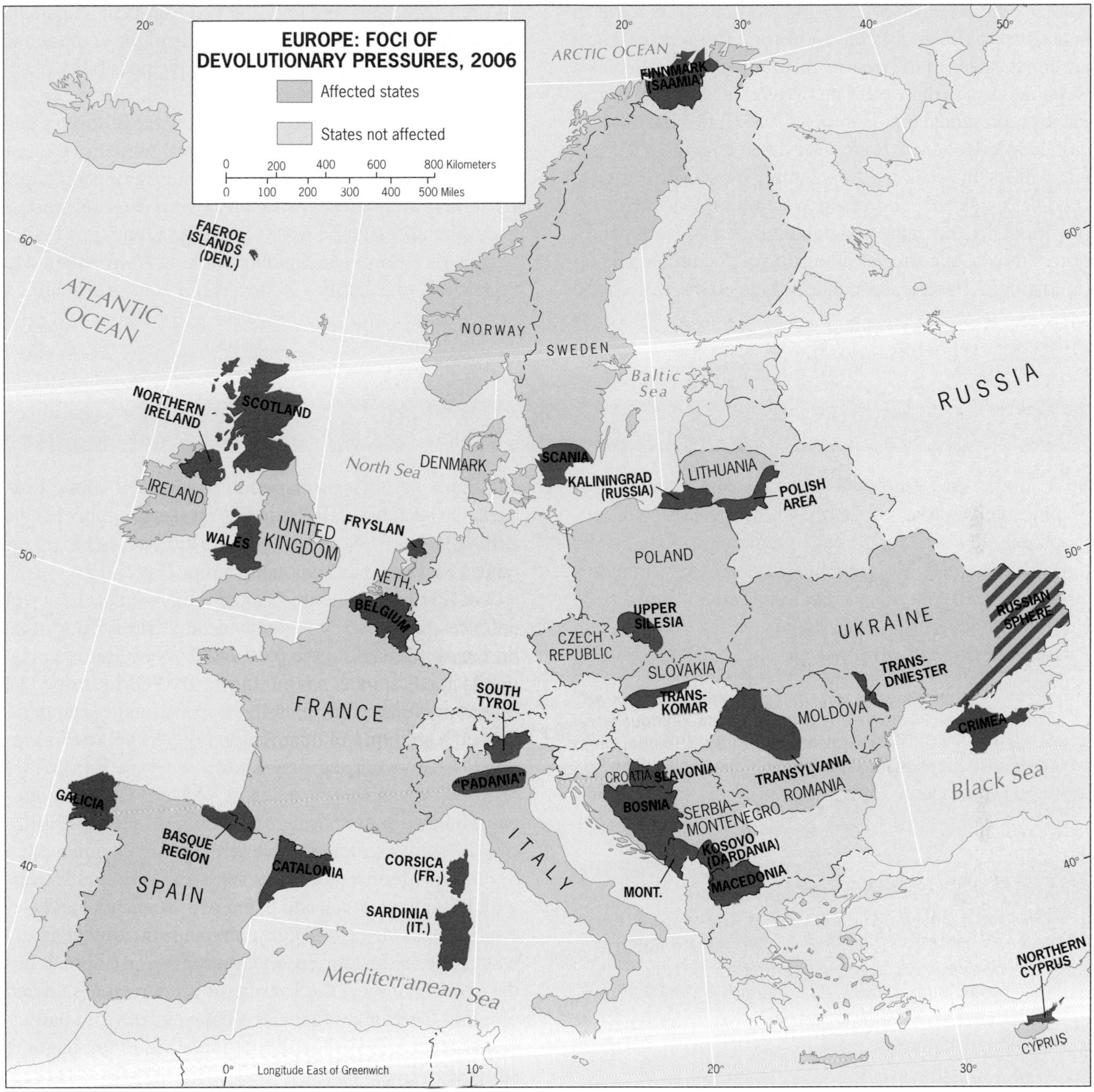

FIGURE 1-12 © H. J. de Blij, P. O. Muller, and John Wiley & Sons, Inc.

the United States and even to a future superpower China, but in 2005 the prospects were not favorable. Even as democracy, free markets, and human rights on the European model were progressing in Bulgaria, Romania, Croatia, and even Turkey, Russia was losing the gains it had made in the previous decade. Government in Russia was becoming more authoritarian, economic controls were tightening, the media were losing their freedom and independence, and the human rights issue raised growing concern in Europe. Russia's president, Vladimir Putin, stated repeatedly that he wished Russia to become, as the Soviet Union had been before its collapse, a global political, economic, and military power. Clearly he had in mind a Russian, not a Russian-European, power. He left no doubt that Russia's continuing influence in former colonies of the Soviet Empire, including parts of Eastern Europe, the Transcaucasus, and Central Asia, were part of this plan. Surveys indicated that he was accurately reflecting public opinion in his country.

In late 2004 a presidential election in Ukraine, Europe's largest country territorially now wedged between the European Union and Russia, became a stage for

Russia's geopolitical strategy. As we note in more detail later, Russian influences (and a historic Russian presence) are far stronger in the east of this former Soviet Socialist Republic than in the west. The two candidates for the presidency represented the "European" west and the "Russian" east, respectively, and Putin intervened directly in Ukrainian politics by openly campaigning for the pro-Russian candidate. The election result was close (in favor of the pro-Russian candidate), and hundreds of thousands of protesters took to the streets, protesting alleged fraud and intimidation. After a tense standoff the election was voided and repeated, and this time the pro-European candidate prevailed—but not before eastern parts of the country had proclaimed their intent to investigate the possibility of seceding should the result be reversed.

Ukraine was a clear contest between European and Russian spheres of interest. In general, western Ukrainians want to take their country into the European Union. Easterners want closer association with Moscow and its geopolitical designs. Clearly, it will be a long time before visions of a European Union stretching from Atlantic to Pacific become reality—if they ever materialize at all.

REGIONS OF THE REALM

So complex are Europe's physical and human geographies that it proves to be one of the most difficult geographic realms when it comes to regionalization. Europe may not be large territorially, but it incorporates enormous environmental, cultural, political, and economic diversity. Europe extends from moist Atlantic shores to continental Russian borders and from Arctic cold to Mediterranean warmth. Small as Europe is, the realm contains nearly 40 countries.

To better comprehend Europe's geographic diversity, we group its countries into five regions based on their proximity, environmental similarities, historical associations, cultural congruities, social parallels, and economic linkages. On such bases, we map (1) Western Europe, (2) the British Isles, (3) Northern (Nordic) Europe, (4) Mediterranean Europe, and (5) Eastern Europe (Fig. 1-13).

A few years ago an irritated American secretary of defense spoke contemptuously of an "Old Europe" that kept harking back to the past and a "New Europe" that would lead the way to the future. By "Old Europe" he meant uncooperative Western Europe, and particularly France and Germany, the region colored yellow in Figure 1-13. Indeed, the map shows the way Europe has been perceived for centuries: a Western Europe dominant, flanked by the United Kingdom offshore; a Northern (Nordic) Europe, remote and sparsely populated; a Southern (Mediterranean) Europe, the *real* Old Europe with its ancient Greek and Roman histories; and an Eastern Europe, always lagging behind in almost every way. Here in Eastern Europe, the secretary implied, lies the "New Europe," the Europe of populous Poland and obliging Romania, of new NATO members and participants in the U.S.-led coalition operating in the ongoing struggle in Iraq.

In fact, this regionalization of Europe still has some merit, but it is being overtaken by events. Northern Europe will always be cold and Mediterranean Europe comparatively warm, and Western Europe will always be moist and Eastern Europe comparatively dry; but human endeavors, as we saw in the first part of this chapter, are changing the functional map. The growing European Union is the most prominent manifestation of this change. As the divisive effects of European boundaries diminish and cross-border interconnections increase, a new regionalization of Europe is emerging. Using this functional-region concept, we would recognize a European core, consisting essentially of regions (1) through

MAJOR CITIES OF THE REALM

City	Population* (in millions)
Amsterdam, Netherlands	1.1
Athens, Greece	3.2
Barcelona, Spain	4.4
Berlin, Germany	3.3
Brussels, Belgium	1.0
Frankfurt, Germany	3.7
London, UK	7.6
Lyon, France	1.4
Madrid, Spain	5.1
Milan, Italy	4.1
Paris, France	9.8
Prague, Czech Republic	1.2
Rome, Italy	2.7
Stuttgart, Germany	2.7
Vienna, Austria	2.2
Warsaw, Poland	2.2

*Based on 2006 estimates.

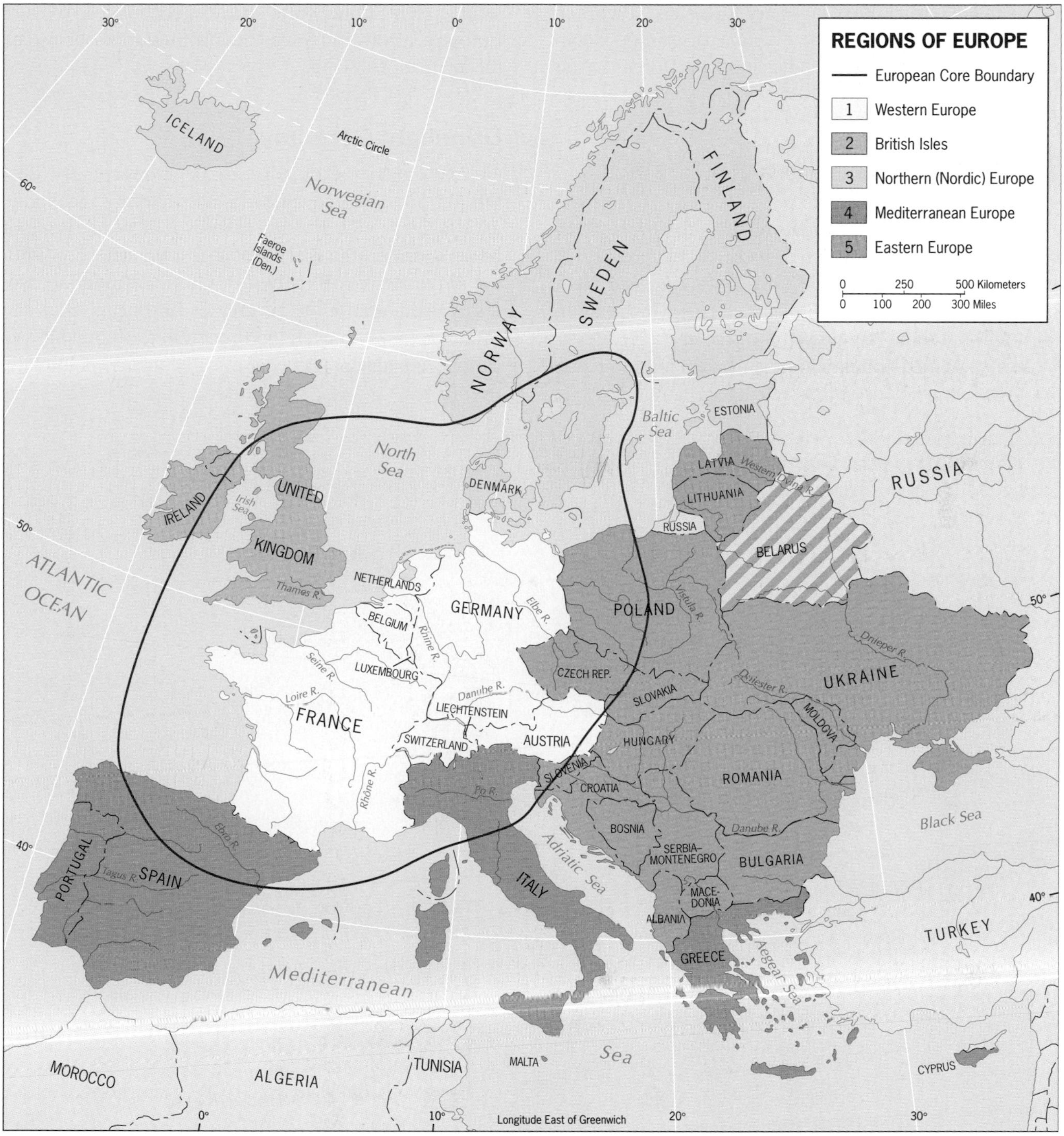

FIGURE 1-13 © H. J. de Blij, P. O. Muller, and John Wiley & Sons, Inc.

(4) on Figure 1-13 and a periphery, mostly in region (5). Embedded in this functional framework are the "Four Motors," "Euroregions," and other spatial expressions of Europe's growing interconnections.

The framework in Figure 1-13 continues to be useful because it provides insight into both Europe's enduring formal regions and its changing functional regions. For all its operational integration, Europe still is a realm whose differences in natural environments, economic opportunities, historical experiences, and relative locations have produced societies and countries with particular, often contrasting, cultural traditions and outlooks. These contrasts make Europe's political map what it is: one of the most tortuously fragmented in the world. And as we noted earlier, devolutionary forces are at work to fracture it even further. The outcome of Europe's contest between centripetal and centrifugal forces is far from certain.

WESTERN EUROPE

"Old" or not, Western Europe remains the heart of the realm and its Union, the hub of its economic power, and the focus of its unifying drive. Germany and France dominate it, flanked by the three Benelux countries in the northwest and the three Alpine states in the east (Fig. 1-14). With a combined population of some 186 million representing eight of the world's richest economies, Western Europe is a powerful force not only in Europe but also on the international stage.

Dominant Germany

Germany today is Europe's most populous country (82.4 million), its most powerful (albeit faltering) economy, and, together with France, its most ardent advocate of union. Germany has been the largest financial contributor to the EU's coffers by far. German leaders ceaselessly promote notions of greater European unity and often try to outdo even the French in their pursuit of a Europe united not only economically but also politically.

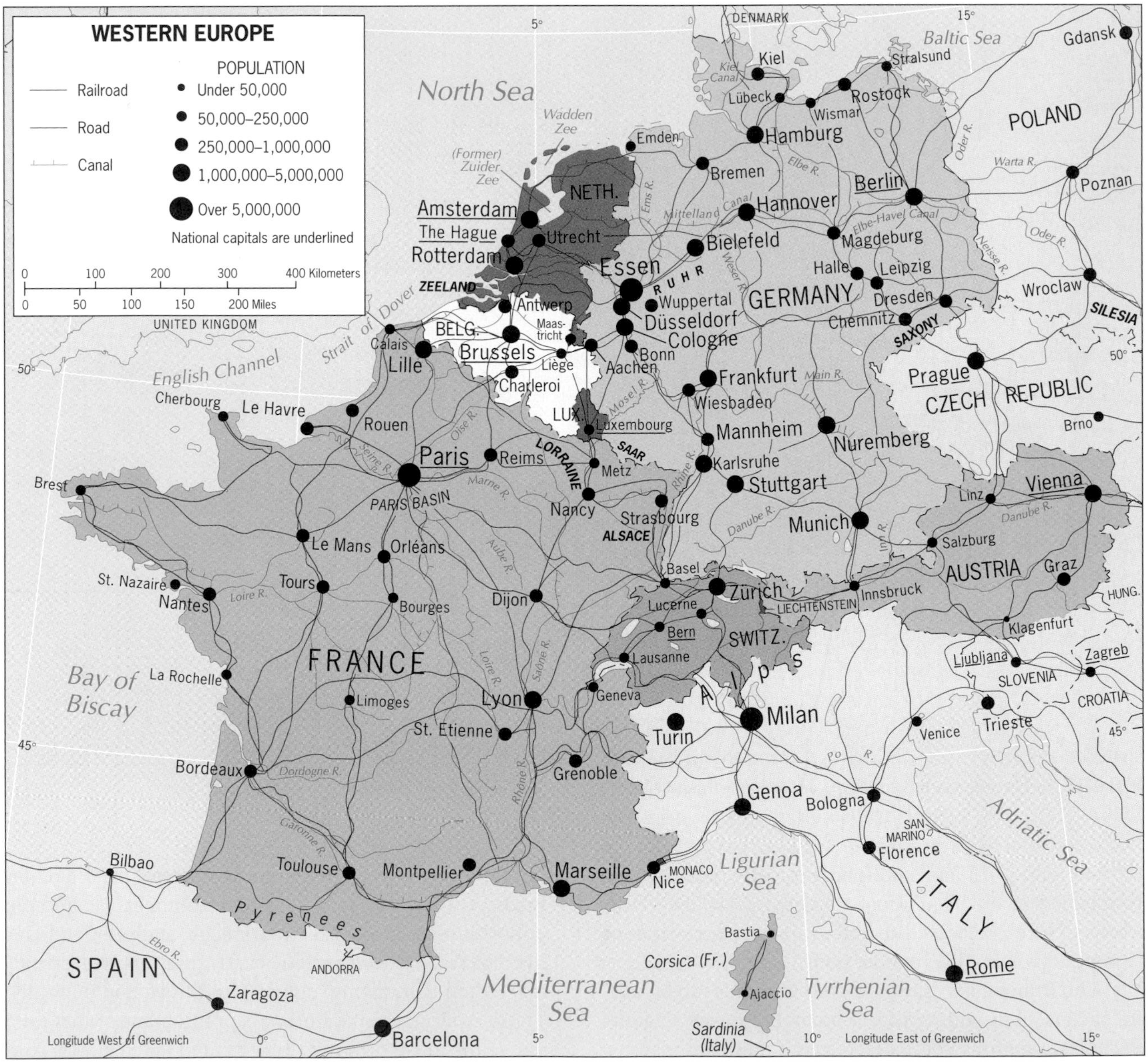

FIGURE 1-14 © H. J. de Blij, P. O. Muller, and John Wiley & Sons, Inc.

Twice during the twentieth century Germany plunged Europe and the world into war, until, finally, in 1945, the defeated and devastated German state was divided into two parts, West and East (see the red map label and line in Fig. 1-15). Its eastern boundaries were also changed, leaving the industrial district of Silesia in newly defined Poland, that of Saxony in communist-ruled and Soviet-controlled East Germany, and the Ruhr in West Germany (Fig. 1-14). Aware that these were the industrial centers that had enabled Nazi Germany to seek world domination through war, the victorious allies laid out this new boundary framework to make sure this would not happen again.

In the aftermath of World War II, Soviet and Allied administration of East and West Germany differed. Soviet rule in East Germany was established on the Russian-communist model and, given the extreme hardships the USSR had suffered at German hands during thc war, harshly punitive. The American-led authority in West Germany was less strict and aimed more at rehabilitation. When the Marshall Plan was instituted, West Germany was included, and its economy recovered rapidly. Meanwhile, West Germany was reorganized politically into a modern federal state along democratic foundations.

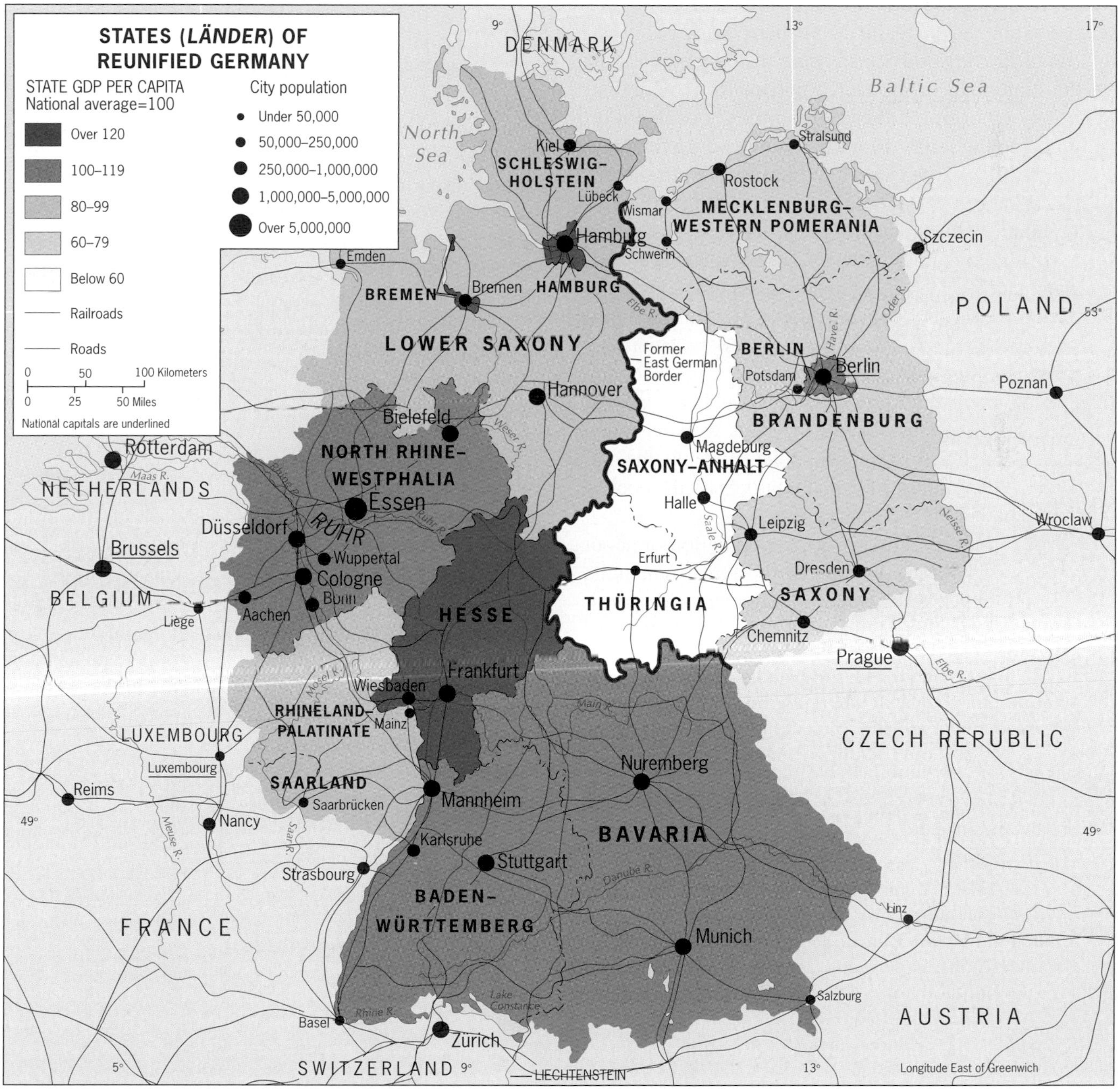

FIGURE 1-15 © H. J. de Blij, P. O. Muller, and John Wiley & Sons, Inc.

West Germany's economy thrived. Between 1949 and 1964, its GNI tripled while industrial output rose 60 percent. It absorbed millions of German-speaking refugees from Eastern Europe (and many escapees from communist East Germany as well). Since unemployment was virtually nonexistent, hundreds of thousands of Turkish and other foreign workers arrived to take jobs Germans could not fill or did not want.

Simultaneously, West Germany's political leaders participated enthusiastically in the OEEC and in the negotiations that led to the six-member Common Market. Geography worked in West Germany's favor: it had common borders with all but one of the EEC member-states. Its transport infrastructure, rapidly rebuilt, was second to none in the realm. More than compensating for its loss of Saxony and Silesia were the expanding Ruhr (in the hinterland of the Dutch port of Rotterdam) and the newly emerging industrial complexes centered on Hamburg in the north, Frankfurt (the leading financial center as well) in the center, and Stuttgart in the south. West Germany exported huge quantities of iron, steel, motor vehicles, machinery, textiles, and farm products.

No economy grows without setbacks and slowdowns, however, and Germany experienced such problems in the 1970s and 1980s, when energy shortages, declining competitiveness on world markets, lagging modernization (notably in the aging Ruhr), and social dilemmas involving rising unemployment, an aging population, high taxation, and a backlash against foreign resident workers roiled West German society. And then, quite suddenly, the collapse of the communist Soviet Union opened the door to reunification with East Germany.

In 1990, West Germany had a population of about 62 million and East Germany 17 million. Communist misrule in the East had yielded outdated factories, crumbling infrastructures, polluted environments, drab cities, inefficient farming, and inadequate legal and other institutions. Reunification was more a rescue than a merger, and the cost to West Germany was enormous. When the West German government imposed sales-tax increases and an income-tax surcharge on its citizens, a sizeable percentage of Westerners doubted the wisdom of reunification. It was projected that it would take decades to reconstruct Virginia-sized East Germany: ten years later, exports from the former East still contributed only about 7 percent of the national total.

Once again, however, the German economy proved its capacity for recovery. Helped by the initial (but, as it turned out, temporary) weakness of the new European currency, the euro, German exports rebounded on world markets. Economic and social conditions in former East Germany showed signs of improvement, helped by the relocation of the national capital from Bonn to Berlin. But by the middle of the current decade Germany's national economy, in need of further reform, had slowed, and the magnitude of the challenge to bring the East up to par with the West was starkly evident.

A Federal Republic

Germany's enthusiasm for the EU should be seen in the context of its own political geography. Following the Allied victory in 1945, the United States organized West Germany into a democratic, federal republic consisting of ten States or *Länder* (Fig. 1-15). Meanwhile, in East Germany the Soviet Union established 15 administrative districts including East Berlin. The smooth functioning of the maturing Federal Republic in the West made possible the absorption of East Germany, which was immediately reorganized into six new States based on its traditional provinces. Political parties active in the West now campaigned for votes in these new States, and the political transformation went remarkably smoothly. But, as Figure 1-15 underscores, the economic picture was darker. Regional disparity, as reflected by per-capita income levels, reflects the East's continuing poverty. Note that five of the East's States (Berlin being the sole exception) are in the lowest income categories, while *none* of the West's States has such low incomes. Similar contrasts exist in other criteria ranging from unemployment (much higher in the East) to productivity per worker (much lower). Such economic differences have significant political implications, and the rebuilding of Germany has a long way to go.

France

German dominance in the European Union is a constant concern in the other leading Western European country. The French and the Germans have been rivals in Europe for centuries. France (population: 60.5 million) is an old state, by most measures the oldest in Western Europe. Germany is a young country, created in 1871 after a loose association of German-speaking states had fought a successful war against . . . the French.

Territorially, France is much larger than Germany, and the map suggests that France has a superior relative location, with coastlines on the Mediterranean Sea, the Atlantic Ocean, and, at Calais, even a window on the North Sea. But France does not have any good natural harbors, and oceangoing ships cannot navigate its rivers and other waterways far inland. France has no equivalent to Rotterdam either internally or externally.

The map of Western Europe (Fig. 1-14) reveals a significant demographic contrast between France and Germany. France has one dominant city, Paris, at the heart of the Paris Basin, France's core area. No other city in France comes close to Paris in terms of population or centrality: Paris has 9.8 million residents, whereas its closest rival, Lyon, has only 1.4 million. Germany has no city to match

AMONG THE REALM'S GREAT CITIES . . . PARIS

If the greatness of a city were to be measured solely by its number of inhabitants, Paris (9.8 million) would not even rank in the world's top 20. But if greatness is measured by a city's historic heritage, cultural content, and international influence, Paris has no peer. Old Paris, near the *Île de la Cité* that housed the original village where Paris began and carries the eight-century-old Notre Dame Cathedral, contains an unparalleled assemblage of architectural and artistic landmarks old and new. The *Arc de Triomphe,* erected by Napoleon in 1806 (though not completed until 1836), commemorates the emperor's victories and stands as a monument to French neoclassical architecture, overlooking one of the world's most famous streets, the *Champs Elysées,* which leads to the grandest of city squares, the *Place de la Concorde,* and on to the magnificent palace-turned-museum, the *Louvre.*

Even the Eiffel Tower, built for the 1889 International Exposition over the objections of Parisians who regarded it as ugly and unsafe, became a treasure. From its beautiful Seine River bridges to its palaces and parks, Paris embodies French culture and tradition. It is perhaps the ultimate primate city in the world.

As the capital of a globe-girdling empire, Paris was the hearth from which radiated the cultural forces of Francophone assimilation, transforming much of North, West, and Equatorial Africa, Madagascar, Indochina, and many smaller colonies into societies on the French model. Distant cities such as Dakar, Abidjan, Brazzaville, and Saigon acquired a Parisian atmosphere. France, meanwhile, spent heavily to keep Paris, especially Old Paris, well maintained—not just as a relic of history, but as a functioning, vibrant center, an example to which other cities can aspire.

Today, Old Paris is ringed by a new and different Paris. Stand on top of the *Arc de Triomphe* and turn around from the *Champs Elysées*, and the tree-lined avenue gives way to *La Défense*, an ultramodern high-rise complex that is one of Europe's leading business districts (see photo, p. 57). But from atop the Eiffel Tower you can see as far as 50 miles (80 km) and discern a Paris visitors rarely experience: grimy, aging industrial quarters, and poor, crowded neighborhoods where discontent and unemployment fester—and where Muslim immigrants cluster in a world apart from the splendor of the old city.

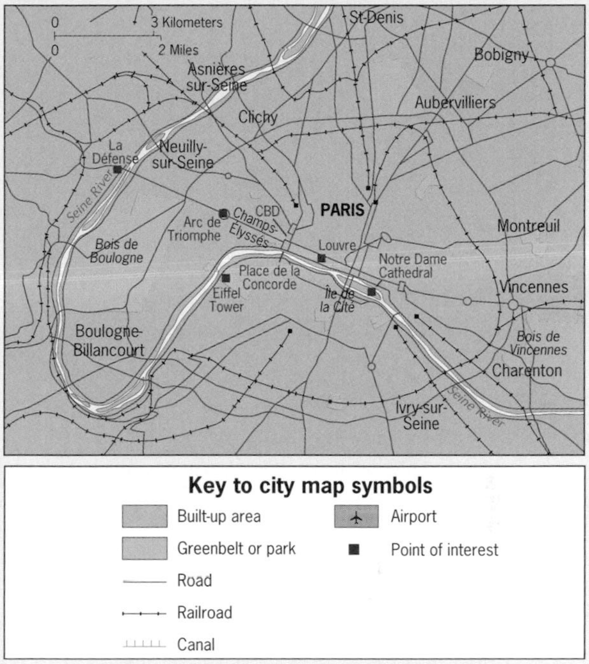

Paris, but it does have a number of cities with populations between 1 and 5 million. And as Table G-1 shows, Germany is much more highly urbanized overall than France.

Why should Paris, without major raw materials nearby, have grown so large? Whenever geographers investigate the evolution of a city, they focus on two impor-
21 tant locational qualities: its **site** (the physical attributes of
22 the place it occupies) and its **situation** (its location relative to surrounding areas of productive capacity, other cities and towns, barriers to access and movement, and other aspects of the greater regional framework in which it lies).

The site of the original settlement at Paris lay on an island in the Seine River, a defensible place where the river was often crossed. This island, the *Île de la Cité*, was a Roman outpost 2000 years ago; for centuries its security ensured continuity. Eventually the island became overcrowded, and the city expanded along the banks of the river (Fig. 1-16, left map).

Soon the settlement's advantageous situation stimulated its growth and prosperity. Its fertile agricultural hinterland thrived, and, as an enlarging market, Paris's focality increased steadily. The Seine River is joined near Paris by several navigable tributaries (the Oise, Marne, and Yonne). When canals extended these waterways even farther, Paris was linked to the Loire Valley, the Rhône-Saône Basin, Lorraine (an industrial area), and the northern border with Belgium. When Napoleon reorganized France and built a radial system of roads—followed later

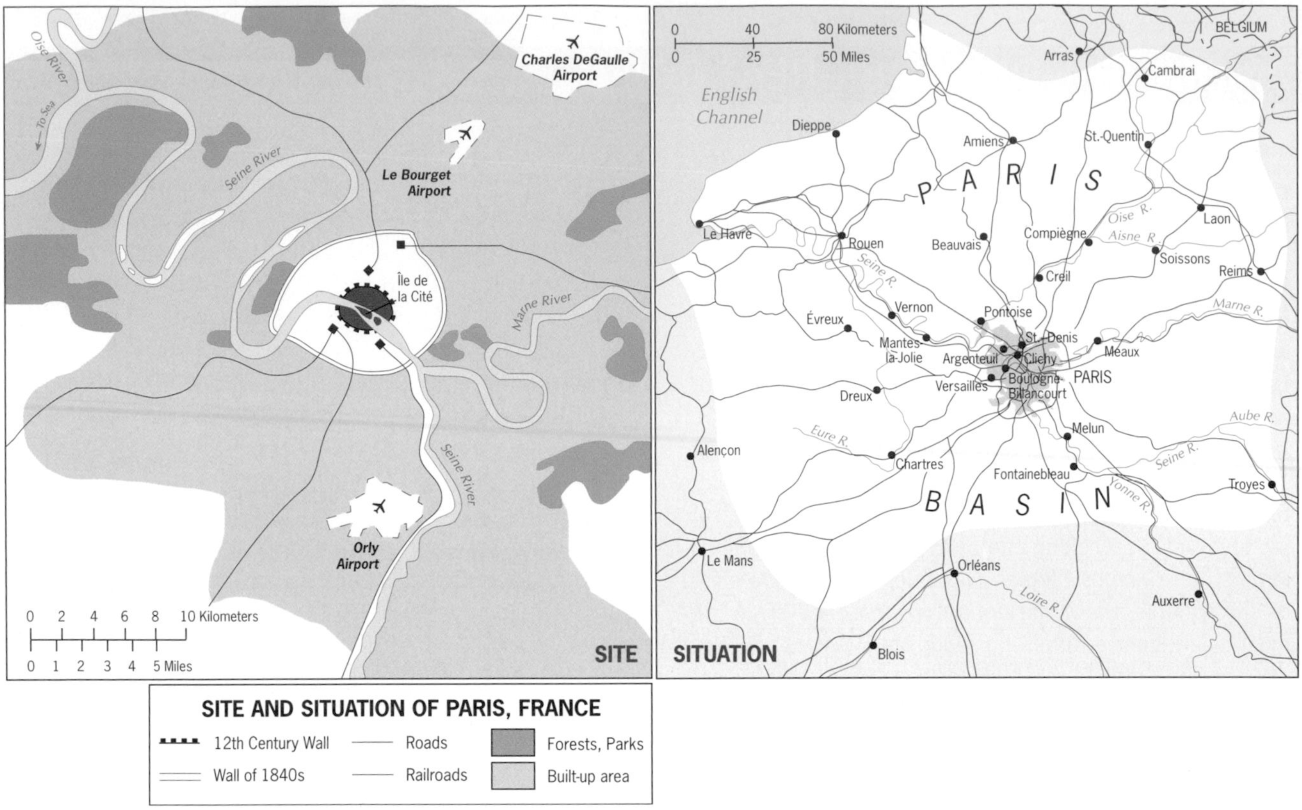

FIGURE 1-16 © H. J. de Blij, P. O. Muller, and John Wiley & Sons, Inc.

FROM THE FIELD NOTES

"Because the Seine has a meandering course in an often steep-sided valley, large ships can sail upstream only as far as Rouen, not to the French capital of Paris. At the river's mouth, the port of Le Havre is no great natural harbor, so that the lower Seine is lined with smaller riverside ports handling such commodities as oil, coal, and (as in the photo at the left) scrap iron . . . Beyond and to the southeast of Paris lies Bourgogne (Burgundy), famous for its vineyards and wines and known also for crops such as wheat, rye, mustard, and sunflowers. From the top of an old church I had a superb view of the rural landscape of Burgundy, with its nucleated settlements and its highly fragmented fields. The diversity of colors signifies the variety of crops even in this small area, but farming here is quite efficient. Farmers, who often own several separate parcels of land, use downsized mechanized equipment; in the Burgundy tradition they live in their clustered villages where they have vegetable gardens and keep their few livestock." (*left*) © H. J. de Blij, (*right*) © Barbara A. Weightman.

by railroads—that focused on Paris from all parts of the country, the city's primacy was assured (Fig. 1-16, right map). The only disadvantage in Paris's situation lies in its seaward access: oceangoing ships can sail up the Seine River only as far as Rouen.

Paris, in accordance with Weber's agglomeration principle, grew into one of Europe's greatest cities. French industrial development was less spectacular, but French agriculture remained Europe's most productive and varied, exploiting the country's wide range of soils and climates. Today France's economic geography is marked by new high-tech industries. It is a leading producer of high-speed trains, aircraft, fiber-optic communications systems, and space-related technologies. It also is the world leader in nuclear power, which currently supplies more than 75 percent of its electricity and thereby reduces its dependence on foreign oil.

When Napoleon reorganized France in the early 1800s, he broke up the country's large traditional subregions and established more than 80 small *départements* (additions and subdivisions later increased this number to 96). Each *département* had representation in Paris, but the power was concentrated in the capital, not in the individual *départements*. France became a highly centralized state and remained so for nearly two centuries (see inset map, Fig. 1-17). Only the island *département* of

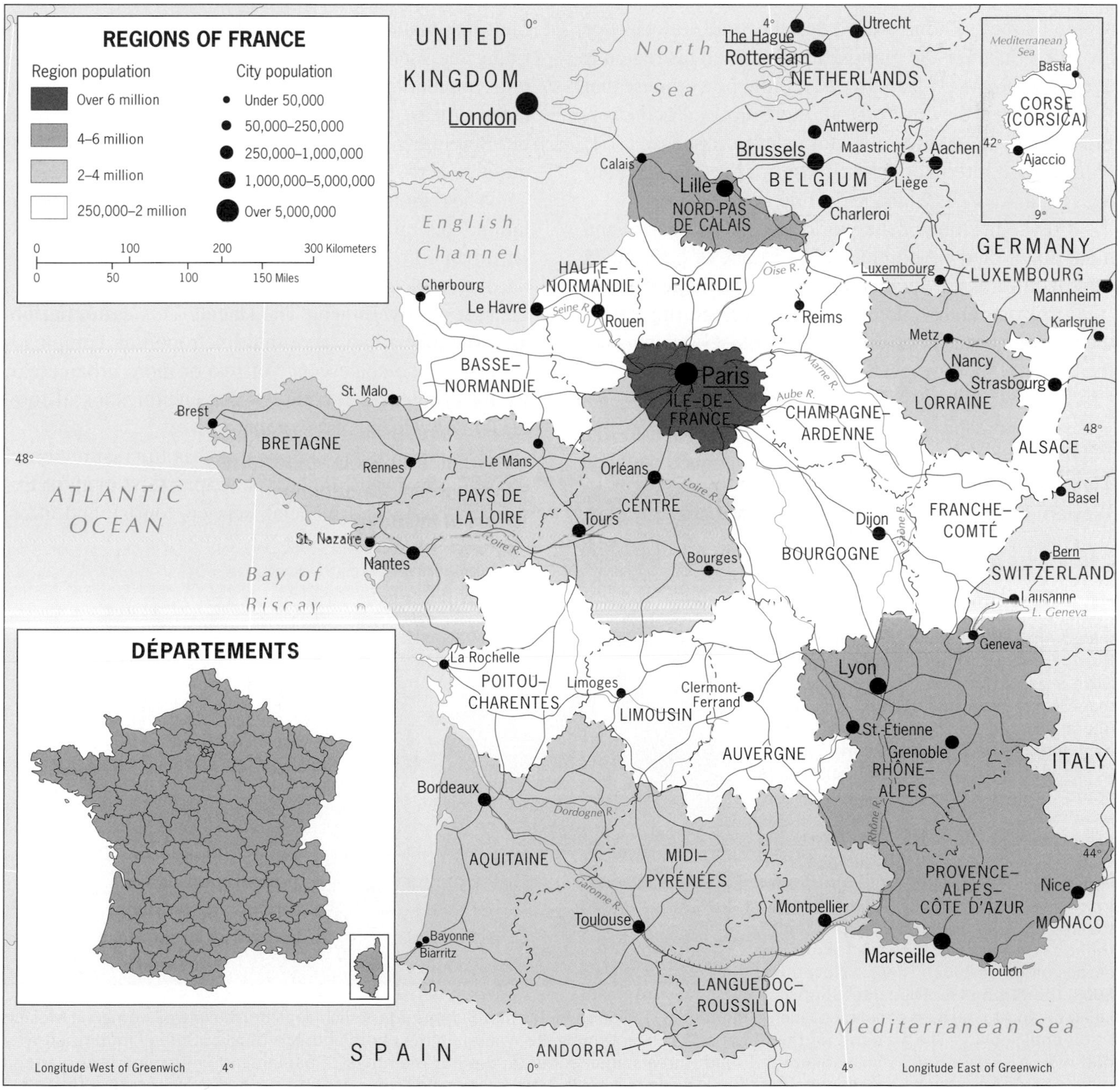

FIGURE 1-17 © H. J. de Blij, P. O. Muller, and John Wiley & Sons, Inc.

Corsica produced a rebel movement, whose violent opposition to French rule continued for decades and even touched the mainland.

Today, France is decentralizing. A new subnational framework of 22 historically significant provinces, groupings of *départements* called *regions* (Fig. 1-17), has been established to accommodate the devolutionary forces felt throughout Europe and, indeed, throughout the world. These regions, though still represented in the Paris government, have substantial autonomy in such areas as taxation, borrowing, and development spending. The cities that anchor them benefit because they are the seats of governing regional councils that can attract investment, not only within France but also from abroad.

Lyon, France's second city and headquarters of the region named Rhône-Alpes, has become a focus for growth industries and multinational firms. This region is evolving into a self-standing economic powerhouse that is becoming a driving force in the European economy. Indeed, it is one of the Four Motors of Europe (see p. 64) with its own international business connections to countries as far away as China and Chile.

France has one of the world's most productive and most diversified economies, based in one of humanity's richest cultures. Protecting that cultural heritage in a unifying Europe and a globalizing world is an imperative that causes occasional differences with Germany over the future of the European Union.

Benelux

Three countries are crowded into the northwest corner of Western Europe: Belgium, the Netherlands, and tiny Luxembourg, collectively referred to by their first syllables (*Be-Ne-Lux*). The major differences that evolved between the agriculturally productive Netherlands and the industrially developed Belgium yielded a double complementarity that led to the 1944 Benelux customs union.

The Benelux countries are among the most densely populated on Earth. For centuries the Dutch have been expanding their living space—not by warring with their neighbors but by wresting land from the sea. The greatest project so far, the draining of almost the entire Zuider Zee (Southern Sea), began in 1932 and continues. In the province of Zeeland, islands are being connected by dikes and the water is being pumped out, adding more *polders* (reclaimed lands) to the total national territory.

Geographically, **Belgium** is marked by a cultural fault line that extends diagonally across the state, separating the Flemish in the northwest (58 percent of the total population of 10.4 million) from the Walloons in the southeast (31 percent). Devolution threatens here, but Belgium's great asset is its capital, Brussels, a headquarters of the European Union.

The regional geography of the **Netherlands** (population: 16.4 million) is noted for the *Randstad*, a triangular urban core area anchored by the constitutional capital, Amsterdam, Europe's largest port, Rotterdam, and
the seat of government, The Hague. This **conurbation**, 23
as geographers call large multimetropolitan complexes formed by the coalescence of two or more urban areas, forms a ring-shaped complex that encircles a still-rural center (*rand* means edge or margin).

Luxembourg, small as it is, has the distinction of recording by far the highest per-capita GNI in all of Europe (Table G-1). Financial, service, tourist, and infor-

For many years, the "Dutch model" of immigrant accommodation and integration was seen as a beacon for the rest of Europe. Welcoming immigrants from Suriname, Indonesia, North Africa, and elsewhere, generous to asylum seekers, liberal with aid, and patient with offenders even in the face of objections from locals, Dutch national and local governments could point to relative harmony and peaceful coexistence as proof that the model worked. An early sign of trouble was the popularity of a politician, Pim Fortuyn, who wanted to limit further immigration and enforce Dutch norms on the immigrant community of more than 1 million (his career was cut short by assassination in 2002). Then, in November 2004, the filmmaker Theo van Gogh was killed on an Amsterdam street by a Muslim militant in retaliation for his film *Submission*, which depicted the fate of women in Muslim traditional society. The scenario for that film had been written by the woman in the photo above, a Muslim Somali immigrant who had won a seat in the Dutch parliament. Death threats against her (and other members of parliament) compelled her to attend meetings surrounded by a team of bodyguards; she and her colleagues lived on an army base and in a prison for security. The Dutch model was, to say the least, tarnished.

mation-technology industries make this country, still run by a "Grand Duke," the most prosperous ministate in the realm.

The Alpine States

Switzerland, Austria, and the *microstate* of Liechtenstein on their border share an absence of coasts and the mountainous topography of the Alps—and little else. Austria speaks one language; the Swiss speak German in the north, French in the west, Italian in the southeast, and even a bit of Rhaeto-Romansch in the remote central highlands (Fig. 1-8). Austria has a large primate city; multicultural Switzerland does not. Austria has a substantial range of domestic raw materials; Switzerland does not. Austria is twice the size of Switzerland and has a larger population, but far more trade crosses the Swiss Alps between Western and Mediterranean Europe than crosses Austria.

Switzerland, not Austria, is in most ways the leading state in Alpine Western Europe (Table G-1). Moun-
24 tainous terrain and **landlocked location** can constitute crucial barriers to economic development, tending to inhibit the dissemination of ideas and innovations, obstruct circulation, constrain farming, and divide cultures. That is why Switzerland is such an important lesson in human geography. Through the skillful maximization of their opportunities (including the transfer needs of their neighbors), the Swiss have transformed their seemingly restrictive environment into a prosperous state. They used the waters cascading from their mountains to generate hydroelectric power in order to develop highly specialized industries. Swiss farmers perfected ways to optimize the productivity of mountain pastures and valley soils. Swiss leaders converted their country's isolation into stability, security, and neutrality, making it a world banking giant, a global magnet for money. Zürich, in the German sector, is the financial center; Geneva, in the French sector, is one of the world's most international cities. The Swiss feel that they do not need to join the EU, and they have not done so.

Austria, which joined the EU in 1995, is a remnant of the Austro-Hungarian Empire and has a historical geography that is far more reminiscent of unstable Eastern Europe than of Switzerland. Even Austria's physical geography seems to demand that the country look eastward: it is at its widest, lowest, and most productive in the east, where the Danube links it to Hungary, its old ally in the anti-Muslim wars of the past.

Vienna, by far the Alpine subregion's largest city, also lies on the country's eastern perimeter. One of the world's most expressive primate cities with magnificent architecture and monumental art, Vienna today is Western Europe's easternmost city, situated on the doorstep of fast-changing Eastern Europe. But Austria's cultural landscapes and economic and political standards are those of Western Europe, a nation-state of the European core.

THE BRITISH ISLES

Off the coast of mainland Western Europe lie two major islands, surrounded by a constellation of tiny ones, that constitute the British Isles, a discrete region of the European realm (Fig. 1-18). The larger of the two major islands, which also lies nearest to the mainland (a mere 21 miles, or 34 km, at the closest point), is the island called *Britain*; its smaller neighbor to the west is *Ireland*.

The names attached to these islands and the countries they encompass are the source of some confusion. They still are called the British Isles, even though British dominance over most of Ireland ended in 1921. The nation-state that occupies Britain and a corner of northeastern Ireland is officially called the United Kingdom of Great Britain and Northern Ireland—United Kingdom for short and UK by abbreviation. But this country often is referred to simply as Britain, and its people are known as the British. The nation-state of Ireland officially is the Republic of Ireland (*Eire* in Irish Gaelic), but it does not include the whole island of Ireland.

How convenient it would be if physical and political geography coincided! Unfortunately, the two do not. During the long British occupation of Ireland, which is overwhelmingly Catholic, many Protestants from northern Britain settled in northeastern Ireland. In 1921, when British domination ended, the Irish were set free—except in that corner in the north, where London kept control to protect the area's Protestant settlers. That is why the country to this day is officially known as the United Kingdom of Great Britain and Northern Ireland.

Northern Ireland (Fig. 1-18) was home not only to Protestants from Britain, but also to a substantial population of Irish Catholics who found themselves on the wrong side of the border when Ireland was liberated. Ever since, conflict has intermittently engulfed Northern Ireland and spilled over into Britain and even into Western Europe. It has been and remains one of Europe's costliest struggles.

Although all of Britain lies in the United Kingdom, political divisions exist here as well. England is the largest of these units, the center of power from which the rest of the region was originally brought under unified control. The English conquered Wales in the Middle Ages, and Scotland's link to England, cemented when a Scottish king ascended the English throne in 1603, was ratified by the Act of Union of 1707. Thus England, Wales, Scotland, and Northern Ireland became the United Kingdom.

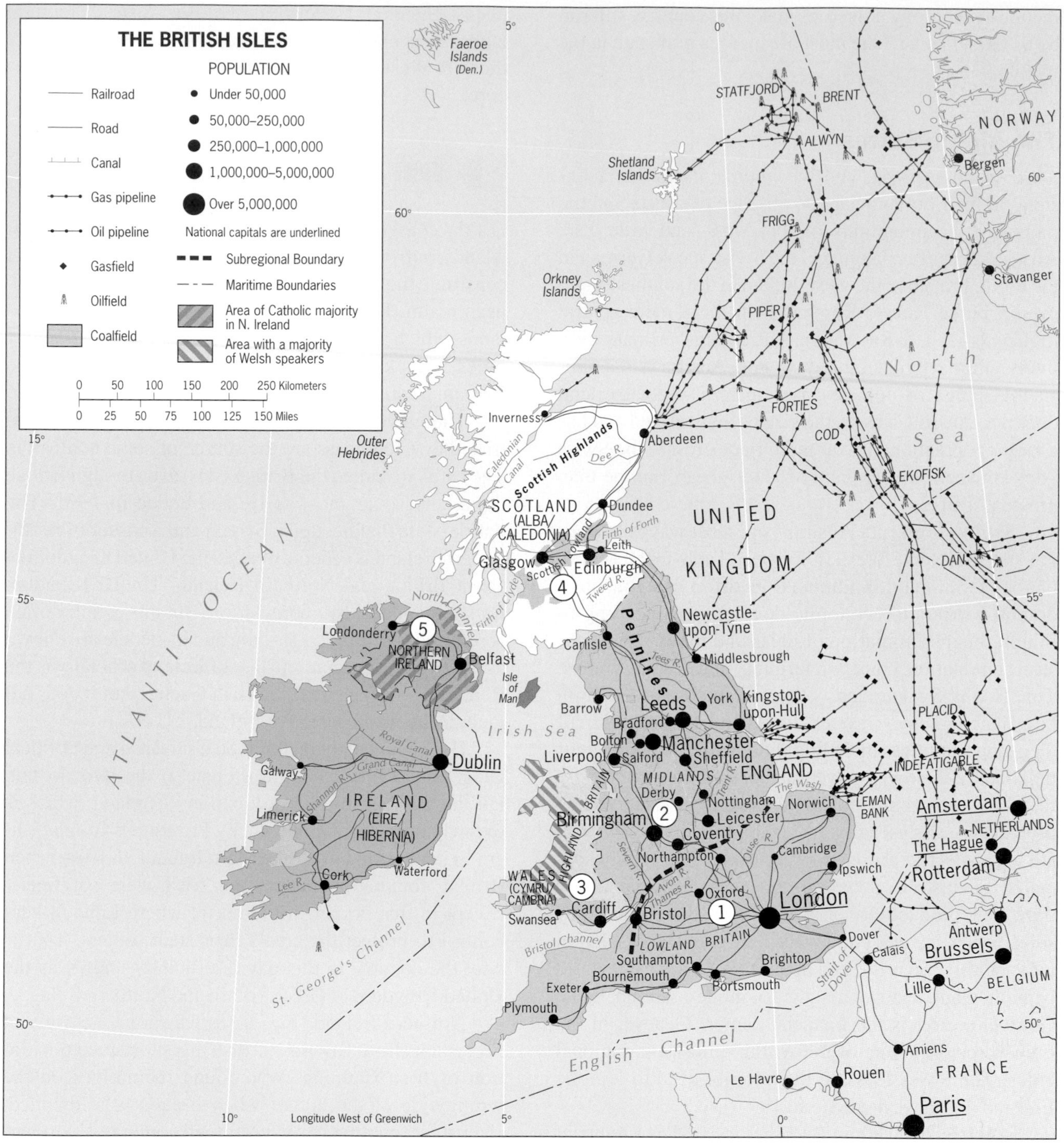

FIGURE 1-18 © H. J. de Blij, P. O. Muller, and John Wiley & Sons, Inc.

The British Isles form a distinct region of Europe for several reasons. Britain's insularity provided centuries of security from turbulent Europe, protecting the evolving British nation as it achieved a system of parliamentary government that had no peer in the Western world. Having united the Welsh, Scots, and Irish, the British set out to forge what would become the world's largest colonial empire. An era of mercantilism and domestic manufacturing (the latter based on water power from streams flowing off the Pennines, Britain's mountain backbone) foreshadowed the momentous Industrial Revolution, which transformed Britain—and much of the world. British cities became synonyms for specialized products as the smokestacks of factories rose like forests over the urban scene. London on the Thames River anchored an English core area that mushroomed into the headquarters

AMONG THE REALM'S GREAT CITIES . . . LONDON

Sail westward up the meandering Thames River toward the heart of London, and be prepared to be disappointed. London does not overpower or overwhelm with spectacular skylines or beckoning beauty. It is, rather, an amalgam of towns—Chelsea, Chiswick, Dulwich, Hampstead, Islington—each with its own social character and urban landscape. Some of these towns come into view from the same river the Romans sailed 2000 years ago: Silvertown and its waterfront urban renewal; Greenwich with its famed Observatory; Thamesmead, the model modern commuter community. Others somehow retain their identity in the vast metropolis that remains (Fig. 1-9), in many ways Europe's most civilized and cosmopolitan city. And each contributes to the whole in its own way: every part of London, it seems, has its memories of empire, its memorials to heroes, its monuments to wartime courage.

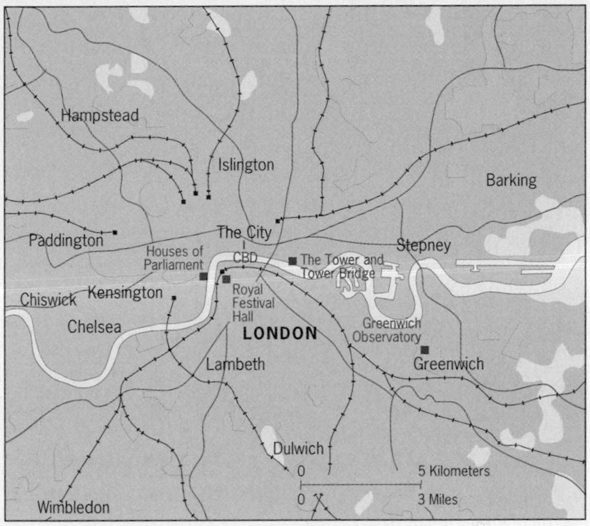

© H. J. de Blij, P. O. Muller, and John Wiley & Sons, Inc.

Along the banks of the Thames, London displays the heritage of state and empire: the Tower and the Tower Bridge, the Houses of Parliament (officially known as the Palace of Westminster), the Royal Festival Hall. Step ashore, and you find London to be a memorable mix of the historic and (often architecturally ugly) modern, of the obsolete and the efficient, of the poor and the prosperous. Public transportation, by world standards, is excellent, even though facilities are aging rapidly; traffic, however, often is chaotic, gridlocked by narrow streets. Recreational and cultural amenities are second to none. London seems to stand with one foot in the twenty-first century and the other in the nineteenth. Its airports are ultramodern. But when the Eurostar TGV train from Paris emerges from the Channel Tunnel, it has to slow down for the final hour in its three-hour run because the tracks into London's Waterloo Station still have not been upgraded to carry it at its normal high speed (see photo caption, p. 55).

London remains one of the world's most livable metropolises for its size (7.6 million) largely because of farsighted urban planning that created and maintained, around the central city, a so-called Greenbelt set aside for recreation, farming, and other nonresidential, noncommercial uses (Fig. 1-9). Although London's growth eroded this Greenbelt, in places leaving only "green wedges," the design preserved crucial open space in and around the city, channeling suburbanization toward a zone at least 25 miles (40 km) from the center. The map reveals the design's continuing impact.

of a global political, financial, and cultural empire. As recently as World War II, the narrow English Channel ensured the United Kingdom's impregnability against German invasion, giving the British time to organize their war machine. When the United Kingdom emerged from that conflict as a leading power among the victorious allies, it seemed that its superpower role in the postwar era was assured.

Two unanticipated developments changed that prospect: the worldwide collapse of colonial empires and the rapid resurgence of mainland Europe. Always ambivalent about the EC and EU, and with its first membership application vetoed by the French in 1963, Britain (admitted in 1973) has worked to restrain moves toward tighter integration. When most member-states adopted the new euro in favor of their national currencies, the British kept their pound sterling and delayed their participation in the European Monetary Union (EMU). As for a federalized Europe, to Britain this is out of the question. In this as in other respects, Britain's historic, insular standoffishness continues.

The United Kingdom

As we noted earlier, the British Isles as a region consists of two political entities: the United Kingdom and Ireland. The UK, with an area about the size of Oregon and a population of 59.9 million, is by European standards quite a large country. Based on a combination of physiographic, historical, cultural, economic, and political criteria, the

United Kingdom can be divided into five subregions (numbered in Fig. 1-18):

1. *Southern England.* Centered on the gigantic London metropolitan area, this is the UK's most affluent subregion, with one-third of the country's population. Financial, communications, engineering, and energy-related industries cluster in this economically and politically dominant area. London exemplifies the momentum of long-term agglomeration. Today this subregion benefits anew not only from its superior links to the European mainland but also as one of the three leading "world cities" on the global scene.
2. *Northern England.* As the map suggests, this subregion should really be called Northern, Central, and Western England. Its center, appropriately called the Midlands, was the focus of the Industrial Revolution and the spectacular rise of Manchester, the source; Liverpool, the great port; Birmingham, unmatched industrial city; and many other manufacturing cities and towns. But here obsolescence has overtaken what was once ultramodern, and the North now suffers from Rustbelt conditions and high unemployment among immigrants from South Asia, Africa, and the Caribbean.
3. *Wales.* This nearly rectangular, rugged territory was a refuge for ancient Celtic peoples, and in its western counties more than half the inhabitants still speak Welsh. Because of the high-quality coal reserves in its southern tier, Wales too was engulfed by the Industrial Revolution, and Cardiff, the capital, was once the world's leading coal exporter. But the fortunes of Wales also declined, and many Welsh emigrated. Among the nearly 3 million who remained, however, the flame of Welsh nationalism survived, and in 1997 the voters approved the establishment of a Welsh Assembly to administer public services in Wales, a first devolutionary step.
4. *Scotland.* Nearly twice as large as the Netherlands and with a population about the size of Denmark's, Scotland is a major component of the United Kingdom. As Figure 1-18 suggests, most of Scotland's 5.1 million people live in the Scottish Lowlands anchored by

FROM THE FIELD NOTES

"From a ferry on the Mersey River, the skyline of downtown Liverpool seemed to mirror the troubles of this city, the nucleus of a metropolitan area with almost one million inhabitants. The dominant buildings in the CBD remain those built during Liverpool's heyday a century ago, when Liverpool's docks outnumbered even London's, and the city thrived on its trade with America and Africa as Merseyside's manufacturing industries prospered. After World War II, Liverpool declined as its port functions dwindled, its industries weakened, and unemployment soared; this view reveals no modern, glass-sided buildings towering over the older cityscape, symbols of investment and confidence in the future. Once the apex of what was called the *Liverpool Triangle* linking this port to West Africa and eastern America, the Merseyside area now lies remote from the centers of British and European action. Some investment has been made in waterfront redevelopment, but the city's stagnation is evident throughout its historic center (where memories of the famed rock group, the Beatles, have developed into a minor industry). 'There may be ambivalence about the European Union in other parts of Britain,' said my colleague of the Department of Geography at the University of Liverpool as we walked the streets, 'but not here. At least we can see more tangible results of EU investment, for which Liverpool is eligible.' But what will lift Merseyside out of its doldrums?" © H. J. de Blij

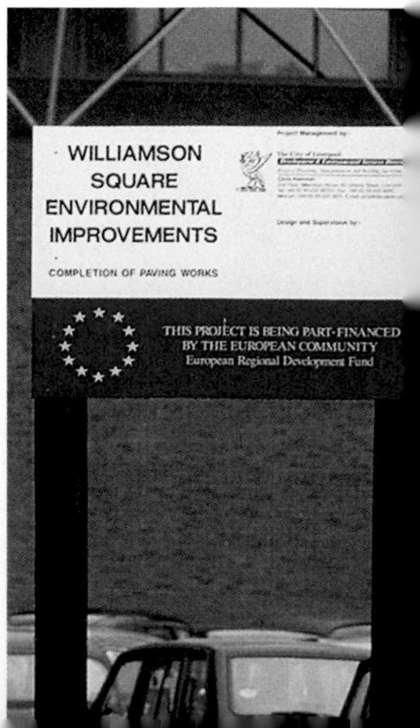
WILLIAMSON
SQUARE
ENVIRONMENTAL
IMPROVEMENTS
COMPLETION OF PAVING WORKS
THIS PROJECT IS BEING PART-FINANCED
BY THE EUROPEAN COMMUNITY
European Regional Development Fund

Edinburgh, the capital, in the east and Glasgow in the west. Attracted there by the labor demands of the Industrial Revolution (coal and iron reserves lay in the area), the Scots developed a world-class shipbuilding industry. Decline and obsolescence were followed by high-tech development, notably in the hinterland of Glasgow, and by Scottish participation in the exploitation of oil and gas reserves under the North Sea (Fig. 1-18), which transformed the eastern ports of Aberdeen and Leith (Edinburgh). But many Scots feel that they are disadvantaged within the UK and should play a major role in the EU. Therefore, when the British government put the option of a Scottish parliament before the voters in 1997, 74 percent approved. Many Scots still hope that total independence lies in their future.

5. *Northern Ireland.* Prospects of devolution in Scotland pale before the devastation caused by political and sectarian conflict in Northern Ireland. With a population of 1.7 million occupying the northeastern one-sixth of the island of Ireland, this area represents the troubled legacy of British colonial rule. A declining majority, now about 54 percent of the people in Northern Ireland, trace their ancestry to Scotland or England and are Protestants; a growing minority, currently around 45 percent, are Roman Catholics, who share their Catholicism with virtually the entire population of the Irish Republic on the other side of the border. Although Figure 1-18 suggests that there are majority areas of Protestants and Catholics in Northern Ireland, no clear separation exists; mostly they live in clusters throughout the territory, including walled-off neighborhoods in the major cities of Belfast (photo below) and Londonderry. Partition is no solution to a conflict that has raged for over three decades at a cost of thousands of lives; Catholics accuse London as well as the local Protestant-dominated administration of discrimination, whereas Protestants accuse Catholics of seeking union with the Republic of Ireland. Repeated efforts to achieve a settlement have ended in failure; the most recent mediation produced a Northern Ireland Assembly to which powers would be devolved from London. But in 2003 this political program failed, and the British government was forced to resume direct rule.

Republic of Ireland

What Northern Ireland is missing through its conflicts is shown by the Irish Republic itself: a growing, booming service-based economy, the fastest-growing in all of Europe for a time around the turn of this century, when Ireland was dubbed the "Celtic Tiger." Burgeoning cities and towns (and rising real estate prices), mushrooming industrial parks, bustling traffic, and construction everywhere reflect this new era. For the first time ever, workers of Irish descent returned from foreign places to take jobs at home; non-Irish immigrants also arrived, posing some new social problems for a closely knit, long-isolated society.

The Republic of Ireland fought itself free from British colonial rule just three generations ago. Its cool, moist climate had earlier led to the adoption of the

"The Troubles" between Protestants and Catholics, pro- and anti-British factions that have torn Northern Ireland apart for decades at a cost of more than 3000 lives, are etched in the cultural landscape. The so-called "Peace Wall" across West Belfast, shown here separating Catholic and Protestant neighborhoods, is a tragic monument to the failure of accommodation and compromise, a physical manifestation of the emotional divide that runs as deep as ever—as reflected by the results of the May 5, 2005 British elections. Hard-line parties won the day after several years of slow movement toward concession and consent. "Good fences make good neighbors," said one analyst the following day, "the Peace Wall will have to stay up for a few more decades yet." © Geray Sweeney/Corbis Images.

American potato as the staple crop, but excessive rain and a blight in the late 1840s caused famine and cost over 1 million lives. Another 2 million Irish emigrated.

Independence from Britain in 1921 did not bring economic prosperity, and Ireland's economy stagnated even after it joined the European Union in 1973 and EU subsidies helped improve infrastructure. But in the 1990s European telecommunications service industries, and soon American and other foreign high-tech industries as well, saw the advantages in Ireland's well-educated but not highly-paid labor force. Suddenly Ireland boomed, its economy expanding faster than any other in Europe, its real estate values skyrocketing, its roads clogged with new cars. Not only did people of Irish ancestry begin returning to Ireland—a first in its history—but the job market attracted other immigrants, diversifying one of Europe's most homogeneous populations. Economic geographers took to calling Ireland the Celtic Tiger as the Irish government set about modernizing laws (on such matters as divorce and abortion), a process long blocked by the Roman Catholic Church.

Inevitably, rapid growth of the kind Ireland experienced created some economic instability and brought certain complications. Ireland, unlike its United Kingdom neighbor, joined the EMU and made the euro its currency. Labor costs rose and some companies closed their operations. The housing market suffered a setback. Foreign immigrants and locals had difficulty adjusting to each other. But after a slowdown in the early 2000s growth resumed, and Ireland continues to be an EU success story.

NORTHERN (NORDIC) EUROPE

North of Europe's Western European core area lies a disconnected group of six countries that exemplify core-periphery contrasts. Northern Europe is a region of difficult environments: generally cold climates, poorly developed soils, limited mineral resources, long distances. Together, the six countries in this region—Sweden, Norway, Denmark, Finland, Estonia, and Iceland—contain just under 26 million inhabitants, which is a lower total population than that of Benelux and only one-seventh that of Western Europe. The overall land area, on the other hand, is almost the size of the entire European core. Here in peripheral *Norden*, as the people call their northerly domain, national core areas lie in the south: note the location of the capitals of Helsinki, Stockholm, and Oslo at approximately the same latitude (Fig. 1-19).

Northern Europe's peripheral situation is more than environmental. As viewed from Europe's core area, Norden is on the way to nowhere. No major shipping lanes lead from Western Europe past Norden to other productive areas of the world, limiting interaction of the sort that ties the British Isles to the mainland. Moreover, except for relatively small Denmark and Estonia, all of Norden lies separated from the European mainland by water. At all levels of spatial generalization, isolation is pervasive in Norden.

Northern Europe's remoteness, isolation, and environmental severity also have had positive effects for this region. The countries of the Scandinavian Peninsula lay removed from the wars of mainland Europe (although Norway was overrun by Nazi Germany during World War II). The three major languages—Danish, Swedish, and Norwegian—are mutually intelligible, which creates one of the criteria delimiting this region. Another regional criterion is the overwhelming adherence to the same Lutheran church in each of the Scandinavian countries (Norway, Sweden, and Denmark), Iceland, and Finland. Furthermore, democratic and representative governments emerged early, and individual rights and social welfare have long been carefully protected. Women participate more fully in government and politics here than in any other region of the world.

Sweden is the largest Nordic country in terms of both population (9.0 million) and territory. Most Swedes live south of 60° North latitude (which passes through Uppsala), in what is climatically the most moderate part of the country (Fig. 1-19). Here lie the capital, core area, and, as Figure 1-7 shows, the main industrial districts; here, too, are the main agricultural areas that benefit from the lower relief, better soils, and milder climate.

Sweden long exported raw or semifinished materials to industrial countries, but today the Swedes are making finished products themselves, including automobiles, electronics, stainless steel, furniture, and glassware. Much of this production is based on local resources, including a major iron ore reserve at Kiruna in the far north (there is a steel mill at Luleå). Swedish manufacturing, in contrast to that of several Western European countries, is based in dozens of small and medium-sized towns specializing in particular products. Energy-poor Sweden was a pioneer in the development of nuclear power, but a national debate over the risks involved has reversed that course.

Norway does not need a nuclear power industry to supply its energy needs. It has found its economic opportunities on, in, and beneath the sea. Norway's fishing industry, now augmented by highly efficient fish farms, long has been a cornerstone of the economy, and its merchant marine spans the world. But since the 1970s, Norway's economic life has been transformed by the bounty of oil and natural gas discovered in its sector of the North Sea.

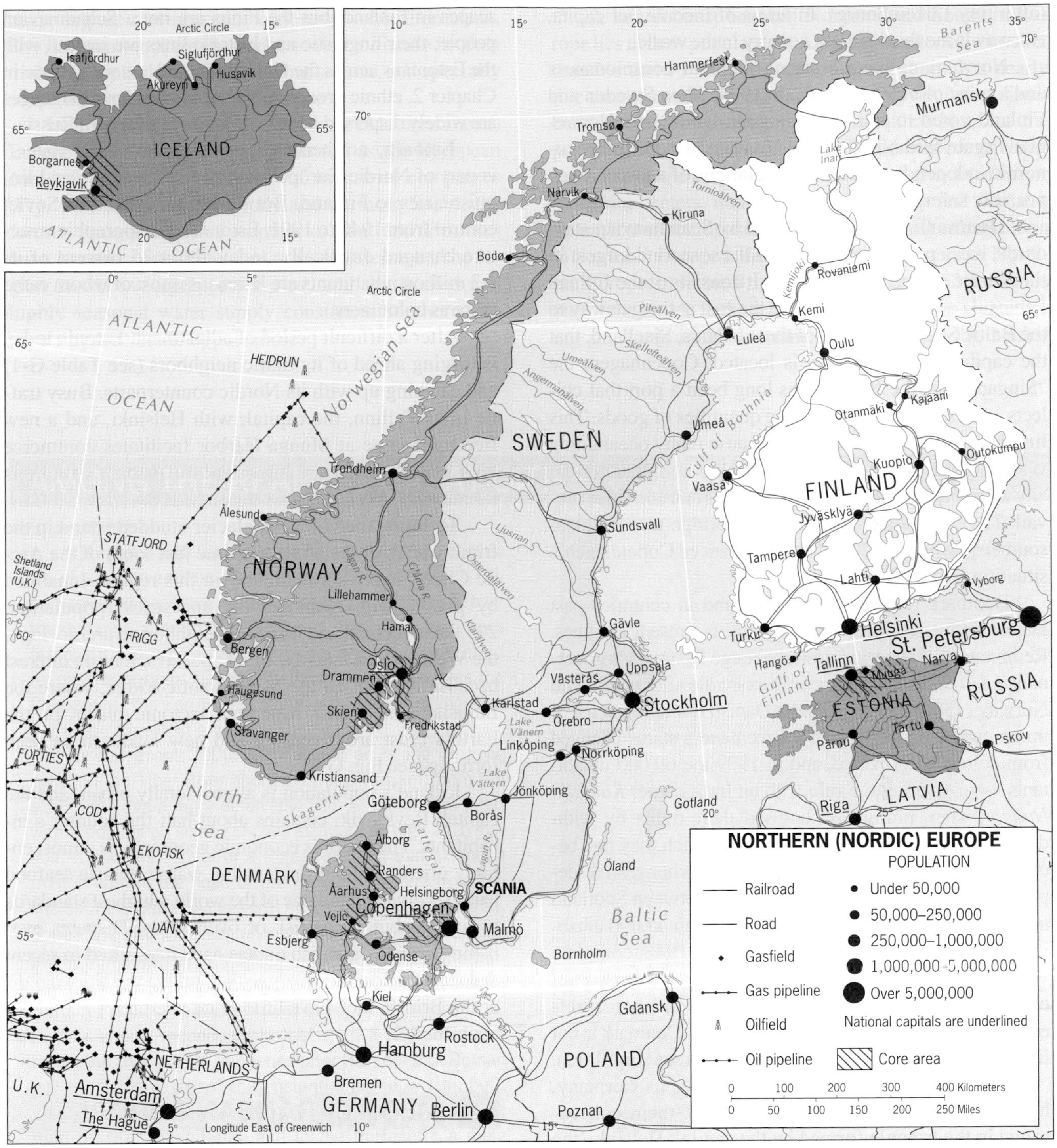

FIGURE 1-19 © H. J. de Blij, P. O. Muller, and John Wiley & Sons, Inc.

With its limited patches of cultivable soil, high relief, extensive forests, frigid north, and spectacularly fjorded coastline, Norway has nothing to compare to Sweden's agricultural or industrial development. Its cities, from the capital Oslo and the North Sea port of Bergen to the historic national focus of Trondheim as well as Arctic Hammerfest, lie on the coast and have difficult overland connections. The isolated northern province of Finnmark has even become the scene of an autonomy movement among the reindeer-herding indigenous Saami (Fig. 1-12). Norway (population 4.6 million) has been described as a necklace, its beads linked by the thinnest of strands. But this has not constrained national development. Norway in 2003 had the second-lowest unemployment rate in Europe

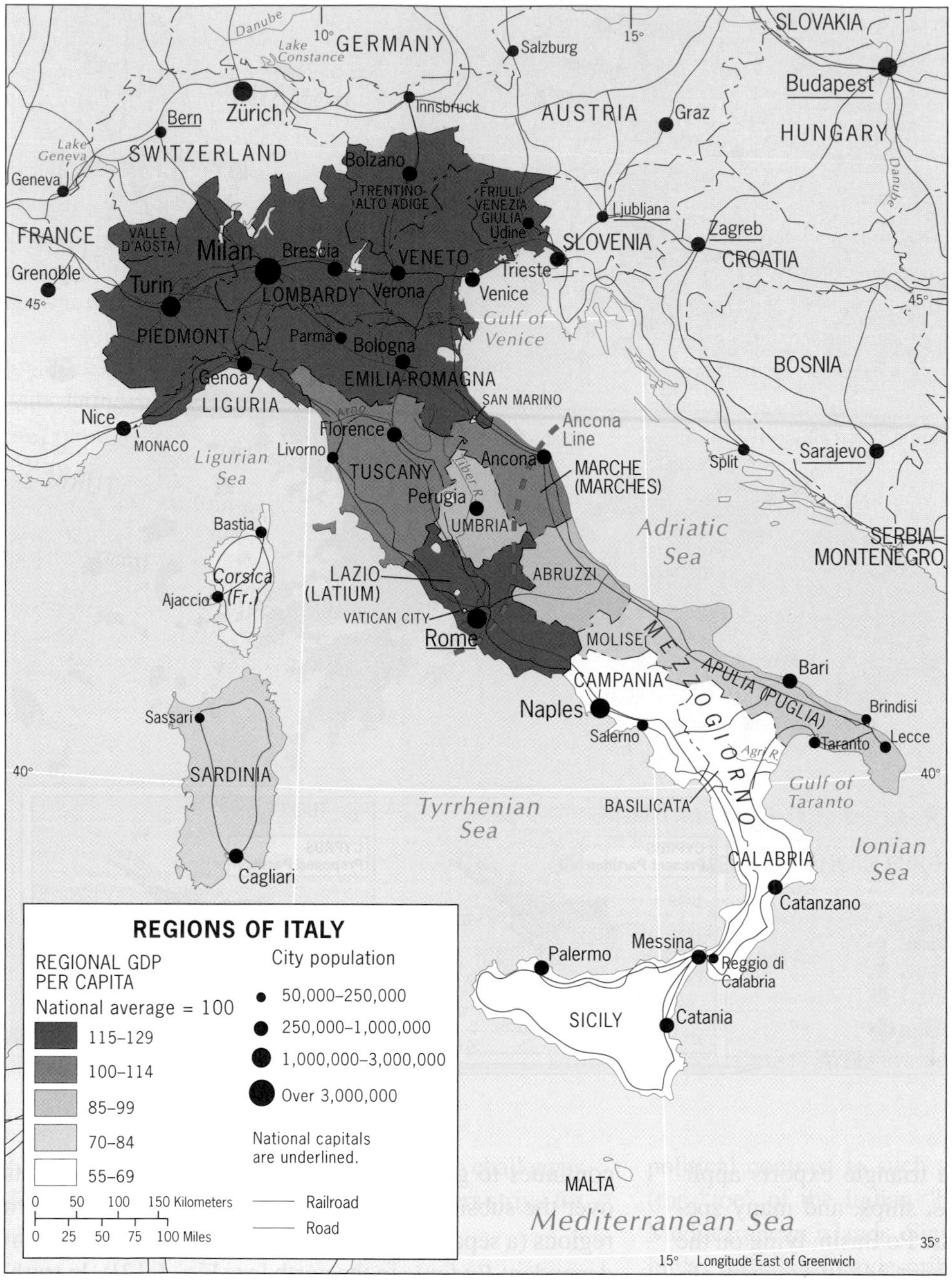

FIGURE 1-21 © H. J. de Blij, P. O. Muller, and John Wiley & Sons, Inc.

Imperial Romans, Muslim Moors, and Catholic kings left their imprints on Iberia, notably the boundary between Spain and Portugal dating from the twelfth century. The golden age of colonialism was followed by dictatorial rule and economic stagnation.

Today both Spain and Portugal are democracies, and their economies are doing well, helped in the case of Portugal by EU subsidies. As Figure 1-22 shows, **Spain** followed the leads of Germany and France and decentralized its administrative structure, creating 17 regions called Autonomous Communities (ACs). Each AC has its own parliament and administration that control planning, public works, cultural affairs, education, environmental matters, and even, to some extent, international commerce. Each AC can negotiate its own degree of autonomy with the central government in Madrid. This new system, however, has not been enough to defuse Spain's most problematic devolutionary issue, which involves the Basques in the AC mapped as Basque Country (Fig. 1-22, especially the top inset map).

Particularly noteworthy is Catalonia, the triangular AC in Spain's northeastern corner, adjacent to France. Centered on prosperous, productive Barcelona, Catalonia is Spain's leading industrial area, an AC with a popu-

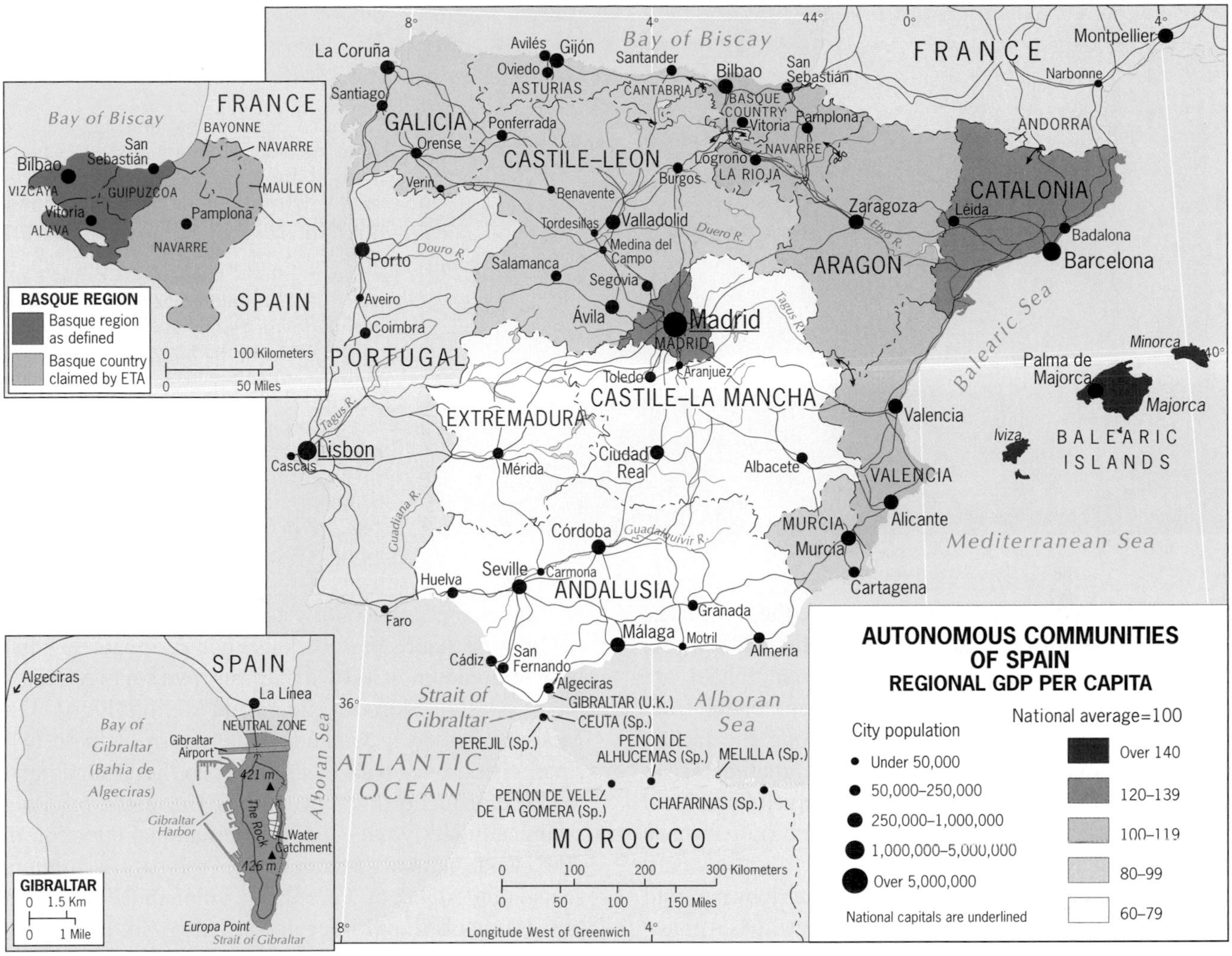

FIGURE 1-22 Adapted and updated with permission from *The Economist*, April 25, 1992, special insert p. 17. © 1992, The Economist Newspaper Group, Inc. Further reproduction prohibited.

lation of 6.4 million. It is imbued with a fierce nationalism, endowed with its own language and culture, and economically it is one of the Four Motors of Europe. Note that while most of Spain's industrial raw materials lie in the northwest, its major industrial development has taken place in the northeast, where innovations and skills propel a high-technology-driven regional economy. In recent years, Catalonia—with 6 percent of Spain's territory and 15 percent of its population—has annually produced 25 percent of all Spanish exports and nearly 40 percent of its industrial exports. Such economic strength translates into political power, and in Spain the issue of Catalonian separatism is never far from the surface.

Ironically, Catalan nationalism was spurred by the very EU subsidies that were intended to redress imbalances in member-states. Spain received hundreds of millions of euros from Brussels, and the central government decided to spend a substantial part of this fund to build a high-speed train link between Madrid and Seville in Andalusia, the poor, southern AC on the Mediterranean coast (Fig. 1-22). This, the government hoped, would stimulate development where it was most needed. But public opinion in Catalonia held that Barcelona, not Seville, should have been rewarded with that modern rail link, and Catalan nationalism got another boost.

As Figure 1-22 indicates, Spain's capital and largest city, Madrid, lies near the geographic center of the state. It also lies along an economic-geographic divide. Catalonia and Madrid are Spain's most prosperous ACs; the contiguous group of five ACs between the Basque Country and Valencia rank next. Tourism (notably along the Mediterranean coast) and winegrowing (especially in La Rioja) contribute importantly here. Ranking below this cluster of ACs are the four such units of the northwest: Galicia, Castile-Leon, and industrialized Asturias and Cantabria. Incomes in these regions are below the national average because industrial obsolescence, dwindling raw material sources, and emigration have plagued

On the morning of March 11, 2004, a series of near-simultaneous explosions aboard commuter trains heading for Madrid killed more than 170 passengers and wounded over 500 in Spain's worst terrorist attack ever. Although the government initially suspected Basque separatist (ETA) involvement, the assault was planned and executed by Islamic extremists and had major social and political consequences. The governing party lost the election a few days later; the new government withdrew Spain's forces from the campaign in Iraq; and Madrid set a new course in several social arenas, including the separation of religion and state. © AP/Wide World Photos.

their economies. Worst off, however, are the three large ACs to the south of Madrid: Extremadura, Castile-La Mancha, and especially Andalusia. Drought, inadequate land reform, scarce resources, and remoteness from Spain's fast-growing northeast are among the factors that inhibit development here. Even as EU members seek to reduce the differences among themselves, individual countries face the geographic consequences of focused growth that deepens internal divisions.

Although Spain and the United Kingdom are both EU members committed to cooperative resolution of territorial problems, the two countries are embroiled in a dispute over a sliver of land at the southern tip of Iberia: legendary Gibraltar (see lower inset map, Fig. 1-22). "The Rock" was ceded by Spain to the British in perpetuity in 1713 (though not the neck of the peninsula linking it to the mainland, which the British later occupied) and has been a British colony ever since. The 30,000 residents are used to British institutions, legal rules, and schools. Successive Spanish governments have demanded that Gibraltar be returned to Madrid's rule, but the colony's residents are against it. And under their 1969 constitution, Gibraltarians have the right to vote on any transfer of sovereignty.

Now the British and Spanish governments are trying to reach an agreement under which they would share the administration of Gibraltar for an indefinite time; both are EU members, so little would have to change. The advantages would be many: economic development is being slowed by the ongoing dispute, border checks would be lifted, EU benefits would flow. But Gibraltarians have their doubts and want to put the issue to a referendum. Spain will not accept the idea of a referendum, and the matter is far from resolved.

Spain's refusal to allow, and abide by, a referendum in Gibraltar stands in remarkable contrast to its demand for a referendum in its own outposts, two small exclaves on the coast of Morocco, Ceuta and Melilla (Fig. 1-22). Morocco has been demanding the return of these two small cities, but Spain has refused on the grounds that the local residents do not want this. The matter has long simmered quietly but came to international attention in 2002 when a small detachment of Moroccan soldiers seized the island of Perejil, an uninhabited, Spanish possession off the Moroccan coast also wanted by Morocco. In the diplomatic tensions that followed, the entire question of Spain's holdings in North Africa (and its anti-Moroccan stance in the larger matter of Western Sahara) exposed some contradictions Madrid had preferred to conceal.

The state of **Portugal** (population: 10.5 million), a comparatively poor country that has benefited enormously from its admission to the EU, occupies the southwestern corner of the Iberian Peninsula. One rule of EU membership is that the richer members assist the poorer ones; Portugal shows the results in a massive renovation project in its capital, Lisbon, as well as in the modernization of surface transport routes.

Unlike Spain, which has major population clusters on its interior plateau as well as its coastal lowlands, Portugal has a population that is concentrated along and near the Atlantic coast. Lisbon and the second city, Porto, are coastal cities; the best farmlands lie in the moister western and northern zones of the country. But the farms are small and inefficient, and although Portugal remains dominantly rural, it must import as much as half of its foodstuffs. Exporting textiles, wines, cork, and fish, and running up an annual deficit, the indebted Portuguese economy remains a far cry from those of other European countries of similar dimensions.

Greece and Cyprus

The eastern segment of the region we define as Southern Europe is dominated by **Greece**, an outlier of both Mediterranean Europe and the European Union. Greece's land boundaries are with Turkey, Bulgaria, Macedonia, and Albania; as Figure 1-23 reveals, it also owns islands just offshore from mainland Turkey. Altogether, the Greek archipelago (island chain) numbers some 2000 islands ranging in size from Crete (3218 square miles [8335 sq km]) to small specks of land in the Cyclades. In addition, Greeks represent the great majority on the now-divided island of Cyprus.

Ancient Greece was a cradle of Western civilization, and later it was absorbed by the expanding Roman Empire. For some 350 years beginning in the mid-fifteenth century, Greece was under the sway of the Ottoman Turks. Greece regained independence in 1827, but not until nearly a century later, through a series of Balkan wars, did it acquire its present boundaries. During World War II, Nazi Germany occupied and ravaged the country, and in the postwar period the Greeks have quarreled with the Turks, the Albanians, and the newly independent Macedonians. Today, Greece finds itself between the Muslim world of Southwest Asia and the Muslim communities of Eastern Europe; it is still the only noncontiguous mainland EU member even after the accession of ten more states in 2004.

Volatile as Greece's surroundings are, Greece itself is a country on the move, an EU success story to rival Ireland's. Political upheavals in the 1970s and economic stagnation in the 1980s are all but forgotten in the new century: now Greece, its economy booming, is described as the locomotive for the Balkans, a beacon for the EU in a crucial part of the world. Infrastructure improvements focused on the Athens urban area, where more than one-third of the population is concentrated, include new subways, a new beltway, and a new airport.

Greece will confront serious challenges in the post–2004 period. With the accession of ten Eastern European members to the EU, Greece will lose much of its EU subsidy to these poorer countries. Greece's democratic institutions still need strengthening, and corruption remains a serious problem. Educational institutions need modernization. And while Greece is on better terms

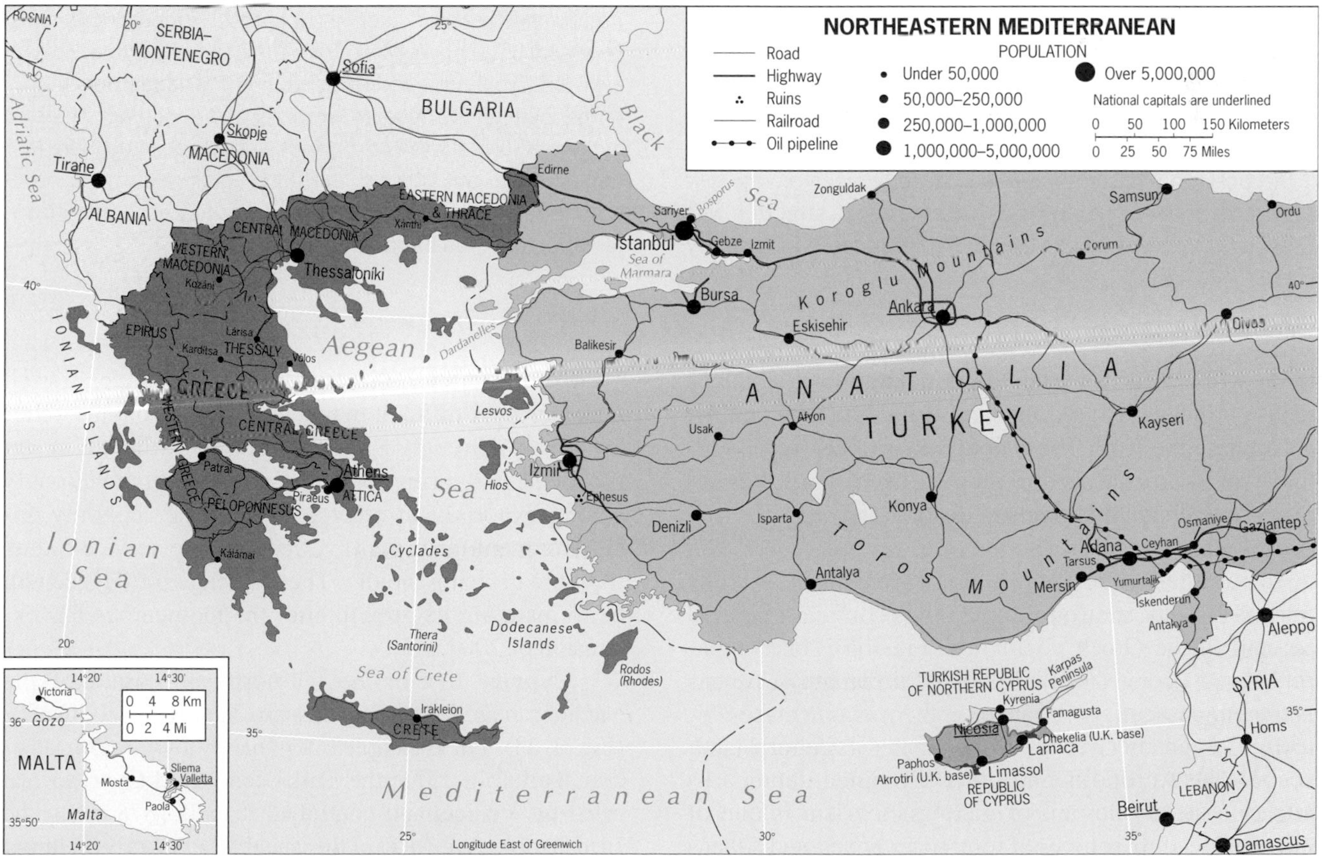

FIGURE 1-23 © H. J. de Blij, P. O. Muller, and John Wiley & Sons, Inc.

AMONG THE REALM'S GREAT CITIES . . . ATHENS

Take a map of Greece. Draw a line to encompass its land area as well as all the islands. Now find the geographic center of this terrestrial and maritime territory, and you will find a major city nearby. That city is Athens, capital of modern Greece, culture hearth of ancient Greece, source of Western civilization.

Athens today is only part of a greater metropolis that sprawls across, and extends beyond, a mountain-encircled, arid basin that opens southward to the Bay of Phaleron, an inlet of the Aegean Sea. Here lies Piraeus, Greece's largest port, linked by rail and road to the adjacent capital. Other towns, from Keratsinion in the west to Agios Dimitrios in the east, form part of an urban area that not only lies at, but *is*, the heart of modern Greece. With a population of 3.2 million, most of the country's major industries, its densest transport network, and a multitude of service functions ranging from government to tourism, Greater Athens is a well-defined national core area.

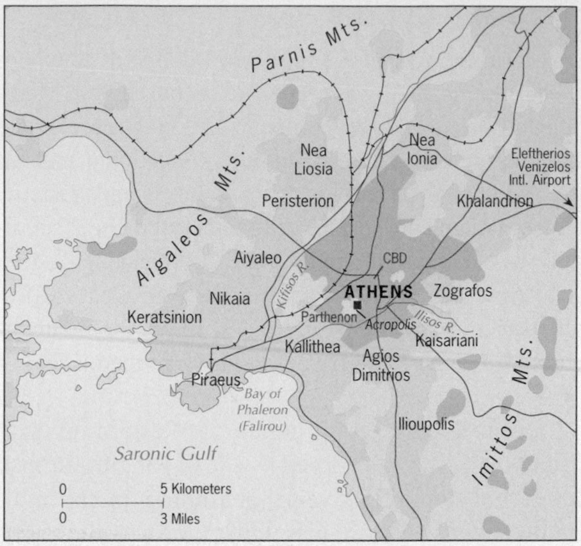

Arrive at the new Eleftherios Venizelos Airport east of Athens and drive into the city, and you can navigate on the famed Acropolis, the 500-foot (150-m) high hill that was ancient Athens' sanctuary and citadel, crowned by one of humanity's greatest historic treasures, the Parthenon, temple to the goddess Athena. It is only one among many relics of the grandeur of Athens 25 centuries ago, now engulfed by mainly low-rise urbanization and all too often obscured by smog generated by factories and vehicles and trapped in the basin. Conservationists decry the damage air pollution has inflicted on the city's remaining historic structures, but in truth the greatest destruction was wrought by the Ottoman Turks, who plundered the country during their occupation of it, and the British and Germans, who carted away priceless treasures to museums and markets in Western Europe.

Travel westward from the Middle East or Turkey, and Athens will present itself as the first European city on your route. Travel eastward from the heart of Europe, however, and your impression of its cultural landscape may differ: behind the modern avenues and shopping streets are bazaar-like alleys and markets where the atmosphere is unlike that of any other in this geographic realm. And beyond the margins of Greater Athens lies a Greece that lags far behind this bustling, productive coreland.

today with its fractious neighbors and invests in development projects in Macedonia and Bulgaria (for example, an oil pipeline from Thessaloniki to a Greek-owned refinery in Skopje, Macedonia), it will take skillful diplomacy to navigate the shoals of historic discord.

Modern Greece is a nation of 11 million centered on historic Athens, one of the realm's great cities. With its port of Piraeus, metropolitan Athens contains about 30 percent of the Greek population, making it one of Europe's most congested and polluted urban areas. Athens is the quintessential primate city; the monumental architecture of ancient Greece still dominates its cultural landscape. The Acropolis and other prominent landmarks attract a steady stream of visitors; tourism is one of Greece's leading sources of foreign revenues, and Athens is only the beginning of what the country has to offer.

Deforestation, soil erosion, and variable rainfall make farming difficult in much of Greece, but the country remains strongly agrarian. It is self-sufficient in staple foods, and farm products continue to figure strongly among exports. But other sectors of the economy, including manufacturing (textiles) and the service industries, are growing rapidly. The challenge for Greece will be to maintain its growth after the momentous EU expansion of 2004.

Cyprus lies in the far northeast corner of the Mediterranean Sea, much closer to Turkey than to Greece (Fig. 1-23), but is peopled dominantly by Greeks rather than Turks. In 1571, the Turks conquered Cyprus, then ruled by Venice, and controlled it until 1878 when the British took over. Most of the island's Turks arrived during the Ottoman period; the Greeks have been there longest.

FROM THE FIELD NOTES

"For a few days in early 2003, the island ministate of Malta held the attention of Europe and, indeed, the world. It was the first of the ten states joining the European Union in 2004 to perform one of the requirements for entry: a referendum on the question among the voters. The multicultural Maltese are very active politically, and vigorous campaigns for as well as against EU membership crested with large and vocal meetings. Polls suggested that the vote would be close, and EU leaders worried: what if the very first country to test its will were to reject the initiative? Might this lead other candidates to do the same, threatening the whole EU expansion plan? But Malta's government, which favored incorporation, persuaded an albeit slender majority that Malta would benefit from EU membership, and the yes vote carried the day. Valletta, the capital (shown here) could celebrate." © H. J. de Blij

When the British were ready to give Cyprus independence after World War II, the 80 percent Greek majority mostly preferred union with Greece. Ethnic conflict followed, but in 1960 the British granted Cyprus independence under a constitution that prescribed majority rule but guaranteed minority rights.

This fragile order broke down in 1974, and civil war engulfed the island. Turkey sent in troops and massive dislocation followed, resulting in the partition of Cyprus into northern Turkish and southern Greek sectors (Fig. 1-23; inset Fig. 1-20). In 1983, the 40 percent of Cyprus under Turkish control, with about 100,000 inhabitants (and some 30,000 Turkish soldiers), declared itself the independent Turkish Republic of Northern Cyprus. Only Turkey recognizes this ministate (which now contains a population of nearly 200,000); the international community recognizes the government on the Greek side as legitimate. With about 875,000 residents, a relatively prosperous agricultural and tourist economy, and strong links with Europe, the Greek-side government qualified the south for EU membership in 2004.

The potential for serious conflict over Cyprus has not disappeared. In effect, the "Green Line" that separates the Turkish and Greek communities constitutes not just a regional border but a boundary between geographic realms.

The Mediterranean region contains one other ministate, **Malta**, located south of Sicily. Malta is a small archipelago of three inhabited and two uninhabited islands with a population under 400,000 (Fig. 1-23, inset).

An ancient crossroads and culturally rich with Arab, Phoenician, Italian, and British infusions, Malta became a British dependency and served British shipping and its military. It suffered terribly during World War II bombings, but despite limited natural resources recovered strongly during the postwar period. Today Malta has a booming tourist industry and a relatively high standard of living, and is a new member of the European Union.

EASTERN EUROPE

As Figure 1-13 shows, Eastern Europe is not only territorially the largest region in the European realm: it also contains more countries (17) than any other European region. Almost all of Eastern Europe lies outside the core, and the problems of the periphery affect many of its countries. From the North European Lowland in Poland to the rugged highlands of the south, this is a region of physiographic, cultural, and political fragmentation. Open plains, major rivers, strategic mountains, isolated valleys, and crucial corridors all have influenced Eastern Europe's tumultuous migrations, epic battles, foreign invasions, and imperial episodes. Illyrians, Slavs, Turks, Hungarians, and other peoples converged on this region from near and far. Ethnic and cultural differences have kept them in chronic conflict.

Geographers call this region a **shatter belt**, a zone of 27
persistent splintering and fracturing. Geographic terminology uses several expressions to describe the breakup

Chapter 2

Russia

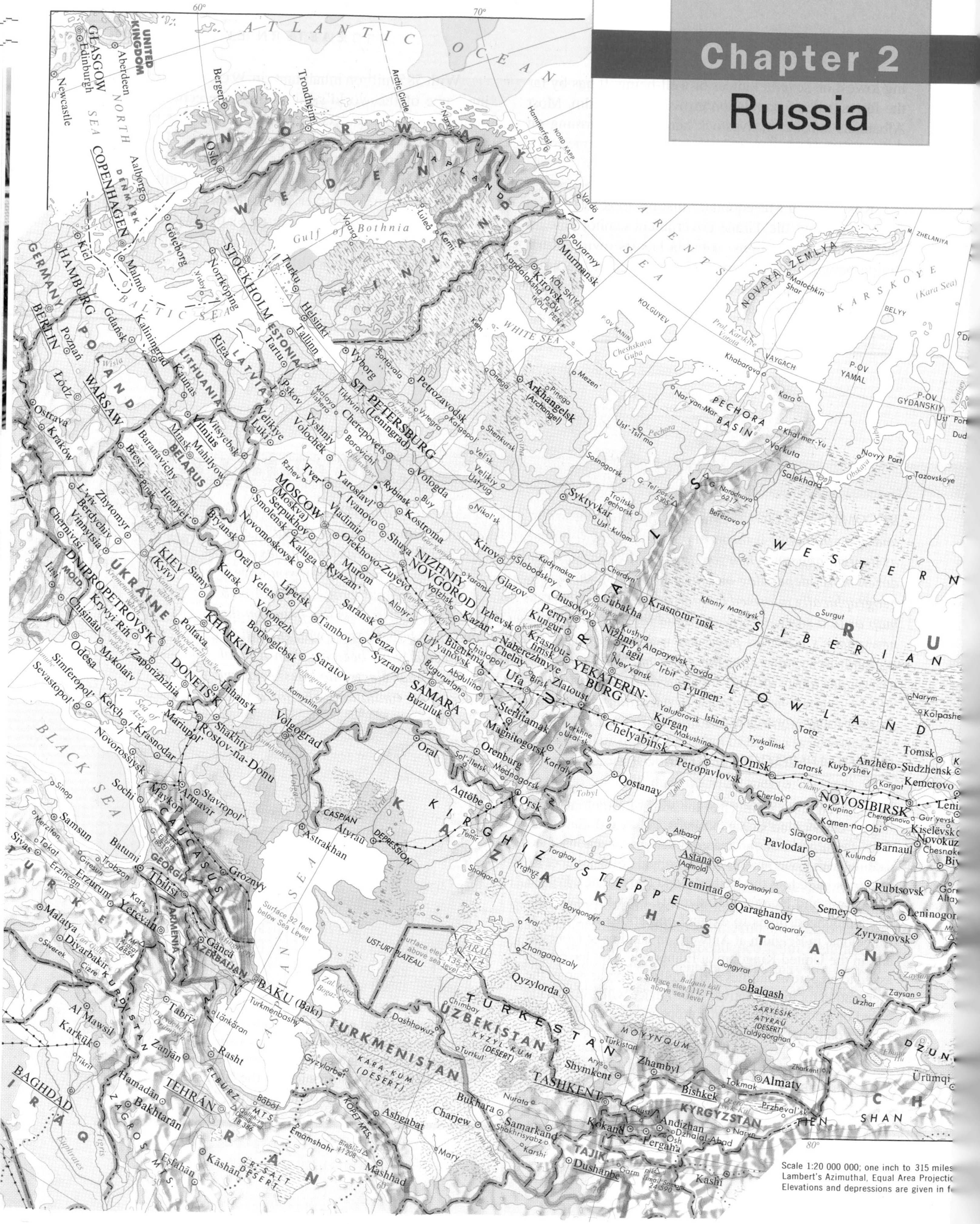

FIGURE 2-1 Reprinted with permission from *Goode's World Atlas*, 21st edition, pp. 178–179. © Rand McNally, 2005. License R.L. 05-S-64.

CONCEPTS, IDEAS, AND TERMS

1 Near Abroad	11 Imperialism
2 Oligarch	12 Russification
3 Climatology	13 Federation
4 Continentality	14 Collectivization
5 Weather	15 Command economy
6 Tundra	16 Unitary state system
7 Taiga	17 Distance decay
8 Permafrost	18 Population decline
9 Forward capital	19 Heartland theory
10 Colonialism	20 Core area

REGIONS

- RUSSIAN CORE AND PERIPHERIES
- EASTERN FRONTIER
- SIBERIA
- RUSSIAN FAR EAST

IS THE RUSSIAN state a geographic realm? A look at a globe yields an affirmative answer. Russia is the world's largest state territorially, it has a substantial population, it has common borders with countries from Norway to North Korea, and it extends from the Arctic Ocean to the Black and Caspian seas. More than twice as large as the conterminous United States and stretching across 11 time zones, Russia is the planet's giant.

For most of the twentieth century, Russia was the cornerstone of the Soviet Union, one of the world's two superpowers with a mighty military and a large arsenal of nuclear weapons. With a communist ideology and a global alliance that stretched from Havana to Hanoi, the Soviets built a political and economic system that extended far beyond Russia's boundaries.

Even now, there is more to Russia than lies within its political borders. Russia, like France and Spain and other former colonial powers, lay at the heart of a colonial sphere that continues to this day to carry the imprints of empire. In the centuries of the conquering czars, Russian armies drove far into Russia's periphery and established a permanent Russian presence from Transcaucasia to Central Asia. During the seven decades of communist colonization, Russian settlers by the millions moved into non-Russian territories from Kaliningrad to Kazakhstan. When the Soviet Union collapsed in 1991, just as in the case of British, French, Portuguese, and other empires earlier, its colonial settlers were left stranded, in many cases far from home in newly independent countries such as Kyrgyzstan and Tajikistan. The new leaders in post-Soviet Moscow had many problems, and one of these was the well-being of those Russians beyond its borders. Several million ethnic Russians returned home, but many millions more remained—and continue to remain—under the jurisdiction of non-Russian governments (see Table 2-1).

Travel from Rostov in the southwest across the nearby Ukrainian border to Donetsk, and you will have left Russia but not Russia's cultural landscapes (Fig. 2-1). Travel from Kurgan east of the Urals across the Kazakhstan border to Petropavlovsk, and don't expect to see anything different: northern Kazakhstan is as Russian as southern Russia itself. Russian minorities in Estonia, Latvia, Moldova, and other former colonies continue to look to Moscow for support when their privileges or rights are abridged. In Moscow, the geographic concept of a **Near Abroad** took hold, a sphere of influence in the 1
former Soviet periphery where Russia would reserve the right to protect the interests of its kin. As time went on, and Russian expatriates either returned home or adjusted to their new situation, that notion lost its urgency. Still Russia projected its power onto small neighbors when it perceived the need and opportunity, as it did in the Georgian province of Abkhazia, where its actions fomented secessionist ideas and had the effect of destabilizing the government in Tbilisi, Georgia's capital.

During the 1990s, postcommunist, post-Soviet Russia was a reorganizing state, struggling to achieve democratic government, to reform its economy, to deal with minority issues, and to define a new role in the world. It was a time of chaos and setbacks, of corruption and opportunism, of administrative weakness, of cultural breakdown and military failure. Russia's first elected president, the amiable, unpredictable, ineffectual, pliable Boris Yeltsin, in some ways personified the rudderless state. Surveys indicated that a majority of Russian

MAJOR GEOGRAPHIC QUALITIES OF Russia

1. Russia is the largest territorial state in the world. Its area is nearly twice as large as that of the next-ranking country (Canada).
2. Russia is the northernmost large and populous country in the world; much of it is cold and/or dry. Extensive rugged mountain zones separate Russia from warmer subtropical air, and the country lies open to Arctic air masses.
3. Russia was one of the world's major colonial powers. Under the czars, the Russians forged the world's largest contiguous empire; the Soviet rulers who succeeded the czars took over and expanded this empire.
4. For so large an area, Russia's population of under 142 million is comparatively small. The population remains heavily concentrated in the westernmost one-fifth of the country.
5. Development in Russia is concentrated west of the Ural Mountains; here lie the major cities, leading industrial regions, densest transport networks, and most productive farming areas. National integration and economic development east of the Urals extend mainly along a narrow corridor that stretches from the southern Urals region to the southern Far East around Vladivostok.
6. Russia is a multicultural state with a complex domestic political geography. Twenty-one internal Republics, originally based on ethnic clusters, function as politico-geographical entities.
7. Its large territorial size notwithstanding, Russia suffers from land encirclement within Eurasia; it has few good and suitably located ports.
8. Regions long part of the Russian and Soviet empires are realigning themselves in the postcommunist era. Eastern Europe and the heavily Muslim Southwest Asia realm are encroaching on Russia's imperial borders.
9. The failure of the Soviet communist system left Russia in economic disarray. Many of the long-term components described in this chapter (food-producing areas, railroad links, pipeline connections) broke down in the transition to the postcommunist order.
10. Russia long has been a source of raw materials but not a manufacturer of export products, except weaponry. Few Russian (or Soviet) automobiles, televisions, cameras, or other consumer goods reach world markets.

citizens yearned for a stronger government, more security (Islamic terrorism in Moscow alone took hundreds
2 of lives), a harnessing of the opportunists, known as **oligarchs**, who had used their ties to government to enrich themselves, and much more stability than the new administration was able to provide.

And that is what Russia got when President Vladimir Putin succeeded Yeltsin and was subsequently reelected. Putin left no doubt: he intended to restore Russia's military power, reinforce its economy, weaken political opposition, imprison oligarchs, muzzle the media, defeat the terrorists in their home base, protect Russian interests beyond the country's borders, and reassert Russia's global geopolitical role. In 2004, when a close presidential election in neighboring Ukraine pitted a pro-Moscow candidate against a pro-Europe opponent, Putin actually campaigned for the pro-Russian candidate. When the election, narrowly won by the candidate he favored, was voided because of widespread fraud, he angrily denounced that action. Europe and Russia are drifting apart, not converging, and the geographic-realm boundary in Figure G-2 is as divisive as ever.

DEFINING THE REALM

As we noted in the Introduction, geographic realms usually are bounded not by sharply etched borders, but by zones of transition. In the case of the realm dominated by Russia, this certainly is the case, and Russia's imperial history has much to do with it. In Chapter 1 we noted how *Belarus* is a component of neither Europe nor Russia, a malfunctioning state that was part of the Soviet Union and today remains a throwback to the days of authoritarian politics and stagnant economics. This country fits neither the European nor the Russian realm criteria, although a convergence with Russia appeared more likely in 2005 than any opening toward Europe. A second transitional area is the *Transcaucasus* between the Black and Caspian seas. Here, as we note later, Russia has an internal problem with a set of ethnic "republics" of which one, Chechnya, has caused a still-unfolding disaster by rejecting Moscow's

Table 2-1
SOCIAL CHARACTERISTICS OF THE OTHER 14 EX-SOVIET REPUBLICS

Name	Population (Millions)	Percent Russians	Religion	Official Language
Armenia	3.2	1	Armenian Orthodox (Christian) 65%	Armenian
Azerbaijan	8.4	3	Shi'ite Muslim 93%	Azerbaijani
Belarus	9.7	11	Belarussian Orthodox, Roman Catholic 49%	Belarussian Russian
Estonia	1.3	26	Estonian Orthodox, Estonian Lutheran 34%	Estonian
Georgia	4.5	6	Georgian, Russian Orthodox 40%; Sunni Muslim 11%	Georgian
Kazakhstan	15.2	30	Sunni Muslim 47%; Russian Orthodox 8%	Kazakh
Kyrgyzstan	5.2	13	Sunni Muslim 75%	Kyrgyz; Russian
Latvia	2.3	29	Christian Churches 40%	Latvian
Lithuania	3.4	6	Roman Catholic 79%	Lithuanian
Moldova	4.2	13	Eastern Orthodox 46%	Romanian
Tajikistan	6.9	1	Sunni Muslim 85%	Tajik
Turkmenistan	5.9	6	Sunni Muslim 87%	Turkmen
Ukraine	46.7	17	Ukrainian Orthodox 30%	Ukrainian
Uzbekistan	27.3	6	Sunni Muslim 76%	Uzbek

rule with armed resistance. Here, too, lie three former Soviet colonies, now independent countries, that are nonetheless drawn into the Russian orbit: Georgia, Armenia, and, to a lesser degree, Azerbaijan. And still another transitory area lies, as we noted, in Russian-populated northern *Kazakhstan*. There is indeed more to the Russian realm than lies within Russia itself.

And within its borders. Russia still is an empire. It was the Soviet Empire that disintegrated, not the Russian one: about 20 percent of Russia's more than 140 million inhabitants are non-Russians, including numerous minorities large and small whose presence is reflected by the administrative map. The Chechens are but one of those minorities, many of which look back upon a history of mistreatment but most of which have accommodated themselves to the reality of Russian dominance. Russia's failure in Chechnya should not obscure what was in fact a remarkable success of the Yeltsin administration: avoiding the collapse of administrative order and the secession of minority-based "republics" throughout the country. For a time, it seemed that Tatarstan, the historic home of Muslim Tatars, would declare independence along with other minorities long unhappy under Soviet-Russian rule. But Yeltsin's government gave these peoples ways to express their demands and to negotiate their terms, and in the end the Russian map stayed intact—except in the Transcaucasus. Still, one of the risks inherent in a return to "strong" government is a reversal of this good fortune. Chechnya's rebellion and Muslim Chechens' terrorist actions have already made life much more difficult for other minorities on the streets of Moscow and other cities, where many ethnic Russians equate minority appearance and dress with rebellion and risk. Alienating Russia's other minorities risks the future of the state itself.

ROOTS OF THE RUSSIAN REALM

The name *Russia* evokes cultural-geographic images of a stormy past: terrifying czars, conquering Cossacks, Byzantine bishops, rousing revolutionaries, clashing cultures. Russians repulsed the Tatar (Mongol) hordes, forged a powerful state, colonized a vast contiguous empire, defeated Napoleon, adopted communism, and, when the communist system failed, lost most of their imperial domain. Today, Russia seeks a new place in the globalizing world.

Precommunist Russia may be described as a culture of extremes. It was a culture of strong nationalism, resistance to change, and despotic rule. Enormous wealth was

In the 1960s visitors to the (then) Soviet Union, one of the authors of this book among them, were shocked by the fate of virtually all of the country's Christian churches, whose condition ranged from bad repair to outright ruin. Architecturally ornate and historic church buildings with superb murals and exquisite glasswork stood empty, subject to the elements and vandalism. Doors and furniture had been taken for firewood; some churches served as barns, others as storage. In his atheist zeal, Josef Stalin ordered the demolition of one of Moscow's religious treasures, the Christ-the-Savior Cathedral (other Moscow cathedrals fared better and were turned into museums). This photograph symbolizes the reversal of fortunes following the collapse of the Soviet Union: religion is again a part of Russian life, the Russian Orthodox Church again dominates the religious map, and a reconstructed Cathedral of the Savior again stands where it did before the communist era.

concentrated in a small elite. Powerful rulers and bejeweled aristocrats perpetuated their privileges at the expense of millions of peasants and serfs who lived in dreadful poverty. The Industrial Revolution arrived late in Russia, and a middle class was slow to develop. Yet the Russian nation gained the loyalty of many of its citizens, and its writers and artists were among the world's greatest. Authors such as Tolstoy and Dostoyevsky chronicled the plight of the poor; composers celebrated the indomitable Russian people and, as Tchaikovsky did in his *1812 Overture*, commemorated their victories over foreign foes.

Under the czars, Russia grew from nation into empire. The czars' insatiable demands for wealth, territory, and power sent Russian armies across the plains of Siberia, through the deserts of interior Asia, and into the mountains along Russia's rim. Russian pioneers ventured even farther, entering Alaska, traveling down the Pacific coast of North America, and planting the Russian flag near San Francisco in 1812. But as Russia's empire expanded, its internal weaknesses gnawed at the power of the czars. Peasants rebelled. Unpaid (and poorly fed) armies mutinied. When the czars tried to initiate reforms, the aristocracy objected. The empire at the beginning of the twentieth century was ripe for revolution, which began in 1905 (see box titled "The Soviet Union, 1924–1991").

The last czar, Nicholas II, was overthrown in 1917, and civil war followed. The victorious communists led by V. I. Lenin soon swept away much of the Russia of the past. The Russian flag disappeared. The czar and his family were executed. The old capital of Russia, St. Petersburg, was renamed Leningrad in honor of the revolutionary leader. Moscow, in the interior of the country, was chosen as the new capital for a country with a new name, the *Soviet Union*. Eventually, this Union consisted of 15 political entities, each a Soviet Socialist Republic (SSR). Russia was just one of these republics, and the name *Russia* disappeared from the international map.

But on the Soviet map, Russia was the giant, the dominant republic, the Slavic center. Not for nothing was the communist revolution known as the Russian Revolution. The other republics of the Soviet Union were for minorities the czars had colonized or for countries that fell under Soviet sway later, but none could begin to match Russia. The Soviet Empire was the legacy of czarist expansionism, and the new communist rulers were Russian first and foremost (with the exception of Josef Stalin, who was a Georgian).

Officially, the Union of Soviet Socialist Republics endured from 1924 to 1991. When the USSR collapsed, the republics were set free and world maps had to be redrawn. Now Russia and its former colonies had to adjust to the realities of a changing world order. It proved to be a difficult, sometimes desperate journey.

RUSSIA'S PHYSICAL ENVIRONMENTS

Russia's physiography is dominated by vast plains and plateaus rimmed by rugged mountains (Fig. 2-2). Only the Ural Mountains break an expanse of low relief that extends from the Polish border to eastern Siberia. In the western part of this huge plain, where the northern pine

The Soviet Union, 1924–1991

For 67 years Russia was the cornerstone of the *Soviet Union*, the so-called Union of Soviet Socialist Republics (USSR). The Soviet Union was the product of the Revolution of 1917, when more than a decade of rebellion against the rule of Czar Nicholas II led to his abdication. Russian revolutionary groups were called soviets ("councils"), and they had been active since the first workers' uprising in 1905. In that year, thousands of Russian workers marched on the czar's palace in St. Petersburg in protest, and soldiers fired on them. Hundreds were killed and wounded. Russia descended into chaos.

A coalition of military and professional men forced the czar's abdication in 1917. Then Russia was ruled briefly by a provisional government. In November 1917, the country held its first democratic election—and, as it turned out, its last for more than 70 years.

The provisional government allowed the exiled activists in the Bolshevik camp to return to Russia (there were divisions among the revolutionaries): Lenin from Switzerland, Trotsky from New York, and Stalin from internal exile in Siberia. In the ensuing political struggle, Lenin's Bolsheviks gained control over the revolutionary soviets, and this ushered in the era of communism. In 1924, the new communist empire was formally renamed the Union of Soviet Socialist Republics, or Soviet Union in shorthand.

Lenin the organizer died in 1924 and was succeeded by Stalin the tyrant, and many of the peoples under Moscow's control suffered unimaginably. In pursuit of communist reconstruction, Stalin and his adherents starved millions of Ukrainian peasants to death, forcibly relocated entire ethnic groups (including a people known as the Chechens), and exterminated "uncooperative" or "disloyal" peoples. The full extent of these horrors may never be known.

On December 25, 1991, the inevitable occurred: the Soviet Union ceased to exist, its economy a shambles, its political system shattered, the communist experiment a failure. The last Soviet president, Mikhail Gorbachev, resigned, and the Soviet hammer-and-sickle flag flying atop the Kremlin was lowered for the last time and immediately replaced by the white, red, and blue Russian tricolor. A new and turbulent era had begun, but Soviet structures and systems will long cast their shadows over transforming Russia.

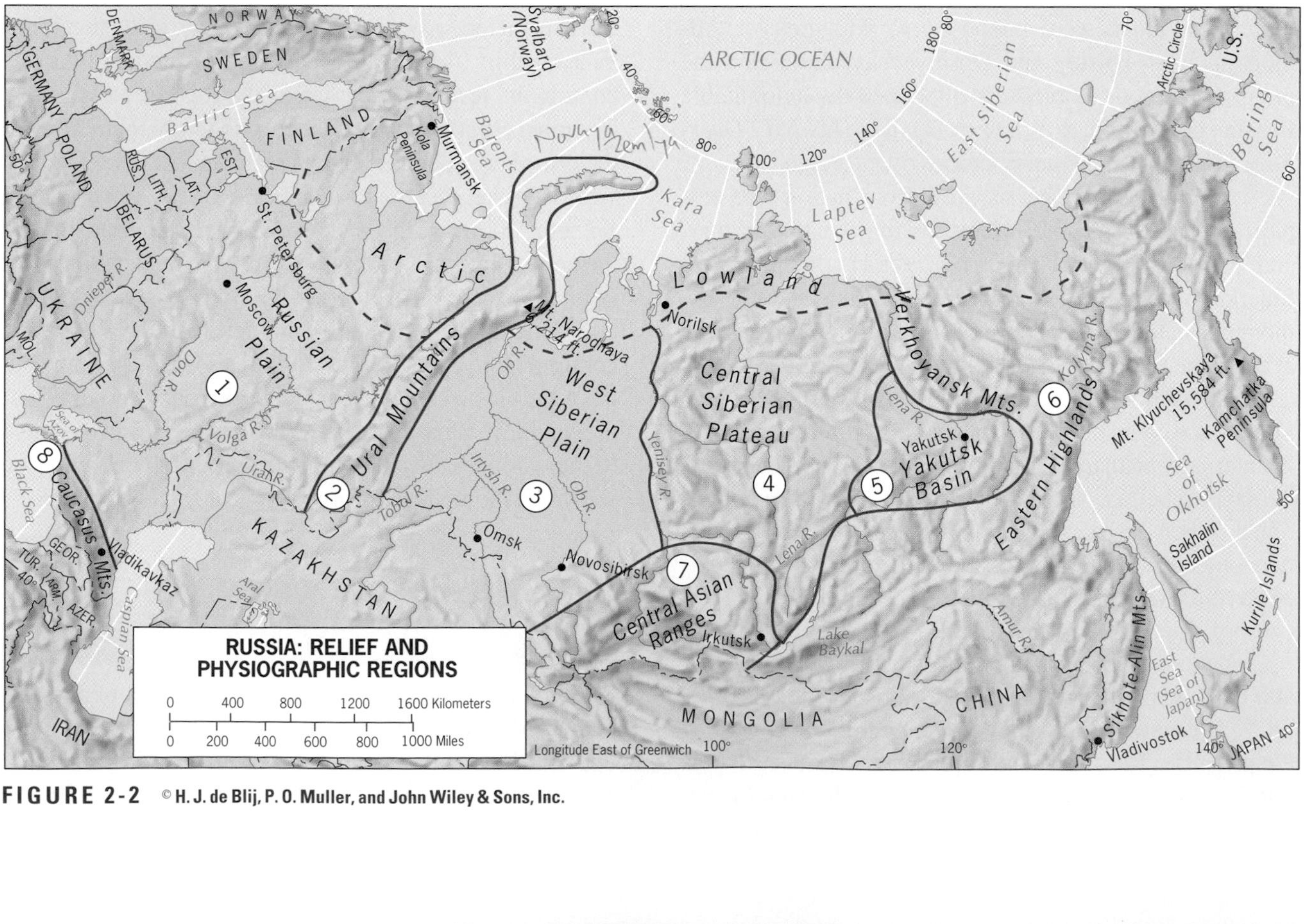

FIGURE 2-2 © H. J. de Blij, P. O. Muller, and John Wiley & Sons, Inc.

forests meet the mid-latitude grasslands, the Slavs established their domain.

The historical geography of Russia is the story of Slavic expansion from its populous western heartland across interior Eurasia to the east and into the mountains and deserts of the south. This eastward march was hampered not only by vast distances but also by harsh natural conditions. As the northernmost populous country on Earth, Russia has virtually no natural barriers against the onslaught of Arctic air. Moscow lies farther north than Edmonton, Canada, and St. Petersburg lies at latitude 60° North—the latitude of the southern tip of Greenland. Winters are long, dark, and bitterly cold in most of Russia; summers are short and growing seasons limited. Many a Siberian frontier outpost was doomed by cold, snow, and hunger.

It is therefore useful to view Russia's past, present,
3 and future in the context of its **climatology**. This field of geography investigates not only the distribution of climatic conditions over the Earth's surface but also the processes that generate this spatial arrangement. The Earth's atmosphere traps heat received as radiation from the Sun, but this *greenhouse effect* varies over planetary space—and over time. As we noted in the introductory chapter, much of what is today Russia was in the grip of a glaciation until the onset of the warmer Holocene. But even today, with a natural warming cycle in progress augmented by human activity, Russia still suffers from severe cold and associated drought. If the enhanced global warming cycle continues, Russia (in contrast to low-lying countries faced by rising sea levels) may benefit. But that will take generations.

Currently precipitation totals, even in western Russia, range from modest to minimal because the warm, moist air carried across Europe from the North Atlantic Ocean loses much of its warmth and moisture by the time it reaches Russia. Figures G-7, G-8, and 2-3 (below) reveal the consequences. Russia's climatic **continentality** (inland climatic 4
environment remote from moderating and moistening maritime influences) is expressed by its prevailing *Dfb* and *Dfc* conditions. Compare the Russian map to that of North America (Fig. G-8), and you note that, except for a small corner off the Black Sea, Russia's climatic conditions resemble those of the U.S. Upper Midwest and Canada. Along its entire north, Russia has a zone of *E* climates, the most frigid on the planet. In these Arctic latitudes originate the polar air masses that dominate its environments.

By studying Russia's climates we can begin to understand what the map of population distribution (Fig. G-9) shows. The overwhelming majority of the country's

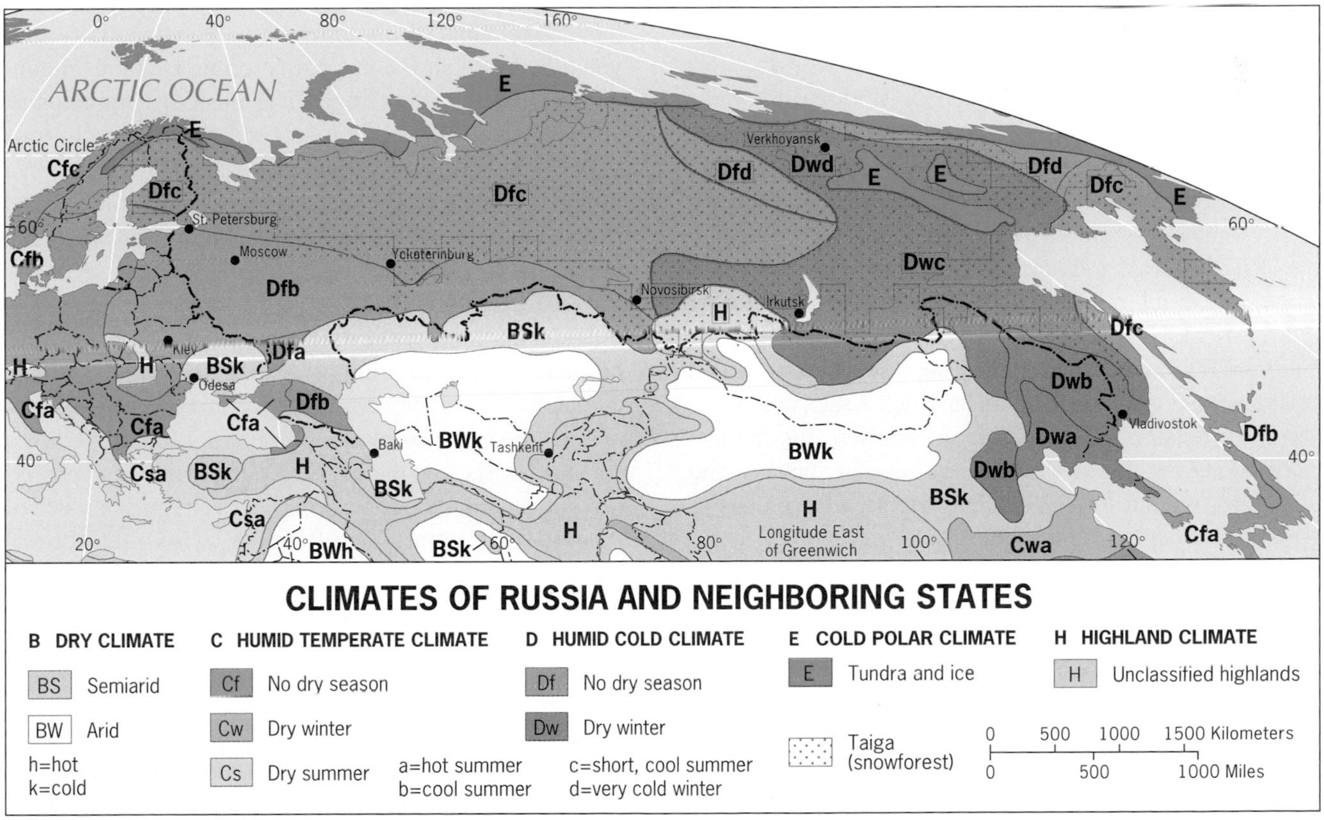

FIGURE 2-3 © H. J. de Blij, P. O. Muller, and John Wiley & Sons, Inc.

142 million people are concentrated in the west and southwest, where climatic conditions were least difficult at a time when farming was the mainstay for most of the population. This map is a legacy that will mark Russia's living space for generations to come.

Climate and weather (there is a distinction: *climate* is
5 a long-term average, whereas **weather** refers to existing atmospheric conditions at a given place and time) have always challenged Russia's farmers. Conditions are most favorable in the west, but even there temperature extremes, variable and undependable rainfall, and short growing seasons make farming difficult. During the Soviet period, fertile and productive Ukraine supplied much of Russia's food needs, but even then Russia often had to import grain. Soviet rulers wanted to reduce their country's dependence on imported food, and their communist planners built major irrigation projects to increase crop yields in the colonized republics of Central Asia. As we will see in Chapter 7, some of these attempts to overcome nature's limitations spelled disaster for the local people.

Physiographic Regions

To assess the physiography of this vast country, refer again to Figure 2-2. Note how mountains and deserts encircle Russia: the Caucasus in the southwest ⑧; the Central Asian Ranges in the center ⑦; the Eastern Highlands facing the Pacific from the Bering Sea to the East Sea (Sea of Japan) ⑥. The Kamchatka Peninsula has a string of active volcanoes in one of the world's most earthquake-prone zones (Fig. G-5). Warm subtropical air thus has little opportunity to penetrate Russia, while cold Arctic air sweeps southward without impediment. Russia's Arctic north is a gently sloping lowland broken only by the Urals and Eastern Highlands.

Russia's vast and complex physical stage can be divided into eight physiographic regions, each of which, at a larger scale, can be subdivided into smaller units. In the Siberian region, one criterion for such subdivision would be the vegetation. The Russian language has given us two
6 terms to describe this vegetation: **tundra**, the treeless plain along the Arctic shore where mosses, lichens, and
7 some grasses survive, and **taiga**, the mostly coniferous forests that begin to the south of where the tundra ends, and extend over vast reaches of Siberia, which means the "sleeping land" in Russian.

The Russian Plain ① is the eastward continuation of the North European Lowland, and here the Russian state formed its first core area. Travel north from Moscow at its heart, and the countryside soon is covered by needleleaf forests like those of Canada; to the south lie the grain fields of southern Russia and, beyond, those of Ukraine. Note the Kola Peninsula and Barents Sea in the far north: warm water from the North Atlantic comes around northern Norway and keeps the port of Murmansk ice free most of the year. The Russian Plain is bounded on the east by the Ural Mountains ②; though not a high range, it is topographically prominent because it separates two extensive plains. The range of the Urals is more than 2000 miles (3200 km) long and reaches from the shores of the Kara Sea to the border with Kazakhstan. It is not a barrier to east-west transportation, and its southern end is densely populated. Here the Urals yield minerals and fossil fuels.

East of the Urals lies Siberia. The West Siberian Plain ③ has been described as the world's largest unbroken lowland; this is the vast basin of the Ob and Irtysh rivers. Over the last 1000 miles (1600 km) of its course to the Arctic Ocean, the Ob falls less than 300 feet (90 m). In Figure 2-2, note the dashed line that extends from the Finnish border to the East Siberian Sea and offsets the Arctic Lowland. North of the line, water in the ground is
permanently frozen; this **permafrost** creates another 8
obstacle to permanent settlement. Looking again at the West Siberian Plain, we see that the north is permafrost-ridden and the central zone is marshy. The south, however, has such major cities as Omsk and Novosibirsk, within the corridor of the Trans-Siberian Railroad.

East of the West Siberian Plain the country begins to rise, first into the central Siberian Plateau ④, another sparsely settled, remote, permafrost-affected region. Here winters are long and cold, and summers are short; the area remains barely touched by human activity. Beyond the Yakutsk Basin ⑤, the terrain becomes mountainous and the relief high. The Eastern Highlands ⑥ are a jumbled mass of ranges and ridges, precipitous valleys, and volcanic mountains. Lake Baykal lies in a trough that is over 5000 feet (1500 m) deep—the deepest rift lake in the world (see Chapter 6). On the Kamchatka Peninsula, volcanic Mount Klyuchevskaya reaches nearly 15,600 feet (4750 m).

The northern part of region ⑥ is Russia's most inhospitable zone, but southward along the Pacific coast the climate is less severe. Nonetheless, this is a true frontier region. The forests provide opportunities for lumbering, a fur trade exists, and there are gold and diamond deposits.

Mountains also mark the southern margins of much of Russia. The Central Asian Ranges ⑦, from the Kazakh border in the west to Lake Baykal in the east, contain many glaciers whose annual meltwaters send alluvium-laden streams to enrich farmlands at lower elevations. The Caucasus ⑧, in the land corridor between the Black and Caspian seas, forms an extension of Europe's Alpine Mountains and exhibits a similarly high *relief* (range of elevations). Here Russia's southern border is sharply defined by *topography* (surface configuration).

As our physiographic map suggests, the more habitable terrain in Russia becomes latitudinally narrower from west to east. Beyond the southern Urals, the zone of settlement in Russia becomes discontinuous (in Soviet times it extended into northern Kazakhstan, where a large Russian population still lives). Isolated towns did develop in Russia's vast eastern reaches, even in Siberia, but the tenuous ribbon of settlement does not widen again until it reaches the country's Far Eastern Pacific Rim.

EVOLUTION OF THE RUSSIAN STATE

Four centuries ago, when European kingdoms were sending fleets to distant shores to search for riches and capture colonies, there was little indication that the largest of all empires would one day center on a city in the forests halfway between Sweden and the Black Sea. The plains of Eurasia south of the taiga had seen waves of migrants sweep from east to west: Scythians, Sarmatians, Goths, Huns, and others came, fought, settled, and were absorbed or driven off. Eventually, the Slavs, heirs to these Eurasian infusions, emerged as the dominant culture in what is today Ukraine, south of Russia and north of the Black Sea.

From this base on fertile, productive soils, the Slavs expanded their domain, making Kiev (Kyyiv) their southern headquarters and establishing Novgorod on Lake Ilmen in the north. Each was the center of a state known as a *Rus*, and both formed key links on a trade route between the German-speaking Hanseatic ports of the Baltic Sea and the trading centers of the Mediterranean. During the eleventh and twelfth centuries, the Kievan Rus and the Novgorod Rus united to form a large and comparatively prosperous state astride forest and steppe (short-grass prairie).

The Mongol Invasion

Prosperity attracts attention, and far to the east, north of China, the Mongol Empire had been building under Genghis Khan. In the thirteenth century, the Mongol (Tatar) "hordes" rode their horses into the Kievan Rus, and the state fell. By mid-century, Slavs were fleeing into the forests, where the Mongol horsemen had less success than on the open plains of Ukraine. Here the Slavs reorganized, setting up several new Russes. Still threatened by the Tatars, the ruling princes paid tribute to the Mongols in exchange for peace.

Moscow, deep in the forest on the Moscow River, was one of these Russes. Its site was remote and defensible. Over time, Moscow established trade and political links with Novgorod and became the focus of a growing area of Slavic settlement. When the Mongols, worried about Moscow's expanding influence, attacked near the end of the fourteenth century and were repulsed, Moscow emerged as the unchallenged center of the Slavic Russes. Soon its ruler, now called a Grand Duke, took control of Novgorod, and the stage was set for further growth.

Grand Duchy of Muscovy

Moscow's geographic advantages again influenced events. Its centrality, defensibility, links to Slavic settlements including Novgorod, and open frontiers to the north and west, where no enemies threatened, gave Moscow the potential to expand at the cost of its old enemies, the Mongols. Many of these invaders had settled in the basin of the Volga River and elsewhere, where the city of Kazan had become a major center of Islam, the Tatars' dominant religion. During the rule of Ivan the Terrible (1547–1584), the Grand Duchy of Muscovy became a major military power and imperial state. Its rulers called themselves czars and claimed to be the heirs of the Byzantine emperors. Now began the expansion of the Russian domain (Fig. 2-4), marked by the defeat of the Tatars at Kazan and the destruction of hundreds of mosques.

The Cossacks

This eastward expansion of Russia was spearheaded by a relatively small group of seminomadic peoples who came to be known as Cossacks and whose original home was in present-day Ukraine. Opportunists and pioneers, they sought the riches of the eastern frontier, chiefly furbearing animals, as early as the sixteenth century. By the mid-seventeenth century they reached the Pacific Ocean, defeating Tatars in their path and consolidating their gains by constructing *ostrogs* (strategic fortified waystations) along river courses. Before the eastward expansion halted in 1812 (Fig. 2-4), the Russians had moved across the Bering Strait to Alaska and down the western coast of North America into what is now northern California (see box titled "Russians in North America").

Czar Peter the Great

When Peter the Great became czar (he ruled from 1682 to 1725), Moscow already lay at the center of a great empire—great, at least, in terms of the territories it controlled. The Islamic threat had been ended with the defeat of the Tatars. The influence of the Russian Orthodox Church was represented by its distinctive religious architecture and powerful bishops. Peter consolidated Russia's gains and hoped to make a modern, European-style state out of his loosely knit country. He built St. Petersburg as a **forward capital** on the doorstep of Swedish- 9
held Finland, fortified it with major military installations, and made it Russia's leading port.

Czar Peter the Great, an extraordinary leader, was in many ways the founder of modern Russia. In his desire to

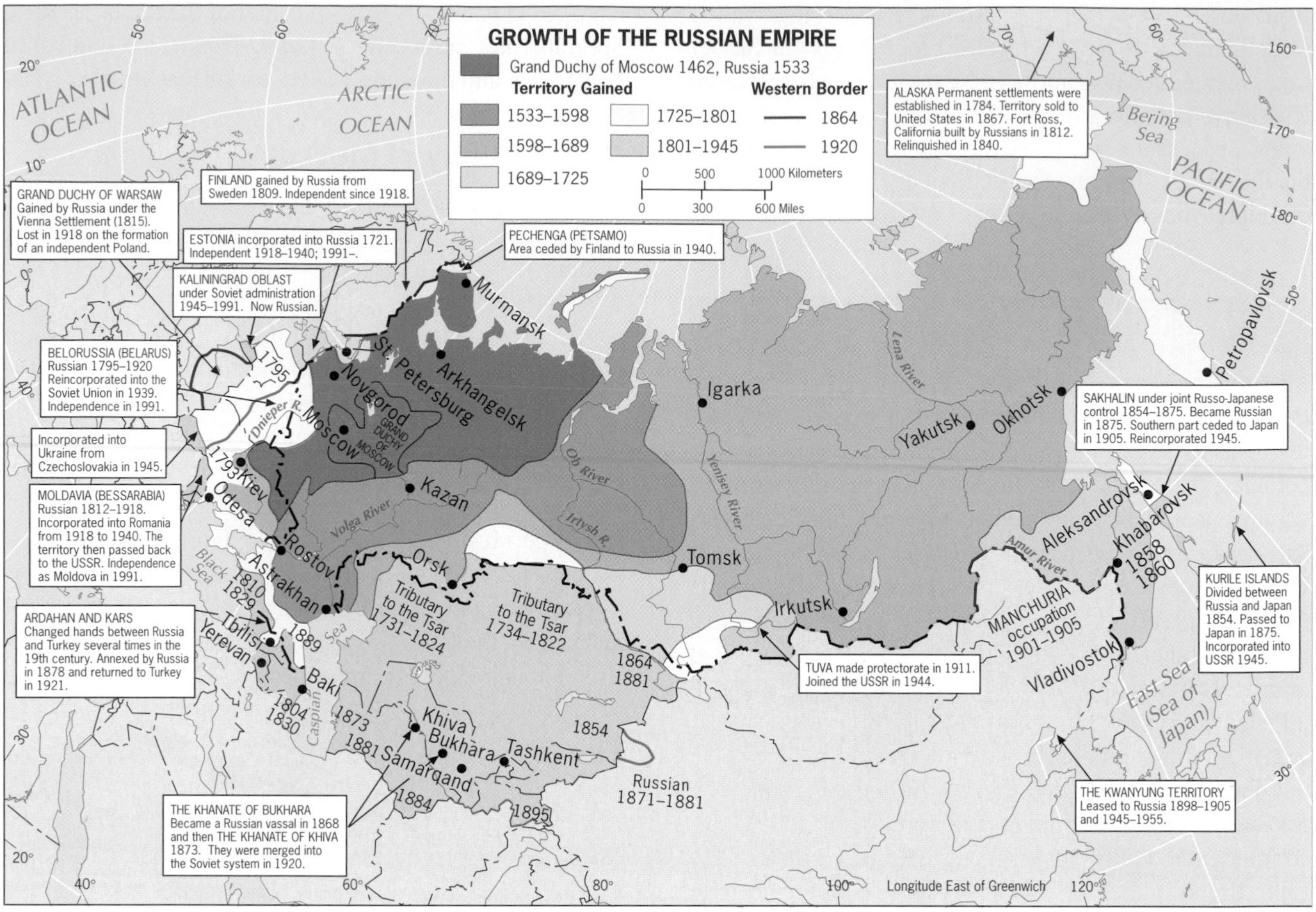

FIGURE 2-4 © H. J. de Blij, P. O. Muller, and John Wiley & Sons, Inc.

remake Russia—to pull it from the forests of the interior to the waters of the west, to open it to outside influences, and to relocate its population—he left no stone unturned. Prominent merchant families were forced to move from other cities to St. Petersburg. Ships and wagons entering the city had to bring building stones as an entry toll. The czar himself, aware that to become a major power Russia had to be strong at sea as well as on land, went to the Netherlands to work as a laborer in the famed Dutch shipyards to learn the most efficient method for building

Russians in North America

The first white settlers in Alaska were Russians, not Western Europeans, and they came across Siberia and the Bering Strait, not across the Atlantic and North America. Russian hunters of the sea otter, valued for its high-priced pelt, established their first Alaskan settlement at Kodiak Island in 1784. Moving south along the North American coast, the Russians founded additional villages and forts to protect their tenuous holdings until they reached as far as the area just north of San Francisco Bay, where they built Fort Ross in 1812.

But the Russian settlements were isolated and vulnerable. European fur traders began to put pressure on their Russian competitors, and St. Petersburg found the distant settlements a burden and a risk. In any case, American, British, and Canadian hunters were decimating the sea otter population, and profits declined. When U.S. Secretary of State William Seward offered to purchase Russia's North American holdings in 1867, St. Petersburg quickly agreed—for $7.2 million. Thus Alaska, including its lengthy southward coastal extension—the Alaskan "panhandle"—became U.S. territory and, eventually, the forty-ninth State. Although Seward was ridiculed for his decision—Alaska was called "Seward's Folly" and "Seward's Icebox"—he was vindicated when gold was discovered there in the 1890s. The twentieth century proved the wisdom of Seward's action, strategically as well as economically. At Prudhoe Bay off Alaska's northern Arctic slope, large oil reserves were tapped in the 1970s and are still being exploited. And like Siberia, Alaska probably contains yet-unknown riches.

ships. Meanwhile, the czar's forces continued to conquer people and territory: Estonia was incorporated in 1721, widening Russia's window to the west, and major expansion soon occurred south of the city of Tomsk (Fig. 2-4).

Czarina Catherine the Great

Under Czarina Catherine the Great, who ruled from 1760 to 1796, Russia's empire in the Black Sea area grew at the expense of the Ottoman Turks. The Crimea Peninsula, the port city of Odesa (Odessa), and the whole northern coastal zone of the Black Sea fell under Russian control. Also during this period, the Russians made a fateful move: they penetrated the area between the Black and Caspian seas, the mountainous Caucasus with dozens of ethnic and cultural groups, many of which were Islamized. The cities of Tbilisi (now in Georgia), Baki (Baku) in Azerbaijan, and Yerevan (Armenia) were captured. Eventually, the Russian push toward an Indian Ocean outlet was halted by the British, who held sway in Persia (modern Iran), and also by the Turks.

Meanwhile, Russian colonists entered Alaska, founding their first American settlement at Kodiak Island in 1784. As they moved southward, numerous forts were built to protect their tenuous holdings against indigenous peoples. Eventually, they reached nearly as far south as San Francisco Bay where, in the fateful year 1812, they erected Fort Ross.

Catherine the Great had made Russia a colonial power, but the Russians eventually gave up on their North American outposts. The sea-otter pelts that had attracted the early pioneers were running out, European and white American hunters were cutting into the profits, and Native American resistance was growing. When U.S. Secretary of State William Seward offered to purchase Russia's Alaskan holdings in 1867, the Russian government quickly agreed—for $7.2 million. Thus Alaska and its Russian-held panhandle became U.S. territory and, ultimately, the forty-ninth State.

A Russian Empire

Although Russia withdrew from North America, Russian expansionism during the nineteenth century continued in Eurasia. While extending their empire southward, the Russians also took on the Poles, old enemies to the west, and succeeded in taking most of what is today the Polish state, including the capital of Warsaw. To the northwest, Russia took over Finland from the Swedes in 1809. During most of the nineteenth century, however, the Russians were preoccupied with Central Asia—the region between the Caspian Sea and western China—where Tashkent and Samarqand (Samarkand) came under St. Petersburg's control (Fig. 2-4). The Russians here were still bothered by raids of nomadic horsemen, and they sought to establish their authority over the Central Asian steppe

FROM THE FIELD NOTES

"Not only the city of St. Petersburg itself, but also its surrounding suburbs display the architectural and artistic splendor of czarist Russia. The czars built opulent palaces in these outlying districts (then some distance from the built-up center), among which the Catherine Palace, begun in 1717 and completed in 1723 followed by several expansions, was especially majestic. During my first visit in 1994 the palace, parts of which had been deliberately destroyed by the Germans during World War II, was still being restored; large black and white photographs in the hallways showed what the Nazis had done and chronicled the progress of the repairs during communist and post-communist years. A return visit in 2000 revealed the wealth of sculptural decoration on the magnificent exterior (left) and the interior detail of a set of rooms called the "golden suite" of which the ballroom (right) exemplifies eighteenth-century Russian Baroque at its height." © H. J. de Blij

country as far as the edges of the high mountains that lay to the south. Thus Russia gained many Muslim subjects, for this was Islamic Asia they were penetrating. Under czarist rule, these people retained some autonomy. Much farther to the east, a combination of Japanese expansionism and a decline of Chinese influence led Russia to annex from China several provinces east of the Amur River. Soon thereafter, in 1860, the port of Vladivostok on the Pacific was founded.

Now began the events that were to lead to the first involuntary halt to the Russian drive for territory. As Figure 2-4 shows, the most direct route from western Russia to the port of Vladivostok lay across northeastern China, the territory then still called Manchuria. The Russians had begun construction of the Trans-Siberian Railroad in 1892, and they wanted China to permit the track to cross Manchuria. But the Chinese resisted. Then, taking advantage of the Boxer Rebellion in China in 1900 (see Chapter 9), Russian forces occupied Manchuria so that railroad construction might proceed.

This move, however, threatened Japanese interests in this area, and the Japanese confronted the Russians in the Russo-Japanese War of 1904–1905. Not only was Russia defeated and forced out of Manchuria: Japan even took possession of the southern part of Sakhalin Island, which they named Karafuto and retained until the end of World War II.

THE COLONIAL LEGACY

Thus Russia, like Britain, France, and other European
10 powers, expanded through **colonialism**. Yet whereas the other European powers expanded overseas, Russian influence traveled overland into Central Asia, Siberia, China, and the Pacific coastlands of the Far East. What emerged was not the greatest empire but the largest territorially contiguous empire in the world. At the time of the Japanese war, the Russian czar controlled more than 8.5 million square miles (22 million sq km), just a tiny fraction less than the area of the Soviet Union after the 1917 Revolution. Thus the communist empire, to a large extent, was the legacy of St. Petersburg and European Russia, not the product of Moscow and the socialist revolution.

The czars embarked on their imperial conquests in part because of Russia's relative location: Russia always lacked warm-water ports. Had the Revolution not intervened, their southward push might have reached the Persian Gulf or even the Mediterranean Sea. Czar Peter the Great envisaged a Russia open to trading with the entire world; he developed St. Petersburg on the Baltic Sea into Russia's leading port. But in truth, Russia's historical geography is one of remoteness from the mainstreams of change and progress, as well as one of self-imposed isolation.

An Imperial, Multinational State

Centuries of Russian expansionism did not confine itself to empty land or unclaimed frontiers. The Russian state became an imperial power that annexed and incorporated many nationalities and cultures. This was done by employing force of arms, by overthrowing uncooperative rulers, by annexing territory, and by stoking the fires of ethnic conflict. By the time the ruthless Russian regime began to face revolution among its own people, czarist Russia was a
hearth of **imperialism**, and its empire contained peoples 11
representing more than 100 nationalities. The winners in the ensuing revolutionary struggle—the communists who forged the Soviet Union—did not liberate these subjugated peoples. Rather, they changed the empire's framework, binding the peoples colonized by the czars into a new system that would in theory give them autonomy and identity. In practice, it doomed those peoples to bondage and, in some cases, extinction.

When the Soviet system failed and the Soviet Socialist Republics became independent states, Russia was left without the empire that had taken centuries to build and consolidate—and that contained crucial agricultural and mineral resources. No longer did Moscow control the farms of Ukraine and the oil and natural gas reserves of Central Asia. But look again at Figure 2-4 and you will see that, even without its European and Central Asian colonies, Russia remains an empire. Russia lost the "republics" on its periphery, but Moscow still rules over a domain that extends from the borders of Finland to North Korea. Inside that domain Russians are in the overwhelming majority, but many subjugated nationalities, from Tatars to Yakuts, still inhabit ancestral homelands. Accommodating these many indigenous peoples is one of the challenges facing the Russian Federation today.

On the basis of physiographic, ethnic, historic, and cultural criteria, therefore, Russia constitutes a geographic realm, although transition zones rather than sharp boundaries mark its limits in some areas. Encircled by mountains and deserts, ruled by climatic continentality, unified by a dominant culture, and unmatched on Earth in terms of dimensions, Russia stands apart—from Europe to the west, China to the east, Central Asia to the south. But Russia, as we will see, is a society in transition. The realm's boundaries are unstable, still changing. To the west, Belarus at the opening of the twenty-first century was redirecting its interests from Europe toward Moscow. In the Caucasus, Armenia and Georgia effectively were in the Russian realm's orbit. In Central Asia, millions of Russians still live across the border in northern Kazakhstan.

In the 1990s, Russia began to reorganize in the aftermath of the collapse of the Soviet Union. This reorganization cannot be understood without reference to the seven decades of Soviet communist rule that went before. We turn next to this crucial topic.

THE SOVIET LEGACY

The era of communism may have ended in the Soviet Empire, but its effects on Russia's political and economic geography will long remain. Seventy years of centralized planning and implementation cannot be erased overnight; regional reorganization toward a market economy cannot be accomplished in a day.

While the world of capitalism celebrates the failure of the communist system in the former Soviet realm, it should remember why communism found such fertile ground in the Russia of the 1910s and 1920s. In those days Russia was infamous for the wretched serfdom of its peasants, the cruel exploitation of its workers, the excesses of its nobility, and the ostentatious palaces and riches of the czars. Ripples from the Western European Industrial Revolution introduced a new age of misery for those laboring in factories. There were workers' strikes and ugly retributions, but when the czars finally tried to better the lot of the poor, it was too little too late. There was no democracy, and the people had no way to express or channel their grievances. Europe's democratic revolution passed Russia by, and its economic revolution touched the czars' domain only slightly. Most Russians, and tens of millions of non-Russians under the czars' control, faced exploitation, corruption, starvation, and harsh subjugation. When the people began to rebel in 1905, there was no hint of what lay in store; even after the full-scale Revolution of 1917, Russia's political future hung in the balance.

The Russian Revolution was no unified uprising. There were factions and cliques; the Bolsheviks ("Majority") took their ideological lead from Lenin, while the Mensheviks ("Minority") saw a different, more liberal future for their country. The so-called Red army factions fought against the Whites, while both battled the forces of the czar. The country stopped functioning; the people suffered terrible deprivations in the countryside as well as the cities. Most Russians (and other nationalities within the empire as well) were ready for radical change.

That change came when the Revolution succeeded and the Bolsheviks bested the Mensheviks, most of

FROM THE FIELD NOTES

"June 25, 1964. We left Moscow this morning after our first week in the Soviet Union, on the road to Kiev. Much has been seen and learned after we arrived in Leningrad from Helsinki (also by bus) last Sunday. Some interesting contradictions: in a country that launched sputnik nearly a decade ago, nothing mechanical seems to work, from elevators to vehicles. The arts are vibrant, from the ballet in the Kirov Theatre to the Moscow Symphony. But much of what is historic here seems to be in disrepair, and in every city and town you pass, rows and rows of drab, gray apartment buildings stand amid uncleared rubble, without green spaces, playgrounds, trees. My colleagues here talk about the good fortune of the residents of these buildings: as part of the planning of the "socialist city," their apartments are always close to their place of work, so that they can walk or ride the bus a short distance rather than having to commute for hours. Our bus broke down as we entered the old part of Tula, and soon we learned that it would be several hours before we could expect to reboard. We had stopped right next to a row of what must have been attractive shops and apartments a half century ago; everything decorative seems to have been removed, and nothing has been done to paint, replace rotting wood, or fix roofs. I walked up the street a mile or so, and there was the "new" Tula—the Soviet imprint on the architecture of the place. This scene could have been photographed on the outskirts of Leningrad, Moscow, or any other urban center we had seen. What will life be like in the Soviet city of the distant future?" © H. J. de Blij

whom were exiled. In 1918, the capital was moved from Petrograd (as St. Petersburg had been renamed in 1914, to remove its German appellation) to Moscow. This was a symbolic move, the opposite of the forward-capital principle: Moscow lay deep in the Russian interior, not even on a major navigable waterway (let alone a coast), amid the same forests that much earlier had protected the Russians from their enemies. The new Soviet Union would look inward, and the communist system would achieve with Soviet resources and labor the goals that had for so long eluded the country. The chief political and economic architect of this effort was the revolutionary leader who prevailed in the power struggle: V. I. Lenin (born Vladimir Ilyich Ulyanov).

The Political Framework

Russia's great expansion had brought many nationalities under czarist control; now the revolutionary government sought to organize this heterogeneous ethnic mosaic into a smoothly functioning state. The czars had conquered, but they had done little to bring Russian culture to the peoples they ruled. The Georgians, Armenians, Tatars, and residents of the Muslim states of Central Asia were among dozens of individual cultural, linguistic, and religious groups that had not been "Russified." In 1917, however, the Russians themselves constituted only about one-half of the population of the entire realm. Thus it was impossible to establish a Russian state instantly over this vast political region, and these diverse national groups had to be accommodated.

The question of the nationalities became a major issue in the young Soviet state after 1917. Lenin, who brought the philosophy of Karl Marx to Russia, talked from the beginning about the "right of self-determination for the nationalities." The first response by many of Russia's subject peoples was to proclaim independent republics, as they did in Ukraine, Georgia, Armenia, Azerbaijan, and even in Central Asia. But Lenin had no intention of permitting the Russian state to break up. In 1923, when his blueprint for the new Soviet Union went into effect, the last of these briefly independent units was fully absorbed into the sphere of the Moscow regime. Ukraine, for example, declared itself independent in 1917 and managed to sustain this initiative until 1919. But in that year the Bolsheviks set up a provisional government in Kiev, the Ukrainian capital, thereby ensuring the incorporation of the country into Lenin's Soviet framework.

The Bolsheviks' political framework for the Soviet Union was based on the ethnic identities of its many incorporated peoples. Given the size and cultural complexity of the empire, it was impossible to allocate territory of equal political standing to all the nationalities; the communists controlled the destinies of well over 100

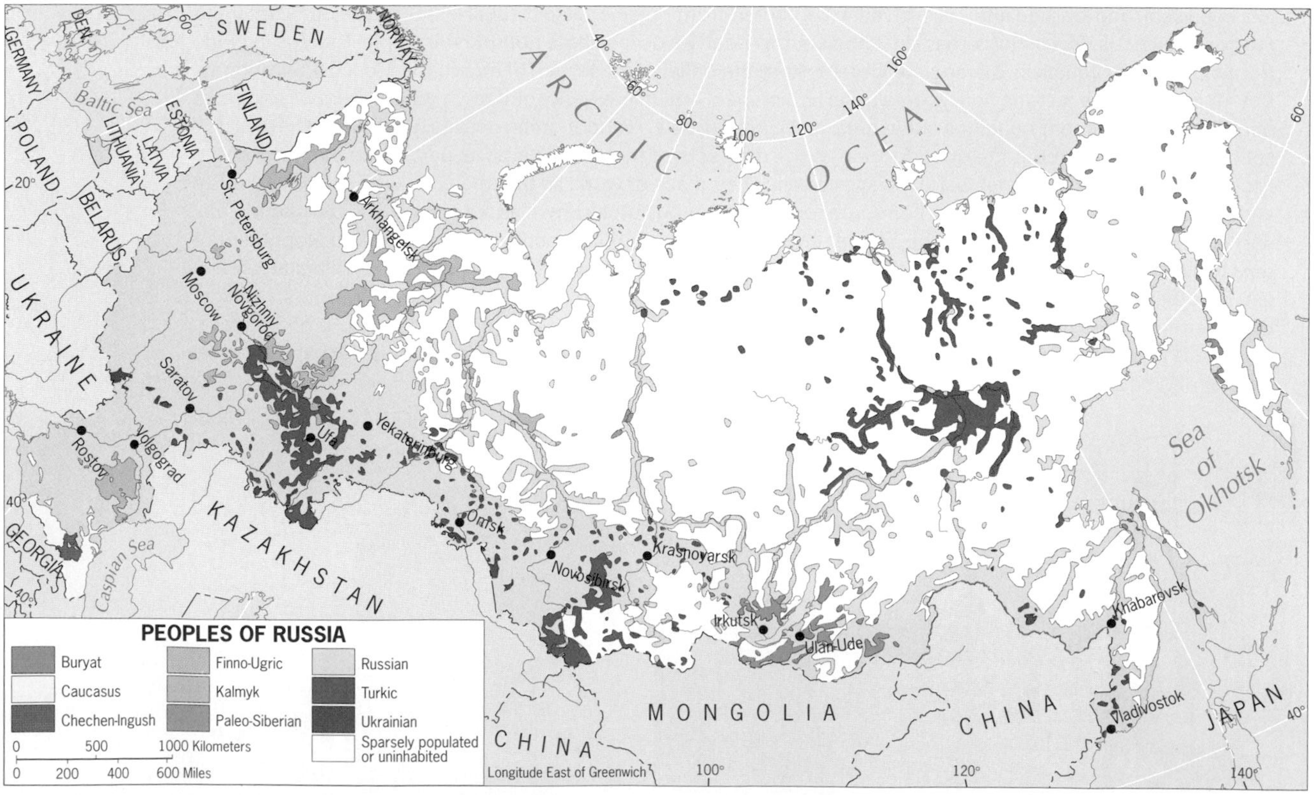

FIGURE 2-5 © H. J. de Blij, P. O. Muller, and John Wiley & Sons, Inc.

peoples, both large nations and small isolated groups. It was decided to divide the vast realm into Soviet Socialist Republics (SSRs), each of which was delimited to correspond broadly to one of the major nationalities. At the time, Russians constituted about half of the developing Soviet Union's population, and, as Figure 2-5 shows, they also were (and still are) the most widely dispersed ethnic group in the realm. The Russian Republic, therefore, was by far the largest designated SSR, comprising just under 77 percent of total Soviet territory.

Within the SSRs, smaller minorities were assigned political units of lesser rank. These were called Autonomous Soviet Socialist Republics (ASSRs), which in effect were republics within republics; other areas were designated Autonomous Regions or other nationality-based units. It was a complicated, cumbersome, often poorly designed framework, but in 1924 it was launched officially under the banner of the Union of Soviet Socialist Republics (USSR).

Eventually, the Soviet Union came to consist of 15 SSRs (shown in Fig. 2-6), including not only the original republics of 1924 but also such later acquisitions as Moldova (formerly Moldavia in Romania), Estonia, Latvia, and Lithuania. The internal political layout often was changed, sometimes at the whim of the communist empire's dictators. But no communist apartheid-like system of segregation could accommodate the shifting multinational mosaic of the Soviet realm. The republics quarreled among themselves over boundaries and territory. Demographic changes, migrations, war, and economic factors soon made much of the layout of the 1920s obsolete. Moreover, the communist planners made it Soviet policy to relocate entire peoples from their homelands in order to better fit the grand design, and to reward or punish—sometimes capriciously. The overall effect, however, was to move minority peoples eastward and to
replace them with Russians. This **Russification** of the 12
Soviet Empire produced substantial ethnic Russian minorities in all the non-Russian republics.

The Soviet planners called their system a **federation**. 13
We focus in more detail on this geographic concept in Chapter 11, but we note here that federalism involves the sharing of power between a country's central government and its political subdivisions (provinces, States, or, in the Soviet case, "Socialist Republics"). Study the map of the former Soviet Union (Fig. 2-6), and an interesting geographic corollary emerges: every one of the 15 Soviet Republics had a boundary with a non-Soviet neighbor. Not

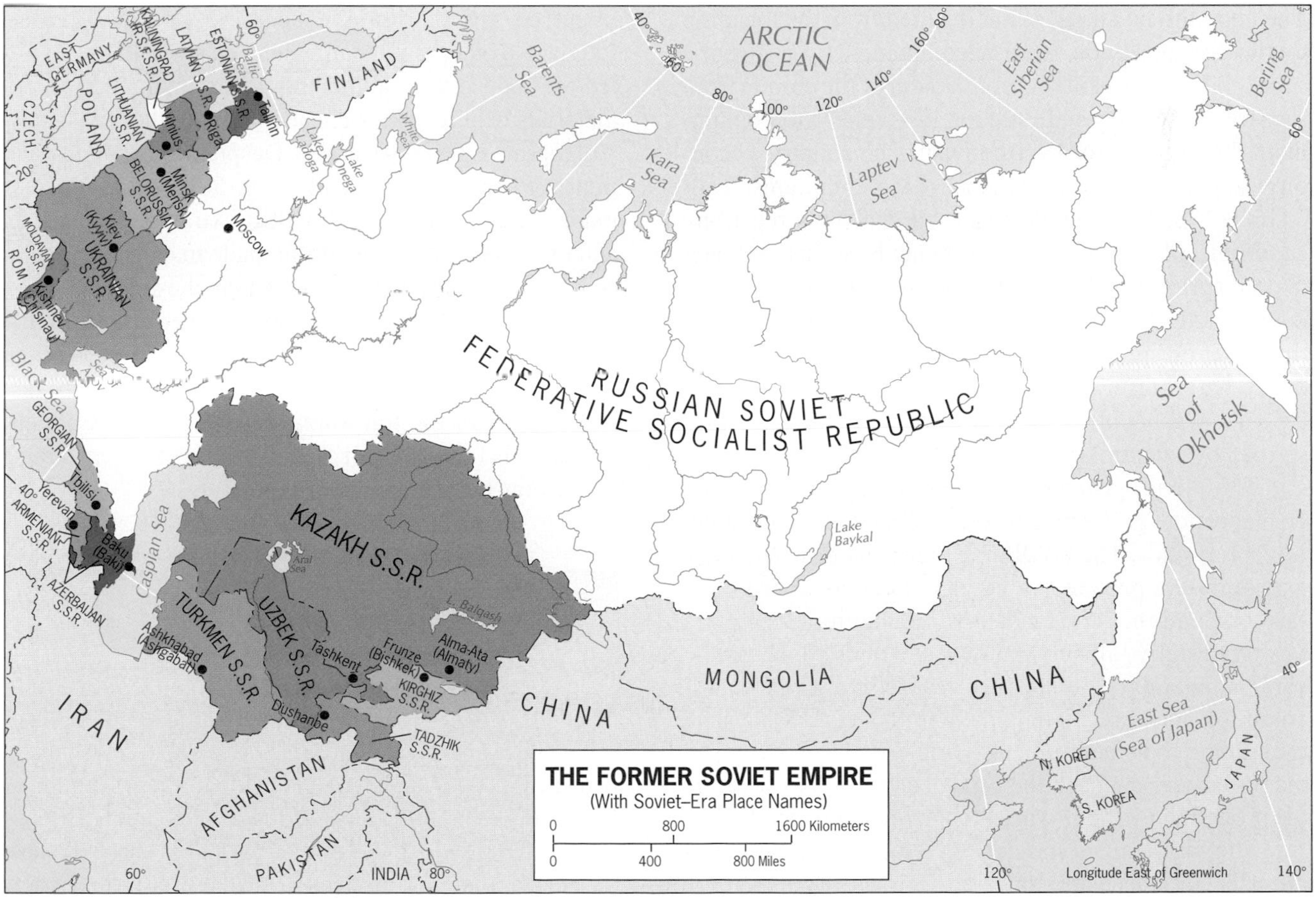

FIGURE 2-6 © H. J. de Blij, P. O. Muller, and John Wiley & Sons, Inc.

one was spatially locked within the others. This seemed to give geographic substance to the notion that any Republic was free to leave the USSR if it so desired. Reality, of course, was different. Moscow's control over the republics made the Soviet Union a federation in theory only.

The centerpiece of the tightly controlled Soviet "federation" was the Russian Republic. With half the vast state's population, the capital city, the realm's core area, and over three-quarters of the Soviet Union's territory, Russia was the empire's nucleus. In other republics, "Soviet" often was simply equated with "Russian"—it was the reality with which the lesser republics lived. Russians came to the other republics to teach (Russian was taught in the colonial schools), to organize (and often dominate) the local Communist Party, and to implement Moscow's economic decisions. This was colonialism, but somehow the communist disguise—how could socialists, as the communists called themselves, be colonialists?—and the contiguous spatial nature of the empire made it appear to the rest of the world as something else. Indeed, on the world stage the Soviet Union became a champion of oppressed peoples, a force in the decolonization process. It was an astonishing contradiction that would, in time, be fully exposed.

The Soviet Economic Framework

The geopolitical changes that resulted from the founding of the Soviet Union were accompanied by a gigantic economic experiment: the conversion of the empire from a czarist autocracy with a capitalist veneer to communism. From the early 1920s onward, the country's economy would be centrally planned—the communist leadership in Moscow would make all decisions regarding economic planning and development. Soviet planners had two principal objectives: (1) to accelerate industrial-
14 ization and (2) to **collectivize** agriculture. For the first time ever on such a scale, and for the first time in accordance with Marxist-Leninist principles, an entire country was organized to work toward national goals prescribed by a central government.

The Soviet planners believed that agriculture could be made more productive by organizing it into huge state-run enterprises. The holdings of large landowners were expropriated, private farms were taken away from the farmers, and the land was consolidated into collective farms. Initially, all such land was meant to be part of a *sovkhoz*, literally a grain-and-meat factory in which agricultural efficiency, through maximum mechanization and minimum labor requirements, would be at its peak. But many farmers opposed the Soviets and tried to sabotage the program in various ways, hoping to retain their land.

The farmers and peasants who obstructed the communists' grand design suffered a dreadful fate. In the 1930s, for instance, Stalin confiscated Ukraine's agricultural output and then ordered part of the border between the Russian and Ukrainian republics sealed—thereby leading to a famine that killed millions of farmers and their families. In the Soviet Union under communist totalitarianism, the ends justified the means, and untold hardship came to millions who had already suffered under the czars. In his book *Lenin's Tomb*, David Remnick estimates that between 30 and 60 million people lost their lives from imposed starvation, political purges, Siberian exile, and forced relocation. It was an incalculable human tragedy, but the secretive character of Soviet officialdom made it possible to hide it from the world.

The Soviet planners hoped that collectivized and mechanized farming would free hundreds of thousands of workers to labor in factories. Industrialization was the prime objective of the regime, and here the results were superior. Productivity rose rapidly, and when World War II engulfed the empire in 1941, the Soviet manufacturing sector was able to produce the equipment and weapons needed to repel the German invaders.

Yet even in this context, the Soviet grand design entailed liabilities for the future. The USSR practiced a
command economy, in which state planners assigned 15
the production of particular manufactures to particular places, often disregarding the rules of economic geography. For example, the manufacture of railroad cars might be assigned (as indeed it was) to a factory in Latvia. No other factory anywhere else would be permitted to produce this equipment—even if supplies of raw materials would make it cheaper to build them near, say, Volgograd 1200 miles away. Despite an expanded and improved transport network (Fig. 2-7), such practices made manufacturing in the USSR extremely expensive, and the absence of competition made managers complacent and workers less productive than they could be.

Of course, the Soviet planners never imagined that their experiment would fail and that a market-driven economy would replace their command economy. When that happened, the transition was predictably difficult; indeed, it is far from over, and it is putting severe strains on the now more democratic state.

RUSSIA'S CHANGING POLITICAL GEOGRAPHY

When the USSR dissolved in 1991, Russia's former empire devolved into 14 independent countries, and Russia itself was a changed nation. Russians now made up about 83 percent of the population of under 150 million, a far higher proportion than in the days of the Soviet Union. But numerous minority peoples remained under Moscow's new flag, and millions of Russians found themselves under new governments in the former Republics.

Soviet planners had created a complicated administrative structure for their "Russian Soviet Federative Socialist Republic," and Russia's postcommunist leaders had to use this framework to make their country function. In 1992, most of Russia's internal "republics," autonomous regions, oblasts, and krays (all of them components of the administrative hierarchy) signed a document known as the Russian Federation Treaty, committing them to cooperate in the new federal system. At first a few units refused to sign, including Tatarstan, scene of Ivan the Terrible's brutal conquest more than four centuries ago, and a republic in the Caucasus periphery, then known as Chechenya-Ingushetia, where Muslim rebels waged a campaign for independence. As the map shows, Chechnya-Ingushetiya split into two separate republics whose names at the time were spelled Chechenya and Ingushetia (Fig. 2-8). Eventually, only Chechnya refused to sign the Russian Federation Treaty, and subsequent Russian military intervention led to a prolonged and violent conflict, with disastrous consequences for Chechnya's people and infrastructure (the capital, Groznyy, was completely destroyed). The Chechnya war continues today and is a disaster for Russia's government as well (see the Issue Box titled "Chechnya").

The Federal Framework of Russia

The spatial framework of the still-evolving Russian Federation is as complex as that of the Russian Federative Socialist Republic of communist times. As the twenty-first century opened, the Federation consisted of 89 entities: 2 Autonomous Federal Cities, 21 Republics, 11 Autonomous Regions (Okrugs), 49 Provinces (Oblasts), and 6 Territories (Krays). Moscow and St. Petersburg are the two Autonomous Federal Cities. The 21 Republics, recognized to accommodate substantial ethnic minorities in the population, lie in several clusters (Fig. 2-8).

Russia's post–1991 governments have faced complicated administrative problems. A multinational, multicultural state that had been accustomed to authoritarian rule and government control over virtually everything—from factory production to everyday life—now had to be governed in a new way. Democratization of the political system, transition to a market economy, sale of state-owned industries (privatization), and liberation of the press and other media must all take place in an orderly way, or the country would, literally, fall apart.

Russia's leaders knew that their options were limited. They could continue to hold as much power as possible at the center, making decisions in Moscow that would apply to all the Republics, Regions, and other subdivisions of
16 the state. Such a **unitary state system**, with its centralized government and administration, marked authoritarian kingdoms of the past and serves totalitarian dictatorships of the present. Or they could share power with the Republics and Regions, allowing elected regional leaders to come to Moscow to represent the interests of their people. This is the federal system Russia chose as the only way to accommodate the country's economic and cultural diversity.

In a *federal system*, the national government usually is responsible for matters such as defense, foreign policy, and foreign trade. The Regions (or provinces, States, or other subdivisions) retain authority over affairs ranging from education to transportation. A federal system does not create unity out of diversity, but it does allow diverse components of the state to coexist, their common interests represented by the national government and their regional interests by their local administrations. Some countries owe their survival as coherent states to their federal frameworks. India and Australia are cases in point.

But to maintain a generally acceptable balance of power between the center and the Regions (or States) is difficult. Disputes over "States' rights" continue to roil the American political scene more than two centuries after the Constitution was adopted. Russia, which also owes its continuity to a grand federal experiment, is barely more than ten years old. Power shifts continue, and the framework of the Russian Federation will undoubtedly change over time.

An additional problem arises from Russia's sheer
size. Geographers refer to the principle of **distance** 17
decay to explain how increasing distances between places tend to reduce interactions among them. Because Russia is the world's largest country, distance is a significant factor in the relationships between the capital and outlying areas. Furthermore, Moscow lies in the far west of the giant country, half a world away from the shores of the Pacific. Not surprisingly, one of the most obstreperous Regions has been remote Primorskiy, the Region of Vladivostok.

Still another problem is reflected in Figure 2-8: the enormous size variation among the Republics (and to a lesser extent, the Regions). Whereas the (territorially) smallest Republics are concentrated in the Russian core area, the largest lie far to the east, where Sakha is nearly a thousand times as large as Ingushetiya. On the other hand, the populations of the huge eastern Republics are tiny compared to those of the smaller ones in the core. Such diversity spells administrative difficulty.

But perhaps the most serious current problem for Russia is the growing social disharmony between Moscow and the subnational political entities. Almost wherever one goes in Russia today, Moscow is disliked and often berated by angry locals. The capital is seen as the privileged playground for those who have benefited most from the post-Soviet transition—bureaucrats and hangers-on whose economic policies have driven down standards of living, whose greed and corruption have hurt the economy,

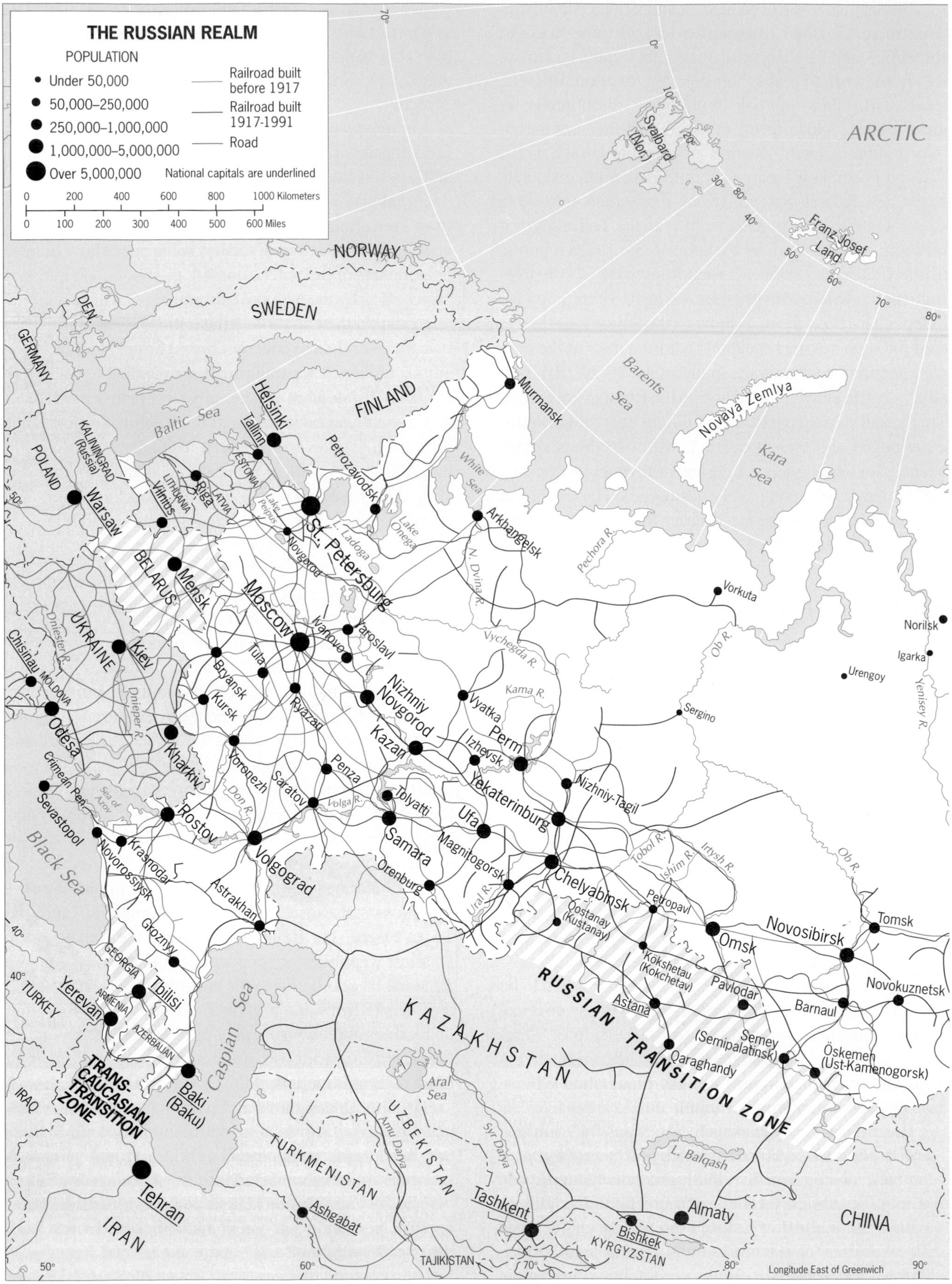

FIGURE 2-7 © H. J. de Blij, P. O. Muller, and John Wiley & Sons, Inc.

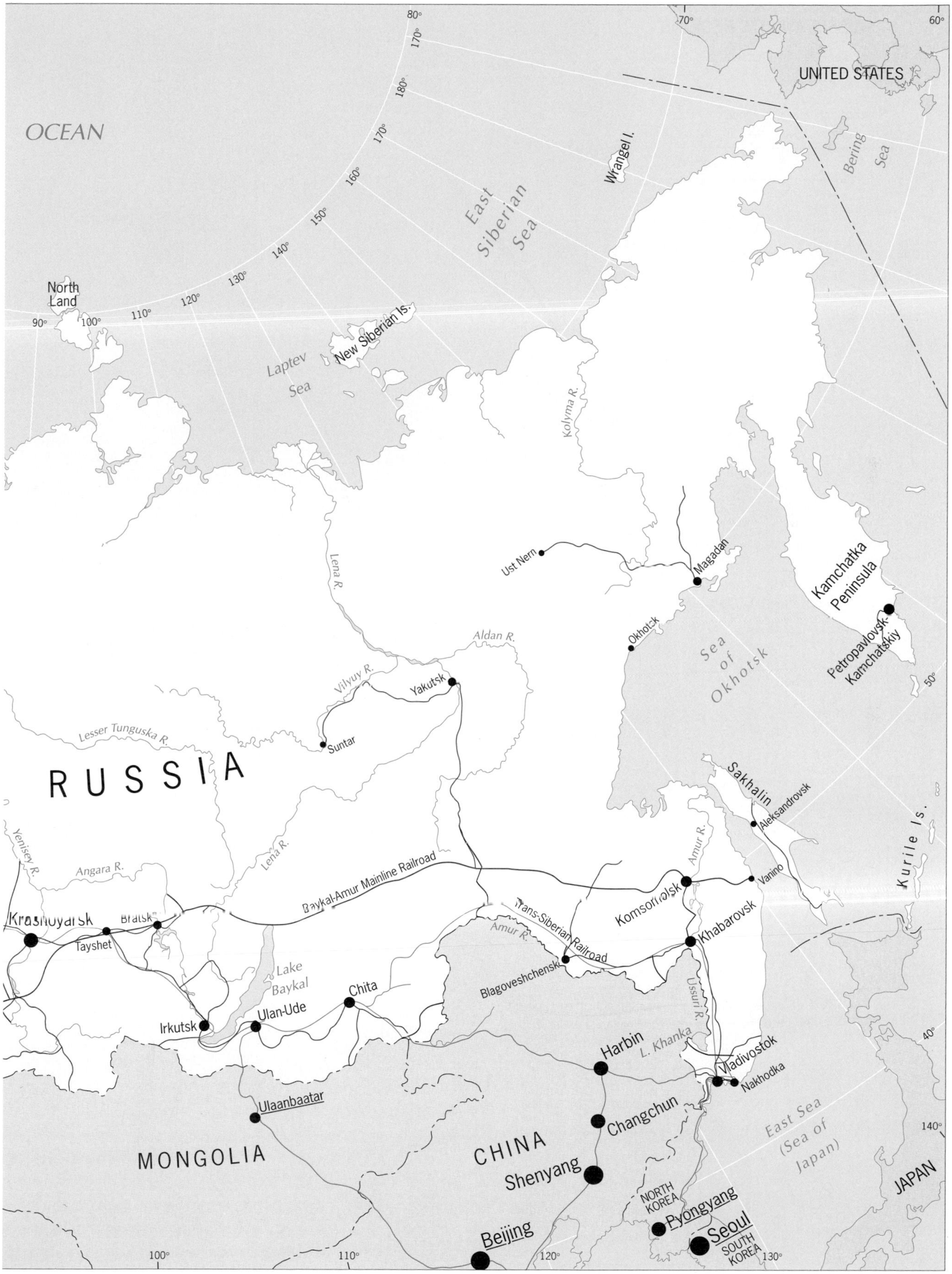
OCEAN
UNITED STATES
Bering Sea
Wrangel I.
East Siberian Sea
North Land
New Siberian Is.
Laptev Sea
Kolyma R.
Lena R.
Ust Nern
Magadan
Kamchatka Peninsula
Petropavlovsk-Kamchatskiy
Okhotsk
Sea of Okhotsk
Aldan R.
Vilyuy R.
Yakutsk
Lesser Tunguska R.
Suntar
RUSSIA
Sakhalin
Aleksandrovsk
Kurile Is.
Yenisey R.
Angara R.
Lena R.
Amur R.
Baykal-Amur Mainline Railroad
Vanino
Komsomolsk
Krasnoyarsk
Bratsk
Tayshet
Trans-Siberian Railroad
Khabarovsk
Amur R.
Blagoveshchensk
Lake Baykal
Chita
Ussuri R.
Irkutsk
Ulan-Ude
Harbin
L. Khanka
Vladivostok
Nakhodka
Ulaanbaatar
Changchun
East Sea (Sea of Japan)
MONGOLIA
CHINA
Shenyang
NORTH KOREA
Pyongyang
JAPAN
Beijing
Seoul
SOUTH KOREA
90°
100°
110°
120°
130°
140°
150°
160°
170°
180°
170°
80°
70°
60°
50°
40°

RUSSIA AND ITS REGIONS

Ethnic Republics and Autonomous Regions

Russian regions

National capital is underlined

0 200 400 600 800 Kilometers
0 100 200 300 400 500 Miles

RUSSIAN ADMINISTRATIVE UNITS
(named after their capitals)

1. Astrakhan 19. Penza
2. Belgorod 20. Pskov
3. Bryansk 21. Rostov
4. Chelyabinsk 22. Ryazan
5. Ivanovo 23. St. Petersburg
6. Kaluga 24. Samara
7. Kemerovo 25. Saratov
8. Kostroma 26. Smolensk
9. Krasnodar 27. Stavropol
10. Kurgan 28. Tambov
11. Kursk 29. Tula
12. Lipetsk 30. Tver
13. Mariy-el 31. Ulyanovsk
14. Moscow 32. Vladimir
15. Novosibirsk 33. Volgograd
16. Omsk 34. Voronezh
17. Orenburg 35. Yaroslavl
18. Orel

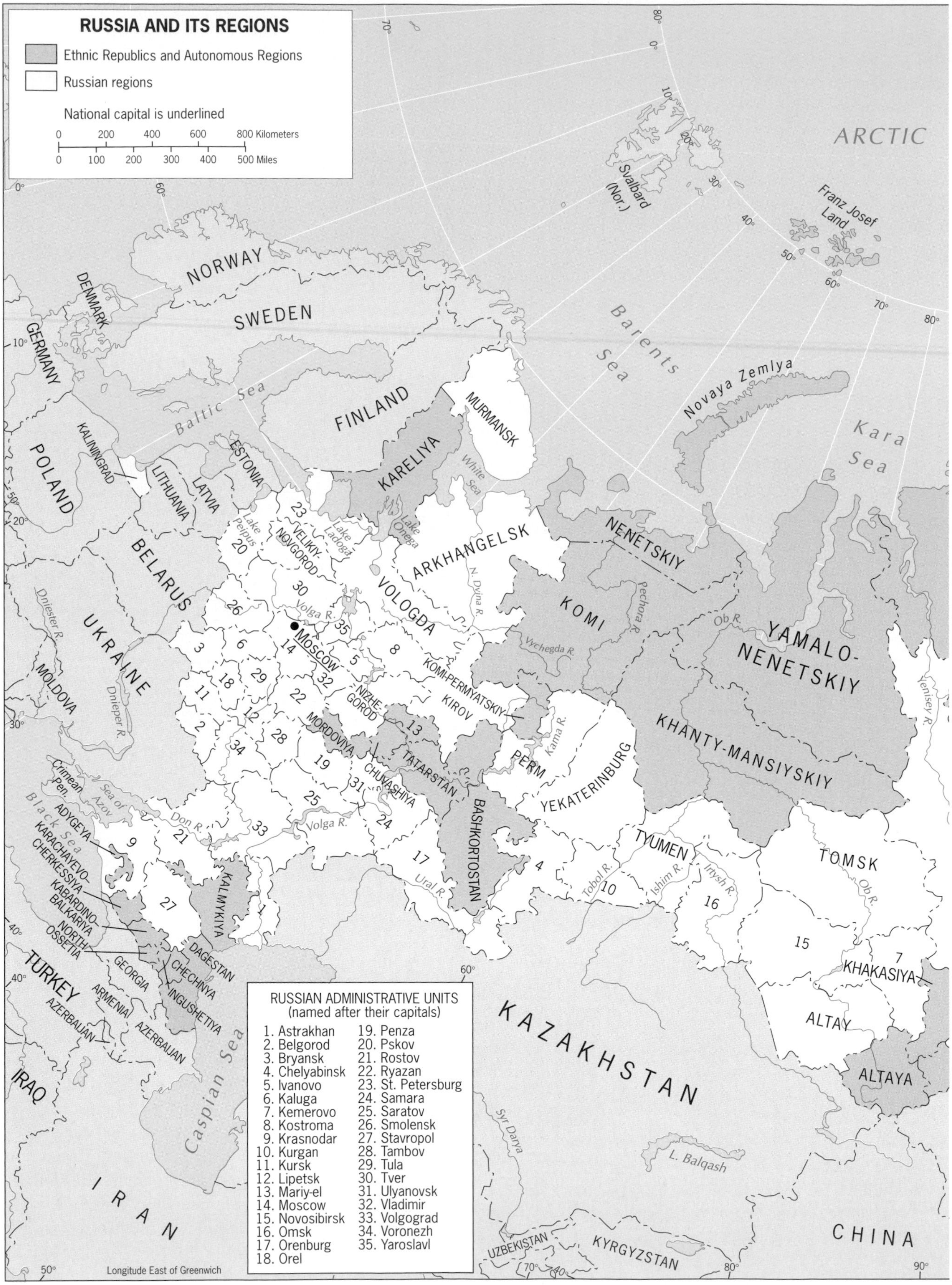

FIGURE 2-8 © H. J. de Blij, P. O. Muller, and John Wiley & Sons, Inc.

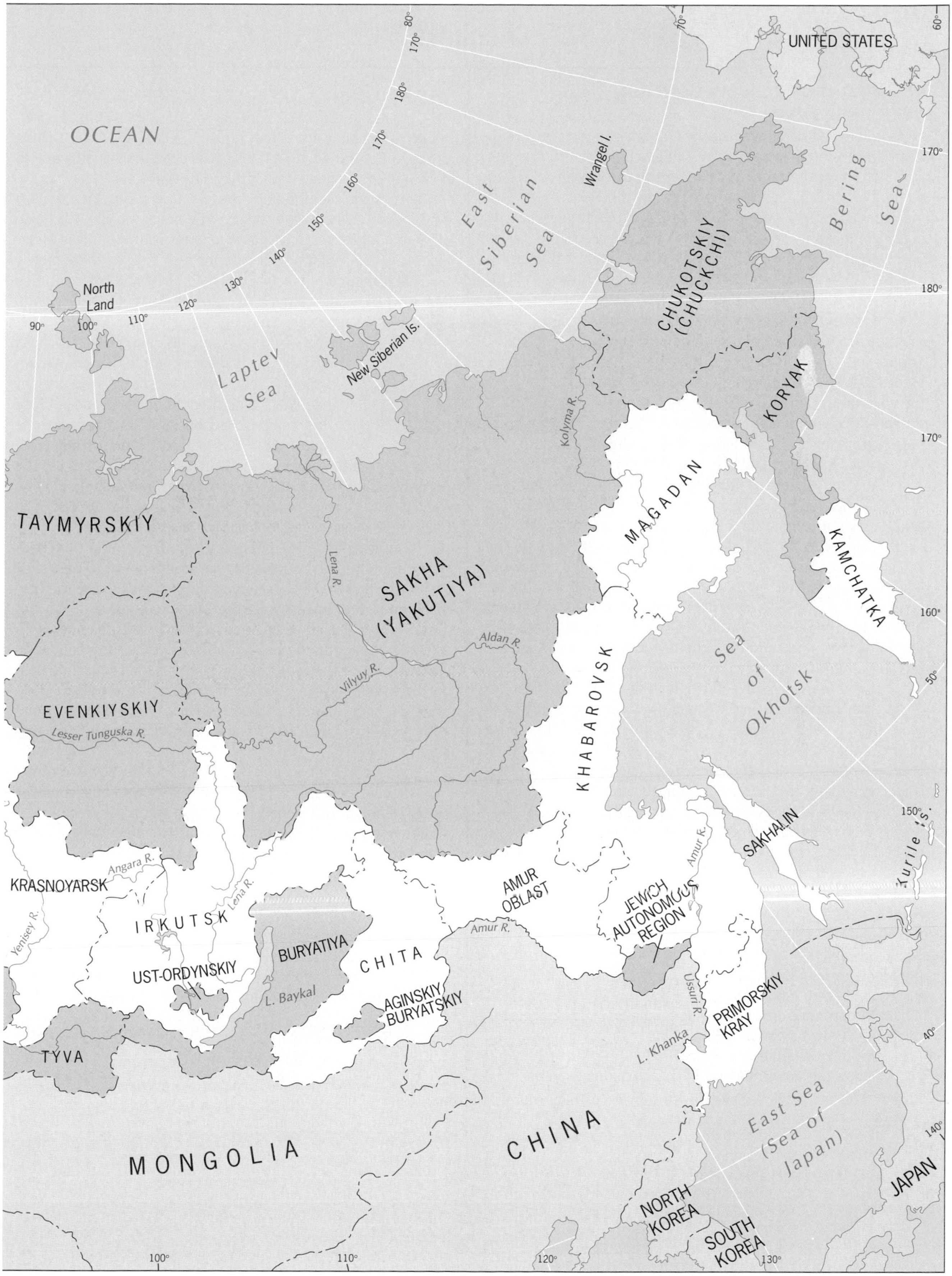
OCEAN
UNITED STATES
Wrangel I.
East Siberian Sea
Bering Sea
CHUKOTSKIY (CHUCKCHI)
North Land
Laptev Sea
New Siberian Is.
KORYAK
Kolyma R.
MAGADAN
KAMCHATKA
TAYMYRSKIY
Lena R.
SAKHA (YAKUTIYA)
Aldan R.
Vilyuy R.
Sea of Okhotsk
KHABAROVSK
EVENKIYSKIY
Lesser Tunguska R.
SAKHALIN
Kurile Is.
Angara R.
KRASNOYARSK
Amur R.
AMUR OBLAST
JEWISH AUTONOMOUS REGION
Yenisey R.
IRKUTSK
Lena R.
BURYATIYA
CHITA
UST-ORDYNSKIY
L. Baykal
AGINSKIY BURYATSKIY
Ussuri R.
PRIMORSKIY KRAY
TYVA
L. Khanka
East Sea (Sea of Japan)
MONGOLIA
CHINA
NORTH KOREA
SOUTH KOREA
JAPAN

Chechnya

Regional ISSUE

IN SUPPORT OF RUSSIAN CONTROL OVER CHECHNYA

"As a policeman here in Moscow, I strongly support the government and the army in their efforts to establish control over the criminal elements trying to take control of Chechnya. Let us remember what happened there. When the Russian government back in 1991 had to take charge of our country and its many components, there was lots of opposition. The Tatars (also Muslims, like the Chechens) talked about establishing an independent state in their republic. The Kalmyks, the Udmurts, the Bashkirs, the Chuvash, and many others proclaimed that they wanted everything ranging from autonomy to independence. There were even Russians demanding it in places like Yekaterinburg and Primorskiy. But the government under President Yeltsin negotiated all these claims and gave those people reasons to want to stay under the Russian flag. Except the Chechens. Nothing was going to satisfy them.

"And the Chechens were even given their own republic after the changeover from Soviet to Russian administration. In 1991 they still shared their territory with the Ingush, who are also Muslims, in the so-called Chechen-Ingush Republic. So what did the Chechens do? They installed their separatist leader as ruler of the republic and started fighting with the Ingush minority. To help solve this crisis, Russia's government divided the republic's territory into two, the larger, richer part including the capital and the oilfields for the Chechens and the remainder for the Ingush.

"But it wasn't good enough for the Chechens. They attacked the Russians living in Chechnya, causing the army to move in to protect Russians and Russian interests. They broke the truce that followed. Then they mounted a full-scale war from their mountain hideouts and caused us hideous casualties. They didn't care that their capital was totally destroyed; these Muslim warlords fought among themselves even as they fought us. Next they started using terror to get their way. Remember the autumn of 1999, when they blew up apartment buildings in Moscow, causing more than 300 dead? And how about what they did in Dagestan, where they took over a bunch of villages and declared an 'independent Chechen republic'? Not to mention that hospital full of doctors, nurses, and patients, whom they took hostage, eventually killing hundreds of innocent victims. These are people to whom we should entrust the government of an independent country on our borders? Never.

"So don't criticize us when it comes to our strategies to deal with these barbarians. Foreign governments say that the Russian army does terrible things to captured Chechens, but what about the treatment of Russian soldiers by Chechen rebels? War is brutal, and this is a war for the soil of Mother Russia.

"If we give up in Chechnya, other minorities will get ideas about independence too, and that would be the end of the nation. There is nothing to negotiate. Russia must and will prevail."

WHY CHECHNYA DESERVES INDEPENDENCE

"My grandparents were born in what is today the Russian colony of Chechnya, and they died a horrible death somewhere in Soviet Central Asia. I am here to avenge their deaths and to punish the Russians for what they did to my people.

"Russians seem to think that we are fighting for independence just because we want it. Why aren't we like all those other minorities that have come to terms with their lives under Moscow's heel? Well, in our case there is more to it. We fought the Russian imperialists to a standstill less than two centuries ago, when the czars' armies colonized Islamic peoples from the Caucasus to Central Asia. Yes, we were eventually defeated, but those Russians never really penetrated our mountain hideouts. Then came the Soviets, who thought they did us a big favor by creating one of those 'Autonomous Republics' for us along with the poor Ingush. But look at the map. They combined our traditional homeland with a stretch of flatland to the north, which was full of Russian farmers and oilmen. Do you think Groznyy was a Chechen town? Think again. Or look for a mosque on those photographs of the 'good old days.'

"But we might have put up with it all except for what happened during the war between the Soviets and the Germans, World War II. Some of us Chechens were happy to see the Germans do to the Soviets what the Soviets had done to us, but Josef Stalin accused us all of collaborating with the enemy. In a few short months he packed all 500,000 of my people on trains and sent them to exile in Kazakhstan. Thousands died on the trains and were simply thrown onto the tracks. Many more, probably about 125,000, perished in the desert. I don't know where or when my grandparents died, but my parents survived.

"We got our homeland back because the Soviet dictator Khrushchev reinstated our Republic in 1957, and the survivors straggled back. But nobody ever forgot what was done to us. We're not talking of a few hundred hostages here. We're talking about the killing of a quarter of a nation. Still, we tried to restart our society under those atheist Soviets, keeping our Islamic practices quiet and private. But then the Soviet Empire collapsed and we got our chance. In 1991 we declared our independence from the new Russia, and at first it seemed that the Russians would be sensible and approve.

"But before long the Russians changed their minds and tried to intervene in our internal affairs. We managed to defeat the Russian army once again, and the war devastated Groznyy, but the Russians would not even consider independence. We got assistance from Muslims elsewhere and money from many sources, but we can see that Russia will not give us what we want. So we turn to any means we can to wound the Russian bear, and we will chase it out of Chechnya with the help of Allah."

Vote your opinion at www.wiley.com/college/debli

Until September 3, 2004, Beslan was an unremarkable small town in the Russian border republic of North Ossetia. But on that fateful day, North Ossetia's relative location coupled with its people's pro-Russian leanings in the Transcaucasian struggle between Moscow and Muslim separatism made it a target for 32 ruthless Chechen militants. In the early morning they seized control of a school where students, teachers, and parents were gathering for the season's opening day, took more than 1,100 people hostage, and rigged the building with explosives. After a 54-hour siege during which the terrorists killed children as well as adults, Russian antiterrorist forces stormed the building and the terrorists set off their bombs, resulting in the deaths of more than 150 youngsters and nearly 200 adults. In the photograph, a desolate Ossetian woman holds her grandchild near the burned-out school a week after the attack, when relatives were still searching in the rubble for missing loved ones and authorities worked to defuse anger and control retaliation. © Maxim Marmur/Getty Images.

whose actions in Chechnya have been a disaster, who have allowed foreigners (prominently the United States) to erode Russian power and prestige, who fail to pay the wages of workers toiling in industries still owned by the state, and who do not represent the Russian people. Complaints about the capital are not uncommon in free countries, but in Russia the mistrust between capital and subordinate areas has become so serious that it constitutes a key challenge for the government.

In 2000, the Putin administration moved to diminish the influence of the Regions by creating a new spatial framework (Fig. 2-9) that combines the 89 Regions, Republics, and other entities into seven new administrative units—not to enhance their influence in Moscow, but to

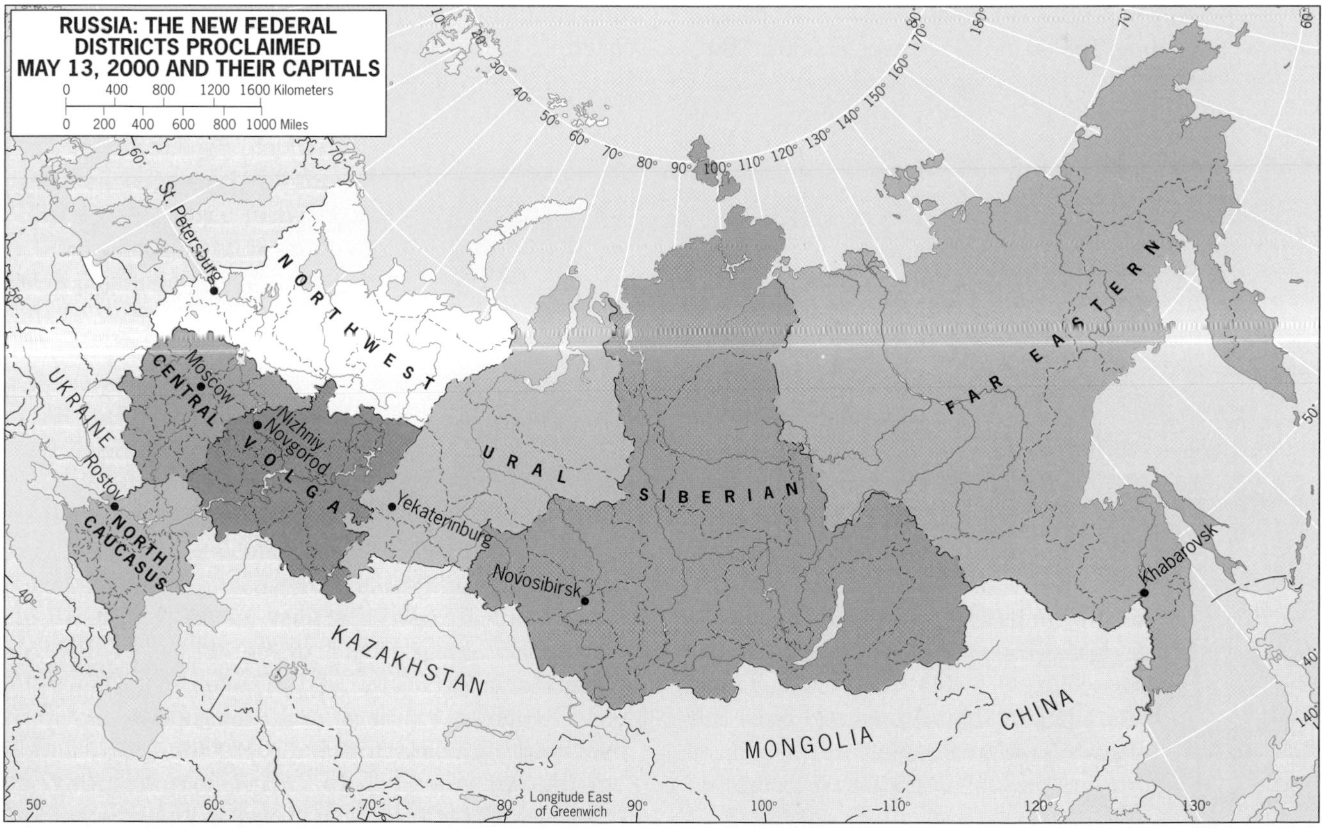

FIGURE 2-9 © H. J. de Blij, P. O. Muller, and John Wiley & Sons, Inc.

increase Moscow's authority over them. Each of these new federal administrative districts has its capital city, which will become the conduit for Moscow's "guidance," as the official plan puts it. This framework will counter some powerful vested interests in the Republics and Regions, however, and it will reverse a trend toward ever-greater local self-determination. It is far from certain that the system will work.

CHANGING SOCIAL GEOGRAPHIES

The political-geographical machinations discussed above give no indication of what is happening in the daily lives of most of Russia's citizens. But when you see statistics on the country's population, its general health, financial problems, life expectancy, and related conditions, the extent of the country's post-Soviet dislocation becomes clear. When the Soviet Union broke apart, the Russian Republic had a population of almost 150 million, still growing albeit very slowly. By 2006 it was down to just over 142 million. If
18 this **population decline** continues, there will be barely 100 million Russian citizens by 2050. Life expectancy for males, over 64 ten years ago, is now 58, and for females it has declined from over 74 to under 72. Among men, alcoholism, suicide, and other manifestations of social disorder drive down life expectancy. Deaths from alcohol poisoning rose from about 16,000 in 1991 to over 41,000 in 2001. In the general population, drug abuse, heavy smoking, and poor diets are to blame.

The incidence of diseases is staggering for a country that not long ago was a world power. Tuberculosis may now affect more than 200,000 people (World Health Organization and Russian sources vary widely), and deaths in just one recent year were nearly 30,000 compared to less than 1000 in the United States with double Russia's population. Rates of heart disease are rising. HIV cases can only be estimated because Russian sources are unreliable. In 2003 Russia officially reported 240,000 people infected, but UN estimates hovered around 1 million, extrapolated to 10 million by 2020. Deaths attributed to AIDS were less than 1000 by mid-2003, but UN projections warned of 100,000 deaths annually by 2010 and 500,000 by 2020.

Coupled with this distress is the decline of national institutions. The armed forces are in disrepair. Major universities lack modern facilities. Leading industries are failing; the energy industry is in the grip of powerful magnates who flout the rules. Legal systems, banking operations, and property regulations are not sufficiently developed to ensure orderly procedures. Corruption and cronyism are rife, and some observers describe organized crime as Russia's most successful enterprise.

Yet Russia remains a formidable nuclear power with a vast arsenal of weapons. Russia's quest for social and political recovery and stability is a matter of global interest.

Symbol of the new era in Russia: a Coca-Cola sign and a bank logo tower over traffic and pedestrians on Novi Arbat Street near the Kremlin in Moscow. The city's skyline still does not reflect its political or economic power or its rank as one of the world's most populous centers, but Moscow's ambience has been transformed since 1990. © Oleg Nikishin/Newsmakers/Getty Images News and Sport Services.

RUSSIA'S PROSPECTS

When Peter the Great began to reorient his country toward Europe and the outside world, he envisaged a Russia with warm-water ports, a nation no longer encircled by Swedes, Lithuanians, Poles, and Turks, and a force in European affairs. Core-periphery relationships always have been crucial to Russia, and they remain so today.

Russia's relative location has long been the subject of study and conjecture by geographers. Just over a century ago, the British geographer Sir Halford Mackinder (1861–1947) argued in a still-discussed article entitled "The Geographical Pivot of History" that western Russia and Eastern Europe enjoyed a combination of natural protection and resource wealth that would someday propel its occupants to world power. The protected core area, he reasoned, overshadowed the exposed periphery. Eventually, this pivotal interior region of Eurasia, which he later called the *heartland*, would become a stage for world domination.

19 Mackinder's **heartland theory** was published when Russia was a weak, economically backward society ripe for revolution, but Mackinder stuck to his guns. When Russia, in control of Eastern Europe and with a vast colonial empire, emerged from World War II as a superpower, Mackinder's conjectures seemed prophetic.

But not all geographers agreed. Probably the first scholar to use the term *rimland* (today *Pacific Rim* is part of everyday language) was Nicholas Spykman, who in 1944 countered Mackinder by calculating that Eurasia's periphery, not its core, held the key to global power. Spykman foresaw the rise of rimland states as superpowers and viewed Japan's emergence to wealth and power as just the beginning of that process. In that perspective, Russia's twentieth-century superpower status was just a temporary phenomenon.

Russia, Europe, and the World

The expansion of the European Union in 2004 has reached Russia's borders. Former components of the Soviet Empire now are members of NATO. How will Russia's relationship with Europe evolve, and what are the prospects for Russia in the wider world?

To Europeans, Russia has always been the enigmatic colossus to the east, alternately bungling and threatening but never in tandem with the West. When the Berlin Wall came down and the Soviet banner folded, there were hopeful predictions of a "European Russia" finally joining its erstwhile ideological adversaries in a regional partnership that would extend from the Atlantic to the Pacific. Such forecasts were based on real as well as idealized evidence: Russians' ethnic ties with Eastern Europe; the revival of Christian churches after decades of communist atheism; the sprouting of Russian democracy; the budding of a market economy; the long-term dependence of Europe on Russian energy supplies.

Reality, however, was rather different. Russian democracy remained tinged by an authoritarian streak that scared investors. The media continued to face government interference. The Russian Orthodox Church suppressed efforts by other churches to become reestablished. Russian leaders objected vigorously, then acquiesced as NATO expanded toward its borders. Russia was unhelpful during efforts to mitigate the collapse of Yugoslavia. Russia's treatment of minorities and migrants seemed to violate EU standards. And in 2004, a major schism emerged over Iran's efforts to develop its nuclear capacity beyond mere electrical-power generation: it seemed to leading European governments that the Iranians were trying to achieve the capacity to make nuclear weapons. The Russians were of no help in the difficult negotiations that followed, creating difficulties not only between Russia and Europe but also between Russia and the United States. Already, Russia had taken a position against the U.S.-led intervention in Iraq, creating still greater distance between the present and former superpowers.

While it is true that Russia has more in common with Europe than with any other geographic realm, Russian leaders presently do not appear to see their country as a potential partner in a greater European sphere. The "Europeanization of Russia" may be an ultimate goal for Europeans seeking a European Union extending from the Atlantic to the Pacific, but in the first decade of the twenty-first century the Russian government and its increasingly autocratic President Putin appeared determined to establish the new Russia as a counterweight, not a partner, to Europe. In the immediate aftermath of the 2001 terrorist attacks on the United States, there was a search for common ground against a threat faced by the United States, Europe, and Russia alike, and for some time afterward European and American criticism of Russia's war against Chechnya's Islamic rebels was muted. But when the cycle of Chechen terrorism and Russian retaliation resumed, Europeans as well as Americans censured Russia for its military actions as well as its manipulation of elections in the rebel republic. This angered Russia's leaders as well as Russians generally, and reversed the convergence that had marked Russian-European and Russian-American relations for more than a decade after 1991.

These significant developments overshadow other aspects of Russia's interactions with its neighbors. Earlier in this chapter we noted the continuing presence of Russian minorities in the former "republics" of the Soviet Union. In Russia's Far East, the declining Russian population and the growing in-migration of Chinese are causing concerns about the future of that region. And while a number of boundary disputes between Russia and China have been resolved, the larger issue of the "lost lands" east of the Amur and Ussuri rivers, taken from China more than a century ago, lingers in the background. There is also concern over the growing links between China and Mongolia, once firmly in Moscow's orbit (a topic we revisit in Chapter 9).

Other external problems Russia confronts include: (1) the ownership of oil and gas reserves in the Caspian Basin, where maritime boundaries in the Caspian Sea have not been satisfactorily established; (2) settlement of a long-standing issue involving four small groups of islands in the Kurile archipelago off northern Japan, a legacy of World War II still not resolved (a matter we consider in Chapter 9); (3) Russia's relations with Georgia, Armenia, and Azerbaijan in the Transcaucasus, where Moscow's interventions and pressures have created serious geopolitical problems; and (4) Russia's relations with the neighboring states on its western flank that separate it from the European Union, particularly Ukraine and Belarus, where Russian actions reflect a new and troubling posture.

Russia's greatest challenge, according to its president, is to recover its superpower status. To much of the rest of the world, Russia's main objectives should be to advance—not reverse—its democratic reforms, strengthen the rule of law, reduce its dependence on the sale of oil and gas, diversify its economy, improve its human-rights record, and nurture an open society. These are not just internal matters: what happens to and in Russia is crucial to the entire planet.

REGIONS OF THE REALM

So vast is Russia, so varied its physiography, and so diverse its cultural landscape that regionalization requires a small-scale perspective and a high level of generalization. Figure 2-10 outlines a four-region framework: the Russian Core and its Peripheries west of the Urals, the Eastern Frontier, Siberia, and the Far East. As we will see, each of these massive regions contains major subregions.

RUSSIAN CORE AND PERIPHERIES

20 The heartland of a state is its **core area**. Here much of the population is concentrated, and here lie its leading cities, major industries, densest transport networks, most intensively cultivated lands, and other key components of the country. Core areas of long standing strongly reflect the imprints of culture and history. The Russian core area, broadly defined, extends from the western border of the Russian realm to the Ural Mountains in the east (Fig. 2-10). This is the Russia of Moscow and St. Petersburg, of the Volga River and its industrial cities.

Central Industrial Region

At the heart of the Russian Core lies the Central Industrial Region (Fig. 2-11). The precise definition of this subregion varies, for all regional definitions are subject to debate. Some geographers prefer to call this the Moscow Region, thereby emphasizing that for over 250 miles (400 km) in all directions from the capital, everything is oriented toward this historic focus of the state. As both Figures 2-1 and 2-7 show, Moscow has maintained its decisive centrality: roads and railroads converge in all directions from Ukraine in the south; from Mensk (Belarus) and the rest of Eastern Europe in the west; from St. Petersburg and the Baltic coast in the northwest; from Nizhniy Novgorod (formerly Gorkiy) and the Urals in the east; from the cities and waterways of the Volga Basin in the southeast (a canal links Moscow to the Volga, Russia's most important navigable river); and even from the subarctic northern periphery that faces the Barents Sea.

Moscow (population: 10.6 million) is the focus of an area that includes some 50 million inhabitants (more than

MAJOR CITIES OF THE REALM

City	Population* (in millions)
Baki (Baku), Azerbaijan	1.8
Irkutsk, Russia	0.6
Kazan, Russia	1.1
Moscow, Russia	10.6
Nizhniy Novgorod, Russia	1.3
Novosibirsk, Russia	1.3
St. Petersburg, Russia	5.3
Tbilisi, Georgia	1.1
Vladivostok, Russia	0.6
Volgograd, Russia	1.0
Yekaterinburg, Russia	1.3
Yerevan, Armenia	1.1

*Based on 2006 estimates.

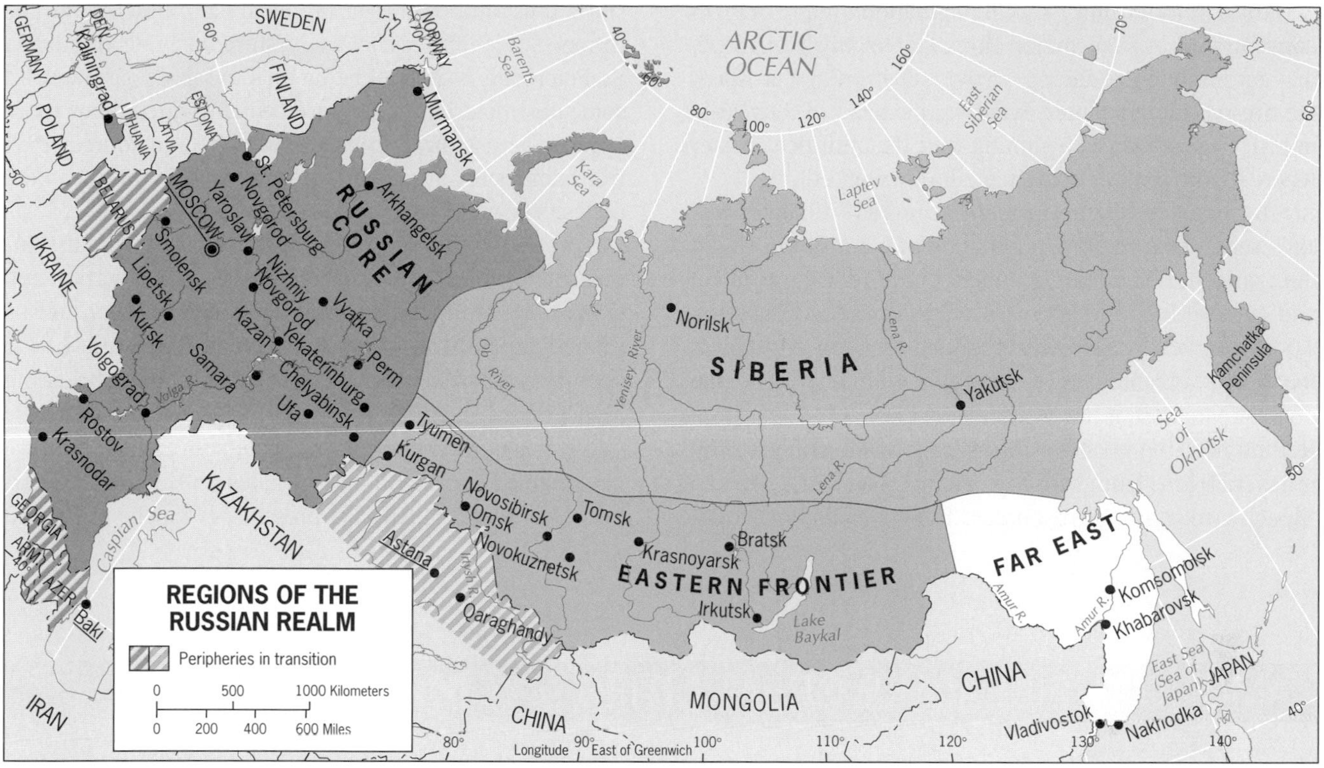

FIGURE 2-10 © H. J. de Blij, P. O. Muller, and John Wiley & Sons, Inc.

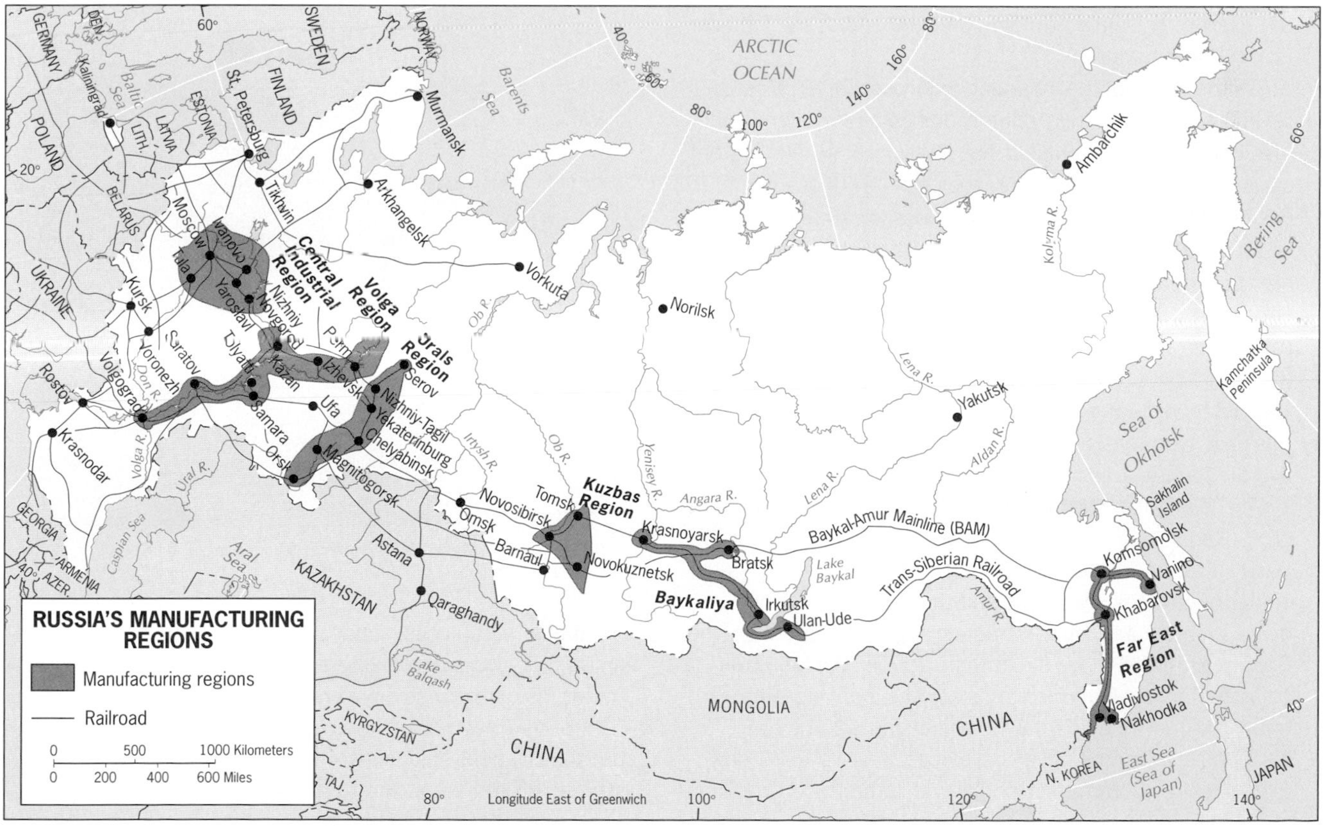

FIGURE 2-11 © H. J. de Blij, P. O. Muller, and John Wiley & Sons, Inc.

one-third of the country's total population), many of them concentrated in such major cities as Nizhniy Novgorod, the automobile-producing "Soviet Detroit"; Yaroslavl, the tire-producing center; Ivanovo, the heart of the textile industry; and Tula, the mining and metallurgical center where lignite (brown coal) deposits are worked.

St. Petersburg (the former Leningrad) remains Russia's second city, with a population of 5.3 million. Under the czars, St. Petersburg was the focus of Russian political and cultural life, and Moscow was a distant second city. Today, however, St. Petersburg has none of Moscow's locational advantages, at least not with respect to the domestic market. It lies well outside the Central Industrial Region near the northwestern corner of the country, 400 miles (650 km) from Moscow. Neither is it better off than Moscow in terms of resources: fuels, metals, and foodstuffs must all be brought in, mostly from far away. The former Soviet emphasis on self-sufficiency even reduced St. Petersburg's asset of being on the Baltic coast because some raw materials could have been imported much more cheaply across the Baltic Sea from foreign sources than from domestic sites in distant Central Asia (only bauxite deposits lie nearby, at Tikhvin).

Yet St. Petersburg was at the vanguard of the Industrial Revolution in Russia, and its specialization and skills have remained important. Today, the city and its immediate environs contribute about 10 percent of the country's manufacturing, much of it through fabricating high-quality machinery.

North of the latitude of St. Petersburg, the region we have named the Russian Core takes on Siberian properties. However, the Russian presence is much stronger in

AMONG THE REALM'S GREAT CITIES . . . MOSCOW

In the vastness of Russia's territory, Moscow, capital of the Federation, seems to lie far from the center, close to its western margin. But in terms of Russia's population distribution, Moscow's centrality is second to none among the country's cities. Moscow (10.6 million) lies at the heart of Russia's primary core area and at the focus of its circulation systems.

On the banks of the meandering Moscow River, Moscow's skyline of onion-shaped church domes and modern buildings rises from the forested flat Russian Plain like a giant oasis in a verdant setting. Archeological evidence points to ancient settlement of the site, but Moscow enters recorded history only in the middle of the twelfth century. Forest and river provided defenses against Tatar raids, and when a Muscovy force defeated a Tatar army in the late fourteenth century, Moscow's primacy was assured. A huge brick Kremlin (citadel, fortress), with walls more than a mile in length and with 18 towers, was built to ensure the city's security. From this base Ivan the Terrible expanded Muscovy's domain and laid the foundations for Russia's vast empire.

The triangular Kremlin and the enormous square in front of it (Red Square of revolutionary times), flanked by St. Basil's Cathedral and overlooking the Moscow River, is the center of old Moscow and is still the heart of the city. From here, avenues and streets radiate in all directions to the Garden Ring and beyond. Until the 1970s, the Moscow Circular Motorway encircled most of the built-up area, but today the metropolis sprawls far outside this beltway. For all its size, Moscow never developed a world-class downtown skyline. Communist policy was to create a "socialist city" in which neighborhoods—consisting of apartment

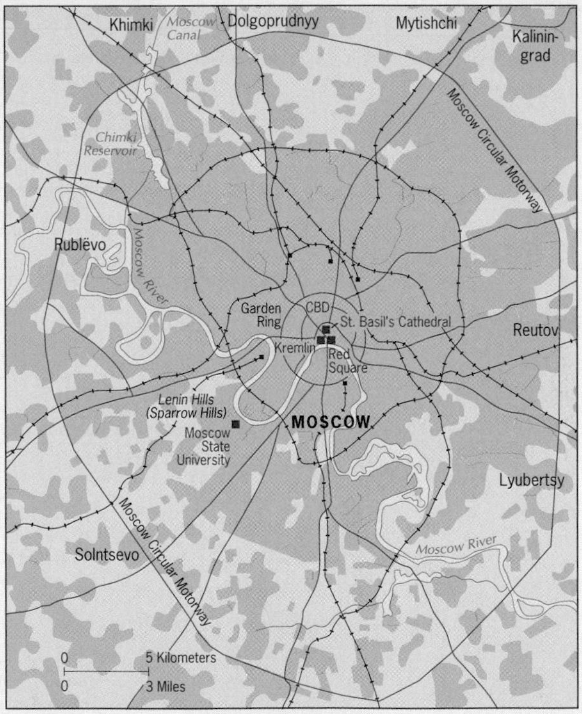

© H. J. de Blij, P. O. Muller, and John Wiley & Sons, Inc.

buildings, workplaces, schools, hospitals, shops, and other amenities—would be clustered so as to obviate long commutes into a high-rise city center.

Moscow may not be known for its architectural appeal, but the city contains noteworthy historic as well as modern structures, including the Cathedral of the Archangel and the towers of Moscow State University.

this remote northern periphery than it is in Siberia proper. Two substantial cities, Murmansk and Arkhangelsk, are road- and rail-connected outposts in the shadow of the Arctic Circle.

Murmansk lies on the Kola Peninsula not far from the border with Finland (Fig. 2-7). In its hinterland lie a variety of mineral deposits, but Murmansk is particularly important as a naval base. During World War II, Allied ships brought supplies to Murmansk; the city's remoteness shielded it from German occupation. After the war, it became a base for nuclear submarines. This city is also an important fishing port and a container facility for cargo ships.

Arkhangelsk is located near the mouth of the Northern Dvina River where it reaches an arm of the White Sea (Fig. 2-7). Ivan the Terrible chose its site during Muscovy's early expansion; the czar wanted to make this the key port on a route to maritime Europe. Yet Arkhangelsk, mainly a port for lumber shipments, has a shorter ice-free season than Murmansk, whose port can be kept open with the help of the warm North Atlantic Drift ocean current (and by icebreakers if necessary).

Nothing in Siberia east of the Urals rivals either of these cities—yet. But their existence and growth prove that Siberian barriers to settlement can be overcome.

Povolzhye: The Volga Region

A second region lying within the Russian Core is the *Povolzhye*, the Russian name for an area that extends along the middle and lower valley of the Volga River. It would be appropriate to call this the Volga Region, for that greatest of Russia's rivers is its lifeline and most of the cities that lie

AMONG THE REALM'S GREAT CITIES . . . ST. PETERSBURG

Czar Peter the Great and his architects transformed the islands of the Neva Delta, at the head of the Gulf of Finland, into the Venice of the North, its palaces, churches, waterfront façades, bridges, and monuments giving St. Petersburg a European look unlike that of any other city in Russia. Having driven out the Swedes, Peter laid the foundations of the Peter and Paul Fortress on the bank of the wide Neva River in 1703, and the city was declared the capital of Russia in 1712.

Peter's "window on Europe," St. Petersburg (named after the saint, not the czar) was to become a capital to match Paris, Rome, and London. During the eighteenth century, St. Petersburg acquired a magnificent skyline dominated by tall, thin spires and ornate cupolas, and graced by baroque and classical architecture. The Imperial Winter Palace and the adjoining Hermitage Museum at the heart of the city are among a host of surviving architectural treasures.

Revolution and war lay in St. Petersburg's future. In 1917, the Russian Revolution began in the city (then named Petrograd), and following the communist victory it lost its functions as a capital to Moscow and its name to Lenin. As Leningrad, it suffered through the 872-day Nazi siege during World War II, holding out heroically through endless bombardment and starvation that took nearly 1 million lives and severely damaged many of its buildings.

The communist period witnessed the neglect and destruction of some of Leningrad's most beautiful churches, the intrusion and addition of crude monuments and bleak apartment complexes, and the rapid industrialization and growth of the city (which now totals 5.3 million). Immediately after the collapse of the Soviet Union, it regained its original name and a new era opened. The Russian Orthodox Church revived, building restoration went forward, and tourism boomed. But the transformation from communist to capitalist ways has been accompanied by social problems that include crime and poverty.

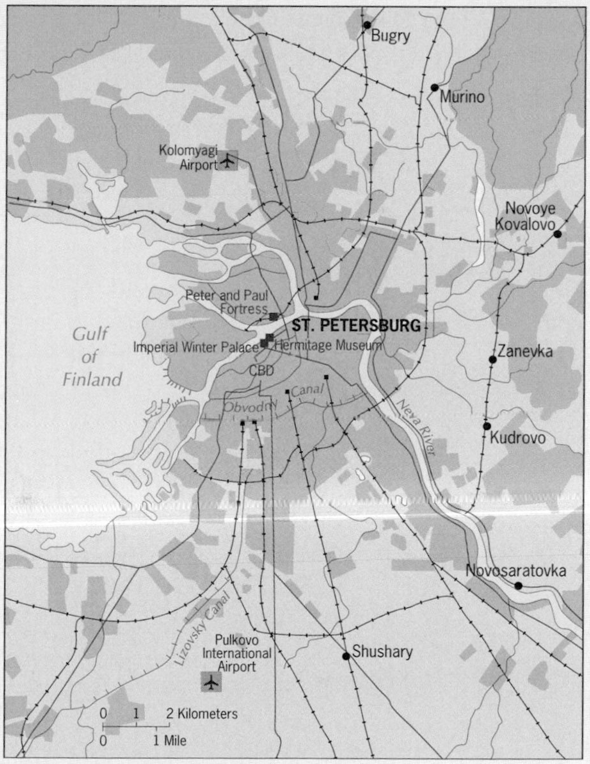

in the Povolzhye are on its banks (Fig. 2-11). In the 1950s, a canal was completed to link the lower Volga with the lower Don River (and thereby the Black Sea).

The Volga River was an important historic route in old Russia, but for a long time neighboring regions overshadowed it. The Moscow area and Ukraine were far ahead in industry and agriculture. The Industrial Revolution came late in the nineteenth century to the Moscow Region and did not have much effect in the Povolzhye. Its major function remained the transit of foodstuffs and raw materials to and from other regions.

This transport function is still important, but the Povolzhye has changed. First, World War II brought furious development because the Volga River, located east of Ukraine, was far from the German armies that invaded from the west. Second, in the postwar era the Volga-Urals Region for some time was the largest known source of petroleum and natural gas in the entire Soviet Union. From near Volgograd (formerly Stalingrad) in the southwest to Perm on the Urals' flank in the northeast lies a belt of major oilfields (Fig. 2-12).

Third, the transport system has been greatly expanded. The Volga-Don Canal directly connects the Volga waterway to the Black Sea; the Moscow Canal extends the northern navigability of this river system into the heart of the Central Industrial Region; and the Mariinsk canals provide a link to the Baltic Sea. Today, the Volga Region's population exceeds 25 million, and the cities of Samara (formerly Kuybyshev), Volgograd, Kazan, and Saratov all have populations between 1.0 and 1.3 million. Manufacturing has also expanded into the middle Volga Basin, emphasizing more specialized engineering industries. The huge Fiat-built auto assembly plant in Tolyatti, for example, is one of the world's largest of its kind.

The Internal Southern Periphery

We turn now to an area that has generated some of the most difficult politico-geographical problems the new Russia has faced since the collapse of the Soviet Union. Between the Black Sea to the west and the Caspian Sea to the east, and with the Caucasus Mountains to the south, Moscow seeks to stabilize and assert its authority over a tier of minority "republics," holdovers from the Soviet era (Fig. 2-13).

A combination of geographic factors causes the problems Moscow confronts in this peripheral zone. Here the Russians (and later the Soviets) met and stalled

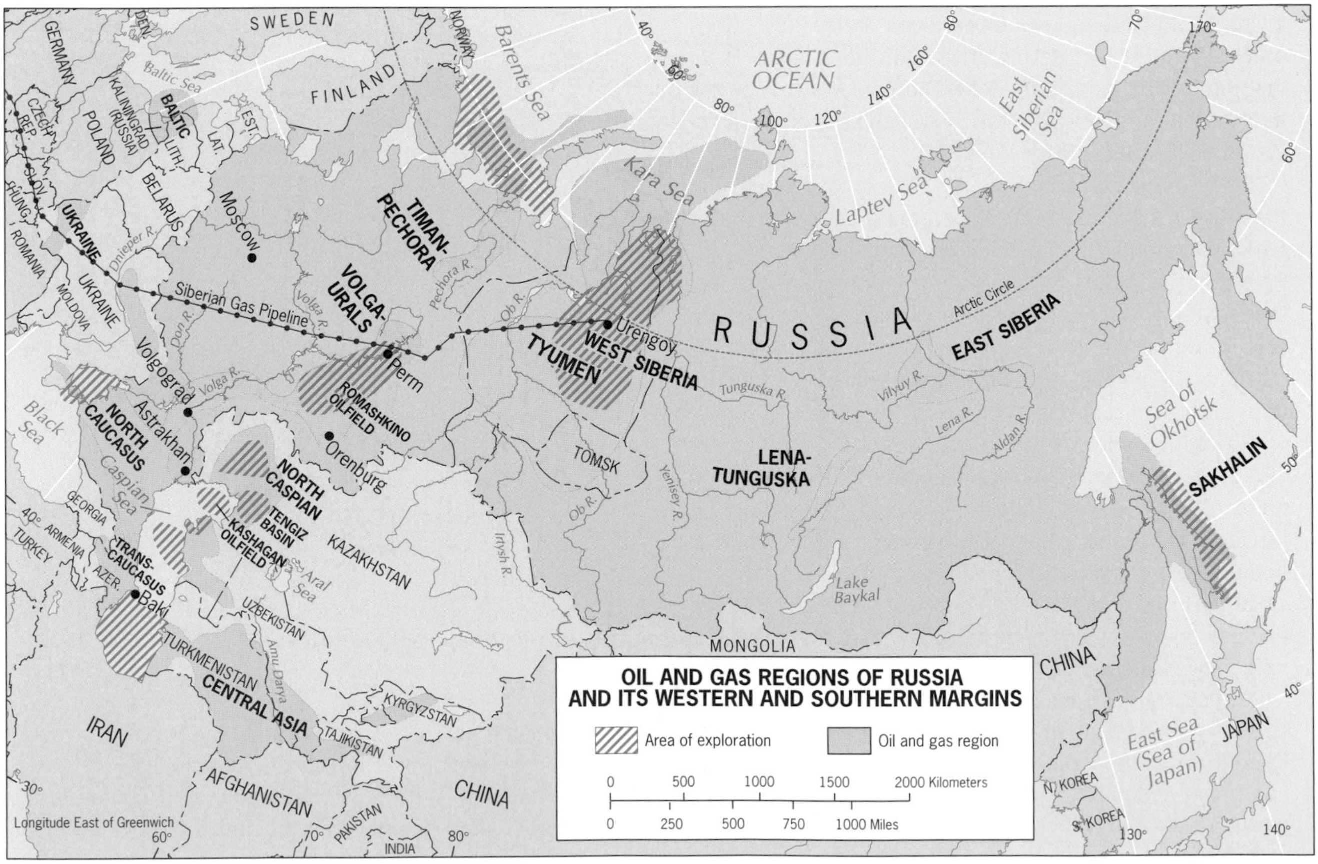

FIGURE 2-12

SOUTHERN RUSSIA:
INTERIOR AND EXTERIOR PERIPHERIES
POPULATION
Under 50,000
50,000–250,000
250,000–1,000,000
1,000,000–5,000,000
Over 5,000,000
National capitals are underlined
Railroad
Road
Canal
Oil pipeline
Proposed oil pipeline
0 50 100 150 200 250 300 350 Kilometers
0 50 100 150 200 Miles
RUSSIA
KAZAKHSTAN
KALMYKIYA
Volgograd
Astrakhan
Elista
Rostov
Mariupol
Sea of Azov
Tikhoretsk
Kropotkin
Krasnodar
Novorossiysk
Armavir
Stavropol
Budennovsk
Maykop
ADYGEYA
Cherkessk
Mineral'nyye Vody
Pyatigorsk
KARACHAYEVO-CHERKESSIYA
KABARDINO-BALKARIYA
Nalchik
INGUSHETIYA
Groznyy
CHECHNYA
Makhachkala
DAGESTAN
Vladikavkaz
Nazran
NORTH OSSETIA
Sochi
ABKHAZIA
Sokhumi
Black Sea
Caspian Sea
CAUCASUS MOUNTAINS
Pankisi Gorge
SOUTH OSSETIA
Tskhinval
Samtredia
Kutaisi
Supsa
Rioni R.
Gori
Tbilisi
Rustavi
Batumi
AJARIA
GEORGIA
Borcka
TURKEY
Trabzon
AZERBAIJAN
Sumgait
Baki (Baku)
Gänjä
Gyumri
ARMENIA
L. Sevan
Yerevan
NAGORNO-KARABAKH
Xankändi (Stepanakert)
Kura R.
Aras R.
Ceyhan
Agri
NAXCIVAN AZER.
Armenian controlled
AZERBAIJAN PROVINCE
Marand
Tabriz
IRAN
Ardabil
Lake Urmia
Lake Van
Tatvan
Van
Don R.
Volga River
Tsimlyansk Res.
Longitude East of Greenwich
CHECHNYA
Kargalinskaja
Russian North (Plains)
Iscerskaja
Selkovskaja
Cervlënnaja
Terek R.
INGUSHETIYA
Groznyy
Gudermes
Argun
Sunzha R.
Shali
Urban-Industrial Middle Zone
Vedeno
NORTH OSSETIA
Chechen South (Mountains)
Shatoy
CAUCASUS MOUNTAINS
DAGESTAN
Argun R.
0 20 40 Kilometers
0 10 20 30 Miles

FIGURE 2-13 © H. J. de Blij, P. O. Muller, and John Wiley & Sons, Inc.

the advance of Islam. Here the cultural mosaic is as intricate as anywhere in the Federation: Dagestan, the southernmost republic fronting the Caspian Sea, has 2.6 million inhabitants comprising some 30 ethnic groups speaking about 80 languages. And beneath the land as well as under the waters of the Caspian Sea lie oil and gas reserves that make this a crucial corner of the vast country. Indeed, Russia has a stake even in the oilfields belonging to neighboring Azerbaijan: as the map shows, the Russians control the pipelines leading from those reserves across Russian territory to the Black Sea oil terminal at Novorossiysk. They watch with great concern as new pipelines are under construction from Azerbaijan across Georgia and Turkey to alternate outlets.

Failure in Chechnya

The presence of oil and oil installations endow this Southern Periphery with far greater significance in the Russian Federation than would otherwise be the case. Note again the route of the pipeline to Novorossiysk: it crosses the republic called Chechnya whose now-devastated capital, Groznyy, was a major oil-industry center and service hub during the Soviet era (Fig. 2-13, including inset).

But Chechnya also contains a sizeable Muslim population, and fiercely independent Muslim Chechens used Caucasus mountain hideouts to resist Russian colonization during the nineteenth century. Accused of collaboration with the Nazis during World War II, on Stalin's orders the Soviets exiled the entire Chechen population to Central Asia, with much loss of life. Rehabilitated and allowed to return by Premier Nikita Khrushchev in the 1950s, the Chechens seized their opportunity in 1991 after the collapse of the Soviet Union and fought the Russian army to a stalemate. But Moscow never granted Chechnya independence (one quarter of the population of 1.2 million was Russian), and an uneasy standoff continued. Meanwhile, Chechen hit-and-run attacks on targets in neighboring Republics continued, and in 1999 three apartment buildings in Moscow were bombed, resulting in 230 deaths. Russian authorities blamed Chechnyan terrorists for these bombings and ordered a full-scale attack on Chechnya's Muslim holdouts. Groznyy, already severely damaged in the earlier conflict, was totally devastated and the rebels were driven into the mountains, but the Russian armed forces, still taking substantial losses, were unable to establish unchallenged control of all of the Republic.

The cycle of violence resumed in 2002 when a band of 41 terrorists took more than 700 performers and theatergoers hostage in a Moscow auditorium, demanding an immediate end to Russian military action in their homeland and the withdrawal of all Russian forces from Chechnya. After a three-day standoff, a bungled rescue effort led to the deaths of nearly 130 of the hostages and all of the terrorists. In 2004 an attack by Islamic militants from Chechnya and neighboring (also partly Muslim) Ingushetiya on a school in Beslan, a small town in North Ossetia (see Fig. 2-13 and photo p. 125), killed as many as 350, nearly half of them schoolchildren. And only a week before the Beslan attack, suicide bombers had struck simultaneously to down two airliners shortly after taking off from a Moscow airport. Whatever Moscow has tried in Chechnya, from accommodation to manipulation to repression, has not worked. Thus a "republic" the size of Connecticut with a population barely over 1 million threatens security and stability in what is territorially the largest state in the world with a population of more than 140 million. It is a mirror not only on Russia but on the world of the present.

It is important to familiarize yourself with the layout of Russia's Internal Southern Periphery, because the problems here not only afflict Russia but spill over into its neighbors. The southern republics, colored pink in Figure 2-13, all present their particular (sometimes intertwined) problems. Half the population in Kalmykia, north of Dagestan, is Buddhist, and local leaders presented Moscow with a dilemma when they decided in 2004 to invite Tibet's Dalai Lama to visit (China, which controls Tibet and calls it Xizang, does not recognize the Dalai Lama's position and denigrates his role). Also bordering the Caspian Sea is Dagestan which, as we noted, contains more than two dozen ethnic groups. Ingushetiya, which was once part of the Chechnyan republic, has a Muslim majority but has not joined the Chechnyan campaign although individual Islamic fighters have done so. North Ossetia is called *North* Ossetia because its kinspeople live outside Russia in *South* Ossetia—which is part of neighboring Georgia. Russian leaders argue that terrorists pass through South Ossetia into North Ossetia and on into Chechnya, and that the government in Georgia is not doing enough to prevent this movement. The Muslim Balkar people of Kabardino-Balkariya, like the Chechens, were accused by Stalin of collaboration with the Nazis and were forcibly exiled with great loss of life; today they constitute only about 12 percent of the republic's population. The same thing happened to the Muslim Karachay of Karachayevo-Cherkessiya: many were deported, and today they constitute a minority in their republic, where Russians and Christian Cherkess form the majority. Only in the separate republic of Adygeya is there no significant Muslim element: Cherkess form the non-Russian population sector here.

This summation of the regional geography of the Internal Southern Periphery helps explain why Russia faces such severe difficulties here. Even after a solution to the Chechnyan problem is found, the challenge will continue: stability in this physiographically and culturally fractured subregion has always been elusive.

The External Southern Periphery

Instability marks not only Russia's interior south but the realm's exterior periphery as well. As Figure 2-13 shows, the area between Russia to the north and Turkey and Iran to the south is occupied by three countries, two of which border Russia: Georgia on the Black Sea, Azerbaijan on the Caspian Sea, and landlocked Armenia. This part of Transcaucasia has historically been a battleground for Christians and Muslims, Russians and Turks, Armenians and Persians. Today the three states on the map, all former Soviet Socialist Republics of communist vintage, suffer from political and administrative mismanagement and economic stagnation despite, at least in the case of Georgia and Azerbaijan, considerable potential. As we will see, Russian involvement has not helped, especially not in Georgia. In addition, there are boundary disputes and ethnic quarrels, and many people here, especially in Armenia, will tell you that life was better during Soviet times than it is today.

Georgia

Of the three former Soviet Republics in Transcaucasia, only Georgia has a Black Sea coast and thereby an outlet to the wider world. Smaller than South Carolina, Georgia is a country of high mountains and fertile valleys. Its social and political geographies are complicated. The population of 4.5 million is more than 65 percent Georgian but also includes Armenians (8 percent), Russians (6 percent), Ossetians (3 percent), and Abkhazians (2 percent). The Georgian Orthodox Church dominates the religious community, but about 10 percent of the people are Muslims, most of them concentrated in Ajaria in the southwest.

As the map shows, Georgia actually has four regions: the main territory centered on the capital, Tbilisi; the Muslim-tinged southwest, known as Ajaria; the Russian-favored northwest, named Abkhazia; and the small interior entity of South Ossetia. *Sakartvelos*, as the Georgians call their country, has a long and turbulent history. Tbilisi, the capital for 15 centuries, lay at the core of an empire around the turn of the thirteenth century, but the Mongol invasion ended this era. Next, the Christian Georgians found themselves in the path of wars between Islamic Turks and Persians. Turning northward for protection, the Georgians were annexed by the Russians, who were looking for warm-water ports, in 1800. Like other peoples overpowered by the czars, the Georgians took advantage of the Russian Revolution to reassert their independence; but the Soviets reincorporated Georgia in 1921 and proclaimed a Georgian Soviet Socialist Republic in 1936. Josef Stalin, the communist dictator who succeeded Lenin, was a Georgian.

Georgia is renowned for its scenic beauty, warm and favorable climates, agriculture (especially tea), timber, manganese, and other products. Georgian wines, tobacco, and citrus fruits are much in demand. The diversified economy could support a viable state.

Unfortunately, Georgia's political geography is loaded with centrifugal forces. After Georgia declared its independence in 1991, factional fighting destroyed its first elected government. But worse was to come. In the Autonomous Region of South Ossetia, a movement for union with (Russian) North Ossetia created havoc; Abkhazia proclaimed independence; and Muslim-infused Ajaria asserted itself by ignoring a government preoccupied with problems elsewhere. The price of all this instability was Russian intervention, including the establishment of Russian military bases; a former Soviet communist foreign minister born in Georgia became president. He tolerated Russia's virtual colonization of Abkhazia and allowed Russian troops to patrol the Caucasus passes in South Ossetia. But in 2003 a political transformation occurred when, following a fraudulent election, thousands of unarmed supporters invaded the parliament and ousted the Russian puppet, and, following negotiations, a new election was scheduled in 2004. The newly-elected leader, a Western-educated economist, started the massive task of reorganizing the country by regaining control over South Ossetia and Ajaria, although the reintegration of secession-minded Abkhazia proved a more elusive goal; after drawing Abkhazia tightly into the Russian economic orbit, the Russians abruptly closed the border, causing economic havoc in the already chaotic province. Meanwhile, Georgia's new leaders dreamed of a time when Georgia might apply for membership in the European Union.

Armenia

As Figure 2-1 shows, landlocked Armenia (population: 3.2 million) occupies some of the most rugged and mountainous terrain in the earthquake-prone Transcaucasus. The Armenians are an embattled people who adopted Christianity 17 centuries ago and for more than a millennium sought to secure their ancient homeland here on the margins of the Muslim world. During World War I, the Ottoman Turks massacred much of the Christian Armenian minority and drove the survivors from eastern Anatolia and what is now Iraq into the Transcaucasus. At the end of that war in 1918, an independent Armenia arose, but its autonomy lasted only two years. In 1920, Armenia was taken over by the Soviets; in 1936, it became one of the 15 constituent republics of the Soviet Union. The collapse of the Soviet Empire gave Armenia what it had lost three generations earlier: independence.

Or so it seemed. Soon afterward, the Armenians found themselves at war with neighboring Azerbaijan over the fate of some 150,000 Armenians living in Nagorno-Karabakh, a pocket of territory surrounded by Azerbaijan. Such a separated territory is called an *exclave*, and this one had been created by Soviet sociopolitical planners who, while acknowledging the cultural (Christian) distinctiveness of this cluster of Armenians, nevertheless gave (Muslim) Azerbaijan jurisdiction over it.

That was a recipe for trouble: the arrangement was made to work under authoritarian Soviet rule, but once this rule ended, the Christian Armenians encircled by Muslim Azerbaijan felt insecure and appealed to Armenia for help. In the ensuing conflict, Armenian troops entered Azerbaijan and gained control over the exclave, even ousting Azerbaijanis from the zone between the main body of Armenia and Nagorno-Karabakh (Fig. 2-13). The international community, however, has not recognized Armenia's occupation, and officially the territory remains a part of Azerbaijan. In 2005, the cultural landscape of Nagorno-Karabakh and its orderly capital stood in ever-starker contrast to that of Azerbaijan, but the fundamental issue remained unresolved.

Azerbaijan

Azerbaijan is the name of an independent state *and* of a province in neighboring Iran. The Azeris (short for Azerbaijanis) on both sides of the border have the same ancestry: they are a Turkish people divided between the (then) Russian and Persian empires by a treaty signed in 1828. By that time, the Azeris had become Shi'ite Muslims, and when the Soviet communists laid out their grand design for the USSR, they awarded the Azeris their own republic. On the Persian side, the Azeris were assimilated into the Persian Empire, and their domain became a province. Today the former Soviet Socialist Republic is the independent state of Azerbaijan (population: 8.4 million), and the 10 million Azeris to the south live in the northwesternmost Iranian province.

During the brief transition to independence and at the height of their war with the Armenians, the dominantly Muslim Azeris tended to look southward, toward Iran. But geographic realities dictate a more practical orientation. Azerbaijan possesses huge reserves of oil and natural gas; under the Soviets it was one of Moscow's chief regional sources of fuels. The center of the oil industry is Baki (Baku), the capital on the shore of the Caspian Sea—but the Caspian Sea is a lake. To export its oil, Azerbaijan needs pipelines, but those of Soviet vintage link Baki to Russia's Black Sea terminal of Novorossiysk. During the mid-1990s, when Azerbaijan's leaders announced plans to build new pipelines via Georgia or Iran, Russia objected and even meddled in Azeri politics to stymie such plans.

A look at the map (Fig. 2-13) shows Azerbaijan's options. A pipeline could be routed from Baki through Georgia to the Black Sea coast near Batumi, avoiding Russian territory altogether. Another route could run through Georgia and across Turkey to the Mediterranean terminal at Ceyhan. Still another option would run the pipeline across Iran to a terminal on the Persian Gulf. Moscow objects to such alternatives, but Azerbaijan now has powerful international allies. American, French, British, and Japanese oil companies are developing Azerbaijan's offshore Caspian reserves, and the U.S. government preferred the Turkish route. That is what materialized, and the first oil began to flow through the new Baki-Ceyhan pipeline in mid-2005.

For all its income from oil, Azerbaijan remains mired in poverty, poorly governed, and bedeviled by corruption (it ranks seventh-worst in the world in the Corruption Index in Table G-1); in 2004 an election described by international observers as fraudulent, and by the United States as "falling short of democratic standards," anointed the son of the long-time ruler as president. None of this, however, stopped Azerbaijan from receiving World Bank and European Bank loans to build the strategic pipeline across Georgia and Turkey to the Mediterranean coast. In this energy-dependent world, oil talks. Given its Muslim roots and its location, the time may come when Azerbaijan turns toward the Islamic world. For the moment, however, this country is inextricably bound up with its neighbors in Transcaucasia and to the north.

The Urals Region

The Ural Mountains form the eastern limit of the Russian Core. They are not particularly high; in the north they consist of a single range, but southward they broaden into a hilly zone. Nowhere are they an obstacle to east-west transportation. An enormous storehouse of metallic mineral resources located in and near the Urals has made this area a natural place for industrial development. Today, the Urals Region, well connected to the Volga and Central Industrial Region, extends from Serov in the north to Orsk in the south (Fig. 2-11).

The Central Industrial, Volga, and Urals regions form the anchors of the Russian core area. For decades they have been spatially expanding toward one another, their interactions ever more intensive. These regions of the Russian Core stand in sharp contrast to the comparatively less developed, forested, Arctic north and the remote upland of the Southern Periphery lying to the southwest between the Black and Caspian seas. Thus even within this Russian coreland, frontiers still await growth and development.

As noted in Chapter 1, because the Ural Mountains form a prominent north-south divide between Russia's western heartland and its eastern expanses, some geographers use this to separate a "European" Russia from its presumably non-European, "Asian" east. (The National Geographic Society on its maps and in its atlases even draws a green line along the crest of the Urals marking this transition.) But geographically this makes no sense. Take the train or drive eastwards across the southern end of the Urals, and you will see no changes in cultural landscapes, no geographic evidence on which to base such a partition. Indeed, as we will see, Russia remains Russia all the way to the end of its transcontinental railroad at Vladivostok on the Pacific coast.

THE EASTERN FRONTIER

From the eastern flanks of the Ural Mountains to the headwaters of the Amur River, and from the latitude of Tyumen to the northern zone of neighboring Kazakhstan, lies Russia's vast Eastern Frontier Region, product of a gigantic experiment in the eastward extension of the Russian Core (Fig. 2-10). As the maps of cities and surface communications suggest, this eastern frontier is more densely peopled and more fully developed in the west than in the east. At the longitude of Lake Baykal, settlement has become linear, marked by ribbons and clusters along the east-west railroads. Two subregions dominate the geography: the Kuznetsk Basin in the west and the Lake Baykal area in the east.

The Kuznetsk Basin (Kuzbas)

Some 900 miles (1450 km) east of the Urals lies another of Russia's primary regions of heavy manufacturing resulting from the communist period's national planning: the Kuznetsk Basin, or *Kuzbas* (Fig. 2-11). In the 1930s, it was opened up as a supplier of raw materials (especially coal) to the Urals, but that function became less important as local industrialization accelerated. The original plan was to move coal from the Kuzbas west to the Urals and allow the returning trains to carry iron ore east to the coalfields. However, good-quality iron ore deposits were subsequently discovered near the Kuznetsk Basin itself. As the new resource-based Kuzbas industries grew, so did its urban centers. The leading city, located just outside the region, is Novosibirsk, which stands at the intersection of the Trans-Siberian Railroad and the Ob River as the symbol of Russian enterprise in the vast eastern interior. To the northeast lies Tomsk, one of the oldest Russian towns in all of Siberia, founded in the seventeenth century and now caught up in the modern development of the Kuzbas Region. Southeast of Novosibirsk lies Novokuznetsk, a city that produces steel for the region's machine and metal-working plants and aluminum products from Urals bauxite.

The Lake Baykal Area (Baykaliya)

East of the Kuzbas, development becomes more insular, and distance becomes a stronger adversary. North of the Tyva Republic and eastward around Lake Baykal, larger and smaller settlements cluster along the two railroads to the Pacific coast (Fig. 2-11). West of the lake, these rail corridors lie in the headwater zone of the Yenisey River and its tributaries. A number of dams and hydroelectric projects serve the valley of the Angara River, particularly the city of Bratsk. Mining, lumbering, and some farming sustain life here, but isolation dominates it. The city of Irkutsk, near the southern end of Lake Baykal, is the principal service center for a vast Siberian region to the north and for a lengthy east-west stretch of southeastern Russia.

Beyond Lake Baykal, the Eastern Frontier really lives up to its name: this is southern Russia's most rugged, remote, forbidding country. Settlements are rare, many being mere camps. The Buryat Republic (Fig. 2-8) is part of this zone; the territory bordering it to the east was taken from China by the czars and may become an issue in the future. Where the Russian-Chinese boundary turns southward, along the Amur River, the region called the Eastern Frontier ends and Russia's Far East begins.

SIBERIA

Before we assess the potential of Russia's Pacific Rim, we should remember that the ribbons of settlement just discussed hug the southern perimeter of this giant country, avoiding the vast Siberian region to the north (Fig. 2-10). Siberia extends from the Ural Mountains to the Kamchatka Peninsula—a vast, bleak, frigid, forbidding land. Larger than the conterminous United States but inhabited by only an estimated 15 million people, Siberia quintessentially symbolizes the Russian environmental plight: vast distances, cold temperatures worsened by strong Arctic winds, difficult terrain, poor soils, and limited options for survival.

But Siberia also has resources. From the days of the first Russian explorers and Cossack adventurers, Siberia's riches have beckoned. Gold, diamonds, and other precious minerals were found there, and later, metallic ores including iron and bauxite were discovered. Still more recently, the Siberian interior proved to contain sizeable quantities of oil and natural gas (Fig. 2-12) and began to contribute significantly to Russia's energy supply.

As the physiographic map (Fig. 2-2) shows, major rivers—the Ob, Yenisey, and Lena—flow gently northward across Siberia and the Arctic Lowland into the Arctic Ocean. Hydroelectric power development in the basins of these rivers has generated electricity used to extract and refine local ores, and run the lumber mills that have been set up to exploit the vast Siberian forests.

The human geography of Siberia is fragmented, and much of the region is virtually uninhabited (Fig. 2-5). Ribbons of Russian settlement have developed; the Yenisey River, for instance, can be traced on this map of Soviet peoples (a series of small settlements north of Krasnoyarsk), and the upper Lena Valley is similarly fringed by ethnic Russian settlement. Yet hundreds of miles of empty territory separate these ribbons and other islands of habitation.

The political geography of eastern Siberia is marked by the growing identity of Sakha (the Yakut Republic). As additional resources are discovered here (including oil and natural gas), this Republic, centered on the capital, Yakutsk, will become more important.

Siberia, Russia's freezer, is stocked with goods that may become mainstays of future national development. Already, precious metals and mineral fuels are bolstering the Russian economy. In time, we may expect Siberian resources to play a growing role in the economic development of the Eastern Frontier and the Russian Far East as well. One step in that process was already taken during Soviet times: the completion of the BAM (Baykal-Amur Mainline) Railroad in the 1980s. This route, lying north of and parallel to the old Trans-Siberian Railroad, extends 2200 miles (3540 km) eastward from Tayshet (near the important center of Krasnoyarsk) directly to the Far East city of Komsomolsk (Fig. 2-7). In the post-Soviet era, the BAM Railroad has been beset by equipment breakdowns and workers' strikes. Nonetheless, it is a key element of the infrastructure that will serve the Eastern Frontier's economic growth in the twenty-first century.

THE RUSSIAN FAR EAST

Imagine this: a country with 5000 miles (8000 km) of Pacific coastline, two major ports, interior cities nearby, huge reserves of resources ranging from minerals to fuels to timber, directly across from one of the world's largest economies—all this at a time when the Asian Pacific Rim was the world's fastest-growing economic region. Would not that country have burgeoning cities, busy harbors, growing industries, and expanding trade?

In the Russian Far East (Fig. 2-10), the answer is—no. Activity in the port of Vladivostok is a shadow of what it was during the Soviet era, when it was the communists' key naval base. The nearby container terminal at Nakhodka suffers from breakdowns and inefficiencies. The railroad to western Russia carries just a fraction of the trade it did during the 1970s and 1980s. Cross-border trade with China is minimal. Trade with Japan is inconsequential. The region's cities are grimy, drab, moribund. Utilities are shut off for hours at a time because of fuel shortages and system breakdowns. Outdated factories are closed down, their workers dismissed. Political relations with Moscow are poor. There is potential here, but little of it has been realized.

As a region, the Russian Far East consists of two parts: the mainland area extending from Vladivostok to the Stanovoy Mountains and the large island of Sakhalin (Figs. 2-1, 2-14). This is cold country: icebreakers have

Russia's oil and natural gas reserves are key to the overall health of the Russian economy. Oil and gas are Russia's most valuable exports by far, and Europe's dependence on Russian energy supplies grows continuously. This photograph shows a small segment of what has been described as the biggest work site in the world: the laying of the Trans-Siberian gas pipeline from near the shore of the Arctic Ocean to the border of the Czech Republic, where it is being connected to the gas pipeline network leading to Western Europe. Here are some statistics: the pipeline shown here is nearly five feet (1.42 m) in diameter. It will lie across permafrost and through trenches for a total of almost 2800 miles (4450 km) from the Arctic to Eastern Europe. It will be capable of sending 34 billion cubic meters of gas from its source to consumers. © Fabian Cevallos/Corbis Sygma.

to keep the ports of Vladivostok and Nakhodka open throughout the winter. Winters here are long and bitterly cold; summers are brief and cool. Although the population is small (about 7 million), food must be imported because not much can be grown. Most of the region is rugged, forested, and remote. Vladivostok, Khabarovsk, and Komsomolsk are the only cities of any size. Nakhodka and the newer railroad terminal at Vanino are smaller towns; the population of the whole island of Sakhalin is about 600,000 (on an island the size of Caribbean Hispaniola [17.5 million]). Offshore lie productive fishing grounds. Most important of all are the huge oil and gas reserves recently discovered on and around Sakhalin: a new era may be dawning as Russian-partnered foreign companies move in to exploit them.

The Soviet regime rewarded people willing to move to this region with housing and subsidies. The communists, like the czars before them, realized the importance of this frontier (Vladivostok means "We Own the East"), and they used every possible incentive to develop it and link it ever closer to Russia's distant western core. Freight rates on the Trans-Siberian Railroad, for example, were about 10 percent of their real costs; the trains were always loaded in both directions. Vladivostok was a military base and a city closed to foreigners, and Moscow invested heavily in its infrastructure. Komsomolsk in the north and Khabarovsk near the region's center were endowed with state-owned industries using local resources: iron ore from Komsomolsk, oil from Sakhalin, timber from the ubiquitous forests. The steel, chemical, and furniture industries sent their products westward by train, and they received food and other consumer goods from the Russian heartland.

For several reasons, the post-Soviet transition has been especially difficult here in the Far East. The new economic order has canceled the region's communist-era

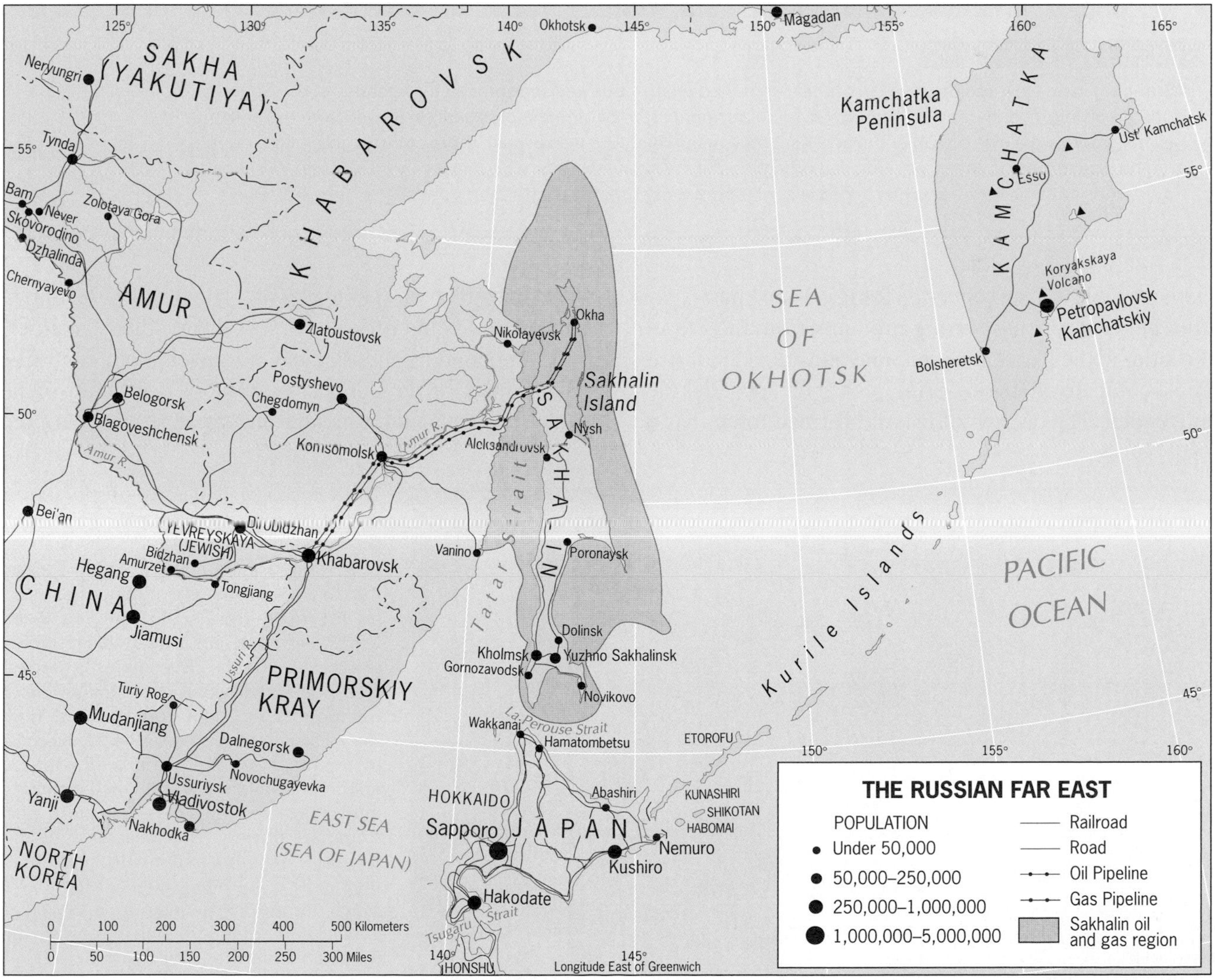

FIGURE 2-14 © H. J. de Blij, P. O. Muller, and John Wiley & Sons, Inc.

FROM THE FIELD NOTES

"Standing in an elevated doorway in the center of Vladivostok, you can see some of the vestiges of the Soviet period: the omnipresent, once-dominant, GUM department store behind the blue seal on the left, and the communist hammer-and-sickle on top of the defunct hotel on the right. But in the much more colorful garb of the people, and the private cars in the street, you see reflections of the new era. Once a city closed to foreigners, Vladivostok now throbs with visitors—and has become a major point of entry for contraband goods." © George W. Moore.

advantages: the Trans-Siberian Railroad now must charge the real cost of transporting products from the Eastern Frontier to the Russian Core. State-subsidized industries must compete on market principles, their subsidies having ended. The decline of Russia's armed forces has hit Vladivostok hard (photo above). The fleet lies rusting in port; service industries have lost their military markets; the shipbuilding industry has no government contracts. Coal miners in the Bureya River Valley (a tributary of the Amur) go unpaid for months and go on strike; coal-fired

The Kamchatka Peninsula constitutes one of the Pacific Rim's most geologically active zones, with more than 125 volcanoes of which more than 20 are active. Sudden massive eruptions can endanger airliners on the northern Pacific route between North America and East Asia, which skirts Kamchatka. The region's largest city, Petropavlovsk-Kamchatskiy, lies at the foot of the 11,400-foot (3456-m) high Koryakskaya Volcano near the Pacific coast in southeastern Kamchatka. Soviet-era apartment buildings dominate the cultural landscape of this bleak fishing port founded during Russia's eighteenth-century eastward expansion. Note that, in this *Dfc* climate, the countryside is much greener (with deciduous as well as coniferous vegetation) than it is toward the north and the Siberian interior. © Mark Newman/Photo Researchers.

power plants do not receive fuel shipments, and cities and towns go dark.

Locals put much of the blame for their region's failure on Moscow, and with reason. As Figure 2-8 shows, the Far East contains only five administrative regions: Primorskiy Kray, Khabarovsk Kray, Amur Oblast, Sakhalin Oblast, and Yevreyskaya, originally the Jewish Autonomous Region. This does not add up to much political clout, which is the way Moscow appears to want it. Compare Figures 2-9 and 2-14 and you will note that the city of Khabarovsk, not the Primorskiy Kray capital of Vladivostok, has been made the headquarters of the Far Eastern Region as defined by Moscow.

For all its stagnation, Russia's Far East will figure prominently in Russian (and probably world) affairs. Here Russia meets China on land and Japan at sea (an unresolved issue between Russia and Japan involves several Kurile islands). Here lie vast resources ranging from Sakhalin's fuels to Siberia's lumber, Sakha's gold to Khabarovsk's metals. Here Russia has a foothold on the Pacific Rim, a window on the ocean on whose shores the world is being transformed.

WHAT YOU CAN DO

RECOMMENDATION: Do you have a good-sized political globe and/or a large wall map of the world in your home or room? If not, we recommend that you do so (we are aware of how crowded desktop and table space can become in a room, especially a shared one). There even are self-standing globes you can place on the floor. We predict that you will be referring to this useful addition to your scholarly arsenal more than you can imagine, and you will find the map equally useful. National Geographic Maps, the Society's cartographic division, produces world maps on a sensible projection (see Appendix A) at several scales, so you can select one that will fit on your wall. In general, the bigger the better, and you will soon be challenging guests to find places you have discovered in this course.

GEOGRAPHIC CONNECTIONS

1. The American Geographical Society in New York, the country's oldest geographical association, presents travelers the opportunity not only to visit remote parts of the world, but to do so accompanied by a professional geographer who specializes in the region or area being visited. Other organizations also appoint geographers as lecturers and guides on ships, planes, and even trains and buses. Imagine that you have been asked to accompany a group of your fellow students taking a riverboat down the Volga and Don rivers in July from Moscow to Rostov near the Sea of Azov, an arm of the Black Sea. In your first briefing, what will you tell them about the weather they should expect, about the minority "republics" of which they will get a glimpse, the crops and land uses they will encounter, the cities and towns they will see, and how they will get from the Volga to the Don? What might be the most interesting side trips along this itinerary?

2. President Vladimir Putin of Russia, who succeeded President Boris Yeltsin, soon after his reelection proclaimed his intention to restore Russia to superpower status, to recapture much of the global might the Soviet Union once possessed. How will Russia's leaders use their vast country's natural resources to advance this plan? Is any Soviet-era legacy likely to be helpful in achieving it? In what ways will this plan affect future relations between the European Union and Russia? Might the boundaries of the Russian realm as mapped in this chapter change as a result, and if so, where and how?

3. Russia has experienced the impact of Islamic terrorism principally in two locales: the Moscow urban area and the Interior Southern Periphery. But Muslim minorities are far more widely distributed throughout this vast country beyond those two areas. Why is Islamic terrorism in Russia so prevalent in these locales and virtually (though not totally) absent elsewhere? What role does physiography play in the persistence of terrorism in the Interior Southern Periphery?

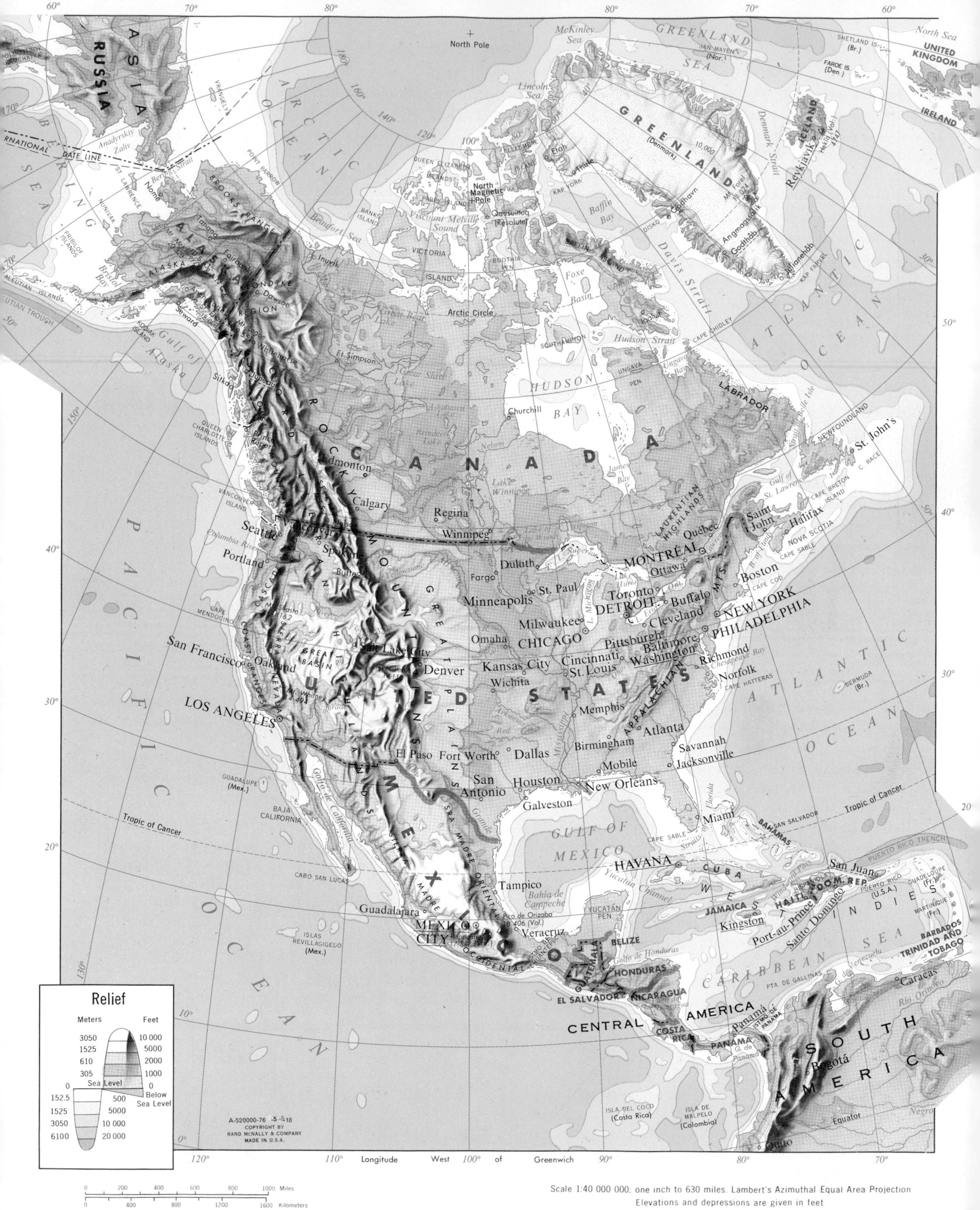

FIGURE 3-1 Reprinted with permission from *Goode's World Atlas*, 21st edition, pp. 89.

Chapter 3

North America

CONCEPTS, IDEAS, AND TERMS

1 Cultural pluralism
2 Physiographic province
3 Rain shadow effect
4 Sunbelt
5 Migration
6 Push/pull factors
7 American Manufacturing Belt
8 Ghetto
9 Outer city
10 Suburban downtown
11 Urban realms model
12 Mosaic culture
13 Productive activities
14 Fossil fuels
15 Economies of scale
16 Postindustrialism
17 Technopole
18 Ecumene
19 Regional state
20 Pacific Rim

REGIONS

- NORTH AMERICAN CORE
- MARITIME NORTHEAST
- FRENCH CANADA
- CONTINENTAL INTERIOR
- SOUTH
- SOUTHWEST
- WESTERN FRONTIER
- NORTHERN FRONTIER
- PACIFIC HINGE

NORTH AMERICA AS a geographic name means different things to different people. To physical geographers, North America refers to the landmass that extends from Alaska to Panama. As Figure G-4 shows, the North American continent is the dominant feature of the North American tectonic plate, which reaches almost—but not quite—to Panama. To human geographers, North America (sometimes inappropriately called "Anglo" America) signifies a geographic realm consisting of two large countries with much in common, Canada and the United States. We therefore recognize three realms in the Americas: (1) *North America*, propelled by the United States; (2) *Middle America*, dominated by Mexico; and (3) *South America*, anchored by Brazil.

DEFINING THE REALM

Defined in the context of human geography, North America is constituted by two of the world's most highly advanced countries by virtually every measure of human development. Blessed by an almost endless range of natural resources and bonded by trade as well as culture, Canada and the United States are locked in a mutually productive embrace that is reflected by the statistics. In an average year of the recent past, 87 percent of Canadian exports went to the United States and 64 percent of Canada's imports came from its southern neighbor. For

The destruction of the twin towers of the World Trade Center on September 11, 2001, left a gaping hole not only in the skyline of New York City but in the ground as well. In a remarkable operation that began before the smoke had cleared, the rubble of terrorism's aftermath was removed in less than nine months, as shown in this photograph taken on Memorial Day 2002. But the success of the terrorists' deadly onslaught set in motion a chain of events that is changing the world as well as the United States. American forces invaded Afghanistan and ousted the regime that had given refuge to Islamic militants. Claiming a terrorist connection among other justifications, the United States led an armed attack on Iraq and overthrew its dictator. American soldiers found themselves headed for Pakistan, Djibouti, the Philippines, and other destinations in a new "War on Terror." Meanwhile, U.S. relations with long-term allies such as France and Canada were strained by disagreements over this campaign even as economic stresses attributable to post–9/11 uncertainties buffeted America as a whole and New York City in particular. Those who watched the dreadful events at this site unfold, and who predicted that the world and the nation would never be the same, were right. © AP/Wide World Photos.

the United States, Canada is its leading export market and its number one source of imports.

Both countries also rank among the world's most highly urbanized, and in each 79 percent of the population is concentrated in cities that have evolved into metropolitan agglomerations. The old core cities are now encircled by rings of suburban development, which draw an ever-greater proportion of the metropolitan population and activity base into their newly urbanized environments. Nothing symbolizes the North American city as strongly as the skyscrapered panoramas of New York, Chicago, and Toronto or the vast, beltway-connected suburban expanses of Los Angeles, Washington, and Houston.

North Americans also are the most mobile people in the world. Commuters stream into and out of suburban business centers and central-city downtowns by the millions each working day (see photos p. 146); most of them drive cars, whose numbers have multiplied more than six times faster than the human population since 1970. Moreover, each year nearly one out of every six individuals changes his or her residence.

Although Canada and the United States share many historical, cultural, and economic qualities, they also differ in significant ways, as can be seen on the map. The United States, somewhat smaller territorially than Canada, occupies the heart of the North American continent and, as a re-

MAJOR GEOGRAPHIC QUALITIES OF

North America

1. North America encompasses two of the world's biggest states territorially (Canada is the second largest in size; the United States is third).
2. Both Canada and the United States are federal states, but their systems differ. Canada's is adapted from the British parliamentary system and is divided into ten provinces and three territories. The United States separates its executive and legislative branches of government, and it consists of 50 States, the Commonwealth of Puerto Rico, and a number of island territories under U.S. jurisdiction in the Caribbean Sea and the Pacific Ocean.
3. Both Canada and the United States are plural societies. Although ethnicity is increasingly important, Canada's pluralism is most strongly expressed in regional bilingualism. In the United States, major divisions occur along racial lines.
4. A substantial number of Quebec's French-speaking citizens supports a movement that seeks independence for the province. The movement's high-water mark was reached in the 1995 referendum in which (minority) non-French-speakers were the difference in the narrow defeat of separation. Prospects for a break-up of the Canadian state have diminished since 2000.
5. North America's population, not large by international standards, is the world's most highly urbanized and mobile. Largely propelled by a continuing wave of immigration, the realm's population total is expected to grow by more than 40 percent over the next half-century.
6. By world standards, this is a rich realm where high incomes and high rates of consumption prevail. North America possesses a highly diversified resource base, but nonrenewable fuel and mineral deposits are consumed prodigiously.
7. North America is home to one of the world's great manufacturing complexes. The realm's industrialization generated its unparalleled urban growth, but today a new postindustrial society and economy are rapidly maturing in both countries.
8. The two countries heavily depend on each other for supplies of critical raw materials (e.g., Canada is the leading source of U.S. energy imports) and have long been each other's chief trading partners. Today, the North American Free Trade Agreement (NAFTA), which also includes Mexico, is linking all three economies ever more tightly as the last barriers to international trade and investment flow are dismantled.
9. North Americans are the world's most mobile people. Although plagued by recurrent congestion problems, the realm's networks of highways, commercial air routes, and cutting-edge telecommunications are still the most efficient on Earth.

sult, encompasses a greater environmental range. The U.S. population is dispersed across most of the country, forming major concentrations along both the (north-south-trending) Atlantic and Pacific coasts. The overwhelming majority of Canadians, however, live in an interrupted east-west corridor that lies across southern Canada, mainly within 200 miles (320 km) of the U.S. border. The United States also encompasses North America's northwestern extension, Alaska. (Offshore Hawai'i, however, belongs in the Pacific Realm.)

Demographic contrasts between the United States and Canada are substantial. The United States is the world's third most populous country with 297.1 million in 2006; Canada ranks 36th with 32.2 million. Comparatively small as its population may be, however, Canada has varied cultures and traditions ranging from multilingualism (French has equal status with English) to ethnic diversity that, as we will note later, have strong spatial expression. Thus Canada shares with the United States a characteristic feature of this realm: the prevalence of **cultural pluralism**. In 1
the United States, this is exhibited by a range of ethnic backgrounds from European and African to Asian and Native American. And while the civil rights movement and other initiatives have eliminated *de jure* (legal) segregation in the United States, ethnic clustering and *de facto* (real-world) self-segregation continue to mark the social landscape. In part, this is due to a troubling aspect of this wealthiest of all realms: poverty and deprivation still afflict a substantial minority as income gaps between the successful rich and the struggling poor continue to widen.

NORTH AMERICA'S PHYSICAL GEOGRAPHY

Before we examine the human geography of the United States and Canada more closely, we need to consider the physical setting in which they are rooted. The North

The photo at the left shows the heart of the central business district (CBD) of Houston, Texas, the fourth-largest U.S. city and one of the leading centers in the nation's energy industry.The central city's skyscraper-dominated skyline originated during the era of industrial urbanization that peaked in the mid-twentieth century. . . . But we now live in the twenty-first century, and the postindustrial transformation of urban America is all but complete. The landscape of the latter is symbolized by the low-rise world headquarters of IBM, completed in 1997, the centerpiece of a forested suburban office campus located in Armonk, New York about 30 miles north of Times Square. This is actually IBM's second headquarters facility on the site, the company having first moved out of Manhattan to this suburban Westchester County community nearly 50 years ago. These scenes represent the most far-reaching change in the history of the American city, shaped by the forces of the past four decades that turned it inside-out as well as replacing its single-centered with a multi-centered spatial structure. *left*: © Alan Schein Photography/Corbis Images. *right*: © Kohn Pedersen Fox Associates.

American continent extends from the Arctic Ocean to Panama (Fig. 3-1), but we will confine ourselves here to the territory north of Mexico—a geographic realm that still stretches from the near-tropical latitudes of southern Florida and Texas to subpolar Alaska and Canada's far-flung northern periphery. The remainder of the North American continent comprises a separate realm, *Middle America*, which is covered in Chapter 4.

Physiography

North America's physiography is characterized by its clear, well-defined division into physically homogeneous
2 regions called **physiographic provinces**. Each region is marked by considerable uniformity in relief, climate, vegetation, soils, and other environmental conditions, resulting in a scenic sameness that comes readily to mind. For example, we identify such regions when we refer to the Rocky Mountains, the Great Plains, and the Appalachian Highlands. However, not all the physiographic provinces of North America are so easily delineated.

Figure 3-2 maps the complete layout of the continent's physiography and includes a cross-sectional terrain profile along the 40th parallel. The most obvious aspect of this map of North America's physiographic provinces is the north-south alignment of the continent's great mountain backbone, the Rocky Mountains, whose rugged topography dominates the western segment of the continent from Alaska to New Mexico. The major feature of eastern North America is another, much lower chain of mountain ranges called the Appalachian Highlands, which also trend approximately north-south and extend from Canada's Atlantic Provinces to Alabama. The orientation of the Rockies and Appalachians is important because, unlike Europe's Alps, they do not form a topographic barrier to polar or tropical air masses flowing southward or northward, respectively, across the continent's interior.

Between the Rocky Mountains and the Appalachians lie North America's vast interior plains, which extend from the Mackenzie Delta on the Arctic Ocean to the Gulf of Mexico. We can subdivide these into several provinces: (1) the great Canadian Shield, which is the geologic core area containing North America's oldest rocks; (2) the Interior Lowlands, covered largely by glacial debris laid down by ice, meltwater, and wind during the Pleistocene glaciation; and (3) the Great Plains, the extensive sedimentary surface that rises gently westward toward the Rocky Mountains. Along the southern margin, these interior plainlands merge into the Gulf-Atlantic Coastal Plain, which extends from southern Texas along the seaward margin of the Appalachian Highlands and the neighboring Piedmont until it ends at Long Island just to the east of New York City.

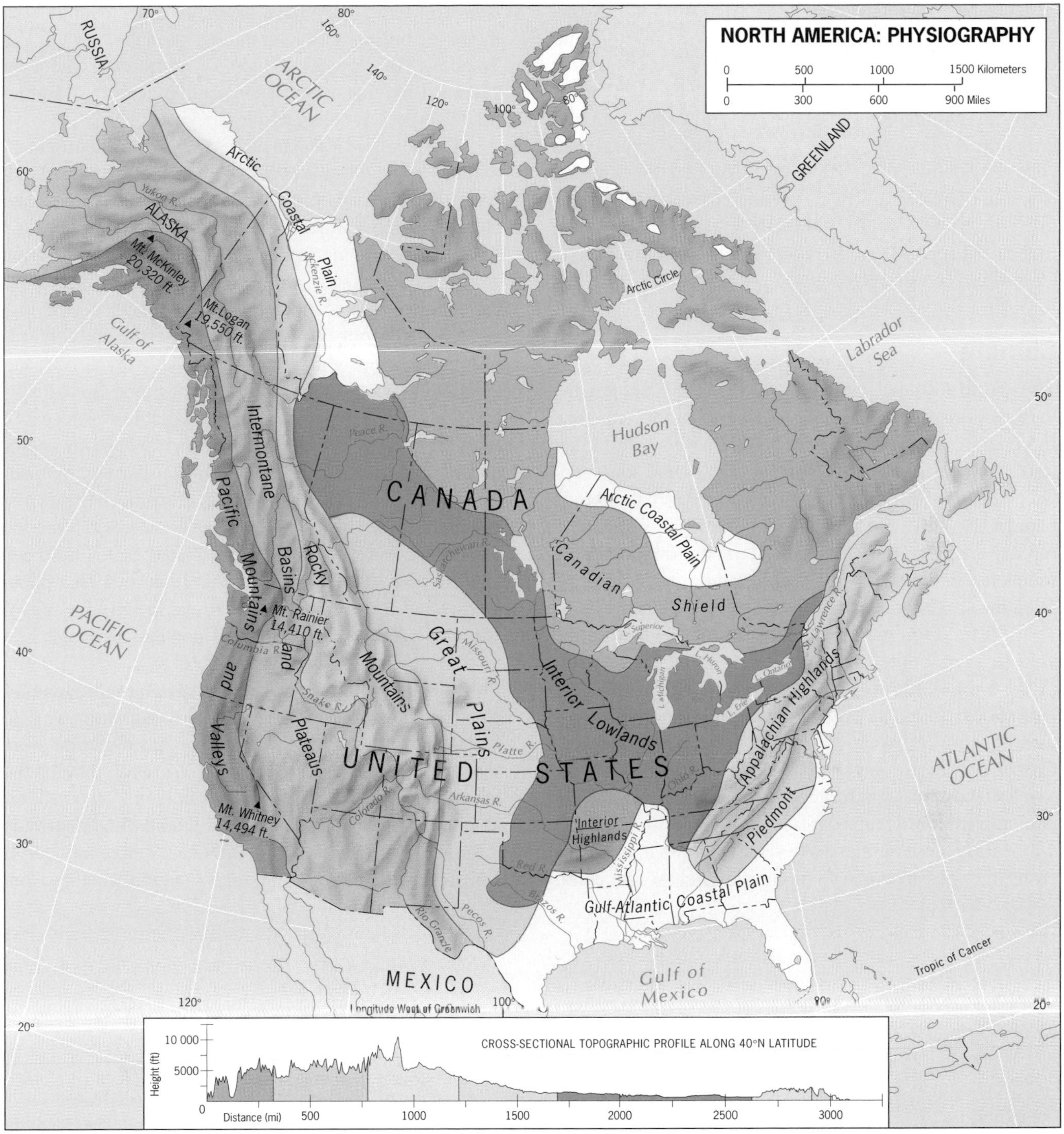

FIGURE 3-2 © H. J. de Blij, P. O. Muller, and John Wiley & Sons, Inc.

On the western side of the Rocky Mountains lies the zone of Intermontane Basins and Plateaus. Within the conterminous United States, this physiographic province includes: (1) the Colorado Plateau in the south, with its thick sediments and spectacular Grand Canyon; (2) the lava-covered Columbia Plateau in the north, which forms the watershed of the Columbia River; and (3) the central Basin-and-Range country (Great Basin) of Nevada and Utah, which contains several extinct lakes from the glacial period as well as the surviving Great Salt Lake. This province is called *intermontane* because of its position between the Rocky Mountains to the east and the Pacific coast mountain system to the west.

From the Alaskan Peninsula to Southern California, the west coast of North America is dominated by an almost unbroken corridor of high mountain ranges that originated from the contact between the North American and Pacific Plates (Fig. G-4). The major components of this coastal mountain belt include California's Sierra Nevada, the Cascades of Oregon and Washington, and the long chain of highland massifs that line the British Columbia and southern Alaska coasts. Three broad valleys—which contain dense populations—are the only noteworthy interruptions: California's Central (San Joaquin-Sacramento) Valley; the Cowlitz-Puget Sound lowland of Washington State, which extends southward into western Oregon's Willamette Valley; and the lower Fraser Valley, which slices through southern British Columbia's coast range.

Climate

The world climate map (Fig. G-8) clearly depicts the various climatic regimes and regions of North America. In general, temperature varies latitudinally—the farther north one goes, the cooler it gets. Regional land-and-water-heating differentials, however, distort this broad pattern. Because land surfaces heat and cool far more rapidly than water bodies, yearly temperature ranges are much larger where *continentality* (interior remoteness from the sea) is greatest.

Precipitation generally tends to decline toward the west (except for the Pacific coastal strip itself) as a result
3 of the **rain shadow effect**. This occurs because Pacific air masses, driven by prevailing winds, carry their moisture onshore but soon collide with the Sierra Nevada-Cascades wall, forcing them to rise—and cool—in order to crest those mountain ranges. Such cooling is accompanied by major condensation and precipitation, so that by the time these air masses descend and warm along the eastern slopes to begin their journey across the continent's interior, they have already deposited much of their moisture. Thus the mountains produce a downwind "shadow" of dryness, which is reinforced whenever eastward-moving air must surmount other ranges farther inland, especially the massive Rockies.

This semiarid—and in places truly arid—environment extends so deeply into the central United States that a broad division can be made between Arid (western) and Humid (eastern) America, which face each other along a fuzzy boundary that is best viewed as a wide transition zone. Although the separating criterion of 20 inches (50 cm) of annual precipitation is easily mapped (see Fig. G-7), that generally north-south *isohyet* (the line connecting all places receiving exactly 20 inches per year) can and does swing widely across the drought-prone Great Plains from year to year because highly variable warm-season rains from the Gulf of Mexico come and go in unpredictable fashion. Drought is indeed a constant environmental hazard, and Arid America today remains in the grip of one of the most severe dry periods ever recorded. Since the late 1990s, most of the western United States has been challenged by the shrinkage of its reservoirs (see photo below), the dessication of its soils, and the proliferation of wildfires—all at a time when its population growth and land development continue to mushroom.

Precipitation in Humid America is far more regular. The prevailing westerly winds (blowing from west to east—winds are always named for the direction *from* which they come), which usually come up dry for the large zone west of the 100th meridian, pick up considerable moisture over the Interior Lowlands, and distribute it throughout eastern North America. A large number of storms develop here on the highly active weather front

This stretch of the Colorado River lies on the Colorado Plateau in southeastern Utah, and is known as Lake Powell. This elongated lake, named after the geographer-explorer who led the first expedition through the nearby Grand Canyon in 1869, formed during the 1960s as the reservoir behind newly completed Glen Canyon Dam. What is unusual about this photo (taken in early 2005) is that it shows Lake Powell at its lowest level in 40 years, dozens of feet below its usual surface elevation. This is the local expression of Arid America's prolonged drought, which began in the mid-1990s. Although the winter of 2004-2005 produced a massive snow pack in the central Rockies where the Colorado River rises, its seasonal melting does not guarantee an end to the long dry spell. Droughts are multi-year phenomena, and eliminating the water deficit will take at least one more season of ample precipitation in the upper Colorado Basin.

between tropical Gulf air to the south and polar air to the north. Even if major storms do not materialize, local weather disturbances created by sharply contrasting temperature differences are always a danger. There are more tornadoes (nature's most violent weather) in the central United States each year than anywhere else on Earth. And in winter, the northern half of this region receives large amounts of snow, particularly around the Great Lakes.

Figure G-8 shows the absence of humid temperate (*C*) climates from Canada (except along the narrow Pacific coastal zone) and the prevalence of cold in Canadian environments. East of the Rocky Mountains, Canada's most *moderate* climates correspond to the *coldest* of the United States. Nonetheless, southern Canada shares the environmental conditions that mark the Upper Midwest and Great Lakes areas of the United States, so that agricultural productivity in the Prairie Provinces and in Ontario is substantial. Canada is a leading food exporter (chiefly wheat), as is the United States, despite its comparatively short growing season.

The broad environmental partitioning into Humid and Arid America is also reflected in the distribution of the realm's soils and vegetation. For farming purposes there is usually sufficient soil moisture to support crops where annual precipitation exceeds the critical 20 inches; where the yearly total is less, soils may still be fertile (especially in the Great Plains), but irrigation is often necessary to achieve their full agricultural potential. As for vegetation, the Humid/Arid America dichotomy is again a valid generalization: the natural vegetation of areas receiving more than 20 inches of water annually is *forest*, whereas the drier climates give rise to a *grassland* cover.

Hydrography (Surface Water)

Surface water patterns in North America are dominated by the two major drainage systems that lie between the Rockies and the Appalachians: (1) the five Great Lakes (Superior, Michigan, Huron, Erie, and Ontario) that drain into the St. Lawrence River, and (2) the mighty Mississippi-Missouri river network, fed by such major tributaries as the Ohio, Tennessee, and Arkansas rivers. Both are products of the last episode of Pleistocene glaciation, and together they amount to nothing less than the best natural inland waterway system in the world. Human intervention has further enhanced this network of navigability, mainly through the building of canals that link the two systems as well as the St. Lawrence Seaway.

Elsewhere, the northern east coast of the continent is well served by a number of short rivers leading inland from the Atlantic. In fact, many of the major northeastern seaboard cities of the United States—such as Washington, D.C., Baltimore, and Philadelphia—are located at the waterfalls that marked the limit to tidewater navigation (hence their designation as *Fall Line cities*). Rivers in the Southeast and west of the Rockies at first offered little practical value because of their orientation and the difficulty of navigating them. In the far west, however, the Colorado and Columbia rivers have become supremely important as suppliers of drinking and irrigation water as well as hydroelectric power.

INDIGENOUS NORTH AMERICA

When the first Europeans set foot on North American soil, the continent was occupied by millions of people whose ancestors had reached the Americas from Asia, via Alaska and probably also across the Pacific, more than 13,000 years before (and possibly as long as 30,000 years ago). In search of Asia, the Europeans misnamed them "Indians," but the historic affinities of these earliest Americans were with the peoples of eastern and northeastern Asia, not India. In North America these *Native Americans* or *First Nations*—as they are now called in the United States and Canada, respectively—had organized themselves into hundreds of nations with a rich mosaic of languages and a great diversity of cultures (Fig. 3-3). Farmers grew crops that the Europeans had never seen; other nations depended chiefly on fishing, herding, hunting, or some combination of these. Elaborate houses, efficient watercraft, effective weaponry, decorative clothing, and wide-ranging art forms distinguished the aboriginal nations. Certain nations had formulated sophisticated health and medical practices; ceremonial life was complex and highly developed; and political institutions were mature and elaborate.

The eastern nations were the first to bear the brunt of the European invasion. By the end of the eighteenth century, ruthless, land-hungry settlers had driven most of the Native American peoples living along the Atlantic and Gulf coasts from their homes and lands, beginning a westward push that was to devastate indigenous society. The U.S. Congress in 1789 proclaimed that "Indian . . . land and property shall never be taken from them without their consent," but in fact this is just what happened. One of the sorriest episodes in American history involved the removal of the eastern Cherokee, Chickasaw, Choctaw, Creek, and Seminole from their homelands in forced marches a thousand miles westward to Oklahoma. One-fourth of the entire Cherokee population died along the way from exposure, starvation, and disease, and the others fared little better. Again, Congress approved treaties that would at least protect the native peoples of the

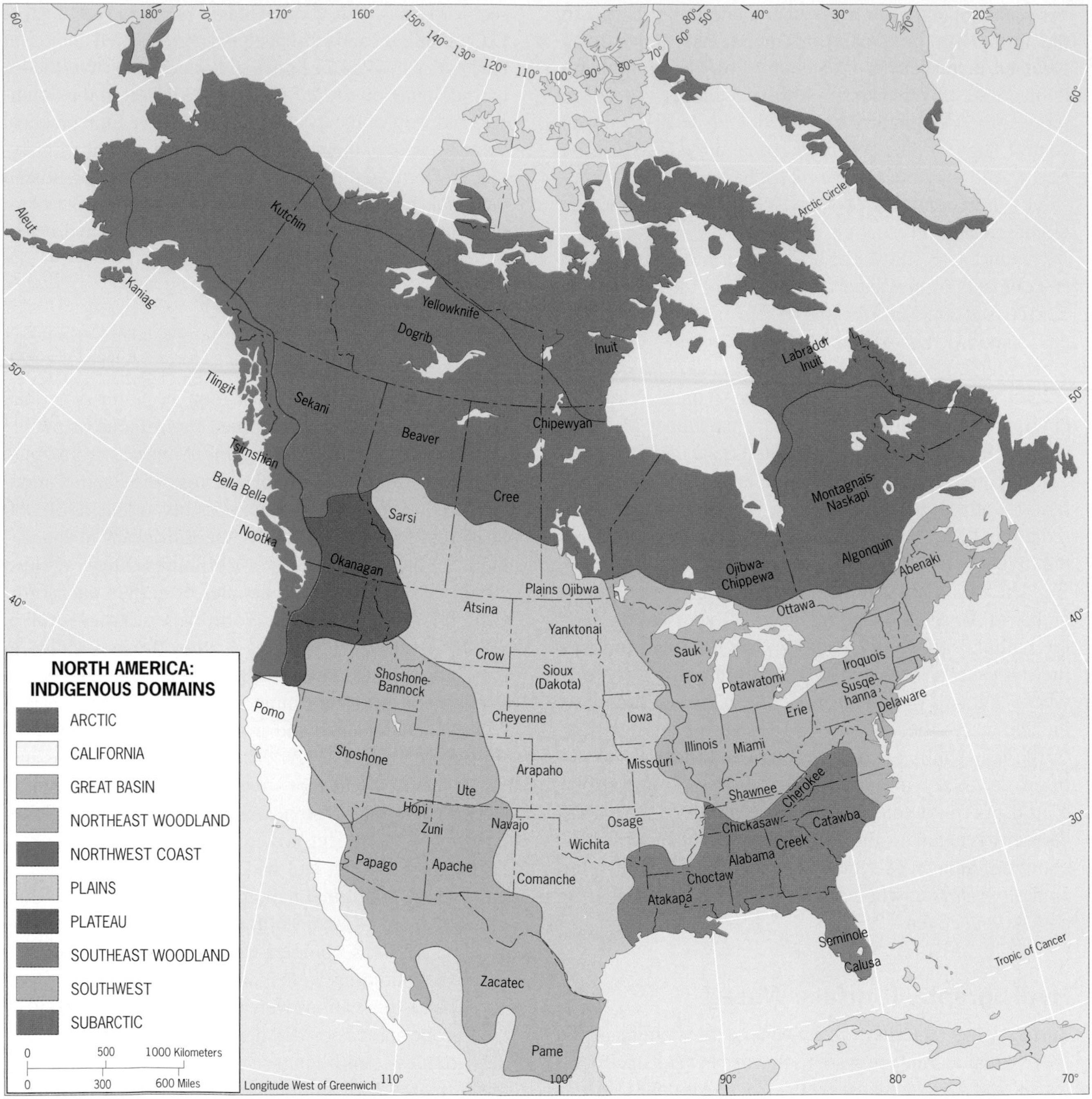

FIGURE 3-3 © H. J. de Blij, P. O. Muller, and John Wiley & Sons, Inc.

Plains (Fig. 3-3) and those farther to the west, but after the mid-nineteenth century the white settlers ignored those guarantees as well. A half-century of war left what remained of North America's nations with about 4 percent of U.S. territory in the form of mostly impoverished reservations.

In what is today Canada, too, the comparatively small First Nations population was overwhelmed by the numbers and power of European settlers, and decimated by the diseases they introduced. Efforts at restitution and recognition of First Nations rights, however, have gone farther in Canada than in the United States.

As we noted earlier, two large countries, with geographic similarities as well as differences, constitute the North American realm. Canada and the United States share physiographic provinces as well as border-straddling economic and cultural subregions (for example, large-scale grain cultivation in the Great Plains). It would be possible to base the regionalization of this realm on contrasts between the two states; but as we will see in the second part of this chapter, quite a different set of regions emerges from our geographic analysis. To fully appreciate this regional framework, we must examine in some detail the changing human geography of the United States and Canada individually.

THE UNITED STATES

As we note throughout this book, the administrative components of states are assuming ever-greater importance in the scheme of things. In Europe, all the major states have reorganized their political-economic systems, forged regions from smaller units, and devolved power to the provinces. As a result, it is increasingly important not just to know Spain but also Catalonia, not just France but also Rhône-Alpes, not just Italy but also Lombardy. So it is in the United States. We should have a clear mental map of the functional layout of this federation, which is why it may be helpful to take a few moments to review Figure 3-4.

Population in Time and Space

The current population distribution of the United States is shown in Figure 3-5. It is important to note that this map is the latest "still" in a motion picture, one that has been unreeling for four centuries since the founding of the first permanent European settlement on the northeastern coast. Slowly at first, then with accelerating speed after 1800, as one major transportation breakthrough followed another,

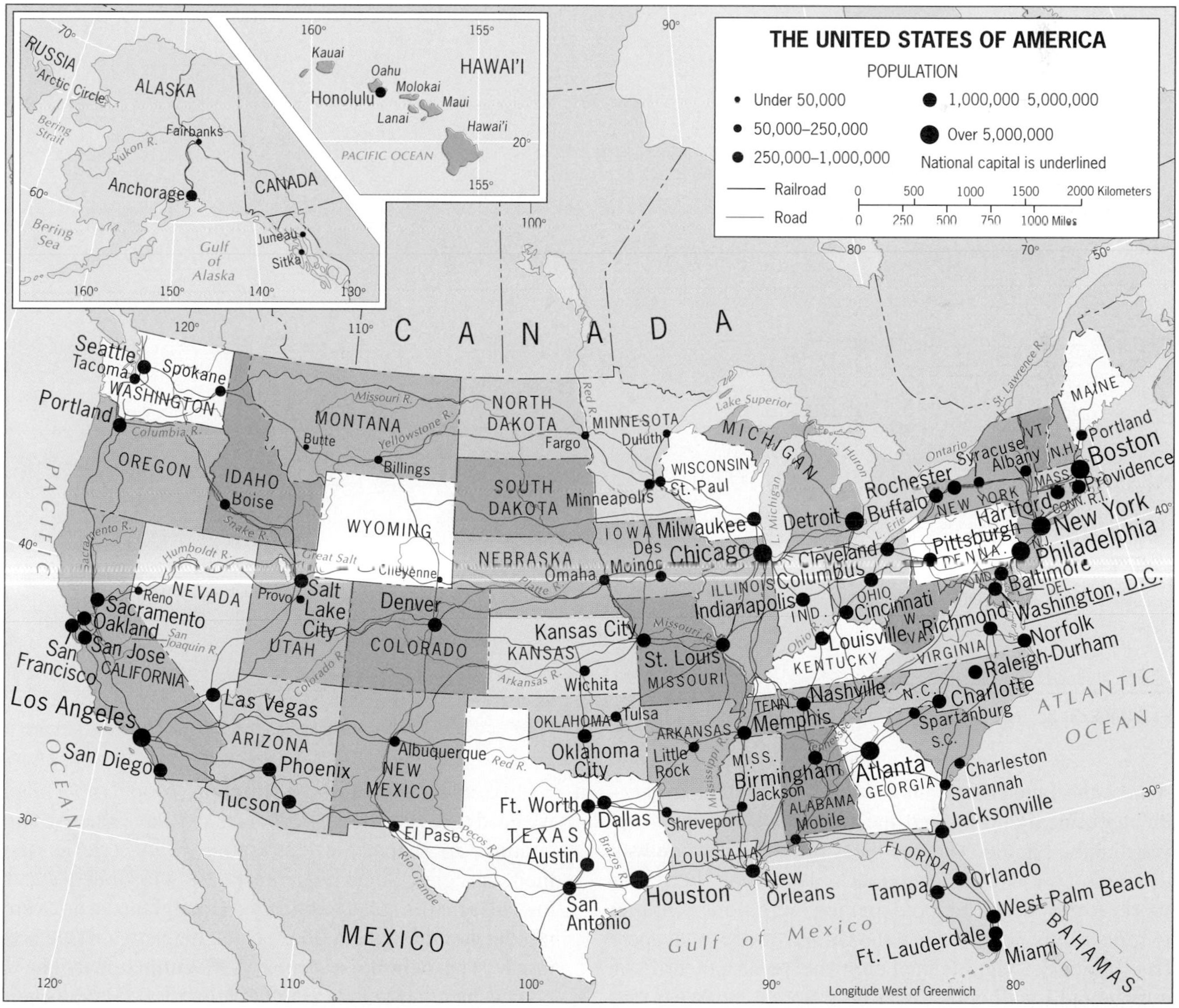

FIGURE 3-4 © H. J. de Blij, P. O. Muller, and John Wiley & Sons, Inc.

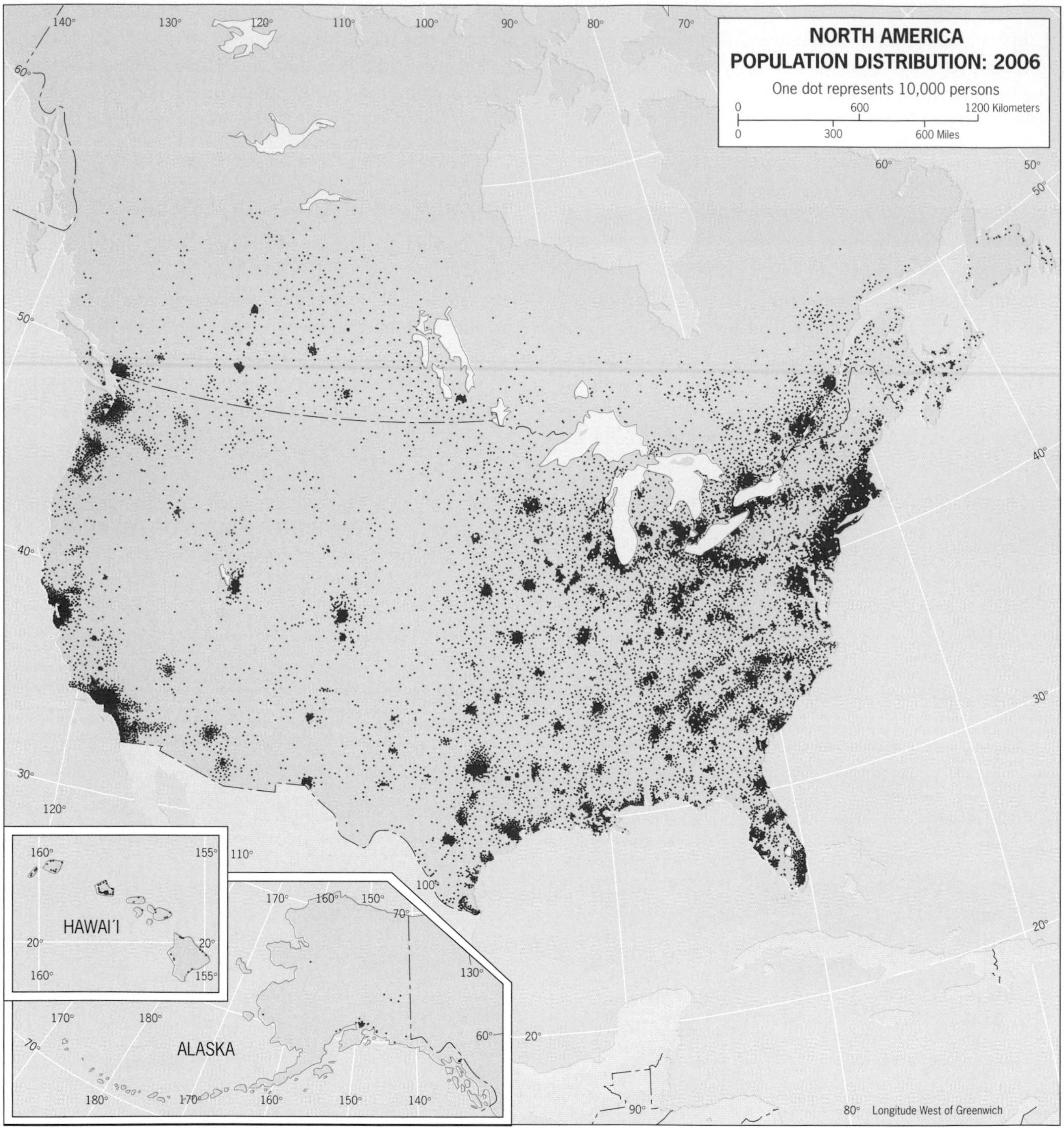

FIGURE 3-5 © H. J. de Blij, P. O. Muller, and John Wiley & Sons, Inc.

Americans (and Canadians) took charge of their remarkable continent and pushed the settlement frontier westward to the Pacific. The swiftness of this expansion was dramatic, but North Americans have long been the world's most mobile people. In fact, migrations continue to redistribute population in the United States today, perhaps the most significant being the persistent drift of people and livelihoods toward the South and West (the
4 so-called **Sunbelt**).

To understand the contemporary population map, we need to review the major forces that have shaped, and continue to shape, the distribution of Americans and their activities. Since its earliest days, the United States has attracted a steady influx of immigrants who were rapidly assimilated into the societal mainstream (see box titled "The Migration Process"). Within the country, people have sorted themselves out to maximize their proximity to existing economic opportunities, and they

The Migration Process

The United States as well as Canada is the product of
5 **migration** (a change in residential location intended to be permanent). After tens of thousands of years of native settlement, European explorers reached the shores of North America beginning with Columbus in 1492 (and probably centuries before that). The first permanent colonies were established in the early 1600s, and from them evolved the modern United States of America. The Europeanization of North America doomed the continent's aboriginal societies, but this was only one of many areas around the world where local cultures and foreign invaders came face to face. Between 1835 and 1935, perhaps as many as 75 million Europeans departed for distant shores—most of them bound for the Americas (Fig. 3-6). Some sought religious freedom; others escaped poverty and famine; still others simply hoped for a better life.

Studies of the *migration decision* indicate that migration flows vary in size with: (1) the perceived degree of difference between one's home, or source, and the destination; (2) the effectiveness of the information flow, that is, the news about the destination that migrants sent to those who stayed behind waiting to decide; and (3) the distance between the source and the destination (shorter moves attract many more migrants than longer ones). More than a century ago, the British social scientist Ernst Georg Ravenstein studied the migration process, and many of his conclusions remain valid today. For example, every migration stream from source to destination produces a counter-stream of returning migrants who cannot adjust, are unsuccessful, or are otherwise persuaded to go back home. Studies of migration also conclude that
several factors are at work in the migration process. **Push** 6
factors motivate people to move away; **pull factors** attract them to new destinations. To those early Europeans, the United States was a new frontier, a place where one might acquire a piece of land and some livestock. The opportunities were reported to be unlimited.

That perception has never changed, and immigration continues to significantly shape the human-geographic complexion of the United States. Today's inmigrants account for more than one-third of the country's annual population growth. Important details about this latest stage of the international migration process are discussed on pages 160–161.

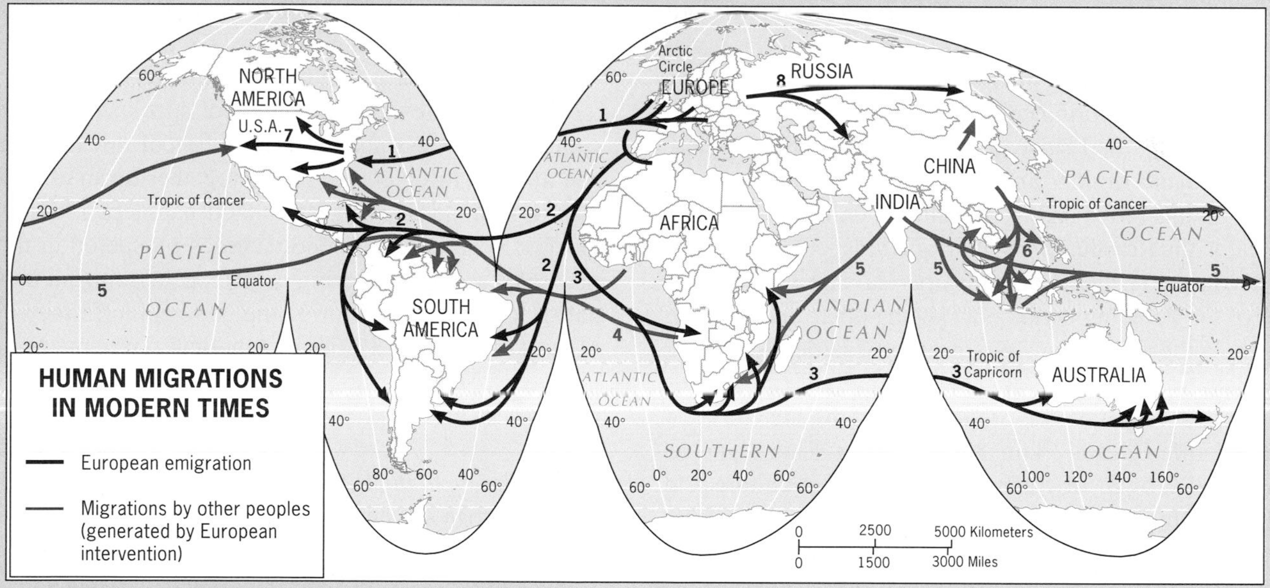

FIGURE 3-6 © H. J. de Blij, P. O. Muller, and John Wiley & Sons, Inc.

have shown little resistance to relocating as the nation's evolving economic geography has successively favored different sets of places over time.

During the past century these transformations spawned a number of major migrations: (1) the still-continuing westward shift but now with that southerly, Sunbelt deflection; (2) the rapid growth of metropolitan areas, first triggered by the late-nineteenth-century Industrial Revolution, which since the 1960s has been largely rechanneled from the central cities to the suburban ring; (3) the movement of African Americans from the rural South to the urban North, which since the 1970s has become a stronger return flow, particularly among middle-income blacks; and (4) the influx of immigrants from outside the United States, mostly European before 1960 but now dominated by a new wave from Middle America and eastern Asia that is directed, respectively, toward the States near the southern border and the urban areas of the Pacific coast.

Let us now look more closely at the historical geography of these changing population patterns, considering first the initial rural influence and then the decisive impacts of industrial urbanization.

Colonial and Nineteenth-Century Development

The current spatial distribution of the U.S. population is rooted in the colonial era of the seventeenth and eighteenth centuries that was dominated by England and France. The French sought mainly to organize a lucrative fur-trading network, while the English established settlements along the coast of what is today the northeastern U.S. seaboard (the oldest, Virginia's Jamestown, was founded almost exactly 400 years ago). These British colonies quickly became differentiated in their local economies, a diversity that was to endure and later shape American cultural geography. The northern colony of New England (Massachusetts Bay and environs) specialized in commerce; the southern Chesapeake Bay colony (Tidewater Virginia and Maryland) emphasized the plantation farming of tobacco; the Middle Atlantic area lying in between (southeastern New York, New Jersey, eastern Pennsylvania) was home to a number of smaller, independent-farmer colonies.

These neighboring colonies soon thrived and yearned to expand, but the British government responded by closing the inland frontier and tightening economic controls. By 1783 this move had led to colonial unification, British defeat in the Revolutionary War, and independence for the newly formed United States of America. The western frontier of the fledgling nation now swung open, and the zone north of the Ohio River was promptly settled following the discovery that the soils (and climate) of the Interior Lowlands were more favorable for farming than those of the Atlantic Coastal Plain and Piedmont. This triggered the rapid growth of trans-Appalachian agriculture and the widening of seaboard-interior trading ties. The new interregional complementarities signified that U.S. spatial organization was assuming national-scale proportions.

By the time the westward-moving frontier swept across the Mississippi Valley in the 1820s, the three former seaboard colonies (**A**, **B**, and **C** in Fig. 3-7) had become separate *culture hearths*—primary source areas and innovation centers from which migrants carried cultural traditions into the central United States (as the arrows in Fig. 3-7 indicate). The northern half of this vast interior space soon became well unified as its infrastructure steadily improved following the introduction of the railroad in the 1830s. The American South, however, did not wish to integrate itself economically with the North, preferring to export tobacco and cotton from its plantations to overseas markets; its insistence on preserving slavery to support this system soon led the South into secession, disastrous Civil War (1861–1865), and a dismal aftermath that took a full century to overcome.

The second half of the nineteenth century saw the frontier cross the western United States, and by 1869 agriculturally booming California was linked to the rest of the nation by transcontinental railroad (these same steel tracks also opened up the bypassed, semiarid Great Plains). When the American frontier closed in the 1890s, today's rural settlement pattern was firmly in place, anchored to a set of enduring national agricultural regions (discussed later in this chapter). By then, however, the exodus of rural Americans toward the burgeoning cities had begun in response to the Industrial Revolution that had taken hold after 1870.

Post–1900 Industrial Urbanization

The Industrial Revolution occurred almost a century later in the United States than in Europe, but when it finally did cross the Atlantic in the 1870s, it took hold so successfully and advanced so robustly that only 50 years later America was surpassing Europe as the world's mightiest industrial power. The impact of industrial urbanization occurred simultaneously at two levels of spatial generalization. At the national or *macroscale*, a system of new cities swiftly emerged, specializing in the collection, processing, and distribution of raw materials and manufactured goods, linked together by an efficient web of railroad lines. Within that urban network, at the local or *microscale*, individual cities prospered in their new roles as manufacturing centers, generating an internal structure that still forms the geographic framework of most of the central cities of America's large metropolitan areas. We now examine the urban trend at both of these scales.

Evolution of the U.S. Urban System. The rise of the national urban system in the late nineteenth century was based on the traditional external role of cities: providing goods and services for their hinterlands in exchange for raw materials. Because people (both as laborers and as consumers), commercial activities, investment capital, and transport facilities were already agglomerated in existing (preindustrial) cities, the emerging industrialization movement tended to favor such locations. Their growing incomes, in turn, permitted industrially intensifying cities to invest in a bigger local infrastructure of private and public services as well as housing—and thereby convert each round of industrial expansion into a new stage of urban development.

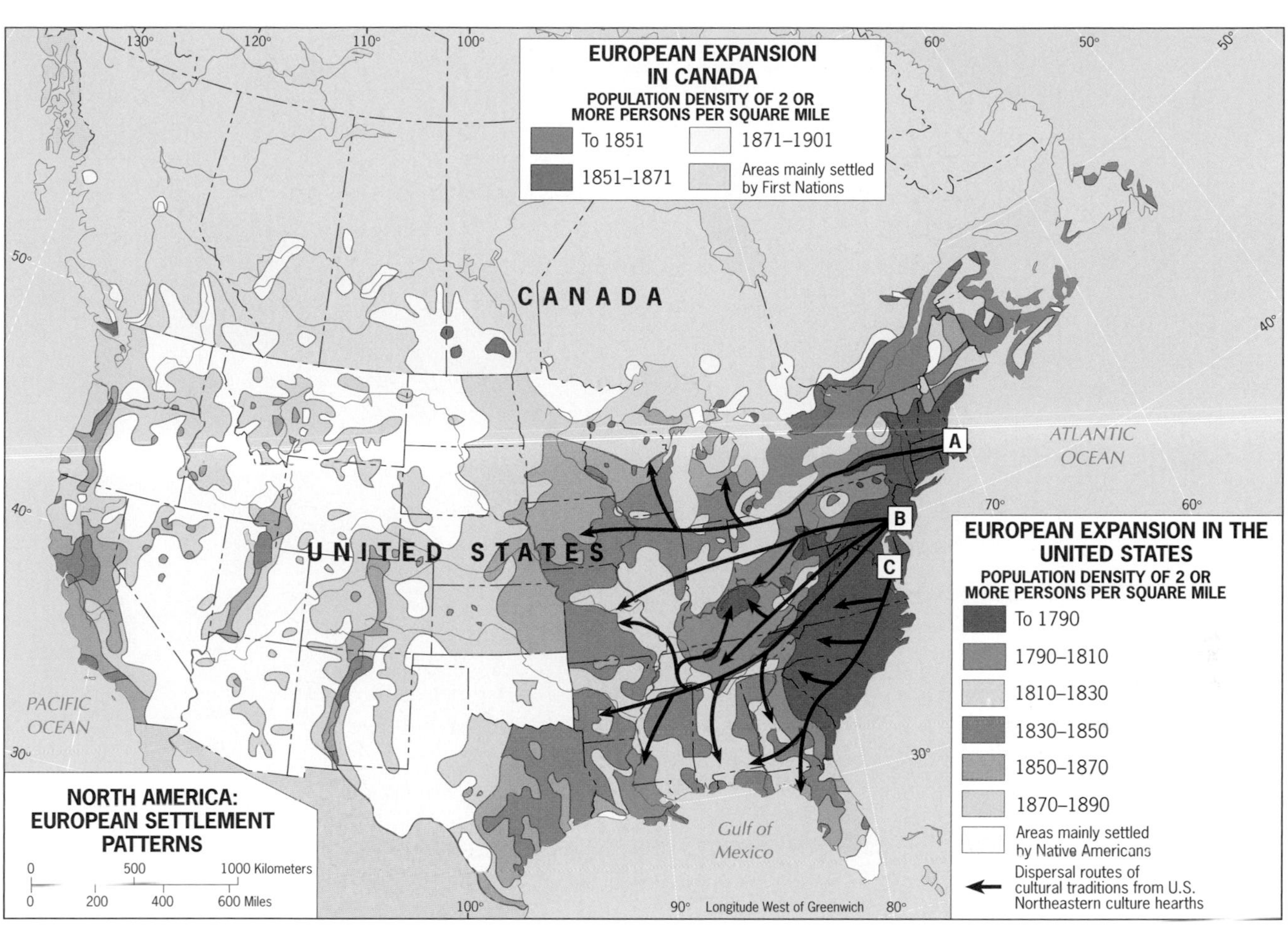

FIGURE 3-7 © H. J. de Blij, P. O. Muller, and John Wiley & Sons, Inc.

The national urban system had actually been in the process of formation long before its flowering after 1870. Its evolution over the past two centuries has been studied by John Borchert, who identified five epochs of metropolitan evolution based on transportation technology and industrial energy. (1) *The Sail-Wagon Epoch* (1790–1830), marked by primitive overland and waterway circulation; the leading cities were the northeastern ports, which were more heavily oriented to the European overseas trade than to their barely accessible western hinterlands. (2) *The Iron Horse Epoch* (1830–1870), dominated by the arrival and spread of the steam-powered railroad; a nationwide transport system had now been forged, and the national urban system began to take shape, with New York emerging as the primate city by 1850. (3) *The Steel-Rail Epoch* (1870–1920), which spanned the Industrial Revolution and saw the full establishment of the national metropolitan system; the new forces shaping growth were the increasing scale of manufacturing, the rise of the steel and auto industries in Midwestern cities, and the introduction of steel rails that enabled trains to travel faster and haul heavier cargoes. (4) *The Auto-Air-Amenity Epoch* (1920–1970), which encompassed the later stage of U.S. industrial urbanization and the maturation of the national urban hierarchy; its key elements were the automobile and the airplane, the expansion of white-collar services jobs, and the growing locational pull of *amenities* (pleasant environments) that increasingly stimulated the urbanization of the suburbs and selected Sunbelt locales. (5) *The Satellite-Electronic-Jet Propulsion Epoch* (1970–), shaped by the newest advancements in information management, computer technologies, global communications, and intercontinental travel; it favors globally oriented metropolises (now often called *world cities*), particularly those along the Pacific and Atlantic coasts that function as international gateways.

Industrialization and the accompanying growth of the urban system reconfigured the realm's economic landscape. The most notable regional transformation was the emergence of the North American Core, or **American Manufacturing Belt**, which contained the lion's share of industrial activity in both the United States and Canada. 7
As Figure 3-8 shows, the geographic form of the Core Region—which includes southern Ontario—was a near-rectangle whose four corners were Boston, Milwaukee,

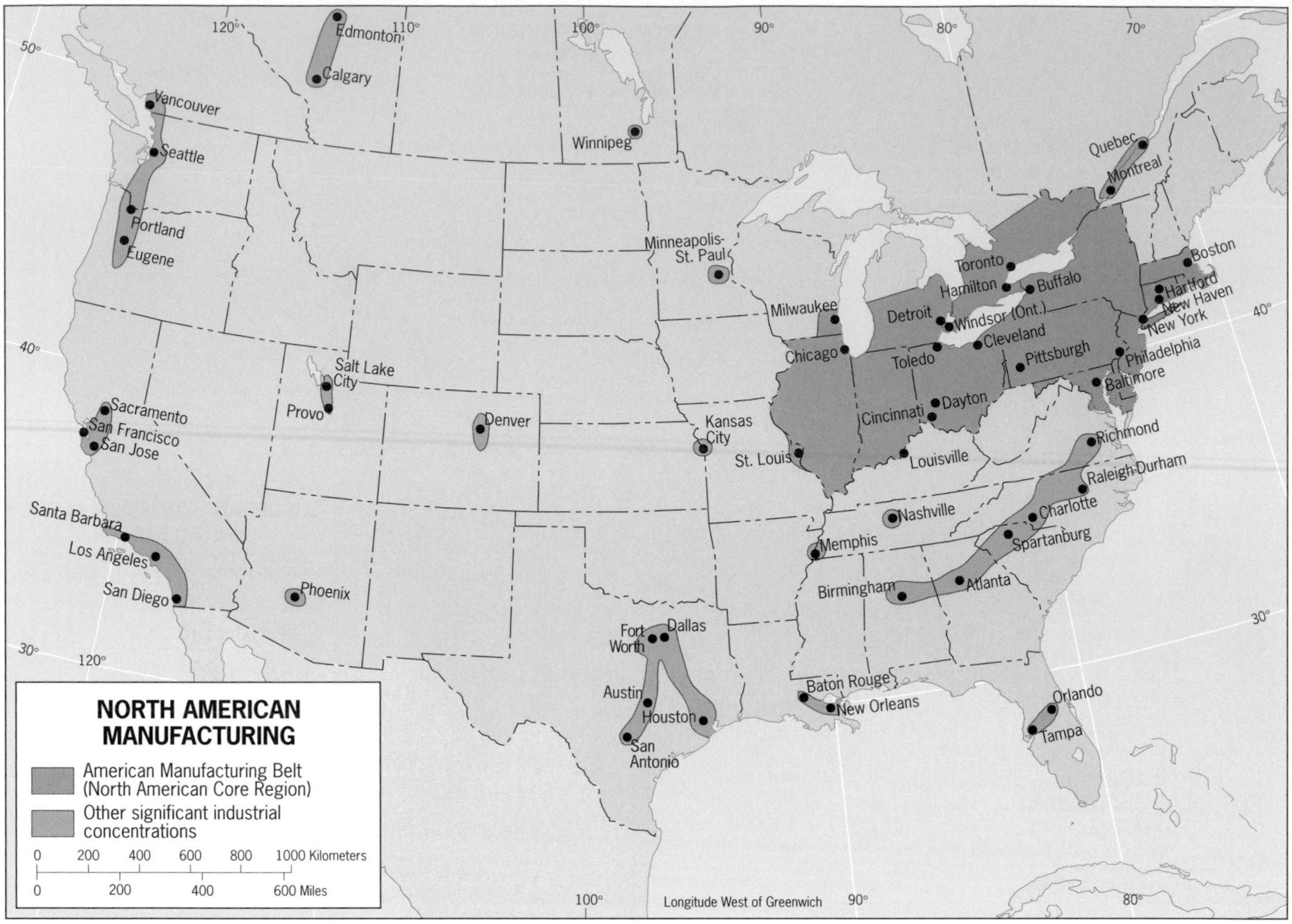

FIGURE 3-8 © H. J. de Blij, P. O. Muller, and John Wiley & Sons, Inc.

St. Louis, and Baltimore. However, because manufacturing is such a spatially concentrated activity, less than 1 percent of the territory of the Belt is devoted to industrial land use: most of its factories are tightly clustered into a dozen districts centered on the cities mapped in Figure 3-8.

At the subregional scale, as transportation breakthroughs permitted progressive urban decentralization and *megalopolitan growth*, the expanding peripheries of major cities soon coalesced to form a number of conurbations. The most important of these by far is the *Atlantic Seaboard Megalopolis* (Fig. 3-9), the 600-mile (1000-km) urbanized northeastern coastal strip extending from southern Maine to Virginia that contains metropolitan Boston, New York, Philadelphia, Baltimore, and Washington. This was the economic heartland of the Core; the seat of U.S. government, business, and culture; and the trans-Atlantic trading interface between much of North America and Europe. Six other primary conurbations have also emerged: *Lower Great Lakes* (Chicago–Detroit–Cleveland–Pittsburgh), *Piedmont* (Atlanta–Charlotte–Raleigh/Durham), *Florida* (Jacksonville–Tampa–Orlando–Miami), *Texas* (Houston–Dallas/Ft. Worth–San Antonio), *California* (San Diego–Los Angeles–San Francisco), and the *Pacific Northwest* (Portland–Seattle–Vancouver). Note that the last spills across the border into Canada, which has also spawned its own nationally predominant conurbation—*Main Street* (Windsor–Toronto–Montreal–Quebec City).

The Changing Structure of the U.S. Metropolis The internal structure of the metropolis reflected the same mixture of forces that shaped the national urban system, especially transportation technology. Rails—in this case, lighter street-rail lines—once again shaped spatial organization as horse-drawn trolleys were succeeded by electric streetcars in the late nineteenth century. The mass introduction of the automobile after World War I changed all that, and America steadily turned from building compact cities to the widely dispersed metropolises of the post–World War II highway era. By 1970, the new intraurban expressway network had equalized location costs throughout the metropolis, setting the stage for suburbia to swiftly transform itself from a residential preserve into a complete outer city

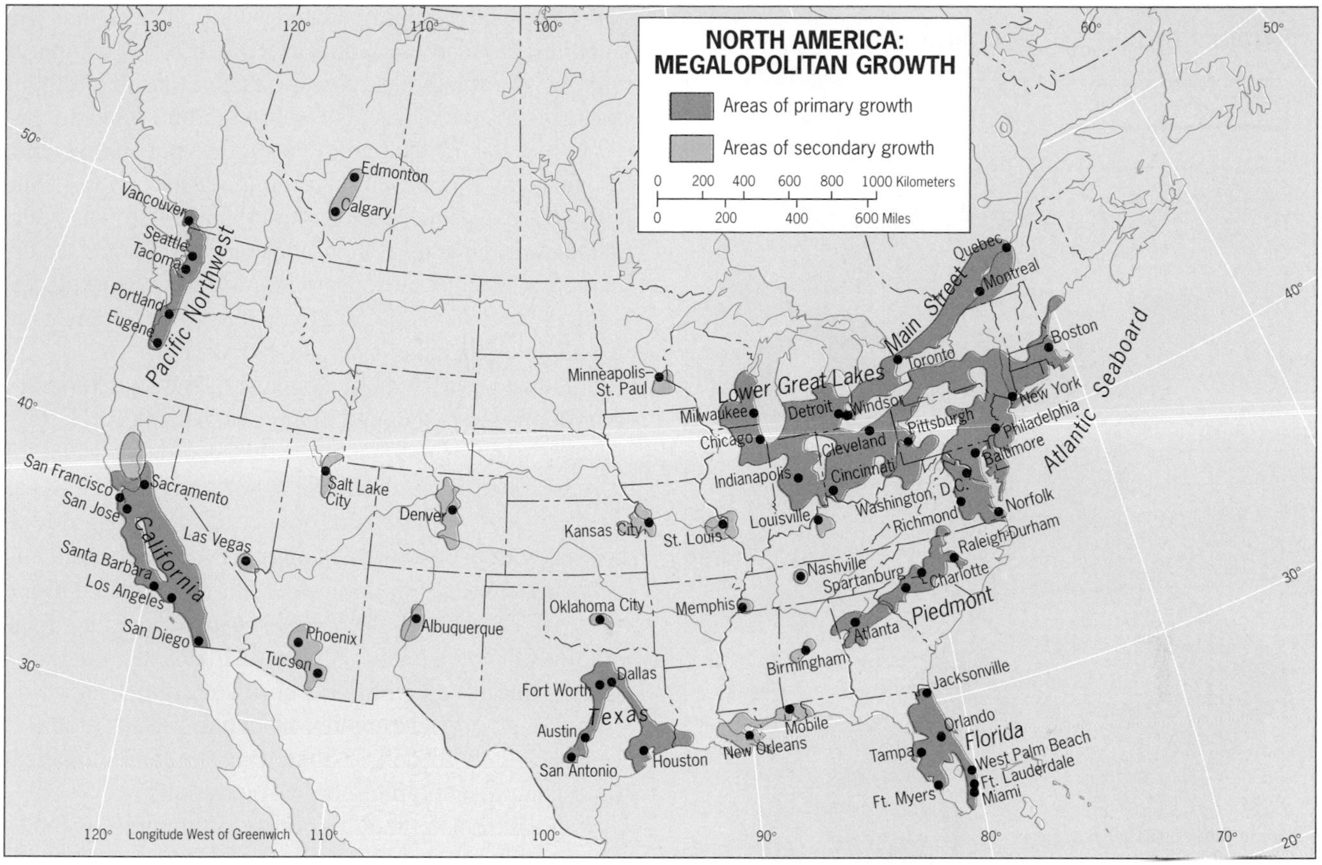

FIGURE 3-9 © H. J. de Blij, P. O. Muller, and John Wiley & Sons, Inc.

with amenities and new prestige that proved highly attractive to the business world. As the newly urbanized suburbs increasingly captured major economic activities, many large cities saw their status diminish to that of coequal. Their once thriving central business districts (CBDs) were now all but reduced to serving the less affluent populations that increasingly dominated the central city's neighborhoods.

This growth process was conceptualized into a four-stage model by John Adams, who identified the four **eras of intraurban structural evolution** (see Fig. 3-10). Stage I, prior to 1888, was the *Walking-Horsecar Era*, which produced a compact pedestrian city in which everything had to be within a 30-minute walk, a layout only slightly augmented when horse-drawn trolleys began to operate after 1850. The 1888 invention of the electric traction motor launched Stage II, the *Electric Streetcar Era* (1888–1920); higher speeds enabled the 30-minute travel radius and the urbanized area to expand considerably along new outlying trolley corridors; in the older core city, the CBD, industrial, and residential land uses differentiated into their modern form. Stage III, the *Recreational Automobile Era* (1920–1945), was marked by the initial impact of cars and highways that steadily improved the accessibility of the outer metropolitan ring, thereby launching a wave of mass suburbanization that further extended the urban frontier. During this era, the still-dominant central city experienced its economic peak and the partitioning of its residential space into neighborhoods sharply defined by income, ethnicity, and race. Stage IV, the *Freeway Era* (under way since 1945), saw the full impact of automobiles, with the metropolis turning inside-out as expressways pushed suburban development more than 30 miles (50 km) from the CBD.

The social geography of the evolving industrial metropolis has been marked by the development of a residential mosaic that exhibited the congregating of ever-more-specialized groups. The electric streetcar, which introduced "mass" transit that every urbanite could afford, allowed the heterogeneous, immigrant-dominated city population to sort itself into ethnically uniform neighborhoods. When the United States sharply curtailed foreign immigration in the 1920s, industrial managers discovered the large African American population of the rural South—increasingly unemployed there as cotton-related agriculture declined—and began to recruit these workers by the thousands for the factories of Manufacturing Belt cities. This influx had an immediate impact on the social

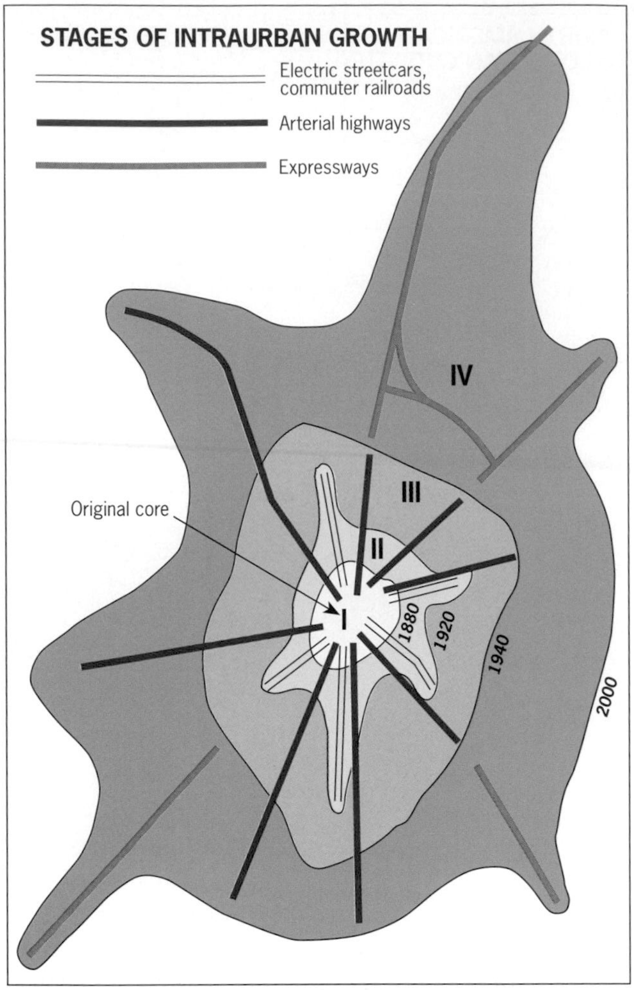

FIGURE 3-10 Adapted with permission from John S. Adams, "Residential Structure of Midwestern Cities," *Annals of the AAG*, Vol. 60 (1970): 56. © Association of American Geographers, 1970.

geography of the industrial city because whites were unwilling to share their living space with the racially different newcomers. The result was the involuntary segregation of these newest migrants, who were channeled into geographically separate, all-black areas. By the 1950s, these
8 mostly inner-city areas became large expanding **ghettos**, speeding the departure of many white central-city communities and reinforcing the trend toward a racially divided urban society.

The huge suburban component of the intraurban residential mosaic is not home just to the more affluent residents of the metropolis: over the past three decades it received so massive an infusion of nonresidential activities
9 that it was transformed into a full-fledged **outer city**. As its ties to the central city loosened, the outer city's growing independence was accelerated by the rise of major new suburban nuclei (particularly near key freeway interchanges) to serve the new local economies. These multipurpose activity nodes often developed around large regional shopping centers, whose prestigious images attracted scores of industrial parks, office campuses and high-rises, hotels, restaurants, entertainment facilities, and even major league sports stadiums and arenas, which together formed burgeoning new **suburban** 10
downtowns that are an automobile-age version of the CBD (see photo, p. 159). As suburban downtowns flourish, they attract tens of thousands of local residents to organize their lives around them—offering workplaces, shopping, leisure activities, and all the other elements of a complete urban environment.

These newest spatial elements of the contemporary metropolis are assembled in the model displayed in Figure 3-11. The rise of the outer city has produced a *multicentered* metropolis consisting of the traditional CBD as well as a set of increasingly coequal suburban downtowns, with each activity center serving a discrete and self-sufficient surrounding area. James Vance defined these tributary areas as *urban realms*, recognizing in his studies that each such realm maintains a separate, distinct economic, social, and political significance and
strength. Figure 3-12 applies the **urban realms model** to 11
Los Angeles; we could easily draw a similar regionalization scheme for other large U.S. metropolises.

The position of the central city within the new multinodal metropolis of realms is eroding. No longer the dominant metropolitanwide center for goods and services, the CBD increasingly serves the less affluent residents of the innermost realm and those working downtown. As manufacturing employment declined precipitously, many large cities adapted successfully by shifting toward service industries. Accompanying this switch is downtown commercial revitalization, but in many cities for each shining new skyscraper that goes up several old commercial buildings

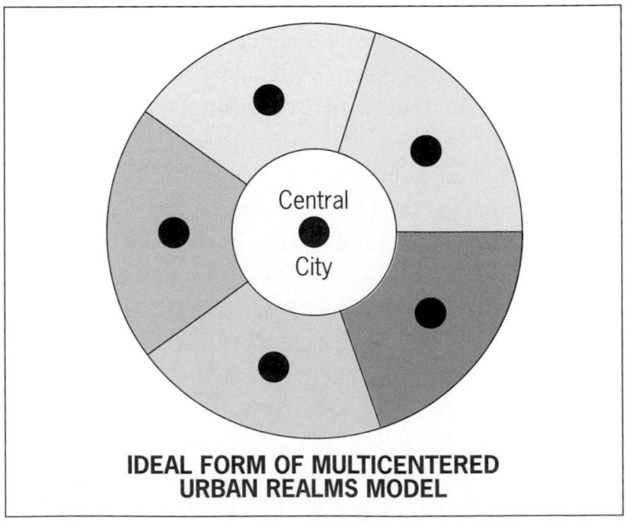

FIGURE 3-11 © H. J. de Blij, P. O. Muller, and John Wiley & Sons, Inc.

FROM THE FIELD NOTES

"Monitoring the urbanization of U.S. suburbs for the past three decades has brought us to Tyson's Corner, Virginia on many a field trip and data-gathering foray. It is now hard to recall from this late-1990s view that only 40 years ago this place was merely a near-rural crossroads. But as nearby Washington, D.C. steadily decentralized, *Tyson's* capitalized on its unparalleled regional accessibility (its Capital Beltway location at the intersection with the radial Dulles Airport Toll Road) to attract a seemingly endless parade of high-level retail facilities, office complexes, and a plethora of supporting commercial services. We would rank it today among the three largest and most influential suburban downtowns (or 'edge cities') in North America."

are abandoned. Residential reinvestment has also occurred in many downtown-area neighborhoods but usually requires the displacement of established lower-income residents, an emotional issue that has sparked many conflicts. Beyond the CBD zone, the vast inner city remains the problem-ridden domain of low- and moderate-income people, with most forced to reside in ghettos.

Cultural Geography

In the United States, over the past two centuries, the contributions of a wide spectrum of immigrant groups have shaped, and continue to shape, a rich and varied cultural complex. Great numbers of these newcomers were willing to set aside their original cultural baggage in favor of

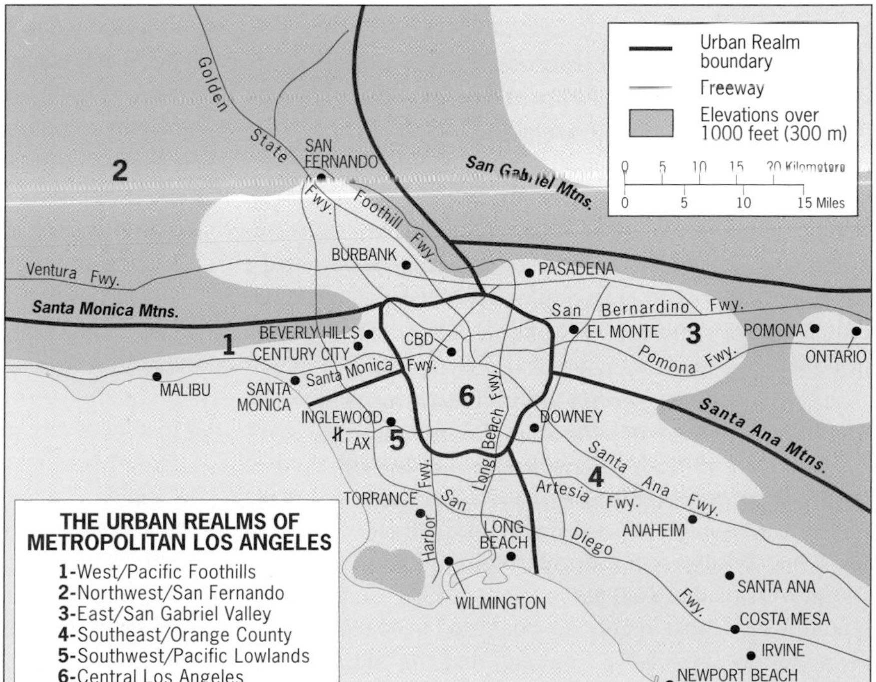

FIGURE 3-12 Adapted with permission from Peter O. Muller, *Contemporary Suburban America*, p. 10. © Prentice Hall, Inc., 1981.

assimilation into the emerging culture of their adopted homeland, which itself was a hybrid nurtured by constant infusions of new influences. For most upwardly mobile immigrants, this plunge into the much-touted "melting pot" promised a ticket for acceptance into mainstream American society.

American Cultural Bases

As the hybrid culture of the United States matured, it built on a set of powerful values and beliefs: (1) love of newness; (2) desire to be near nature; (3) freedom to move; (4) individualism; (5) societal acceptance; (6) aggressive pursuit of goals; and (7) a firm sense of destiny. Brian Berry has discerned these cultural traits in the behavior of people throughout the evolution of urban America. A "rural ideal" has prevailed throughout U.S. history and is still expressed in a strong bias against residing in cities. When industrialization made urban living unavoidable, those able to afford it soon moved to the emerging suburbs (*newness*) where a form of country life (*close to nature*) was possible in a semiurban setting. The fragmented metropolitan residential mosaic, composed of myriad slightly different neighborhoods, encouraged frequent *unencumbered mobility* as middle-class life revolved around the *individual* nuclear family's *aggressive pursuit* of its aspirations for *acceptance into the next higher stratum of society*. To most Americans these accomplishments confirmed that they could attain their goals through hard work and perseverance and that they had the ability to realize their *destiny* by achieving the "American Dream" of homeownership, affluence, and total satisfaction.

Language and Religion

Although linguistic variations play a far more important role in Canada, more than one-eighth of the U.S. population speaks a primary language other than English. Differences in English usage are also evident at the subnational level in the United States, where regional variations (*dialects*) can still be noted despite the recent trend toward a truly national society. The South and New England immediately come to mind as areas that still possess distinctive accents.

North America's Christian-dominated kaleidoscope of religious faiths contains important spatial variations. Many major Protestant denominations are clustered in particular regions, with Southern Baptists localized in the southeastern quadrant of the United States, Lutherans in the Upper Midwest and northern Great Plains, and Mormons west of the Rockies but focused on Utah. Roman Catholics are most visibly concentrated in Manufacturing Belt metropolises, New England, and the Mexican borderland zone. Judaism is the nation's most highly agglomerated major religious group, whose largest congregations are clustered in the suburbs of Megalopolis, Southern California, South Florida, and the Midwest.

Ethnic Patterns

Ethnicity (national ancestry) has always played a key role in American cultural geography. Today, whites of European background no longer dominate the increasingly diverse U.S. ethnic tapestry, with ethnics of color and non-European origin comprising more than one-third of the population—a proportion that demographers predict will rise to 50 percent by mid-century. In the late 1990s, Hispanic Americans surpassed African Americans to become the nation's largest minority, and today they account for 15 percent of the U.S. total.

The spatial distribution of the four largest ethnic minorities is mapped in Figure 3-13. Hispanics are regionally clustered, with about 60 percent residing in California, Texas, New Mexico, and South Florida; nonetheless, population geographers expect this rapidly expanding minority to continue to disperse across the country, especially in the West, the interior South, and along the eastern seaboard. African Americans are regionally concentrated as well, most notably in the South, reflecting the legacy of slavery and the plantation economy; blacks are also a major presence in the central cities of the Manufacturing Belt and west coast (most metropolitan areas do not register prominently on these small-scale maps). Asian Americans are the most agglomerated of the leading ethnic minorities, their distribution dominated by urban-based clusters along the Pacific coast (entry points for the large number who are recent immigrants). The main concentrations of Native Americans are found in the West, where they largely occupy tribal lands on reservations ceded by the federal government; many also reside in communities widely scattered across the nation.

Immigration

The changing ethnic complexion of the United States has long been influenced by immigration. That trend continues strongly, and during the first half of this decade at least 700,000 legal immigrants annually entered the country. The source areas, however, have changed dramatically over the past half-century. During the 1950s, just over 50 percent came from Europe, 25 percent from Middle and South America, and 15 percent from Canada. Today fully 50 percent come from Middle and South America, more than 25 percent from Asia, and only about 15 percent from Europe and Canada. Given this ongoing influx of immigrants, it is not surprising that in 2004 the Census Bureau reported that more than 33 million residents (11.5 percent of the national population) had been born in other countries—the highest total

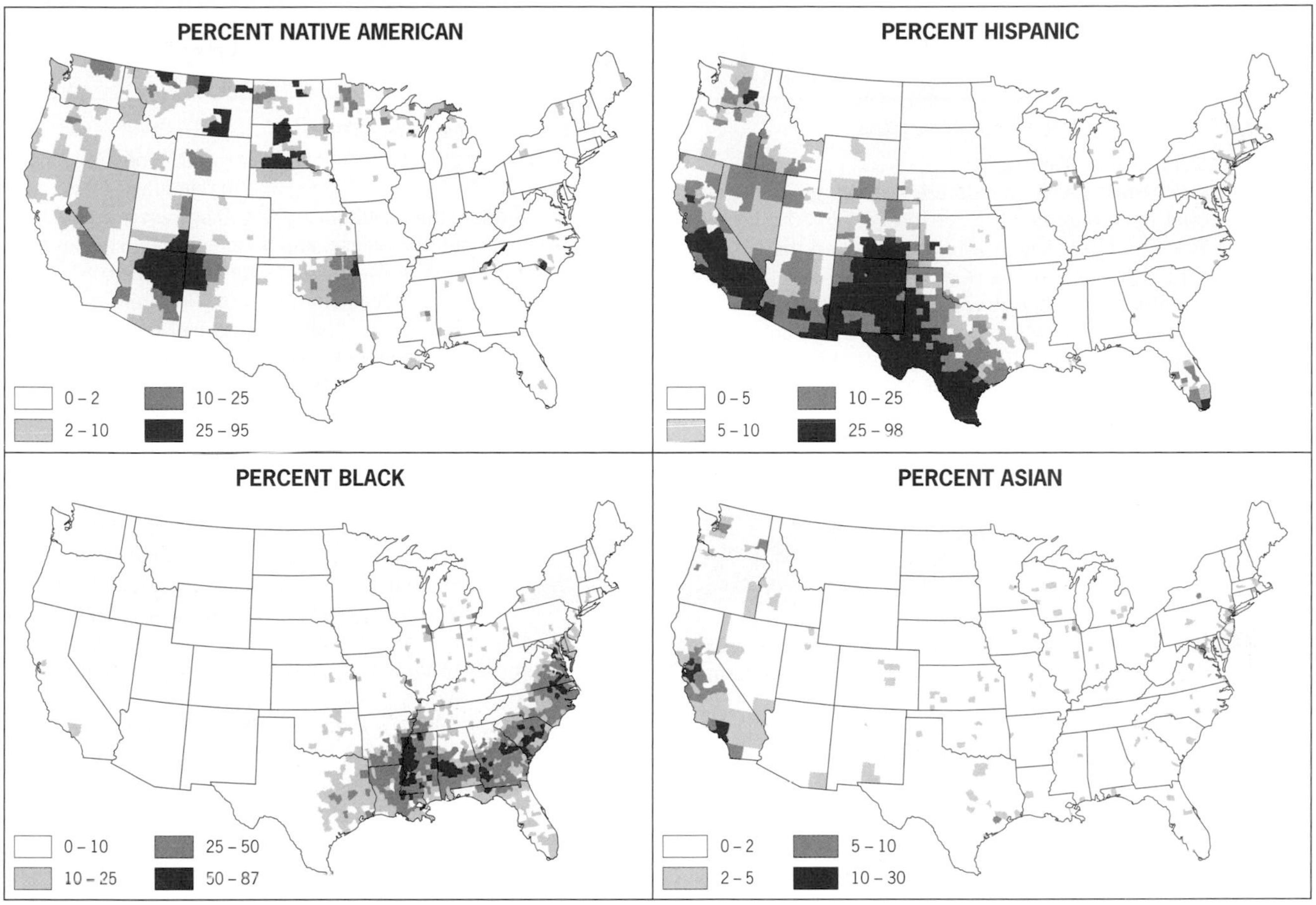

FIGURE 3-13 © H. J. de Blij, P. O. Muller, and John Wiley & Sons, Inc.

in U.S. history (see Issue Box titled "Immigrants: How Many Can North America Accommodate?").

Although still overwhelmingly directed at urban areas (led by Miami, North America now contains the world's five largest immigrant cities), the geography of immigrant destinations is changing. The once-dominant Manufacturing Belt States have slipped to less than 25 percent, with most of the newcomers now heading for either metropolitan New York or Chicago. Not unexpectedly, with the ongoing growth of Asian and Hispanic American populations (Fig. 3-13), the Sunbelt States have taken up the slack, with California, Texas, and Florida attracting more than 50 percent of all U.S. immigrants.

The Emerging Mosaic Culture

American cultural geography continues to evolve. What is now taking place is a new fragmentation into the
12 emerging nationwide **mosaic culture**, an increasingly heterogeneous complex of separate, uniform "tiles" that cater to more specialized groups than ever before. No longer based solely on such broad divisions as income, race, and ethnicity, today's residential communities of interest are also forming along the dimensions of age, occupational status, and especially lifestyle. Their success and steady proliferation reflect an obvious satisfaction on the part of most Americans. Yet such balkanization—fueled by people choosing to interact only with others exactly like themselves—threatens the survival of important democratic values that have prevailed throughout the evolution of U.S. society.

The Changing Geography of Economic Activity

The economic geography of the United States today is the product of all of the foregoing, as bountiful environmental, human, and technological resources have cumulatively blended together to create one of the world's most advanced economies. Perhaps the greatest triumph was overcoming the tyranny of distance, so that people and activities could be organized into a continentwide spatial economy that took maximum advantage of agricultural, industrial, and urban development opportunities. Yet, despite these past achievements, American economic geography at the outset of the twenty-first century is once again in the throes of restructuring as the

Immigrants: How Many Can North America Accommodate?

Regional ISSUE

IMMIGRATION BRINGS BENEFITS—THE MORE THE MERRIER!

"The United States and Canada are nations of immigrants. What would have happened if our forebears had closed the door to America after they arrived and stopped the Irish, the Italians, the Eastern Europeans, and so many other nationalities from entering this country? Now we're arguing over Latinos, Asians, Russians, Muslims, you name it. Fact is, newcomers have always been viewed negatively by most of those who came before them. When Irish Catholics began arriving in the 1830s the Protestants already here accused them of assigning their loyalty to some Italian pope rather than to their new country, but Irish Catholics soon proved to be pretty good Americans. Sound familiar? Muslims can be very good Americans too. It just takes time, longer for some immigrant groups than others. But don't you see that America's immigrants have always been the engine of growth? They become part of the world's most dynamic economy and make it more dynamic still.

"My ancestors came from Holland in the 1800s, and the head of the family was an architect from Rotterdam. I work here in western Michigan as an urban planner. People who want to limit immigration seem to think that only the least-educated workers flood into the United States and Canada, depriving the less-skilled among us of jobs and causing hardship for citizens. But in fact America attracts skilled and highly educated as well as unskilled immigrants, and they all make contributions. The highly educated foreigners, including doctors and technologically skilled workers, are quickly absorbed into the workforce; you're very likely to have been treated by a physician from India or a dentist from South Africa. The unskilled workers take jobs we're not willing to perform at the wages offered. Things have changed! A few decades ago, American youngsters on summer break flooded the job market in search of temporary employment in hotels, department stores, and restaurants. Now they're vacationing in Europe or trekking in Costa Rica, and the managers of those establishments bring in temporary workers from Jamaica and Romania.

"And our own population is aging, which is why we need the infusion of younger people immigration brings with it. We don't want to become like Japan or some European countries, where they won't have the younger working people to pay the taxes needed to support the social security system. I agree with opponents of immigration on only one point: what we need is legal immigration, so that the new arrivals will get housed and schooled, and illegal immigration must be curbed. Otherwise, we need more, not fewer, immigrants."

LIMIT IMMIGRATION NOW

"The percentage of recent immigrants in the U.S. population is the highest it has been in 70 years, and in Canada in 60 years. America is adding the population of San Diego every year, over and above the natural increase, and not counting illegal immigration. This can't go on. By 2010, 15 percent of the U.S. population will consist of recent immigrants. One-third of them will not have a high-school diploma. They will need housing, education, medical treatment, and other social services that put a huge strain on the budgets of the States they enter. The jobs they're looking for often aren't there, and then they start displacing working Americans by accepting lower wages. It's easy for the elite to pontificate about how great immigration is for the American melting pot, but they're not the ones affected on a daily basis. Immigration is a problem for the working people. We see company jobs disappearing across the border to Mexico, and at the same time we have Mexicans arriving here by the hundreds of thousands, legally and illegally, and more jobs are taken away.

"And don't talk to me about how immigration now will pay social security bills later. I know a thing or two about this because I'm an accountant here in Los Angeles, and I can calculate as well as the next guy in Washington. Those fiscal planners seem to forget that immigrants grow older and will need social security too. And as for that supposed slowdown in the aging of our population because immigrants are so young and have so many children, over the past 20 years the average age in the United States has dropped by four months. So much for that nonsense. What's needed is a revamping of the tax structure, so those fat cats who rob corporations and then let them go under will at least have paid their fair share into the national kitty. There'll be plenty of money to fund social services for the aged. We don't need unskilled immigrants to pay those bills.

"And I'm against this notion of amnesty for illegal immigrants being talked about these days. All that would do is to attract more people to try to make it across our borders. I heard President Fox of Mexico propose opening the U.S.-Mexican border the way they're opening borders in the European Union. Can you imagine what would happen? What our two countries really need is a policy that deters illegal movement across that border, which will save lives as well as jobs, and a system that will confine immigration to legal channels. These days, that's not just a social or economic issue; it's a security matter as well."

Vote your opinion at www.wiley.com/college/deblij

transition is completed from industrial to postindustrial society.

Major Components of the Spatial Economy

Economic geography is mainly (though not exclusively)
13 concerned with the locational analysis of **productive activities**. Four major sets may be identified:

- **Primary activity:** the extractive sector of the economy in which workers and the environment come into direct contact, especially in *mining* and *agriculture*.
- **Secondary activity:** the *manufacturing* sector in which raw materials are transformed into finished industrial products.
- **Tertiary activity:** the *services* sector, including a wide range of activities from retailing to finance to education to routine office-based jobs.
- **Quaternary activity:** today's dominant sector, involving the collection, processing, and manipulation of *information*; a subset, sometimes referred to as **quinary activity**, is the managerial activity associated with decision-making in large organizations.

Historically, each of these activities has successively dominated the American labor force for a time over the past 200 years, with the quaternary sector now dominant. Agriculture dominated until the late nineteenth century, giving way to manufacturing by 1900. The steady growth of services after 1920 finally surpassed manufacturing in the 1950s but now shares a dwindling portion of the limelight with the still-rising quaternary sector. The approximate breakdown by major sector of employment in the U.S. labor force today is agriculture, 2 percent; manufacturing, 15 percent; services, 18 percent; and quaternary, 65 percent (with about 10 percent in the quinary sector). We now proceed to review these major productive components of the spatial economy in the following coverage of resource use, agriculture, manufacturing, and the postindustrial revolution.

Resource Use

The United States (and Canada) was blessed with abundant deposits of mineral and energy resources. Fortunately, these resources were usually concentrated in sufficient quantities to make long-term extraction an economically feasible proposition, and most of the richest raw material sites are still the scene of major drilling or mining operations. Moreover, the continental and offshore mineral/fuel storehouse may yet contain significant undiscovered resources for future exploitation.

Mineral Resources. North America's rich mineral deposits are localized in three zones: the Canadian Shield north of the Great Lakes, the Appalachian Highlands, and scattered areas throughout the mountain ranges of the West. The Shield's most noteworthy minerals are iron ore, nickel, gold, uranium, and copper. Besides vast deposits of soft (bituminous) coal, the Appalachian region also contains hard (anthracite) coal in northeastern Pennsylvania and iron ore in central Alabama. The western mountain zone contains significant deposits of coal, copper, lead, zinc, molybdenum, uranium, silver, and gold.

Fossil Fuel Energy Resources. The realm's most strategically important resources are its petroleum (oil), natural gas, and coal supplies—the **fossil fuels**, so named 14
because they were formed by the geologic compression and transformation of plant and tiny animal organisms that lived hundreds of millions of years ago. These energy supplies are mapped in Figure 3-14, which reveals abundant deposits and far-flung distribution networks.

The leading *oil*-production areas of the United States are located along and offshore from the Texas–Louisiana Gulf Coast; in the Midcontinent district, extending through western Texas–Oklahoma–eastern Kansas; and along Alaska's central North Slope facing the Arctic Ocean. (Canada's major oilfields lie in a wide crescent curving southeastward from northern Alberta to southern Manitoba and beneath the waters around Newfoundland in the extreme east.) The distribution of *natural gas* deposits generally resembles the geography of oilfields because petroleum and energy gas are usually found in similar geologic formations (the floors of ancient shallow seas); natural gas is now considered the most glamorous fossil fuel because of its growing dominance in the generation of North America's electrical power supply. The realm's *coal* reserves rank among the greatest on Earth, and the U.S. portion alone contains at least a 400-year supply; the main producing coalfields are found in Appalachia, the northern U.S. Great Plains/southern Alberta, and southern Illinois/western Kentucky.

Agriculture

Despite the post–1900 emphasis on developing the nonprimary sectors of the spatial economy, agriculture remains an important element in America's human geography. Because it is the most space-consuming economic activity, vast expanses of the U.S. (and Canadian) landscape are clothed with fields of grain. In addition, great herds of livestock are sustained by pastures and fodder crops because this wealthy realm can afford the luxury of feeding animals from its farmlands to meet huge demands for red meat in its diet. The growing application of high-technology mechanization to farming has steadily increased both the volume and value of total agricultural production. It also has been accompanied by a sharp reduction in the number of those

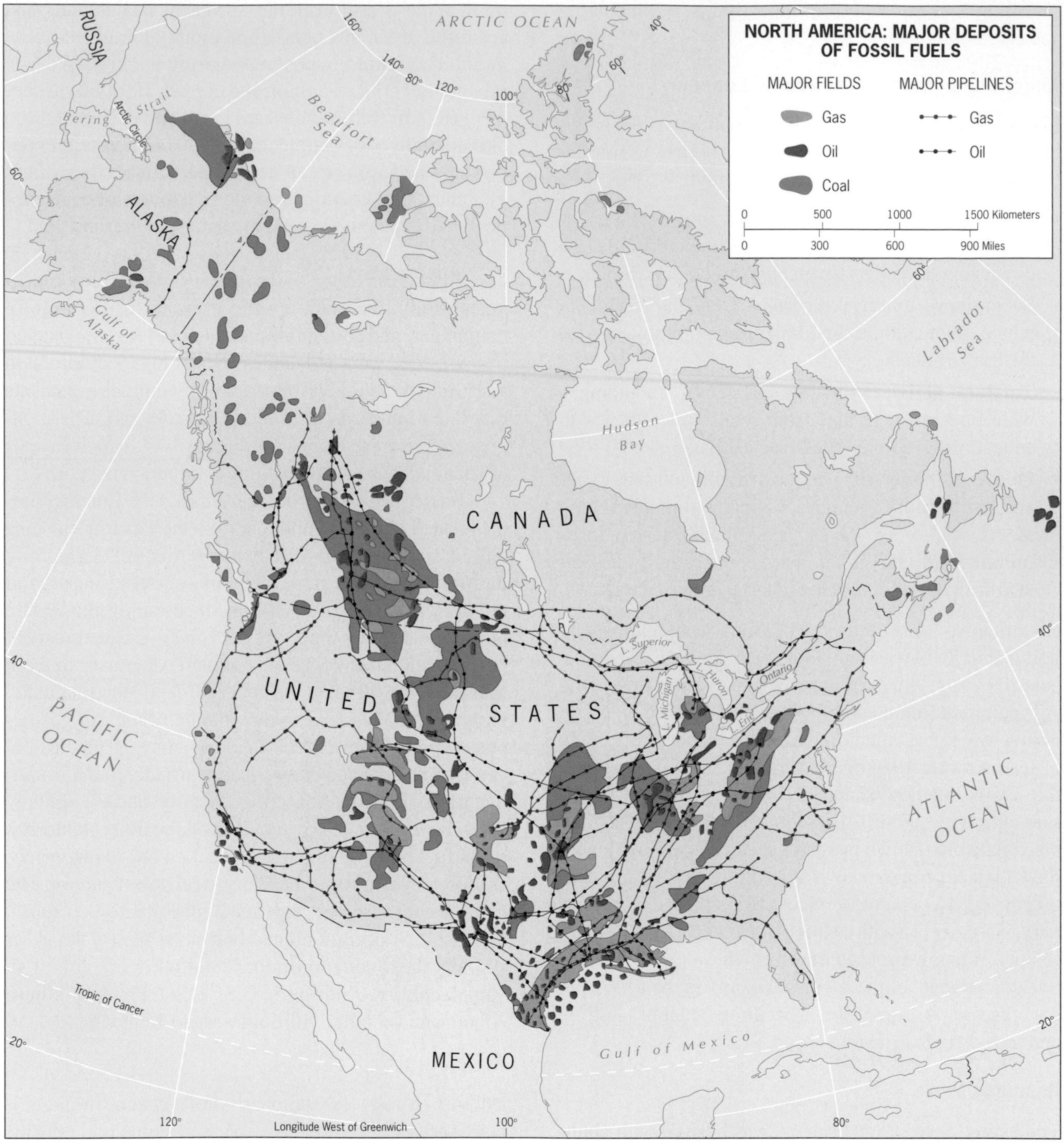

FIGURE 3-14

actively engaged in agriculture, and today only slightly more than 1 percent of the U.S. population still lives on farms.

The regionalization of U.S. agricultural production is shown in Figure 3-15, its spatial organization developed largely within the framework of the *von Thünen model* (see p. 49). As in Europe (Fig. 1-6), the early-nineteenth-century, original-scale model of town and hinterland expanded outward (driven by constantly improving transportation technology) from a locally "isolated state" to encompass the entire continent by 1900. As this macrogeographic structure formed, the greatly enlarged original Thünian production zones (Fig. 1-5) were modified: (1) the first ring was now differentiated into an inner fruit/vegetable zone and a surrounding belt of dairying; (2) the forestry ring was displaced to the out-

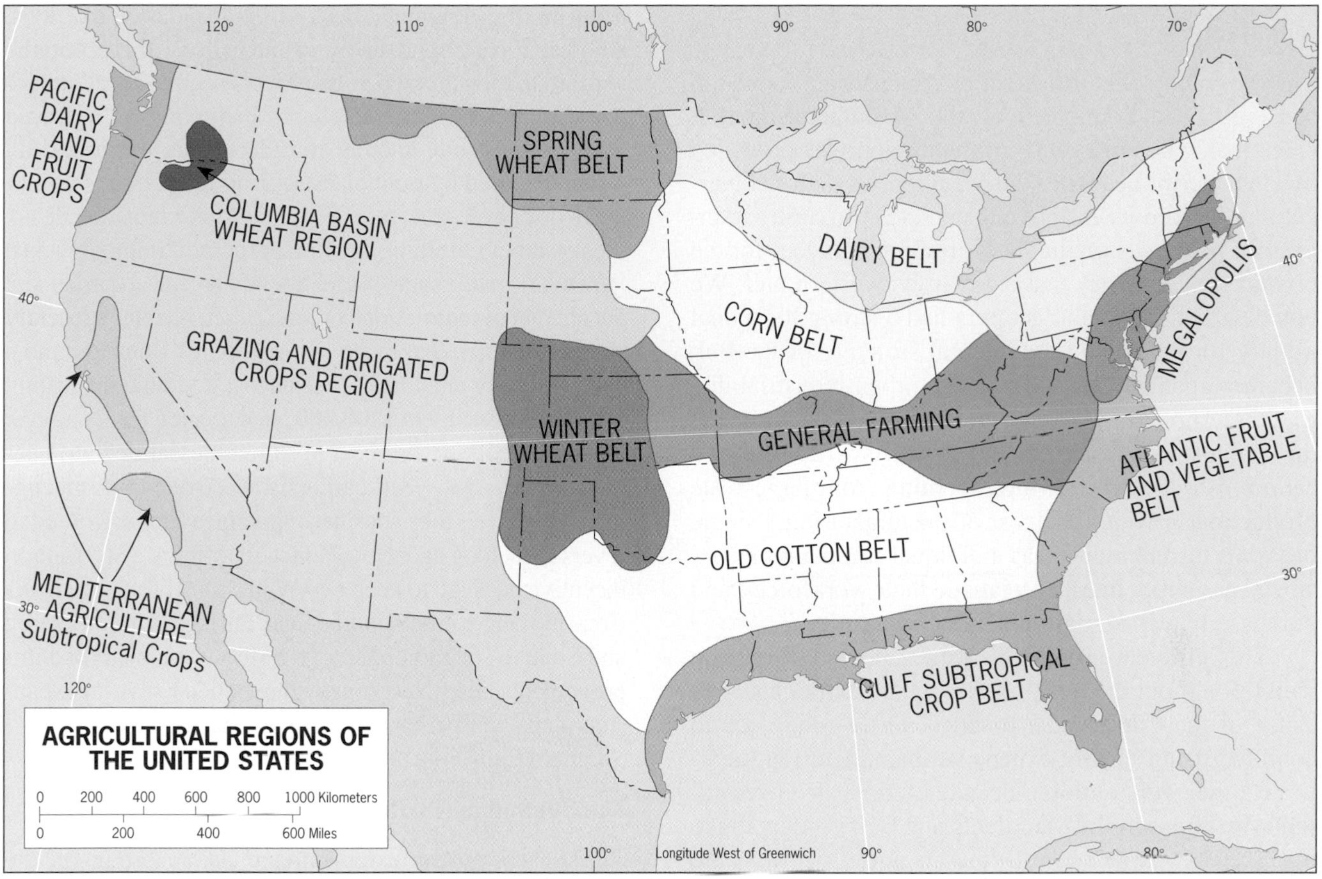

FIGURE 3-15 Adapted with permission from James O. Wheeler et al., *Economic Geography*, 3 rev. ed., p. 319. © John Wiley & Sons, Inc., 1998.

ermost limits of the regional system because railroads could now transport wood quite cheaply; (3) the field crops ring subdivided into an inner mixed crop-and-livestock ring to produce meat (the Corn Belt, as it came to be called) and an outer zone that specialized in the mass production of wheat grains (the Wheat Belts); and (4) the ranching area remained in its outermost position, a grazing zone that supplied young animals to be fattened in the meat-producing Corn Belt as well as supporting an indigenous sheep-raising industry. The "supercity" anchoring this macro-Thünian regional system was the northeastern Megalopolis, well on its way toward coalescence and already the dominant food market and transport focus of the entire country.

Although the circular rings of the model are not apparent in Figure 3-15, many spatial regularities can be observed (remember that von Thünen, too, applied his model to the real world and thereby distorted the theoretically ideal pattern). Most significant is the sequence of farming regions as distance from the national market increased, especially westward from Megalopolis toward central California, which was the main directional thrust of the historic inland penetration of the United States. The Atlantic Fruit and Vegetable Belt, Dairy Belt, Corn Belt, Wheat Belts, and Grazing Region are indeed consistent with the model's logical structure, each zone successively farther inland astride the main transcontinental routeway. We can attribute deviations from the scheme to irregularities in the environment or to unique conditions. For example, the nearly year-round growing seasons of California and the Gulf Coast-Florida region permit those distant areas (with the help of efficient refrigerated transport) to produce fruits and vegetables in competition with the regional system's (winter-dormant) innermost zone.

Today, driven by globalization forces that require U.S. producers to become ever more competitive on the international food market, the agricultural spatial economy continues to evolve. In his studies of the changing scale of American agriculture, John Fraser Hart has identified an emerging framework that constitutes a new tripartite macrogeography: a Midwestern Corn Belt heartland, increasingly devoted to cash-grain farming; a constellation of livestock-raising zones along the heartland's wide periphery; and a coastal "rimland" of specialized crop-producing areas.

Manufacturing

The geography of North America's industrial production has long been dominated by the Manufacturing Belt (Fig. 3-8). The emergence of this region was propelled by (1) superior access to the Megalopolis national market that formed its eastern edge and (2) proximity to industrial resources, particularly iron ore and coal for the pivotal steel industry that arose in its western half. We noted earlier that manufacturers had a strong locational affinity for cities, and the internal structure of the Belt became organized around a dozen urban-industrial districts interlinked by a dense transportation network. As these industrial centers expanded, they swiftly achieved
15 **economies of scale**, savings accruing from large-scale production in which the cost of manufacturing a single item was further reduced as individual companies mechanized assembly lines, specialized their workforces, and purchased raw materials in massive quantities.

This efficient production pattern served the nation well throughout the remainder of the industrial age. Because of the effects of *historical inertia*—the need to continue using hugely expensive manufacturing facilities for their full, multiple-decade lifetimes to cover initial investments—the Manufacturing Belt is not going to disappear anytime soon. But as its aging facilities are retired, a process well under way, the distribution of American industry will change. As transportation costs equalize among U.S. regions, as energy costs now favor the south-central oil- and gas-producing States, as high-technology manufacturing advances reduce the need for less skilled labor, and as locational decision-making intensifies its attachment to noneconomic factors, industrial management has increasingly demonstrated its willingness to relocate to regions it perceives as more desirable in the South and West.

Parts of the Manufacturing Belt have been resisting this trend, particularly the industrial Midwest, whose prospects have greatly improved since the "Rustbelt" days of the 1970s and 1980s. Shedding that label in recent years, many Midwestern manufacturers are successfully reinventing their operations by rooting out inefficiencies, investing in cutting-edge factories and technologies, and not only beating back foreign competition in the United States but significantly expanding their exports. Nonetheless, there is a downside to this progress because as high-tech manufacturing proliferates, older factory workers discover that the demand for their lower-tech skills shrinks accordingly.

As the U.S. manufacturing map continues to change, the economic lives of countless communities are affected. If the past 15 years were a sign of things of come, entrepreneurial restructuring will also play an increasingly important role because corporate mergers and acquisitions often result in the geographic reshuffling of thousands of jobs. Another recent trend likely to intensify is the economic warfare that occurs when employers seek to locate major new facilities and deliberately pit cities, counties, and even States against one another in order to get the best deal. Often exhorted by politicians seeking reelection, communities offer ever greater incentive packages that include tax breaks, labor retraining, and infrastructural improvements. However, many such competitions can fail to produce a decent rate of return. For example, when Alabama recently went to extreme lengths to lure Mercedes-Benz to build a new assembly plant near Tuscaloosa, it wound up costing the State government $200,000 per expected job. (The cost to Tennessee's government to attract the Saturn automaking complex a few years earlier had been one-tenth that figure.) Moreover, these business strategies can be worked in reverse; not long ago, New York City offered a $60 million incentive package to keep two Wall Street brokerage firms from moving to the suburbs, and Disneyland got its city and State to spend hundreds of millions on local road improvements after the company threatened to cancel expansion plans and perhaps even move the theme park out of Southern California.

The Postindustrial Revolution

The signs of **postindustrialism** are visible throughout the 16
United States (and much of Canada) today; they are popularly grouped under such labels as "the computer age" or "the new economy." The term *postindustrial* by itself, of course, tells us mainly what the theme of the American economy no longer is. Yet many social scientists also use the term to refer to a set of societal traits that signal an historic break with the past (for example, work experiences now focus mainly on person-to-person interaction rather than person-product or person-environment contact). Many of the urban spatial expressions of the ongoing societal transformation have already been highlighted; here we focus on some broader economic-geographical patterns.

High-technology, white-collar, office-based activities are the leading growth industries of the postindustrial economy. Most are relatively footloose and are therefore responsive to such noneconomic locational forces as geographic prestige, local amenities, and proximity to recreational opportunities. Northern California's *Silicon Valley*—the world's leading center for computer research and development and the headquarters of the U.S. microprocessor industry—epitomizes the blend of locational qualities that attract a critical mass of high-tech companies to a given locality. These include: (1) a world-class research university (Stanford); (2) technological know-how in the form of a large pool of highly educated, highly skilled labor; (3) close proximity to a cosmopolitan urban center (San Francisco); (4) abundant venture capital; (5) a local economic climate and entrepreneurial culture that supports risk-taking and forgives failure by both smaller and big companies; (6) a locally based network of global business linkages; and (7) a high-amenity environment in the form of top-quality housing, pleasant weather, scenic countrysides, and year-round recreational opportunities.

The development of Silicon Valley is so significant to the new postindustrial era that regional-planning scholars Manuel Castells and Peter Hall have conceptualized it as
17 the first **technopole**. Technopoles are planned techno-industrial complexes that innovate, promote, and manufacture the hardware and software products of the new informational economy. On the landscape of the outer suburban city, where almost all of these complexes are located, the signature of a technopole is a low-density cluster of ultramodern, low-rise buildings laid out as a campus—an image as symbolic of today's spatial economy as the smoke-belching factory was of the industrial age a century ago. From Silicon Valley, technopoles have spread in all directions—from San Diego in southwesternmost California to the Route 128 corridor around Boston and from North Carolina's Research Triangle to the lakeside suburbs of Seattle; many of these technopoles will be noted in the regional section that concludes this chapter. These high-tech complexes are also becoming a global phenomenon as they spring up in other geographic realms. They often arise as joint ventures between government and the private sector; among the local names they go by are science city (Japan; Russia), technopolis (France; Japan), science park (Taiwan)—and even Silicon Glen (Scotland).

One of the hallmarks of the postindustrial revolution is the emergence of the Internet, which has already been called the defining technology of the twenty-first century. The recent growth of the Internet has been driven by the need to transmit ever greater quantities of data at ever higher speeds. Fiber-optic cables are best able to meet these needs, and have given rise to a network of backbones or linkages that connect information-rich centers across the planet. The leading component of this global infrastructure is its United States network, which handles up to 75 percent of the world's Internet traffic and is dominated by a set of major backbone cities and their interconnections (Fig. 3-16). Interestingly, the three leading metropolitan areas in terms of connected backbone capacity are Washington, D.C. (federal government operations coupled with a booming technopole

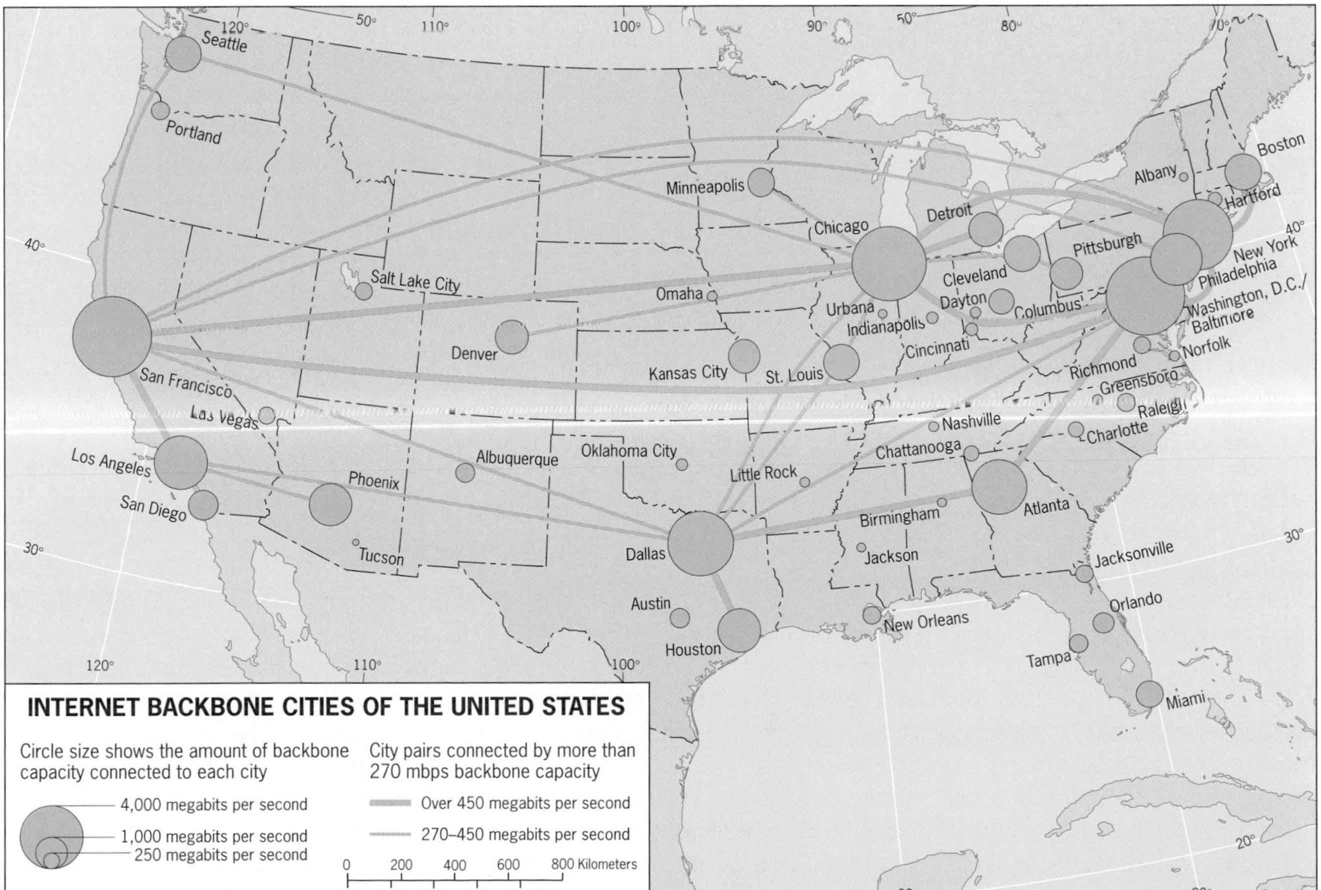

FIGURE 3-16 Adapted with permission from H.J. de Blij, ed., *Oxford Atlas of North America*, p. 41. © Oxford University Press, 2005.

in suburban northern Virginia), Chicago (central location on the network, which facilitates wide access to the continental interior), and San Francisco (preeminent influence of Silicon Valley). Thus, at least in its initial stage of development, the hierarchy of backbone capacity does not match the urban population hierarchy. The primate city, New York, continues to rank in the second tier of Internet backbone cities together with another major technopole (Dallas's "Silicon Prairie") and a leading regional hub (Atlanta, which serves the South).

CANADA

Like the United States, Canada is a federal state, but it is organized differently. Canada is divided administratively into ten provinces and three territories (Fig. 3-17). The provinces—where 99.7 percent of all Canadians live—range in territorial size from tiny, Delaware-sized Prince Edward Island to sprawling Quebec, more than twice the area of Texas. Beginning in the east, the four Atlantic Provinces are Nova Scotia, New Brunswick, Prince Edward Island, and Newfoundland and Labrador. To their west lie Quebec and Ontario, Canada's two biggest provinces. Most of western Canada (which is what the Canadians call everything west of Lake Superior) is covered by the three Prairie Provinces—Manitoba, Saskatchewan, and Alberta. In the far west, beyond the Canadian Rockies and facing the Pacific, lies the tenth province, British Columbia.

The three territories—Yukon, the Northwest Territories, and Nunavut—together occupy a massive area half the size of Australia but are inhabited by only about 100,000 people. Nunavut is the newest addition to Canada's political map and deserves special mention. Created in 1999, this new territory is the outcome of a major aborig-

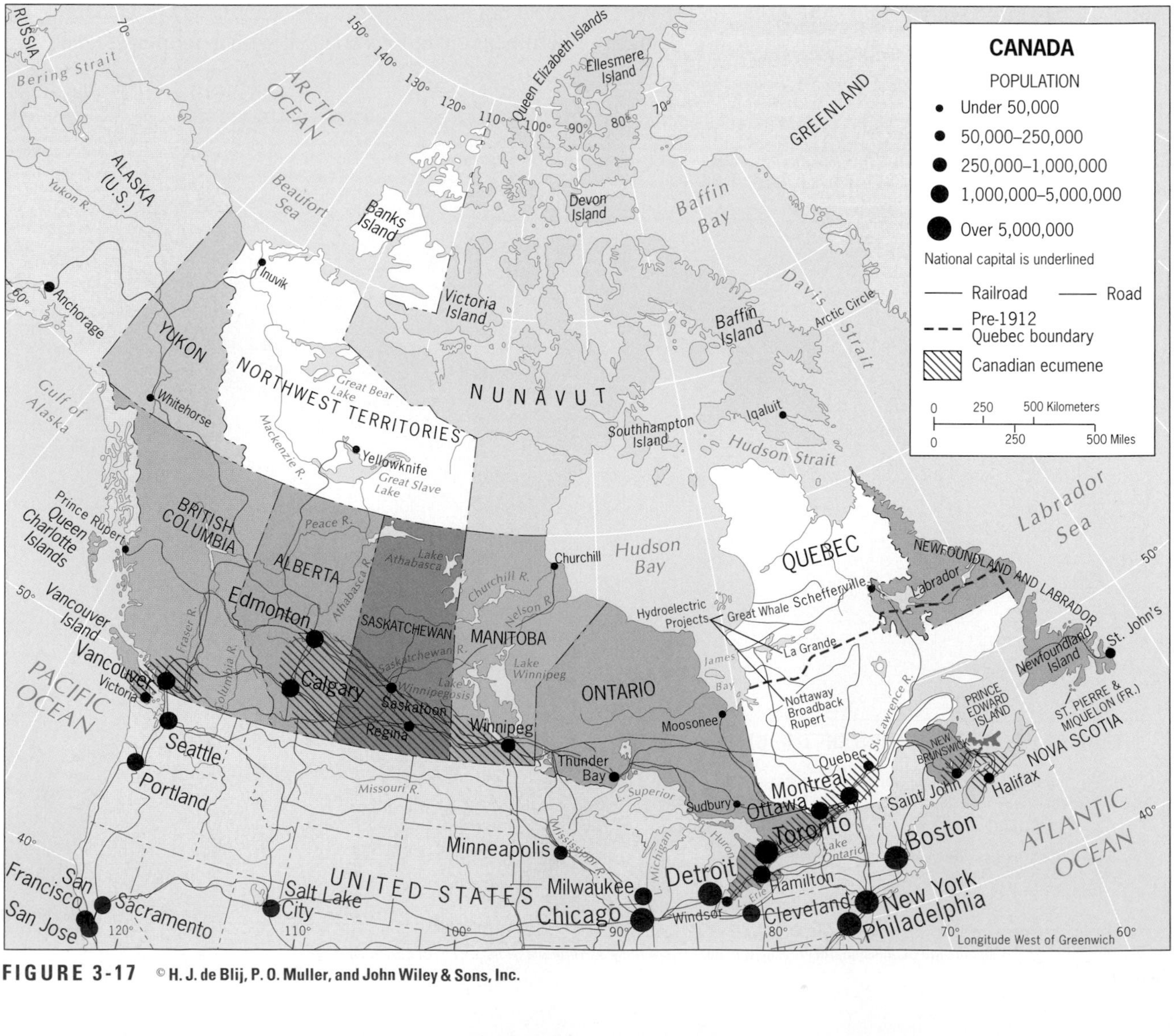

FIGURE 3-17 © H. J. de Blij, P. O. Muller, and John Wiley & Sons, Inc.

inal land claim agreement between the *Inuit* people (formerly called Eskimos) and the federal government, and encompasses all of Canada's eastern Arctic. Nunavut covers fully one-fifth of Canada's land, an area larger than that of any other province or territory. Since 80 percent of its 31,000 residents are Inuit, Nunavut—which means "our land"—is the first territory to have a substantial degree of native self-government (in addition to outright ownership of an area half the size of Texas).

In population size (in descending order), Ontario (12.4 million) and Quebec (7.6 million) are again the leaders; British Columbia ranks third with 4.3 million; next come the three Prairie Provinces with a combined total of 5.5 million; the Atlantic Provinces are the four smallest, together containing 2.3 million. The sparsely populated territories in the far north contain barely 100,000 residents. Canada's total population of 32.2 million is only slightly larger than one-tenth the size of the U.S. population. Spatially, as noted at the outset of the chapter, Canadians overwhelmingly reside within 200 miles (320 km) of the U.S. border.

Though comparatively small, Canada's population is divided by culture and tradition, and this division has a pronounced regional expression. Just under 60 percent of Canada's citizens speak English as their mother tongue, 23 percent speak French, and the remaining 17 percent other languages; about 18 percent of the population speak both English and French (of these, 43 percent of those who speak French as a first language are bilingual, but only 9 percent of the English speakers). The spatial clustering of more than 90 percent of the country's *Francophones* (French-speakers) in Quebec accentuates this Canadian social division along ethnic and linguistic lines (Fig. 3-18). More than 80 percent of this province's population is French Canadian, and Quebec is the historic, traditional, and emotional focus of French

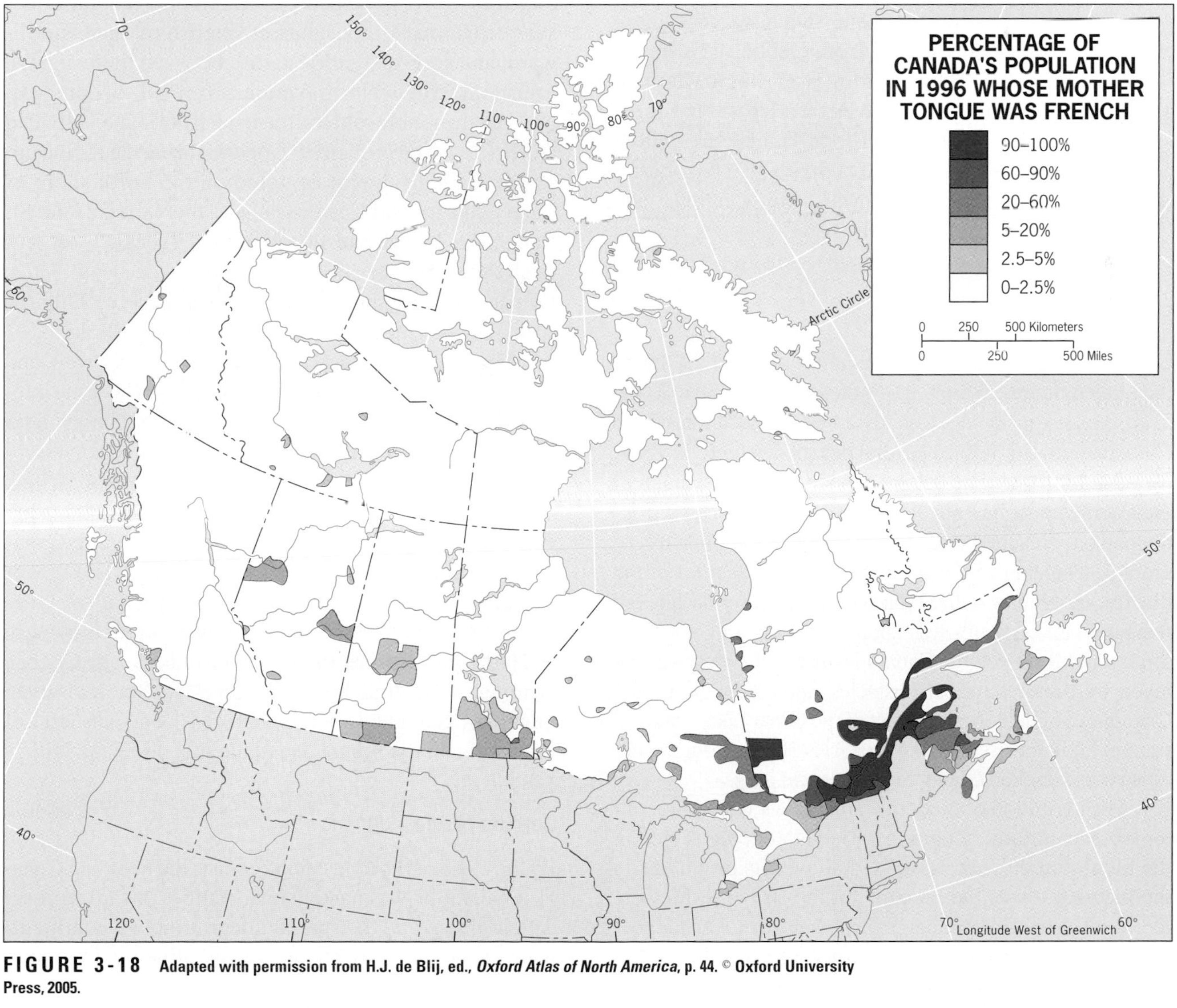

FIGURE 3-18 Adapted with permission from H.J. de Blij, ed., *Oxford Atlas of North America*, p. 44. © Oxford University Press, 2005.

culture in Canada. Over the past few decades a strong nationalist movement emerged in Quebec to demand outright separation from the rest of Canada. That movement crested in 1995, when independence was narrowly defeated in a landmark referendum. Even though support for secession has waned since then, this ethnolinguistic division continues to be the cornerstone of Canada's cultural pluralism, and the issue could still pose a threat to national unity in the future.

Population in Time and Space

The map showing the distribution of Canada's population (Fig. 3-5, p. 152) reveals that only about one-eighth of this
18 enormous country can be classified as its **ecumene**—the inhabitable zone of permanent settlement. As Figure 3-17 indicates, the Canadian ecumene is dominated by a discontinuous strip of population clusters that lines the southern border. We can identify four such clusters on the map, the largest by far being Main Street (home to more than six out of every ten Canadians). As noted earlier (Fig. 3-9), *Main Street* is the conurbation that stretches across southernmost Quebec and Ontario, from Quebec City on the lower St. Lawrence River southwest through Montreal and Toronto to Windsor on the Detroit River. The three lesser clusters are: (1) the Saint John-Halifax crescent in central New Brunswick and Nova Scotia; (2) the prairies of southern Alberta, Saskatchewan, and Manitoba; and (3) the southwestern corner of British Columbia, focused on Canada's third-largest metropolis, Vancouver.

Pre-Twentieth-Century Canada

Canada's population map evolved more slowly than that of the United States. As Figure 3-7 shows, compared to European expansion in the United States, penetration of the Canadian interior lagged several decades behind. In 1850, for example, when the eastern United States and much of the Pacific coast had already been settled, Canada's frontier had only reached Lake Huron. In terms of political geography, Canada did not unify before the last third of the nineteenth century—and then only reluctantly because of fears the United States was about to expand in a northerly direction. Moreover, the Canadian push to the west was delayed by major physical obstacles, including the barren Canadian Shield north and west of the Great Lakes, and the rugged highlands separating the Great Plains from the Pacific (which lacked convenient mountain passes).

The evolution of modern Canada—as well as its contemporary cultural geography—is also deeply rooted in the bicultural division discussed above. Its origin lies in the fact that it was the French, not the British, who were the first European colonizers of present-day Canada, beginning in the 1530s. *New France*, during the seventeenth century, grew to encompass the St. Lawrence Basin, the Great Lakes region, and the Mississippi Valley. A series of wars between the English and French began in the 1680s, ending with France's defeat and the cession of New France to Britain in 1763. By the time London took control of its new possession, the French had made considerable progress in their North American domain. French laws, the French land-tenure system, and the Roman Catholic Church prevailed, and substantial settlements (including Montreal on the St. Lawrence) had been established. The British, anxious to avoid a war of suppression and preoccupied with problems in their other American colonies, gave former French Quebec—the territory extending from the Great Lakes to the mouth of the St. Lawrence—the right to retain its legal and land-tenure systems as well as freedom of religion.

After the American War for Independence, London was left with a region it called British North America (the name *Canada*—derived from an aboriginal word meaning settlement—was not yet in use), but whose cultural imprint still was decidedly French. The Revolutionary War drove many thousands of English refugees northward, and soon difficulties arose between them and the French in British North America. In 1791, heeding appeals by these new settlers, the British Parliament divided Quebec into two provinces: Upper Canada, the region upstream from Montreal centered on the north shore of Lake Ontario, and Lower Canada, the valley of the St. Lawrence. Upper and Lower Canada became, respectively, the provinces of Ontario and Quebec (Fig. 3-17). By parliamentary plan, Ontario would become English-speaking and Quebec would remain French-speaking.

This earliest cultural division did not work well, and in 1840 the British Parliament tried again. This time it reunited the two provinces through the Act of Union under which Upper and Lower Canada would have equal representation in the provincial legislature. That, too, failed, and efforts to find a better system finally led to the 1867 British North America Act that established the Canadian federation (consisting initially of Upper and Lower Canada, New Brunswick, and Nova Scotia, and later, between 1870 and 1999, to be joined by the other provinces and territories). Under the 1867 Act, Ontario and Quebec were again separated, but this time Quebec was given important guarantees: the French civil code was left unchanged, and the French language was protected in Parliament and in the courts.

Canada Since 1900

By the close of the nineteenth century, the fledgling Canadian federation was making major strides toward regional development and the spatial integration of a continent-wide economy. In 1886, the transcontinental Canadian

Pacific Railway had finally been completed to Vancouver, a condition of British Columbia's entry into the federation 15 years earlier. This new lifeline soon spawned the settlement of not only the far west but also the fertile Prairie Provinces, whose wheat-raising economy expanded steadily as immigrants from the east and abroad poured in during the early twentieth century. Industrialization also began to stir, and by 1920 Canadian manufacturing had surpassed agriculture as the leading source of national income. As noted earlier (Fig. 3-8), the dominant zone of industrial activity is the Toronto–Hamilton–Windsor corridor of southern Ontario (which also functions as one of the 12 districts of the American Manufacturing Belt). That has been the case since the Industrial Revolution took hold in Canada during World War I (1914–1918), when the country heavily supported Britain's war effort on the European mainland.

As industrial intensification proceeded, it was accompanied by the expected parallel maturation of the national urban system. Along the lines of Borchert's five epochs of American metropolitan evolution, Maurice Yeates has constructed a similar multistage model of urban-system development that divides the telescoped Canadian experience into three eras. The initial *Frontier-Staples Era* (prior to 1935) encompasses the century-long transition from a frontier-mercantile economy to one oriented to staples (production of raw materials and agricultural goods for export), with increasing manufacturing activity in the budding industrial heartland. By 1930, Montreal and Toronto, reflecting their different cultural constituencies, had emerged as the two leading cities atop the national urban hierarchy (thus Canada has no single primate city).

Toronto's central business district, which fronts the northern shore of Lake Ontario, reflects a mixture of building styles from the traditional to the ultramodern. The density of these skyscrapers is remarkable, and underscores that the downtown skyline of Canada's largest city compares with anything on the continent, including midtown Manhattan and Chicago's Loop. © Corbis Images.

Next came the *Era of Industrial Capitalism* (1935–1975), during which Canada achieved U.S.-style prosperity. However, most of this development—involving the massive growth of manufacturing, tertiary activities, and urbanization—took place after 1950 because the Great Depression lasted until World War II, which was followed by a lengthy recession. A major stimulus was the investment of U.S. corporations in Canadian branch-plant construction, especially in the automobile industry in Ontario near the automakers' Detroit-area headquarters. In western Canada, the rapid growth of oil and natural gas production fueled Alberta's urban development, but new agricultural technologies reduced farm labor needs and sparked a rural outmigration in neighboring Saskatchewan and Manitoba. The postwar period also saw the ascent of Main Street, which on less than 2 percent of Canada's land quickly came to contain more than 60 percent of its people, contributed two-thirds of its national income, and had nearly 75 percent of its manufacturing jobs.

The third stage, ongoing since 1975, is the *Era of Global Capitalism*, signifying the rise of additional foreign investment from the Asian Pacific Rim and Europe. This, of course, is also the era of transformation into a postindustrial economy and society, and in the process Canada is experiencing many of the same upheavals as the United States. Most development today is occurring in the form of new suburbanization. This represents a departure from the recent past because the pre-1990 Canadian metropolis had experienced far less automobile-generated deconcentration than the United States, with the central city retaining much of its middle-income population and economic base. But many large Canadian cities are now turning inside-out, and the new intraurban geography is increasingly symbolized by the suburban downtowns that anchor the ultramodern business complexes along the Highway 401 freeway north of Toronto and Alberta's West Edmonton mega-mall, the world's biggest shopping center.

Cultural/Political Geography

The historic cleavage between Canada's French- and English-speakers has resurfaced over the past four decades to dominate the country's cultural and political geography. By the time the Canadian federation observed its centennial in 1967, it had become evident that Quebecers regarded themselves as second-class citizens; they believed that bilingualism meant that French-speakers had to learn English but not vice versa; and they perceived that Quebec was not getting its fair share of the country's wealth. Since the 1960s, the intensity of ethnic feelings in Quebec has risen in surges despite the federal government's efforts to satisfy the province's demands. During the 1970s, while a separatist political party came to power in Quebec, a new federal constitution was drawn up in Ottawa. In 1980, Quebec's voters solidly rejected independence when given that choice in a referendum. But the new constitution did *not* satisfy the Quebecers, and throughout the 1980s and early 1990s the Ottawa government struggled unsuccessfully to devise a plan, acceptable to all the provinces, that would keep Quebec in the Canadian federation.

By 1995, with Canada's interest in constitutional reform exhausted, a second referendum on Quebec's sovereignty could no longer be put off. With the reenergized separatist party again leading the way, the Francophone-dominated electorate very nearly approved independence (which attracted an astonishing 49.4 percent of the vote), an outcome now regarded by Canadians as their country's "near-death experience." Subsequently, despite calls by many separatists for a follow-up vote that might turn narrow defeat into victory, plans for a third referendum were shelved because opinion polls in Quebec have shown an erosion in public support for secession. Several reasons lie behind that shift in attitude: (1) the implementation of provincial laws that firmly established the use of French and the primacy of Québécois culture; (2) the increased bilinguality of Quebec's Anglophones (English speakers), which has calmed Francophone fears concerning assimilation into Canada's English-speaking mainstream; (3) the arrival of a new wave of immigrants (some Francophone, some Anglophone) from Asia, Africa, Eastern Europe, and the Caribbean, who have accelerated the weakening of language barriers by settling in both French-speaking and English-speaking neighborhoods; and (4) the landslide defeat of the ruling separatist Parti Québécois in the provincial election of 2003.

Although the issue of Quebec's separation from the Canadian federation has been pushed into the background, it has not been buried and could resurface in the foreseeable future. The federalist opposition party that came to power in the 2003 provincial election has not fared well, and current polls show a persistent level of voter support for independence within striking distance at around 45 percent. Meanwhile, the Parti Québécois—which has lost previous elections only to come back strong—was engaged in a leadership struggle during 2005 that would determine how aggressively it will renew its campaign for sovereignty.

On the wider Canadian scene, the culture-based Quebec crisis has stirred the ethnic feelings of the country's 1.3 million native (First Nations and Inuit) peoples. Their assertions have also received a sympathetic hearing in Ottawa, and in recent years breakthroughs were achieved with the creation of Nunavut and a treaty for limited tribal self-government in northern British Columbia. A leading concern among these peoples is that their aboriginal rights be protected by the federal government against the provinces. This is especially true for the First Nations of Quebec's northern frontier, the Cree, whose historic domain covers more than half of the province of Quebec as it appears on current maps. Administration of the Cree was assigned to Quebec's government in 1912, a responsibility that a move toward independence might well invalidate. In any case, the Cree would also likely be empowered to seek independence. This would leave the French-speaking remnant of Quebec with only about 45 percent of the province's present territory. As Figure 3-17 shows, the territory of the Cree is no unproductive wilderness: it contains vital facilities of the James Bay Hydroelectric Project, a massive scheme of dikes, dams, and artificial lakes that has transformed much of northwestern Quebec and generates electrical power for a huge market within and outside the province. In 2002, after its attempts to block construction of the Project were supported by the federal courts, the Cree negotiated a groundbreaking treaty with the Quebec government whereby this First Nation dropped its opposition in return for a portion of the income earned from electricity sales as well as the granting of power to control its own economic and community development.

Finally, the events of the past quarter-century have profoundly impacted Canada's national political landscape, and not only in Quebec. Regionalism has also intensified in the west, whose leaders opposed federal concessions to Quebec and insisted that the equal treatment of all ten provinces is a basic principle that precludes the designation of special status for anyone. The most recent federal elections have clearly revealed the emerging fault lines that surround Quebec and set the western provinces (British Columbia, Alberta, and Saskatchewan) off from the rest of the country. Even the remaining Ontario-led center and the eastern bloc of Atlantic Provinces voted divergently, and today the politico-geographical hypothesis of "Four Canadas" may be on its way toward becoming reality. Thus, with its national unity under pressure, Canada today confronts the coalescing forces of *devolution* that threaten to transform the new fault lines into permanent fractures.

Economic Geography

As in the United States, the growth of Canada's spatial economy has been supported by a diversified, high-quality *resource base*. We noted earlier that the Canadian Shield is

AMONG THE REALM'S GREAT CITIES . . . TORONTO

Toronto, the capital of Ontario and Canada's largest metropolis (5.1 million), is the historic heart of English-speaking Canada. The landscape of much of its center is dominated by exquisite Victorian-era architecture and surrounds a healthy downtown that is one of Canada's leading economic centers. Landmarks abound in this CBD, including the SkyDome stadium, the famous City Hall with its facing pair of curved high-rises, and mast-like CN Tower, which at 1815 feet (553 m) is the world's tallest free-standing structure. Toronto also is a major port and industrial complex, with miles of facilities lining the shoreline of Lake Ontario.

Livability is one of the first labels Torontonians apply to their city, which has retained more of its middle class than central cities of its size in the United States. *Diversity* is another leading characteristic because this is North America's richest urban ethnic mosaic. Among the largest of more than a dozen thriving ethnic communities are those dominated by Italians, Portuguese, Chinese, Greeks, and Ukrainians; overall, Toronto now includes residents from 169 countries who speak more than 100 languages; and the immigrant inflow continues strongly, with those born outside Canada constituting 44 percent of the city's population (in all the world only Miami exhibits a higher percentage). *Vibrancy* is yet another hallmark of this remarkable city, whose resiliency was tested but not broken by the SARS virus outbreak of 2003.

Toronto has worked well in recent decades, thanks to a metropolitan government structure that fostered central city–suburban cooperation. But that relationship is increasingly stressed as the outer city gains a critical mass of population, economic activity, and political clout, while the central city was recently forced to accommodate an unwanted amalgamation with five of its nearest suburban neighbors.

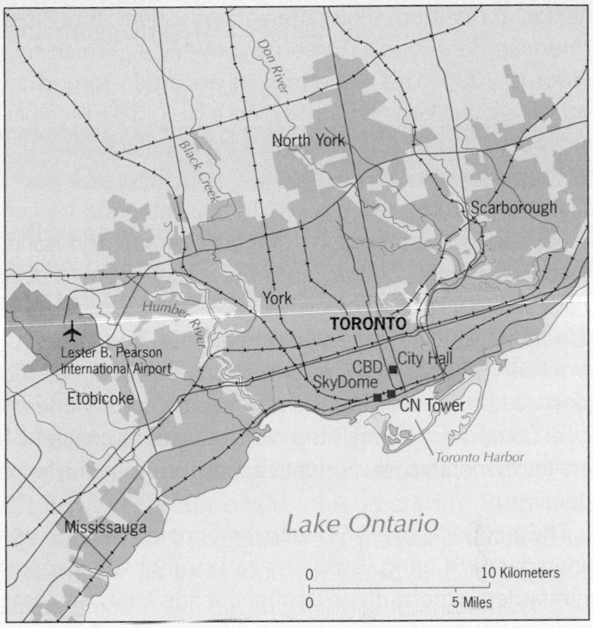

© H. J. de Blij, P. O. Muller, and John Wiley & Sons, Inc.

endowed with rich mineral deposits and that oil and natural gas are extracted in sizeable quantities in Alberta to the southwest of the Shield. Canada has long been a leading *agricultural* producer and exporter, especially of wheat and other grains from its breadbasket in the Prairie Provinces; new technologies keep productivity high, but labor requirements continue to diminish and farm workers now account for less than 2 percent of the national workforce. Postindustrialization has caused substantial employment decline in the *manufacturing sector*, with Southern Ontario's industrial heartland being the most adversely affected area. On the other hand, Canada's robust *tertiary and quaternary sectors* (which today employ more than 70 percent of the total workforce) are creating a host of new economic opportunities. Fortunately, Southern Ontario is benefiting from this postindustrial transformation because the country's leading high-technology, research-and-development complex has grown up around Waterloo and Guelph, a pair of university towns about 60 miles (100 km) west of Toronto.

Canada's economic future is also going to be strongly affected by the continuing development of its trading relationships. The landmark United States–Canada Free Trade Agreement, signed in 1989, initiated a now-completed phasing out of all tariffs and investment restrictions between the two countries, whose annual cross-border flow of goods and services is the largest in the international trade arena. (In the early 2000s more than four-fifths of Canada's exports went to the United States, from which it also derived about two-thirds of its imports.) In 1994 these economic linkages were further tightened by the implementation of the North American Free Trade Agreement (NAFTA), which consolidated the gains of the 1989 pact and opened major new opportunities for both countries by adding Mexico to the trading partnership (see box titled "North American Free Trade Agreement [NAFTA]").

agriculture is the predominant feature of the landscape. Whereas the innermost fruit-and-vegetable and dairying belts in the U.S. macro-Thünian regional system (Fig. 3-15) are in competition with other economic activities in the North American Core to the east, by the time one reaches the Mississippi Valley, meat and grain production prevail for much of the next thousand miles (1600 km) west to the base of the Rocky Mountains. Because the eastern half of the Continental Interior lies in Humid America and closer to the national food market on the northeastern seaboard, mixed crop-and-livestock farming won out over less competitive wheat raising, which was relegated to the fertile but semiarid environment of the central and western Great Plains on the dry side of the 100th meridian. The latter area also contains Canada's agricultural heartland north of the 49th-parallel border. Although these fertile Prairie Provinces are less subject to serious drought than the U.S. high plains to the south, they are situated at a more northerly latitude and thus have a shorter growing season.

The distribution of North American corn and wheat farming is mapped in Figure 3-20, and the edges of the most productive zones of these two leading Continental Interior crops coincide with several of the region's boundaries. As we saw earlier in Figure 3-15, the Corn Belt is most heavily focused on Iowa and part of neighboring Illinois, while the Spring and Winter Wheat Belts are respectively centered on North Dakota and western Kansas. This separation of the Wheat Belts is also the basis for subdividing the Continental Interior region, as the dashed line in Figure 3-19 indicates. The North subregion encompasses the Upper Midwest, the northern Great Plains, and the cultivable southern portions of Canada's three Prairie Provinces; the Spring Wheat Belt is its leading agricultural component and is so named because crops are planted in the spring and harvested in late summer or early autumn. The South subregion contains a major part of the Corn Belt in its eastern half and all of the Winter Wheat Belt in its western. The Corn Belt should probably be renamed the *Corn-Soybean Belt* because it also produces huge soybean crops that are grown in rotation with corn (see box titled "Geography of the Mighty Soybean"). The Winter Wheat Belt is centered in the southern Great Plains, and its name is derived from the practice of planting wheat in the fall and harvesting it by late spring; during its infancy, winter wheat can withstand the milder winter temperatures south of 40°N latitude. The great advantage of this growth cycle is that the crop is harvested before the hot, dry summer takes hold in this semiarid environment.

Throughout the Continental Interior, economic activity is markedly oriented toward farming. Its leading metropolises—Kansas City, Minneapolis–St. Paul, Winnipeg, Omaha, and even Denver—are major processing and marketing points for pork and beef packing, flour milling, and soybean, sunflower, and canola oil production (all of which are increasingly exported). Although family farms are still distributed widely across the rural landscape, the unrelenting, aggressive incursion of large-scale corporate farming threatens that way of life. This is particularly true for hog raising, where big companies are taking control of the entire production process in their ultramodern, factory-like complexes that gobble up huge swaths of farmland and disgorge appalling quantities of animal wastes that pollute air and groundwater for miles around.

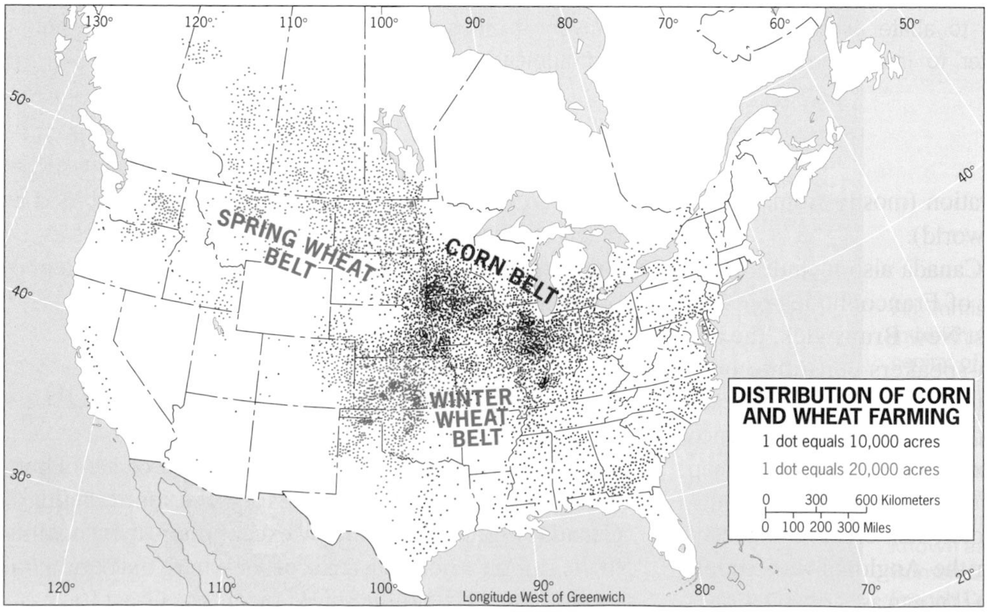

FIGURE 3-20 © H. J. de Blij, P. O. Muller, and John Wiley & Sons, Inc.

Geography of the Mighty Soybean

Over the past century, no crop in this realm has expanded faster than the soybean. This hardy legume was one of the earliest plants domesticated, and for nearly 5000 years the Chinese have revered it as a grain vital to life itself. The advantages of soybeans are numerous. They are the world's cheapest and most efficient source of protein, with a level of concentration double that of meat or fish. They are easy to cultivate, adapting to a wide range of environmental conditions, and they are less vulnerable to pests and diseases than other temperate-climate food crops. Economically, crop yields are comparatively high, and the demand for soybeans has steadily risen across the world in recent decades.

The dramatic increase of soybean production in the United States has been propelled not only by these advantages but also by this crop's particular suitability to the Corn Belt. It was quickly discovered that soybeans were an ideal crop to rotate with corn because they recondition the soil with nitrogen and other nutrients needed by corn. In terms of cultivation, soybeans also proved highly profitable to grow on the large, flat fields of the Corn Belt (which continues to account for about 80 percent of all U.S. soybean production) where their farm-machinery requirements were identical to those already in place for raising corn. Moreover, clever marketing constantly widened the versatility of soybeans beyond their use as a nutritious livestock feed, and today they are used in a variety of human foods (ranging from vegetable oil to a plethora of low-calorie products) as well as such industrial goods as fertilizers, paints, and insect sprays.

All of the above also contributed to steadily rising foreign demand, and in many recent years soybeans have ranked as the leading U.S. export commodity. Other producing countries have recognized this opportunity as well, and intensifying global competition since 1990 has resulted in a decline of America's share of the world market. The United States today produces roughly 35 percent of the world's annual soybean output (down from more than 80 percent in 1980, and 50 percent as recently as the late 1990s). As we note in Chapter 5, most of its competition—a skyrocketing 50-plus percent of the global supply in 2004—comes from Brazil and Argentina in mid-latitude South America; China, the largest single market country, is also seeking to expand its own acreage in a bid to become a major producer, but the unavailability of sufficient agricultural land has kept its production to just under 10 percent of the world supply. Soybeans will continue to play a key role in North American and international agriculture as the twenty-first century unfolds, buoyed in no small part by this grain's important potential in the struggle to ease the global food shortage.

As individual agricultural opportunities decline, especially in the less intensively farmed Great Plains to the west, the exodus of younger people accelerates. Left behind is an ever more elderly and poorer population, thinly dispersed across a constellation of stressed rural communities, struggling to survive. The Great Plains depopulation trend, in fact, is rapidly plummeting toward demographic collapse: fewer than 12 million residents remained on the Plains in 2005, a population total equal to only about half that of the New York metropolitan area. As Great Plains society unravels, a policy debate intensifies as to whether or not the best solution is planned abandonment and the return of at least a portion of these grasslands to their natural state.

Despite the key role that agriculture plays in the regional economy, the nonfarming sectors of the Continental Interior continue to develop and diversify, especially in and around major metropolitan centers. Minnesota has been the most successful in promoting such growth, capitalizing on the reputations of that State's stable economy, highly educated workforce, track record as a product innovator, and "north woods" amenities. In the urban areas of the eastern Great Plains, telemarketing has become a major pursuit, with Omaha, Nebraska particularly well known as a toll-free call center thanks to the many credit-card, reservations, and catalog-order companies that have concentrated there. Elsewhere, Gateway 2000, the computer sales giant, has singlehandedly put North Sioux City, South Dakota, on the map, and Sprint, a global telecommunications leader, has built its headquarters just southwest of Kansas City. In addition, the development of energy resources has propelled the growth of the region's northwestern corner, most notably southern Alberta's burgeoning Calgary-Edmonton corridor.

THE SOUTH

The American South occupies the realm's southeastern corner, extending from the lush Bluegrass Basin of northern Kentucky to the swampy bayous of Louisiana's Gulf Coast, and from the knobby hills of West Virginia to the flat sandy islands off the southernmost extremity of peninsular Florida (Fig. 3-19). Of the realm's nine regions, none has undergone more overall change during the past half-century. Choosing to pursue its own sectional interests practically from the advent of nationhood, the South experienced an even deeper economic and cultural isolation from the rest of the United States during the long and bitter aftermath of the ruinous Civil War. For over a century it languished in economic stagnation, but the 1970s finally witnessed a reassessment of the nation's perception of the region that launched a still-ongoing wave of growth and change unprecedented in Southern history.

Propelled by the forces that created the Sunbelt phenomenon, people and activities streamed into the urban South. Cities such as Atlanta, Charlotte, Miami, and Tampa became booming metropolises practically overnight, and conurbations swiftly formed in such places as southern and central Florida, the Carolina Piedmont, and the western Gulf Coast from Houston (just across the Texas border) through New Orleans to Mobile, Alabama. This "bulldozer revolution" was matched in the most favored rural areas by an agricultural renaissance that stressed such higher-value commodities as soybeans, winter wheat, poultry, and wood products. On the social front, institutionalized racial segregation was dismantled more than three decades ago.

Yet for all the growth that has taken place since 1970, the South remains a region beset by many economic problems because the geography of its development has been decidedly uneven. Although several metropolitan and certain farming areas have benefited, many others containing sizeable populations have not, and the juxtaposition of progress and backwardness is frequently encountered in the Southern landscape. Not surprisingly, the gap between rich and poor is wider here than in any other U.S. region. This economic disparity also carries racial dimensions: compared to whites, blacks exhibit higher poverty rates and rank well behind in median family income and achievement of middle-

A port is an interface between the land behind it and the water body in front of it, and some of the world's most productive ports lie at the mouths of major rivers—such as Rotterdam at the North Sea outlet of Western Europe's Rhine. New Orleans, located at the apex of the Mississippi Delta, has benefited from such a favored location since the early nineteenth century. Today, in the era of NAFTA, as the products of the U.S. heartland increasingly enter the international trade arena, southern Louisiana senses a heightened opportunity to tap this goods flow by convincing shippers to export by water via the Mississippi. In order to attract this business, huge investments have recently been made to modernize and expand the Port of New Orleans. Some of the results are visible here along this stretch of the great river just above the Crescent City itself, part of a huge new port complex that will extend more than 20 miles (32 km) upstream from downtown New Orleans (seen in the middle distance). © Courtesy of the Port of New Orleans. Photo by Henry Clay.

class status. Within this checkerboard spatial framework of development, the South's new affluence is largely concentrated in its central-city CBDs and burgeoning suburban rings. At the other end of the income spectrum, its far more widely dispersed struggling areas include less favored agricultural zones (such as the chronically depressed bottomlands that line the lower Mississippi) and myriad rural towns whose jobs are tied mainly to small-scale plants in locally stagnating industries such as food processing and garmentmaking. And at the regional scale, yet another geographic pattern of inequality emerges: except for centrally located Atlanta and the Interstate-85 "Boombelt" corridor extending from there northeastward into central North Carolina, a significant proportion of the South's recent economic growth has taken place on its periphery in the Washington suburbs of northern Virginia, on the western Gulf Coast, and along Florida's central (Interstate-4) and southern-coast corridors.

The transformation of the South is also changing the complexion of its demographic mosaic. Racially and economically, the composition of the region's population continues to be reshaped by the in-migration of whites and blacks from the North, with both groups exhibiting higher income levels than their resident Southern counterparts. Ethnically, the steadily increasing presence of Hispanics and Asians is intensifying the South's cultural heterogeneity. The region's Hispanic population growth has been particularly strong, more than doubling since 1990 as largely Mexican and Central American migrants have streamed in to fill agricultural and unskilled industrial jobs that others would not take. In age structure, too, the region is diversifying as an influx of retirees continues, not just in Florida but also in Georgia, Virginia, and the Carolinas. Finally, we should remember that the massive construction of new urban landscapes since 1980 has provided ample opportunity for creating fragmented residential environments that typify today's mosaic culture. Thus the filling of the tiles of the Southern population mosaic by an ever more diverse array of social and lifestyle groups is another signal that this once outcast region has all but completed its convergence with the rest of the country.

This 30-year transition period has ended with the South not only fully absorbed into the *national* economy and society, but also trying to become a player on the *international* stage. During the 1990s, European and Japanese manufacturers were active investors in the South, but industrial jobs are now proving hard to hang onto as factories increasingly relocate outside the United States to take advantage of cheaper foreign labor. Thus the overall impact of globalization has so far been modest, and the South today is the least successful exporter of any U.S. region. As for the future, in terms of relative location the most promising new foreign business opportunities to develop in the NAFTA era lie to the south in Middle and South America. South Florida has been in the vanguard here, with Miami appropriately billing itself as "the gateway to Latin America." Lately, other cities of the South have indicated an interest in undercutting this monopoly, most notably Atlanta (expanded nonstop air routes), New Orleans (seaport enlargement; photo p. 184), and Memphis (global distribution hub for both Federal Express and United Parcel Service).

THE SOUTHWEST

Before the late twentieth century, most geographers did not even identify a distinct Southwestern region, classifying the U.S.-Mexican borderland zone into separate southward extensions of the Continental Interior and the mountain/plateau country to its west. Today, however, this steadily developing area—constituted by the States of Texas, New Mexico, and Arizona—has earned its place on the regional map of North America (Fig. 3-19). The still-emerging Southwest is also unique in the United States because it is a bicultural regional complex. Atop the crest of the Sunbelt wave, in-migrating Anglo-Americans joined a rapidly expanding Mexican American population anchored to the sizeable long-time resident Hispanic population, which traces its roots to the Spanish colonial territory that once extended from Texas's Gulf Coast to central California. In fact, when we add the large Native American population, it is not inappropriate to recognize the Southwest as a *tricultural* region.

Recent rapid development in Texas, Arizona, and New Mexico is built on a three-pronged foundation: (1) availability of huge amounts of *electricity* to power air conditioners through the long, brutally hot summers; (2) sufficient *water* to supply everything from swimming pools to irrigated crops, so that large numbers may reside in this dry environment; and (3) the *automobile*, so that affluent newcomers may spread themselves out at much-desired low densities. The first and third of these have been rather easily attained because the eastern half of the Southwest is abundantly endowed with oil and natural gas (Fig. 3-14). But the future of water supplies, especially in this decade of record drought throughout the western U.S., is far more problematical. For instance, Phoenix and Tucson could not keep growing at their current swift pace without the massive new Central Arizona Project (CAP) Canal, which transports Colorado River water across hundreds of miles of desert.

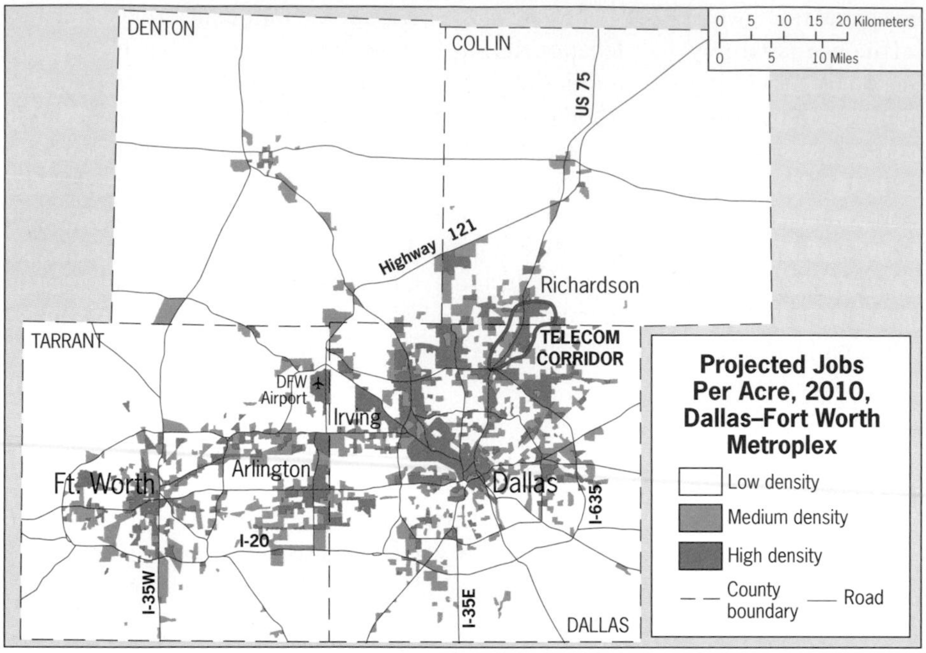

FIGURE 3-21 Adapted with permission from *North Central Texas: Regional Profile, 1990 to 2010.* © North Central Texas Council of Governments, 1994.

The economic growth of the Southwest in recent years has been led by Texas, which during the 1990s weaned itself away from dependence on its oil and gas industries (whose share of the State's annual income is now only about one-quarter of the 1980 level). Following a serious oil-induced recession in the 1980s, the Lone Star State's mix of economic activities was restructured and diversified. Success was not long in coming, and today the eastern Texas triangle formed by Dallas–Ft. Worth, Houston, and San Antonio has become one of the world's most productive postindustrial complexes, specializing in business-information and health-care services and high-technology manufacturing. The metropolis centered on the State capital—Austin, which lies near the heart of the triangle—has become a technopole that is second only to Silicon Valley as a leader in computer research and development as well as a producer of semiconductors and PCs. To the north at the triangle's apex, the Dallas–Ft. Worth (DFW) Metroplex is keeping pace with its own expanding cluster of high-tech industries focused on the *Telecom Corridor* technopole just north of Dallas (the world's largest concentration of telecommunications firms). As Figure 3-21 shows, the geography of employment is increasingly oriented to booming suburban freeway corridors, a spatial pattern intensifying as well in metropolitan Austin, Houston, and San Antonio. The DFW Metroplex is also the leading Texas command post for international trade and NAFTA-spawned development, and it is becoming the northern anchor of a new transnational growth corridor that extends for more than 500 miles (800 km) southward from Dallas through Austin and San Antonio to Monterrey 150 miles (250 km) inside Mexico.

The western flank of the Southwest—whose development is concentrated mostly in southern Arizona—is unique. Equaled only by parts of southern Israel's Negev, this is the most technologically transformed desert environment on Earth. But while the Negev is devoted to agriculture, Arizona's far more intensive development is marked by vast urbanized landscapes focused on two of the largest western U.S. cities, Phoenix and Tucson. During most of the 1990s, metropolitan Phoenix was expanding into the surrounding desert at the rate of an acre per hour, and it still grows at an astonishing yearly rate of roughly 4 percent. Although the forces of the slow-growth movement are gaining momentum, Arizona's conservative political and business leaders have been loath to restrict development. This debate is becoming moot, however, because everyone agrees that retrievable water supplies—the State's lifeblood and key to its future—have just about reached their limit.

THE WESTERN FRONTIER

To the north of the western half of the Southwest lies the realm's newest region: the swiftly growing, economically booming Western Frontier. In its east-west extent, this mountain-studded plateau region stretches between (and includes) the Rocky Mountains and the Sierra Nevada–Cascades chain; its longer north-south axis reaches from northern Arizona's Grand Canyon country

There is something quite remarkable about the ongoing urbanization of arid valleys in the high desert of the Western Frontier, and nowhere is this phenomenon more starkly on view than in the Las Vegas Valley in Nevada's southernmost extremity. This particular municipality is Henderson, not only Las Vegas's largest suburb but also the fastest-growing urban settlement in the entire United States between 1990 and 2000. Many affluent retirees have also been attracted to this immediate area, and together with their younger counterparts have shaped a classic mosaic-culture urban fabric dominated by a low-density checkerboard of self-contained, socially homogeneous communities. Architectural critics and environmentalists, of course, would have no trouble in declaring the filling of this landscape to be one of the most blatant examples of urban sprawl on view anywhere in North America. © Mike Yamashita/Woodfin Camp & Associates.

to the edge of the subarctic snowforest west of the Canadian Rockies (Fig. 3-19). The region's heartland encompasses Utah, Nevada, Idaho, and western Colorado, and it is here that the old intermountain West is most actively reinventing itself. Until recently, such an opportunity would have been unthinkable because of this region's remoteness, dryness, and sparse population. But the Western Frontier's geographic prospects are changing significantly as the postindustrial national economy matures, advances in communications and transportation technologies eliminate interregional differences in the costs of doing business, and amenities become the leading locational preference of employers. Undoubtedly, proximity to the increasingly expensive and overcrowded Pacific coast is also a reinforcing factor: well over 1 million disenchanted Californians have moved into the Western Frontier since 1995, lured by its sunny climate, wide-open spaces, spectacular scenery, proliferating jobs, and lower cost of living.

The label *Western Frontier* further emphasizes that the transformation under way is being driven by the influx of people and activities from outside the region. Indeed, the populations of the four States of its heartland all rank among the fastest-growing in the United States since 1990, with the more than 2.5 million migrants who arrived during the past decade already constituting more than 25 percent of the Western Frontier's population. In the economic-geographic sphere, the region's identity is being reforged as traditional resource-based industries (mining, timber, livestock grazing) decline, while in-migrating and new indigenous companies create thousands of high-tech manufacturing and specialized services jobs. The spatial pattern of this development is strongly focused on the region's mushrooming urban areas and creates an unusual frontier: instead of a single line of advancing settlement, a dispersed set of population clusters is simultaneously expanding around large- and middle-sized cities, high-amenity zones, and interstate freeway corridors.

The hottest growth area of all in the Western Frontier region—and the fastest-growing large metropolis in the United States—is the Las Vegas Valley of southernmost Nevada (photo above). This boom was triggered by Las Vegas's burgeoning recreation industry (which now attracts about 40 million visitors annually), but it is also the product of new economic development dominated by the influx of high-tech and professional services firms that specialize in computers, finance, and health care. Business entrepreneurs have flocked here, and the new employment opportunities sparked a local population explosion, spearheaded by those relocating Californians. Relatively cheap land, modest housing prices, low taxes, and the sunny climate further fueled this migration.

Other urban areas are also experiencing accelerated growth, particularly the region's two largest metropolises, centered on Denver and Salt Lake City (which have spawned conurbations extending more than 50 miles north and south of the central city). In both cases the leading amenity is the Rocky Mountains, which soar majestically above the urbanizing flatlands and foothills. Metropolitan

Denver has expanded rapidly along (and into) central Colorado's Front Range, building a diverse new, globally oriented economic base dominated by telecommunications, computer software, and financial services. On a somewhat smaller scale, Salt Lake City has followed in Denver's footsteps on the Rockies' western flank and now centers a thriving conurbation that lines northern Utah's Wasatch Front from Ogden in the north to Provo in the south. In fact, the budding technopole of *Software Valley*, the 40-mile (65-km) corridor linking Salt Lake City and Provo, contains one of the world's largest concentrations of information technology companies, built up around two universities, a highly educated workforce, and a postindustrial complex specializing in the production of word-processing and computer-simulation software.

As much as 90 percent of the Western Frontier's development is confined to its widely dispersed urban areas, but the effects of this rapid growth are rippling deeply into the region's vast rural zones. Some of the problems—such as the steadily rising incidence of wildfires as zones bordering forests become ever more populated—are caused by the spillover effects of urbanization on adjacent areas. But many others result from the collision of lifestyles as higher-income, recreation-seeking city dwellers invade a countryside whose long-time, nonaffluent residents are struggling to preserve a way of life based on declining activities such as mining and extensive agriculture. In the valleys of the Colorado Rockies, ranchers are further squeezed by the upwardly spiraling land values set off by the massive growth of ski resorts—and usually are forced to sell and relocate their operations to less favorable land in even more remote areas. In the small towns of northern Idaho and western Montana, social tensions are inflamed as local right-wing militias encounter more liberal transplanted Californians. And seemingly everywhere, the plans of developers are blocked by environmentalists. The rapid emergence of a new preservation ethic, which increasingly threatens environmentally insensitive land and resource developers, is part of a larger antigrowth movement that is certain to intensify. Throughout this region today, the buzzwords are "managing growth without restricting economic opportunity"—which usually translates into bringing opposing interests together to work out compromises.

Whatever our new century brings, the growth boom of the 1990s that forged the realm's newest region has forever altered the economic-geographic and demographic complexion of much of the old intermountain West. Now reborn as the Western Frontier, this no-longer-remote region is putting the often disastrous boom/bust cycles of its mining- and grazing-dominated past behind it and is making major strides toward establishing itself as a globally connected, technological powerhouse on the inland doorstep of North America's burgeoning sector of the Pacific Rim (see box on p. 190).

THE NORTHERN FRONTIER

The northern half of the realm, lying poleward of roughly 52°N latitude, constitutes the Northern Frontier region (Fig. 3-19). This is by far North America's biggest regional subdivision, covering almost 90 percent of Canada and all of the U.S. State of Alaska. Indeed, if this huge territory were a separate country, it would rank as the world's fifth-largest in size. The latitudinal extent of this region leaves no doubt as to why it is designated as *northern*. In fact, after northern Russia (whose dimensions of continentality are even more pronounced), this is our planet's second greatest expanse of harsh cold environment affecting an inhabited area. As with the Arctic fringes of Eurasia, very few physical barriers in northern North America ameliorate the masses of supercold air, which throughout the long winter plunge directly southward from the ice-covered Arctic Ocean. Not surprisingly, most of the Northern Frontier's people and activities are concentrated in the milder, heavily forested southern half of the region.

Unquestionably, this region is also a *frontier*—and far more classically so than the Western Frontier just discussed. The latter's newly activated, dispersed, urbanized development pattern and dramatic population influx are largely attributable to its improved economic-geographic position astride a national axis, within which significant east-west interaction was already taking place. The Northern Frontier, on the other hand, represents a thrust into undeveloped country in an entirely new direction, one marked by the leisurely, long-term advance of a line across parts of Canada's near- and middle-northern periphery that absorbs the areas it crosses into the national spatial economy. Boom and bust cycles since the late nineteenth century have controlled the rate of movement of this development line, which frequently has ebbed backward as much as it has flowed forward. In general, the still sparsely populated Northern Frontier is mostly a region of recent European settlement based on the extraction of newly discovered resources. As we shall see, however, First Nations and other indigenous peoples, who have always regarded this region as a homeland, are today playing a growing role in the affairs of the Northern Frontier and will have much to say about shaping its future.

As noted earlier, the Canadian Shield, which covers the eastern half of the Northern Frontier (Fig. 3-2), is one of the world's richest storehouses of mineral resources. Most of the Shield's developing areas are focused on mining complexes that extract such high-demand metallic ores as nickel, uranium, copper, gold, silver, lead, and zinc. The leading zones of production are located along the Shield's southern margins, but since 1980 the search for additional deposits, the expansion of road and rail networks, and the opening of new mines have all advanced steadily northward. The Yukon and Northwest Territories have proven bountiful, with gold- and diamond-mining (Canada is now the world's fourth largest producer) paving the way; and at the opposite, eastern edge of the Shield, near Voisey's Bay on the central coast of Labrador, the largest body of high-grade nickel ores ever discovered is being opened up for full-scale production.

Besides metallic minerals, the Northern Frontier is endowed with an abundance of other resources, which include fossil fuels, hydroelectric power opportunities, timber, and fisheries. These are dominated by the substantial oil and gas reserves that are concentrated in a wide zone east of the Canadian Rockies between the Yukon and the U.S. border (Fig. 3-14). Most intriguing are the Athabasca Tar Sands of northeastern Alberta, which contain more crude oil than all of Saudi Arabia but are deposited as a tar-like substance (bitumen) in near-surface sand formations. Recent technological breakthroughs are enabling this energy supply to be extracted ever more profitably, output is steadily rising, and over the next decade the Tar Sands are expected to become one of Canada's leading oil-producing zones.

The productive activities of the Northern Frontier are linked together within a far-flung network of mines, oil- and gasfields, pulp mills, and hydropower stations that have spawned hundreds of settlements and thousands of miles of interconnecting transport and communications facilities. Though dominant, this commercial economy constitutes just one component of a dual regional economy. The other component, as Robert Bone points out in his book *The Geography of the Canadian North*, consists of the native economy that "evolved from a subsistence hunting economy to one in which wage employment, transfer payments, and trapping provide cash income while hunting provides country food." This native economy, which adheres to the traditional values of the indigenous peoples (First Nations), has long been tied to the commercial economy as a source of wages. It has also been heavily land-based, and in recent years First Nations throughout the region have become more assertive and have filed legal actions to recover lands that Europeans took in the past without negotiating treaties. Both the courts and the federal government have been sympathetic, and the creation of Nunavut in 1999 (see map, p. 168) has been the most dramatic breakthrough to date. It now seems clear that First Nations will increasingly be consulted on future resource development in their traditional homelands; however, because vast tracts of northern lands are likely to be involved, this represents a potential threat to national unity should related issues of self-government become enmeshed in the process.

The Northern Frontier also contains Alaska, whose geographic opportunities and challenges differ somewhat from those of the Canadian North. Besides being the largest U.S. State, Alaska earns a sizeable income from oil production on its northeastern Arctic slope, and the south-central coast around Anchorage increasingly benefits from its near-midpoint location on the Pacific Rim air route between the "Lower 48" States and East Asia. These advantages have propelled a growth spurt of more than 100 percent since 1970, and today Alaska's total of just over 680,000 accounts for nearly one-third of the Northern Frontier's entire population on less than one-sixth of its land area. Here, too, metallic minerals have long drawn migrants northward. But, unlike Canada to the east, Alaska's recent boom was triggered by the opening of its vast oilfields on the North Slope, one of the hemisphere's leading energy sources since the 800-mile Trans-Alaska Pipeline began operating in 1977 (Fig. 3-14). Dwindling supplies at the main field at Prudhoe Bay are now compelling producers to turn to the huge additional reserves of the Arctic slope, but strong opposition from nature preservationist groups has slowed their plans. To the east lies the Arctic National Wildlife Refuge, where the Bush administration plans to begin drilling. To the west lies the National Petroleum Reserve-Alaska, which the federal government partially opened in 1999 to oil leasing by drillers, who insist that their newly gained sensitivity paired with the latest technologies will minimize the impact of production on the tundra's fragile ecosystems. Another major project that faces a similar battle is the proposal to construct a natural-gas pipeline to parallel the oil pipeline (huge supplies of liquified gas are a byproduct of oil drilling). As noted earlier, demand for clean-burning natural gas by electricity producers is rising all across North America.

THE PACIFIC HINGE

The Pacific coastlands of the conterminous United States and southwesternmost Canada, which comprise the realm's ninth region (Fig. 3-19), have been a powerful lure to migrants since the Oregon Trail was pioneered more than 160 years ago. Unlike the remainder of western North America south of 50°N latitude, the strip of land between the Sierra Nevada–Cascade mountain wall and the sea receives adequate moisture. It also possesses a far more hospitable environment, with generally delightful weather south of San Francisco, highly productive

FROM THE FIELD NOTES

"A short climb to an overlook near the Oregon-California border provides an instructive perspective over a geologically active segment of the Pacific coastline. This is quite a contrast to the smooth-sloping shores of most of eastern North America: deep water, unobstructed (wind) fetch, and pounding waves drive back this shore along steep cliffs. The owners of those apartments must have dramatic ocean views, but future generations will face the reality of nature's onslaught." © H. J. de Blij

farmlands in California's Central Valley, and such scenic glories as the Big Sur coast, Washington State's Olympic Mountains, and the spectacular waters surrounding San Francisco, San Diego, Seattle, and Vancouver. Most of the major development here took place during the post–World War II era, accommodating enormous population and economic growth along this outermost edge of the conterminous United States. But with the arrival of the twenty-first century, it is clear that, in terms of its economic geography, the West Coast no longer represents an end but rather a beginning—a gateway to an abundance of growing opportunities that in recent times have blossomed on many of the distant shores encircling the Pacific Basin. That is why we use the term *hinge*, because now this region increasingly forms an interface between North America and the booming, still-emerging *Pacific Rim* (see box titled "The Pacific Rim Connection").

Sixty years of unrelenting growth have taken their toll, and whereas the West Coast States and British Columbia are now savoring the prospects of their Pacific Rim location, they must also confront the less pleasant consequences of regional maturity. This is especially true in California, where the massive development of America's most populated and multiethnic State (37 million in 2006, 15 percent larger than all of Canada) has been overwhelmingly concentrated in the teeming conurbation extending south from San Francisco through San Jose, the San

The Pacific Rim Connection

Today we continue to witness the rise of a far-flung region born of a string of economic miracles on the shores of the Pacific Ocean. It is a still-discontinuous region, led by its foci on the Pacific's western margins: Japan, coastal China, South Korea, Taiwan, Thailand, Malaysia, and Singapore.
20 The regional term **Pacific Rim** has come into use to describe this dramatic development, which over the past quarter-century has redrawn the map of the Pacific periphery—not only in East and Southeast Asia but also in the United States and Canada, and even in such Southern Hemisphere locales as Australia and South America's Chile.

The Pacific Rim also is a superb example of a functional region, with economic activity in the form of capital flows, raw-material movements, and trade linkages generating urbanization, industrialization, and labor migration. In the process, human landscapes from Sydney to Santiago are being transformed within the 20,000-mile (32,000-km) corridor that girdles the globe's largest body of water. We will continue to discuss this important regional development and its spatial impacts, in later chapters, wherever the Pacific Rim intersects Pacific-bordering geographic realms.

FROM THE FIELD NOTES

"Minutes after takeoff from the Seattle-Tacoma International Airport on the way to Tokyo we got a superb view of central Seattle from the ferry terminals on Elliott Bay to Interstate-5 linking California to Canada and beyond. Like Sydney, Australia, Seattle has suburbs across the water reached most easily by boat; this is one of America's most scenic and vibrant cities. Seattle has transformed itself repeatedly through economic cycles reflected by the urban landscape: the dominance of precision-equipment manufacture, long dominated by Boeing Aircraft, is ending in favor of a new era of computer technology (Microsoft has its headquarters in a suburb) and related industries of the 'New Economy.' " (There is something, though, that dates this photograph . . . what is it?) © H. J. de Blij

Joaquin Valley, the Los Angeles Basin, and the southwestern coast into San Diego at the Mexican border. Environmental hazards bedevil this entire corridor, including inland droughts, coastal-zone flooding, mudslides, brushfires, and particularly earthquakes—with the ominous San Andreas Fault practically the axis of megalopolitan coalescence. To all this, humans have added their own abuses of California's fragile habitat, from overuse of water supplies (requiring vast aqueduct systems to import water from hundreds of miles away) to the chronic air pollution of the Los Angeles–San Diego coastal strip.

These challenges notwithstanding, Californians over the past half-century have built one of the realm's most productive economic machines. Indeed, if the Golden State were an independent country, its economy would today rank as the world's sixth largest. The road to success, however, has often proven to be a bumpy one, as was the case in Southern California during the 1990s. This subregion began the decade struggling as the national recession combined with the unexpected end of the Cold War to deliver a serious blow to the Los Angeles-area economy, which at the time was heavily dependent on its aerospace industry driven by federal government defense expenditures. By the late 1990s, a turnaround was underway as Southern California successfully diversified its economy and developed such new growth sectors as the entertainment industry, foreign trade, and the nation's largest single metropolitan manufacturing complex—whose products range from high-tech medical equipment to the clothing churned out by the low-tech factories of a burgeoning, immigrant-staffed garment industry that continues to outpace New York City's.

The northern portion of the region encompasses Northern California and the Pacific Northwest, the latter extending northward from Oregon's Willamette Valley through the Cowlitz-Puget Sound lowland of western Washington State into the adjoining coastal zone of British Columbia beyond the Canadian border. Since 1990, Northern California, focused on the San Francisco Bay Area, has experienced a smoother path than the southern part of the State. Here an unprecedented period of economic expansion continues, led by Silicon Valley. Within that technopole, no community stands to gain more from this boom than Palo Alto, which is situated at the largest switching point on the Internet that handles more than half of its global traffic (see Fig. 3-16). As Internet commerce grows explosively, companies are flocking to Palo Alto to take advantage of its unparalleled access to cyberspace. This is further facilitated by the city's state-of-the-art local fiber-optic network, and what is taking shape here is nothing less than an infrastructural hub that will become one of the cornerstones of the so-called New Economy.

Continuing up the coast, the Pacific Northwest is also deeply involved in the pioneering and production of computer technology, especially in the technopole centered on suburban Seattle's Redmond (Microsoft's headquarters). Originally built on timber and fishing—primary activities that still survive here despite human pressures on the environment—this subregion found its impetus for

AMONG THE REALM'S GREAT CITIES . . . LOS ANGELES

It is hard to think of another city that has been in more headlines in the past several years than Los Angeles. Much of this publicity has been negative. But despite its earthquakes, mudslides, brushfires, pollution alerts, showcase trials, and riots, L.A.'s glamorous image has hardly been dented. This is compelling testimony to the city's resilience and vibrancy, and to its unchallenged position as the Western world's entertainment capital.

The plane-window view during the descent into Los Angeles International Airport, almost always across the heart of the metropolis, gives a good feel for the immensity of this urban landscape. It not only fills the huge natural amphitheater known as the Los Angeles Basin, but it also oozes into adjoining coastal strips, mountain-fringed valleys and foothills, and even the margins of the Mojave Desert more than 50 miles (80 km) inland. This quintessential spread city, of course, could only have materialized in the automobile age. In fact, most of it was built rapidly over the past 60 years, propelled by the swift expansion of a high-speed freeway network unsurpassed anywhere in metropolitan America. In the process, Greater Los Angeles became so widely dispersed that today it has reorganized within a geographic framework of six sprawling urban realms (see Fig. 3-12).

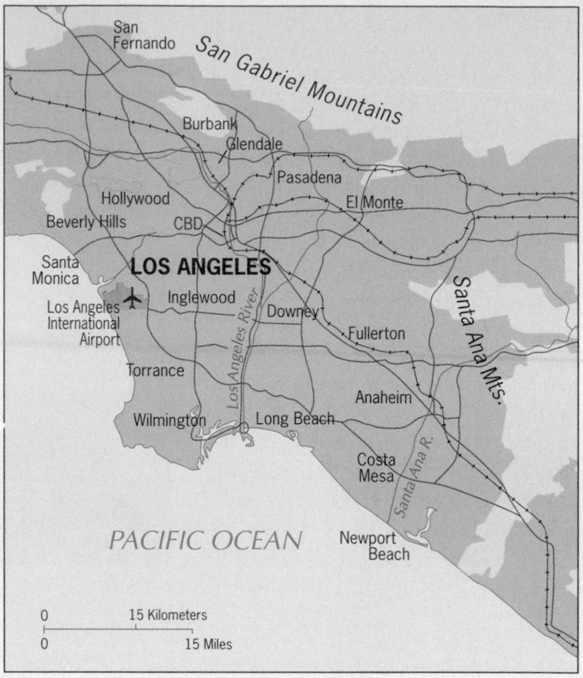

The metropolis as a whole is home to 17.4 million Southern Californians, constitutes North America's second largest urban agglomeration, and forms the southern anchor of the huge California conurbation that lines the Pacific coast from San Francisco southward to the Mexican border (Fig. 3-9). It also is the Pacific Hinge's leading trade, manufacturing, and financial center. In the global arena, Los Angeles is the eastern Pacific Rim's biggest city and the origin of the largest number of transoceanic flights and sailings to Asia.

industrialization in the massive Columbia River dam projects of the 1930s and 1950s, which generated the cheap hydroelectricity that attracted aluminum and aircraft manufacturers. Although Boeing is still a major employer, the recent influx of affluent newcomers is fueling the growth of software and Internet companies that is spearheading Seattle's transformation into the prototype metropolis of the New Economy. Nonetheless, what appear to be short-term economic problems have arisen in the early twenty-first century as a result of the downturn in the global computer and airline industries.

North of the Canadian border, the Pacific Hinge's northernmost subregion stands out from the others not only because of its British Columbia location but also because its economic core, Vancouver, has become the most Asianized metropolis in North America. Vancouver lies closer to East Asia on the air and sea routes than the cities to its south, and it got a head start in forging trade and investment linkages to the western Pacific Rim thanks to its large and growing Asian population. Today, ethnic Chinese residents constitute fully 20 percent of the metropolitan population, whose total Asian component now exceeds a remarkable 35 percent. Although a new east-west hybrid culture is trying to establish itself, frictions intensified during the late 1990s as wealthy Asians (many of them expatriates of now-Chinese Hong Kong) took over neighborhoods and replaced traditional homes with much larger houses, surrounded them with walls and fences where none had existed, and cut down most of the trees. More serious is the growing belief by the majority Anglos that the newcomers do not really want to assimilate, preferring instead to avoid the use of English and live, shop, and transact business exclusively within their own communities.

With North America ever more tightly enmeshed in the global economy, nowhere are the realm's international linkages more apparent today than in the Pacific Hinge. From the northern end of this region to the Mexican border, its geographic advantages as a gateway to the Pacific Rim are providing opportunities undreamed of barely a decade ago. Asianizing British Columbia now exports more than 40 percent of its goods to the countries of the western Pacific Rim (a proportion far higher than in any other Canadian province or U.S. State). The rest of the Vancouver–Seattle–Portland corridor is following suit, led by its high-technology industries. In California, the foreign trade sector of the State's gigantic economy has tripled since 1990, and today more than 15 percent of its jobs are export-dependent. Both Northern and Southern California are expected to thrive as these trends continue, with metropolitan Los Angeles becoming the financial, manufacturing, and trading capital of the eastern Pacific Rim as well as its transport hub (L.A. overtook New York in the mid-1990s as the leading U.S. seaport in the value of goods passing through it). Thus it is abundantly clear today that the mid-latitude West Coast no longer serves as North America's back door, but forms the hinges of a new front door to the Pacific arena thrown wide open to foster international interactions of every kind.

WHAT YOU CAN DO

RECOMMENDATION: Get involved! Help a Geographic Alliance! Because "geographic illiteracy" still prevails in the United States, the National Geographic Society in the 1980s launched a project to help teachers prepare themselves for geography instruction. Teachers from every State in the Union were invited to participate in seminars at NGS headquarters in Washington, D.C., and these teachers subsequently presented similar seminars for colleagues in their home States. Thus a State-based network of Geographic Alliances was created and still functions today. These Geographic Alliances are always looking for assistance, resources, ideas, and other contributions. It would be an interesting research project to discover whether geography is taught in the schools in the town where your college is located, and whether the teachers teaching it were part of the Alliance network.

GEOGRAPHIC CONNECTIONS

1 Imagine that you are the chief executive of a successful young company that manufactures software for the newest high-speed computers. Where would you locate your company and why? Establish a "short list" of three possible sites, and choose one by a process of elimination in which your decision is built upon the most important locational variables for your plant and its highly skilled workforce.

2 Immigration has long been a potent force in shaping the population mosaic of North America. Why is the "geography of demography" in the United States today so different from that of stagnant Europe? How does that difference express itself regionally? Does the immigration process affect your daily life, and how might it affect your competitive position when you are ready to enter the job market following completion of your education? Enhance your experience in considering these questions by surfing the Internet and consulting appropriate websites. As an extra endeavor, ponder these same questions as they would apply to the contemporary Canadian scene, keeping an eye out for similarities and differences between the two countries of this realm.

a

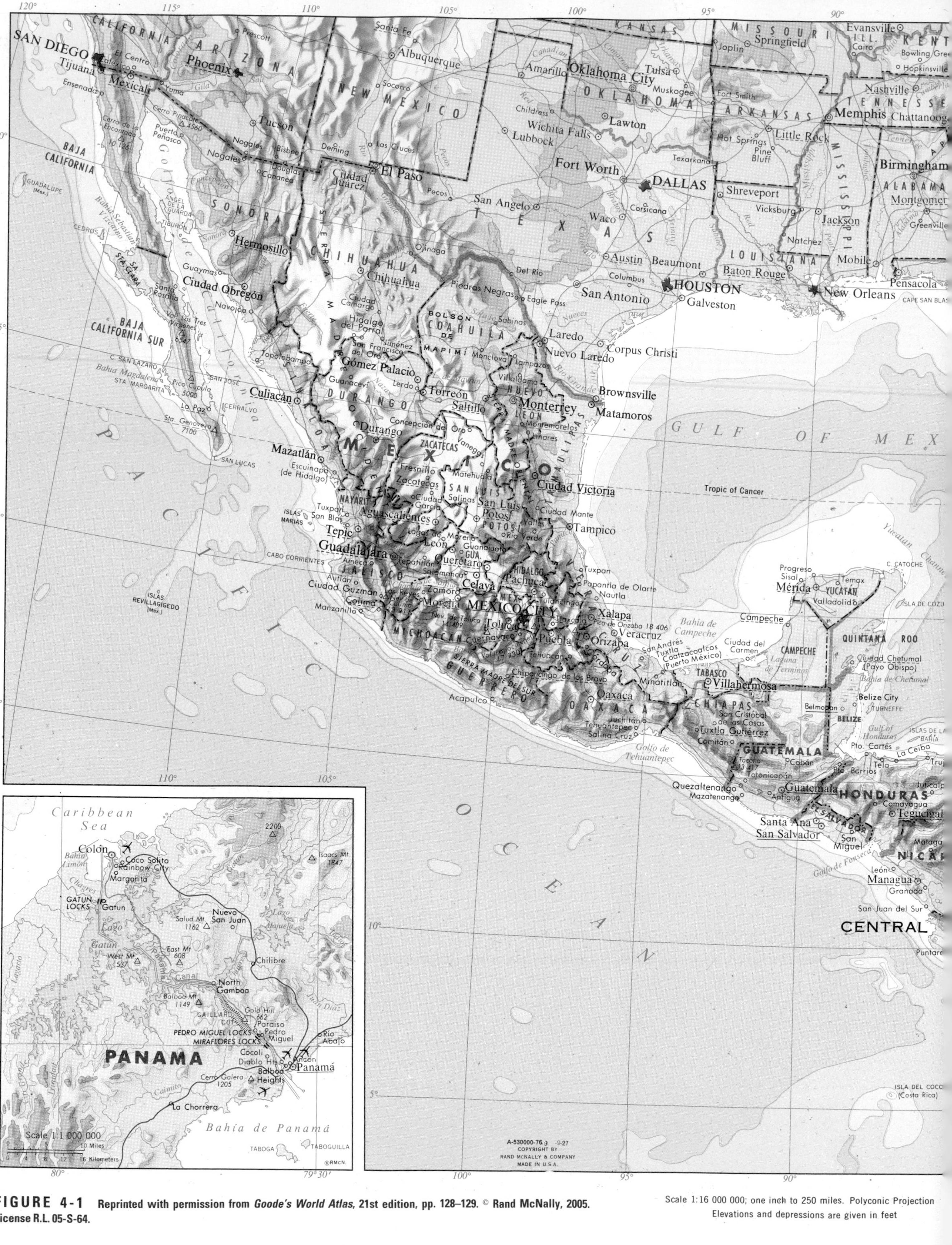

FIGURE 4-1 Reprinted with permission from *Goode's World Atlas*, 21st edition, pp. 128–129. © Rand McNally, 2005. License R.L. 05-S-64.

Scale 1:16 000 000; one inch to 250 miles. Polyconic Projection
Elevations and depressions are given in feet

Chapter 4

Middle America

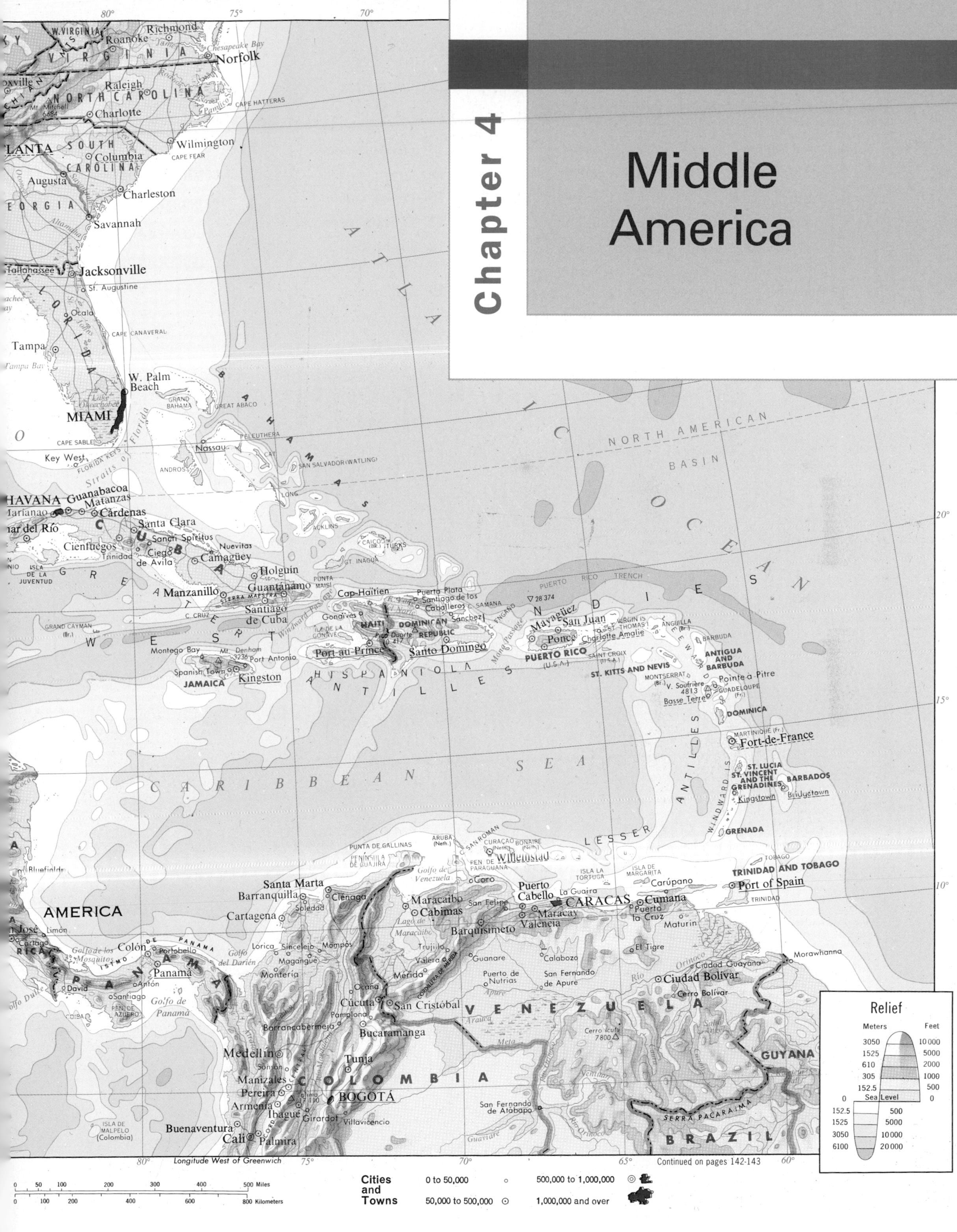

CONCEPTS, IDEAS, AND TERMS

1. Land bridge
2. Archipelago
3. Culture hearth
4. Mainland-Rimland framework
5. Mestizo
6. Hacienda
7. Plantation
8. Acculturation
9. Transculturation
10. Maquiladora
11. Dry canal
12. Altitudinal zonation
13. *Tierra caliente*
14. *Tierra templada*
15. *Tierra fría*
16. *Tierra helada*
17. *Tierra nevada*
18. Tropical deforestation
19. Mulatto

REGIONS

- MEXICO
- CENTRAL AMERICA
- CARIBBEAN BASIN
  - GREATER ANTILLES
  - LESSER ANTILLES

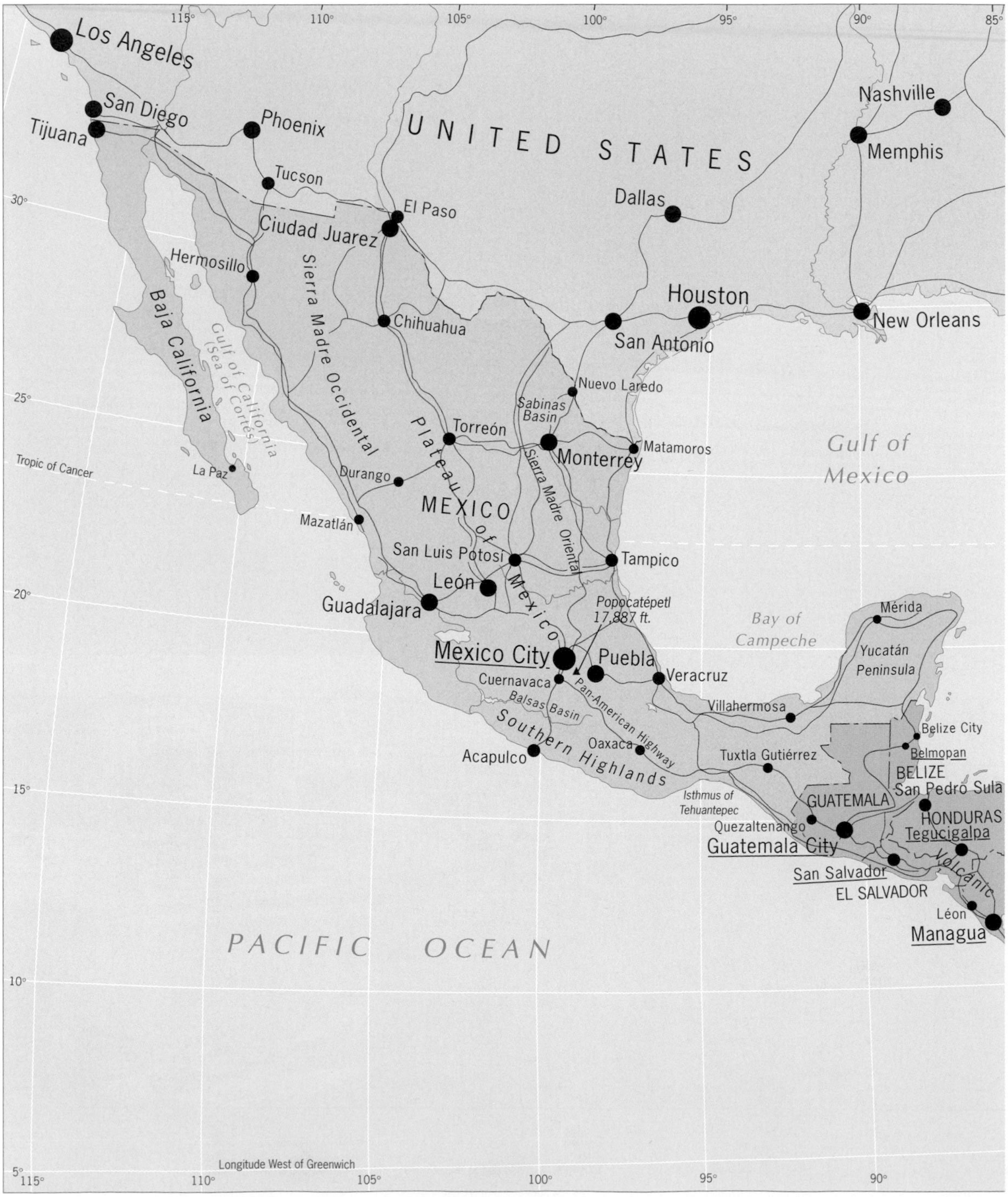

Middle America is a realm of vivid contrasts, turbulent history, continuing political turmoil, and an uncertain future. The realm's diversity, visible in Figure 4-1, encompasses all the lands and islands between the United States to the north and South America to the south. Its regions are mapped in Figure 4-2 and include: (1) the substantial landmass of *Mexico*; (2) the narrowing strip of land to its southeast that constitutes *Central America*; and (3) the many large and small islands—respectively known as the *Greater Antilles* and *Lesser Antilles*—of the Caribbean Sea to the east.

Middle America is a realm of soaring volcanoes and forested plains, of mountainous islands and flat coral cays. Moist tropical winds sweep in from the east, watering windward (wind-facing) coasts while leaving leeward (wind-protected) areas dry. Soils vary from fertile volcanic to desert barren. Spectacular scenery abounds, and tourism is one of the realm's leading industries.

FIGURE 4-2 © H. J. de Blij, P. O. Muller, and John Wiley & Sons, Inc.

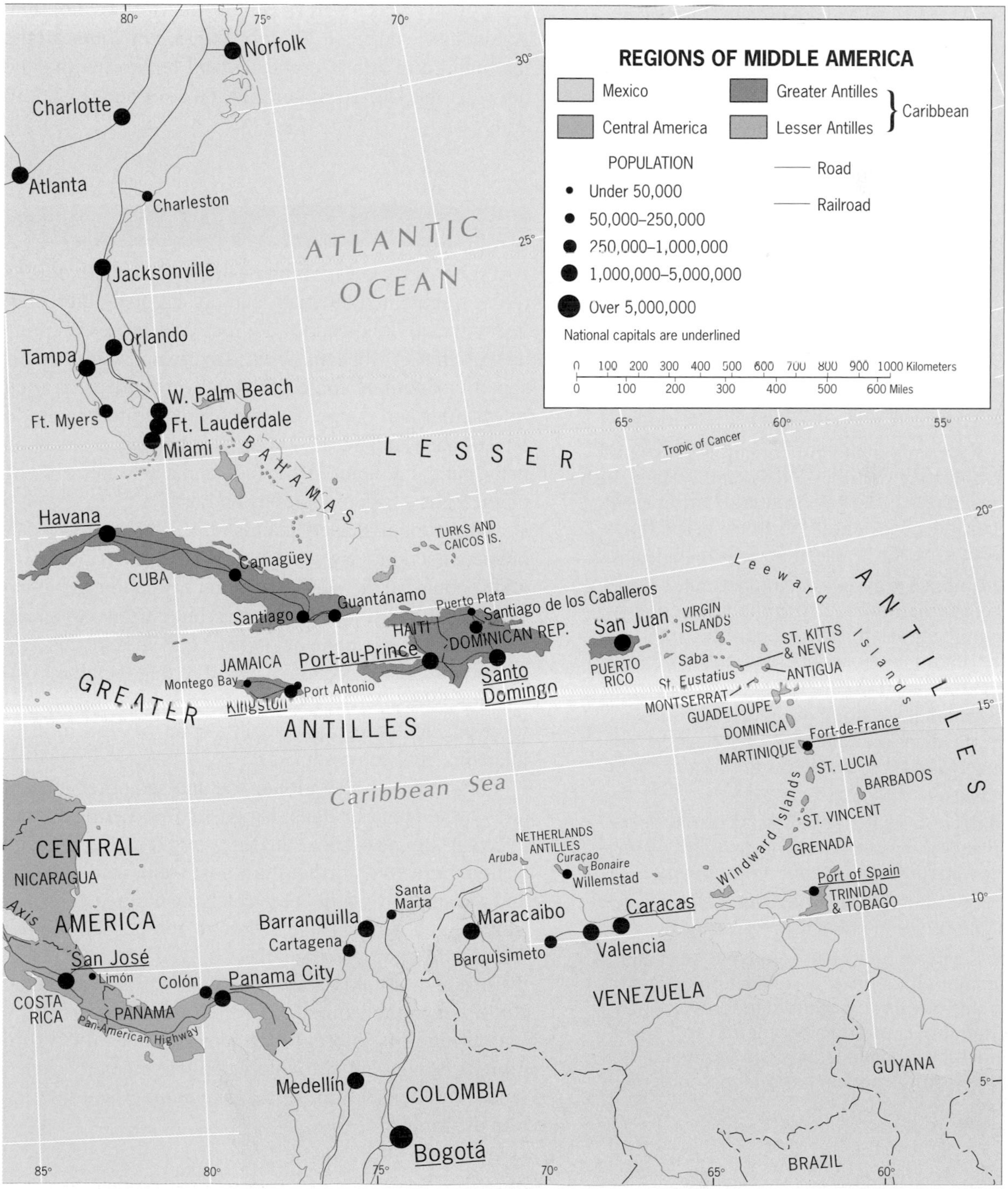

DEFINING THE REALM

Is Middle America a discrete geographic realm? Some geographers combine Middle and South America into a realm they call "Latin" America, citing the dominant Spanish (and in Brazil, Portuguese) heritage and the prevalence of Roman Catholicism. But these criteria apply more strongly to South America. In Middle America, large populations exhibit African and Asian as well as European ancestries. And nowhere in South America has native, Amerindian culture contributed to modern civilization as strongly as it has in Mexico. The Caribbean Basin is a patchwork of independent states, territories in political transition, and residual colonial dependencies. The Dominican Republic speaks Spanish, adjacent Haiti uses French, Dutch is spoken in Aruba, while English is spoken in Jamaica. Middle America thus gives vivid definition to concepts of cultural-geographical pluralism.

Middle America is occasionally called *Central America*, but that name actually refers to a region within the Middle American realm (Fig. 4-2). Central America comprises the republics that occupy the strip of mainland between Mexico and Panama: Guatemala, Belize, Honduras, El Salvador, Nicaragua, and Costa Rica. Although Panama itself is regarded here as belonging to Central America, it should be noted that many Central Americans still do not consider Panama to be part of their region because for most of its history that country was a part of South America's Colombia. Again, as we define it, *Middle America* includes all the mainland and island countries and territories that lie between the United States and the continent of South America.

MAJOR GEOGRAPHIC QUALITIES OF Middle America

1. Middle America is a fragmented realm that consists of all the mainland countries from Mexico to Panama and all the islands of the Caribbean Basin to the east.
2. Middle America's mainland constitutes a crucial barrier between Atlantic and Pacific waters. In physiographic terms, this is a land bridge that connects the continental landmasses of North and South America.
3. Middle America is a realm of intense cultural and political fragmentation. The political geography defies unification efforts, but countries and regions are beginning to work together to solve mutual problems.
4. Middle America's cultural geography is complex. African influences dominate the Caribbean, whereas Spanish and Amerindian traditions survive on the mainland.
5. The realm contains the Americas' least-developed territories. New economic opportunities may help alleviate Middle America's endemic poverty.
6. In terms of area, population, and economic potential, Mexico dominates the realm.
7. Mexico is reforming its economy and has experienced major industrial growth. Its hopes for continuing this development are tied to overcoming its remaining economic problems and to expanding trade with the United States and Canada under the North American Free Trade Agreement (NAFTA).

PHYSIOGRAPHY

As Figure 4-1 shows, Middle America is a realm of high relief, fragmented territory, and crustal instability studded with active volcanoes. Figures G-4 and G-5 remind us why: the Caribbean, North American, South American, and Cocos tectonic plates converge here, creating spectacular and dangerous landscapes subject to earthquakes and landslides. Add to this, the realm's location in the path of Atlantic hurricanes, and it amounts to some of the highest-risk topography on Earth.

The funnel-shaped mainland, a 3800-mile (6000-km) connection between North and South America, is wide enough in the north to contain two major mountain chains and a vast interior plateau, but narrows to a slim 40-mile (65-km) ribbon of land in Panama. Here this strip of land, or *isthmus*, bends eastward so that Panama's orientation is east-west. Thus mainland Middle America is what physical geographers call a **land bridge**, an isth- 1
mian link between continents.

If you examine a globe, you can see other present and former land bridges: Egypt's Sinai Peninsula between Asia and Africa, the (now-broken) Bering land bridge between northeasternmost Asia and Alaska, and the shallow waters between New Guinea and Australia. Such land bridges, though temporary surface features in geologic time, have played crucial roles in the dispersal of animals and humans across the planet. But even though mainland Middle America forms a land bridge, its internal fragmentation has always inhibited movement. Mountain ranges, swampy coastlands, and dense rainforests make contact and interaction difficult.

In terms of natural hazards, Middle America is the most dangerous realm of all. Hurricanes, volcanic eruptions, earthquakes, and landslides threaten virtually every corner of this part of the world. In January 2001, two of these hazards lethally combined to devastate the Las Colinas neighborhood of Santa Tecla, a suburb of the city of San Salvador, capital of the Central American country of El Salvador. A severe earthquake (magnitude 7.6 on the Richter Scale) rocked much of that country, causing only moderate damage overall, but a few places took the brunt of the seismic forces unleashed by the temblor. One was the geologically unstable hillside above Las Colinas, which in a matter of seconds produced this massive landslide that buried hundreds of homes and claimed more than half of the 687 Salvadorians who perished in the quake that day. © Lopez/El Diario de Hoy/Gamma-Presse, Inc.

Island Chains

As shown in Figure 4-2, the approximately 7000 islands of the Caribbean Sea stretch in a lengthy arc from Cuba and the Bahamas east and south to Trinidad, with numerous outliers outside (such as Barbados) and inside (e.g., the Cayman Islands) the main chain. The four large islands—Cuba, Hispaniola (containing Haiti and the Dominican Republic), Puerto Rico, and Jamaica—are called the *Greater Antilles*. All the remaining smaller islands are called the *Lesser Antilles*.

2 The entire Antillean **archipelago** (island chain) consists of the crests and tops of mountain chains that rise from the floor of the Caribbean, the result of collisions between the Caribbean Plate and its neighbors (Fig. G-4). Some of these crests are relatively stable, but elsewhere they contain active volcanoes, and almost everywhere in this realm earthquakes are an ever-present danger—in the islands as well as on the mainland (Fig. G-5).

LEGACY OF MESOAMERICA

Mainland Middle America was the scene of the emergence of a major ancient civilization. Here lay one of the
3 world's true **culture hearths** (see Fig. 7-3), a source area from which new ideas radiated and whose population could expand and make significant material and intellectual progress. Agricultural specialization, urbanization, and transport networks developed, and writing, science, art, and other spheres of achievement saw major advances. Anthropologists refer to the Middle American culture hearth as *Mesoamerica*, which extended southeast from the vicinity of present-day Mexico City to central Nicaragua. Its development is especially remarkable because it occurred in very different geographic environments, each presenting obstacles that had to be overcome in order to unify and integrate large areas. First, in the low-lying tropical plains of what is now northern Guatemala, Belize, and Mexico's Yucatán Peninsula, and perhaps simultaneously in Guatemala's highlands to the south, the Maya civilization arose more than 3000 years ago. Later, far to the northwest on the high plateau in central Mexico, the Aztecs founded a major civilization centered on the largest city ever to exist in pre-Columbian times.

The Lowland Maya

The Maya civilization is the only one in the world that ever arose in the lowland tropics. Its great cities, with their stone pyramids and massive temples, still yield archeological information today. Maya culture reached its zenith from the third to the tenth centuries A.D.

The Maya civilization, anchored by a series of city-states, unified an area larger than any of the modern Middle American countries except Mexico. Its population probably totaled between 2 and 3 million; certain Maya languages are still used in the area to this day. The Maya city-states were marked by dynastic rule that functioned alongside a powerful religious hierarchy, and the great cities that now lie in ruins were primarily ceremonial centers. We also know that Maya culture produced skilled

FROM THE FIELD NOTES

"We spent Monday and Tuesday upriver at Lamanai, a huge, still mostly overgrown Maya site deep in the forest of Belize. On Wednesday we drove from Belize City to Altun Ha, which represents a very different picture. Settled around 200 B.C., Altun Ha flourished as a Classic Period center between A.D. 300 and 900, when it was a thriving trade and redistribution center for the Caribbean merchant canoe traffic and served as an entrepôt for the interior land trails, some of them leading all the way to Teotihuacán. Altun Ha has an area of about 2.5 square miles (6+ sq km), with the main structures, one of which is shown here, arranged around two plazas at its core. I climbed to the top of this one to get a perspective, and sat down to have my sandwich lunch, imagining what this place must have looked like as a bustling trade and ceremonial center when the Roman Empire still thrived, but a more urgent matter intruded. A five-inch tarantula emerged from a wide crack in the sun-baked platform, and I noticed it only when it was about two feet away, apparently attracted by the crumbs and a small piece of salami. A somewhat hurried departure put an end to my historical-geographical ruminations." © H. J. de Blij.

artists and scientists and that these people achieved a great deal in agriculture and trade. They grew cotton, created a rudimentary textile industry, and exported cotton cloth by seagoing canoes to other parts of Middle America in return for valuable raw materials. They domesticated the turkey and grew cacao, developed writing systems, and studied astronomy.

The Highland Aztecs

In what is today the intermontane highland zone of Mexico, significant cultural developments were also taking place. Here, just north of present-day Mexico City, lay Teotihuacán, the first true urban center in the Western Hemisphere, which prospered for nearly seven centuries after its founding around the beginning of the Christian era. Later the Toltecs migrated into this area from the north, conquered and absorbed the local Amerindian peoples, and formed a powerful state. When, about 300 years later, the Toltec state itself was penetrated by new elements from the north, it was already in decay; but its technology was readily adopted and developed by the conquering Aztecs.

The Aztec state, the pinnacle of organization and power in pre-Columbian Middle America, is thought to have originated in the early fourteenth century with the founding of a settlement on an island in a lake that lay in the *Valley of Mexico* (the area surrounding what is now Mexico City). This urban complex, a functioning city as well as a ceremonial center, named Tenochtitlán, was soon to become the greatest city in the Americas and the capital of a large powerful state. The Aztecs soon gained control over the entire Valley of Mexico, a pivotal 30-by-40-mile (50-by-65-km) mountain-encircled basin that is still the heart of the modern state of Mexico. Both elevation and interior location moderate its climate; for a tropical area, it is quite dry and very cool. The basin's lakes formed a valuable means of internal communication, and the Aztecs built canals to connect several of them. This fostered a busy canoe traffic, bringing agricultural produce to the cities, and tribute was paid by their many subjects to the headquarters of the ruling nobility.

With its heartland consolidated, the new Aztec state soon began to conquer territories to the east and south. The Aztecs' expansion of their empire was driven by their desire to subjugate peoples and towns in order to extract taxes and tribute. As Aztec influence spread throughout Middle America, the state grew ever richer, its population mushroomed, and its cities thrived and expanded.

The Aztecs produced a wide range of impressive accomplishments, although they were better borrowers and refiners than they were innovators. They developed irrigation systems, and they built elaborate walls to terrace slopes where soil erosion threatened. Indeed, the greatest contributions of Mesoamerica's Amerindians surely came from the agricultural sphere. Corn (maize), the sweet potato, various kinds of beans, the tomato, squash, cacao beans (the raw material of chocolate), and tobacco are just a few of the crops that grew in Mesoamerica when the Europeans first made contact.

COLLISION OF CULTURES

We in the Western world all too often believe that history began when the Europeans arrived in some area of the world and that the Europeans brought such superior power to the other continents that whatever existed there previously had little significance. Middle America confirms this misperception: the great, feared Aztec state apparently fell before a relatively small band of Spanish invaders in an incredibly short period of time (1519–1521). But let us not lose sight of a few realities. At first, the Aztecs believed the Spaniards were "White Gods" whose arrival had been predicted by Aztec prophecy. Hernán Cortés, for all his 508 soldiers, did not single-handedly overthrow this powerful empire: he ignited a rebellion by Amerindian peoples who had fallen under Aztec domination and had seen their relatives carried off for human sacrifice to Aztec gods. Led by Cortés with his horses and guns, these peoples rose against their Aztec oppressors and joined the band of Spaniards headed toward Tenochtitlán (where thousands of them would die in combat against Aztec defenders).

Effects of the Conquest

Spain's defeat of Middle America's dominant indigenous state opened the door to Spanish penetration and supremacy. Throughout the realm, the confrontation between Hispanic and native cultures spelled disaster for the Amerindians: a catastrophic decline in population (perhaps as high as 90 percent), rapid deforestation, pressure on vegetation from grazing animals, substitution of Spanish wheat for maize (corn) on cropland, and the concentration of Amerindians into newly built towns.

The Spaniards were ruthless colonizers but not more so than other European powers that subjugated other cultures. True, the Spaniards first enslaved the Amerindians and were determined to destroy the strength of indigenous society. But biology accomplished what ruthlessness could not have achieved in so short a time: diseases introduced by the Spaniards and slaves imported from Africa decimated millions of Amerindians.

Middle America's cultural landscape—its great cities, its terraced fields, its dispersed aboriginal villages—was thus drastically modified. Unlike the Amerindians, who had used stone as their main building material, the Spaniards employed great quantities of wood and used charcoal for heating, cooking, and smelting metal. The onslaught on the forests was immediate, and rings of deforestation swiftly expanded around the colonizers' towns. The Spaniards also introduced large numbers of cattle and sheep, and people and livestock now had to compete for available food (requiring the opening of vast areas of marginal land that further disrupted the region's food-production balance). Moreover, the Spaniards introduced their own crops (notably wheat) and farming equipment, and soon large fields of wheat began to encroach upon the small plots of corn that the natives cultivated.

The Spaniards' most far-reaching cultural changes derived from their traditions as town dwellers. To facilitate domination, the Amerindians were moved off their land into nucleated villages and towns that the Spaniards established and laid out. In these settlements, the Spaniards could exercise the kind of rule and administration to which they were accustomed (Fig. 4-3). The internal focus of each Spanish town was the central *plaza* or market square, around which both the local church and government buildings were located. The surrounding street pattern was deliberately laid out in *gridiron* form, so that any insurrections by the resettled Amerindians could be contained by

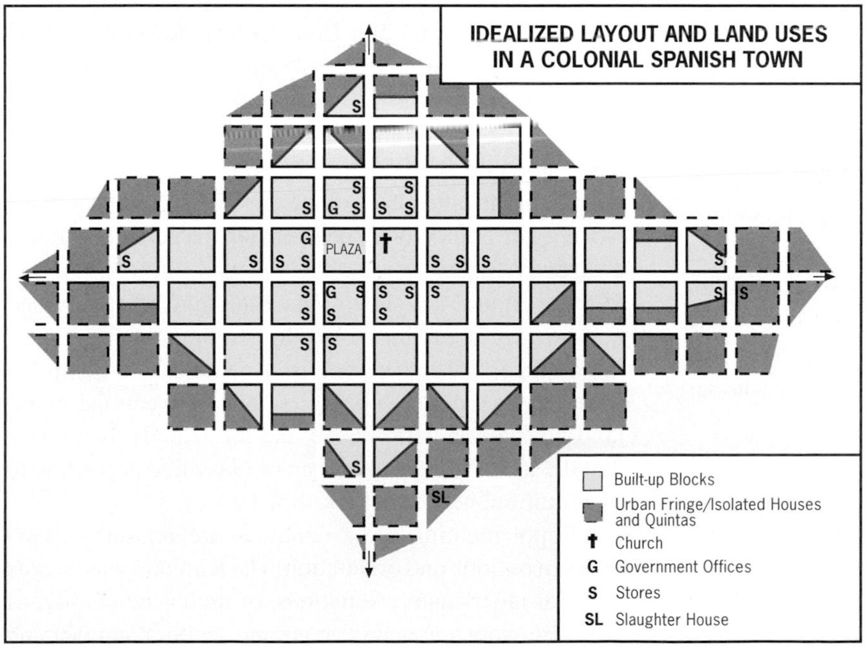

FIGURE 4-3 Adapted with permission from Charles S. Sargent, Jr., "The Latin American City," in Brian & Olwyn Blouet, eds., *Latin America and the Caribbean: A Systematic and Regional Geography*, 4 rev. ed., p. 167. © John Wiley & Sons, Inc., 2002.

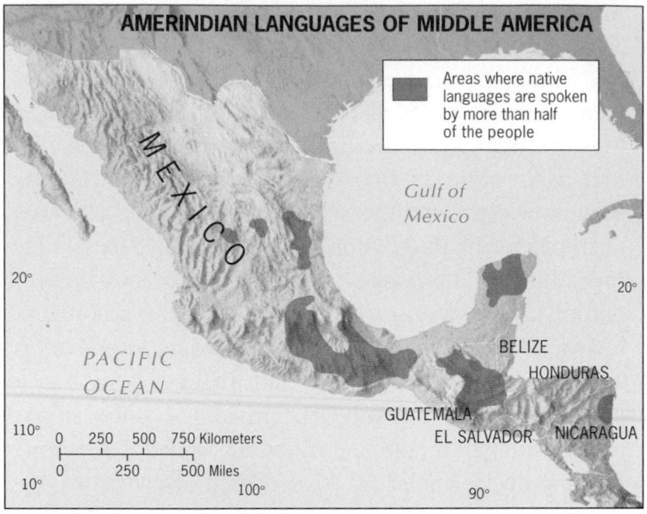

FIGURE 4-4 © H. J. de Blij, P. O. Muller, and John Wiley & Sons, Inc.

having a small military force seal off the affected blocks and then root out the troublemakers. Each town was located near what was thought to be good agricultural land (which was often not so good), so that the Amerindians could go out each day and work in the fields. Packed tightly into these towns and villages, they came face to face with Spanish culture. Here they learned the Europeans' Roman Catholic religion and Spanish language, and they paid their taxes and tribute to a new master. Nonetheless, the nucleated indigenous village survived under colonial (and later postcolonial) administration and is still a key feature of remote Amerindian areas in southeastern Mexico and inner Guatemala, where to this day native languages prevail over Spanish (Fig. 4-4).

Once the indigenous population was conquered and resettled, the Spaniards were able to pursue another primary goal in their New World territory: the exploitation of its wealth for their own benefit. Lucrative trade, commercial agriculture, livestock ranching, and especially mining were avenues to affluence. And wherever the Spaniards ruled—in towns, farms, mines, or indigenous villages—the Roman Catholic Church was the supreme cultural force transforming Amerindian society. With Jesuits and soldiers working together to advance the frontiers of New Spain, in the wake of conquest it was the church that controlled, pacified, organized, and acculturated the Amerindian peoples.

MAINLAND AND RIMLAND

In Middle America outside Mexico, only Panama, with its twin attractions of interoceanic transit and gold deposits, became an early focus of Spanish activity. From there, following the Pacific side of the isthmus, Spanish influence radiated northwestward through Central America and into Mexico. The major arena of international competition in Middle America, however, lay not on the Pacific side but on the islands and coasts of the Caribbean Sea. Here the British gained a foothold on the mainland, controlling a narrow coastal strip that extended southeast from Yucatán to what is now Costa Rica. As the colonial-era map (Fig. 4-5) shows, in the Caribbean the Spaniards faced the British, French, and Dutch, all interested in the lucrative sugar trade, all searching for instant wealth, and all seeking to expand their empires.

Later, after centuries of European colonial rivalry in the Caribbean Basin, the United States entered the picture and made its influence felt in the coastal areas of the mainland, not through colonial conquest but through the introduction of widespread, large-scale, banana plantation agriculture. The effects of these plantations were as far-reaching as the impact of colonialism on the Caribbean islands. Because the diseases the Europeans had introduced were most rampant in these hot, humid lowlands (as well as the Caribbean islands to the east), the Amerindian population that survived was too small to provide a sufficient workforce. This labor shortage was remedied through the generation of the trans-Atlantic slave trade from Africa that transformed the Caribbean Basin's demography (see map, p. 289).

These contrasts between the Middle American highlands and the coastal areas and Caribbean islands were
conceptualized by John Augelli into the **Mainland-** 4
Rimland framework (Fig. 4-6). Augelli recognized (1) a Euro-Amerindian **Mainland**, which consisted of continental Middle America from Mexico to Panama, with the exception of the Caribbean coastal belt from mid-Yucatán southeastward; and (2) a Euro-African **Rimland**, which included this coastal zone as well as the islands of the Caribbean. The terms *Euro-Amerindian* and *Euro-African* underscore the cultural heritage of each region. On the Mainland, European (Spanish) and Amerindian influences
are paramount and also include **mestizo** sectors where the 5
two ancestries mixed. In the Rimland, the heritage is European and African.

As Figure 4-6 shows, the Mainland is subdivided into several areas based on the strength of the Amerindian legacy. The Rimland is also subdivided, with the most obvious division being the one between the mainland-coastal plantation zone and the islands. Note, too, that the islands themselves can be classified according to their cultural heritage (Figs. 4-5, 4-6).

Supplementing these contrasts are regional differences in outlook and orientation. The Rimland was an area of sugar and banana plantations, of high accessibility, of seaward exposure, and of maximum cultural contact and

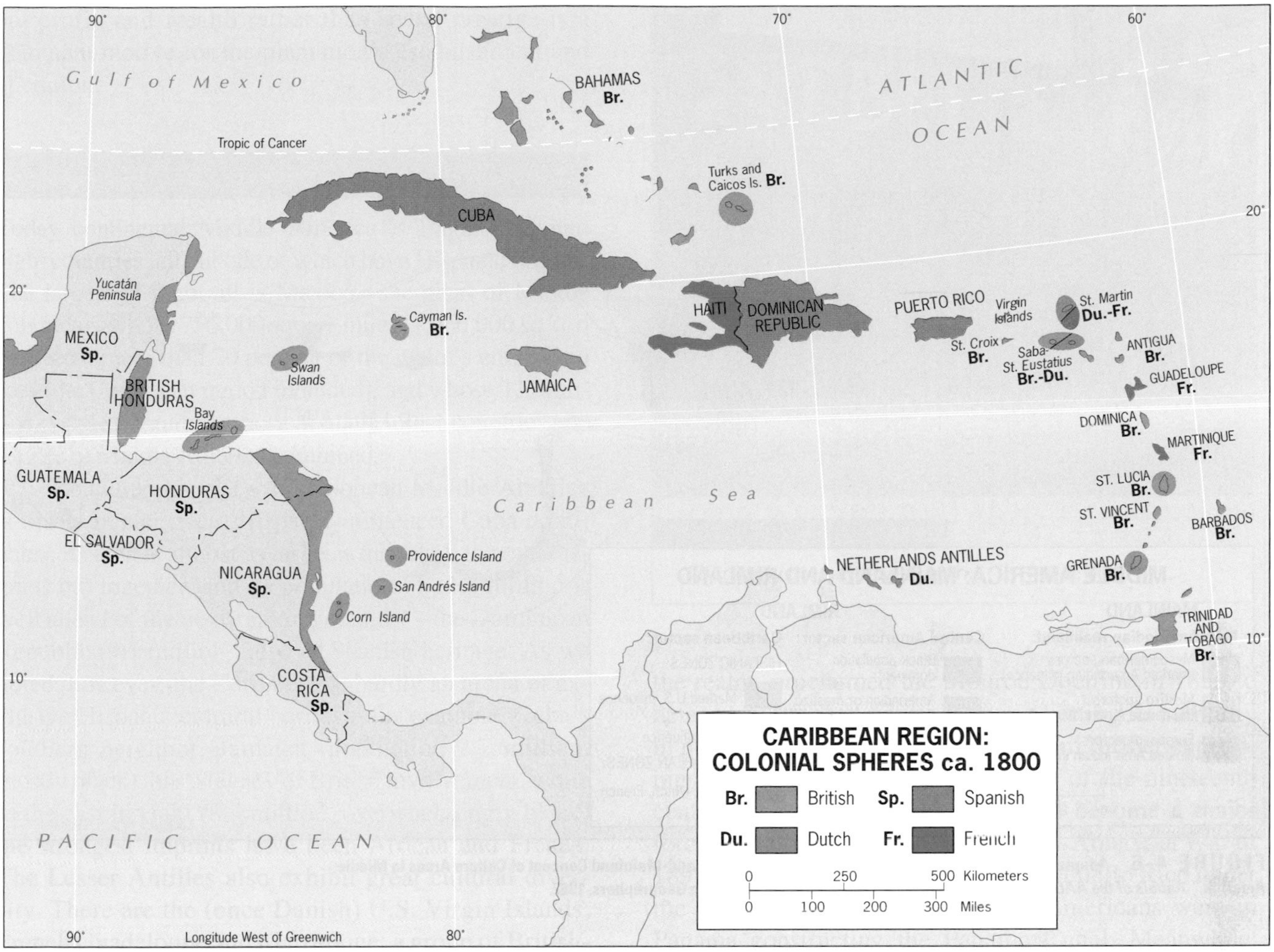

FIGURE 4-5 © H. J. de Blij, P. O. Muller, and John Wiley & Sons, Inc.

mixture. The Mainland, being farther removed from these contacts, was an area of greater isolation. The Rimland was the region of the great *plantation*, and its commercial economy was therefore susceptible to fluctuating world markets and tied to overseas investment capital. The Mainland was the region of the *hacienda*, which was more self-sufficient and less dependent on external markets.

The Hacienda

This contrast between plantation and hacienda land tenure in itself constitutes strong evidence for the Rimland-Mainland division. The hacienda was a Spanish institution, but the modern plantation, Augelli argued, was the concept of Europeans of more northerly origin. In the
6 **hacienda**, Spanish landowners possessed a domain whose productivity they might never push to its limits: the very possession of such a vast estate brought with it social prestige and a comfortable lifestyle. Native workers lived on the land—which may once have been *their* land—and had plots where they could grow their own subsistence crops. All this is written as though it is mostly in the past, but the legacy of the hacienda, with its inefficient use of land and labor, is still visible throughout mainland Middle America.

The Plantation

The **plantation** was conceived as something entirely dif- 7
ferent from the hacienda. Robert West and John Augelli list five characteristics of Middle American plantations that illustrate the differences between hacienda and plantation: (1) plantations are located in the humid tropical coastal lowlands; (2) plantations produce for export almost exclusively—usually a single crop; (3) capital and skills are often imported so that foreign ownership and an outflow of profits occur; (4) labor is seasonal—needed in large numbers mainly during the harvest period—and such labor has been imported because of the scarcity of Amerindian workers; and (5) with its "factory-in-the-field" operation, the plantation is more efficient in its use of land and labor than the hacienda. The objective was not self-sufficiency

Likened to a glowing hot bar of freshly cast steel plucked from a furnace, the *Faro del Comercio* (Lighthouse of Commerce) shoots skyward in the historic heart of Monterrey. Saddle Mountain in the background is a reminder that this northeastern Mexican metropolis is situated at a strategic point where the desert meets the great Sierra Madre Oriental range, in the heart of an area so well endowed with raw materials that Mexico's iron and steel industry was born here a century ago. Ever since, Monterrey has been an economic leader, and its new landmark signifies that no city more boldly confronts the Mexican future. Today, building on the opportunities offered by NAFTA, Monterrey is busy forging a role for itself on the international circuit. Capitalizing on its function as the southern anchor of the booming growth corridor that stretches northward to Texas' Dallas–Ft. Worth Metroplex, Monterrey has built a new downtown business complex that includes a convention center booked solidly with trade shows at least two years ahead. And in 1999, the city even hosted the National League's season-opening baseball game. © Trygve Bolstad/Panos Pictures.

different picture. Mexico's overall economy may have benefited in the early days of NAFTA, especially when expanding trade with North America helped it overcome the financial crisis of 1994, but since then growth in Mexico has been very slow, averaging about 1 percent per year (compare this to China's more than 7 percent and South Korea's more than 4 percent). Despite predictions that NAFTA would alleviate poverty in Mexico, consider this: real wages have been dropping there every year over the past ten, and income disparities between the United States and Mexico have risen by more than 10 percent during the same period.

Meanwhile, unemployment in the once-booming maquiladora cities is growing as jobs are being lost when corporations move their factories to East Asia. Visit the Mexican "twins" on the opposite side of American cities on the Rio Bravo (Grande), and you will be reminded of conditions in the harsh, concrete-jungle manufacturing complexes on China's Pacific Rim—except that workers in Mexico, unable to pay the rent for decent apartments, huddle in even more squalid shacks in vast slums that encircle the factory clusters. Unemployment, crime (especially against women), insecurity, and poverty cloud the NAFTA picture in Mexico's maquiladora cities, which have grown far beyond capacity and cannot cope with the resulting needs.

Notions that the string of maquiladora cities with their job opportunities would stem the tide of illegal migration from Mexico across the U.S. border faded quickly. In 2004 U.S. President Bush proposed a program that would begin to substitute a form of legalization for the failed interdiction effort (an estimated 10 million Mexican citizens are in the United States today). What would really help stem this tide, however, would be rapid economic growth of the Asian Pacific Rim variety in Mexico itself. Some positive steps have indeed been taken—democracy and open government have progressed, for example—but Mexico, as we noted earlier, needs other advances ranging from fairer and more comprehensive tax collection (which would be a byproduct of a national attack on corruption) to better education. In the first decade of this new century Mexico remains at a crossroads, the NAFTA promise fading and the alternatives beckoning, but as yet beyond reach.

THE CENTRAL AMERICAN REPUBLICS

Crowded onto the narrow segment of the Middle American land bridge between Mexico and the South American continent are the seven countries of Central America (Fig. 4-12). Territorially, they are all quite small; their population sizes range from Guatemala's 13.4 million down to Belize's 303,000. Physiographically, the land bridge here consists of a highland belt flanked by coastal lowlands on both the Caribbean and Pacific sides (Fig. 4-1). These highlands are studded with volcanoes, and local areas of fertile volcanic soils are scattered throughout them. From earliest times, the region's inhabitants have been concentrated in this upland zone, where tropical temperatures are moderated by elevation and rainfall is sufficient to support a variety of crops.

Altitudinal Zonation of Environments

Continental Middle America and the western margin of South America are areas of high relief and strong environmental contrasts. Even though settlers have always

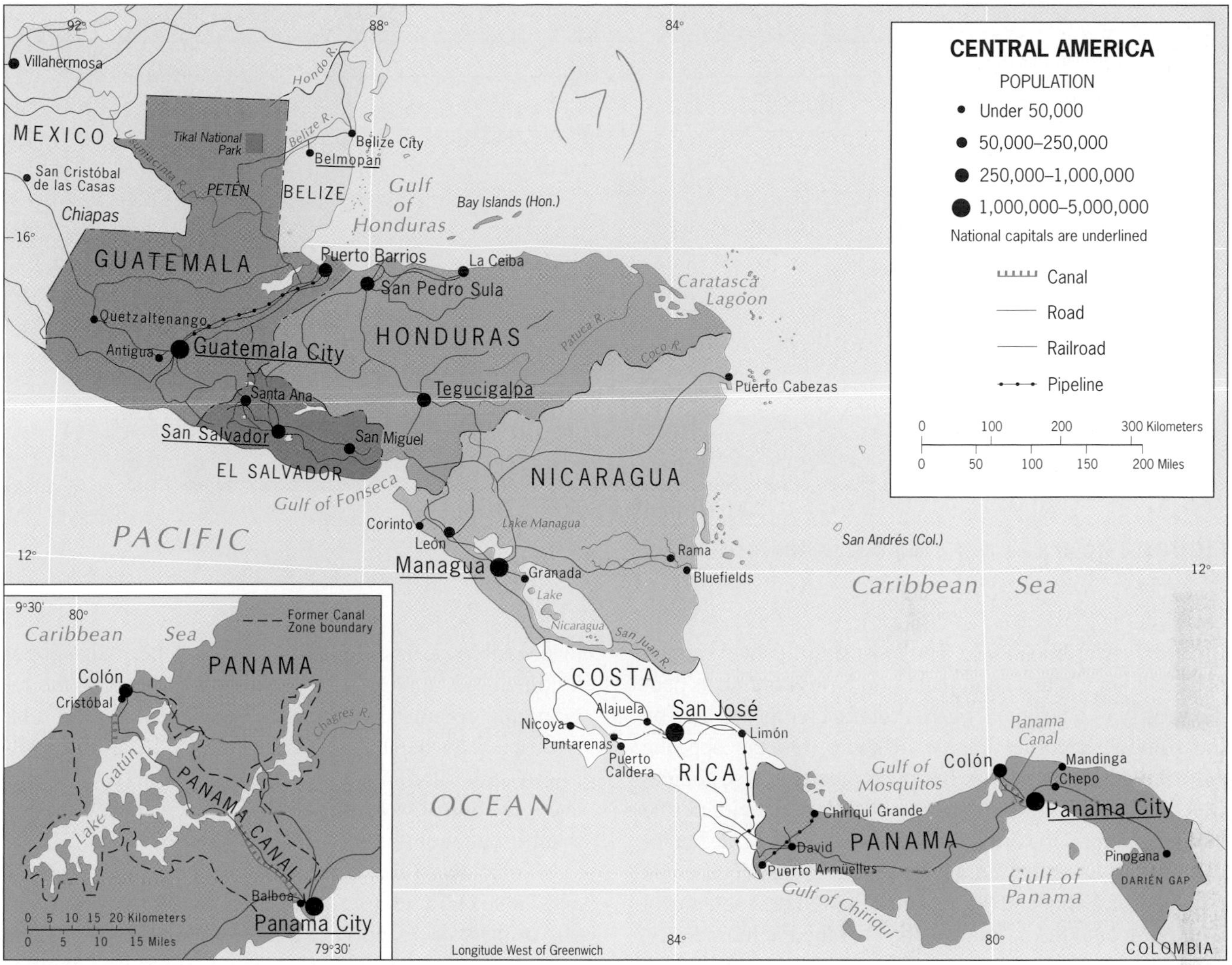

FIGURE 4-12 © H. J. de Blij, P. O. Muller, and John Wiley & Sons, Inc.

favored temperate intermontane basins and valleys, people also cluster in hot tropical lowlands as well as high plateaus just below the snow line in South America's Andes Mountains. In each of these zones, distinct local climates, soils, vegetation, crops, domestic animals, and
12 modes of life prevail. Such **altitudinal zones** (diagrammed in Fig. 4-13) are known by specific names as if they were regions with distinguishing properties—as, in reality, they are.

The lowest of these vertical zones, from sea level to
13 2500 feet (about 750 m), is known as the ***tierra caliente***, the "hot land" of the coastal plains and low-lying interior basins where tropical agriculture predominates. Above this zone lie the tropical highlands containing Middle and
14 South America's largest population clusters, the ***tierra templada*** of temperate land reaching up to about 6000 feet (1800 m). Temperatures here are cooler; prominent among the commercial crops is coffee, while corn (maize) and wheat are the staple grains. Still higher, from about 6000 feet to nearly 12,000 feet (3600 m), is the ***tierra fría***, the 15
cold country of the higher Andes where hardy crops such as potatoes and barley are mainstays. Above the tree line, which marks the upper limit of the *tierra fría*, lies the ***tierra*** 16
helada; this fourth altitudinal zone, extending from about 12,000 to 15,000 feet (3600 to 4500 m), is so cold and barren that it can support only the grazing of sheep and other hardy livestock. The highest zone of all is the ***tierra*** 17
nevada, a zone of permanent snow and ice associated with the loftiest Andean peaks. As we will see, the varied cultural geography of Middle and western South America is closely related to these diverse environments.

Population Patterns

Even at its comparatively small scale, Figure G-9 indicates that Central America's population tends to concentrate in the uplands of the *tierra templada* and that population densities are generally greater toward the Pacific than

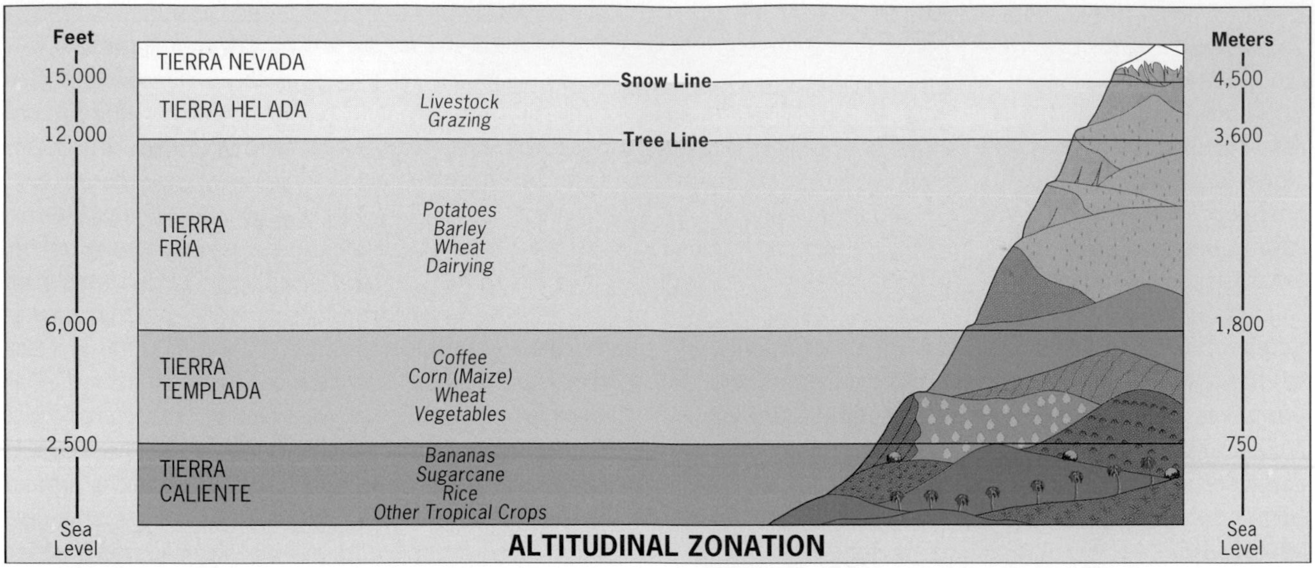

FIGURE 4-13 © H. J. de Blij, P. O. Muller, and John Wiley & Sons, Inc.

toward the Caribbean side. The most significant exception is El Salvador, whose political boundaries confine its people mostly to its tropical *tierra caliente*, tempered here by the somewhat cooler Pacific offshore. On the opposite side of the isthmus, Belize has the typically sparse population of the hot, wet Caribbean coastlands and their infertile soils. Panama is the only other exception to the rule but for different reasons. Economic development has focused on the Panama Canal and the coasts, although new settlement is moving onto the mountain slopes of the Pacific side.

Central America, we noted earlier, actually begins within Mexico, in Chiapas and in Yucatán, and the region's republics face many of the same problems as less-developed parts of Mexico. Population pressure is one of these problems: a population explosion began in the mid-twentieth century, increasing the region's human inhabitants from 9 million to nearly 40 million by the start of the twenty-first. Unlike Mexico, Central America's population growth is not yet slowing down except in Costa Rica and Panama, and population geographers talk of a demographic catastrophe in the making.

Emergence from a Turbulent Era

Devastating inequities, repressive governments, external interference, and the frequent unleashing of armed forces have destabilized Central America for much of its modern history. The roots of these upheavals are old and deep, and today the region continues its struggle to emerge from a period of turmoil that lasted through the 1980s into the mid-1990s.

Central America is not a large region, but because of its physiography it contains many isolated, comparatively inaccessible locales. Conflicts between Amerindian population clusters and mestizo groups are endemic to the region, and contrasts between the privileged and the poor are especially harsh. Dictatorial rule by local elites followed authoritarian rule by Spanish colonizers, and the latest episode of violent confrontation was simply another manifestation of this persistent polarization.

Today there are signs of progress but, as Figure G-11 and Table G-1 underscore, this region as a whole remains one of the world's poorest, a legacy of disadvantage that typifies the global periphery. Central American countries produce little the world needs (except Panama, whose Canal income gives it the region's second-highest GNI). Its farmers face subsidized competition on world markets. There is little or no money to maintain, let alone improve, infrastructure. Intraregional trade is minimal; the Pan American Highway is in disrepair and does not even reach Panama's eastern border. The turbulence of decades past may be in abeyance, but its causes remain.

The Seven Republics

Guatemala, the westernmost of Central America's republics, has more land neighbors than any other. Straight-line boundaries across the tropical forest mark much of the border with Mexico, creating the box-like region of Petén between Chiapas State on the west and Belize on the east; also to the east lie Honduras and El Salvador (Fig. 4-12). This heart of the ancient Maya Empire, which remains strongly infused by Amerindian culture and tradition, has just a small window on the Caribbean but a longer Pacific coastline. Guatemala was still part of Mexico when the Mexicans threw off the Spanish yoke, and although independent from Spain after 1821, it did not become a separate republic until 1838. Mestizos, not the Amerindian majority, secured the country's independence.

Most populous of the seven republics with 13.4 million inhabitants (mestizos are in the majority with 64 percent, Amerindians 33 percent), Guatemala has seen much conflict. Repressive regimes made deals with U.S. and other foreign economic interests that stimulated development but at a high social cost. Over the past half-century, military regimes have dominated political life. The deepening split between the wretchedly poor Amerindians and the better-off mestizos, who here call themselves *ladinos*, generated a civil war that started in 1960 and has since claimed more than 200,000 lives as well as 50,000 "disappearances." An overwhelming number of the victims have been of Mayan descent; the mestizos control the government, army, and land-tenure system.

The tragedy of Guatemala is that its economic geography has considerable potential but has long been shackled by the unending internal conflicts that have kept the income of 80 percent of the population below the poverty line. The country's mineral wealth includes nickel in the highlands and oil in the lower-lying north. Agriculturally, soils are fertile and moisture is ample over highland areas large enough to produce a wide range of crops including excellent coffee.

Belize, strictly speaking, is not a Central American republic in the same tradition as the other six. Until 1981, this country, a wedge of land between northern Guatemala, Mexico's Yucatán Peninsula, and the Caribbean (Fig. 4-12), was a dependency of the United Kingdom known as British Honduras. Slightly larger than Massachusetts and with a minuscule population of only 303,000 (many of African descent), Belize has been more reminiscent of a Caribbean island than of a continental Middle American state. Today, all that is changing as the demographic complexion of Belize is being reshaped. Thousands of residents of African descent have recently emigrated (many went to the United States) and were replaced by tens of thousands of Spanish-speaking immigrants. Most of the latter are refugees from strife in nearby Guatemala, El Salvador, and Honduras, and their proportion of the Belizean population has risen from 33 to nearly 50 percent since 1980. Within the next few years the newcomers will be in the majority, Spanish will become the *lingua franca*, and Belize's cultural geography will exhibit an expansion of the Mainland at the expense of the Rimland.

The Belizean transformation extends to the economic sphere as well. No longer just an exporter of sugar and bananas, Belize is producing new commercial crops, and its seafood-processing and clothing industries have become major revenue earners. Also important is tourism, which annually lures more than 150,000 vacationers to the country's Mayan ruins, resorts, and newly legalized casinos; a growing speciality is ecotourism, based on the natural attractions of the country's near-pristine environment. Belize is also known as a center for *offshore banking*—a financial haven for foreign companies and individuals who want to avoid paying taxes in their home countries.

Honduras is a country on hold as it struggles to rebuild its battered infrastructure and economy. In 1998 Hurricane Mitch struck Honduras, and the consequences were catastrophic as massive floods and mudslides were unleashed across the country, killing 9200 people, demolishing more than 150,000 homes, destroying 21,000 miles of roadway and 335 bridges, and rendering 2 million homeless. Also devastated was the critical agricultural sector that employed two-thirds of Honduras's labor force, accounted for nearly a third of its gross national income, and earned more than 70 percent of its foreign revenues.

With 7.4 million inhabitants, about 87 percent mestizo, bedeviled Honduras still has years to go even to restore what was already the third poorest economy in the Americas (after Haiti and Nicaragua). Agriculture, livestock, forestry, and limited mining formed the mainstays of the pre-1998 economy, with the familiar Central American products—bananas, coffee, shellfish, apparel—earning most of the external income.

Honduras, in direct contrast to Guatemala, has a lengthy Caribbean coastline and a small window on the Pacific (Fig. 4-12). The country also occupies a critical place in the political geography of Central America, flanked as it is by Nicaragua, El Salvador, and Guatemala—all continuing to grapple with the aftermath of years of internal conflict and, most recently, natural disaster. The road back to economic viability is an arduous one, but once traversed will still leave four out of five Hondurans deeply mired in poverty and the country with little overall improvement in its development prospects.

El Salvador is Central America's smallest country territorially, smaller even than Belize, but with a population almost 25 times as large (7.0 million) it is the most densely peopled. Again, like Belize, it is one of only two continental republics that lack coastlines on both the Caribbean and Pacific sides (Fig. 4-12). El Salvador adjoins the Pacific in a narrow coastal plain backed by a chain of volcanic mountains, behind which lies the country's heartland. Unlike neighboring Guatemala, El Salvador has a quite homogeneous population (88 percent mestizo and just 9 percent Amerindian). Yet ethnic homogeneity has not translated into social or economic equality or even opportunity. Whereas other Central American countries were called banana republics, El Salvador was a coffee republic, and the coffee was produced on the huge landholdings of a few landowners and on the backs of a subjugated peasant labor force. The military supported this system and repeatedly suppressed violent and desperate peasant uprisings.

From 1980 to 1992, El Salvador was torn by a devastating civil war that was worsened by outside arms supplies from the United States (supporting the government) and Nicaragua (aiding the Marxist rebel forces). But ever since the negotiated end to that war, efforts have been under way to prevent a recurrence because El Salvador is having difficulty overcoming its legacy of searing inequality. The civil war did have one positive result: affluent citizens who left the country and did well in the United States and elsewhere send substantial funds back home, which now provide the largest single source of foreign revenues. This has helped stimulate such industries as apparel and footwear manufacturing, as well as food processing. But a major stumbling block to revitalization of the agricultural sector has again been land reform, and El Salvador's future still hangs in the balance.

Nicaragua is best approached by reexamining the map (Fig. 4-12), which underscores the country's pivotal position in the heart of Central America. The Pacific coast follows a southeasterly direction, but the Caribbean coast is oriented north-south so that Nicaragua forms a triangle of land with its lakeside capital, Managua (photo below), located in a valley on the mountainous, earthquake-prone, Pacific side (the country's core area has always been located here). The Caribbean side, where the uplands yield to a coastal plain of rainforest, savanna, and swampland, has for centuries been home to Amerindian peoples such as the Miskito, who have been remote from the focus of national life.

Until the end of the 1970s, Nicaragua was the typical Central American republic, ruled by a dictatorial government and exploited by a wealthy land-owning minority, its export agriculture dominated by huge plantations owned by foreign corporations. It was a situation ripe for insurgency, and in 1979, leftist rebels overthrew the government; but the new regime quickly produced its own excesses, resulting in civil war through most of the 1980s. That conflict ended in 1990, and more democratic governments have since been voted into office.

Nicaragua's economy has been a leading casualty of this turmoil, and for the past two decades it has ranked as continental Middle America's poorest. Hurricane Mitch struck here too in 1998, devastating the country's farms and driving tens of thousands into the impoverished towns just at a time when the agricultural sector was recovering and land reform promised a better life for some 200,000 peasant families.

Nicaragua's options are limited: none of the billions of aid dollars headed for Bosnia, Iraq, and Subsaharan Africa will be matched here. For years there has been talk of a *dry canal* transit role for this country, but other potential overland routes are more promising. Meanwhile, Nicaragua's population explosion continues (2006 total: 5.9 million), dooming hopes for a rise in standards of living.

Costa Rica underscores what was said about Middle America's endless variety and diversity because it differs significantly from its neighbors and from the norms of Central America as well. Bordered by two volatile countries (Nicaragua to the north and Panama to the east), Costa Rica is a nation with an old democratic tradition and, in this cauldron, no standing army for the

A country's capital may be its primate city; it may also reflect the economic state of the nation. Nicaragua is Central America's poorest republic, its capital leveled by catastrophic earthquakes in 1931 and 1972 and damaged by civil war in 1978–1979, its already weak infrastructure devastated by Hurricane Mitch and its national aspiration constrained by the power of the United States. Poverty of the kind experienced by Nicaraguans leaves little in the way of resources for maintenance and upkeep, and Managua reflects this state of affairs. © D. Donne Bryant Stock Photography.

past half-century! Although the country's Hispanic imprint is similar to that found elsewhere on the Mainland, its early independence, its good fortune to lie remote from regional strife, and its leisurely pace of settlement allowed Costa Rica the luxury of concentrating on its economic development. Perhaps most important, internal political stability has prevailed over much of the past 175 years.

Like its neighbors, Costa Rica is divided into environmental zones that parallel the coasts. The most densely settled is the central highland zone, lying in the cooler *tierra templada*, whose heartland is the Valle Central (Central Valley), a fertile basin that contains the country's main coffee-growing area and the leading population cluster focused on San José (Fig. 4-12)—the most cosmopolitan urban center between Mexico City and the primate cities of northern South America. To the east of the highlands are the hot and rainy Caribbean lowlands, a sparsely populated segment of Rimland where many plantations have been abandoned and replaced by subsistence farming. Between 1930 and 1960, the U.S.-based United Fruit Company shifted most of the country's banana plantations from this crop-disease-ridden coastal plain to Costa Rica's third zone—the plains and gentle slopes of the Pacific coastlands. This move gave the Pacific zone a major boost in economic growth, and it is now an area of diversifying and expanding commercial agriculture.

The long-term development of Costa Rica's economy has given it the region's highest standard of living, literacy rate, and life expectancy. Agriculture continues to dominate (with bananas, coffee, seafood, and tropical fruits the leading exports), and tourism is expanding steadily. Costa Rica is widely known for its superb scenery and for its efforts to protect what is left of its diverse tropical flora and fauna. But rainforest destruction is the regionwide price of human population growth and woodland exploitation, and even in Costa Rica as much as 70 percent of the original forest has vanished (see box titled "Tropical Deforestation").

Still, the country's veneer of development cannot mask serious problems. In terms of social structure, about one-quarter of its population of 4.3 million is trapped in an unending cycle of poverty, and the huge gap between the poor and the affluent is constantly widening. With volatile neighbors and an economy that remains insufficiently diversified against risk, Costa Rica is only one step ahead of its regional partners.

Panama owes its existence to the idea of a canal connecting the Atlantic and Pacific oceans to avoid the lengthy circumnavigation of South America. In the 1880s, when Panama was still an extension of neighboring Colombia, a French company tried and failed to build such a waterway here. By the turn of the twentieth century, U.S. interest in a Panama canal rose sharply, and the United States in 1903 proposed a treaty that would permit a renewed effort at construction across Colombia's Panamanian isthmus. When the Colombian Senate refused to go along, Panamanians rebelled and the United States supported this uprising by preventing Colombian forces from intervening. The Panamanians, at the behest of the United States, declared their independence from Colombia, and the new republic immediately granted the United States rights to the Canal Zone, averaging about 10 miles (16 km) in width and just over 50 miles (80 km) in length.

Soon canal construction commenced, and this time the project succeeded as American technology and medical advances triumphed over a formidable set of obstacles. The Panama Canal (see inset map, Fig. 4-12) was opened in 1914, a symbol of U.S. power and influence in Middle America. The Canal Zone was held by the United States under a treaty that granted it "all the rights, powers, and authority" in the area "as if it were the sovereign of the territory." Such language might suggest that the United States held rights over the Canal Zone in perpetuity, but the treaty nowhere stated specifically that Panama permanently yielded its own sovereignty in that transit corridor. In the 1970s, as the canal was transferring more than 14,000 ships per year (that number is now only slightly lower, but the cargo tonnage is up significantly) and generating hundreds of millions of dollars in tolls, Panama sought to terminate U.S. control in the Canal Zone. Delicate negotiations began. In 1977, an agreement was reached on a staged withdrawal by the United States from the territory, first from the Canal Zone and then from the Panama Canal itself (a process completed on December 31, 1999).

Panama today reflects some of the usual geographic features of the Central American republics. Its population of 3.3 million is nearly 60 percent mestizo and also contains substantial Amerindian, white, and black minorities. Spanish is the official language, but English is also widely used. Ribbon-like and oriented east-west, Panama's topography is mountainous and hilly. Eastern Panama, especially Darien Province adjoining Colombia, is densely forested, and here is the only remaining gap in the intercontinental Pan American Highway. Most of the rural population lives in the uplands west of the canal; there, Panama produces bananas, shrimps and other seafood, sugarcane, coffee, and rice. Much of the urban population is concentrated in the vicinity of the waterway, anchored by the cities at each end of the canal.

Near the northern end of the Panama Canal lies the city of Colón, site of the Colón Free Zone, a huge trading entrepôt designed to transfer and distribute goods bound for South America. It is augmented by the Manzanillo

Tropical Deforestation

Before the Europeans arrived, two-thirds of continental Middle America was covered by tropical rainforests. The clearing and destruction of this precious woodland resource to make way for expanding settlement frontiers and the exploitation of new economic opportunities began in the sixteenth century during the Spanish colonial era, and the practice has continued systematically ever since.
18 In recent decades, however, the pace of **tropical deforestation** in Central America has accelerated alarmingly, and since 1950 almost 90 percent of the region's forests have been decimated. Now, about 3 million acres of Central American and Mexican woodland disappear each year, an area equivalent to one-third the size of Belgium. El Salvador has already lost 99 percent of its forests, and most of the six other republics will soon reach that stage.

The causes of tropical deforestation are related to the persistent economic and demographic problems of disadvantaged countries. In Central America, the leading cause has been the need to clear rural lands for cattle pasture as many countries, especially Costa Rica, became meat producers and exporters. The price of this environmental degradation has been enormous, although some gains have been recorded. Because tropical soils are so nutrient-poor, newly deforested areas can function as pastures for only a few years at most. These fields are then abandoned for other freshly cut lands and quickly become the ravaged landscape seen in the photo in this box. Without the protection of trees, local soil erosion and flooding immediately become problems, affecting still-productive nearby areas (a sequence of events that reached catastrophic dimensions all across Honduras and northern Nicaragua when Hurricane Mitch struck in 1998). A second cause of deforestation is the rapid logging of tropical woodlands as the timber industry increasingly turns from the exhausted forests of the midlatitudes to harvest the rich tree resources of the equatorial zones, responding to accelerating global demands for housing, paper, and furniture. The third major contributing factor is related to the region's population explosion: as more and more peasants are required to extract a subsistence from inferior lands, they have no choice but to cut down the remaining forest for both firewood and additional crop-raising space, and their intrusion prevents the trees from regenerating (Haiti is the extreme example of this denudation process).

Although deforestation is a depressing event, tropical pastoralists, farmers, and timber producers do not consider it to be life-threatening, and perhaps it even seems to offer some short-term economic advantages. Why, then, should there be such an outcry from the scientific community? And why should the World Resources Institute call this "the world's most pressing land-use problem"? The answer is that unless immediate large-scale action is taken, by the middle of this century the world's tropical rainforests will be reduced to two disappearing patches—the western Amazon Basin of northern South America and the middle Congo Basin of Equatorial Africa.

The tropical forest, therefore, must be a very important part of our natural world, and indeed it is. Biologically, the rainforest is by far the richest, most diversified arena of life on our planet: even though it covers only about (a shrinking) 3 percent of the Earth's land area, it contains about three-quarters of all plant and animal species. Its loss would cause not only the extinction of millions of species, but also what ecologist Norman Myers calls "the death of birth" because the evolutionary process that produces new species would be terminated. Because tropical rainforests already yield countless valuable medicinal, food, and industrial products, many potential disease-combatting drugs or new crop varieties to feed undernourished millions will be irretrievably lost if they are allowed to disappear.

This scene in Costa Rica shows how badly the land can be scarred in the wake of deforestation. Without roots to bind the soil, tropical rains swiftly erode the unprotected topsoil. © Peter Poulides.

FROM THE FIELD NOTES

"The Panama Canal remains an engineering marvel 90 years after it opened in August 1914. The parallel lock chambers each are 1000 feet long and 110 feet wide, permitting vessels as large as the Queen Elizabeth II to cross the isthmus. Ships are raised by a series of locks to Gatún Lake, 85 feet above sea level. We watched as tugs helped guide the QEII into the Gatún Locks, a series of three locks leading to Gatún Lake, on the Atlantic side. A container ship behind the QEII is sailing up the dredged channel leading from the Limón Bay entrance. The lock gates are 65 feet wide and seven feet thick, and range in height from 47 to 82 feet. The motors that move them are recessed in the walls of the lock chambers. Once inside the locks, the ships are pulled by powerful locomotives called *mules* that ride on rails that ascend and descend the system. It was still early morning, and a major fire, probably a forest fire, was burning near the city of Colón, where land clearing was in progress. This was the beginning of one of the most fascinating days ever." © H. J. de Blij

International Terminal, an ultramodern port facility capable of transshipping more than 1000 containers a day. By 2002 China had become the third-largest user of the Canal (after the United States and Japan) and accounted for more than 20 percent of the cargo entering the Colón Free Zone.

The Panama Canal has political as well as economic implications for Panama. For many years, Panama has recognized Taiwan as an independent entity and has sponsored resolutions to get Taiwan readmitted to the United Nations (where it was ousted in favor of China in 1971). In return, Taiwan has spent hundreds of millions of dollars in Panama in the form of investment and direct aid. But now, China's growing presence in the Canal and the Colón Free Zone are causing a conundrum. Panama has awarded a contract to a Hong Kong-based (and thus Chinese) firm to operate and modernize the ports at both ends of the Canal, resulting in investments approaching U.S. $500 million. China is lobbying Panama for a switch in recognition, using this commitment as leverage. But the Panamanians are cautious. Having just ousted the United States, they are not eager to be dominated by another giant. It is more profitable to play the "two Chinas" off against each other.

Panama also feels the effects of its location adjacent to rebel-plagued, drug-infested, and unstable Colombia. For many years Panama City, Central America's only coastal capital city, was a financial center that handled more than just the funds generated by the Canal. The link between the Colón Free Zone and Panama City also had a money-laundering dimension, and the Miami-like skyline of the capital reflected that. In the 1990s, Colombia was the Colón Free Zone's biggest customer, but when Panama began a crackdown on money laundering, its share of trade dropped significantly.

Fortunately for Panama, its border with Colombia in the Darien is short (170 mi/270 km) and densely forested, occasionally providing hideouts for rebel bands and traffickers but creating a buffer between the main theaters of rebel activity and the Panamanian core area. Colombia's campaign against drug production, as we note in Chapter 5, focuses on other parts of the country and plagues its other neighbors far more. But Colombia's chronic difficulties are having another effect: well-to-do Colombians, looking for a relatively safe place to live and invest, are moving to Panama in growing numbers. Panama welcomes them, an ironic revival of linkage to a country on which the Panamanians turned their backs a century ago.

THE CARIBBEAN BASIN

As Figure 4-2 reveals, the Caribbean Basin, Middle America's island region, consists of a broad arc of numerous islands extending from the western tip of Cuba to the southern coast of Trinidad. The four larger islands or *Greater Antilles* (Cuba, Hispaniola, Jamaica, and Puerto Rico) are clustered in the western half of this arc. The smaller islands or *Lesser Antilles* extend to the east in a crescent-shaped zone from the Virgin Islands to Trinidad and Tobago. (Breaking this tectonic-plate-related regularity are the Bahamas and the Turks and Caicos, north of the Greater

Antilles, and numerous other islands too small to appear on a map at the scale of Fig. 4-2.)

On these islands, whose combined land area constitutes only 9 percent of Middle America, lie 33 states and several other entities. (Europe's colonial flags have not totally disappeared from this region, and the U.S. flag flies over Puerto Rico.) The populations of these states and territories, however, comprise 21 percent of the entire geographic realm, making this the most densely peopled part of the Americas.

Economic and Social Patterns

The Caribbean is also one of the poorest regions of the world. Environmental, demographic, political, and economic circumstances combine to impede development in the region at almost every turn. The Caribbean Basin is scenically beautiful, but it is a difficult place to make a living, and poverty is the norm.

It was not always thus. Untold wealth flowed to the European colonists who invaded this region, enslaved and eventually obliterated the indigenous Amerindian (Carib) population, and brought to these islands in bondage the Africans to work on the sugar plantations that made them rich. By the time competition from elsewhere ended the near-monopoly Caribbean sugar had enjoyed on European markets, the seeds of disaster had been sowed.

As the sugar trade collapsed, millions were pushed into a life of subsistence, malnutrition, and even hunger. Population growth now created ever-greater pressure on the land, but Caribbean islanders had few emigration options in their fragmented, mountainous habitats. The American market for sugar allowed some island countries to revive their sugar exports, including the Dominican Republic and Jamaica (Cuba's exports went to the Soviet Union, but when the USSR disintegrated Cuba's sugar industry was doomed). Some agricultural diversification occurred in the Lesser Antilles where bananas, spices, and sea-island cotton replaced sugar, but farming is a risky profession in the Caribbean. Foreign markets are undependable, and producers are trapped in a disadvantageous international economic system they cannot change.

Not surprisingly, many farming families simply abandon their land and leave for towns and cities, hoping to do better there. Although the Caribbean region is less urbanized, overall, than other parts of the Americas, it is far more urbanized than, say, Africa or the Pacific Realm. About two-thirds of the Caribbean's population now live in cities such as Santo Domingo in the Dominican Republic, Havana in Cuba, Port-au-Prince in Haiti, and San Juan in Puerto Rico. These cities reflect the poverty of the citizens who were driven to seek refuge there. Port-au-Prince has some of the world's worst slums (photo below), and desolate squatter settlements ring cities such as Kingston, Jamaica, and smaller towns.

Ethnicity and Advantage

The human geography of the Caribbean region still carries imprints of the cultures of Subsaharan Africa, and the legacy of European domination also lingers. The historical geography of Cuba, Hispaniola, and Puerto Rico is suffused with Hispanic culture; Haiti and Jamaica carry stronger African legacies. But the reality of this ethnic diversity is that European lineages still hold the

The Caribbean is a region of sharp, often searing contrasts. Haiti is the Western Hemisphere's poorest country, its city slums (here in the capital, Port-au-Prince) the most desperate of human environments. Tourism and wealthy waterfront living create another face of the region, generating landscapes of conspicuous consumption—but producing some of the limited economic opportunities in the Caribbean Basin. © Wesley Boyce-Upon/Photo Researchers.

advantage. Hispanics tend to be in the best positions in the Greater Antilles; people who have mixed European-
19 African ancestries, and who are described as **mulatto**, rank next. The largest part of this social pyramid is also the least advantaged: the Afro-Caribbean majority. In virtually all societies of the Caribbean, the minorities hold disproportionate power and exert overriding influence. In Haiti, the mulatto minority accounts for barely 5 percent of the population but has long held most of the power. In the adjacent Dominican Republic, the pyramid of power puts Hispanics (17 percent) at the top, the mixed sector (70 percent) in the middle, and the Afro-Caribbean minority (12 percent) at the bottom. Historic advantage has a way of perpetuating itself.

The composition of the population of the islands is further complicated by the presence of Asians from both China and India. During the nineteenth century, the emancipation of slaves and ensuing local labor shortages brought some far-reaching solutions. Some 100,000 Chinese emigrated to Cuba as indentured laborers, and Jamaica, Guadeloupe, and especially Trinidad saw nearly 250,000 South Asians arrive for similar purposes. To the African-modified forms of English and French heard in the Caribbean, therefore, can be added several Asian languages. The ethnic and cultural variety of the plural societies of Caribbean America is indeed endless.

Tourism: Promising Alternative?

Given the Caribbean region's limited economic options, does the tourist industry offer better opportunities? Opinions on this question are divided (see the Issue Box entitled "The Role of the Tourist Industry in Middle American Economies"). The resort areas, scenic treasures, and historic locales of Caribbean America attract well over 20 million visitors annually, with about half of these tourists traveling on Florida-based cruise ships. Certainly, Caribbean tourism is a prospective moneymaker for many islands. In Jamaica alone, this industry now accounts for about one-sixth of the gross national income and employs more than one-third of the labor force.

But Caribbean tourism also has serious drawbacks. The invasion of poor communities by affluent tourists contributes to rising local resentment, which is further fueled by the glaring contrasts of shiny new hotels towering over substandard housing and luxury liners gliding past poverty-stricken villages. At the same time, tourism can have the effect of debasing local culture, which often is adapted to suit the visitors' tastes at hotel-staged "culture" shows. And while tourism does generate income in the Caribbean, the intervention of island governments and multinational corporations removes opportunities from local entrepreneurs in favor of large operators and major resorts.

The Greater Antilles

The four islands of the Greater Antilles contain five political entities: Cuba, Haiti, the Dominican Republic, Jamaica, and Puerto Rico (Fig. 4-2). Haiti and the Dominican Republic share the island of Hispaniola.

Cuba, the largest Caribbean island-state in terms of both territory (43,000 square miles/111,000 sq km) and population (11.4 million), lies only 90 miles (145 km) from the southern tip of Florida (Fig. 4-14). Havana, the now-dilapidated capital, lies almost directly across from the Florida Keys on the northwest coast of the elongated island. Cuba was a Spanish possession until the late 1890s when, with American help in the Spanish-American War, it attained independence. Fifty years later, a U.S.-backed dictator was in control, and by the 1950s Havana had become an American playground. The island was ripe for revolution, and in 1959 Fidel Castro's insurgents gained control, thereby making Cuba a communist dictatorship and a Soviet client. Castro's rule survived the collapse of the Soviet Empire despite the loss of subsidies and sugar markets on which it had long relied.

As Fig. 4-14 shows, sugar was Cuba's economic mainstay for many years; the plantations, once the property of rich landowners, extend all across the territory. But as this map also shows, sugarcane is losing its position as the leading Cuban foreign-exchange earner. Mills are being closed down, and the cane fields are being cleared for other crops and for pastures. Cuba has other economic opportunities, however, especially in its highlands. There are three mountainous areas, of which the southeastern chain, the Sierra Maestra, is the highest and most extensive. These highlands create considerable environmental diversity as reflected by extensive, timber-producing tropical forests and varied soils on which crops ranging from tobacco to subtropical and tropical fruits are grown. Rice and beans are the staples, but Cuba cannot meet its needs and so must import food. The savannas of the center and west support livestock. Cuba has limited mineral reserves and no oil, but its nickel deposits are extensive and have been mined for a century.

Poverty, crumbling infrastructure, crowded slums, and unemployment mark the Cuban cultural landscape, but Cuba's regime still has support among the general population. During the Castro period, much has been done to bring the Afro-Cuban population into the mainstream through education and health provision. Cubans point to Guatemala, Nicaragua, and El Salvador and ask whether those countries are better off than they are under

The Role of the Tourist Industry in Middle American Economies

Regional ISSUE

IN SUPPORT OF THE TOURIST INDUSTRY

"As the general manager of a small hotel in St. Maarten, on the Dutch side of the island, I can tell you that without tourists, we would be in deep trouble economically. Mass tourism, plain and simple, has come to the rescue in the Caribbean and in other countries of Middle America. Look at the numbers. Here it's the only industry that is growing, and it already is the largest worldwide industry. For some of the smaller countries of the Caribbean, this is not just the leading industry but the only one producing external revenues. Whether it's hotel patrons or cruise-ship passengers, tourists spend money, create jobs, fill airplanes that give us a link to the outside world, require infrastructure that's good not just for them but for locals too. We've got better roads, better telephone service, more items in our stores. All this comes from tourism. I employ 24 people, most of whom would be looking for nonexistent jobs if it weren't for the tourist industry.

"And it isn't just us here in the Caribbean. Look at Belize. I just read that tourism there, based on their coral reefs, Mayan ruins, and inland waterways, brought in U.S. $100 million last year, in a country with a total population of only about 300,000! That sure beats sugar and bananas. In Jamaica, they tell me, one in every three workers has a job in tourism. And the truth is, there's still plenty of room for the tourist industry to expand in Middle America. Those Americans and Europeans can close off their markets against our products, but they can't stop their citizens from getting away from their awful weather by coming to this tropical paradise.

"Here's another good thing about tourism. It's a clean industry. It digs no mine shafts, doesn't pollute the atmosphere, doesn't cause diseases, doesn't poison villagers, isn't subject to graft and corruption the way some other industries are.

"Last but not least, tourism is educational. Travel heightens knowledge and awareness. There's always a minority of tourists who just come to lie on the beach or spend all their time in some cruise-ship bar, but most of the travelers we see in my hotel are interested in the place they're visiting. They want to know why this island is divided between the Dutch and the French, they ask about coral reefs and volcanoes, and some even want to practice their French on the other side of the border (don't worry, no formalities, just drive across and start talking). Tourism's the best thing that happened to this part of the world, and other parts too, and I hope we'll never see a slowdown."

CRITICAL OF THE TOURIST INDUSTRY

"You won't get much support for tourism from some of us teaching at this college in Puerto Rico, no matter how important some economists say tourism is for the Caribbean. Yes, tourism is an important source of income for some countries, like Kenya with its wildlife and Nepal with its mountains, but for many countries that income from tourism does not constitute a real and fundamental benefit to the local economies. Much of it may in fact result from the diversion to tourist consumption of scarce commodities such as food, clean water, and electricity. More of it has to be reinvested in the construction of airport, cruise-port, overland transport, and other tourist-serving amenities. And as for items in demand by tourists, have you noticed that places with many tourists are also places where prices are high?

"Sure, our government people like tourism. Some of them have a stake in those gleaming hotels where they can share the pleasures of the wealthy. But what those glass-enclosed towers represent is globalization, powerful multinational corporations colluding with the government to limit the opportunities of local entrepreneurs. Planeloads and busloads of tourists come through on prearranged (and prepaid) tour promotions that isolate those visitors from local society.

"You geographers talk about cultural landscapes. Well, picture this: luxury liners gliding past poverty-stricken villages, luxury hotels towering over muddy slums, restaurants serving caviar when, down the street, children suffer from malnutrition. If the tourist industry offered real prospects for economic progress in poorer countries, such circumstances might be viewed as the temporary, unfortunate byproducts of the upward struggle. Unfortunately, the evidence indicates otherwise. Name me a tourism-dependent economy where the gap between the rich and poor has narrowed.

"As for the educational effect of tourism, spare me the argument. Have you sat through any of those 'culture' shows staged by the big hotels? What you see there is the debasing of local culture as it adapts to visitors' tastes. Ask hotel workers how they really feel about their jobs, and you'll hear many say that they find their work dehumanizing because expatriate managers demand displays of friendliness and servitude that locals find insulting to sustain.

"I've heard it said that tourism doesn't pollute. Well, the Alaskans certainly don't agree—they sued a major cruise line on that issue and won. Not very long ago, cruise-ship crews routinely threw garbage-filled plastic bags overboard. That seems to have stopped, but I'm sure you've heard of the trash left by mountain-climbers in Nepal, the damage done by off-road vehicles in the wildlife parks of Kenya, the coral reefs injured by divers off Florida. Tourism is here to stay, but it is no panacea."

Vote your opinion at www.wiley.com/college/deblij

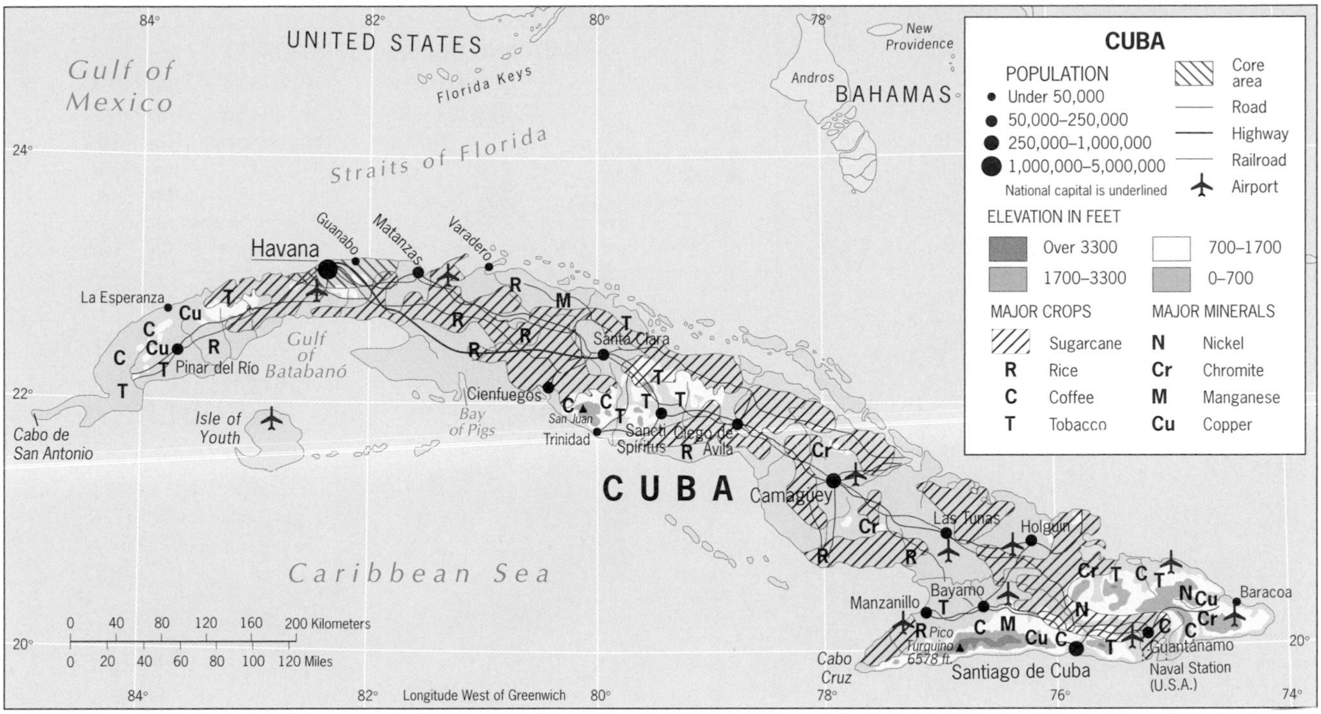

FIGURE 4-14 © H. J. de Blij, P. O. Muller, and John Wiley & Sons, Inc.

Castro. But in the United States, the view is different: Cuba, exiles and locals agree, could be the shining star of the Caribbean, its people free, its tourist economy booming, its products flowing to American markets. "The future Ireland of the Caribbean," suggested a Cuban geographer recently. Much will have to change for that prediction to come true.

Jamaica lies across the deep Cayman Trench from southern Cuba, and a cultural gulf separates these two countries as well. Jamaica, a former British dependency, has an almost entirely Afro-Caribbean population. As a member of the British Commonwealth, Jamaica still recognizes the British monarch as the chief of state, represented by a governor-general. The effective head of government in this democratic country, however, is the prime minister. English remains the official language in Jamaica, and British customs still linger.

Smaller than Connecticut and with 2.7 million people, Jamaica has experienced a steadily declining GNI over the past few decades despite its relatively slow population growth. Tourism has become the largest source of income, but the markets for bauxite (aluminum ore), of which Jamaica is a major exporter, have dwindled. And like other Caribbean countries, Jamaica has trouble making money from its sugar exports. Jamaican farmers also produce crops ranging from bananas to tobacco, but the country faces the disadvantages on world markets common to those in the periphery. Meanwhile, Jamaica must import all of its oil and much of its food because the densely populated coastal flatlands suffer from overuse and shrinking harvests.

The capital, Kingston on the south coast, reflects Jamaica's economic struggle. Almost none of the hundreds of thousands of tourists who visit the country's beaches, explore its Cockpit Country of (karst) limestone towers and caverns, or populate the cruise ships calling at Montego Bay or other points along the north coast get even a glimpse of what life is like for the ordinary Jamaican.

Haiti, the poorest state in the Western Hemisphere by virtually any measure, occupies the western part of the island of Hispaniola, directly across the Windward Passage from eastern Cuba (Fig. 4-15). Arawaks, not Caribs, formed the dominant indigenous population here, but the Spanish colonists who first took Hispaniola killed many, worked most of the survivors to death on their plantations, and left the others to die of the diseases they brought with them. French pirates established themselves in coastal coves along Hispaniola's west coast even as Spanish activity focused on the east, and just before the eighteenth century opened, France formalized the colonial status of "Saint Dominique." During the following century, French colonists established vast plantations in the valleys of the northern mountains and in the central plain, laid out elaborate irrigation systems, built dams, and brought in a large number of Africans in bondage to work the fields. Prosperity made the colonists rich, but the African workers suffered terribly.

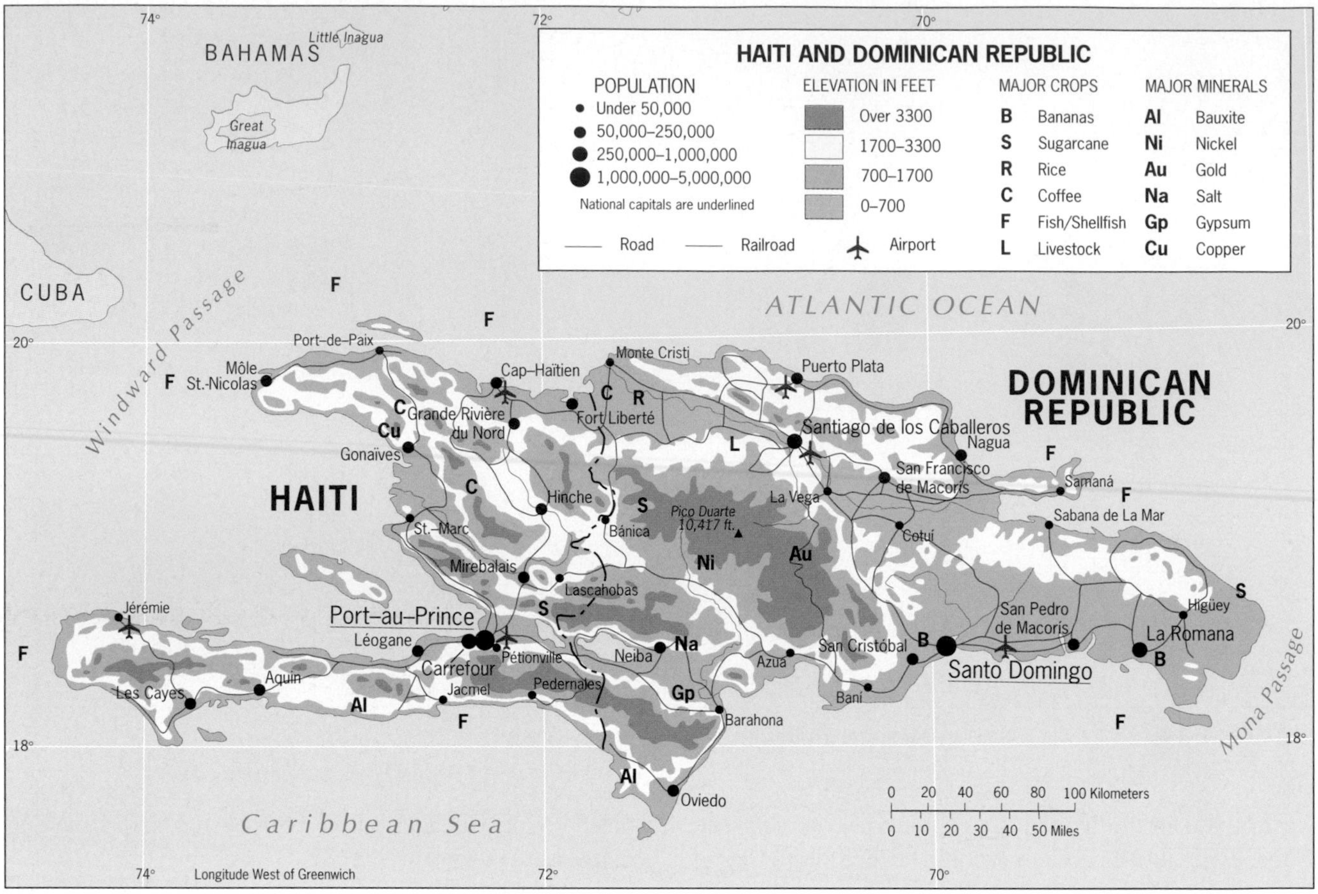

FIGURE 4-15 © H. J. de Blij, P. O. Muller, and John Wiley & Sons, Inc.

They rebelled and, in a momentous victory over their European oppressors, established the independent republic of Haiti (resurrecting the original Arawak name for it) in 1804.

Strife among Haitian groups, American intervention, mismanagement, and dictatorship doomed the fortunes of the republic. In the 1990s, the United States attempted to help move Haiti toward more representative government, but the country was in economic and social collapse. By 2003, Haiti's GNI per capita had fallen to less than half that of Jamaica, a level lower than that of many poor African countries; foreign aid makes possible most of the country's limited public expenditures. Health conditions are dreadful: malnutrition is common, AIDS is rampant, diseases ranging from malaria to tuberculosis are rife, but hospital beds and doctors are in short supply. Conditions in and around the capital, Port-au-Prince, are among the worst in the world—600 miles (less than 1000 km) from the United States.

The **Dominican Republic** has a larger share of the island of Hispaniola than Haiti (Fig. 4-15) in terms of both territory and population. Fly along the north-south border between the two countries, and you see a crucial difference: to the west, Haiti's hills and plains are treeless and gulleyed, its soils eroded and its streams silt-laden. To the east, forests drape the countryside and streams run clear.

Indigenous Caribs inhabited this eastern part of Hispaniola, and when the European colonists arrived on the island, they were in the process of driving the Arawaks westward. Spanish colonists made this a prosperous colony, but then Mexico and Peru diverted Spanish attention from Hispaniola and the territory was ceded to France. But Hispanic culture lingered, and after the revolution in Haiti the French gave eastern Hispaniola back to Spain. After the Dominican Republic declared its independence in 1821, Haitian forces invaded it and occupied the republic until 1844, creating an historic animosity that persists today.

The mountainous Dominican Republic has a wide range of natural environments and a far stronger resource base than Haiti. Nickel, gold, and silver have long been exported along with sugar, tobacco, coffee, and cocoa, but tourism (the great opportunity lost to Haiti) is the leading industry. A long period of dictatorial rule punctuated by revolutions and U.S. military intervention ended in 1978 with the first peaceful transfer of power following a democratic election.

Political stability brought the Dominican Republic rich rewards, and during the late 1990s the economy, based on manufacturing, high-tech industries, and remittances from Dominicans abroad as well as tourism, grew at an average 7 percent per year. But in the early 2000s the economy collapsed, not only because of the downturn in the world economy but also because of bank fraud and corruption in government. Suddenly the Dominican peso collapsed, inflation skyrocketed, jobs were lost, and blackouts prevailed. As the people protested, lives were lost and the self-enriched elite blamed foreign financial institutions that were unwilling to lend the government more money. Once again the hopes of ordinary citizens were dashed by greed and corruption among those in power.

Puerto Rico is the largest U.S. domain in Middle America, the easternmost and smallest island of the Greater Antilles (Fig. 4-16). This 3500-square-mile (8950-sq-km) island, with a population of 4.0 million, is larger than Delaware and more populous than Oregon. It fell to the United States more than a century ago during the Spanish-American War of 1898. Since the Puerto Ricans had been struggling for some time to free themselves from Spanish control, this transfer of power was in their view only a change from one colonial power to another. As a result, the first half-century of U.S. administration was difficult, and it was not until 1948 that Puerto Ricans were permitted to elect their own governor.

When the island's voters approved the creation of a Commonwealth in a 1952 referendum, Washington D.C. and San Juan, the two seats of government, entered into a complicated arrangement. Puerto Ricans are U.S. citizens but pay no federal taxes on local incomes. The Puerto Rican Federal Relations Act governs the island under the terms of its own constitution and awards it considerable autonomy. Puerto Rico also receives a large annual subsidy from Washington, totaling about U.S. $9 billion per year.

Despite these apparent advantages in the poverty-mired Caribbean, Puerto Rico has not thrived under U.S. administration. Long dependent on a single-crop economy (sugar), the island's industrialization during the 1950s and 1960s was based on its comparatively cheap labor, tax breaks for corporations, political stability, and special access to the U.S. market. As a result, pharmaceuticals, electronic equipment, and apparel top today's list of exports, not sugar or bananas. But this industrialization

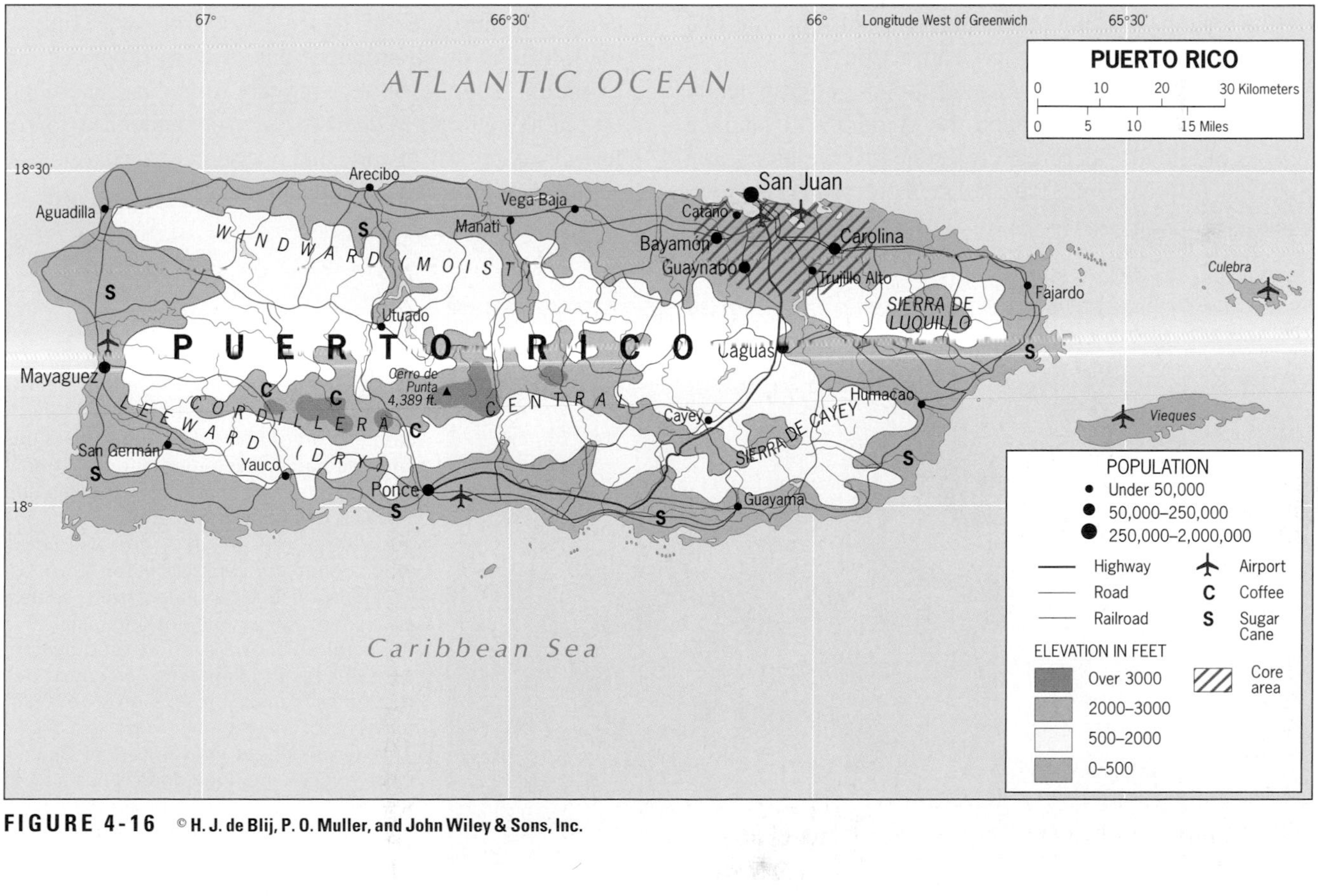

FIGURE 4-16 © H. J. de Blij, P. O. Muller, and John Wiley & Sons, Inc.

failed to stem a tide of emigration that carried more than 1 million Puerto Ricans to New York City alone. The same wages that favored corporations kept many Puerto Ricans poor or unemployed. Behind the impressive waterfront skyline of San Juan with its high-rise hotels and tourist attractions lies a landscape of economic malaise. Today, approximately 60 percent of all Puerto Ricans continue to live below the poverty line.

Not surprisingly, Puerto Ricans have been vocal in demanding political change. Nonetheless, successive referendums during the 1990s resulted in retention of the status quo—continuation of Commonwealth status rather than either Statehood or independence. The issue will no doubt continue to pose a formidable challenge to American statecraft in the years ahead.

The Lesser Antilles

As Figure 4-2 shows, the Greater Antilles are flanked by two clusters of islands: the extensive Bahamas-Turks/Caicos archipelago to the north and the Lesser Antilles to the east and south. The Bahamas, the former British colony that is now the closest Caribbean neighbor to the United States, alone consists of nearly 3000 coral islands, most of them rocky, barren, and uninhabited, but about 700 carrying vegetation, of which some 30 are inhabited. Centrally positioned New Providence Island carries most of the country's 330,000 inhabitants and contains the capital, Nassau, a leading tourist attraction.

The Lesser Antilles are grouped geographically into the Leeward Islands and the Windward Islands, a (climatologically incorrect) reference to the prevailing airflows in this tropical area. The Leeward Islands extend from the U.S. Virgin Islands to the French dependencies of Guadeloupe and Martinique, and the Windward Islands from St. Lucia to the Netherlands Antilles off the Venezuelan coast (Fig. 4-2). It would be impractical to detail the individual geographic characteristics of each of the Lesser Antilles, but we should note that these countries and territories share the environmental risks of this region: earthquakes, volcanic eruptions, and hurricanes; that they face, to varying degrees, similar economic problems in the form of limited domestic resources, overpopulation, soil deterioration, land fragmentation, and market limitations; that tourism has become the leading industry for many; that political status ranges from complete sovereignty to continuing dependency; and that cultural diversity is strong not only between islands but within them as well. In economic terms, the GNI figures given (where available) in Table G-1 may look encouraging, but these numbers conceal the social reality in virtually every entity: the gap between the fortunate few who are well-off and the great majority who are poor remains enormous.

Under such circumstances one looks for hopeful signs, and one of these signs comes from **Trinidad and Tobago**, the two-island republic at the southern end of the Lesser Antilles. This country (population: 1.3 million) has embarked on a natural-gas-driven industrialization boom that could turn this two-island country into an economic tiger (see photo below). Trinidad has long been an oil producer, but lower world prices and dwindling supplies in recent years forced a reexamination of its natural gas deposits to help counter the downturn. That quickly resulted in the discovery of major new supplies, and this cheap and abundant fuel has sparked a

Many Caribbean towns depend primarily on the tourist industry but not Trinidad's capital, Port of Spain. For many years this city has had a diversified and expanding industrial base ranging from textiles and plastics to building materials and foods (next time you are in a grocery, check where Angostura® bitters come from!). A key location on Caribbean trade routes, efficient container and docking facilities, a major airport, and a good road system combine with domestic oil and gas resources to make the economy of "T&T" into one of the region's most successful. This recent photo shows Port of Spain's modern container-loading facilities and, in the background, some of the contemporary architecture in this bustling city. © Don Kelin/D. Donne Bryant Stock Photographers.

local gas-production boom as well as an influx of energy, chemical, and steel companies from Western Europe, Canada, and even India. Many of the new industrial facilities have agglomerated at the ultramodern Point Lisas Industrial Estate outside the capital, Port of Spain, and they have propelled Trinidad to become the world's leading exporter of ammonia and methanol. Natural gas is also an efficient fuel for the manufacturing of metals, and steelmakers as well as aluminum refiners have been attracted to locate here. With Trinidad lying only a few miles from the Venezuelan coast of South America, it is also a sea-lane crossroads that is well connected to the vast, near-coastal supplies of iron ore and bauxite that are mined in nearby countries, particularly the Brazilian Amazon.

Middle America is a physically, culturally, and economically fragmented and diverse geographic realm that defies generalization. Given its strong Amerindian presence, its North American and African infusions, and its lingering Western European traditions, this certainly is not "Latin" America.

WHAT YOU CAN DO

RECOMMENDATION: Take a "serious" vacation! Hundreds of thousands of hard-working students take off for warmer weather between semesters or during spring break. Most go to relax in Florida or another "Sunbelt" State. But Middle America is nearby, and you could combine some sun and warmth with a valuable geographic field experience—if you prepare yourself and are ready for some challenges. Take some good, large-scale maps and a notebook and, when you spend a week in Puerto Rico, ask the people you meet how they feel about their island's political status. Visit Martinique and learn on the spot what happened to the "Paris of the Caribbean," St. Pierre, just over a century ago: the evidence is still visible. Go to Curaçao and find out how the island's economy survived the decline of its oil-refinery era. Or if you're in the West, cross the Mexican border and get your own take on the impact of NAFTA. The opportunities are right on our doorstep, and when you use your geographic perspective, you'll see the world in new ways of lasting value.

GEOGRAPHIC CONNECTIONS

1 When we studied the expansion of the European Union (EU), the issue of Turkey's application for membership loomed large (Chapter 1). Turkish leaders want admission and are working to meet the conditions of membership, but Turkey still fails to meet these criteria in important respects. European Union members' views range across the spectrum, from enthusiastic to ambivalent to negative. Now imagine a North American Union of which Canada and the United States are members, and which Mexico wishes to join (NAFTA is a far less comprehensive union than the EU, in effect just a first step toward it). Use your geographic insights to analyze the prospects of such a merger, and predict reactions in Canada and the United States. In what ways would Mexico qualify for membership, and in what respects would it not? Do you see any similarities between current Turkey–Europe and Mexico–North America relationships? In what ways are the two situations crucially different?

2 For many years, Caribbean islands such as St. Lucia have been exporting bananas to markets in North America and Europe, with these markets protected by preferential trade agreements. Now those agreements are ending as trade barriers fall all over the world, and the single-crop economies of the Caribbean are in trouble. What are the options for the farmers of St. Lucia and its neighbors? Relate your answer to core-periphery contrasts in the modern world.

Scale 1:40 000 000; one inch to 630 miles. Lambert's Azimuthal, Equal Area Projection
Elevations and depressions are given in feet

FIGURE 5-1 Reprinted with permission from *Goode's World Atlas*, 21st edition, p. 139. © Rand McNally, 2005. License R.L. 05-S-64.

Chapter 5

South America

CONCEPTS, IDEAS, AND TERMS

1. *Altiplano*
2. Land alienation
3. Plural society
4. Commercial agriculture
5. Subsistence agriculture
6. South American culture spheres
7. Free Trade Area of the Americas (FTAA)
8. Urbanization
9. Rural-to-urban migration
10. Megacity
11. "Latin" American City model
12. Informal sector
13. *Barrio* (*favela*)
14. Insurgent state
15. Failed state
16. Von Thünen model
17. El Niño
18. Forward capital
19. Growth-pole concept

REGIONS

- THE NORTH
- THE WEST
- THE SOUTH
- BRAZIL

Of all the continents, South America has the most familiar shape—a giant triangle connected by mainland Middle America's tenuous land bridge to its sister continent in the north. South America also lies not only south but mostly east of its northern counterpart. Lima, the capital of Peru—one of the continent's westernmost cities—lies farther east than Miami, Florida. Thus South America juts out much more prominently into the Atlantic Ocean toward southern Europe and Africa than does North America. But lying so far eastward means that South America's western flank faces a much wider Pacific Ocean, with the distance from Peru to Australia nearly twice that from California to Japan.

As if to reaffirm South America's northward and eastward orientation, the western margins of the continent are rimmed by one of the world's longest and highest mountain ranges, the Andes, a gigantic wall that extends unbroken from Tierra del Fuego near the southern tip of the triangle to Venezuela in the far north (Fig. 5-1). The other major physiographic feature of South America dominates its central north—the Amazon Basin; this vast humid-tropical amphitheater is drained by the mighty Amazon, which is fed by several major tributaries. Much of the remainder of the continent can be classified as plateau, with the most important components being the Brazilian Highlands that cover most of Brazil southeast of the Amazon Basin, the Guiana Highlands located north of the lower Amazon Basin, and the cold Patagonian plateau that blankets the southern third of Argentina. Figure 5-1 also reveals two other noteworthy river basins beyond Amazonia: the Paraná-Paraguay Basin of south-central South America, and the Orinoco Basin in the far north that drains interior Colombia and Venezuela.

DEFINING THE REALM

During most of the twentieth century, South America was a realm of political turmoil and dictatorial regimes, strong and damaging regional disparities, poor internal connections and limited international circulation, and development inertia. Toward the end of the century, however, things appeared to be improving. The major countries, previously accustomed to going their separate ways, saw the benefits of forging closer multinational ties. New transport routes crossed international borders and opened settlement frontiers. Democratic governments replaced authoritarian regimes. Economies were growing, and an era of stability and progress seemed to have arrived.

But in this first decade of the twenty-first century, South America's promise is again in doubt. Corruption and mismanagement, endemic failures in this part of the world, damaged economies from Argentina, whose economy imploded, to Ecuador, where the political framework faltered. Instability afflicted poverty-stricken Bolivia, where an elected government was ousted by violent protests, and oil-rich Venezuela, where deadly turmoil accompanied a drive to recall the president. A costly drug war continued in Colombia, confidence in the government was at a low ebb in Peru, and isolated Paraguay remained South America's version of a failed state. So what remains of the hopes of a decade ago? A democratic election in Brazil, representing a considerable ideological shift from right to left, was a defining moment for this realm's giant state. The economic success of Chile can be cited as an example for the entire continent. And the stability of Uruguay in a turbulent area (Argentina is its neighbor) is a beacon of hope.

South America thus remains a realm of boom-bust cycles that erode confidence and deter long-term investment. Infrastructure remains weak, inefficiency rampant, and corruption endemic. And the gulf between rich and poor is widening, not narrowing: overall, the richest 20 percent of the population control 70 percent of South America's wealth, while the poorest one-fifth own only 2 percent (in some individual countries, the gap is even wider). By some measures this disparity is greater in South America than in any other world realm. The prospects of this continent, therefore, remain uncertain, but its needs, as we will see, are clear.

MAJOR GEOGRAPHIC QUALITIES OF South America

1. South America's physiography is dominated by the Andes Mountains in the west and the Amazon Basin in the central north. Much of the remainder is plateau country.
2. Half of the realm's area and half of its population are concentrated in one country—Brazil.
3. South America's population remains concentrated along the continent's periphery. Most of the interior is sparsely peopled, but sections of it are now undergoing significant development.
4. Interconnections among the states of the realm are improving rapidly. Economic integration has become a major force, particularly in southern South America.
5. Regional economic contrasts and disparities, both in the realm as a whole and within individual countries, are strong.
6. Cultural pluralism exists in almost all of the realm's countries and is often expressed regionally.
7. Rapid urban growth continues to mark much of the South American realm, and the urbanization level overall is today on a par with the levels in the United States and Europe.

THE HUMAN SEQUENCE

Although modern South America's largest populations are situated in the east and north, during the height of the Inca Empire the Andes Mountains contained the most densely peopled and best organized state on the continent. Although the origins of Inca civilization are still shrouded in mystery, it has become generally accepted that the Incas were descendants of ancient peoples who came to South America via the Middle American land bridge. Thus for thousands of years before the Europeans arrived in the sixteenth century, indigenous Amerindian societies had been developing in South America.

The Inca Empire

About one thousand years ago, a number of regional cultures thrived in Andean valleys and basins and at places along the Pacific coast. By A.D. 1300, the Incas had established themselves in the intermontane basin of Cuzco (Fig. 5-2). With their hearth consolidated, they were now ready to begin forging the greatest pre-European empire in the Americas by steadily conquering and extending their authority over the peoples of coastal Peru and other Andean basins. When the Inca civilization is compared to that of ancient Mesopotamia, Egypt, and the Mexican Aztec Empire, it quickly becomes clear that this civiliza-

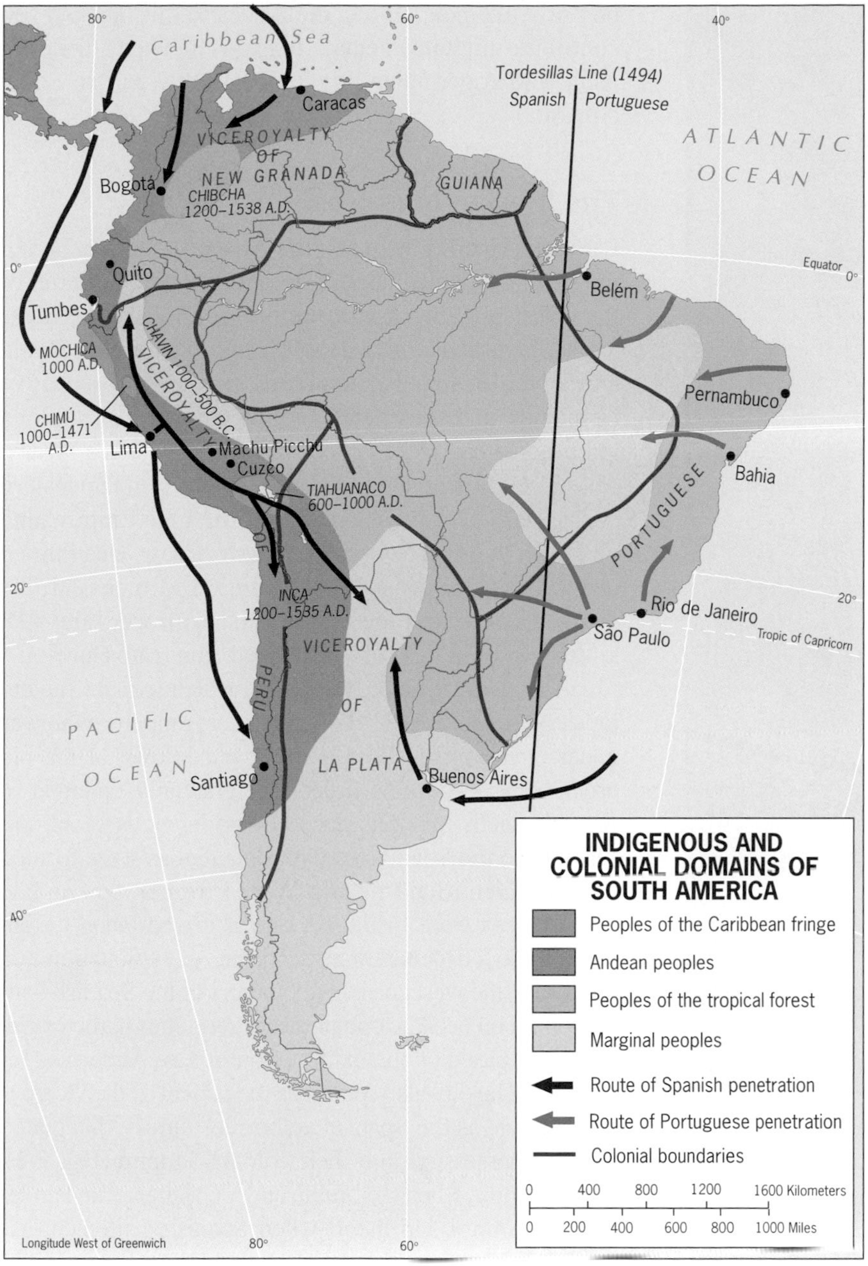

FIGURE 5-2 © H. J. de Blij, P. O. Muller and John Wiley & Sons, Inc.

tion was an unusual achievement. Everywhere else, rivers and waterways provided avenues for the circulation of goods and ideas. Here, however, an empire was forged
1 from a series of elongated basins (called ***altiplanos***) in the high Andes, created when mountain valleys between parallel and converging ranges filled with erosional materials from surrounding uplands. These *altiplanos* are often separated by some of the world's most rugged terrain, with high snowcapped mountains alternating with precipitous canyons.

More impressive than the Incas' military victories was their subsequent capacity to integrate the peoples and regions of the Andean domain into a stable and efficiently functioning state. The odds would seem to have been against them because as they progressed their domain became ever more elongated, making effective control much more difficult. The Incas, however, were expert road and bridge builders, colonizers, and administrators, and in an incredibly short time they unified their new territories that stretched from Colombia southward to central Chile (brown zone, Fig. 5-2).

The Incas themselves were always in a minority in this huge state, and their position became one of a ruling elite in a rigidly class-structured society. A bureaucracy of Inca administrators strictly controlled the life of the empire's subjects, and the state was so highly centralized

FROM THE FIELD NOTES

"From this high vantage point I got a good perspective of a valley near Pisac in the Peruvian Andes, not far from Cuzco. This was part of the Incan domain when the Spaniards arrived to overthrow the empire, but the terraces you can see actually predate the Inca period. Human occupation in these rugged mountains is very old, and undoubtedly the physiography here changed over time. Today these slopes are arid and barren, and only a few hardy trees survive; the stream in the valley bottom is all the water in sight. But when the terrace builders transformed these slopes, the climate may have been more moist, the countryside greener." © Philip L. Keating.

that a takeover at the top was enough to gain power over the entire empire—as the Spaniards quickly proved in the 1530s. The Inca Empire disintegrated abruptly under the impact of the Spanish invaders, but it left behind spectacular ruins such as those at Peru's Machu Picchu. It also bequeathed a legacy of social values that have remained a part of Amerindian life in the Andes to this day and still contribute to fundamental divisions between the Hispanic and Amerindian population in this part of South America.

The Iberian Invaders

In South America as in Middle America, the location of indigenous peoples largely determined the direction of the thrusts of European invasion. The Incas, like Mexico's Maya and Aztec peoples, had accumulated gold and silver at their headquarters, possessed productive farmlands, and constituted a ready labor force. Not long after the defeat of the Aztecs in 1521, Francisco Pizarro sailed southward along the continent's northwestern coast, learned of the existence of the Inca Empire, and withdrew to Spain to organize its overthrow. He returned to the Peruvian coast in 1531 with 183 men and two dozen horses, and the events that followed are well known. In 1533, his party rode victorious into Cuzco.

At first, the Spaniards kept the Incan imperial structure intact by permitting the crowning of an emperor who was under their control. But soon the breakdown of the old order began. The new order that gradually emerged in western South America placed the indigenous peoples in serfdom to the Spaniards. Great haciendas were formed by **land alienation** (the takeover of former Amerindian lands), taxes were instituted, and a forced-labor system was introduced to maximize the profits of exploitation. 2

Lima, the west coast headquarters of the Spanish conquerors, soon became one of the richest cities in the world, its wealth based on the exploitation of vast Andean silver deposits. The city also served as the capital of the viceroyalty of Peru, as the Spanish authorities quickly integrated the new possession into their colonial empire (Fig. 5-2). Subsequently, when Colombia and Venezuela came under Spanish control and, later, when Spanish settlement expanded in what is now Argentina and Uruguay, two additional viceroyalties were added to the map: New Granada and La Plata.

Meanwhile, another vanguard of the Iberian invasion was penetrating the east-central part of the continent, the coastlands of present-day Brazil. This area had become a Portuguese sphere of influence because Spain and Portugal had signed a treaty in 1494 to recognize a north-south line 370 leagues west of the Cape Verde Islands as the boundary between their New World spheres of influence. This border ran approximately along the meridian of 50°W longitude, thereby cutting off a sizeable triangle of eastern South America for Portugal's exploitation (Fig. 5-2). But a brief look at the political map of South America (Fig. 5-3) shows that this treaty did not limit Portuguese colonial territory to the east of the 50th meridian. Instead, Brazil's

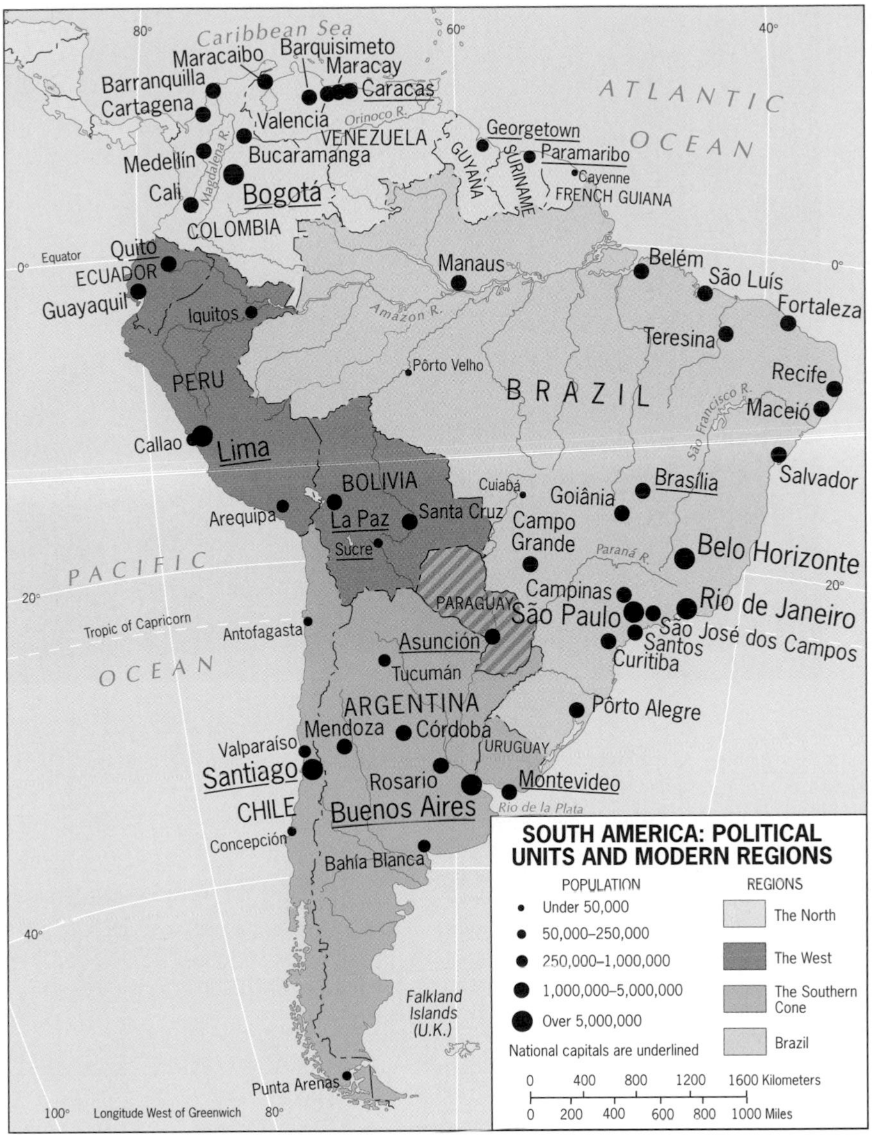

FIGURE 5-3 © H. J. de Blij, P. O. Muller, and John Wiley & Sons, Inc.

boundaries were bent far inland to include almost the entire Amazon Basin, and the country came to be only slightly smaller in territorial size than all the other South American countries combined. This westward thrust was the work of many Brazilian elements, particularly the *Paulistas*, the settlers of São Paulo who needed Amerindian slave labor to run their plantations.

The Africans

As Figure 5-2 shows, the Spaniards initially got very much the better of the territorial partitioning of South America—not just in land quality but also in the size of the aboriginal labor force. When the Portuguese began to develop their territory, they turned to the same lucrative activity that their Spanish rivals had pursued in the Caribbean—the plantation cultivation of sugar for the European market. And they, too, found their labor force in the same source region, as millions of Africans were brought in slavery to the tropical Brazilian coast north of Rio de Janeiro (see map, p. 289). Not surprisingly, Brazil now has South America's largest black population, which is still heavily concentrated in the country's poverty-stricken northeastern States. With blacks of "pure" and mixed ancestry today accounting for 45 percent of Brazil's population of 184 million, the Africans decidedly constitute the third major immigration of foreign peoples into South America.

Longstanding Isolation

Despite their adjacent location on the same continent, their common language and cultural heritage, and their shared national problems, the countries that arose out of South America's Spanish viceroyalties (as well as Brazil) have existed in a considerable degree of isolation from one another. Distance and physiographic barriers reinforced this

FROM THE FIELD NOTES

"Salvador is one of Brazil's oldest and most vibrant cities. Magnificent churches, public buildings, and mansions were built during the time when this was, by many measures, the most important city in the Southern Hemisphere. Long the capital of Brazil, Salvador was the point of entry for tens of thousands of Africans, and the fortune-making plantation economy, augmented by the whaling industry, concentrated enormous wealth here, some of which went into the construction of an opulent city center. But fortunes change, Salvador lost its political as well as economic advantages, and the city fell into disrepair. Walking the streets of the old town in 1982, I noted the state of decay of much of Salvador's architectural heritage and wondered about its survival: weathering in this tropical environment was destroying woodwork, façades, and roofs. But then the United Nations proclaimed Salvador's old town a World Heritage site, and massive restoration began (left). By the late 1990s, much of the district had been revived (right), and tourism's contribution to the local economy was on the rise." © H. J. de Blij.

separation, and the realm's major population agglomerations still adhere to the coast, mainly the eastern and northern coasts (Fig. G-9). The viceroyalties existed primarily to extract riches and fill Spanish coffers. In Iberia there was little interest in developing the American lands for their own sake. Only after those who had made Spanish and Portuguese America their home and who had a stake there rebelled against Iberian authority did things begin to change, and then very slowly. South America was saddled with the values, economic outlook, and social attitudes of eighteenth-century Iberia—not the best tradition from which to begin the task of forging modern nation-states.

Independence

Certain isolating factors had their effect even during the wars for independence. Spanish military strength was always concentrated at Lima, and those territories that lay farthest from their center of power—Argentina and Chile—were the first to establish their independence from Spain (in 1816 and 1818, respectively). In the north, Simón Bolívar led the burgeoning independence movement, and in 1824 two decisive military defeats there spelled the end of Spanish power in South America.

This joint struggle, however, did not produce unity because no fewer than nine countries emerged from the three former viceroyalties. It is not difficult to understand why this fragmentation took place. With the Andes intervening between Argentina and Chile and the Atacama Desert between Chile and Peru, overland distances seem even greater than they really were, and these obstacles to contact proved quite effective. Hence, from their outset the new countries of South America began to grow apart, and friction and even wars have been frequent. Only within the past two decades have the countries of this realm finally begun to recognize the mutual advantages of increasing co-operation and to make lasting efforts to steer their relationships in this direction.

CULTURAL FRAGMENTATION

When we speak of the "interaction" of South American countries, it is important to keep in mind just who does the interacting. The fragmentation of colonial South America into ten individual republics, and the subsequent postures of each of these states, was the work of a small minority that constituted the landholding, upper-class elite. Thus in every country a vast majority—be they Amerindians in Peru or people of African descent in Brazil—could only watch as their European masters struggled with each other for supremacy.

South America, then, is a realm where Amerindians of diverse cultures, Europeans from Iberia and elsewhere, Africans from West and Equatorial Africa, and Asians from Japan, India, and Indonesia have
3 forged **plural societies** exhibiting various degrees of cultural integration. South America's human geography is a cultural kaleidoscope of almost endless variety, whose enduring divisions and modern mixtures combine to create images of past and future, which are also reflected in the economic landscape. On the map of agriculture, the realm's dominant livelihood, the
4 persistence of divisions is all too clear: **commercial**
5 (for-profit) and **subsistence** (minimum life-sustaining) forms of farming exist side by side in a pattern more pervasive than it is anywhere else in the world (Fig. 5-4). The geography of commercial agriculture tends to match the distribution of landholders of European background, whereas subsistence farming prevails in areas historically occupied by indigenous peoples. As in Middle America and, indeed, wherever Europeans and indigenous peoples made contact, the European colonists took the best land and left the less productive land to the natives. In South America, the survival of the large Amerindian region of traditional farming that extends along the Andes from Colombia to Bolivia reflects indigenous achievements in adapting to a highland environment the Europeans could not exploit.

Four decades ago, the cultural geographer John
6 Augelli drew a map that showed the status of **South America's culture spheres** (as he called them) in the mid-twentieth century (Fig. 5-5). *Tropical plantation* farming of the Middle American type took root along the north and east coast, where the forced labor of African slaves made the profits. *Commercial* life of the European variety prevailed in the three southernmost countries—Argentina, Chile, and Uruguay—as well as southernmost Brazil. *Subsistence* farming, as long practiced by the Amerind peoples of the Andes, still prevailed there. In an area he mapped as *mestizo-transitional*, Augelli recognized the integration of South American lifeways and the changes taking place in interior Brazil, northern Venezuela, and Colombia, and parts of Peru and Argentina. Today, the fifth zone he mapped as *undifferentiated*—that is, mostly the (then-unchanged) Amazonian interior—is shrinking rapidly as the *transitional* culture sphere has penetrated the rainforest

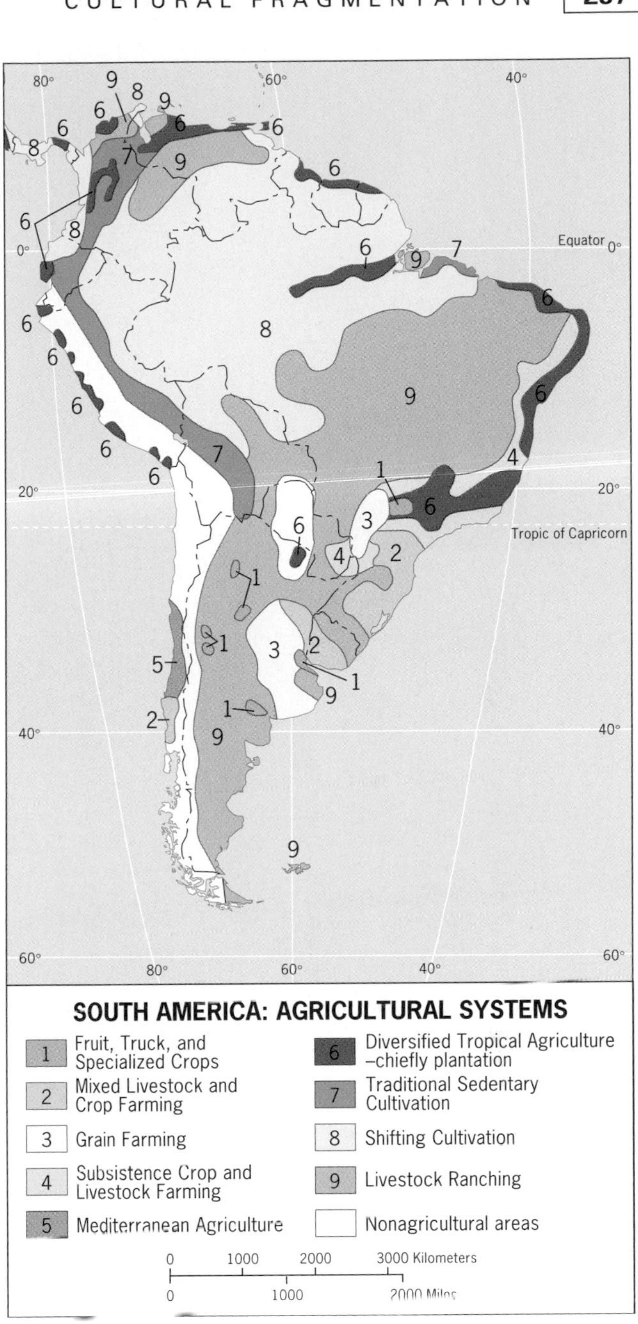

FIGURE 5-4 © H. J. de Blij, P. O. Muller, and John Wiley & Sons, Inc.

FIGURE 5-5 Adapted with permission from John P. Augelli, "The Controversial Image of Latin America: A Geographer's View," *Journal of Geography*, Vol. 62 (1963): 103–112. © National Council for Geographic Education, 1963.

that is being destroyed to make way for roads, farms, and towns.

As we noted earlier, it is inappropriate to refer to Middle *or* South America as "Latin" America. Many millions of this realm's inhabitants are Amerindians, the successors to great indigenous civilizations that are no more "Latin" than North America's Native Americans are "Anglo." Tens of millions more are of African descent, also without any ancestral "Latin" connections. Millions of others are Asians (Brazil alone is home to more than 1 million ethnic Japanese). About one-sixth of Argentina's population has Arab ancestry. And, as any visitor to South America quickly recognizes, many of those who can indeed claim southern European lineage are likely to have African, Amerindian, or other non-"Latin" blood in their veins. Thus the indisputable geographic name "South" is the best one to use for this part of the Americas, although, as the map reminds us, "Southeast" would be even more accurate.

ECONOMIC INTEGRATION

If you were to look at an atlas map of South America from the mid-twentieth century, you would be struck by the scarcity of roads or railroads linking neighboring countries. North Americans may not think twice before driving from Toronto to Mexico City, but in those days you did not just take off from Buenos Aires on your way to Lima. In fact, few people (and not many goods) crossed the borders between South American countries. These countries had more effective links with European countries than they did with each other. They looked overseas for commercial, cultural, educational, and other connections.

Today that picture is quite different. Maps of surface communications (see the regional maps later in this chapter) show an expanding network with extensions that are penetrating long-isolated frontiers. Cross-border rail, road, and pipeline projects are underway. These maps reflect a new South American era: the states of the realm have come to realize that local and regional partnerships are beneficial. Increasingly, South America's countries are trading with each other, lowering or eliminating tariff barriers, and banding together in the interest of free trade. One key multinational project that symbolizes this new age is the *Hidrovia* (water-route) system of river locks in the Paraná-Paraguay Basin, where five formerly contentious nations are jointly investing to facilitate barge traffic all the way to interior Brazil.

But it will take a long time to overcome the realm's fragmented historical geography. True, border disputes no longer tend to erupt into violent confrontations between neighbors, but true regional integration has a long way to go. South American governments, noting the benefits of supranational cooperation in Europe (EU) and North America (NAFTA), are also seeking ways to enhance trade among themselves. One result of this new look at regional integration was the formation in 1994 of an organization called *Mercosur*, consisting of Argentina, Uruguay, Paraguay, and Brazil (where it is called *Mercosul*). The idea was to reduce trade barriers among the members, to create a common external tariff, and to negotiate jointly with other trade blocs such as NATFA and the EU.

Reality, however, has been tough on Mercosur. One problem is Brazil's enormous size and powerful economy compared to crisis-prone Argentina and tiny Uruguay and Paraguay. Yet Brazil is not rich enough to help its neighbors in any significant way, nor is Brazil willing to yield sovereignty the way European states have done in order to make the EU work. For example, in 2004 Brazil blocked rice imports from Uruguay and Argentina; the Argentinians in turn put up barriers against certain imports from Brazil such as televisions and leather goods. Meanwhile, Mercosur members are accusing one another of making individual deals with overseas countries—as was the practice before Mercosur was formed to change things.

Mercosur's troubles have not stopped South Americans from dreaming of a wider regional union. In December 2004, leaders of eight countries met in Cuzco, Peru, to proclaim the founding of the *South American Community of Nations*, which is planned to include all 12 of the realm's independent states, encompassing a population of 365 million and a combined output of U.S. $1 trillion. Brazil, which would be the dominant member, as it is of Mercosur, announced in Cuzco an agreement to construct a highway linking Brazil to Peru and hence the Pacific coast. The leaders present spoke hopefully of a union similar to the EU, with a single currency, a South American passport, and a parliament and common market. But four of the twelve putative members did not send leaders to attend, and members of another regional group, the *Andean Community*, continued to trade individually with the United States far more than they do with Mercosur. The history of South America's regional integration is brief; its future is likely to be prolonged and difficult.

Meanwhile, another vision of Pan American integration, this one not originating in South America but projected to include it, is also in trouble. The success of NAFTA led North American economic planners to propose the formation of a **Free Trade Area of the Ameri- 7
cas (FTAA)** that would extend from Alaska to Tierra del Fuego and would combine all existing regional multinational agreements, including NAFTA and Mercosur, into a single gigantic trading bloc. But Chile, once South America's most enthusiastic proponent of this project, voted against the U.S.-led military intervention in Iraq, and the perception that U.S.-planned initiatives are designed to entrench American domination has put FTAA on the back burner in much of South America.

URBANIZATION

As in most other less advantaged realms, South Americans are leaving the land and moving to the cities. This
urbanization process intensified throughout the twenti- 8
eth century, and it persists so strongly that South America's urban population percentage (79 percent, or 288 million of the continent's 365 million inhabitants) now exceeds both those of Europe and the United States. And nowhere is South America's population increasing faster than in the towns and cities. We usually assume that the populations of rural areas grow more rapidly than urban areas because farm families traditionally have more children than city dwellers. Yet, overall, the urban population of South America has grown annually by about 5 percent since 1950, while the increase in rural areas was less than 2 percent. These numbers underscore not only
the dimensions but also the durability of the **rural-to- 9
urban migration** from the countryside to the cities.

The generalized spatial pattern of South America's urban transformation is displayed in Figure 5-6, which shows a *cartogram* of the continent's population.* Here

*A cartogram is a specially transformed map in which countries and cities are represented in proportion to their populations. Those containing large numbers are "blown up" in population-space, while those containing lesser numbers are "shrunk" in size accordingly. This technique is used to map the countries of the world in Figure G-10.

we see not only the realm's 13 countries in population-space relative to each other, but also the proportionate sizes of individual large cities within their total national populations.

Regionally, the realm's highest urban percentages are found in southern South America. Today in Argentina, Chile, and Uruguay, at least 87 percent of the people reside in cities and towns (statistically on a par with Europe's most urbanized countries). Ranking next among the most heavily urbanized populations is Brazil at 81 percent. The next highest group of countries, averaging 79 percent urban, border the Caribbean in the north. Not surprisingly, the Amerind-subsistence-dominated Andean countries constitute the realm's least

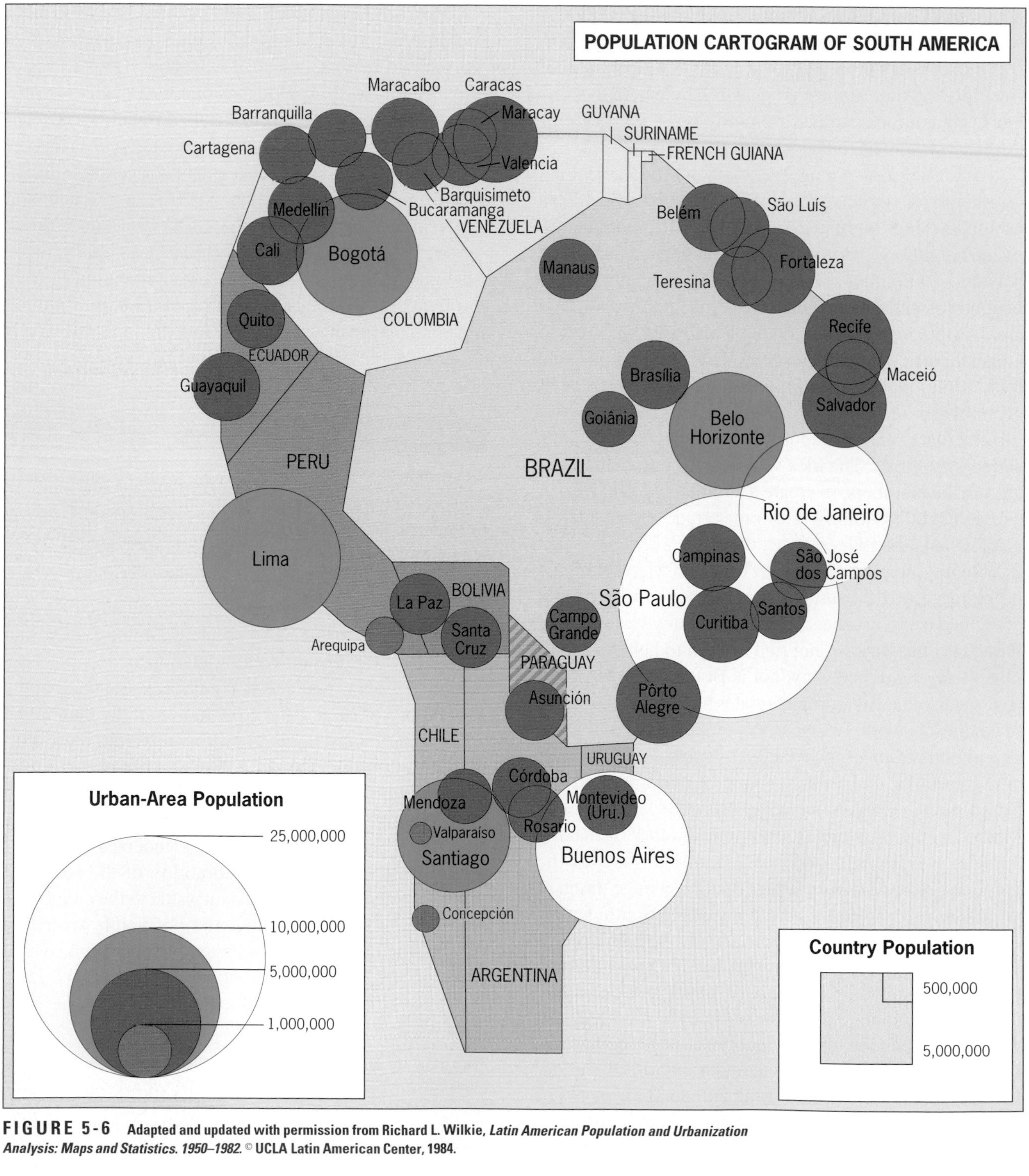

FIGURE 5-6 Adapted and updated with permission from Richard L. Wilkie, *Latin American Population and Urbanization Analysis: Maps and Statistics. 1950–1982.* © UCLA Latin American Center, 1984.

urbanized zone (Peru, whose population is 72 percent urbanized, is the region's leader). Figure 5-6 also tells us a great deal about the relative positions of major metropolises in their countries. Three of them—Brazil's São Paulo and Rio de Janeiro, and Argentina's Buenos
10 Aires—rank among the world's **megacities** (whose populations exceed 10 million).

In South America, as in Middle America, Africa, and Asia, people are attracted to the cities and driven from the poverty of the rural areas. Both *pull* and *push factors* are at work. Rural land reform has been slow in coming, and every year tens of thousands of farmers simply give up and leave, seeing little or no possibility for economic advancement. The cities lure them because they are perceived to provide opportunity—the chance to earn a regular wage. Visions of education for their children, better medical care, upward social mobility, and the excitement of life in a big city draw hordes to places such as São Paulo and Caracas. But the actual move can be traumatic. Cities in developing countries are surrounded and often invaded by squalid slums, and this is where the urban immigrant most often finds a first—and frequently permanent—abode in a makeshift shack without even the most basic amenities and sanitary facilities. And unemployment is persistently high, often exceeding 25 percent of the available labor force. But still the people come, the overcrowding in the shantytowns worsens, and the threat of epidemic-scale disease (and other disasters) rises.

The "Latin" American City Model

Although the urban experience in the South and Middle American realms has varied because of diverse historical, cultural, and economic influences, there are many common threads that have prompted geographers to search for useful generalizations. One is the
model of the intraurban spatial structure of the **"Latin"** 11
American city proposed by Ernst Griffin and Larry Ford (Fig. 5-7), which may have wider application to cities of developing countries in other geographic realms as well.

FIGURE 5-7 Adapted with permission from Ernst C. Griffin & Larry R. Ford, "A Model of Latin American City Structure," *Geographical Review*, Vol. 70 (1980): 406. © American Geographical Society, 1980.

FROM THE FIELD NOTES

"Two unusual perspectives of Rio de Janeiro form a reminder that here the wealthy live near the water in luxury high-rises, such as these overlooking Ipanema Beach, while the poor have million-dollar views from their hillslope *favelas*, such as Rocinho." © H. J. de Blij.

AMONG THE REALM'S GREAT CITIES . . . LIMA

Even in a realm marked by an abundance of primate cities, Lima (8.4 million) stands out. Here reside 30 percent of the Peruvian population—who produce over 70 percent of the country's gross national income, 90 percent of its collected taxes, and 98 percent of its private investments. Economically, at least, Lima *is* Peru.

Lima began as a modest oasis in a narrow coastal desert squeezed between the cold waters of the Pacific and the soaring heights of the nearby Andes. Near here the Spanish conquistadors discovered one of the best natural harbors on the western shoreline of South America, where they founded the port of Callao; but they built their city on a site 7 miles (11 km) inland where soils and water supplies proved more beneficial. This city was named Lima and quickly became the Spaniards' headquarters for all their South American territories.

Since independence, Lima has continued to dominate Peruvian national life. The city's population growth was manageable through the end of the 1970s, but the past quarter-century has witnessed a disastrous doubling in its size. As usual, the worst problems are localized in the peripheral squatter shantytowns that now house close to one-half of the metropolitan population. Lima's CBD, however, is worlds removed from this squalor, its historic landscape dotted with five-star hotels, new office towers, and prestigious shops as foreign investments have poured in since 1990.

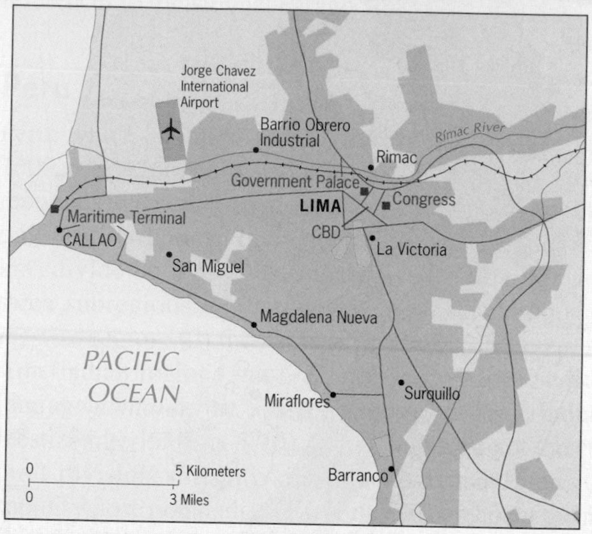

© H. J. de Blij, P. O. Muller, and John Wiley & Sons, Inc.

discussed and illustrated on pp. 215–216). The major mineral products from the Sierra are copper, zinc, lead, and several other metallic minerals, with the largest mining complex centered on Cerro de Pasco.

Of Peru's three subregions, the *Oriente*, or East—the inland slopes of the Andes and the Amazon-drained, rainforest-covered *montaña*—is the most isolated. The focus of the eastern subregion, in fact, is Iquitos, a city that looks east rather than west and can be reached by oceangoing vessels sailing 2300 miles (3700 km) up the Amazon River across northern Brazil. Iquitos grew rapidly during the Amazon wild-rubber boom of a century ago and then declined; now it is finally growing again and reflects Peruvian plans to open up the eastern interior.

Petroleum was discovered west of Iquitos in the 1970s, and since 1977 oil has flowed through a pipeline built across the Andes to the Pacific port of Bayóvar. In the 1990s, a major natural gas reserve, the Camisea field, was discovered in the east, and Peru hopes to transfer this supply via a pipeline to a conversion plant at Paracas in the western desert. From there some of this energy gas would be sent to Lima through a connecting pipeline, but more of it would be sent to tankers at an offshore terminal for export to California.

These are ambitious plans, but environmentalists worry. The Camisea field lies between two of Peru's major conservation areas, and already wells are being drilled in one of them, a reserve established long ago to protect three small tribal peoples living traditional lives. Furthermore, the Paracas plant, already under construction, is just 4.3 miles (7 km) from the only marine preserve along the entire Peruvian coast. Recent reports suggest that major oil as well as gas discoveries are in the offing in what remains of Peru's pristine interior. Everyone here knows about what happened in neighboring Ecuador, where the oil industry caused ecological disaster. Will Peru fare better?

Peru today appears to have put behind it the lengthy period of instability during which its government was threatened by well-organized guerrilla movements. The most serious threat, which for a time during the 1980s seemed on the verge of achieving an insurgent state in its Andean base, was the so-called *Sendero Luminoso* (Shining Path) which, at its height, even threatened Lima. But forceful action under the later-disgraced president of Japanese ancestry, Alberto Fujimori, decimated the rebel movement and returned Peru to political stability and economic growth. In the process, the government also tackled the problem of cocaine production in the Huallaga Valley and made progress against that scourge as well.

In 2000 Peru had to cross a political minefield when President Fujimori attempted to hold on to power, and the issue led to the exposure of political fraud, evidence of bribery, gunrunning, and worse. Having succeeded in getting himself elected for an illegal third term, Fujimori was confronted by large-scale riots in the capital. He finally left for Japan, announced that he would not return to Peru, and thereby cleared the way for the installation of Alejandro Toledo as president.

This episode underscored the weaknesses of Peru's democratic institutions, and while investors and tourists returned to the country, its underlying disabilities were painfully exposed. Even as the economic statistics turned more favorable, another measure of the condition of the country looked worse: the percentage of the people living in poverty increased from 46 in 1995 to more than 50 in 2005. The challenges facing Peru's leaders are daunting indeed.

Ecuador

On the map, Ecuador, smallest of the three Andean West republics, appears to be just a corner of Peru. But that would be a misrepresentation because Ecuador possesses a full range of regional contrasts (Fig. 5-10). It has a coastal belt; an Andean zone that may be narrow (under 150 miles or 250 km) but by no means of lower elevation than elsewhere; and an *Oriente*—an eastern subregion that is as sparsely settled and as economically marginalized as that of Peru. As in Peru, nearly half of Ecuador's population (which totals 14.0 million) is concentrated in the Andean intermontane basins and valleys, and the most productive region is the coastal strip. Here, however, the similarities end.

Ecuador's Pacific coastal zone consists of a belt of hills interrupted by lowlands, of which the most important lies in the south between the hills and the Andes, drained by the Guayas River. Guayaquil—the country's largest city, main port, and leading commercial center—forms the focus of this subregion. Unlike Peru's coastal strip, Ecuador's is not a desert: it consists of fertile tropical plains not afflicted by excessive rainfall. Seafood (especially shrimp) is a leading product, and these lowlands support a thriving commercial agricultural economy built around bananas, cacao, cattle-raising, and coffee on the hillsides. Moreover, Ecuador's western subregion is also far less Europeanized than Peru's because its white component of the national population is only about 10 percent.

A greater proportion of whites are engaged in administration and hacienda ownership in the central Andean zone, where most of the 41 percent of the Ecuadorians who are Amerindian also reside—and, not surprisingly, where land-tenure reform is an explosive issue. The differing interests of the Guayaquil-dominated coastal lowland and the Andean-highland subregion focused on the capital (Quito) have long fostered a deep regional cleavage between the two. This schism has intensified in recent years, and autonomy and other devolutionary remedies are now being openly discussed in the coastlands.

FROM THE FIELD NOTES

"I walked through the street market in Otavalo, Ecuador, on the Andean road north from Quito to the Colombian border. This is Inca country, and at these elevations (Mount Cayambe at nearly 19,000 feet [5800 m] lies nearby) the staple is the potato. Beets and turnips are also for sale, and many other crops—common and unusual—are cultivated in this area. The people here, both sellers and buyers, seem quite prosperous. They were well dressed against the cool mountain weather and business was brisk. Comparing this scene to what I had seen in the Ecuadorian port city of Guayaquil helped me understand this small country's intense regionalism better. Mestizo coast and Amerindian interior are worlds apart." © Philip L. Keating.

In the rainforests of the *Oriente* subregion, oil production is expanding as a result of the discovery of additional reserves. Some analysts are predicting that interior Ecuador as well as Peru will prove to contain reserves comparable to those of Venezuela and Colombia, and that an "oil era" will soon transform their economies. As it is, oil already tops Ecuador's export list, and the industry's infrastructure is being modernized. This is essential because much ecological damage has already been done: the trans-Andean pipeline to the port of Esmeraldas, constructed in 1972, was the source of numerous oil spills, and toxic waste was dumped along its route in a series of dreadful environmental disasters. On the positive side, a longstanding boundary dispute between Ecuador and Peru was settled in 1998, giving the Ecuadorians less territory than they wanted but awarding Ecuador navigation rights on the Amazon and its northern Peruvian tributaries.

These and other growth opportunities notwithstanding, most development is now on hold because Ecuador is in the vanguard of a regionwide social movement, led by indigenous peoples, which is threatening to destabilize the Andean West. This movement, grounded in the persistent exclusion of Amerindians within each of the region's three national societies, has spawned a number of uprisings as frustrations can no longer be contained over repeatedly failed promises by governments that have raised expectations for social and economic improvement. Ongoing geographic change further fuels this unrest because the poorest people increasingly urbanize and come to witness close-up how their second-class standing and lack of education bar them from making progress. In Ecuador, the widely supported Confederation of Indigenous Nationalities of Ecuador (CONAIE) spearheads the protest movement, demanding a multiethnic state in which the Amerindians are guaranteed greater political involvement, better access to landownership, and an end to government neglect, which ensures that indigenous communities end up with inadequate educational systems and infrastructural linkages.

Bolivia

Bolivia is one of South America's troubled countries, a malfunctioning one, if not a failed state. To grasp the fundamental reason, have a look at Figure 5-10. Bolivia is part Andean (the west) and part Amazonian lowland (the east or *Oriente*). The west and northwest are traditional Andean Amerindian domains; the east and southeast are mestizo-dominated. The European conquerors invaded the Amerindian highlands, forced the indigenous people to work in their mines, and established their headquarters there, the gritty capital of La Paz. But Bolivia had another capital, where the money made in the highlands was spent by those it enriched: culturally elegant Sucre. Today Bolivia has a third center, almost an economic capital: Santa Cruz, the hub of the energy industry that is transforming Bolivia's economy—and shaking up its social geography.

Ethnic, cultural, and economic divisions like these spell trouble, and that is what Bolivia has. The Andes broaden into the 450-mile- (720-km-) wide *Altiplano*, a plateau so high that it lies just below the snow line, its local climate moderated by the waters of Lake Titicaca, 12,500 feet (3800 m) above sea level. Aymara and Quechua subsistence farmers have cultivated the thin soils for centuries, dating to pre-Inca times. Today, Amerindians still constitute almost two-thirds of the country's population of 9.1 million, but they lost much of their land to the European invaders and never gained political power consistent with their numbers. What made the richest Europeans in Bolivia wealthy was not land, however, but minerals. The town of Potosí in the eastern Andes became legendary—and infamous—for the immense deposits of silver in its vicinity, as well as tin, zinc, copper, and several ferroalloys. The conditions under which debt-ridden Amerindian workers labored, in the bitter cold and for pitiful wages, made the name Potosí synonymous with serfdom. Eventually the reserves were depleted and world market prices for metals fell, and workers lost their jobs, their landless families now destitute.

Meanwhile, the Europeans fought among themselves over territory, and Bolivia, which during the nineteenth century had a coastline on the Pacific, lost it in a war with Chile during the 1880s. Bolivia also lost territory to Brazil and Paraguay, but it was the loss of its maritime outlet that would damage it the most. To this day, even with rail links and transit rights through Chile's Arica and Antofagasta, Bolivian politicians press for a boundary change that would give them a corridor to the ocean.

Bolivia's political geography was galvanized during the 1990s and early 2000s by the expansion of its energy industries. The government, already exporting oil to Brazil and Argentina (see the pipelines in Fig. 5-10), wanted to construct a gas pipeline across the Andes to a terminal near Arica, from where gas would be exported to the United States. Many Amerindian citizens of Bolivia, seeing little potential benefit in this project, objected. Already, other causes—poverty, neglect, remoteness (less than half the people in this "Latin" country speak Spanish as a first language, and an estimated one-third do not speak Spanish at all)—had begun to roil the social order, and in 2000 a week-long protest featuring blockaded roads and wildcat strikes brought the country to a standstill. But the key issue was the plan to send gas to the United States via Chile. In 2003, clashes in the capital between Amerindian protesters and security forces caused more than 70 deaths, and the country's president was driven from office.

Another factor in Bolivia's now-chronic instability is its production of coca, long a traditional palliative for Amerindians but now part of the international cocaine network. Amerindian farmers deeply resented American pressure on the Bolivian government to eradicate the "illegal" fields, and in 2005 renewed protests again forced a presidential resignation.

Landlocked, ethnically split, poverty-mired, and very poorly governed, Bolivia is a country of limited options and limitless problems.

Paraguay

As Figure 5-3 indicates, Paraguay is one of those countries (Kazakhstan is another) that lies across a regional transition zone. In terms of economic development, Paraguay's per-capita GNI is more typical of the Andean West than the Southern Cone. Ethnically, too, the pattern is more characteristic of the West: more than 85 percent of Paraguay's 6.3 million people are mestizo, but with so pervasive an Amerindian influence that any white ancestry is almost totally submerged. As for languages, Amerindian Guaraní is so widely spoken alongside Spanish that the country is surely one of the world's most completely bilingual. The physiography, however, is decidedly non-Andean because all of Paraguay lies to the east of the mountains in the center of the Paraná-Paraguay Basin; from here, the outlet to the sea has always been to the south (Fig. 5-11). As the Basin's waterways are improved, that regional orientation is steadily strengthening.

Paraguay's landlocked position has much to do with its modest economic development to date, but political instability contributes greatly to the problem. Among South American states, only Guyana is less urbanized than Paraguay, and the vast majority of the people living on the land are mired in an inescapable cycle of poverty. In 2005, nearly two-thirds of the people, in the slums around the capital, Asunción, and on the land from the moist plains in the east to the dry, scrub-forested Chaco in the west, lived at or below the poverty level.

As Figure 5-11 shows, Paraguay is flanked to the east by a narrow corridor called Entre Rios ("between the rivers") belonging to Argentina, as well as Brazil's southern State of Paraná. In recent years, this complicated border area has taken on regional and even global significance because it lies remote from the capitals and core areas of all three of the countries that converge here, and has become a haven not only for smugglers but also for people with terrorist ties. A substantial number of Arab immigrants has settled in what representatives of national governments refer to as this *Triple Frontier*, where law enforcement is lax, money is made illegally, and dangerous causes are given financial support. The great majority of these Arab immigrants, who live on all three sides of the two borders but concentrate especially in Foz do Iguaçu, Brazil, and Ciudad del Este, Paraguay, are not involved in terrorist activities and attend their mosques (there are dozens in the two towns) for religious, not political, purposes. But Argentinian authorities have long been concerned over the small minority that has links with the terrorist organization Hizbollah in Lebanon, and over the suspected roles this organization may have played in the two deadly attacks on Jewish targets in the Argentine capital, Buenos Aires, in 1992 (the Israeli embassy) and in 1994 (the Jewish Information Center, in which 87 persons perished). When it was reported that maps of the Triple Frontier had been found in an al-Qaeda hideout in Afghanistan, following the fall of the Taliban regime there, this concern led to an investigation that exposed financial ties involving millions of dollars between certain Muslim businesspeople in the Triple Frontier and Hizbollah in Lebanon. What once may have been a flow of funds from the Middle East to the Triple Frontier now appears to have a reverse direction, suggesting to investigators that the prevailing commercial chaos in the area has become a boon to supporters of what Western governments designate as terrorist organizations and cells, not only in the Middle East but also in other locales in South America as well as North America. A political geographer in Argentina described landlocked and lawless eastern Paraguay as an Afghanistan in South America, dangerously neglected in the War on Terror. Whether Paraguay's fragile political system will be able to deal with this emerging challenge is a key question for the future.

THE SOUTH: MID-LATITUDE SOUTH AMERICA

South America's three southern countries—Argentina, Chile, and Uruguay—constitute a region sometimes referred to as the *Southern Cone* because of its pointed, ice-cream-cone shape (Fig. 5-11). As noted earlier, Paraguay has strong links to this region and in some ways forms part of it, although social contrasts between Paraguay and those in the Southern Cone remain sharp.

Since 1995, the countries of this region have been drawing closer together in an economic union named Mercosur, the hemisphere's second-largest trading bloc after NAFTA. Despite setbacks and disputes, Mercosur

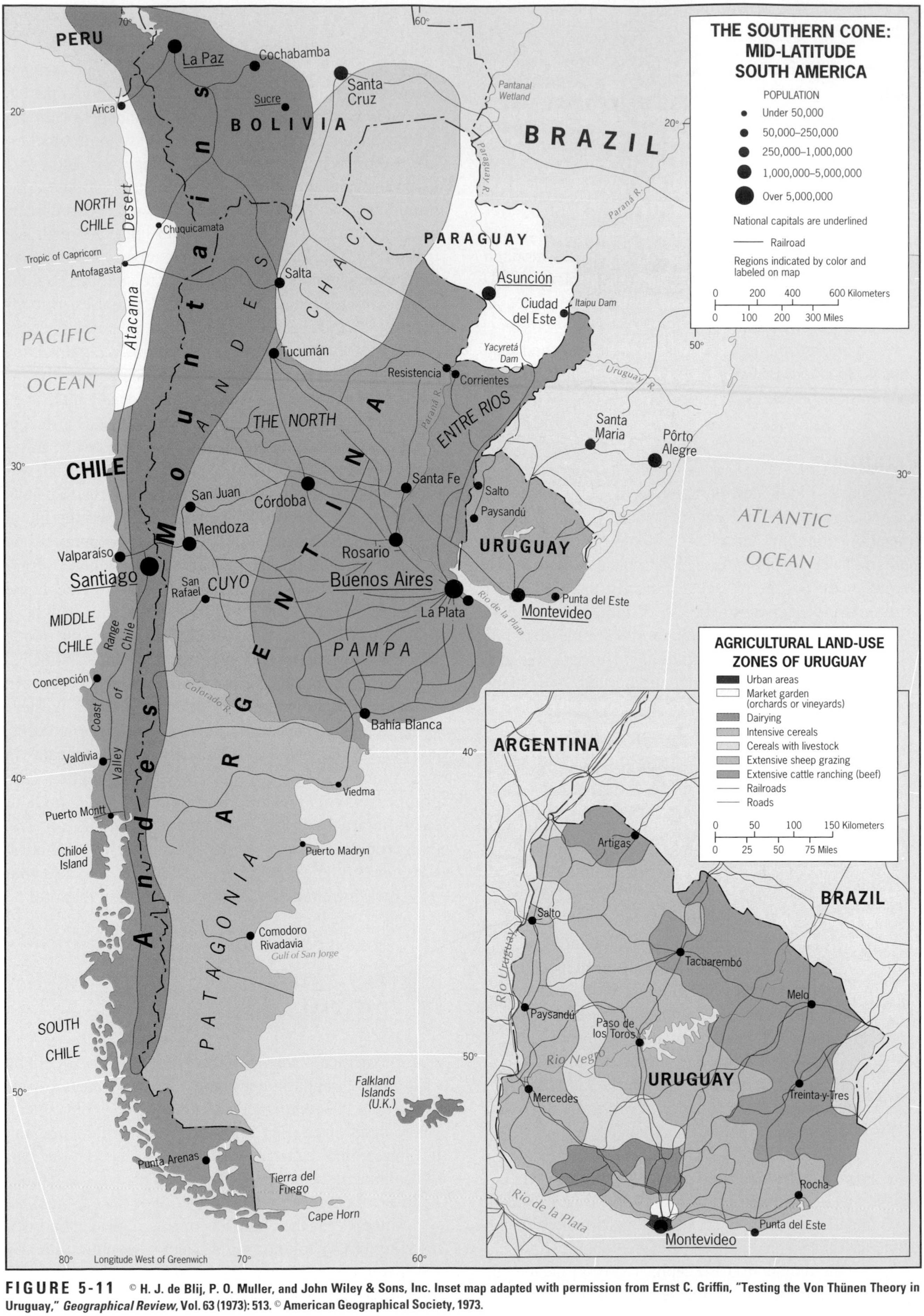

FIGURE 5-11 © H. J. de Blij, P. O. Muller, and John Wiley & Sons, Inc. Inset map adapted with permission from Ernst C. Griffin, "Testing the Von Thünen Theory in Uruguay," *Geographical Review*, Vol. 63 (1973): 513. © American Geographical Society, 1973.

has expanded and today encompasses Argentina, Uruguay, Paraguay, and Brazil, with Chile and Bolivia participating as associate members.

The Southern Cone represents South America's European-commercial culture sphere (Fig. 5-5). Mercosur has had the effect of extending these cultural and economic norms not only into Brazil but also into Paraguay, which lies in the vast basin of the Paraguay-Paraná river system that flows into the Rio de la Plata (Fig. 5-11). Two of the world's largest dams, the Yacyretá and the Itaipu, were constructed on the Paraná River along Paraguay's eastern border, and one of Mercosur's biggest infrastructure programs envisages the completion of the 2150-mile (3450-km) *Hidrovia* waterway that would open most of the Paraguay-Paraná river basin to barge transport. Other Mercosur initiatives, including roads through the Andes, bridges, highways, and railroads, are intended to strengthen intraregional ties.

Argentina

The largest Southern Cone country by far is Argentina, whose territorial size ranks second only to Brazil in this geographic realm; its population of 38.7 million ranks third after Brazil and Colombia. Argentina exhibits a great deal of physical-environmental variety within its boundaries, and the vast majority of the Argentines are concentrated in the physiographic subregion known as the *Pampa* (a word meaning "plain"). Figure G-9 underscores the degree of clustering of Argentina's inhabitants on the land and in the cities of the Pampa. It also shows the relative emptiness of the other six subregions (mapped in Fig. 5-11): the scrub-forest *Chaco* in the northwest; the mountainous *Andes* in the west, along whose crestline lies the boundary with Chile; the arid plateaus of *Patagonia* south of the Rio Colorado; and the undulating transitional terrain of intermediate *Cuyo*, *Entre Rios* (also known as "Mesopotamia" because it lies between the Paraná and Uruguay rivers), and the *North*.

The Argentine Pampa is the product of the past 150 years. During the second half of the nineteenth century, when the great grasslands of the world were being opened up (including those of the interior United States, Russia, and Australia), the economy of the long-dormant Pampa began to emerge. The food needs of industrializing Europe grew by leaps and bounds, and the advances of the Industrial Revolution—railroads, more efficient ocean transport, refrigerated ships, and agricultural machinery—helped make large-scale commercial meat and grain production in the Pampa not only feasible but also highly profitable. Large haciendas were laid out and farmed by tenant workers; railroads radiated ever farther outward from the booming capital of Buenos Aires and brought the entire Pampa into production.

Argentina once was one of the richest countries in the world. Its historic affluence is still reflected in its architecturally splendid cities whose plazas and avenues are flanked by ornate public buildings and private mansions. This is true not only of the capital, Buenos Aires, at the head of the Rio de la Plata estuary; it also applies

FROM THE FIELD NOTES

"Arrived in Mendoza (Argentina) yesterday to begin a week's study of the wine industry here. Stayed at an old hotel still showing cracks in the walls from the 1861 earthquake. Was invited to the local opera performing *Carmen*, starting time: 11:30 PM. Mendoza is so hot in the daytime that people here use the night to work as well as play, and I was told that the city is 'dead' from about 1 to 6 PM. This morning's guided excursion produced some remarkable viticultural vistas, but nothing topped the drama of the mountains and the famous 'Christ of the Andes' statue at the foot of what is actually the Andean range called the *Sierra de Los Paramillos*. Beyond the mountains in this westward view lies Chile, and during the movement for independence from Spain, General José de San Martin planned and executed an expedition to liberate Chile in 1817 from his headquarters in Mendoza." © H. J. de Blij.

AMONG THE REALM'S GREAT CITIES . . . BUENOS AIRES

Its name means "fair winds," which first attracted European mariners to the site of Buenos Aires alongside the broad estuary of the muddy Rio de la Plata. The shipping function has remained paramount, and to this day the city's residents are known throughout Argentina as the *porteños* (the "port dwellers"). Modern Buenos Aires was built on the back of the nearby Pampa's grain and beef industry. It is often likened to Chicago and the Corn Belt in the United States because both cities have thrived as interfaces between their immensely productive agricultural hinterlands and the rest of the world.

Buenos Aires (13.4 million) is yet another classic South American primate metropolis, housing more than one-third of all Argentines, serving as the capital since 1880, and functioning as the country's economic core. Moreover, Buenos Aires is a cultural center of global standing, a monument-studded city that contains the world's widest boulevard (*Avenida 9 de Julio*).

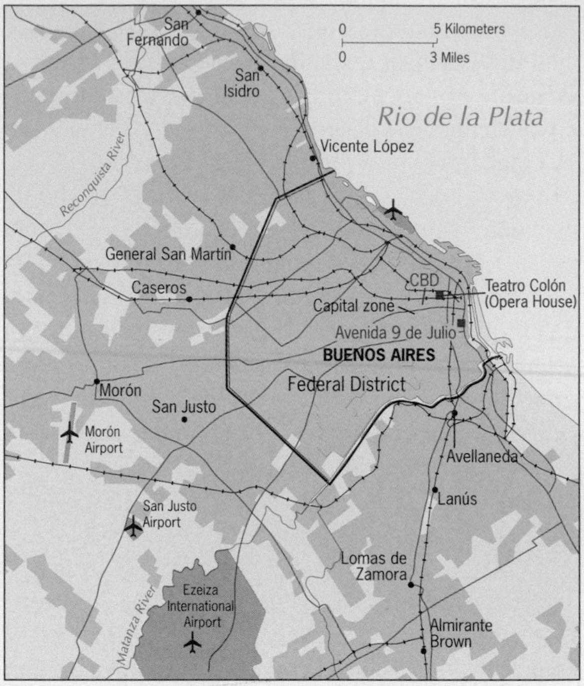

During the half-century between 1890 and 1940, the city was known as the "Paris of the South" for its architecture, fashion leadership, book publishing, and performing arts activities (it still has the world's biggest opera house, the *Teatro Colón*). With the recent restoration of democracy, Buenos Aires is now trying to recapture its golden years. Besides reviving these cultural functions, the city has added a new one: the leading base of the hemisphere's motion picture and television industry for Spanish-speaking audiences.

to interior cities such as Córdoba, Rosario, and Mendoza. The cultural imprint is dominantly Spanish, but the cultural landscape was diversified by a massive influx of Italians and smaller but influential numbers of British, French, and German immigrants. A sizeable immigration from Lebanon resulted in the diffusion of Arab ancestry to more than 12 percent of the Argentinian population.

Argentina has long been one of the realm's most urbanized countries: 89 percent of its population is concentrated in cities and towns, a higher percentage even than Western Europe or the United States. About one-third of all Argentinians live in metropolitan Buenos Aires, also by far the leading industrial complex where processing Pampa products dominates. Córdoba has become the second-ranking industrial center and was chosen by foreign automobile manufacturers as the car-assembly center for the expanding Mercosur market. One in four Argentinian wage earners is engaged in manufacturing, another indication of the country's economic progress. But what concentrates the urban populations is the processing of products from the vast, sparsely peopled interior: Tucumán (sugar), Mendoza (wines), Santa Fe (forest products), Salta (livestock). Argentina's product range is enormous. There is even an oil reserve near Comodoro Rivadávia on the coast of Patagonia.

Despite all these riches, Argentina's economic history is one of boom and bust. With fewer than 40 million inhabitants, a vast territory with diverse natural resources, adequate infrastructure, and good international linkages, Argentina should still be one of the world's wealthiest countries, as it once was. But political infighting and economic mismanagement have combined to ruin a vibrant and varied economy. What began as a severe recession toward the end of the 1990s became an economic collapse in the first years of the new century.

To understand how Argentina finds itself in this situation, a map of its administrative structure is useful. On paper, Argentina is a federal state consisting of the Buenos Aires Federal District and 23 provinces (Fig. 5-12). As would be expected from what we have just learned, the urbanized provinces are populous while the mainly rural

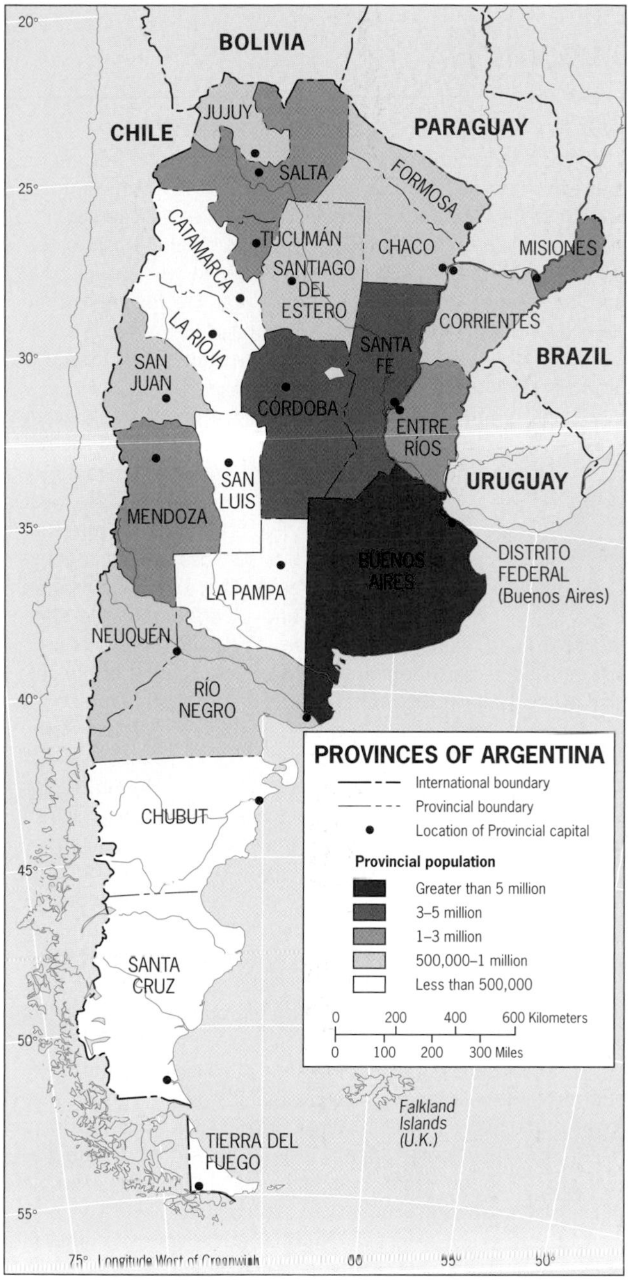

FIGURE 5-12 © H. J. de Blij, P. O. Muller, and John Wiley & Sons, Inc.

ones have smaller populations—but the gap between Buenos Aires Province (whose capital is La Plata), with nearly 15 million, and Tierra del Fuego (capital: Ushuaia), with barely over 100,000, is wide indeed. Several other provinces have under 500,000 inhabitants, so that the larger ones in addition to dominant Buenos Aires are also disproportionately influential in domestic politics, especially Córdoba and Santa Fe, both with capitals of the same name.

The never-ending problem for Argentina has been corrupt politics and associated mismanagement. Following an army coup in 1946, Juan Perón got himself elected president, bankrupted the country, and was succeeded by a military junta that plunged the country into its darkest days, culminating in the "Dirty War" of 1976–1983, which saw more than 10,000 Argentinians disappear without a trace. In 1982, this ruthless military clique launched an invasion of the British-held Malvinas (Falkland Islands), resulting in a major defeat for Argentina. By the time civilian government replaced the discredited junta, inflation was soaring and the national debt had become staggering (see the Issue Box entitled "Who Needs Democracy?"). Economic revival during the 1990s was followed by another severe downturn that exposed the flaws in Argentina's fiscal system, including scandalously inefficient tax collection and unconditional federal handouts to the politically powerful provinces.

In 2003, a new administration took office, led by President Néstor Kirchner, who hails from a small Patagonian province, not from one of the country's traditional power centers. He inherited a country in debt, an economy in difficulty, a nation still at war with itself over the "disappeared," and a cadre of corrupt military, police, and judicial officials, some of whom had secured immunity from prosecution for misdeeds during military rule. Kirchner, with considerable popular support, took on these challenges, from negotiating debt rescheduling to sacking judges. Argentina's recovery, however, will require long-term political stability, not just one presidential term. On that score, the future is still uncertain.

Chile

For 2500 miles (4000 km) between the crestline of the Andes and the coastline of the Pacific lies the narrow strip of land that is the Republic of Chile. On average just 90 miles (150 km) wide (and only rarely over 150 miles or 250 km in width), Chile is the world's quintessential example of what *elongation* means to the functioning of a state. Accentuated by its north-south orientation, this severe territorial attentuation not only results in Chile extending across numerous environmental zones; it has also contributed to the country's external political, internal administrative, and general economic problems. Nonetheless, throughout most of their modern history, the Chileans have made the best of this potentially disastrous centrifugal force: from the beginning, the sea has constituted an avenue of longitudinal communication; the Andes Mountains continue to form a barrier to encroachment from the east; and when confrontations loomed at the far ends of the country, Chile proved to be quite capable of coping with its northern rivals, Bolivia and Peru, as well as Argentina in the extreme south.

As Figures 5-11 and 5-13 underscore, Chile is a three-subregion country. About 90 percent of its people

Who Needs Democracy?

SOUTH AMERICA DOES!

"I'm old enough to remember the good as well as the bad old days here in Argentina. As a young man I worked for the national railroad company when that guy Perón came on the scene. I guess he learned his craft from the fascists in Italy, because he excelled in election by intimidation. He divided this country as never before, spending money on the workers and the poor while curbing freedoms and even abolishing freedoms guaranteed under our constitution. To tell you the truth, my salary actually went up a bit, and I even voted for him the second time around—largely, I think, because I wanted to do something for his beautiful wife Evita, who was loved by the whole country (except the military).

"Anyway, if you looked around South America in those days, we could have done worse. Wall to wall dictators. The military and police, secret and otherwise, put the fear of God into the populace, helped by the church. At least we had a guy who went through the motions of election. While those others divided the wealth among their cronies, Perón tried to spread it around. Even wealth we didn't have! Yes, he certainly put us into debt. And he had his cronies too.

"When they finally deposed Perón, I thought that we might see democratic elections here. (In retrospect, I don't know why I thought that—Spain and Portugal were ruled by two of the worst, Franco and Salazar; that's our "Latin" heritage). Well, you already know how wrong I was. Instead of more democracy, we got military repression. We lived in terror, not just fear. You've probably heard of those nuns who asked questions about the fate of a baby whose mother had 'disappeared.' They loaded them into a helicopter, in their outfits, and threw them out over the Rio de la Plata. Later I heard some soldiers joke about "the flying nuns." Rumors of what could happen to you were rife. Everyone knew someone who knew someone who was a victim.

"Our cloud of military terror lifted after they made their mistake in the Malvinas, 20 years ago, and since then we've had a taste of real democracy. It hasn't been easy, but let me tell you, it's better than 'strongman' rule so many of us South Americans are familiar with. Most of all, there's openness. You can express your views without fear. Political parties can argue their positions without military intimidation. Corrupt public officials can be found out by reporters, and they can't send machine-gun-toting colonels to kill their pursuers. Yes, there are downsides—that openness applies to the economy too, and as we've discovered, outsiders can interfere in our financial affairs. But I'd rather have democracy and open disorder than dictatorship and isolated order, and I've known them both."

Regional ISSUE

SAY NO TO THE EXCESSES OF DEMOCRACY!

"You can't help noticing that the Great Democratic Revolution that was supposed to be sweeping Middle and South America is not exactly a success. Here in Brazil we've had democratic government of a sort since 1989, when our man Fernando Collor de Mello was allowed by the military to win an election. You'll remember what happened to him. He resigned three years later after being implicated in a corruption and influence-peddling scheme. I'm sure the poor benefited hugely from this return to democracy.

"South American countries have a history of being led by men who captured the imagination of the nation and whose vision forged the character of the state. These men knew that the state must serve the people. The state must ensure that its railroads and bus services are available at low cost to all citizens. The state must provide electricity, fuel, and clean water. The state must staff and maintain the schools. The state and the church are indivisible, and the networks of the church are in the service of the state. If the military is needed to maintain order in the interest of stability, so be it. Call them strongmen if you like, but they personified their nations and did so for centuries.

"Democracy is a luxury for rich countries. The Americans like to teach us about democracy, but as a geography high-school teacher here in Santa Catarina, let me ask you this: how can the U.S. call itself a democracy when, in their Senate, a few hundred thousand people in Wyoming have the same representation as over 30 million in California? Oh, I see. You call yourself a 'representative republic.' Well, then don't lecture us about democracy. Even Juan Perón was 'democratically' elected, you know. In fact, most of those allegedly dictatorial rulers you so harshly criticize would probably win at the ballot box.

"Anyway, look at what democracy has done for Argentina and for Venezuela before the Venezuelans had the good sense to throw the rascals out and elect Chavez. Look what it is doing for Peru, where democratically elected President Toledo was trying to sell off the electric company serving Arequipa before the citizens stopped him. That 'openness' you hear pro-democracy advocates talk about just means that outsiders can come in, put your country in debt, demand 'privatization' of public companies, buy up corporations and banks and haciendas, fire thousands of workers to increase profits and share prices back home, and make the poor even poorer. What we need in this part of the world is a strong hand, not only to guide the nation but to protect the state."

Vote your opinion at www.wiley.com/college/deblij

are concentrated in what is called Middle Chile, where Santiago, the capital and largest city, and Valparaíso, the chief port, are located. North of Middle Chile lies the Atacama Desert, which is wider, drier, and colder than the coastal desert of Peru.

In the southern part of Middle Chile, the invading Europeans met the local Amerindian peoples, mainly the Mapuche, who constitute about 80 percent of the 1 million indigenous people in Chile's population of 16.4 million. The Mapuche fought off the Spanish conquerors

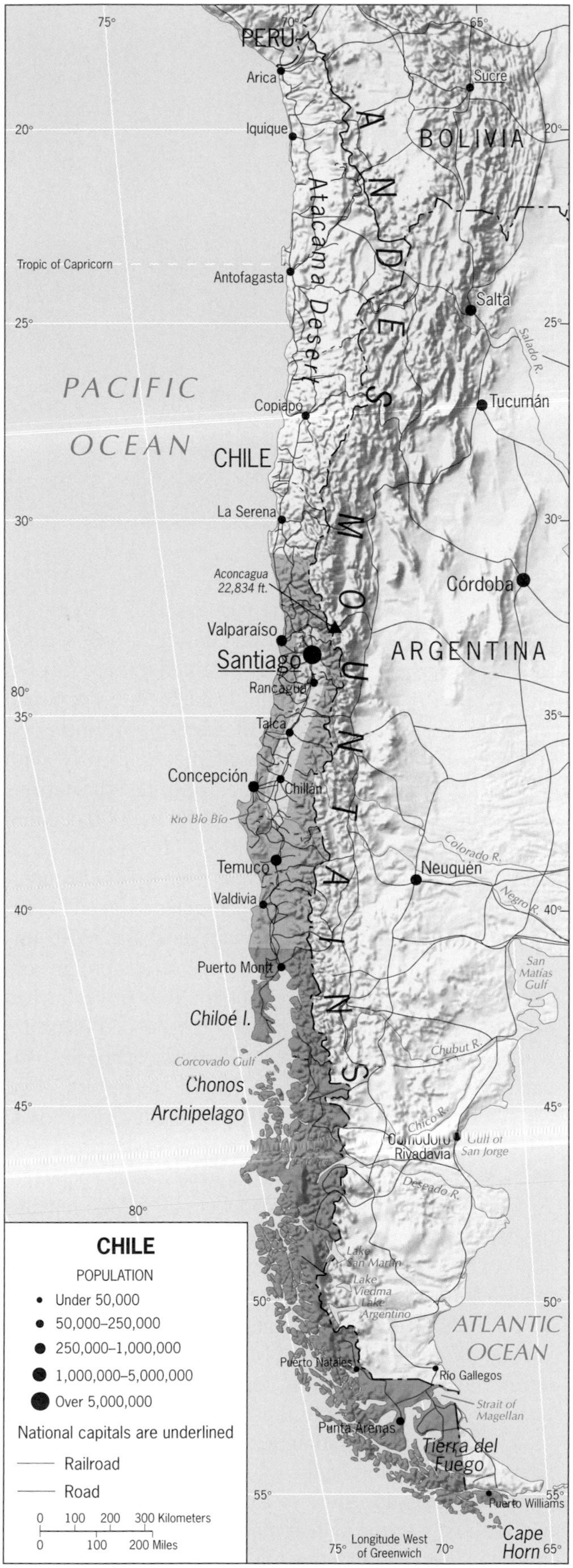

FIGURE 5-13 © H. J. de Blij, P. O. Muller, and John Wiley & Sons, Inc.

from the sixteenth century on, slowly retreating in their forest domain until they had all been pushed south of the Bío Bío River (Fig. 5-13). In the nineteenth century the Mapuche even succeeded in having their independence in Southern Chile recognized through treaties, but as happened to so many indigenous peoples, their treaties were violated by military action and they were soon relocated into small reserves. Meanwhile the Europeans cut down their forests and exported lumber, planting tree farms where natural forests had stood. In the late twentieth century, following the repression of the Pinochet regime, the Mapuche began a campaign of destruction by fire, burning planted trees and even the houses of European "settlers." They took to calling their homeland "Araucania," and their leaders talked of secession. In the climate of the early twenty-first century, such actions were equated with terrorism, and public support for the Mapuche cause declined. Still, the issue continues to roil Chilean politics, and the injustices done to the Mapuche must be addressed.

South of Middle Chile, the coast is broken by a plethora of fjords and islands, the topography is mountainous, and the climate—wet and cool near the Pacific—soon turns drier and colder against the Andean interior. South of the latitude of Chiloé Island, there are no permanent overland routes and hardly any settlements. These three subregions are also clearly apparent on the map of culture spheres (Fig. 5-5), which displays a mestizo north, a European-commercial zone in Middle Chile, and an undifferentiated south. In addition, a small Amerind-subsistence zone in northern Chile's Andes is shared with Argentina and Bolivia.

Some intraregional differences exist between northern and southern Middle Chile, the country's core area. Northern Middle Chile, the land of the hacienda and of Mediterranean climate with its dry summer season, is an area of (usually irrigated) crops that include wheat, corn, grapes, and other Mediterranean products. Livestock raising and fodder crops also take up much of the productive land, but continue to give way to the more efficient and profitable cultivation of fruits for export. Southern Middle Chile, into which immigrants from both the north and Europe (especially Germany) have pushed, is a better-watered area where raising cattle has predominated. But here, too, more lucrative fruits, vegetables, and grains are changing the area's agricultural specializations.

Prior to the 1990s, the arid Atacama region in the north accounted for more than half of Chile's foreign revenues. The Atacama Desert contains the world's largest

FROM THE FIELD NOTES

"A sense of impermanence pervades the physical as well as the cultural geography here in southern Chile. On the bus ride south along the Pan-American Highway, you see the landscape change as the ever-present backdrop of the tall Andes to the east gives way, south of the town of Osorno, to nearby lakes and snowcapped volcanoes obviously active and menacing. Moreover, the ground—even the urban site of Puerto Montt on the shore of the Gulf of Ancud—reveals the effects of recent earthquakes and tremors, Steep, fault-generated slopes and narrow, deep valleys divide the city's living and traveling space into congested districts that seem to warn of the hazards of life here. Look up and you see Mount Calbuco towering over the townscape. This is world-renowned scenery that attracts a steady stream of tourists on their way to Patagonia, but for the locals, it spells constant danger." © H. J. de Blij.

exploitable deposits of nitrates, which was the country's economic mainstay before the discovery of methods of synthetic nitrate production a century ago. Subsequently, copper became the chief export (Chile again possesses the world's largest reserves). It is found in several places, but the main concentration lies on the eastern margin of the Atacama near Chuquicamata, not far from the port of Antofagasta.

Chile today is emerging from a development boom that transformed its economic geography during the 1990s, a growth spurt that established its reputation as South America's greatest success story. Following the withdrawal of its brutal military dictatorship in 1990, Chile embarked on a program of free-market economic reform that brought stable growth, lowered inflation and unemployment, reduced poverty, and attracted massive foreign investment. The last is of particular significance because these new international connections enabled the export-led Chilean economy to diversify and develop in some badly needed new directions. Copper remains the single leading export, but many other mining ventures have been launched. In the agricultural sphere, fruit and vegetable production for export has soared because Chile's harvests coincide with the winter farming lull in the affluent countries of the Northern Hemisphere. Industrial expansion is occurring as well, though at a more leisurely pace, and new manufactures include a modest array of goods that range from basic chemicals to computer software.

Chile's newly internationalized economy has propelled the country to forge a prominent role for itself on the global trading scene. Even the Japanese expressed interest in building ties, and for a time during the 1990s Japan was its leading trading partner. This foothold in the Pacific Rim arena notwithstanding, Chile's trade linkages are now more heavily oriented toward its own hemisphere. The Chileans quickly realized that regional economic integration offered the best opportunities for sustained progress. Chile has affiliated locally with Mercosur as an associate member; despite exhortations from its Southern Cone partners, Chile in 2000 decided against seeking full membership.

Chile ended the twentieth century on a less encouraging note as its still-dominant copper sector dragged the national economy down to a growth standstill. With this commodity earning more than a third of all foreign revenues, there was little the Chileans could do in 1998 when the Asian economic crisis stifled demand and caused a worldwide collapse in the already low price of copper. This was followed by a period of painful readjustment as Chile's planners discovered that the country's economic diversification away from its reliance on copper had been neither as comprehensive nor as rapid as was first believed. Among the other lessons learned were that Chile remains too dependent on its natural resources, needs to expand the skills of its labor force, and must work harder to develop its manufacturing, services, and quaternary sectors.

Chile's recent transformation is widely touted as the "economic model" for all of Middle and South America to emulate. That is almost certainly an overstatement because few countries in these geographic realms can match the Chilean combination of natural and human resources and development circumstances. Yet all of those countries strongly aspire to follow in Chile's footsteps and for the first time in their modern histories are cooperating toward making that a reality. Their goal is the economic integration of all the Americas within the next few years—an accomplishment that could lead to a better life for every inhabitant of the Western Hemisphere.

Uruguay

Uruguay, unlike Argentina or Chile, is compact, small, and rather densely populated. This buffer state of old became a fairly prosperous agricultural country, in effect a smaller-scale Pampa (though possessing less favorable soils and topography). Figures 5-1 and G-8 show the similarity of physical conditions on the two sides of the Plata estuary. Montevideo, the coastal capital, contains more than 35 percent of the country's population of 3.4 million; from here, railroads and roads radiate outward into the productive agricultural interior (Fig. 5-11). In the immediate vicinity of Montevideo lies Uruguay's major farming area, which produces vegetables and fruits for the metropolis as well as wheat and fodder crops. Most of the rest of the country is used for grazing cattle and sheep, with beef products, wool and textile manufactures, and hides dominating the export trade. Tourism is another major economic activity as Argentines, Brazilians, and other visitors increasingly flock to the Atlantic beaches at Punta del Este and other thriving resort towns.

16 In Chapters 1 and 3, we described the **von Thünen model**'s layout of agricultural zones. Here in Uruguay we can discern another real-world example of that scheme (inset map, Fig. 5-11). Market gardens and dairying cluster nearest the urban area of Montevideo, with increasingly extensive agriculture ringing this national market in concentric land-use zones extending to the Brazilian border.

Uruguay's area of 68,500 square miles (177,000 sq km)—less territory even than Guyana—does not leave much room for population clustering. Nevertheless, a special quality of the land area of Uruguay is that it is rather evenly peopled right up to the boundaries with Brazil and Argentina. Of all the countries in South America, Uruguay is the most truly European. As for future development prospects as Mercosur matures (Montevideo is the bloc's administrative capital), Uruguay is in an excellent position to capitalize on its location between the Pampa to the south and Brazil's most dynamic areas to the north.

BRAZIL: GIANT OF SOUTH AMERICA

The next time you board an airplane in the United States, don't be surprised if the aircraft you enter was built in Brazil. When you go to the supermarket to buy provisions, take a look at their sources. Chances are some of them will come from Brazil. When you listen to your car radio, some of the best music you hear is likely to have originated in Brazil. The emergence of Brazil as the regional superpower of South America and an economic superpower in the world at large is going to be the story of the twenty-first century. In 2003 Brazil's newspapers carried quotes from national leaders who asked why Brazil, for all its territorial size, population numbers, and energy needs, did not have a nuclear-energy program. Why, asked one editorial, should countries like North Korea and Pakistan have nuclear weapons when Brazil does not even possess nuclear technology? Certainly not because Brazil lacks the capacity.

Why is Brazil so upward-bound today? In large measure it is because, after a long period of dictatorial rule by a minority elite that used the military to stay in power, Brazil embraced democratic government in 1989 (its first democratically elected president was ousted on charges of corruption three years later) and has not looked back since. The era of military coups and repeated quashings of civil liberties is over, and in Brazil's presidential election of 2002 the leader of the Workers Party, Luiz Inacio da Silva, popularly known as Lula, was elected in a runoff that marked a watershed in Brazilian democratic politics. (In a show of displeasure at the result, the United States did not send a single high-ranking representative to Lula's inauguration.) A champion of the poor and dispossessed, Lula came into office amid a rising tide of expectations but with the reality of a faltering economy confronting his administration. Subsequently, his skillful management of Brazil's numerous countervailing forces has begun to unleash the country's regional and global potential.

The Realm's Giant

By any measure, Brazil is South America's colossus. Territorially, the country is so large that it has common boundaries with all the realm's other states except Ecuador and Chile, occupying only slightly less than half of the entire landmass (Fig. 5-3). Only Russia, Canada, the United States, and China are larger. Its population of 183.8 million, just over half the realm's total, also ranks fifth in the world. By some measures, Brazil's economy is seventh-largest in the world. And Brazil's agricultural production, according to some analysts, is growing so fast that it may, within a decade, exceed that of the country that has for generations been the world's agricultural superpower—the United States.

Population Patterns

Brazil's population of 184 million is as diverse as that of the United States. In a pattern quite familiar in the Americas, the indigenous inhabitants of the country

China alone. Indeed, Africa does have some areas of good soils, ample water, and high productivity: the volcanic soils of Mount Kilimanjaro and those of the highlands around the Western Rift Valley, the soils in the Ethiopian Highlands, and those in the moister areas of higher-latitude South Africa and parts of West Africa yield good crops when social conditions do not disrupt the farming communities. But such areas are small in the vastness of Africa. This realm has nothing to compare to the huge alluvial basins of India and China, not even a lower Nile Valley and Delta, where comparatively small areas of fertile, irrigated soils can support tens of millions of people. Except for small (in some cases experimental) patches of rice and wheat, Africa is the land of corn (maize), millet, and root crops, far less able to provide high per-acre yields. Africa's natural environment poses a formidable challenge to the millions who depend directly on it.

ENVIRONMENT AND HEALTH

From birth, Africans (especially rural people living in *A* climates) are exposed to a wide range of diseases spread by insects and other organisms. The study of human
4 health in spatial context is the field of **medical geography**, and medical geographers employ modern methods of analysis (including geographic information systems) to track disease outbreaks, identify their sources, detect their carriers, and prevent their repetition. Alliances between doctors and geographers have already yielded significant results. Doctors know how a disease ravages the body; geographers know how climatic conditions such as wind direction or variations in river flow can affect the distribution and effectiveness of disease carriers. This collaboration helps protect vulnerable populations.

Tropical Africa, the source of many serious illnesses, is the focus of much of medical geography's work. Not only the carriers (*vectors*) of infectious diseases but also cultural traditions that facilitate transmission, such as sexual practices, food selection and preparation, and personal hygiene, play their role—and all can be mapped. Comparing medical, environmental, and cultural maps can lead to crucial evidence that helps combat the scourge.

In Africa today, hundreds of millions of people carry one or more maladies, often without knowing exactly what ails them. A disease that infects many people (the *hosts*) in a kind of equilibrium, without causing
5 rapid and widespread deaths, is said to be **endemic** to the population. People affected may not die suddenly or dramatically, but their health deteriorates, energy levels fall, and the quality of life declines. In tropical Africa hepatitis, venereal diseases, and hookworm are among the public health threats in this category.

When a disease outbreak has local or regional dimensions, it is called **epidemic**. It may claim thousands, 6
even tens of thousands of lives, but it remains confined to a certain area, perhaps one defined by the range of its vector. In tropical Africa, trypanosomiasis, the disease known as sleeping sickness and vectored by the tsetse fly, has regional proportions. The great herds of savanna wildlife form the *reservoir* of this disease, and the tsetse fly transmits it to livestock and people. It is endemic to the wildlife, but it kills cattle, so Africa's herders try to keep their animals in tsetse-free zones. African sleeping sickness appears to have originated in a West African source area during the fifteenth century, and it spread throughout much of tropical Africa (Fig. 6-4). Its epidemic range was limited by that of the tsetse fly: where there are no tsetse flies, there is no sleeping sickness. More than anything else, the tsetse fly has kept Subsaharan Africa's savannalands largely free of livestock and open to wildlife. Should a remedy be found, livestock would replace the great herds on the grasslands.

When a disease spreads worldwide, it is described as
pandemic. Africa's and the world's most deadly vec- 7

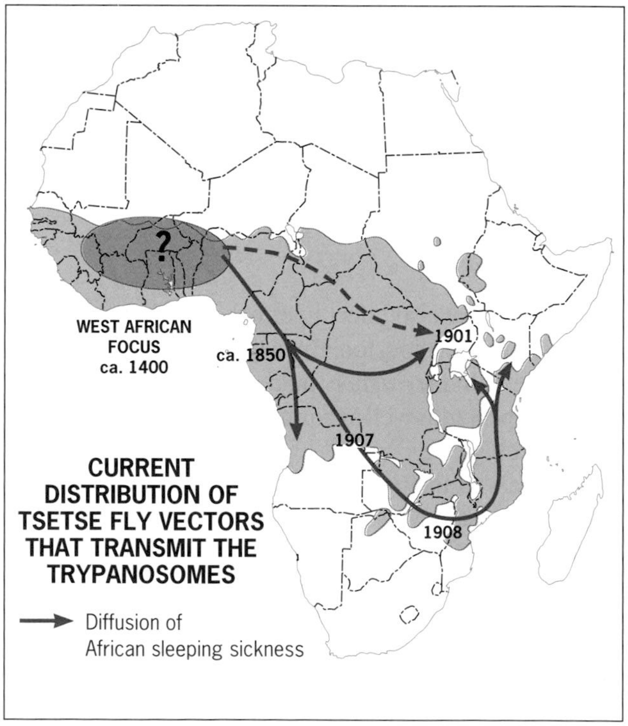

FIGURE 6-4 Adapted with permission from (1) J. Ford, "The Distribution of the Vectors of African Pathogenic Trypanosomes," *Bulletin of the World Health Organization*, Vol. 28 (1963): 655; and (2) K. C. Willett, "Trypanosomiasis and the Tsetse Fly Problem in Africa," *Annual Review of Entomology*, Vol. 8 (1963): 197.

tored disease is malaria, transmitted by a mosquito and killer of as many as 1 million children each year. Whether malaria has an African origin is not known, but it is an ancient affliction. Hippocrates, the Greek physician of the fifth century B.C., mentions it in his writings. Apes, monkeys, and several other species also suffer from it. Fever attacks, anemia, and enlargement of the spleen are its symptoms. Malaria has diffused around the world and prevails not only in tropical but also in temperate areas. Eradication campaigns against the mosquito vector have had some success, but always the carrier has come back with renewed vigor. At present, as many as 300 million people annually are recorded as having malaria, but the actual number probably is much higher because in much of tropical Africa malaria is simply an omnipresent malady affecting entire populations. The short life expectancies for tropical Africa reported in Table G-1 in part reflect infant and child mortality from malarial infection.

Another pandemic disease, also vectored by a mosquito and with African origins, is yellow fever. It crossed the Atlantic, reached South and Middle America, and even penetrated the United States during the nineteenth century; its defeat in Panama exactly a century ago made possible the construction of the Panama Canal. In Africa, however, it remains a threat: Senegal had an outbreak during the 1960s that claimed more than 20,000 lives.

During the 1970s, Africa was afflicted by yet another malady that appears to have started in the equatorial forest, became an epidemic, and grew into a pandemic: AIDS (see box titled "AIDS in Subsaharan Africa"). Monkeys may have activated this disease in humans, but whatever the source, Africa has been the most severely affected by it. In 2005, there were an estimated 41 million AIDS cases in the world, and at least 27 million of them were in Africa. Again, these official calculations probably conceal a much worse crisis, and the population growth rates for many African countries are declining as a result. During the early part of the twenty-first century, some countries will experience actual population decreases. Life expectancies are dropping to levels not seen for centuries.

The great incubator of vectored and nonvectored diseases in humid equatorial Africa continues to threaten local populations. Sudden outbreaks of Ebola fever in Sudan during the 1970s, in The Congo* in the 1990s, and in Uganda in 2000–2001 projected the risks that many Africans live with onto the world's television screens. But Africa's woes are soon forgotten. Only the fear that an African epidemic might evolve into a global pandemic mobilizes world attention. Africans cope with the planet's most difficult environments, but they are least capable of combating its hazards because their resources are so limited.

Even when international efforts are made to assist Africa, the unintended results may be catastrophic. Among Africa's endemic diseases is schistosomiasis, also called bilharzia. The vector is a snail, and the transmitted parasites enter via body openings when people swim or wash in slow-moving water where the snails thrive (fast-moving streams are relatively safe). When development projects designed to help African farmers dammed rivers and streams and sent water into irrigation ditches, farm production rose—but so did the incidence of schistosomiasis, which causes internal bleeding and fatigue, although it is rarely fatal. Medical geographers' maps showed the spread of schistosomiasis around the dam projects, and soon the cause was clear: by slowing down the water flow, the engineers had created an ideal environment for the snail vector.

Poverty is a powerful barrier to improving health, and major interventions in the natural environment carry costs as well as benefits. Africa needs improved medical services, more effective child-immunization programs, mobile clinics for remote areas, and similar remedies. While 67 percent of the realm's people continue to live in rural areas, the natural environment will weigh heavily on a vulnerable population.

LAND AND FARMING

In their penetrating book, *Geography of Sub-Saharan Africa*, editor Samuel Aryeetey-Attoh and his colleagues focus on the issue of land tenure, crucial in Africa because most Africans are farmers. **Land tenure** refers to 0
the way people own, occupy, and use land. African traditions of land tenure are different from those of Europe or the Americas. In most of Subsaharan Africa, communities, not individuals, customarily hold land. Occupants of the land have temporary, custodial rights to it and cannot sell it. Land may be held by large (extended) families, by a village community, or even by a traditional chief who holds the land in trust for the people. His subjects may house themselves on it and farm it, but in return they must follow his rules.

When the European colonizers took control of much of Subsaharan Africa, their land ownership practices clashed head-on with those of Africa. Africans believed that their land belonged to their ancestors, the living, and the yet-unborn; Europeans saw unclaimed space and

*Two countries in Africa have the same short-form name: Congo. In this book, we use *The Congo* for the larger Democratic Republic of the Congo, and *Congo* for the smaller, neighboring Republic of Congo.

AIDS in Subsaharan Africa

Acquired Immune Deficiency Syndrome—AIDS—spread in Africa during the 1980s and became a pandemic during the 1990s. Its impact on Africa has been devastating. In 2004, more than 2.3 million people died of AIDS in Subsaharan Africa, 75 percent of the global toll.

Persons infected by HIV (human immunodeficiency virus) do not immediately or even soon display symptoms of AIDS. In the early stages, only a blood test will reveal infection, and then only by indicating that the body is mobilizing antibodies to fight HIV. People can carry the virus for years without being aware of it; during that period they can unwittingly transmit it to others. For that reason, official statistics of AIDS cases lag far behind the reservoir of those infected, especially in countries where populations cannot be thoroughly tested. Since AIDS was identified in the early 1980s, nearly 25 million people have died of it.

By 2005, more than 40 million people were known to be infected by AIDS worldwide, and of these about 27 million lived in 34 tropical African countries according to United Nations sources. In the early 1990s, the worst-hit countries were in Equatorial Africa in what was called the AIDS Belt from (then) Zaïre to Kenya. But a decade later, the most severely afflicted countries lay in Southern Africa. In Zimbabwe, Botswana, Lesotho, and Swaziland, more than 25 percent of all persons 15–49 were infected with HIV; in South Africa, the region's most populous country, the proportion is 22 percent (10.5 million). These are the official numbers; medical geographers estimate that 20 to 25 percent of the *entire* population of several tropical African countries are infected.

As the number of deaths from AIDS rises, African countries' vital statistics are showing the demographic impact. In Botswana, life expectancy in 1994 was 60 years; by 2005 it had declined to 36. In South Africa it declined from 66 to 52. In Zimbabwe, population growth was 3.3 percent per year in the early 1980s; today it is just over 1 percent. Meanwhile, AIDS is killing parents at such a rate that by 2010, it is projected by the United Nations, it will be responsible for 20 million African orphans.

Geographer Peter Gould, in his classic book *The Slow Plague*, called Africa a "continent in catastrophe," pointing out that the official data on which these depressing reports are based understate the magnitude of the calamity. African governments, Gould wrote, conceal the real dimensions of the tragedy for various reasons: a sense of shame, concern that the real figures will drive investors out and keep tourists away, and the arrogance of military rulers used to denying reality. Many doctors in Africa, he added, say that AIDS cases are 80 to 90 percent underreported.

No part of tropical Africa has been spared. West African countries, too, are reporting growing numbers of AIDS cases. Ivory Coast, reporting that a rising 7 percent of adults are infected there, may provide the most accurate data. In Nigeria, where the full onslaught may not yet be known, AIDS incidence among adults is more than 5 percent (which, in that populous country, means that about 8 million people are ill with AIDS).

The misery the AIDS pandemic has created is beyond measure, but its economic impact is beyond doubt. Workers are sick on the job; but jobs are scarce and medical help is unavailable or unaffordable for many, so the workers stay on the job until they collapse. In a realm where unemployment is high, new workers immediately fill the vacancies, but they, too, are unwell. Corporations report skyrocketing AIDS-related costs. In Botswana, whose economy has been prospering due to increased diamond production, highly trained diamond sorters are being lost to AIDS, and training replacement workers is depressing profits. Medical expenses, multiplying death benefits, funeral payments, and recruiting and training costs are reducing corporate incomes from Kenya's sugar plantations to South Africa's gold mines. Thus the impact of AIDS is recorded in the continent's economic statistics. The pattern shown in Figure G-11, where Africa is concerned, is due in substantial part to the AIDS pandemic.

In countries of the global core, progress has been made in containing AIDS and prolonging the lives of those afflicted. This has taken AIDS off the medical center stage, but such a turnaround is not in prospect for Africa. The cost of the medications that combat AIDS is too high, and private drug companies will not dispense their products at prices affordable to African patients. Thus the best short-term strategy is for governments to attempt to alter public attitudes and behavior, notably by making condoms available as widely and freely as possible. Uganda, once one of tropical Africa's worst-afflicted countries, managed to slow the rate of AIDS dispersal by these methods—and through a vigorous advertising campaign on billboards, in newspapers, and on radio and television.

But what Africa needs is a coordinated attack on AIDS through the provision of medicines. In 2003 President George W. Bush announced an initiative to provide Subsaharan Africa with the financial means to combat the disease: a (U.S.) $10 billion fund to facilitate treatment as well as prevention. The urgency of the matter is underscored by estimates that, in the absence of a massive campaign, some of Africa's countries may see population declines of as much as 10 to 20 percent, a disaster comparable only to Europe's fourteenth-century bubonic plague and the collapse of Native American populations after the introduction of smallpox by the European invaders in the sixteenth. Once again environmental, cultural, and historical factors combine to afflict Africa's peoples.

9 felt justified in claiming it. What Africans called **land alienation**—the expropriation of (often the best) land by Europeans—changed the pattern of land tenure in Africa. By the time the Europeans withdrew, private land ownership was widely dispersed and could not be reversed. Postcolonial African states tried to deal with this legacy by nationalizing all the land and doing away with private ownership, reverting in theory to the role of the traditional chief. But this policy has not worked well. In the rural areas, the government in the capital is a remote authority often seen as unsympathetic to the plight of farmers. Governments keep the price of farm products low on urban markets, pleasing their supporters but frustrating farmers. Large landholdings once owned by Europeans seem now to be occupied by officials or those the government favors. The colonial period's approach to land tenure left Africa with a huge social problem.

Rapid population growth makes the problem worse. Traditional systems of land tenure, which involve subsistence farming in various forms ranging from shifting cultivation to pastoralism, work best when population is fairly stable. Land must be left fallow to recover from cultivation, and pastures must be kept free of livestock so the grasses can revive. A population explosion of the kind Africa experienced during the mid-twentieth century destroys this equilibrium. Soils cannot rest, and pastures are overgrazed. As land becomes degraded, yields decline.

This is the situation in much of Subsaharan Africa, where two-thirds of the people depend on farming for their livelihood. That percentage is slowly declining, but it remains the highest for any realm in the global periphery. Subsaharan Africa is sometimes thought of as a storehouse of minerals and fuels and great underground riches, but farming is the key to its economy. And not just subsistence farming: agricultural exports also produce foreign exchange revenues needed to pay for essential imports. The European colonizers identified suitable environments and laid out plantations to grow cocoa, coffee, tea, and other luxury crops for the markets of the global core. Irrigation schemes enabled the export of cotton, groundnuts (peanuts), and sesame seeds. Today, national governments and some private owners continue to operate these projects.

But most African farmers are subsistence farmers who grow grain crops (corn, millet, sorghum) in the drier areas and root crops (yams, cassava, sweet potatoes) where moisture is ample. Shifting cultivation still occurs in remote parts of The Congo and neighboring locales in Equatorial Africa, but this practice is dwindling as the forest shrinks. Others are pastoralists, driving their cattle and goats along migratory routes to find water and pasture, sometimes clashing with sedentary farmers whose fields they invade. Whatever their principal mode of farming, however, all African farmers today feel the effects of population pressure and must cope with dwindling space and damaged ecologies. By *intercropping*, planting several types of crops in one cleared field, some of which resemble those in the forest, shifting cultivators extend the life of their plots. In *compound farming*, usually near a market center, farmers combine the use of compound (village) and household waste as fertilizer with intensive care of the crops to produce vegetables, fruits, and root crops as well as eggs and chickens for urban consumers. Other subsistence farmers use *diversification* to cope with such problems; they combine their cultivation of subsistence crops with nonfarm work, including full-time jobs in nearby towns.

Notwithstanding all these adaptations, African farm production is declining for many reasons, including government mismanagement and corruption. When Nigeria became independent in 1960, its government embarked on costly industrialization, and agriculture was given a lower priority. In that year, Nigeria was the world's largest producer and leading exporter of palm oil; today, Nigeria must import palm oil to meet its domestic demand. Inadequate rural infrastructures create another obstacle. Roads from the rural areas to the market centers are in bad and worsening shape, often impassable during the rainy season and rutted when dry. Ineffective links between producers and consumers create economic losses. Storage facilities in the villages tend to be inadequate, so that untold quantities of grains, roots, and vegetables are lost to rot, pests, and theft. Electricity supply in rural areas also requires upgrading: without power, pumps cannot be run for water supply, and the processing of produce cannot be mechanized. Rural farmers have inadequate access to credit: banks would rather lend to urban homeowners than to distant villagers. This is especially true for women who, according to current estimates, produce 75 percent of the food in Subsaharan Africa.

Along the banks of Africa's rivers and on the shores of the Great Lakes fishing augments local food supplies, but Africa's fishing industries are still underdeveloped and contribute only modestly to overall nutrition. During the postcolonial period, Subsaharan Africa's population has more than doubled; food production has declined. Agricultural exports have also decreased, but food imports have mushroomed. The situation is especially clear in terms of per-capita output: in 1999 it was 10 percent below that in 1989. Even advances in biotechnology that closed the global gap between food production and demand largely passed Africa by (see box titled "A Green Revolution for Africa?")

A Green Revolution for Africa?

10 The **Green Revolution**—the development of more productive, higher-yielding types of grains—has narrowed the gap between world population and food production. Where people depend mainly on rice and wheat for their staples, the Green Revolution has pushed back the specter of hunger. The Green Revolution has had less impact in Subsaharan Africa, however. In part, this relates to the realm's high rate of population growth (which is substantially higher than that of India or China). Other reasons have to do with Africa's staples: rice and wheat support only a small part of the realm's population. Corn (maize) supports many more, along with sorghum, millet, and other grains. In moister areas, root crops, such as the yam and cassava, as well as the plantain (similar to the banana) supply most calories. These crops were not priorities in Green Revolution research.

Lately there have been a few signs of hope. Even as terrible famines struck several parts of the continent and people went hungry in places as widely scattered as Moçambique and Mali, scientists worked toward two goals: first, to develop strains of corn and other crops that would be more resistant to Africa's virulent crop diseases, and second, to increase the productivity of those strains. But these efforts faced serious problems. An average African acre planted with corn, for example, yields only about half a ton of corn, whereas the world average is 1.3 tons. A virus-resistant corn variety has now been developed that also yields more than the old variety, and it has been introduced throughout Subsaharan Africa. In Nigeria, where it was first distributed, yields rose significantly, contributing to a near doubling of farm production in that country since 1980 according to UN data. Hardier types of root crops also raised yields significantly.

But Africa needs even more than this to reverse the cycle of food deficiency. (Nor is the Green Revolution an unqualified remedy: the poorest farmers, who need the help most, can least afford the more expensive higher-yielding seeds and also cannot pay for the pesticides that may be required.) It has been estimated that, for the realm as a whole, food production has fallen about 1 percent annually even as population grew by more than 2 percent. Lack of capital, inefficient farming methods, inadequate equipment, soil exhaustion, male dominance, apathy, and devastating droughts contributed to this decline. The seemingly endless series of civil conflicts (Uganda, The Congo, Liberia, Angola, Rwanda) also reduced farm output. The Green Revolution may narrow the gap between production and need, but the battle for food sufficiency in Africa is far from won.

AFRICA'S HISTORICAL GEOGRAPHY

Africa is the cradle of humanity. Archeological research has chronicled 7 million years of transition from australopithecenes to hominids to *Homo sapiens*. It is therefore ironic that we know comparatively little about Subsaharan Africa from 5000 to 500 years ago—that is, before the onset of European colonialism. This is partly due to the colonial period itself, during which African history was neglected, many African traditions and artifacts were destroyed, and many misconceptions about African cultures and institutions became entrenched. It is also a result of the absence of a written history over most of Africa south of the Sahara until the sixteenth century—and over a large part of it until much later than that. The best records are those of the savanna belt immediately south of the Sahara, where contact with North African peoples was greatest and where Islam achieved a major penetration.

The absence of a written record does not mean, as some scholars have suggested, that Africa does not have a history as such prior to the coming of Islam and Christianity. Nor does it mean that there were no rules of social behavior, no codes of law, no organized economies. Modern historians, encouraged by the intense interest shown by Africans generally, are now trying to reconstruct the African past, not only from the meager written record but also from folklore, poetry, art objects, buildings, and other such sources. Much has been lost forever, though. Almost nothing is known of the farming peoples who built well-laid terraces on the hillsides of northeastern Nigeria and East Africa or of the communities that laid irrigation canals and constructed stone-lined wells in Kenya; and very little is known about the people who, perhaps a thousand years ago, built the great walls of Zimbabwe (photos at right). Porcelain and coins from China, beads from India, and other goods from distant sources have been found in Zimbabwe and other points in East and Southern Africa, but the trade routes within Africa itself—let alone the products that circulated and the people who handled them—still remain the subject of guesswork.

African Genesis

Africa on the eve of the colonial period was a continent in transition. For several centuries, the habitat in and near one of the continent's most culturally and economically productive areas—West Africa—had been changing. For 2000 years, probably more, Africa had been innovating as well as adopting ideas from outside. In West Africa, cities were developing on an impressive scale; in central and Southern Africa, peoples were moving, readjusting, sometimes struggling with each other for territorial supremacy. The Romans had penetrated to southern Sudan, North African peoples were trading with West Africans, and Arab *dhows* were sailing the waters along the eastern coasts, bringing Asian goods in exchange for gold, copper, and a comparatively small number of slaves.

Consider the environmental situation in West Africa as it relates to the past. As Figures G-7 and G-8 indicate, the environmental regions in this part of the continent exhibit a decidedly east-west orientation. The isohyets (lines of equal rainfall totals) run parallel to the southern coast (Fig. G-7); the climatic regions, now positioned somewhat differently from where they were two millennia ago, still trend strongly east-west (Fig. G-8); the vegetation pattern also reflects this situation, with a coastal forest belt yielding to savanna (tall grass with scattered trees in the south, shorter grass in the north) that gives way, in turn, to steppe and desert.

We know that African cultures had been established in all these environmental settings for thousands of years. One of these, the Nok culture, endured for over eight centuries on the Benue Plateau (north of the Niger-Benue confluence in modern Nigeria) from about 500 B.C. to the third century A.D. The Nok people made stone as well as iron tools, and they left behind a treasure of art in the form of clay figurines representing humans and animals. But we have no evidence that they traded with distant peoples. The opportunities created by environments and technologies still lay ahead.

FROM THE FIELD NOTES

"Got up before dawn this date in the hope of being the first visitor to the Great Zimbabwe ruins, a place I have wanted to see ever since I learned about it in a historical geography class. Climbed the hill and watched the sun rise over the great elliptical 'temple' below, then explored the maze of structures of the so-called fortification on the hilltop above. Next, for an incredible 90 minutes I was alone in the interior of the great oval walls of the temple. What history this site has seen—it was settled for perhaps six centuries before the first stones were hewn and laid here around A.D. 750, and from the 11th to the 15th centuries this may have been a ceremonial or religious center of a vast empire whose citizens smelted gold and copper and traded them via Sofala and other ports on the Indian Ocean for goods from South Asia and even China. No cement was used: these walls are built with stones cut to fit closely together, and in the valley they rise over 30 feet (nearly 10 m) high. And what secrets Great Zimbabwe still conceals—where are the plans, the tools, the quarries? The remnants of an elaborate water-supply system suggest a practical function for this center, but a decorated conical tower and adjacent platform imply a religious role. Whatever the answers, history hangs heavily, almost tangibly, over this African site." © H. J. de Blij

Early Trade

West Africa, over a north-south span of a few hundred miles, displayed an enormous contrast in environments, economic opportunities, modes of life, and products. The peoples of the tropical forest produced and needed goods that were different from the products and requirements of the peoples of the dry, distant north. For example, salt is a prized commodity in the forest, where the humidity precludes its formation, but it is plentiful in the desert and steppe. This enabled the desert peoples to sell salt to the forest peoples in exchange for ivory, spices, and dried foods. Thus there evolved a degree of *regional complementarity* between the peoples of the forest and those of the dryland. And the savanna peoples—those located in between—found themselves in a position to channel and handle the trade (which is always economically profitable).

The markets in which these goods were exchanged prospered and grew, and cities arose in the savanna belt of West Africa. One of these old cities, now an epitome of isolation, was once a thriving center of commerce and learning and one of the leading urban places in the world—Timbuktu. Others, predecessors as well as successors of Timbuktu, have declined, some of them into oblivion. Still other savanna cities, such as Kano in the northern part of Nigeria, remain important.

Early States

Strong and durable states arose in the West African culture hearth (see Fig. 7-3). The oldest state we know anything about is Ghana. Ancient Ghana was located to the northwest of the modern country of Ghana, and covered parts of present-day Mali, Mauritania, and adjacent territory. Ghana lay astride the upper Niger River and included gold-rich streams flowing off the Futa Jallon Highlands, where the Niger has its origins. For a thousand years, perhaps longer, old Ghana managed to weld various groups of people into a stable state. The country had a large capital city complete with markets, suburbs for foreign merchants, religious shrines, and, some miles from the city center, a fortified royal retreat. Taxes were collected from the citizens, and tribute was extracted from subjugated peoples on Ghana's periphery; tolls were levied on goods entering Ghana, and an army maintained control. Muslims from the northern drylands invaded Ghana in about A.D. 1062, when it may already have been in decline. Even so, the capital was protected for 14 years. However, the invaders had ruined the farmlands and destroyed the trade links with the north. Ghana could not survive. It finally broke into smaller units.

In the centuries that followed, the focus of politico-territorial organization in the West African culture hearth shifted almost continuously eastward—first to ancient Ghana's successor state of Mali, which was centered on Timbuktu and the middle Niger River Valley, and then to the state of Songhai, whose focus was Gao, a city on the Niger that still exists. This eastward movement may have been the result of the growing influence and power of Islam. Traditional religions prevailed in ancient Ghana, but Mali and its successor states sent huge, gold-laden pilgrimages to Mecca along the savanna corridor south of the Sahara, passing through present-day Khartoum and Cairo. Of the tens of thousands who participated in these pilgrimages, some remained behind. Today, many Sudanese trace their ancestry to the West Africa savanna kingdoms.

West Africa's savanna region undoubtedly witnessed momentous cultural, technological, and economic developments, but other parts of Africa also progressed. Early states emerged in present-day Sudan, Eritrea, and Ethiopia. Influenced by innovations from the Egyptian culture hearth, these kingdoms were stable and durable: the oldest, Kush, lasted 23 centuries (Fig. 6-5). The Kushites built elaborate irrigation systems, forged iron tools, and built impressive structures as the ruins of their long-term capital and industrial center, Meroe, reveal. Nubia, to the southeast of Kush, was Christianized until the Muslim wave overtook it in the eighth century. And Axum was the richest market in northeastern Africa, a powerful kingdom that controlled Red Sea trade and endured for six centuries. Axum, too, was a Christian state that confronted Islam, but Axum's rulers deflected the Muslim advance and gave rise to the Christian dynasty that eventually shaped modern Ethiopia.

The process of **state formation** spread throughout 11
Africa and was still in progress when the first European contacts occurred in the late fifteenth century. Large and effectively organized states developed on the equatorial west coast (notably Kongo) and on the southern plateau from the southern part of The Congo to Zimbabwe. East Africa had several city-states, including Mogadishu, Kilwa, Mombasa, and Sofala.

A crucial event affected virtually all of Equatorial, West, and Southern Africa: the great Bantu migration from present-day Nigeria-Cameroon southward and eastward across the continent. This migration appears to have occurred in waves starting as long as 5000 years ago, populating the Great Lakes area and penetrating South Africa, where it resulted in the formation of the powerful Zulu Empire in the nineteenth century (Fig. 6-5).

All this reminds us that, before European colonization, Africa was a realm of rich and varied cultures, diverse lifestyles, technological progress, and external trade (Fig. 6-6). But Europe's intervention would forever change its evolving political map.

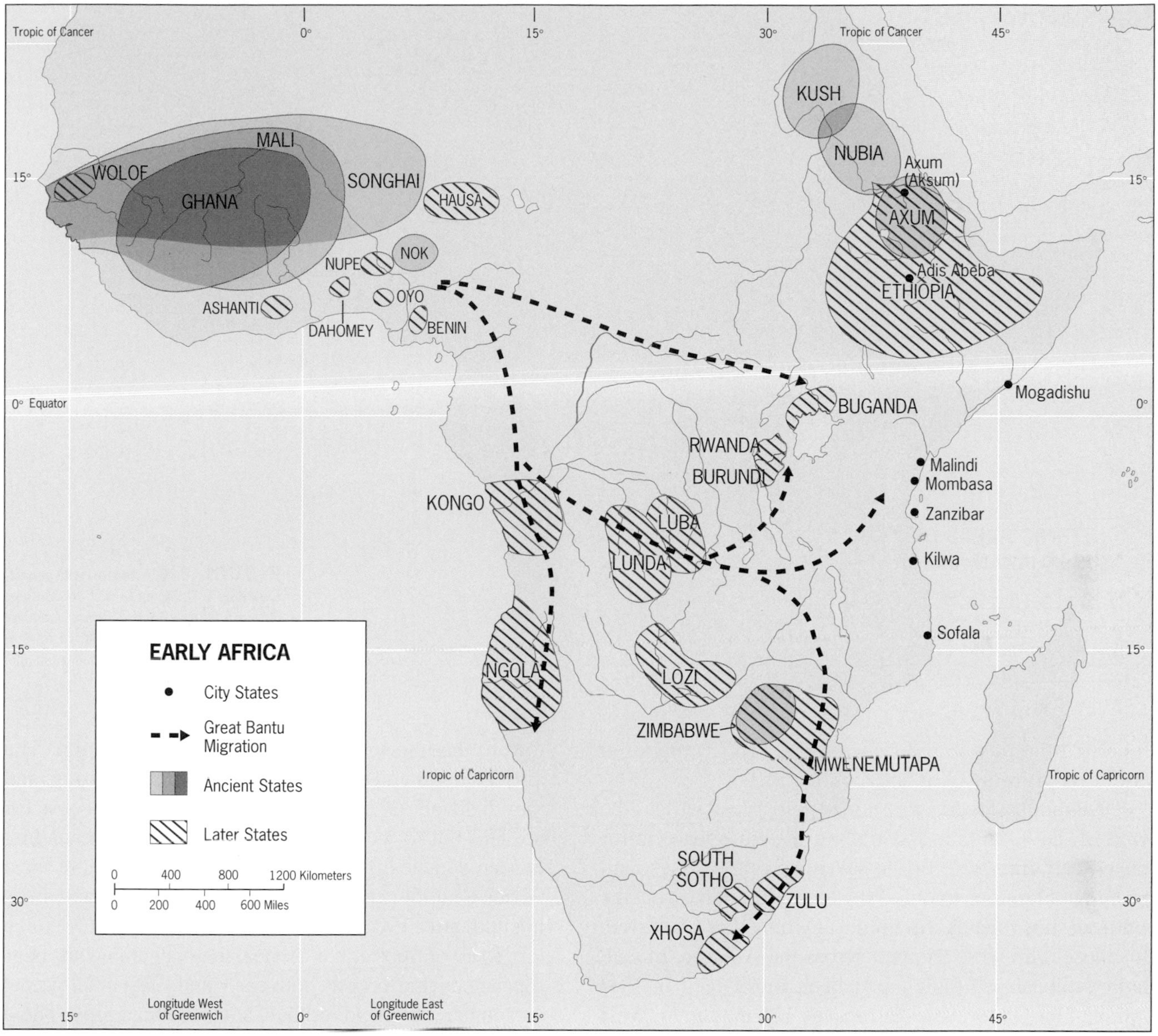

FIGURE 6-5 © H. J. de Blij, P. O. Muller, and John Wiley & Sons, Inc.

The Colonial Transformation

European involvement in Subsaharan Africa began in the fifteenth century. It would interrupt the path of indigenous African development and irreversibly alter the entire cultural, economic, political, and social makeup of the continent. It started quietly in the late fifteenth century, with Portuguese ships groping their way along the west coast and rounding the Cape of Good Hope. Their goal was to find a sea route to the spices and riches of the Orient. Soon other European countries were sending their vessels to African waters, and a string of coastal stations and forts sprang up. In West Africa, the nearest part of the continent to European spheres in Middle and South America, the initial impact was strongest. At their coastal control points, the Europeans traded with African middlemen for the slaves who were needed to work New World plantations, for the gold that had been flowing northward across the desert, and for ivory and spices.

Suddenly, the centers of activity lay not with the cities of the savanna but in the foreign stations on the Atlantic coast. As the interior declined, the coastal peoples thrived. Small forest states gained unprecedented wealth, transferring and selling slaves captured in the interior to the European traders on the coast. Dahomey (now called Benin) and Benin (now part of neighboring Nigeria) were states built on the slave trade. When slavery eventually came under attack in Europe, those who had

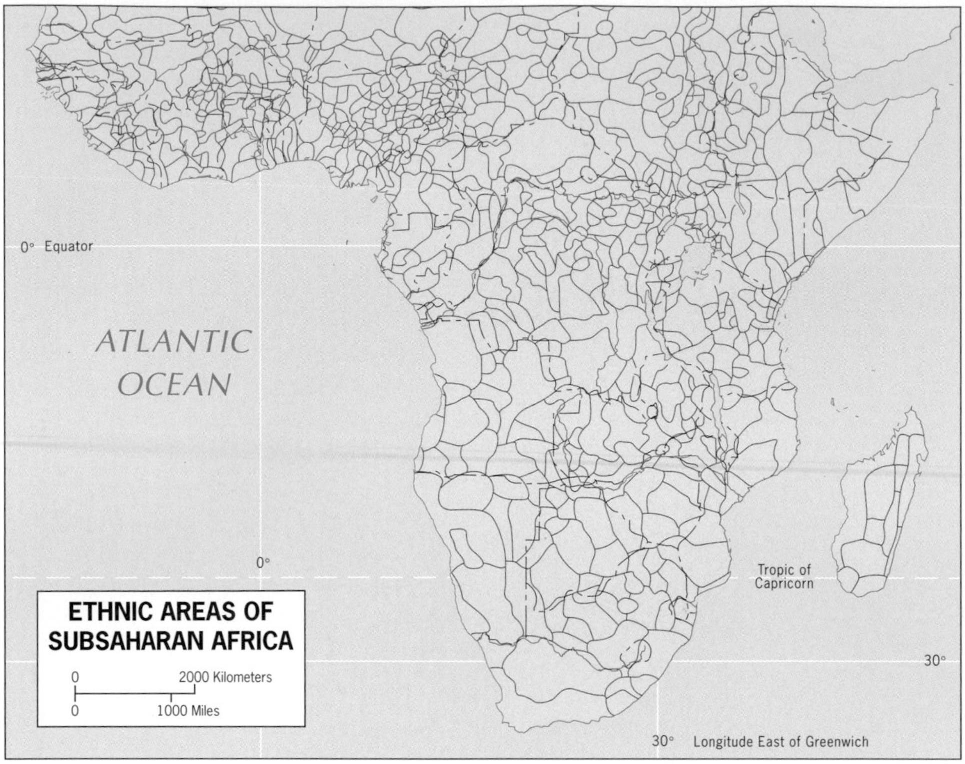

FIGURE 6-6 Adapted with permission from L. D. Stamp & W. T. W. Morgan, *Africa: A Study in Tropical Development*, 3rd rev. ed., after G. P. Murdock, *Africa: Its Peoples and Their Cultures.*

inherited the power and riches it had brought opposed abolition vigorously in both continents.

Although slavery was not new to West Africa, the *kind* of slave raiding and trading the Europeans introduced certainly was. In the savanna, kings, chiefs, and prominent families traditionally took a few slaves, but the status of those slaves was unlike anything that lay in store for those who were shipped across the Atlantic. In fact, large-scale slave trading had been introduced in East Africa long before the Europeans brought it to West Africa. African middlemen from the coast raided the interior for able-bodied men and women and marched them in chains to the Arab markets on the coast (Zanzibar was a notorious market). There, packed in specially built *dhows*, they were carried off to Arabia, Persia, and India. When the European slave trade took hold in West Africa, however, its volume was far greater. Europeans, Arabs, and collaborating Africans ravaged the continent, forcing perhaps as many as 30 million persons away from their homelands in bondage (Fig. 6-7). Families were destroyed, as were whole villages and cultures; those who survived their exile suffered unfathomable misery.

The European presence on the West African coast completely reoriented its trade routes, for it initiated the decline of the interior savanna states and strengthened the coastal forest states. Moreover, the Europeans' insatiable demand for slaves ravaged the population of the interior. But it did not lead to any major European thrust toward the interior or produce colonies overnight. The African middlemen were well organized and strong, and they held off their European competitors, not just for decades but for centuries. Although the Europeans first appeared in the fifteenth century, they did not carve West Africa up until nearly 400 years later, and in many areas not until after 1900.

In all of Subsaharan Africa, the only area where European penetration was both early and substantial was at its southernmost end, at the Cape of Good Hope. There the Dutch founded Cape Town, a waystation to their developing empire in the East Indies. They settled in the town's hinterland and migrated deeper into the interior. They brought thousands of slaves from Southeast Asia to the Cape, and intermarriage was common. Today, people of mixed ancestry form the largest part of Cape Town's population. Later the British took control not only of the Cape but also of the expanding frontier and brought in tens of thousands of South Asians to work on their plantations. Multiracial South Africa was in the making.

Elsewhere in the realm, the European presence remained confined almost entirely to the coastal trading stations, whose economic influence was strong. No real frontiers of penetration developed. Individual travelers, missionaries, explorers, and traders went into the interior, but nowhere else in Africa south of the Sahara was there an invasion of white settlers comparable to that of southernmost Africa's.

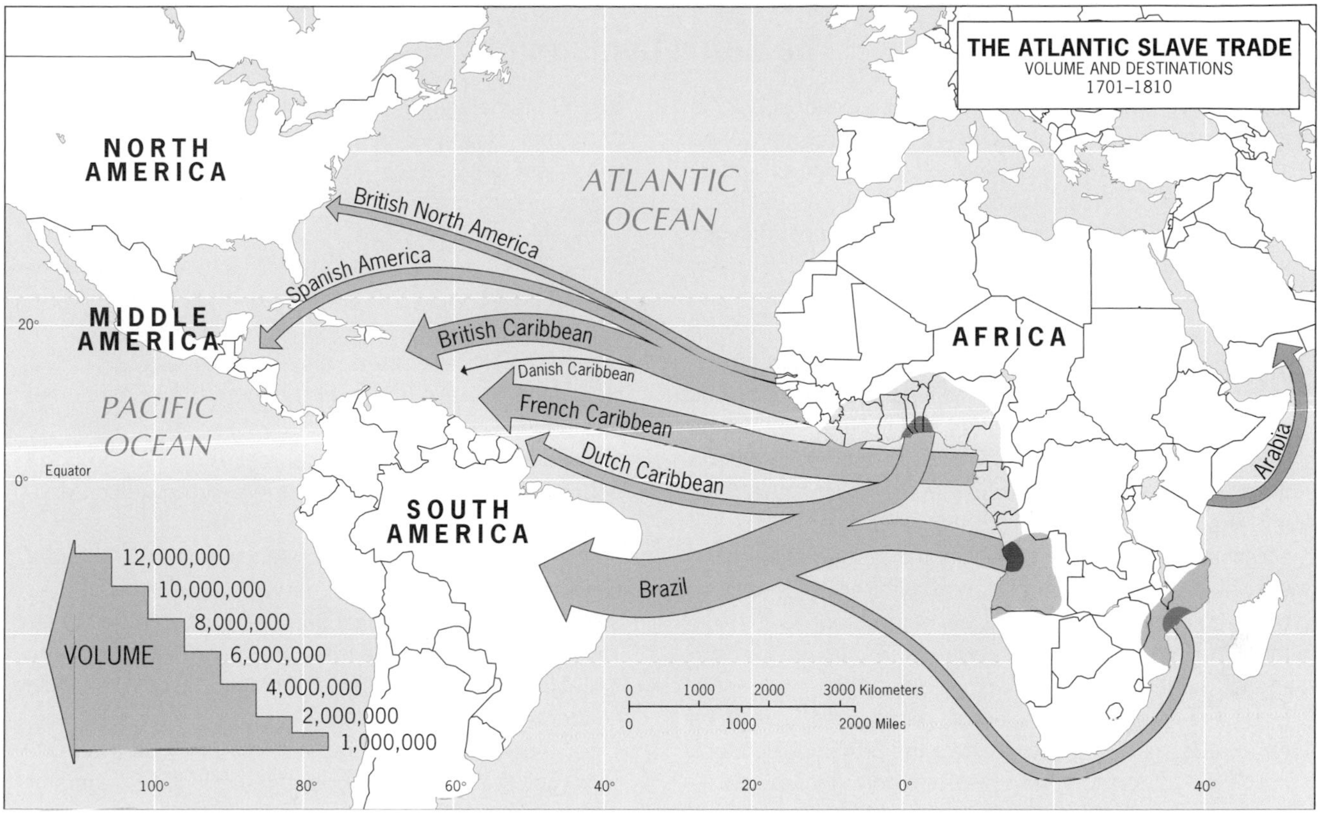

FIGURE 6-7 Adapted with permission from (1) Philip D. Curtin, *The Atlantic Slave Trade*, p. 57, © University of Wisconsin Press, 1969; and (2) Donald K. Fellows, *Geography*, p. 121, © John Wiley & Sons, Inc., 1967.

Colonization

In the second half of the nineteenth century, after more than four centuries of contact, the European powers finally laid claim to virtually all of Africa. Parts of the continent had been "explored," but now representatives of European governments and rulers arrived to create or expand African spheres of influence for their patrons. Competition was intense. Spheres of influence began to crowd each other. It was time for negotiation, and in late 1884 a conference was convened in Berlin to sort things out. This conference laid the groundwork for the now familiar politico-geographical map of Africa (see box titled "The Berlin Conference").

Figure 6-8 shows the result. The French dominated most of West Africa, and the British East and Southern Africa. The Belgians acquired the vast territory that became The Congo. The Germans held four colonies, one in each of the realm's regions. The Portuguese held a small colony in West Africa and two large ones in Southern Africa (see the map dated 1910).

The European colonial powers shared one objective in their African colonies: exploitation. But in the way they governed their dependencies, they reflected their differences. Some colonial powers were themselves democracies (the United Kingdom and France); others were dictatorships (Portugal, Spain). The British established a system of indirect rule over much of their domain, leaving indigenous power structures in place and making local rulers representatives of the British Crown. This was unthinkable in the Portuguese colonies, where harsh, direct control was the rule. The French sought to create culturally assimilated elites that would represent French ideals in the colonies. In the Belgian Congo, however, King Leopold II, who had financed the expeditions that staked Belgium's claim in Berlin, embarked on a campaign of ruthless exploitation. His enforcers mobilized almost the entire Congolese population to gather rubber, kill elephants for their ivory, and build public works to improve export routes. For failing to meet production quotas, entire communities were massacred. Killing and maiming became routine in a colony in which horror was the only common denominator. After the impact of the slave trade, King Leopold's reign of terror was Africa's most severe demographic disaster. By the time it ended, after a growing outcry around the world, as many as 10 million Congolese had been murdered. In 1908 the Belgian government took over, and slowly its Congo began to mirror Belgium's own internal divisions: corporations, government administrators, and the Roman Catholic Church each pursued their sometimes competing interests. But no

The Berlin Conference

In November 1884, the imperial chancellor and architect of the German Empire, Otto von Bismarck, convened a conference of 14 states (including the United States) to settle the political partitioning of Africa. Bismarck wanted not only to expand German spheres of influence in Africa but also to play off Germany's colonial rivals against one another to the Germans' advantage. The major colonial contestants in Africa were: (1) the British, who held beachheads along the West, South, and East African coasts; (2) the French, whose main sphere of activity was in the area of the Senegal River and north of the Congo Basin; (3) the Portuguese, who now desired to extend their coastal stations in Angola and Moçambique deep into the interior; (4) King Leopold II of Belgium, who was amassing a personal domain in the Congo Basin; and (5) Germany itself, active in areas where the designs of other colonial powers might be obstructed, as in Togo (between British holdings), Cameroon (a wedge into French spheres), South West Africa (taken from under British noses in a swift strategic move), and East Africa (where German Tanganyika broke the British design for a solid block of territory from the Cape north to Cairo).

When the conference convened in Berlin, more than 80 percent of Africa was still under traditional African rule. Nonetheless, the colonial powers' representatives drew their boundary lines across the entire map. These lines were drawn through known as well as unknown regions, pieces of territory were haggled over, boundaries were erased and redrawn, and African real estate was exchanged among European governments. In the process, African peoples were divided, unified regions were ripped apart, hostile societies were thrown together, hinterlands were disrupted, and migration routes were closed off. Not all of this was felt immediately, of course, but these were some of the effects when the colonial powers began to consolidate their holdings and the boundaries on paper became barriers on the African landscape (Fig. 6-8).

The Berlin Conference was Africa's undoing in more ways than one. The colonial powers superimposed their domains on the African continent. By the time Africa regained its independence after the late 1950s, the realm had acquired a legacy of political fragmentation that could neither be eliminated nor made to operate satisfactorily. The African politico-geographical map is thus a permanent liability that resulted from three months of ignorant, greedy acquisitiveness during a period when Europe's search for minerals and markets had become insatiable.

one thought to change the name of the colonial capital: it was Leopoldville until the Belgian Congo achieved independence in 1960.

12 **Colonialism** transformed Africa, but in its post-Berlin form it lasted less than a century. In Ghana, for example, the Ashanti (Asante) Kingdom still was fighting the British in the early years of the twentieth century; by 1957, Ghana was independent again. In a few years, much of Subsaharan Africa will have been independent for half a century, and the colonial period is becoming an interlude rather than a paramount chapter in modern African history.

Legacies

Nevertheless, some colonial legacies will remain on the map for centuries to come. The boundary framework has been modified in a few places, but it is a politico-geographical axiom that boundaries, once established, tend to become entrenched and immovable. The transport systems were laid out to facilitate the movement of raw materials from interior sources to coastal outlets. Internal circulation was only a secondary objective, and most African countries still are not well connected to each other. Moreover, the colonizers founded many of Subsaharan Africa's cities or built them on the sites of small towns or villages.

In many countries, the elites that gained advantage and prominence during the colonial period also have retained their preeminence. This has led to authoritarianism in some postcolonial African states (for details, see the regional discussion to come), and to violence and civil conflict in others. Military takeovers of national governments have been a byproduct of decolonization, and hopes for democracy in the struggle against colonialism have too often been dashed. Yet a growing number of African countries, including Ghana, Tanzania, Botswana, South Africa, and, most recently, Nigeria, have overcome the odds and achieved representative government under dauntingly difficult circumstances.

Although not strictly a colonial legacy, the impact of the Cold War (1945–1990) on African states should be noted. In three countries—Ethiopia, Somalia, and Angola—great-power competition, weapons, military advisers, and, in Angola, foreign armed forces magnified civil wars that would perhaps have been inevitable in any case. In other countries, ideological adherence to foreign dogmas led to political and social experiments (such as one-party Marxist regimes and costly farm collectivization) that cost Africa dearly. Today, debt-ridden Africa is again being told what to do, this time by foreign financial institutions. The cycle of poverty that followed colonial-

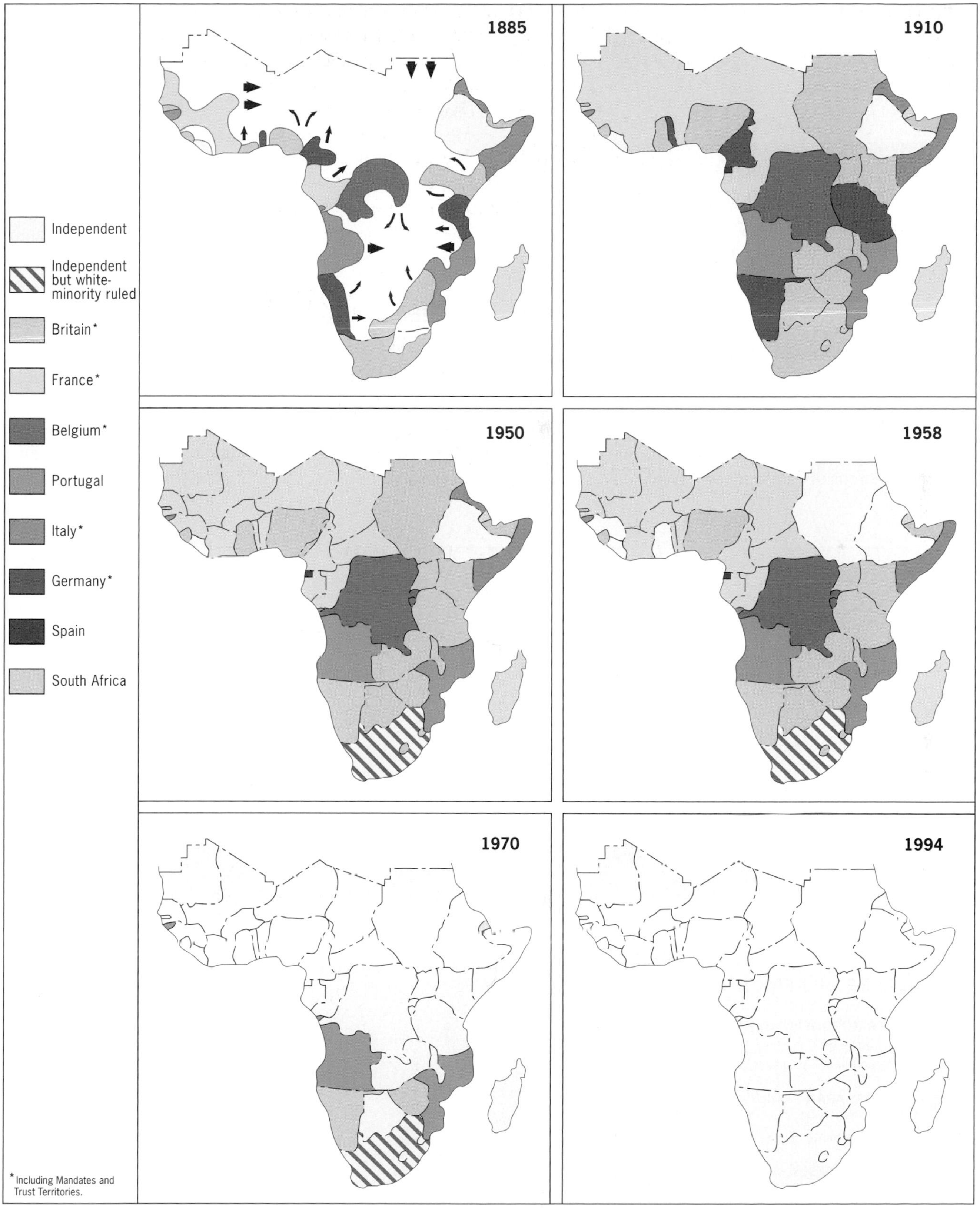

COLONIZATION AND DECOLONIZATION SINCE 1885

FIGURE 6-8

ism exacts a high price from African societies. As for future development prospects, the disadvantages of peripheral location vis-à-vis the world's core areas continue to handicap Subsaharan Africa.

CULTURAL PATTERNS

We may tend to think of Africa in terms of its prominent countries and famous cities, its development problems and political dilemmas, but Africans themselves have another perspective. The colonial period created states and capitals, introduced foreign languages to serve as the *lingua franca*, and brought railroads and roads. The colonizers stimulated labor movements to the mines they opened, and they disrupted other migrations that had been part of African life for many centuries. But they did not change the ways of life of most of the people. No less than 67 percent of the realm's population still live in, and work near, Africa's hundreds of thousands of villages. They speak one of more than a thousand languages in use in the realm. The villagers' concerns are local; they focus on subsistence, health, and safety. They worry that the conflicts over regional power or political ideology will engulf them, as has happened to millions in Liberia, Sierra Leone, Ethiopia, Rwanda, The Congo, Moçambique, and Angola since the 1970s. Africa's largest peoples are major nations, such as the Yoruba of Nigeria and the Zulu of South Africa. Africa's smallest peoples number just a few thousand. As a geographic realm, Subsaharan Africa has the most complex cultural mosaic on Earth (see Fig. 6-6).

African Languages

Africa's linguistic geography is a key component of that cultural intricacy. Most of Subsaharan Africa's more than 1000 languages do not have a written tradition, making classification and mapping difficult. Scholars have attempted to delimit an African language map, and Figure 6-9 is a composite of their efforts. One feature is common

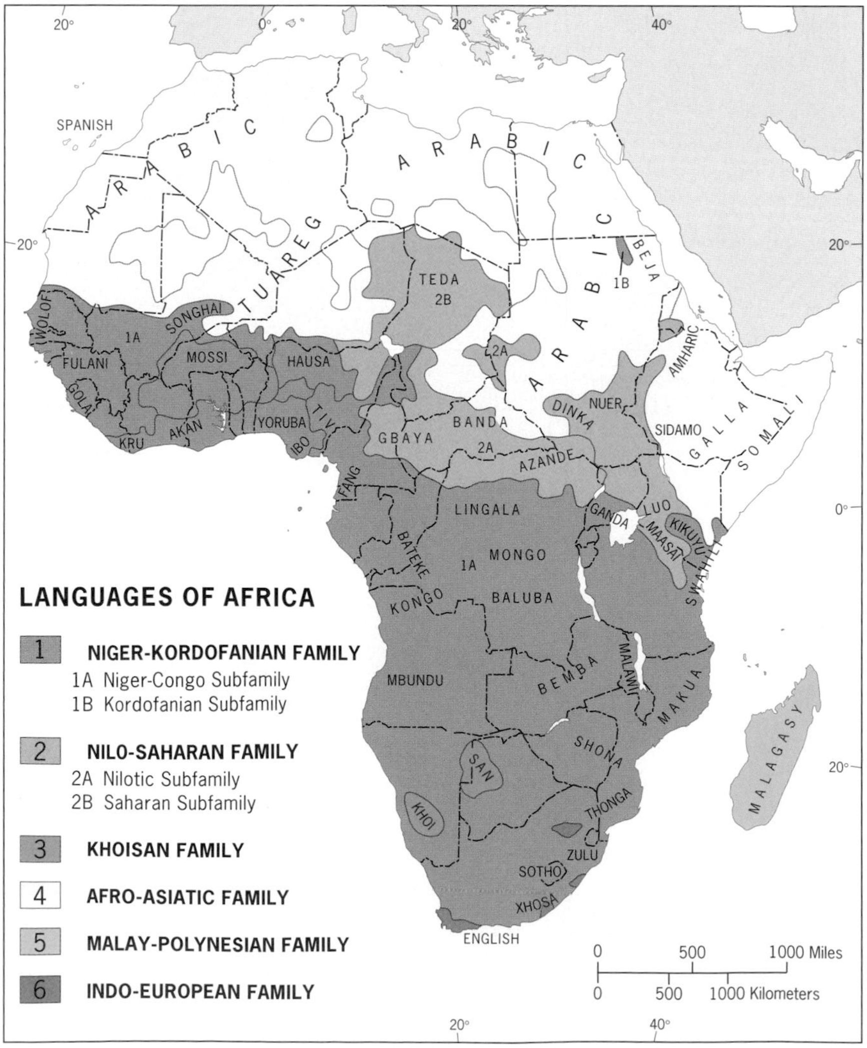

FIGURE 6-9 © H. J. de Blij, P. O. Muller, and John Wiley & Sons, Inc.

to all language maps of Africa: the geographic realm begins approximately where the Afro-Asiatic language family (mapped in yellow in Fig. 6-9) ends, although the correlation is sharper in West Africa than to the east.

In Subsaharan Africa, the dominant language family is the Niger-Kordofanian family. It consists of two subfamilies, the tiny Kordofanian concentration in northeastern Sudan and the pervasive Niger-Congo subfamily that extends across most of the realm from West Africa to East and Southern Africa. The Bantu language forms the largest branch in this subfamily, but Niger-Congo languages in West Africa, such as Yoruba and Akan, also have millions of speakers. Another important language family is the Nilo-Saharan family, extending from Maasai in Kenya northwest to Teda in Chad. No other language families are of similar extent or importance: the Khoisan family, of ancient origins, now survives among the dwindling Khoi and San peoples of the Kalahari; the small white minority in South Africa speak Indo-European languages; and Malay-Polynesian languages prevail in Madagascar, which was peopled from Southeast Asia before Africans reached it.

About 40 African languages are spoken by 1 million people or more, and a half-dozen by about 10 million or more: Hausa (50 million), Yoruba (23 million), Ibo, Swahili, Lingala, and Zulu. Although English and French have become important *linguae francae* in multilingual countries such as Nigeria and Côte d'Ivoire (where officials even insist on spelling the name of their country—Ivory Coast—in the Francophone way), African languages also serve this purpose. Hausa is a common language across the West African savanna; Swahili is widely used in East Africa. And pidgin languages, mixtures of African and European tongues, are spreading along West Africa's coast. Millions of Pidgin English (called *Wes Kos*) speakers use this medium in Nigeria and Ghana.

Multilingualism can be a powerful centrifugal force 13
in society, and African governments have tried with varying success to establish "national" alongside local languages. Nigeria, for example, made English its official language because none of its 250 languages, not even Hausa, had sufficient internal interregional use. But using a European, colonial language as an official medium invites criticism, and Nigeria remains divided on the issue. On the other hand, making a dominant local language official would invite negative reactions from ethnic minorities. Language remains a potent force in Africa's cultural life.

Religion in Africa

Africans had their own religious belief systems long before Christians and Muslims arrived to convert them. And for all of Subsaharan Africa's cultural diversity, Africans had a consistent view of their place in nature. Spiritual forces, according to African tradition, are manifest everywhere in the natural environment, not in a supreme deity that exists in some remote place. Thus gods and spirits affect people's

The faithful kneel during Friday prayers at a mosque in Kano, northern Nigeria. The survival of Nigeria as a unified state is an African success story; the Nigerians have overcome strong centrifugal forces in a multi-ethnic country that is dominantly Muslim in the north, Christian in the south. In the 1990s, some Muslim clerics began calling for an Islamic Republic in Nigeria, and after the death of the dictator Abacha and the election of a non-Muslim president, the Islamic drive intensified. A number of Nigeria's northern States adopted Sharia (strict Islamic) law, which led to destructive riots between the majority Muslims and minority Christians who felt threatened by this turn of events. Can Nigeria avoid the fate of Sudan (see pp. 349–350)? © M. & E. Bernheim/Woodfin Camp & Associates.

daily lives, witnessing every move, rewarding the virtuous, and punishing (through injury or crop failure, for example) those who misbehave. Ancestral spirits can inflict misfortune on the living. They are everywhere: in the forest, rivers, mountains.

As with land tenure, the religious views of Africans clashed fundamentally with those of outsiders. Monotheistic Christianity first touched Africa in the northeast when Nubia and Axum were converted, and Ethiopia has been a Coptic Christian stronghold since the fourth century A.D. But the Christian churches' real invasion did not begin until the onset of colonialism after the turn of the sixteenth century. Christianity's various denominations made inroads in different areas: Roman Catholicism in much of Equatorial Africa mainly at the behest of the Belgians, the Anglican Church in British colonies, and Presbyterians and others elsewhere. But almost everywhere, Christianity's penetration led to a blending of traditional and Christian beliefs, so that much of Subsaharan Africa is nominally, though not exclusively, Christian. Go to a church in Gabon or Uganda or Zambia, and you may hear drums instead of church bells, sing African music rather than hymns, and see African carvings alongside the usual statuary.

Islam had a different arrival and impact. Long before the colonial invasion, Islam advanced out of Arabia, across the desert, and down the east coast. Muslim clerics converted the rulers of African states and commanded them to convert their subjects. They Islamized the savanna states and penetrated into present-day northern Nigeria, Ghana, and Ivory Coast. They encircled and isolated Ethiopia's Coptic Christians and Islamized the Somali people in Africa's Horn. They established beachheads on the Kenya coast and took over Zanzibar. On the map, the African Transition Zone defines the Islamic Front (see Fig. 6-10). In the field, Arabizing Islam and European Christianity competed for African minds, and Islam proved to be a far more pervasive force. From Senegal to Somalia, the population is virtually 100 percent Muslim, and Islam's rules dominate everyday life. The Sunni *mullahs* would never allow the kind of marriage between traditional and Christian beliefs seen in much of formerly colonial Africa. This fundamental contradiction between Islamic dogma and Christian accommodation creates a potential for conflict in countries where both religions have adherents.

MODERN MAP AND TRADITIONAL SOCIETY

The political map of Subsaharan Africa has 45 states but no nation-states (apart from some microstates and ministates in the islands and in the south). Centrifugal forces are powerful, and outside interventions during the Cold War, when communist and anticommunist foreigners took sides in local civil wars, worsened conflict within African states. Colonialism's economic legacy was not much better. In tropical Africa, core areas, capitals, port cities, and transport systems were laid out to maximize profit and facilitate exploitation of minerals and soils; the colonial mosaic inhibited interregional communications except where cooperation enhanced efficiency. Colonial Zambia and Zimbabwe, for example (then called Northern and Southern Rhodesia), were landlocked and needed outlets, so railroads were built to Portuguese-owned ports. But such routes did little to create intra-African linkages. The modern map reveals the results: in West Africa you can travel from the coast into the interior of all the coastal states along railways or adequate roads. But no high-standard roadway was ever built to link these coastal neighbors to each other.

To overcome such disadvantages, African states must cooperate internationally, continentwide as well as regionally. The Organization of African Unity (OAU) was established for this purpose in 1963 and in 2001 was superseded by the African Union. In 1975 the Economic Community of West African States (ECOWAS) was founded by 15 countries to promote trade, transportation, industry, and social affairs in the region. And in the early 1990s another important step was taken when 12 countries joined in the Southern African Development Community (SADC), organized to facilitate regional commerce, intercountry transport networks, and political interaction.

Population and Urbanization

Subsaharan Africa is the second least urbanized world realm, but it is urbanizing at a fast pace. The percentage of urban dwellers today stands at 33 percent. This means that almost 240 million people now live in cities and towns, many of which were founded and developed by the colonial powers.

African cities became centers of embryonic national core areas, and of course they served as government headquarters. This *formal sector* of the city used to be the dominant one, with government control and regulations affecting civil service, business, industry, and workers. Today, however, African cities look different. From a distance, the skyline still resembles that of a modern center. But in the streets, on the sidewalks right below the shopwindows, there are hawkers, basket weavers, jewelry sellers, garment makers, wood carvers—a second economy, most of it beyond government control. This *informal sector* now dominates many African cities. It is peopled by the rural immigrants, who also work as servants, apprentices, construction workers, and in countless other menial jobs.

Millions of urban immigrants, however, cannot find work, at least not for months or even years at a time. They live in squalid circumstances, in desperate poverty, and governments cannot assist them. As a result, the squatter rings around (and also within) many of Africa's cities are unsafe—uncomfortable, unhealthy slums without adequate shelter, water supply, or basic sanitation. Garbage-strewn (no solid-waste removal here), muddy and insect-infested during the rainy season, and stifling and smelly during the dry period, they are incubators of disease. Yet very few residents return to their villages. Every new day brings hope.

In our regional discussion we focus on some of Subsaharan Africa's cities, all of which, to varying degrees, are stressed by the rate of population influx. Despite the plight of the urban poor and the poverty of Africa's rural areas, some of Africa's capitals remain the strongholds of privileged elites who, dominant in governments, fail to address the needs of other ethnic groups. Discriminatory policies and artificially low food prices disadvantage farmers and create even greater urban-rural disparities than the colonial period saw. But today the prospect of democracy brings hope that Africa's rural majorities will be heard and heeded in the capitals.

ECONOMIC PROBLEMS

As the world map of economic development (Fig. G-11) shows, Subsaharan Africa as a geographic realm is the weakest link in the international economy. Proportionately more countries are poor and debt-ridden in Subsaharan Africa than anywhere else. Annual income per person is low, infrastructure is inadequate, linkages with the rest of the world are weak, farmers face obstacles to profit at home as well as abroad. Champions of globalization say that Africa's poverty is proof that a lack of globalizing interaction spells disadvantage for any country. Opponents of globalization argue that Africa's condition is proof that globalization is making the poor poorer and the already-rich richer.

From Mauritania to Madagascar, Africa's low-income economies reflect a combination of difficulties so daunting that overcoming them should be a global, not just an African, objective. If the world is a village of neighborhoods, as is so often said, this is the neighborhood that needs the most help—ranging from medical assistance against AIDS to fair market conditions for farmers. Lifting tens of millions of subsistence farmers from the vagaries of individual survival to the security of collective prosperity will take more than a Green Revolution. Africa's vast expanse of savanna soil may not be any less fertile than that of Brazil's booming *cerrado*, but if there is no capital to provide fertilizers and no incentive to produce, no secure financial system and no roads to markets, no government policies to secure fair domestic prices and no equal access to the international marketplace, Africa's key industry will not make progress. That is the crucial problem: farming will form Africa's dominant economic activity for decades to come. True, Africa continues to sell commodities ranging from asbestos to zinc to foreign buyers in a pattern established during the colonial era, but the gauge of Africa's well-being lies in its agriculture.

Figure G-11 may appear to contradict this prospect, but consider how few the exceptions are. Of Subsaharan Africa's nearly 50 countries, all but five have low-income economies. The only well-balanced economy among these five is South Africa, for reasons we discuss in the regional section of this chapter: South Africa combines industrial production, commodity sales, and commercial farming. Among the other four, Botswana's prosperity is based on commodity (diamond) sales, Namibia's on metals, and that of Gabon on oil and forest exploitation. But even the export of oil or metals cannot lift Nigeria, Angola, Zimbabwe, or any other African economy out of its low-income status.

The remnants of colonial economic ties are mostly gone, and profits from commodities, from gold to manganese, have dropped as production costs rise and market prices fluctuate. But a new buyer has arrived on the African scene: China. The Chinese demand for metals, especially iron ore but also ferroalloys and copper, is burgeoning, and Africa has what China wants—except that Subsaharan Africa's railroad system, even in comparatively efficient South Africa, cannot get the cargoes to the coast in sufficient quantity or, in many areas, at all. China is offering to pay for infrastructure improvements, and there is hardly a country in Africa today without Chinese business representatives negotiating with the government. If China's interest in commodities were to help link farmers as well as mines to cities and ports, Africa's economic condition might improve.

REGIONS OF THE REALM

On the face of it, Africa seems to be so massive, compact, and unbroken that any attempt to justify a contemporary regional breakdown is doomed to fail. No deeply penetrating bays or seas create peninsular fragments as they do in Europe. No major islands (other than Madagascar) provide the broad regional contrasts we see in

Middle America. Nor does Africa really taper southward to the peninsular proportions of South America. And Africa is not cut by an Andean or a Himalayan mountain barrier. Given Africa's colonial fragmentation and cultural mosaic, is regionalization possible? Indeed it is.

Maps of environmental distributions, ethnic patterns, cultural landscapes, historic culture hearths, and other spatial data yield a four-region structure complicated by a fifth, overlapping zone as shown in Figure 6-10. Beginning in the south, we identify the following regions:

1. *Southern Africa*, extending from the southern tip of the continent to the northern borders of Angola, Zambia, Malawi, and Moçambique. Ten countries constitute this region, which extends beyond the tropics and whose giant is South Africa. The island state of Madagascar, with Southeast Asian influences, is neither Southern nor East African.
2. *East Africa*, where natural (equatorial) environments are moderated by elevation and where plateaus, lakes, and mountains, some carrying permanent snow, define the countryside. Six countries, including the highland part of Ethiopia, comprise this region.
3. *Equatorial Africa*, much of it defined by the basin of the Congo River, where elevations are lower than in East Africa, temperatures are higher and moisture more ample, and where most of Africa's surviving rainforests remain. Among the eight countries that form this region, The Congo dominates territorially and demographically, but others are much better off economically and politically.

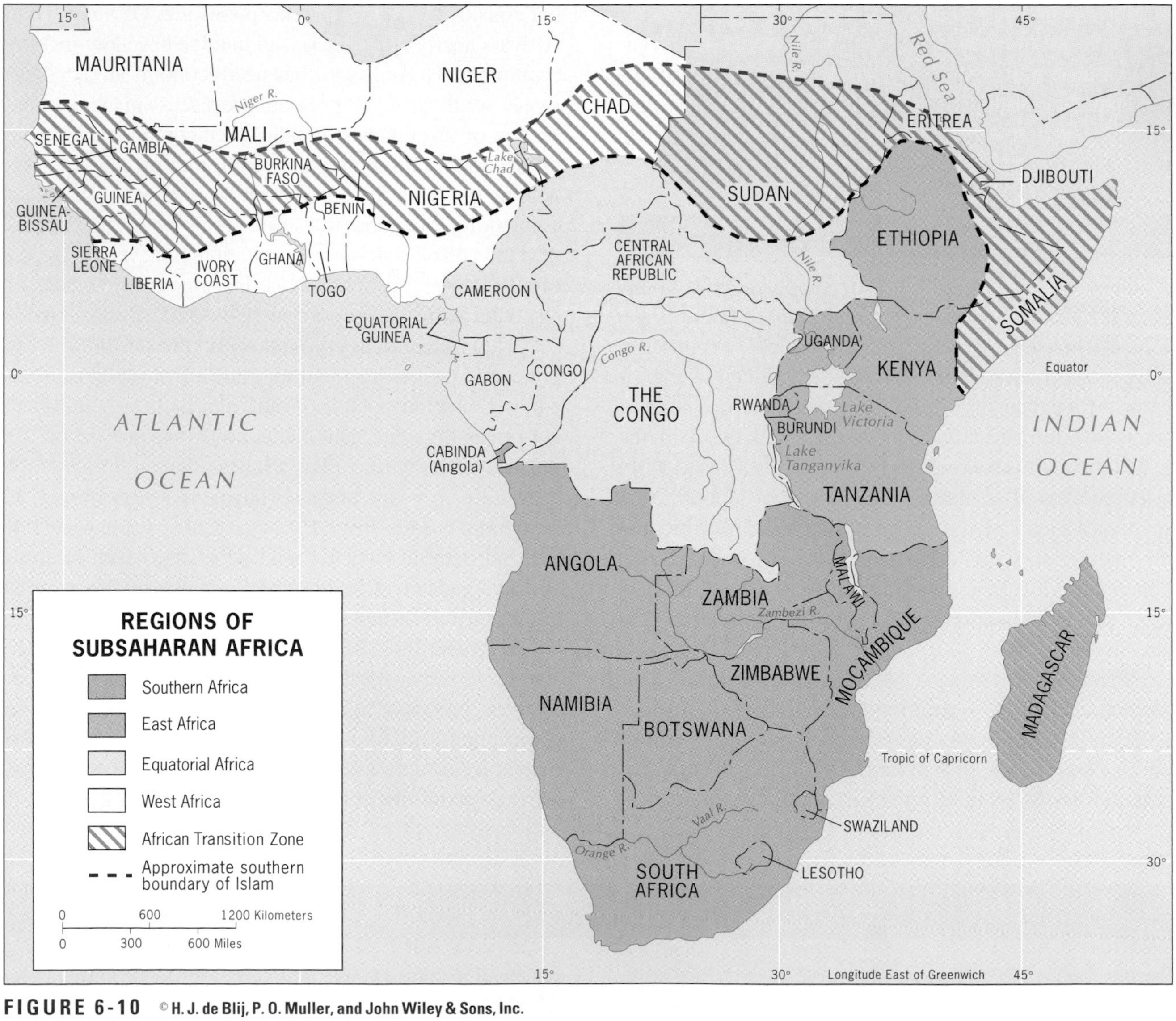

FIGURE 6-10 © H. J. de Blij, P. O. Muller, and John Wiley & Sons, Inc.

MAJOR CITIES OF THE REALM

City	Population* (in millions)
Abidjan, Ivory Coast	3.6
Accra, Ghana	2.0
Adis Abeba, Ethiopia	3.1
Cape Town, South Africa	3.0
Dakar, Senegal	2.5
Dar es Salaam, Tanzania	2.8
Durban, South Africa	2.6
Harare, Zimbabwe	1.6
Ibadan, Nigeria	2.5
Johannesburg, South Africa	3.3
Kinshasa, The Congo	6.3
Lagos, Nigeria	12.3
Lusaka, Zambia	1.5
Mombasa, Kenya	0.8
Nairobi, Kenya	3.0

*Based on 2006 estimates.

4. *West Africa*, which includes the countries of the western coast and those on the margins of the Sahara in the interior, a populous region anchored in the southeast by Africa's demographic giant, Nigeria. Fifteen countries form this crucial African region.
5. *The African Transition Zone*, the complicating factor on the regional map of Africa. In Figure 6-10, note that this zone of increasing Islamic influence completely dominates some countries (e.g., Somalia in the east and Senegal in the west) while cutting across others, thereby creating Islamized northern areas and non-Islamic southern zones (Nigeria, Chad, Sudan).

SOUTHERN AFRICA

Southern Africa, as a geographic region, consists of all the countries and territories lying south of Equatorial Africa's The Congo and East Africa's Tanzania (Fig. 6-11). Thus defined, the region extends from Angola and Moçambique (on the Atlantic and Indian Ocean coasts, respectively) to South Africa and includes a half-dozen landlocked states. Also marking the northern limit of the region are Zambia and Malawi. Zambia is nearly cut in half by a long land extension from The Congo, and Malawi penetrates deeply into Moçambique. The colonial boundary framework, here as elsewhere, produced many liabilities.

Southern Africa constitutes a geographic region in both physiographic and human terms. Its northern zone marks the southern limit of the Congo Basin in a broad upland that stretches across Angola and into Zambia (the tan corridor extending eastward from the Bihe Plateau in Fig. 6-2). Lake Malawi is the southernmost of the East African rift-valley lakes; Southern Africa has none of East Africa's volcanic and earthquake activity. Most of the region is plateau country, and the Great Escarpment is much in evidence here. There are two pivotal river systems: the Zambezi (which forms the border between Zambia and Zimbabwe) and the Orange-Vaal (South African rivers that combine to demarcate southern Namibia from South Africa).

Southern Africa is the continent's richest region materially. A great zone of mineral deposits extends through the heart of the region from Zambia's Copperbelt through Zimbabwe's Great Dyke and South Africa's Bushveld Basin and Witwatersrand to the goldfields and diamond mines of the Orange Free State and northern Cape Province in the heart of South Africa. Ever since these minerals began to be exploited in colonial times, many migrant laborers have come to work in the mines.

Southern Africa's agricultural diversity matches its mineral wealth. Vineyards drape the slopes of South Africa's Cape Ranges; tea plantations hug the eastern escarpment slopes of Zimbabwe. Before civil war destroyed its economy, Angola was one of the world's leading coffee producers. South Africa's relatively high latitudes and its range of altitudes create environments for apple orchards, citrus groves, banana plantations, pineapple farms, and many other crops.

Despite this considerable wealth and potential, the countries of Southern Africa have not prospered. As Figure G-11 shows, most remain mired in the low-income category (Malawi, with a per-capita GNI of only $600, is one of the world's poorest states); only South Africa and its western neighbors, Namibia and Botswana, are in the middle-income rank, with the latter desert state being the realm's second most sparsely populated country. A period of rapid population growth, followed by the devastating onslaught of AIDS, civil conflict, political instability, incompetent government, widespread corruption, unfair competition on foreign markets, and environmental problems have constrained economic development. In 2003, countries in this region needed emergency food supplies to stave off malnutrition and hunger; one of them (Zimbabwe) failed to distribute the provisions where they were most needed, and another (Zambia) refused genetically modified (GM) grain because it was thought to be unhealthy.

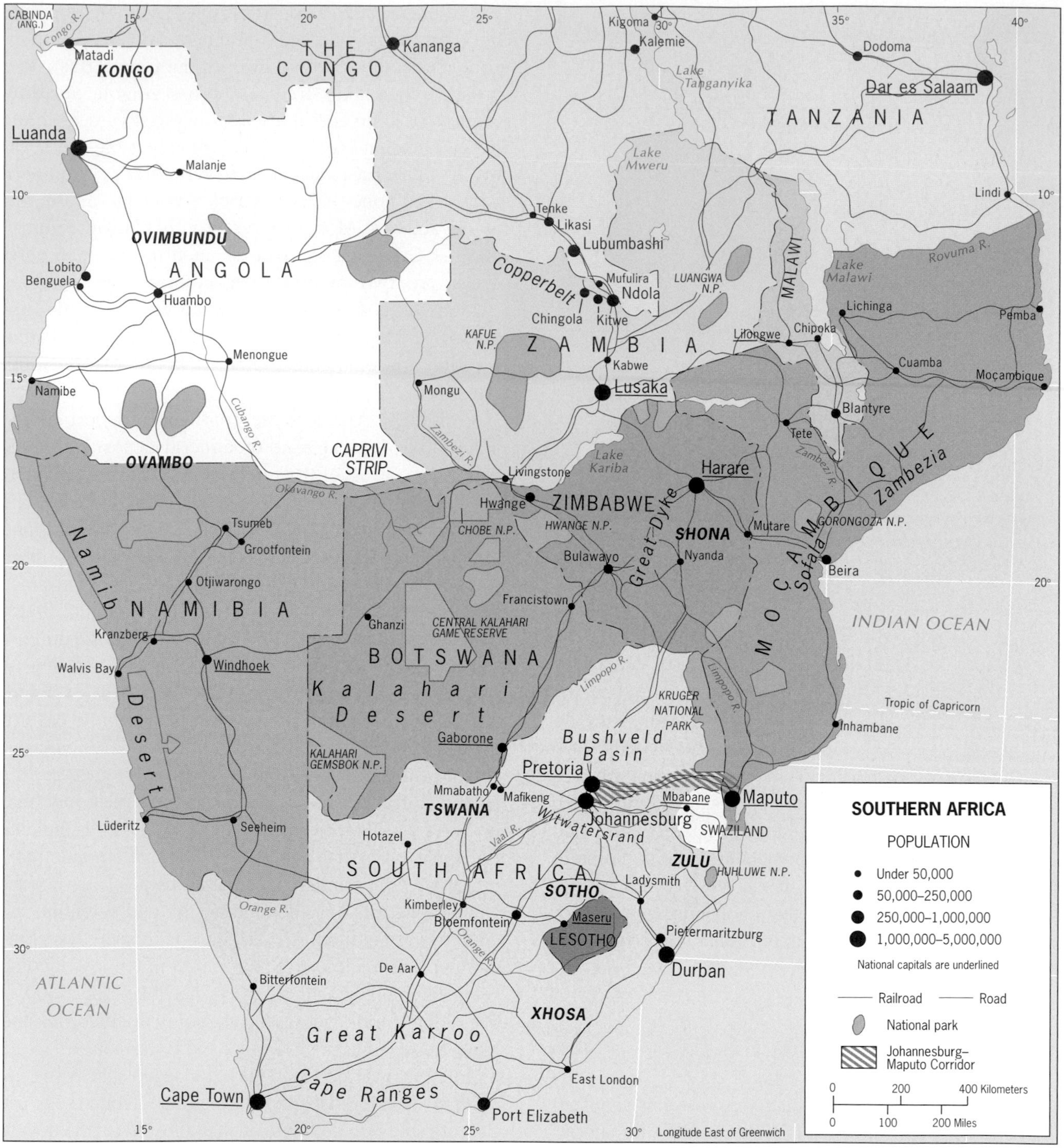

FIGURE 6-11 © H. J. de Blij, P. O. Muller, and John Wiley & Sons, Inc.

Still, Southern Africa as a region is better off than any other in Subsaharan Africa, with four of its countries having risen above the lowest-income rank among world states. Some international cooperation including a regional association and a customs union are emerging. And South Africa, the realm's most important country by many measures, gives hope for a better future.

South Africa

The Republic of South Africa is the giant of Southern Africa, an African country at the center of world attention, a bright ray of hope not only for Africa but for all humankind.

Long in the grip of one of the world's most notorious racial policies (**apartheid**, or "apartness," and its deriva- 14

15 tive, **separate development**), South Africa today is shedding its past and is building a new future. That virtually all parties to the earlier debacle are now working cooperatively to restructure the country under a new flag, a new national anthem, and a new leadership was one of the great events of the late twentieth century. Now, with a new century opened, South Africa is poised to take its long-awaited role as the economic engine for the region—and perhaps beyond.

South Africa stretches from the warm subtropics in the north to Antarctic-chilled waters in the south. With a land area in excess of 470,000 square miles (1.2 million sq km) and a heterogeneous population of 47.9 million, South Africa is the dominant state in Southern Africa. It contains the bulk of the region's minerals, most of its good farmlands, its largest cities, best ports, most productive factories, and most developed transport networks. Mineral exports from Zambia and Zimbabwe move through South African ports. Workers from as far away as Malawi and as nearby as Lesotho work in South Africa's mines, factories, and fields.

Historical Geography

South Africa's location has much to do with its human geography. On the African continent, peoples migrated southward—first the Khoisan-speakers and then the Bantu peoples—into the South African cul-de-sac. On the oceans, the Europeans arrived to claim the southernmost Cape as one of the most strategic places on Earth, the gateway from the Atlantic to the Indian Ocean, a waystation on the route to Asia's riches. The Dutch East India Company founded Cape Town as early as 1652, and soon the Hollanders began to bring Southeast Asians to the Cape to serve as domestics and laborers. When the British took over about 150 years later, Cape Town had a substantial population of mixed ancestry, the source of today's so-called Coloured sector of the country's citizenry.

The British also altered the demographic mosaic by bringing tens of thousands of indentured laborers from their South Asian domain to work on the sugar plantations of east-coast Natal. Most of these laborers stayed after their period of indenture was over, and today South Africa counts about 1 million Indians among its people. Most are still concentrated in Natal and prominently so in metropolitan Durban.

As we noted earlier, South Africa had been occupied by Europeans long before the colonial "scramble for Africa" gained momentum. The Dutch, after the British took control of the Cape, trekked into the South African interior and, on the high plateau they called the *highveld*, founded their own republics. When diamonds and gold were discovered there, the British challenged the Boers (descendants of the Dutch settlers) for these prizes. In 1899–1902, the British and the Boers fought the Boer War. The British won, and British capitalists took control of South Africa's economic and political life. But the Boers negotiated a power-sharing arrangement and eventually achieved hegemony. Having long since shed their European links, they now called themselves *Afrikaners*, their word for Africans, and proceeded to erect the system known as apartheid.

These foreign immigrations and struggles took place on lands that Africans had already entered and fought over. When the Europeans reached the Cape, Bantu nations were driving the weaker Khoi and San peoples into less hospitable territory or forcing them to work in bondage. One great contest was taking place in the east and southeast, below the Great Escarpment. The Xhosa nation was moving toward the Cape along this natural corridor. Behind them, in Natal (Fig. 6-11), the Zulu Empire became the region's most powerful entity in the nineteenth century. On the highveld, the North and South Sotho, the Tswana, and other peoples could not stem the tide of European aggrandizement, but their numbers ensured survival.

In the process, South Africa became Africa's most pluralistic and heterogeneous society. People had converged on the country from Western Europe, Southeast Asia, South Asia, and other parts of Africa itself. At the end of the twentieth century, Africans outnumbered non-Africans by just about 4 to 1 (Table 6-1).

Table 6-1
DEMOGRAPHIC DATA FOR SOUTH AFRICA

Population Groups	Estimates 2006 (in millions)
African nations	37.8
Zulu	11.4
Xhosa	8.4
Tswana	3.9
Sotho (N and S)	3.8
Others (6)	10.3
Whites	4.6
Afrikaners	2.9
English-speakers	1.6
Others	0.1
Mixed (Coloureds)	4.3
African/White	4.0
Malayan	0.3
South Asian	1.2
Hindus	0.8
Muslims	0.4
TOTAL 47.9	

Social Geography

Heterogeneity also marks the spatial demography of South Africa. Despite centuries of migration and (at the Cape) intermarriage, labor movement (to the mines, farms, and factories), and massive urbanization, regionalism pervades the human mosaic. The Zulu nation still is largely concentrated in the province the Europeans called Natal. The Xhosa still cluster in the Eastern Cape, from the city of East London to the Natal border and below the Great Escarpment. The Tswana still occupy ancestral lands along the Botswana border. Cape Town still is the core area of the Coloured population; Durban still has the strongest Indian imprint. Travel through South Africa, and you will recognize the diversity of rural cultural landscapes as they change from Swazi to Ndebele to Venda.

This historic regionalism was among the factors that led the Afrikaner government to institute its "separate development" scheme, but it could not stem the tide of urbanization. Millions of workers, job-seekers, and illegal migrants converged on the cities, creating vast shantytowns on their margins. In the legal African "townships" such as Johannesburg's Soweto ("SOuth WEstern TOwnships") and in these squatter settlements the anti-apartheid movement burgeoned, and the strength of the African National Congress (ANC) movement grew. In February 1990 Nelson Mandela, imprisoned for 28 years on Robben Island, South Africa's Alcatraz, became a free man, and following the momentous first democratic election of 1994 he became president of an ANC-dominated government in Cape Town.

Before this election could take place, however, South Africa's administrative geography had to be radically changed. The republic's political geography was a legacy of the Union period, when it was divided into four provinces: the Cape, by far the largest and centered on the legislative capital of Cape Town; Natal, anchored by the port city of Durban; the Transvaal ("across the Vaal River"), focused on the administrative capital of Pretoria and including the great Johannesburg metropolis; and the Orange Free State, the Boer stronghold with Bloemfontein as its headquarters.

This structure was replaced by a new map of nine provinces (Fig. 6-12), creating a federal arrangement in which each province would have its own administration while being represented in the central government. The boundaries of Natal and the Orange Free State remained essentially unchanged, but Natal's name was changed to Kwazulu-Natal, recognizing the Zulu presence there, and the Orange Free State became simply the Free State. But the Cape Province was divided into four new provinces, one of them overlapping into the Transvaal, and the rest of the Transvaal was divided into three, one of which is Gauteng, which includes the Johannesburg–Witwatersrand–Pretoria megalopolis.

The 1994 election, based on this new layout, produced an ANC victory in seven of these nine provinces, excepting only Kwazulu-Natal, where the Zulu vote went to a local political movement called Inkatha, and the Western Cape, where a combination of white and Coloured voters outpolled the ANC. This was a very fortunate result, proving that the ANC was not all-powerful and giving minorities, including Coloureds and whites, reason to trust and participate in the new system.

The transition from Afrikaner domination to democratic government still was complicated by numerous circumstances. As it turned out, it was not white opposition that most threatened majority rule. The white extremist political parties collapsed relatively soon, and a brief terrorist episode was quickly halted. More consequential was the uncertain role of the largest single nation in South Africa, the Zulu, whose historic domain lies in Kwazulu-Natal, and whose prominent leader, Chief Buthelezi, at times hinted at secession if the new South Africa was not acceptable. In the runup to the election, violence killed thousands and threatened to precipitate a civil war. But that danger, too, was overcome, and the Zulu nation and its leaders remained in the fold.

In 1999, ANC leader Thabo Mbeki, who had served as President Mandela's deputy, became the country's second popularly elected president. In other African states, the succession from heroic founder-of-the-nation to political inheritor-of-the-presidency often has not gone well, but South Africa's new constitution proved its worth. The new president faced several disadvantages: inevitable comparisons to the incomparable Mandela; the rising tide of AIDS, which Mbeki controversially attributed to causes other then HIV; and the farm invasions in Zimbabwe (see p. 303) on which his leadership seemed to falter. Nevertheless, Mbeki's economic policies, social programs, and foreign initiatives (The Congo, Rwanda) have served his country well.

Economic Geography

Undoubtedly the most serious immediate problems for the new South Africa are economic. Ever since diamonds were discovered at Kimberley in the 1860s, South Africa has been synonymous with minerals. The Kimberley finds, made in a remote corner of what was then the Orange Free State (the British soon annexed it to the Cape), set into motion a new economic geography. Rail lines were laid from the coast to the "diamond capital" even as fortune seekers, capitalists, and tens of thousands of African workers, many from as far afield as Lesotho, streamed to the site. One of the capitalists was Cecil Rhodes, of Rhodes Scholarship fame,

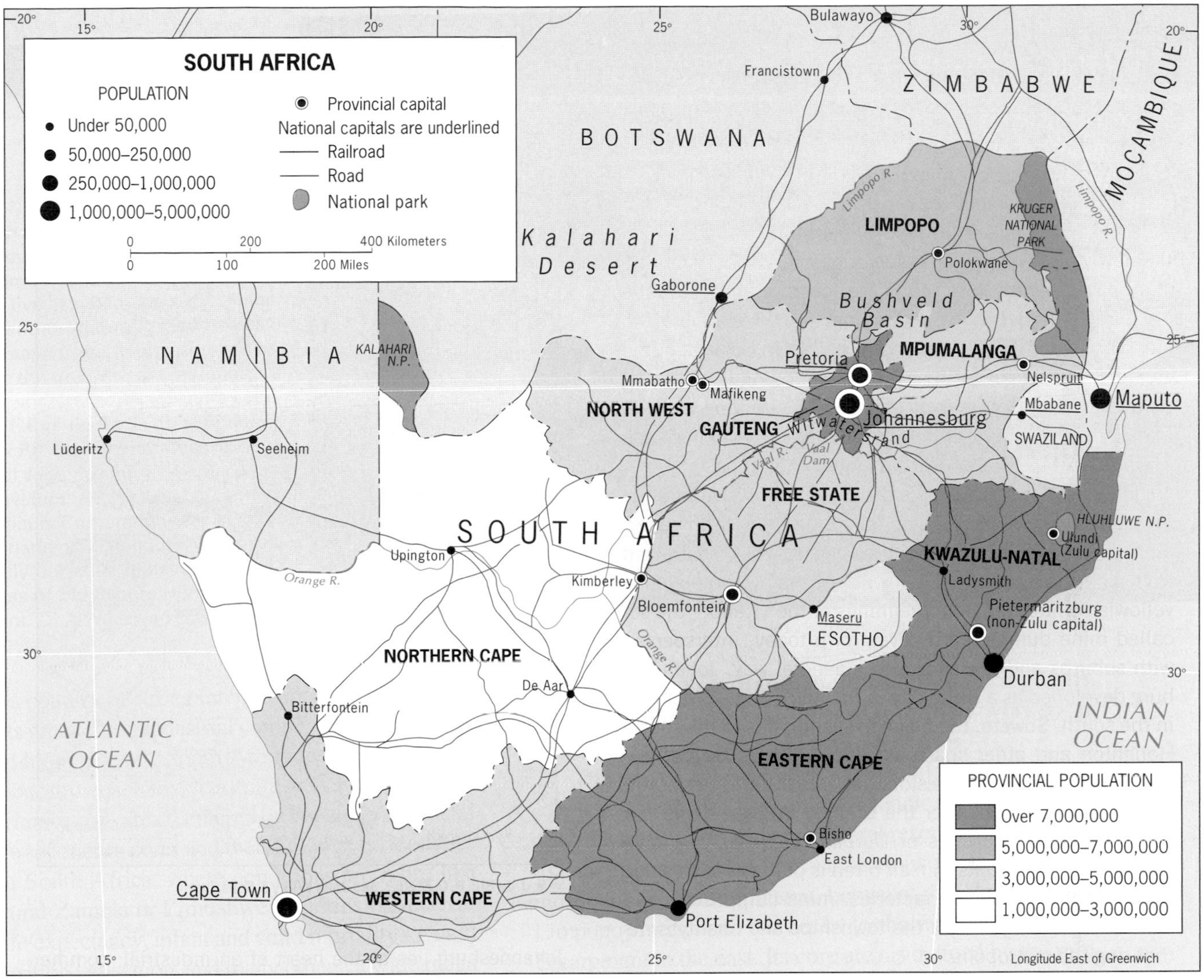

FIGURE 6-12 © H. J. de Blij, P. O. Muller, and John Wiley & Sons, Inc.

who used his fortune to help Britain dominate Southern Africa.

Just 25 years after the diamond discoveries, prospectors found what was long to be the world's greatest goldfield on a ridge called the Witwatersrand (Fig. 6-12). This time the site lay in the so-called South African Republic (the Transvaal), and again the Boers were unable to hold the prize. Johannesburg became the gold capital of the world, and a new and even larger stream of foreigners arrived, along with a huge influx of African workers. Cheap labor enlarged the profits. Johannesburg grew explosively, satellite towns developed, and black townships mushroomed. The Boer War was only an interlude here on the mineral-rich Witwatersrand.

During the twentieth century, South Africa proved to be richer than had been foreseen. Additional goldfields were discovered in the Orange Free State. Coal and iron ore were found in abundance, which gave rise to a major iron and steel industry. Other metallic minerals, including chromium and platinum, yielded large revenues on world markets. Asbestos, manganese, copper, nickel, antimony, and tin were mined and sold; a thriving metallurgical industry developed in South Africa itself. Capital flowed into the country, white immigration grew, farms and ranches were laid out, and markets multiplied.

South Africa's cities grew apace. Johannesburg was no longer just a mining town: it became an industrial complex and a financial center as well. The old Boer capital, Pretoria, just 30 miles (50 km) north of the Witwatersrand, became the country's administrative center during apartheid's days. In the Orange Free State, major industrial growth (including oil-from-coal technology) matched the expansion of mining. While the core area developed megalopolitan characteristics, coastal cities expanded as well. Durban's port served not only the Witwatersrand but a wider regional hinterland as well. Cape

The humanitarian disaster that has gripped the western extremity of East Africa and the adjoining eastern margin of Equatorial Africa defies imagination. As many as 3.5 million people may have perished (only the roughest estimates of the carnage exist), and millions more have been dislocated and dispossessed. Marauding armed gangs raid crowded refugee camps already ravaged by hunger and disease. This photo shows part of a refugee camp near the place where the borders of Rwanda, Burundi, and The Congo meet, and the situation here is actually better than most—some tents have been provided to the displaced families, some trees have not yet been cut down for firewood, and there are a few stalks of corn. But an entire interregional zone has been destabilized, and only a massive aid effort can begin to restore what has been lost. ©Ian Berry/ Magnum Photos, Inc.

Rwanda (8.7 million) and Burundi (6.5 million) are physiographically part of East Africa, but their cultural geography is linked to the north and west. Here, Tutsi pastoralists from the north subjugated Hutu farmers (who had themselves made serfs of the local Twa [pygmy] population), setting up a conflict that was originally ethnic but became cultural. Certain Hutu were able to advance in the Tutsi-dominated society, becoming to some extent converted to Tutsi ways, leaving subsistence farming behind, and rising in the social hierarchy. These so-called moderate Hutu were—and are—often targeted by other Hutus, who resent their position in society. This longstanding discord, worsened by colonial policies, had repeatedly devastated both countries and, in the 1990s, spilled over into The Congo, generating the first interregional war in Subsaharan Africa.

An estimated 3.5 million people have perished as Hutu, Tutsi, Ugandan, and Congolese rebel forces have fought for control over areas of the eastern Congo, unleashing longstanding local animosities (such as those between Hema and Lendu around Bunia) that worsened the death toll. Only massive international intervention could stabilize the situation, but the world has turned a blind eye to the region's woes—again.

Highland Ethiopia

As Figure 6-14 shows, the East African region also encompasses the highland zone of Ethiopia, including the capital, Adis Abeba; the source of the Blue Nile, Lake Tana; and the Amharic core area that was the base of the empire that lost its independence only from 1935 to 1941. Ethiopia, mountain fortress of the Coptic Christians who held their own here, eventually became a colonizer itself. Its forces came down the slopes of the highlands and conquered much of the Islamic part of Africa's Horn, including present-day Eritrea and the Ogaden area, a Somali territory. (Geographically, these are parts of what we have mapped as the African Transition Zone [Fig. 6-10], which will be discussed in the final section of this chapter.)

Physiographically and culturally, highland Ethiopia is part of the East African region. But because Ethiopia was not colonized and because its natural outlets are to the Red Sea, not southward to Mombasa, effective interconnections between former British East Africa and highland Ethiopia never developed. But the Amhara and Oromo peoples of Ethiopia are Africans, not Arabs; nor have they been Arabized or Islamized as in northern Sudan and Somalia. The independence and secession of Eritrea in 1993 effectively landlocked Ethiopia, but for a few years there was cooperation and Ethiopia used Eritrean Red Sea ports. In 1998, however, a boundary dispute led to a bitter and costly war, and Ethiopia was forced to turn to Djibouti for a maritime outlet. This border conflict cost thousands of lives and did not end until UN mediation in 2000 and the arrival of a peacekeeping force; as recently as mid-2005, the matter was still not closed. Meanwhile, there were complaints from Somalia that Ethiopian forces were violating their joint border. With adversaries on three sides, Ethiopia is likely to turn increasingly toward East Africa, however tenuous its surface links with Kenya are today.

EQUATORIAL AFRICA

The term *equatorial* is not just locational but also environmental. The equator bisects Africa, but only the western part of central Africa features the conditions associated with the low-elevation tropics: intense heat, high rainfall and extreme humidity, little seasonal variation, rainforest and monsoon-forest vegetation, and enormous biodiversity. To the east, beyond the Western Rift Valley, elevations rise, and cooler, more seasonal climatic regimes prevail. As a result, we recognize two regions in these lowest latitudes: Equatorial Africa to the west and (just discussed) East Africa to the east.

Equatorial Africa is physiographically dominated by the gigantic Congo Basin. The Adamawa Highlands separate this region from West Africa; rising elevations and climatic change mark its southern limits (see the *Cwa* boundary in Fig. G-8). Its political geography consists of eight states, of which The Congo (formerly Zaïre) is by far the largest in both territory and population (Fig. 6-15).

Five of the other seven states—Gabon, Cameroon, São Tomé and Príncipe, Congo, and Equatorial Guinea—all have coastlines on the Atlantic Ocean. The Central African Republic and Chad, the south of which is part of this region, are landlocked. In certain respects, the physical and human characteristics of Equatorial Africa extend even into southern Sudan. This vast and complex region is in many ways the most troubled in the entire Subsaharan Africa realm.

The Congo

As the map shows, The Congo has but a tiny window (23 miles; 37 km) on the Atlantic Ocean, just enough to accommodate the mouth of the Congo River. Oceangoing ships can reach the port of Matadi, inland from which falls and rapids make it necessary to move goods by road or rail to the capital, Kinshasa. This is not the only place where the Congo River fails as a transport route. Follow it upstream on Figure 6-15, and you note that other transshipments are necessary between Kisangani and Ubundu, and at Kindu. Follow the railroad south from Kindu, and you reach another narrow corridor of territory at the city of Lubumbashi. That vital part of Katanga Province contains most of The Congo's major mineral resources, including copper and cobalt.

With a territory not much smaller than the United States east of the Mississippi, a population of 62.0 million, a rich and varied mineral base, and much good agricultural land, The Congo would seem to have all the ingredients needed to lead this region and, indeed, Africa. But strong centrifugal forces, arising from its physiography and cultural geography, pull The Congo apart. The immense forested heart of the basin-shaped country creates communication barriers between east and west, north and south. Many of The Congo's productive areas lie along its periphery, separated by enormous distances. These areas tend to look across the border, to one or more of The Congo's nine neighbors, for outlets, markets, and often ethnic kinship as well.

The Congo's civil wars of the 1990s started in one such neighbor, Rwanda, and spilled over into what was then still known as Zaïre. Rwanda, Africa's most densely populated country, has for centuries been the scene of conflict between sedentary Hutu farmers and invading Tutsi pastoralists. Colonial borders and practices worsened the situation (see the Issue Box titled "The Impact of Colonialism on Subsaharan Africa"), and after independence a series of terrible crises followed. In the mid-1990s, the latest of these crises generated one of the largest refugee streams ever seen in the world, and the conflict engulfed eastern (and later northern and western) Congo. The death toll will never be known, but estimates range from 3 to 4 million, a calamity that was not enough to propel the "international community" into concerted peacemaking action. By 2005, a combination of power transfer in the capital, Kinshasa, negotiation among the rebel groups and the African states involved in various ways in the conflict, UN assistance, and exhaustion had produced a semblance of stability in all but some eastern areas of The Congo (Fig. 6-15).

Across the River

To the west and north of the Congo and Ubangi rivers lie Equatorial Africa's other seven countries (Fig. 6-15). Two of these are landlocked. **Chad**, straddling the African Transition Zone as well as the regional boundary with West Africa, is one of Africa's most remote countries, although recent oil discoveries in the south are likely to make it less isolated. The **Central African Republic**, chronically unstable and poverty-stricken, never was able to convert its agricultural potential and mineral resources (diamonds, uranium) into real progress. And one country consists of two small, densely forested volcanic islands: **São Tomé and Príncipe**, a ministate whose small economy based on some cocoa and coconut trees was recently transformed by the discovery of oil beneath the waters between it and Nigeria.

The four coastal states present a different picture. All possess oil reserves and share the Congo Basin's equatorial forests; oil and timber, therefore, rank prominently among their exports. In **Gabon**, this combination has produced Equatorial Africa's only upper-middle-income economy. Of the four coastal states, Gabon also has the largest proven mineral resources, including manganese, uranium, and iron ore. Its capital, Libreville (the only coastal capital in the region), reflects all this in its

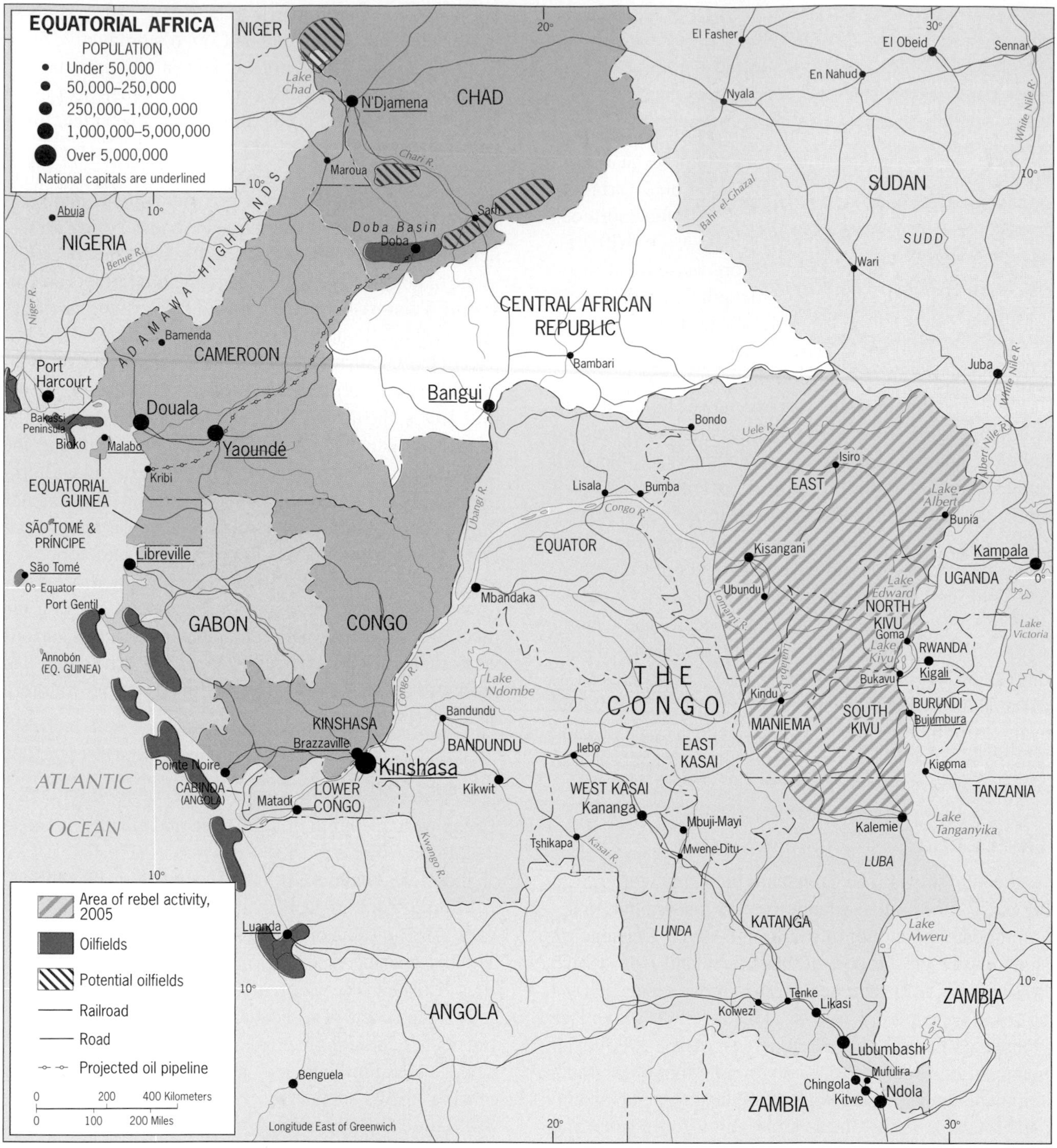

FIGURE 6-15 © H. J. de Blij, P. O. Muller, and John Wiley & Sons, Inc.

high-rise downtown, bustling port, and fast-growing squatter settlements.

Cameroon, less well endowed with oil or other raw materials, has the region's strongest agricultural sector by virtue of its higher-latitude location and high-relief topography. Western Cameroon is one of the more developed parts of Equatorial Africa and includes the capital, Yaoundé, and the port of Douala.

With five neighbors, **Congo** could be a major transit hub for this region, especially for The Congo if it recovers from civil war. Its capital, Brazzaville, lies across the Congo River from Kinshasa and is linked to the port of Pointe Noire by road and rail. But devastating power struggles have negated Congo's geographic advantages.

As Figure 6-15 shows, **Equatorial Guinea** consists of a rectangle of mainland territory and the island of

The Impact of Colonialism on Subsaharan Africa

Regional ISSUE

COLONIALISM IS THE CULPRIT!

"If you want to see and feel the way colonialism ruined Africa, you don't have to go beyond the borders of my own country, the so-called Democratic Republic of the Congo, for the evidence. I'm a school teacher here in Kikwit, east of Kinshasa, and my whole life I've lived under the dictatorship left us by the Belgians and in the chaos before and after it. All you have to do is look at the map. Who can say that the political map of Africa isn't a terrible burden? Imagine some outside power, say the Chinese, coming into Europe and throwing the Germans and the French together in one country. There would never be peace! Well, that's what they did here. All over Africa they put old enemies together and parceled up our great nations. And now they tell us to get over our animosities and 'tribalism' and live in democratic peace.

"And let me tell you something about tribalism. Before the conquest, we had ethnic groups and clans like all other peoples have, from the Scots to the Siamese. And we certainly fought with each other. But mostly we got along, and there was lots of gray area between and among us. You weren't immediately labeled a Hutu or a Luba or a Bemba. The colonial powers changed all that. Suddenly we were tribalized. Whether you liked it or not, whether you filled the 'profile' or not, you were designated a Tutsi or a Hutu, a Ganda or a Toro. This enabled the European rulers to use the strong to dominate the weak, to use the rich to tax the poor, to police migrants and assemble labor. The tribalization of Africa was a colonial invention. It worsened our divisions and destroyed our common ground. And now we just have to get over it?

"Of course the colonialists exploited our mineral resources—I tell my pupils how The Congo enriched Europeans and other 'civilized' peoples outside of Africa. But they also grabbed our best land, turning food-producing areas into commercial plantations. Are you surprised that in Zimbabwe Mugabe wants to chase those white farmers off 'their' land? I'm not, and I'm with him all the way. If the colonialists hadn't ruined traditional agriculture, there wouldn't be any food shortages today.

"Look at my country. The war between the Hutu and the Tutsi that started during colonial times has now spread deep into The Congo. We've got so used to taking sides in 'tribal' conflicts that the whole east is out of control. And who are the losers? The poorest of the poor, who get robbed and killed by the Mai, the Interahamwe, and the Tutsis and their friends. What are the ex-colonial powers doing now to help us with the mess they left behind? Nothing! So don't complain to me about poor African leadership and our failure to carry out economic reforms dictated by some foreign bank. The poisonous legacy of colonialism will hold Africa back for generations to come."

COLONIALISM IS A SCAPEGOAT!

"We Africans have many factors to blame for our troubles, but colonialism isn't one of them. Many African countries have been independent for nearly two generations; most Africans were born long after the colonial era ended. Are we better off today than we were at the end of the colonial period? Are those military dictatorships that ran so many countries into the ground any better than the colonizers who preceded them? Are the ghosts of colonialists past rigging our elections, stifling the media? Is our endemic corruption the fault of lingering colonialism? Did the colonial governors set up Swiss bank accounts to which to divert development funds? Can extinct colonial regimes be blamed for the environmental problems we face? Did the bandits of interior Sierra Leone or eastern areas of The Congo learn their mutilation practices from colonialism? As a nurse in a Uganda hospital for AIDS victims, I say that it's time to stop blaming ancient history for our current failures.

"It's not that we Africans aren't at a disadvantage in this unforgiving world. Our postcolonial population explosion was followed by an HIV-induced implosion. Apartheid South Africa spread its malign influence far beyond its borders. The Cold War pitted regimes against rebels and, to use a colonial term, tribe against tribe. Droughts and other plagues of biblical proportions ravaged our continent when we were more vulnerable than ever. Prices for our raw materials on international markets fell when we most needed a boost. Foreign governments protected their farmers against competition from African producers. Those are legitimate complaints. But colonialism?

"Forget about blaming boundaries, white farmers, tribalism, religion, and other residues of colonialism for our present troubles. Look at what happened here in Uganda. A decade ago we were at the heart of the 'AIDS Belt' in tropical Africa, with one of the worst infection rates on the continent. Then came President Museveni, whose government launched a vigorous self-help campaign, educating people to the risks, urging restraint, distributing condoms, encouraging women to resist unwanted sex. Now Uganda's AIDS incidence is dropping and we are held up as an example for all of Africa to follow. While South Africa's President Mbeki reveals his confusion about the causes of AIDS, Zimbabwe's Mugabe is too busy chasing white farmers, and Kenya's Moi is trying to rig his succession, we in Uganda have leadership in an area where we need it desperately. That certainly beats blaming colonialism for everything.

"We have an African Union. If those colonial boundaries are so bad, let's change them, or do what the European Union is doing and begin to erase them. But first, let's isolate tyrants, nurture open democracy, and combat corruption—and stop blaming colonialism."

Vote your opinion at www.wiley.com/college/deblij

Substantial areas of The Congo remain beyond the control of the capital and are the scene of costly fighting among rebel armies with devastating impact on local populations. These heavily armed troops of an outfit called the Union of Congolese Patriots are shown having taken over the northeastern town of Bunia in advance of still another round of talks to end more than four years of civil war (the talks failed, and strife continues). © AP/WideWorld Photos.

Bioko, where the capital of Malabo is located. A former Spanish colony that remained one of Africa's least developed territories, Equatorial Guinea, too, has been affected by the oil business in this area. Petroleum products now dominate its exports, but, as in so many other oil-rich countries, this bounty has not significantly raised incomes for most of the people.

One other territory would seem to be a part of Equatorial Africa: *Cabinda*, wedged between the two Congos just to the north of the Congo River's mouth. But Cabinda is one of those colonial legacies on the African map—it belonged to the Portuguese and was administered as part
17 of Angola. Today it is an **exclave** of independent Angola, and a valuable one: it contains major oil reserves.

WEST AFRICA

West Africa occupies most of Africa's Bulge, extending south from the margins of the Sahara to the Gulf of Guinea coast and from Lake Chad west to Senegal (Fig. 6-16). Politically, the broadest definition of this region includes all those states that lie to the south of Western Sahara, Algeria, and Libya and to the west of Chad (itself sometimes included) and Cameroon. Within West Africa, a rough division is sometimes made between the large, mostly steppe and desert states that extend across the southern Sahara (Chad could also be included here) and the smaller, better-watered coastal states.

Apart from once-Portuguese Guinea-Bissau and long-independent Liberia, West Africa comprises four former British and nine former French dependencies. The British-influenced countries (Nigeria, Ghana, Sierra Leone, and Gambia) lie separated from one another, whereas Francophone West Africa is contiguous. As Figure 6-16 shows, political boundaries extend from the coast into the interior, so that from Mauritania to Nigeria, the West African habitat is parceled out among parallel, coast-oriented states. Across these boundaries, especially across those between former British and former French territories, there is only limited interaction. For example, in terms of value, Nigeria's trade with Britain is about 100 times as great as its trade with nearby Ghana. The countries of West Africa are not interdependent economically, and their incomes are largely derived from the sale of their products on the non-African international market.

Given these cross-currents of subdivision within West Africa, why are we justified in speaking of a single West African region? First, this part of the realm has remarkable cultural and historical momentum. The colonial interlude failed to extinguish West African vitality, expressed not only by the old states and empires of the savanna and the cities of the forest, but also by the vigor and entrepreneurship, the achievements in sculpture, music, and dance, of peoples from Senegal to Nigeria's southeastern Iboland. Second, West Africa contains a set of parallel east-west ecological belts, clearly reflected in Figures G-7 and G-8, whose role in the development of the region is pervasive. As the transport-route pattern on the map of West Africa indicates, overland connections within each of these belts, from country to country, are poor; no coastal or interior railroad ever connected this tier of countries. Yet spatial interaction is stronger across these

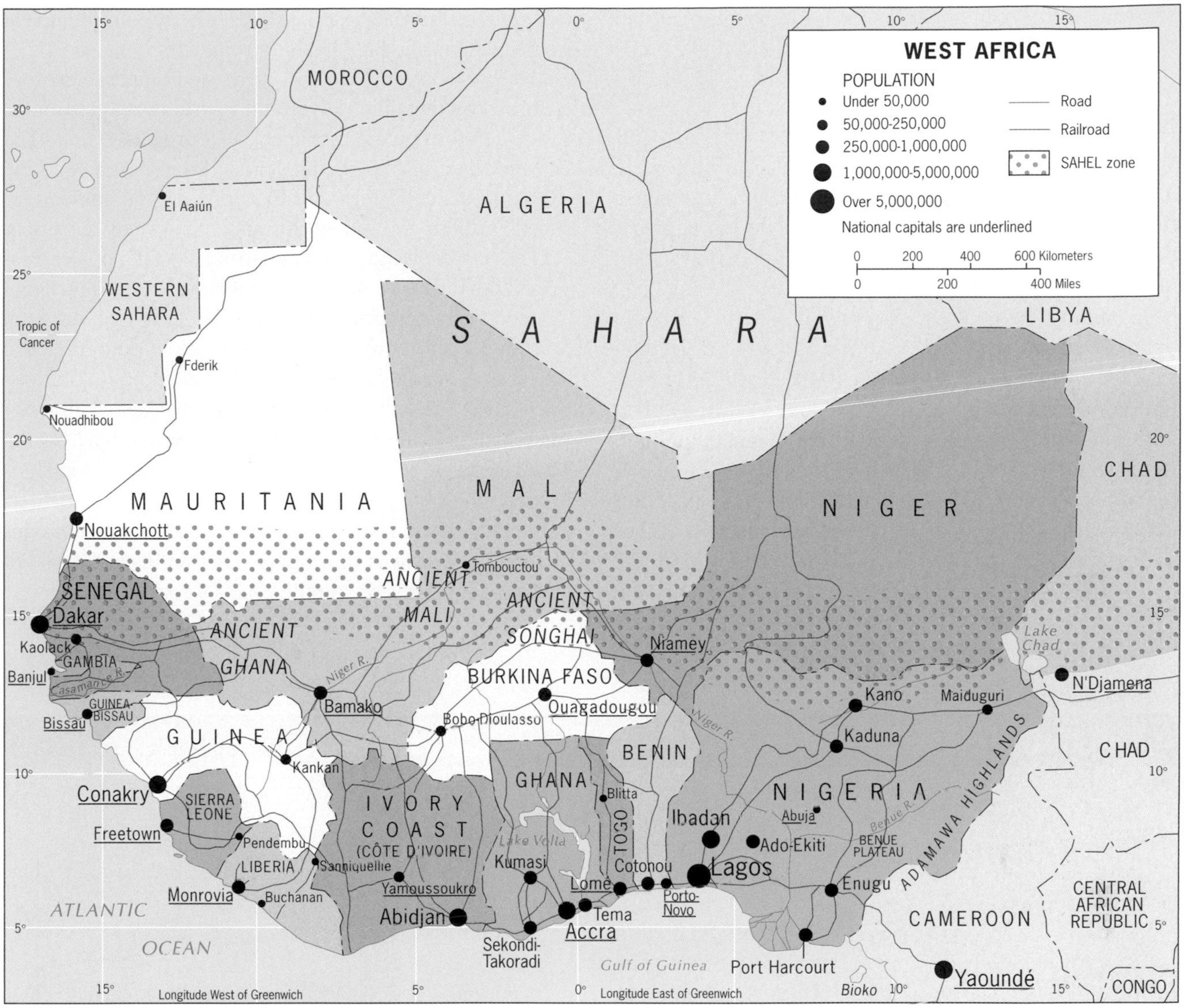

FIGURE 6-16

belts, and some north-south economic exchange does take place, notably in the coastal consumption of meat from cattle raised in the northern savannas. And third, West Africa received an early and crucial imprint from European colonialism, which—with its maritime commerce and slave trade—transformed the region from one end to the other. This impact reached into the heart of the Sahara and set the stage for the reorientation of the whole area, from which emerged the present patchwork of states.

Despite the effects of the slave trade, West Africa today is Subsaharan Africa's most populous region (Fig. G-9). In these terms, Nigeria (whose census results are in doubt, but with an estimated population of 145 million) is Africa's largest state; Ghana (22.4 million) and Ivory Coast (17.6 million) rank high as well. The southern half of the region, understandably, is home to most of the people. Mauritania, Mali, and Niger include too much of the unproductive Sahel's steppe and the arid Sahara to sustain populations comparable to those of Nigeria, Ghana, or Ivory Coast.

The peoples along the coast reflect the modern era that the colonial powers introduced: they prospered in their newfound roles as middlemen in the coastward trade. Later, they experienced the changes of the colonial period; in education, religion, urbanization, agriculture, politics, health, and many other endeavors, they adopted new ways. In contrast, the peoples of the interior retained their ties with a different era in African history. Distant and aloof from the main theater of European colonial activity and often drawn into the Islamic orbit, they experienced a significantly different kind of change. But the map reminds us that Africa's boundaries were not drawn to accommodate such contrasts. Both Nigeria and Ghana possess population clusters representing the interior as well as the coastal peoples, and in both countries the wide cultural gap between north and south has produced political problems.

Nigeria: West Africa's Cornerstone

When Nigeria achieved full independence from Britain in 1960, its government faced the task of administering a European political creation containing three major nations and nearly 250 other peoples ranging from several million to a few thousand in number. The country had been endowed with a federal political system consisting of three regions, each focused on one of the three leading nations. In the southwest lay the Western Region, the traditional home of the Yoruba, a people with long urban traditions and a complex and highly developed culture. In the southeast lay the Eastern Region, historic home of the Ibo people, less affected than the Yoruba by the colonial impact, less urbanized, but widely dispersed to other parts of Nigeria and even beyond. And in the north, the Northern Region was the largest and most populous, the domain of the Hausa-Fulani cluster, a region of ancient cities, Islam, and Sahelian environments.

The map of Nigeria shows a compact territory crossed by the Niger River and its tributary, the Benue, together forming a Y-shaped system (Fig. 6-17). The lower Niger and its delta separated the Yoruba and the Ibo and their Western and Eastern Regions; the core of the Northern Region was situated between the upper Niger and the Benue. The departing colonial power, recognizing the new state's regional diversity, believed that a federal framework would help Nigeria's government to cope with the resulting stresses. Some progress had been made in the construction of road and railroad links across the "Middle Zone" between north and south, English had been introduced as the common tongue for the educated elite (it is still one of the four official languages today along with Yoruba, Ibo, and Hausa), a considerable civil service had been built up, and in terms of education Nigeria was in a better position than most other African countries or colonies.

AMONG THE REALM'S GREAT CITIES . . . LAGOS

In a realm that is only 33 percent urbanized, Lagos, former capital of federal Nigeria, is the exception: a teeming metropolis of 12.3 million, sometimes called the Calcutta of Africa.

Lagos evolved over the past three centuries from a Yoruba fishing village, Portuguese slaving center, and British colonial headquarters into Nigeria's largest city, major port, leading industrial center, and first capital. Situated on the country's southwestern coast, it consists of a group of low-lying islands and sand spits between the swampy shoreline and Lagos Lagoon. The center of the city still lies on Lagos Island, where the highrises adjoining the Marina overlook Lagos Harbor and, across the water, Apapa Wharf and the Apapa industrial area. The city expanded southward onto Ikoyi Island and Victoria Island, but after the 1970s most urban sprawl took place to the north, on the western side of Lagos Lagoon.

Lagos's cityscape is a mixture of modern high-rises, dilapidated residential areas, and squalid slums. From the top of a high-rise one sees a seemingly endless vista of rusting corrugated roofs, the houses built of cement or mud in irregular blocks separated by narrow alleys. On the outskirts lie the shantytowns of the less fortunate, where shelters are made of plywood and cardboard and lack even the most basic facilities.

By world standards, Lagos ranks among the most severely polluted, congested, and disorderly cities. Mismanagement and official corruption are endemic. Laws, rules, and regulations, from zoning to traffic, are flouted. The international airport is notorious for its inadequate security and for extortion by immigration and customs officers. In many ways, Lagos is a city out of control.

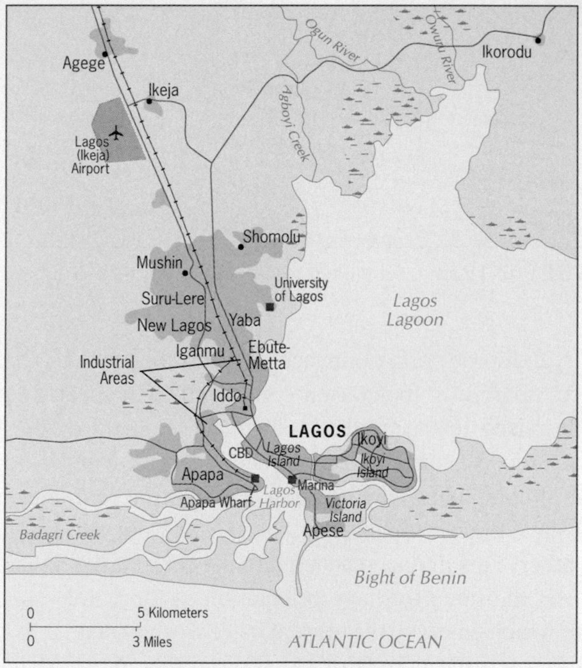

© H. J. de Blij, P. O. Muller, and John Wiley & Sons, Inc.

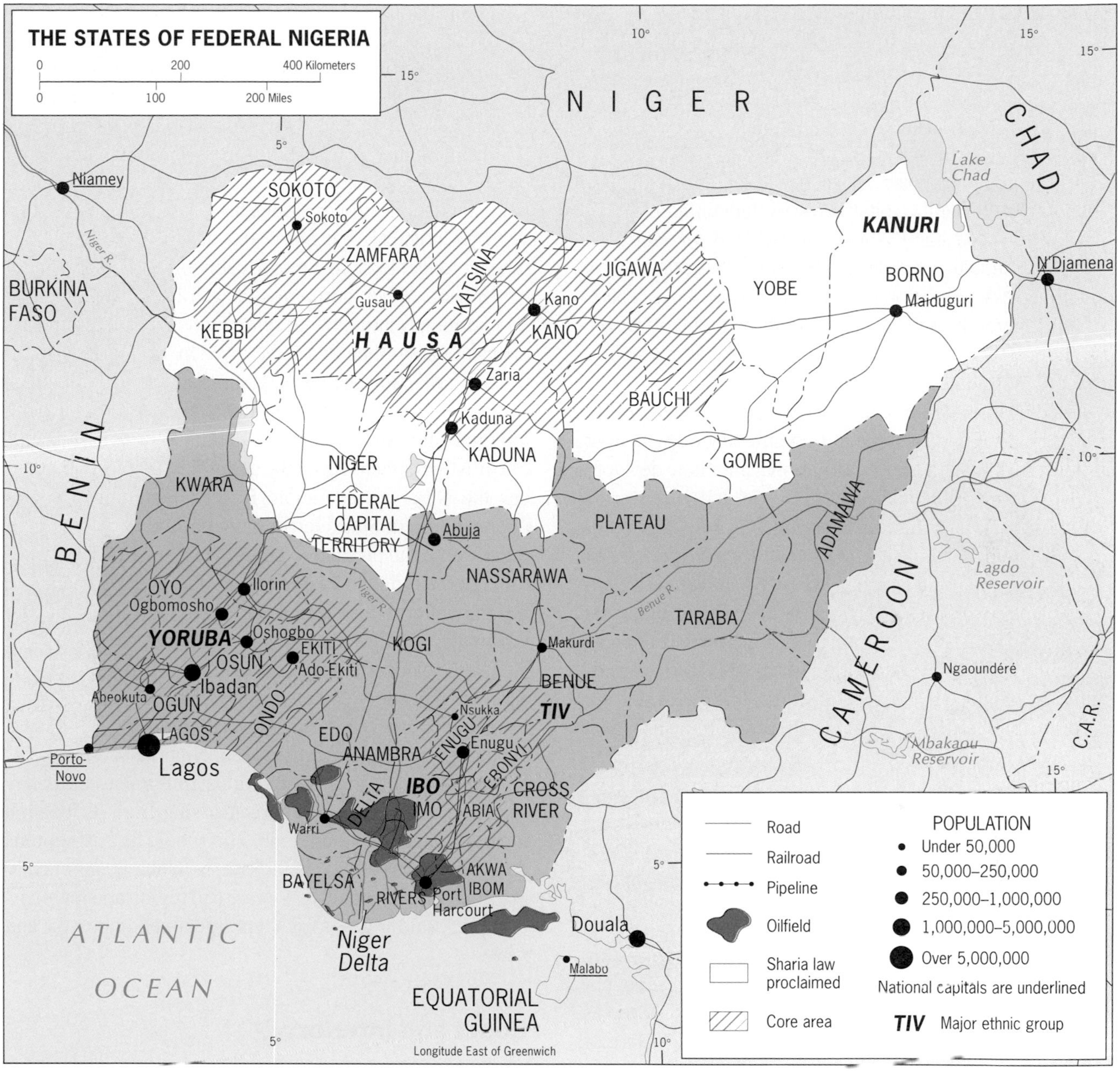

FIGURE 6-17 © H. J. de Blij, P. O. Muller, and John Wiley & Sons, Inc.

For reasons obvious from the map, Britain's colonial imprint always was stronger in the two southern regions than in the north. Christianity became the dominant faith in the south, and southerners, especially Yoruba, took a lead role in the transition from colony to independent state. The choice of Lagos, the port of the Western Region, as the federal capital (and not one of the cities in the more populous north) reflected British hopes for the country's future.

Nigeria is a large country territorially (about the size of Texas plus Oklahoma), and as Figure 6-16 shows, it extends farther northward than any of the other coastal West African states as far west as Guinea. This has the effect of incorporating much larger populations within its northern borders than its neighbors have. It was an objective of Nigeria's original three-region federal system that the two smaller southern regions would balance the single, larger one in the Islamic north.

But Nigeria's three-region federation did not last long. Initially, it was not the Northern Region but the Eastern Region that mounted a campaign to secede from the federation, although the north did play a role. Many Ibo had moved to northern cities to take jobs northerners were unwilling to do; some became successful and prominent entrepreneurs. These energetic Christians aroused animosities among the majority Muslims, and in September and October 1966 a series of massacres occurred. Surviving Ibos fled from the north in huge numbers, and in May 1967 the government of the Eastern Region declared its independence as the Republic of Biafra. This led to a civil war that lasted three years and cost an estimated 1 million lives. To ensure that such a secession attempt would not happen again, successive Nigerian governments and regimes repeatedly modified the federal map. Today Nigeria has 36 States as well as a new Federal Capital Territory containing the more centrally located capital, Abuja (Fig. 6-17). But during most of its existence as an independent state, Nigeria has not functioned as a federation. Civilian government was repeatedly abrogated by military dictatorships that centralized power, enriched the rulers, and impoverished the country.

Large oilfields were discovered under the Niger Delta during the 1950s, when Nigeria's agricultural sector produced most of its exports (peanuts, palm oil, cocoa, cotton) and farming still had priority in national and State development plans. Soon, revenues from oil production dwarfed all other sources, bringing the country a brief period of prosperity and promise. But before long Nigeria's oil wealth brought more bust than boom. Misguided development plans now focused on grand, ill-founded industrial schemes and costly luxuries such as a national airline; the continuing mainstay of the vast majority of Nigerians, agriculture, fell into neglect. Worse, poor management, corruption, outright theft of oil revenues during military misrule, and excessive borrowing against future oil income led to economic disaster. The country's infrastructure collapsed. In the cities, basic services broke down. In the rural areas, clinics, schools, water supplies, and roads to markets crumbled. In the Niger Delta area, local people beneath whose land the oil was being exploited demanded a share of the revenues and reparations for ecological damage; the military regime under General Abacha responded by arresting and executing nine of their leaders. On global indices of national well-being, Nigeria sank to the lowest rungs even as its production ranked it as high as the world's tenth-largest oil producer, with the United States its chief customer.

In 1999 Nigeria's hopes were raised when, for the first time since 1983, a democratically elected president was sworn into office. But even as President Obasanjo confronted the country's tough realities—idle, dated factories; abandoned farms and plantations; countless unemployed; ingrained habits of corruption and nonpayment of taxes; a fast-growing population that now nears 150 million; the deepening AIDS crisis—a cultural imbroglio with catastrophic potential loomed. The return to democracy brought with it a loss of power among the northern elite that had long prevailed in the military, and now northern States, predominantly Muslim, began flexing their muscles in other ways. The State of Zamfara, whose politicians had fallen out of favor in the government of President Obasanjo, proclaimed that *Sharia* (strict Islamic) law would henceforth apply not only to matters such as marriage and inheritance, but to all aspects of life. Sharia criminal punishments are severe and include amputations for theft. But it also affects daily life: men and women cannot ride in taxis together, schools are segregated by gender, many jobs are reserved for males. Muslim leaders in Zamfara insisted that Sharia law would affect only Muslims, not Christians and others in the State, but this was not so. Christian women, for example, had to find public taxis driven by Christians, but most taxis are driven by Muslims who no longer pick up either Muslim or Christian women.

Several other northern States followed Zamfara in proclaiming Sharia law, despite the federal government's appeal to slow their initiatives. When Kaduna State imposed Sharia, riots between Muslims and Christians devastated the old capital city of Kaduna. There and elsewhere, the imposition of Sharia led to the departure of thousands of Christians, intensifying the cultural fault line that threatens the Nigerian federation. By the beginning of 2003, twelve northern States territorially comprising nearly half the country had proclaimed Sharia law (Fig. 6-17). In October 2003, radical Muslim clerics succeeded in persuading the governments of three of those northern States to terminate, with disastrous consequences, a UN World Health Organization polio vaccination campaign. They asserted that the vaccines, from Western countries, had been made to kill or sterilize the Muslims.

All this raises the prospect that Nigeria, West Africa's cornerstone and one of Africa's most important states, may succumb to devolutionary forces arising from its location in the African Transition Zone. This would be a calamitous development. A stable, well-governed, and economically growing Nigeria would be a beacon to region and realm; its collapse could infect Africa's political geography far and wide.

Coast and Interior

Nigeria is one of 17 states (counting Chad and offshore Cape Verde [not shown in Fig. 6-16]) that constitute the region of West Africa. Four of these countries, comprising a huge territory on the Sahara's margins but containing small populations, are landlocked: Mali, Burkina Faso, Niger, and Chad. Figure G-8 shows clearly how steppe and desert conditions dominate the natural environments of these four interior states. Figure G-9 reveals the concentration of population in the steppe zone and along the ribbon of water provided by the Niger River. Scattered oases form the remaining settlements and anchor regional trade.

But even the coastal states do not escape the dominance of the desert over West Africa. Mauritania's environment is almost entirely desert. Senegal, as Figure 6-16 shows, is mostly a Sahel country; and not only northern Nigeria but also northern Benin, Togo, and Ghana have interior steppe zones. The loss of pastures to **desertifica-** 18

tion (human-induced desert expansion) is a constant worry for the livestock herders there.

West Africa's states share the effects of the environmental zonation depicted in Figures G-7 and G-8, but they also have distinct regional geographies. **Benin**, Nigeria's neighbor, has a growing cultural and economic link with the Brazilian State of Bahía, where many of its people were taken in bondage and where elements of West African culture have survived. **Ghana**, once known as the Gold Coast, was the first West African state to achieve independence (1957), with a sound economy based on cocoa exports. Two grandiose postindependence schemes can be seen on its map: the port of Tema, which was to serve a vast West African hinterland, and Lake Volta, which resulted from the region's largest dam project. When neither fulfilled expectations, Ghana's economy collapsed. But in the 1990s, Ghana's military regime was replaced by democratic and stable government, and over the past decade Ghana has become a model for Africa. Even ethnic strife in the northern region in 2002 did not disrupt the country's progress, and its agriculture-dominated export economy, based on cocoa, received an unexpected boost when chaos descended on its chief West African competitor, neighboring Ivory Coast.

Ivory Coast (which, as we have stated, officially still goes by its French name, *Côte d'Ivoire*) translated three decades of autocratic but stable rule into economic progress that gave it lower-middle-income status based mainly on cocoa and coffee sales. Continued French involvement in the country's affairs contributed to this prosperity; the capital, Abidjan, reflected its comparative well-being. But in a familiar pattern, Ivory Coast's president-for-life first engineered the transfer of the capital to his home village, Yamoussoukro, and then spent tens of millions of dollars building a Roman Catholic basilica there to rival that of St. Peter's in Rome. It was dedicated just as the country's economy was slowing and its social conditions worsened.

By the late 1990s, Ivory Coast's economy had reverted to the low-income category (Fig. G-11). Worse, regional strife intensified during the run-up to a presidential election in 2000, when southern politicians objected to the ethnic origins and alleged Muslim sympathies of a northern candidate. In September 2002, a series of coordinated attacks by northern rebels failed to take the largest city and former capital, Abidjan, but succeeded in capturing major northern centers. The attacks cost hundreds of lives and confirmed the worst fears of citizens in this long-stable and comparatively prosperous country, where immigrants have long contributed importantly to the economy and society. At the turn of this century, the population included about 2 million Burkinabes (immigrants from Burkina Faso to the north), as many as 2 million Nigerians, about 1 million Malians, 500,000 Senegalese, and smaller groups of Ghanaians, Guineans, and Liberians. This cultural mix had been kept stable by strong central government and a growing economy, but when both these mainstays failed, disorder followed. In 2004 French troops, sent to Ivory Coast to stabilize the situation and to facilitate negotiations as well as protect the French expatriate community, suffered a deadly attack by the Ivoirian Air Force. This resulted in retaliatory action and led to assaults on French citizens in Abidjan and elsewhere (see photo below). Most of the expatriate community departed, leaving Ivory Coast, its cocoa-based economy devastated, in dire straits, a West African success story with a tragic turn.

All of West Africa has been affected by what has happened in Liberia and Sierra Leone. **Liberia**, a country founded in 1822 by freed slaves who returned to Africa

The Islamic Front cuts across Ivory Coast (see Fig. 6-18), creating a regional division long kept in check by an authoritarian government and economic prosperity. But the country's political stability failed as southerners, long accustomed to dominating the country, faced challenges from the north, and the economy suffered. In 2004, violence broke out and France, Ivory Coast's former ruler, sent an expeditionary force to impose order. It was unable to do so immediately, and in the continuing strife thousands of expatriates hastily departed from the country, leaving their possessions behind and further undermining the economy. On November 9, 2004, a large group of expatriates, most of them French, sought safety in Abidjan's Hôtel d'Ivoire, which was soon surrounded by protesting crowds. French soldiers used force to keep the hotel from being overrun, and at least seven people were killed. In April 2005, a tentative agreement between southern and northern politicians was reached, but mutual distrust remained and economic conditions continued to worsen.

with the help of American colonization societies, was ruled by their descendants for more than six generations. Rubber plantations and iron mines made life comfortable for the "Americo-Liberians," but among the local peoples, resentment simmered. A military coup in 1980, in which the president was killed, was followed in 1989 by full-scale civil war that pitted ethnic groups against each other and drove hundreds of thousands of refugees into neighboring countries, including Ivory Coast, Guinea, and Sierra Leone. Monrovia, the capital (named after U.S. president James Monroe) was devastated; an estimated 230,000 people, almost 10 percent of the population, perished. In 1997, one of the rebel leaders, Charles Taylor, became president, but he was indicted by a UN war-crimes tribunal in 2003, just as he was negotiating a truce between his regime and other rebel groups. Nigerian forces and UN peacekeeping troops entered Monrovia and in 2005 were continuing to try to bring stability back to this blighted country.

One of the countries affected by these events was **Sierra Leone**, Liberia's coastal neighbor, also founded as a haven for freed slaves, in this case by the British in 1787. Independent since 1961, Sierra Leone went the all-too-familiar route from self-governing Commonwealth member to republic to one-party state to military dictatorship. But in the 1990s a civil war brought untold horror to this small country even as refugees from Liberia's war were arriving. Rebels enabled by diamond sales fought supporters of the legitimate government in a struggle that devastated town and countryside alike and killed and mutilated tens of thousands. Eventually a combination of West African, United Nations, and British forces intervened in the conflict and resurrected a semblance of representative government, but not before Sierra Leone had sunk to dead last on the world's list of national well-being.

Senegal had the advantage of lying separated from Sierra Leone by Guinea, another Francophone country along the West African coast. Its Sahelian environmental problems notwithstanding, Senegal managed to convert its colonial advantage (its capital, Dakar, was the headquarters for France's West African empire) into lasting progress. By no means a rich country, Senegal depends for foreign income on farm exports (chiefly peanuts), fishing (the leading source), phosphate sales, and iron ore production. But its most valuable asset has been its democratic tradition, which has now lasted more than four decades. This has enabled Senegal to weather some storms, including a failed union with its English-speaking enclave, Gambia, and a secessionist movement in the southwestern Casamance District (Fig. 6-16).

Senegal is over 90 percent Muslim, but its population of 11.5 million is dominated by the Wolof, the largest ethnic group constituting about 35 percent of the total, the Fulani (17 percent), and the Serer (12 percent). The Wolof are concentrated in and around the capital and wield most of the power. Senegal's leaders have maintained close relationships with France, which is still Senegal's leading trading partner and financial supporter in times of need. Without oil, diamonds, or other lucrative income sources and with an overwhelmingly subsistence-farming population, Senegal nevertheless managed to achieve GNI levels that ranked among the region's highest (see Table G-1). Here is proof that reasonably representative government and stability are greater assets than gold, liquid or otherwise.

Populous, multicultural, divided West Africa remains a region of farmers and herders (the herders range over the savannas and steppes of the north) along a tier of fast-changing environments between ocean and desert. Local village markets drive the traditional economy, and **periodic markets**—not open every day but operating 19
every three, four, or more days—ensure that all villages participate in the exchange network. Traditions of this kind endure here, even as the cities beckon the farmers and burst at the seams. The region's great challenges are economic survival and nation-building, constrained by a boundary framework that is as burdensome as any in Africa.

THE AFRICAN TRANSITION ZONE

From our discussion of East, Equatorial, and West Africa, it is obvious that the northern margin of the realm we define as Subsaharan Africa is in turmoil. Along a zone from Ethiopia in the east to Guinea in the west, cultural and ethnic tensions tend to erupt into conflict, and often the conflict spreads to engulf large areas across international borders. It is also clear that much of this turmoil has to do with the religious transition from Islamic and Arabized Africa to the Africa where Christianity and traditional beliefs hold sway. This zone, called the *African Transition Zone*, epitomizes the way one geographic realm yields to another. But in terms of its instability, variability, and volatility, this particular transition between realms is unique.

It is useful to look again at Figure 6-10, which shows that some entire countries lie within the African Transition Zone, such as Senegal, Burkina Faso, Eritrea, Djibouti, and Somalia. Others are bisected by it, including Ivory Coast, Nigeria, Chad, Sudan, and Ethiopia. Africa's colonial boundary framework was laid out even as Islam was on the march and paid little heed to the consequences. But it is Islam that defines the southern periphery of the African Transition Zone along a religious frontier sometimes referred to as the **Islamic Front** (Fig. 6-18). 20

As we noted earlier in this chapter, Islam arrived across the Sahara by caravan and up the Nile by boat, thus reaching and converting the peoples of the interior steppes and savannas on the southern side of the great desert. (This wide stretch of semiarid Africa is known today as the *Sahel*, an Arabic word meaning "border" or "margin.") Then came Europe's colonial powers, and the modern politico-geographical map of Subsaharan Africa took shape. Muslims and non-Muslims were thrown together in countries not of their making. Vigorous Christian proselytism slowed the march of Islam, resulting in the distribution shown in Figure 7-2. The north-to-south transition is especially clear in the heart of the continent: Libya is 97 percent Muslim, Chad 54, the Central African Republic 16, and The Congo 1.4.

Conflict is the hallmark of much of the Islamic Front across Africa. In Sudan, a 30-year war between the Arab, Islamic north and the African, Christian-animist south, costing hundreds of thousands of lives and displacing millions, was finally settled in late 2004. Even so, a new and deathly conflict had arisen in Sudan's far-western Darfur Province even as the southern war ended. In Ivory Coast, livestock owned by Muslim cattle herders from Burkina Faso had been trampling African farms for years before the north-south schism broke into open conflict in 2002. In Nigeria, Islamic revivalism in the North is clouding the entire country's prospects. But perhaps the most conflict-prone part of the African Transition Zone lies in Africa's Horn, involving the historic Christian state of Ethiopia and its neighbors (Fig. 6-19).

Earlier we noted that highland **Ethiopia**, where the country's core area, capital (Adis Abeba), and Christian heartland lie, is virtually encircled by dominantly Muslim societies. One poor-quality road, along which bandits rob bus passengers and law enforcement is virtually nonexistent, forms the only direct surface link between Adis Abeba and Nairobi, the capital of Kenya to the south (Fig. 6-19). But, as Figure 6-19 reminds us, Ethiopia's eastern Ogaden is traditional Somali country and almost entirely Islamic, a vestige of a time when the rulers in Adis Abeba extended their power from their highland fortress over the plains below. Today, 33 percent of Ethiopia's population of 75.8 million is Muslim, and the Islamic Front is quite sharply defined (Fig. 6-18).

In 1998, Ethiopia and neighboring **Eritrea** fought a senseless border war over a few villages and a small strip of mainly desert territory, an aftermath of Eritrea's secession from Ethiopia. Once part of what may be called the Ethiopian Empire, Eritrea was under British administration during World War II and was made part of an

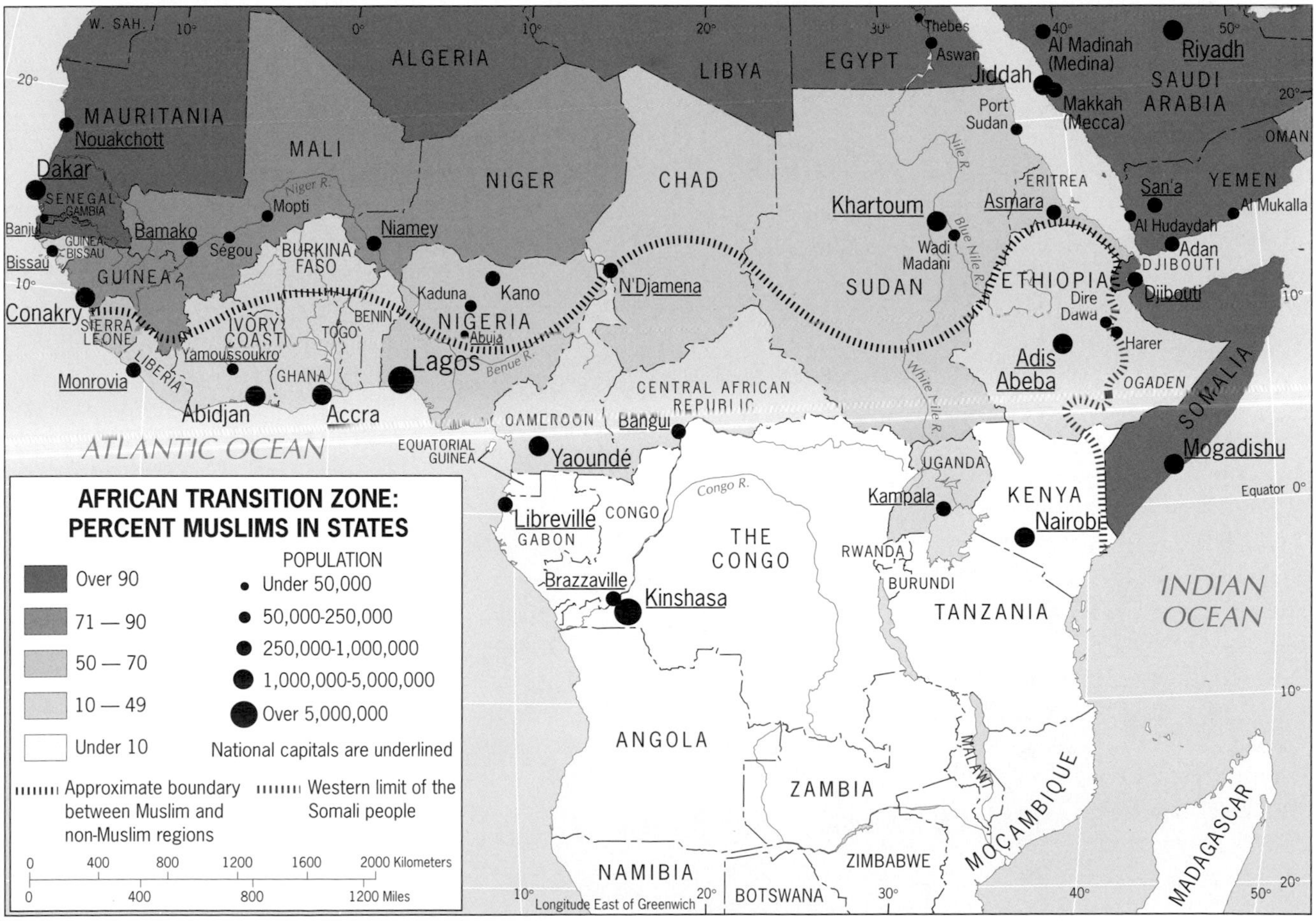

FIGURE 6-18

Ethiopian Federation in 1952. After Ethiopia dissolved this federal arrangement in 1962 and absorbed Eritrea, Muslim as well as Christian Eritreans fought a war of secession. In 1993 their efforts were rewarded by international recognition of Eritrea's independence. This had the effect of landlocking Ethiopia, which had made Assab its key Red Sea port (Fig. 6-19). With the Muslim component of Eritrea's population of 4.6 million growing faster than the Christian sector, relationships between Ethiopia and Eritrea were tense, and the 1998 border war in fact was more than that. After spending U.S. $3 billion on the conflict and accepting foreign mediation in 2000, Ethiopia and Eritrea needed international food assistance to ward off starvation as a combined 16 million people suffered from hunger.

In 2002, a ministate in this part of the African Transition Zone, **Djibouti**, came to international attention as a result of the so-called War on Terror declared by the United States. Formerly a French colony and still host

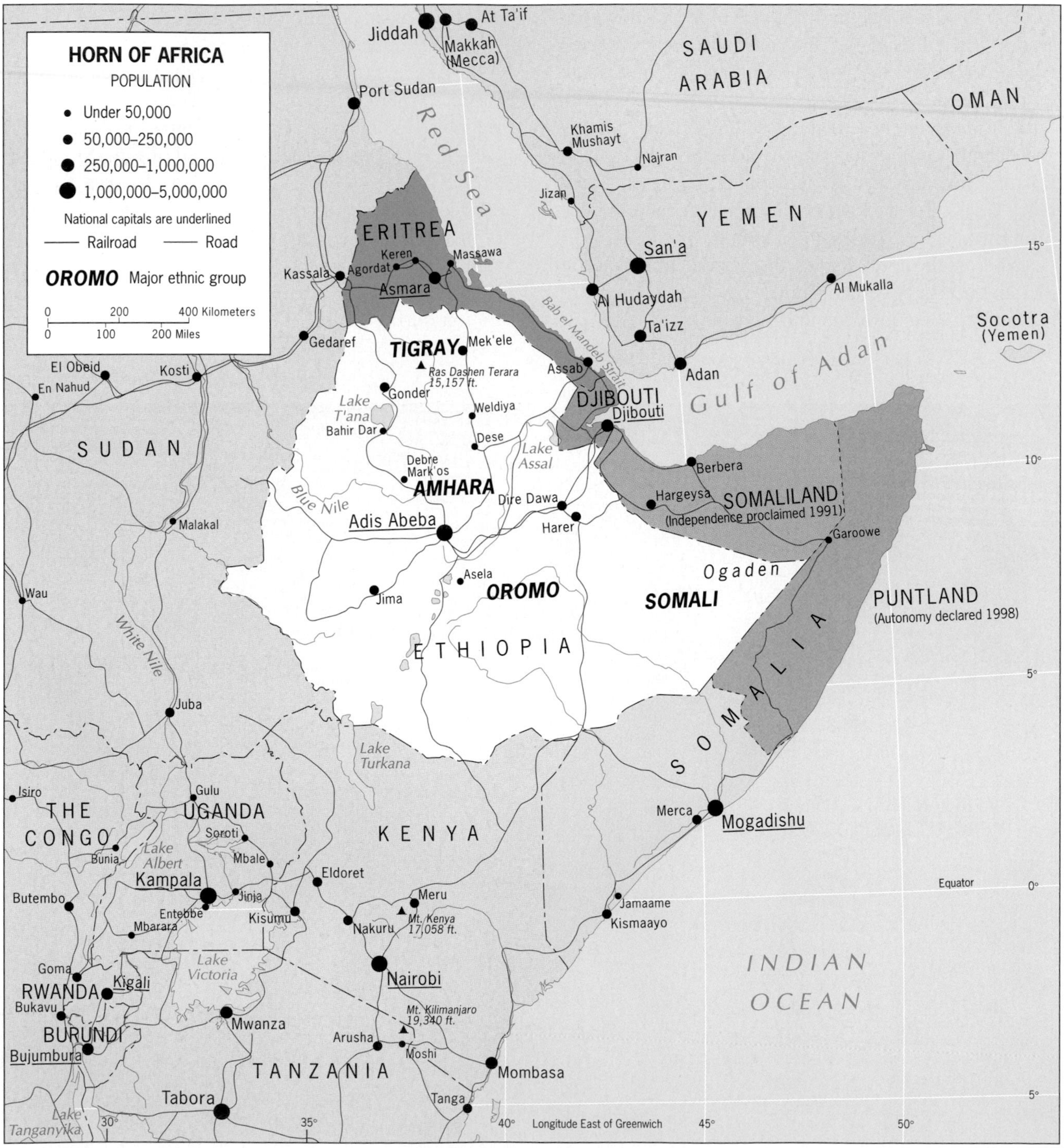

FIGURE 6-19 © H. J. de Blij, P. O. Muller, and John Wiley & Sons, Inc.

to a French military base, coastal Djibouti, ethnically divided between Afars and Issas but politically stable, proved to have a favorable relative location between troubled Somalia, vulnerable Eritrea, and crucial Ethiopia—and directly across the narrow Bab el Mandeb Strait (which separates the Red Sea and the Gulf of Adan) from Yemen, the ancestral home of Usama bin Laden.

But the country whose name is etched most vividly in American and Canadian minds is **Somalia**, where nearly 9 million people, virtually all Muslim, live pastoral and farming lives at the mercy of a desert-dominated climate. As Figures 6-18 and 6-9 show, Somali ethnic clans also inhabit most of adjoining eastern Ethiopia, so that the Somali "nation" exceeds 10 million. But "nation" is not a term that applies to Somalia, where five major ethnic groups and hundreds of clans are engaged in an unending contest for power as well as survival. In the early 1990s, media coverage of the latest of Somalia's death-dealing droughts led to U.S.-led United Nations intervention with the aim of food provision as well as nation-building, but the donors were immediately caught up in Somalia's relentless strife and a number of U.S. and other forces were killed, their bodies dragged through the streets of the capital, Mogadishu. Clan warfare destroyed what little infrastructure Somalia possessed, and the state essentially ceased to exist.

The fragmentation of Somalia, a merger of former British and Italian dependencies along the Horn's coast soon controlled by a harsh military dictatorship, actually began long before this debacle. In 1991, the former British sector declared its unilateral independence as the Republic of Somaliland (Fig. 6-19). In sharp contrast to the rest of the war-torn country, Somaliland has thrived; its capital, Hargeysa, bustles with commerce; and its political system, based on participation by elders of all clans, has a modicum of democracy. But the international community has not recognized the state. In 1998 another part of Somalia, named Puntland, also proclaimed its autonomy, although it left open the possibility of future participation in a federal Somalia. Meanwhile, the warring factions continue their struggles in the more populous south, vying for control of Mogadishu and the southern port of Kismaayo. In the process, interior southern Somalia has become a hotbed of Islamic extremism, a feature of the African Transition Zone that will be felt from the Bulge to the Horn.

WHAT YOU CAN DO

RECOMMENDATION: Write an op-ed piece! This is not merely a matter of writing a longer letter to the editor. An op-ed (opinion-editorial) essay is more challenging because you cannot assume the reader's familiarity with the topic. Therefore, you will want to write an introductory paragraph setting the stage and explaining what issue you plan to address, leading into a concisely argued position on the matter citing the necessary evidence, followed by a logical and judicious conclusion. Give yourself 1000 to 1200 words, and make every word count! It's great practice, and if the newspaper or magazine editor accepts your article, it will have much more impact than a letter to the editor. And, unlike a letter, you are able to title your op-ed piece. For example, you might try one under the headline Give African Farmers a Break! *or another asking* What has Globalization Done for Africa? *And here's a switch: readers are likely to write letters to the editor responding to* your *argument. They may even ask the editor to invite you to write again on another topic.*

GEOGRAPHIC CONNECTIONS

1 Tourism is one of Africa's important industries, in some countries contributing substantially to the government's income and creating jobs where unemployment is high. But if you were a member of an organized tour group, you would soon discover that visitors to Subsaharan Africa tend to be more interested in the realm's remaining wildlife than in the people and their cultures. How would you go about adding some human interest to the itinerary? What aspects of the cultural geography of Subsaharan Africa would you like to see and why? In what ways are villages and small towns usually more evocative of regional cultures than large cities?

2 You have just been appointed to a key U.S.-government commission charged with the following responsibility: sufficient funds are available to give substantial assistance to just four countries in Subsaharan Africa, and the government hopes that the beneficial effects of this assistance will radiate out into neighboring states. The first task of the commission of which you are a member is to identify the four countries. If you had this decision to make, which four African states would you nominate, and what criteria, including geographic factors, would you employ to arrive at that selection?

Chapter 7

North Africa/Southwest Asia

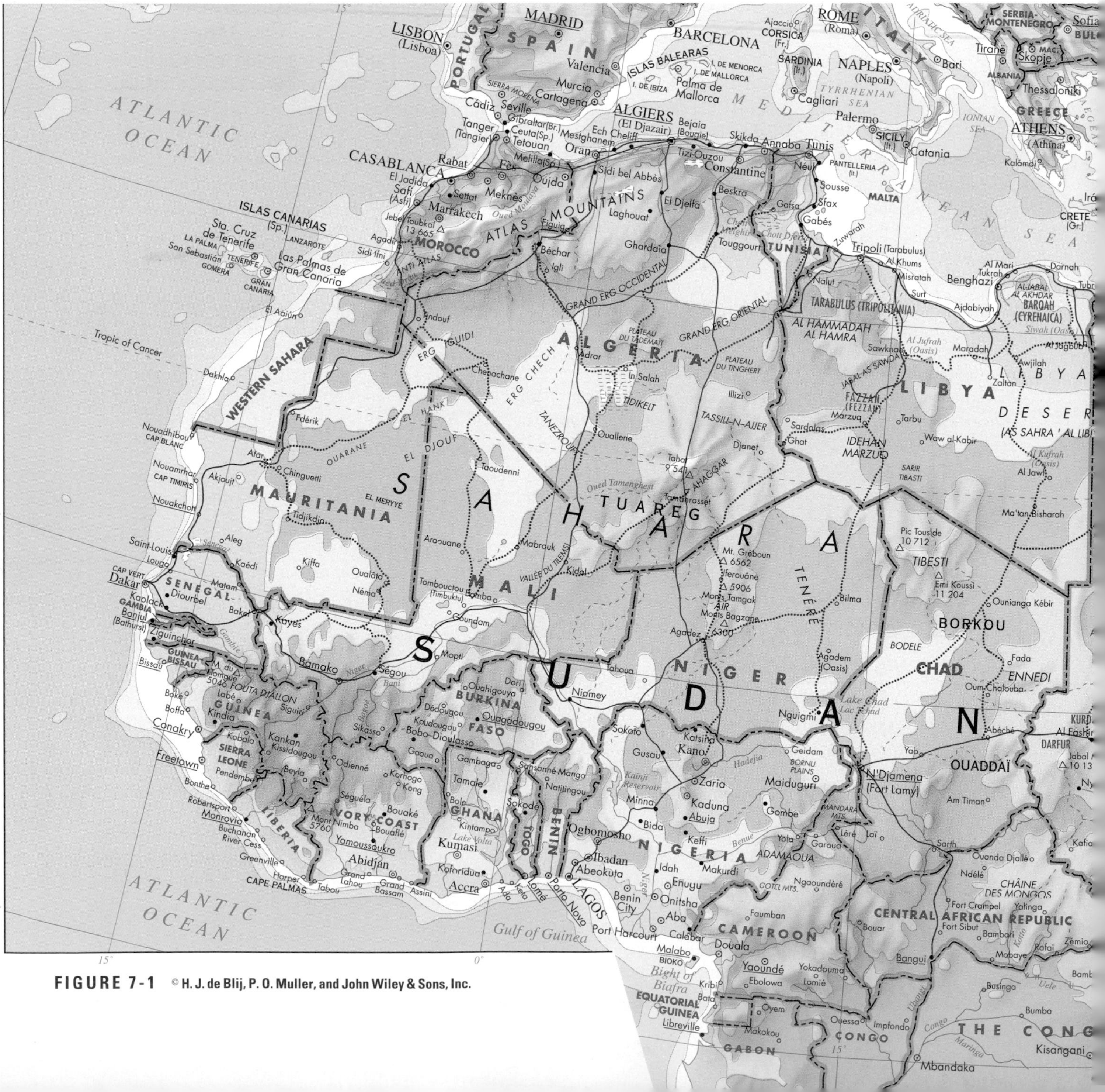

FIGURE 7-1 © H. J. de Blij, P. O. Muller, and John Wiley & Sons, Inc.

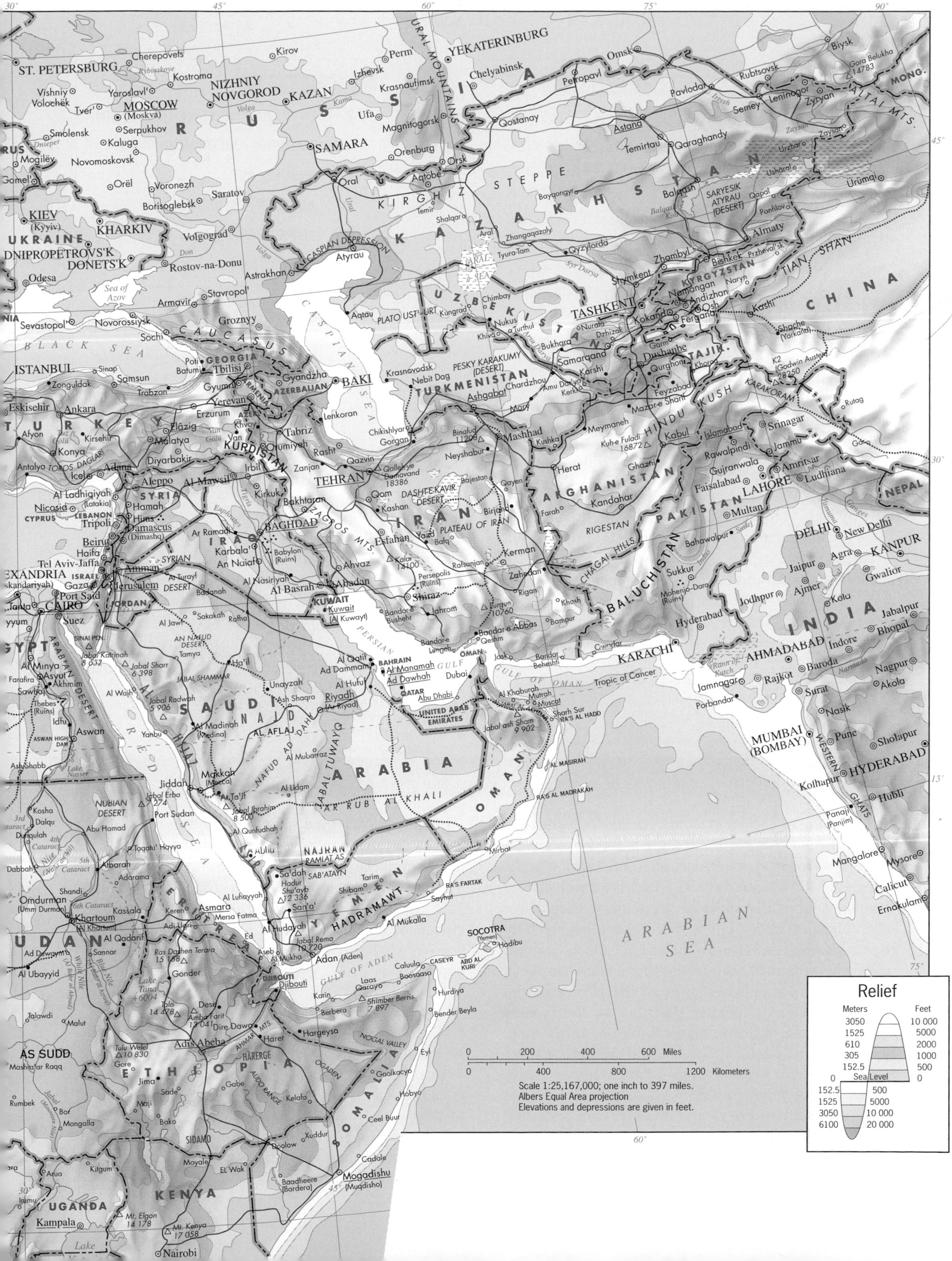

30°
45°
60°
75°
90°
ST. PETERSBURG
Cherepovets
Kirov
Perm'
YEKATERINBURG
URAL MOUNTAINS
Omsk
Biysk
Gora Belukha 14783
MONG.
Vishniy Volochëk
Yaroslavl'
Kostroma
NIZHNIY NOVGOROD
KAZAN
Izhevsk
Krasnoufimsk
Chelyabinsk
Petropavl
Pavlodar
Rubtsovsk
Leninogor
Zyryan
ALTAI MTS.
Tver'
MOSCOW
(Moskva)
Ufa
Semey
Smolensk
Serpukhov
R U S S I A
Magnitogorsk
Qostanay
Astana
Kaluga
SAMARA
Orenburg
Temirtau
Qaraghandy
RUS
Mogilëv
Novomoskovsk
Orsk
Zaysan
Urzhar
Gomel'
Orël
Voronezh
Oral
Aqtöbe
STEPPE
Usharal
Ürümqi
Saratov
KIRGHIZ
Bayqongyr
Balqash
SARYESIK ATYRAU (DESERT)
Qapal
Panfilov
KIEV
(Kyyiv)
Borisoglebsk
Temir
Shalqar
K A Z A K H S T A N
UKRAINE
KHARKIV
Volgograd
Aral
Zhangaqazaly
Almaty
DNIPROPETROVS'K
DONETS'K
CASPIAN DEPRESSION
Atyrau
Tyura-Tam
Qyzylorda
Zhambyl
Bishkek
Przheval'sk
TIAN SHAN
Odesa
MOSCOW
Rostov-na-Donu
Astrakhan
ARAL SEA
Syr Darya
KYRGYZSTAN
Naryn
Sea of Azov
Stavropol'
Shymkent
Namangan
Andizhan
CHINA
Armavir
U Z B E K I S T A N
Chimbay
TASHKENT
Kokand
Osh
Kashi
NIA
Sevastopol'
Novorossiysk
Groznyy
Aqtau
PLATO UST'-URT
Kungrad
Nukus
Turthul
Nurata
Dzhizak
Fergana
Shache (Yarkand)
BLACK SEA
Sochi
CAUCASUS
CASPIAN SEA
Khiva
Bukhara
Garm
Dushanbe
TAJIK.
Poti
GEORGIA
Tbilisi
Samarqand
Qurghonteppa
Khorog
K2 (Godwin Austen) 28250
ISTANBUL
Sinop
Batumi
Krasnovodsk
PESKY KARAKUMY (DESERT)
Karshi
Zonguldak
Samsun
ARMENIA
Gyandzha
BAKI
Nebit Dag
TURKMENISTAN
Chardzhou
Amu Dar'ya
Kerki
Feyzabad
KARAKORAM
Trabzon
Gyumri
AZERBAIJAN
Ashgabat
Mary
Mazar-e Sharif
HINDU KUSH
Eskisehir
Ankara
Yerevan
Lenkoran
Rutag
T U R K E Y
Erzurum
Elâzig
Khvoy
Tabriz
Chikishlyar
Gorgan
Binalud 11208
Mashhad
Meymaneh
Srinagar
Afyon
Kirsehir
Malatya
KURDISTAN
Orumiyeh
Kuh-e Fuladi 16872
Kabul
Islamabad
Jammu
Konya
Diyarbakir
Rasht
Neyshabur
Kushka
Rawalpindi
Antalya
TOROS DAGLARI
Adana
Icel
Qazvin
Zanjan
Qolleh-ye Damavand 18386
Ghazni
Gujranwala
Amritsar
Aleppo
Al Mawsil
Irbil
TEHRAN
Bajestan
Qayen
Herat
Faisalabad
LAHORE
Ludhiana
Al Ladhiqiyah (Latakia)
SYRIA
Kirkuk
Qom
DASHT-E KAVIR DESERT
A F G H A N I S T A N
NEPAL
Nicosia
CYPRUS
LEBANON
Hamah
Bakhtaran
Kashan
Birjand
Kandahar
PAKISTAN
Tripoli
Hims
Damascus
(Dimashq)
BAGHDAD
ZAGROS MTS.
I R A N
Farah
Multan
Beirut
Haifa
Ar Ramadi
IRAQ
Esfahan
Yazd
Bafq
PLATEAU OF IRAN
RIGESTAN
Bahawalpur
DELHI
New Delhi
Tel Aviv-Jaffa
Karbala
An Najaf
Babylon (Ruins)
Kerman
CHAGAI HILLS
Agra
KANPUR
EXANDRIA
ISRAEL
Amman
SYRIAN DESERT
Al Turayf
Ahvaz
Kalai 14100
Rafsanjan
BALUCHISTAN
Jaipur
Gwalior
Port Said
Gaza
Jerusalem
Badanah
Al Nasiriyah
Abadan
Persepolis (Ruins)
Zahedan
Sukkur
Ajmer
Tanta
CAIRO
JORDAN
Al Basrah
KUWAIT
Kuwait
(Al Kuwayt)
Shiraz
Rigan
Mohenjo-Daro (Ruins)
Jodhpur
Kota
Suez
Sakakah
Rafha
Bandar-e Bushehr
Jahrom
Kuh-e Furgun 10760
Khash
Hyderabad
I N D I A
Jabalpur
Al Jawf
Bandar-e Abbas
Bampur
Bhopal
SINAI PEN.
AN NAFUD DESERT
Tamya
PERSIAN GULF
Bandar-e Lengeh
Qeshm
Gwadar
KARACHI
AHMADABAD
Indore
Jabal Katrinah 8652
Jabal Sharr 6398
Ha'il
Al Qatif
BAHRAIN
Ad Dammam
Al Manamah
Ad Dawhah
OMAN
Jask
Bandar Beheshti
Rann of Kutch
Baroda
Nagpur
Al Minya
Asyut
JABAL SHAMMAR
Al Hufuf
QATAR
Dubai
GULF OF OMAN
Tropic of Cancer
Jamnagar
Rajkot
Akola
Farafra
Akhmim
Unayzah
Al Khaburah
Mutrah
Muscat
Sawhaj
Al Wajh
Jabal Radwah 5906
Ash Shaqra
Riyadh
(Ar Riyad)
Abu Dhabi
Porbandar
Surat
Thebes (Ruins)
S A U D I
NAJD
UNITED ARAB EMIRATES
Sharh Sur
RA'S AL HADD
Nasik
Idfu
Aswan
Al Madinah (Medina)
AL AFLAJ
AD DAHL
Jabal ash Sham 9902
ASWAN HIGH DAM
RED SEA
Yanbu
HIJAZ
NAFUD
Al Mubarraz
JABAL TUWAYQ
MUMBAI (BOMBAY)
Pune
Sholapur
Ash Shabb
Lake Nasser
Makkah (Mecca)
A R A B I A
OMAN
AL MASIRAH
HYDERABAD
Kolhapur
Jiddah
Jabal Erba 7274
At Ta'if
Al Lidam
RA'S AL MADRAKAH
Hubli
NUBIAN DESERT
Jabal Ibrahim 8500
AR RUB' AL KHALI
Kosha
Dalqu
Abu Hamad
Port Sudan
Panaji (Panjim)
WESTERN GHATS
Dunqulah
4th Cataract
Al Qunfudhah
Togatu' Hayya
Abha
NAJRAN
RAMLAT AS SAB'ATAYN
Mirbat
Mangalore
Mysore
Dabbah
5th Cataract
Atbarah
Sa'dah
Tarim
Adarama
Hadur Shu'ayb 12336
Shibam
Y E M E N
RA'S FARTAK
Calicut
Omdurman
(Umm Durman)
Shandi
6th Cataract
Kassala
Al Luhayyah
Sayhut
Ernakulam
Khartoum
(Al Khartum)
Keren
Asmara
Mersa Fatma
San'a'
HADRAMAWT
Al Mukalla
ARABIAN SEA
UDAN
Al Qadarif
Adi Ugri
ERITREA
Al Hudaydah
Jabal Rema 10720
SOCOTRA
(Yemen)
Hadibu
Ad Duwaym
Sannar
Ras Dashen Terara 15158
Ed
Aseb
Al Mukha
Adan (Aden)
GULF OF ADEN
Caluula
CASEYR
ABD AL-KURI
Al Ubayyid
White Nile
Blue Nile
Gonder
DJIBOUTI
Djibouti
Laas Qoray
Boosaaso
Lake Tana 6004
Karin
Hurdiya
Talawdi
Malut
Tala 14478
Dese
Amba Farit 13041
Berbera
Shimber Berris 7897
Bender Beyla
Dire Dawa
Hargeysa
AS SUDD
Tulu Welel 10830
Adis Abeba
Harer
NOGAL VALLEY
Mashra'ar Raqq
Gore
E T H I O P I A
HARERGE
OGADEN
Eyl
Jima
Gaalkacyo
Sade
Gabe
Kelafo
Hobyo
Rumbek
Bor
Maji
Mongalla
Bako
SOMALIA
Ceel Buur
SIDAMO
Xuddur
Doolow
Moyale
El Wak
Cadale
Arua
Kitgum
Baadheere (Bardera)
Mogadishu
(Muqdisho)
UGANDA
KENYA
Mt. Elgon 14178
Kampala
Mt. Kenya 17058
Lake
Nairobi
0 200 400 600 Miles
0 400 800 1200 Kilometers
Scale 1:25,167,000; one inch to 397 miles.
Albers Equal Area projection
Elevations and depressions are given in feet.
Relief
Meters
Feet
3050 10 000
1525 5000
610 2000
305 1000
152.5 500
0 Sea Level 0
152.5 500
1525 5000
3050 10 000
6100 20 000

CONCEPTS, IDEAS, AND TERMS

1 Cultural geography
2 Culture hearth
3 Cultural diffusion
4 Cultural environment
5 Cultural ecology
6 Hydraulic civilization theory
7 Climate change
8 Spatial diffusion
9 Expansion diffusion
10 Relocation diffusion
11 Contagious diffusion
12 Hierarchical diffusion
13 Islamization
14 Culture region
15 Religious revivalism
16 Wahhabism
17 Cultural revival
18 Stateless nation
19 Choke point
20 Nomadism
21 Buffer state

REGIONS

- EGYPT AND THE LOWER NILE BASIN
- MAGHREB AND ITS NEIGHBORS
- MIDDLE EAST
- ARABIAN PENINSULA
- EMPIRE STATES
- TURKESTAN

FROM MOROCCO ON the shores of the Atlantic to the mountains of Afghanistan, and from the Horn of Africa to the steppes of inner Asia, lies a vast geographic realm of enormous cultural complexity. It stands at the crossroads where Europe, Asia, and Africa meet, and it is part of all three (Fig. 7-1). Throughout history, its influences have radiated to these continents and to practically every other part of the world as well. This is one of humankind's primary source areas. On the Mesopotamian Plain between the Tigris and Euphrates rivers (in modern-day Iraq) and on the banks of the Egyptian Nile arose several of the world's earliest civilizations. In its soils, plants were domesticated that are now grown from the Americas to Australia. Along its paths walked prophets whose religious teachings are still followed by hundreds of millions. And at the opening of the twenty-first century, the heart of this realm is beset by some of the most bitter and dangerous conflicts on Earth.

DEFINING THE REALM

It is tempting to characterize this geographic realm in a few words and to stress one or more of its dominant features. It is, for instance, often called the "Dry World," containing as it does the vast Sahara as well as the Arabian Desert. But most of the realm's people live where there is water—in the Nile Delta, along the hilly Mediterranean coastal strip (the *tell*, meaning "mound" in Arabic) of northwesternmost Africa, along the Asian eastern and northeastern shores of the Mediterranean Sea, in the Tigris-Euphrates Basin, in far-flung desert oases, and along the lower mountain slopes of Iran south of the Caspian Sea and of Turkestan to the northeast. We know this world region as one where water is almost always at a premium, where peasants often struggle to make soil and moisture yield a small harvest, where nomadic peoples and their animals circulate across dust-blown flatlands, where oases are islands of sedentary farming and trade in a sea of aridity. But it also is the land of the Nile, the lifeline of Egypt, the crop-covered *tell* of northern Algeria, the verdant shores of western Turkey, the meltwater-fed valleys of Central Asia. Compare Figure 7-1 to Figure G-8, and the dominance of *B* climates becomes evident. Also consult Figure G-9, and you will see how water-dependent this realm's clustered population is.

An "Arab World"?

North Africa/Southwest Asia is also often referred to as the Arab World. This term implies a uniformity that does not actually exist. First, the name *Arab* is applied loosely to the peoples of this area who speak Arabic and related languages, but ethnologists normally restrict it to certain occupants of the Arabian Peninsula—

MAJOR GEOGRAPHIC QUALITIES OF

North Africa/ Southwest Asia

1. North Africa and Southwest Asia were the scene of several of the world's great ancient civilizations, based in its river valleys and basins.
2. From this realm's culture hearths diffused ideas, innovations, and technologies that changed the world.
3. The North Africa/Southwest Asia realm is the source of three world religions: Judaism, Christianity, and Islam.
4. Islam, the last of the major religions to arise in this realm, transformed, unified, and energized a vast domain extending from Europe to Southeast Asia and from Russia to East Africa.
5. Drought and unreliable precipitation dominate natural environments in this realm. Population clusters exist where water supply is adequate to marginal.
6. Certain countries of this realm have enormous reserves of oil and natural gas, creating great wealth for some but doing little to raise the living standards of the majority.
7. The boundaries of the North Africa/Southwest Asia realm consist of volatile transition zones in several places in Africa and Asia.
8. Conflict over water sources and supplies is a constant threat in this realm, where population growth rates are high by world standards.
9. The Middle East, as a region, lies at the heart of this realm; and Israel lies at the center of the Middle East conflict.
10. Religious, ethnic, and cultural discord frequently cause instability and strife in this realm.

the Arab "source." In any case, the Turks are not Arabs, and neither are most Iranians or Israelis. Moreover, although the Arabic language prevails from Mauritania in the west across all of North Africa to the Arabian Peninsula, Syria, and Iraq in the east, it is not spoken in other parts of this realm. In Turkey, for example, Turkish is the major language, and it has Ural-Altaic rather than Arabic's Semitic or Hamitic roots. The Iranian language belongs to the Indo-European linguistic family. Other "Arab World" languages that have separate ethnological identities are spoken by the Jews of Israel, the Tuareg people of the Sahara, the Berbers of northwestern Africa, and the peoples of the transition zone between North Africa and Subsaharan Africa to the south.

An "Islamic World"?

Yet another name given to this realm is the World of Islam. The prophet Muhammad (Mohammed) was born in Arabia in A.D. 571, and in the centuries after his death in 632, Islam spread into Africa, Asia, and Europe. This was the age of Arab conquest and expansion. Their armies penetrated Southern Europe, their caravans crossed the deserts, and their ships plied the coasts of Asia and Africa. Along these routes they carried the Muslim (Islamic) faith, converting the ruling classes of the states of the West African savanna, threatening the Christian stronghold in the highlands of Ethiopia, penetrating the deserts of inner Asia, and pushing into India and even the island extremities of Southeast Asia. Today, the Islamic faith extends far beyond the limits of the realm under discussion (Fig. 7-2). Nor is the World of Islam entirely Muslim. Judaism, Christianity (notably in Egypt and Lebanon), and other faiths survive in the heartland of the Islamic World. So this connotation is not satisfactory either.

"Middle East"?

Finally, this realm is frequently called the Middle East. That must sound odd to someone in, say, India, who might think of a Middle West rather than a Middle East! The name, of course, reflects the biases of its source: the "Western" world, which saw a "Near" East in Turkey, a "Middle" East in Egypt, Arabia, and Iraq, and a "Far" East in China and Japan. Still, the term has taken hold, and it can be seen and heard in everyday usage by scholars, journalists, and members of the United Nations. Even so, it should be applied to only one of the regions of this vast realm, not to the realm as a whole.

HEARTHS OF CULTURE

This geographic realm occupies a pivotal part of the world: here Eurasia, crucible of human cultures, meets Africa, source of humanity itself. A million years ago, the ancestors of our species walked from East Africa into North Africa and Arabia and spread from one end of Asia to the other. More than one hundred thousand years ago, *Homo sapiens* crossed these lands on the way to Europe, Australia, and, eventually, the Americas. Ten thousand years ago, human communities in what we now call the Middle East began to domesticate plants and animals, learned to irrigate their fields, enlarged their settlements into towns, and formed the earliest states. One thousand years ago, the heart of the realm was stirred and mobilized by the teachings of Muhammad and the Quran (Koran), and Islam was on the march from North Africa

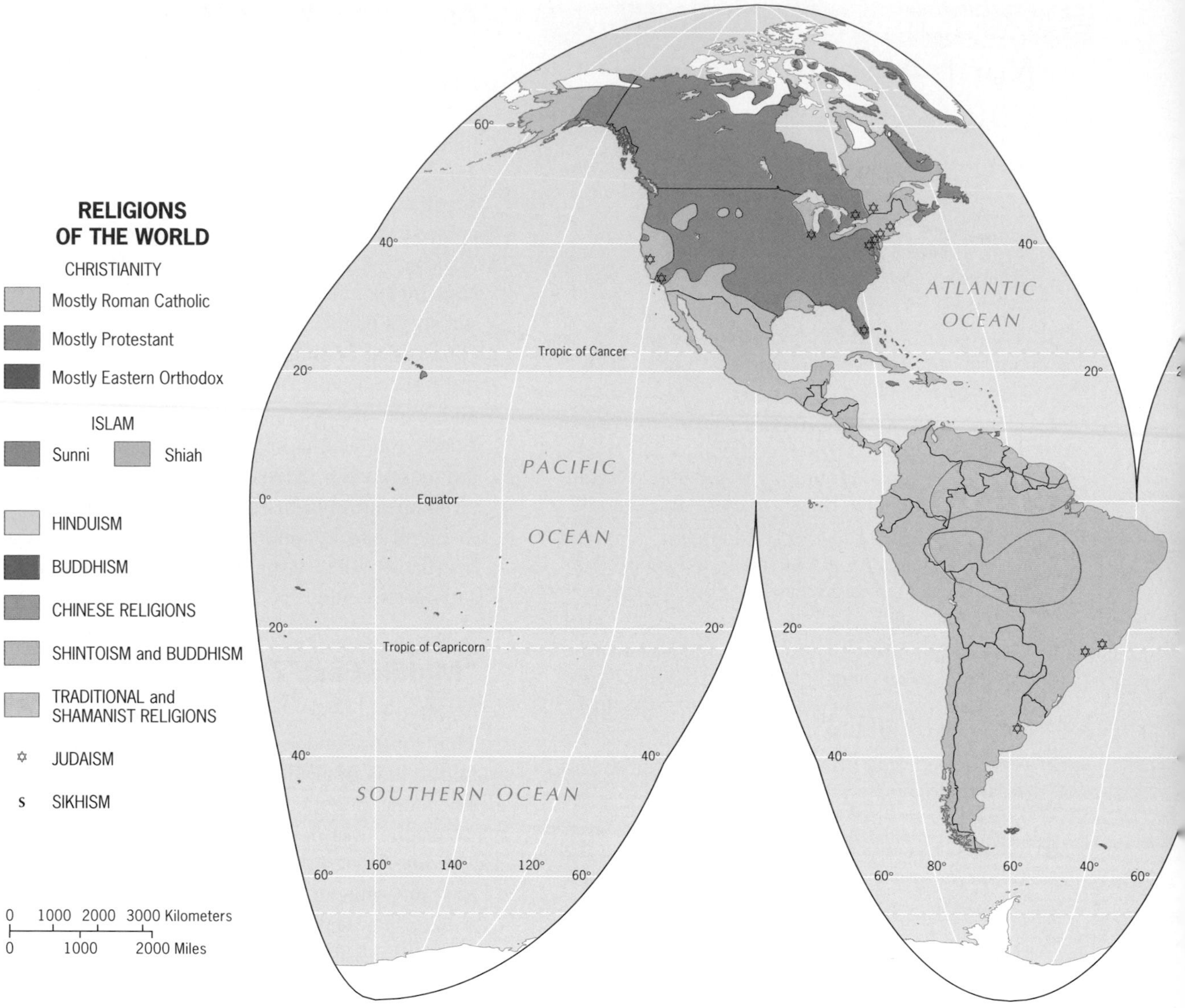

FIGURE 7-2 © H. J. de Blij, P. O. Muller, and John Wiley & Sons, Inc.

to India. Today this realm is a cauldron of religious and political activity, weakened by conflict but empowered by oil, plagued by poverty but fired by a wave of religious fundamentalism (Muslims prefer the term religious *revivalism*).

In the Introduction (p. 16) we discussed the concept of culture and its regional expression in the cultural land-
1 scape. **Cultural geography**, we noted, is a wide-ranging and comprehensive field that studies spatial aspects of human cultures, focusing not only on cultural landscapes
2 but also on **culture hearths**—the crucibles of civilization, the sources of ideas, innovations, and ideologies that changed regions and realms. Those ideas and innovations spread far and wide through a set of processes that we
3 study under the rubric of **cultural diffusion**. Because we understand these processes better today, we can reconstruct ancient routes by which the knowledge and achievements of culture hearths spread (that is, diffused) to other areas. Another topic of cultural geography, also relevant in the context of the North Africa/Southwest Asia realm, is the **cultural environment** that a dominant culture creates. 4
Human cultures exist in long-term accommodation with (and adaptation to) their natural environments, exploiting opportunities that these environments present and coping with the extremes they can impose. The study of the relationship between human societies and natural environments has become a separate branch of cultural geography called **cultural ecology**. As we will see, the North Africa/ 5
Southwest Asia realm presents many opportunities to investigate cultural geography in regional settings.

Mesopotamia and the Nile

In the basins of the major rivers of this realm (the Tigris and Euphrates of modern-day Turkey, Syria, and Iraq, and the Nile of Egypt) lay two of the world's earliest

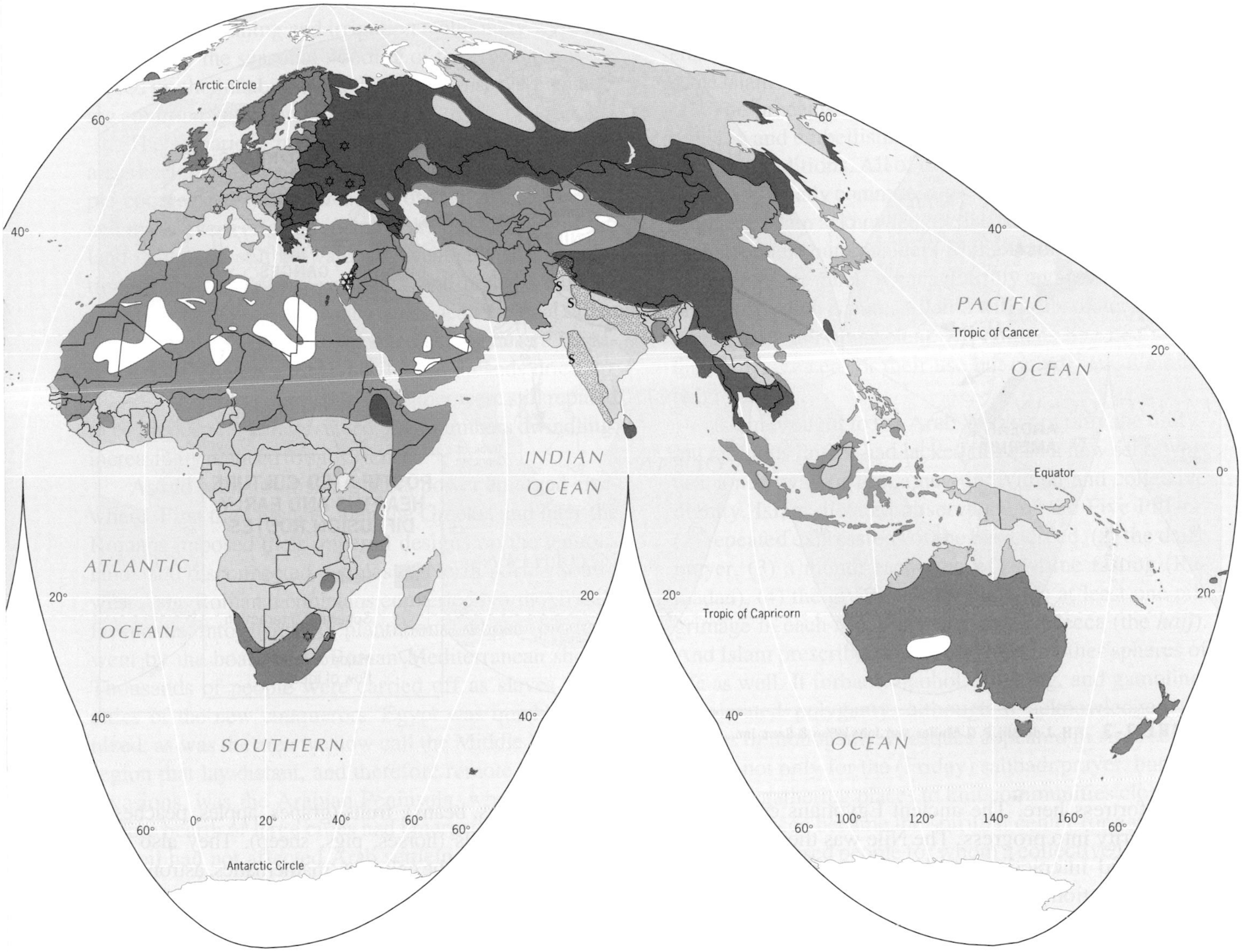

culture hearths (Fig. 7-3). Mesopotamia, "land amidst the rivers," had fertile alluvial soils, abundant sunshine, ample water, and animals and plants that could be domesticated. Here, in the Tigris-Euphrates lowland between the Persian Gulf and the uplands of present-day Turkey, arose one of humanity's first culture hearths, a cluster of communities that grew into larger societies and, eventually, into the world's first states. (Early state development probably was going on simultaneously in East Asia's river basins as well.) Mesopotamians were innovative farmers who knew when to sow and harvest crops, water their fields, and store their surplus. Their knowledge diffused to villages near and far, and a *Fertile Crescent* evolved extending from Mesopotamia across southern Turkey into Syria and the Mediterranean coast beyond (Fig. 7-3).

Irrigation was the key to prosperity and power in Mesopotamia, and urbanization was its reward. Among many settlements in the Fertile Crescent, some thrived, grew, enlarged their hinterlands, and diversified socially and occupationally; others failed. What determined success? One theory, the **hydraulic civilization theory**, 6
holds that cities that could control irrigated farming over large hinterlands held power over others, used food as a weapon, and thrived. One such city, Babylon on the Euphrates River, endured for nearly 4000 years (from 4100 B.C.). A busy port, its walled and fortified center endowed with temples, towers, and palaces, Babylon for a time was the world's largest city.

Egypt's cultural evolution may have started even earlier than Mesopotamia's, and its focus lay upstream from (south of) the Nile Delta and downstream from (north of) the first of the Nile's series of rapids, or cataracts (Fig. 7-3). This part of the Nile Valley is surrounded by inhospitable desert, and unlike Mesopotamia (which lay open to all comers), the Nile provided a

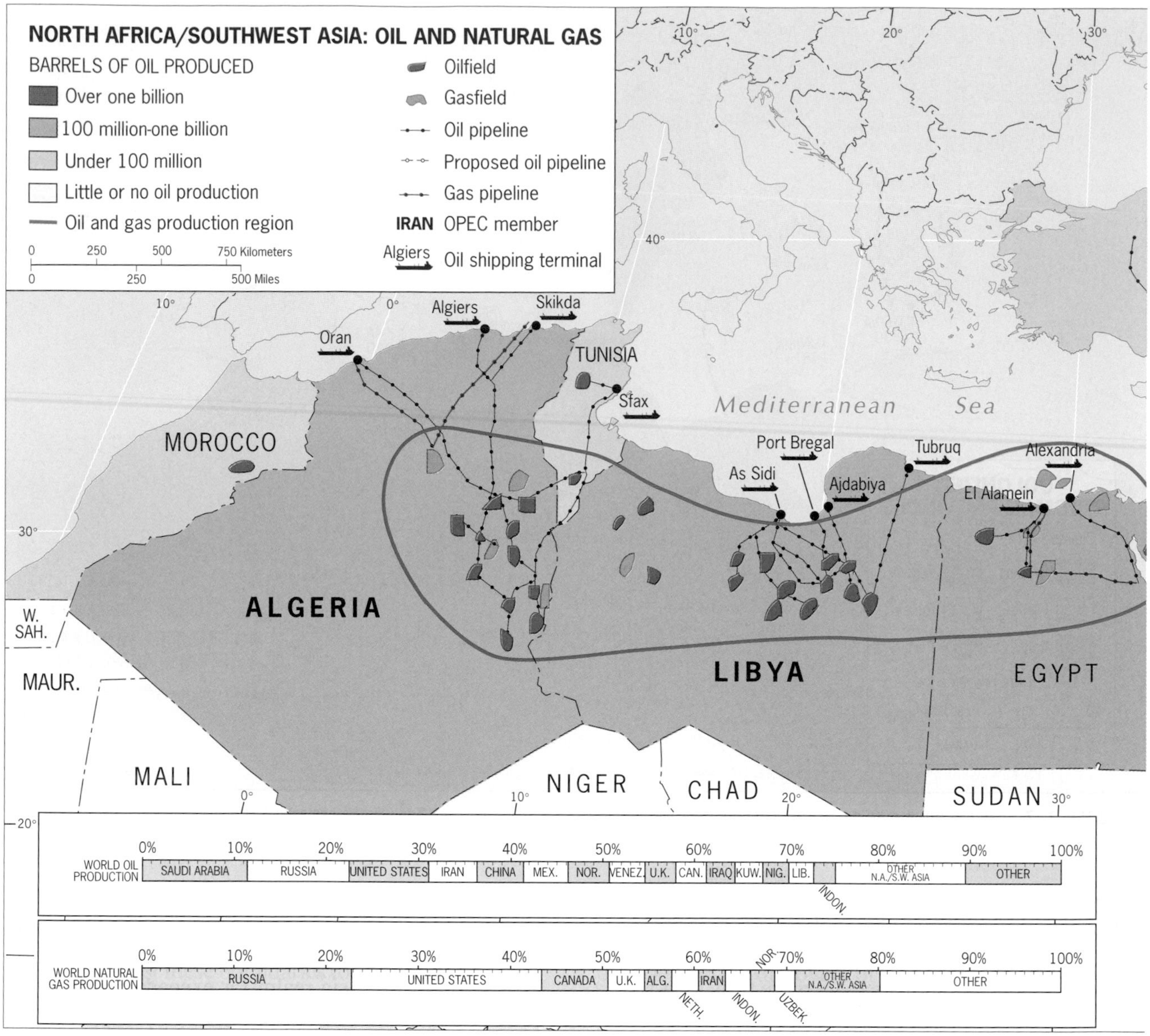

FIGURE 7-8 © H. J. de Blij, P. O. Muller, and John Wiley & Sons, Inc.

Turkish Ottoman Empire collapsed. As Figure 7-8 shows, however, others were less fortunate. A few countries had (and still have) potential. The smaller, weaker emirates and sheikdoms on the Arabian Peninsula always feared that powerful neighbors would try to annex them (Kuwait faced this prospect in 1990 when Iraq invaded it). The unevenly distributed oil wealth, therefore, created another source of division and distrust among Islamic neighbors.

A Foreign Invasion

The oil-rich countries of the realm found themselves with a coveted energy source, but without the skills, capital, or equipment to exploit it. These had to come from the Western world and entailed what many tradition-bound Muslims feared most: a strong foreign presence on Islamic soil, foreign intervention in political as well as economic affairs, and penetration of Islamic societies by the vulgarities of Western ways. Moreover, the advent of the oil industry created a veneer of modernity in the cultural landscape (where gleaming skyscrapers towered over ornate and historic minarets) as well as among the elites who benefited most directly from the energy riches. This set up tensions and even clashes between the modern and the traditional that continue to plague oil-rich Islamic countries today. It also created contrasts between countries that took different paths as their cultures changed. In the United Arab Emirates (UAE), it is no surprise to see a woman emerge from a supermarket, get into her car, and drive home with the groceries. That would be unthink-

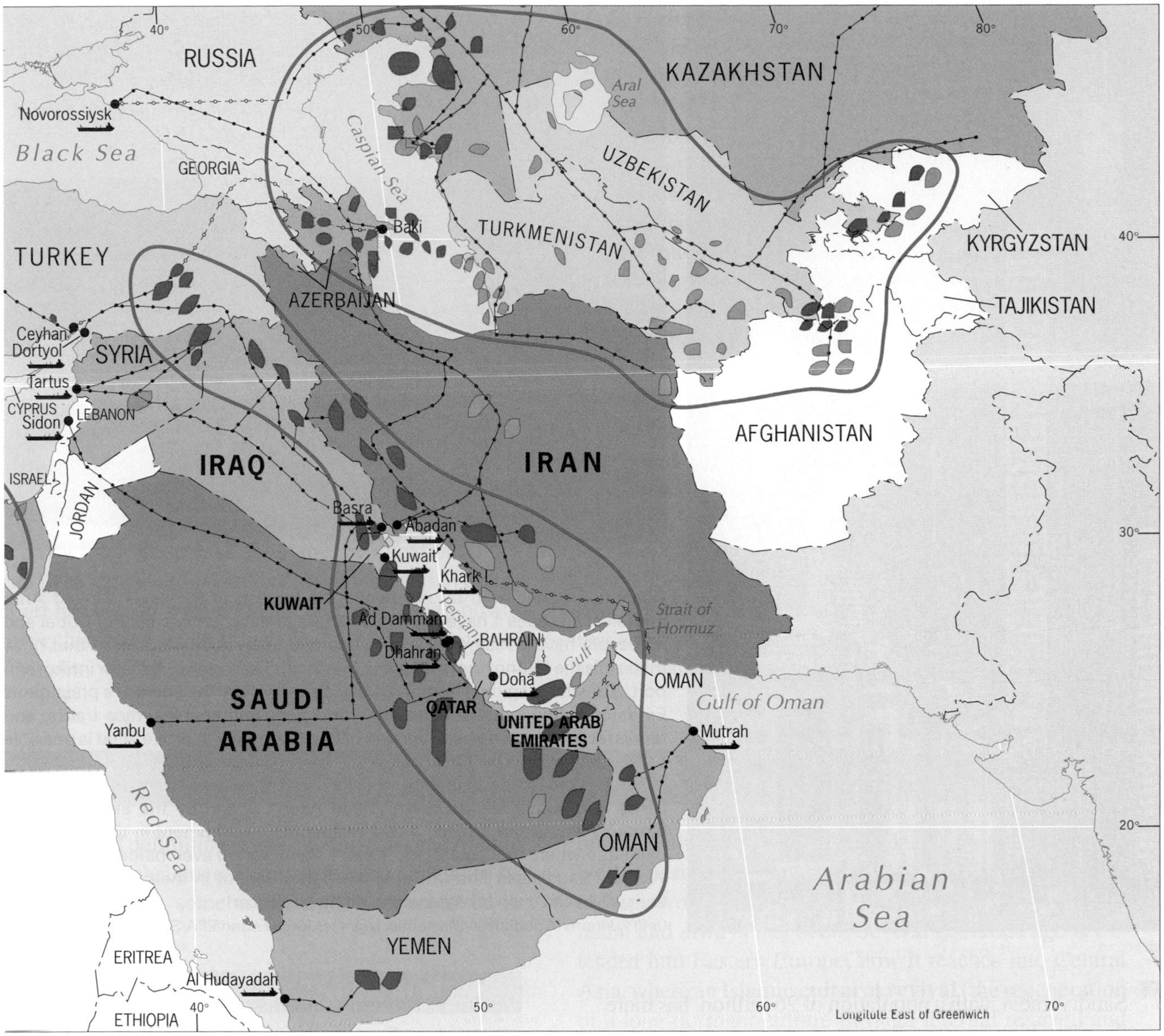

able in neighboring Saudi Arabia, where women are prohibited from driving cars.

In geographic terms, the impact of oil, its production and sale, in the exporting countries of this realm can be summarized as follows:

1. *High Incomes.* When oil prices on international markets were high, several countries in this realm ranked as the highest-income societies in the world. Even when oil prices declined, virtually all the petroleum-exporting states remained at least in the upper-middle-income category (Fig. G-11).
2. *Modernization.* Notably in Saudi Arabia but also in Kuwait and UAE, huge oil revenues have been used to modernize infrastructure. Futuristic city skyscrapers are only one manifestation of this (photo p. 342); modern ports, airports, and superhighways are others.
3. *Industrialization.* Farsighted governments among those with oil wealth, realizing that petroleum reserves will not last forever, have invested some of their income in industrial plants that will outlast the era of oil. Petrochemical industries, plastics fabrication, and desalinization plants are among these facilities.
4. *Intra-Realm Migration.* The oil wealth has attracted millions of workers from less favored parts of the realm to work in the oilfields, in the ports, and in many other, mostly menial capacities. This has brought many Shi'ites to the countries of eastern Arabia; hundreds of thousands of Palestinians also work there as laborers.

of northern Africa merges into Subsaharan Africa (a region we discussed on pp. 320–323).

3. **The Middle East.** This region includes Israel, Jordan, Lebanon, Syria, and Iraq. In effect, it is the crescent-like zone of countries that extends from the eastern Mediterranean coast to the head of the Persian Gulf.
4. **The Arabian Peninsula.** Dominated by the large territory of Saudi Arabia, the Arabian Peninsula also includes the United Arab Emirates, Kuwait, Bahrain, Qatar, Oman, and Yemen. Here lies the source and focus of Islam, the holy city of Mecca; here, too, lie many of the world's greatest oil deposits.
5. **The Empire States.** We refer to this region as the Empire States because two of the realm's giants, both the centers of historic empires, dominate its geography. Turkey, heart of the Ottoman Empire, now is the realm's most secular state and aspires to joining the European Union. Shi'ite Iran, the core of the former Persian Empire, has become an Islamic Republic. To the north, Azerbaijan, once a part of the Persian Empire, lies in the turbulent, Muslim-infused Transcaucasian Transition Zone. To the south, the island of Cyprus is divided between Turkish and Greek spheres, the Turkish sector being another remnant of imperial times.
6. **Turkestan.** Turkish influence ranged far and wide in Southwest Asia, and following the Soviet collapse in 1991 that influence proved to be durable and strong. In the five former Soviet Central Asian republics the strength and potency of Islam vary, and the new governments deal warily (sometimes forcefully) with Islamic revivalists. Russian influence continues, democratic institutions remain weak, and Western (chiefly American) involvement has increased since the War on Terror began and Afghanistan—which also forms part of this region—became the major target of that campaign.

EGYPT AND THE LOWER NILE BASIN

Egypt occupies a pivotal location in the heart of a realm that extends over 6000 miles (9600 km) longitudinally and some 4000 miles (6400 km) latitudinally. At the northern end of the Nile and of the Red Sea, at the eastern end of the Mediterranean Sea, in the northeastern corner of Africa across from Turkey to the north and Saudi Arabia to the east, adjacent to Israel, to Islamic Sudan, and to militant Libya, Egypt lies in the crucible of this realm. Because it owns the Sinai Peninsula (recently lost to and regained from Israel), Egypt, alone among states on the African continent, has a foothold in Asia, a foothold that gives it a coast overlooking the strategic Gulf of Aqaba (the northeasternmost arm of the Red Sea). Egypt also controls the Suez Canal, vital link between the Indian and Atlantic oceans and lifeline of Europe. The capital, Cairo (El Qahira), is the realm's largest city and a leading center of Islamic civilization. We hardly need to further justify Egypt's designation (together with northern Sudan) as a discrete region.

Egypt's Nile is the aggregate of two great branches upstream: the White Nile, which originates in the streams that feed Lake Victoria in East Africa, and the Blue Nile, whose source lies in Lake Tana in Ethiopia's highlands. The two Niles converge at Khartoum in modern-day Sudan. About 95 percent of Egypt's 76.4 million people live within a dozen miles (20 km) of the great river's banks or in its delta (Fig. 7-10). It has always been this way: the Nile rises and falls seasonally, watering and replenishing soils and crops on its banks.

The ancient Egyptians used *basin irrigation*, building fields with earthen ridges and trapping the floodwaters with their fertile silt, to grow their crops. That practice continued for thousands of years until, during the nineteenth century, the construction of permanent dams made it possible to irrigate Egypt's farmlands year round. These dams, with locks for navigation, controlled floods, expanded the country's cultivable area, and allowed the farmers to harvest more than one crop per year on the same field. In a single century, all of Egypt's farmland was brought under *perennial irrigation*.

The greatest of all Nile projects, the Aswan High Dam (which began operating in 1968), creates Lake Nasser, one of the world's largest artificial lakes (Fig. 7-10). As the map shows, the lake extends into Sudan, where 50,000 people had to be resettled to make way for it. The Aswan High Dam increased Egypt's irrigable land by nearly 50 percent and today provides the country with about 40 percent of its electricity. But, as is so often the case with megaprojects of this kind, the dam also produced serious problems. Snail-carried schistosomiasis and mosquito-transmitted malaria thrived in the dam's standing water, afflicting hundreds of thousands of people living nearby. By blocking most of the natural fertilizers in the Nile's annual floodwaters, the dam necessitated the widespread use of artificial fertilizers, very costly to small farmers and damaging to the natural environment. And the now fertilizer- and pesticide-laden Nile no longer supports the fish fauna offshore, reducing the catch and depriving coastal populations of badly needed proteins.

Egypt's elongated oasis along the Nile, just 3 to 15 miles (5 to 25 km) wide, broadens north of Cairo across a delta anchored in the west by the great city of Alexandria and in the east by Port Said, gateway to the Suez Canal.

AMONG THE REALM'S GREAT CITIES . . . CAIRO

Stand on the roof of one of the high-rise hotels in the heart of Cairo, and in the distance you can see the great pyramids, monumental proof of the longevity of human settlement in this area. But the present city was not founded until Muslim Arabs chose the site as the center of their new empire in A.D. 969. Cairo became and remains Egypt's primate city, situated where the Nile River opens into its delta, home today to more than one-seventh of the entire country's population.

Cairo at 11.4 million ranks among the world's 20 largest urban agglomerations, and it shares with other cities of the poorer world the staggering problems of crowding, inadequate sanitation, crumbling infrastructure, and substandard housing. But even among such cities, Cairo is noteworthy for its stunning social contrasts. Along the Nile waterfront, elegant skyscrapers rise above carefully manicured, Parisian-looking surroundings (photo below). But look eastward, and the urban landscape extends gray, dusty, almost featureless as far as the eye can see. Not far away, more than a million squatters live in the sprawling cemetery known as the City of the Dead. On the outskirts, millions more survive in overcrowded shantytowns of mud huts and hovels.

And yet Cairo is the dominant city not only of Egypt but for a wider sphere; it is the cultural capital of the Arab World, with centers of higher learning, splendid museums, world-class theater and music, and magnificent mosques and Islamic learning centers. Although Cairo always has been primarily a center of government, administration, and religion, it also is a river port and an industrial complex, a commercial center, and, as it sometimes seems, one giant bazaar. Cairo is the heart of the Arab World, a creation of its geography, and a repository of its history.

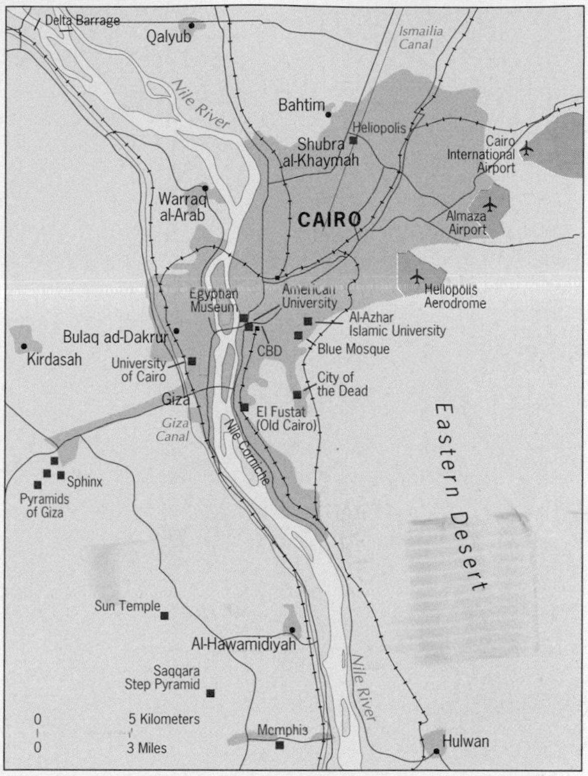

Looking eastward from El Duqqi, high over the west bank of the Nile, you see the heart of the great city of Cairo, where such landmarks as the Egyptian Museum, the headquarters of the Arab League, the Ministry of Foreign Affairs and the Supreme Constitutional Court are located. Beyond the modern buildings flanking the river and the city's crowded high-rise center lie vast, poverty-stricken settlements where the majority of its more than 11 million inhabitants live. © Stone/Getty Images.

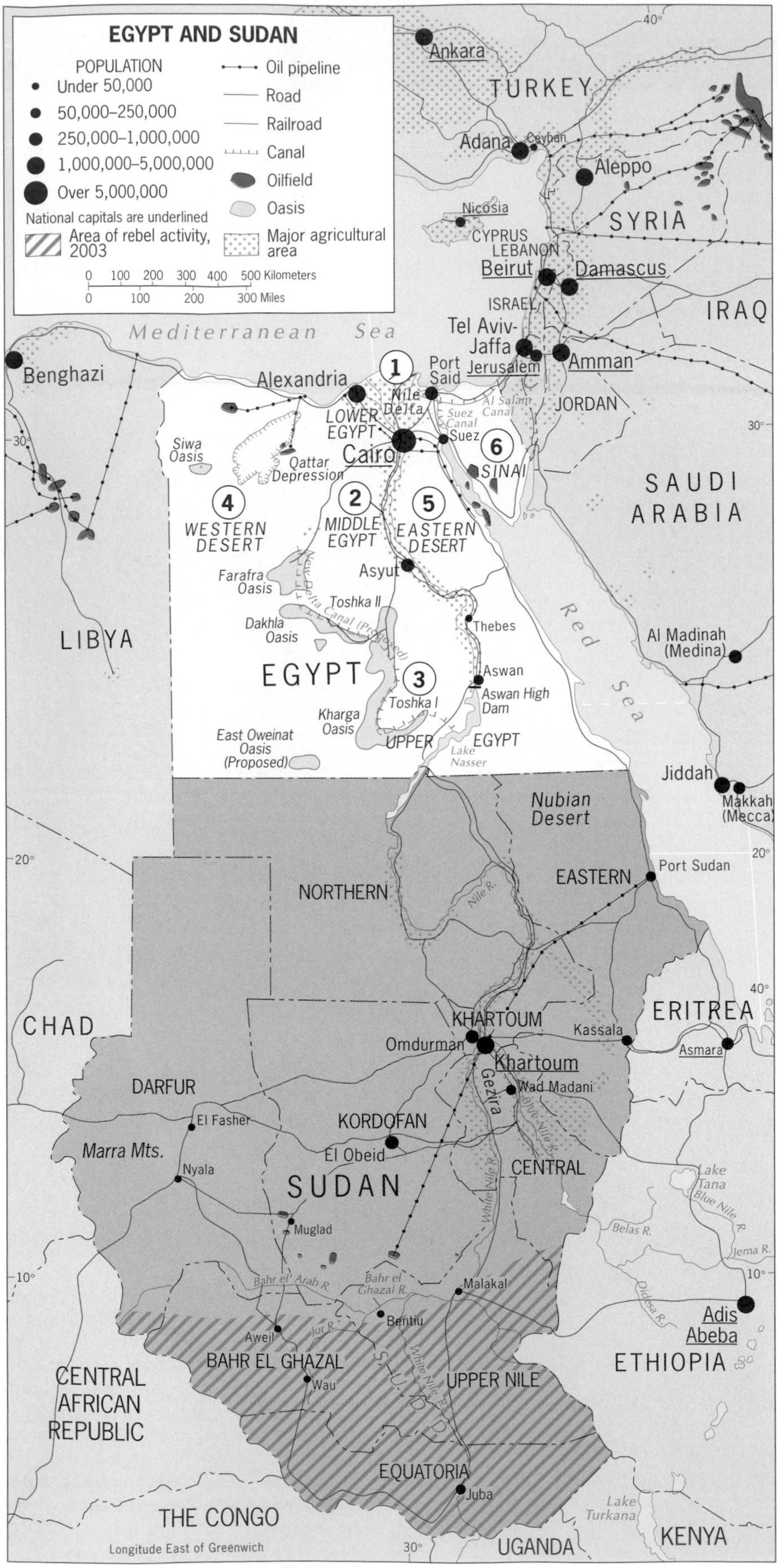

FIGURE 7-10 ©H. J. de Blij, P. O. Muller, and John Wiley & Sons, Inc.

The delta contains extensive farmlands, but it is a troubled area today. The ever more intensive use of the Nile's water and silt upstream is depriving the delta of much needed replenishment. And the low-lying delta is geologically subsiding, raising fears of salt-water invasion from the Mediterranean Sea that would damage soils here.

Egypt's millions of subsistence farmers, the *fellaheen*, still struggle to make their living off the land, as did the peasants of the Egypt of five millennia past. Rural landscapes seem barely to have changed; ancient tools are still used, and dwellings remain rudimentary. Poverty, disease, high infant mortality rates, and low incomes prevail. The Egyptian government is embarked on grandiose plans to expand its irrigated acreage, but without modernization and greater efficiency in the existing farmlands, the future is dim.

Egypt's Regional Geography

Egypt has six subregions, mapped in Figure 7-10. Most Egyptians live and work in Lower (i.e., northern) and Middle Egypt (regions ① and ②), the country's core area anchored by Cairo and flanked by the leading port and second industrial center, Alexandria. The economy has benefited from further oil discoveries in the Sinai (region ⑥) and in the Western Desert (region ④), so that Egypt now is self-sufficient and even exports some petroleum. Cotton and textiles are the other major source of external income, but the important tourist industry has repeatedly been hurt by Islamic extremists. As the population mushrooms, the gap between food supply and demand widens, and Egypt must import grain. Since the late 1970s, Egypt has been a major recipient of U.S. foreign aid.

Egypt today is at a crossroads in more ways than one. Its planners know that reducing the high birth rate would improve the demographic situation, but revivalist Muslims object to any programs that promote family planning. Its accommodation with Israel helps ensure foreign aid but divides the people. Its government faces a fundamentalist challenge. Egypt's future, in this crucial corner of the realm, is uncertain.

Divided Sudan

As Figure 7-10 shows, Egypt is flanked by two countries that have posed challenges to its leadership: Sudan to the south and Libya to the west. Sudan, more than twice as large as Egypt and with 41.3 million people, lies centered on the confluence of the White Nile (from Uganda) and the Blue Nile (from Ethiopia). Here the twin capital, Khartoum-Omdurman, anchors a large agricultural area where cotton was planted during colonial times. The British administration combined northern Sudan, which was Arabized and Islamized, with a large area to the south, which was African and where many villagers had been Christianized. After the British left, the regime in Khartoum wanted to impose its Islamic rule on this portion of the African Transition Zone that formed southern Sudan, and a bitter civil war ensued.

The cost in human lives and dislocation of this 30-year war is incalculable. The United Nations estimates that as many as 2 million people died, and as recently as 2003 as many as 4 million people were refugees in their own country. But international mediation finally brought relief in the form of a peace agreement whereby the leader of the rebellious South would become vice president of a federal Sudan and the people of the South would, six years hence, be able to vote on independence. In mid-2005, shortly after the agreement had been signed, hundreds of thousands of displaced Southerners were moving back to their homeland. A UN peacekeeping force was being deployed, and the town of Juba, the southern headquarters on the Nile River long cut off from the rest of the world, was coming to life. The peace agreement stipulates that the South will be governed by Southerners and that it will receive a share of Sudan's newfound oil revenues.

The discovery of significant oil reserves in southern Kordofan Province in the 1990s produced assets as well as liabilities for Sudan (Fig. 7-10). They transformed the country from an importer to a seller of oil, but Sudan's armed forces reportedly drove tens of thousands of villagers in the oil-rich areas from their homes and lands, killing those who resisted and burning their villages to the ground. The foreign companies involved in oil production came under heavy international pressure to withdraw, and this prospect was a factor in the government's willingness to negotiate an end to the conflict with the South.

But no sooner was one terrible conflict near the point of resolution than another erupted, also along the Islamic Front (mapped in Fig. 6-18): a war in Darfur (Arabic for "House of the Fur," the African inhabitants of the southern part of this western province). Although Islamized, the Fur are ethnically African, and they share the province with Arabs who occupy its center and north (Fig. 7–10). Apparently, the centuries-old Islamization of the African Fur was not enough for the regime in Khartoum, which blamed these sedentary farmers for sympathizing with anti-Khartoum rebels. Soon the province's camel- and cattle-herding Arabs were invading the Africans' lands, supported by the army and by an informal militia named the *janjaweed* that burned villages, destroyed farms, and killed at will (see photo next page). By mid-2005, more than 2 million Africans had been driven from their lands, more than 180,000 had been killed, and fields lay untilled, ensuring

Not much is left standing in Um Hashab village in northern Darfur province, Sudan, after an air force plane and several helicopters attacked and left almost all its houses burning. A farmer stands where his house stood, the walls outlined by mounds of ash. The Arabic-speaking Islamic regime in Khartoum denied responsibility, but the United Nations took note of the violence in Darfur and cited evidence that ethnic African villagers, who are also Muslims, were being driven from their land by armed bands empowered by the government. By mid-2005, estimates of the death toll in Khartoum's Darfur campaign passed 180,000 with more than 2 million people uprooted from their homes. Amr Nabil/© AP/Wide World Photos.

widespread hunger and starvation later in the year. The UN Security Council threatened the Khartoum regime with sanctions unless it disarmed the militias, which it (understandably) ignored. The tragedy of Sudan forms a reminder of the capacity of a single rapacious regime to destroy the lives and hopes of millions under its territorial sway.

THE MAGHREB AND ITS NEIGHBORS

The countries of northwestern Africa are collectively called the *Maghreb*, but the Arab name for them is more elaborate than that: *Djezira-al-Maghreb*, or "Isle of the West," in recognition of the great Atlas Mountain range rising like a huge island from the Mediterranean Sea to the north and the sandy flatlands of the immense Sahara to the south.

The countries of the Maghreb (sometimes spelled *Maghrib*) are Morocco, last of the North African kingdoms; Algeria, a secular republic beset by the religious-political problems we noted earlier; and Tunisia, smallest and most Westernized of the three (Fig. 7-11). Libya, facing the Mediterranean between the Maghreb and Egypt, is unlike any other North African country: an oil-rich desert state whose population is almost entirely clustered in settlements along the coast.

Whereas Egypt is the gift of the Nile, the Atlas Mountains facilitate the settled Maghreb. These high ranges wrest from the rising air enough orographic rainfall to sustain life in the intervening valleys, where good soils support productive farming. From the vicinity of Algiers eastward along the coast into Tunisia, annual rainfall averages more than 30 inches (75 cm), a total more than three times as high as that recorded for Alexandria in Egypt's delta. Even 150 miles (240 km) inland, the slopes of the Atlas still receive over 10 inches (25 cm) of rainfall. The effect of the topography can be read on the world map of precipitation (Fig. G-7): where the highlands of the Atlas terminate, desert conditions immediately begin.

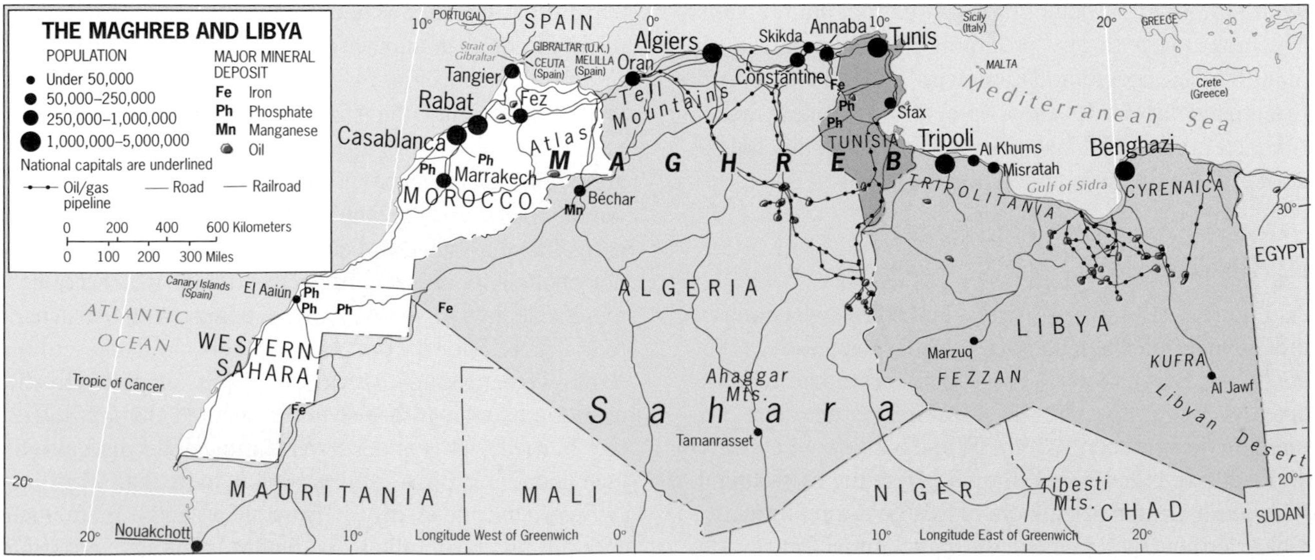

FIGURE 7-11 © H. J. de Blij, P. O. Muller, and John Wiley & Sons, Inc.

FROM THE FIELD NOTES

"Having seen the *Giralda* in Seville, Spain (photo p. 333), I was interested to view the tower of the Koutoubia Mosque in Marrakech, Morocco, also built in the twelfth century and closely resembling its Andalusian counterpart. Some 220 feet (67 m) tall, the Koutoubia minaret was built by Spanish slaves and is a monument to the heyday of Islamic culture of the time. I soon discovered that it is not as easy as it looks to get a clear view of the tower, and a boy offered his help, guiding me around the block on which the mosque is situated. He was well informed on the building's history and on the varying fortunes of his home town. It was European mispronunciation of the name 'Marrakech' that led to Morocco being called Morocco, he reported. And Marrakech was Morocco's capital for a very long time, and still should be. 'But back to the mosque,' I said. 'What does the sign say?' He smiled. 'You'll just have to learn some Arabic, just as we know French,' he said. Obviously a teacher in the making, I thought." © H. J. de Blij

The Atlas Mountains trend southwest-northeast and begin in Morocco as the High Atlas, with elevations close to 13,000 feet (4000 m). Eastward, two major ranges dominate the landscapes of Algeria proper: the Tell Atlas to the north, facing the Mediterranean, and the Saharan Atlas to the south, overlooking the great desert. Between these two mountain chains, each consisting of several parallel ranges and foothills, lies a series of intermontane basins (analogous to South America's Andean *altiplanos* but at lower elevations), markedly drier than the northward-facing slopes of the Tell Atlas. In these valleys, the rain shadow effect of the Tell Atlas is reflected not only in the steppe-like natural vegetation but also in land-use patterns: pastoralism replaces cultivation, and stands of short grass and bushes blanket the countryside.

During the colonial era, which began in Algeria in 1830 and lasted until the early 1960s, well over a million Europeans came to settle in North Africa—most of them French, and a large majority bound for Algeria—and these immigrants soon dominated commercial life. They stimulated the renewed growth of the region's towns; Casablanca, Algiers, Oran, and Tunis became the urban foci of the colonized territories. Although the Europeans dominated trade and commerce and integrated the North African countries with France and the European Mediterranean world, they did not confine themselves to the cities and towns. They recognized the agricultural possibilities of the favored parts of the tell (the lower Tell Atlas slopes and narrow coastal plains that face the Mediterranean) and established thriving farms. Not surprisingly, agriculture here is Mediterranean. Algeria soon became known for its vineyards and wines, citrus groves, and dates; Tunisia has long been one of the world's leading exporters of olive oil; and Moroccan oranges went to many European markets. Following independence in the 1950s and 1960s (and after a bitter war in the case of Algeria), economic circumstances changed as French expatriates left the region and Islamic regulations, which included the prohibition of alcoholic-beverage production, altered cultural and commercial landscapes.

The Countries

Between desert and sea, the Maghreb states display considerable geographic diversity. **Morocco**, a conservative kingdom in a revolutionary region, is tradition-bound and weak economically. Its core area lies in the north, anchored by four major cities; but the Moroccans' attention is focused on the south, where the government is seeking to absorb its neighbor, **Western Sahara** (a former Spanish dependency with about 300,000 inhabitants, many of them immigrants from Morocco). Even if this campaign is successful, it will do little to improve the lives of most of Morocco's 31.5 million people. Hundreds of thousands have emigrated to Europe, many of them via Spain's two tiny exclaves on the Moroccan coast, Ceuta and Melilla (Fig. 7-11).

Algeria, rich in oil and natural gas concentrated in its eastern interior, fought a bitter war of independence against the French colonizers. Sovereignty, however, did not bring stability or prosperity. A long-term domestic conflict pitted the country's governing modernizers, supported by the army, against Islamic fundamentalists who wanted to make Algeria an Islamic republic along the lines of Sudan or Iran. When the initial vote in a two-round election indicated that the fundamentalists would win, the second round was canceled and the stage was set for all-out war. Eventually, more than 100,000 Algerians lost their lives in this tragic conflict, which abated only recently and has not been completely resolved. Meanwhile, a recent cultural dispute arose in the Kabylia area east of the capital, Algiers, where the Berber minority is concentrated and rebelled against alleged mistreatment by the government.

Algeria's capital, Algiers, is home to 3.4 million of the country's 33.3 million inhabitants, but these numbers should be weighed against the massive emigration of Algerians to Europe, mainly to France. There are now several times as many Algerian immigrants in France than there were French colonists in Algeria at the height of the colonial period, which accounts for France's prominence on the Muslim population map of Western Europe (Fig. 1-10).

Almost rectangular in shape, **Libya** (5.9 million) lies on the Mediterranean Sea between the Maghreb states and Egypt (Fig. 7-11). What limited agricultural possibilities exist lie in the district known as Tripolitania in the northwest, centered on the capital, Tripoli, and in the northeast in Cyrenaica, where Benghazi is the urban focus. But it is oil, not farming, that drives the economy of this highly urbanized country. The oilfields lie well inland from the Gulf of Sidra, linked by pipelines to coastal terminals. Libya's two interior corners, the desert Fezzan district in the southwest and the Kufra oasis in the southeast, are connected to the coast by two-lane roads subject to sandstorms.

Muammar Qaddafi, an army colonel, overthrew King Idris of Libya in 1969 and in the 1970s burnished his Islamic credentials by expelling Libya's Jewish community, supporting Palestinian causes, and promoting "Allah's language" throughout North Africa and the Middle East. In the 1980s Libya became involved in state-sponsored terrorism, and in the early 1990s the United Nations imposed sanctions because of Qaddafi's refusal to turn over information or suspects in the bombing of French and American civilian airliners.

Since the late 1990s, Libya's role in the world has taken some dramatic turns. First, Qaddafi turned his interest from Middle Eastern and Palestinian causes to Africa, using oil revenues to make large payments to African leaders and earning praise even from South Africa's Nelson Mandela. Next, Qaddafi cooperated with an international tribunal that charged two Libyans with the 1988 downing of a Pan American aircraft over Scotland, and agreed to make substantial payments to the families of the victims. And in 2004 Qaddafi invited UN inspectors to investigate Libya's nuclear weapons program and equipment, later permitting the removal of materials the country had purchased for this purpose. After 35 years in power, it appeared that the unpredictable ruler had decided to seek an end to the isolation his policies had imposed on his country. By late 2004, tourists were already returning to visit Libya's fabled Roman ruins, but for others these developments were less welcome. Libya's West African workers found themselves less needed and jobless in the newly opened Libya, and were leaving the country by the thousands.

As Figure 7-11 shows, the Maghreb countries and Libya adjoin several desert-dominated states in the African Transition Zone. These are African countries with strong Islamic imprints, but as Figure 6-18 shows, they lie north of the Islamic Front with only one exception, Chad. Vast and sparsely populated **Mauritania** is the most Islamized of these countries, with about half its population of 3.2 million concentrated in and near the capital, Nouakchott, base of a small fishing fleet. Neighboring **Mali** depends on the waters of the upper Niger River, on which its capital of Bamako lies. A multiparty, democratic state, Mali is also multicultural, with a sizeable minority adhering to traditional beliefs and a small Christian minority. To the east of Mali lies **Niger**, which shares boundaries with Algeria and Libya; like Mali, this is one of the world's least urbanized countries. Unlike Mali, however, Niger has only a short stretch of the middle course of the Niger River, where the capital, Niamey, is situated. And **Chad**, directly south of Libya, has the lowest percentage of Muslim inhabitants but the strongest division between Islamized north and animist/Christian south. As Figure 6-18 reminds us, the capital, N'Djamena, lies on the southern edge of the African Transition Zone, directly on the Islamic Front. Since early 2003, even as Chad was experiencing a (still-ongoing) oil boom in the south, a cultural conflict in Sudan's neighboring Darfur Province has sent hundreds of thousands of refugees across its border.

THE MIDDLE EAST

The regional term *Middle East*, we noted earlier, is not satisfactory, but it is so common and so generally used that avoiding it creates more problems than it solves. It originated when Europe was the world's dominant realm and when places were "near," "middle," and "far" from Europe: hence a Near East (Turkey), a Far East (China, Japan, Korea, and other countries of East Asia), and a Middle East (Egypt, Arabia, Iraq). If you check definitions used in the past, you will see that the terms were applied inconsistently: Syria, Lebanon, Palestine, even Jordan sometimes were included in the "Near" East, and Persia and Afghanistan in the "Middle" East.

Today, the geographic designation *Middle East* has a more specific meaning. And at least half of it has merit: this region, more than any other, lies at the middle of the vast Islamic realm (Fig. 7-9). To the north and east of it, respectively, lie Turkey and Iran, with Muslim Turkestan beyond the latter. To the south lies the Arabian Peninsula. And to the west lie the Mediterranean Sea and Egypt, and the rest of North Africa. This, then, is the pivotal region of the realm, its very heart.

Five countries form the Middle East (Fig. 7-12): Iraq, largest in population and territorial size, facing the Persian Gulf; Syria, next in both categories and fronting the Mediterranean; Jordan, linked by the narrow Gulf of Aqaba to the Red Sea; Lebanon, whose survival as a unified state has come into question; and Israel, Jewish nation in the crucible of the Muslim world. Because of the extraordinary importance of this region in world affairs, we focus in some detail on issues of cultural, economic, and political geography in the following discussion.

Iraq

In the immediate aftermath of the 9/11 terrorist attacks in the United States in 2001, American forces overthrew the ruling revivalist Taliban regime in Afghanistan that

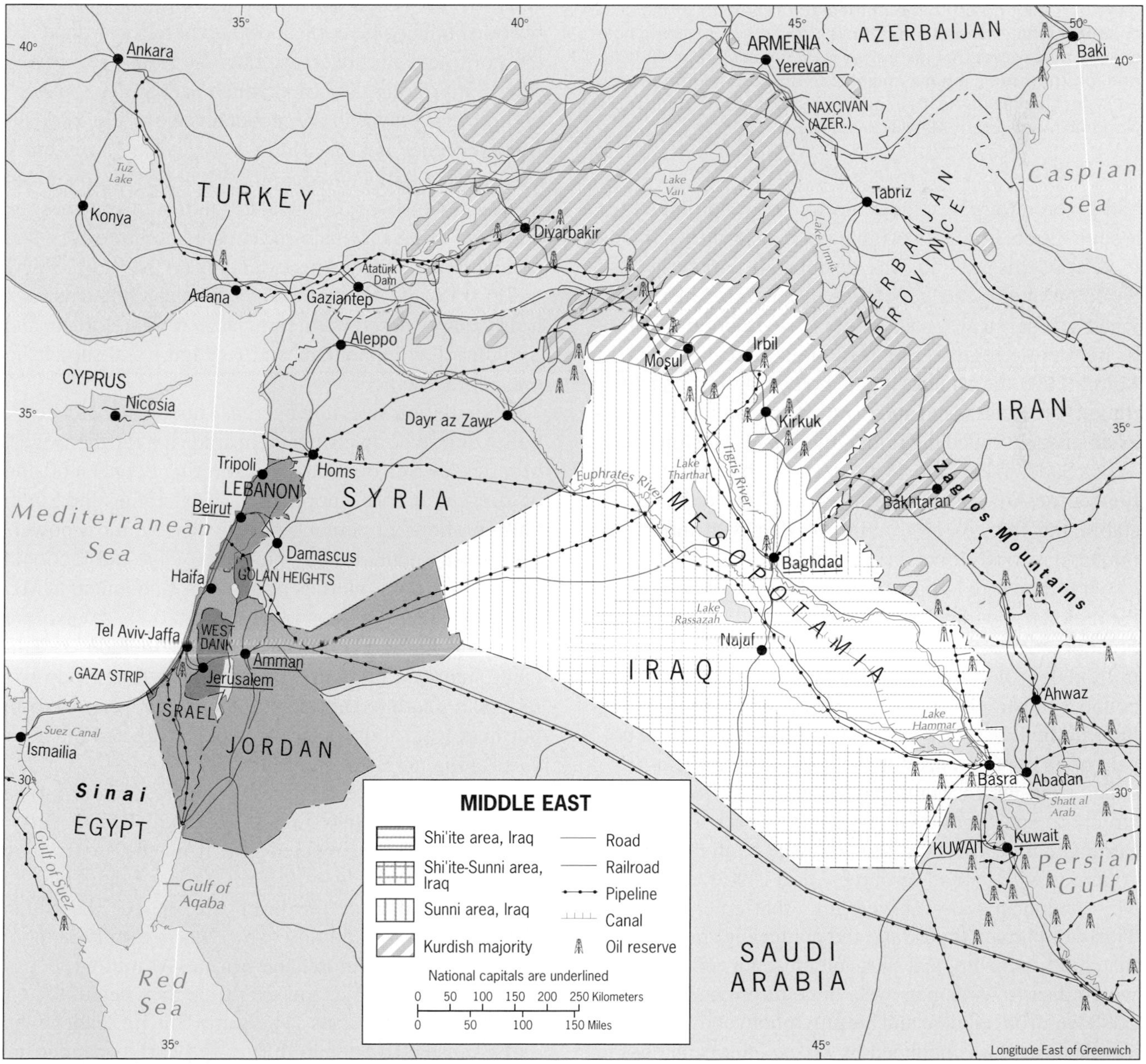

FIGURE 7-12 © H. J. de Blij, P. O. Muller, and John Wiley & Sons, Inc.

One of the tragedies arising from the U.S.-led intervention in Iraq was the collateral damage it inflicted on the country's cultural heritage. Museums and archeological sites were damaged and looted in the immediate aftermath of the U.S. conquest when law and order broke down. Here a gaping hole caused by a rocket that just missed a relief sculpture showing one of the area's many ancient civilizations affords easy entry for thieves in the main archway leading into Baghdad's National Museum. © AP/Wide World Photos.

had given refuge to Usama bin Laden, who plotted the assault. Soon, however, Afghanistan lost priority among U.S. concerns as Iraq was designated a security risk based on intelligence reports of its terrorist links, its alleged arsenal of weapons of mass destruction (WMD), its murderous regime, and the regional threat it posed. In March 2003 the United States, helped by UK and Australian forces, attacked and invaded Iraq and following a brief struggle ended the rule of the Sunni-Muslim clique led by Saddam Hussein. While the search for WMD continued, the American government now faced the task of stabilizing Iraq, reconstructing its damaged infrastructure, and reconstituting its government.

It is therefore important to understand key aspects of the regional geography of California-sized Iraq and its official population of 27.3 million (which may in fact be 2 million less because of unregistered emigration). The country constitutes nearly 60 percent of the total area of the Middle East as we define it and has just over 40 percent of its population. With its world-class oilfields (revenues from which would pay for much of Iraq's reconstruction), major gas reserves, and large areas of irrigable farmland, Iraq also is the best endowed of all the region's countries with natural resources. Iraq is heir to the early Mesopotamian states and empires that emerged in the Tigris-Euphrates Basin, and the country is studded with matchless archeological sites and museum collections, to which disastrous damage was done during and after the 2003 war from combat and looting (photo above).

Today, Iraq is bounded by six neighbors and has in recent times had adversarial relations with most of them. To the north lies Turkey, source of both Iraq's vital rivers. To the east lies Iran, target of a destructive war during the 1980s. Kuwait, at the head of the Persian Gulf, was invaded by Iraq's armies in 1990. To the south, Saudi Arabia long was a friend of Iraq's enemies. And to the west, Iraq is adjoined by Jordan, which gave Iraq some help during the 1991 Gulf War but later turned its back on the Saddam regime, and Syria, functionally most remote from Iraq but also a country ruled by a strongman and his minority clique. As Figure 7-13 shows, pipelines across Syria link Iraq's oilfields with Mediterranean terminals to the west.

This map also displays Iraq's regional components. The capital and core area, Baghdad (6.1 million) and its environs, lie about midway between the Persian Gulf and the Turkish border astride the Tigris River. Three historic domains lie north and south of this heartland. To the north and northwest extends the Sunni stronghold (the "Sunni Triangle") from which Saddam's repressive regime was drawn. In the mountains along the Turkish-Iranian border lies the Kurdish zone that forms but a small part of a much larger ethnic Kurdish region. From the eastern suburbs of Baghdad southward stretches the third and largest domain in terms of both territory and population: the Shiah area extending all the way to Basra (Al Basrah) and the shores of the Persian Gulf. The vast, nearly empty desert bordering Saudi Arabia, Jordan, and Syria has few inhabitants and, as Figure 7-8 shows, no oil to attract settlers.

Population size, however, has not been matched by political power or influence in Iraq. By far the largest group is the Shi'ite sector, constituting between 60 and 65 percent of Iraq's population. Next comes the Sunni cluster, about 22 percent, historically the most powerful but fragmented along numerous clan and factional lines. The Kurds, victims of Saddam's chemical WMD in 1990, make up around 15 percent. There also are numerous smaller minorities, which include the ancient, mainly Christian, non-Arab Assyrian group living today scattered among the Sunnis and the Kurds in northern Iraq, and the Turkmen in Kirkuk and other locales along the Kurd-Sunni frontier.

The Shi'ites were ruthlessly suppressed during the rule by Saddam Hussein and his Sunni Ba'ath Party. When U.S. forces invaded Iraq from the south, they faced comparatively little resistance on their way to Baghdad, and after the brief war ended, the Shiah south remained relatively calm. But in a truly democratic Iraq of the future, the Shi'ites would expect to play a dominant role, and over time they began to agitate for early elections. The same Shi'ite leader who had counseled patience during the early occupation, the cleric Ali al-Sistani, in early 2004 started to press

FIGURE 7-13 © H. J. de Blij, P. O. Muller, and John Wiley & Sons, Inc.

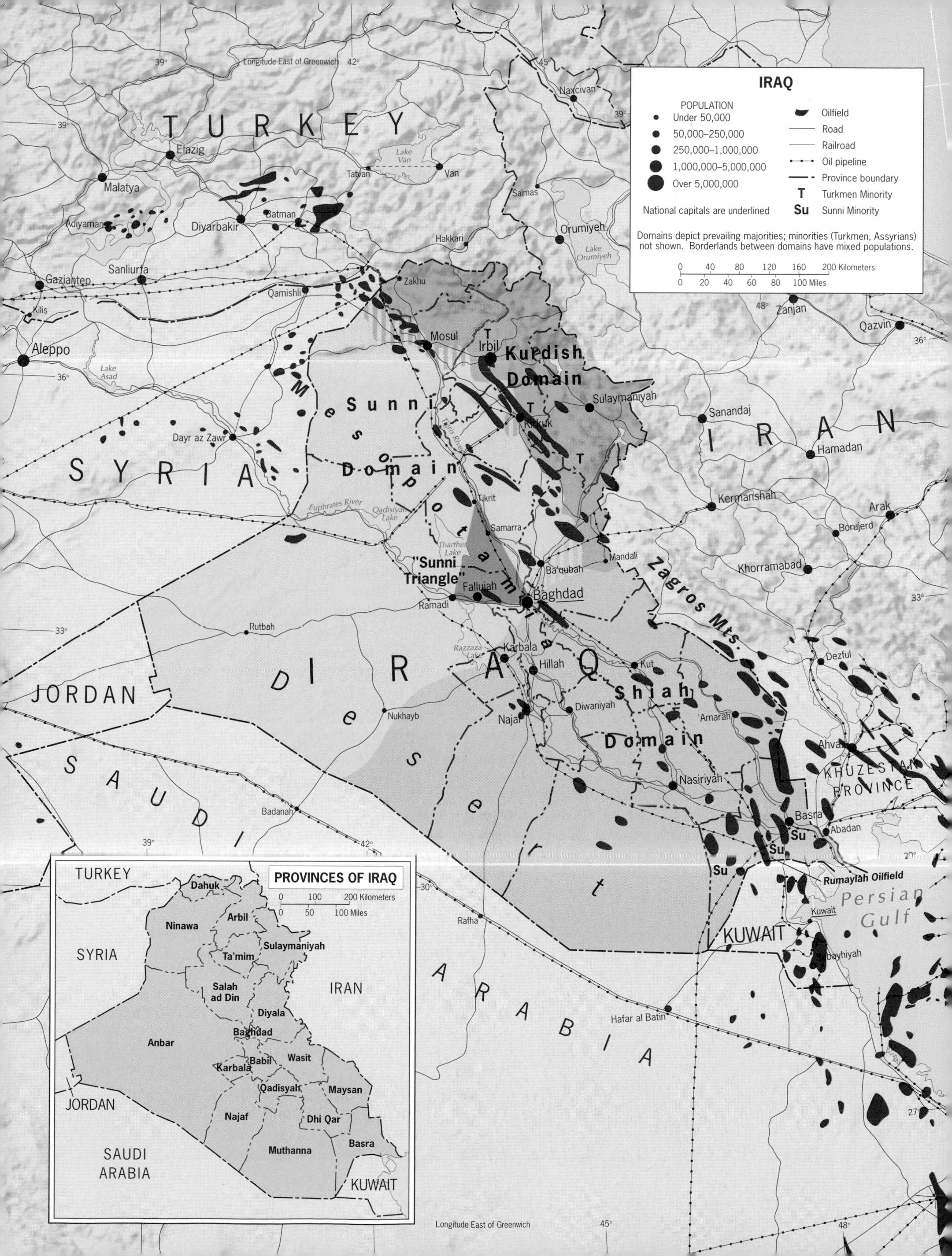

IRAQ
POPULATION
Under 50,000
50,000–250,000
250,000–1,000,000
1,000,000–5,000,000
Over 5,000,000
National capitals are underlined
Oilfield
Road
Railroad
Oil pipeline
Province boundary
T Turkmen Minority
Su Sunni Minority
Domains depict prevailing majorities; minorities (Turkmen, Assyrians) not shown. Borderlands between domains have mixed populations.
0 40 80 120 160 200 Kilometers
0 20 40 60 80 100 Miles
Longitude East of Greenwich
TURKEY
SYRIA
IRAN
JORDAN
SAUDI ARABIA
KUWAIT
IRAQ
Kurdish Domain
Sunni Domain
Shiah Domain
"Sunni Triangle"
Mesopotamia
Desert
Zagros Mts.
KHUZESTAN PROVINCE
Persian Gulf
Rumaylah Oilfield
Elazig
Malatya
Adiyaman
Diyarbakir
Batman
Tatvan
Van
Lake Van
Naxcivan
Salmas
Orumiyeh
Lake Orumiyeh
Hakkari
Sanliurfa
Gaziantep
Kilis
Aleppo
Lake Asad
Qamishli
Zakhu
Mosul
Irbil
Sulaymaniyah
Kirkuk
Zanjan
Qazvin
Sanandaj
Hamadan
Kermanshah
Arak
Borujerd
Khorramabad
Dayr az Zawr
Euphrates River
Qadisiyah Lake
Tigris River
Tikrit
Samarra
Tharthar Lake
Ba'qubah
Mandali
Fallujah
Ramadi
Baghdad
Rutbah
Razzaza Lake
Karbala
Hillah
Kut
Diwaniyah
Najaf
Nukhayb
'Amarah
Nasiriyah
Dezful
Ahvaz
Basra
Abadan
Badanah
Rafha
Kuwait
Subayhiyah
Hafar al Batin
PROVINCES OF IRAQ
0 100 200 Kilometers
0 50 100 Miles
Dahuk
Arbil
Ninawa
Sulaymaniyah
Ta'mim
Salah ad Din
Diyala
Baghdad
Anbar
Babil
Wasit
Karbala
Qadisiyah
Maysan
Najaf
Dhi Qar
Muthanna
Basra

for quick resolution, and mobilized large demonstrations to that end.

The situation in the Sunni domain (Fig. 7-13) was quite different. Sunnis known to have been part of Saddam's regime through membership in the Ba'ath Party or armed forces were arrested or dismissed from their positions, a practice that appeared to create more problems than it solved (even university professors who were party members lost their jobs, although many had joined as a matter of survival, not ideological commitment). While much of the Sunni area remained fairly calm, a major and costly insurrection arose in the so-called Sunni Triangle, within which Saddam Hussein's clan and that of others in his inner circle had their homes. This insurrection reached into Baghdad and other areas beyond the zone mapped in Figure 7-13 and to some extent destabilized the heart of the country.

The Kurds suffered severely in the old Iraq, but they benefited from the special protection they had during the period following the earlier Gulf War. They converted their security and their productive domain into stability and progress, and many feel that they would have a great deal to lose in a future Iraq dominated by the majority Shi'ites. Iraq's 4 million Kurds, we should note, are part of a greater Kurdish nation that extends into Turkey, Iran, and other neighboring countries (see box titled "A Future Kurdistan?").

American administrators hope to weld these disparate geographies into one democratic (possibly federal) Iraq, which would require enormous investment and much time—time that may not be available in the face of a specter of rising expectations.

Syria

Like Iraq, Syria is no stranger to minority rule. Although Syria's population of 18.8 million is about 75 percent Sunni Muslim, the ruling elite comes from a smaller Islamic sect based there, the Alawites. Leaders of this powerful minority have retained control over the country for decades, at times by ruthless suppression of dissent. In 2000, president-for-life Hafez al-Assad died and was succeeded by his son Bashar, signaling a continuation of the political status quo.

Like Lebanon and Israel, Syria has a Mediterranean coastline where crops can be raised without irrigation. Behind this densely populated coastal zone, Syria has a much larger interior than its neighbors, but its areas of productive capacity are widely dispersed. Damascus, in the southwest corner of the country, was built on an oasis and is considered to be the world's oldest continuously inhabited city. It is now the capital of Syria, with a population of 2.4 million.

The far northwest is anchored by Aleppo, the focus of cotton- and wheat-growing areas in the shadow of the Turkish border. Here the Orontes River is the chief source of irrigation water, but in the eastern part of the country the Euphrates Valley is the crucial lifeline. It is in Syria's interest to develop its eastern provinces, and recent discoveries of oil there will speed that process.

A Future Kurdistan?

Political maps of the region do not show it, but where Turkey, Iraq, and Iran meet, the cultural landscape is not Turkish, Iraqi, or Iranian. Here live the Kurds, a fractious and fragmented nation of at least 25 million (their numbers are uncertain). More Kurds live in Turkey than in any other country (perhaps as many as 14 million); possibly as many as 8 million in Iran; about half that number in Iraq; and smaller numbers in Syria, Armenia, and even Azerbaijan (see Fig. 7-12).

The Kurds have occupied this isolated, mountainous, frontier zone for over 3000 years. They are a nation, but they have no state; nor do they enjoy the international attention that peoples of other **stateless nations** (such as
18
the Palestinians) receive. Turkish and Iraqi repression of the Kurds, and Iranian betrayal of their aspirations, briefly make the news but are soon forgotten. Relative location has much to do with this: spatial remoteness and the obstacles created by the ruling regimes inhibit access to their landlocked domain.

Many Kurds dream of a day when their fractured homeland will be a nation-state. Most would agree that the city of Diyarbakir, now in southeastern Turkey, would become the capital. It is the closest any Kurdish town comes to a primate city, although the largest urban concentration of Kurds today is in the shantytowns of Istanbul, where more than 3 million have migrated. In their heartland, meanwhile, the Kurds, their shared goals notwithstanding, are a divided people whose intense disunity has thwarted their objectives; and, as in the case of Europe's Basques, a small minority of extremists has tended to use violence in pursuit of their aims.

The Kurds—without oil reserves, without a seacoast, without a powerful patron, without a global public relations machine—are victims of a conspiracy between history and geography. One of the world's largest stateless nations hopes for a deliverance that is unlikely to come.

A milestone was reached in Iraq on Sunday, January 30, 2005, when voters in all 18 of Iraq's provinces went to the polls to elect a 275-member National Assembly. Turnout was substantial in the Shiah and Kurdish provinces, but very low in the Sunni domain, where religious leaders had urged a boycott that would serve to invalidate the effort. Whatever the outcome, however, this was a massive organizational effort that ranged from securing streets and voting places (as U.S. troops are shown doing in the road in front of the interim prime minister's residence) to preparing and checking voters' rolls against ID cards on election day as millions of Iraqis defied threats of violence by casting their ballots—including women on an equal basis with men. (*left and right*) © AP/Wide World Photos.

Jordan

Syria's southern neighbor is Jordan, a classic example (and victim) of changing relative location. This desert kingdom was a product of the Ottoman collapse, but when Israel was created it lost its window on the Mediterranean Sea, the (now Israeli) port of Haifa, previously in the British-administered Mandate of Palestine. Following its independence in 1946 with about 400,000 inhabitants, the creation of Israel bequeathed Jordan some 500,000 West Bank Palestinians and, later, a huge inflow of refugees. Today Palestinians outnumber original residents by more than two to one in the population of 5.9 million. It may be said that the 47-year rule by King Hussein, ending in 1999, was the key centripetal force that held the country together.

With an impoverished capital, Amman, without oil reserves, and possessing only a small and remote outlet to the Gulf of Aqaba, Jordan has survived with U.S., British, and other aid. It lost its West Bank territory in the 1967 war with Israel, including its sector of Jerusalem (then the kingdom's second-largest city). No third country has a greater stake in a settlement between Israel and the Palestinians than Jordan.

Lebanon

The map suggests that Lebanon has significant geographic advantages in this region: a lengthy coastline on the Mediterranean Sea; a world-class capital, Beirut, on its shoreline; oil terminals on its coast; and the Syrian capital in its hinterland. The map at the scale of Figure 7-12 cannot reveal still another asset: the fertile, agriculturally productive Bekaa Valley in the eastern interior.

French colonialism in the post-Ottoman era created a territory which in 1930 was about equally Muslim and Christian. Beirut became known as the "Paris of the Middle East." After independence in 1946, Lebanon functioned as a democracy and did well economically as the region's leading banking and commercial center. But the Muslim sector of the population grew much faster than the Christian one, and in the late 1950s the Arabs launched their first rebellion against the established order. Following the first of several waves of influx by Palestinian refugees, Lebanon fell apart in 1975 in a civil war that wrecked Beirut, devastated the economy, and left the country at the mercy of Syrian forces which entered to stabilize the situation.

Today Lebanon (population: 4.6 million) has restabilized and Beirut is being rebuilt, but it will take generations to overcome the losses sustained during a quarter-century of dislocation.

Israel and the Palestinian Territories

Between the Mediterranean coast to the west, the tip of the Gulf of Aqaba to the south, and the Jordan River and Dead Sea to the east, lies a strip of land that has been bitterly contested throughout human history. Its ground is holy to three great religions; its soil is vital to two peoples; its fate is in the hands of one state—the state of Israel. This is the land once owned by ancient Egypt of the Pharaohs, settled by Jews more than 3000 years ago, fought over by Islamic rulers and Christian crusaders, annexed by the Ottoman Turks, colonized and partitioned by European colonists, and settled in modern times by Jews and Arabs in uneasy accommodation. The Jews call it Israel. The Arabs call it Palestine.

Figure 7-14 helps us understand the complex issues involved here. In 1946 the British, who had administered this area in post-Ottoman times, granted independence to what was then called "Transjordan," the kingdom east of the Jordan River. In 1948, the orange-colored area became the U.N.-sponsored state of Israel—including, of course, land that had long belonged to Arabs in this territory called Palestine.

As soon as Israel proclaimed its independence, neighboring Arab states attacked it. Israel, however, not only held its own but pushed the Arab forces back beyond its borders, gaining the green areas shown in Figure 7-14. Meanwhile, Transjordanian armies crossed the Jordan River and annexed the yellow-colored area named the

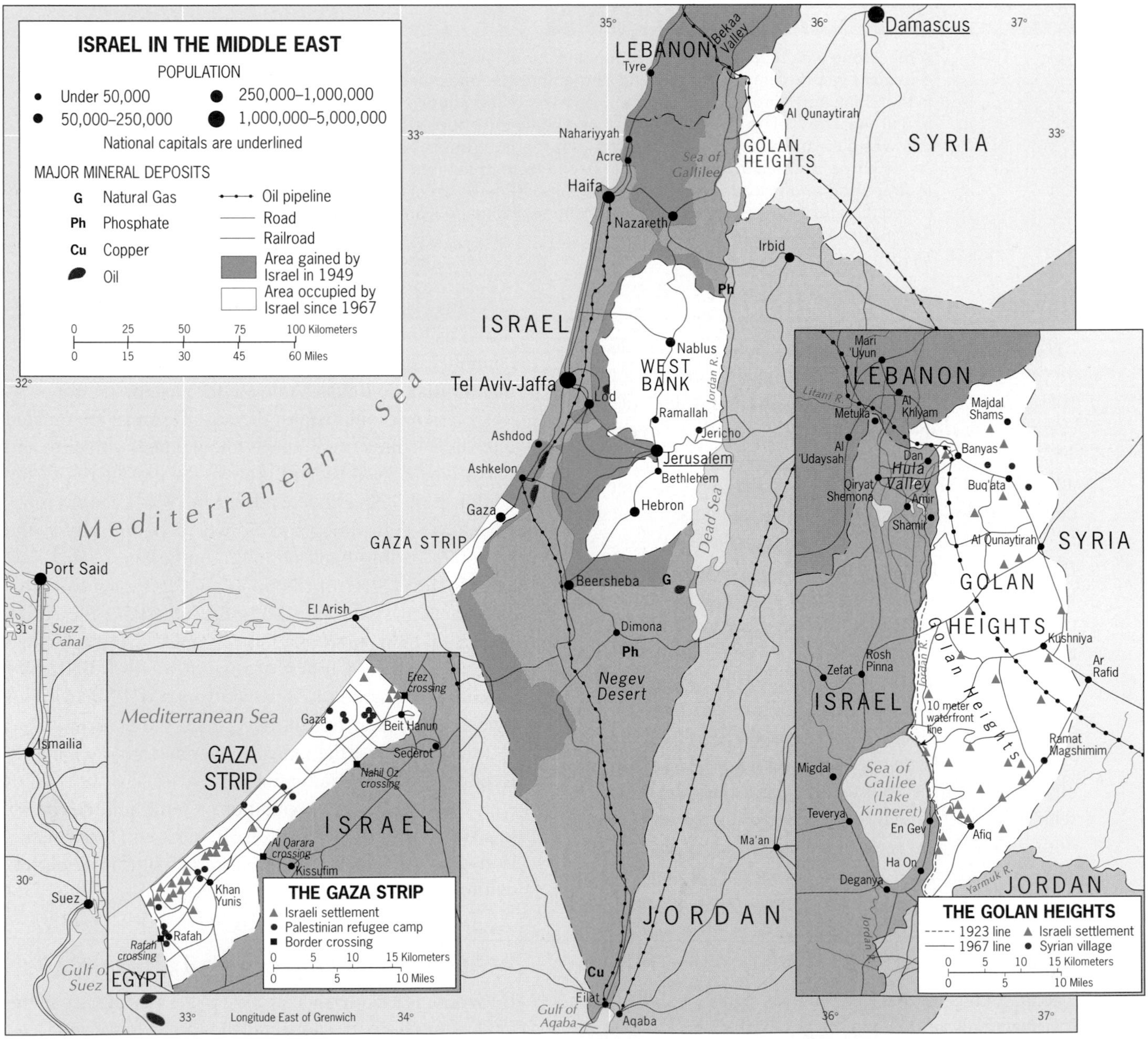

FIGURE 7-14 © H. J. de Blij, P. O. Muller, and John Wiley & Sons, Inc.

West Bank, including part of the city of Jerusalem. The king called his newly enlarged country Jordan.

More conflict followed. In 1967 a week-long war produced a major Israeli victory: Israel took the Golan Heights from Syria, the West Bank from Jordan, and the Gaza Strip from Egypt, and conquered the whole Sinai Peninsula all the way to the Suez Canal. In later peace agreements, Israel returned the Sinai but not the Gaza Strip.

All this strife produced a huge outflow of Palestinian Arab refugees and displaced persons. The Palestinians have been a stateless nation; about 1.2 million continue to live as Israeli citizens within the borders of Israel, but more than 2.5 million are in the West Bank and another 1.5 million in the Gaza Strip (the Golan Heights population is comparatively insignificant). An even larger number, however, live in neighboring and nearby countries including Jordan (2.8 million), Syria (440,000), Lebanon (415,000), and Saudi Arabia (310,000); another quarter-million reside in Iraq, Egypt, Kuwait, and Libya. Many have been assimilated into the local societies, but tens of thousands of others continue to live in refugee camps. In 2005 the scattered Palestinian population was estimated to total 9.5 million.

Israel is about the size of Massachusetts and has a population of 7.0 million (including 1.5 million Arabs who hold Israeli citizenship). But because of its location amid adversaries and its strong international links, notably with the United States, these data do not reflect Israel's importance in the region. Israel has built a powerful military, is widely believed to possess weapons of mass destruction, pursues policies that arouse Arab passions, and is the target of Arab hostility near and far. In recent times, Israel's regional neighbors have grown stronger as well, and Palestinian demands have been punctuated by a campaign of terror inside Israel that has cost hundreds of lives and has led to an unending cycle of even more deadly retaliation. It is a cycle of violence that could spin out of control with consequences far beyond the Middle East.

As a democracy with strong American ties, Israel is a recipient of massive U.S. assistance, and it has been the policy of the United States to seek an accommodation between Jews and Palestinians as well as between Israel and its Arab neighbors. The latter has resulted in peace agreements between Israel and its neighbors Egypt and Jordan, but the Palestinian issue has defied solution. To some Palestinians as well as Arabs abroad, Israel has no right to exist. To others more moderate and realistic, Israel must alter its policies to secure recognition and peace. United States governments, most recently under President Clinton, have tried to mediate, but failures of Palestinian as well as Israeli leadership doomed these efforts. During the Bush Administration, the president for the first time mentioned publicly the outlines of the geography that must form the basis for a long-term solution: the establishment of a Palestinian state alongside Israel. But much compromise will be required if such a solution is to be reached. Here are the key issues:

1. *The West Bank.* Even after its capture by Israel in 1967, the West Bank might have become a Palestinian homeland (and possibly a state), but Jewish immigration to the area made such a future difficult. In 1977 only 5000 Jews lived in the West Bank; by 2005 there were over 200,000, making up close to 10 percent of the population and creating a seemingly inextricable jigsaw of Jewish and Arab settlements (Fig. 7-15). Palestinians insist that Israel cease its building of West Bank settlements and dismantle those built without Israeli authorization. In any future Palestinian state, Jewish settlers who decide not to return to Israel will become citizens of Palestine just as Palestinian Arabs living in Israel are citizens of Israel.
2. *The Security Fence.* In response to the infiltration of suicide terrorists, Israel has walled off most of the West Bank along a border shown in Figure 7-15. This wall does not conform to the *de facto* boundary of the West Bank, cutting off about 10 percent of the West Bank's territory and in effect annexing this to Israel. The Security Fence in places also separates Palestinians from their land and cuts roads to schools and hospitals (photo p. 362). The Palestinians demand that this barrier be taken down; the Israelis reply that the Palestinian administration must control the terrorists.
3. *The Golan Heights.* The right inset map in Figure 7-14 suggests how difficult the Golan Heights issue is. The "heights" overlook a large area of northern Israel, and they flank the Jordan River and crucial Lake Kinneret (the Sea of Galilee), the water reservoir for Israel. Relations with Syria are not likely to become normal again until the Golan Heights are returned, but in democratic Israel the political climate may make ceding this territory impossible.
4. *The Gaza Strip.* As the left inset map in Figure 7-14 shows, the small and barren but populous Gaza Strip is a mosaic of Arab towns, small Israeli settlements, and Palestinian refugee camps. Of the 1.5 million Palestinian Arabs, many need jobs across the border in Israel to survive; others depend on international refugee relief. Israel's tight control is resented, and the comparatively neat and prosperous Israeli settlements form an inviting target for terrorists. In 2004, Israel's president began a domestic campaign to close down the Israeli settlements and bring their residents home, a proposal

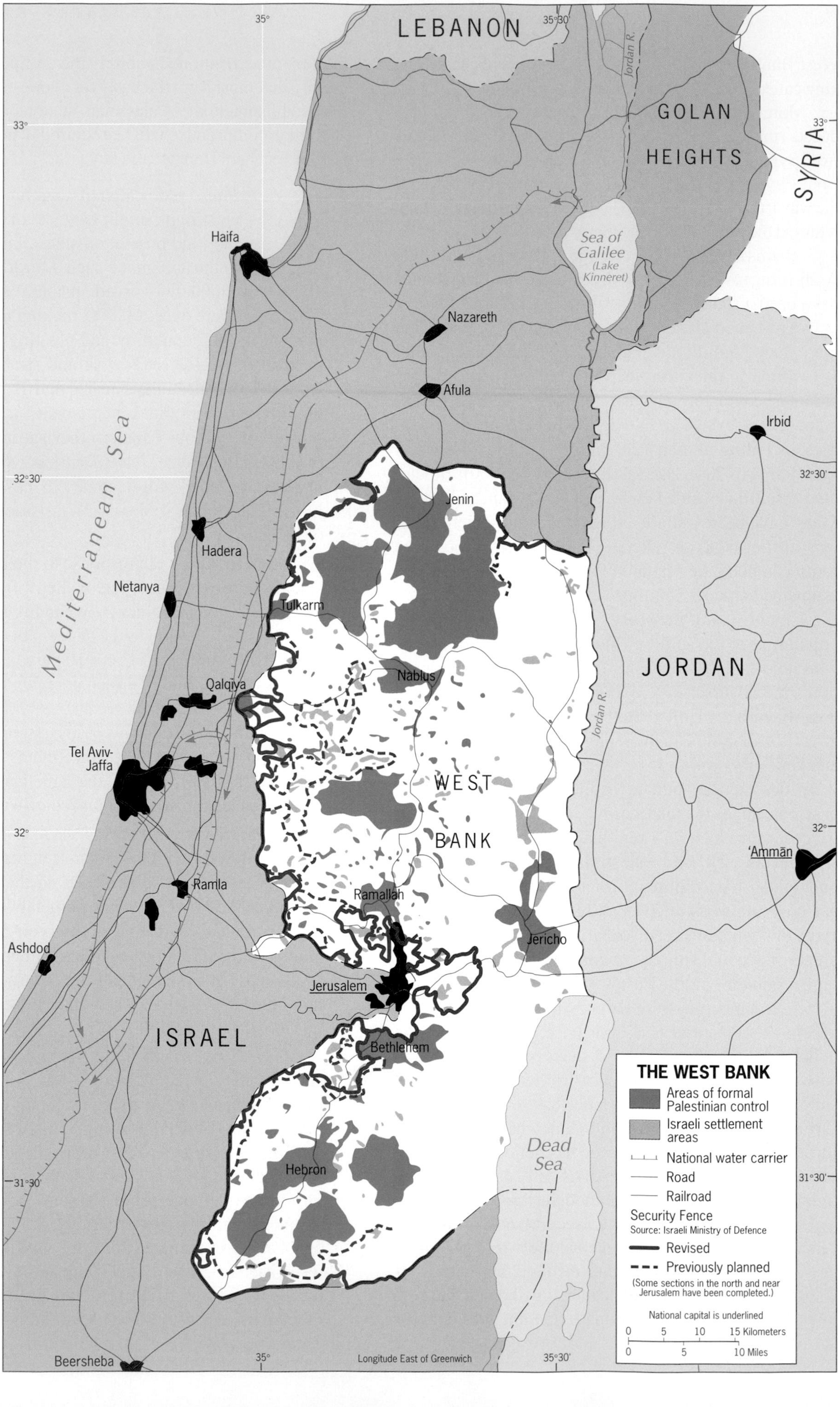

LEBANON
GOLAN
HEIGHTS
SYRIA
Sea of
Galilee
(Lake
Kinneret)
Jordan R.
Haifa
Nazareth
Afula
Irbid
Mediterranean Sea
Jenin
Hadera
Netanya
Tulkarm
Nablus
JORDAN
Qalqiya
Tel Aviv-
Jaffa
WEST
BANK
'Ammān
Ramla
Ramallah
Ashdod
Jericho
Jerusalem
ISRAEL
Bethlehem
Dead
Sea
Hebron
Beersheba
35°
35°30'
33°
32°30'
32°
31°30'
Longitude East of Greenwich
THE WEST BANK
Areas of formal Palestinian control
Israeli settlement areas
National water carrier
Road
Railroad
Security Fence
Source: Israeli Ministry of Defence
Revised
Previously planned
(Some sections in the north and near Jerusalem have been completed.)
National capital is underlined
0 5 10 15 Kilometers
0 5 10 Miles

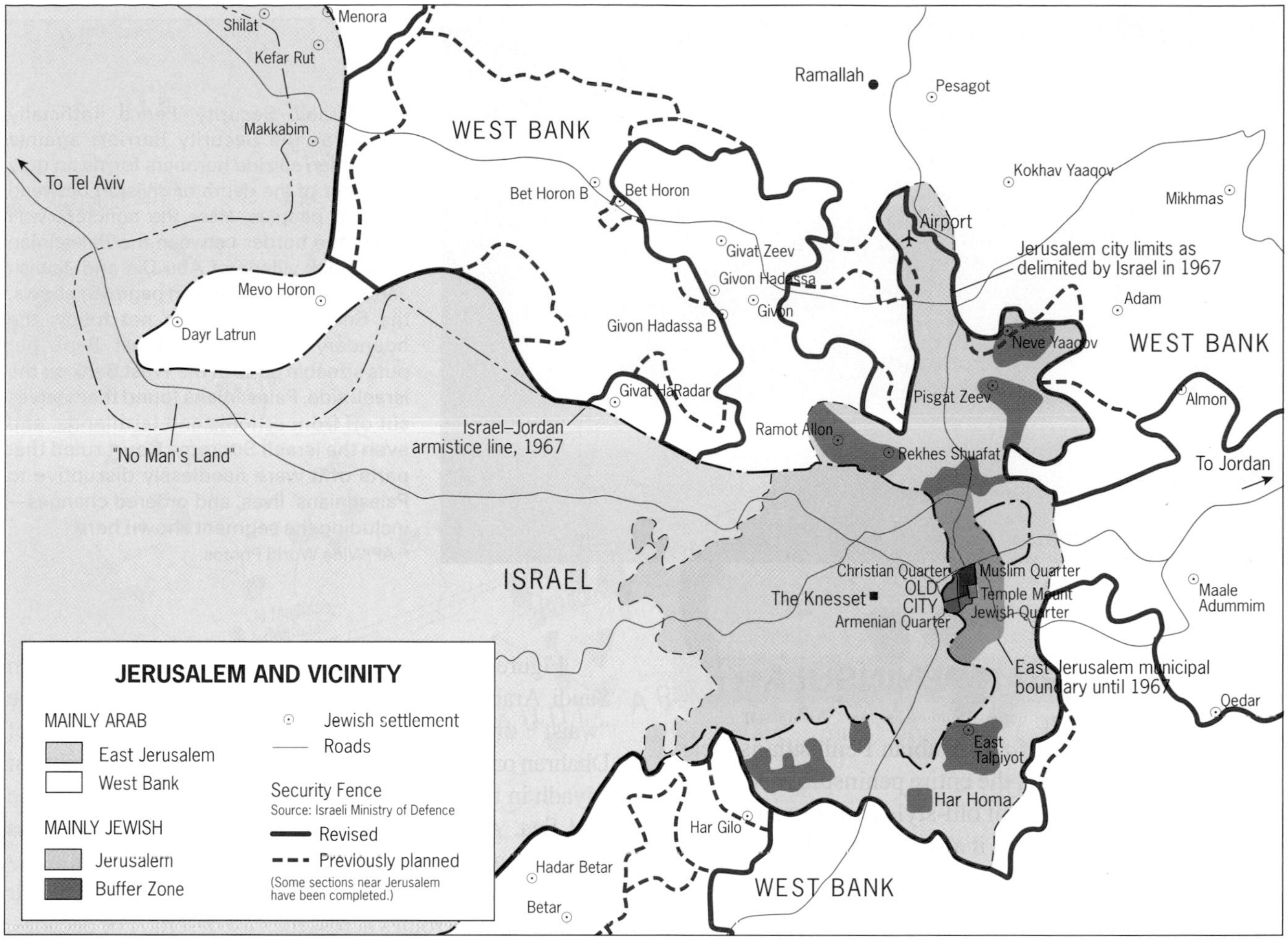

FIGURE 7-16 © H. J. de Blij, P. O. Muller, and John Wiley & Sons, Inc.

that had majority public support in Israel but elicited acrimonious opposition from conservative parties in the government.

5. *Jerusalem.* The United Nations intended Tel Aviv to be Israel's capital and Jerusalem an international city. But the Arab attack and the 1948–1949 war allowed Israel to drive toward Jerusalem (see Fig. 7-14, the green wedge into the West Bank). By the time a cease-fire was arranged, Israel held the western part of the city, and Arab forces the eastern sector. But in this eastern sector lay major Jewish historic sites, including the Western Wall of the holy Temple. Still, in 1950 Israel declared the western sector of Jerusalem its capital, making this, in effect, a *forward capital*. Figure 7-16 shows the position of the armistice line, leaving most of the Old City in Jordanian hands. But then, in the 1967 war, Israel conquered all of the West Bank, including East Jerusalem, and in 1980 the Jewish state reaffirmed Jerusalem's status as capital, calling on all nations to move their embassies from Tel Aviv. Meanwhile, the government redrew the map of the ancient city, building Jewish settlements in a ring around East Jerusalem that would end the old distinction between Jewish west and Arab east. This enraged Palestinian leaders, who still view Jerusalem as the eventual headquarters of their hoped-for Palestinian state. The segment of the Security Fence being planned for the Jerusalem area in 2005 (and partially completed) will have the effect of annexing virtually the entire Jerusalem urban region to Israel, creating a potentially insurmountable problem on the road to a two-state solution.

Israel lies at the center of a fast-moving geopolitical storm; the five issues raised here are only part of the overall problem (others include water rights, compensation for land expropriation, and the Palestinians' "right to return" to pre-refugee abodes). Now, in the age of nuclear, chemical, and biological weapons and longer-range missiles, Israel's search for an accommodation with its Arab neighbors is a race against time.

FIGURE 7-15 Adapted from map originally drawn by C.B. Williams & C.T. Elsworth, *The New York Times*, November 17, 1995, p. A-6. © 1995 by The New York Times Company. Reused by permission.

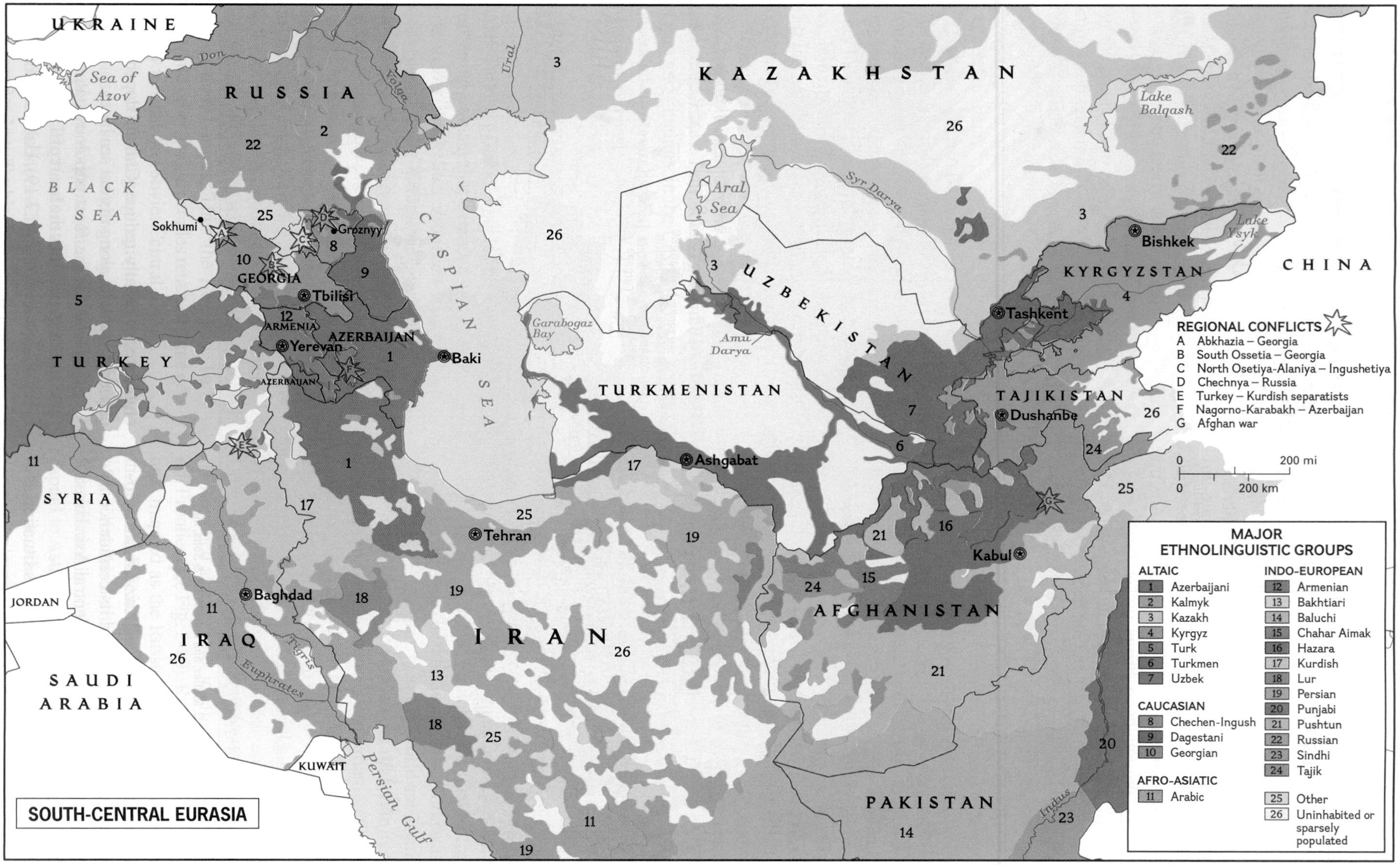
SOUTH-CENTRAL EURASIA
REGIONAL CONFLICTS
A Abkhazia – Georgia
B South Ossetia – Georgia
C North Osetiya-Alaniya – Ingushetiya
D Chechnya – Russia
E Turkey – Kurdish separatists
F Nagorno-Karabakh – Azerbaijan
G Afghan war
0 200 mi
0 200 km
MAJOR ETHNOLINGUISTIC GROUPS
ALTAIC
1 Azerbaijani
2 Kalmyk
3 Kazakh
4 Kyrgyz
5 Turk
6 Turkmen
7 Uzbek
CAUCASIAN
8 Chechen-Ingush
9 Dagestani
10 Georgian
AFRO-ASIATIC
11 Arabic
INDO-EUROPEAN
12 Armenian
13 Bakhtiari
14 Baluchi
15 Chahar Aimak
16 Hazara
17 Kurdish
18 Lur
19 Persian
20 Punjabi
21 Pushtun
22 Russian
23 Sindhi
24 Tajik
25 Other
26 Uninhabited or sparsely populated
UKRAINE
RUSSIA
KAZAKHSTAN
CHINA
KYRGYZSTAN
UZBEKISTAN
TAJIKISTAN
TURKMENISTAN
AFGHANISTAN
PAKISTAN
IRAN
IRAQ
SYRIA
JORDAN
SAUDI ARABIA
KUWAIT
TURKEY
GEORGIA
ARMENIA
AZERBAIJAN
BLACK SEA
Sea of Azov
CASPIAN SEA
Aral Sea
Garabogaz Bay
Lake Balqash
Lake Ysyk
Persian Gulf
Don
Volga
Ural
Syr Darya
Amu Darya
Tigris
Euphrates
Indus
Sokhumi
Groznyy
Tbilisi
Yerevan
Baki
Tehran
Ashgabat
Tashkent
Bishkek
Dushanbe
Kabul
Baghdad

FIGURE 7-20 Reproduced from double map supplement (Caspian Sea), *National Geographic*, May 1999.

Tashkent, lies in the far eastern core area of the country, where most of the people live in towns and farm villages, and the crowded Fergana Valley is the focus. In the west lies the shrinking Aral Sea (photos below), whose feeder streams were diverted into cotton fields and croplands during the Soviet occupation; heavy use of pesticides contaminated the groundwater, and countless thousands of local people suffered severe medical problems as a result.

In the post-Soviet period, Islamic (Sunni) revivalism has become a problem for Uzbekistan. Wahhabism, the particularly virulent form of Sunni fundamentalism, took root in eastern Uzbekistan; government efforts to suppress it have only increased its strength in the countryside.

Turkmenistan, a desert republic that extends all the way from the Caspian Sea to the borders of Afghanistan, has a population of just 5.9 million, of which about three-quarters are Turkmen. The Turkmen (or Turkomen, as they are called elsewhere) are a traditionally nomadic Muslim people, many of whom were forced into sedentary farming during the country's days as a Soviet Socialist Republic, when a dictatorial regime controlled them. After independence, a former ruler reinvented himself as a Turkmen patriot, banned all opposition, and outdid his erstwhile Soviet masters in his own control over education and religion.

During Soviet times, Turkmenistan's communist planners began work on a massive project: the Garagum (Kara Kum) Canal, designed to bring water from Turkestan's eastern mountains into the heart of the desert. Today the canal is 700 miles (1100 km) long, and it has enabled the cultivation of some 3 million acres of cotton, vegetables, and fruits. The plan is to extend the canal all the way to the Caspian Sea, but meanwhile Turkestan has hopes of greatly expanding its oil and gas output

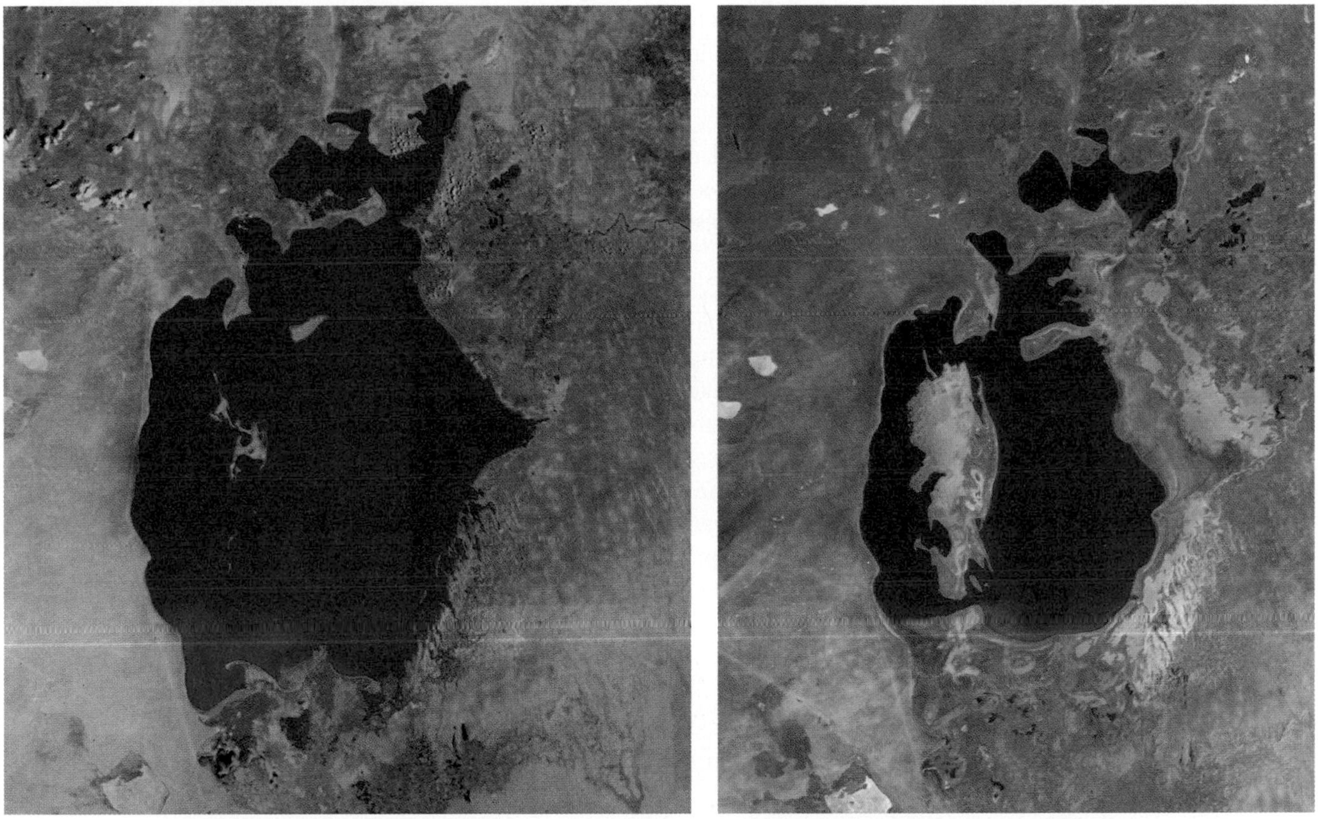

The Aral Sea is a saltwater lake on the boundary between Kazakhstan and Uzbekistan, both former Soviet Socialist Republics. During the communist period, the lake's two inflowing rivers were diverted for irrigation, resulting in the dramatic shrinkage illustrated by these two Landsat images. The image on the left was taken in the 1960s, when the Aral Sea covered an area of about 26,300 square miles (68,000 sq km). Thirty years later, the lake had been cut in two and the total water area combined was less than 13,000 square miles (33,700 sq km). The surface had dropped about 50 feet (15 m) and a large island in what is now known as the "Greater Sea" (as opposed to the "Lesser Sea" to the north) threatened to become another land bridge. The fate of the Aral Sea is but one chapter in Soviet-era mismanagement of the environment in the headlong rush to increase agricultural productivity. The Kazakh and Uzbek governments, with assistance from international agencies, are now in the process of restoring the flow of the two feeder rivers to arrest the Aral Sea's decline, but it is unlikely that a significant rise in the water level can be achieved.

from Caspian Basin reserves. But look again at Figure 7-19: Turkmenistan's relative location is not advantageous for export routes.

Kyrgyzstan's topography and political geography are reminiscent of the Caucasus. Shown in yellow in Figure 7-19, Kyrgyzstan lies intertwined with Uzbekistan and Tajikistan to the point of exclaves and enclaves along its borders. The Kyrgyz, for whom the Soviets established this "republic," constitute barely 60 percent of the population of 5.2 million. Uzbeks and other minorities form a complex cultural geography; mountains and valleys isolate communities and make nation-building difficult. The agricultural economy is weak, consisting of pastoralism in the mountains and farming in the valleys. About 75 percent of the people profess allegiance to Islam, and Wahhabism has gained a strong foothold here. The town of Osh is often referred to as the headquarters of the movement in Turkestan.

As if to confirm the comparison to the Caucasus, Kyrgyzstan has become a staging area in the War on Terror. In the aftermath of America's invasion of Afghanistan in 2001, the Kyrgyz government permitted the United States to establish a military base near the capital, Bishkek. This impelled Moscow to start negotiations to be allowed to build its own military base nearby, ostensibly against terrorism but undoubtedly also to create a counterweight against the American presence in this former Russian colony.

Tajikistan is the poorest of the five states of Turkestan; as many as 800,000 of its workers (out of a total population of 6.9 million) cross into Russia every year looking for temporary employment. The Tajiks, who constitute about 80 percent of the population, are of Persian, not Turkic origin and speak a Persian, and thus Indo-European, language. Most Tajiks, despite their Persian affinities, are Sunni Muslims, not Shi'ites. Tajikistan's terrain is even higher than that of neighboring Kyrgyzstan, creating similar barriers to the integration of a multicultural society. Small though the country is, regionalism plagues this state: the government in Dushanbe often is at odds with the barely connected north (see Fig. 7-19), a hotbed not only of Islamic revivalism but also of anti-Tajik, Uzbek activism in Tajikistan's portion of the Fergana Valley. Tajikistan also is troubled by its location relative to the heroin-producing industry of Afghanistan and the major Russian market for this drug. Several tons are confiscated every year, but many more cross the country on their way north, sustaining a vast international criminal enterprise.

Afghanistan

Turkestan's southernmost country, Afghanistan, exists because the British and Russians, competing for hegemony in this area during the nineteenth century, agreed to tolerate it as a cushion—or **buffer state**—between them. This is 21
how Afghanistan acquired the narrow extension leading from the main territory eastward to the Chinese border—the Vakhan Corridor (Fig. 7-21). As the colonialists delimited it, Afghanistan adjoined the domains of the Turkmen, Uzbeks, and Tajiks to the north, Persia (now Iran) to the west, and the western flank of British India (now Pakistan) to the east.

Geography and history seem to have conspired to divide Afghanistan. As Figure 7-21 shows, the towering Hindu Kush range dominates the center of the country, creating three broad environmental zones: the relatively well-watered and fertile northern plains and basins; the rugged, earthquake-prone central highlands; and the desert-dominated southern plateaus. Kabol (Kabul), the capital, lies on the southeastern slope of the Hindu Kush, linked by narrow passes to the northern plains and by the Khyber Pass to Pakistan.

Across this variegated landscape moved countless peoples: Greeks, Turks, Arabs, Mongols, and others. Some settled here, their descendants today speaking Persian, Turkic, and other languages. Others left archeological remains or no trace at all. The present population of Afghanistan (30.1 million) has no ethnic majority. This is a country of minorities in which the Pushtuns (or Pathans) of the east are the most numerous but make up barely 40 percent of the total. The second-largest minority are the Tajiks, a world away across the Hindu Kush, concentrated in the zone near Afghanistan's border with Tajikistan. The Hazara of the central highlands and the south, the Uzbeks and Turkmen in the northern border areas, the Baluchi of the southern deserts, and other, smaller groups scattered across this riven country create one of the world's most complex cultural mosaics (Fig. 7-20).

Episodes of conflict have marked the history of Afghanistan, but none was as costly as its involvement in the Cold War. Following the Soviet intervention of 1979, the United States supported the Muslim opposition, the *Mujahideen* ("strugglers"), with modern weapons and money, and the Soviets were forced to withdraw. Soon the factions that had been united during the anti-Soviet campaign were in conflict, delaying the return of some 4 million refugees who had fled to Pakistan and Iran. The situation resembled the pre-Soviet past: a feudal country with a weak and ineffectual government in Kabol.

In 1994, what at first seemed to be just another warring faction appeared on the scene: the so-called Taliban ("students of religion") from religious schools in Pakistan. Their avowed aim was to end Afghanistan's chronic factionalism by instituting strict Islamic law. Popular support in the war-weary country, especially among the Pushtuns, led to a series of successes, and by 1996 the Taliban had taken Kabol. Afghanistan's physical geography and

FIGURE 7-21 © H. J. de Blij, P. O. Muller, and John Wiley & Sons, Inc.

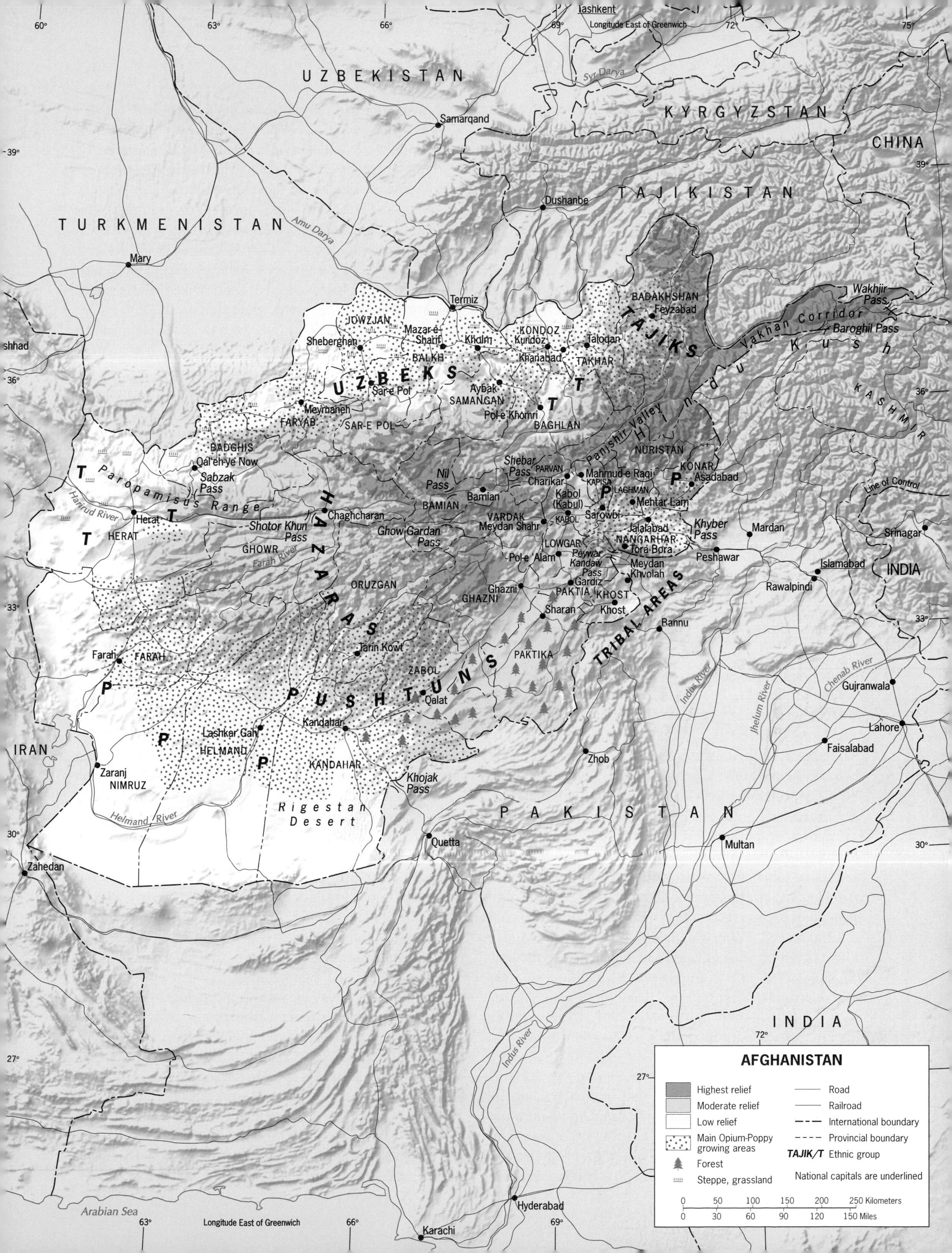
AFGHANISTAN
Highest relief
Moderate relief
Low relief
Main Opium-Poppy growing areas
Forest
Steppe, grassland
Road
Railroad
International boundary
Provincial boundary
TAJIK/T Ethnic group
National capitals are underlined
0 50 100 150 200 250 Kilometers
0 30 60 90 120 150 Miles
Longitude East of Greenwich
UZBEKISTAN
TURKMENISTAN
KYRGYZSTAN
TAJIKISTAN
CHINA
IRAN
PAKISTAN
INDIA
Tashkent
Samarqand
Dushanbe
Mary
Termiz
Mashhad
Amu Darya
Syr Darya
UZBEKS
TAJIKS
HAZARAS
PUSHTUNS
TRIBAL AREAS
JOWZJAN
BALKH
KONDOZ
TAKHAR
BADAKHSHAN
SAMANGAN
BAGHLAN
FARYAB
SAR-E POL
BADGHIS
HERAT
GHOWR
ORUZGAN
FARAH
HELMAND
NIMRUZ
KANDAHAR
ZABOL
GHAZNI
PAKTIKA
PAKTIA
KHOST
LOWGAR
VARDAK
BAMIAN
PARVAN
KAPISA
KABOL
LAGHMAN
NANGARHAR
NURISTAN
KONAR
Sheberghan
Mazar-e Sharif
Kholm
Kundoz
Khanabad
Taloqan
Feyzabad
Sar-e Pol
Aybak
Pol-e Khomri
Meymaneh
Qal'eh-ye Now
Sabzak Pass
Paropamisus Range
Harirud River
Herat
Chaghcharan
Shotor Khun Pass
Farah River
Ghow Gardan Pass
Bamian
Nil Pass
Shebar Pass
Charikar
Mahmud-e Raqi
Asadabad
Kabol (Kabul)
Sarowbi
Mehtar Lam
Jalalabad
Tora Bora
Khyber Pass
Meydan Shahr
Pol-e 'Alam
Peywar Kandaw Pass
Gardiz
Meydan Khvolah
Khost
Ghazni
Sharan
Tarin Kowt
Qalat
Kandahar
Lashkar Gah
Farah
Zaranj
Khojak Pass
Rigestan Desert
Helmand River
Panjshir Valley
Hindu Kush
Vakhan Corridor
Wakhjir Pass
Baroghil Pass
KASHMIR
Line of Control
Srinagar
Mardan
Peshawar
Islamabad
Rawalpindi
Bannu
Zhob
Quetta
Multan
Gujranwala
Lahore
Faisalabad
Indus River
Jhelum River
Chenab River
Zahedan
Hyderabad
Karachi
Arabian Sea

northern opposition then slowed the Taliban's drive, but in 1998 they penetrated the Uzbek and Tajik strongholds and captured Mazar-e-Sharif, a key northern city.

The Taliban's imposition of Islamic law was so strict and severe that Islamic as well as non-Islamic countries objected. Restrictions on the activities of women ended their professional education, employment, and freedom of movement, and had a devastating impact on children as well. Public amputations and stonings enforced the Taliban's code. In the process, Afghanistan became a haven for groups of revolutionaries whose goals went far beyond those of the Taliban: they plotted attacks on Western interests throughout the realm and threatened Arab regimes they deemed compliant with Western priorities. Taliban-ruled, cave-riddled, remote, and isolated Afghanistan was an ideal locale for these outlaws. Already in possession of arms and ammunition (Soviet as well as American) left over from the Cold War, they also benefited from Afghanistan's huge, illicit opium trade. In 2000, according to UN data, Afghanistan produced between 70 and 79 percent of the world's opium. Much of the revenue found its way into the coffers of the conspirators.

With such resources, the militants were able to organize and launch attacks against several Western targets in the realm and elsewhere, but events took a fateful turn in 1996 that would empower them as never before. During the conflict with the Soviets, the Mujahideen cause had been supported not only by the United States but also by the Saudi extremist referred to earlier in this chapter, Usama bin Laden. The son of a Saudi construction billionaire who had more than 50 children, Usama graduated from a university in Jiddah in 1979 and headed for Afghanistan with an inherited fortune estimated at about U.S. $300 million to help the anti-Soviet campaign. Well-connected in Saudi Arabia and now a pivotal figure in Afghanistan, bin Laden saw his fame as well as his war chest grow as the Soviet intervention collapsed. Following the withdrawal of the communist forces, he returned to Saudi Arabia, where he denounced his government for allowing U.S. troops on Saudi soil during the Gulf War. The Saudi regime responded by stripping him of his citizenship and expelling him, and bin Laden fled to Sudan. There he set up several legitimate businesses to facilitate his now-global financial transactions, but he also established terrorist training camps. Under international pressure, the Khartoum regime ousted him in 1996, and bin Laden returned to a country he knew would welcome him again—Afghanistan.

Bin Laden used his fortune to support the Taliban's drive to conquer Afghanistan, and in close association with the movement's leader, Mullah Omar, he established the headquarters of the terrorist organization al-Qaeda in the country's mountainous east. From this base he coordinated the global network that had already committed numerous terrorist acts, including the first (1993) attack on New York's World Trade Center. Al-Qaeda's exploits were funded by numerous Muslim sources, including many in Saudi Arabia. This made possible well-coordinated attacks on the U.S. embassies in Nairobi, Kenya, and Dar es Salaam, Tanzania, the bombing of the *USS Cole* in Adan, Yemen, and other strikes. There was no effective response from U.S. or other Western sources. After the embassy bombings, President Clinton ordered an inconsequential missile strike on a suspected terrorist camp in eastern Afghanistan.

Meanwhile, the suicide attacks of September 11, 2001 were being planned as operatives traveled the world and moved hundreds of thousands of dollars through secret channels to enable al-Qaeda's teams to prepare. Three of the four planned attacks succeeded, killing thousands and causing billions of dollars in damage. Several weeks later (having given al-Qaeda leaders time to plan their escape), U.S. and British forces, with the acquiescence of the president of Pakistan, attacked both the Taliban regime and the al-Qaeda infrastructure in Afghanistan, the first stage in the proclaimed "War on Terror" that would leave virtually no country unaffected.

The allied assault on Afghanistan quickly led to the collapse and destruction of the Taliban regime, many of its soldiers and supporters killed by warlords settling old scores. But neither the leader of the Taliban nor Usama bin Laden was captured, and only a few of al-Qaeda's top 40 lieutenants were apprehended or eliminated. Usama bin Laden, whose cameraman had taped a celebratory dinner in a restaurant in Kandahar after the 9/11 attacks, continued to distribute recordings of polemics and threats.

Meanwhile, Allied forces led by the United States ended the terror the Taliban had imposed on Afghan society and stabilized the political situation insofar as possible, focusing their efforts on the capital Kabol and its environs. Remote areas remained beyond effective control, but reconstruction of the devastated country got started where possible. An interim Afghan administration saw to it that women were allowed once again to attend schools, serve as teachers and nurses and enter other professions, and move about with some freedom. But what was accomplished in and around Kabol could not soon be repeated in the outlying provinces and their capitals such as Herat, Mazar-e-Sharif, or Feyzabad, where corrupt regional governors or rapacious warlords remained in power. Still, the central government made headway, relieving some of the governors of their jobs, coopting others to join the government, and extending Kabol's reach to the borders. Only the south and the mountainous, forested areas adjacent to Pakistan's Tribal Areas (Fig. 7-21) remained largely uncontrolled, and here remnants of al-Qaeda continued to operate.

In the autumn of 2004 Afghanistan achieved a remarkable feat: a democratic election that validated the presi-

President Hamid Karzai, the first elected post-Taliban leader of Afghanistan, walks the red carpet at the presidential palace in Kabol on his way to a meeting with tribal leaders, accompanied by heavily armed bodyguards. The president's cabinet includes a plurality of Pushtuns but has Uzbek, Tajik, and other representatives of Afghanistan's diverse cultural makeup. © Paula Bronstein/Getty Images News and Sports Services.

dency of Hamid Karzai and produced a consensus government including some of his former adversaries (photo above). Ballots were cast in even the most remote areas of the country, turnout was high, and disruption was minimal. On another front, however, Afghanistan faced trouble: the widespread revival of opium-poppy cultivation and the rapid growth of heroin exports. The Taliban regime had outlawed opium-poppy cultivation in 2000 and, through its draconian punishments, had reduced the active acreage to about 4000 (1600 hectares). But the new Kabol government did not have the policing powers the Taliban had, and soon cultivation was on the upswing; by 2003, more than 80,000 acres of opium-poppy fields were once again making Afghanistan the world's leading exporter of heroin. This presented the new government as well as its American and other coalition allies with a difficult challenge, because an eradication campaign would alienate farmers throughout Afghanistan at a time when stability in the countryside was key to national progress.

The events of the past several years have greatly altered the regional geography of this part of the world, including the functioning of the Afghanistan–Pakistan boundary, the movement of people and goods, the diffusion of religious doctrines, and the operation of political systems. There was a time when the transition from Southwest Asia to South Asia was evident to anyone emerging from the Khyber Pass onto the plain at Peshawar, but those days are over. Pakistan now lies directly in the path of the winds of change coming from the southern flank of Turkestan.

WHAT YOU CAN DO

RECOMMENDATION: Seize a campus opportunity! If you are taking this course while studying at a college or university (rather than online), you may be living and working in the most multicultural social environment you will ever encounter. Every campus administration provides ways to meet and interact with foreign students, and using this chance to do so can be as valuable an educational experience as any you will have, and may lead to lasting friendships, perhaps even visits in years to come. Foreign students tend to be pleased when their hosts express an interest in their home countries, and their reactions to what they encounter here can be interesting and enlightening. Ask them if they've ever taken any geography courses back home!

GEOGRAPHIC CONNECTIONS

1 If you were to travel through the Muslim world, visiting universities and schools, you would frequently see a version of Figure 7-5 (Areas Under Muslim Rule at Certain Times) on office walls and in classrooms. Using your knowledge of spatial diffusion processes (covered on pp. 331–332), explain how this map came about. What areas on the map has Islam "lost"? Where does Islam appear to be gaining? Why do some of Islam's most significant gains not appear on this map?

2 The United States and its allies intervened in Iraq in 2003 for several reasons, including the overthrow of the country's dictatorial regime. Some geographers argued that, among the contingency plans for the country following the military campaign, consideration should have been given to the "Bosnia Model"—the option to temporarily divide Iraq into three parts or perhaps four (the fourth being the capital, Baghdad) to allow for variable rates of adjustment to the occupation, followed eventually by reunification. With the support of the maps in Chapter 7, briefly characterize the cultural geography of Iraq and comment on its traditional regionalism in this context. What might be the responses of Iraq's neighbors to such a plan?

KYRGYZSTAN
UZBEKISTAN
TASHKENT
TURKMENISTAN
TAJIKISTAN
AFGHANISTAN
Kābol
Herāt
Kandahār
Mazār-e Sharīf
PAKISTAN
Islāmābād
Peshawar
Rawalpindi
Lahore
Faisalabad
Multan
Quetta
Hyderābād
KARACHI
BALUCHISTAN
SIND
IRAN
XINJIANG UYGUR (SINKIANG)
TAKLA MAKAN (DESERT)
KUNLUN SHAN
ALTUN SHAN
QINGHAI
CHINA
PLATEAU OF TIBET
XIZANG (TIBET)
Lhasa
GANGDISE SHAN
NYAINQENTANGLHA SHAN
TANGGULA SHAN
KARAKORAM
HINDU KUSH
JAMMU AND KASHMIR
Srinagar
K2 (Godwin Austen) 28 250
HIMALAYA
NEPAL
Kathmandu
Mt. Everest 29 035
BHUTAN
Thimphu
SIKKIM
ARUNACHAL PRADESH
ASSAM
NAGALAND
MEGHALAYA
MANIPUR
MIZORAM
TRIPURA
BANGLADESH
Dhaka
Chittagong
MYANMAR (BURMA)
INDIA
DELHI
New Delhi
PUNJAB
HARYANA
UTTARANCHAL
HIMACHAL PRADESH
UTTAR PRADESH
Lucknow
KANPUR
Ganges
RĀJASTHĀN
THAR DESERT
Jaipur
Jodhpur
Ajmer
GUJARAT
AHMADĀBĀD
RANN OF KUTCH
KATHIAWAR PENINSULA
Tropic of Cancer
MADHYA PRADESH
Bhopāl
Indore
Jabalpur
BIHAR
Patna
JHARKHAND
WEST BENGAL
KOLKATA (CALCUTTA)
ORISSA
Cuttack
Bhubaneshwar
CHHATTISGARH
Raipur
Nāgpur
MAHĀRASHTRA
MUMBAI (BOMBAY)
Pune
KONKAN COAST
WESTERN GHATS
EASTERN GHATS
DECCAN PLATEAU
HYDERABAD
ANDHRA PRADESH
Vishākhapatnam
GOLCONDA COAST
GOA
Panaji
KARNATAKA
BANGALORE
Mysore
CHENNAI (MADRAS)
COROMANDEL COAST
TAMIL NADU
Madurai
KERALA
MALABAR COAST
Calicut
Coimbatore
Trivandrum
LAKSHADWEEP (LACCADIVE IS.) (India)
LAKSHADWEEP
MALDIVES
SRI LANKA (CEYLON)
Colombo
Kandy
Jaffna
Trincomalee
Galle
ARABIAN SEA
LACCADIVE SEA
BAY OF BENGAL
INDIAN OCEAN
ANDAMAN ISLANDS (India)
NICOBAR ISLANDS (India)
A Area occupied by Pakistan and claimed by India
B Area claimed and occupied by India; status disputed by Pakistan.
C Area occupied by China and claimed by India.
D Area occupied by India and claimed by China.
Relief
Meters
3050
1525
610
305
152.5
Sea Level
152.5 500
1525 5000
3050 10 000
6100 20 000
0 50 100 200 300 400 500 Miles
0 100 200 400 600 800 Kilometers
Scale 1:16,896,000; one inch to 266 miles
Polyconic projection
Elevations and depressions are given in feet
Longitude East of Greenwich

Chapter 8

South Asia

CONCEPTS, IDEAS, AND TERMS

1 **Wet monsoon**
2 **Social stratification**
3 **Refugees**
4 **Population geography**
5 **Population distribution**
6 **Population density**
7 **Physiologic density**
8 **Rate of natural population change**
9 **Demographic transition**
10 **Population explosion**
11 **Forward capital**
12 **Irredentism**
13 **Caste system**
14 **Intervening opportunity**
15 **Natural hazards**
16 **Failed state**
17 **Insurgent state**

REGIONS

- PAKISTAN
- INDIA
- BANGLADESH
- MOUNTAINOUS NORTH
- SOUTHERN ISLANDS

THE GREAT LANDMASS of Eurasia projects numerous peninsulas into the seas that surround it—Scandinavia, Iberia, Arabia, Malaya, Korea, Kamchatka—but none matches the great triangle that points like a dagger into the ocean: India. So vast is India that it divides the northern Indian Ocean into two seas, the Arabian Sea to the west and the Bay of Bengal to the east (Fig. 8-1). So populous is the geographic realm that has India at its heart that it is poised to overtake China centered East Asia during the present century. South Asia, from Pakistan to Bangladesh and from Nepal to Sri Lanka, is a world unto itself.

DEFINING THE REALM

Mountains, deserts, and coastlines combine to make South Asia one of the world's most vividly defined physiographic realms. To the north, the Himalaya Mountains create a natural wall between South Asia and China. To the east, mountain ranges and dense forests mark the boundary between South and Southeast Asia. To the west, rugged highlands and expansive deserts separate South Asia from its neighbors. Within these confines lies a geographic realm that is more densely populated than any other. If current population trends continue, it will soon be the most populous realm on Earth as well.

South Asia consists of five regions (Fig. 8-2). Its keystone is India, whose population passed the 1 billion mark in 1999. In the west lies Pakistan. South Asia's eastern

FIGURE 8-1

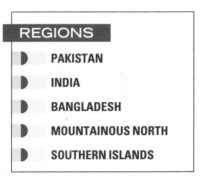

MAJOR GEOGRAPHIC QUALITIES OF

South Asia

1. South Asia is clearly defined physiographically and is bounded by mountains, deserts, and ocean; the Indian peninsula is Eurasia's largest.
2. South Asia is the world's second most poverty-filled realm (ranking just above Subsaharan Africa), with low average incomes, low levels of education, poorly balanced diets, and poor overall health.
3. With only 3 percent of the world's land area but 23 percent of its population, more than half of it engaged in subsistence farming, South Asia's economic prospects are bleak.
4. Population growth rates in South Asian countries exceed the global average; India's population surpassed the 1 billion mark in 1999.
5. The North Indian Plain, the lower basin of the Ganges River, contains the heart of the world's second largest population cluster.
6. Despite encircling mountain barriers, invaders from ancient Greeks to later Muslims penetrated South Asia and complicated its cultural mosaic.
7. British colonialism unified South Asia under a single flag, but the empire fragmented into several countries along cultural lines after Britain's withdrawal in 1947.
8. Pakistan, South Asia's western region, lies on the flanks of two realms: largely Muslim North Africa/Southwest Asia and dominantly Hindu South Asia.
9. India is the world's largest federation and most populous democracy, but its political achievements have not been matched by enlightened economic policies.
10. Religion remains a powerful force in South Asia. Hinduism in India, Islam in Pakistan, and Buddhism in Sri Lanka all show tendencies toward fundamentalism and nationalism.
11. Active and potential boundary problems involve internal areas (notably between India and Pakistan in Kashmir) as well as external locales (between India and China in the northern mountains).

flank is centered on Bangladesh. The northern region consists of the mountainous lands of Kashmir, Nepal, and Bhutan. And the southern region includes the islands of Sri Lanka and the Maldives. As the map shows, India divides into several subregions.

South Asia's physiographic boundaries are formidable barriers, but they have not prevented conquerors or proselytizers from penetrating it. As a result, today this realm is a patchwork of religions, languages, traditions, and cultural landscapes. So complex is this mosaic that, remarkably, its political geography numbers just seven states (and only five on the mainland).

Among the cultural infusions was Islam. Today, Pakistan is an Islamic republic, and Islam provides the cement for the state. Pakistan's eastern border with India is a cultural divide in more ways than one: dominantly Hindu India is a secular, not a theocratic, state. Why, then, do we include Pakistan in the South Asian rather than the Southwest Asian/North African realm? One criterion is ethnic continuity, which links Pakistan to India rather than to Afghanistan or Iran. Another is historical geography. Pakistan was part of Britain's South Asian Empire, and it originated from the partition of that domain between Muslim and Hindu majorities. Although Urdu is the official national language of Pakistan, English is the *lingua franca*, as it is in India. Furthermore, the border between India and Pakistan does not signify the eastern frontier of Islam in Asia. About 150 million of India's more than 1 billion citizens are Muslims, and in South Asia's eastern region, Bangladesh (population: 147 million) is more than 85 percent Muslim. Finally, Pakistan and India are locked in a struggle to control a vital mountainous area in the far north, where the British withdrawal in 1947 left the boundary between them unresolved. Even as Indian and Pakistani cricket teams play each other on sun-baked pitches, their armies face off in a deadly conflict—a conflict that has the potential to unleash a nuclear war.

South Asia and the War on Terror

The western boundary of South Asia has long had the characteristics of a frontier—a zone of transition and contention not all of which is fully integrated into the political systems of the adjoining states. The fateful Soviet invasion and occupation of Afghanistan (1979–1989), when Moscow tried to stabilize that fractious country and establish a secular regime in Kabol (Kabul), caused a bitter conflict in which Afghans fought each other even as they resisted their joint enemy. That war drove several million Pushtun and other Afghan refugees across the Pakistani border. Already along this same border, inside Pakistan but effectively not under governmental control, lay a mountainous, forbidding, inaccessible strip of land called the *Tribal Areas* (see the map on page 375)—where ethnicity, traditions, clan systems, and cultural landscapes were more representative of remote parts of Turkestan than of mainstream Pakistan. Here no one, not even a Pakistani official, goes without the consent of chief and mullah.

In the proxy-war context of the Cold War, the United States saw an opportunity to damage Soviet interests and

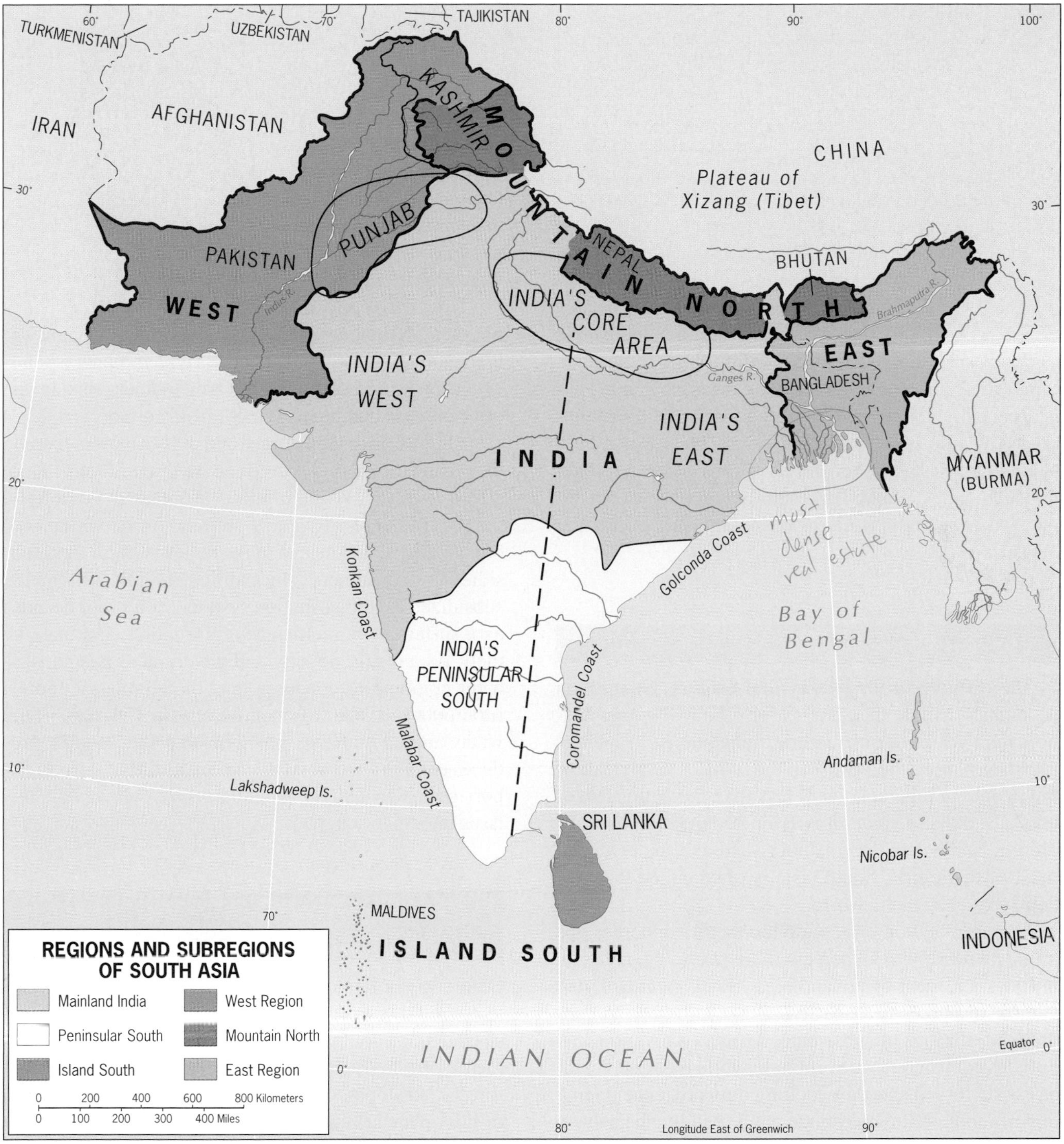

FIGURE 8-2 © H. J. de Blij, P. O. Muller, and John Wiley & Sons, Inc.

so dispatched an enormous quantity of weapons to the Afghan resistance fighters via Pakistan. This infusion, in addition to the impact of the ever-growing refugee population, changed the ideological landscape in Pakistan itself. Even as the Soviets were defeated (Usama bin Laden was on the Afghan scene and financed the campaign as well), Islamic revivalism mushroomed in neighboring Pakistan. Islamic schools teaching strict Islamic doctrines, approved by Pakistan and supported financially by Saudi Arabians, gave rise to the fundamentalist Taliban movement that swept into power in chaotic Afghanistan during the 1990s.

The Taliban regime brought Afghanistan stability at a dreadful social price; it also created a safe base for the al-Qaeda terrorist movement that was to strike at Western interests the world over, including, ultimately, the United States itself. As noted in Chapter 7, the attacks of

September 11, 2001 provoked massive U.S. retaliation in Afghanistan, where the Taliban regime was eliminated and al-Qaeda members were killed or arrested (most of the leadership escaped and went into hiding, possibly in the Tribal Areas). But for Pakistan this campaign had momentous consequences. The leader of its military regime, General Pervez Musharraf, decided to side with the United States in its war against terrorism—a complete reversal for a country that had spawned the Taliban, recognized the Taliban regime, and sympathized with the movement's Islamist aspirations.

We defer the implications of all this for Pakistan's political geography until our discussion of this realm's regions, but in terms of regional definitions it is clear that things are changing on South Asia's western flank. Pakistan and India continue their deadly South Asian embrace over Kashmir and other issues. Meanwhile, Pakistan looks westward, toward its porous western frontier with Afghanistan and Turkestan beyond, as it contemplates its role in the world of Islam in the era of the War on Terror.

A REALM OF POVERTY

At the opening of the twenty-first century, South Asia accounts for more than one-fifth of the world's population and two-thirds of its poorest inhabitants. Its literacy rates are among the lowest in the world. Nearly half of the people in this realm earn less than the equivalent of one U.S. dollar per day. It is estimated that half the children in South Asia are malnourished and underweight, most of them girls. South Asia is often called the most deprived realm in the world.

A combination of geographic factors underlies this tragic picture. With 23 percent of the world's population but just 3 percent of its land area, South Asia lacks the natural resources to raise living standards for its hundreds of millions of subsistence farmers. Governmental policies contribute to the problem: while East and Southeast Asia forged ahead by looking outward, encouraging exports and foreign investment, and spending heavily on literacy and technical education, health care, and land reform, South Asian governments tended to adopt bureaucratic controls and elaborate forms of state planning. Cultural traditions also play their role. Resistance to change and reluctance by the privileged to open doors of opportunity to the less advantaged inhibit economic advancement for all.

Here is a grim reality: we will note repeatedly in this chapter that South Asia is self-sufficient in food and that India produces enough grain not only to be able to feed its own population but also to fill the needs of its neighbors. In 2003, official statistics showed that India had a stored wheat surplus of more than 50 million metric tons and a rice surplus of 5 million tons, enough to feed the entire realm's population for more than a year, with another good harvest in sight. But this apparent triumph of the Green Revolution did not end malnourishment in India. Despite growing annual surpluses, India's children remain among the most malnourished in the world: at the turn of the millennium, 47 percent were reported to be either severely malnourished or moderately malnourished. While food rots in government warehouses in Punjab, children die in villages in the neighboring State of Rajasthan.

How can a government hoard food in huge quantities while letting the poor die of starvation? The answer is economics—and politics. India's government responds to the pressures of powerful farm lobbies, raising the price of grain and then buying it from commercial farmers at higher prices for storage and redistribution, thereby increasing the market cost for poor villagers who cannot afford to buy it. While millions of villagers go hungry, the government is also under pressure from international organizations, which lend India money, to limit food subsidies to "consumers." In addition, India's nationwide subsidized food distribution system, set up four decades ago, suffers from inefficiency, corruption, and massive theft. As a result, poverty and government mismanagement combine to condemn tens of millions of India's rural poor to malnutrition and even starvation at a time when official statistics proclaim that the country produces enough food not only to feed itself but even to export grains to hungry neighbors. We will revisit this issue later in the chapter.

PHYSIOGRAPHIC REGIONS OF SOUTH ASIA

Before we look into South Asia's complex and fascinating cultural geography, we need to discuss the physical stage of this populous realm. South Asia is a realm of immense physiographic variety, of snowcapped peaks and forest-clad slopes, of vast deserts and broad river basins, of high plateaus and spectacular shores (Fig. 8-3). The collision of two of the Earth's great tectonic plates created the world's highest mountain ranges, their icy crests yielding meltwaters for great rivers below. The workings of the Earth's atmosphere put South Asia in the path of tropical cyclones and produce reversing seasonal windflows known as *monsoons*. This is a realm of almost infinite variety, a world unto itself.

In general terms, we can recognize three clearly defined physiographic zones in South Asia: the northern mountains, the southern peninsular plateaus, and between them a belt of river lowlands. Superimposed on this configuration is an east-west precipitation gradient

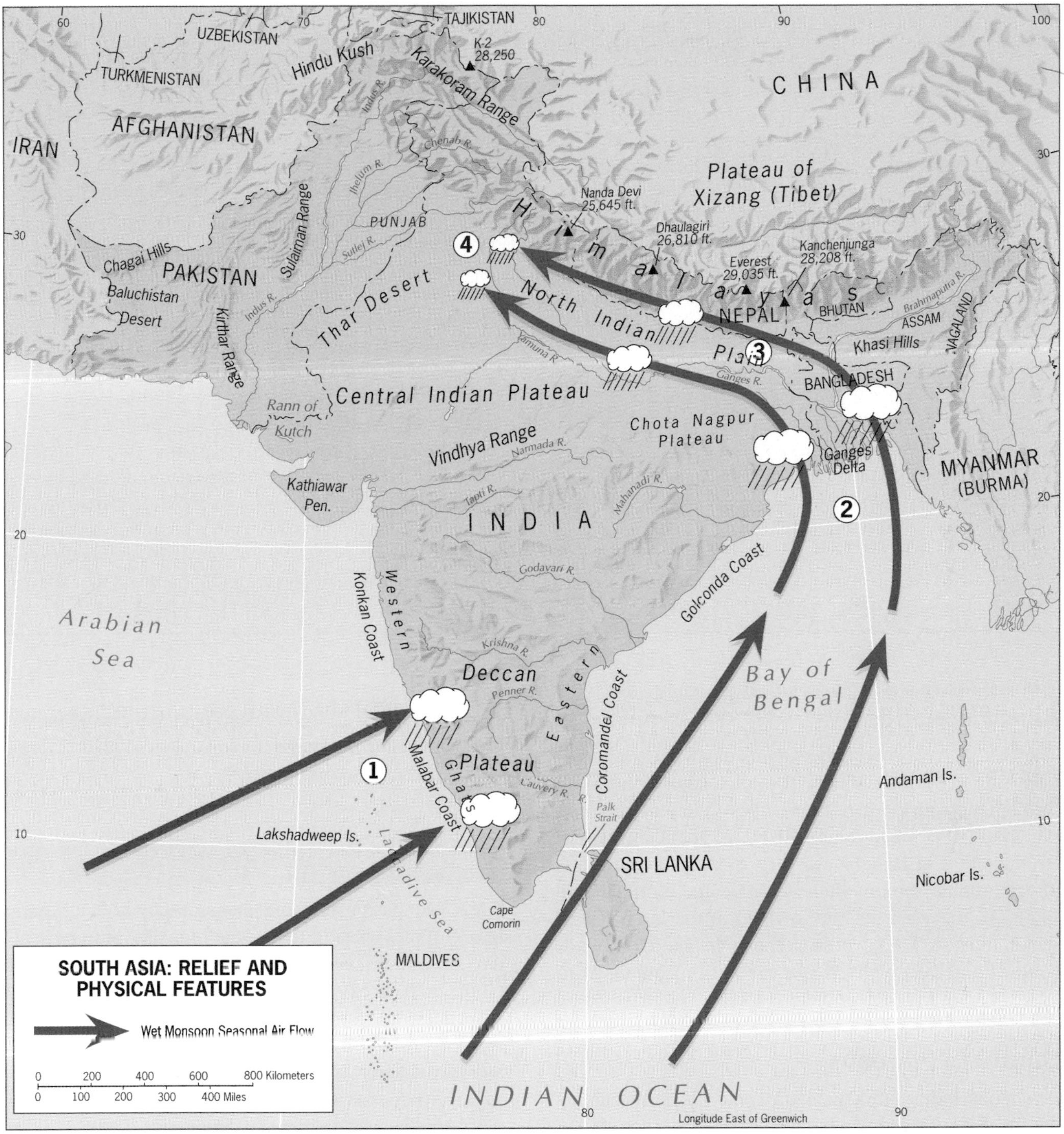

FIGURE 8-3 © H. J. de Blij, P. O. Muller, and John Wiley & Sons, Inc.

from wet (Bangladesh) to dry (western Pakistan) that is clearly visible in Figures G-7 and G-8, broken only by the strip of high moisture along India's southwestern Malabar Coast (Fig. 8-1).

Northern Mountains

The northern mountains extend from the Hindu Kush and Karakoram ranges in the northwest through the Himalayas in the center (Mount Everest, the world's tallest peak, lies in Nepal) to the ranges of Bhutan and the Indian State of Arunachal Pradesh in the east. Dry and barren in the west on the Afghanistan border, the ranges become green and tree-studded in Kashmir, forested in the lower sections of Nepal, and even more densely vegetated in Arunachal Pradesh. Transitional foothills, with many deeply eroded valleys cut by rushing meltwaters, lead to the river basins below.

South Asia's Life-Giving Monsoon

The dynamics of daily summer sea breezes are well known. The hot sun heats up the land, the air over this surface rises, and cool air from over the water blows in across the beach, replacing it. Teeming beaches and high-cost seafront apartments attest to the efficacy of this natural air conditioning.

Under certain circumstances, this kind of circulation can affect whole regions. When an entire landmass heats up, a huge low-pressure system forms over it. This system can draw vast volumes of air from over the oceans onto the land. It is not just a local, daily phenomenon. It takes months to develop, but once it is in place, it also persists for months. When the inflow of moist oceanic air
1 starts over South Asia, the **wet monsoon** has arrived. It may rain for 60 days or more. The countryside turns green, the paddies fill, and another dry season's dust and dirt are washed away. The region is reborn (see photo pair, p. 413).

Not all continents or coasts experience monsoons. A particular combination of topographic and atmospheric circumstances is needed. Figure 8-3 shows how it works in South Asia. It is June. For months, a low-pressure system has been building over northern India. Now, this cell has become so powerful that it draws air from over a large area of the warm, tropical Indian Ocean toward the interior. Some of that moisture-laden air is forced upward against the Western Ghats ①, cooling as it rises and condensing large amounts of rainfall. Other streams of air flow across the Bay of Bengal and get caught up in the convection over northeastern India and Bangladesh ②. Seemingly endless rain now inundates a much larger area, including the whole North Indian Plain. The mountain wall of the Himalayas stops the air from spreading and the rain from dissipating ③. The air moves westward, drying out as it flows toward Pakistan ④.

After persisting for weeks, the system finally breaks down and the wet monsoon gives way to periodic rains and, eventually, another dry season. Then the anxious wait begins for the next year's monsoon, for without it India would face disaster. In recent years, the monsoon has shown signs of irregularity and sometimes partial failure. In India, life hangs by a meteorological thread.

River Lowlands

The belt of river lowlands extends eastward from Pakistan's lower Indus Valley (the area known as Sind) through the wide plain of the Ganges Valley of India and on across the great double delta of the Ganges and Brahmaputra in Bangladesh (Fig. 8-1). In the east, this physiographic region often is called the North Indian Plain. To the west lies the lowland of the Indus River, which rises in Tibet, crosses Kashmir, and then bends southward to receive its major tributaries from the Punjab ("Land of Five Rivers").

Southern Plateaus

Peninsular India is mostly plateau country, dominated by the massive Deccan, a tableland built of basalt that poured out when India separated from Africa during the breakup of Gondwana (see Fig. 6-3). The Deccan (meaning "South") tilts to the east, so that its highest areas are in the west and the major rivers flow into the Bay of Bengal. North of the Deccan lie two other plateaus, the Central Indian Plateau to the west and the Chota-Nagpur Plateau to the east (Fig. 8-1). On the map, note the Eastern and Western Ghats ("hills") that descend from Deccan plateau elevations to the narrow coastal plains below. Onshore winds of the annual wet monsoon bring ample rain to the Western Ghats (see box titled "South Asia's Life-Giving Monsoon" and Fig. 8-3). As a result, here lies one of India's most productive farming areas and one of southern India's largest population concentrations.

THE HUMAN SEQUENCE

Great river basins mark the physiography of South Asia; in one of these basins, that of the Indus River in present-day Pakistan, lies evidence of the realm's oldest major civilization. It existed at the same time, and interacted with, ancient Mesopotamia, and it was centered on large, well-organized cities (Fig. 8-4). From here, influences and innovations diffused into India. In fact, India's very name is believed to derive from the ancient Sanskrit word *sindhu*, used to identify the ancient civilizations in the Indus Valley.

Eventually, such cities as Harappa and Mohenjo-Daro seem to have experienced the same fate as those of Mesopotamia, perhaps also because of environmental change. About 3500 years ago Aryan (Indo-European)-speaking peoples migrated into the region from the northwest, penetrating India and starting the process of welding the Ganges Basin's isolated tribes and villages into an organized system. Having absorbed much of the culture of the Indus Valley, they brought a new order to this region. A new belief system, *Hinduism*, arose and

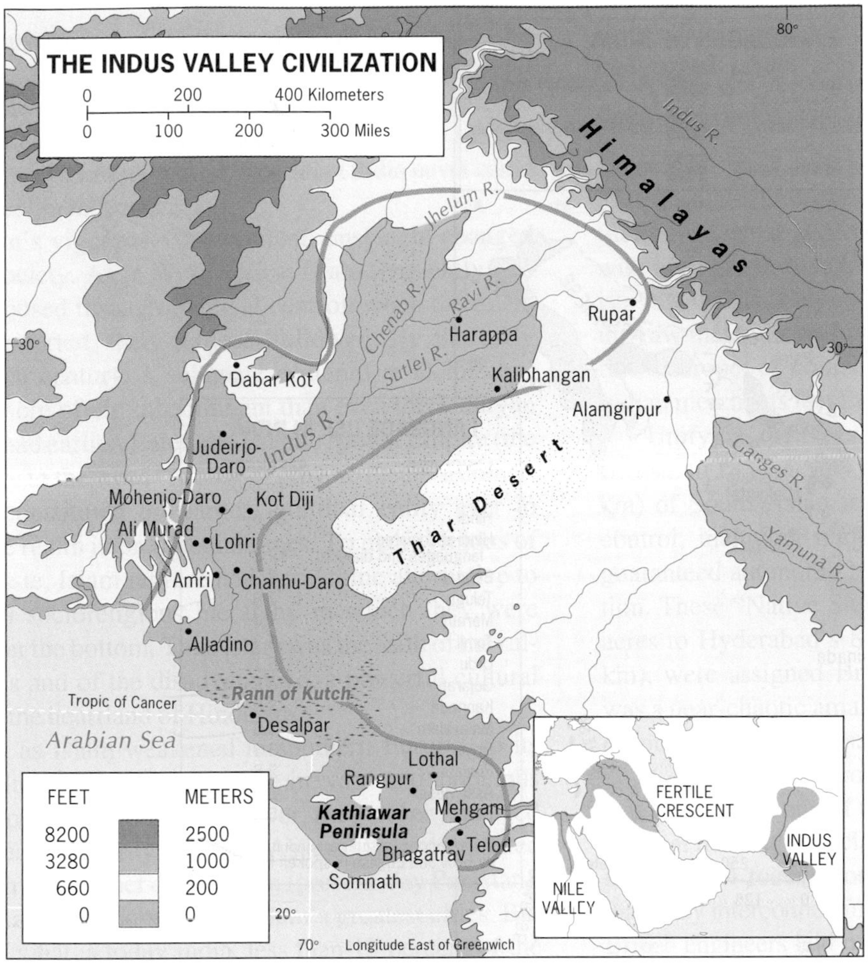

FIGURE 8-4 © H. J. de Blij, P. O. Muller, and John Wiley & Sons, Inc.

2 with it a new way of life. A complex **social stratification** developed in which Brahmans, powerful priests, stood at the head of a caste system in which soldiers, merchants, artists, peasants, and all others had their place.

Hinduism was restrictive, especially for those of the lower castes, and in the sixth century B.C. (more than 2500 years ago) a prince born into one of northern India's kingdoms sought a better way. Prince Siddhartha, better known as Buddha, gave up his royal position to teach religious salvation through meditation, the rejection of Earthly desires, and a reverence for all forms of life. His teachings did not have a major impact on the Hindu-dominated society of his time, but what he taught was not forgotten. Centuries later, the ruler of a powerful Indian state decided to make *Buddhism* the state religion, following which the faith spread far and wide.

But first South Asia was penetrated by other foreign influences: the Persians were followed by the Greeks under Alexander the Great in the late fourth century B.C. This is why, on a world map of languages, Hindi (the major domestic language of India) lies at the eastern end of the region of Indo-European languages, in the same family as Persian, Italian, and English. But the map of India's languages (Fig. 8-5) shows that southern India's languages are *not* Indo-European. Indeed, while northern India was a subregion of cultural infusion and turmoil, the south lay remote and isolated. This part of the peninsula had apparently been settled long before the Indus and Ganges civilizations arose, and southern India developed into a distinctive subregion with its own cultures. The *Dravidian* languages spoken here—Telugu, Tamil, Kanarese (Kannada), and Malayalam—have long literary histories.

Aśoka's Mauryan Empire

When the Greeks withdrew from the Ganges Basin and the Hindu heartland was once again free, a powerful empire arose there—the first true empire in the realm. This, the Mauryan Empire, extended its influence over India as far west as the Indus Valley (thus incorporating the populous Punjab) and as far east as Bengal (the double delta of the Ganges and Brahmaputra); it reached as far south as the modern city of Bangalore.

FROM THE FIELD NOTES

"More than a half-century after the end of British rule, the centers of India's great cities continue to be dominated by the Victorian-Gothic buildings the colonizers constructed here. Here is evidence of a previous era of globalization, when European imprints transformed urban landscapes. Walking the streets of Mumbai (the British called it Bombay) you can turn a corner and be forgiven for mistaking the scene for London, double-deckered buses and all. One of the British planners' major achievements was the construction of a nationwide railroad system, and railway stations were given great prominence in the urban architecture. I had walked up Naoroji Road, having learned to dodge the wild traffic around the circles in the Fort area, and watched the throngs passing through Victoria Station. Inside, the facility is badly worn, but the trains continue to run, bulging with passengers hanging out of doors and windows." © H. J. de Blij.

political entity. As early as the 1930s, the idea of a separate state (that would become Pakistan) was being promoted by Muslim activists. They circulated pamphlets arguing that British India's Muslims were a nation distinct from the Hindus and that a separate state consisting of Sind, Punjab, Baluchistan, Kashmir, and a portion of Afghanistan should be created from the British South Asian Empire in this area. The first formal demand for such partitioning was made in 1940, and, as later elections proved, the idea had almost universal support among the realm's Muslims.

As the colony moved toward independence, a political crisis developed: India's majority Congress Party would not even consider partition, and the minority Muslims refused to participate in any future unitary government. But partition would not be a simple matter. True, Muslims were in the majority in the western and eastern sectors of British India, but Islamic clusters were scattered throughout the realm (Fig. 8-6A). Any new boundaries between Hindus and Muslims to create an Islamic Pakistan and a Hindu India would have to be drawn right through areas where both sides coexisted. People by the millions would be displaced.

The consequences of this migration for the social geography of India were far-reaching, especially in the northwest (Fig. 8-6B). Comparing the 1931 and 1951 distributions in Figure 8-6, you can see the dimensions of the losses in the Indian Punjab and in what is today the State of Rajasthan. (Because Kashmir was mapped as three entities before the partition and as one afterward, the change there represents an administrative, not a major numerical alteration.) Even in the east a Muslim exodus occurred, as reflected on the map by the State of West Bengal, adjacent to Bangladesh, where the Muslim component in the population declined substantially.

The world has seen many **refugee** migrations but none 3
involving so many people in so short a time as the one resulting from British India's partition (which occurred on independence day, August 15, 1947). Scholars who study the refugee phenomenon differentiate between "forced" and "voluntary" migrations but, as this case underscores, it is not always possible to separate the two. Many Muslims, fearing they had no choice, feared for their future in India and joined the stampede. Others had the means and the ability to make a decision to stay or leave, but even these better-off migrants undoubtedly sensed a threat. In any event, massive as the refugee movement was, it left far more Muslims in India than departed. As Figure 8-6 shows, the percentage of Muslims in India's total population (then about 360 million) declined even below what it had been in 1931, but India's Muslim minority has always had a high rate of population growth. By 2005, it exceeded 150 million, approaching 15 percent of the total population—the largest cultural minority in the world and almost as large as Pakistan's entire population.

And what about Hindus who found themselves on the "wrong" side of the border at the time of partition? The great majority of Hindus preferred to live in a secular democracy than under Islamic rule, so they, too, crossed the border (photo p. 390). The Hindu component of present-day Pakistan may have neared 20 percent in 1947 but is only about 1 percent today; in Bangladesh (which was East Pakistan at the time of partition) it declined from

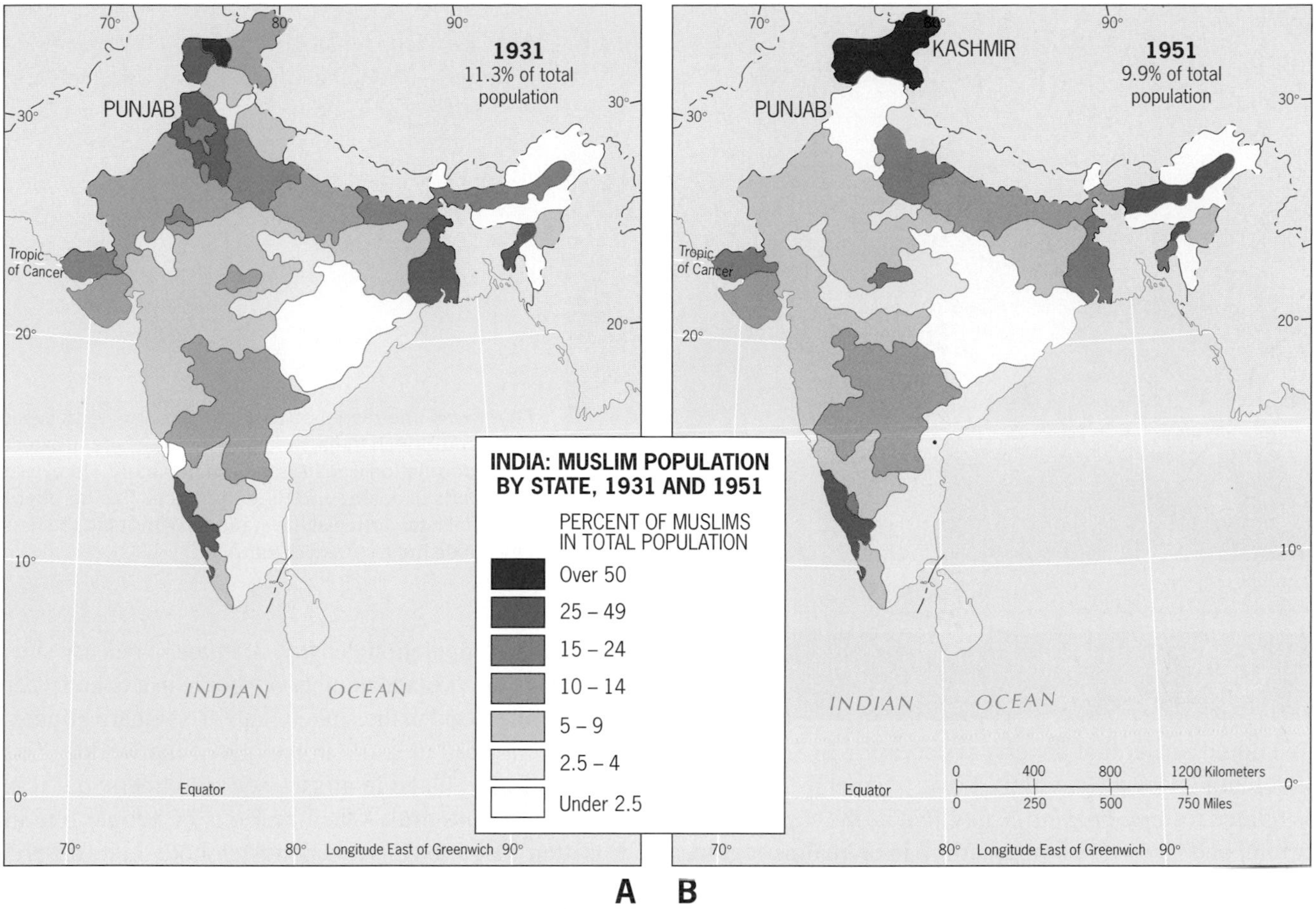

FIGURE 8-6 © H. J. de Blij, P. O. Muller, and John Wiley & Sons, Inc.

30 percent to about 9 percent today. The end of British rule in South Asia changed the realm's cultural geography forever.

Twenty-First-Century Regional Flashpoints

Ethnically complex, multicultural South Asia is a realm of remarkable accommodation as well as persistent conflict. The conflicts have ancient, recent, and current causes. The Islamic invasion of Hindu society is an ancient source of continuing strife; the imperial partition and colonial restructuring of South Asia created enduring causes of conflict; the Cold War produced hotbeds of strife still smoldering; and now the War on Terror is spawning still another basis for discord. No country in South Asia has been left unaffected. India's huge Muslim minority constitutes only one of its social and political challenges. Sri Lanka's Tamil minority, brought to the island under colonial auspices, generated a revolt that has destabilized the country for decades. In Nepal, Maoist rebels struggle to overthrow the government in Cold War fashion.

But the most serious flashpoint in South Asia is Kashmir, where India and Pakistan meet in the mountains of the far north, because this is an international, not a domestic, problem. When the boundary between Hindu India and Muslim Pakistan was hastily delimited in 1947, it stopped short of the northern territory of Jammu and Kashmir, one of the 562 "Native States" recognized by the British colonial administration. Kashmir (the short form of the territory's name) had a Hindu maharajah but a majority Muslim population, and both Pakistan and India wanted it. When British rule ended, Pakistani and Indian armies were soon at war in Kashmir. After repeated bouts of conflict, Kashmir today is divided along the latest armistice line, but neither side is prepared to yield. When this protracted and costly struggle started, India and Pakistan were armed with conventional weapons. Today they are both nuclear powers, which transforms Kashmir from another of the world's problem frontiers into a potentially catastrophic flashpoint for nuclear war. We focus on the background and implications in the regional section of this chapter (especially in the box on pp. 396–397).

Flight was one response to the 1947 partition of what had been British India, resulting in one of the greatest mass population transfers in human history. Here, two trainloads of eastbound Hindu refugees fleeing (then) West Pakistan arrive at the station in Amritsar, the first city inside India. © Corbis-Bettmann.

SOUTH ASIA'S POPULATION DILEMMA

We noted earlier that poverty is pervasive in South Asia and that hunger goes with it. Poverty's persistence also is related to *demographic issues*, that is, the size, distribution, and growth of population in a realm, region, country, or province and factors associated with these
4 conditions. The field of **population geography**, the spatial expressions of demography, focuses on such matters. It is useful to refer back to Figure G-9, which displays the continuing concentration of population in South Asia's major river valleys, underscoring the enduring dependence of the majority of the people on these ribbons of life.

Figure G-9 is an example of the various ways maps
5 can reveal **population distribution**. There we use dots representing quantity (that is, numbers of people) to create an overall image of the demographic situation in the world in general and, if you look in more detail, in the realms in particular. Even at the small scale used in Figure G-9, it is clear that even populous India has sparsely peopled areas as well as densely populated river basins and coastal strips.

But this general impression of population distribution does not tell us much about more specific measures that can be useful in understanding regional and local economic problems. These measures report a region's,
6 country's, or State's **population density**—the number of people who reside in an average unit area such as a square mile or a square kilometer. As you note in Table G-1 (pp. 35–41), we report two kinds of population density for each country: arithmetic population density and physiologic population density. *Arithmetic* density simply means the average number of people in a country per square mile (used in the United States) or square kilometer (the standard measure in the rest of the world). You see this figure listed in atlases and gazetteers, but it is not very meaningful. When millions of people live in Pakistan's river basins and almost nobody in its deserts and high mountains, what does its arithmetic average of 543 per square mile really mean? Not much—which
is why geographers prefer to use the **physiologic density** measure, which reports the number of people in a 7
country per unit of land suitable for farming or grazing. According to this measure, note that Sri Lanka and Bangladesh, not India, are the most densely populated countries in this realm (the very high figure for the tiny islands of the Maldives reflects the virtual absence of farmland there).

Another key index used in population geography is a
realm's or country's **rate of natural population change**, 8
the number of births minus the number of deaths, usually reported as a percentage (as we do in Table G-1) or per thousand in the population. During the twentieth century, it was routine to talk of the rate of natural "increase" because almost every country's population was growing. But today, as we noted in the chapters on Europe and Russia, an ever-larger number of countries report more deaths than births, so that subtracting the deaths from the births produces a negative number—and signals "negative population change." Again during the twentieth century, demographers studying the then-rapid population increase in many countries calculated how long it would take for those populations to double in size, and "doubling time" became a warning flag for fast-expanding na-

tions. But today this measure has lost most of its utility. A few countries still have fast-growing populations by world standards, but almost everywhere the growth rate is declining, even if it is still comparatively high.

Dynamics of Population Growth

In South Asia, we can still talk about rates of natural population increase because in all of this realm's countries the population still is growing faster than the world average. This is a matter of concern not just to the governments of these countries, but to the world as a whole, because rapid population growth in countries with large populations, even at a declining rate, produces economic, social, and political problems that extend beyond the realm's borders.

Three-quarters of South Asia's population live in India, so we should consider this global giant in the context of its population geography and its impact on the realm as a whole. Comparing the map of world economies (Fig. G-11) with the list of population growth rates in Table G-1, we note a clear pattern: the bulk of population growth is occurring in the lower-income economies. In many of the high-income economies, population growth is small, has leveled off, or is even negative. These higher-
9 income economies have gone through the so-called **demographic transition**, a four-stage sequence that took them from high birth rates and high death rates in preindustrial times to very low birth rates and very low death rates today (Fig. 8-7). Stages 2 and 3 in this model constitute the
10 **population explosion**, a hallmark of the twentieth century: death rates in the industrializing and urbanizing countries dropped, but birth rates took longer to decline. In 1900 the world's population was about 1.5 billion; by 2000 it had surpassed 6 billion.

When the British ruled India during the nineteenth century, the country still was in the first stage, with high birth rates and high death rates; the high death rates were caused not only by a high incidence of infant and child mortality but also by famines and epidemics. As Figure 8-7 indicates, the population during Stage 1 does not grow or decline much, but it is not stable. Famines and disease outbreaks kept erasing the gains made during better times. But then India entered the second stage. Birth rates remained high, but death rates declined because medical services improved (soap came into widespread use), food distribution networks became more effective, farm production expanded, and urbanization developed. In the 1920s, India's population still was growing at a rate of only 1.04 percent, but by the 1970s, that rate had shot up to 2.22 percent per year (Fig. 8-8). Note that India gained 28 million people during the 1920s but a staggering 135 million during the 1970s.

Has India entered the third stage, when the death rate begins to level off and birth rates decline substantially, narrowing the gap and slowing the annual increase? The rate of increase suggests that it has: from 2.22 percent during the 1970s, it dropped to 2.11 percent in the 1980s and then to 1.88 percent during the 1990s (Fig. 8-8). But India has another problem. During its population explosion, its numbers grew so large that even a declining rate of natural change continues to add ever greater numbers to its total. In Figure 8-8, we see that while the decadal

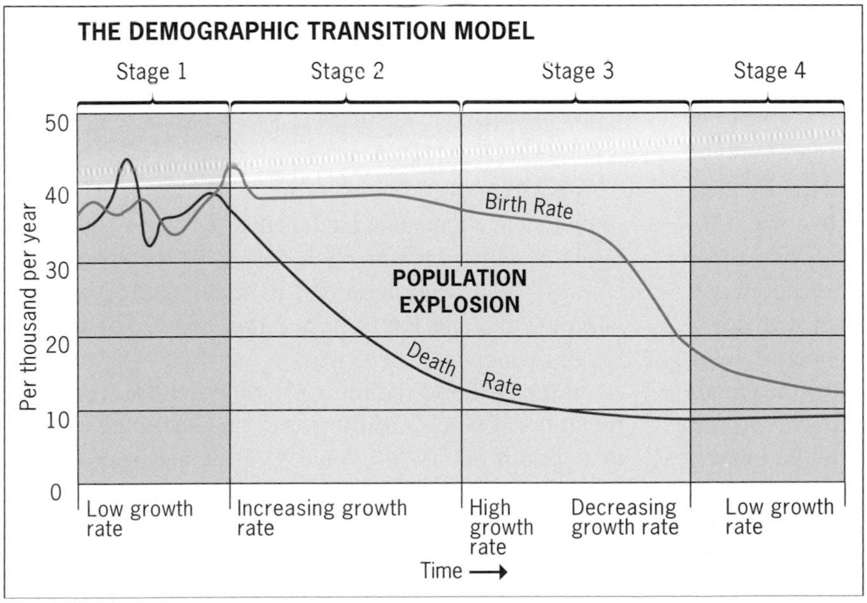

FIGURE 8-7 ©H. J. de Blij, P. O. Muller, and John Wiley & Sons, Inc.

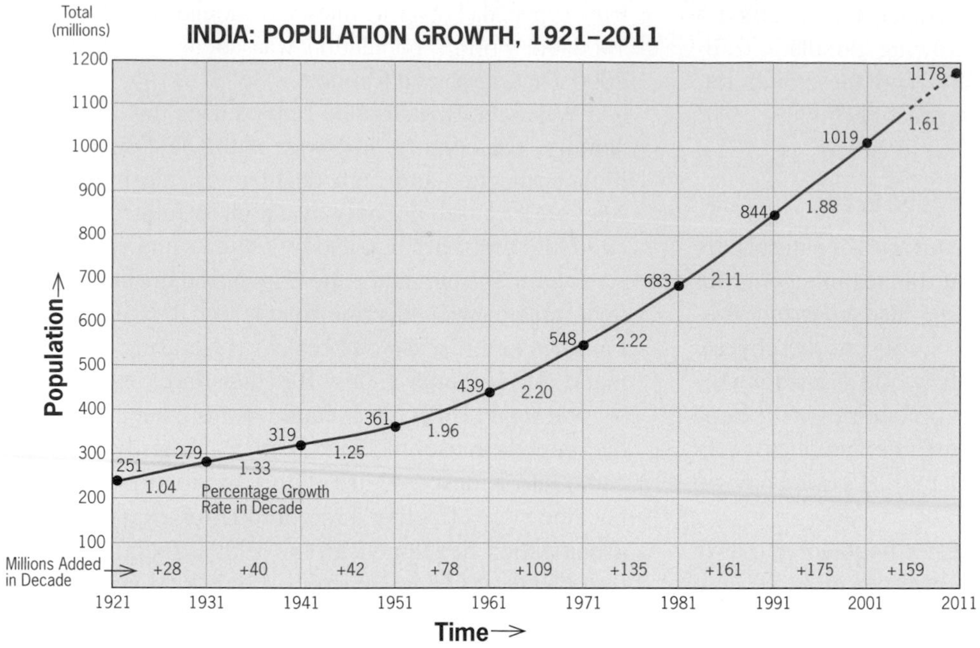

FIGURE 8-8 © H. J. de Blij, P. O. Muller, and John Wiley & Sons, Inc.

rate of increase dropped from 2.22 to 1.88 percent between 1971 and 2001, the millions added grew from 135 in the 1970s to 161 in the 1980s to 175 during the 1990s—taking the total past 1 billion during 1999. If India has indeed entered the third stage of the demographic transition, it will not feel its effects for some time.

Some population geographers theorize that the populations of all countries will eventually stabilize at some level, just as Europe's did. Certain governments, notably China's, have instituted regulations to limit family size, but this policy is more easily implemented by dictatorships than democracies. And even if such stabilization were just one doubling of its current population away, India still would have an astronomical total of more than 2 billion residents.

Geography of Demography

Statistics for a country as large as India tend to lose their usefulness unless they are put in geographic context. In its demographic as in so many of its other aspects, there is not just one India but several regionally different and distinct Indias. Figure 8-9 takes population growth down to the State level and provides a comparison between the census periods of 1981–1991 and 1991–2001 (the names of India's States are shown in Fig. 8-12).

During the period from 1981 to 1991, the highest growth rates were recorded in the States of the northeast, and only eight States had growth rates below 2 percent. But between 1991 and 2001, no fewer than 11 States, including Goa, had growth rates below 2 percent. Comparing these two maps tells us that little has changed in India's heartland, where the populous States of West Bengal and Madhya Pradesh show slight decreases but Uttar Pradesh and Bihar display similarly slight increases. The most important reductions in rates of natural increase are recorded in parts of the northeast, the east, and the south. In the tip of the peninsula, Kerala and neighboring Tamil Nadu have growth rates lower than that of the region's leading country, Sri Lanka.

One reason for these contrasts lies in India's federal system: individual States pursued population-control policies to reduce population growth, including mass sterilizations. Another reason, as we note in more detail in the regional section, relates to spatial differences in India's development. As in the world at large, India's economically-better-off States have lower rates of population growth.

India is a Hindu-dominated, officially secular country in which population policies can be debated and implemented. Pakistan, on the other hand, is a strictly Islamic state in which no such options exist. Population-control policies are regarded as incompatible with Islamic tenets, and Pakistan remains one of the world's fast-growing nations.

Among world realms, only Subsaharan Africa suffers more severely from poverty and social dislocation than South Asia. As Table G-1 indicates, however, South Asian conditions are only marginally better than Africa's, and South Asia has more than twice as many people as Subsaharan Africa. We turn next to South Asia's regional components.

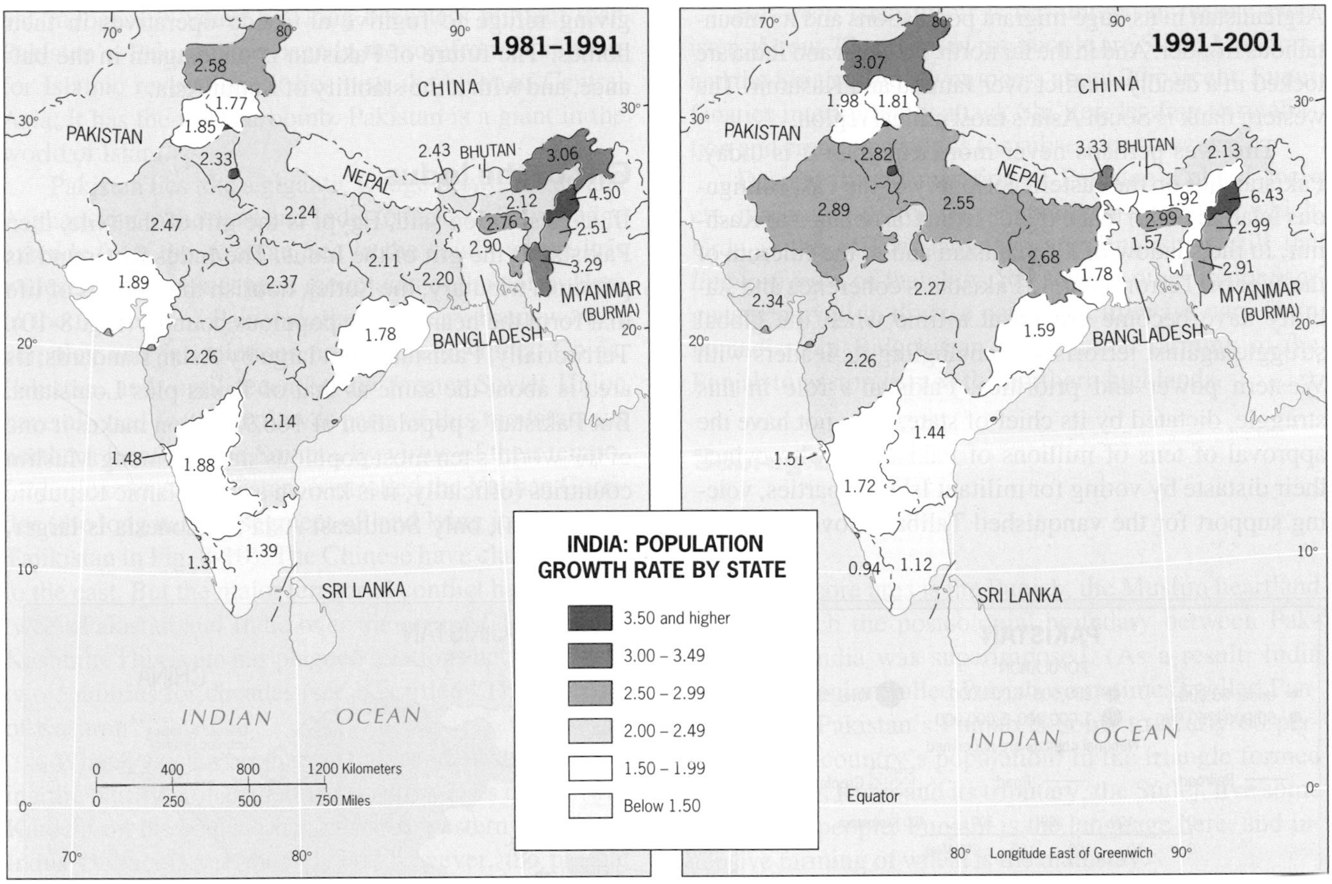

FIGURE 8-9 © H. J. de Blij, P. O. Muller, and John Wiley & Sons, Inc.

REGIONS OF THE REALM

PAKISTAN: ON SOUTH ASIA'S WESTERN FLANK

If India is the dominant entity in South Asia, why focus first on Pakistan? There are several reasons, both historic and geographic. Here lay South Asia's earliest urban civilizations, whose innovations radiated into the great peninsula. Here lies South Asia's Muslim frontier, contiguous to the great Islamic realm to the west and irrevocably linked to the enormous Muslim minority to its east. Pakistan's cultural landscapes bear witness to its transitional location. Teeming, disorderly Karachi is the typical South Asian city; as in India, the largest urban center lies on the coast. Historic, architecturally Islamic Lahore is reminiscent of the scholarly centers of Muslim Southwest Asia. In Pakistan's east, the partition boundary divides a Punjab that is otherwise continuous, a land of villages, wheat fields, and irrigation ditches. In the northwest, Pakistan resembles

MAJOR CITIES OF THE REALM

City	Population* (in millions)
Ahmadabad, India	5.4
Bangalore, India	6.7
Chennai (Madras), India	7.1
Colombo, Sri Lanka	0.7
Delhi–New Delhi, India	16.2
Dhaka, Bangladesh	13.3
Hyderabad, India	6.4
Karachi, Pakistan	12.5
Kathmandu, Nepal	0.9
Kolkata (Calcutta), India	14.6
Lahore, Pakistan	6.7
Mumbai (Bombay), India	18.9
Varanasi, India	1.3

*Based on 2006 estimates.

FROM THE FIELD NOTES

"Driving through the Khyber Pass that links Afghanistan and Pakistan was a riveting experience. From the dry and dusty foothills on the Afghanistan side, with numerous ruins marking historic battles, I traversed the rugged Hindu Kush via multiple hairpin turns and tunnels into Pakistan. This is one of the most strategic passes in the world, but for invading armies it was no easy passage: defensive forces had the advantage in these treacherous valleys. But now the roads and tunnels facilitate the movement of refugees, drugs, and arms, and they are used by militant separatist forces from both sides of the border. Pakistan's North West Frontier Province, and especially the city of Peshawar, are hotbeds of these activities." © Barbara A. Weightman.

stated that they would lead all of Pakistan in a new direction commanded by the faith. Included in the alliance were representatives from the Tribal Areas, nominally but not functionally a part of the province (Fig. 8-10).

The North West Frontier's physical geography is dominated by mountain ranges, mountain-encircled basins, and strategic passes, of which the most famous is the Khyber Pass (see photo above). The Turks invaded the upper Indus Valley via this route; later, the Mongols (Moguls) streamed through it on their way to India; recently, millions of Pushtun and other refugees used it to escape the carnage in Afghanistan. Almost directly across from the Khyber Pass lies Peshawar in an alluvium-filled, fertile valley where wheat and corn cover the countryside. After Baluchistan, this remains Pakistan's least-urbanized and poorest province.

Peshawar also is home to some of Pakistan's most militant Islamic groups, who forged an uneasy alliance with the several million refugees who fled Afghanistan and settled in United Nations-assisted refugee camps in the province. In pre-Taliban times these refugees were
subject to the pressures of **irredentism** from their kin 12
in Afghanistan, who urged them to demand autonomy within Pakistan. The government noted the risks involved and improved infrastructural linkages between the North West Frontier and the Punjab, strengthening the integration of the aptly named frontier province into the national framework. As the 2002 election showed, however, the North West Frontier still goes its own way, resisting national policies and obstructing the War on Terror.

Livelihoods

For all its size and growing regional influence, Pakistan remains a low-income economy. The level of urbanization currently stands at only 34 percent, and population growth continues at a high annual rate of 2.4 percent. Subsistence farming of food crops still occupies most of the people. Life expectancy for a child born today is just 61 years. Infant mortality is significantly higher than in any of the realm's other countries.

Nevertheless, Pakistan has made significant economic progress during its five decades-plus of independence. Irrigation has expanded enormously; land reform has progressed; and the cotton-based textile industry has generated important revenues. Despite a limited mineral resource base, manufacturing has grown, including a steel mill at Port Qasim near Karachi.

To the rest of the world, Pakistan sells textiles, carpets, tapestries, leather goods, and rice. Despite its large and growing population, Pakistan during the 1990s was able to export rice (although it did import some wheat), which is a measure of the effect of the Green Revolution as well as its program of farm expansion. In recent years, Pakistan has also become one of the largest producers of heroin in southern Asia, and its opium and hashish enter the international drug trade despite government efforts to interdict it. Afghanistan is the leading producer of heroin in this part of the world; sales were the biggest source of income for the Taliban and continue to support warlords and regional "commanders" in the post-Taliban era.

Pakistan's legitimate economy has shown substantial growth in recent years, and the country's huge public debt has been partially forgiven in return for its support in the War on Terror. As Table G-1 shows, Pakistan ranks second in per-capita GNI among the major states of mainland South Asia, and a reduction in the rate of population growth would further improve its prospects. As we will see, these prospects are now endangered by a new threat: a nuclear arms race with its giant neighbor.

Emerging Regional Power

A country's underdevelopment is not necessarily a barrier against its emergence as a power to be reckoned with in international affairs. When Pakistan became independent in 1947, the country was weak, disorganized, and divided. Just five and a half decades later, Pakistan is a major military force and a nuclear power.

To say that Pakistan in 1947 was a divided country is no exaggeration. In fact, upon independence, present-day Pakistan was united with present-day Bangladesh, and the two countries were called West Pakistan and East Pakistan, respectively. The basis for this scheme was Islam: in Bangladesh, too, Islam is the state religion. Between the two Islamic wings of Pakistan lay Hindu India. But there was little else to unify the easterners and westerners, and their union lasted less than 25 years. In 1971, a costly war of secession led to independence for East Pakistan, which took the name Bangladesh; at the same time, West Pakistan became Pakistan.

The loss of Bangladesh was no disaster for Pakistan: in virtually every respect, Bangladesh was (and remains) even more severely impoverished than Pakistan (see Table G-1). Pakistan's challenges lay closer to home—in Kashmir, in the North West Frontier, and, most importantly, to the east in India. India had supported independence for Bangladesh and did not resume diplomatic relations with Pakistan until five years later. But the political relationship between India and Pakistan has remained tense. Apart from the conflict in Kashmir (elaborated further in the Issue Box entitled "Who Should Govern Kashmir?"), their other differences are numerous, notably, India remained a democracy while Pakistan became a military dictatorship; India's treatment of its Muslim minorities frequently riled Pakistan; and India developed a close relationship with the same Soviet Union seen as a threat by Islamabad. During the Cold War, India tilted toward Moscow, whereas Pakistan, despite frequent disputes, was favored by Washington in order to curb Soviet influence in Afghanistan.

Pakistan was confronted by new challenges across its eastern border in the 1990s when India exploded several nuclear bombs at its test site. Worldwide consternation at this provocation put intense pressure on Pakistan not to follow suit, but Pakistan had little choice. Days later, it set off nuclear explosions at its own atomic test site. The "Islamic Bomb" was now reality, and the prospect of nuclear war, which had seemed to be receding, once again reasserted itself.

No longer merely a newly decolonized, economically disadvantaged country trying to survive, Pakistan has taken a crucial place in the political geography of two neighboring realms in turbulent transition.

INDIA: GIANT OF THE REALM

Just over three-quarters of the great land triangle of South Asia is occupied by a single country—India, the world's most populous democracy and, in terms of human numbers, the world's largest federation. Consider this: India (population: 1.124 billion) has nearly as many inhabitants as live in all the countries of Subsaharan Africa *plus* North Africa/Southwest Asia combined—75 of them.

That India has endured as a unified country is a politico-geographical miracle. India is a cultural mosaic of immense ethnic, religious, linguistic, and economic diversity and contrast; it is a state of many nations. The period of British colonialism gave India the underpinnings of unity: a single capital, an interregional transport network, a *lingua franca*, a civil service. Upon independence in 1947, India adopted a federal system of government, giving regions and peoples some autonomy and identity, and allowing others to aspire to such status. Unlike Africa, where federal systems failed and where military dictatorships replaced them, India remained essentially democratic and retained a federal framework in which States have considerable local authority.

This political, democratic success has been achieved despite the presence of powerful centrifugal forces in this vast, culturally diverse country. Relations between the Hindu majority and the enormous Muslim minority, better in some States than in others, have at times threatened to destabilize the entire federation. Local rebellions, demands by some minorities for their own States, frontier wars, even involvement in a foreign but nearby civil war (in Sri Lanka) have buffeted the system—which has bent but not collapsed. India has succeeded where others in the postcolonial world have failed.

This success has not been matched in the field of economics, however. After more than 50 years of independence India remains a very poor country, and not all of this can be blamed on the colonial period or on population growth, although overpopulation remains a strong impediment to improvement of living standards. Much of it results from poor and inconsistent economic planning, too much state ownership of inefficient industries, excessive government control over economic activities, bureaucratic suppression of initiative, corruption, and restraints against foreign investment. As we shall presently see, a

Who Should Govern Kashmir?

Regional ISSUE

KASHMIR SHOULD BE PART OF PAKISTAN!

"I don't know why we're even debating this. Kashmir should and would have been made part of Pakistan in 1947 if that colonial commission hadn't stopped mapping the Pakistan–India boundary before they got to the Chinese border. And the reason they stopped was clear to everybody then and there: instead of carrying on according to their own rules, separating Muslims from Hindus, they reverted to that old colonial habit of recognizing "traditional" States. And what was more traditional than some Hindu potentate and his minority clique ruling over a powerless majority of Muslims? It happened all over India, and when they saw it here in the mountains they couldn't bring themselves to do the right thing. So India gets Jammu and Kashmir and its several million Muslims, and Pakistan loses again. The whole boundary scheme was rigged in favor of the Hindus anyway, so what do you expect?

"Here's the key question the Indians won't answer. Why not have a referendum to test the will of all the people in Kashmir? India claims to be such a democratic example to the world. Doesn't that mean that the will of the majority prevails? But India has never allowed the will of the majority even to be expressed in Kashmir. We all know why. About two-thirds of the voters would favor union with Pakistan. Muslims want to live under an Islamic government. So people like me, a Muslim carpenter here in Srinagar, can vote for a Muslim collaborator in the Kashmir government, but we can't vote against the whole idea of Indian occupation.

"Life isn't easy here in Srinagar. It used to be a peaceful place with boats full of tourists floating on beautiful lakes. But now it's a violent place with shootings and bombings. Of course we Muslims get the blame, but what do you expect when the wishes of a religious majority are ignored? So don't be surprised at the support our cause gets from Pakistan across the border. The Indians call them terrorists and they accuse them of causing the 60,000 deaths this dispute has already cost, but here's a question: why does it take an Indian army of 600,000 to keep control of a territory in which they claim the people prefer Indian rule?

"Now this so-called War on Terror has made things even worse for us. Pakistan has been forced into the American camp, and of course you can't be against 'terrorism' in Afghanistan while supporting it in Kashmir. So our compatriots on the Pakistani side of the Line of Control have to stay quiet and bide their time. But don't underestimate the power of Islam. The people of Pakistan will free themselves of collaborators and infidels, and then they will be back to defend our cause in Kashmir."

KASHMIR BELONGS TO INDIA!

"Let's get something straight. This stuff about that British boundary commission giving up and yielding to a maharajah is nonsense. Kashmir (all of it, the Pakistani as well as the Indian side) had been governed by a maharajah for a century prior to partition. What the maharajah in 1947 wanted was to be ruled by neither India nor Pakistan. He wanted independence, and he might have gotten it if Pakistanis hadn't invaded and forced him to join India in return for military help. As a matter of fact, our Prime Minister Nehru prevailed on the United Nations to call on Pakistan to withdraw its forces, which of course it never did. As to a referendum, let me remind you that a Kashmir-wide referendum was (and still is) contingent on Pakistan's withdrawal from the area of Kashmir it grabbed. And as for Muslim 'collaborators,' in the 1950s the pre-eminent leader on the Indian side of Kashmir was Sheikh Muhammad Abdullah (get it?), a Muslim who disliked Pakistan's Muslim extremism even more than he disliked the maharajah's rule. What he wanted, and many on the Indian side still do, is autonomy for Kashmir, not incorporation.

"In any case, Muslim states do not do well by their minorities, and we in India generally do. As far as I am concerned, Pakistan is disqualified from ruling Kashmir by the failure of its democracy and the extremism of its Islamic ideology. Let me remind you that Indian Kashmir is not just a population of Hindus and Muslims. There are other minorities, for example the Ladakh Buddhists, who are very satisfied with India's administration but who are terrified at the prospect of incorporation into Islamic Pakistan. You already know what Sunnis do to Shi'ites in Pakistan. You are aware of what happened to ancient Buddhist monuments in Taliban Afghanistan (and let's not forget where the Taliban came from). Can you imagine the takeover of multicultural Kashmir by Islamabad?

"To the Muslim citizens of Indian Kashmir, I, as a civil servant in the Srinagar government, say this: look around you, look at the country of which you are a subject. Muslims in India are more free, have more opportunities to participate in all spheres of life, are better educated, have more political power and influence than Muslims do in Islamic states. Traditional law in India accommodates Muslim needs. Women in Muslim-Indian society are far better off than they are in many Islamic states. Is it worth three wars, 60,000 lives, and a possible nuclear conflict to reject participation in one of the world's greatest democratic experiments?

"Kashmir belongs to India. All inhabitants of Kashmir benefit from Indian governance. What is good for all of India is good for Kashmir."

Vote your opinion at www.wiley.com/college/deblij

few bright spots in some of India's States contrast sharply to the overwhelming poverty of hundreds of millions.

States and Peoples

The map of India's political geography shows a federation of 28 States, 6 Union Territories (UTs), and 1 National Capital Territory (NCT) (Fig. 8-12). The federal government retains direct authority over the UTs, all of which are small in both territory and population. The NCT, however, includes Delhi and the capital, New Delhi, and has more than 16 million inhabitants.

The political spatial organization shown in Figure 8-12 is mainly the product of India's restructuring following independence from Britain. Its State boundaries reflect the broad outlines of the country's cultural mosaic: as far as

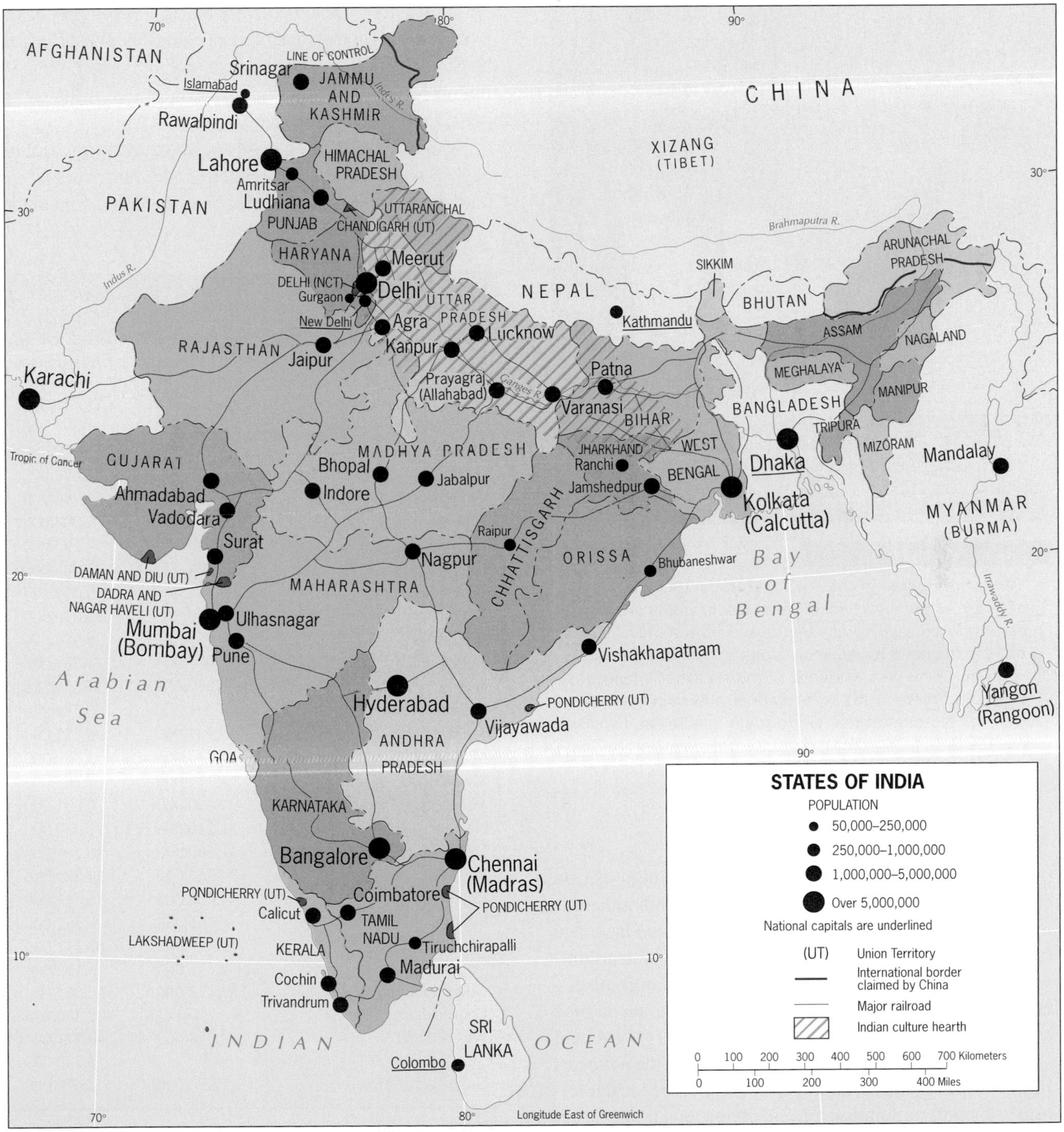

FIGURE 8-12 © H. J. de Blij, P. O. Muller, and John Wiley & Sons, Inc.

FROM THE FIELD NOTES

"We were docked in the port of Mumbai (Bombay), literally surrounded by a fleet of warships including an aircraft carrier (foreground) and a half-dozen submarines. India is often in the news because of its ongoing conflict with Pakistan, but here was evidence of India's other agenda: its projection of strength as a regional power in the Indian Ocean." Note: after this photograph was taken, India as well as Pakistan successfully tested nuclear bombs, and the potential for devastating conflict grows. © H. J. de Blij.

possible the system recognizes languages, religions, and cultural traditions. Indians speak 14 major and numerous minor languages, and while Hindi is the official language (and English is the *lingua franca*), it is by no means universal. The map is the product of endless compromise—endless because demands for modifications of it continue to this day. As recently as late 2000, the federal government authorized the creation of three new States. In the northeast lie very small States established to protect the local traditions of small populations; minority groups in the larger States ask why they should not receive similar recognition.

With only 28 States for a national population of 1.124 billion, several of India's States contain more people than many countries of the world (Table 8-1). As Figure 8-12 shows, the (territorially) largest States lie in the heart of the country and on the great southward-pointing peninsula. Uttar Pradesh (166 million, according to the 2001 Census) and Bihar (about 83 million) constitute much of the Ganges River Basin and are the core area of modern India (see box titled "Solace and Sickness from the Holy Ganges"). Maharashtra (almost 97 million), anchored by the great coastal city of Bombay (renamed Mumbai in 1996), also has a population larger than that of most countries. West Bengal, the State that adjoins Bangladesh, contains more than 80 million residents, 14.6 million of whom live in its urban focus, Calcutta (renamed Kolkata in 2000).

These are staggering numbers, and they do not decline much toward the south. Southern India consists of four States linked by a discrete history and by their distinct Dra-

Table 8-1
INDIA: POPULATION BY STATE

State	1991 Census	2001 Census
Andhra Pradesh	66,508,000	75,728,000
Arunachal Pradesh	865,000	1,091,000
Assam	22,414,000	26,638,000
Bihar	86,374,000	82,879,000
Chhattisgarh*		20,796,000
Goa	1,170,000	1,344,000
Gujarat	41,310,000	50,597,000
Haryana	16,464,000	21,083,000
Himachal Pradesh	5,171,000	6,077,000
Jammu and Kashmir	7,719,000	10,070,000
Jharkhand*		26,909,000
Karnataka	44,977,000	52,734,000
Kerala	29,099,000	31,839,000
Madhya Pradesh	66,181,000	60,385,000
Maharashtra	78,937,000	96,752,000
Manipur	1,837,000	2,389,000
Meghalaya	1,775,000	2,306,000
Mizoram	690,000	891,000
Nagaland	1,210,000	1,989,000
Orissa	31,660,000	36,707,000
Punjab	20,282,000	24,289,000
Rajasthan	44,006,000	56,473,000
Sikkim	406,000	541,000
Tamil Nadu	55,859,000	62,111,000
Tripura	2,757,000	3,191,000
Uttar Pradesh	139,112,000	166,053,000
Uttaranchal*		8,480,000
West Bengal	68,078,000	80,221,000
National Capital Territory**		13,783,000
Union Territories	1,998,000	2,670,000

*Established 2000.
**Established 1993.

Solace and Sickness from the Holy Ganges

Stand on the banks of the Ganges River in Varanasi, Hinduism's holiest city, and you will see people bathing in holy water, drinking it, and praying as they stand in it—while the city's sewage flows into it nearby, and the partially cremated corpses of people and animals float past. It is one of the world's most compelling—and disturbing—sights.

The Ganges (Ganga, as the Indians call it) is Hinduism's sacred river. Its ceaseless flow and spiritual healing power are earthly manifestations of the Almighty. Therefore, tradition has it, the river's water is immaculate, and no amount of human (or other) waste can pollute it. On the contrary: just touching the water can wash away a believer's sins.

At Varanasi, Prayagraj (Allahabad), and other cities and towns along the Ganges, the river banks are lined with Hindu temples, decaying ornate palaces, and dozens of wide stone staircases called ghats. These stepped platforms lead down to the water, enabling thousands of bathers to enter the river. They come from the city and from afar, many of them pilgrims in need of the healing and spiritual powers of the water. It is estimated that more than a million people enter the river somewhere along its 1600-mile (2600-km) course every day. During religious festivals, the number may be ten times as large.

By any standards, the Ganges is one of the world's most severely polluted streams, and thousands among those who enter it become ill with diarrhea or other diseases; many die. In 1986, then-Prime Minister Rajiv Gandhi launched a major scheme to reduce the level of pollution in the river, a decade-long construction program of sewage treatment plants and other facilities. In the mid-1980s, Gandhi was told, nearly 400 million gallons of sewage and other wastes were being disgorged into the Ganges every day. The plan called for the construction of nearly 40 sewage treatment plants in riverfront cities and towns.

Many Hindus, however, opposed this costly program to clean up the Ganges. To them, the holy river's spiritual purity is all that matters. Getting physically ill is merely incidental to the spiritual healing power contained in a drop of Ganga's water.

The stone steps leading into the Ganges' waters in Varanasi, India. Varanasi is India's holiest city, and millions descend these *ghats* every year. © David Zimmerman/Masterfile.

vidian languages. Facing the Bay of Bengal are Andhra Pradesh (just under 76 million) and Tamil Nadu (62 million), both part of the hinterland of the city of Madras (renamed Chennai in 1997) and located on the coast near their joint border. Facing the Arabian Sea are Karnataka (almost 53 million) and Kerala (32 million). Kerala, often at odds with the federal government in New Delhi, has long had the highest literacy rate in India and one of the lowest rates of population growth owing to strong local government and strictly enforced policies. "It's a matter of geography," explained a teacher in the Kerala city of Cochin. "We are here about as far away as you can get from the capital, and we make our own rules."

As Figure 8-12 shows, India's smaller States lie mainly in the northeast, on the far side of Bangladesh, and in the northwest, toward Jammu and Kashmir. North of Delhi, India is flanked by China and Pakistan, and physical as well as cultural landscapes change from the flatlands of the Ganges to the hills and mountains of spurs of the Himalayas. In the State of Himachal Pradesh, forests cover the hillslopes and relief reduces living space; only 6 million people live here, many in small, comparatively isolated clusters. Before independence and political consolidation, the colonial government called this area the "Hill States."

The map becomes even more complex in the distant northeast, beyond the narrow corridor between Bhutan and Bangladesh. The dominant State here is Assam (26.6 million), famed for its tea plantations and important because its oil and gas production amounts to more than 40 percent of India's total.

In the Brahmaputra Valley, Assam resembles the India of the Ganges. But in almost all directions from Assam, things change. To the north, in sparsely populated Arunachal Pradesh (1.1 million), we are in the Himalayan offshoots again. To the east, in Nagaland (2 million), Manipur (2.4 million), and Mizoram (0.9 million), lie the forested and terraced hillslopes that separate India from Myanmar (Burma). This is an area of numerous ethnic groups (more than a dozen in Nagaland alone) and of frequent rebellion against Delhi's government. And to the south, the States of Meghalaya (2.3 million) and Tripura (3.2 million), hilly and still wooded, border the teeming floodplains of Bangladesh. Here in the country's northeast, where peoples are always restive and where population growth is still soaring, India faces one of its strongest regional challenges.

India's Changing Map

After independence, the Indian government began phasing out the privileged "Princely States" which the British had protected during the colonial period. Next, the government reorganized the country on the basis of its major regional languages (see Fig. 8-5). Hindi, then spoken by more than one-third of the population, was designated the country's official language, but the Indian constitution gave national status to 13 other major languages, including the four Dravidian languages of the south. English, it was anticipated, would become India's common language, its *lingua franca* at government, administrative, and business levels. Indeed, English not only remained the language of national administration but also became the chief medium of commerce in growing urban India. English was the key to better jobs, financial success, and personal advancement, and the language constituted a common ground in higher education.

The newly devised framework based on the major regional languages, however, proved to be unsatisfactory to many communities in India. In the first place, many more languages are in use than the 14 that had been officially recognized. Demands for additional States soon arose. As early as 1960, the State of Bombay was divided into two language-based States, Gujarat and Maharashtra.

This devolutionary pressure has continued throughout India's existence as an independent country. In 2000, the three newest States were recognized: Jharkhand, carved from southern Bihar State on behalf of 18 poverty-stricken districts there; Chhattisgarh, where tribal peoples had been agitating since the 1930s for separation from the State of Madhya Pradesh; and Uttaranchal, which split from India's most populous, Ganges Basin, core-area State of Uttar Pradesh on the basis of its highland character and lifeways (Fig. 8-12).

For many years India has faced quite a different set of cultural-geographic problems in its northeast, where numerous ethnic groups occupy their own niches in a varied, forest-clad topography. The Naga, a cluster of peoples whose domain had been incorporated into Assam State, rebelled soon after India's independence. A protracted war brought federal troops into the area; after a truce and lengthy negotiations, Nagaland was proclaimed a State in 1961. That led the way for other politico-geographical changes in India's problematic northeastern wing.

The Sikhs

A further dilemma involves India's Sikh population. The Sikhs (the word means "disciples") adhere to a religion that was created about five centuries ago to unite warring Hindus and Muslims into a single faith. This faith's principles rejected the negative aspects of Hinduism and Islam, and it gained millions of followers in the Punjab and adjacent areas. During the colonial period, many Sikhs supported British administration of India, and by doing so they won the respect and trust of the British, who employed tens of thousands of Sikhs as soldiers and policemen. By 1947, there was a large Sikh middle class in the Punjab. When independence came, many left their rural homes and moved to the cities to enter urban professions. Today, they still exert a strong influence over Indian affairs, far in excess of the less than 2 percent of the population (about 19 million) they constitute.

After independence, the Sikhs demanded that the original Indian State of Panjab (Punjab) be divided into a Sikh-dominated northwest and a Hindu-majority southeast. The government agreed, so that Punjab as now constituted (Fig. 8-12) is India's Sikh stronghold, whereas neighboring Haryana State is mainly Hindu.

The Muslims

When India became a sovereign state and the great population shifts across its borders had ended, the country was left with a Muslim minority of about 35 million widely dispersed throughout the country. By 1991, that minority had grown to nearly 99 million, representing 11.7 percent of the total population (Fig. 8-13A). The 2001 Census of India reported a Muslim population of more than 138 million, constituting 13.4 percent and growing at a rate of nearly 4 million per year. This indicated a rate of increase in the Muslim sector of about 2.8 percent compared to India's overall rate of 1.7 percent, that is, two-thirds higher than the national average. As Table G-1 indicates, this is one of the world's fastest rates of natural increase; only a few countries grow as rapidly, and none in South Asia ranks higher than 2.5 percent.

The map displaying the 2001 distribution of Muslims in India (Fig. 8-13B) reveals actual numbers as well as percentages and indicates that the core-area State of Uttar Pradesh and the Bangladesh-bordering State of West Bengal each contain more than 20 million Muslim inhabitants, while another two States, core-area Bihar and west-coast Maharashtra, each contain more than 10 million. In terms of percentage of the total State population, however, Jammu and Kashmir leads with 67 percent, followed by Assam in northeastern India with 31 percent. Also noteworthy is Kerala in the far south, with nearly 25 percent of its population Muslim. The great geographic advantage India has is that its Muslim minority is not regionally concentrated, precluding the kind of secession movement that often develops when communities with different cultures are spatially divided (as we shall see, Sri Lanka in this realm suffers from exactly that kind of problem).

Relations between the Hindu majority and the Muslim minority are complex. What makes the news is conflict—for example, in 2002 when Muslims attacked a train carrying Hindus to a contested holy site in Ayodhya, killing dozens, which was followed by retaliation in several towns (but not others) in Gujarat. What does not make the news is that a Muslim population of more than 150 million lives and participates in the kind of democracy that is all but unknown in the Muslim world itself. In the first decade of the twenty-first century, a Hindu nationalist party was the leading force in the federal government, but a Muslim scientist, the man who developed India's first nuclear weapons, was the country's president. On the other hand, many Muslims have been angered by the Hindu-inspired name changes appearing on India's map, most recently the substitution of Prayagraj for the venerable Muslim name of Allahabad in the State of Uttar Pradesh.

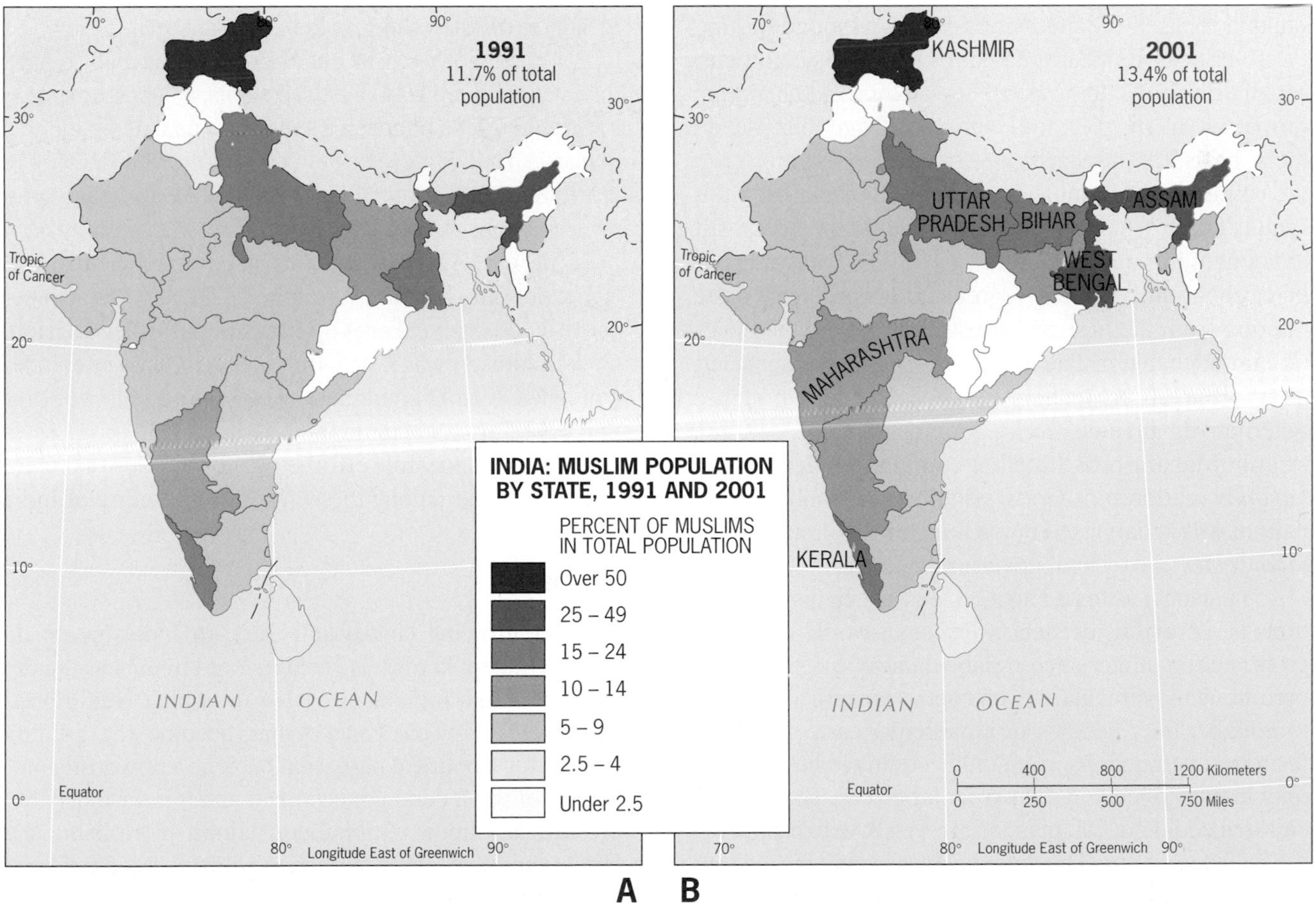

FIGURE 8-13

Concern over links between local Muslim subversives and foreign terrorists is changing Hindu attitudes, as is reflected by growing cooperation between Israeli and Indian intelligence operations and official statements linking a recent terrorist attack on the Indian Parliament not only to Pakistan (the usual suspect) but also to domestic collaborators. Pakistan's prominent role in the War on Terror is seen as a contradiction in terms by Indians facing continuous terrorism in their sector of Kashmir. Achieving a lasting accommodation of the Muslim minority will be a key challenge for India in the decades ahead.

Centrifugal Forces: From India to Hindustan?

In Chapter 1 we introduced the concept of centrifugal and centripetal forces affecting the fabric of the state. No country in the world exhibits greater cultural diversity than India, and variety in India comes on a scale unmatched anywhere else on Earth. Such diversity spells strong centrifugal forces, although, as we will see, India also has powerful consolidating bonds.

Among the centrifugal forces, Hinduism's stratification of society into castes remains pervasive. Under Hindu dogma, *castes* are fixed layers in society whose ranks are based on ancestries, family ties, and occupa-
13 tions. The **caste system** may have its origins in the early social divisions into priests and warriors, merchants and farmers, craftspeople and servants; it may also have a racial basis, for the Sanskrit term for caste is color. Over the centuries, its complexity grew until India had thousands of castes, some with a few hundred members, others containing millions. Thus, in city as well as in village, communities were segregated according to caste, ranging from the highest (priests, princes) to the lowest (the untouchables). The term *untouchable* has such negative connotations that some scholars object to its use. Alternatives include *dalits* (oppressed), the common term in Maharashtra State but coming into general use; *harijans* (children of God), which was Gandhi's designation, still widely used in the State of Bihar; and *Scheduled Castes*, the official government label.

A person was born into a caste based on his or her actions in a previous existence. Hence, it would not be appropriate to counter such ordained caste assignments by permitting movement (or even contact) from a lower caste to a higher one. Persons of a particular caste could perform only certain jobs, wear only certain clothes, worship only in prescribed ways at particular places. They or their children could not eat, play, or even walk with people of a higher social status. The untouchables occupying the lowest tier were the most debased, wretched members of this rigidly structured social system. Although the British ended the worst excesses of the caste system, and postcolonial Indian leaders—including Mohandas (Mahatma) Gandhi (the great spiritual leader who sparked the independence movement) and Jawaharlal Nehru (the first prime minister)—worked to modify it, a few decades cannot erase centuries of class consciousness. In traditional India, caste provided stability and continuity; in modernizing India, it constitutes an often painful and difficult legacy.

Today we can discern a geography of caste—a degree of spatial variation in its severity. Cultural geographers estimate that about 15 percent of all Indians are of lower caste, about 40 percent of backward caste (one important rank above the lower caste), and some 18 percent of upper caste, at the top of which are the Brahmans, men in the priesthood. (The caste system does not extend to the Muslims or Sikhs, although converted Christians are affected, which is why these percentages do not add up to 100.) The colonial government and successive Indian governments have tried to help the lowest castes. This effort has had more effect in the urban than in the rural areas of India. In the isolated villages of the countryside, the untouchables often are made to sit on the floor of their classroom (if they go to school at all); they are not allowed to draw water from the village well because they might pollute it; and they must take off their shoes, if they wear any, when they pass higher-caste houses. But in the cities, untouchables have reserved for them places in the schools, a fixed percentage of State and federal government jobs, and a quota of seats in national and State legislatures. Gandhi, who took a special interest in the fate of the untouchables in Indian society, accomplished much of this reform.

The caste system remains a powerful centrifugal force, not only because it fragments society but also because efforts to weaken it often result in further division. Gandhi himself was killed, only a few months after independence, by a Hindu fanatic who opposed his work for the least fortunate in Indian society. But progress is being made, and while efforts to help the poorest are not always popular among the better-off, the future of India depends on it.

Hindutva

Another growing centrifugal force in India has to do with a concept known as *Hindutva* or Hinduness—a desire to remake India as a society in which Hindu principles prevail. This concept has become the guiding agenda for a political party that became a powerful component of the federal government, and it is variously expressed as Hindu nationalism, Hindu patriotism, and Hindu heritage. This naturally worries Muslims and other minorities, but it also concerns those who understand that India's secularism, its separation of religion and state, is indispensable to the survival of its democracy. *Hindutva* enthusiasts want to impose a Hindu curriculum on schools, change the flexible family law in ways that would make it unacceptable to Muslims, inhibit the activities of non-Hindu religious proselytizers, and forge an India in which non-Hindus are essentially outsiders. Moderate Hindus and non-Hindus in India oppose such notions, which are as divisive as any India has faced. They nonetheless acknowledged the appeal of the Bharatiya Janata Party (BJP), which swept to power on a Hindu-nationalist platform, and realized that only the constraints of a coalition government kept the party from pursuing its more extreme goals. Recent State and national elections have seen a decline in the fortunes of the BJP, reflecting an internal party struggle between moderates and *hindutva* hardliners—and also confirming the continuing strength and vitality of Indian democracy.

The radicalization of Hinduism and the infusion of Hindu nationalism into federal politics might be seen by outsiders as threats to India's unity, but Indian voters have not rushed to embrace these initiatives. During the most recent election campaign, the opposition Congress Party accused the BJP of polarizing India, citing the recent spate of name changes on India's map as one example of this. The BJP narrowly lost its majority in the 2004 elections and its prime minister resigned, yielding to a successor from the opposition Congress Party. But the Congress Party does not have a stellar record of governance, so the pendulum could swing back again. Increasingly, however, the key issues are likely to revolve around the management of India's diverse and growing economy, and the *hindutva* campaign may well take a back seat.

FROM THE FIELD NOTES

"The streets of India's cities often seem to be one continuous market, with people doing business in open storefronts, against building walls, or simply on the sidewalk. Here the formal and informal sectors of India's economy intermingle. I walked this way in Delhi every morning, and the store selling mattresses and pillows was always open. But the women in the foreground, selling handkerchiefs and other small items from a portable iron rack, were sometimes here, and sometimes not; one time I ran into them about a half-mile down the road. As I learned one tumultuous day, every time the government tries to exercise some control over the street hawkers and sidewalk sellers, massive opposition results and chaos can ensue. A few days later, everything is as it was. Change comes very slowly here." © H. J. de Blij.

Centripetal Forces

In the face of all these divisive forces, what bonds have kept India unified for so long? Without question, the dominant binding force in India is the cultural strength of Hinduism, its sacred writings, holy rivers, and influence over Indian life. For most Indians, Hinduism is a way of life as much as it is a faith, and its diffusion over virtually the entire country (regardless of the Muslim, Sikh, and Christian minorities) brings with it a national coherence that constitutes a powerful antidote to regional divisiveness. Over the long term, however, the key ingredients of this Hinduism have been its gentility and introspection, radical outbursts notwithstanding.

Another centripetal force lies in India's democratic institutions. In a country as culturally diverse and as populous as India, reliance on democratic institutions has been a birthright ever since independence, and democracy's survival—raucous, often corrupt, always free—has been a crucial unifier.

Furthermore, communications are better in much of India than in many other countries in the global periphery, and the continuous circulation of people, ideas, and goods helps bind the disparate state together. Before independence, opposition to British rule was a shared philosophy, a strong centripetal force. After independence, the preservation of the union was a common objective, and national planning made this possible.

India's capacity for accommodating major changes and its flexibility in the face of regional and local demands are also a centripetal force. Boundaries have been shifted; internal political entities have been created, relocated, or otherwise modified; and secessionist demands have been handled with a mixture of federal power and cooperative negotiation. Indians in South Asia have accomplished what Europeans in former Yugoslavia could not, and India's history of success is itself a centripetal force.

Still another centripetal force in India is education. The country takes great pride in its high literacy rates, which exceed 96 percent for both males and females; these numbers are substantially higher than those of neighboring mainland countries and reflect Indians' determination to avail themselves of every educational opportunity, another colonial legacy. Private institutes teaching English abound in the cities, and when opportunities for the outsourcing of service jobs arose on the global economic scene, India had the educated workforce to seize them.

Finally, no discussion of India's binding forces would be complete without mentioning the country's strong leadership. Gandhi, Nehru, and their successors did much to unify India by the strength of their compelling personalities. For many years, leadership was a family affair: Nehru's daughter, Indira Gandhi, twice took decisive control (in 1966 and 1980) after weak governments, and her son, Rajiv Gandhi (who, in 1991, like his mother seven years earlier, also was assassinated), was prime minister in the late 1980s. Since then, political leadership of India has been less dynastic—and also less cohesive.

Urbanization

India is famous for its great and teeming cities, but India is not yet an urbanized society. Only 28 percent of the population lived in cities and towns in 2005—but in terms of sheer numbers, that 28 percent amounts to over 315 million people, more than the entire population of the United States.

India's rate of urbanization is on the upswing. People by the hundreds of thousands are arriving in the already-overcrowded cities, swelling urban India by about 5 percent annually, almost three times as fast as the overall population growth. Not only do the cities attract as they do everywhere; many villagers are driven off the land by the desperate conditions in the countryside. As villagers manage to establish themselves in Mumbai or Kolkata or Chennai, they help their relatives and friends

FROM THE FIELD NOTES

"Searing social contrasts abound in India's overcrowded cities. Even in Mumbai (Bombay), India's most prosperous large city, hundreds of thousands of people live like this, in the shadow of modern apartment buildings. Within seconds we were surrounded by a crowd of people asking for help of any kind, their ages ranging from the very young to the very old. Somehow this scene was more troubling here in well-off Mumbai than in Kolkata (Calcutta) or Chennai (Madras), but it typified India's urban problems everywhere." © H. J. de Blij.

FROM THE FIELD NOTES

"Negotiating the traffic in India's chaotic cities is always a challenge as creaking buses, vintage cars, taxis large and small, scooters, and bicycles mingle in a mass of movement that sometimes makes streets look like rivers of humanity. I watched the scene from a vantage point on Nungambakkam Road in Chennai (formerly Madras), impressed that so much high-speed congestion produced no accidents. 'You have to have a sense of humor to be part of all that,' said a man who stopped to chat. 'Just down the street, make a left and then a right, and you'll see that even the authorities do.' He was referring to the sign appealing to drivers to obey the seemingly nonexistent rules. India is in dire need of highway and road improvement; in cities like Chennai, the road system is pretty much the way the British left it in the 1940s when India became independent." © H. J. de Blij.

to join them in squatter settlements that often are populated by newcomers from the same area, bringing their language and customs with them and cushioning the stress of the move.

As a result, India's cities display staggering social contrasts. Squatter shacks without any amenities at all crowd against the walls of modern high-rise apartments and condominiums (photo at left). Hundreds of thousands of homeless roam the streets and sleep in parks, under bridges, on sidewalks. As crowding intensifies, social stresses multiply. Disorder never seems far from the surface; sporadic rioting, often attributable to rootless urban youths unable to find employment, has become commonplace in India's cities.

India's modern urbanization has its roots in the colonial period, when the British selected Calcutta (Kolkata), Bombay (Mumbai), and Madras (Chennai) as regional trading centers and fortified ports. Madras was fortified as early as 1640; Bombay (1664) had the situational advantage of being the closest of all Indian ports to Britain; and Calcutta (1690) lay on the margin of India's largest population cluster and had the most productive hinterland, to which the Ganges Delta's countless channels connected it. This natural transport network made Calcutta an ideal colonial headquarters, but the population of Bengal was often rebellious. In 1912 the British moved their colonial government from Calcutta to the safer interior city of New Delhi, built adjacent to the old Mogul capital of Delhi.

Figure 8-12 displays the distribution of major urban centers in India. Except for Delhi–New Delhi, the largest cities have coastal locations: Kolkata dominates the east, Mumbai the west, and Chennai the south. But urbanization also has expanded in the interior, notably in the core area. The surface interconnections among India's cities remain inadequate (notably the road network), but an Indian urban system is emerging.

Economic Geography

If India has faced problems in its great effort to achieve political stability and national cohesion, these problems are more than matched by the difficulties that lie in the way of economic growth and development. The large-scale factories and power-driven machinery of the colonial rulers wiped out a good part of India's indigenous industrial base. Indian trade routes were taken over. European innovations in health and medicine sent the rate of population growth soaring, without introducing solutions for the many problems this spawned. Surface communications improved and food distribution systems became more efficient, but it would be a mistake to assume that this adequately protected Indian peasants from starvation. Local and regional food shortages occurred as El Niño-

AMONG THE REALM'S GREAT CITIES . . . MUMBAI (BOMBAY)

Another historic name is disappearing from the map: Bombay. In precolonial times, fishing folk living on the seven small islands at the entrance to this harbor named the place after their local Hindu goddess, *Mumbai*. The Portuguese, first to colonize it, called it Bom Bahia, "Beautiful Bay." The British, who came next, corrupted both to *Bombay*, and so it remained for more than three centuries. Now local politics has taken its turn. Leaders of a Hindu nationalist party that control the government of the State of which Bombay is the capital, Maharashtra, passed legislation to change its name back to Mumbai. In 1996, India's federal government approved this change.

Mumbai's 18.9 million people make this India's largest city. Maharashtra is India's economic powerhouse, the State that leads the country in virtually every respect. Locals dream of an Indian Ocean Rim of which Mumbai will be the anchor.

As such, Mumbai is a microcosm of India, a burgeoning, crowded, chaotic, fast-moving agglomeration of humanity (see photo, p. 408). The gothic-Victorian architecture of the city center is a legacy of the British colonial period (see photo, p. 388). Shrines, mosques, temples, and churches evince the pervasive power of religion in this multicultural society. Street signs come in a bewildering variety of scripts and alphabets. Creaking double-decker buses compete with oxwagons and handcarts on the congested roadways. The throngs on the sidewalks—businesspeople, holy men, sari-clad women, beggars, clerks, homeless wanderers—spill over into the streets.

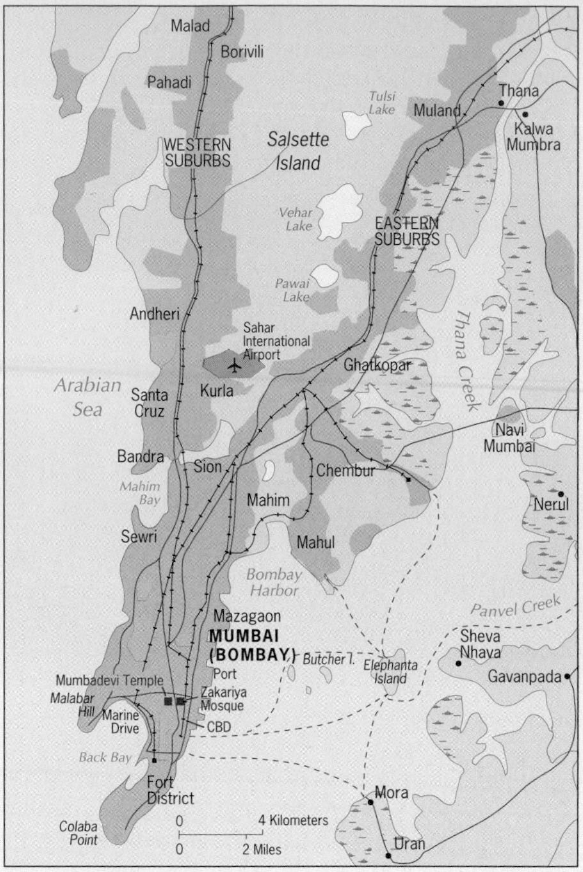

Mumbai is an urban agglomeration of city-sized neighborhoods, each with its own cultural landscape. The seven islands have been connected by bridges and causeways, and the resulting Fort District still is the center of the city, with many of its monuments and architectural landmarks. Marine Drive leads to the wealthy Malabar Hill area, across Back Bay. Northward lie the large Muslim districts, the Sikh neighborhoods, and other ethnic and cultural enclaves. Beyond are some of the world's largest and poorest squatter settlements. It is situation, not site, that gave Bombay primacy in South Asia. The opening of the Suez Canal in 1869 made Bombay the nearest Indian port to Europe. Today, Mumbai is the focus of India's fastest-growing economic zone—not yet a tiger but on the move.

generated droughts struck India repeatedly, grain prices rose beyond what villagers could afford, and the colonial (and later sovereign) governments failed to intervene effectively. Today, more than 70 percent of Indians still live on the land, where, as we noted earlier, their social, economic, and political disadvantages sometimes spell disaster. Although tens of millions of Indians have been lifted above the poverty line over the past decade, an estimated 26 percent of the population still lives below it.

India's economy in the 2000s is on the move, but its heavy dependence on agriculture continues. Farming still accounts for nearly 70 percent of employment and yields 25 percent of India's GNI, which makes the entire economy vulnerable to the whim of the annual wet monsoon. When weather conditions are favorable, however, India thrives, and such was the case in 2003, when an unprecedented harvest gave India a financial boost that radiated across every economic sector; another good year in 2004 again boosted India's fortunes. The overall economy grew at 6.5 percent, and investments in infrastructure increased, including a nationwide roadbuilding program to link the metropolitan areas of New Delhi, Mumbai, Chennai, and Kolkata with four-lane highways.

AMONG THE REALM'S GREAT CITIES . . . KOLKATA (CALCUTTA)

Calcutta is synonymous with all that can go wrong in large cities: poverty, dislocation, disease, pollution, crime, corruption. To call a city the Calcutta of its region is to summarize urban catastrophe. West Bengal's leaders had this in mind when they proposed that the city's local name, *Kolkata*, be restored. The city got its dreadful reputation (as well as its Anglicized name) during colonial times, when plague, malaria, and other diseases claimed countless thousands in legendary epidemics. The British chose the site, 80 miles (130 km) up the Hooghly River from the Bay of Bengal and less than 30 feet (9 m) above sea level, not far from some unhealthy marshes but well placed for commerce and defense. When the British East India Company was granted freedom of trade in the populous hinterland, Calcutta's heyday began; when (in 1772) the British made it their colonial capital, the city prospered. The British sector of the city was drained and raised, and so much wealth accumulated here that Calcutta became known as the "city of palaces." Outside the British town, rich Indian merchants built magnificent mansions. Beyond lay neighborhoods that were often based on occupational caste, whose names are still on the map today (such as Kumartuli, the potters' district). Almost everywhere, on both banks of the Hooghly, lay the huts and hovels of the poorest of the poor. Searing social contrasts characterized Calcutta.

The twentieth century was not kind to Calcutta. In 1912 the British moved their colonial capital to New Delhi. The 1947 partition that created Pakistan also created then-East Pakistan (now Bangladesh), cutting off a large part of Calcutta's hinterland and burdening the city with a flood of refugees. The Indian part of the city had arisen virtually without any urban planning, and the influx created almost unimaginable conditions. Today, Kolkata counts 14.6 million residents, including as many as 500,000 homeless. Beyond the façade of the downtown lies what may be the sickest city of all.

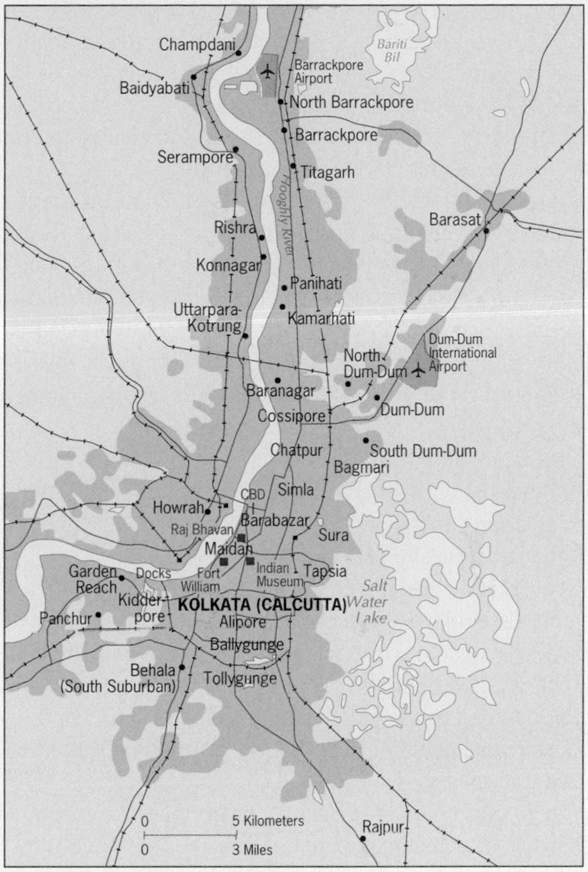

Agriculture

Growing contradictions mark agriculture in India. In Punjab, Haryana, and western Uttar Pradesh, farming remains relatively large-scale and comparatively efficient; from these areas come much of the huge grain surplus that India has been accruing over the last several years. In Gujarat, Chhattisgarh, and Bihar, some children still starve and tens of millions are malnourished as staples do not reach them or are unaffordable when they do.

Traditional farming methods persist as they do in Subsaharan Africa, and yields per acre are low by world standards for virtually every crop grown this way. Transport systems are inadequate, hampering the movement of farm produce. In 2005, only about half of India's 600,000 villages were accessible by motorable road, and in this era of modern transportation animal-drawn carts still outnumber vehicles nationwide.

As the total population grows, the amount of cultivated land per person declines. Today, this physiologic density is 1707 per square mile (659 per sq km). However, this is nowhere near as high as the physiologic density in neighboring Bangladesh, where the figure is more than twice as great (3913 and 1511, respectively). But India's farming is so inefficient that this comparison is deceptive. Fully two-thirds of India's huge working population depends directly on the land for its livelihood, but the great majority of Indian farmers are poor and cannot

AMONG THE REALM'S GREAT CITIES . . . DELHI NEW AND OLD

Fly directly over the Delhi–New Delhi conurbation into the regional airport, and you may not see the place at all. A combination of smog and dust creates an atmospheric soup that can limit visibility to a few hundred feet for weeks on end. Relief comes when the rains arrive, but Delhi's climate is mostly dry. The tail-end of the wet monsoon reaches here during late June or July, but altogether the city only gets about 25 inches (60 cm) of rain a year. When the British colonial government decided to leave Calcutta and build a new capital city adjacent to Delhi, conditions were different. South of the old city lay a hill about 50 feet (15 m) above the surrounding countryside, on the right bank of the southward-flowing Yamuna tributary of the Ganges. Compared to Calcutta's hot, swampy environment, Delhi's was agreeable. In 1912 it was not yet a megacity. Skies were mostly clear. Raisina Hill became the site of a New Delhi.

This was not the first time rulers chose Delhi as the seat of empire. Ruins of numerous palaces mark the passing of powerful kingdoms. But none brought to the Delhi area the transformation the British did. In 1947 the Indian government decided to keep its headquarters here. In 1970 the metropolitan population exceeded 4 million. By 2006 it was 16.2 million.

Delhi is popular as a seat of government for the same reason as its present expansion: the city has a fortuitous relative location. The regional topography creates a narrow corridor through which all land routes from northwestern India to the North Indian Plain must pass, and Delhi lies in this gateway. Thus the twin cities not only contain the government functions; they also anchor the core area of this populous country.

Old Delhi was once a small, traditional, homogeneous town. Today Old and New Delhi form a multicultural, multifunctional urban giant.

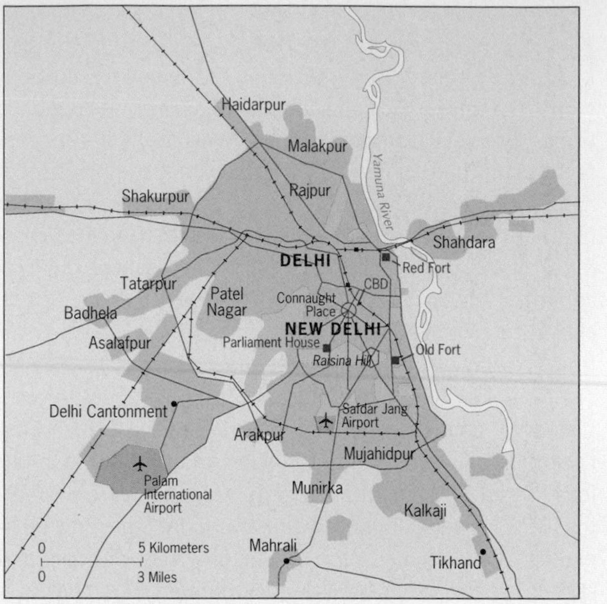

improve their soils, equipment, or yields. Those areas in which India has substantially modernized its agriculture (as in Punjab's wheat zone) remain islands in a sea of agrarian stagnation.

This stagnation has persisted in large measure because India, after independence, failed to implement a much-needed nationwide land reform program. Roughly one-quarter of India's entire cultivated area is still owned by less than 5 percent of the country's farming families, and little land redistribution was taking place. Perhaps half of all rural families own either as little as an acre or no land at all. Independent India inherited inequities from the British colonial period, but the individual States of the federation would have had to cooperate in any national land reform program. As always, the large landowners retained considerable political influence, so the program never got off the ground.

To make matters worse, much of India's farmland is badly fragmented as a result of local rules of inheritance, thereby inhibiting cooperative farming, mechanization, shared irrigation, and other opportunities for progress (top photo p. 413). Not surprisingly, land consolidation efforts have had only limited success except in the States of Punjab, Haryana, and parts of Uttar Pradesh, where modernization has gone farthest. Official agricultural development policy, at the federal and State levels, has also contributed to India's agricultural malaise and the uneven distribution of progress. Unclear priorities, poor coordination, inadequate information dissemination, and other failures have been reflected in the country's disappointing output.

It is instructive to compare Figure 8-14, showing the distribution of crop regions in India, with Figure G-7, which shows mean annual precipitation in India and the world. In the comparatively dry northwest, notably in the Punjab and neighboring areas of the upper Ganges Basin, wheat is the leading cereal crop. Here, India has made major gains in annual production through the in-

FROM THE FIELD NOTES

"Travel into rural India, and you soon grasp the realities of Indian agriculture. Human and animal labor predominate; farming methods and equipment are antiquated. At this village I watched the women feed scrawny sticks of sugarcane into a rotating press turned slowly by a pair of bullocks prodded by a boy. In this family enterprise, the grandfather is responsible for boiling the liquid sugar in a huge pan over the fire pit (upper right). The dehydrated sugar is then molded into large round cakes to be sold at the local market. As I observed this slow and inefficient process, I understood better why India is among the world's leaders in terms of sugarcane *acreage*—and among the world's last in terms of *yields*."
© Barbara A. Weightman.

troduction of high-yielding grain varieties developed under the banner of the Green Revolution, the international research program that played a key role in overcoming the food crises of the 1960s. These "miracle crops" led to the expansion of cultivated areas, the construction of new irrigation systems, and the more intensive use of fertilizer (a mixed blessing, for fertilizers tend to be expensive and the "miracle" crops are more heavily dependent on them).

Toward the moister east, and especially in the wet-monsoon-drenched areas (Fig. 8-14), rice becomes the dominant staple. About one-fourth of India's total farmland lies under rice cultivation, most of it in the States of Assam, West Bengal, Bihar, Jharkhand, Orissa, and eastern Uttar Pradesh and along the Malabar-Konkan coastal strip facing the Arabian Sea. These areas receive over 40 inches (100 cm) of rainfall annually, and irrigation supplements precipitation where necessary.

India devotes more land to rice cultivation than any other country, but yields per acre remain among the world's lowest, despite the introduction of "miracle rice." Nevertheless, when monsoon conditions are good, India now is capable of producing far more grain than it needs (when people are malnourished, the cause is cost not availability), and it is able to export food to needy countries in Africa. But India's dependence on its wet monsoon remains precarious, and the country is not much more than a single poor-harvest year away from a potential food crisis.

Subsistence remains the fate of tens of millions of Indian villagers who cannot afford fertilizers, cannot cultivate the new and more productive strains of rice or

The arrival of the annual rains of the wet monsoon transforms the Indian countryside. By the end of May, the paddies lie parched and brown, dust chokes the air and it seems that nothing will revive the land. Then the rains begin, and blankets of dust turn into layers of mud. Soon the first patches of green appear on the soil, and by the time the monsoon ends all is green. The photograph on the left, taken just before the onset of the wet monsoon in the State of Goa, shows the paddies before the rains begin; three months later the countryside looks as on the right.
©Steve McCurry/Magnum Photos, Inc.

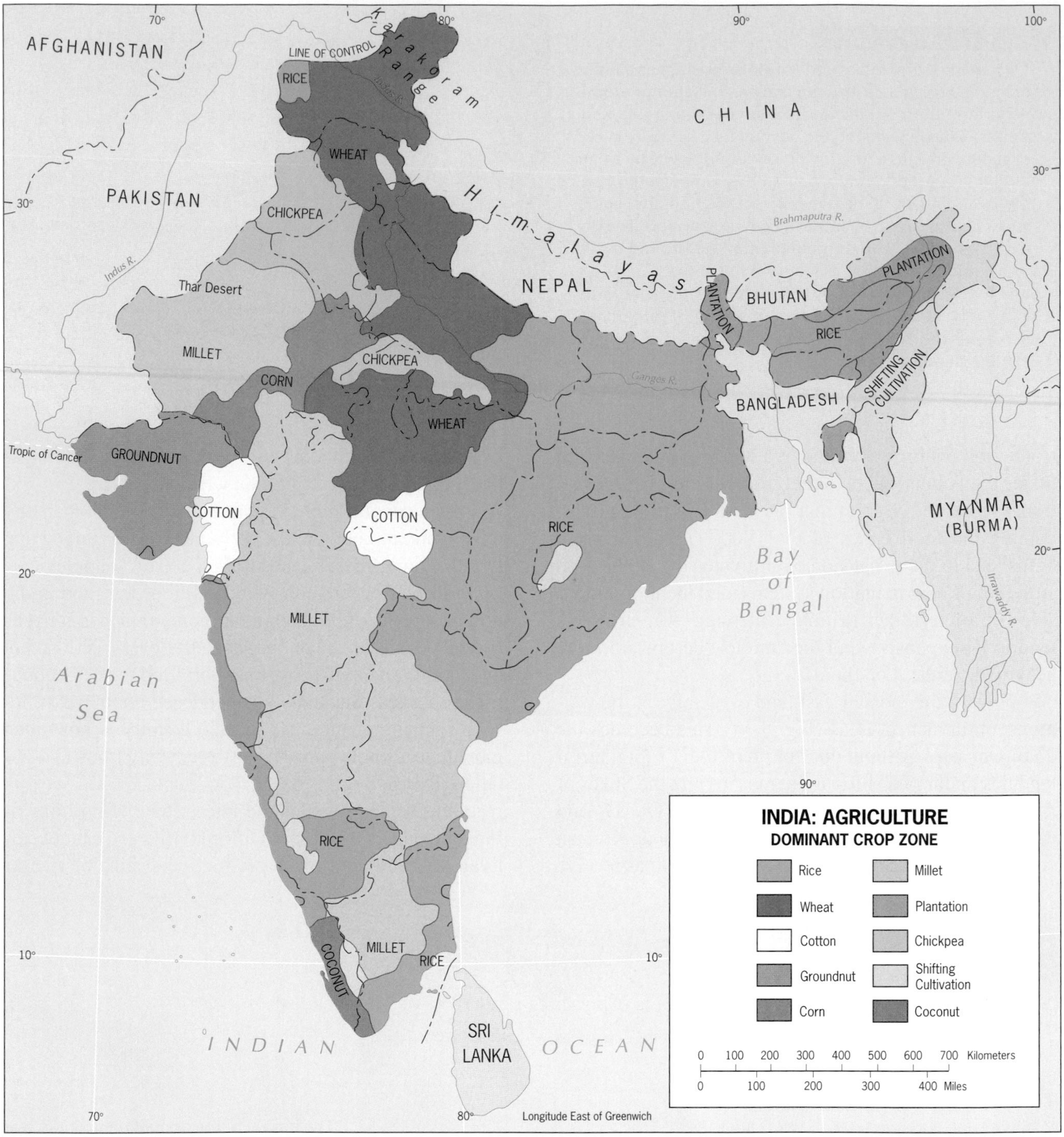

FIGURE 8-14 © H. J. de Blij, P. O. Muller, and John Wiley & Sons, Inc.

wheat, and cannot escape the cycle of poverty. Perhaps as many as 175 million of these people do not even own a plot of land and must live as tenants, always uncertain of their fate. This is the enduring reality against which optimistic predictions of improved nutrition in India must be weighed. True, rice and wheat yields have increased at slightly more than the rate of population growth since the Green Revolution. But food security remains elusive, and India continues to face the risks inherent in the ever-growing needs of its burgeoning population.

Industrialization

In 1947, India inherited the mere rudiments of an industrial framework. After more than a century of British control over the economy, only 2 percent of India's workers were engaged in industry, and manufacturing and mining combined produced only about 6 percent of the national income. Textile and food processing were the dominant industries. Although India's first iron-making plant opened in 1911 and the first steel mill began operating in 1921, the initial major stimulus for heavy industrialization did not come until after the outbreak of World War II. Manufacturing was concentrated in the largest cities: Kolkata led, Mumbai was next, and Chennai ranked third.

The geography of manufacturing still reflects those beginnings, and industrialization in India has proceeded slowly, even after independence (Fig. 8-15). Kolkata now anchors India's eastern industrial region—the Bihar-Bengal District—where jute manufactures dominate, but cotton, engineering, and chemical industries also operate. On the nearby Chota-Nagpur Plateau to the west, coal-mining and iron and steel manufacturing have developed.

On the opposite side of the subcontinent, two industrial areas dominate the western manufacturing region: one is centered on Mumbai and the other on Ahmadabad. This dual region, lying in Maharashtra and Gujarat states, specializes in cotton and chemicals, with some engineering and food processing. Cotton textiles have long been an industrial mainstay in India, and it was one of the few industries to benefit from the nineteenth-century economic order the British imposed. With the local cotton harvest, the availability of cheap yarn, abundant and inexpensive labor, and the power supply from the Western Ghats' hydroelectric stations, the industry thrived.

The southern industrial region consists chiefly of a set of linear, city-linking corridors focused on Chennai, specializing in textile production and light engineering activities. Today, all of India's manufacturing regions are increasing their output of ready-to-wear garments—another legacy of the early development of cotton textiles. Clothing has become India's second-leading export by value; the production of gems and jewelry, another growing specialization, ranks first. As for the future, Chennai in 2003 caught a glimpse of it when a South Korean automobile manufacturer began exporting cars to the European market from its newly built plant in the city.

A very important development in the south is occurring in and around Bangalore in the State of Karnataka, India's "Silicon Plateau." Several hundred software

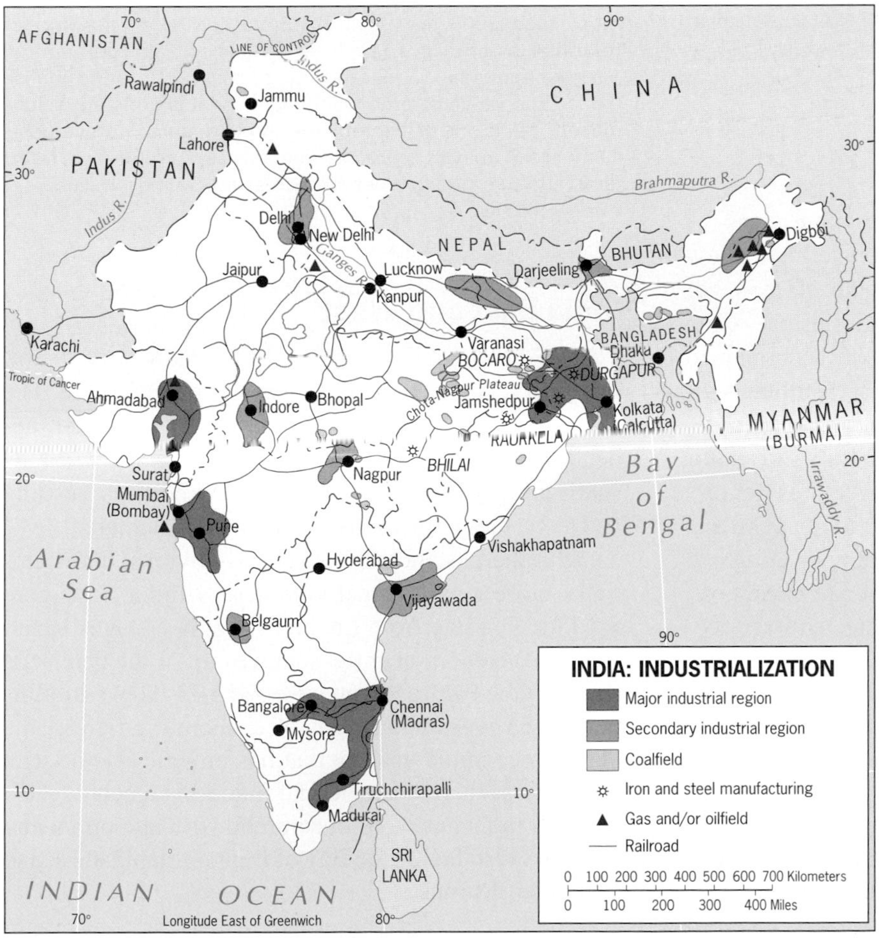

FIGURE 8-15 © H. J. de Blij, P. O. Muller, and John Wiley & Sons, Inc.

companies are based here, many of them foreign with names such as IBM, Texas Instruments, and Motorola (photo at right). What attracts them, and makes them profitable, are the modest salaries of India's software engineers, who earn about one-fifth of what their foreign colleagues earn. The emigration of technicians is becoming a problem, but replacements are still plentiful. Bangalore is proof of India's potential in the modern world.

Despite some imbalances and inefficiencies, India's industrial resource base is well endowed. Limited high-quality coal deposits are exploited in the Chota-Nagpur area. In combination with large lower-grade coalfields elsewhere, the country's total output is high enough to rank it among the world's ten leading coal producers. With no known major petroleum reserves (some oil comes from Assam, Gujarat, Punjab, and offshore from Mumbai), India must spend heavily on fuel imports every year. Major investments have been made in hydroelectric plants, especially multipurpose dams that provide electricity, enhance irrigation, and facilitate flood control. India's iron ore deposits in Bihar (northwest of Kolkata) and Karnataka (in the heart of the Deccan) may rank among the largest in the world. Jamshedpur, located west of Kolkata in the eastern industrial region, has become India's steel-making and metals-fabrication center. Yet India still exports iron ore as a raw material to the higher-income industrialized countries, mainly Japan. For low-income, revenue-needy countries, entrenched practices are difficult to break.

Improving Prospects

Nonetheless, India's economy is showing signs of breaking out of its long-term mold. Despite India's low current GNI of U.S. $2880 per person, the sheer size of its economically active population makes this one of the world's largest economies, ranking sixth according to the latest data. A middle class of nearly 300 million is fueling demand for goods ranging from mobile phones (recent sales were about 1 million per month!) to motor bikes (more than 10,000 a day according to press reports in 2004). India's enormous market is attracting more foreign investment each year, and government efforts to constrain corruption and reduce red tape are paying off. High-tech industries exporting software, and service-industry outsourcing, are transforming places like Bangalore and Gurgaon into ultramodern, globally connected centers. Until 1999, India did not have a single modern shopping mall. By 2004 it had more than 100, with dozens of others under construction. India's economy is on the move (recording a soaring growth rate of 10.4 percent in the final quarter of 2003), with huge implications for the global economic picture.

Bangalore, near the meeting point of the states of Karnataka, Andhra Pradesh, and Tamil Nadu, is the center of what has become known as India's "Silicon Plateau." This technopole is the heart of the country's IT development, and from here many skilled workers have made their way to Europe and the U.S., where their abilities earn them more than at home. But many more workers choose to stay here and benefit from the higher local incomes the high-tech IT industry provides. This scene could be in any modern city—except for the dress worn by these workers. Bangalore attracts the best and the brightest, has its own fine educational institutions, and lies at more than 3000 feet (900m) above sea level, making for tolerable summers. The city is a meeting (and competing) place for Kannada-, Tamil-, and Telugu-speakers, making English the indispensable common language, and it is growing rapidly. Indeed, that is one of its problems: infrastructure improvements cannot keep up with needs and demands, and congestion, traffic jams, and commuting times are increasing. Sound familiar?

India East and West

The most commonly cited, and most clearly evident, regional division of India is between north and south. The north is India's heartland, the south its Dravidian appendage; the north speaks Hindi as its *lingua franca*, the south prefers English over Hindi; the north is bustling and testy, the south seems slower and less agitated.

But there is another, as yet less obvious, but potentially more significant divide across India. In Figure 8-15, draw a line from Lucknow, on the Ganges River, south to Madurai, near the southern tip of the peninsula (also review Figure 8-2 on page 381). To the west of this line, India is showing signs of economic progress, the kind of economic activity that has brought Pacific Rim countries such as Thailand and Indonesia a new life. To the east, India has more in common with less promising countries also facing the Bay of Bengal: Bangladesh and Myanmar (Burma).

As with other regional divides, there are exceptions to our east-west delineation. Indeed, our map seems to suggest that much of India's industrial strength lies in the east. But what the map cannot reveal is the profitability of those industries. True, the east is rich in iron and coal, but the heavy industries built by the state in the 1950s are now outdated, uncompetitive, and in decline. The hinterland of Kolkata contains India's Rustbelt. The government keeps many industries going but at a high cost. Old industries, such as carpetmaking and cotton-weaving, continue to use child labor to remain viable. The State of Bihar represents the stagnation that afflicts much of India east of our line: by several measures it ranks among the poorest of the 28 States.

Compare this to western India. The State of Maharashtra, the hinterland of Mumbai, leads India in many categories, and Mumbai leads Maharashtra. Many smaller, private industries have emerged here, manufacturing goods ranging from umbrellas to satellite dishes and from toys to textiles. Across the Arabian Sea lie the oil-rich economies of the Arabian Peninsula. Hundreds of thousands of workers from western India have found jobs there, sending money back to families from Punjab to Kerala. More importantly, many have used their foreign incomes to establish service industries back home. Outward-looking western India, in contrast to the inward-looking east, has begun to establish other ties to the outside world. Satellite links have enabled Bangalore to become the center of a growing software-producing complex reaching world markets. The beaches of Goa, the small State immediately to the south of Maharashtra, appeal to the tourist markets of Europe. This is, in fact, a
14 classic case of **intervening opportunity** because resorts have sprung up along Goa's coast, and European tourists who once went to the more distant Maldives and Seychelles are coming to Goa. Maharashtra's economic success also has spilled over into Gujarat to the north, and even landlocked Rajasthan (the next State to the north) is experiencing the beginnings of what, by Indian standards, is a boom.

The boom has created political problems, however. Not only is Maharashtra State a rising economic power; it also is the base of a strong Hindu nationalist political movement whose leaders object to foreign intrusions and have blocked major development projects and other enterprises. They halted a huge industrial scheme about halfway through and closed a fast-food operation that they deemed incompatible with local culture. Such clashes between foreign interests and domestic traditions continue even as India's economy forges ahead.

Nevertheless, India's east-west divide shows a growing contrast that puts the west far ahead. The hope is that Maharashtra's success will spread northward and southward along the Arabian Sea coast and will ultimately diffuse eastward as well. But for this to happen, India will have to bring its population spiral under control.

BANGLADESH: CHALLENGES OLD AND NEW

On the map of South Asia, Bangladesh looks like another State of India: the country occupies the area of the double delta of India's great Ganges and Brahmaputra rivers, and India almost completely surrounds it on its landward side (Fig. 8-16). But Bangladesh is an independent country, born in 1971 after its brief war for independence against Pakistan, with a territory about the size of Wisconsin. Today it remains one of the poorest and least developed countries on Earth, with a population of 147.3 million that is growing at an annual rate of 2.1 percent.

Not only is Bangladesh a poor country; it also is highly susceptible to damage from **natural hazards**. 15
During the twentieth century, eight of the ten deadliest natural disasters in the entire world struck this single country. Most recently, a 1991 cyclone (as hurricanes are called in this part of the world) killed over 150,000 people.

The reasons for Bangladesh's vulnerability can be deduced from Figures 8-16 and 8-1. Southern Bangladesh is the deltaic plain of the Ganges–Brahmaputra river system, combining fertile alluvial soils that attract farmers with the low elevations that endanger them when water rises. The shape of the Bay of Bengal forms a funnel that sends cyclones and their storm surges of wind-whipped water barreling into the delta coast. Without money to build seawalls, floodgates, elevated shelters in sufficient numbers, or adequate escape routes, hundreds of thousands of people are at continuous risk, with deadly consequences. And as if this is not enough, millions of people have been found to be exposed to excessive (natural) arsenic in the drinking water from their wells.

Bangladesh remains a nation of subsistence farmers; urbanization is at only 23 percent, and Dhaka, the megacity capital, and the southeastern port of Chittagong are the only urban centers of consequence. Moreover, Bangladesh has one of the highest physiologic densities in the world (3913 people per square mile/1511 per sq km), and only higher-yielding varieties of rice and the introduction of wheat in the crop rotation (where climate allows) have improved diets and food security. But diets remain poorly balanced, and, overall, the people's

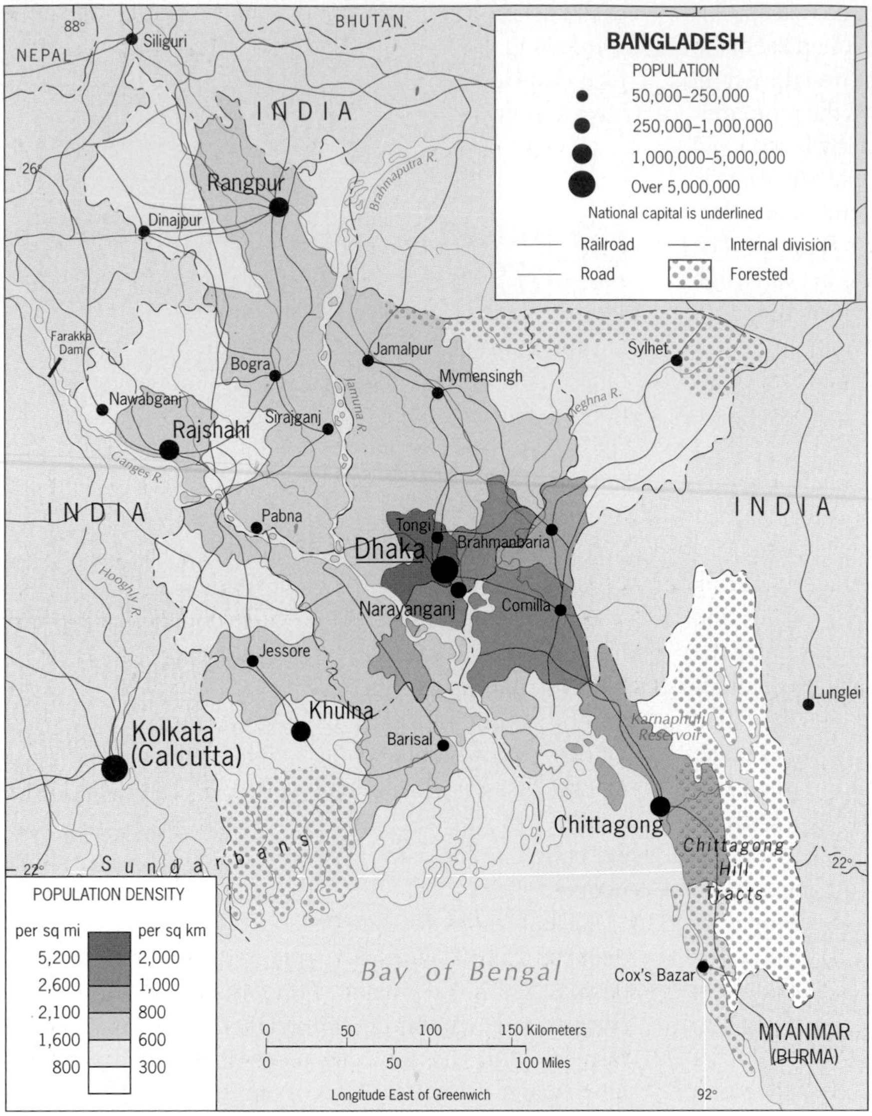

FIGURE 8-16 © H. J. de Blij, P. O. Muller, and John Wiley & Sons, Inc.

nutrition is unsatisfactory. The textile industry provides most of Bangladesh's foreign revenues, but the once-thriving jute industry continues its decline. The discovery of a natural gas reserve is now the subject of a national debate: home consumption or money-making export?

Bangladesh is a dominantly Muslim society, but not one dominated by revivalists. For example, 30 seats in the national legislature are reserved for women. Its relations with neighboring India have at times been strained over water resources (India's control over the Ganges that is Bangladesh's lifeline), cross-border migration (10 percent of the population is Hindu), and transit between parts of India across Bangladesh's north (refer to Figure 8-15 to see the reason). All the disadvantages of the global periphery afflict this populous, powerless country where survival is the leading industry and all else is luxury.

THE MOUNTAINOUS NORTH

As Figures 8-1 and 8-2 show, a tier of landlocked countries and territories lies across the mountainous zone that walls India off from China. One of them, Kashmir, is in a state of near-war. Another, Sikkim, was absorbed by India in 1975 and made into one of its federal States. But Nepal and Bhutan retain their independence.

Nepal, northeast of India's Hindu core, has a population of 25.9 million and is the size of Illinois. It has three geographic zones (Fig. 8-17): a southern, subtropical, fertile lowland called the Terai; a central belt of Himalayan foothills with swiftly flowing streams and deep valleys; and the spectacular high Himalayas themselves (topped by Mount Everest) in the north. The capital, Kathmandu, lies in the east-central part of the country in an open valley of the central hill zone.

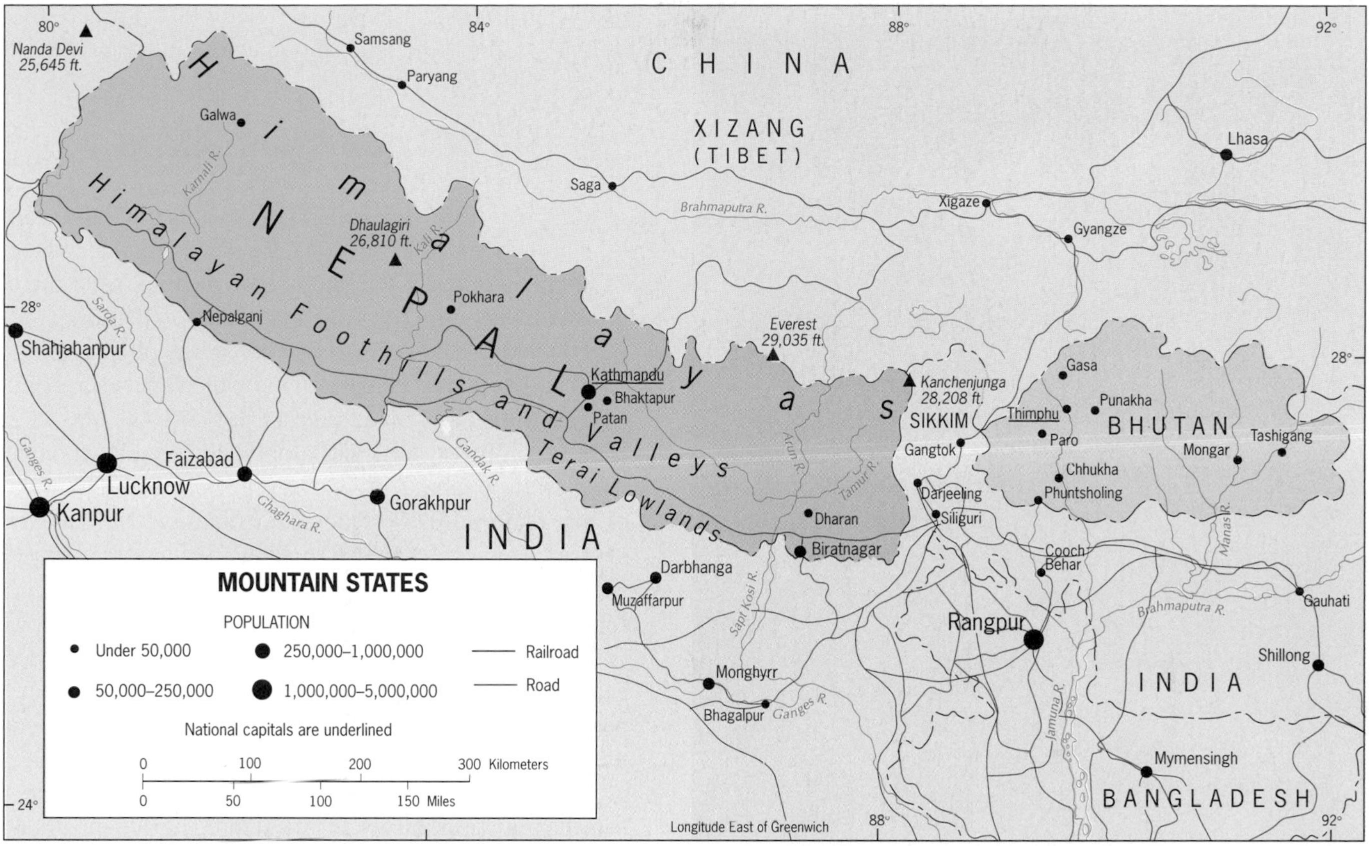

FIGURE 8-17 © H. J. de Blij, P. O. Muller, and John Wiley & Sons, Inc.

Nepal is materially poor but culturally rich. The Nepalese are a people of many sources, including India, Tibet, and interior Asia. About 80 percent are Hindu, and Hinduism is the country's official religion; but Nepal's Hinduism is a unique blend of Hindu and Buddhist ideals. Thousands of temples and pagodas ranging from the simple to the ornate grace the cultural landscape, especially in the valley of Kathmandu, the country's core area. Although over a dozen languages are spoken, 90 percent of the people also speak Nepali, a language related to Indian Hindi.

As the data in Table G-1 suggest, Nepal is a troubled country suffering from severe underdevelopment, with a GNI near the lowest in the entire realm. It also faces strong centrifugal social and political forces. Environmental degradation, crowded farmlands and soil erosion, and deforestation scar the countryside. The Himalayan peaks form a world-renowned tourist attraction, but tourist spending in Nepal, always relatively modest, has been cut back because of the current revival of Maoist-communist terrorism, not only in the western hills where it has long been active but also in the capital itself.

Nepal's political geography has long been troubled. The end of absolute monarchy in 1991 and the advent of democracy did not end the country's regional divisions: the southern Terai with its tropical lowlands is a world apart from the hills of central Nepal, and the peoples of the west have origins and traditions different from those in the east. Moreover, in 2001 the assassination of Nepal's king threatened the disintegration of the state, narrowly averted by the leadership of the Nepal Congress Party. Then in 2002 the country's stability was imperiled again as the Maoist insurgency spread. Three years later, Nepal was in effect a **failed state**, its government unable to control the insurgency, its economy collapsed, and its future hanging in the balance. 16

Mountainous **Bhutan**, wedged between India and China's Tibet (Fig. 8-17), is the only other buffer between Asia's giants. In landlocked, fortress-like Bhutan, time seems to have stood still. Bhutan is officially a constitutional monarchy, but its king rules the country with virtually absolute power; economic subsistence and political allegiance are the norms of life for most of the population of just over 1 million. Thimphu, the capital, has about 60,000 inhabitants. The symbols of Buddhism, the state religion, dominate its cultural landscape. Social tensions arise from the large but diminishing Nepalese minority, most of whom are Hindus and some of whom have been persecuted by the dominant Bhutia.

Forestry, hydroelectric power, and tourism all have potential here, and Bhutan has considerable mineral

Human-induced environmental degradation is vividly displayed on Nepal's highest inhabitable slopes. Against the spectacular backdrop of the main Himalaya range north of Kathmandu, population pressures in the teeming valleys have forced farming and fuel-wood extraction to ever higher elevations. The inevitable result is deforestation and when the dense vegetation and binding roots are removed, copious summer-monsoon rainwater is free to cascade down the steep, denuded mountainsides. This massive runoff soon produces myriad gullies that further destabilize slopes by triggering mass movements of soil and rock. The lower half of this photo exhibits the evidence of this erosional process, which affects the more level terrain in the foreground as well. Ominously, we can also discern human dwellings creeping upslope, which portends a final onslaught on the single remaining forested ridge. © Galen Rowell/Mountain Light Photography, Inc.

resources. But isolation and inaccessibility preserve traditional ways of life in this mountainous buffer state.

THE SOUTHERN ISLANDS

As Figure 8-1 shows, South Asia's continental landmass is flanked by several sets of islands: Sri Lanka off the southern tip of India, the Maldives in the Indian Ocean to the southwest, and the Andaman Islands (belonging to India) marking the eastern edge of the Bay of Bengal.

The **Maldives** consists of more than a thousand tiny islands whose combined area is just 115 square miles (less than 300 sq km) and whose highest elevation is barely over 6 feet (2 m) above sea level. Its population of just under 300,000 from Dravidian and Sri Lankan sources is now 100 percent Muslim, one-quarter of which is concentrated on the capital island named Maale. The Maldives might be unremarkable, except that, as Table G-1 shows, this country has the realm's highest GNI per capita. The locals have translated their palm-studded, beach-fringed islands into a tourist mecca that attracts tens of thousands of mainly European visitors annually.

The Maldives' low elevation has been mentioned repeatedly in assessments of the future impact of global warming, which would produce rising sea levels (top photo p. 421). Those risks became sudden reality on December 26, 2004 when the Indian Ocean tsunami generated off Indonesia swept over the islands, killing more than 100 residents as well as tourists and destroying resort facilities along the shore as well as inland. In mid-2005 it was too early to gauge the impact on the Maldives' realm-leading economy, but the tourist industry was badly damaged.

Sri Lanka: South Asian Tragedy

Sri Lanka (known as Ceylon before 1972), the compact, pear-shaped island located just 22 miles (35 km) across the Palk Strait from India, became independent from Britain in 1948 (Fig. 8-18). There were good reasons to create a separate sovereignty for Sri Lanka. This is neither a Hindu nor a Muslim country: the majority of its more than 20 million people—about 75 percent—are Buddhists. Furthermore, unlike India or Pakistan, Sri Lanka is a plantation country, and commercial farming still is the mainstay of the agricultural economy.

The great majority of Sri Lanka's people are descended from migrants who came to this island from northwest India beginning about 2500 years ago. Those migrants introduced the advanced culture of their source area, building towns and irrigation systems and bringing Buddhism. Today, their descendants, known as the Sinhalese, speak a language (Sinhala) that belongs to the Indo-European language family of northern India.

FROM THE FIELD NOTES

"Some countries have to take warnings of global warming more seriously than others. As we approached the Maldives, the islands lay like lilypads on the surface of a pond. No part of this country's natural surface lies more than 6 feet (less than 2 meters) above sea level. The upper floors of the buildings in the capital, Maale, form the Maldives's highest points. Almost any rise in sea level would threaten this Indian Ocean outpost of South Asia."

The Dravidians who lived on the mainland, just across the Palk Strait, came later and in far smaller numbers—until the British colonialists intervened. During the nineteenth century the British brought hundreds of thousands of Tamils to work on their tea plantations, and soon a small minority became a substantial segment of Ceylonese society. The Tamils brought their Dravidian tongue to the island and introduced their Hindu faith. At the time of independence, they constituted more than 15 percent of the population; today they total about 18 percent.

When Ceylon became independent, it was one of the great hopes of the postcolonial world. The country had a sound economy and a democratic government, and it was renowned for its tropical beauty. Its reputation soared when a massive campaign succeeded in eradicating malaria and when family-planning campaigns reduced population growth while the rest of the realm was experiencing a population explosion. Rivers from the cool, forested interior highlands fed the paddies that provided ample rice; crops from the moist southwest paid the bills, and the capital, Colombo, grew to reflect the optimism that prevailed.

In the midst of this glowing scenario, the seeds of disaster were already being sown. Sri Lanka's Tamil minority soon began proclaiming its sense of exclusion, demanding better treatment from the Sinhalese majority.

FROM THE FIELD NOTES

"The long civil war involving Tamil demands for a separate state has severely damaged Sri Lanka's economy. Its impact can be seen in the capital, Colombo, where the tourist industry collapsed and new investment dwindled. In the mid-1990s, with hopes for a resolution of the crisis rising, some new building finally began in Colombo: the twin towers seen rising here above the cityscape were being built by Singaporean investors anticipating a resumption of economic progress in this embattled country. But even the capture of Jaffna by government forces did not lower the tension in the capital. We saw armed military checkpoints everywhere." Note: The towers were completed, but badly damaged later by a massive bomb set off by the Tamil Tigers in the business district. A suicide bomber made an attempt on the life of the Prime Minister; the Temple of the Tooth, Sri Lankan Buddhism's holiest shrine, was damaged by a car bomb that killed many. Sri Lanka's future is threatened by a conflict from which there seems to be no escaping.

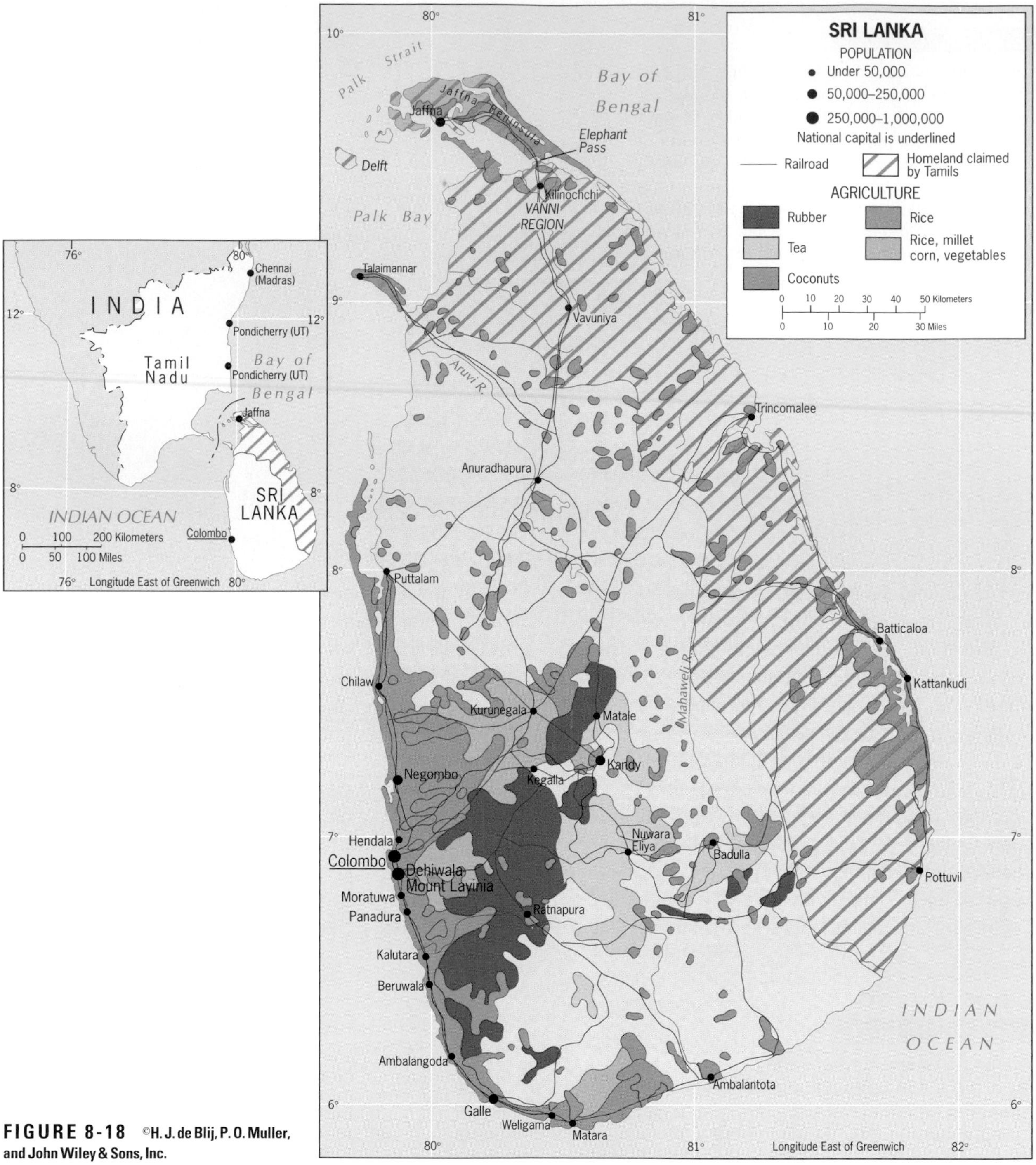

FIGURE 8-18 ©H. J. de Blij, P. O. Muller, and John Wiley & Sons, Inc.

Although the government recognized Tamil as a "national language" in 1978, sporadic violence marked the Tamil campaign, and in 1983 full-scale civil war began. Now many in the Tamil community demanded a separate Tamil state to encompass the north and east of the country (see Fig. 8-18), and a rebel army called the Tamil Tigers confronted Sri Lanka's national forces.

The sequence of events fits the model of the evolution
17 of the **insurgent state** discussed in Chapter 5. In Sri Lanka, the *equilibrium* stage was reached in the early 1990s, when the Tamil Tigers claimed the Jaffna Peninsula and set up their headquarters there. The *counteroffensive* stage is now in progress, but the Tamil forces strike at will (their attacks on Colombo are detailed in the photo caption on p. 421) and a negotiated settlement, now being pursued with the help of Norwegian mediators, appears inevitable. Even if the Tamils do not succeed in securing the independent state they call Eelam, they will likely force the government to make some territorial concessions. And even the devastation from the 2004 Indian Ocean tsunami, which struck Sri

FROM THE FIELD NOTES

"Sri Lanka is a dominantly Buddhist country, the religion of the majority Sinhalese. Most areas of the capital, Colombo, have numerous reminders of this in the form of architecture and statuary: shrines to the Buddha, large and small, abound. But walk into the Tamil parts of town, and the cultural landscape changes drastically. This might as well be a street in Chennai or Madurai: elaborate Hindu shrines vie for space with storefronts and Buddhist symbols are absent. The people here seemed to be less than enthused about the Tamil Tigers' campaign for an independent state. 'This would never be a part of it anyway,' said the fellow walking toward me as I took this photograph. 'We're here for better or worse, and for us the situation up north makes it worse.' But, he added, Sri Lankan governments of the past had helped create the situation by discriminating against Tamils."

© H. J. de Blij.

Lanka especially hard and killed an estimated 35,000 people, most of them along the exposed southern and eastern coast, could not dampen the hostility between the two sides. Accusations of theft of relief supplies and intimidation of survivors occurred even as both parties faced a common enemy.

Once-promising Sri Lanka is paying dearly for its politicians' failures. The cost of the civil war is enormous. The economic consequences are incalculable, with the tourist industry having been reduced to a fraction of what it could be. And South Asia has lost a beacon of opportunity and progress.

WHAT YOU CAN DO

RECOMMENDATION: Seek an internship! If you are willing to trade your time for some experience and perhaps even education and training, consider looking for an internship where you can use your knowledge of geography while learning more about it. Should you decide to become a geography major or take a minor, you may be eligible for one of the National Geographic Society's internships in Washington, D.C. The Society places interns in many of its operations, from the Geography Bee to National Geographic Maps. Inquire about this and other internship opportunities at the office of the Geography Department at your college or university.

GEOGRAPHIC CONNECTIONS

1 One of the current issues in the United States involves the so-called outsourcing of jobs: the transfer of salaried employment from American to foreign locations. Many economists say that this is a logical outgrowth of globalization, that it involves comparatively few jobs, and that it helps U.S. companies survive competition to which they might otherwise succumb—costing more jobs than the outsourcing did in the first place. India has become a major player in the outsourcing of certain work. What are these jobs? Where in India do they cluster and why? Do language and/or time-zone location have anything to do with India's success, and if so, how?

2 Six decades after India and Pakistan ceased to be part of the British colonial empire, the contrasts between India and Pakistan—demographic, economic, social, political—could hardly be greater despite the fact that they are neighbors. What, in your view, are the key differences between the two South Asian states in these contexts? Can you identify crucial geographic factors that would account for such enormous dissimilarities? Given their shared histories, do you see any common ground between Pakistan and India today that might form the basis for future convergence and cooperation?

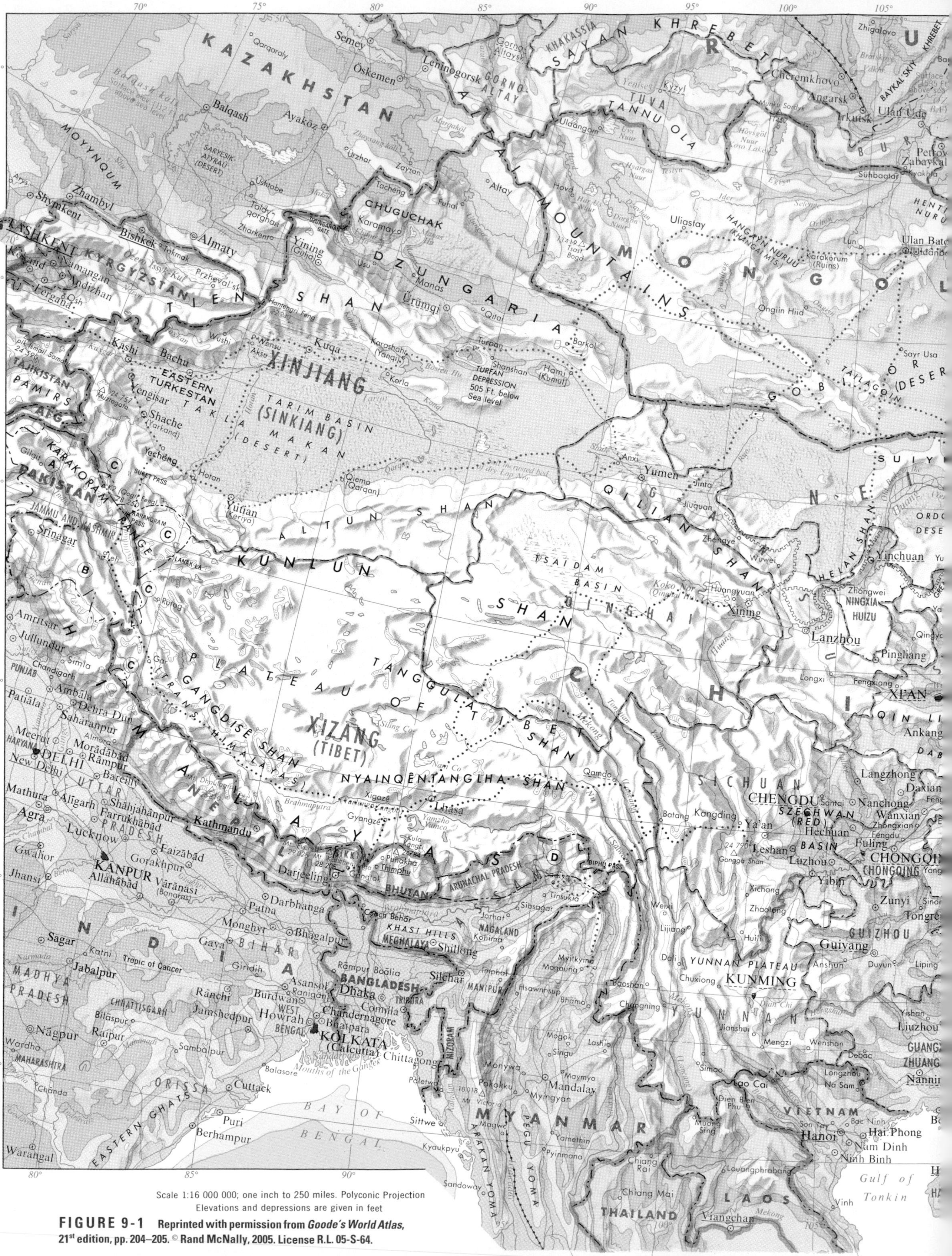

Scale 1:16 000 000; one inch to 250 miles. Polyconic Projection
Elevations and depressions are given in feet

FIGURE 9-1 Reprinted with permission from *Goode's World Atlas*, 21st edition, pp. 204–205. © Rand McNally, 2005. License R.L. 05-S-64.

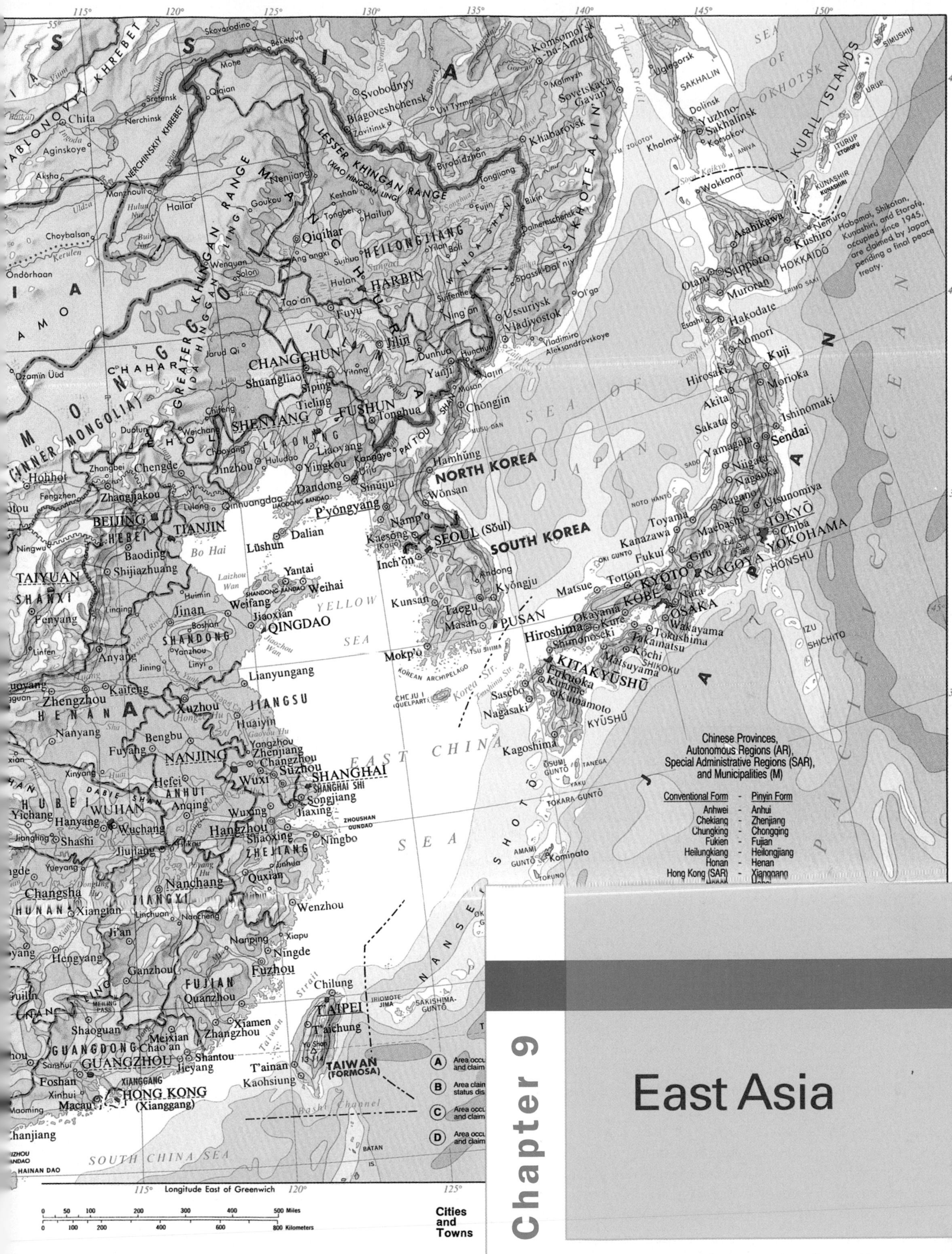

Chapter 9

East Asia

CONCEPTS, IDEAS, AND TERMS

1. Pacific Rim
2. Confucius
3. Sinicization
4. Extraterritoriality
5. Special Administrative Region (SAR)
6. Restrictive population policies
7. Economic restructuring
8. Core area
9. Geography of development
10. Overseas Chinese
11. Special Economic Zone (SEZ)
12. Regional state
13. Economic tiger
14. Buffer state
15. Jakota Triangle
16. Modernization
17. Relative location
18. Areal functional organization
19. Regional complementarity
20. State capitalism

REGIONS

- CHINA PROPER
- XIZANG (TIBET)
- XINJIANG
- MONGOLIA
- JAKOTA TRIANGLE (JAPAN-KOREA-TAIWAN)

East Asia is a geographic realm like no other. At its heart lies the world's most populous country. On its periphery lies one of the globe's most powerful national economies. Along its coastline, on its peninsulas, and on its islands an economic boom has transformed cities and countrysides. Its interior contains the world's highest mountains and vast deserts. It is a storehouse of raw materials. The basins of its great rivers produce food that can sustain more than a billion people.

DEFINING THE REALM

The East Asian geographic realm consists of six political entities: China, Mongolia, North Korea, South Korea, Japan, and Taiwan. Note that we refer here to "political entities" rather than "states." In changing East Asia, the distinction is significant. Taiwan, which its government officially calls the Republic of China, functions as a state but is regarded by mainland China (the People's Republic of China) as a temporarily wayward province. North Korea is not a full member of the United Nations, and the division of the Korean Peninsula may be temporary.

As defined here, East Asia lies between the vast expanses of Russia to the north and the populous countries of South and Southeast Asia to the south. This geographic realm extends from the deserts of Central Asia to the Pacific islands of Japan and Taiwan. Environmental diversity is one of its hallmarks.

East Asia also is the hub of the evolving regional
1 phenomenon called the **Pacific Rim**. From Japan to Taiwan and from South Korea to Hong Kong (Xianggang), the Pacific frontage of East Asia is being transformed (see box titled "Names and Places"). Skyscrapers tower over Haikou, capital of the once-dormant southern Chinese island of Hainan. Luxury automobiles from Europe and America ply the streets of Dalian, long the drab port for China's Northeast. A forest of construction cranes marks the emergence of Pudong, a huge commercial and industrial zone where Shanghai stakes its entry into the Pacific Rim's booming economy. Millions of people are on the move, abandoning their farms and villages and seeking work in such projects.

Japan was the leader in East Asia's Pacific Rim development. Long before this regional term even came into general use, Japan had built a giant economy with global connections and was the only highly developed country in Asia not plagued by stagnation. Although China lay isolated and South Korea struggled in the aftermath of its terrible war against communism (1950–1953), Japan built a society unlike that of any other in eastern Eurasia and merited recognition as a discrete geographic realm. But Pacific Rim developments are diminishing the contrasts be-

MAJOR GEOGRAPHIC QUALITIES OF East Asia

1. East Asia is encircled by snowcapped mountains, vast deserts, cold climates, and Pacific waters.
2. East Asia was one of the world's earliest culture hearths, and China is one of the world's oldest continuous civilizations.
3. East Asia is the world's most populous geographic realm, but its population remains strongly concentrated in its eastern regions.
4. China, the world's largest nation-state demographically, is the current rendition of an empire that has expanded and contracted, fragmented and unified many times during its long existence.
5. China today remains a mainly rural society, and its vast eastern river basins feed hundreds of millions in a historic pattern that continues today.
6. China's sparsely peopled western regions are strategically important to the state, but they lie exposed to minority pressures and Islamic influences.
7. Along China's Pacific frontage an economic transformation is taking place, affecting all the coastal provinces and creating an emerging Pacific Rim region.
8. Increasing regional disparities and fast-changing cultural landscapes are straining East Asian societies.
9. Japan, the economic giant of the East Asian realm, has a history of colonial expansion and wartime conduct that still affects international relations here.
10. East Asia may witness the rise of the world's next superpower as China's economic and military strength and influence grow—and if China avoids the devolutionary forces that fractured the Soviet Union.
11. The political geography of East Asia contains a number of flashpoints that can generate conflict, including Taiwan, North Korea, and several island groups in the realm's seas.

tween Japan and its East Asian neighbors. South Korea today challenges Japan on international markets, selling goods ranging from electronics to automobiles. In southeastern China, Guangdong Province is an economic juggernaut that includes Shenzhen, recently the world's fastest growing city. Japan's own investments in Pacific Rim economies have helped not only to lessen the contrasts, but also to enhance the linkages between Tokyo and the mainland. In short, Japan has again become part of a functional region within the East Asian realm.

Names and Places

In 1958, the government of the People's Republic of China adopted the so-called *pinyin* system of standard Chinese, which replaced the Wade-Giles system used since colonial times. Pinyin was adopted not to teach foreigners how to spell and pronounce Chinese names and words but to establish a standard form of the Chinese language throughout China. The pinyin system is based on the pronunciation of Chinese characters in Northern Mandarin, the Chinese spoken in the region of the capital and the north in general.

The new linguistic standard caught on slowly outside China, but today it is in general use. The old name of the capital, Peking, has become Beijing. Canton is now Guangzhou. The Yangtze Kiang (River) is now the Chang Jiang/Yangzi. Tientsin, Beijing's port, is now Tianjin. A few of the old names persist, however. The Chinese call their colony Xizang, but many maps still carry the name Tibet.

Pinyin usage also affected personal names. China's long-time ruler, Mao Tse-tung, is now called Mao Zedong. His eventual successor, Teng Hsiao-ping, was more simply Deng Xiaoping. And remember: when the Chinese write their names, they use the last name first. The current president, Hu Jintao, is Jintao to friends and Mr. Hu to others. To the Chinese, it is Bush George and Cheney Dick, not the other way around.

No matter how powerful Japan's economy became during the second half of the twentieth century, China remains the colossus of East Asia. One of the world's oldest continuous civilizations, China was a major culture hearth when Japan was an isolated frontier inhabited by the Ainu. Migrations that originated in China advanced through the Korean Peninsula and reached the islands. Philosophies, religions, and cultural traditions (including, for example, urban design and architecture) diffused from China to Korea and Japan, and to other locales on the margins of the Chinese hearth. Over the past millennium, China repeatedly expanded to acquire empires, only to collapse in chaos when its center failed to maintain control. European colonial powers took such opportunities to partition China among themselves. The Russians and the Japanese pushed their empires deep into China. But always China recovered, and today China may again be poised to expand its sphere of influence and, indeed, to take its place as a world power in the twenty-first century.

NATURAL ENVIRONMENTS

Figure 9-1 dramatically illustrates the complex physical geography of the East Asian realm. In the southwest lie ice-covered mountains and plateaus, the Earth's crust in this region crumpled up like the folds of an accordion. A gigantic collision of tectonic plates is creating this landscape as the Indian Plate pushes northward into the underbelly of the Eurasian Plate (Fig. G-4). The result is some of the world's most spectacular scenery, but snow, ice, and cold are not the only dangers to human life here. Earthquakes and tremors occur almost continuously, causing landslides and avalanches. As the map shows, the high mountains and plateaus widen from a relatively narrow belt in the Karakoram to form Xizang's (Tibet's) vast plateau, flanked by the Himalayas to the south. Then, east of Tibet, the mountain ranges converge again and bend southward into Southeast Asia, where they lose their high relief.

FROM THE FIELD NOTES

"From the train, traveling from Beijing to Xian, we had a memorable view of the Loess Plateau. Loess is a fine-grained dust formed from rocks pulverized by glacial action, blown away by persistent winds, and deposited in sometimes well-defined locales. Loess covers much of the North China Plain, which is what makes it so fertile, but there it is not as thick as in the Loess Plateau in the middle basin of the Huang He (Yellow River). Here the loess averages 250 feet (75 m) in thickness and in places reaches as much as 600 feet (180 m). Because loess has some very distinctive physical properties, it tends to create unusual landscapes. Through a complicated physical process following deposition, loess develops the capacity to stand upright in walls and columns, and resists collapse when it is excavated. As a result the landscape looks terraced: streams cut deep and steep-sided valleys. The Loess Plateau is a physiographic region, but it is a cultural region too. Hundreds of thousands of people have literally dug their homes into vertical faces of the kind shown here, creating cave-like, multiroom dwellings with wooden exterior doors." © H. J. de Blij.

As Figure G-9 shows, this Asian interior is one of the world's most sparsely populated areas, but it is nevertheless critical to the lives of hundreds of millions of people. In these high mountains, fed by the melting ice and snow, rise the great rivers that flow eastward across China and southward across Southeast and South Asia. Throughout the Holocene, these rivers have been eroding the uplands and depositing their sediments in the lowlands, in effect creating the alluvium-filled basins that now sustain huge populations. Fertile alluvial soils and adequate growing seasons, combined with ample water and millions of hands to sow the wheat and plant the rice, have allowed the emergence of one of the great population concentrations on Earth.

Figure G-8 records this high-elevation zone of East Asia as category *H*, highland climate, which is by far the largest area of its kind in the world. Here one criterion overpowers the rules of climatic classification: altitude. The mountains wrest virtually all moisture from air masses moving northward and block their path into interior Asia. As Figure G-8 shows, desert conditions prevail in East Asia's northern interior. Figure 9-2 names two of the more famous ones: the Takla Makan of far western China and the Gobi of Mongolia.

Physiography, therefore, has much to do with East Asia's population distribution, but even the more habitable and agriculturally productive east has its limitations. The northeast suffers from severe continentality, with long and bitterly cold winters. High relief encircles the river basins north of the Yellow Sea and dominates much of the northern part of the Korean Peninsula, creating strong local environmental contrasts. South Korea, as Figure G-8 shows, experiences relatively moderate conditions comparable to those of the U.S. Southeast; North Korea has a harsh continental climate like that of North Dakota.

In general, coastal, peninsular, and insular East Asia possess more moderate climates than the interior. Like South Korea, southern Japan and southeastern China have humid temperate climates, and southernmost Taiwan and Hainan Island even have areas of tropical (*A*) conditions. Proximity to the ocean tends to moderate climatic environments, and coastal East Asia proves the point.

Before we focus on East Asia's human geography, it is useful to look at a map of this realm's complex physical stage (Fig. 9-2). From the high-relief interior come

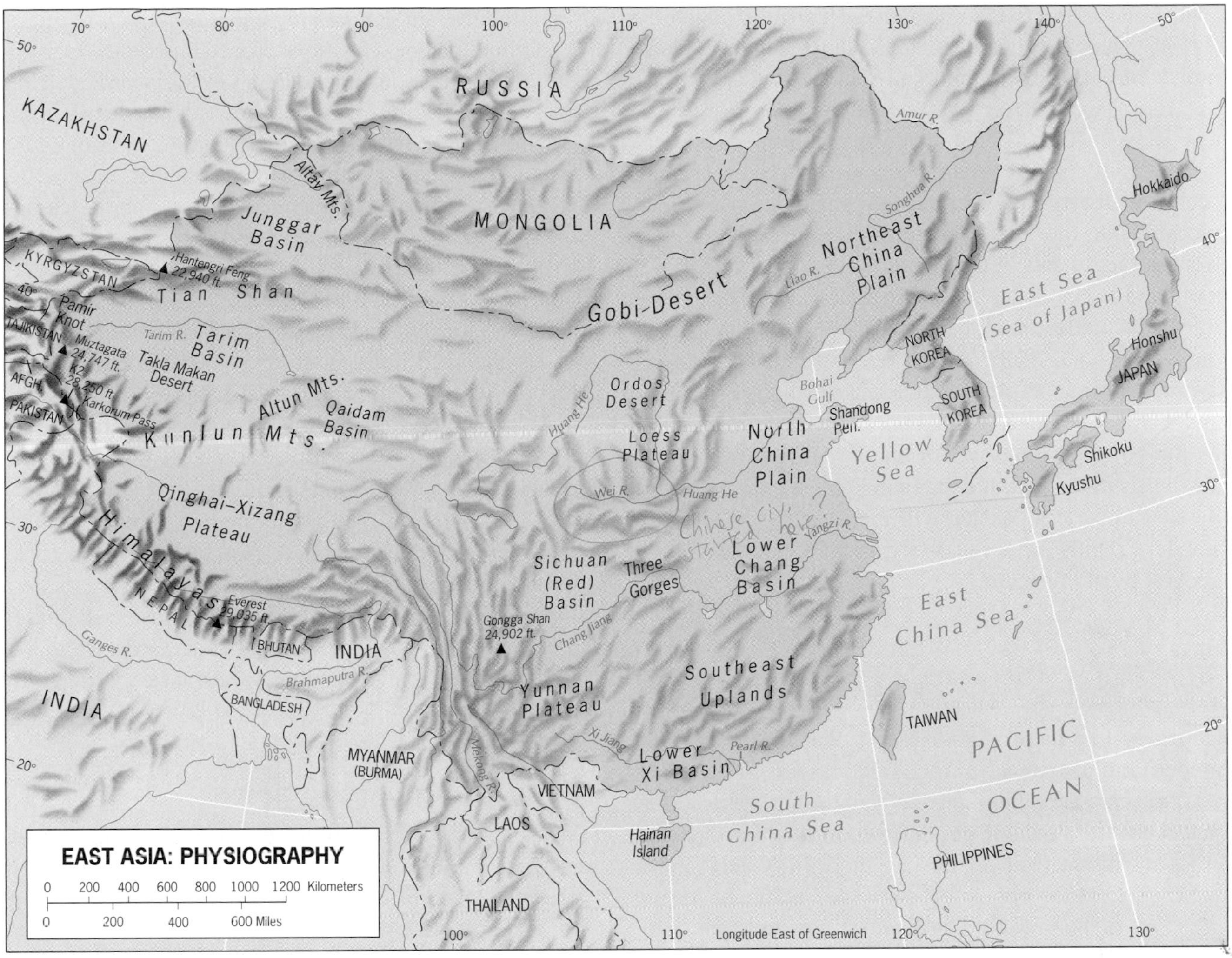

FIGURE 9-2 © H.J. de Blij, P.O. Muller, and John Wiley & Sons, Inc.

three major river systems that have played crucial roles in the human drama. In the north, the Huang He (Yellow River) arises deep in the high mountains, crosses the Ordos Desert and the Loess Plateau, and deposits its fertile sediments in the vast North China Plain, where East Asia's earliest states emerged. In the center, the Chang Jiang (Long River), called the Yangzi downstream, crosses the Sichuan Basin and the Three Gorges, where a huge dam project is under way, and waters extensive ricefields in the Lower Chang Basin. And in the south, the Xi Jiang (West River) originates on the Yunnan Plateau and becomes the Pearl River in its lowest course. Its estuary, flanked by several of China's largest urban-industrial complexes, has become one of the hubs of the evolving Pacific Rim.

Further scrutiny of Figure 9-2 indicates that a fourth river system plays a role in China: the Liao River in the northeast and its basin, the Northeast China Plain. As the map suggests, however, the Liao is not comparable to the great rivers to its south, its course being shorter and its basin, in this higher-latitude area, much smaller. As we will see, China's Northeast was for some time the country's industrial, not agricultural, heartland.

Looking again toward the interior, note the Loess Plateau south of the Ordos Desert, where the Huang He makes its giant loop. Loess is a fertile, windblown deposit composed of rock pulverized by glaciers (see photo on page at left). Add water (in this case the middle Huang and its tributaries) and an adequate growing season, and a sizeable population will arise. To the south, deep in the interior, lies the Sichuan (Red) Basin, crossed by the Chang Jiang. This basin has supported human communities for a long time, and you can actually make out its current population cluster on the world population map (Fig. G-9). The Sichuan Basin, encircled as it is by mountains, is one of the world's most clearly defined physiographic regions, and the concentration of its approximately 120 million inhabitants reflects that definition.

Still farther to the south lies the Yunnan Plateau, source of the tributaries that feed the Xi River. Much of southeastern China has comparatively high relief; it is hilly and in places mountainous. This high relief has helped limit contacts between China and Southeast Asia.

East Asia's Pacific margin is a jumble of peninsulas and islands. The Korean Peninsula looks like a near-bridge from Asia to Japan, and indeed it has served as such in the past. The Liaodong and Shandong peninsulas protrude into the Yellow Sea, which continues to silt up from the sediments of the Huang and Liao rivers. Off the mainland lie the islands that have played such a crucial role in the modern human geography of Asia and, indeed, the world: Japan, Taiwan, and Hainan. Japan's environmental range is expressed by cold northern Hokkaido and warm southern Kyushu, but Japan's core area lies on its main island, Honshu. As Figure 9-1 shows, myriad smaller islands flank the mainland and dot the East and South China seas. As we will discover, some of these smaller islands have major significance in the human geography of this realm.

HISTORICAL GEOGRAPHY

Consider this: there is no evidence that *Homo sapiens*, modern humans, reached the Americas any earlier than 40,000 years ago, and many paleoanthropologists argue that the only available evidence indicates that humans crossed from Eurasia to Alaska a mere 14,000 to 13,000 years ago. On the opposite side of the Pacific, however, the story of hominid and human settlement spans hundreds of thousands of years—perhaps 1 million.

What took so long? If hominids like *Homo erectus* left their ancestral African homelands and migrated as far away as eastern Asia more than half a million years ago, and if modern humans followed them and reached Australia some 50,000 years ago, what kept the hominids from the Americas altogether, and what delayed the humans?

No satisfactory answers have yet been found to questions like these. The width of the Pacific Ocean may have doomed maritime migrants, although trans-Pacific migration may eventually have occurred. The frigid temperatures along the northern route from Siberia to Alaska may have stopped hominids and humans alike. But the Pleistocene was punctuated by warm interglaciations; why did it take until the early Holocene for humans to make the Bering Strait crossing?

All this implies that for hundreds of thousands of years East Asia was a cul-de-sac, a dead-end for migrants out of Africa. Many archeological sites in the realm have yielded evidence of *Homo erectus*, including what is perhaps the most famous of all: Peking Man. In a cave near the Chinese capital, Beijing, an archeologist in 1927 found a single tooth he recognized as representing a hominid. Later excavations proved him right as parts of more than three dozen skeletons were unearthed. They proved that Peking Man and his associates made stone and bone tools, had a communal culture, controlled the use of fire, hunted wildlife, and cooked their meat. Later arrivals undoubtedly challenged older communities, and the frontier of settlement must have expanded across the river basins of the east. But when *Homo sapiens*, modern humans, arrived in East Asia, the hominids could not compete. Current anthropological theory holds that the better-equipped humans eliminated the hominids, perhaps between 60,000 and 40,000 years ago, after which the human migration into the Americas could begin.

In the context of East Asia we should mention a minority view. Some anthropologists state that the evidence supports the notion that modern humanity developed *not* from one stock in Africa, but from four stocks in widely separated parts of the world: Africa, western Eurasia (the Caucasoids), Australia (the Australoids), and eastern Eurasia (the Mongoloids). According to this idea, today's East Asians trace their ancestry to Peking Man and beyond.

Early Cultural Geography

Whatever the outcome of the debate over human origins in East Asia, it is clear that humans have inhabited the plains and river basins, foothills, and islands of this realm for a very long time. Hunting sustained both the hominids and the early human communities; fishing drew them to the coasts and onto the islands. The first crossing into Japan may have occurred as long as 10,000 to 12,000 years ago, possibly much longer, when the Jomon people, a Caucasoid population of uncertain geographic origins, entered the islands; their modern descendants, the Ainu, spread throughout the Japanese archipelago. Today, only about 20,000 persons living in northernmost Hokkaido trace their ancestry to Ainu sources.

About 2300 years ago the Yayoi people, rice farmers who had settled in Korea, appear to have crossed by boat to Kyushu, Japan's southernmost island, from where they advanced northward. The Ainu, who subsisted by fishing, trapping, and hunting, were driven back, but gene-pool studies show that much mixing of the groups took place; they also show that the Yayoi invasion was followed by other incursions from the Asian mainland. By then, powerful dynastic states had already arisen in what is today China, and early Chinese culture traits thus found their way into Japan through the process we know as relocation diffusion.

On the Asian mainland, plant and animal domestication had begun as early as anywhere on Earth. We in the

Western world take it for granted that these momentous processes began in what we now call the Middle East and diffused from the Fertile Crescent to other parts of Eurasia and the rest of the world. But the taming of animals and the selective farming of plants may have begun as early, or earlier, here in East Asia. As in Southwest Asia, the fertile alluvial soils of the great river basins and the ebb and flow of stream water created an environment of opportunity, and millet and rice were being harvested between 7000 and 8000 years ago.

Even during this Neolithic period of increasingly sophisticated stone tools, East Asia was a mosaic of regional cultures. Their differences are revealed by the tools they made and the decorations on their bowls, pots, and other utensils. An especially important discovery of two 8000-year-old pots in the form of a silkworm cocoon, from China's Hebei Province, suggests a very ancient origin for one of the region's leading historic industries.

As noted earlier, plant and animal domestication produced surpluses and food storage, enabling population growth and requiring wider regional organization. Here as elsewhere during the Neolithic, settlements expanded, human communities grew more complex, and power became concentrated in a small group, an *elite*.

This process of *state formation* is known to have occurred in only a half-dozen regions of the world, and China was one of these. But evidence about China's earliest states has long been scarce. Today, however, archeologists are focusing on the lower Yi-Luo River Valley in the western part of Henan Province, where the first documented Chinese dynasty, the Xia Dynasty (2200–1770 B.C.) existed. The capital of this ancient state, Erlitou, has been found, and archeologists now refer to the Xia Dynasty as the Erlitou culture. Secondary centers are being discovered, and Erlitou tools and implements in a wider area prove that the Xia Dynasty represents a substantial state.

All early states were ruled by elites, but China's political history is chronicled in *dynasties* because here the succession of rulers came from the same line of descent, sometimes enduring for centuries. In the transfer of power, family ties counted for more than anything else. Dynasties were overthrown, but the victors did not change this system. Dynastic rule lasted into the twentieth century.

The Xia Dynasty may have been the earliest Chinese state, but it lay in the area where, later, more powerful dynastic states arose: the North China Plain. Here the tenets of what was to become Chinese society were implanted early and proved to be extremely durable. From this culture hearth ideas, innovations, and practices diffused far and wide. From agriculture to architecture, poetry to porcelain-making, influences radiated southward into Southeast Asia, westward into interior Asia, and eastward into Korea and Japan. In the North China Plain lay the origins of what was to become the Middle Kingdom, which its citizens considered the center of the world.

Table 9-1 summarizes the sequence of dynasties that may have begun with Xia more than four millennia ago and ended in 1911 when the last emperor of the Qing (Manchu) Dynasty, a boy 6 years old, was forced to abdicate the Chinese throne. As the table shows, every dynasty contributed importantly to the development of Chinese society, sometimes progressively through enlightened policies and efficient administration, at other times regressively when capricious cruelty, authoritarian excesses, and xenophobia prevailed. During those 4000 years, China evolved into the world's most populous nation, always rebounding after disastrous famines and floods, its territory covering more than 80 percent of the East Asian realm (Fig. 9-3).

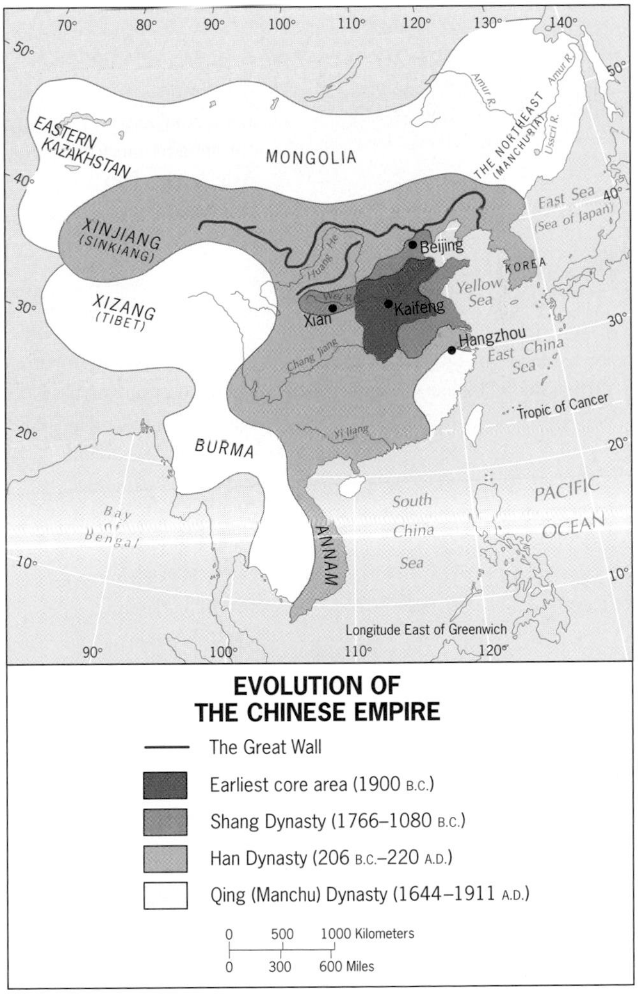

FIGURE 9-3 © H.J. de Blij, P.O. Muller, and John Wiley & Sons, Inc.

Table 9-1
THE CHINESE DYNASTIES, 2200 B.C. TO A.D. 1911

Dynasty Name(s)	Date(s)	Areas Governed	Major Features	Geographical Impact
Xia	c2200–c1770 B.C.	Small part of Huang River Basin, centered in Yi-Luo tributary.	First dynastic state (?); Neolithic technology, sophisticated stone tools.	Beginnings of stream diversion, irrigation.
Shang (latter part Yin)	c1766–c1080 B.C.	Huang-Wei River confluence across Shandong and Henan Provinces. Anyang capital.	Neolithic to Bronze Age transition. Timber houses with thatched roofs, beginnings of Chinese writing, moon-cycle-based calendar. Superb bronze vases and other implements.	Unifies large region in lower Huang River Basin.
Zhou	c1027–221 B.C. Two periods: Spring and Autumn c1027–481 B.C.; Warring States 475–221 B.C.	A people centered in Shaanxi Province to the west of the Shang area overthrow the Shang rulers and expand their domain eastward. Warring States is a period of feudalism.	Formative period for China. Irrigation systems, iron smelting, horses, expanded farm production. Major towns develop. Writing system is established. Chopsticks come into use. Daoism and Confucianism arise; Daoism centers on mysticism, individualism, the personal "way," and Confucianism on duties, community standards, respect for government.	Taoism diffuses widely throughout East Asia, reaches South Asia, and influences Buddhism. Confucianism becomes China's guiding philosophy for more than 2000 years. Building of *Great Wall* begins.
Qin (Ch'in)	221–206 B.C.	Most of North China Plain, south into Lower Chang Basin; first consolidation of large Chinese state.	Time of bureaucratic dictatorship after ferment of Zhou: Confucian Classics burned, Taoism combated. Much of Great Wall is built at huge cost in lives. Cruel despotism, excesses. Emperor's tomb discovered at Xian in 1976 with 6000 terracotta life-sized men and horses.	Ruler Ch'in (Qin)'s name immortalized as *China*.
Han	206 B.C.–A.D. 220	Major territorial expansion adds Xinjiang in interior Asia, Vietnam in Southeast Asia.	Second formative dynasty for China; restoration of Confucian principles and a flowering of culture. First use of eunuchs by authoritarian government, planting a long-term weakness in administration. Xian becomes one of the greatest cities of the ancient world. Chinese refer to themselves as the *People of Han*.	About the same time, same size as the Roman Empire. *Silk Road* carries goods from China across inner Asia to Syria and on to Rome.
Period from AD 220–580 witnessed disunion and division. Important developments included the arrival of Buddhism in East Asia and the diffusion of Chinese influence and ideas into Korea and across Korea into Japan.				
Sui	A.D. 581–618	North and South China reunified, Xian rebuilt and greatly expanded. Conflict with Turks in Central Asia and with Koreans in east.	Brief but important dynasty with revival of Confucian rituals and practices in education. Brutal but modernizing regime conducts a census, establishes a penal code, and engages in massive public works.	Construction of key segments of Grand Canal links Huang and Yangzi Rivers and thus connects northern and southern economies.
Tang	A.D. 618–907	Defeat of Turks in Central Asia. Campaign to put down rebellions in the south succeeds, though northern frontier remains unstable.	A golden age for China and a third formative dynasty. Enormous and efficient bureaucracy dictates virtually all aspects of life. Xian now the cultural capital and largest city in the world. *Buddhism* thrives, pagodas arise everywhere. The arts, Chinese and foreign, flourish.	Arab and Persian seafarers visit Chinese ports, the Silk Route is loaded with trade, and China seems set for an international era. It is not to be: *Islam* arrives in Central Asia and Tang armies are defeated by Arabs in 751. The Silk Route breaks down. In the east, however, Chinese influences permeate Korea and penetrate Japan.

Table 9-1
(Continued)

Dynasty Name(s)	Date(s)	Areas Governed	Major Features	Geographical Impact
The 907–960 period is known as the Five Dynasties because five leaders tried to establish dynasties in northern China in quick succession, each failing in turn. In southern China it is known as the time of the Ten Kingdoms, for ten regimes ruled various parts of the region. Cultural continuity prevails despite political instability.				
Song (Northern (Southern	A.D. 960–1279 A.D. 960–1127) A.D. 1127–1279)	Rulers consolidate the Five (northern) Dynasties, then begin capturing the Ten (southern) Kingdoms. They are unable to control the Liao state north of the Great Wall. Mongols invade, drive Song rulers southward, and end dynasty in 1279.	Preoccupied with organization and administration, Song rulers create a competent, efficient *civil service*. A defensive rather than an expansionist period. An era of rich cultural and intellectual achievement in mathematics, astronomy, mapmaking. Paper and movable type are invented, as is *gunpowder*. Large ships are built. The arts, from porcelain-making to poetry, continue to thrive. Practice of footbinding begins. Improved rice varieties feed estimated 100 million inhabitants.	Capital is moved from northern Kaifeng on Huang River to Hangzhou in the south as Song rulers lose control over the north. Mongol invasion is followed by Kublai Khan's choice of Beijing as his capital (1272). Mongols use Chinese authoritarian and bureaucratic traditions to forge their own dynastic rule. Several cities have populations exceeding 1 million.
Yuan (Mongol)	A.D. 1264–1368	Controls eastern Asia from Siberia in the north to the Vietnam border in the south, including present-day Russian Far East and most of Korea.	Dynasty begins in north before Song ends in south. Repressive regime, social chasms between Mongol rulers and Chinese ruled. Deep and enduring hostility between Chinese and Mongols. Cultural *isolationism* results; Mongols are acculturated to Chinese norms. Marco Polo visits East Asia.	Most of present-day Mongolia is part of the Chinese sphere for the first time. Mongol effort to enter Japan fails. Mongol power over much of Eurasia stimulates trade and flow of information. Beijing grows into major city.
Ming	A.D. 1368–1644	Rules over all eastern China from Amur River in north to Red River (Vietnam) in south, as far west as Inner Mongolia and Yunnan. At various times incorporates North Korea, most of Mongolia, even Myanmar (Burma).	Chinese rule again after Mongol Yuan dynasty. Stable but autocratic and inward-looking regime. Major advances in science and technology. In first half of fifteenth century, Chinese oceangoing vessels, larger than any in Europe, explore Pacific and Indian Ocean waters and reach East Africa. Farming expands, silk and cotton industries improve, printing plants multiply. Defensive walls are built in the north and around cities, including the Forbidden City in Beijing, the Ming capital. But mismanagement, infighting, eunuch influence in the palaces, and anticommercialism weaken Ming rule.	Early Ming ships and fleets reach Southeast Asian and Indian Ocean shores long before smaller European vessels do, but the Chinese advantage is lost when Ming rulers turn isolationist and order maritime ventures halted. Population growth causes serious problems of food supply and political control.
Qing (Manchu)	A.D. 1644–1911	*Largest* China-centered empire ever incorporates Mongolia, much of Turkestan, Xizang (Tibet), Myanmar, Indochina, Korea, Taiwan.	Foreign rule again: Manchu, a people with Tatar links living in present-day Northeast China, seize an opportunity and take control of Beijing. Improbably, a group numbering about 1 million rule a nation of several hundred million. It is done by keeping Ming systems of administration and retaining (and rewarding) Ming officials. Expansion creates an empire. Population pressure and the concentration of land ownership in fewer hands create problems worsened by floods and famines. European powers and Japan force concessions on weakening Qing rulers; defeats in war lead to revolution and collapse.	Manchu (Qing) rulers create the map that today forms the justification for China's claims to Xizang (Tibet), Taiwan, and the central Eurasian interior; at the height of its power, the Qing empire forces the Russians to recognize Manchu authority over China's Northeast as far north as the Argun River. Latent claims to Russia's Far East are based on Manchu predominance there.

Dominant as China is in its regional sphere, there is more to East Asia than the Chinese giant. To the north, the landlocked state of Mongolia is what remains of the time when Mongol armies conquered much of Eurasia (as Table 9-1 shows, even China fell under the Mongol sway). To the east, the Korean Peninsula escaped incorporation into China. Offshore, Japan first surpassed China as a military power, colonizing Manchuria (as it was then called) and penetrating deep into China's heartland. Later, Japan outclassed China as an economic power, achieving world status while China languished under Maoist communism. And as the twenty-first century dawned, Taiwan remained a separate political entity, its detachment from China resulting from the struggle that brought the communists to power in Beijing.

REGIONS OF THE REALM

East Asia presents us with an opportunity to illustrate the changeable nature of regional geography. Our regional delimitation is based on current circumstances, and it predicts ways the framework may change. It is anything but static. At the beginning of the twenty-first century, we can identify five geographic regions in the East Asian realm (Fig. 9-4). These are:

1. *China Proper.* Almost any map of China's human geography—population distribution, urban centers,

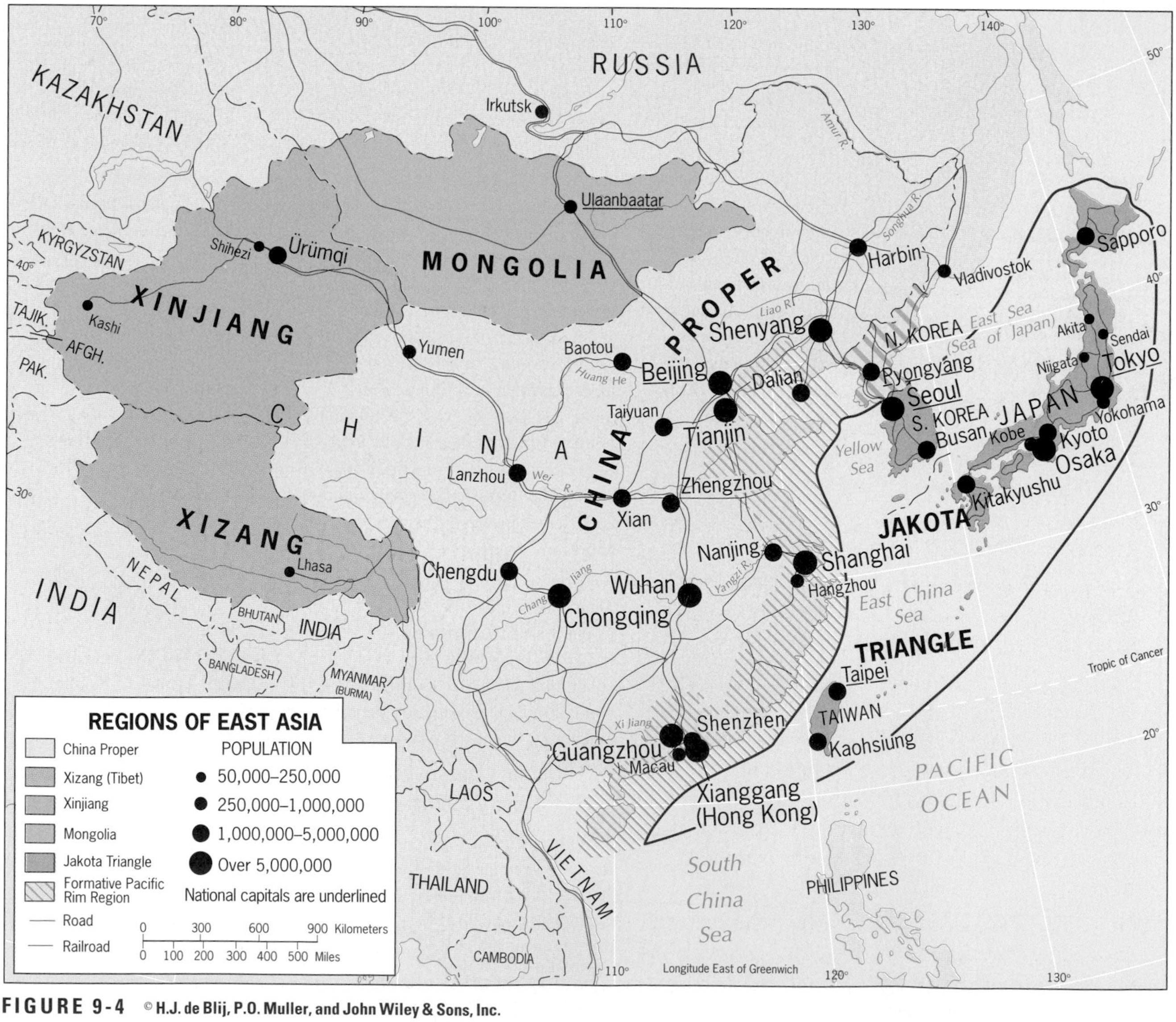

FIGURE 9-4 © H.J. de Blij, P.O. Muller, and John Wiley & Sons, Inc.

MAJOR CITIES OF THE REALM

City	Population* (in millions)
Beijing (Shi), China	10.9
Busan, South Korea	3.8
Chongqing (Shi), China	5.1
Guangzhou, China	3.9
Hong Kong (Xianggang) SAR, China	7.3
Macau SAR, China	0.5
Nanjing, China	2.8
Osaka, Japan	11.2
Seoul, South Korea	9.6
Shanghai (Shi), China	12.8
Shenyang, China	5.0
Shenzhen, China	3.8
Taipei, Taiwan	2.5
Tianjin (Shi), China	9.4
Tokyo, Japan	26.8
Wuhan, China	6.3
Xian, China	3.3

*Based on 2006 estimates.

surface communications, agriculture, industry—emphasizes the strong concentration of Chinese activity in the country's eastern sector. This is the "real" China, where its great cities, populous farmlands, and historic sources are located. Long ago, scholars called this *China Proper*, and it is a good regional designation. But China is a large and complex country, and a number of subregions are nested within China Proper. Some of these, such as the North China Plain and the Sichuan Basin, are old and well-established geographic units. One in particular is new: China's Pacific Rim, still growing, yet poorly defined, and shown as "formative" in Figure 9-4.

2. *Xizang (Tibet)*. The high mountains and plateaus of Xizang, ruled by China but still widely known by its older name of Tibet, form a stark contrast to teeming China Proper. Here, next to one of the world's largest and most populous regions, lies one of the emptiest and, in terms of inhabited space, smallest regions.
3. *Xinjiang*. The vast desert basins and encircling mountains of Xinjiang form a third East Asian region. Again, physical as well as human geographic criteria come into play: here China meets Islamic Central Asia.
4. *Mongolia*. The desert state of Mongolia forms East Asia's fourth region. Like Tibet, landlocked Mongolia, vast but sparsely peopled, stands in stark contrast to populous China Proper.
5. *Jakota Triangle*. East Asia's fifth region is defined by its economic geography. *Ja*pan, South *Ko*rea, and *Ta*iwan (the name "Jakota" derives from the first two letters of each) were transformed by Pacific Rim economic developments during the second half of the twentieth century. Its recent emergence foreshadows further changes in the decades ahead as Korean unification becomes a possibility and contrasts between the Jakota Triangle and Pacific Rim China diminish.

CHINA PROPER

When we in the Western world chronicle the rise of civilization, we tend to focus on the historical geography of Southwest Asia, the Mediterranean, and Western Europe. Ancient Greece and Rome were the crucibles of culture; Mediterranean and Atlantic waters were the avenues of its diffusion. China lay remote, so we believe, barely connected to this Western realm of achievement and progress. When an Italian adventurer named Marco Polo visited China during the thirteenth century and described the marvels he saw there, his work did little to change European minds. Europe was and would always be the center of civilization.

The Chinese, naturally, take a different view. Events on the western edge of the great Eurasian landmass were deemed irrelevant to theirs, the most advanced and refined culture on Earth. Roman emperors were rumored to be powerful, and Rome was a great city, but nothing could match the omnipotence of China's rulers. Certainly the Chinese city of Xian (which had a different name at the time) far eclipsed Rome as a center of sophistication. Chinese civilization existed long before ancient Greece and Rome emerged, and it was still there long after they collapsed. China, the Chinese teach themselves, is eternal. It was, and always will be, the center of the civilized world.

We should remember this notion when we study China's regional geography because 4000 years of Chinese culture and perception will not change overnight—not even in a generation. Time and again, China overcame the invasions and depredations of foreign intruders, and afterward the Chinese would close off their vast country against the outside world. Just 35 years ago, in the early 1970s, there were just a few *dozen* foreigners in the entire country with its (then) nearly 1 billion inhabitants. The institutionalization of communism required this insularity, and even the Soviet advisors had been thrown out. But by the early 1970s, China's rulers decided that an opening to the Western world would be advantageous, and so U.S. President Richard Nixon was invited to visit Beijing. That historic occasion, in 1972, ended this latest period of isolation—as always, on China's terms. Since then, China has been open to tourists and businesspeople, teachers, and investors. Tens of thousands of Chinese students have been sent to study at American and other Western institutions. Long-suppressed ideas flowed into China, and a pro-democracy movement arose and climaxed in 1989. China's rulers knew that their violent repression of this movement would anger the world, but that did not matter because they deemed foreign condemnation irrelevant. Foreigners in China had done much worse. Moreover, Westerners had no business interfering in China's domestic affairs.

RELATIVE LOCATION

Throughout their nation's history, the Chinese have at times decided to close their country to any and all foreign influences; the most recent episode of exclusion occurred just a few decades ago. Exclusion is one of China's recurrent traditions, made possible by China's relative location and Asia's physiography. In other words, China's "splendid isolation" was made possible by geography.

Earlier we noted the role of relief and desert in encircling the culture hearth of East Asia, but equally telling is the factor of distance. Until recently, China lay far from the modern source areas of innovation and change. True, China—as the Chinese emphasize—was itself such a hearth, but China's contributions to the outside world remained limited, essentially, to finely made arts and crafts. China did interact with Korea, Japan, Taiwan, and parts of Southeast Asia, and eventually millions of Chinese emigrated to neighboring countries. But compare these regional links to those of the Arabs, who ranged worldwide and who brought their knowledge, religion, and political influence to areas from Mediterranean Europe to Bangladesh and from West Africa to Indonesia. Later, when Europe became the center of intellectual and material innovation, China found itself farther removed, by land or sea, than almost any other part of the world.

Today, modern communications notwithstanding, China still is distant from almost anywhere else on Earth. Going by rail from Beijing to Moscow, the capital of China's Eurasian neighbor, involves a tedious journey that takes the better part of a week. Direct surface connections with India are practically nonexistent. Overland linkages with Southeast Asian countries, though improving, remain tenuous.

But for the first time in its history, China now lies near a world-class hearth of technological innovation and financial power: Japan. This proximity to an industrial and financial giant is critical for the momentous economic developments taking place in China's coastal provinces. Japanese investments and business partnerships have transformed the economic landscape of Pacific-coast China. American and European trade links also are important, but Japan's role has been crucial. Japan's economic success set the Pacific Rim engine in motion, and Japan's best financial years happened to coincide with China's reopening to foreigners in the 1970s. Geographic and economic circumstances combined to transform the map and made "Pacific Rim" a household word around the world.

EXTENT AND ENVIRONMENT

China's total area is slightly smaller than that of the United States including Alaska: each country has about 3.7 million square miles (9.6 million sq km). As Figure 9-5 reveals, the longitudinal extent of China and the 48 contiguous U.S. States also is similar. Latitudinally, however, China is considerably wider. Miami, near the southern limit of the United States, lies halfway between Shanghai and Guangzhou. Thus China's lower-latitude southern region takes on characteristics of tropical Asia. In the Northeast, too, China incorporates much of what in North America would be Quebec and Ontario. Westward, China's land area becomes narrower and physiographic similarities increase. But, of course, China has no west coast.

Now compare the climate maps of China and the United States in Figure 9-6 (which are enlargements of the appropriate portions of the world climate map in Fig. G-8). Note that both have a large southeastern climatic region marked *Cfa* (that is, humid, temperate, warm-summer), flanked in China by a zone of *Cwa* (where winters become drier). Westward in both countries, the *C* climates yield to colder, drier climes. In the United

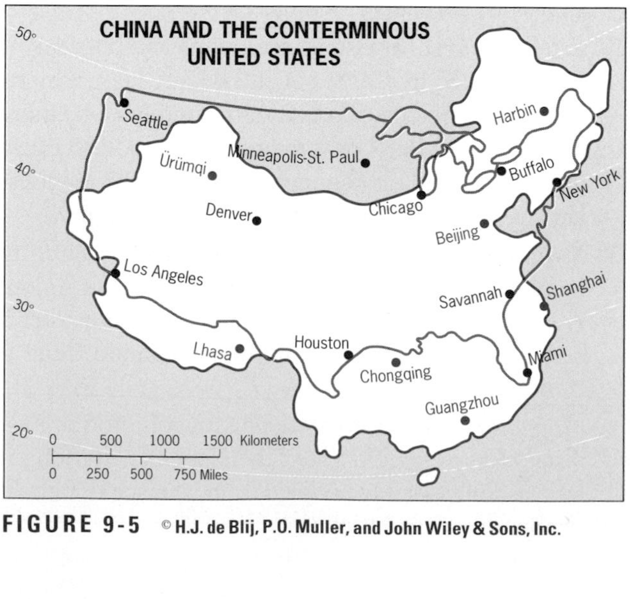

FIGURE 9-5 © H.J. de Blij, P.O. Muller, and John Wiley & Sons, Inc.

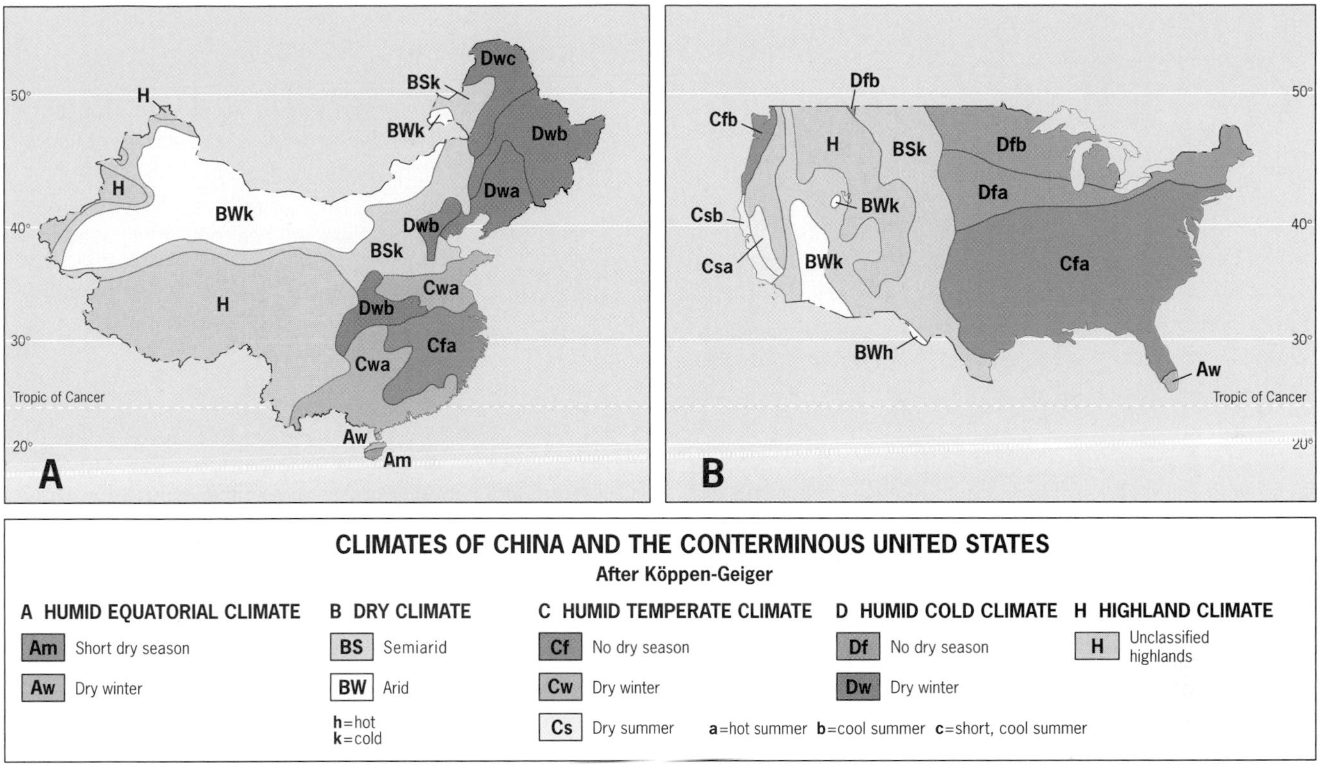

FIGURE 9-6 © H.J. de Blij, P.O. Muller, and John Wiley & Sons, Inc.

States, moderate *C* climates develop again along the Pacific coast. China, however, stays dry and cold as well as high in elevation at equivalent longitudes.

Note especially the comparative location of the U.S. and Chinese *Cfa* areas in Figure 9-6. China's lies much farther to the south. In the United States, the *Cfa* climate extends beyond 40° North latitude, but in China, cold and generally winter-dry *D* climates take over at the latitude of Virginia. Beijing has a warm summer but a bitterly cold and long winter. Northeast China, in the general latitudinal range of Canada's lower Quebec and Newfoundland, is much more severe than its North American equivalent. Harsh environments prevail over vast regions of China, but, as we will see, nature compensates in spectacular fashion. From the climatic zone marked *H* (for highlands) in the west come the great life-giving rivers whose wide basins contain enormous expanses of fertile soils. Without these waters, China would not have a population more than four times that of the United States.

EVOLVING CHINA

Even if China is not the world's longest continuous civilization (Egypt may claim this distinction), no other state on Earth can trace its cultural heritage as far back as China can. China's fortunes rose and fell, but over more than 40 centuries its people created a society with strong traditions, values, and philosophies. Kongfuzi (Confucius), whose teachings and writing still influence China, lived during the Zhou Dynasty, 2500 years ago (see box titled "Kongfuzi [Confucius]"). The Chinese still refer to themselves as the "People of Han," the dynasty that marks the breakdown of the old feudal order, the rise of military power, the unification of a large empire, the institution of property rights, the flourishing of architecture, the arts, and the sciences, and the development of trade along the Silk Route (Table 9-1). The Han Dynasty reigned 2000 years ago. The walled city of Xian, then called Ch'angan, was the early Han capital and one of the greatest cities of the ancient world.

As we try to gain a better understanding of China's present political, economic, and social geography, we should not lose sight of the China that endured for thousands of years under a sequence of imperial dynasties as a single entity. Larger and more populous than Europe, China had its divisive feudal periods, but always it came together again, ruled—mostly dictatorially—from a strong center as a unitary state. China was, and remains, a predominantly rural society, a powerless and subservient population controlled by an often ruthless bureaucracy. Retributions, famines, epidemics, and occasional uprisings took heavy tolls in the countryside. But in the cities, Chinese

Kongfuzi (Confucius)

2 **Confucius** (*Kongfuzi* or *Kongzi* in pinyin) was China's most influential philosopher and teacher. His ideas dominated Chinese life and thought for over 20 centuries.

Kongfuzi was born in 551 B.C. and died in 479 B.C. . Appalled at the suffering of ordinary people during the Zhou Dynasty, he urged the poor to assert themselves and demand explanations for their harsh treatment by the feudal lords. He tutored the indigent as well as the privileged, giving the poor an education that had hitherto been denied them and ending the aristocracy's exclusive access to the knowledge that constituted power.

Kongfuzi's revolutionary ideas extended to the rulers as well as the ruled. He abhorred supernatural mysticism and cast doubt on the divine ancestries of China's aristocratic rulers. Human virtues, not godly connections, should determine a person's place in society, he taught. Accordingly, he proposed that the dynastic rulers turn over the reins of state to ministers chosen for their competence and merit. This was another Kongzi heresy, but in time this idea came to be accepted and practiced.

His Earthly philosophies notwithstanding, Kongfuzi took on the mantle of a spiritual leader after his death. His thoughts, distilled from the mass of philosophical writing (including Daoism) that poured forth during his lifetime, became the guiding principles of the formative Han Dynasty. The state, he said, should not exist just for the power and pleasure of the elite; it should be a cooperative system for the well-being and happiness of the people.

With time, a mass of writings evolved, much of which Kongfuzi never wrote. At the heart of this body of literature lay the Confucian Classics, 13 texts that became the basis for education in China for 2000 years. From government to morality and from law to religion, the Classics were Chinese civilization's guide. The entire national system of education (including the state examinations through which everyone, poor or privileged, could enter the civil service and achieve political power) was based on the Classics. Kongfuzi championed the family as the foundation of Chinese culture, and the Classics prescribe a respect for the aged that was a hallmark of Chinese society.

But Kongfuzi's philosophies also were conservative and rigid, and when the colonial powers penetrated China, Kongzi's Classics came face to face with practical Western education. For the first time, some Chinese leaders began to call for reform and modernization, especially of teaching. Kongzi principles, they said, could guide an isolated China, but not China in the new age of competition. But the Manchu rulers resisted this call, and the Nationalists tried to combine Kongzi and Western knowledge into a neo-Kongzi philosophy, which was ultimately an unworkable plan.

The communists who took power in 1949 attacked Kongzi thought on all fronts. The Classics were abandoned, indoctrination pervaded education, and, for a time, even the family was viewed as an institution of the past. Here the communists miscalculated. It proved impossible to eradicate two millennia of cultural conditioning in a few decades. When China entered its post-Mao period, public interest in Kongfuzi surged, and the shelves of bookstores again sag under the weight of the *Analects* and other Confucian and post-Kongfuzi writings. The spirit of Kongfuzi will pervade physical and mental landscapes in China for generations to come.

culture flourished, subsidized by the taxes and tribute extracted from the hinterlands. As China grew, it incorporated minorities ranging from Koreans and Mongols to Uyghurs and Tibetans; in the far south, it now includes several peoples with Southeast Asian affinities. Like the Chinese citizenry itself, these minorities have experienced both benevolent government and brutal subjugation. But
3 **sinicization** has always been China's wish: to endow them with the elements of Chinese culture.

A Century of Convulsion

When the European colonialists appeared in East Asia, China long withstood them with a self-assured superiority based on the strength of its culture and the reassuring continuity of the state. There was no market for the British East India Company's rough textiles in a country long used to finely fabricated silks and cottons. There was little interest in the toys and trinkets the Europeans produced in the hope of barter for Chinese tea and porcelain. Even key European inventions, such as the mechanical clock, though considered amusing and entertaining, were ignored and even deprecated as irrelevant to Chinese culture.

The self-confident Ming emperors even beat the Europeans at their own game. In the fourteenth century, great oceangoing vessels sailed the South China Sea; Chinese fleets carrying as many as 20,000 men reached Southeast and South Asia and even East Africa. Several times as large and far more sophisticated than anything built in Europe, China's ships, together with their sailors and the goods they carried, demonstrated the potential of the Middle Kingdom. But suddenly China faced a crisis at home: the impact of the onset of the Little Ice Age that

AMONG THE REALM'S GREAT CITIES . . . XIAN

The city known today as Xian is the site of one of the world's oldest urban centers. It may have been a settlement during the Shang-Yin Dynasty more than 3000 years ago; it was a town during the Zhou Dynasty, and the Qin emperor was buried here along with 6000 life-sized terracotta soldiers and horses, reflecting the city's importance. During the Han Dynasty the city, then called Ch'angan, was one of the greatest centers of the ancient world, the Rome of ancient China. Ch'angan formed the eastern terminus of the Silk Route, a storehouse of enormous wealth. Its architecture was unrivalled, from its ornamental defensive wall with elaborately sculpted gates to the magnificent public buildings and gardens at its center.

Situated on the fertile loess plain of the upper Wei River, Ch'angan was the focus of ancient China during crucial formative periods. After two centuries of Han rule, political strife led to a period of decline, but the Sui emperors rebuilt and expanded Ch'angan when they made it their capital. During the Tang Dynasty, Ch'angan again became a magnificent city with three districts: the ornate Palace City; the impressive Imperial City, which housed the national administration; and the busy Outer City containing the homes and markets of artisans and merchants.

After its Tang heyday the city again declined, although it remained a bustling trade center. During the Ming Dynasty it was endowed with some of its architectural landmarks, including the Great Mosque marking the arrival of Islam; the older Big Wild Goose Pagoda dates from the influx of Buddhism. After the Ming period, Ch'angan's name was changed to Xian (meaning "Western Peace"), then to Siking, and in 1943 back to Xian again.

Having been a gateway for Buddhism and Islam, Xian in the 1920s became a center of Soviet communist ideology.

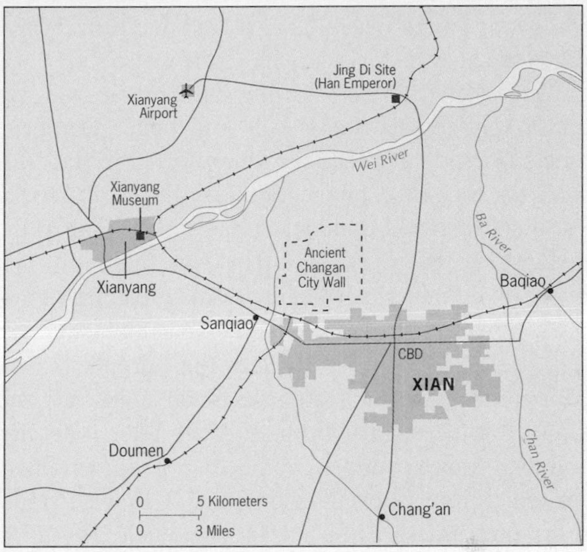

The Nationalists, during the struggle against the Japanese, moved industries from the vulnerable east to Xian, and when the communists took power, they enlarged Xian's industrial base still further. The present city (population: 3.3 million) lies southwest of the famed tombs, its cultural landscape now dominated by a large industrial complex that includes a steel mill, textile factories, chemical plants, and machine-making facilities. Little remains (other than some prominent historic landmarks) of the splendor of times past, but Xian's location on the railroad to the vast western frontier of China sustains its long-term role as one of the country's key gateways.

FROM THE FIELD NOTES

"To pass through the gate from Tiananmen Square and enter the Manchu Emperors' Forbidden City was a riveting experience. Look back now toward the gate, and you see throngs of Chinese visiting what less than a century ago was the exclusive domain of the Qing Dynasty's absolute rulers—so absolute that unauthorized entry was punished by instant execution. The vast, walled complex at the heart of Beijing not only served as the imperial palace; it was a repository of an immense collection of Chinese technological and artistic achievements. Some of this is still here, but most of all you are awed by the lingering atmosphere, the sense of what happened here in these exquisitely constructed buildings that epitomized what high Chinese culture could accomplish. Inevitably you think of the misery of millions whose taxes and tribute paid for what stands here."

also affected Europe. Cold and drought decimated the wheat harvest on the North China Plain, and boats were needed by the thousands to carry rice from the Yangzi/Chang Basin to the hungry north. That spelled the end of China's long-range maritime explorations; it may have changed the historical geography of the world.

Even when Europe's sailing ships made way for steam-driven vessels and newer and better European products (including weapons) were offered in trade for China's tea and silk, China continued to reject European imports and resisted commerce in general. (When the Manchus took control of Beijing in 1644, their bows and arrows proved superior to Chinese-manufactured muskets which were so heavy and hard to load that they were almost useless.) The Chinese kept the Europeans confined to small peninsular outposts, such as Macau, and minimized interaction with them. Long after India had succumbed to mercantilism and economic imperialism, China maintained its established order. This was no surprise to the Chinese. After all, they had held a position of undisputed superiority in their Celestial Kingdom as long as could be remembered, and they had dealt with foreign invaders before.

A (Lost) War on Drugs

All this confidence was shattered during the Manchu (Qing) Dynasty, China's last imperial regime. The Manchus, who had taken control in Beijing as a small minority and who had grafted their culture onto China's, had the misfortune of reigning when the standoff with Europeans ended and the balance of power shifted in favor of the colonialists.

On two fronts in particular, the economic and the political, the European powers destroyed China's invincibility. Economically, they succeeded in lowering the cost and improving the quality of manufactured goods, especially textiles, and the handicraft industries of China began to collapse in the face of unbeatable competition. Politically, the demands of the British merchants and the growing English presence in China led to conflicts. In the early part of the nineteenth century, the central issue was the importation into China from British India of opium, a dangerous and addictive intoxicant. Opium was destroying the very fabric of Chinese culture, weakening the society, and rendering China easy prey for colonial profiteers. As the Manchu government moved to stamp out the opium trade in 1839, armed hostilities broke out, and soon the Chinese found themselves losing a war on their own territory. The First Opium War (1839–1842) ended in disaster: China's rulers were forced to yield to British demands, and the breakdown of Chinese sovereignty was under way.

British forces penetrated up the Chang Jiang and controlled several areas south of it (Fig. 9-7); Beijing hurriedly sought a peace treaty by which it granted leases and concessions to foreign merchants. In addition, China ceded Hong Kong Island to the British and opened five ports, including Guangzhou (Canton) and Shanghai, to foreign commerce. No longer did the British have to accept a status that was inferior to the Chinese in order to do business; henceforth, negotiations would be pursued on equal terms. Opium now flooded into China, and its impact on Chinese society became even more devastating. Fifteen years after the First Opium War, the Chinese again tried to stem the disastrous narcotic tide, and the foreigners who had attached themselves to their country again defeated them. Now the government legalized cultivation of the opium poppy in China itself. Chinese society was disintegrating; the scourge of this drug abuse was not defeated until after the revival of Chinese power in the twentieth century.

But before China could reassert itself, much of what remained of its independence steadily eroded (see box titled "Extraterritoriality"). In 1898 the Germans obtained a lease on Qingdao on the Shandong Peninsula, and the French acquired a sphere of influence in the far south at Zhanjiang (Fig. 9-7). The Portuguese took Macau; the Russians obtained a lease on Liaodong in the Northeast as well as railway concessions there; even Japan got into the act by annexing the Ryukyu Islands and, more importantly, Formosa (Taiwan) in 1895.

After four millennia of recurrent cultural cohesion, economic security, and political continuity, the Chinese world lay open to the aggressions of foreigners whose innovative capacities China had negated to the end. Now the ships flying European flags lay in the ports of China's coasts and rivers, but China had not learned to manufacture the cannons to blast them out of the water. The smokestacks of foreign factories rose over the townscapes of its great cities, and no Chinese force could dislodge them. The Japanese, seeing China's weakness, invaded Korea. The Russians entered Manchuria, the home base of the Manchus. The foreign invaders even took to fighting among themselves, as Japan and Russia did in Manchuria in 1904. (Today this part of China is called the *Northeast*, the name Manchuria having been—understandably—rejected.)

A New China Rises

In the meantime, organized opposition to the foreign presence in China was gathering strength, and the twentieth century opened with a large-scale revolt against all outside elements. Bands of revolutionaries roamed both cities and countryside, attacking not only the hated foreigners but

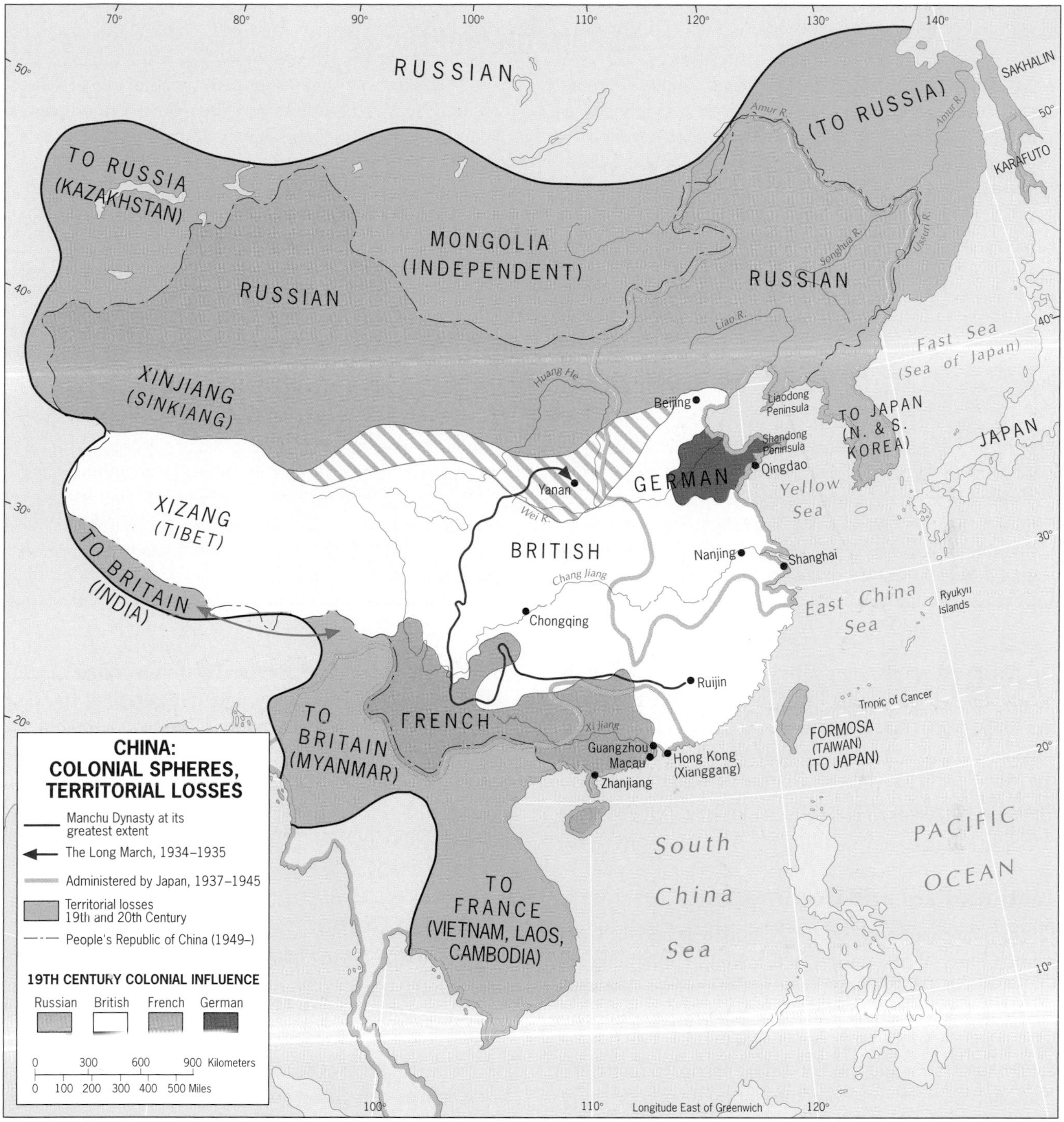

FIGURE 9-7 © H.J. de Blij, P.O. Muller, and John Wiley & Sons, Inc.

also Chinese who had adopted Western cultural traits. Known as the Boxer Rebellion (after a loose translation of the Chinese name for these revolutionary groups), this 1900 uprising was put down with much bloodshed by an international force consisting of British, Russian, French, Italian, German, Japanese, and American soldiers. Simultaneously, another revolutionary movement was gaining support, aimed against the Manchu leadership itself. In 1911, the emperor's garrisons were attacked all over China, and in a few months the 267-year-old Qing Dynasty was overthrown. Indirectly, it too was a casualty of the foreign intrusion, and it left China divided and disorganized.

The fall of the Manchus and the proclamation of a republican government in China did little to improve the country's overall position. The Japanese captured Germany's holdings on the Shandong Peninsula, including the city of Qingdao, during World War I. When the victorious European powers met at Versailles in 1919 to divide

Extraterritoriality

During the nineteenth century, as China weakened and European colonial invaders entered China's coastal cities and sailed up its rivers, the Europeans forced China to accept a European doctrine of international law—
4 **extraterritoriality**. Under this doctrine, foreign states and their representatives are immune from the jurisdiction of the country in which they are based. Today, this applies to embassies and diplomatic personnel. But in Qing (Manchu) China, it went far beyond that.

The European, Russian, and Japanese invaders established as many as 90 *treaty ports*—extraterritorial enclaves in China's cities under unequal treaties enforced by gunboat diplomacy. In their "concessions," diplomats and traders were exempt from Chinese law. Not only port areas but also the best residential suburbs of large cities were declared to be "extraterritorial" and made inaccessible to Chinese citizens. In the city of Guangzhou (Canton in colonial times), Sha Mian Island in the Pearl River was a favorite extraterritorial enclave. A sign at the only bridge to the island stated, in English and Cantonese, "No Dogs or Chinese."

Christian missionaries fanned out into China, their residences and churches fortified with extraterritorial security. In many places, Chinese found themselves unable to enter parks and buildings without permission from foreigners. This involved a loss of face that contributed to bitter opposition to the presence of foreigners—a resentment that exploded in the Boxer Rebellion of 1900.

After the collapse of the Qing Dynasty in 1911, the Chinese Nationalists negotiated an end to all commercial extraterritoriality in China Proper; the Russians, however, would not yield in then-Manchuria. Only Hong Kong and Macau retained their status as colonies.

When China's government in 1980 embarked on a new economic policy that gave major privileges and exemptions to foreign firms in certain coastal areas and cities, opponents argued that this policy revived the practice of extraterritoriality in a new guise. This issue remains a sensitive one in a China that has not forgotten the indignities of the colonial era.

the territorial spoils, they affirmed Japan's rights in the area. This led to yet another Chinese effort to counter the foreign scourge. Nationwide protests and boycotts of Japanese goods were organized in what became known as the May Fourth Movement. One participant in these demonstrations was a charismatic young man named Mao Zedong.

Nationalists and Communists

During the chaotic 1920s, the Nationalists and the Communist Party at first cooperated, with the remaining foreign presence their joint target. After Sun Yat-sen's death in 1925, Chiang Kai-shek became the Nationalists' leader, and by 1927 the foreigners were on the run, escaping by boat and train or falling victim to rampaging Nationalist forces. But soon the Nationalists began purging communists even as they pursued foreigners, and in 1928, when Chiang established his Nationalist capital in the city of Nanjing, it appeared that the Nationalists would emerge victorious from their campaigns. They had driven the communists ever deeper into the interior, and by 1933 the Nationalist armies were on the verge of encircling the last communist stronghold in the area of Ruijin in Jiangxi Province.

This led to a momentous event in Chinese history: the *Long March*. Nearly 100,000 people—soldiers, peasants, leaders—marched westward from Ruijin in 1934, a communist column that included Mao Zedong and Zhou Enlai. The Nationalist forces rained attack after attack on the marchers, and of the original 100,000, about three-quarters were killed. But new sympathizers joined along the way (see the route marked on Fig. 9-7), and the 20,000 survivors found a refuge in the mountainous interior of Shaanxi Province, 2000 miles (3200 km) away. There, they prepared for a renewed campaign that would bring them to power.

Japan in China

Although many foreigners fled China during the 1920s and 1930s, others seized the opportunity presented by the contest between the Nationalists and communists. The Japanese took control over the Northeast, and when the Nationalists proved unable to dislodge them, they set up a puppet state there, appointed a Manchu ruler to represent them, and called their possession Manchukuo.

The inevitable full-scale war between the Chinese and the Japanese broke out in 1937, with the Nationalists bearing the brunt of it (which gave the communists an opportunity to regroup). The gray boundary on Figure 9-7 shows how much of China the Japanese conquered. The Nationalists moved their capital to Chongqing, and the communists controlled the area centered on Yanan. China had been broken into three pieces.

The Japanese committed unspeakable atrocities in their campaign in China. Millions of Chinese citizens were shot, burned, drowned, subjected to gruesome chemical and biological experiments, and otherwise wantonly victimized. The "Rape of Nanking," an orgy of murder, rape, torture, and pillage following the fall of Nanjing to the Japanese Army in late 1937, was perhaps the most heinous of Japan's war crimes during this period. Years later, when China's economic reforms of the 1980s and 1990s led to a renewed Japanese presence in China, the Chinese public and its leaders called for Japan to acknowledge and apologize for these wartime abuses. In Japan, this pitted apologists against strident nationalists, a dispute that continues today. The Chinese remain dissatisfied with Japan's failure to apologize unconditionally for its actions, and the book is not yet closed on this sensitive issue.

Communist China Arises

After the U.S.-led Western powers defeated Japan in 1945, the civil war in China quickly resumed. The United States, hoping for a stable and friendly government in China, sought to mediate the conflict but at the same time recognized the Nationalists as the legitimate government. The United States also aided the Nationalists militarily, destroying any chance of genuine and impartial mediation. By 1948, it was clear that Mao Zedong's well-organized militias would defeat Chiang Kai-shek. Chiang kept moving his capital—back to Guangzhou, seat of Sun Yat-sen's first Nationalist government, then back to Chongqing. Late in 1949, after a series of disastrous defeats in which hundreds of thousands of Nationalist forces were killed, the remnants of Chiang's faction gathered Chinese treasures and valuables and fled to the island of Taiwan. There, they took control of the government and proclaimed their own Republic of China.

Meanwhile, on October 1, 1949, standing in front of the assembled masses at the Gate of Heavenly Peace on Beijing's Tiananmen Square, Mao Zedong proclaimed the birth of the People's Republic of China.

CHINA'S HUMAN GEOGRAPHY

After more than a half-century of communist rule, China is a society transformed. It has been said that the year 1949 actually marked the beginning of a new dynasty not so different from the old, an autocratic system that dictated from the top. In that view, Mao Zedong simply bore the mantle of his dynastic predecessors. Only the family lineage had fallen away; now communist "comrades" would succeed each other.

And certainly some of China's old traditions continued during the communist era, but in many other ways Chinese society was totally overhauled. Benevolent or otherwise, the dynastic rulers of old China headed a country in which—for all its splendor, strength, and cultural richness—the fate of landless people and of serfs often was undescribably miserable; in which floods, famines, and diseases could decimate the populations of entire regions without any help from the state; in which local lords could (and often did) repress the people with impunity; in which children were sold and brides were bought. The European intrusion made things even worse, bringing slums, starvation, and deprivation to millions who had moved to the cities.

The communist regime, dictatorial though it was, attacked China's weaknesses on many fronts, mobilizing virtually every able-bodied citizen in the process. Land was taken from the wealthy; farms were collectivized; dams and levees were built with the hands of thousands; the threat of hunger for millions receded; health conditions improved; child labor was reduced. But China's communist planners also made terrible mistakes. The *Great Leap Forward*, requiring the reorganization of the peasantry into communal brigade teams to speed industrialization and make farming more productive, had the opposite effect and so disrupted agriculture that between 20 and 30 million people died of starvation between 1958, when the program was implemented, and 1962, when it was abandoned.

Mao ruled China from 1949 to 1976, long enough to leave lasting marks on the state. Another of his communist dictums had to do with population. Like the Soviets (and influenced by a horde of Soviet advisors and planners), Mao refused to impose or even recommend any population policy, arguing that such a policy would represent a capitalist plot to constrain China's human resources. As a result, China's population grew explosively during his rule.

Yet another costly episode of Mao's rule was the so-called *Great Proletarian Cultural Revolution*, launched by Mao Zedong during his last decade in power (1966–1976). Fearful that Maoist communism was being contaminated by Soviet "deviationism" and worried about his own stature as its revolutionary architect, Mao unleashed a campaign against what he viewed as emerging elitism in society. He mobilized young people living in cities and towns into cadres known as Red Guards and ordered them to attack "bourgeois" elements throughout China, criticize Communist Party officials, and root out "opponents" of the system. He shut down all of China's schools, persecuted untrustworthy intellectuals, and encouraged the Red Guards to engage in what he called a renewed "revolutionary experience." The results were disastrous: Red Guard factions took to fighting among themselves, and anarchy, terror, and economic paralysis followed. Thousands of China's leading intellectuals died, moderate leaders were purged, and teachers, elderly citizens, and older revolutionaries were tortured to make them confess to "crimes" they did not commit. As the economy suffered, food and industrial production declined. Violence and famine killed as many as 30 million people as the Cultural Revolution spun out of control. One of those who survived was a Communist Party leader who had himself been purged and then been reinstated—Deng Xiaoping. Deng was destined to lead the country in the post-Mao period of economic transformation.

Political and Administrative Divisions

Before we investigate the emerging human geography of contemporary China, we should acquaint ourselves with the country's political and administrative framework (Fig. 9-8). For administrative purposes, China is divided into the following units:

4 Central-Government-Controlled Municipalities (*Shi's*)
5 Autonomous Regions
22 Provinces
2 Special Administrative Regions

The four central-government-controlled municipalities are the capital, Beijing; its nearby port city, Tianjin; China's largest metropolis, Shanghai; and the Chang River port of Chongqing, in the interior. These *shi's* form the cores of China's most populous and important subregions, and direct control over them from the capital entrenches the central government's power.

We should note that the administrative map of China continues to change—and to pose problems for geographers. The city of Chongqing was made a *shi* in 1996, and its "municipal" area was enlarged to incorporate not only the central urban area but a huge hinterland covering

FIGURE 9-8 © H.J. de Blij, P.O. Muller, and John Wiley & Sons, Inc.

all of eastern Sichuan Province. As a result, the "urban" population of Chongqing is officially 30 million, making this the world's largest metropolis—but in truth, as shown in the cities population table on page 435, the central urban area contains only 5.1 million inhabitants. And because Chongqing's population is officially *not* part of the province that borders it to the west (Fig. 9-8), the official population of Sichuan declined by 30 million when the *Chongqing Shi* was created. So despite what Table 9-2 suggests, the population within Sichuan Province's outer borders remains the largest of any of China's administrative divisions.

The five Autonomous Regions were established to recognize the non-Han minorities living there. Some laws that apply to Han Chinese do not apply to certain minorities. As we saw in the case of the former Soviet Union, however, demographic changes and population movements affect such regions, and the policies of the 1940s may not work in the twenty-first century. Han Chinese immigrants now outnumber several minorities in their own regions. The five Autonomous Regions (A.R.'s) are: (1) Nei Mongol A.R. (Inner Mongolia); (2) Ningxia Hui A.R. (adjacent to Inner Mongolia); (3) Xinjiang Uyghur A.R. (China's northwest corner); (4) Guangxi Zhuang A.R. (far south, bordering Vietnam); and (5) Xizang A.R. (Tibet).

China's 22 Provinces, like U.S. States, tend to be smallest in the east and largest toward the west. The territorially smallest are the three easternmost provinces on China's coastal bulge: Zhcjiang, Jiangsu, and Fujian. The two largest are Qinghai, flanked by Tibet, and Sichuan, China's Midwest.

As with all large countries, some provinces are more important than others. The Province of Hebei nearly surrounds Beijing and occupies much of the core of the country. The Province of Shaanxi is centered on the great ancient city of Xian. In the southeast, momentous economic developments are occurring in the Province of Guangdong, whose urban focus is Guangzhou. When, in the pages that follow, we refer to a particular province or region, Figure 9-8 is a useful locational guide.

In 1997, the British dependency of Hong Kong (Xianggang) was taken over by China and became the country's first **Special Administrative Region (SAR)**. 5
In 1999, Portugal similarly transferred Macau, opposite Hong Kong on the Pearl River Estuary, to Chinese control, creating the second SAR under Beijing's administration.

Table 9-2
CHINA: POPULATION BY MAJOR ADMINISTRATIVE DIVISIONS*
(in Millions)

Provinces	1990 Population	2001 Population
Anhui	56.2	63.3
Fujian	30.0	34.4
Gansu	22.4	25.8
Guangdong	62.8	77.8
Guizhou	32.4	38.0
Hainan	6.6	8.0
Hebei	61.1	67.0
Heilongjiang	35.2	38.1
Henan	85.5	95.6
Hubei	54.0	59.8
Hunan	60.7	66.0
Jiangsu	67.1	73.6
Jiangxi	37.7	41.7
Jilin	24.7	26.9
Liaoning	39.5	41.9
Qinghai	4.5	5.2
Shaanxi	32.9	36.6
Shandong	84.4	90.4
Shanxi	28.8	32.7
Sichuan	107.2	86.4
Yunnan	37.0	42.9
Zhejiang	41.5	46.1
Autonomous Regions		
Guangxi Zhuang	42.2	47.9
Nei Mongol	21.5	23.8
Ningxia Hui	4.7	5.6
Xinjiang Uyghur	15.2	18.8
Xizang	2.2	2.6

*Population data for the four centrally administered municipalities and the Hong Kong (Xianggang) and Macau Special Administrative Regions are shown in the table on page 435.

Population Issues

Table 9-2 underscores the enormity of China's population: many Chinese provinces have more inhabitants than do most of the world's countries. With a 2006 population of 1.323 billion, China remains the largest nation on Earth. We should view these huge numbers in the context of China's population growth rate. High rates of population growth dim national development prospects in many countries in the periphery, but China has a special problem. Even a modest growth rate—such as today's 0.6 percent annually—still adds nearly 8 million to the total every year.

Mao Zedong, as we have noted, refused to institute policies that would reduce the rate of population increase because he believed (encouraged by Soviet advisors) that this would play into the hands of his capitalist enemies. Fast-growing populations, he said, constituted the only asset many poor, communist, or socialist countries had. If the capitalist world wanted to reduce growth rates, it must be an anticommunist plot.

But other Chinese leaders, realizing that rapid population growth would stymie their country's progress, tried to stem the tide through local propaganda and education. Some of these leaders lost their positions as "revisionists," but their point was not lost on China's future leaders. After Mao's death, China embarked on a vigorous population-control program. In the early 1970s, the annual rate of natural increase was about 3 percent; by the mid-1980s, it was down to 1.2 percent. Families were ordered to have one child only, and those who violated the policy were penalized by losing tax advantages, educational opportunities, and even housing privileges. As we noted, China's census bureau now officially reports a moderate growth rate of 0.6 percent.

The one-child policy was vigorously policed throughout China by Communist Party members, leading to serious excesses. The number of abortions, sometimes forced, rose sharply; female infanticide increased. During the 1980s, when reports of these byproducts of the policy reached the outside world, the regime began to institute certain exceptions (minorities were not subject to the one-child limit). Couples living in rural areas were allowed to have a second child if their firstborn was female. Certain farming and fishing families could apply for permission to raise a second child in view of future labor needs. Since 2001, urban couples could have two children if both the father and mother came from one-child families.

6 Chinese demographers report that these **restrictive population policies** have produced 300 million "nonbirths" over the past 30 years; that is, China's population would be more than 1.6 billion today if these policies had not been instituted. They estimate that the fertility rate declined from 2.30 children per woman in 1960 to 1.69 in 2004, which would mean that China's population will level off at mid-century and then begin a steady decline. This will entail a rapid increase in the proportion of the population over 65 years old, with consequences similar to those in Europe—a fast reduction in the number of workers to support these old-age pensioners. And this is not some remote, distant worry: China is already having problems meeting its pension obligations today.

China's one-child policy may have prevented a population explosion, but because families preferred male heirs it also led to an excess of males over females in the population pyramid. The official ratio at birth over the past 30 years is 118 boys for every 100 girls; this is a troubling statistic because it implies selective abortion, female infanticide, and child neglect. Another concern arising from this ratio imperils the future: millions of young men will be unable to find wives, and geographers who study this phenomenon, and its precedents elsewhere, report that such situations tend to induce governments to expand their armed forces and, sometimes, to find reasons to give those forces a purpose.

As a result, China in 2005 was engaged in an internal debate over the one-child policy, and Chinese newspapers and magazines carried articles proposing a substitute two-child policy, with no exceptions, even in urban areas. Such a change would raise China's fertility rate and keep the population growing beyond 2050, but the advantages would include a larger number of younger workers to support the aged and a reduction in the skewed male-female ratio that threatens to undermine this vast country's future social fabric.

The Minorities

We asserted earlier that China remains an empire; its government controls territories and peoples that are in effect colonized. The ethnolinguistic map (Fig. 9-9) apparently confirms this proposition. This map should be seen in context, however. It does not show local-area majority populations but instead reveals where minorities are concentrated. For example, the Mongolian population in China is shown to be clustered along the southeastern border of Mongolia in the Autonomous Region called Nei Mongol (see Fig. 9-8). But even in that A.R., the Han Chinese, not the Mongols, are now in the majority.

Nonetheless, the map gives definition to the term *China Proper* as the home of the People of Han, the ethnic (Mandarin-speaking) Chinese depicted in tan and light orange in Figure 9-9. From the upper Northeast to the border with Vietnam and from the Pacific coast to the margins of Xinjiang, this Chinese majority dominates. When you compare this map to that of population distribution (to be discussed shortly), it will be clear that the minorities constitute only a small percentage of the country's total. The Han Chinese form the largest and densest clusters.

In any case, China controls non-Chinese areas that are vast, if not populous. The Tibetan group numbers under 3 million, but it extends over all of settled Xizang. Turkic peoples inhabit large areas of Xinjiang. Thai, Vietnamese, and Korean minorities also occupy areas on the margins of Han China. As we will see later, the Southeast Asian minorities in China have participated strongly in the Pacific Rim developments on their doorsteps. Hundreds of thousands have migrated from their Autonomous Regions to the economic opportunities along the coast.

Numerically, Chinese dominate in China to a far greater degree than Russians dominated their Soviet Empire. But territorially, China's minorities extend over a proportionately larger area. The Ming and Manchu rulers bequeathed the People of Han an empire.

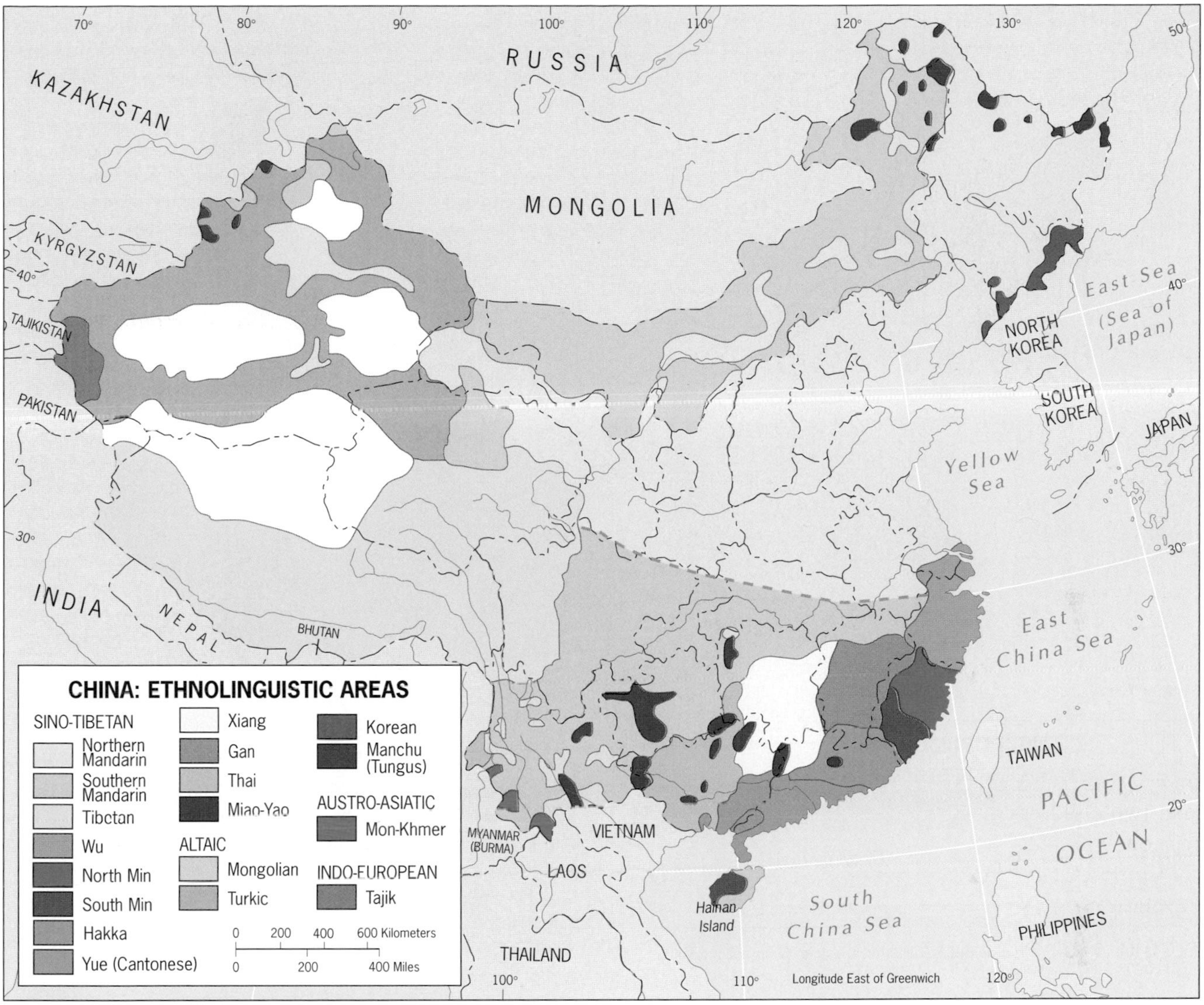

FIGURE 9-9 © H.J. de Blij, P.O. Muller, and John Wiley & Sons, Inc.

PEOPLE AND PLACES OF CHINA PROPER

A map of China's population distribution (Fig. 9-10) reveals the continuing relationship between the physical environment and its human occupants. In technologically advanced countries, we have noted, people shake off their dependence on what the land can provide; they cluster in cities and in other areas of economic opportunity. This depopulates rural areas that may once have been densely inhabited. In China, that stage has not yet been reached. Although China has large cities, as in India a sizeable majority of the people (59 percent) still live on—and from—the land. Thus the map of population distribution reflects the livability and productivity of China's basins, lowlands, and plains. Compare Figures 9-6A and 9-10, and China's continuing dependence on soil, water, and warmth will be evident.

The population map also suggests that in certain areas environmental limitations are being overcome. Industrialization in the Northeast, irrigation in the Inner Mongolia Autonomous Region, and oil-well drilling in Xinjiang enabled millions of Chinese to migrate from China Proper into these frontier zones, where they now outnumber the indigenous minorities.

Nonetheless, physiography and demography remain closely related in China. To grasp this relationship, it is useful to compare Figures 9-2 (physiography) and 9-10 (population) as our discussion proceeds. On the population map, the darker the color, the denser the population: in places we can follow the courses of major rivers through this scheme. Look, for example, at China's Northeast. A ribbon of population follows the Liao and

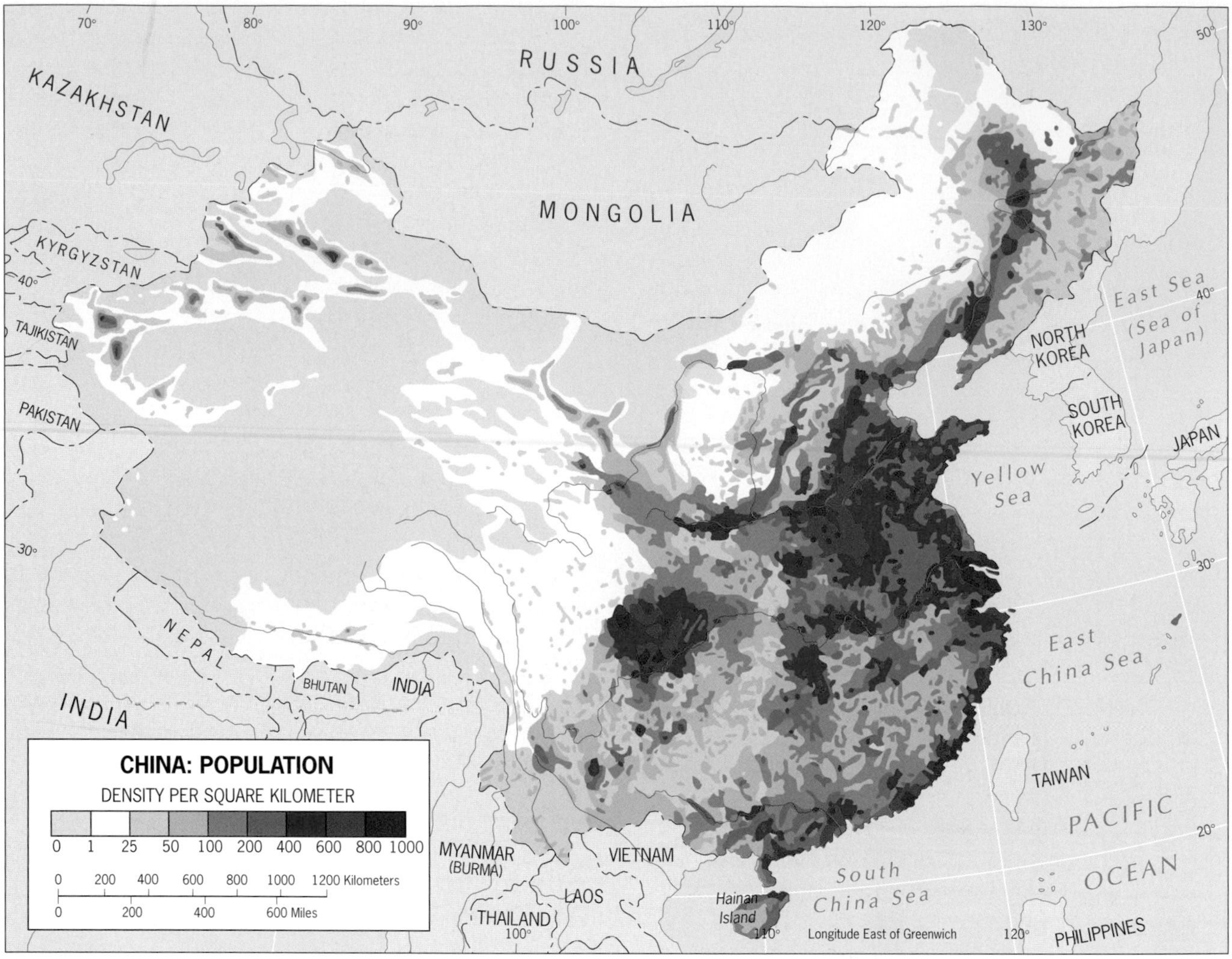

FIGURE 9-10 © H.J. de Blij, P.O. Muller, and John Wiley & Sons, Inc.

the Songhua rivers. Also note the huge, nearly circular population concentration on the western edge of the red zone; this is the Sichuan Basin, in the upper course of the Chang Jiang. Four major river basins contain more than three-quarters of China's 1.3 billion people:

1. The Liao-Songhua Basin or the Northeast China Plain
2. The Lower Huang He (Yellow River) Basin, known as the North China Plain
3. The Upper and Lower Basins of the Chang Jiang (Yangzi River)
4. The Basins of the Xi (West) and Pearl River

Not all the people living in these river plains are farmers, of course. China's great cities also have developed in these populous areas, from the industrial centers of the Northeast to the Pacific Rim upstarts of the South. Both Beijing and Shanghai lie in major river basins. And, as the map shows, the hilly areas between the river basins are not exactly sparsely populated either. Note that the hill country south of the Chang Basin (opposite Taiwan) still has a density of 130 to 260 people per square mile (50 to 100 per sq km).

Northeast China Plain

The Northeast China Plain is the heartland of China's Northeast, ancestral home of the Manchus who founded China's last dynasty, battleground between Japanese and Russian invaders, Japanese colony, heartland of communist industrial development, and now losing ground to the industries of the Pacific Rim. This region used to be called Manchuria. As the administrative map (Fig. 9-8) shows, there are three provinces here: Liaoning in the south, facing the Yellow Sea; Jilin in the center; and Heilongjiang, by far the largest, in the north.

When you are familiar with the vast farmlands of the North China Plain, the southernmost part of the Northeast China Plain looks so similar that it seems to be a continuation. Near the coast, farms look as they do near the mouth of the great Huang He; to the east, the Liaodong Peninsula looks like the Shandong Peninsula across the water. That impression soon disappears, however. Travel northward, and the Northeast China Plain reveals the effects of cold climate and thin soils. Farmlands are patchy; smokestacks seem to rise everywhere. This is industrial, not primarily farming, country. But the landscape looks like a Rustbelt: equipment often is old and outdated, transport systems are inadequate, buildings are in disrepair. Stream, land, and air pollution is dreadful. The Northeast in some ways resembles industrial zones in former communist Eastern Europe.

The Northeast has experienced many ups and downs. Although Japanese colonialism was ruthless and exploitive, Japan did build railroads, roads, bridges, factories, and other components of the regional infrastructure. (At one time, half of the entire railroad mileage in China was in the Northeast.) After the Japanese were ousted, the Soviets looted the area of machinery, equipment, and other goods. During the late 1940s, the Northeast was a ravaged frontier. But then the communists took power and made the industrial development of the Northeast a priority. From the 1950s until the 1970s, the Northeast led the nation in manufacturing growth. Its population, just a few million in the 1940s, mushroomed to more than 100 million. Towns and cities grew exponentially.

All this growth was based on the Northeast's considerable mineral wealth (Fig. 9-11). Iron ore deposits and coalfields lie concentrated in the Liao Basin; aluminum ore, ferroalloys, lead, zinc, and other metals are plentiful, too. As the map shows, the region also is well endowed with oil reserves; the largest lie in the Daqing Reserve between Harbin and Qiqihar. The communist planners made Changchun capital of Jilin Province, the Northeast's automobile manufacturing center. Shenyang, capital of Liaoning Province, became the leading steelmaking complex. Fushun and Anshan were assigned other functions, and the communists encouraged Han Chinese to migrate to the Northeast to find employment there. To attract them, the government built apartments, schools, hospitals, recreational facilities, even old-age homes—all to speed the region's industrial development.

For a time, it worked. During the 1970s, the Northeast was responsible for fully one-quarter of the entire country's industrial output. Stimulated by the needs of the growing cities and towns, farms expanded and crops diversified. The Songhua River Basin in the hinterland of Harbin, capital of Heilongjiang Province, became a productive agricultural zone despite the environmental limitations here. The Northeast was a shining example of what communist planning could accomplish.

7 Then came the **economic restructuring** of the post-Mao era. Under the new, market-driven economic order, the state companies that had led China's industrial resurgence were unable to adapt. Suddenly they were aging, inefficient, obsolete. Official policy encouraged restructuring and privatization, but the huge state factories could not (and much of their labor force would not) comply. Today, the Northeast's contribution to China's industrial output has fallen below 10 percent as the production of factories in the Southeast, from Shanghai to Guangdong, rises annually.

Economic geographers, however, will not write off the prospects of this northeastern corner of China; this area, as we noted, has had its downturns before. The Northeast still contains a storehouse of resources, including extensive stands of oak and other hardwood forests on the mountain slopes that encircle the Northeast China Plain. The locational advantages of the major centers are undiminished. Harbin, for example, lies at the convergence of five railroads and is situated at the head of navigation on the Songhua River, providing a link to the northeast corner of the region and, beyond the Amur River, to Russia's Khabarovsk. Moreover, when the Koreas reunite, producing what may become the western Pacific Rim's second Japan, and when the Russian Far East takes off (Vladivostok and Nakhodka are the natural ports for the upper Northeast), this northern frontier of China may yet reverse its decline and enter still another era of economic success.

North China Plain

Take the new superhighway between the port of Tianjin and the capital, Beijing, or the train from Beijing southward to Anyang, and you see a physical landscape almost entirely devoid of relief and a cultural landscape that repeats itself endlessly. Villages, many showing evidence of the collectivization program of the early communist era, pass at such regular intervals that some kind of geometry seems to control the cultural landscape. Carefully diked canals stretch in all directions. Rows of trees mark fields and farms. Depending on the season, the air is clear, as it is during the moist summer when the wheat crop covers the countryside like a carpet, or choked with dust, which can happen almost any other time of the year. The soil here is a fertile mixture of alluvium and loess (deposits of fine, windborne silt or dust of glacial origin), and the slightest breeze stirs it up. In the early spring the dust hangs like a fog over the landscape, covering everything with a fine powder. Add this to urban pollution and the cities become as unhealthy as any in the world.

For all we see, there are geographic explanations. The geometry of the settlement pattern results from the flatness of the uniformly fertile surface and the distance

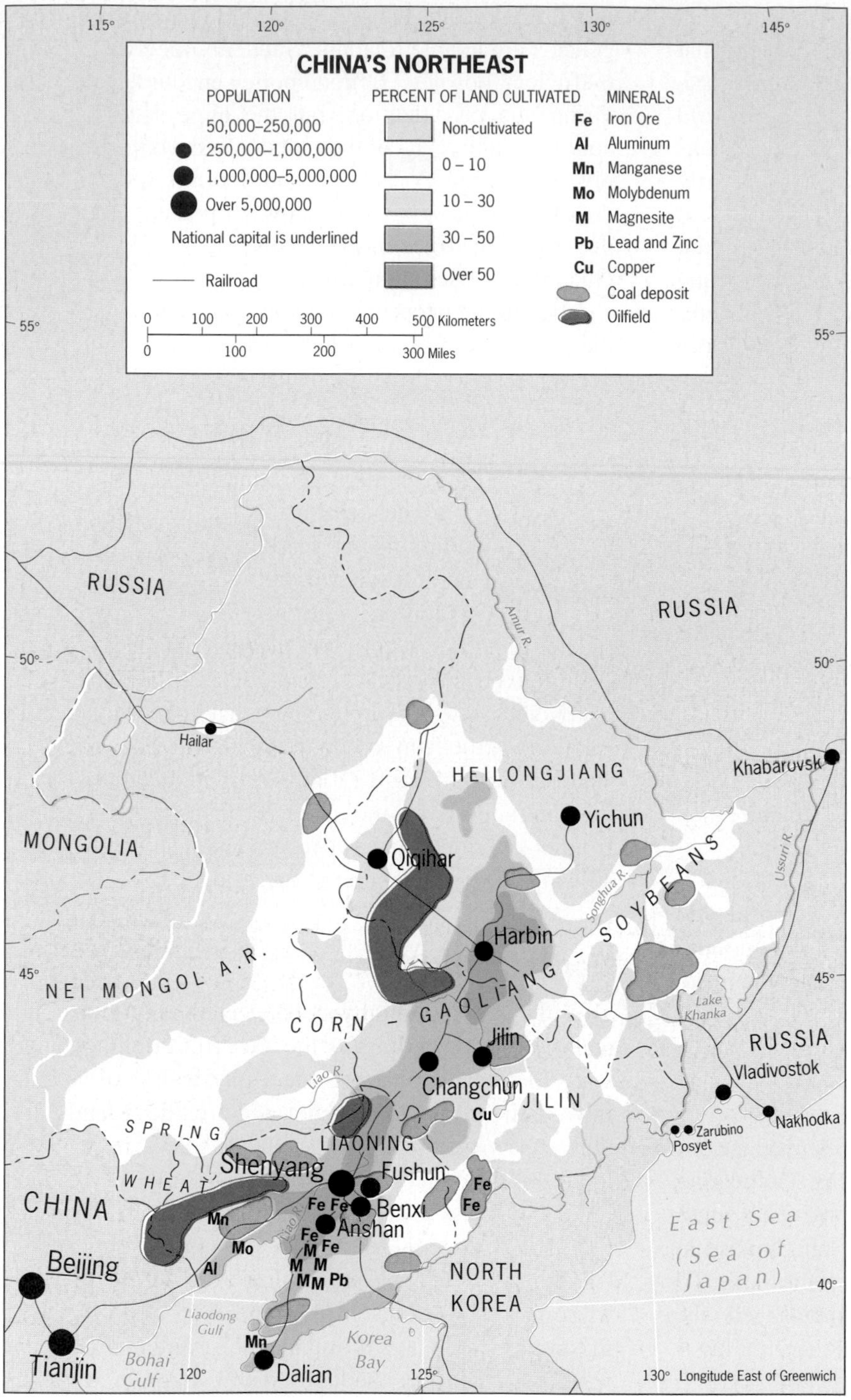

FIGURE 9-11 © H.J. de Blij, P.O. Muller, and John Wiley & Sons, Inc.

villagers could walk to neighboring villages and back during daylight (photo at right). The canals form part of a vast system of irrigation channels designed to control the waters of the Huang He (Yellow River). The rows of trees serve as windbreaks. And while wheat dominates the crops of the North China Plain, you will also see millet, sorghum, soybeans, cotton for the textile industry, tobacco (the Chinese are heavy smokers), and fruits and vegetables for the urban markets.

The North China Plain is one of the world's most heavily populated agricultural areas. Figure 9-10 shows that most of it has a density of more than 1000 people per square mile (400 per sq km), and in some parts the density is twice as high. Here, the ultimate hope of the Beijing government lay less in land redistribution than in raising yields through improved fertilization, expanded irrigation facilities, and the more intensive use of labor. A series of dams on the Huang River, including the Xiaolangdi Dam

AMONG THE REALM'S GREAT CITIES . . . BEIJING

Beijing (10.9 million), capital of China, lies at the northern apex of the roughly triangular North China Plain, just over 100 miles (160 km) from its port, Tianjin, on the Bohai Gulf. Urban sprawl has reached the hills and mountains that bound the Plain to the north, a defensive barrier fortified by the builders of the Great Wall. You can reach the Great Wall from central Beijing by road in about an hour.

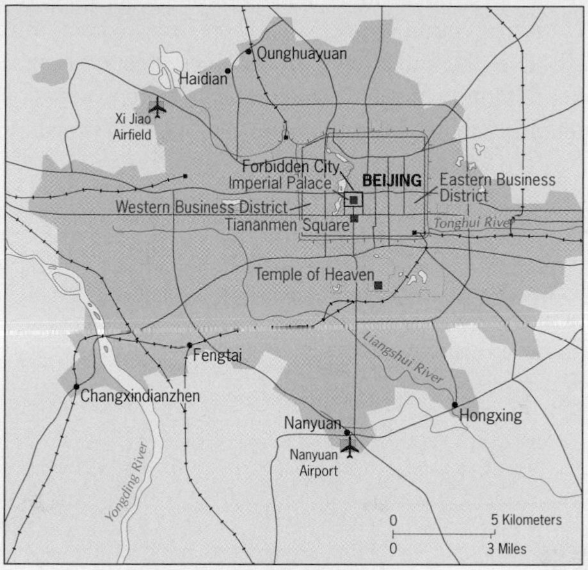

© H. J. de Blij, P. O. Muller, and John Wiley & Sons, Inc.

Although settlement on the site of Beijing began thousands of years ago, the city's rise to national prominence began during the Mongol (Yuan) Dynasty, more than seven centuries ago. The Mongols, preferring a capital close to their historic heartland, endowed the city with walls and palaces. Following the Mongols, later rulers at times moved their capital southward, but the government always returned to Beijing. From the third Ming emperor onward, Beijing was China's focus; it was ideally situated for the Manchus when they took over in 1644. During the twentieth century, China's Nationalists again chose a southern capital, but when the communists prevailed in 1949 they reestablished Beijing as their (and China's) headquarters.

Ruthless destruction of historic monuments carried out from the time of the Mongols to the communists (and now during China's modernization) has diminished, but not destroyed, Beijing's heritage. From the Forbidden City of the Manchu emperors (photo p. 439) to the fifteenth-century Temple of Heaven, the capital remains an open-air museum and the cultural focus of China. Successive emperors and aristocrats, moreover, bequeathed the city numerous parks and additional recreational spaces that other cities lack. Today Beijing is being transformed as a result of China's new economic policies (photo p. 453). A forest of high-rises towers above the retreating traditional cityscape, avenues have been widened, and a beltway has been built. A new era dawns in this historic capital.

FROM THE FIELD NOTES

"We flew over the North China Plain for an hour, and one could not miss the remarkably regular spacing of the villages on this seemingly table-flat surface. Many of these villages became communes during the communist reorganization of China's social order. The elongated buildings (many added since the 1950s) each were designed to contain one or more extended families. After the demise of Mao and his clique the collectivization effort was reversed and the emphasis returned to the small nuclear family. In this village and others we overflew, the long sides of the buildings almost always lay east-west, providing maximum sun exposure to windows and doors, and enhancing summer ventilation from longitudinal breezes. The more distant view looks hazy, because the North China Plain during the spring is wafted by loess-carrying breezes from the northwest. On the ground, a yellow-gray hue shrouds the atmosphere until a passing weather front temporarily clears the air." © H. J. de Blij.

upstream in Henan Province, now reduce the flood danger, but outside the irrigated areas the ever-present problem of rainfall variability and drought persists. The North China Plain has not produced any substantial food surplus even under normal circumstances; thus when the weather turns unfavorable, the situation soon becomes precarious. The specter of famine may have receded, but the food situation is still uncertain in this very critical part of China Proper.

The North China Plain is an excellent example of the concept of a national **core area**. Not only is it a densely 8
populated, highly productive agricultural zone, but it also is the site of the capital and other major cities, a substantial industrial complex, and several ports, among which Tianjin ranks as one of China's largest (Fig. 9-12). Tianjin, on the Bohai Gulf, is linked by rail and highway (less than a two-hour drive) to Beijing. Like many of China's

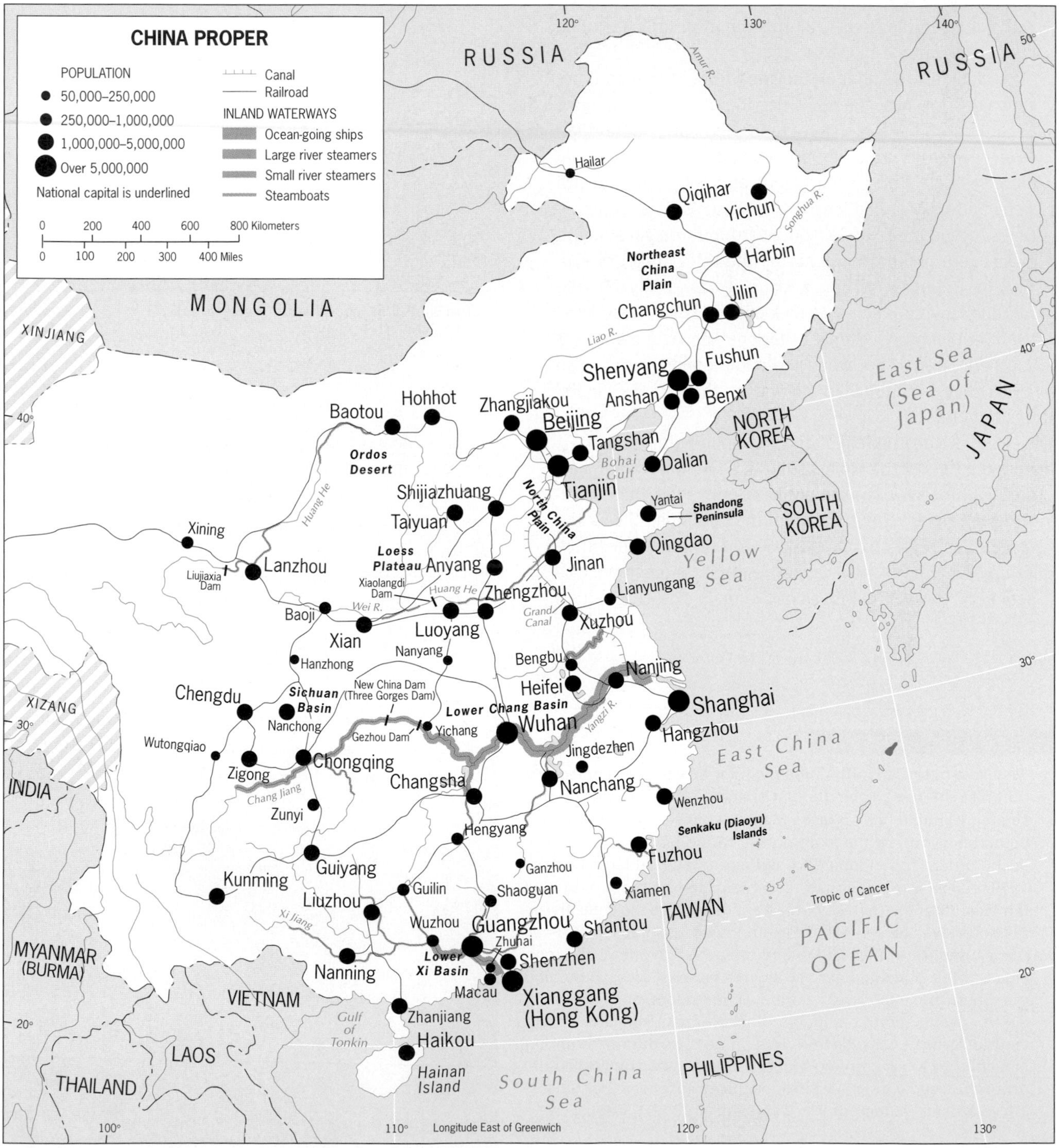

FIGURE 9-12 © H.J. de Blij, P.O. Muller, and John Wiley & Sons, Inc.

FROM THE FIELD NOTES

"Beijing seems to be one gigantic construction project, with six-lane ring roads and arteries connecting clusters of high-rises still sprouting from the ground. But there is another side to this. Old, historic Beijing's neighborhoods are being swept away, often without any consideration either for the heritage being lost or for the families being displaced. Here, near the old city wall in the southwest, part of a neighborhood already has been bulldozed, and locals salvage what they can (including reusable bricks) from the rubble. Laundry still hangs from lines in front of the still-occupied houses in the background, but any day they, too, will be gone." © H. J. de Blij.

harbors, that of Tianjin's river port is not particularly good; but Tianjin is well situated to serve not only the northern sector of the North China Plain and the capital, but also the Upper Huang Basin and Inner Mongolia beyond (see Fig. 9-1). Tianjin, again like several other Chinese ports, had its modern start as a treaty port, but the city's major growth awaited communist rule. For decades it was a center for light industry and a flood-prone harbor, but after 1949 the communists constructed a new artificial port and flood canals. They also chose Tianjin as a site for major industrial development and made large investments in the chemical industry (in which Tianjin still leads China), iron and steel production, heavy machine manufacturing, and textiles. Today, with a population of 9.4 million, Tianjin is China's third-largest metropolis and the center of one of its leading industrial complexes.

Northeast of Tianjin lies the smaller city of Tangshan. In 1976, a devastating earthquake, with its epicenter between Tangshan and Tianjin, killed an estimated 750,000 people, the deadliest natural disaster of the twentieth century. So inept was the government's relief effort in nearby Beijing that the widespread resentment it provoked may have helped end the Maoist period of communist rule in China.

Beijing, unlike Tianjin, is China's political, cultural, and educational center. Its industrial development has not matched Tianjin's. The communist administration did, however, greatly expand the municipal area of Beijing, which (as we noted) is not controlled by the Province of Hebei but is directly under the central government's authority. In one direction, Beijing was enlarged all the way to the Great Wall—30 miles (50 km) to the north—so that the "urban" area includes hundreds of thousands of farmers. Not surprisingly, this has circumscribed an enormous total population, enough to rank Beijing, with its 10.9 million inhabitants, among the world's megacities.

Inner Mongolia

Northwest of the core area as defined by the North China Plain, along the border with the state of Mongolia, lies Inner Mongolia, administratively defined as the Nei Mongol Autonomous Region (see Fig. 9-8, p. 444). Originally established to protect the rights of the approximately 5 million Mongols who live outside the Mongolian state, Inner Mongolia has been the scene of massive immigration by Han Chinese. Today Chinese outnumber Mongols here by nearly four to one. Near the Mongolian border, Mongols still traverse the steppes with their tents and herds, but elsewhere irrigation and industry have created an essentially Chinese landscape. The A.R.'s capital, Hohhot, has been eclipsed by Baotou on the Huang He, which supports a corridor of farm settlements as it crosses the dry land here on the margins of the Gobi and Ordos deserts. With about 25 million inhabitants, Inner Mongolia still cannot compare to the huge numbers that crowd the North China Plain, but recent mineral discoveries have boosted industry in Baotou and livestock herding is expanding. Nei Mongol may retain its special administrative status, but it functions as part of Han China in all but name.

Basins of the Chang/Yangzi

In contrast to the contiguous, flat agricultural-urban-industrial North China Plain defined by the sediment-laden Huang River, the basins and valleys of the Chang Jiang (Long River) display variation in elevation and relief. The Chang River, whose lower-course name becomes the Yangzi, is an artery like no other in China. Near

AMONG THE REALM'S GREAT CITIES . . . SHANGHAI

Sail into the mouth of the great Yangzi River, and you see little to prepare you for your encounter with China's largest city (population: 12.8 million). For that, you turn left into the narrow Huangpu River, and for the next several hours you will be spellbound. To starboard lies a fleet of Chinese warships. On the port side you pass oil refineries, factories, and neighborhoods. Next you see an ultramodern container facility, white buildings, and rust-free cranes, built by Singapore. Soon rusty tankers and freighters line both sides of the stream. High-rise tenements tower behind the cluttered, chaotic waterfront where cranes, sheds, boatyards, piles of rusting scrap iron, mounds of coal, and stacks of cargo vie for space. In the river, your boat competes with ferries, barge trains, and cargo ships. Large vessels, anchored in midstream, are being offloaded by dozens of lighters tied up to them in rows. The air is acrid with pollution. The noise—bells, horns, whistles—is deafening.

What strikes you is the vastness and sameness of Shanghai's cityscape, until you pass beneath the first of two gigantic suspension bridges. Suddenly, everything changes. To the left, or east, lies *Pudong*, an ultramodern district with the space-age Oriental Pearl Television Tower rising like a rocket on its launchpad (photo p. 469) above a forest of modern, glass-and-chrome skyscrapers that make the Huangpu look like Hong Kong's Victoria Harbor. To the right, along the Bund, stand the remnants of Victorian buildings, monuments to the British colonialists who made Shanghai a treaty port and started the city on its way to greatness. Everywhere, construction cranes rise above the cityscape. Shanghai looks like one vast construction zone—complete with a 1510-foot office tower (called the World Financial Center) that is one of the world's three tallest buildings.

The Chinese spent heavily to improve infrastructure in the Shanghai area, including brand-new Pudong International Airport, connected to the city by the world's first *mag-lev* (magnetic levitation) train—the fastest anywhere at 267 mph (430 kph). Pudong has become a magnet for foreign investment, attracting as much as 10 percent of the annual total for the entire country. Shanghai's income is rising much faster than China's, and the Yangzi River Delta is becoming a counterweight to the massive Pearl River Delta development in China's South. Among other things, Shanghai is becoming China's "motor city," complete with a Formula One race track seating 200,000 at the center of an automobile complex where all levels of the industry, from manufacturing to sales, are being concentrated.

In 2010 Shanghai will host World Expo on the banks of the Huangpu River (the mag-lev track will be extended to the site), and the city's planners, who have already transformed the place from just another Open City to a symbol of the New China, will have an opportunity to show the world what has been accomplished here in just one generation.

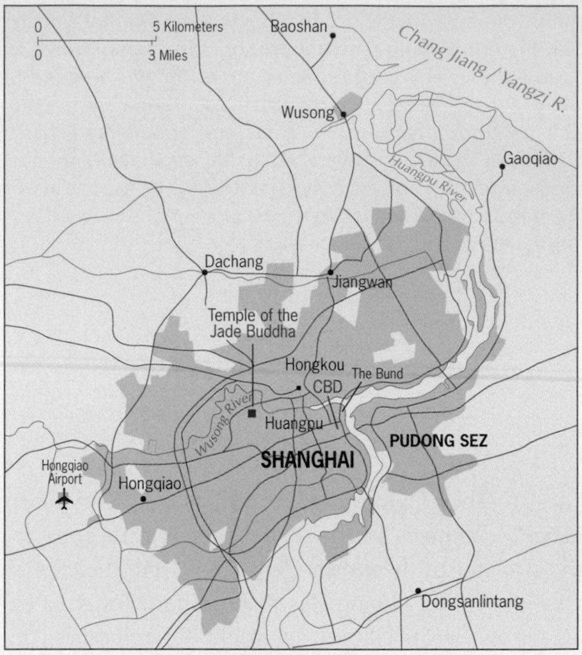

its mouth lies the country's largest city, Shanghai. Part of its middle course is now being transformed by a gigantic engineering project. Farther upstream, the Chang crosses the populous, productive Sichuan Basin. And unlike the Huang, the Chang Jiang is navigable to oceangoing ships for over 600 miles (1000 km) from the coast all the way to Wuhan (Fig. 9-12). Smaller ships can reach Chongqing, even after dam construction. Several of the Chang River's tributaries also are navigable, so that 18,500 miles (30,000 km) of water transport routes serve its drainage basin. Thus the Chang Jiang constitutes one of China's leading transit corridors. With its tributaries it handles the trade of a vast area, including nearly all of middle China and sizeable parts of the north and south. The North China Plain may be the core area of China, but in many ways the Lower Chang Basin is its heart.

Early in Chinese history, when the Chang's basin was being opened up and rice and wheat cultivation began, a

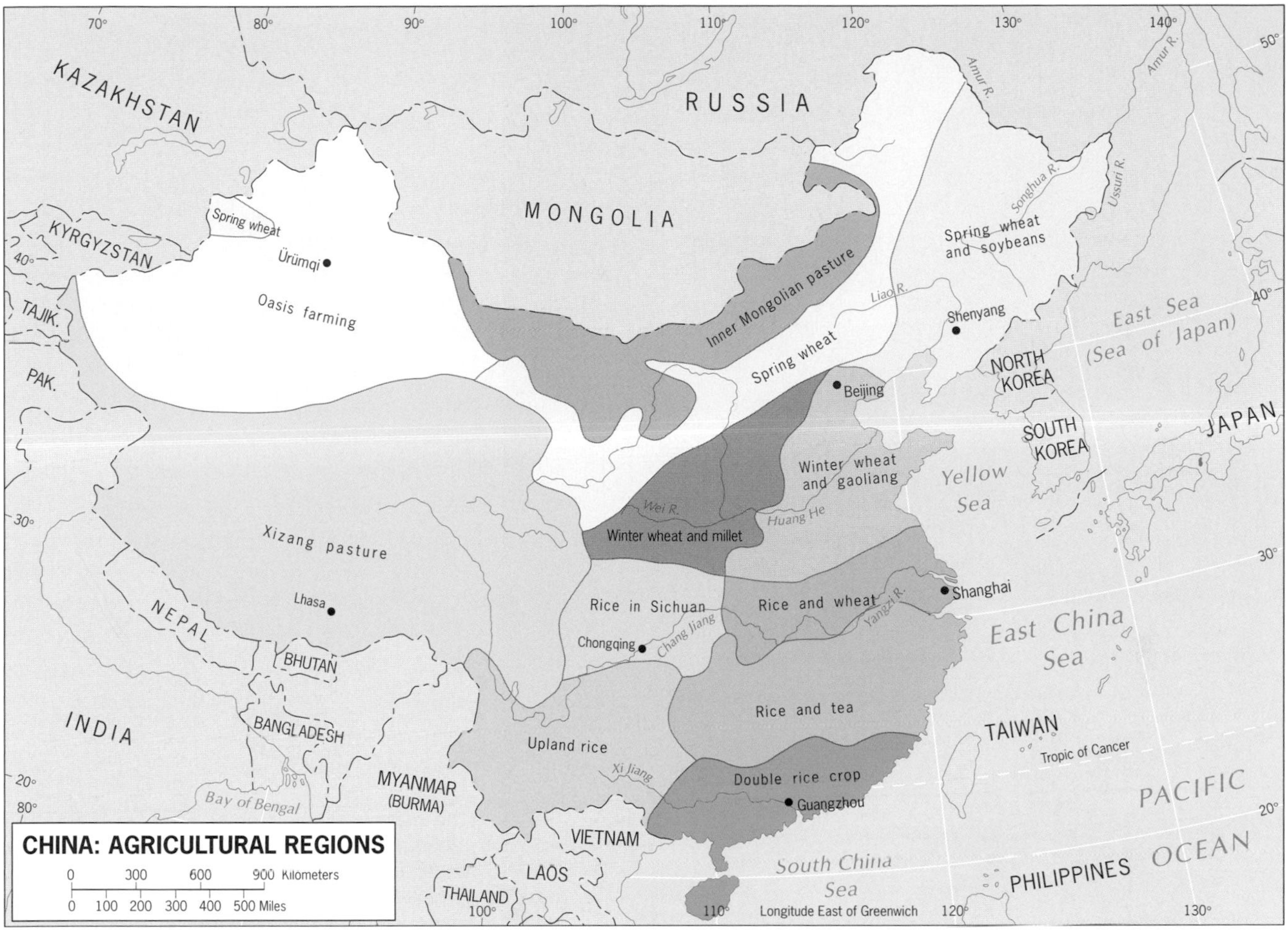

FIGURE 9-13 © H.J. de Blij, P.O. Muller, and John Wiley & Sons, Inc.

FROM THE FIELD NOTES

"China's new consumer culture is perhaps nowhere more evident than in burgeoning, modernizing, globalizing Shanghai. I walked today down Nanjing Road, exactly 20 years after I first saw it in 1982, but only four years after my last excursion here. The rate of transformation is increasing. Thousands of locals have taken up the fast-food habit and are eating at Kentucky Fried Chicken (the colonel's face is everywhere) or McDonald's (I counted five of the latter along my route). A section of Nanjing Road has been made into a pedestrian mall, and here too is evidence of a new habit: private automobile transportation. Car sales lined the street, and potential buyers thronged the displays. Given the existing congestion on local roads, where will these cars fit?" © H. J. de Blij.

The massive Three Gorges Dam on the Yangzi (Chang) River above Yichang in Hubei Province under construction in 2003. This is the world's largest-ever hydroelectric project, yielding its first power supplies in 2003 with capacity expanding until the entire project is at full production in 2009. © AP/Wide World Photo.

canal was built to link this granary to the northern core of old China. Over 1000 miles (1600 km) long, this was the longest artificial waterway in the world, but during the nineteenth century it fell into disrepair. Known as the Grand Canal, it was dredged and rebuilt when the Nationalists controlled eastern China. After 1949, the communist regime continued this restoration effort, and much of the canal is now again open to barge traffic, supplementing the huge fleet of vessels that hauls domestic interregional trade along the east coast (Fig. 9-12).

As Figure 9-13 shows, the Lower Chang Basin is an area of both rice and wheat farming, offering further proof of its pivotal situation between south and north in the heart of China Proper. Shanghai lies at the coastal gateway to this productive region on a small tributary of the Chang (now Yangzi) River, the Huangpu. The city has an immediate hinterland of some 20,000 square miles (50,000 sq km)—smaller than West Virginia—containing more than 50 million people. About two-thirds of this population are farmers who produce food, silk filaments, and cotton for the city's industries.

Travel upriver along the Yangzi/Chang Jiang, and you meet an unending stream of vessels large and small, including numerous "barge trains"—as many as six or more barges pulled by a single tug in an effort to save fuel (they are slow, but time is not the primary concern). The traffic between Shanghai and Wuhan—Wuhan is short for Wuchang, Hanyang, and Hankou, a coalescing conurbation of three cities—makes the Yangzi one of the world's busiest waterways. Smoke-belching boat engines create a more or less permanent plume of pollution here, worsening the regional smog created by the factories of Nanjing and others perched on the waterfront. Here China's Industrial Revolution is in full gear, with all its environmental consequences.

Above Wuhan the river traffic dwindles because the depth of the Chang reduces the size of vessels that can reach Yichang. River boats carry coal, rice, building materials, barrels of fuel, and many other items of trade. But the middle course of the Chang River is becoming more than a trade route. In the northward bend of the great river between Yichang and Chongqing, where the Chang has cut deep troughs on its way to the coast, one of the world's largest engineering projects is in progress (photo at left). It has several names: the Sanxia Project, the Chang Jiang Water Transfer Project, the Three Gorges Dam, and the New China Dam. It is most commonly called the Three Gorges Dam because it is here that the great river flows through a 150-mile (240-km) series of steep-walled valleys less than 360 feet (110 m) wide. Near the lower end of this natural trough, a dam now rises to over 600 feet (180 m) above the valley floor to a width of 1.3 miles (2.1 km), creating a reservoir that inundates the historic Three Gorges and extends more than 380 miles (over 600 km) upstream. China's project engineers say that the dam will end the river's rampaging flood cycle, enhance navigation, stimulate development along the new lake perimeter, and provide at least one-tenth, and perhaps as much as one-eighth, of China's electrical power supply. As such it will transform the heart of China, which is why many Chinese like to compare this gigantic project to the Great Wall and the Grand Canal, and call it the New China Dam.

The Sichuan Basin

As Figure 9-1 shows, all this development is taking place downstream from Chongqing, the city near the eastern margin of the great and populous Sichuan Basin that was made, along with its hinterland, a *shi* in the hierarchy of China's administrative units. Obviously this city is strongly affected by the dam's construction, which disrupted its river trade (boats up to 1000 tons could reach Chongqing previously). A diversion channel around the Three Gorges Dam will restore and enhance Chongqing's port and transit functions, and the city figures prominently in an important economic design for China: the expansion of its coastal economic success into the deep interior. We return to this topic later, but consider Chongqing's relative location on the western margin of the Yangzi's great economic corridor and on the eastern edge of the vast and productive Sichuan Basin, now upstream from the country's greatest engineering project and about midway between Kunming to the south and Xian to the north. This city is likely to become the most important growth pole in China's interior.

Sichuan, its basin crossed by the Chang (called the Yangzi downstream) after the river emerges from the mountains of the west, frequently was a problem province for both China's dynastic rulers and the communist regime. Its thriving capital, Chengdu, was one of China's most active centers of dissent during the 1989 pro-democracy movement. With about 120 million inhabitants (including Chongqing) today, Sichuan Province would be one of the world's ten largest states, but it does not even contain 10 percent of China's population. Sichuan's fertile, well-watered soils yield a huge variety of crops: grains (rice, wheat, corn), soybeans, tea, sugarcane, and many fruits and vegetables.

In many ways Sichuan is a country unto itself, isolated by topography and distance from the government in Beijing. The people here are Han Chinese, but cultural landscapes reveal the area's proximity to Southeast Asia. Along the great riverine axis of the Chang/Yangzi, China changes character time and again.

Basins of the Xi (West) and Pearl River

As Figure 9-12 shows, the Xi River and its basins are no match for the Chang or Huang, not even for the Liao. This southernmost river even seems to have the wrong name: Xi means West! It reaches the coast in a complex area where it forms a delta immediately adjacent to the estuary of the Pearl River. In this subtropical part of China, local relief is higher than in the lowlands and basins of the center and north, so that farmlands are more confined. Especially in the interior areas, water supply is a recurrent problem. On the other hand, its warm climate permits the double-cropping of rice (see photo below). Regional food

FROM THE FIELD NOTES

"It may not look like much, but this home-made, powered thresher is doing the work of several people in this paddy near Guilin in the Guangxi Zhuang Autonomous Region of southern China. Three people do the bunching, feeding, and bagging! This rural scene led me to wonder: what will happen when China's agriculture becomes more mechanized? How will millions of redundant farmers be employed?"
© Barbara A. Weightman.

production, however, has never approached that of the North China Plain or the Lower Chang Basin. And, as Figure 9-10 shows, the population in this southern part of China is smaller than that in the more northerly heartlands.

Again unlike the Huang and Chang rivers, which rise in snow-covered interior mountains, the Xi is a shorter stream whose source is on the Yunnan Plateau (see Fig. 9-2). Just up the coast from its delta, however, lies one of the most important areas of modern China. As we will note later, the mouth of the Pearl River is flanked by some of China's fastest-growing economic complexes, including Guangzhou, capital of Guangdong Province, Hong Kong, the former British dependency, and adjoining Shenzhen, one of the world's fastest-growing cities.

This is all the more remarkable because, as the maps emphasize, South China is not especially well endowed with geographic advantages, locational or otherwise. The Pearl River is navigable for larger ships only to Guangzhou, and the Xi for small river boats no farther than Wuzhou. Guangdong Province, the key province of southern China, lies far from China's heartlands. Nor is South China blessed with the resources that propelled the industrial development of the Northeast and other parts of China. Indeed, one of the principal objectives of the Three Gorges Dam project is to provide South China with ample electricity. As Figure 9-14 shows, northern, central, and western China have most of the coal, oil, and natural gas resources, not the South.

China Proper, with its populous and powerful subregions, is in many ways the dominant region in the East Asian geographic realm. But China Proper is only one part of China. To the west lie two vast regions with relatively sparse populations: Xizang (Tibet) and Xinjiang.

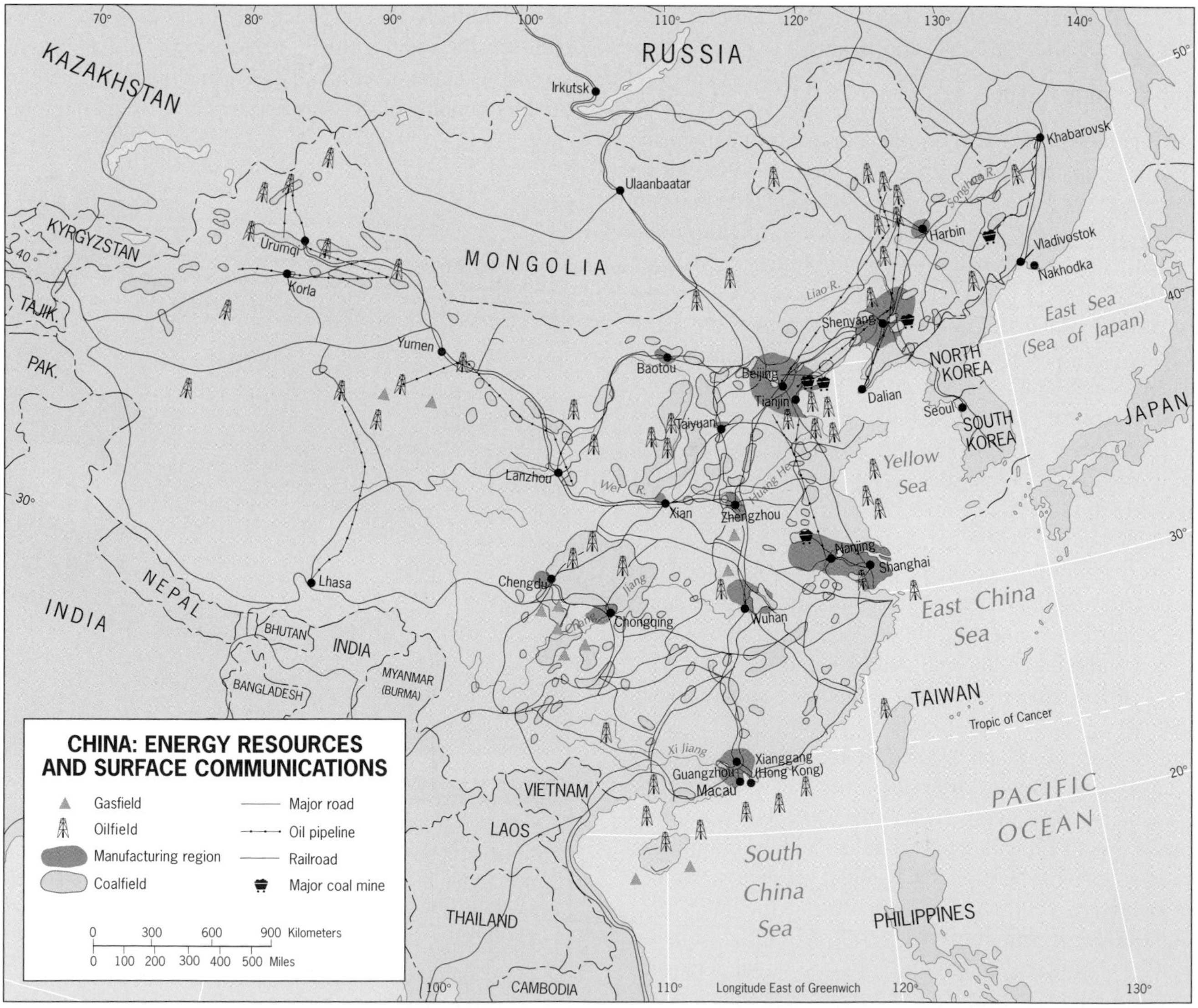

FIGURE 9-14 © H.J. de Blij, P.O. Muller, and John Wiley & Sons, Inc.

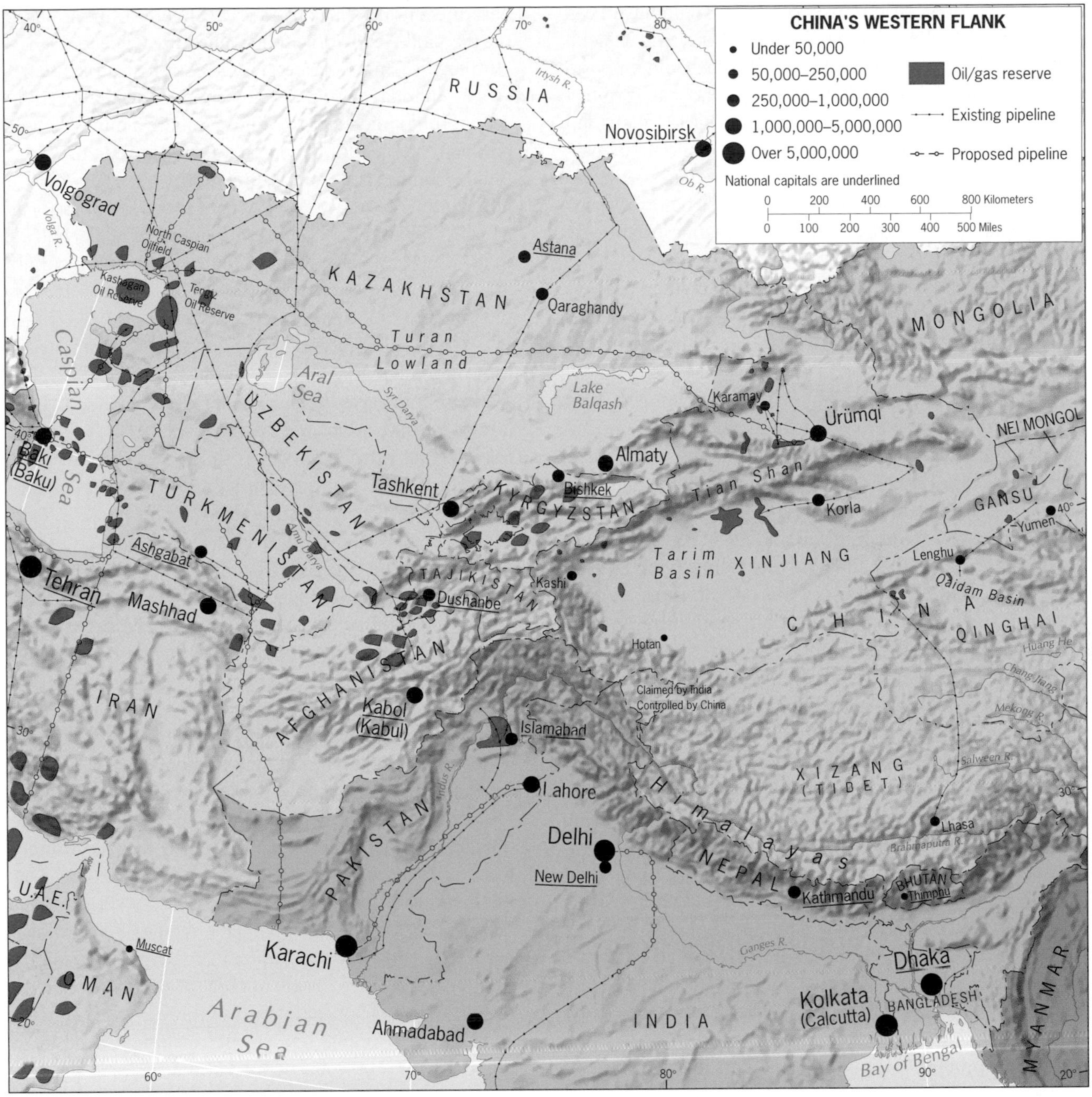

FIGURE 9-15 © H.J. de Blij, P.O. Muller, and John Wiley & Sons, Inc.

And here, in this remote western periphery adjoining Turkestan, China meets Muslim Central Asia (Fig. 9-15).

XIZANG (TIBET)

As the climate and physiographic maps demonstrate, harsh physical environments dominate this colony of China. It is a vast, sparsely peopled region designated as an Autonomous Region, but in fact Xizang (Tibet) is an occupied society. In terms of its human geography, we should take note of two subregions.

The first is the core area of Tibetan culture, between the Himalayas to the south and the Trans-Himalayas lying not far to the north. In this area, some valleys lie below 7000 feet (2130 m); the climate is comparatively mild, and some cultivation is possible. Here lies Tibet's main population cluster, including the crossroads capital of Lhasa. The Chinese government has made investments to develop these valleys, which contain excellent sites for hydroelectric power projects (some have been used for a few light industries) and promising mineral deposits. The second area of interest is the Qaidam Basin in the north (Fig. 9-15). This basin lies thousands of feet below the surrounding Kunlun and Altun Mountains and has always contained a concentration of nomadic pastoralists. Recently, however, exploration has revealed the presence of oilfields and coal reserves below the surface of the Qaidam Basin, and these resources are now being developed (Fig. 9-16).

As Figure 9-3 reminds us, Tibet came under Chinese domination during the Manchu (Qing) Dynasty in 1720, but it regained its separate status in the late nineteenth century. China's communist regime took control after the invasion of 1950; in 1959, China crushed an uprising after Tibetan villagers tried to resist the Chinese presence. Tibetan society had been organized around the fortress-like monasteries of Buddhist monks who paid allegiance to their supreme leader, the Dalai Lama. The Chinese wanted to modernize this feudal system, but the Tibetans clung to their traditions. In 1959, they proved no match for the Chinese armed forces; the Dalai Lama was ousted, and the monasteries were emptied. The Chinese destroyed much of Tibet's cultural heritage, looting its religious treasures and works of art. Their harsh rule devastated Tibetan society, but after Mao's death in 1976 the Chinese relaxed their tight control. Although amends were made (religious treasures were returned to Xizang, monastery reconstruction was permitted, and Buddhist religious life resumed), pro-independence rioting has been frequent since 1987. As a result, the Chinese have again tightened their grip. Since its formal annexation in 1965, Xizang has been administered as an Autonomous Region. Although it is large in area, its population still totals less than 3 million.

Visiting Chinese-occupied Tibet is a depressing experience. Tibetan domestic architecture has beauty and expressiveness, but the Chinese have built ugly, gray structures to house the occupiers—often directly in front of Tibetan monasteries and shrines. This juxtaposition of Chinese banality and Tibetan civility seems calculated to perpetuate hostility, an anachronism in this supposedly postcolonial world. During the rule of Mao Zedong, the Chinese officially regarded Tibet as a territorial buffer against India (then allied with the Soviet Union). The breakup of the Soviet Empire has not softened Beijing's stand.

XINJIANG

China's northwestern corner consists of the giant Xinjiang-Uyghur Autonomous Region, constituting over one-sixth of the country's land area and situated in a strategic part of Central Asia (Fig. 9-15). Linked to China Proper via the vital Hexi Corridor in Gansu Province to the east, Xinjiang is China's largest single administrative area.

Even today, however, after more than a half-century of vigorous sinicization, Chinese remain a minority here. About 40 percent of the population of more than 20 million is Chinese (up from 5 percent in 1949), and the Chinese control virtually all aspects of life. The majority are Muslim Uyghurs, who make up approximately half the total, and Kazakhs, Kyrgyz, and others with cultural affinities across the western border to all five republics of former Soviet Central Asia.

The physical geography of Xinjiang is dominated by high mountain ranges and vast basins (Fig. 9-2). The southern margin is defined by the Kunlun Shan (Mountains) beyond which lies Xizang (Tibet). Across the north-center of the region lies the mountain wall of the Tian Shan. Between the Kunlun and Tian Shan extends the vast, dry Tarim Basin within which lies the Takla Makan Desert, and north of the Tian Shan lies another depression, the Junggar Basin. Rivers rising on the mountain slopes disappear beneath the sediments of the basins, sustaining a ring of oases where their waters come near the surface or where wells and *qanats* make irrigation possible (Fig. 9-16).

This would not seem to be the most favorable geography, but Xinjiang has other assets that China exploits. The Qing (Manchu) armies that fought the nomadic Muslim warlords here in the 1870s saw the agricultural possibilities and planted the first crops. The communist rulers after 1949 exiled political opponents as well as criminals to notorious prison camps in this remote frontier. Then Xinjiang turned out to have extensive reserves of oil and natural gas (Figs. 9-15, 9-16). Here, too, China could experiment with space technology, rocketry, and nuclear weaponry far from densely populated areas (and shielded from foreign spies). And now, as Figure 9-15 reminds us, China's westernmost borders lie relatively close to the great energy reservoirs of the Caspian Sea Basin—and only one neighbor lies in the way. Pipeline construction across Kazakhstan to Xinjiang is in progress.

Xinjiang's evolving human geography is revealed by Figure 9-16. Most of the Han Chinese are concentrated in the north-central cluster of cities and towns between the capital, Ürümqi, and the oil center of Karamay. Of special interest here is the city of Shihezi, founded by the Chinese army and now China's model for the region. Over 90 percent of its population of 400,000 are ethnic Chinese; Shihezi's comparative modernity and orderly life stand in sharp contrast to the poorer, often squalid towns and settlements outside the Chinese heartland. Also centered on the Ürümqi-Karamay area is Xinjiang's largest agricultural zone, where cotton and fruits yield export revenues. As the

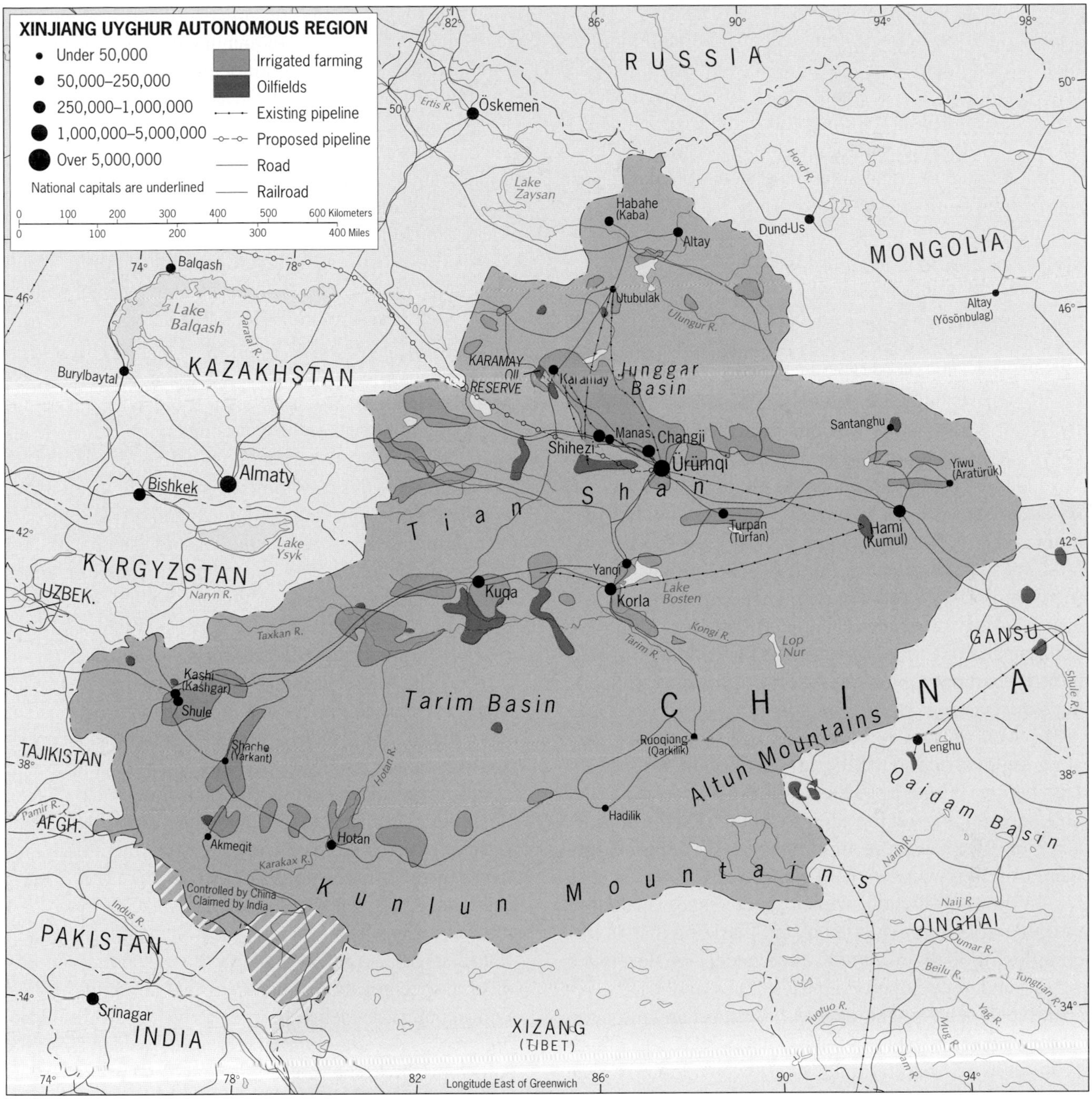

FIGURE 9-16 © H.J. de Blij, P.O. Muller, and John Wiley & Sons, Inc.

map shows, a ring of oases and clusters of irrigated agriculture encircles both the Tarim and Junggar basins.

Beijing has made extensive efforts to integrate Xinjiang into the Chinese state. A railroad and highway link Ürümqi and Lanzhou via the Hexi Corridor. Westward from Ürümqi, a railroad crosses the border with Kazakhstan and connects to the Öskemen-Almaty line; another railroad, opened in 2000, connects the capital to Kashi (Kashgar) in the distant southwest (Fig. 9-16). But vast spaces in Xinjiang remain distant and inaccessible. As the map shows, just one road, recently upgraded, loops around the vast southern and eastern flank of the Tarim Basin from Kashi via Hotan to Korla.

China's western frontier region, therefore, is a place of challenges as well as opportunities. Here a regime that from time to time displays intolerance of "foreign" belief systems meets a majority Muslim population that resists incorporation and whose leaders sometimes call for secession. Here China is having its brush with the War on Terror, accusing some of these Muslims of al-Qaeda sympathies, if not activism. Here the gap between comparatively well-off Chinese and poorer minorities is large—and spatially vivid. Here, too, China has unresolved boundary disputes with neighbors, including Tajikistan and India. On the other hand, Xinjiang's energy resources (also including some coal), agricultural and industrial potential, and relative location make this a promising frontier, a symbol for the rest of Central Asia of what Chinese administration can accomplish.

CHINA'S PACIFIC RIM: EMERGING REGION

From Dalian on the doorstep of the Northeast to Hainan Island off the country's southernmost point, China is booming. In the cities, old neighborhoods are being bulldozed to make way for modern skyscrapers. In the countryside of the coastal provinces, thousands of factories employing millions of workers who used to farm the land are changing life in the towns and villages. New roads, airports, power plants, and dams are being built. It is an uneven, sometimes chaotic process that is affecting certain parts of China's coast more than others. It is penetrating parts of China's interior, but meanwhile social gaps are widening: between advantaged and disadvantaged, well-off and poor, urbanite and villager. It is creating regional economic disparities that China has never seen before. It is also producing what may become a separate geographic region along China's Pacific littoral, a region that has more in common with Japan, South Korea, and Taiwan than with Yunnan or Gansu.

In the Introduction, we discussed issues relating to the geography of development, emphasizing that, in this complex, globalizing world, the concept of developed versus underdeveloped countries (DCs and UDCs, respectively) is no longer tenable. By almost any measure, China remains what used to be called a UDC. As Figure G-11 shows, it barely ranks as a lower-middle-income economy. But China now has its own core-periphery duality: in its Pacific coast provinces, per-capita incomes today are considerably above the national average GNI of U.S. $4990, while hundreds of millions of villagers in China's remote interior earn far less. In its burgeoning Pacific Rim, China resembles a modernizing economy with "developed" characteristics. In much of its interior, it does not. This contrast is one reason a discrete Pacific Rim region appears to be forming in China Proper.

Geography of Development

Many yardsticks are used to gauge a country's development, including, as we noted, GNI per capita; percentage of workers in farms, factories, or other kinds of employment; amount of energy consumed; efficiency of transport and communications; use of manufactured metals like aluminum and steel in the economy; productivity of the labor force; and social measures such as literacy, nutrition, medical services, and savings rates. All these data are calculated per person, yielding an overall measure that ranks the country among all those providing statistics. These data do not, however, tell us *why* some countries, or parts of countries, exhibit the level of development they do.

The **geography of development** leads us into the study of raw-material distributions, environmental conditions, cultural traditions, historic factors (such as the lingering effects of colonialism), and forces of location. In the emergence of China's Pacific Rim, relative location played a major role, as we will see. 9

Geographers tend to view the development process spatially, whereas other scholars focus on structural aspects. In the 1960s, economist Walt Rostow formulated a global model of the development process that is still being discussed today. This model suggested that all developing countries follow an essentially similar path through five interrelated growth stages. In the earliest of these stages, a *traditional society* engages mainly in subsistence farming, is locked in a rigid social structure, and resists technological change. When it reaches the second stage, *preconditions for takeoff*, progressive leaders move the country toward greater flexibility, openness, and diversification. Old ways are abandoned, workers move from farming into manufacturing, and transport improves. This leads to the third stage, *takeoff*, when the country experiences a type of industrial revolution. Sustained growth takes hold, industrial urbanization proceeds, and technological and mass-production breakthroughs occur. If the economy continues to expand, the fourth stage, *drive to maturity*, brings sophisticated industrial specialization and increasing international trade. Some countries reach a still more advanced fifth stage of *high mass consumption* marked by high incomes, widespread production of consumer goods and services, and most workers employed in the tertiary and quaternary economic sectors.

This model takes little account of core-periphery contrasts within individual countries. In China, takeoff conditions (stage 3) exist in much of the Pacific Rim, but other major areas of this vast country remain in stage 1. China has passed through its transition from tradition-bound to progressive leadership, but the effect of that leadership's reformist policies has not been the same in all parts of the country. The most fertile ground lies along the Pacific, where China has a history of contact with foreign enterprises, where the country is most open to the world, and from where many Chinese departed for Southeast Asian (and other) countries. These **Overseas** 10
Chinese were positioned to play a crucial role in China's

Pacific Rim development: many had left kin and community there, and those links facilitated the investment of hundreds of millions of dollars when the opportunity arose. Many of the Pacific Rim's factories were built by means of this repatriated money.

Transforming the Economic Map

China's communist era has witnessed the metamorphosis of the world's most populous nation. Despite the dislocations of collectivization, communization, the Cultural Revolution, power struggles, policy reversals, and civil conflicts, China as a communist state in the aftermath of Mao managed to stave off famine and hunger, control its population growth rate, strengthen its military capacity, improve its infrastructure, and enhance its world position. It took over Hong Kong from the British and Macau from the Portuguese, closing the chapter on colonialism. When economic turbulence struck its Pacific Rim and Southeast Asian neighbors, China withstood the onslaught. Its trade with neighbors near and far is growing. Foreign investment in China has mushroomed. And still its government espouses communist dogma.

How could the government manipulate the coexistence of communist politics and market economics? That was the key question confronting Deng Xiaoping and his comrades when they took power in 1979. In terms of ideology, this objective would seem unattainable; an economic "open-door" policy would surely lead to rising pressures for political democracy.

But Deng thought otherwise. If China's economic experiments could be spatially separated from the bulk of the country, the political impact would be kept at bay. At the outset, the new economic policies would apply mainly to China's bridgehead on the Pacific Rim, leaving most of the vast country comparatively unaffected. Accordingly, the government introduced a complicated system of **Special Economic Zones (SEZs)**, so-called *Open* 11
Cities, and *Open Coastal Areas*, which would attract technologies and investments from abroad and transform the economic geography of eastern China (Fig. 9-17).

FIGURE 9-17 © H.J. de Blij, P.O. Muller, and John Wiley & Sons, Inc.

In these economic zones, investors are offered many incentives. Taxes are low. Import and export regulations are eased. Land leases are simplified. The hiring of labor under contract is allowed. Products made in the economic zones may be sold on foreign markets and, under some restrictions, in China as well. Even Taiwanese enterprises may operate here. And profits made may be sent back to the investors' home countries.

When Deng's government made the decisions that would reorient China's economic geography, location was a prime consideration. Beijing wanted China to participate in the global market economy, but it also wanted to cause as little impact on interior China as possible—at least in the first stages. The obvious answer was to position the Special Economic Zones along the coast. Initially, the government established four SEZs, all with particular locational properties (Fig. 9-17):

1. *Shenzhen*, adjacent to then-booming British Hong Kong on the Pearl River Estuary in Guangdong Province
2. *Zhuhai*, across from then still-Portuguese Macau, also on the Pearl River Estuary in Guangdong Province
3. *Shantou*, opposite southern Taiwan, a colonial treaty port, also in Guangdong Province, source of many Chinese now living in Thailand
4. *Xiamen*, on the Taiwan Strait, also a colonial treaty port (then known as Amoy in the local dialect), in Fujian Province, source of many Chinese now based in Singapore, Indonesia, and Malaysia

In 1988 and 1990, respectively, two additional SEZs were proclaimed:

5. *Hainan Island*, declared an SEZ in its entirety, its potential success linked to its location near Southeast Asia
6. *Pudong*, across the river from Shanghai, China's largest city, different from other SEZs because it was a giant state-financed project designed to attract large multinational companies

China also opened 14 other coastal cities for preferential treatment of foreign investors (Fig. 9-17). Again, most of these cities had been treaty ports in the colonial extraterritoriality period. They were chosen for their size, overseas trading history, links to emigrated Chinese, level of industrialization, and pool of local talent and labor.

When all these Pacific Rim initiatives were implemented, China's market-conscious communist leaders had reason to expect an economic boom in the coastal provinces. (On the map, note that the new economic policy affected every coastal province, as well as the single coastal A.R.) Japan already was a highly developed economy (at the final stage of the Rostow model). South Korea and Taiwan were well beyond takeoff. Hong Kong proved what could be done on this side of the Pacific. Singapore, Southeast Asia's most successful state, is 77 percent Chinese. In truth, China's pragmatic planners were unsure of just what forces they might unleash, which is why foreign investment rules were stricter in the Open Cities and Open Coastal Areas than in the free-wheeling SEZs.

CHINA'S PACIFIC RIM TODAY

When China's planners laid out the SEZs, they held the highest hopes for Shenzhen, immediately adjacent to the thriving British dependency of Hong Kong. They knew that the kinds of attractions the SEZs offered to foreign investors would entice many factory owners paying Hong Kong wages to move across the border to Shenzhen. Successful Chinese living in Southeast Asia could use Hong Kong as a base for investment in Shenzhen; Taiwanese, even during the political standoff between Beijing and Taipei, could channel investments into this favored SEZ.

Those hopes were not disappointed. In the late 1970s, Shenzhen was just a sleepy fishing and duck-farming village of about 20,000. By 2006, its population neared 4 million—the fastest growth of any urban center on Earth in human history. Thousands of factories, large and small, moved into the SEZ from Hong Kong. Workers by the hundreds of thousands came from Guangdong Province, Guangxi Zhuang A.R., and beyond to look for jobs. High-rise buildings housing banks, corporate offices, hotels, and other service facilities made Shenzhen look like a new Hong Kong (top photo p. 465).

Shenzhen's economic impact was far-reaching. Persuading Hong Kong industries to cross the border precipitated a sharp modification of Hong Kong's own economy; attracting investments from Overseas Chinese, mainly in Southeast Asia, opened financial coffers long closed to China; and enabling Taiwanese investors to participate in China's Pacific Rim boom created links with the "wayward province" that have economic as well as political benefit. By its very success (in its 1990s heyday, it produced 15 percent of all of China's exports by value), Shenzhen signaled the way for the other SEZs and Open Cities.

Guangdong Province and the Pearl River Estuary

As Figure 9-17 shows, eight provinces (including Hainan Island) and one Autonomous Region face the Pacific and form the basis for recognizing an emerging Pacific Rim

Just 30 years ago, Shenzhen was a town of 20,000 with pig pens and fish ponds across the ironclad border with Hong Kong. Today it is an SEZ city of almost 4 million with numerous links to the adjacent SAR—the fastest-growing urban agglomeration in human history. And Shenzhen is only one growth pole on China's burgeoning Pacific Rim. To be sure, the skyscrapers seen here include hotels, offices, and banks—only a fortunate minority among the city's workers can afford to live in apartments such as those in the left foreground. But this scene symbolizes China's headlong economic rush.
© Gareth Brown/Corbis Images.

region in eastern China. But among these provinces, one stands out by making the largest contribution by far to the national economy: Guangdong Province. Although Hong Kong and Macau, both SARs, are not administratively a part of Guangdong Province, they are functionally crucial parts of the vast economic complex that fringes the estuary of the Pearl River (Fig. 9-18). Indeed, without Hong Kong, there would not have been a Shenzhen of the kind described above.

The economic core of Guangdong Province, the *Pearl River Delta* (*PRD*), consists of Guangzhou, the province's capital; the Hong Kong SAR; the Shenzen SEZ; the Zhuhai SEZ; the Macau SAR; and the latest juggernaut, fast-growing Zhongshan (Fig. 9-18). In a few years the PRD will have three cities with more than 5 million inhabitants, five with more than 1 million, and six with more than a half-million. Per-capita incomes here are estimated to be ten times the national average.

But this Pearl River Delta hub suffers from growing problems of inadequate infrastructure, worsening pollution, power outages, labor shortages (yes, labor shortages in China!), and resulting increases in labor costs. This is making the PRD less attractive to investors, who are looking at the Shanghai-centered Yangzi River Delta as an alternative. Paradoxically, this may stimulate a process China needs: the expansion of its coastal economic "miracle" into the

The Pearl River Delta has become the hub of China's Pacific Rim, a vast complex of dynamic, coalescing cities producing a huge range of exports. It began with Shenzhen, but today Shenzhen is only one of these expanding urban centers; other cities and towns are booming throughout this evolving subregion. Fly over the area, and you see expanses of factories and apartment buildings stretching to the horizon, linked by roads and bridges leading to well-equipped ports. Commercial fleets deliver enormous quantities of raw materials; container ships carry the finished goods to overseas markets. This view over Huizhou, located to the northeast of Shenzhen, is typical: a rapidly transforming landscape, factories the size of several football fields, basic apartments, a polluted canal, a new bridge. Millions of workers flooded to the area to fill the low-paying jobs available in new factories of the kind seen here, their shifts operating around the clock. Huizhou's products range from the traditional ones, such as textiles and garments, to petrochemicals and electronics; today this city is the world's largest producer of laser diodes and a global leader in the manufacture of DVD and CVD players. © Invest Hong Kong.

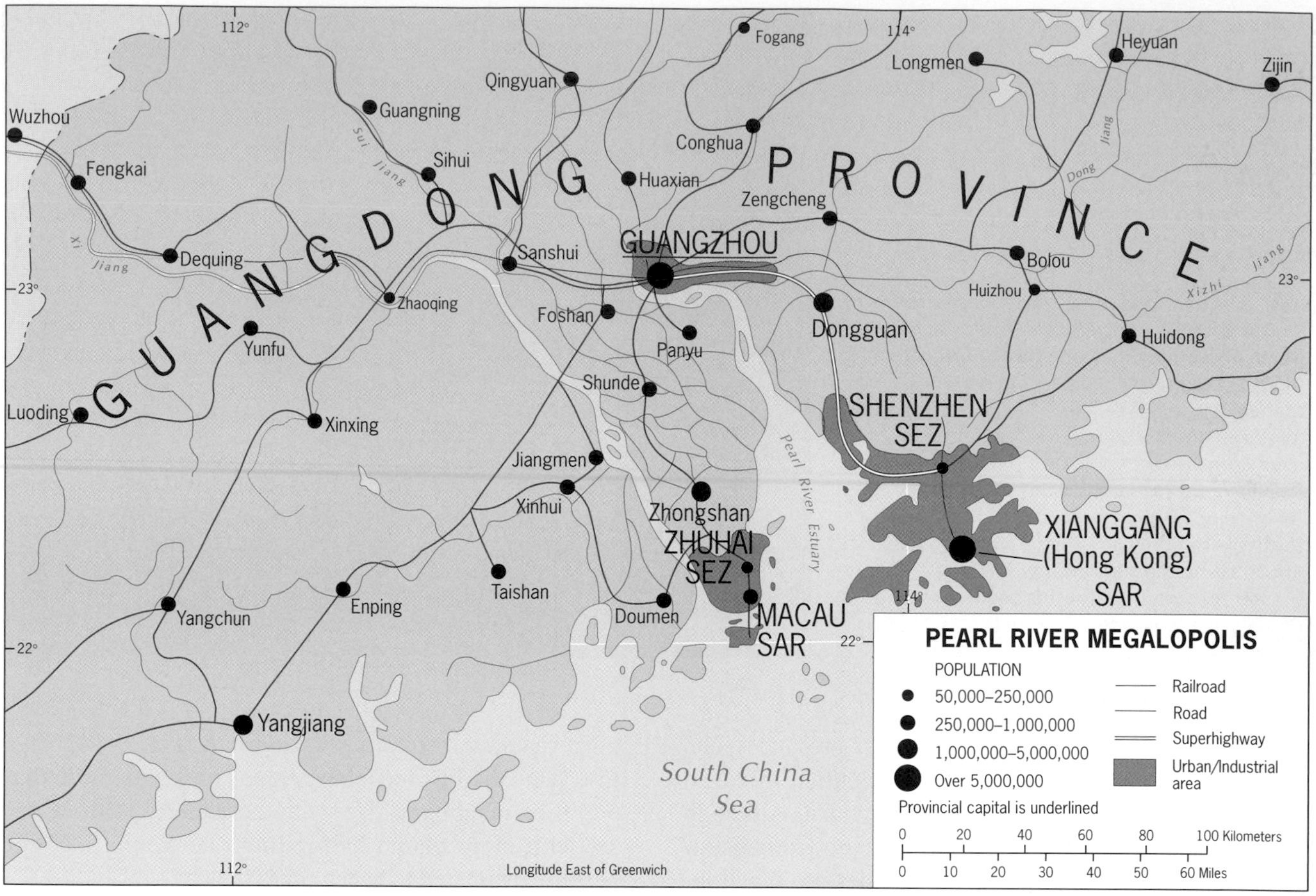

FIGURE 9-18 © H.J. de Blij, P.O. Muller, and John Wiley & Sons, Inc.

deep interior. Leaders in Guangdong Province propose the extension of the PRD's model of economic cooperation into nine provinces of China's South, all the way west to Yunnan and Sichuan. A network of road and railroad links among these provinces would allow raw materials and labor to circulate far more efficiently (there are no labor shortages in the interior) and move finished products not only to ports but also to the huge domestic market. A recent report in *The Economist* suggests that such an integrated region would encompass about 460 million people (one-third of the national population) on one-fifth of China's land producing nearly 40 percent of China's economic output including one-third of all its exports.

The PRD as an economic heartland will need Beijing's help in making its regional aspirations come true, and all signs indicate that such support will be forthcoming. Already, the government has begun to grant SEZ-like privileges to parts of Yunnan Province around the capital, Kunming. Similar plans also exist for areas in Sichuan: the rapid expansion of Shanghai and the Yangzi River Delta will feed into the burgeoning economic corridor that leads through Wuhan toward Chongqing, providing further impetus to the westward spread of Pacific Rim productivity.

The Pearl River Delta as well as the Yangzi River Delta fit the model of the **regional state** as defined by the 12
Japanese economist Kenichi Ohmae. These "natural economic zones," as he called them, defy old political borders and are shaped by the global economy of which they are parts; their representatives deal directly with foreign partners and negotiate the best terms they can with the national governments under which they operate (Europe, we noted, has several examples of this kind). Both the Pearl River Delta and the Yangzi River Delta have now moved so far ahead of China as a whole that they have become political as well as economic forces in the country.

Hong Kong

The process we have been discussing started with the economic success of Hong Kong while it was still a British dependency, empowered by a unique set of geographic circumstances. Here, on the left bank of the Pearl River's estuary, more than 7 million people (97 percent of Chinese descent) crowded onto 400 square miles (1000 sq km) of fragmented, hilly territory under the tropical sun and created an economy larger than that of a hundred countries.

Hong Kong (the name means "Fragrant Harbor") was ceded by the Chinese to Britain during the colonial period in three parts: the islands in 1841, the Kowloon Peninsula in 1860, and the New Territories (a lease) in 1898 (Fig. 9-19). Until the 1950s, Hong Kong was just another trading colony, but then the Korean War and the United Nations embargo on communist China cut Hong Kong off from its hinterland. With a huge labor force and no domestic resources, the colony needed an alternative—and it came in the form of a textile industry based on imported raw materials. Soon Hong Kong textiles were selling on worldwide markets; capital was invested in light industries; and the industrial base diversified into electrical equipment, appliances, and countless other consumer goods. Meanwhile, Hong Kong quietly served as a back door to China, which rewarded the colony by providing fresh water and staple foods. The
13 colony had become an **economic tiger**.

Hong Kong also became one of the world's leading financial centers. When China changed course and embarked on its "Open Door" policy in the 1980s, Hong Kong was ready. Hong Kong factory owners moved their plants to Shenzhen, where wages were even lower; Hong Kong banks financed Shenzhen industries; and Hong Kong's economy adjusted once again to a new era. Hong Kong's port, however, continued to transfer Shenzhen's products. During the 1990s, about 25 percent of all of China's trade passed through Hong Kong harbor.

When the British lease on Hong Kong expired in 1997 and the British agreed to yield authority to China, the Chinese promised to allow Hong Kong's freewheeling economy to continue, in effect creating "one country, two systems." At midnight on June 30, Hong Kong became the Xianggang Special Administrative Region (SAR). True to the letter (if not always the spirit) of this commitment, the name "Hong Kong" has continued in general use. And on December 20, 1999 China took control over the last remaining European colony—Portuguese Macau—and made it a SAR as well, guaranteeing (as in Hong Kong) the continuation of its existing social and economic system for 50 years.

FIGURE 9-19 © H.J. de Blij, P.O. Muller, and John Wiley & Sons, Inc.

Pedestrians on a downtown Hong Kong street in May 2003 wear surgical masks to protect against SARS (severe acute respiratory syndrome). The SARS virus originated in adjacent southern China and had first been detected in Hong Kong during mid-February. Within days SARS reached pandemic proportions as international air travelers unknowingly dispersed the disease across much of the world (Toronto, Canada was particularly hard hit). During the 100 days that Hong Kong grappled with its local outbreak, 1755 confirmed cases of SARS were recorded here, resulting in 298 deaths. Health authorities responded by imposing stringent quarantine measures to isolate Hong Kong's disease victims. These soon proved to be effective, and on June 23 the UN's World Health Organization (WHO) sounded the all-clear. © AP/Wide World Photos.

The Other SEZs

Not only did China grant economic advantages to corporations doing business in the coastal SEZs; China made enormous infrastructure investments to enhance their attractiveness and efficiency. In Shanghai, a massive development project transformed the right bank of the Huangpu River (across from the old downtown) into an ultramodern complex whose skyline rivals that of Hong Kong (photo p. 469). *Pudong*, as this area is called, was designed to offer competition to the great Pearl River hub as part of the Beijing government's effort to reduce the imbalance between north and south along China's Pacific Rim. And in this, the central government succeeded, perhaps beyond its expectations. Shanghai has truly become a world city, foreign investment in the Yangzi River Delta now exceeds that in the PRD, and in 2010 Shanghai will be host to World Expo, just two years after Beijing hosts the 29th Summer Olympics.

None of the other SEZs has come close to what has happened in Shenzhen or Pudong. Xiamen was established especially to attract Taiwanese investment; Shantou has benefited greatly from investment by Overseas Chinese in Southeast Asia (notably Thailand); Zhuhai is benefiting from the fact that its wages are even lower than Shenzhen's, causing the transfer of many industries from Shenzhen; and Hainan lags because of relative location and poor administration. The SEZs are no workers' paradise: thousands of factories producing goods ranging from toys to textiles create a pollution-choked landscape, working conditions often are dreadful, and wages are marginal. Unfortunately, China's new industrial revolution has some of the features of the old one.

CHINA: GLOBAL SUPERPOWER?

China in the early twenty-first century appears likely to become more than an economic force of world proportions: China also appears on course to achieve global superpower stature. As long as it retains its autocratic form of government (which allowed it to impose draconian population policies and comprehensive economic experiments without having to consult the electorate), China will be able to practice the kind of state capitalism that—in another guise—made South Korea an economic power. Unlike Japan or South Korea, however, China has no constraints on its military power. Its People's Liberation Army served as security forces during the pro-democracy turbulence of 1989; China's military is the largest standing army in the world, with about 3 million soldiers and some 1.2 million reserves. Its weaponry is being updated, and the government has made a major investment in its military by raising its armed forces budget significantly since 2000. Already, China is a nuclear power and has a growing arsenal of medium-range and intercontinental ballistic missiles.

During the twentieth century, the United States and the Soviet Union were locked in a 45-year Cold War that repeatedly risked nuclear conflict. That fatal exchange never happened, in part because it was a struggle between superpowers who understood each other comparatively well. While the politicians and military strategists were plotting, the cultural doors never closed: American audiences listened to Prokofiev and Shostakovich, watched Russian ballet, and read Tolstoy and Pasternak even as the Soviets cheered Van Cliburn, read Hemingway, and lion-

Across the Huangpu River from the colonial-era Bund, the Lujiazui Development Zone in the heart of Shanghai's spectacular Pudong District reflects China's new business and commercial era. The futuristic Oriental Pearl Television Tower (left) and the 88-story Mao Observation Tower, containing a hotel with the world's highest atrium (right), will soon be outranked by an even taller structure. There was a time when American cities' skyscrapers symbolized the New World's power and modernity; now it is Asia's turn to show the world that a new era is in the making.

ized American dissidents. In short, it was an intracultural Cold War, which reduced the threat of mutual destruction.

The twenty-first century may witness a far more dangerous geopolitical struggle in which the adversaries may well be the United States and China. U.S. power and influence still prevail in the western Pacific, but it is easy to discern areas where Chinese and American interests will diverge (Taiwan is only one example). American bases in Japan, thousands of American troops in South Korea, and American warships in the East and South China Seas are potential grounds for dispute. All this might generate the world's first *inter*cultural Cold War, in which the risk of fatal misunderstanding is incalculably greater than it was during the last.

How can such a Cold War be averted? Trade, scientific and educational links, and cultural exchanges are obvious remedies: the stronger our interconnections, the less likely is a deepening conflict. We Americans should learn as much about China as we can, to understand it better, to appreciate its cultural characteristics, and to recognize the historico-geographical factors underlying China's views of the West.

For more than 40 centuries, China has known authoritarianism of both the brutal and the benevolent kind, has been fractured by regionalism only to unify again, and has depended on communalism to survive environmental and despotic depredations. For nearly 25 centuries Confucianism has guided it. Time and again, China has been opened to, and ruled by, foreigners, and time and again it has retreated into isolationism when things went wrong. Such a reversal may no longer be possible, given what we have seen in this chapter. But now a renewed force is rising in modernizing China: nationalism. A spate of recent books published in China, including one by Song Qiang et al. titled *China Can Say No*, reflects the frustrations of many Chinese over what they view as American arrogance and insensitivity to China's traditions and interests. The "inadvertent" bombing of the Chinese embassy in Belgrade by American aircraft operating under NATO command in 1999, and the mid-air collision between what the Chinese regarded as a U.S. "spy plane" and one of its aircraft near the island of Hainan in 2001, are the kinds of incidents that contribute to this view and that energize Chinese nationalism. American scorn for China's human rights failings are another source of irritation.

China today is on the world stage, globally engaged, and internationally involved. Constraining the forces that tend to lead to world-power competition between China and the United States will be the geopolitical challenge of the twenty-first century.

MONGOLIA

Between China's Inner Mongolia and Xinjiang to the south and Russia's Eastern Frontier region to the north lies a vast, landlocked, isolated country called Mongolia that constitutes a discrete region of East Asia (Fig. 9-4). With only 2.6 million inhabitants in an area larger than Alaska, Mongolia is a steppe- and desert-dominated vacuum between two of the world's most powerful countries.

From the late 1600s until the revolutionary days of 1911, Mongolia was part of the Chinese Empire. With Soviet support the Mongols held off Chinese attempts to regain control, and in the 1920s the country became a People's Republic on the Soviet model. Free elections in 1990 ushered in a new political era, but Mongolia's economy, still largely based on animal products (chiefly cashmere wool), is troubled by mismanagement and environmental problems.

The map reflects Mongolia's difficulties. Despite its historic associations, ethnic affinities, and cultural involvement with China, Mongolia's incipient core area, including the capital of Ulaanbaatar, adjoins Russia's Eastern Frontier. The country's 800,000 herders and their millions of sheep follow nomadic tracks along the fringes of the vast Gobi Desert. When the cold of a Siberian winter descends on the steppes, as happened twice earlier in this decade, human and livestock losses are severe and the economy breaks down.

Over the past five years, China's involvement in Mongolia has intensified as Russia's has declined. Copper, iron, zinc, and coal reserves attract Chinese companies; road, bridge, and railroad building funded by China are in the works; and today more than 1000 Chinese businesses are active in Mongolia. Public opinion polls reflect concern on the part of Mongolians over China's role as the leading foreign investor and trade partner, but Mongolia has accepted what many regard as the inevitable. A 2004 *New York Times* article reported another side of the Chinese coin: a government official is quoted as stating that Mongolia had 4 million deer before 1990, but now very few are left "because Chinese, who like deer antlers, shot them one by one."

Mongolia's Soviet period has yielded to a time of Chinese involvement, and Mongolia's relatively high rate of economic growth is undoubtedly the result of its proximity to China's markets. But Mongolia is a vulner-
14 able **buffer state** wedged between former and future superpowers. Will it escape the fate of other Asian buffers?

THE JAKOTA TRIANGLE REGION

We turn now to a region that exemplifies the future of East Asia. Along the Pacific Rim, from Japan (through South Korea, Taiwan, China's Guangdong) to Singapore, rapid economic development has transformed community and society. In China, as we saw, a Pacific Rim region is in the making, still discontinuous today but likely to extend all along the coast in the future. In Japan, South Korea, and Taiwan we can observe that future—today. That is why we
15 recognize an East Asian region we call the **Jakota Triangle**, consisting of Japan, Korea (South Korea at present but probably a reunited Korea later), and Taiwan (Fig. 9-20). This is a region of great cities, huge consumption of raw materials from all over the world, voluminous exports, and global financial linkages. It also is a region of social problems, political uncertainties, and economic vulnerabilities.

JAPAN

When we assess China's prospects of becoming a superpower, we should remember what happened in Japan in the nineteenth century. In 1868, a group of reform-minded modernizers seized power from an old guard, and by the end of the century Japan was a military and economic force. From the factories in and around Tokyo and from urban-industrial complexes elsewhere poured forth a stream of weapons and equipment the Japanese used to embark on colonial expansion. By the mid-1930s, Japan lay at the center of an empire that included all of the Korean Peninsula, the whole of China's Northeast (which the Japanese called Manchukuo), the Ryukyu Islands, and Taiwan, as well as the southern half of Sakhalin Island (called Karafuto). Not even a disastrous earthquake, which destroyed much of Tokyo in 1923 and killed 143,000 people, could slow the Japanese drive.

World War II saw Japan expand its domain farther than the architects of the 1868 "modernization" could have anticipated. By early December 1941, Japan had conquered large parts of China Proper, all of French Indochina to the south, and most of the small islands in the western Pacific. Then, on December 7, 1941, Japanese-built aircraft carriers moved Tokyo's warplanes within striking range of Hawai'i, and the surprise attack on Pearl Harbor underscored Japan's confidence in its war machine. Soon the Japanese overran the Philippines, the (then) Netherlands East Indies, Thailand, and British Burma and Malaya, and drove a wide corridor through the heart of China to the border with Vietnam.

A few years later, Japan's expansionist era was over. Its armies had been driven from virtually all its possessions, and when American nuclear bombs devastated two Japanese cities in 1945, the country lay in ruins. But once again, Japan, aided this time by an enlightened U.S. postwar administration, surmounted disaster.

Japan's economic recovery and its rise to the status of world economic superpower was the success story of the second half of the twentieth century. Japan lost the war and its empire, but it scored many economic victories in a new global arena. Japan became an industrial giant, a technological pacesetter, a fully urbanized society, a political power, and an affluent nation. No city in the world today is without Japanese cars in its streets; few photography stores lack Japanese cameras and film; laboratories the world over use Japanese optical equipment. From microwave ovens to VCRs, from oceangoing ships to camcorders, Japanese goods flood the world's markets. Even today,

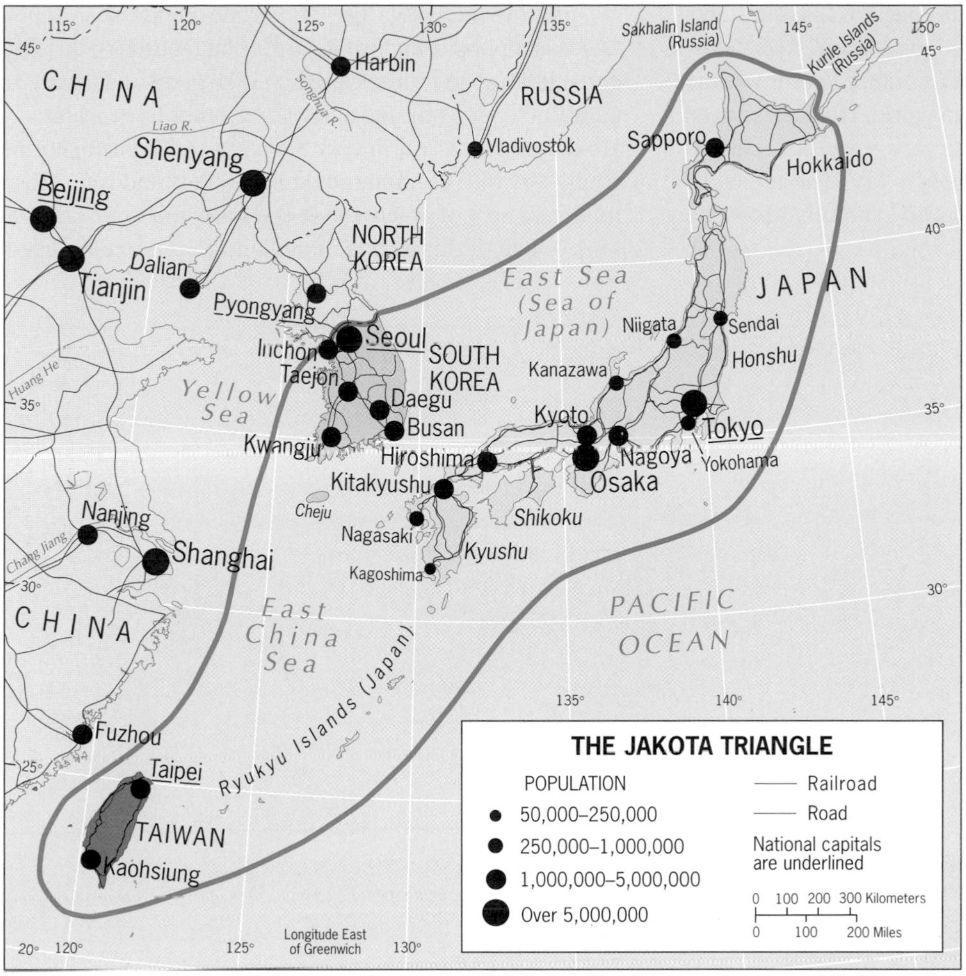

FIGURE 9-20 © H.J. de Blij, P.O. Muller, and John Wiley & Sons, Inc.

while experiencing a severe downturn, the Japanese economy remains the second largest national economy in the world. As we will see later in this chapter, Japan faces several challenges, economic, political, and demographic. But even while in recession, the Japanese economy remains formidable.

Japan's brief colonial adventure helped lay the groundwork for other economic successes along the western Pacific Rim. The Japanese ruthlessly exploited Korean and Formosan (Taiwanese) natural and human resources, but they also installed a new economic order there. After World War II, this infrastructure facilitated an economic transition—and soon made both Taiwan and South Korea competitors on world markets.

Spatial Limitations

In discussing the economic geography of Europe, we noted how the comparatively small island of Britain became the crucible of the Industrial Revolution, which gave it a huge head start and advantage over the continent, across which the Industrial Revolution spread decades later. Britain was not large, but its insular location, its local raw materials, the skills of its engineers, its large labor force, and its social organization combined to endow it not only with industrial strength but also with a vast overseas empire.

After the modernizers took control of Japan in 1868—an event known as the *Meiji Restoration* (the return of "enlightened rule" centered on the Emperor Meiji)—they turned to Britain for guidance in reforming their nation and its economy. In the decades that followed, the British advised the Japanese on the layout of cities and the construction of a railroad network, on the location of industrial plants, and on the organization of education. The British influence is still visible in the Japanese cultural landscape today: the Japanese, like the British, drive on the left side of the road. Consider how this affects the effort to open the Japanese market to U.S. automobiles!

The Japanese reformers of the late nineteenth century undoubtedly saw many geographic similarities between Britain and Japan. At that time, most of what mattered in Japan was concentrated on the country's largest island, Honshu (literally, "mainland"). The ancient capital, Kyoto, lay in the interior, but the modernizers wanted a coastal, outward-looking headquarters. So they chose the town of

Edo, on a large bay where Honshu's eastern coastline turns sharply (Fig. 9-21). They renamed the place *Tokyo* ("eastern capital"), and little more than a century later it was the largest urban agglomeration in the world. Honshu's coasts were near mainland Asia, where raw materials and potential markets for Japanese products lay. The notion of a greater Japanese empire followed naturally from the British example.

But in other ways, the British and Japanese archipelagoes, at opposite ends of the Eurasian landmass, differed considerably. In total area, Japan is larger than Britain. In addition to Honshu, Japan has three other large islands—Hokkaido to the north and Shikoku and Kyushu to the south—as well as numerous small islands and islets, for a total land area of about 146,000 square miles (377,000 sq km). Much of this area is mountainous and steep-sloped,

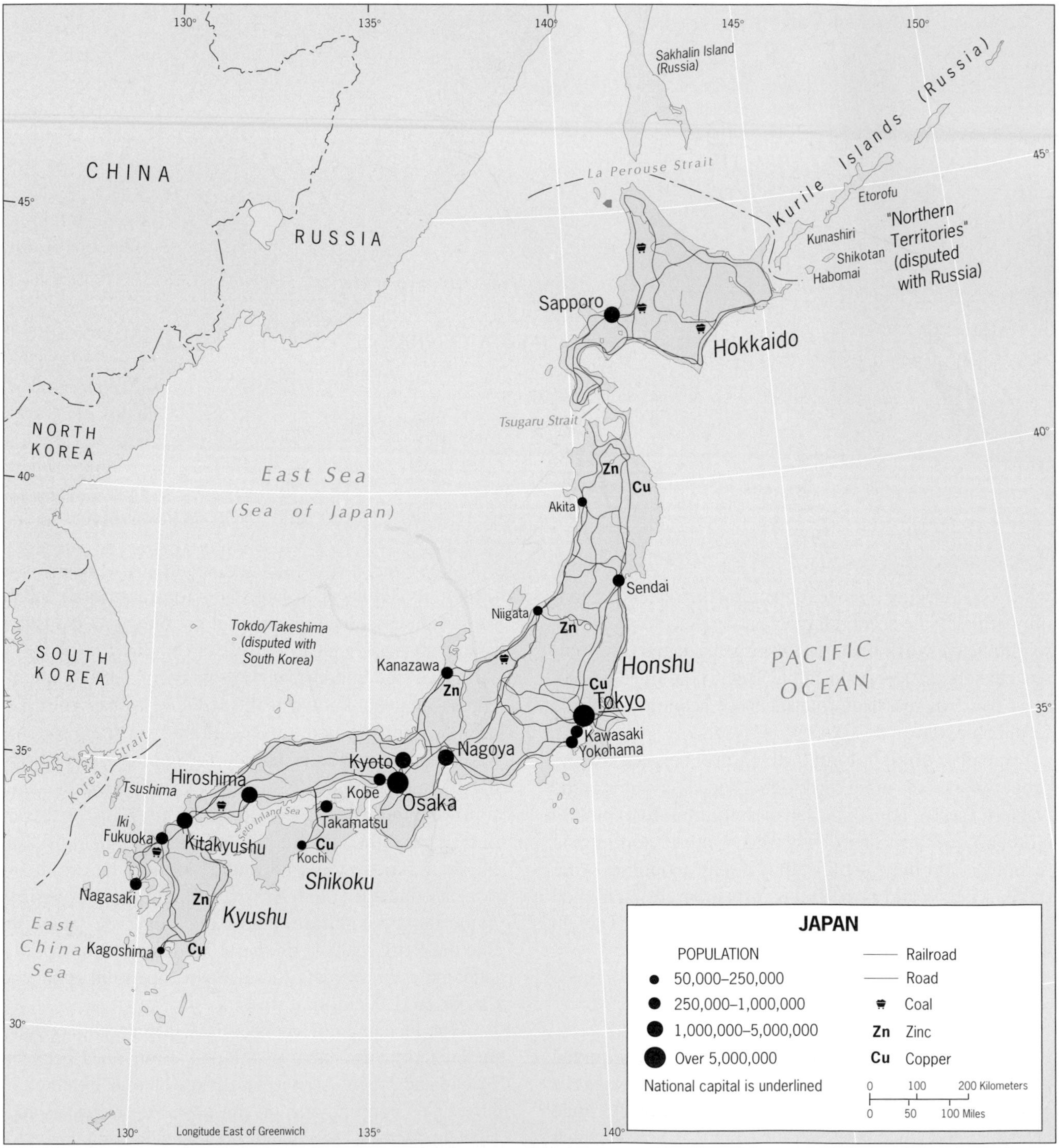

FIGURE 9-21 © H.J. de Blij, P.O. Muller, and John Wiley & Sons, Inc.

geologically young, earthquake-prone, and studded with volcanoes. Britain has lower relief, is older geologically, does not suffer from severe earthquakes, and has no active volcanoes. And in terms of the raw materials for industry, Britain was much better endowed than Japan. Self-sufficiency in iron ore and high-quality coal gave Britain the head start that lasted a century.

Japan's high-relief topography has been an ever-present challenge. All of Japan's major cities, except the ancient capital of Kyoto, are perched along the coast, and virtually all lie partly on artificial land claimed from the sea. Sail into Kobe harbor, and you will pass artificial islands designed for high-volume shipping and connected to the mainland by automatic space-age trains. Enter Tokyo Bay, and the refineries and factories to your east and west stand on huge expanses of landfill that have pushed the bay's shoreline outward. With 127.9 million people, the vast majority (78 percent) living in towns and cities, Japan uses its habitable living space intensively—and expands it wherever possible.

As Figure 9-22 shows, farmland in Japan is both limited and regionally fragmented. Urban sprawl has invaded much cultivable land. In the hinterland of Tokyo lies the Kanto Plain; around Nagoya, the Nobi Plain; and surrounding Osaka, the Kansai District—each a major farming zone under relentless urban pressure. All three of these plains lie within Japan's fragmented but well-defined core area (delimited by the red line on the map), the heart of Japan's prodigious manufacturing complex.

From Realm to Region

For nearly a half-century, Japan's seemingly insurmountable economic lead, its almost unique (for so large a population) ethnic homogeneity, its insular situation, and its cultural uniformity justified its designation as a discrete geographic realm, not a region of a greater East Asia. But rapid change on the rim of the western Pacific—both within Japan and beyond it—has eroded the bases for this distinction. In the mid-1990s, Japan's economic engine faltered while those of its Pacific Rim neighbors accelerated; its social fabric was torn by unprecedented acts of violence, and its foreign relations were soured by disputes over wartime atrocities in China and Korea. Japan stood alone in the vanguard of Pacific Rim developments, but it is no longer discrete.

Early Directions

On the map, the Korean Peninsula seems to extend like an unfinished bridge from the Asian mainland toward southern Japan. Two Japanese-owned islands, Tsushima and Iki, in the Korea Strait almost complete the connection (Fig. 9-21). In fact, Korea was linked to Japan in prehistoric times, when sea levels were lower. Peoples and cultural infusions reached Japan from Asia time and again, even after rising waters inundated the land bridge between them.

The Japanese islands were inhabited even before the ancestors of the modern Japanese arrived from Asia. As we noted earlier, the Ainu, a people of Caucasian ancestry, had established themselves on all four of the major islands thousands of years before. But the Ainu could not hold their ground against the better organized and armed invaders, and they were steadily driven northward until they retained a mere foothold on Hokkaido. Today, the last vestiges of Ainu ancestry and culture are disappearing.

By the sixteenth century, Japan had evolved a highly individualistic culture, characterized by social practices and personal motivations based on the Shinto belief system. People worshiped many gods, venerated ancestors, and glorified their emperor as a divine figure, infallible and omnipotent. The emperor's military ruler, or *shogun*, led the campaign against the Ainu and governed the nation. The capital, Kyoto, became an assemblage of hundreds of magnificent temples, shrines, and gardens (photo p. 475). In their local architecture, modes of dress, festivals, theater, music, and other cultural expressions, the Japanese forged a unique society. Handicraft industries, based on local raw materials, abounded.

Nonetheless, in the mid-nineteenth century, Japan hardly seemed destined to become Asia's leading power. For 300 years the country had been closed to outside influences, and Japanese society was stagnant and tradition-bound. When the European colonizers first appeared on Asia's shores, Japan tolerated, even welcomed, their merchants and missionaries. But as the European presence in East Asia grew stronger, the Japanese began to shut their doors. Toward the end of the sixteenth century, the emperor decreed that all foreign traders and missionaries should be expelled. Christianity (especially Roman Catholicism) had gained a considerable foothold in Japan, but now this was feared as being a prelude to colonial conquest. Early in the seventeenth century, the shogun launched a massive and bloody campaign that practically stamped out Christianity.

Determined not to suffer the same fate as the Philippines, which by then had fallen to Spain, the Japanese henceforth allowed only minimal contact with Europeans. A few Dutch traders, confined to a little island near the city of Nagasaki, were for many decades the sole representatives of Europe in Japan. Thus Japan entered a long period of isolation, lasting past the middle of the nineteenth century.

Japan could maintain its aloofness, while other areas were being enveloped by the colonial tide, because of its strong central government and well-organized military, the

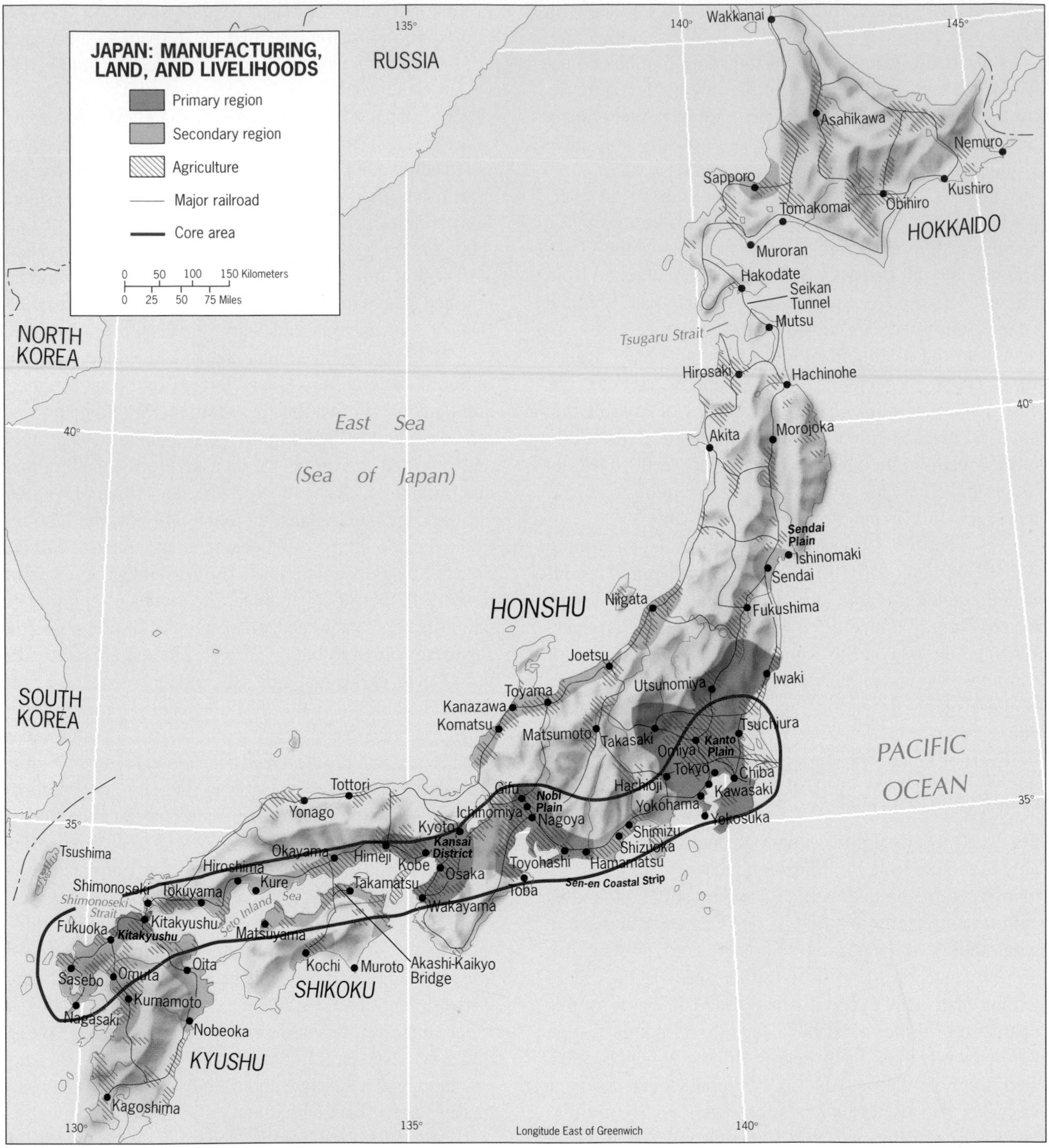

FIGURE 9-22 © H.J. de Blij, P.O. Muller, and John Wiley & Sons, Inc.

natural protection its islands provided, and its remoteness along East Asia's difficult northern coast. Also, Japan's isolation was far less splendid than that of China, whose exquisite silks, prized teas, and skillfully made wares attracted traders and usurpers alike.

When Japan finally came face to face with the new weaponry of its old adversaries, it had no answers. In the 1850s, the steel-hulled "black ships" of the American fleet sailed into Japanese harbors, and the Americans extracted one-sided trade agreements. Soon the British, French, and Dutch were also on the scene, seeking similar treaties. When there was local resistance to their show of strength, the Americans quickly demonstrated their superiority by shelling parts of the

FROM THE FIELD NOTES

"The city of Kyoto, chronologically Japan's second capital (after Nara; before Tokyo), is the country's principal center of culture and religion, education and the arts. Tree-lined streets lead past hundreds of Buddhist temples; tranquil gardens provide solace from the bustle of the city. I rode the bullet train from Tokyo and spent my first day following a walking route recommended by a colleague, but got only part of the way because I felt compelled to enter so many of the temple grounds and gardens. And not only Buddhism, but also Shinto makes its mark on the cultural landscape. I passed under a *torii*—a gateway usually formed by two wooden posts topped by two horizontal beams turned up at their ends—which signals that you have left the secular and entered the sacred, and found this beautiful Shinto shrine with its orange trim and olive-green glazed tiles." © H. J. de Blij.

Japanese coast. By the late 1860s, even as Japan's modernizers were about to overturn the old order, no doubt remained that Japan's protracted isolation had come to an end.

A Japanese Colonial Empire

When the architects of the Meiji Restoration confronted the challenge to build a Japan capable of competing against powerful adversaries in a changing world, they took stock of the country's assets and liabilities. Material assets, they found, were limited. To achieve industrialization, coal and iron ore were needed. In terms of coal, there was enough to support initial industrialization. Coalfields in Hokkaido and Kyushu were located near the coast; since the new industries also were on the coast, cheap water transportation was possible. As a result, the shores of the (Seto) Inland Sea (Fig. 9-22) became the sites of many factories. But there was little iron ore, certainly nothing like what was available domestically in Britain, and not enough to sustain the massive industrialization the reformers had in mind. This commodity would have to be purchased overseas and imported.

On the positive side, manufacturing—light manufacturing of the handicraft type—already was widespread in Japan. In cottage industries and in community workshops, the Japanese produced textiles, porcelain, wood products, and metal goods. The small ore deposits in the country were enough to supply these local industries. Power came from human arms and legs and from wheels driven by water; the chief source of fuel was charcoal. Importantly, Japan did have an industrial tradition and an experienced labor force that possessed appropriate manufacturing skills. All this was not enough to lead directly to industrial modernization, but it did hold promise for capital formation. The community and home workshops were integrated into larger units, hydroelectric plants were built, and some thermal (coal-fired) power stations were constructed in critical areas.

For the first time, Japanese goods (albeit of the light manufactured variety) began to compete with Western products on international markets. The Japanese planners resisted any infusion of Western capital, however. Although Western support would have accelerated industrialization, it would have cost Japan its economic autonomy. Instead, farmers were more heavily taxed, and the money thus earned was poured into the industrialization effort.

Another advantage for Japan's planners lay in the country's military tradition. Although the shogun's forces had been unable to repel the invasions of the 1850s, this was the result of outdated equipment, not lack of manpower or discipline. So while its economic transformation gathered momentum, the military forces, too, were modernized. Barely more than a decade after the Meiji Restoration, Japan laid claim to its first Pacific prize: the Ryukyu Islands (1879). A Japanese colonial empire was in the making.

When Japan gained control over its first major East Asian colonies, Taiwan and Korea, its domestic raw-material problems were essentially solved, and a huge labor pool fell under its sway as well. High-grade coal, high-quality iron ore, and other resources were shipped to the factories of Japan; and in Taiwan as well as in Korea, the Japanese built additional manufacturing plants to augment production. This, in turn, provided the equipment to sustain the subsequent drive into China and Southeast Asia.

Modernization

The reformers who set Japan on a new course in 1868 probably did not anticipate that three generations later their country would lie at the heart of a major empire sustained by massive military might. They set into motion a process
16 of **modernization**, but they managed to build on, not replace, Japanese cultural traditions. We in the Western world tend to equate modernization with Westernization: urbanization, the spread of transport and communications facilities, the establishment of a market (money) economy, the breakdown of local traditional communities, the proliferation of formal schooling, the acceptance and adoption of foreign innovations. In the non-Western world, the process often is viewed differently. There, "modernization" is seen as an outgrowth of colonialism, the perpetuation of a system of wealth accumulation introduced by foreigners driven by greed. In this view, the local elites who replaced the colonizers in the newly independent states only continue the disruption of traditional societies, not their true modernization. Traditional societies, they argue, can be modernized without being Westernized.

In this context, Japan's modernization is unique. Having long resisted foreign intrusion, the Japanese did not achieve the transformation of their society by importing a Trojan horse; it was done by Japanese planners, building on the existing Japanese infrastructure, to fulfill Japanese objectives. Certainly Japan imported foreign technologies and adopted innovations from the British and others, but the Japan that was built, a unique combination of modern and traditional elements, was basically an indigenous achievement.

Relative Location

Japan's changing fortunes over the past century reveal
the influence of **relative location** in the country's devel- 17
opment. When the Meiji Restoration took place, Britain, on the other side of the Eurasian landmass, lay at the center of a global empire. The colonization and Europeanization of the world were in full swing. The United States was still a developing country, and the Pacific Ocean was an avenue for European imperial competition. Japan, even while it was conquering and consolidating its first East Asian colonies (the Ryukyus, Taiwan, Korea), lay remote from the mainstream of global change.

Then Japan became embroiled in World War II and dealt severe blows to the European colonial armies in Asia. The Europeans never recovered: the French lost Indochina, and the Dutch were forced to abandon their East Indies (now Indonesia). When the war ended, Japan was defeated and devastated, but at the same time the Japanese had done much to diminish the European presence in the Pacific Basin. Moreover, the global situation had changed dramatically. The United States, Japan's

FROM THE FIELD NOTES

"Visiting the Peace Memorial Park in Hiroshima is a difficult experience. Over this site on August 6, 1945 began the era of nuclear weapons use, and the horror arising from that moment in history, displayed searingly in the museum, is an object lesson in this time of nuclear proliferation and enhanced risk. In the museum is a model of the city immediately after the explosion (the red ball marks where the detonation occurred), showing the total annihilation of the entire area at an immediate loss of more than 80,000 people and the death from radiation of many more subsequently. In the park outside, the Atomic Bomb Memorial Dome, the only building to partially survive the blast, has become the symbol of Hiroshima's devastation and of the dread of nuclear war." © H. J. de Blij.

trans-Pacific neighbor, had become the world's most powerful and wealthiest country, whereas Britain and its global empire were fading. Suddenly Japan was no longer remote from the mainstream of global action: now the Pacific was becoming the avenue to the world's richest markets. Japan's relative location—its situation relative to the economic and political foci of the world—had changed. Therein lay much of the opportunity the Japanese seized after the postwar rebuilding of their country.

Japan's Spatial Organization

Imagine this. 128 million people crowded into a territory the size of Montana (population: 970,000), most of it mountainous, subject to frequent earthquakes and volcanism, with no domestic oilfields, little coal, few raw materials for industry, and not much level land for farming. If Japan today were an underdeveloped country in need of food relief and foreign aid, explanations would abound: overpopulation, inefficient farming, energy shortages.

True, only an estimated 18 percent of Japan's national territory is designated as habitable. And Japan's large population is crowded into some very big cities. Moreover, Japan's agriculture is not especially efficient. But Japan defeated the odds by calling on old Japanese virtues: organizational efficacy, massive productivity, dedication to quality, and adherence to common goals. Even before the Meiji Restoration, Japan was a tightly organized country of some 30 million citizens. Historians suggest that Edo, even before it became the capital and was renamed Tokyo, may have been the world's largest urban center. The Japanese were no strangers to urban life, they knew manufacturing, and they prized social and economic order.

Areal Functional Organization

All this proved invaluable to the modernizers when they set Japan on its new course. The new industrial growth of the country could be based on the urban and manufacturing development that was already taking place. As we noted, Japan does not possess major domestic raw-material sources, so no substantial internal reorganization was necessary. However, some cities were better sited and enjoyed better situations relative to those limited local resources and, more importantly, to external sources of raw materials than others. As Japan's regional organization took shape, a hierarchy of cities developed; Tokyo took and kept the lead, but other cities grew rapidly into industrial centers.

This process was governed by a geographic principle
18 that Allen Philbrick called **areal functional organization**, a set of five interrelated tenets that help explain the evolution of regional organization, not only in Japan but throughout the world. Human activity, Philbrick reasoned, has spatial focus. It is concentrated in some locale, whether a farm or factory or store. Every one of these establishments occupies a particular location; no two of them can occupy exactly the same spot on the Earth's surface (even in high-rises, there is a vertical form of absolute location). Nor can any human activity proceed in total isolation, so that interconnections develop among these various establishments. This system of interconnections grows more complex as human capacities and demands expand. Each system (for example, farmers sending crops to market and buying equipment at service centers) forms a unit of areal functional organization. In the Introduction, we referred to functional regions as systems of spatial organization; we can map Philbrick's units of areal functional organization as regions. These regions evolve because of what he called "creative imagination" as people apply their total cultural experience and their technological know-how when they organize and rearrange their living space. Finally, Philbrick suggests, we can recognize levels of development in areal functional organization, a ranking of places and regions based on the type, extent, and intensity of exchange.

In the broadest sense, we can divide regions of human organization into three categories: subsistence, transitional, and exchange. Japan's areal organization reflects the exchange category. Within each of these categories, we can also rank individual places on the basis of the number and kinds of activities they generate. Even in regions where subsistence activities dominate, some villages have more interconnections than others. The map of Japan (Fig. 9-22) showing its resources, urban settlements, and surface communications can tell us much about Japan's economy. It looks just like maps of other parts of the world where an exchange type of areal organization has developed—a hierarchy of urban centers ranging from the largest cities to the tiniest hamlets, a dense network of railways and roads connecting these places, and productive agricultural areas near and between the urban centers.

For Japan, however, the map shows us something else: Japan's external orientation, its dependence on foreign trade. All primary and secondary regions lie on the coast. Of all the cities in the million-size class, only Kyoto lies in the interior (Fig. 9-21). If we deduced that Kyoto does not match Tokyo–Yokohama, Osaka–Kobe, or Nagoya in terms of industrial development, that would be correct—the old capital remains a center of small-scale light manufacturing. Actually, Kyoto's ancient character has been deliberately preserved, and large-scale industries have been discouraged. With its old temples and shrines, its magnificent gardens, and its many workshop and cottage industries, Kyoto remains a link with Japan's premodern past.

Leading Economic Regions

As Figure 9-22 shows, Japan's dominant region of urbanization and industry (along with productive agriculture) is the *Kanto Plain*, which contains about one-third of the Japanese population and is focused on the Tokyo–Yokohama–Kawasaki metropolitan area. This gigantic cluster of cities and suburbs (the world's second largest urban agglomeration), interspersed with intensively cultivated farmlands, forms the eastern anchor of the country's elongated and fragmented core area. Besides its flatness, the Kanto Plain possesses other advantages: its fine natural harbor at Yokohama, its relatively mild and moist climate, and its central location with respect to the country as a whole. (The region's only disadvantage is its vulnerability to earthquakes—see box titled "When the Big One Strikes.") It has also benefited enormously from Tokyo's designation as the modern capital, which coincided with Japan's embarkation on its planned course of economic development. Many industries and businesses chose Tokyo as their headquarters in view of the advantages of proximity to the government's decision makers.

The Tokyo–Yokohama–Kawasaki conurbation has become Japan's leading manufacturing complex, producing more than 20 percent of its annual output. The raw materials for all this industry, however, come from far away. For example, the Tokyo area is among the chief steel producers in Japan, using iron ores from the Philippines, Malaysia, Australia, India, and even Africa; most of the coal is imported from Australia and North America, and the petroleum comes from Southwest Asia and Indonesia. The Kanto Plain cannot produce nearly enough food for its massive resident population. Imports must come from Canada, the United States, and Australia as well as other areas in Japan.

AMONG THE REALM'S GREAT CITIES . . . TOKYO

Many urban agglomerations are named after the city that lies at their heart, and so it is with the second largest of all: Tokyo (26.8 million). Even its longer name—Tokyo–Yokohama–Kawasaki—does not begin to describe the congregation of cities and towns that form this massive, crowded metropolis that encircles the head of Tokyo Bay and continues to grow, outward and upward.

Near the waterfront, some of Tokyo's neighborhoods are laid out in a grid pattern. But the urban area has sprawled over hills and valleys, and much of it is a maze of narrow, winding streets and alleys. Circulation is slow, and traffic jams are legendary. The train and subway systems, however, are models of efficiency—although during rush hours you must get used to the *shirioshi* pushing you into the cars to get the doors closed.

At the heart of Tokyo lies the Imperial Palace with its moats and private parks. Across the street, buildings retain a respectful low profile, but farther away Tokyo's skyscrapers seem to ignore the peril of earthquakes. Nearby lies one of the world's most famous avenues, the Ginza, lined by department stores and luxury shops. In the distance you can see an edifice that looks like the Eiffel Tower, only taller: this is the Tokyo Tower (see photo, p. 481), a multipurpose structure designed to test lighter Japanese steel, transmit television and radio signals, detect Earth tremors, monitor air pollution, and attract tourists.

Tokyo is the epitome of modernization, but Buddhist temples, Shinto shrines, historic bridges, and serene gardens still grace this burgeoning urban complex, a cultural landscape that reflects Japan's successful marriage of the modern and the traditional.

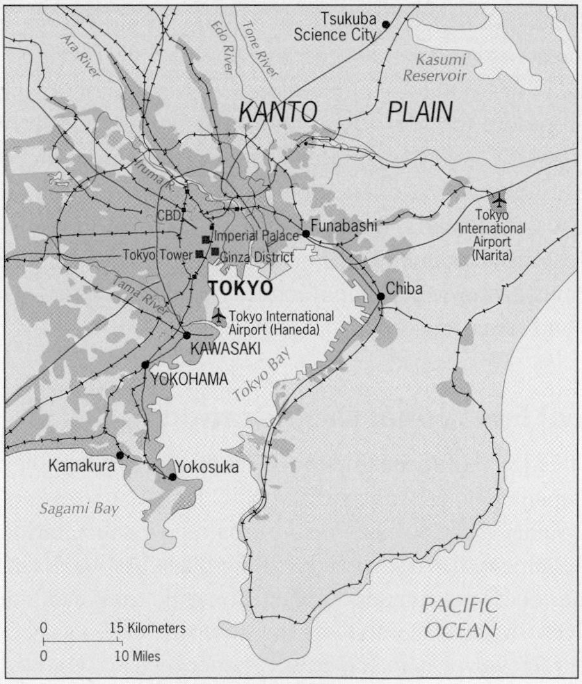

When the Big One Strikes

The Tokyo–Yokohama–Kawasaki urban area is the second biggest metropolis on Earth (only Mexico City is larger). But Tokyo is more than a large city: it constitutes the most densely concentrated financial and industrial complex in the world. Two-thirds of Japan's businesses worth more than U.S. $50 million are clustered here, many with vast overseas holdings. More than half of Japan's huge industrial profits (averaging about U.S. $100 billion annually between the late 1980s and the downturn of the late 1990s) are generated in the factories of this gigantic metropolitan agglomeration.

But Tokyo has a worrisome environmental history because three active tectonic plates are converging here (Fig. 9-23). All Japanese know about the "70-year" rule: over the past three-and-a-half centuries, the Tokyo area has been struck by major earthquakes roughly every 70 years—in 1633, 1703, 1782, 1853, and 1923. The Great Kanto Earthquake of 1923 set off a firestorm that swept over the city and killed an estimated 143,000 people. Tokyo Bay virtually emptied of water; then a *tsunami* (seismic sea wave) roared back in, sweeping everything before it. That Japan could overcome this disaster was evidence of the strength of its economy.

Today, Tokyo is more than a national capital. It is a global financial and manufacturing center in which so much of the world's wealth and productive capacity are concentrated that an earthquake comparable to the one of 1923 would have a calamitous effect worldwide. Ominously, Tokyo at the outset of our new century is a much more vulnerable place than the Tokyo of the 1920s. True, building regulations are stricter, and civilian preparedness is better. But whole expanses of industries have been built on landfill that will liquefy; the city is honeycombed by underground gas lines that will rupture and stoke countless fires; congestion in the area's maze of narrow streets will hamper rescue operations; and many older high-rise buildings do not have the structural integrity that has lately emboldened builders to erect skyscrapers of 50 stories and more. Add to this the burgeoning population of the Kanto Plain—which surpasses 40 million on what may well be the most dangerous 4 percent of Japan's territory—and we realize that the next big earthquake in the Tokyo area will not be a remote, local news story.

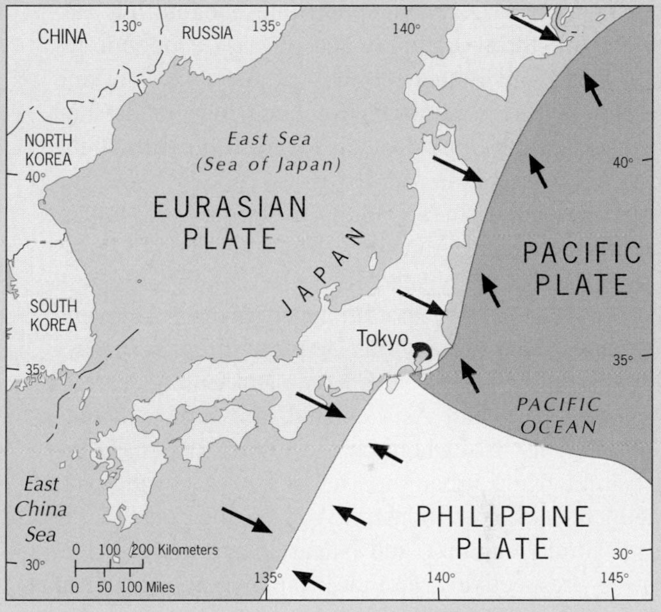

FIGURE 9-23 © H.J. de Blij, P.O. Muller, and John Wiley & Sons, Inc.

Thus Tokyo depends completely on its external trading ties for food, raw materials, and markets for its products, which run the gamut from children's toys to high-precision optical equipment to the world's largest oceangoing ships.

The second-ranking economic region in Japan's core area is the Osaka–Kobe–Kyoto triangle—also known as the *Kansai District*—located at the eastern end of the Inland Sea (Fig. 9-22). The Osaka–Kobe conurbation had an advantageous position with respect to Manchuria (Manchukuo) during the height of Japan's colonial empire. Situated at the head of the Inland Sea, Osaka was the major Japanese base for the China trade and for the exploitation of Manchuria, but it suffered when the empire was destroyed and it lost its trade connections with China after World War II. Kobe (like Yokohama, which is Japan's chief shipbuilding center) has remained one of the country's busiest ports, handling both Inland Sea traffic and extensive overseas linkages. Kyoto, as we noted, remains much as it was before Japan's great leap forward, a city of small workshop industries. The Kansai District is also an important farming area. Rice, of course, is the most intensively grown crop in the warm, moist lowlands, but this agricultural zone is smaller than the one on the Kanto Plain. Here is another huge concentration of people that must import food.

The Kanto Plain and Kansai District, as Figure 9-22 indicates, are the two leading primary regions within Japan's core. Between them lies the *Nobi Plain* (also called the Chubu District), focused on the industrial metropolis of Nagoya, Japan's leading textile producer. The map indicates some of the Nagoya area's advantages and liabilities. The Nobi Plain is larger than the lowlands of the Kansai District; thus its agricultural productivity is greater, though not as great as that of the Kanto Plain. But Nagoya has neither Tokyo's centrality nor Osaka's position on the Inland Sea: its connections to Tokyo are via the tenuous Sen-en Coastal Strip. Its westward connections are better, and the Nagoya area may be coalescing with the Osaka–Kobe conurbation. Still, the quality of Nagoya's port is not nearly as good as Tokyo's Yokohama or Osaka's Kobe, and it has had silting problems.

Westward from the three regions just discussed—which together constitute what is often called the Tokaido megalopolis—extends the Seto Inland Sea, along whose shores the remainder of Japan's core area is continuing to develop. The most impressive growth has occurred around the western entry to the Inland Sea (the Strait of Shimonoseki), where *Kitakyushu*—a conurbation of five cities on northern Kyushu—constitutes the fourth Japanese manufacturing complex and primary economic region. Road and railway tunnels connect Honshu and Kyushu, but the northern Kyushu area does not have an urban-industrial equivalent on the Honshu side of the strait. The Kitakyushu conurbation includes Yawata, site of northwest Kyushu's (rapidly declining) coal mines. The first steel plant in Japan was built there on the basis of this coal; for many years it was the country's largest. The advantages of transportation here at the western end of the Inland Sea are obvious: no place in Japan is better situated to do business with Korea and China. As relations with mainland Asia expand, this area will reap many benefits. Elsewhere on the Inland Sea coast, the Hiroshima–Kure urban area has a manufacturing base that includes heavy industry. And on the coast of the Korea Strait, Fukuoka and Nagasaki are the principal centers—respectively, an industrial city and a center of large shipyards.

Only one major Japanese manufacturing complex lies outside the belt extending from Tokyo in the east to Kitakyushu in the west—the secondary region centered on Toyama on the Sea of Japan (East Sea). The advantage here is cheap power from nearby hydroelectric stations, and the cluster of industries reflects it: paper manufacturing, chemical factories, and textile plants. Figure 9-22 also gives an inadequate picture of the variety and range of industries that exist throughout Japan, many of them oriented to local (and not insignificant) markets. Thousands of manufacturing plants operate in cities and towns other than those shown on this map, even on the cold northern island of Hokkaido, which is connected to Honshu by the Seikan rail tunnel, the world's longest, beneath the treacherous Tsugaru Strait.

The map of Japan's areal organization shows that four primary regions dominate its core area. Each of them is primary because each duplicates to some degree the contents of the others. Each contains iron and steel plants, is served by one major port, and lies in or near a large, productive farming area. What the map does not show is that each also has its own external connections for the overseas acquisition of raw materials and sale of finished products. These linkages may even be stronger than those among the four internal regions of the core area. Only for Kyushu and its coal have domestic raw materials affected the nature and location of heavy manufacturing, and these resources are almost depleted. In the structuring of the country's areal functional organization, therefore, more than just the contents of Japan itself is involved. In this respect Japan is not unique: all countries that have exchange-type organization must adjust their spatial forms and functions to the external interconnections required for progress. But for few countries is this truer than Japan.

East and West, North and South

The previous discussion shows that Japan has its own Pacific Rim: almost all of Honshu's economic expansion is occurring on the east coast. Japanese geographers often call this eastern zone *Omote Nippon* ("Front Japan"), the Pacific-facing, megalopolitan, high-tech, dynamic sector that embodies the modern economic superpower—as opposed to *Ura Nippon* ("Back Japan"), the comparatively underdeveloped periphery of unkempt farms and poor villages bordering the Sea of Japan (East Sea) to the west. In a country that needs food as much as Japan does, the derelict state of farmlands in *Ura Nippon* is of much concern to the country's planners.

The congestion on Honshu is another concern. Efforts to link Hokkaido and Inland-Sea-facing Shikoku to the "mainland" have produced world-record tunnels and bridges such as the Seikan Tunnel and the Akashi Kaikyo Bridge (Fig. 9-22), designed to bring remote areas into the national sphere and to diffuse Honshu's economic primacy beyond the main island. This effort has had limited success, especially in Hokkaido, which constitutes nearly one-quarter of Japan's total area, but contains only about 5 percent of the national population. Farming remains the main pursuit on Hokkaido, producing wheat, dairy products, and meat; fishing and tourism also do well. But Hokkaido has not translated its improved linkages into competitive industries; locals will tell you that there is a "subsidy mentality" here, that Tokyo's investments have led to complacency rather than initiative. Only the Sapporo area, with its automobile parts and electronics factories, shares in the Honshu-led export boom.

Food and Population

Japan's economic modernization so occupies center stage that we can easily forget the country's considerable achievements in agriculture. Japan's planners, who are as interested in closing the food gap as in expanding industries, have created extensive networks of experiment stations to promote mechanization, optimal seed selection and fertilizer use, and information services to distribute knowledge about enhancing crop yields to farmers as rapidly as possible. Although this program has succeeded, Japan faces the unalterable reality of its stubborn topography: it simply lacks sufficient land to farm. Populous Japan has one of the highest physiologic population densities—8101 per square mile (3128 per sq km)—on Earth. (We discuss this concept on pages 390 and 411.)

The Japanese go to extraordinary lengths to maximize their limited farming opportunities, making huge investments in research on higher-yielding rice varieties, mechanization, irrigation, terracing, fertilization, and other practices. More than 90 percent of farmland is assigned to food crops. Hilly terrain is given over to cash crops, such as tea and grapes (Japan has a thriving wine industry). Vegetable gardens ring all the cities.

But all this costs money, and Japan's food prices are high. Millions of Japanese have moved from the farms to the cities, depopulating the countryside which needs farmers. Japan's political system gives underpopulated rural areas disproportionate influence, and produce prices are kept artificially high, all to induce farmers to stay on the land. It has not worked. Between 1920 and 2000, full-time farmers declined from 50 percent to under 4 percent of the labor force. Yet the policies persist. In the early 1990s, American farmers proved that they could provide rice to Japanese consumers at one-sixth the price the Japanese were paying for home-grown rice. But the government resisted pressure to open its markets to imports.

With their national diet so rich in starchy foods like rice, wheat, barley, and potatoes, the Japanese need protein to balance it. Fortunately, they can secure enough of it, not by buying it abroad but by harvesting it from rich fishing grounds near the Japanese islands. With their customary thoroughness, the Japanese have developed a fishing industry that is larger than that of the United States or any of the long-time fishing nations of northwestern Europe, and it now supplies the domestic market with a second staple after rice. Although mention of the Japanese fishing industry brings to mind a fleet of ships scouring the oceans and seas far from Japan, most of this huge catch (about one-seventh of the world's annual total) comes from waters within a few dozen miles of Japan itself. Where the warm Kuroshio and Tsushima currents meet colder water off Japan's coasts, a rich fishing ground yields sardines, herring, tuna, and mackerel in the warmer waters and cod, halibut, and salmon in the seas to the north. Japan's coasts have about 4000 fishing villages, and tens of thousands of small boats ply the waters offshore to bring home catches that are distributed to local and city markets. The Japanese also practice *aquaculture*—the "farming" of freshwater fish in artificial ponds and flooded paddy (rice) fields, of seaweeds in aquariums, and of oysters, prawns, and shrimp in shallow bays. They are even experimenting with cultivating algae for their food potential.

When you walk the streets of Japan's cities, you will notice something that Japanese vital statistics are reporting: the Japanese are getting taller and heavier. Coupled with this change is evidence that heart disease and cancer are rising as causes of mortality. The reason seems to lie

Tokyo, at the center of one of the largest metropolises in the world, continues to change. Land-filling and bridge-building in the bay continue; skyscrapers sprout amid low-rise neighborhoods in this earthquake-prone area; traffic congestion worsens. The red-painted Tokyo Tower, a beacon in this part of the city, was modeled on the Eiffel Tower in Paris but, as a billboard at its base announces, is an improvement over the original: lighter steel, greater strength, less weight. Tokyo Bay, part of which can be seen from this vantage point, was the scene of one of history's most costly environmental disasters during the earthquake of 1923. A giant *tsunami* (seismic sea wave) swept up the bay from the epicenter even as landfills liquefied and buildings sank into the mud. The death toll exceeded 140,000; a much larger population than lived in Tokyo in 1923 is now at risk of a repeat.

in the changing diets of many Japanese, especially the younger people. When rice and fish were the staples for virtually everyone, and the calories available were limited but adequate, the Japanese as a nation were among the world's healthiest and longest-lived. But as fast-food establishments diffused throughout Japan, and Western (especially American) tastes for red meat and fried food spread among children and young adults, their combined impact on the population was soon evident. Coupled with the heavy cigarette smoking that prevails in all the Asian countries of the Pacific Rim, this is changing the region's medical geography.

Japan's Pacific Rim Prospects

As the twenty-first century opens, Japan remains the economic giant on the western Pacific Rim—but a giant with an uncertain future. Japanese products still dominate world markets and Japanese investments span the globe, but in the late 1990s Japan's economy faltered. Its growth rate declined. Many companies scaled back their operations, and some went bankrupt. The banking system proved to be vulnerable. The first Pacific Rim economic tiger was ailing.

Japan's problems stemmed from a combination of circumstances at home and abroad. Overseas, the collapse of Southeast Asian currencies and economies hurt Japanese investments there. From automobile plants to toy factories, long-profitable operations turned into liabilities. In Japan itself, financial mismanagement, long tolerable because of the booming economy, now took its toll. In addition, competition from still-healthy economies, notably from Taiwan but also from struggling South Korea, undercut Japanese goods on world markets.

In the long run, all this may prove to have been a temporary setback, but Japan is unlikely to regain its former economic dominance on the Pacific Rim. Japan's dependence on foreign oil is a potential weakness that could further depress its economy should international supply lines be disrupted. Western competition for Japan's markets is growing. The countless investments Japan made around the world during its time of plenty, from skyscrapers and movie studios to golf courses and hotels, have lost much of their value and must be sold at huge losses. While Japan reorganizes, American and European businesses buy its assets and take advantage of its problems.

Were you to make a field trip to Japan to study its economic geography, none of this might be immediately evident. The cities still bustle with activity, raw materials continue to arrive at the ports, container traffic still stirs the docks. Even in a downturn, this mighty economy outproduces all others on the western Pacific Rim. And, indeed, that observation is relevant to the future. Japan's infrastructure has the capacity to propel an economic rebound. But it needs new opportunities.

One of these opportunities lies in the Russian Far East and beyond, in Siberia. As we noted in Chapter 2, the Russian Far East lies directly across from northern Honshu and Hokkaido and is a storehouse of raw materials awaiting exploitation. For this, the Japanese are ideally located—but political relations with the former Soviet Union, and now with Russia, have stymied all initiatives (see box titled "The Wages of War"). Japan should be one of the major beneficiaries of any substantial oil reserves near Sakhalin (see map, p. 139), but again the necessary connections may not be possible. Japan also has a major potential opportunity in Korea, especially if a unification process takes hold there. But relations with Korea continue to be eroded by Korean memories of Japanese behavior during World War II and by Japan's refusal to acknowledge its misdeeds.

All this must be seen against the backdrop of Japan's changing society. The population of 128 million is aging rapidly, and population geographers expect it to rise only slightly before the end of this decade, when it will begin to decline. This decline will increase quite rapidly, to about 100 million in 2050 and only 67 million in 2100. Already, for some time, Japan has faced a labor shortage (although the recent economic downturn produced rising unemployment, a novelty in the postwar period). Over the long term, Japan will need millions of immigrants to keep its economy going, but for culturally homogeneous, ethnically conscious Nippon, which has historically resisted immigration, this will be a wrenching reality. Still, Japanese leaders will have no choice. The country's "graying" will strain welfare systems and tax burdens.

Modernization, meanwhile, is further stressing Japan's society. The Japanese have a strong tradition of family cohesion and veneration of their oldest relatives, but this tradition is breaking down. Younger people are less willing than their parents and grandparents to accept overcrowded housing, limited comforts, and little privacy. In Japan, for all its material wealth, family homes tend to be small, cramped, often flimsily built, and sometimes lacking the basic amenities other high-income countries consider indispensable. With their disposable income, the Japanese during the boom economy came to rank among the world's most-traveled tourists, and, having seen how Americans and Europeans of similar income levels live, they returned home dissatisfied.

As we just noted, Japan remains one of the most uniform, culturally homogeneous societies in today's interactive world. The Chinese closed their doors to foreigners intermittently; the Japanese have done so far more effectively. Other than the dwindling Ainu, there are no indigenous minorities. Koreans were brought to Japan

The Wages of War

Japan and the former Soviet Union never signed a peace treaty to end their World War II conflict. Why? There are four reasons, and all of them are on the map just to the northeast of Japan's northernmost large island, Hokkaido (Fig. 9-21). Their names are Habomai, Shikotan, Kunashiri, and Etorofu. The Japanese call these rocky specks of the Kurile Island chain their "Northern Territories." The Soviets occupied them late in the war and never gave them back to Japan. Now they are part of Russia, and the Russians have not given them back either.

The islands themselves are no great prize. During World War II, the Japanese brought 40,000 forced laborers, most of them Koreans, to mine the minerals there. When the Red Army overran them in 1945, the Japanese were ordered out, and most of the Koreans fled. Today the population of about 50,000 is mostly Russian, many of them members of the military based on the islands and their families. At their closest point the islands are only 3 miles (5 km) from Japanese soil, a constant and visible reminder of Japan's defeat and loss of land. Moreover, territorial waters bring Russia even closer, so that the islands' geostrategic importance far exceeds their economic potential.

Attempts to settle the issue have failed. In 1956, Moscow offered to return the tiniest two, Shikotan and Habomai, but the Japanese declined, demanding all four islands back. In 1989, then-Soviet President Mikhail Gorbachev visited Tokyo in the hope of securing an agreement. The Japanese, it was widely reported, offered an aid-and-development package worth U.S. $26 billion to develop Russia's eastern zone—its Pacific Rim and the vast resources of the eastern Siberian interior. This would have begun the transformation of Russia's Far East, stimulated the ports of Nakhodka and Vladivostok, and made Russia a participant in the spectacular growth of the western Pacific Rim.

But it was not to be. Subsequently, Russian presidents Yeltsin and Putin also were unable to come to terms with Japan on this issue, facing opposition from the islands' inhabitants and from their own governments in Moscow. And so World War II, more than 60 years after its conclusion, continues to cast a shadow over this northernmost segment of the western Pacific Rim.

during World War II to serve as forced laborers, but otherwise Japanese residency and citizenship imply an almost total ethnic sameness. During periods of labor shortage, the notion to invite foreign workers would not be entertained. When the United States, after the Vietnam War, sought new homes for hundreds of thousands of Vietnamese refugees, Japan accepted just a few hundred after much negotiation. Such jealously guarded ethnic-cultural uniformity is unlikely to be viable in the globalizing world of the twenty-first century.

Japan's future in the geopolitical framework of the western Pacific is also uncertain. Some 60 years after the end of World War II, tens of thousands of American forces still are based on Japanese soil. China and South Korea accept this arrangement, which constrains Japan's rearming. But in Japan, where nationalism is rising, domestic military preparedness is a growing political issue. Over the past ten years, however, Japan's sense of security was shaken by developments in communist North Korea, which in 1998 fired a missile across Honshu into the Pacific Ocean and was suspected of developing nuclear weapons capacity; by admissions from North Korea that it had kidnapped Japanese citizens from streets and beaches and forced them to teach Japanese to North Korean spies; and by the appearance of armed North Korean vessels in Japanese territorial waters. In 2003, the Japanese parliament approved legislation providing for the participation of Japanese troops in the U.S.-led intervention in Iraq, a move that contradicted public opinion as reflected by surveys but strengthened American-Japanese government ties. Japan also joined the United States and three other countries (China, Russia, and South Korea) to negotiate an end to North Korea's nuclear threat.

Japan's economic success and political stability may have been facilitated by enlightened U.S. postwar policy, but the future will be different. In 2002, China overtook the United States as the leading exporter to the Japanese market; today, Japanese companies compete vigorously against American corporations on China's Pacific Rim. Japan continues to have image problems among Chinese and Koreans (a Japanese soccer victory in Beijing in 2004 led to a near-riot), but at the highest levels cooperation is growing. On the other hand, Japan and China will become fierce competitors over critical energy supplies. In 2005, the Japanese were trying to outbid the Chinese over the routing of oil and natural-gas pipelines from Russia: the Chinese want to position these pipelines to link to terminals in its Northeast, whereas Japan wants to route them to terminals on the Sea of Japan (East Sea) at Nakhodka, directly across from northern Honshu. Japan today is the world's second-largest oil consumer, but China's consumption is rising rapidly and heightened competition for Russian exports is inevitable. This will obviously have an effect on Japanese-Chinese relations, political as well as economic, in the decades ahead.

Japan, therefore, is a pivotal nation on the western Pacific Rim, an economic superpower that stands at social and political crossroads whose future will profoundly influence not only East Asia and the Pacific, but the world as a whole.

KOREA

On the Asian mainland, directly across the Sea of Japan (East Sea), lies the peninsula of Korea (Fig. 9-24), a territory about the size of the State of Idaho, much of it mountainous and rugged, and containing a population of 72 million. Unlike Japan, however, Korea has long been a divided country, and the Koreans a divided nation. For uncounted centuries Korea has been a pawn in

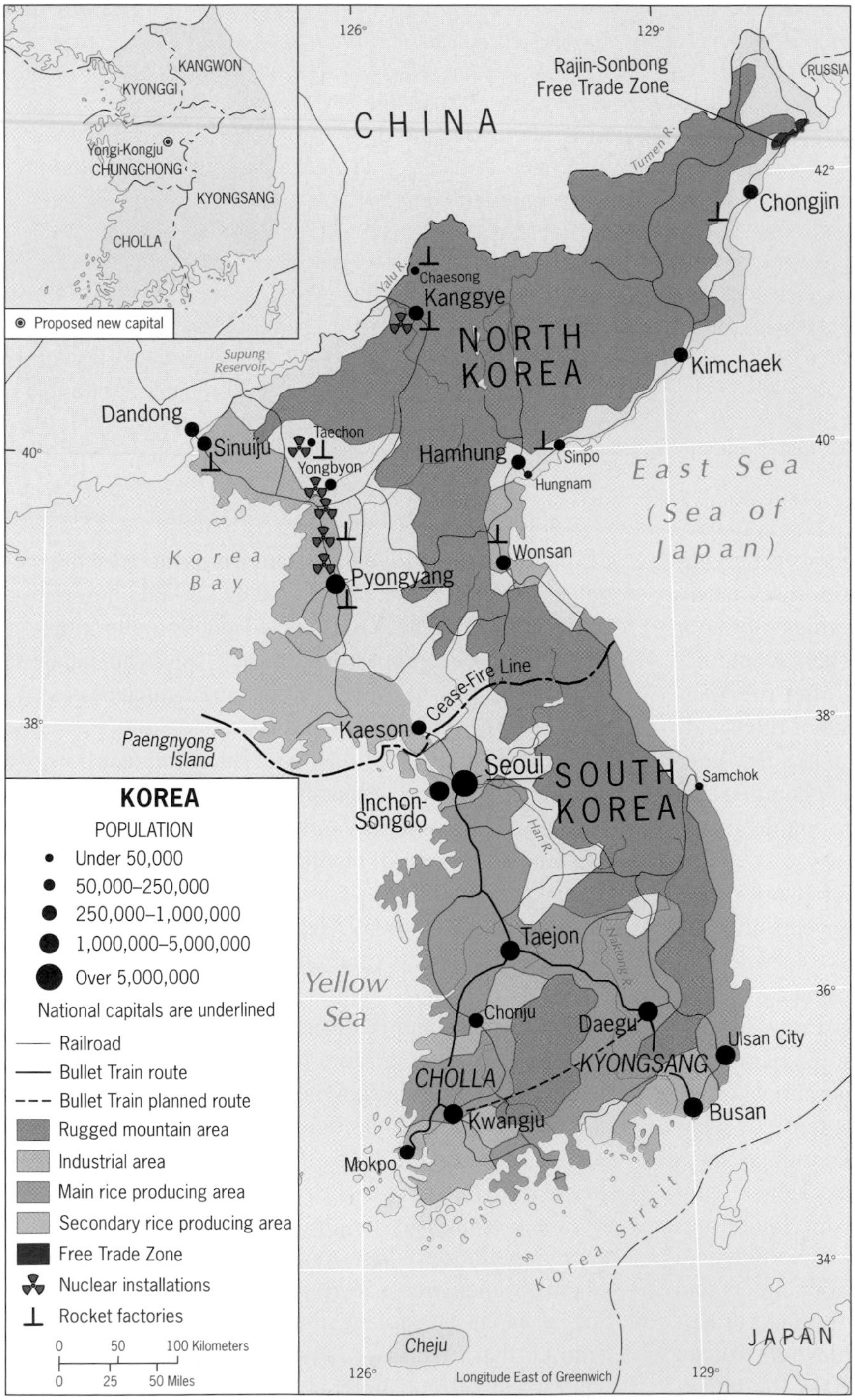

FIGURE 9-24 © H.J. de Blij, P.O. Muller, and John Wiley & Sons, Inc.

AMONG THE REALM'S GREAT CITIES . . . SEOUL

Seoul (9.6 million), located on the Han River, is ideally situated to be the capital of all of Korea, North and South (its name means capital in the Korean language). Indeed, it served as such from the late fourteenth century until the early twentieth, but events in that century changed its role. Today, the city lies in the northwest corner of South Korea, for which it serves as capital; not far to the north lies the tense demilitarized zone (DMZ) that contains the cease-fire line with North Korea. That line cuts across the mouth of the Han, depriving Seoul of its river traffic. Its ocean port, Inchon, has emerged as a result.

Seoul's undisciplined growth, attended by a series of recent accidents including the failure of a major bridge over the Han and the collapse of a six-story department store, reflects the unbridled expansion of the South Korean economy as a whole, as well as the political struggles that carried the country from autocracy to democracy. Central Seoul lies in a basin surrounded by hills to an elevation of about 1000 feet (330 m), and the city has sprawled outward in all directions, even toward the DMZ (see photo p. 3). An urban plan designed in the early 1960s was overwhelmed by the immigrant flow.

During the period of Japanese colonial control, Seoul's surface links to other parts of the Korean Peninsula were improved, and this infrastructure played a role in the city's later success. Seoul is not only the capital but also the leading industrial center of South Korea, exporting sizeable quantities of textiles, clothing, footwear, and (increasingly) electronic goods. South Korea is still an economic tiger on the Pacific Rim, and Seoul is its heart.

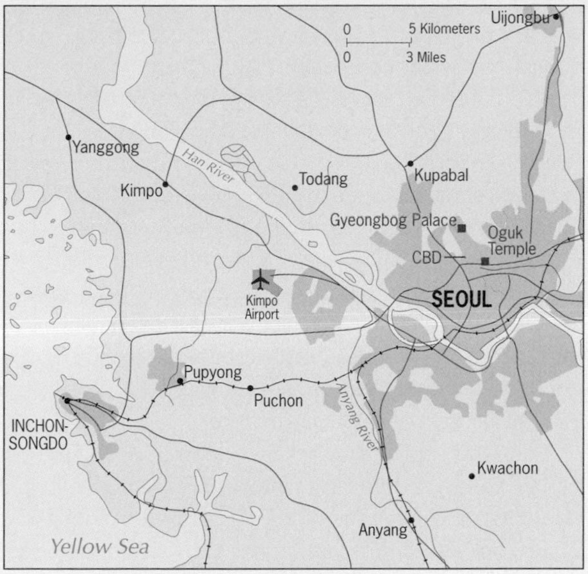

the struggles of more powerful neighbors. It has been a dependency of China and a colony of Japan. When it was freed from Japan's oppressive rule at the end of World War II (1945), the victorious Allied powers divided Korea for administrative purposes. That division gave North Korea (north of the 38th parallel) to the forces of the Soviet Union and South Korea to those of the United States. In effect, Korea traded one master for two new ones. The country was not reunited for the rest of the century because North Korea immediately fell under the communist ideological sphere and became a dictatorship in the familiar (but in this case extreme) pattern. South Korea, with massive American aid, became part of East Asia's capitalist perimeter. Once again, it was the will of external powers that prevailed over the desires of the Korean people.

In 1950, North Korea sought to reunite the country by force and invaded South Korea across the 38th parallel. This attack drew a United Nations military response led by the United States. Thus began the devastating Korean War (1950–1953) in which North Korea's forces pushed far to the south only to be driven back across its own half of Korea almost to the Chinese border. Then China's Red Army entered the war and drove the UN troops southward again. A cease-fire was arranged in 1953, but not before the people and the land had been ravaged in a way that was unprecedented even in Korea's violent past. The cease-fire line shown in Figure 9-24 became a heavily fortified *de facto* boundary. For five decades, until very recently, virtually no contact of any kind occurred across this border, which continues to divide families and economies alike.

As a result, the Jakota Triangle, as a region of East Asia, must include South Korea but exclude North Korea (see box titled "High-Risk North Korea"). Times may change, as they did in formerly divided Germany, but by the turn of this century South Korea, with 45 percent of Korea's land area but with two-thirds of the Korean population, had emerged as one of the economic tigers on the Pacific Rim. As an undivided country, Korea might have achieved even more because North and South Korea exhibit what geographers call **regional complementarity**. 19
This condition arises when two adjacent regions complement each other in economic-geographic terms. In this

High-Risk North Korea

The Korean Peninsula is physiographically well-defined, and its people—the Korean nation—have occupied it for thousands of years. But internal political division, foreign cultural intrusion, and colonial subjugation have created a turbulent history. Now Koreans are divided again, this time by ideology. Even as the capitalist South became one of the Pacific Rim's economic tigers, North Korea remained an isolated, stagnant state on a Stalinist-communist model abandoned long ago even in the former Soviet Union itself. For over a half-century, North Korea has been ruled by two men: Kim Il Sung, the "Great Leader," and his son Kim Jong Il, the "Dear Leader." Total allegiance to the dictators is ensured by an omnipotent police and military. Self-imposed isolation is North Korea's hallmark. The state controls all aspects of life in North Korea.

Economic policies under the communist regime have been disastrous. Per-capita incomes in the North (population: 23.1 million) are less than one-tenth of those in the South; poverty is rampant. Infant mortality is more that 5 times higher than in South Korea. Industrial equipment is 25 years out of date. Exports are negligible. When a combination of failed agricultural policies and environmental extremes created famine in the late 1990s and early 2000s, the regime resisted relief efforts, even from South Korea. As during the Maoist period in China, the world will never know the number of casualties.

North Korea did progress on one front: nuclear capability and associated weaponry. In 1999, it fired a test missile across northern Japan, heightening international concern. The North Koreans used their nuclear threat to negotiate a multibillion-dollar technical aid program to build reactors for peaceful purposes, with Japan and South Korea among the major contributors. But in late 2002, a North Korean source admitted that his country had indeed pursued a nuclear weapons program and was now in possession of "at least one" nuclear bomb. The North Koreans refused to allow visitors to inspect its nuclear facilities at Yongbyon, 60 miles (100 km) due north of the capital, Pyongyang. Other known nuclear installations are shown in Figure 9-24.

President George W. Bush, in his 2002 State of the Union address, called North Korea part of the "axis of evil" to be dealt with as part of the War on Terror. But while Iraq was forced to admit UN weapons inspectors, North Korea, with 37,000 U.S. troops along its border with the South, remained exempt. Meanwhile, North Korea acknowledged having kidnapped Japanese youngsters from public places in Japan and having forced them to teach Japanese language and culture to North Korean spies. In 2002 this admission led to authorized visits to Japan by some of the survivors and a furor among the Japanese people.

Even as the "Dear Leader" was able to reject U.S. and Japanese demands that nuclear operations at Yongbyon be reviewed, he approved some small gestures toward his southern neighbor. Beginning in 2000, a few hundred members of families separated by the 1953 cease-fire line were allowed a few hours to meet each other. He also allowed the preliminary design of a railroad that would link North and South across the demilitarized zone (DMZ). After authorizing the creation of a barbed-wire-enclosed "free-trade zone" in the far northeast at Rajin-Sonbong, where South Korean cruise ships could dock, Kim Jong Il approved another one in the northwest, at Sinuiji, in 2002. But when the Chinese billionaire who was to take charge of this free-trade zone was arrested for corruption in China, that plan fell apart.

Desperate North Koreans flee their country across the Chinese border and on the seas; in foreign countries they invade embassy compounds and plead for asylum. In 2005 the reunification of the two Koreas seemed far off; but so did German reunification in 1985, and then it happened a few years later. But the reunification of the two Koreas will entail far greater difficulties than that of the Germanies. North Korea is far worse off even than former communist East Germany; the cost will be staggering. South Koreans will face far greater sacrifices than West Germans did, with far fewer resources. Still, the boundary superimposed on the Korean nation is a leftover of the Cold War, and its elimination and the reintegration of North Korea into the international community is a laudable goal.

case, North Korea has raw materials that the industries of South Korea need; South Korea produces food the North needs; North Korea produces chemical fertilizers farms in the South need.

After the mid-twentieth century, however, the two Koreas were cut off from each other, and they developed in opposite directions. North Korea, whose large coal and iron ore deposits attracted the Japanese and whose hydroelectric plants produce electricity that the South could use, carried on what limited trade it generated with the Chinese and the Soviets. South Korea's external trade links were with the United States, Japan, and Western Europe.

In the early postwar years, there were few indications that South Korea would emerge as a major economic force on the Pacific Rim and, indeed, on the world stage. Over 70 percent of all workers were farmers; agriculture was inefficient; the country was stagnant. But huge infusions of aid, first from the United States and then Japan, coupled with the reorganization of farming and stimulation of industries (even if they lost money), produced a dramatic turnaround. Large feudal estates were parceled out to farm families in 3-hectare (7.5-acre) plots, and a program of massive fertilizer importation was begun. Production rose to meet domestic needs, and in some years yielded surpluses.

The industrialization program was based on modest local raw materials, plentiful and capable labor, ready overseas (especially American) markets, and continuing foreign assistance. Despite corrupt dictatorial rule, political instability, and social unrest, South Korean regimes managed to sustain a rapid rate of economic growth that placed the country, by the late 1980s, among the world's top ten trading powers. To get the job done, the government borrowed heavily from overseas, and it controlled banks and large industries. South Korea's growth re-
20 sulted from **state capitalism** rather than free-enterprise capitalism, but it had impressive results. South Korea became the world's leading shipbuilding nation; its automobile industry grew rapidly; and its iron and steel and chemical industries began to thrive. It has placed less emphasis, however, on smaller, high-technology industries, which are the strengths of other Pacific Rim economic tigers. So the future remains uncertain.

South Korea (population: 48.7 million) prospers today, and we can see its economic prowess on the map (Fig. 9-24). The capital, Seoul, with about 10 million inhabitants, now ranks among the world's megacities and anchors a huge industrial complex facing the Yellow Sea at the waist of the Korean Peninsula. Hundreds of thousands of farm families migrated to the Seoul area after the end of the Korean War (today only 20 percent of South Koreans remain on the land). They also moved to Busan, the nucleus of the country's second largest manufacturing zone, located on the Korea Strait opposite the western tip of Honshu. And the government-supported, urban industrial drive here continues. Just 30 years ago, Ulsan City, 40 miles (60 km) north of Busan along the coast, was a fishing center with perhaps 50,000 inhabitants; today its population exceeds a million, nearly half of them the families of workers in the Hyundai automobile factories and the local shipyards and docks.

The third industrial area shown in Figure 9-24, anchored by the city of Kwangju, has advantages of relative location that will spur its development, but it faces domestic problems. Kwangju lies at the center of the Cholla region, the southwest corner of South Korea; the southeast is called Kyongsang. For more than three decades beginning in the 1960s, the army generals who ruled South Korea all came from Kyongsang, and through state control of banks and other monopolies they bestowed favors on Kyongsang

FROM THE FIELD NOTES

"South Korea was one of the four 'economic tigers' marking the upsurge of the Western Pacific Rim in the 1980s and 1990s, its products ranging from computers to cars selling on world markets and its GDP rising rapidly. Walking through the prosperous central business district of globalizing Seoul you see every luxury name brand on display amid plentiful evidence of wealth and well-being. But you also see evidence of the survival of traditions long lost elsewhere, in the huge number of bookstores (South Koreans still read voraciously), in the stores selling traditional as well as modern musical instruments, in the school uniforms. And in other ways: when I got on a bus with all seats taken, a boy got up and offered me his place." © H. J. de Blij

companies while denying access to potential competitors in Cholla. The regime in Seoul also thwarted infrastructure development in Cholla, even to the point of impeding plans for improved railroad links between Kwangju and the capital. This regional friction led to protests in Cholla and violent repression by the military rulers, the aftermath of which is fading only slowly in this part of the now-democratic country. These historic discriminations seem to be ending as South Korea's democracy matures and regional differences subside. One reason for this has to do with infrastructural improvements: a new high-speed train system, the Korea Train Express (KTX), now links Seoul with both Mokpo in the Cholla region and Busan in the Kyongsang region (Fig. 9-24), and there are plans to link Kwangju with Daegu, thereby establishing an interregional link in the south. South Koreans know what the bullet-train network did for Japan, and they expect the KTX to accelerate the integration of their country as well.

Another sign of South Korea's political maturation is the government's proposal to move the capital from congested Seoul to an area about 100 miles (160 km) to the south, near Taejon in Chungchong Province (Fig. 9-24). Leaders argue that moving the government to a more centralized location will be costly but will enhance efficiency, but as the plan gains momentum look for politicians and others to raise obstacles. This will ultimately have to be a majority decision; the relocation project is slated to begin in 2007, and time is growing short.

Meanwhile, South Korea has moved far beyond the days when giant corporate conglomerates (known as *chaebol*) conspired to suppress competition, controlled the banks, and dictated economic policy to the government. It was state capitalism at its most extreme, and it proved to have weaknesses similar to planned communist economies. After the system collapsed and South Korea proved its capacity to thrive in the competitive world of the Pacific Rim, the country became one of the region's economic tigers—not without economic problems, but with boundless potential. Seoul's metropolitan economy has become a high-tech hub; the adjacent port city of Inchon is now spawning Songdo, a special economic zone designed to rival Shanghai's Pudong. China has surpassed the United States as South Korea's leading trade partner, and Songdo lies only about 200 miles (320 km) across the Yellow Sea from the Chinese mainland—symbolizing the new century as well as the new Pacific Rim era that is transforming all countries in this realm.

TAIWAN

The third component of the region we have called the Jakota Triangle is another, and more durable, success story on the Pacific Rim: Taiwan. Along with the Penghu (Makung) Islands in the Taiwan Strait and the islands of Jinmen Dao (Quemoy) near Xiamen and Matsu Tao (Matsu) near Fuzhou at the coast of mainland China, Taiwan is regarded by Beijing's communist regime as a Chinese province temporarily "wayward," that is, disobedient. After the reunification of Portuguese Macau with China in 1999, Taiwan remained the only major unresolved territorial issue in China's postcolonial, postwar sphere.

Beijing's claim to Taiwan has strengths as well as weaknesses. Han people settled on Taiwan, but the island had been settled thousands of years earlier by Malay-Polynesian groups who had split into mountain (east) and plains (west) inhabitants. When the Chinese emigrated from famine-stricken Fujian Province to Taiwan in the seventeenth century, they soon displaced and assimilated the plains people, but the mountain aborigines held out until the Japanese colonial invasion. Formal Chinese control began in 1683 when the expansionist Qing (Manchu) Dynasty made Taiwan part of Fujian Province. Rice and sugar became important exports to the mainland, and by the 1840s Taiwan's population was an estimated 2.5 million. In 1886 the island was declared a province of China, but by then the British had established two treaty ports there, the French had blockaded it, and the Japanese had sent expeditions to it. Nine years later, in 1895, China ceded Taiwan to Japan in the treaty settlement of the Sino-Japanese War, and it became a Japanese colony.

When China's aging Manchu Dynasty was overthrown following the rebellion of 1911, a Nationalist government in 1912 proclaimed the Republic of China (ROC), led by Sun Yat-sen. Naturally, this government wanted to oust all colonialists, not only Europeans but also Japanese, and not just from the mainland but from Taiwan as well. On Taiwan, however, the Japanese held on, using the island as a source of food and raw materials as well as a market for Japanese products. To ensure all this, the Japanese launched a prodigious development program involving road and railroad construction, irrigation projects, hydroelectric schemes, mines (mainly for coal), and factories. Farmlands were expanded and farming methods improved. While mainland China was engulfed in conflict, Taiwan remained under Japanese control. Not until 1945, when Japan was defeated in World War II, did Taiwan become an administrative part of China again. But only briefly. As we noted earlier, the Nationalists, under the leadership of Chiang Kai-shek, now faced the rising tide of communism on the mainland, and the ROC government fought a losing war. In 1949, the flag of the Republic of China was taken to a last stronghold: offshore Taiwan. There, Chiang and his followers, with their military might, weapons, and wealth taken from the mainland, established what they (and the world, led by the United States) proclaimed as the legitimate government of all China. On the island, the Nationalists, helped by the United States, began a massive reconstruction of the war-damaged infrastructure. In the international arena, the ROC represented the country; in the still-young United Nations, the Nationalists of Taiwan occupied China's seat. On the huge mainland, however, it was the communists in Beijing who ruled and called their country the People's Republic of China (PRC). Thus was born the two-China dilemma which still confronts the international community. When the ROC was ousted from the United Nations in 1971 to make way for the PRC, legitimacy was also transferred.

All this might have remained an internal dispute but for two major developments: Taiwan's rise as an economic tiger on the Pacific Rim, coupled with remarkable strides toward democracy made in the ROC; and the emergence on the world political and economic stage of communist-ruled China following decades of isolation. Today, the PRC's power is growing rapidly, and Taiwan's heirs to the ROC face an uncertain future; but it is the ROC, not the PRC, that has managed to combine economic success with democratization.

Taiwan, as Figure 9-25 shows, is not a large island. It is smaller than Switzerland but has a population much larger (22.8 million), most of it concentrated in an arc lining the western and northern coasts. The Chungyang Mountains, an area of high elevations (some over 10,000 feet [3000 m]), steep slopes, and dense forests, dominate the eastern half of the island. Westward, these mountains yield to a zone of hilly topography and, facing the Taiwan Strait, a substantial coastal plain. Streams from the mountains irrigate the paddyfields, and farm production has more than doubled since 1950 even as hundreds of thousands of farmers left the fields for work in Taiwan's expanding industries.

Today, the lowland urban-industrial corridor of western Taiwan is anchored by the capital, Taipei (Taibei), at the island's northern end and rapidly growing Kaohsiung (Gaoxiong) in the far south.* The Japanese developed Chilung (Jilong), Taipei's outport, to export nearby coal, but now the raw materials flow the other way. Taiwan imports raw cotton for its textile industry, bauxite (for aluminum) from Indonesia, oil from Brunei, and iron ore from Africa. Taiwan has a developing iron and steel industry, nuclear power plants, shipyards, a large chemical industry, and modern transport networks. Increasingly, however, Taiwan is exporting products of its high-technology industries: personal computers, telecommunications equipment, and precision electronic instruments. Taiwan has enormous brainpower, and many foreign firms join in the research and development carried on in such places as Hsinchu (Xinzhu) in the north, where the government has helped establish a technopole centered on the microelectronics and personal computer industries. In the south, the "science city" of Tainan specializes in microsystems and information technology.

The establishment of China's Special Economic Zones has gone a long way toward solving the economic problems arising from the political discord between China and Taiwan. In the almost-anything-goes environment of the SEZs, Taiwanese business owners are permitted to build or buy factories just as "real" foreigners are, and as we noted earlier, the Xiamen SEZ was situated directly across from Taiwan for just this purpose. As a result, thousands of Taiwanese companies now operate in China even as political tensions continue.

In some ways, Taiwan's emergence as an economic tiger on the Pacific Rim is even more spectacular than Japan's. True, Taiwan received much Western assistance, but it got out of debt faster than anyone expected. Today, annual per-capita income exceeds U.S. $15,000,

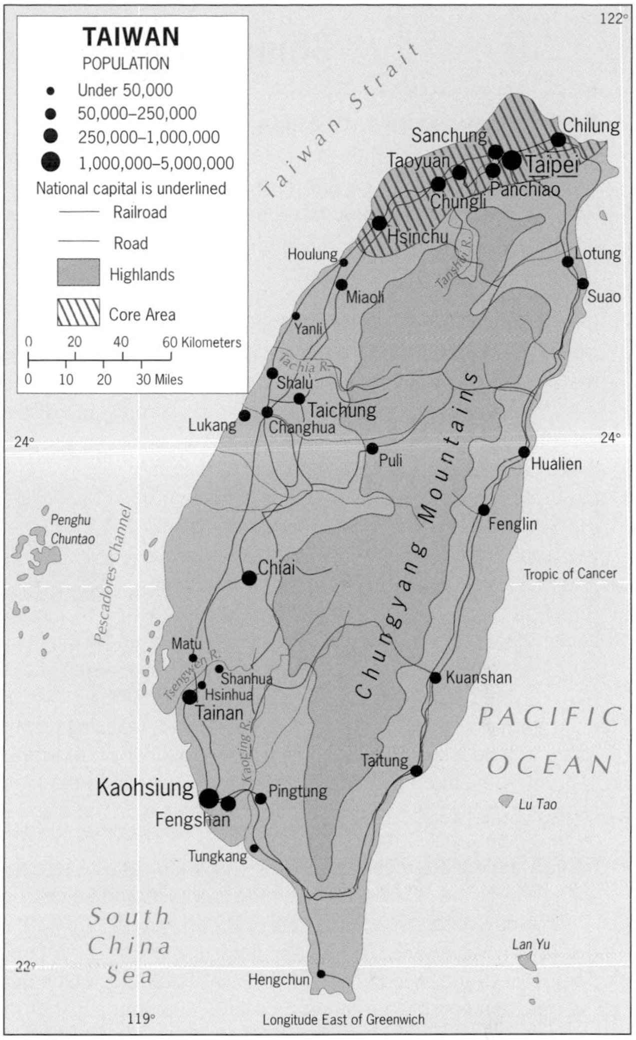

FIGURE 9-25 © H.J. de Blij, P.O. Muller, and John Wiley & Sons, Inc.

which is higher than that in many European countries. Taiwan's trade surplus has yielded tens of billions of dollars in reserves, which is being used to further its own development and to invest overseas. Taiwan, for example, is the largest foreign investor in the economy of Vietnam. But although Taiwan weathered the regional economic crisis of the late 1990s rather well, the new century has been problematic. The U.S. and Japanese markets for high-tech exports, especially computer chips, shrank markedly. Thousands of firms long profitable in Taiwan had to move their operations to China's SEZs, where labor costs were much lower, to stay in business. This reduced economic growth from 6.3 percent in 2000 to 1.1 percent in 2001 and subsequently into negative territory, while unemployment, basically unheard of in Taiwan, shot up to nearly 5 percent. All this happened while mainland China's economy continued to grow at an annual rate that exceeded 5 percent. It was a shock to Taiwan's confidence as an East Asian economic "tiger."

*The Taiwanese have retained the old Wade-Giles spelling of place names; China now uses the *pinyin* system (see box page 427). Names in parentheses are written according to the pinyin system.

Chinese Rule or Independence for Taiwan?

Regional ISSUE

SHOULD CHINA RULE TAIWAN?

"I am a graduate student in political geography here at East China Normal University, and I've been listening to your lecture about Chinese-American relations and the disagreement we have over Taiwan. You talk about this as though you have a right to stand between two parts of China, and with the power you have right now, I guess you can. You said that you hoped that mainland Chinese would become more 'reasonable' in their attitude toward the island province, but let me (as my professor would say) erect a hypothetical.

"Imagine that America wakes up one morning and discovers that a band of thugs have flown several 747s, loaded with weapons of all kinds, including those of mass destruction ready for use, into the San Juan, Puerto Rico airport. After taking the airport amid many casualties, they move into the city and enter administration buildings, threatening mass annihilation unless the local government resigns. By evening, Puerto Rico is under their control, American and Puerto Rican flags have been lowered and replaced by the banner of an 'independent' Puerto Rico.

"In Washington, your Congress meets and, realizing that the entire population of the island is at risk, makes the only move it can under the circumstances: it orders your armed forces to assemble a fleet to head for Puerto Rico, to blockade the island so as to create a basis for negotiation and a platform for possible military action. The fleet assembles near Miami and heads eastward through the channel between the Bahamas and Cuba.

"A day before reaching Puerto Rican waters, the path of the hastily assembled American fleet is blocked by a larger fleet from China, which has sailed from its base at Colón near the northern end of the Panama Canal. The commander of the American fleet communicates with his Chinese counterpart. 'We cannot let you pass,' states the Chinese admiral. 'We have not yet determined whether the invaders of Puerto Rico were invited by the people there to liberate them. We may have common ground with them. Please avoid an unnecessary conflict by going back to your bases.'

"How, sir, do you suppose the American people would feel toward China if a version of this scenario took place? They would be outraged at China's interference in a domestic matter. They would order your navy to attack. Well, we in China have had something like this happen with Taiwan. Thugs took it over. American aircraft carriers appeared in the Taiwan Strait. We've 'avoided conflict' for many decades, but that won't last forever. One way or another, Taiwan will submit to Beijing's rule. The Chinese people demand it."

TAIWAN SHOULD BE INDEPENDENT

"Instead of harping constantly on the 'reunification' issue, why don't Beijing's rulers get their own house in order first? Their political system is straight out of the discarded past, they're apparently allergic to democracy, and they can't seem to accept or accommodate any disobedience or dissent. I represent a Taiwanese high-tech corporation and I fly frequently between Taipei and Hong Kong and visit the so-called People's Republic via that route—and I can tell you, the Chinese could learn a lot about democracy, openness, freedom, transparency, and honesty from the Taiwanese. Politically, that 'One Country Two Systems' promise in Hong Kong is eroding as we speak. Economically, the Chinese ought to acknowledge that a lot of the 'Chinese miracle' on the Pacific Rim is actually a 'Taiwanese miracle.' Without our money, skills, discipline, and rigor there wouldn't be much of a miracle, especially if you also subtract what Southeast Asia's Chinese have contributed.

"Of course, Taiwan is a great rallying point for Beijing's autocrats. Gives them a chance to rail against the Western world, against America, against social excesses (we even have freedom of religion in Taiwan, imagine that!), against international conspiracies. But the truth is that the voters of Taiwan long ago ousted the 'thugs' who took us over, forged a thriving economy and a vibrant democracy, and made this island an example of what might have been across the Taiwan Strait.

"Could I say a few words about the 'historic' justification for China's claim to Taiwan? First of all, this island was peopled by Southeast Asian groups long before any Chinese arrived here, so if it's a matter of who got here first, we're last on the list. Beijing didn't integrate Taiwan politically until the 1680s, after groups of hungry citizens started to cross the Strait. That's pretty late in China's vaunted 5000-year history. Taiwan didn't become a province until 1886, which lasted exactly nine years. From 1895 to 1945 Taiwan belonged to Japan as part of the war settlement, and then Beijing took it back only to lose it again four years later to the 'thugs' under Chiang Kai-shek. That makes 13 years as part of the mainland system and 103 years of something else.

"For a country that has expanded and contracted for millennia, has gained and lost territory from Tibet to Taiwan, you would think that China could live with an ethnically compatible, economically successful, politically stable, democratic and independent neighbor. Those stiff bureaucrats in Beijing should have better things to do than threaten force to 'reincorporate' their exemplary neighbor."

Vote your opinion at www.wiley.com/college/deblij

Taiwan's political future remains clouded. The Nationalists who made the island their base in 1949 established a Republic of China that was authoritarian, but the ROC has evolved into a democratic state. This was not an easy process, but when Taiwan's own democracy movement culminated in 1991, it produced not a massacre, but free elections for a new parliament. Taiwan's leaders now proclaim that the direct and popular election of their president in 1996 was the first such vote not just in Taiwan or the ROC, but in the 4000-year history of the Chinese state. Taiwan thus presents itself as an alternative model to the communist-ruled PRC, proving to the world that democracy is compatible with Chinese culture and tradition. Nevertheless, the international community does not recognize Taiwan as an independent state, and a substantial segment of the Taiwanese electorate itself opposes any move toward outright sovereignty. Beijing dictates that Taiwan is off limits to U.S. cabinet members. In 1995, when the United States permitted the ROC's president to visit the American university of which he is a graduate, so that he might attend a class reunion, relations between Washington and Beijing plummeted. In 2000, when Taiwan's voters elected a president who, during the campaign, had proclaimed Taiwan's "equality" with mainland China, Beijing placed 600 missiles aimed at Taiwan on its shores opposite the island. The president then took to calling his country "Taiwan" rather than the ROC, and urged his fellow Taiwanese to do so as well; he also took a vote on the question of Taiwan's right to hold a referendum on independence.

The political geography of Taiwan, therefore, is at odds with its economic geography. The ROC stood firm against communism in the 1960s and 1970s; but its long-term ally, the United States, was required to publicly commit itself to a one-China policy, the one being the PRC, the communist giant. Even as goods and capital flow between Taiwan and the mainland, Beijing engages in threatening acts (such as nearby missile tests) when democracy in the ROC takes another step forward.

Today, Taiwan's political options are as limited as its economic horizons are endless. The communist Chinese have made it clear that a move toward independence, based on the likely outcome of any referendum, would lead to intervention. Though militarily well prepared, Taiwan could not long hold off the PRC's enormous armed forces. In any case, independence for Taiwan has philosophical disadvantages for the leaders of the ROC, as it would diminish their claim to legitimacy as the alternative model for a reunified China. In practical terms, conflict on the Pacific Rim would be disastrous for both sides, whatever course it were to take.

And so Taiwan's future hangs in the balance, a wayward province with national aspirations, a quasi-state with global connections, a test of Chinese capacity for compromise and accommodation (see the Issue Box entitled "Chinese Rule or Independence for Taiwan?"). In this era of devolutionary forces, the fate of Taiwan will be a harbinger of the world order of the twenty-first century.

WHAT YOU CAN DO

RECOMMENDATION: Join a Research Team! Many of the faculty members of your college or university, the professors you meet in class, have another life: they perform research and publish their findings in their discipline's scholarly literature. Some faculty may be working every summer at a base in some remote locale in a distant land; others will be using the library and the Internet. Often, when you read their publications, you will note that these are joint ventures (some articles in academic journals may have three, four, or even more authors). Such long-term research is likely to require lots of labor. If a professor's course or field is of special interest to you, ask her or him whether she or he is doing research in this particular area, and if so whether you can be of any assistance. Such help may start out in the library, but soon you may be working on interview teams or, who knows, joining a team of scientists overseas. Make your interest known, and an opportunity is likely to arise. You will learn things you would never learn in a classroom!

GEOGRAPHIC CONNECTIONS

1 East Asia is a realm of great cities, including many of the largest in the world. Names like Tokyo, Beijing, and Shanghai represent the megacity phenomenon like few others. But look at East Asia in Table G-1, and you will find that the countries of this realm display a range of urbanization levels from very high (Japan and South Korea are in the lead) to very low (China, for all its burgeoning cities, remains the least urbanized state in the realm). What geographic factors help explain these variable levels of urbanization among East Asia's countries? Why does China have more cities in the one-million-plus population category than any other country in the realm and yet the lowest percentage of urbanization? Do you expect the urban geography of this realm to change, and if so, where and how?

2 Consider core-periphery and globalization concepts (discussed in the Introduction) in the context of the East Asian realm. If you were asked to define what region or regions of East Asia presently form part of the global "core," what criteria would you use and where would this core component appear on the map? In those areas that form part of the periphery, what has been the impact of globalization? Has globalization drawn these areas closer to the core, benefiting them, or has the economic gap widened between core and periphery in East Asia?

Scale 1:16 000 000; one inch to 250 miles. Polyconic Projection

Elevations and depressions are given in feet

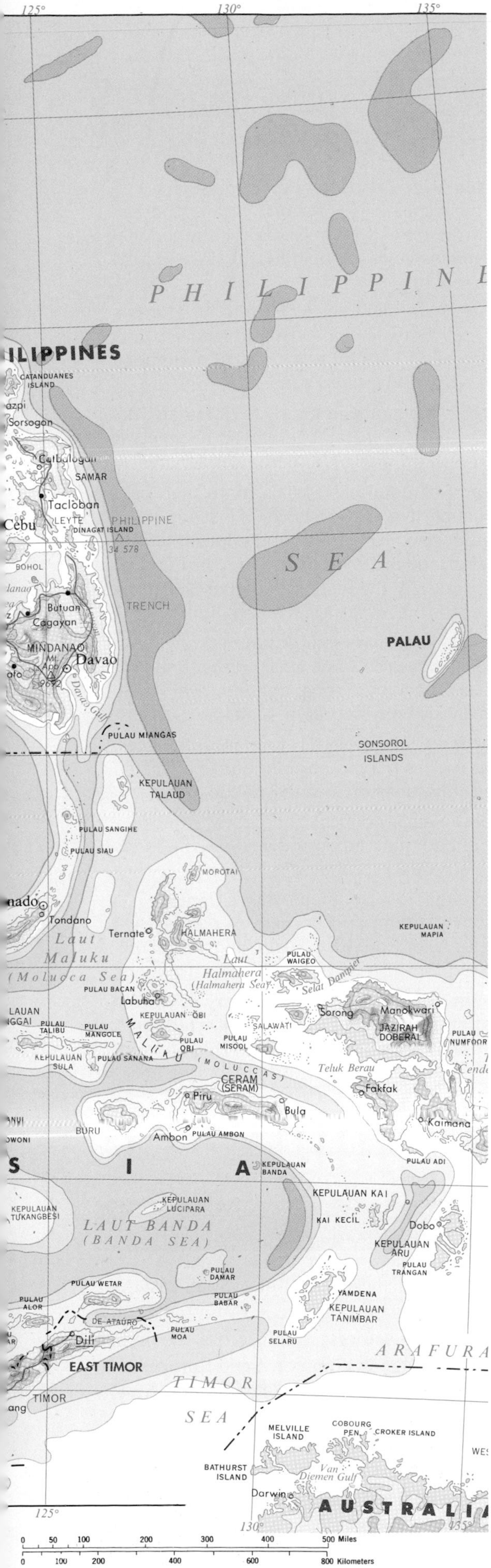

FIGURE 10-1 Reprinted with permission from *Goode's World Atlas*, 21st edition, pp. 212–213.

Chapter 10

Southeast Asia

CONCEPTS, IDEAS, AND TERMS

1 Tsunami
2 Buffer zone
3 Shatter belt
4 Overseas Chinese
5 Organic theory
6 State boundaries
7 Antecedent boundary
8 Subsequent boundary
9 Superimposed boundary
10 Relict boundary
11 State territorial morphology
12 Compact state
13 Protruded state
14 Elongated state
15 Fragmented state
16 Perforated state
17 Domino theory
18 *Entrepôt*
19 Archipelago
20 Transmigration

REGIONS

- MAINLAND SOUTHEAST ASIA
- INSULAR SOUTHEAST ASIA

SOUTHEAST ASIA . . . . the very name roils American emotions. Here the United States owned its only major colony. Here American forces triumphed over Japanese enemies. Here the United States fought the only war it ever lost. Here Washington's worst Cold War fears failed to materialize. Here American companies invested heavily when the Pacific Rim's economic growth transformed dormant economies into potential Pacific tigers. Here American troops and intelligence agents pursue the War on Terror. Southeast Asia, once remote and stagnant, has taken center stage in our globalizing world.

On December 26, 2004, the world was reminded that Southeast Asia's volatility is not limited to its political geography. This realm lies on some of the planet's most mobile geology ranging from unstable subduction zones (Fig. G-4) to active volcanoes (Fig. G-5) and is affected by numerous earthquakes. One of those earthquakes, in the subduction zone west of the Indonesian island of Sumatera* where the Eurasian, Indian, and
1 Australian tectonic plates meet, generated a **tsunami** that thrust walls of water onto coasts near and far, sweeping away entire towns and villages as well as their populations and killing as many as 300,000 people living on the shores of the Indian Ocean from Southeast Asia to East Africa. In 1815, people around the world woke up one April morning to skies that darkened as the day went on, the result of the gigantic eruption of the Indonesian volcano named Tambora. A year later the world was shrouded in a cloak of ash and soot, and there was no summer. Crops would not ripen, and hunger and starvation afflicted millions. The human population of the planet was just 1 billion at the time; consider what would happen today under similar circumstances, with the human population more than six times as large. Southeast Asia's volatility has the potential to disrupt life around the world.

MAJOR GEOGRAPHIC QUALITIES OF

Southeast Asia

1. Southeast Asia extends from the peninsular mainland to the archipelagos offshore. Because Indonesia controls part of New Guinea, its functional region reaches into the neighboring Pacific geographic realm.
2. Southeast Asia, like Eastern Europe, has been a shatter belt between powerful adversaries and has a fractured cultural and political geography shaped by foreign intervention.
3. Southeast Asia's physiography is dominated by high relief, crustal instability marked by volcanic activity and earthquakes, and tropical climates.
4. A majority of Southeast Asia's more than half-billion people live on the islands of just two countries: Indonesia, with the world's fourth-largest population, and the Philippines. The rate of population increase in the Insular region of Southeast Asia exceeds that of the Mainland region.
5. Although the overwhelming majority of Southeast Asians have the same ancestry, cultural divisions and local traditions abound, which the realm's divisive physiography sustains.
6. The legacies of powerful foreign influences, Asian as well as non-Asian, continue to affect the cultural landscapes of Southeast Asia.
7. Southeast Asia's political geography exhibits a variety of boundary types and several categories of state territorial morphology.
8. The Mekong River, Southeast Asia's Danube, has its source in China and borders or crosses five Southeast Asian countries, sustaining tens of millions of farmers, fishing people, and boat owners.
9. The realm's giant in terms of territory as well as population, Indonesia, has not asserted itself as the dominant state because of mismanagement and corruption; but Indonesia has enormous potential.

DEFINING THE REALM

Southeast Asia is a realm of peninsulas and islands, a corner of Asia bounded by India on the northwest and China on the northeast (Fig. 10-1). Its western coasts are washed by the Indian Ocean, and to the east stretches the vast Pacific. From all these directions, Southeast Asia has been penetrated by outside forces. From India came traders; from China, settlers; from across the Indian Ocean, Arabs to engage in commerce and Europeans to build empires; and from across the Pacific, the Americans. Southeast Asia has been the scene of countless contests for power and primacy—the competitors have come from near and far.

Southeast Asia's geography in some ways resembles that of Eastern Europe. It is a mosaic of smaller

*As in Africa, names and spellings have changed with independence. In this chapter, we will use the contemporary spellings, except when we refer to the colonial period. Thus Indonesia's four major islands are Jawa, Sumatera, Kalimantan (the Indonesian part of Borneo), and Sulawesi. The Dutch called them Java, Sumatra, Dutch Borneo, and Celebes, respectively.

countries on the periphery of one of the world's largest
2 states. It has been a **buffer zone** between powerful ad-
3 versaries. It is a **shatter belt** in which stresses and pressures from without and within have fractured the political geography. Like Eastern Europe, Southeast Asia exhibits great cultural diversity. It is a realm of hundreds of cultures, numerous languages and dialects, and several major religions.

LAND AND SEA BORDERS

The Southeast Asian realm borders South Asia and East Asia on land and the Pacific and Austral realms at sea (Fig. 10-2). Note that one political entity that forms part of it, Indonesia, extends beyond the realm's borders. The easternmost province of Indonesia is the western half of the Pacific island of New Guinea, where indigenous cultures are Papuan, not Indonesian. Today Indonesia rules what is in effect a Pacific island colony called Papua. We discuss this territory as part of our assessment of Indonesia, but we will also consider all of New Guinea as a component of the Pacific Realm in Chapter 12.

Because the politico-geographical map (Fig. 10-2) is so complicated, it should be studied attentively. One good way to strengthen your mental map of this realm is to follow the mainland coastline from west to east. The westernmost state in the realm is Myanmar (called Burma before 1989 and still referred to by that name),

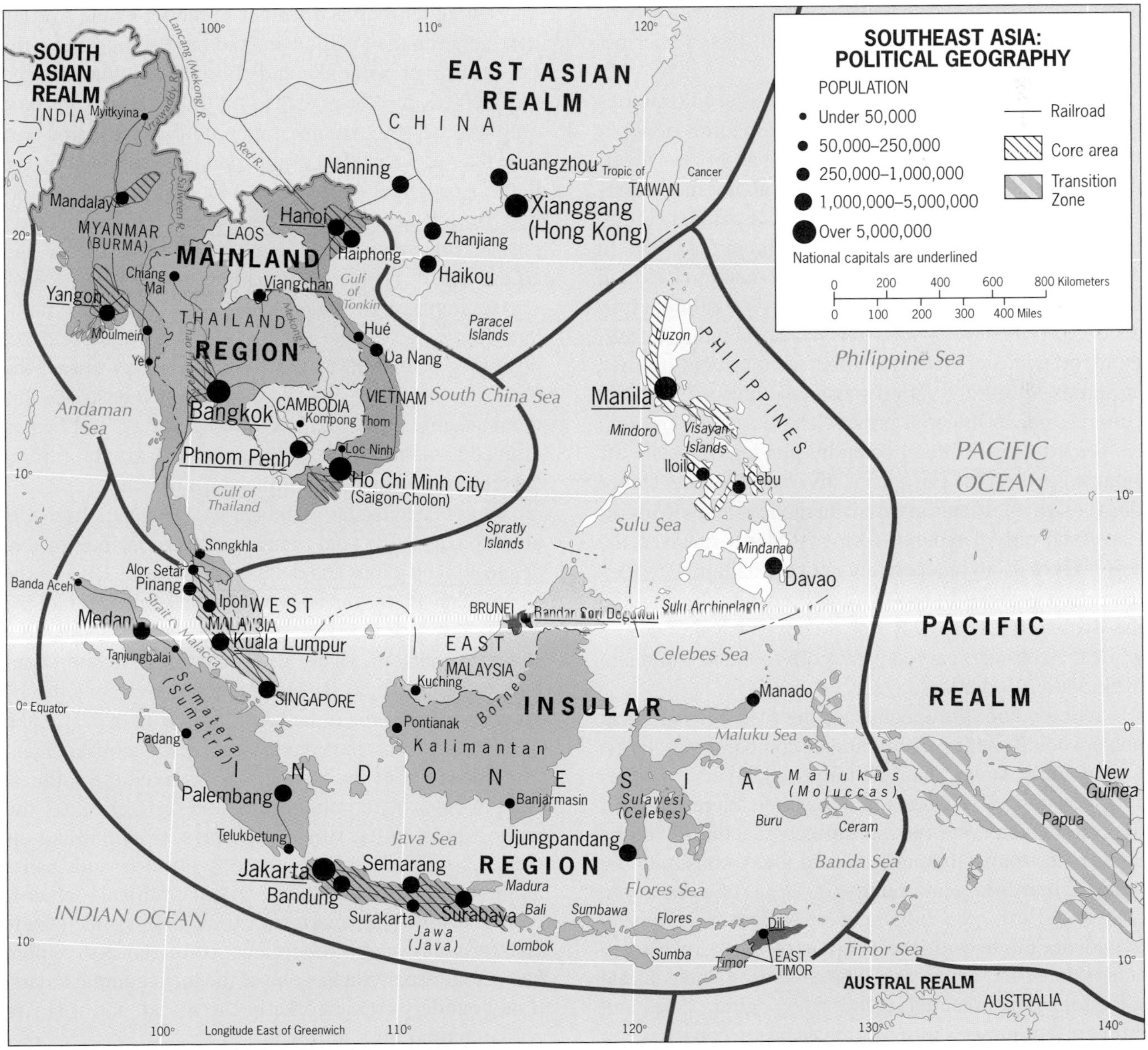

FIGURE 10-2 © H.J. de Blij, P.O. Muller, and John Wiley & Sons, Inc.

the only country in Southeast Asia that borders both India and China. Myanmar shares the "neck" of the Malay Peninsula with Thailand, heart of the *Mainland* region. The south of the peninsula is part of Malaysia—except for Singapore, at the very tip of it. Facing the Gulf of Thailand is Cambodia. Still moving generally eastward, we reach Vietnam, a strip of land that extends all the way to the Chinese border. And surrounded by its neighbors is landlocked Laos, remote and isolated. This leaves the islands that constitute *Insular* Southeast Asia: the Philippines in the north and Indonesia in the south, and between them the offshore portion of Malaysia, situated on the largely Indonesian island of Borneo. Also on Borneo lies the ministate of Brunei, small but, as we will see, important in the regional picture. And finally, a new state is just appearing on the map in the realm's southeastern corner: East Timor, a former Portuguese colony annexed by Indonesia in 1976 and released to United Nations supervision in 1999. After a transition period, East Timor became an independent state in 2002.

No dominant state has arisen among the 11 countries of Southeast Asia. The largest country in terms of population as well as territory, Indonesia, has as-yet unrealized potential to overshadow the rest but not in the foreseeable future. Neither did any single, dominant core of indigenous culture develop here as it did in East Asia. In the river basins and on the plains of the mainland, as well as on the islands offshore, a flowering of cultures produced a diversity of societies whose languages, religions, arts, music, foods, and other achievements formed an almost infinitely varied mosaic—but none of those cultures rose to imperial power. The European colonizers forged empires here, often by playing one state off against another; the Europeans divided and ruled. Out of this foreign intervention came the modern map of Southeast Asia; only Thailand (formerly Siam) survived the colonial era as an independent entity. Thailand was useful to two competing powers, the French to the east and the British to the west. It was a convenient buffer, and while the colonists carved pieces off Thailand's domain, the kingdom endured.

Indeed, the Europeans accomplished what local powers could not: the formation of comparatively large, multicultural states that encompassed diverse peoples and societies and welded them together. Were it not for the colonial intervention, it is unlikely that the 17,000 islands of far-flung Indonesia would today constitute the world's fourth-largest country in terms of population. Nor would the nine sultanates of Malaysia have been united, let alone with the peoples of northern Borneo across the South China Sea. For good or ill, the colonial intrusion consolidated a realm of few culture cores and numerous ministates into less than a dozen countries.

PHYSICAL GEOGRAPHY

As Figure 10-1 shows, Southeast Asia is a realm in which high relief dominates the physiography. From the Arakan Mountains in western Myanmar (Burma) to the glaciers (yes, glaciers!) of the Indonesian part of New Guinea, elevations rise above 10,000 feet (3300 m) in many places. In Myanmar, Mount Victoria is such a peak; in northern Laos, Phu Bia comes close; on the Indonesian island of Sumatera (Sumatra under the old spelling) the tallest volcano is Kerinci; on Jawa (Java) Mounts Slamet and Semeru tower over the countryside; and on Borneo Mount Kinabalu, not a volcano but an erosional remnant, is higher than any other mountain in this realm. In the Philippines, Mount Pinatubo at 4874 feet (1477 m) is not especially high, but this volcano's eruptions have had major impact on the country and, in 1991, on global climate.

The relief map is a further reminder of the overlap here between the Pacific Rim and the Pacific Ring—the Pacific Rim of economic and political development and the Pacific Ring of Fire with its earthquakes and volcanic eruptions. If you examine Figure G-5 closely, you will note that all the islands of this realm are affected by the latter except one, the largely Indonesian island of Borneo. Borneo is a slab of ancient crust, a mini-continent pushed high above sea level by tectonic forces and eroded into its present mountainous topography.

As Figure 10-1 underscores, rivers rise in the highland backbones of the islands and peninsulas, and deposit their sediments as they wind their way toward the coast; the physiography of Sumatera demonstrates this unmistakably. The volcanic hills, plateaus, and better-drained lowlands are fertile and, in the warmth of tropical climates, can yield multiple crops of rice.

On the peninsular mainland we see a pattern that is already familiar: rivers rising in the Asian interior that create alluvial plains and deltas. The Mekong River is the Chang/Yangzi of Southeast Asia: you can trace it all the way from China via Laos, Thailand, and Cambodia into southern Vietnam, where it forms a massive and populous delta. In the west, Myanmar's key river is the Irrawaddy; Thailand's is the Chao Phraya. In the north, the Red River Basin is the breadbasket of northern Vietnam.

No survey of the physical geography of Southeast Asia would be complete without reference to the realm's seas, gulfs, straits, and bays. Irregular and indented coastlines such as these, with thousands of islands near and far, create difficult problems when it comes to drawing maritime boundaries (the islands also form havens for rebels and criminals). As we note later, Southeast Asia has one of the most complex maritime boundary frameworks in the world, and it is rife with potential for conflict.

POPULATION GEOGRAPHY

Compared to the huge population numbers and densities in the habitable regions of South Asia and China, demographic totals for the countries of Southeast Asia, with the exception of Indonesia, seem modest. Again, comparisons with Europe come to mind. Three countries—Thailand, the Philippines, and Vietnam—have populations between 64 and 88 million. Laos, quite a large country territorially (comparable to the United Kingdom), had just 6.1 million inhabitants in 2006. Cambodia, substantially larger than Greece, had 13.7 million.

As Table G-1 shows, arithmetic population densities are not especially high in Southeast Asia except in the urbanized ministate of Singapore. Similarly, physiologic densities are lower than in eastern China or neighboring South Asian countries. When the political situation is stable, several Southeast Asian countries are able to export large quantities of rice. Thailand and, in recent years, Vietnam have ranked among the world's leading exporters in this crucial staple of Asian diets.

Viewed in spatial perspective, one can discern both similarities and differences in the population patterns of Southeast Asia and its giant neighbors. As in East and South Asia, population clusters in the large basins and deltas of major rivers, where densities are high and the countryside is parceled into paddies as far as the eye can see. Numerous small, space-conserving villages dot the landscape in such lowlands as the Red and Mekong basins of Vietnam, a scene reminiscent of South China. But between these alluvial lowlands lies higher ground where the landscape takes on savanna characteristics with reddish, leached, less productive soils, scattered villages, and comparatively sparse population. And in the islands of Indonesia and the Philippines, Southeast Asia has environments that are not present at all in India or China—well-watered volcanic soils supporting luxuriant natural vegetation and capable of sustaining intensive agriculture. Whole countrysides have been meticulously terraced to create paddyfields and irrigate them.

People and Land

Of Southeast Asia's 565 million inhabitants, just over 55 percent live on the islands of Indonesia and the Philippines, leaving the realm's mainland countries with only 44 percent of the population. With such high population pressure in adjacent realms, why has Southeast Asia not been flooded by waves of immigrants?

In fact, Southeast Asia has received its share of immigrants, but overland invasions have been comparatively limited. Several factors have hindered travel along overland routes into Southeast Asia. First, physical obstacles inhibit such movement. In Chapter 8, we noted the barrier effect of the densely forested hills and mountains along the border between northeastern India and northwestern Myanmar (Burma). North of Myanmar lies forbidding Xizang (Tibet), and northeast of Myanmar and north of Laos is the high, rugged Yunnan Plateau. Transit is easier in the east, between southeastern China and northern Vietnam, and this has indeed been an avenue for contact and migration. As the ethnolinguistic map of China (p. 447) shows, migration from Southeast Asia into China also has occurred in this area.

Second—as reflected in the population distribution map (Fig. G-9)—the highly clustered agricultural opportunities in Southeast Asia lie separated by large stretches of less productive land, and these in turn are crossed by high-relief zones that hinder movement. The map of Southeast Asia reveals several concentrated areas of productive capacity but few corridors that would facilitate migration.

As we noted earlier, immigrants to Southeast Asia over the past millennium have come mainly by sea, not by land. That is true not only for Arabs and Europeans but also for neighboring Chinese. Mostly the Chinese immigrants came not as farmers seeking land but as traders and workers looking for opportunities in the cities and towns. "Chinatowns" form a vital part of most Southeast Asian cities today, especially the coastal ones.

The Ethnic Mosaic

Southeast Asia's peoples come from a common stock just as (Caucasian) Europeans do, but this has not prevented the emergence of regionally or locally discrete ethnic or cultural groups. Figure 10-3 displays the broad distribution of ethnolinguistic groups in the realm, but be aware that this is a generalization. At the scale of this map, numerous small groups cannot be depicted.

The map shows the rough spatial coincidence, on the mainland, between major ethnic group and modern political state. The Burman dominate in the country once known as Burma (now Myanmar); the Thai occupy the state once known as Siam (Thailand); the Khmer form the nation of Cambodia and extend northward into Laos; and the Vietnamese inhabit the long strip of territory facing the South China Sea.

Territorially, by far the largest population is classified in Figure 10-3 as Indonesian, the inhabitants of the great archipelago that extends from Sumatera west of the Malay Peninsula to the Malukus (Moluccas) in the east and from the lesser Sunda Islands in the south to the Philippines in the north. Collectively, all these peoples—the Filipinos, Malays, and Indonesians—shown in Figure 10-3 are known as Indonesians, but they have been divided by history and politics. Note, on the map, that the

Indonesians in Indonesia itself include Javanese, Madurans, Sundanese, Balinese, and other large groups; hundreds of smaller ones are not shown. In the Philippines, too, island isolation and contrasting ways of life are reflected in the cultural mosaic. Also part of this Indonesian ethnic-cultural complex are the Malays, whose heartland lies on the Malay Peninsula but who form minorities in other areas as well. Like most Indonesians, the Malays are Muslims, but Islam is a more powerful force in Malay society than, in general, in Indonesian culture.

Figure 10-3 also reminds us that (again like Eastern Europe) Southeast Asia has many ethnic minorities. On the Malay Peninsula, note the South Asian (Hindustani) cluster; Hindu communities with Indian ancestries exist

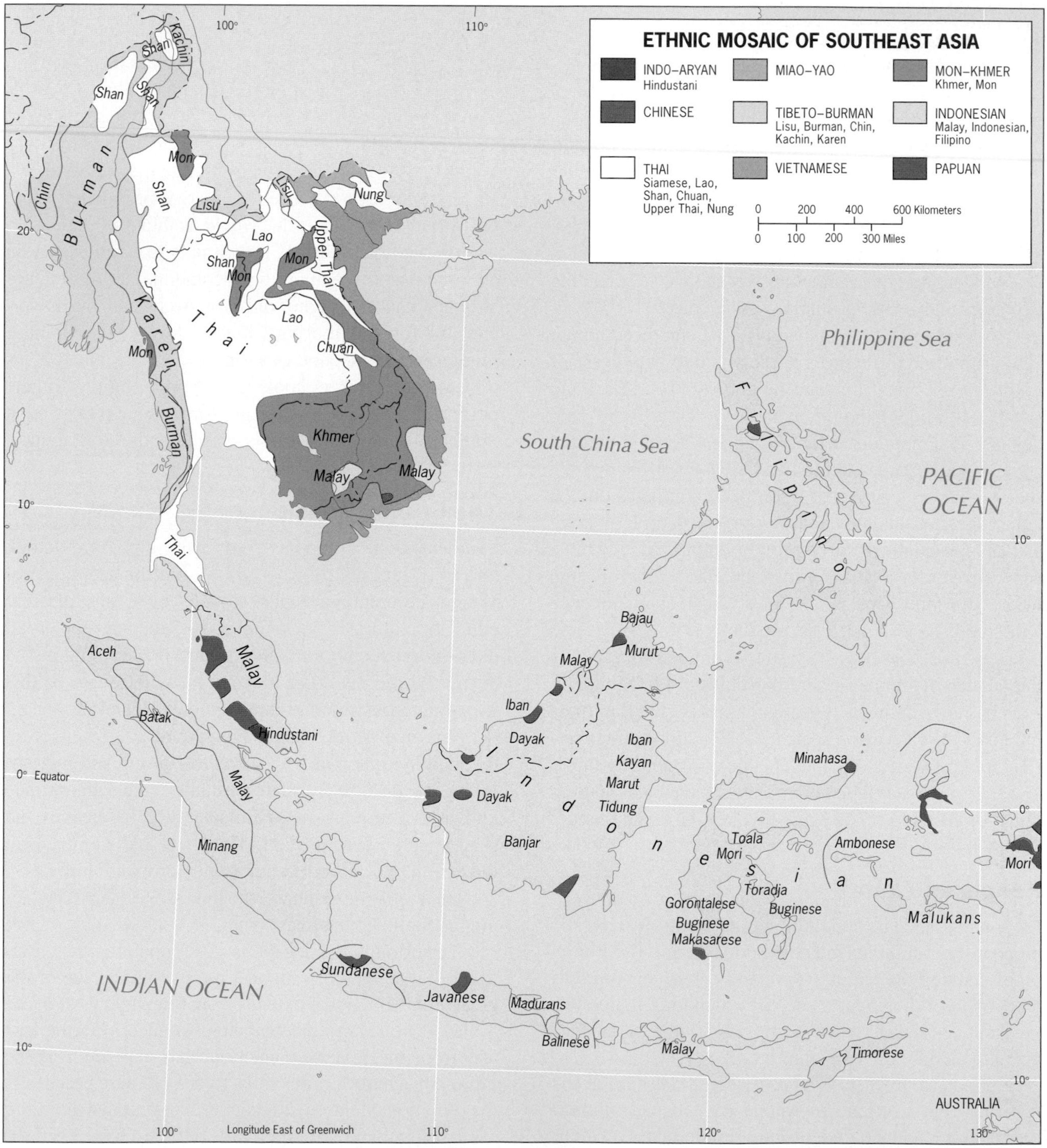

FIGURE 10-3 © H.J. de Blij, P.O. Muller, and John Wiley & Sons, Inc.

FROM THE FIELD NOTES

"It was a relatively cool day in Hanoi, which was just as well because the drive from Haiphong had taken nearly four hours, much of the time waiting behind oxcarts, and the countryside was hot and humid. On my way to an office along Hang Bai Street I ran into this group of youngsters and their teacher on their way to the Ho Chi Minh memorial. He right away realized the level of my broken French, and explained patiently that he taught history, and took his class on a field trip several times a year. 'They get all dressed up and it is a big moment for them,' he said, 'and there is a lot of history to be visited in Hanoi.' The kids had seen me take this picture and now they crowded around. I told them that I noticed that rouge seemed to be a favorite color in their outfits. 'But that should not surprise you, monsieur,' said the teacher, smiling. 'This is Vietnam, communist Vietnam, and red is the color that rallies us!' In the city at the opposite end of the country named after Ho Chi Minh (founder of his country's Communist Party, president from 1945 to 1969, victor over the French colonizers, and leader in the war against the United States), my experiences with teachers and students had been quite different, to the point that most called their city Saigon. But here in Hanoi the ideological flame still burned brightly." © H. J. de Blij.

in many parts of the peninsula, but here in the southwest they form the majority in a small area. Nearby lies the largest single exception to the rule that Chinese minorities in Southeast Asia are urban-based: along the west coast of the Malay Peninsula, Chinese are farmers and villagers as well as city dwellers. In the north of the realm, minorities share the lands in which the Burman, Thai, and Vietnamese dominate. Those minorities, as a comparison between Figures 10-2 and 10-3 proves, tend to occupy zones peripheral to the regional core areas, where forests often are dense and difficult to penetrate, where isolation from the power cores prevails, where remoteness promotes detachment from the national state, and where, time and again, ethnic conflicts have raged as the core-based majority tried to control the frontier. During the war in Vietnam, peoples in the highlands (the Montagnards) tended to be sympathetic to the American campaign because of their longstanding hostility to the dominant (lowland) Vietnamese. Today, the Shan and the Karen in Myanmar are in a state of conflict with the regime based in the core area.

HOW THE POLITICAL MAP EVOLVED

The leading colonial competitors in Southeast Asia were the Dutch, French, British, and Spanish (with the last replaced by the Americans in their stronghold, the Philippines). The Japanese had colonial objectives here as well, but these came and went during the course of World War II.

The Dutch acquired the greatest prize: control over the vast archipelago now called Indonesia (formerly the Netherlands East Indies). France established itself on the eastern flank of the mainland, controlling all territory east of Thailand and south of China. The British conquered the Malay Peninsula, gained power over the northern part of the island of Borneo, and established themselves in Burma as well. Other colonial powers also gained footholds but not for long. The exception was Portugal, which held on to its eastern half of the island of Timor (Indonesia) until after the Dutch had been ousted from their East Indies.

Figure 10-4 shows the colonial framework in the late nineteenth century, before the United States assumed control over the Philippines. Note that while Thailand survived as an independent state, it lost territory to the British in Malaya and Burma and to the French in Cambodia and Laos.

The Colonial Imprint

The colonial powers divided their possessions into administrative units as they did in Africa and elsewhere. Some of these political entities became independent states when the colonial powers withdrew. France, one of the mainland's leading colonial powers, divided its

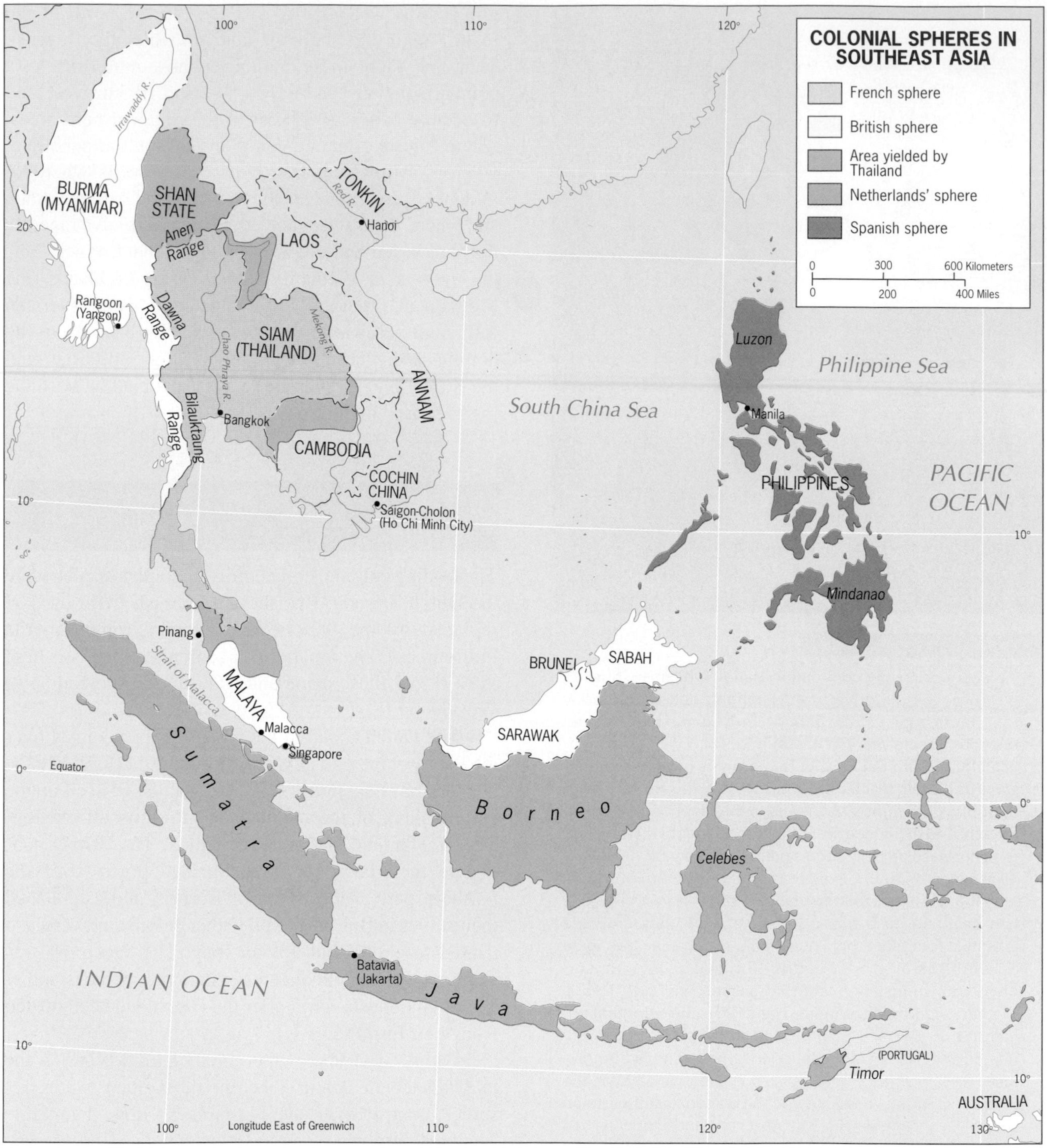

FIGURE 10-4 © H.J. de Blij, P.O. Muller, and John Wiley & Sons, Inc.

Southeast Asian empire into five units. Three of these units lay along the east coast: Tonkin in the north next to China, centered on the basin of the Red River; Cochin China in the south, with the Mekong Delta as its focus; and between these two, Annam. The other two French territories were Cambodia, facing the Gulf of Thailand, and Laos, landlocked in the interior. Out of these five French dependencies there emerged the three states of Indochina. The three east coast territories ultimately became one state, Vietnam; the other two (Cambodia and Laos) each achieved separate independence.

The British ruled two major entities in Southeast Asia (Burma and Malaya) in addition to a large part of northern Borneo and many small islands in the South

China Sea. Burma was attached to Britain's Indian Empire; from 1886 until 1937 it was governed from distant New Delhi. But when British India became independent in 1947 and split into several countries, Burma was not part of the grand design that created West and East Pakistan (the latter now Bangladesh), Ceylon (now Sri Lanka), and India. Instead, in 1948 Burma (now Myanmar) was given the status of a sovereign republic.

In Malaya, the British developed a complicated system of colonies and protectorates that eventually gave rise to the equally complex, far-flung Malaysian Federation. Included were the former Straits Settlements (Singapore was one of these colonies), the nine protectorates on the Malay Peninsula (former sultanates of the Muslim era), the British dependencies of Sarawak and Sabah on the island of Borneo, and numerous islands in the Strait of Malacca and the South China Sea. The original Federation of Malaysia was created in 1963 by the political unification of recently independent mainland Malaya, Singapore, and the former British dependencies on the largely Indonesian island of Borneo. Singapore, however, left the Federation in 1965 to become a sovereign city-state, and the remaining units were later restructured into peninsular Malaysia and, on Borneo, Sarawak and Sabah. Thus the term *Malaya* properly refers to the geographic area of the Malay Peninsula, including Singapore and other nearby islands; the term *Malaysia* identifies the politico-geographical entity of which Kuala Lumpur is the capital city.

The Hollanders took control of the "spice islands" through their Dutch East India Company, and the wealth that they extracted from what is today Indonesia brought the Netherlands its Golden Age. From the mid-seventeenth to the late-eighteenth century, the Dutch could develop their East Indies sphere of influence almost without challenge, for the British and French were preoccupied with the Indian subcontinent. By playing the princes of Indonesia's states against one another in the search for economic concessions and political influence, by placing the Chinese in positions of responsibility, and by imposing systems of forced labor in areas directly under its control, the Company had a ruinous effect on the Indonesian societies it subjugated. Java (Jawa), the most populous and productive island, became the focus of Dutch administration; from its capital at Batavia (now Jakarta), the Company extended its sphere of influence into Sumatra (Sumatera), Dutch Borneo (Kalimantan), Celebes (Sulawesi), and the smaller islands of the East Indies. This was not accomplished overnight, and the struggle for territorial control was carried on long after the Dutch East India Company had yielded its administration to the Netherlands government. Dutch colonialism thus threw a girdle around Indonesia's more than 17,000 islands, paving the way for the creation of the realm's largest and most populous nation-state (226 million today).

In the colonial tutelage of Southeast Asia, the Philippines, long under Spanish domination, had a unique experience. As early as 1571, the islands north of Indonesia were under Spain's control (they were named for Spain's King Philip II). Spanish rule began when Islam was reaching the southern Philippines via northern Borneo. The Spaniards spread their Roman Catholic faith with great zeal, and between them the soldiers and priests consolidated Hispanic dominance over the mostly Malay population. Manila, founded in 1571, became a profitable waystation on the route between southern China and western Mexico (Acapulco usually was the trans-Pacific destination for the galleons leaving Manila's port). There was much profit to be made, but the indigenous people shared little in it. Great landholdings were awarded to loyal Spanish civil servants and to men of the church. Oppression eventually yielded revolution, and Spain was confronted with a major uprising when the Spanish-American War broke out elsewhere in 1898.

As part of the settlement of that war, the United States replaced Spain in Manila. That was not the end of the revolution, however. The Filipinos now took up arms against their new foreign ruler, and not until 1905, after terrible losses of life, did American forces manage to "pacify" their new dominion. Subsequently, U.S. administration in the Philippines was more progressive than Spain's had been. In 1934, Congress passed the Philippine Independence Law, providing for a ten-year transition to sovereignty. But before independence could be arranged, World War II intervened. In 1941, Japan conquered the islands, temporarily ousting the Americans; U.S. forces returned in 1944 and, with strong Filipino support, defeated the Japanese in 1945. The agenda for independence was resumed, and in 1946 the sovereign Republic of the Philippines was proclaimed.

Today, all of Southeast Asia's states are independent, but centuries of colonial rule have left strong cultural imprints. In their urban landscapes, their education systems, their civil service, and countless other ways, this realm still carries the marks of its colonial past.

Cultural-Geographic Legacies

The French, who ruled and exploited a crucial quadrant of Southeast Asia, had a name for their empire: *Indochina*. That name would be appropriate for the rest of the realm as well because it suggests its two leading Asian influences for the past 2000 years. Periodic expansions of the Chinese Empire, notably along the eastern periphery, as well as the arrival of large numbers of Chinese settlers (see box titled "Overseas Chinese"), infused the region

Overseas Chinese

Southeastern China has been the source of Chinese emigrants to Southeast Asia for as long as 2000 years, but the largest movements have occurred over the past six centuries (Fig. 10-5). During the initially outward-looking Ming Dynasty (1368–1644) and early in the following Manchu (Qing) Dynasty, small groups of Chinese emigrants, sometimes pushed by famine at home, sailed southward to establish small communities in Southeast Asian towns. During the early colonial period (1670–1870), growing commercial opportunities formed the pull factors that attracted greater numbers. But the largest exodus came during the late colonial period (1870–1940) when Southeast Asia's expanding job and trade opportunities attracted more than 20 million Chinese, many from Fujian and Guangdong provinces. During the communist period, additional migrants illegally left China, many to or via Hong Kong.

Toward the end of their rule, colonial regimes began to restrict Chinese immigration, as did independent Thailand. The Chinese had become an influential class of middlemen and controlled ever more commerce, and colonial rulers as well as local peoples feared Chinese influence and "expansionism." During World War II and subsequent decolonization, Chinese were persecuted in Southeast Asia, especially in Malaya while under Japanese control, and in Indonesia in the 1960s when Chinese were targeted as communist sympathizers.

Not counting intermarriages, Southeast Asia today may contain as many as 30 million Chinese (more than half the world total of **Overseas Chinese**), and many 4
have become very successful. When China began its new economic policy, Overseas Chinese invested heavily in the SEZs and Open Cities, playing a large role in stimulating the economic growth experienced there.

But success entails risks. Resentment against Chinese prosperity and business practices led to discriminatory legislation in Malaysia. When Indonesia's economy collapsed in 1998, police stood by as gangs looted, burned, murdered, and raped at will in Chinese commercial districts. Wealthier Chinese fled, exacerbating Indonesia's economic crisis.

Chinese everywhere resented Indonesia's failure to protect its Chinese minority, and relations between China and Indonesia were strained. Nonetheless, mutually advantageous trade continued to grow. But in early 2005, when China announced plans to provide limited assistance to Indonesian victims of the December 2004 tsunami, Chinese public opinion strongly opposed this decision and caused a suspension of the program.

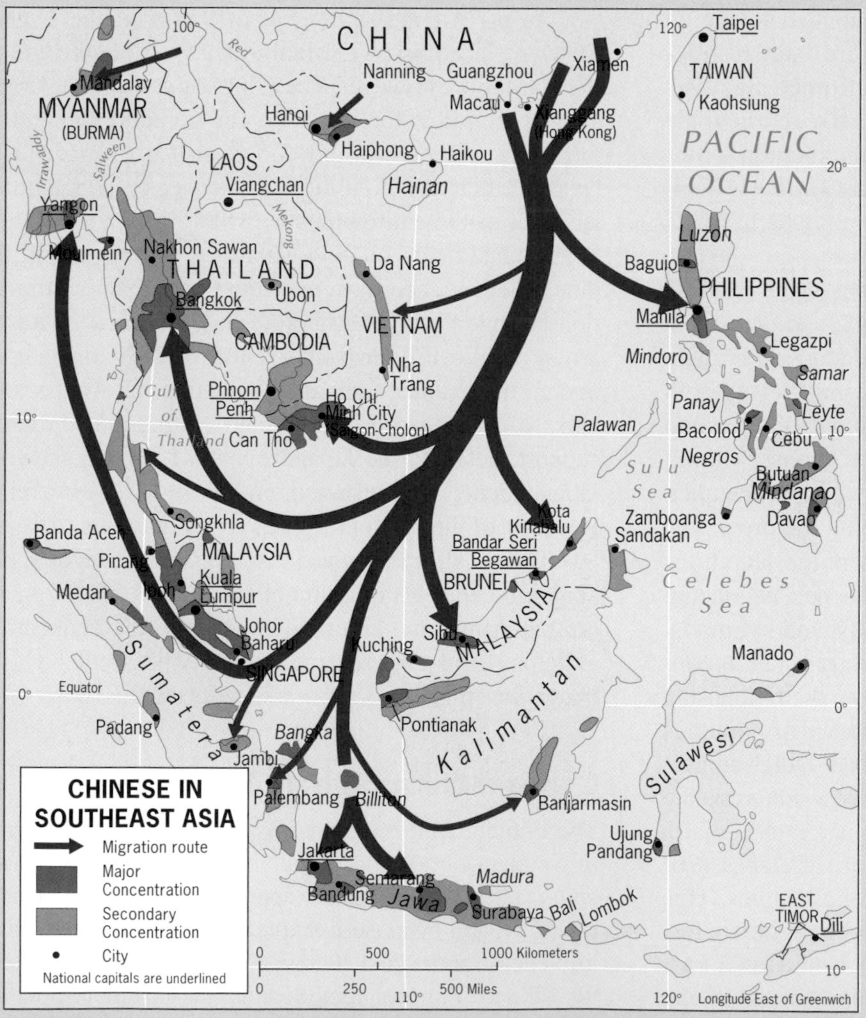

FIGURE 10-5 ©H.J. de Blij, P.O. Muller, and John Wiley & Sons, Inc.

FROM THE FIELD NOTES

"Like most major Southeast Asian cities, Bangkok's urban area includes a large and prosperous Chinese sector. No less than 14 percent of Thailand's population of 64.7 million is of Chinese ancestry, and the great majority of Chinese live in the cities. In Thailand, this large non-Thai population is well integrated into local society, and intermarriage is common. Still, Bangkok's 'Chinatown' is a distinct and discrete part of the great city. There is no mistaking Chinatown's limits: Thai commercial signs change to Chinese, goods offered for sale also change (Chinatown contains a large cluster of shops selling gold, for example), and the urban atmosphere, from street markets to bookshops, is dominantly Chinese. This is a boisterous, noisy, energetic part of multicultural Bangkok, a vivid reminder of the Chinese commercial success in Southeast Asia." © H. J. de Blij.

with cultural norms from the north. The Indians came from the west by way of the sea, as their trading ships plied the coasts, and settlers from India founded colonies on Southeast Asian shores in the Malay Peninsula, on the lower Mekong Plain, on Jawa and Bali, and on Borneo.

With the migrants from the Indian subcontinent came their faiths: first Hinduism and Buddhism, later Islam. The Muslim religion, promoted by the growing number of Arab traders who appeared on the scene, became the dominant religion in Indonesia (where nearly 80 percent of the population adheres to Islam today). But in Myanmar, Thailand, and Cambodia, Buddhism remained supreme, and in all three countries the overwhelming majority of the people are now adherents. In culturally diverse Malaysia, the Malays are Muslims (to be a Malay *is* to be a Muslim), and almost all Chinese are Buddhists; but most Malaysians of Indian ancestry remain Hindus. Although Southeast Asia has generated its own local cultural expressions, most of what remains in tangible form has resulted from the infusion of foreign elements. For instance, the main temple at Angkor Wat, constructed in Cambodia during the twelfth century, remains a monument to the Indian architecture of that time.

The *Indo* part of Indochina, then, refers to the cultural imprints from South Asia: the Hindu presence, the importance of Buddhism (which came to Southeast Asia via Sri Lanka [Ceylon] and its seafaring merchants), the influences of Indian architecture and art (especially sculpture), writing and literature, and social structures and patterns.

The *China* in the name Indochina signifies the role of the Chinese here. Chinese emperors coveted Southeast Asian lands, and China's power reached deep into the realm. Social and political upheavals in China sent millions of sinicized people southward. Chinese traders, pilgrims, seafarers, fishermen, and others sailed from southeastern China to the coasts of Southeast Asia and established settlements there. Over time, those settlements attracted more Chinese emigrants, and Chinese influence in the realm grew (see map at left). Not surprisingly, relations between the Chinese settlers and the earlier inhabitants of Southeast Asia have at times been strained, even violent. The Chinese presence in Southeast Asia is long-term, but the invasion has continued into modern times. The economic power of Chinese minorities and their role in the political life of the realm have led to conflicts.

The Chinese initially profited from the arrival of the Europeans, who stimulated the growth of agriculture, trade, and industries; here the Chinese found opportunities they lacked at home. They harvested rubber, found jobs on the docks and in the mines, cleared the bush, and transported goods in their sampans. They brought useful skills with them, and as tailors, shoemakers, blacksmiths, and fishermen, they prospered. The Chinese were also astute in business; soon, they not only dominated the region's retail trade but also held prominent positions in banking, industry, and shipping. Thus they have always been far more important than their modest numbers in Southeast Asia would suggest. The Europeans used them for their own designs but at times found the Chinese to be stubborn competitors—so much so that eventually they tried to

The Chinese Presence in Southeast Asia

Regional ISSUE

THE CHINESE ARE TOO INFLUENTIAL!

"It's hard to imagine that there was a time when we didn't have Chinese minorities in our midst. I think I understand how Latin Americans [*sic*] feel about their 'giant to the north.' We've got one too, but the difference is that there are a lot more Chinese in Southeast Asia than there are Americans in Latin America. The Chinese even run one country because they are the majority there, Singapore. And if you want to go to a place where you can see what the Chinese would have in mind for this whole region, go there. They're knocking down all the old Malay and Hindu quarters, and they've got more rules and laws than we here in Indonesia could even think of. I'm a doctor here in Bandung and I admire their modernity, but I don't like their philosophy.

"We've had our problems with the Chinese here. The Dutch colonists brought them in to work for them, they got privileges that made them rich, they joined the Christian churches the Europeans built. I'm not sure that a single one of them ever joined a mosque. And then shortly after Sukarno led us to independence they even tried to collaborate with Mao's communists to take over the country. They failed and many of them were killed, but look at our towns and villages today. The richest merchants and the money lenders tend to be Chinese. And they stay aloof from the rest of us.

"We're not the only ones who had trouble with the Chinese. Ask them about it in Malaysia. There the Chinese started a full-scale revolution in the 1950s that took the combined efforts of the British and the Malays to put down. Or Vietnam, where the Chinese weren't any help against the enemy when the war happened there. Meanwhile they get richer and richer, but you'll see that they never forget where they came from. The Chinese in China boast about their coastal economy, but it's coastal because that's where our Chinese came from and where they sent their money when the opportunity arose.

"As a matter of fact, I don't think that China itself cares much for or about Southeast Asia. Have you heard what's happening in Yunnan Province? They're building a series of dams on the Mekong River, our major river, just across the border from Laos and Burma. They talk about the benefits and they offer to destroy the gorges downstream to facilitate navigation, but they won't join the Mekong River Commission. When the annual floods cease, what will happen to Tonle Sap, to fish migration, to seasonal mudflat farming? The Chinese do what they want—they're the ones with the power and the money."

THE CHINESE ARE INDISPENSABLE!

"Minorities, especially successful minorities, have a hard time these days. The media portray them as exploiters who take advantage of the less fortunate in society, and blame them when things go wrong, even when it's clear that the government representing the majority is at fault. We Chinese arrived here long before the Europeans did and well before some other 'indigenous' groups showed up. True, our ancestors seized the opportunity when the European colonizers introduced their commercial economy, but we weren't the only ones to whom that opportunity was available. We banded together, helped each other, saved and shared our money, and established stable and productive communities. Which, by the way, employed millions of locals. I know: my family has owned this shop in Bangkok for five generations. It started as a shed. Now it's a six-floor department store. My family came from Fujian, and if you added it all up we've probably sent more than a million U.S. dollars back home. We're still in touch with our extended family and we've invested in the Xiamen SEZ.

"Here in Thailand we Chinese have done very well. It is sometimes said that we Chinese remain aloof from local society, but that depends on the nature of that society. Thailand has a distinguished history and a rich culture, and we see the Thais as equals. Forgive me, but you can't expect the same relationship in East Malaysia. Or, for that matter, in Indonesia, where we are resented because the 4 percent Chinese run about 60 percent of commerce and trade. But we're always prepared to accommodate and adjust. Look at Malaysia, where some of our misguided ancestors started a rebellion but where we're now appreciated as developers of high-tech industries, professionals, and ordinary workers. This in a Muslim country where Chinese are Christians or followers of their traditional religions, and where Chinese regularly vote for the Malay-dominated majority party. So when it comes to aloofness, it's not us.

"The truth is that the Chinese have made great contributions here, and that without the Chinese this place would resemble South or Southwest Asia. Certainly there would be no Singapore, the richest and most stable of all Southeast Asian countries, where minorities live in peace and security and where incomes are higher than anywhere else. In fact, mainland Chinese officials come to Singapore to learn how so much was achieved there. Imagine a Southeast Asia without Singapore! It's no coincidence that the most Chinese country in Southeast Asia also has the highest standard of living."

Vote your opinion at www.wiley.com/college/deblij

impose restrictions on Chinese immigration. When the United States took control of the Philippines, it also sought to stop the influx of Chinese into those islands.

When the European colonial powers withdrew and Southeast Asia's independent states emerged, Chinese population sectors ranged from nearly 50 percent of the total in Malaysia (in 1963) to barely over 1 percent in Myanmar. In Singapore today, Chinese constitute 77 percent of the population of 4.3 million; when Singapore seceded from Malaysia in 1965, the Chinese component in Malaysia was substantially reduced (it is about 25 percent today). In Indonesia, the percentage of Chinese in the total population is not high (no more than 4 percent), but the Indonesian population is so large that even this small percentage indicates a Chinese sector of more than 9 million. In 1998, this tiny minority reportedly controlled 70 percent of Indonesia's economy. In Thailand, on the other hand, many Chinese have married Thais, and the Chinese minority of about 14 percent has become a cornerstone of Thai society, dominant in trade and commerce.

In general, Southeast Asia's Chinese communities remained aloof and formed their own separate societies in the cities and towns. They kept their culture and language alive through social clubs, schools, and even residential suburbs that, in practice if not by law, were Chinese in character. At one time they were caught in the middle between the Europeans and Southeast Asians when the local people were hostile toward both white people and the Chinese. Since the withdrawal of the Europeans, however, the Chinese have become the main target of this antagonism, which remained strong because of the Chinese involvement in money lending, banking, and trade monopolies (see the Issue Box on the page at left entitled "The Chinese Presence in Southeast Asia"). Moreover, there is the specter of an imagined or real Chinese political imperialism along Southeast Asia's northern flanks.

The *China* in Indochina, therefore, represents a diversity of penetrations. Southern China was the source of most of the old invasions, and Chinese territorial consolidation provided the impetus for successive immigrants. Mongoloid racial features carried southward from East Asia mixed with the preexisting Malay stock to produce a transition from Chinese-like people in the northern mainland to darker-skinned Malay types in the distant Indonesian east. Although Indian cultural influences remained strong, people throughout Southeast Asia adopted Chinese modes of dress, plastic arts, types of houses and boats, and other cultural attributes. During the past century, and especially during the last half-century, renewed Chinese immigration brought skills and energies that propelled these minorities to comparative wealth and influence.

SOUTHEAST ASIA'S POLITICAL GEOGRAPHY

Southeast Asia's diverse and fractured natural landscapes, as well as varied cultural influences, have combined to create a complex political geography. As we noted earlier, *political geography* focuses on the spatial expressions of political behavior; we were introduced to several aspects of it when we examined Western Europe's supranationalism, Eastern Europe's balkanization, Pakistan's irredentism toward Kashmiri Muslims, and the geopolitics of eastern Eurasia. Many political geographers have studied the fundamental causes of the cyclical rise and decline of states; their answers have ranged from environmental changes to biological forces. Friedrich Ratzel (1844–1904) conceptualized the state as a biological organism whose life, from birth through expansion and maturation to eventual senility and collapse, mirrors that of any living thing. Ratzel's **organic** 5 **theory** of state development held that nations, being aggregates of human beings, would over the long term live and die as their citizens did.

Other political geographers sought to measure the strength of the forces that bind states (centripetal forces) and that divide them (centrifugal forces), and thus to assess a state's chances to survive separatism of the kind Canada and Sri Lanka are experiencing today, to name just two examples. Even in today's era of modern warfare, a stretch of water still affects the course of events. Taiwan would not be what it is without the 100 miles (160 km) of water that separates it from mainland China. Singapore's secession from (then) Malaya was facilitated because Singapore is an island; had it been attached to the Malay Peninsula, the centrifugal forces of separatism might not have prevailed. The role of physical geography in political events remains powerful.

In this section we pay particular attention to the boundaries and territories of the component states of Southeast Asia because this realm's political geography displays virtually every possible attribute of both. We focus first on the borders and then on the states' territorial shapes.

The Boundaries

Boundaries are sensitive parts of a state's anatomy: just as people are territorial about their individual properties, so nations and states are sensitive about their territories and borders. The saying that "good fences make good neighbors" certainly applies to states, but, as we know, the boundaries between states are not always "good fences."

Boundaries, in effect, are contracts between states. 6
That contract takes the form of a treaty that contains the *definition* of the boundary in the form of elaborate description. Next, cartographers perform the *delimitation* of the treaty language, drawing the boundary on official, large-scale maps. Throughout human history, states have used those maps to build fences, walls, or other barriers in a process called *demarcation*.

Once established, we can classify boundaries geographically. Some are sinuous, conforming to rivers or mountain crests (*physiographic*) or coinciding with breaks

FROM THE FIELD NOTES

"I stood on the Laotian side of the great Mekong River which, during the dry season, did not look so great! On the opposite side was Thailand, and it was rather easy for people to cross here at this time of the year. But, the locals told me, it is quite another story in the wet season. Then the river inundates the rocks and banks you see here, it rushes past, and makes crossing difficult and even dangerous. The buildings where the canoes are docked are built on floats, and rise and fall with the seasons. The physiographic-political boundary between Thailand and Laos lies in the middle of the valley we see here." © Barbara A. Weightman.

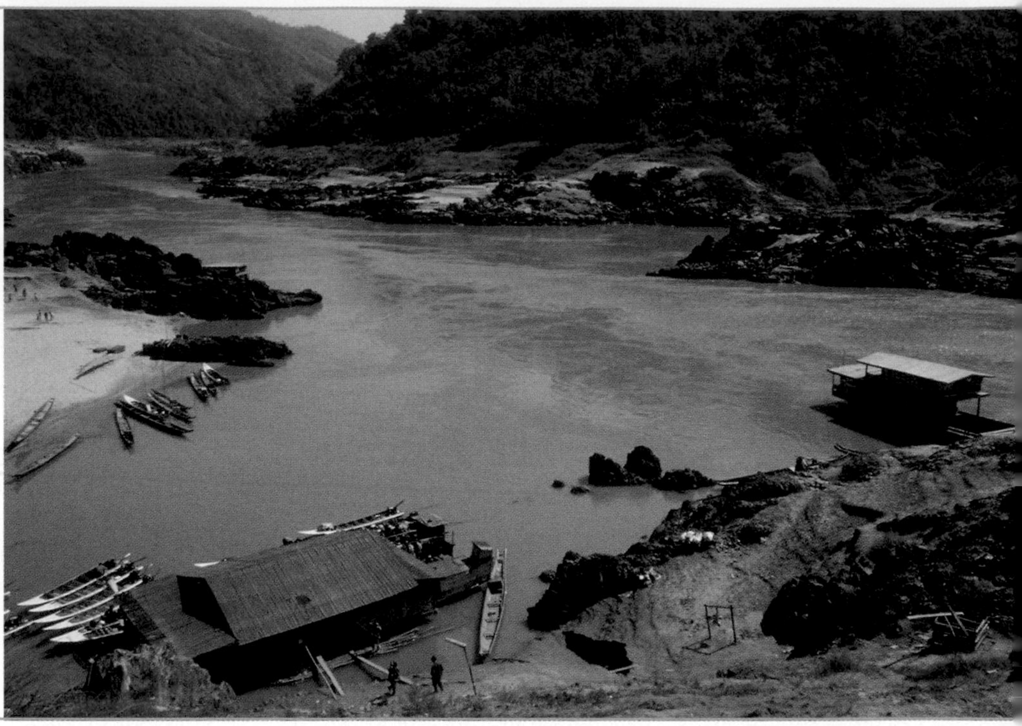

or transitions in the cultural landscape (*anthropogeographic*). As any world political map shows, many boundaries are simply straight lines, delimited without reference to physical or cultural features. These *geometric* boundaries can produce problems when the cultural landscape changes where they exist.

In general, the boundaries of Southeast Asia were better defined than those of several other postcolonial areas of the world, notably Africa, the Arabian Peninsula, and Turkestan. The colonial powers that established the original treaties tried to define boundaries to lie in remote and/or sparsely peopled areas, for example, across interior Borneo. Nevertheless, certain Southeast Asian boundaries have produced problems, among them the geometric boundary between Papua, the portion of New Guinea ruled by Indonesia, and Papua New Guinea, the eastern component of the island.

Even on a small-scale map of the kind we use in this chapter, we can categorize the boundaries of this realm. A comparison between Figures 10-2 and 10-3 reveals that the boundary between Thailand and Myanmar over long segments is anthropogeographic, notably where the name *Karen*, the Myanmar minority, appears in Figure 10-3. Figure 10-1 shows that a large segment of the Vietnam–Laos boundary is physiographic-political, coinciding with the Annamese Cordillera (Mountains).

Boundaries also can be classified genetically, that is, as their evolution relates to the cultural landscapes they traverse. A leading political geographer, Richard Hartshorne (1899–1992), proposed a four-level *genetic boundary classification*. All four of these boundary types can be observed in Southeast Asia.

Certain boundaries, Hartshorne reasoned, were defined and delimited before the present-day human landscape developed. In Figure 10-6 (upper-left map), the boundary between Malaysia and Indonesia on the island of Borneo is an example of the first boundary type, the **antecedent boundary**. Most of this border passes through 7
sparsely inhabited tropical rainforest, and the break in settlement can even be detected on the small-scale world population map (Fig. G-9).

A second category of boundaries evolved as the cultural landscape of an area took shape, part of the ongoing process of accommodation. These **subsequent bound-** 8
aries are represented in Southeast Asia by the map in the upper right of Figure 10-6, which shows in some detail the border between Vietnam and China. This border is the result of a long process of adjustment and modification, the end of which may not yet have come.

The third category involves boundaries drawn forcibly across a unified or at least homogeneous cultural landscape. The colonial powers did this when they divided the island of New Guinea by delimiting a boundary in a nearly straight line (curved in only one place to accommodate a bend in the Fly River), as shown in the lower-left map of Figure 10-6. The **superimposed** 9
boundary they delimited gave the Netherlands the western half of New Guinea. When Indonesia became inde-

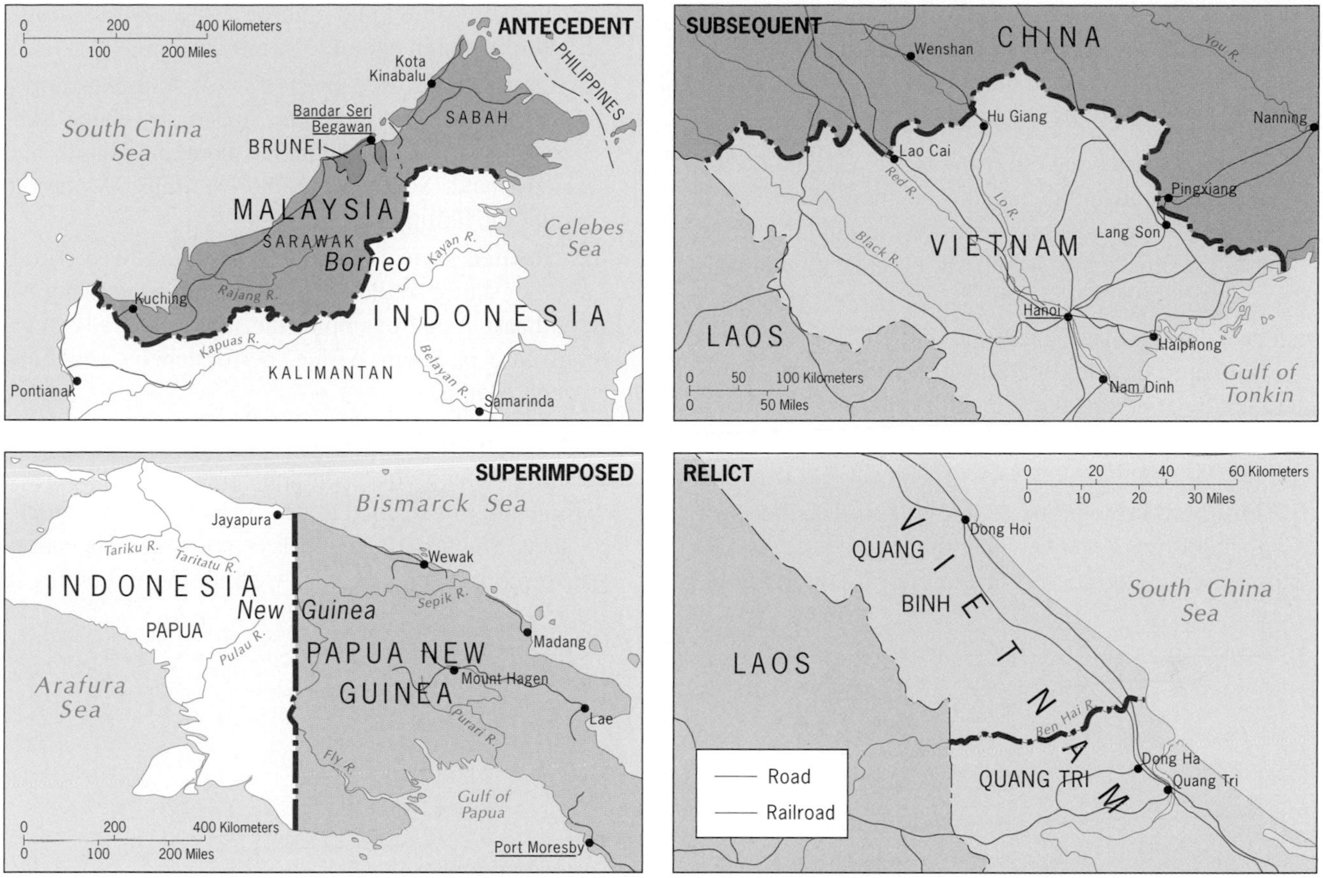

GENETIC POLITICAL BOUNDARY TYPES

FIGURE 10-6 © H.J. de Blij, P.O. Muller, and John Wiley & Sons, Inc.

pendent in 1949, the Dutch did not yield their part of New Guinea, which is peopled mostly by ethnic Papuans, not Indonesians. In 1962 the Indonesians invaded the territory by force of arms, and in 1969 the United Nations recognized its authority there. This made the colonial, superimposed boundary the eastern border of Indonesia and had the effect of extending Indonesia from Southeast Asia into the Pacific Realm. Geographically, all of New Guinea forms part of the Pacific Realm.

The fourth genetic boundary type is the so-called
10 **relict boundary**—a border that has ceased to function but whose imprints (and sometimes influence) are still evident in the cultural landscape. The boundary between the former North and South Vietnam (Fig. 10-6, lower-right map) is a classic example: once demarcated militarily, it has had relict status since 1976 following the reunification of Vietnam in the aftermath of the Indochina War (1964–1975).

Southeast Asia's boundaries have colonial origins, but they have continued to influence the course of events in postcolonial times. Take one instance: the physiographic boundary that separates the main island of Singapore from the rest of the Malay Peninsula, the Johor Strait (see Fig. 10-11). That physiographic-political boundary facilitated, perhaps crucially, Singapore's secession from the state of Malaysia. Without it, Malaysia might have been persuaded to stop the separation process; at the very least, territorial issues would have arisen to slow the sequence of events. As it was, no land boundary needed to be defined. The Johor Strait demarcated Singapore and left no question as to its limits.

State Territorial Morphology

Boundaries define and delimit states; they also create the mosaic of often interlocking territories that give individual countries their shape, also known as their *morphology*. The **territorial morphology** of a state affects its 11
condition, even its survival. Vietnam's extreme elongation has influenced its existence since time immemorial. As we will see, Indonesia has tried to redress its fragmentation into thousands of islands by promoting unity through the "transmigration" of Javanese from the most populous island to many of the others.

Political geographers identify five dominant state territorial configurations, all of which we have encountered in our world regional survey but which we have not categorized until now. All but one of these shapes are represented in Southeast Asia, and Figure 10-7 provides the terminology and examples:

12 • **Compact states** have territories shaped somewhere between round and rectangular, without major indentations. This encloses a maximum amount of territory within a minimum length of boundary. Southeast Asian example: Cambodia.

13 • **Protruded states** (sometimes called *extended*) have a substantial, usually compact territory from which extends a peninsular corridor that may be landlocked or coastal. Southeast Asian examples: Thailand, Myanmar.

14 • **Elongated states** (also called *attenuated*) have territorial dimensions in which the length is at least six times the average width, creating a state that lies astride environmental and/or cultural transitions. Southeast Asian example: Vietnam.

• **Fragmented states** consist of two or more territorial units separated by foreign territory or by water. Subtypes are mainland-mainland, mainland-island, and island-island. Southeast Asian examples: Malaysia, Indonesia, Philippines. 15

• **Perforated states** completely surround the territory of other states, so that they have a "hole" in them. No Southeast Asian example; the most illustrative current case is South Africa, perforated by Belgium-sized Lesotho. 16

In the discussion that follows, we will have frequent occasion to refer to this geographic property of Southeast Asia's states. For so comparatively small a realm with so few states, Southeast Asia displays a considerable variety of state morphologies. When we link these features to

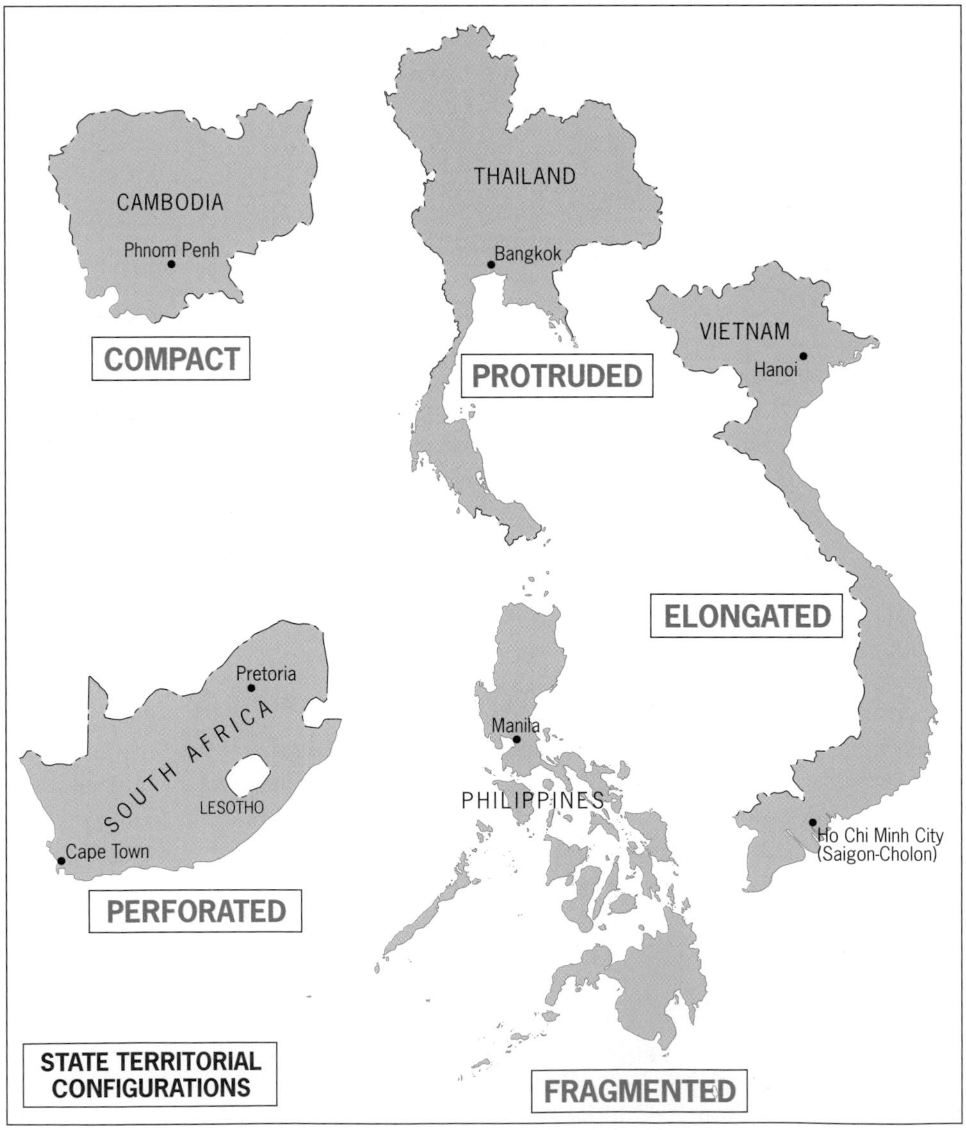

FIGURE 10-7 © H.J. de Blij, P.O. Muller, and John Wiley & Sons, Inc.

other geographic aspects (such as relative location), we obtain useful insights into the regional framework.

One point of caution: states' territorial morphologies do not determine their viability, cohesion, unity, or lack thereof; they can, however, influence these qualities. Cambodia's compactness has not ameliorated its divisive political geography, for example. But as we will find in the pages that follow, shape plays a key role in the still-unfolding political and economic geography of Southeast Asia.

REGIONS OF THE REALM

Southeast Asia's first-order regionalization must be based on its mainland-island fragmentation. But as we have noted there are physiographic, historical, and cultural reasons to include the Malaysian (southern) part of the Malay Peninsula in the Insular region, as shown in Figure 10-2. Using the political framework as our grid, we see that the regions of Southeast Asia are constituted as follows:

Mainland region Vietnam, Cambodia, Laos, Thailand, Myanmar (Burma)

Insular region Malaysia, Singapore, Indonesia, East Timor, Brunei, Philippines

MAINLAND SOUTHEAST ASIA

Five countries form the Mainland region of Southeast Asia: two of them protruded, one compact, one elongated, and one landlocked. Two colonial powers, buffered by Thailand, shaped its modern historical geography. One religion, Buddhism, dominates cultural landscapes, but this is a multicultural, multiethnic region. Although one of the least urbanized regions of the world, it contains several major cities. And as Figure 10-2 shows, two countries (Vietnam and Myanmar) possess more than one core area each. We approach the region from the east.

MAJOR CITIES OF THE REALM

City	Population* (in millions)
Bangkok, Thailand	6.8
Hanoi, Vietnam	4.3
Ho Chi Minh City (Saigon), Vietnam	5.3
Jakarta, Indonesia	13.8
Kuala Lumpur, Malaysia	1.4
Manila, Philippines	10.9
Phnom Penh, Cambodia	1.3
Singapore, Singapore	4.3
Viangchan, Laos	0.8
Yangon, Myanmar (Burma)	4.3

*Based on 2006 estimates.

ELONGATED VIETNAM

Consider this: when the Indochina War ended in 1975, Vietnam had less than half the population it has today. In 2006, Vietnam's population exceeded 83 million, about 60 percent of it under 21 years of age. For the great majority of Vietnamese, therefore, the terrible war of the 1960s and 1970s is history, not memory. What concerns most Vietnamese today is the need to overcome two decades of isolation, to reconnect to the world at large, and to join in the economic boom on the Pacific Rim. In 1995, the United States moved to normalize relations with Hanoi, opening an embassy and authorizing interactions long prohibited in the aftermath of the conflict. Five years later, this was followed by a successful state visit by President Clinton.

Travel up the crowded road from the northern port of Haiphong to the capital, Hanoi, or sail up the river to the southern metropolis officially known as Ho Chi Minh City—but called *Saigon* by almost everyone there—and you are quickly reminded of the cultural effects of Vietnam's elongation (Fig. 10-8). The French colonizers delimited Vietnam as a 1200-mile (2000-km) strip of land extending from the Chinese border in the north to the tip of the Mekong Delta in the south. Substantially smaller than California, this coastal belt was the domain of the Vietnamese (Fig. 10-3). The French recognized that Vietnam, whose average width is under 150 miles (240 km), was not a homogeneous colony, so they divided it into three units: (1) Tonkin, land of the Red River Delta and centered on Hanoi in the north; (2) Cochin China, region of the Mekong Delta and centered on Saigon in the south; and (3) Annam, focused on the ancient city of Hué, in the middle (Fig. 10-4). Today, the Vietnamese prefer to use *Bac Bo*, *Nam Bo*, and *Trung Bo*, respectively, to designate these areas.

The Vietnamese (or Annamese, also Annamites, after their cultural heartland) speak the same language, although the northerners can easily be distinguished from southerners

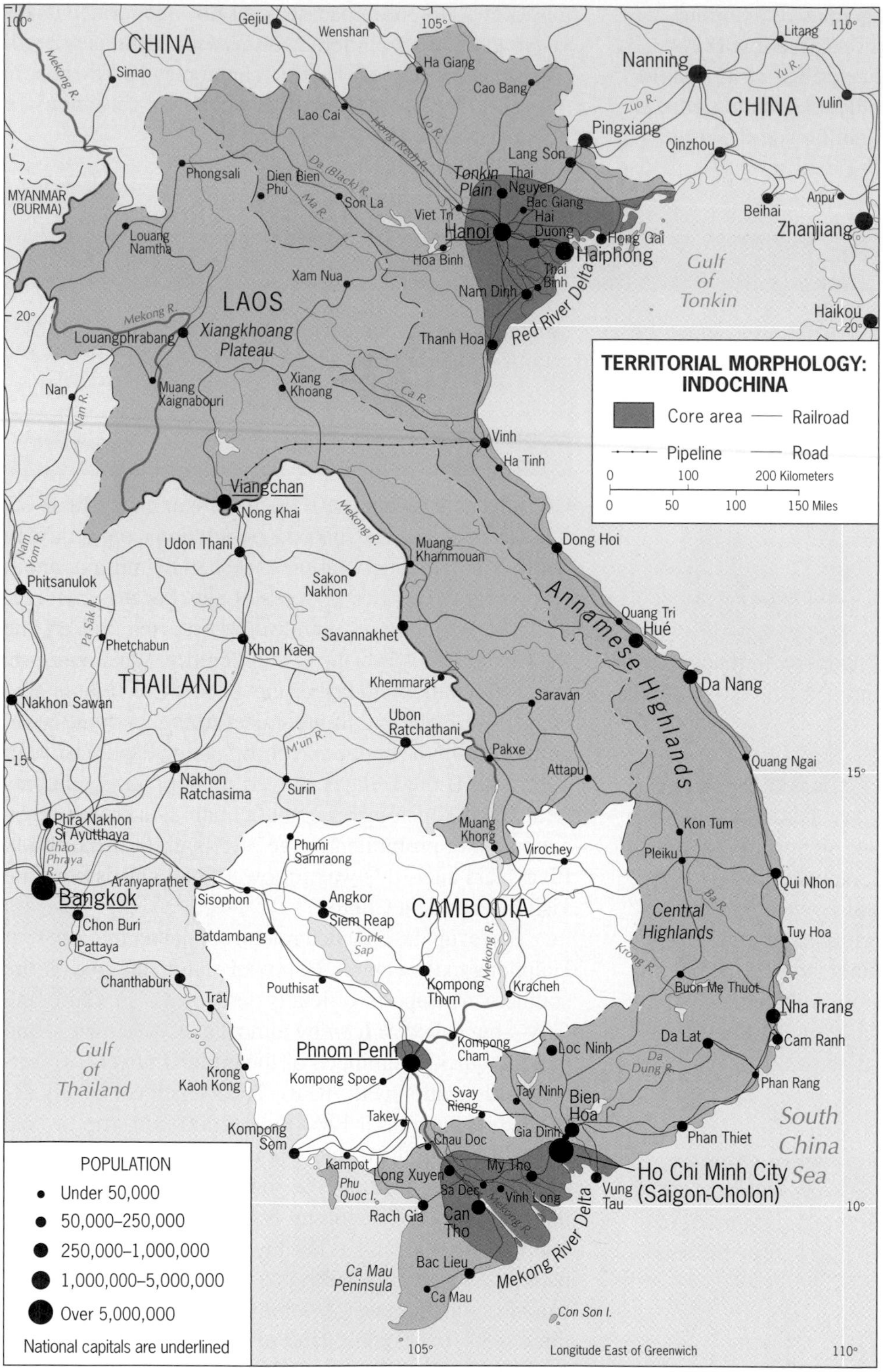

FIGURE 10-8 © H.J. de Blij, P.O. Muller, and John Wiley & Sons, Inc.

by their accent. As elsewhere in their colonial empire, the French made their language the *lingua franca* of Indochina, but their tenure was cut short by the Japanese, who invaded Vietnam in 1940. During the Japanese occupation, Vietnamese nationalism became a powerful force, and after the Japanese defeat in 1945, the French could not regain control. In 1954, the French suffered a disastrous final trouncing on the battlefield at Dien Bien Phu in the far northwest and were ousted from the country.

Even after its forces routed the colonizers, however, Vietnam did not become a unified state. Separate regimes took control: a communist one in Hanoi and a noncommunist counterpart in Saigon. Vietnam's pronounced elongation had made things difficult for the French; now it played its role during the postcolonial period. Note, in Figure 10-8, that Vietnam is widest in the north and south, with a narrow "waist" in its middle zone. North and South Vietnam were worlds apart.

Many Americans still remember how the United States became involved in the inevitable conflict between communists and noncommunists in Vietnam during the 1960s and 1970s. At first, American military advisors were sent to Saigon to help the shaky regime there cope with communist insurgents. When the tide turned against the South, the United States committed military forces and equipment to halt the further spread of communism (see box titled "Domino Theory"); at one time, more than 500,000 U.S. soldiers were in Vietnam.

The conduct of the war, its mounting casualties, and its apparent futility created severe social tensions in the United States. If you had been a student during that period, your college experience would have been radically different from what it is today. Protest rallies, "teach-ins," antidraft demonstrations, "sit-ins," marches, strikes, and even hostage-taking disrupted campus life. The Indochina War threatened the stability of American society. It drove an American president (Lyndon Johnson) from office in 1968 and destroyed the electoral chances of his vice president (Hubert Humphrey), who would not disavow the U.S. role in the war.

In 1975, the Saigon government fell, and the United States was ousted—just a little over two decades after the French defeat at Dien Bien Phu. When the last helicopter left from the roof of the U.S. embassy in Saigon amid scenes of desperation and desertion, it marked the end of a sequence of events that closely conformed to the *insurgent state model* we introduced in the context of contemporary Colombia in Chapter 5 (p. 248).

Vietnam Today

In the aftermath of its devastating war, Vietnam found itself in a territorial conflict with China, and even as countless Vietnamese were boarding boats to escape communist rule, they were joined by thousands of local Chinese. The dimensions of this emigration will never be known; of the estimated 2 million refugee "boat people," more than half perished from storms and sinkings, starvation at sea, and pirate attacks. The majority of those who survived were resettled in America. Meanwhile, the United States placed an embargo on Vietnam, isolating the country and stifling its economy. Only its fertile, rice-yielding river basins allowed it to survive.

At a time when the western Pacific Rim was booming, Vietnam was also held back by its dogmatic communist regime and by its ruined infrastructure. In the early 1990s, the 100-km (60-mile) road trip between the dilapidated capital, Hanoi, and the port of Haiphong could take as long as four hours. In the paddies of the northern (Tonkin) core area, anchored by Hanoi, irrigation water was still raised by bucket. On the roads, animal-drawn carts outnumbered vehicles.

Even then, Saigon-Cholon (officially called Ho Chi Minh City) in the south was far ahead of the northern capital. With about 10 percent of the country's population and a far larger share of its best-educated people, Saigon recovered faster and attracted foreign investment sooner. Cholon, the Chinatown that suffered severely in

Domino Theory

During the Indochina War (1964–1975), it was U.S. policy to contain communist expansion by supporting the efforts of the government of South Vietnam to defeat communist insurgents. Soon, the war engulfed North Vietnam as U.S. bombers attacked targets north of the border between North and South. In the later phases of the war, conflict spilled over into Laos and Cambodia. U.S. warplanes also used bases in Thailand. Like dominoes, one country after another fell to the ravages of the war or was threatened by it.

Some scholars warned that this domino effect could eventually affect not only Thailand but also Malaysia, Indonesia, and Burma (today Myanmar): the whole Southeast Asian realm, they predicted, could be destabilized. But, as we know, that did not happen. The war remained confined to Indochina. And the domino "theory" seemed invalid.

But is the theory totally without merit? Unfortunately, some political geographers to this day mistakenly define this idea in terms of communist activity. Communist insurgency, however, is only one way a country may be destabilized (Peru came close to collapse from it). But right-wing rebellion (Nicaragua's Contras), ethnic conflict (Bosnia), religious extremism (Algeria), and even economic and environmental causes can create havoc in a
country. Properly defined, the **domino theory** holds that 17
destabilization from any cause in one country can result in the collapse of order in a neighboring country, triggering a chain of events that can affect a series of contiguous states in turn.

A recent instance of the domino effect comes from Equatorial Africa, where the collapse of order in Rwanda spilled over into The Congo (then Zaïre) and also affected Burundi, Congo, and Angola.

AMONG THE REALM'S GREAT CITIES . . . SAIGON

Officially, it is known as Ho Chi Minh City, renamed in 1975 after the ouster of American forces, but only bureaucrats and atlases call it that. To the people here, it is still Saigon, city on the Saigon River, served by, it seems, a thousand companies using the name. Saigon Taxi, Saigon Hotel, Saigon Pharmacy—there is hardly a city block without the name Saigon on some billboard.

The phenomenon is significant. Hanoi's northerners ordained the change, but this still is South Vietnam, and many southerners see the North as peopled by slow, dogmatic communists obstructing the hopes and plans for their vibrant city. In truth, Saigon may be at least twice as large as Hanoi, but like the capital it suffers from decades of deterioration, a broken-down infrastructure, and undependable public services. When, in the early 1990s, businesspeople could find no modern accommodations or working telephones, an Australian firm built an eight-story floating hotel with its own power plant, towed it up the Saigon River, and parked it on the downtown waterfront.

Therein lies Saigon's great advantage: large oceangoing vessels can reach the city in a three-hour sail up the meandering river. The Chinese developed its potential: in the 1960s, Chinese merchants controlled more than half of the South's exports and nearly all of its textile factories and foreign exchange. Saigon-Cholon (Cholon means "great market") was a Chinese success story, but after the war the Hanoi regime nationalized all Chinese-owned firms. Tens of

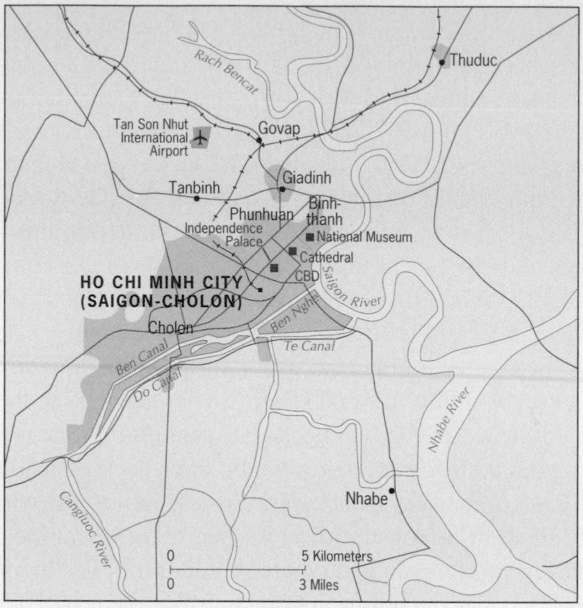

© H. J. de Blij, P. O. Muller, and John Wiley & Sons, Inc.

thousands of middle-class Chinese left the city, its economy devastated.

Today, Saigon (population: 5.3 million) is steadily recovering. A modern skyline is rising above the colonial one. A large Special Economic Zone has been created south of the city. Factories are being built. The Pacific Rim era is under way. Will the South rise again?

the aftermath of the war, thrived again. When, in 1995, the United States moved to normalize relations, Saigon's economy experienced a surge strong enough to trouble the dogmatic bureaucrats in Hanoi. In the city's Special Economic Zone and in its futuristic New Saigon district, signs of a new age are evident.

But Vietnam's communist government has been slow to open the country to foreign investment and involvement. Instead, it has in some ways confirmed the fears of those who resisted communist rule all along: it has harassed Buddhist and Christian groups and detained their leaders, and it has revived old conflicts between minorities in the mountains and ethnic Vietnamese in control. Researchers and visitors report that land is being taken from the hill peoples in large swaths and planted with coffee and other export crops; in 2002, the government announced that Vietnam had become the world's largest exporter of coffee. That may be premature, but the deforestation of the interior highlands is already having an environmental effect as annually worsening floods afflict the once-stable countryside.

By 2003, Vietnam's economy lagged behind those of other countries with much potential on the Pacific Rim. Foreign investment has been limited, and many plans to privatize state industries have been shelved. Like China, Vietnam remains a communist country, but unlike China it has not had the leadership of a Deng Xiaoping. Hanoi's avowed decision to mix communist ideology with market economics never resulted in far-reaching change: the emphasis remained on communist statism, not on economic reform. As a result, Vietnam is experiencing neither the turmoil seen in China nor the huge regional disparities arising there. During the mid-1990s, when Saigon prospered and Hanoi stagnated (photo next page), it appeared that North and South were once again growing apart. Today, Saigon's exuberance is restrained by Hanoi's dogmatism—and with it, Vietnam's economic progress.

FROM THE FIELD NOTES

"Although modernization has changed the skylines of Vietnam's major cities (Saigon more so than Hanoi), most of the country's urban areas look and function much as they have for generations. Main arteries throb with commerce and traffic (still mostly bicycles and mopeds); side streets are quieter and more residential. No high-rises here, but the problems of dense urban populations are nevertheless evident. Every time I turned into a side street, the need for street cleaning and refuse collection was obvious, and drains were clogged more often than not. 'We haven't allowed our older neighborhoods to be destroyed like they have in China,' said my colleague from Hanoi University, 'but we also don't have an economy growing so fast that we can afford to provide the services these people need.' Certainly the Pacific Rim contrast between burgeoning coastal China and slower-growing Vietnam—a matter of communist-government policy—is evident in their cities. 'As you see, nothing much has changed here,' she said. 'Not much evidence of globalization. But we're basically self-sufficient, nobody goes hungry, and the gap between the richest and the poorest here in Vietnam is a fraction of what it is now in China. We decided to keep control.' You would find more agreement on these points here in Hanoi than in Saigon, but the Vietnamese state is stable and progressing." © H. J. de Blij.

COMPACT CAMBODIA

Former French Indochina contained two additional entities—Cambodia and Laos. In Cambodia, the French possessed one of the greatest treasures of Hinduism: the city of Angkor, capital of the ancient Khmer Empire, and the temple complex known as Angkor Wat. The Khmer Empire prevailed here from the ninth to the fifteenth centuries, and King Suryavarman II built Angkor Wat to symbolize the universe in accordance with the precepts of Hindu cosmology. When the French took control of this area during the 1860s, Angkor lay in ruins, sacked long before by Vietnamese and Thai armies. The French created a protectorate, began the restoration of the shrines, established Cambodia's permanent boundaries, and restored the monarchy under their supervision.

Geographically, Cambodia enjoys several advantages, most notably its compact shape. Compact states enclose a maximum of territory within a minimum of boundary and are without peninsulas, islands, or other remote extensions of the national spatial framework. Cambodia had the further advantage of strong ethnic and cultural homogeneity: 85 percent of its 13.7 million inhabitants are Khmers, with the remainder divided between Vietnamese and Chinese. As Figure 10-8 shows, Southeast Asia's greatest river, the Mekong, enters Cambodia from Laos and crosses it from north to south, creating a great bend before flowing into southern Vietnam (see box titled "The Mighty Mekong"). Phnom Penh, the country's capital, lies on the river. The ancient capital of Angkor lies in the northwest, not far from the Tonle Sap ("Great Lake"), a major lake linked to, and filled by, the waters of the Mekong. Cambodia has a coastline on the Gulf of Thailand, but its core area lies in the interior; the Mekong is more important than the Gulf. Kompong Som, the port at the end of the railroad from Phnom Penh, has only limited facilities.

Cambodia's present social geography continues to suffer from the after-effects of the Indochina War. Neither its compactness nor its isolation could overcome the disadvantage of relative location near a conflict that was bound to spill across its borders (Fig. 10-8). Internal disharmony

The Mighty Mekong

From its source among the snowy peaks of China's Qinghai and Xizang, the Mekong River rushes and flows some 2600 miles (4200 km) to its delta in southernmost Vietnam. This "Danube of Southeast Asia" crosses or borders five of the realm's countries (see photo, p. 506), supporting rice farmers and fishing people, forming a transport route where roads are few, and providing electricity from dams upstream. Tens of millions of people depend on the waters of the Mekong, from subsistence farmers in Laos to apartment dwellers in China. The Mekong Delta in southern Vietnam is one of the realm's most densely populated areas and produces enormous harvests of rice.

But problems loom. China is building a series of dams across the Lancang (as the Mekong is called there) to supply Yunnan Province with electricity. Although such hydroelectric dams should not interfere with water flow, countries downstream worry that a severe dry spell in the interior would impel the Chinese to slow the river's flow to keep the reservoirs full. Cambodia is concerned over the future of the Tonle Sap, a large natural lake filled by the Mekong (Fig. 10-8). In Vietnam, farmers worry about salt water invading the paddies should the Mekong's level drop. And the Chinese may not be the only dam builders in the future: Thailand has expressed an interest in building a dam on the Thai-Laos border where it is defined by the Mekong.

In such situations, the upstream states have an advantage over those downstream. Several international organizations have been formed to coordinate development in the Mekong Basin, including the Mekong River Commission (MRC) founded half a century ago. China has offered to sell electricity from its dams to Thailand, Laos, and Myanmar. Coordinated efforts to reduce logging in the Mekong's drainage basin have had some effect. After consultations with the MRC, Australia built a bridge linking Laos and Thailand. There is even a plan to make the Mekong navigable from Yunnan to the coast, creating an alternative outlet for interior China.

Sail the Mekong today, however, and you are struck by the slowness of development along this artery. Wooden boats, thatch-roofed villages, and teeming paddies mark a river still crossed by antiquated ferries and flanked by few towns. Of modern infrastructure, one sees little. And yet the Mekong and its basin form the lifeline of mainland Southeast Asia's dominantly rural societies.

fueled by external interference led in 1970 to the ouster of the last king by the military; in 1975, that regime was itself overthrown by communist revolutionaries, the so-called Khmer Rouge. These new rulers embarked on a course of terror and destruction in order to reconstruct Kampuchea (as they called Cambodia) as a rural society. They drove townspeople into the countryside where they had no place to live or work, emptied hospitals and sent the sick and dying into the streets, outlawed religion and family, and in the process killed as many as 2 million Cambodians (out of

FROM THE FIELD NOTES

"Here near Siem Reap, along a small stream flowing into the Tonle Sap, I had the opportunity to observe one of Cambodia's many Mekong Basin stilt villages. The water rises and falls with the seasons; it is comparatively dry now, and a pool of stagnant water is what is left of the river that will soon form here once again. In the past, stilt houses were built of traditional materials only, but now there are tin roofs and plastic sheeting. Conditions are not sanitary; there was refuse everywhere, waiting to be swept away by the rising waters. Outhouses such as those in the left foreground drained directly into the temporary pond, now mosquito-infested and severely polluted. I wondered how many infants would survive their first year here." © Barbara A. Weightman.

a population of 8 million). In the late 1970s, Vietnam, having won its own war, invaded Cambodia to drive the Khmer Rouge away. But this action led to new terror, and a stream of refugees crossed the border into Thailand. Eventually, remnants of the Khmer Rouge managed to establish a base in the northwest of Cambodia, where their murderous leader, Pol Pot, who was never brought to account for his actions, committed suicide in 1998.

Cambodia's postwar trauma continues today. Once self-sufficient and able to feed others, it now must import food. Rice and beans are the subsistence crops, but the dislocation in the farmlands set production back severely, and continuing strife in the countryside has disrupted supply routes for years.

Cambodians hope that tourism, focused chiefly on the great temples at Angkor, will become a mainstay of the economy. But political stability is a prerequisite for this industry, and it has not yet returned to this ancient kingdom.

LANDLOCKED LAOS

North of Cambodia lies Southeast Asia's only landlocked country, Laos (Fig. 10-8). Interior and isolated, Laos changed little during 60 years of French colonial administration (1893–1953). Then, along with other French-ruled areas, it became an independent state. Soon this well-entrenched, traditional kingdom fell victim to rivalries between traditionalists and the victorious communists, and the old order collapsed.

Laos has no fewer than five neighbors, one of which is the East Asian giant, China. The Mekong River forms a long stretch of its western boundary, and the important sensitive border with Vietnam to the east lies in mountainous terrain. With 6.1 million people (about half of them ethnic Lao, related to the Thai of Thailand), Laos lies surrounded by comparatively powerful states. The country has no railroads, just a few miles of paved roads, and very little industry; it is only 19 percent urbanized (the capital, Viangchan, lies on the Mekong and has an oil pipeline to Vietnam's coast). Laos remains the region's poorest and most vulnerable entity.

FROM THE FIELD NOTES

"Along a road in Xiangkhoang Province, in north-central Laos, I stopped to watch these people operate their gas-powered rice mill. I had some difficulty communicating with the people, but from the young boy I gathered that some of this equipment had been fashioned from U.S. war materiel, including the drum (I wondered what it might have contained) and the funnel, which was made from bomb casings." © Barbara A. Weightman.

PROTRUDED THAILAND

In virtually every way, Thailand is the leading state of the Mainland region. In contrast to its neighbors, Thailand has been a strong participant in the Pacific Rim's economic development. Its capital, Bangkok (population: 6.8 million), is the largest urban center in the Mainland region and one of the world's most prominent primate cities. The country's population, 64.7 million in 2006, is growing at the second slowest rate in this geographic realm (only fully urbanized Singapore grows more slowly). Over the past few decades, only political instability and uncertainty have inhibited economic progress. Thailand is a constitutional monarchy; its moves toward stable democracy have been halting. Corruption remains a major problem here.

Thailand is the textbook example of a protruded state. From a relatively compact heartland, in which lie the core area, capital, and major areas of productive capacity, a 600-mile (1000-km) corridor of land, in places less than 20 miles (32 km) wide, extends southward to the border with Malaysia (Fig. 10-9). The boundary that defines this protrusion runs down the length of the Malay Peninsula to the Kra Isthmus, where neighboring Myanmar peters out and Thailand fronts the Andaman Sea (an arm of the Indian Ocean) as well as the Gulf of Thailand. In the entire country, no place lies farther from the capital than the southern end of this tenuous protrusion.

Consider this spatial situation in the context of Figure 10-3 (p. 498). Note that the Malay ethnic population group extends from Malaysia more than 200 miles (300 km) into Thai territory. In Thailand's five southernmost provinces, 85 percent of the inhabitants are Muslims (the figure for Thailand as a whole is less than 5 percent). The political border between Thailand and Malaysia is porous; in fact,

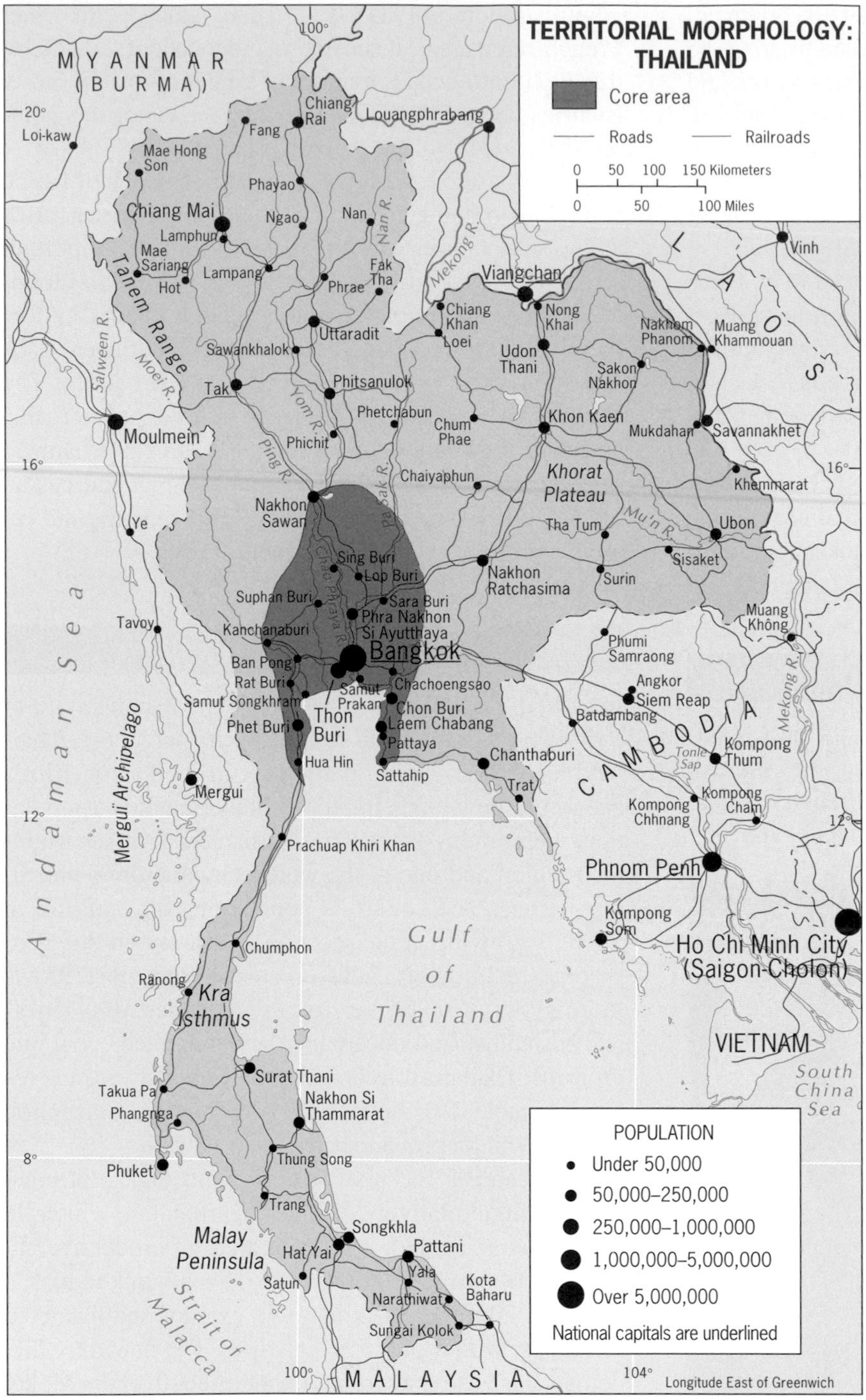

FIGURE 10-9 © H.J. de Blij, P.O. Muller, and John Wiley & Sons, Inc.

you can cross it at will almost anywhere along the Kolok River by canoe, and inland it is a matter of walking a forest trail. For more than a century, the southern provinces have had closer ties with Malaysia across the border than with Bangkok 600 miles (1000 km) away, and the Thai government has not tried to restrict movement or impose onerous rules on the Muslim population here.

But in the new era of the War on Terror and in view of rising violence in this southern frontier zone, Thailand's south is attracting national attention. The Pattani United Liberation Organization, named after the functional capital of the Muslim south, was active in the 1960s and 1970s but had been dormant since. Then a series of bombings at a Pattani hotel, several schools, a Chinese shrine, and a Buddhist temple in 2001 and 2002 raised fears that Muslim-inspired violence was on the rise again. In April 2002, Bangkok newspapers reported that police had found bomb-making materials in the

AMONG THE REALM'S GREAT CITIES . . . BANGKOK

Bangkok (6.8 million), mainland Southeast Asia's largest city, is a massive sprawling metropolis on the banks of the Chao Phraya River, an urban agglomeration without a center, an aggregation of neighborhoods ranging from the immaculate to the impoverished. In this city of great distances and high tropical temperatures, getting around is often difficult because roadways are choked with traffic. A (diminishing) network of waterways affords the easiest way to travel, and life focuses on the busiest waterway of all, the Chao Phraya. Ferries and water taxis carry tens of thousands of commuters and shoppers across and along this bustling artery, flanked by a growing number of high-rise office buildings, luxury hotels, and ultramodern condominiums. Many of these modern structures reflect the Thais' fondness for domes, columns, and small-paned windows, creating a skyline unique in Asia.

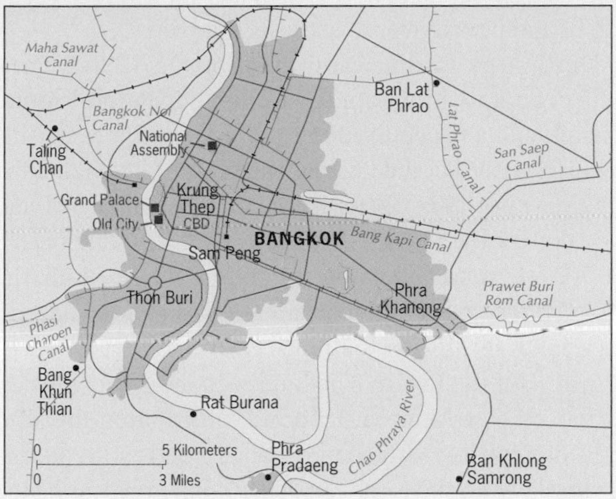

Gold is Buddhist Thailand's symbol, adorning religious and nonreligious architecture alike. From a high vantage point in the city, you can see hundreds of golden spires, pagodas, and façades rising above the townscape. The Grand Palace, where royal, religious, and public buildings are crowded inside a white, crenellated wall more than a mile long (see photo p. 518), is a gold-laminated city within a city embellished by ornate gateways, dragons, and statuary. Across the mall in front of the Grand Palace lie government buildings sometimes targeted by rioters in Bangkok's volatile political atmosphere. Not far away are Chinatown and myriad markets; Bangkok has a throbbing commercial life.

Decentralized, dispersed Bangkok has major environmental problems, ranging from sinking ground (resulting in part from excessive water pumping) to some of the world's worst air pollution. Yet industries, port facilities, and residential areas continue to expand as the city's hinterland grows and prospers with Pacific Rim investment. Bangkok is the pulse of Southeast Asia's Mainland region.

house of a Muslim religious leader. In August 2003, the notorious al-Qaeda terrorist Hambali, who worked to link al-Qaeda with its Southeast Asian counterpart, Jemaah Islamiah, was captured in Thailand. The year 2004, however, brought the most ominous developments yet. A terrorist attack on an army base in January killed four Buddhist soldiers and resulted in the theft of hundreds of rifles. And several dozen people, including Buddhist monks and police officers, were killed in subsequent skirmishes when, in October, counterterrorism forces in the town of Takbai arrested six Muslims suspected of stealing weapons from the Thai armed forces. A riot outside the police station where the suspects were being held then led to a mass arrest and the deaths of nearly 80 Muslims during transfer to a military base. The government responded by declaring martial law in Thailand's three southernmost provinces.

Religious tensions in this increasingly troubled Islamic outpost should be seen in the context of this area's geography. Thailand's pronounced protrusion, its porous border with Muslim Malaysia (the central government now proposes the building of a "security fence" there), its dependence on foreign tourism, and the vulnerable location of its Andaman Sea coastal tourist facilities (which rebounded swiftly from the 2004 tsunami), combine to raise concern over the stability and security of this distant part of the national territory.

As Figure 10-2 shows, Thailand occupies the heart of the Mainland region of Southeast Asia. While Thailand has no Red, Mekong, or Irrawaddy Delta, its central lowland is watered by a set of streams that flow off the northern highlands and the Khorat Plateau in the east. One of these streams, the Chao Phraya, is the Rhine of Thailand. From the head of the Gulf of Thailand to Nakhon Sawan, this river is a highway of traffic. Barge trains loaded with rice head for the coast, ferry boats head upstream, and freighters transport tin and tungsten (of which Thailand is among the world's largest producers). Bangkok sprawls on both sides of the lower Chao Phraya, here flanked by skyscrapers, pagodas, factories, boatsheds, ferry landings, luxury hotels, and modest dwellings in crowded confusion. On the right bank of the

Chao Phraya, Bangkok's west side, lie the city's remaining *klong* (canal) neighborhoods, where waterways and boats form the transport system. Bangkok is still known as the Venice of Asia, although many *klongs* have been filled in and paved over to serve as roadways.

During the Pacific Rim boom of the 1990s, Thailand's economy grew rapidly. Relative location, natural environment, and social conditions all contributed to this growth. Situated at the head of the Gulf of Thailand, the country's core area opens to the South China Sea and hence the Pacific Ocean. The Gulf itself has long been a rich fishing ground and more recently a source of oil (the major refinery lies in the port area of Laem Chabang, an industrial zone just to the north of the tourist resort of Pattaya). But the key to Thailand's economic explosion was the Thai workforce, laboring under often dreadful conditions for low wages to produce goods sold on foreign markets at cheap prices. This labor force attracted major foreign investment that, in turn, stimulated service industries.

For a time, Thailand became a Pacific Rim economic tiger, producing goods ranging from plastic toys to "Japanese" cars. But while the workers toiled, government and financial mismanagement planted seeds of ruin. In 1997, the country's currency, the *baht*, began a slide in value that undermined the economy and led to numerous bank failures. Exports and worker productivity declined, real estate values plunged, and Thailand became the first casualty in a series of economic setbacks that would eventually affect the entire western Pacific Rim.

Although Thailand and the other Pacific Rim economies suffered severe downturns, we should remember that the boom period had yielded many infrastructural improvements not only in the core area but also in distant areas such as the north, centered on Chiang Mai. Thailand, like other Pacific Rim economies, has improved highways, ports, power systems, airports, and housing. Although a period of poor maintenance followed the events of 1997, Thailand's fundamental assets remained to serve the country well when the world economy revived. By 2005, Thailand's GNI per capita had recovered to approach U.S. $7500, its balance of trade had been favorable for seven consecutive years, and the tourist industry, buffeted by the post–2001 religious tensions and more briefly by the 2004 tsunami, was recovering.

Under normal circumstances, Thailand's natural environments and cultural landscapes continue to attract millions of visitors annually. The magnificent architecture of Buddhism (the country's old name, Siam, was a Chinese word meaning golden-yellow), swathed in gold and shining in the tropical sun, is without equal, as the photo on this page underscores. The warm coasts and beaches have long made this country a favorite destination for tourists from Germany to Japan. One of the world's most famous resorts is Phuket, on the distant Andaman coast near the southwestern end of Thailand's Malay Peninsula protrusion.

FROM THE FIELD NOTES

"Thailand exhibits one of the world's most distinctive cultural landscapes, both in its urban and its rural areas. Graceful pagodas, stupas, and spirit houses of Buddhism adorn city and countryside alike. Gold-layered spires rise above, and beautify cities and towns that would otherwise be drab and featureless. The architecture of Buddhism has diffused into public architecture, so that many secular buildings, from businesses to skyscrapers, are embellished by something approaching a national style. The grandest expression of the form undoubtedly is a magnificent assemblage of structures within the walls of the Grand Palace in the capital, Bangkok, of which a small sample is shown here. Climb onto the roof of any tall building in the city's center, however, and the urban scene will display hundreds of such graceful structures, interspersed with the modern and the traditional townscape." © H. J. de Blij.

Tourism is Thailand's leading source of foreign revenues, but there is a dark side to the industry. The Thai people's relaxed attitude toward sex has made Thai tourism in large part a sex industry, candidly advertised in such foreign markets as Japan where "sex tours" attract planeloads of male participants. When the AIDS pandemic reached Thailand in the 1980s, it changed no habits, and by the mid-1990s the country was suffering one of the world's worst outbreaks of the disease. At the turn of this century, more than 1 million of Thailand's population of 60-plus million were infected.

Thailand's territorial morphology creates both problems and opportunities. Problems include the effective integration of the national territory through surface communications, which must cover long distances to reach remote locales; the influx of refugees from neighboring countries dislocated by internal conflict; and control over contraband (opium) drug production and trade in the interior "Golden Triangle" where Myanmar, Thailand, and Laos meet. The government's efforts to speed the development of rural areas, notably the northeast, through megaprojects such as large dams have disrupted local life, damaged ecologies, and deepened the split between city and countryside. On the other hand, Thailand's lengthy Kra Isthmus creates economic opportunities where southernmost Thailand lies close to Malaysia and Indonesia, raising hopes for a three-country development scheme in that area. Given political stability, Thailand's future as a regional leader appears secure.

EXTENDED MYANMAR

Myanmar (formerly Burma) shares with Thailand its protruded morphology but little else. Long languishing under one of the world's most corrupt military regimes, Myanmar is one of the planet's poorest countries where, it seems, time has stood still for centuries. Whereas air, railroad, and highway routes serve Thailand's peninsular protrusion, there is not even an all-weather track to the southern end of Myanmar's. Whereas busy shipping lanes converge on Thailand's coastal core area, Myanmar's Irrawaddy Delta is more reminiscent of Bangladesh.

Myanmar's territorial morphology is complicated by a shift in the Burmese core area that took place during colonial times. Prior to the colonial period, the focus of embryonic Burma lay in the so-called dry zone between the Arakan Mountains and the Shan Plateau. The urban focus of the state was Mandalay, which had a central situation and relative proximity to the non-Burmese highlands all around (Fig. 10-1). Then the British developed the agricultural potential of the Irrawaddy Delta, and Rangoon (now called Yangon) became the hub of the colony. The Irrawaddy waterway links the old and the new core areas, but the center of gravity has shifted to the south.

The political geography of Myanmar constitutes a particularly good example of the role and effect of state morphology on internal state structure. Not only is Myanmar a protruded state; in addition, its core area is surrounded on the west, north, and east by a horseshoe of great mountains—where many of the country's 11 minority peoples had their homelands before the British occupation. The colonial boundaries had the effect of incorporating these peoples into the domain of the Burman people, who today constitute about 55 percent of the population of 51.5 million. When the British departed and Burma became independent in 1948, the less numerous peoples traded one master for another. In 1976, nine of these indigenous peoples formed a union to demand the right to self-determination in their homelands.

As Figure 10-3 shows, the peripheral peoples of Myanmar occupy a significant part of the state. The Shan of the northeast and far north, who are related to the neighboring Thai, account for about 7 percent of the population, or 3.5 million. An even larger minority, the Karen (just under 10 percent, 4.9 million), live in the neck of Myanmar's protrusion and have proclaimed that they wish to create an autonomous territory within a federal Myanmar. Although the powerful military have dealt these aspirations a series of setbacks, centrifugal forces continue to bedevil the central regime. Its response has been to exert power by all available means rather than to accommodate these forces, even to the point of stifling political discourse (let alone opposition) among the Burman themselves.

This is all the more tragic because Myanmar has economic potential far superior to that of Bangladesh, its neighbor to the west, and to other countries in its own region. It can feed itself, and it could export significant quantities of rice. Varied soils and diverse environments can yield a range of other crops. Tin and other metals have been found in the highlands. Today, Myanmar is exploiting the forests of the interior at an alarming rate; teak has risen to first place among exports by value—among legitimate exports, that is. In its corner of the "Golden Triangle," Myanmar also is one of the world's largest illicit producers of opium poppies and harvests other illegal crops for the drug trade as well. For a time, the regime in Yangon tried to control the trade, but its constant conflict with peoples outside the core area diverted its attentions and energies from that campaign.

For decades, the repressive policies of a brutal military regime have kept Myanmar in turmoil and all but halted its development. Infrastructure throughout the country is in dismal condition, but this has not stopped the regime from purchasing a nuclear reactor and jet fighters from Russia and conventional weapons from North Korea. Nor has isolated Myanmar escaped the cycle of violence affecting other countries in Asia: religious riots between Muslims and Buddhists in 2001 claimed numerous victims. About 3 percent of Myanmar's population is Muslim.

In the first years of this new century, the military regime made some overtures toward the opposition led by one of the world's most courageous politicians, Daw

Aung San Suu Kyi. She has been in various forms of detention for more than a decade but has not yielded to threats or enticements. Progress, however, remained minimal. In any case, a change of heart by the regime will not undo the effects of policies that have pushed impoverished and exhausted Myanmar into the ranks of the world's poorest economies.

INSULAR SOUTHEAST ASIA

On the peninsulas and islands of Southeast Asia's southern and eastern periphery lie six of the realm's 11 states (Fig. 10-2). Few regions in the world contain so diverse a set of countries. Malaysia, the former British colony, consists of two major areas separated by hundreds of miles of South China Sea. The realm's southernmost large state, Indonesia, sprawls across thousands of islands from Sumatera in the west to New Guinea in the east. North of the Indonesian archipelago lies the Philippines, a nation that once was a U.S. colony. These are three of the most severely fragmented states on Earth, and each has faced the challenges that such politico-spatial division brings. This Insular region of Southeast Asia also contains two small but important sovereign entities: a city-state and a sultanate. The city-state is Singapore, once a part of Malaysia (and one instance in which internal centrifugal forces were too great to be overcome). The sultanate is Brunei, an oil-rich Muslim territory on the island of Borneo that seems transplanted from the Persian Gulf. In addition, a third small entity, East Timor, achieved independence in May 2002 following three years of UN administration. Few parts of the world are more varied or more interesting geographically.

FROM THE FIELD NOTES

"Walking along Tuanku Abdul Rahman Street in Kuala Lumpur, I had just passed the ultramodern Sultan Abdul Samad skyscraper when this remarkable view appeared: the old and the new in a country that seems to have few postcolonial hang-ups and in which Islam and democracy coexist. The British colonists designed and effected the construction of the Moorish-Victorian buildings in the foreground (now the City Hall and Supreme Court); behind them rises the Bank of Commerce, one of many banks in the capital. Look left, and you see the Bank of Islam, not a contradiction here in economically diversified Malaysia. And just a few hundred yards away stands St. Mary's Cathedral, across the street from still another bank. Several members of the congregation told me that the church was thriving and that there was no sense of insecurity here. 'This is Malaysia, sir,' I was told. 'We're Muslims, Buddhists, Christians. We're Malays, Chinese, Indians. We have to live together. By the way, don't miss the action at the Hard Rock Café on Sultan Ismail Street.' Now there, I thought, was a contradiction as remarkable as this scene." © H. J. de Blij.

MAINLAND-ISLAND MALAYSIA

The state of Malaysia represents one of the three types of fragmented states discussed earlier: the mainland-island type, in which one part of the national territory lies on a continent and the other on an island. Malaysia is a colonial political artifice that combines two quite disparate components into a single state: the southern end of the Malay Peninsula and the northern part of the island of Borneo. These are known, respectively, as West Malaysia and East Malaysia (Fig. 10-2). The name *Malaysia* came into use in 1963, when the original Federation of Malaya, on the Malay Peninsula, was expanded to incorporate the areas of Sarawak and Sabah in Borneo. When the name *Malaya* is used, it refers to the peninsular part of the Federation, whereas Malaysia refers to the total entity.

The Malays of the peninsula, traditionally a rural people, displaced older aboriginal communities there and today make up about 61 percent of the country's population of 26.7 million. They possess a strong cultural identity expressed in adherence to the Muslim faith, a common language, and a sense of territoriality that arises from their perceived Malayan origins and their collective view of Chinese, Indian, European, and other foreign intruders.

The Chinese came to the Malay Peninsula and to northern Borneo in substantial numbers during the colonial period, and today they constitute one-quarter of Malaysia's population (they are the largest single group in Sarawak). During World War II, when the Japanese ruled Malaya, the Chinese were ruthlessly persecuted; many were driven into the forested interior where they founded a communist-inspired resistance movement that long destabilized the area. Today, the commercially active, urbanized Chinese sector still is victimized by discrimination, now by the dominant Malays.

Hindu South Asians were in this area long before the Europeans, and for that matter before the Arabs and Islam arrived on these shores. Today they still form a substantial minority of over 7 percent of the population, clustered, like the Chinese, on the western side of the peninsula (Fig. 10-3).

The populous peninsular part of Malaysia remains the country's dominant sector with 11 of its 13 States and fully 80 percent of its population. Here the Malay-dominated government has strictly controlled economic and social policies while pushing the country's modernization. During the Asian economic boom of the 1990s, Malaysia's planners embraced the notion of symbols: the capital, Kuala Lumpur, was endowed with the world's tallest building; a space-age airport outpaced Malaysia's needs; a high-tech administrative capital began to be built at Putrajaya, and a nearby development was called Cyberjaya—all part of a so-called *Multimedia Supercorridor* to anchor Malaysia's core area (Fig. 10-10).

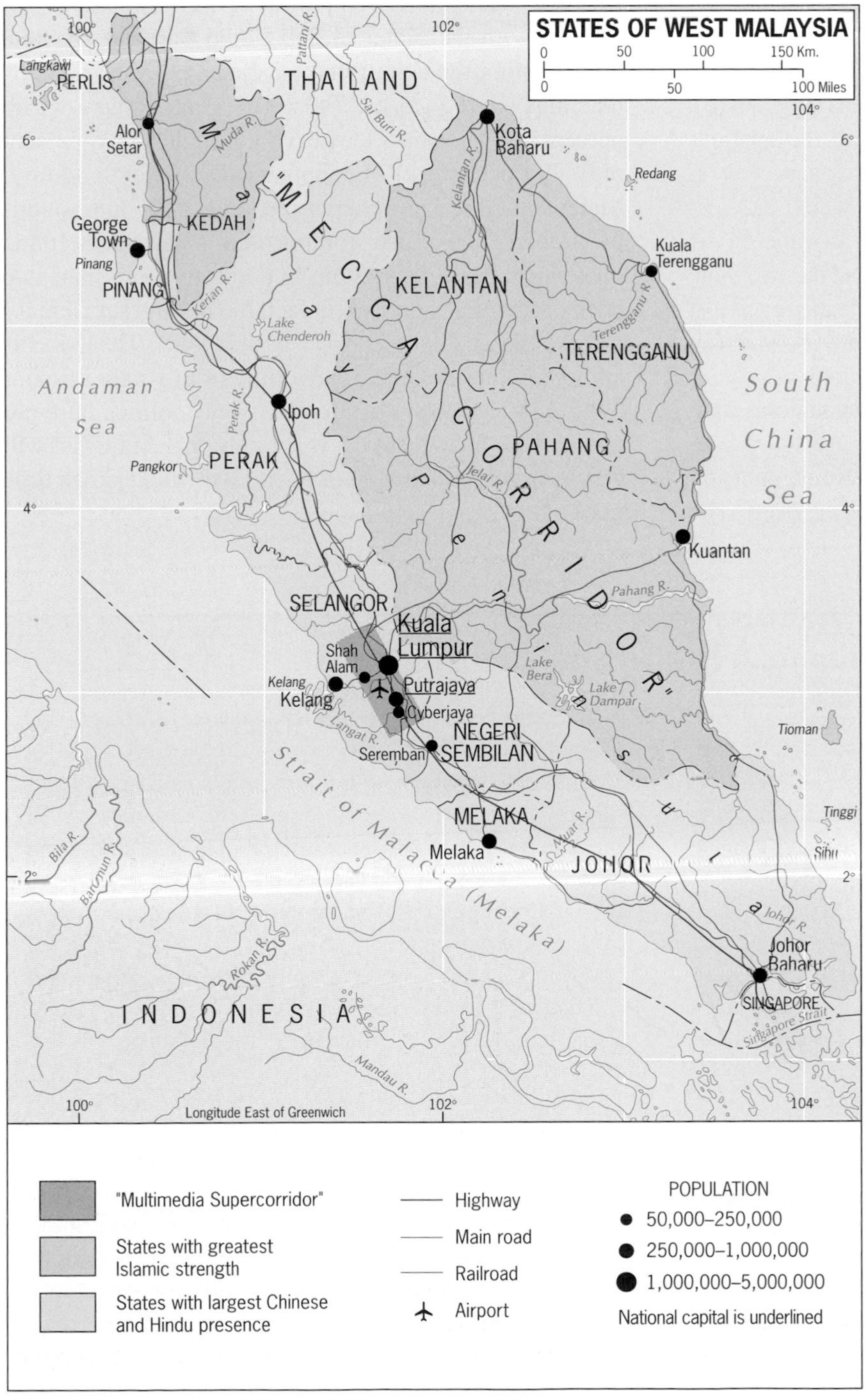

FIGURE 10-10 © H.J. de Blij, P.O. Muller, and John Wiley & Sons, Inc.

The chief architect of this program was Malaysia's long-term and autocratic head of state, Mahathir bin Mohamad, leader of the Malay-dominated party that forms the majority in government. Mahathir not only had the support of the great majority of the country's Malays, but also that of the important and influential ethnic Chinese minority, which saw him as the only acceptable alternative to the more fundamentalist Islamic party challenging his rule. Malaysia's headlong rush to modernize caused a backlash among more conservative Muslims which, in 2001, led to Islamist victories in two States, tin-producing Kelantan and energy-rich but socially poor Terengganu. As the fundamentalist governments in those two States imposed strict religious laws, Malaysians talked of two "corridors" marking their country: the Multimedia Supercorridor in the west and the "Mecca Corridor" in the east (Fig. 10-10). However, after Mahathir's resignation and the appointment of Abdullah Badawi as his more moderate successor, Islamist fervor in the "Mecca Corridor," which had featured calls for a *jihad* in Malaysia, began to wane. In the March 2004 elections, the Islamist party lost the two State legislatures it had gained in 2001, and its leader even lost his seat in the federal parliament. Democracy, flawed as it still may be in this Muslim country, was a victor.

The Malay Peninsula's primacy began long ago, during colonial times, when the British created a substantial economy based on rubber plantations, palm-oil extraction, and mining (tin, bauxite, copper, iron). The Strait of Malacca (Melaka) became one of the world's busiest and most strategic waterways, and Singapore, at the southern end of it, a prized possession (Singapore seceded from the Malaysian Federation in 1965).

Malaysia, despite the loss of Singapore and notwithstanding its recurrent ethnic troubles, became a major player on the burgeoning Pacific Rim. The strong skills and modest wages of the local workforce attracted many companies, and the government capitalized on its opportunities, for example, by encouraging the creation of a high-technology manufacturing complex on the island of Pinang (see box titled "Pinang: A Future Singapore?").

The decision to combine the 11 Sultanates of Malaya with the States of Sabah and Sarawak on Borneo, creating the country now called Malaysia, had far-reaching consequences. These two States make up 60 percent of Malaysia's territory (although they represent only 20 percent of the population). They endowed Malaysia with major energy resources and huge stands of timber. They also complicated Malaysia's ethnic makeup because each State is home to more than two dozen indigenous groups (in fact, the immigrant Chinese form the largest single group in Sarawak). These locals complain that the federal government in Kuala Lumpur treats East Malaysia as a colony, and politics here are contentious and fractious. It is likely that Malaysia will confront devolutionary forces in East Malaysia as time goes on.

Pinang: A Future Singapore?

In the northwestern corner of Malaysia's Malay Peninsula, opposite the northern end of Indonesian Sumatera, where the Strait of Malacca (Melaka) opens into the Indian Ocean, lies an island called Pinang (Fig. 10-10). A thriving outpost during the British colonial period, Pinang (capital: George Town), like Singapore, was one of the so-called Straits Settlements.

Today, Singapore no longer is a part of Malaysia, but Pinang's importance is growing rapidly. Now connected to the mainland by one of Asia's longest bridges (9 miles [14 km]), the 100-square-mile island has a population of over 1 million. Chinese outnumber Malays by 60 to 32 percent, and George Town, the old colonial port, has become the focus of an expanding high-technology manufacturing complex for the international computer industry. The 65-story Tun Abdul Razak Complex, housing government offices, businesses, department stores, and recreational facilities, is just one landmark in an ultramodern skyline. From the top you can see modern industrial parks with neon-lit signs proclaiming the presence of such companies as Intel, SONY, Philips, Motorola, and Hitachi.

The rapid development of Pinang is a calculated strategy by the Malaysian government. The emphasis is on building a highly skilled labor force that is locally nurtured in technical schools and training facilities financed by the multinational corporations themselves. Malaysia makes infrastructure investments ranging from the new toll bridge to a major airport. George Town's colonial and multicultural townscape is attracting a growing tourist trade, and new hotels and resorts are enlarging the service industry.

Malaysia hopes that Pinang will become the leading corner of a major *growth triangle* (a concept that is gaining popularity along the Pacific Rim), together with Indonesian Sumatera to the west and southern Thailand to the north. Meanwhile, Pinang is following in Singapore's economic footsteps. But Kuala Lumpur's politicians surely hope that Chinese-dominated Pinang will not follow Chinese-dominated Singapore's political course.

SINGAPORE

In 1965, a fateful event occurred in Southeast Asia: Singapore, crown jewel of British colonialism in this realm, seceded from the Malaysian Federation and became a sovereign state, albeit a ministate (Fig. 10-11). With its magnificent relative location, human resources, and firm government, Singapore then overcame the limitations of space and the absence of raw materials to become one of the economic tigers on the Pacific Rim.

With a mere 240 square miles (600 sq km) of territory, space is at a premium in Singapore, and this is a constant worry for the government. Singapore's only local spatial advantage over Hong Kong is that its small territory is less fragmented (there are just a few small islands in addition to the compact main island). With a population of 4.3 million and an expanding economy, Singapore must develop space-conserving, high-tech industries. It has succeeded in this effort, and the urban area's high-rise buildings reflect the city-state's prosperity (photo p. 524). You will find no slums in Singapore; the dilapidated streets in the old Chinatown were left merely to conserve a fragment of the city's past. Singapore mostly impresses with its newness and its modernity, and with the order of daily life.

As the map shows, Singapore lies where the Strait of Malacca (Melaka), leading westward to India, opens into the South China Sea and the waters of Indonesia (Fig. 10-11 inset). The port developed as part of the British Southeast Asian empire, and when independence came in 1963, it was made a part of the Malaysian Federation. However, two years later Singapore seceded from Malaysia and became a genuine city-state on the Pacific Rim. No reunification, merger, or annexation looms here.

Benefiting from its relative location, the old port of Singapore had become one of the world's busiest (by numbers of ships served) even before independence. It thrived as an ***entrepôt*** between the Malay Peninsula, 18
Southeast Asia, Japan, and other emerging economic powers on the Pacific Rim and beyond. Crude oil from Southeast Asia still is unloaded and refined at Singapore, then shipped to Asian destinations. Raw rubber from the adjacent Peninsula and from Indonesia's island of Sumatera is shipped to Japan, the United States, China, and other countries. Timber from Malaysia, rice, spices, and other foodstuffs are processed and forwarded via Singapore. In return, automobiles, machinery, and equipment are imported into Southeast Asia through Singapore.

But that is the old pattern. Singapore's leaders want to redirect the city-state's economy and move toward

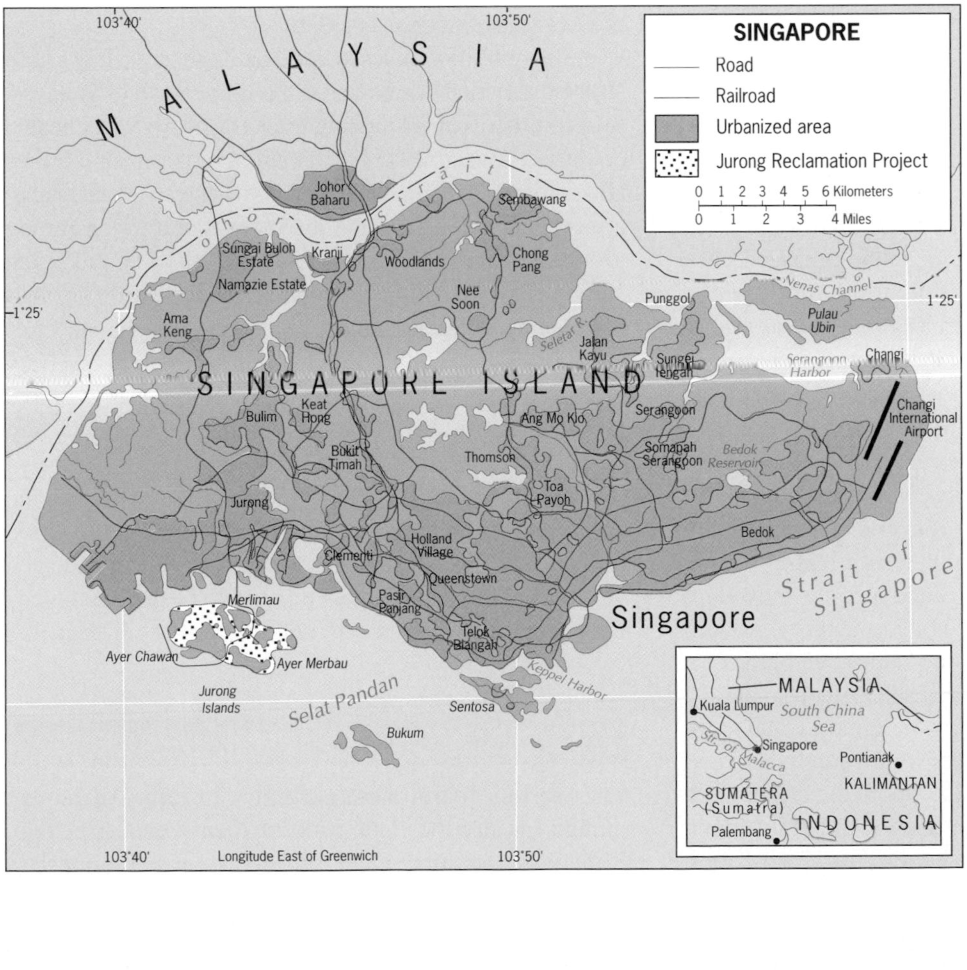

FIGURE 10-11 © H.J. de Blij, P.O. Muller, and John Wiley & Sons, Inc.

FROM THE FIELD NOTES

"My first visit to Singapore was by air, and I got to know the central city, where I was based, rather well. But the second was by freighter from Hong Kong, and this afforded quite a different perspective. Here was one reason for Singapore's success: I have simply never seen a neater, cleaner, better organized, or more modern port facility. Even piles of loose items were carefully stacked. Loading and offloading went quietly and orderly. Singapore is one of the world's leading *entrepôts*, where goods are brought in, stored, and transshipped. Nobody does it better, and Singapore carefully guards its reputation for dependability, on-schedule loading, and lack of corruption." © H. J. de Blij.

high-tech industries for the future. In Singapore the government tightly controls business as well as other aspects of life. (Some newspapers and magazines have been banned for criticizing the regime, and there are even fines for such things as eating on the subway and failing to flush a public toilet.) Its overall success after secession has tended to keep the critics quiet: while GNI per capita from 1965 to 2003 multiplied by a factor of more than 15—to U.S. $24,180—that of neighboring Malaysia reached just U.S. $8940. Among other things, Singapore became (and for many years remained) the world's largest producer of disk drives for small computers.

As multinational corporations settled in, Singapore's political stability on the volatile Pacific Rim was another advantage. Nonetheless, a problem emerged: unlike Taiwan, where brainpower is plentiful, Singapore's technical capacities were limited. As recently as 1995, only 8 percent of the population was university educated. So while the government reduced the incentives for low-technology manufacturers to come here and wages rose, the high-tech industries found themselves thinking about moving to nearby Malaysia and Thailand. That slowed Singapore's boom and challenged the government to find ways to stimulate a revival.

To accomplish this revival, Singapore has moved in several directions. First, it will focus on three growth areas: information technology, automation, and biotechnology. Second, there are notions of a "Growth Triangle" involving Singapore's developing neighbors, Malaysia and Indonesia; those two countries would supply the raw materials and cheap labor, and Singapore the capital and technical know-how. Third, Singapore opened its doors to capitalists of Chinese ancestry who left Hong Kong when China took it over and who wanted to relocate their enterprises here. Singapore's population is 77 percent Chinese, 14 percent Malay, and 8 percent South Asian. The government is Chinese-dominated, and its policies have served to sustain Chinese control. Indeed, China itself often cites Singapore's combination of authoritarianism and economic success as proof that communism and market economics can coexist.

Singapore continues to build on its success. It has constructed a second bridge across the Johor Strait to Malaysia to accommodate the spiraling truck traffic between the two countries. An active land reclamation program already has added over 4 percent more territory to the crowded island; a recent scheme connected the main island to the Jurong group of islands in the southwest (Fig. 10-11). All seven Jurong Islands, in turn, were glued together by dikes, their intervening waters drained to create space for a giant petrochemical complex. Key to Singapore's progress has been its comparative lack of corruption, attracting many companies to establish themselves here as a base from which to do business in notoriously corrupt countries elsewhere in the realm. Still, Singapore is the smallest of the Pacific Rim's economic tigers, and it faces growing competition near and far. Over the long term its relative location, which yielded its early wealth, may turn out to be its most enduring guarantee of survival—if not prosperity.

INDONESIA'S ARCHIPELAGO

The world's fourth-largest country in terms of human numbers is also the globe's most expansive **archipelago**. 19
Spread across more than 17,000 islands, Indonesia's

AMONG THE REALM'S GREAT CITIES . . . JAKARTA

Jakarta, capital of Indonesia, sometimes is called the Kolkata (Calcutta) of Southeast Asia. Stand on the elevated highway linking the port to the city center and see the villages built on top of garbage dumps by scavengers using what they can find in the refuse, and the metaphor seems to fit. There is poverty here unlike that in any other city in the realm.

But there are other sides to Jakarta. Indonesia's economic progress has made its mark here, and the evidence is everywhere. Television antennas and satellite dishes rise like a forest from rusted, corrugated-iron rooftops. Cars (almost all, it seems, late-model), mopeds, and bicycles clog the streets, day and night. A meticulously manicured part of the city center contains a cluster of high-rise hotels, office buildings, and apartments. Billboards advertise planned communities on well-located, freshly cleared land.

Jakarta's population is a cross-section of Indonesia's, and the silver domes of Islam rise above the cityscape alongside Christian churches and Hindu temples. The city always was cosmopolitan, beginning as a conglomerate of villages at the mouth of the Ciliwung River under Islamic rule, becoming a Portuguese stronghold and later the capital of the Dutch East Indies (under the name Batavia). Advantageously situated on the northwestern coast of Jawa, Indonesia's most populous island, Jakarta bursts at the seams with growth. Sail into the port, and hundreds of vessels, carrying flags from Russia to Argentina, await berths. Travel to the outskirts, and huge shantytowns are being expanded by a constant stream of new arrivals. So vast is the human agglomeration—nobody knows how many millions have descended on this place (the official figure is 13.8 million)—that the majority live without adequate (or any) amenities.

© H. J. de Blij, P. O. Muller, and John Wiley & Sons, Inc.

A seemingly permanent dome of metallic-gray air, heavy with pollutants, rests over the city. The social price of economic progress is high, but Jakartans are willing to pay it.

225.8 million people live both separated and clustered separated by water and clustered on islands large and small.

The map of Indonesia requires some attention (Fig. 10-12). Five large islands dominate the archipelago territorially, one of which, New Guinea in the east, is not part of the Indonesian culture sphere, although its western half is under Indonesian control. The other four major islands are collectively known as the Greater Sunda Islands: *Jawa* (Java), smallest but by far the most populous and important; *Sumatera* (Sumatra) in the west, directly across the Strait of Malacca from Malaysia; *Kalimantan*, the Indonesian sector of large, compact, minicontinent Borneo; and wishbone-shaped, distended *Sulawesi* (Celebes) to the east. Extending eastward from Jawa are the Lesser Sunda Islands, including Bali and, near the eastern end, Timor. Another important island chain within Indonesia is the Maluku (Molucca) Islands, between Sulawesi and New Guinea. The central water body of Indonesia is the Java Sea.

As the map shows, Indonesia does not control all of the archipelago. In addition to East Malaysia's territory on Borneo, there also is Brunei (see box titled "Rich and Broken Brunei"). And on the eastern part of the island of Timor, a long struggle against Indonesian rule led, in 1999, to a referendum on independence. The overwhelming majority of East Timorese voted in favor of sovereignty, and after guidance from the United Nations East Timor achieved sovereignty in 2002.

Indonesia is a Dutch colonial creation, and the Dutch chose Jawa as their colonial headquarters, making Batavia (now Jakarta) their capital. Today, **Jawa** remains the core of Indonesia. With about 130 million inhabitants, Jawa is one of the world's most densely peopled places and one of the most agriculturally productive. It also is the most highly urbanized part of a country in which nearly

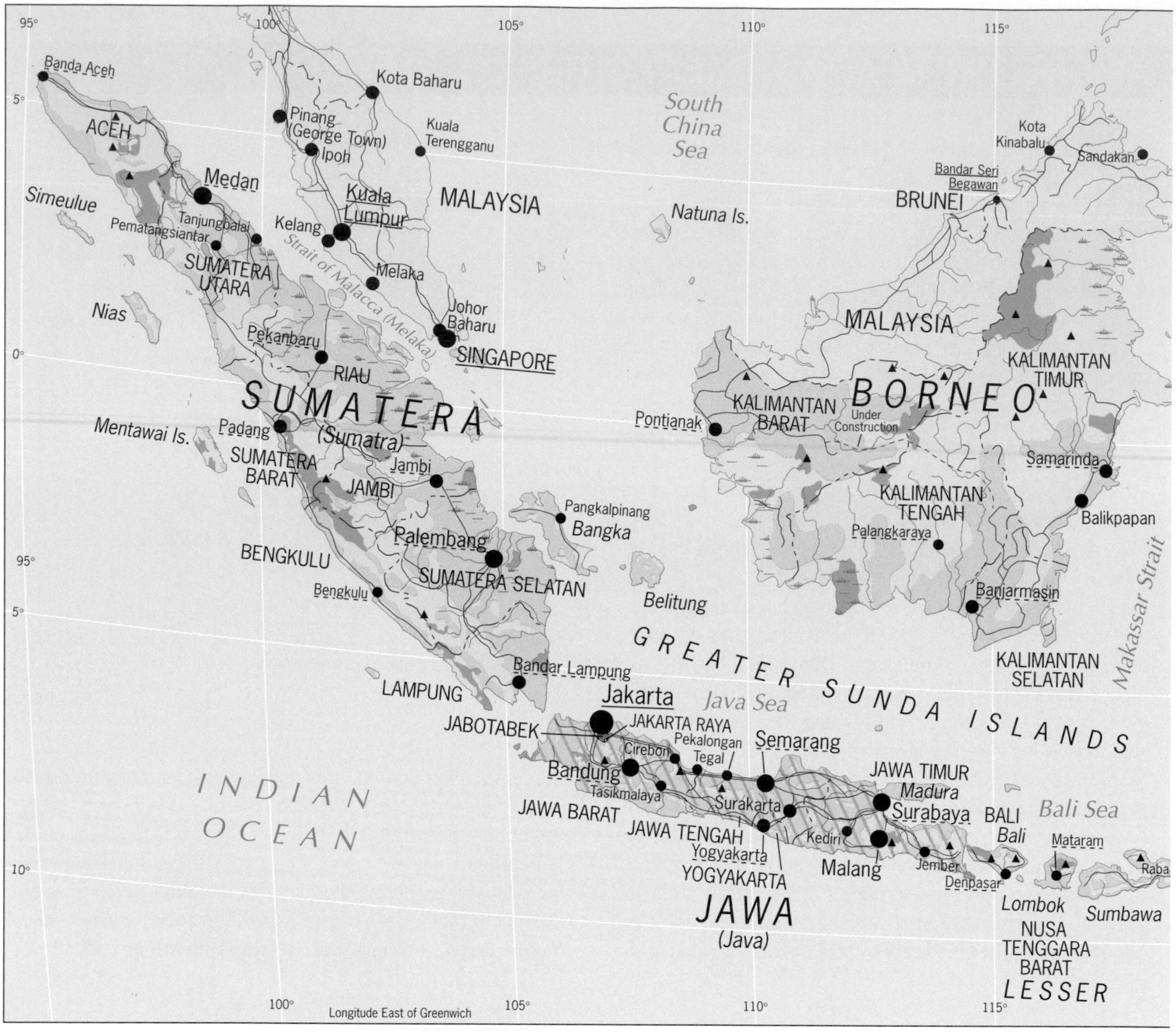

FIGURE 10-12 © H.J. de Blij, P.O. Muller, and John Wiley & Sons, Inc.

60 percent of the people still live on the land, and the Pacific Rim boom of the 1990s had strong impact here. The city of Jakarta, on the northwestern coast, became the heart of a larger conurbation now known as *Jabotabek*, consisting of the capital as well as Bogor, Tangerang, and Bekasi. During the 1990s the population of this megalopolis grew from 15 to 20 million, and it is predicted to reach 30 million by 2010. Already, Jabotabek is home to over 10 percent of Indonesia's entire population and 25 percent of its urban population. Thousands of factories, their owners taking advantage of low prevailing wages, were built in this area, straining its infrastructure and overburdening the port of Jakarta. On an average day, hundreds of ships lie at anchor, awaiting docking space to offload raw materials and take on finished products.

Indonesia's experience with rapid economic growth was tempered by the failures of its government. Suharto, the same president who had ruled the country for 25 years, and who had enriched himself, his family, and numerous friends and cronies, was reelected in 1997 for another five-year term. His authoritarian rule, which some credited for creating the stability that made Indonesia's economic boom possible, tolerated no free opposition or open election campaigns. But when Indonesia's economy was battered by the slump that had earlier struck Thailand and Malaysia, political turmoil forced Suharto to step down. During the rioting that accompanied this transition, the Jawanese, as they had done in the past, turned on the Chinese in their midst.

Always in Indonesia, Jawa is where the power lies. The Jawanese constitute about 55 percent of the country's population. Sumatera, much larger than Jawa, has only about one-third as many people (46 million). Sulawesi has some 17 million, and Kalimantan, Indone-

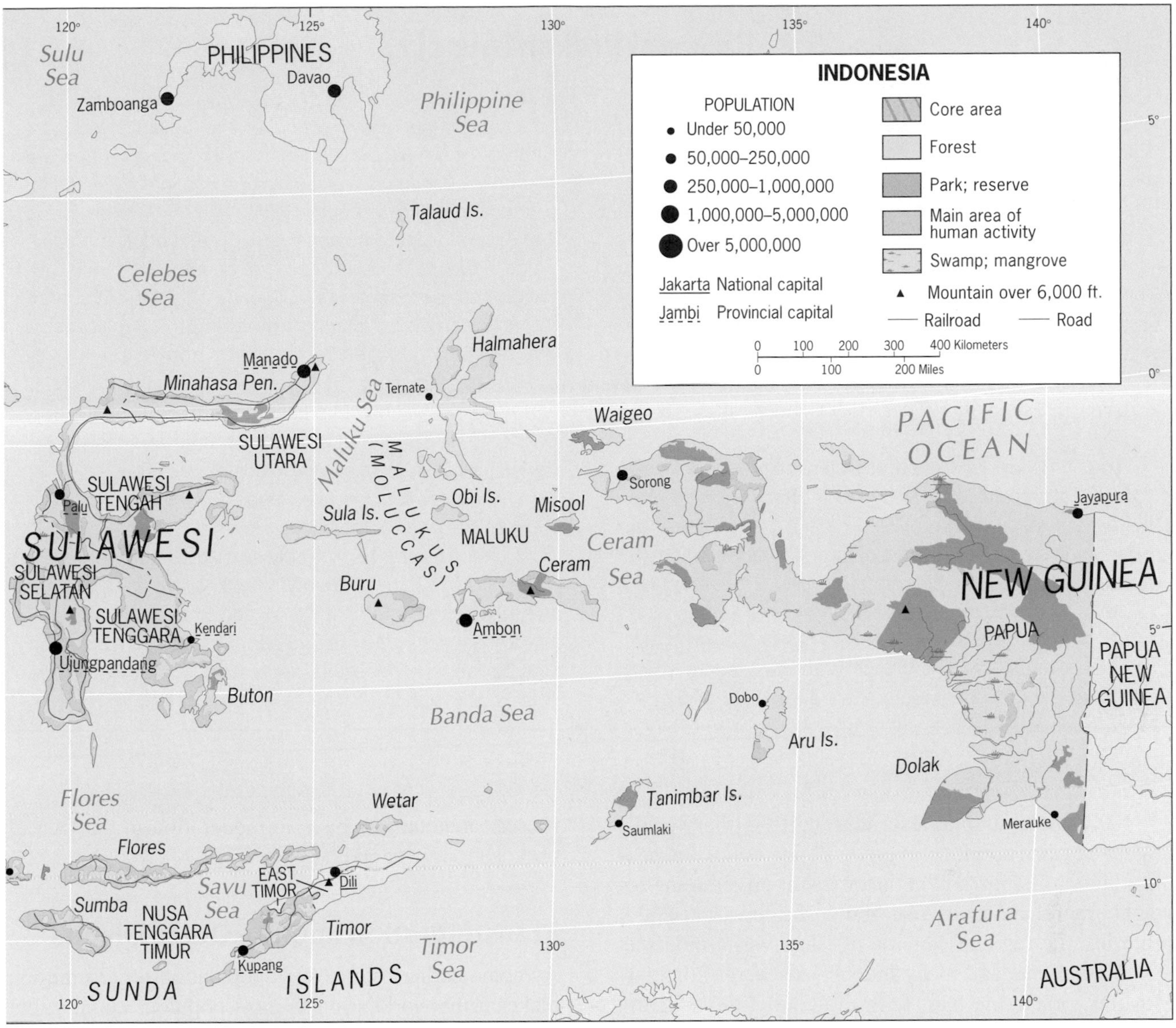

sia's huge territory on Borneo, just 13 million. As for Papua in the far east, which is larger than California, its population is barely 2.5 million.

Diversity in Unity

Indonesia's survival as a unified state is as remarkable as India's and Nigeria's. With more than 300 discrete ethnic clusters, over 250 languages, and just about every religion practiced on Earth (although Islam dominates), actual and potential centrifugal forces are powerful here. Wide waters and high mountains perpetuate cultural distinctions and differences. Indonesia's national motto is *bhinneka tunggal ika*: diversity in unity.

What Indonesia has achieved is etched against the country's continuing cultural complexity. There are dozens of distinct aboriginal cultures; virtually every coastal community has its own roots and traditions. And the majority, the rice-growing Indonesians, include not only the numerous Jawanese—who share an adherence to Islam but have their own ethnic and cultural identity—but also the Sundanese (who constitute 14 percent of Indonesia's population), the Madurese (7 percent), and others. Perhaps the best impression of the cultural mosaic comes from the string of islands that extends eastward from Jawa to Timor (Fig. 10-12). The rice-growers of Bali adhere to a modified version of Hinduism, giving the island a unique cultural atmosphere; the population of Lombok is mainly Muslim, with some Balinese Hinduism; Sumbawa is a Muslim community; Flores is mostly Roman Catholic. In western Timor, Protestant groups dominate; in now-independent East Timor, where the Portuguese ruled, Roman Catholicism prevails.

Rich and Broken Brunei

Brunei is an anomaly in Southeast Asia—an oil-exporting Islamic sultanate far from the Persian Gulf. Located on the north coast of Borneo, sandwiched between Malaysian Sarawak and Sabah (Fig. 10-6, upper-left map), the Brunei sultanate is a former British-protected remnant of a much larger Islamic kingdom that once controlled all of Borneo and areas beyond. Brunei achieved full independence in 1984. With a mere 2228 square miles (5700 sq km)—slightly larger than Delaware—and only 410,000 people, Brunei is dwarfed by the other political entities of Southeast Asia (Fig. 10-1). But the discovery of oil in 1929 (and natural gas in 1965) heralded a new age for this remote territory.

Today, Brunei is one of the largest oil producers in the British Commonwealth, and recent offshore discoveries suggest that production will increase. As a result, the population is growing rapidly through immigration (67 percent of Brunei's residents are Malay, 11 percent Chinese), and the sultanate enjoys one of the highest standards of living in Southeast Asia. In 1998, however, a financial scandal of gigantic proportions was revealed, involving the disappearance of more than U.S. $15 billion and the collapse of Brunei's largest private company, which had been under the direction of the Sultan's youngest brother. This financial calamity continued to buffet the country's economy seven years later and was reflected by a steep drop in its per-capita GNI.

Most of Brunei's inhabitants live and work near the oilfields in the western corner of the country and in the capital, Bandar Seri Begawan. Evidence of profligate spending is everywhere, ranging from sumptuous palaces to magnificent mosques to luxurious hotels. In this respect Brunei offers a stark contrast to neighboring East Malaysia, but even within Brunei there are spatial differences. Brunei's interior remains an area of subsistence agriculture (a small minority of indigenous groups survive here) and rural isolation, its villages a world apart from the modern splendor of the coast.

Brunei's territory is not only small: it is fragmented. A sliver of Malaysian land separates the larger west from the east (where the capital lies), and Brunei Bay provides a water link between the two parts. It has been proposed that Brunei purchase this separating corridor from the Malaysians, but no action has yet been taken.

Nevertheless, Indonesia nominally is the world's largest Muslim country: overall, 77 percent of the people adhere to Islam, and in the cities the silver domes of neighborhood mosques rise above the townscape. For centuries, the regional version of Islam was a moderate and tolerant one, and many Indonesians were "nominal" Muslims, practicing their faith irregularly and adhering only loosely to its precepts. In recent years, however, radical Islamic movements have arisen in Indonesia, treated warily by a government unsure of the right response. As the War on Terror reached into Indonesia in the form of pressure on the Indonesian administration to take firmer action and to investigate possible al-Qaeda links, terrorists responded in 2002 by bombing a nightclub in Bali. More than 200 people, mostly vacationing Westerners, were killed (see photo at right).

Indonesia's economic downturn and its internal political upheavals had already fostered Islamic revivalism in the sprawling country, and not only in Jawa, before the events of September 2001 changed the world. As we will note in some detail, the government's longstanding population policies unintentionally contributed to rising conflict between Muslims and Christians, and until 2001 Jakarta's biggest problem was to contain this strife and to restrain Muslim radicals. Now the question is whether this, the world's largest Muslim-dominated country by far, can maintain its historic moderation in the face of provocations near and far.

Transmigration and the Outer Islands

As noted earlier, Indonesia's population of 226 million makes it the world's fourth most populous country, but Jawa, we also noted, contains approximately 55 percent of it. With about 130 million people on an island the size of Louisiana, the population pressure is enormous here. Moreover, Indonesia's annual rate of population growth remains at 1.6 percent. To deal with this problem, and at the same time to strengthen the core's power over outlying areas, the Indonesian government long pursued a policy known as **transmigration**, inducing Jawanese (and 20
Madurese from the adjacent island of Madura) to relocate to other islands. Several million Jawanese have moved to locales as distant as the Malukus, Sulawesi, and Sumatera; many Madurese were resettled in Kalimantan.

In the government's view, the transmigration policy would serve to counter centrifugal forces arising from Indonesia's geography, spread the official language (Bahasa Indonesia, a modified form of Malay), and bring otherwise unexploited land into production. It has done so, but it also had negative effects. In Kalimantan, the Muslim Madurese found themselves facing armed Dayaks, in-

Indonesia still has vast areas of rainforest from Sumatera to Papua, but the forest is under assault in many parts of this populous country, with disastrous impact on biodiversity, soils, and rivers. High-quality wood fetches high prices in markets such as Japan and China, and for decades Indonesia's corrupt authoritarian regime actively participated in the plunder. More recently, democratically elected governments have sought to curb the rate of destruction, but corruption continues to impede these efforts. Here in Kalimantan, on the Indonesian side of Borneo, about 4000 square miles (10,000 sq km) of rainforest are lost every year. This depressing photograph shows the frontier of clear-cutting in Kalimantan Barat (West Kalimantan). © Wayne G. Lawler/Photo Researchers, Inc.

digenous people who did not want to give up their land, and thousands were killed in what amounted to a regional civil war. Elsewhere, Jawanese often were blamed for introducing methods and practices of administration and business that ran counter to local traditions. In Papua, where Indonesia is in effect a colonial power, aboriginal people have attacked Indonesian newcomers.

In fact, the Dutch colonialists started the Transmigration Program, but Sukarno abandoned it and it was not revived until 1974, when Suharto saw economic and political opportunities in it. Over the next 25 years, nearly 8 million people were relocated; the World Bank gave Indonesia nearly U.S. $1 billion to facilitate the program before it pulled out amid reports of forced migration.

The results were mixed. World Bank project reviews suggested that the vast majority of the transmigrants were happy with the land, schools, medical services, and other amenities with which they were provided, but independent surveys painted a different picture. About half of the migrants never managed to rise above a life of subsistence; many never received even the most basic needs for survival in their new environment; land was simply taken away from indigenous inhabitants, notably in Kalimantan; the newcomers badly damaged the traditional cultures

In the immediate aftermath of the pair of late-Saturday-night bomb blasts that wreaked havoc in the crowded nightclub district of Bali's Kuta Beach on October 13, 2002, foreign tourists carry their baggage through the wreckage as they hastily depart the island. The final death toll came to 202, with many of the victims being young Australian vacationers; in addition, more than 300 people were injured, many of them maimed for life. With the threat of further terrorist attacks by Islamic militants an ever-present one, another casualty has been Bali's tourist boom, which has still not recovered to its pre-attack level. © AP/Wide World Photos.

they invaded; and the program led to massive forest and wetland destruction (photo p. 529). And thousands of migrants died in clashes with local residents.

In 2001, the Transmigration Program was canceled as one of the final acts of the short-lived government of President Abdurrahman Wahid, but the damage it did will trouble Indonesia for generations. From northern Sumatera to western Kalimantan to central Sulawesi to the islands of the Malukus and Papua, cultural strife and its polarizing effect will challenge Indonesian governments of the future.

In September 2004, a significant event occurred that received too little attention in the Western world: the democratic election of a new president and the constitutional succession of government. Those who argue that Islam and democracy are incompatible should note that the obstacles to Indonesia's achievement went far beyond religious ideology: the country's fragmented territory, huge distances, ethnic and cultural diversity, remote corners, and rebellious districts all formed part of the challenge to be overcome, a challenge of which Islamic revivalist movements and their disputes were only part. President Susilo Bambang Yudhoyono took office in October 2004, the first Indonesian president ever to be elected directly by the people. He immediately declared that his priority would be to defuse Indonesia's separatist movements, not by seeking to overpower them but by persuading them that their future would be far brighter within than outside of Indonesia.

The Major Islands

Indonesia is an important state in a crucial geographic location. Each of its four major territorial components, in addition to Jawa, is geographically distinct.

Sumatera, Indonesia's westernmost major island, lies across the narrow Strait of Malacca from West Malaysia. Tropical rainforest still covers enough of it to sustain refuges for the orangutan population, but logging and human population growth are making ever greater inroads. Sharing Malaya's equatorial environments, Sumatera in colonial times became a base for rubber and palm-oil plantations. Its high relief and fertile volcanic soils make possible the cultivation of a wide range of crops, from vegetables in the highlands, to tea and coffee on the slopes, to subsistence root crops in the lowlands. Sumatera and its neighboring small islands contain large coal reserves, petroleum and natural gas, tin, bauxite (for aluminum), and even gold and silver. As the map suggests, intra-island surface communications are still weak, although the north-south Sumatera Highway was completed in the 1980s. The highest population density is around the northern city of Medan; another cluster lies in the middle zone of the island; and a third, augmented by a large contingent of transmigrated Jawanese, focuses on Palembang and Bandar Lampung in the south.

The northern part of the island has long been the most fractious. While the Batak people (see Fig. 10-3) adjusted to colonization and Westernization, and made Medan one of Indonesia's Pacific Rim boom towns, the Aceh fought the Dutch in a 30-year war that ended only in the twentieth century. In the 1990s, while Indonesia, Malaysia, and Thailand talked of an economic growth triangle including northern Sumatera, the Aceh mounted another rebellion—this time against Jakarta. Having just agreed to allow a referendum on independence to take place in East Timor, the new democratic government under then-president Wahid appeared ready to negotiate a settlement. Then, even as the Aceh problem simmered,

FROM THE FIELD NOTES

"I drove from Manado on the Minahasa Peninsula in northeastern Sulawesi to see the ecological crisis at Lake Tondano, where a fast-growing water hyacinth is clogging the water and endangering the local fishing industry. On the way, in the town of Tomolon, I noticed this side street lined with prefabricated stilt houses in various stages of completion. These, I was told, were not primarily for local sale. They were assembled from wood taken from the forests of Sulawesi's northern peninsula, then taken apart again and shipped from Manado to Japan. 'It's a very profitable business for us,' the foreman told me. 'The wood is nearby, the labor is cheap, and the market in Japan is insatiable. We sell as many as we can build, and we haven't even begun to try marketing these houses in Taiwan or China.' At least, I thought, this wood was being converted into a finished product, unlike the mounds of logs and planks I had seen piled up in the ports of Borneo awaiting shipment to East Asia."

the House of Representatives of the central Sumateran Province of Riau approved a "declaration of independence." And the House of East Kalimantan (Kalimantan Timur in Fig. 10-12) proposed that the province be renamed the "Federal State of East Kalimantan." Devolutionary forces are at work throughout Indonesia.

Certainly, the Aceh problem is crucial among these issues. With its population of more than 4 million, its natural resources, its militant history, and its peninsular situation, the province's Free Aceh Movement would seem to have a better chance than most components of the republic to force Jakarta's hand. But in 2003, a negotiated settlement brokered by a Swiss team appeared in prospect, pending the 2004 presidential election. Then nature took a hand in the process when the December 2004 tsunami struck Aceh and its capital, Banda Aceh at the northern tip of the island, harder than any other Indian Ocean coastland, inflicting about 170,000 casualties. Even as soldiers of the Indonesian army helped in the relief effort and political prisoners were released to search for their families, the conflict continued. The government did not lift the state of emergency, and in mid-2005 it appeared that even this environmental calamity would not yield a truce in embattled Aceh.

Kalimantan is the Indonesian part of the minicontinent of Borneo, a slab of the Earth's crystalline crust whose backbone of mountains is of erosional, not volcanic, origin. Compact and massive—at over 290,000 square miles (750,000 sq km) it is larger than Texas—Borneo has a deep, densely forested interior that is a last refuge of some 35,000 orangutans. Elephants, rhinoceroses, and tigers still survive even as the loggers constrict their habitat. Numerous other species of fauna—insects, reptiles, mammals, birds—inhabit the rich mountain and lowland forests. So much of Borneo's Pleistocene heritage has survived principally because of its comparatively small human population. The Indonesian part, Kalimantan, constitutes 28 percent of the national territory, but, with 13 million people, it contains less than 6 percent of Indonesia's population.

Figure 10-12 shows that the main areas of human activity lie in the west, southeast, and east. All towns of any size, including Pontianak in the west and Balikpapan in the east, are on or near the coast; the rivers still form

FROM THE FIELD NOTES

"Getting to Ambon was no easy matter, and my boat trip was enlivened by an undersea earthquake that churned up the waters and shook up all aboard. The next morning our first view of the town of Ambon, provincial capital of Maluku, showed a center dominated by a large mosque with a modern minaret and a large white dome to the right of the main street leading to the port. My host from the local university told me that, as on many of the islands of the Malukus, about half of the people were Christians, not Muslims, and that some large churches were situated in the outskirts. 'But the relationships between Muslims and Christians are worsening,' he said. It had to do with the arrival of Jawanese, some of whom were 'agitators' and Islamic fundamentalists, stirring up religious passions. 'Have you heard of the Taliban?' he asked. 'Well, we have similar so-called religious students here, educated in Islamic schools that teach extremism.' . . . We drove from the town into the countryside to the west, toward Wakisihu and past the old Dutch Fort Rotterdam. 'Let me show you something,' he said as we turned up a dirt road. At its end, in the middle of a field, sat a large single structure called University of Islam, distinguished by a pair of enormous stairways. 'Can you call a single building like this where all they teach is Islam a university?' he asked. 'What they teach here is Islamic fundamentalism and intolerance.' He was prophetic. Weeks later, religious conflict broke out, the mosque in town was burned along with Christian churches; his own university lay in ruins." © H. J. de Blij.

important routes into the interior. Parks and reserves to protect flora and fauna cover but a small portion of the vast territory. Not only the flora and fauna, but also the human population has ancient roots. Indigenous peoples, including some of the Dayak clans, continue their slash-and-burn subsistence in the interior; others have become sedentary farmers. But Indonesia's transmigration policy has brought about half a million Jawanese and Madurese to Kalimantan, and they are laying out farms and digging drainage canals, bringing drastic change to rural areas. Of Kalimantan's four provinces, only the southernmost exports significant raw materials, including oil from offshore reserves, coal, iron ore, and some gold and diamonds; but in the national picture, Indonesian Borneo remains little developed.

Sulawesi is an island that looks like a set of intersecting mountain ranges rising from the sea. Its northern end is a 500-mile (800-km) chain of extinct, dormant, and active volcanoes known as the Minahasa Peninsula, its future northern extension toward the Philippine Sea already marked by a ribbon of seafloor volcanoes just rising above the surface. So rugged is the relief on this island that you cannot take surface transportation from Palu to Ujungpandang. The ethnic mosaic is quite complex, with seven major groups inhabiting more or less isolated parts of the island as well as a sizeable cluster of Jawanese transmigrants who are concentrated in Sulawesi Tengah (Fig. 10-12). Since 1999, the level of strife between local Christians and immigrant Jawanese Muslims has risen and has destabilized this remote province of Sulawesi—forcing Christians who once lived peacefully among Muslim villagers to flee, and culturally segregating settlements.

The two most populous and developed parts of Sulawesi are the southwestern peninsula centered on Ujungpandang and the eastern end of the Minahasa Peninsula, where Manado is the urban focus. Subsistence farming occupies most of the population of 17 million, but logging and wood products, some mining, and fishing augment the cash economy. Cultural landscapes are varied; in Minahasa, Christian churches remain numerous and form reminders of the colonial period, when relations between this province and the Dutch were favorable. During the liberation period, Minahasans sided with the Dutch (as many of their neighbors in the Malukus did) and called their peninsula Holland's "twelfth province." It was an unfulfilled wish, souring relations between Jakarta and Manado for years afterward. But when President Suharto was forced to resign in 1998, his transitional successor, Habibie, was a son of Minahasa, and the past was forgotten. The religious strife in Sulawesi Selatan (Fig. 10-12), however, has been more enduring: by 2004 the Indonesian military had succeeded in pacifying the area, but animosities between Christians and Muslims, migrants and locals run deep, and only time and improving livelihoods are likely to soften them.

Papua, like East Timor, fell to Indonesia well after the Dutch colonial era had ended, but unlike East Timor, the United Nations approved Indonesia's takeover in 1969. Until 2002 this province was known by its Bahasa Indonesian name of *Irian Jaya* (Guinea West), but in that year the government of President Megawati Sukarnoputri bowed to the demands of indigenous leaders of the Papuan communities on the island and changed its name. As the map shows, the eastern half of the island of New Guinea comprises the independent state of Papua New Guinea. The Papuans are divided not only into numerous ethnolinguistic groups but also into two political entities.

Papua is an Indonesian province, but it lies in the Pacific geographic realm, not in Southeast Asia. As such, it has about 22 percent of the country's land area but just over 1 percent of its population, including more than 200,000 ethnic Indonesians, most from Jawa and many of them concentrated in the provincial capital with the non-Papuan name of Jayapura. Forested, high-relief, glacier-peaked Papua, lying as it does in a different geographic realm, is a world apart from teeming Jawa. It has what is reputedly the world's richest gold mine and second-largest open-pit copper mine; its forests not only harbor the indigenous peoples but also sustain a major logging industry.

Sanctioned as Indonesia's rule in Papua may be, local opposition has continued for decades. The *Organisasi Papua Merdeka* (Free Papua Movement, or FPM) and occasional rebel attacks remind the Jakarta government of its role here. In 1999, during the turmoil in the other provinces, the FPM held rallies in Jayapura to promote its cause, and in 2000 a congress was held where independence was demanded and a Papuan flag, the Morning Star, displayed. Later that year, Indonesian troops killed demonstrators in Merauke on the province's south coast. In 2003, the Indonesian government divided Papua into three administrative provinces, and shortly thereafter several executives of a metals corporation accused of polluting Papuan waters and poisoning local residents were arrested, only to be exonerated months later, but not before the world was alerted to the allegations of corporate misdeeds in remote Papua. All this deepened Papuan opposition to Jakarta's rule, and the time may come when Indonesia will rue its acquisition of Papua, potential source of the kinds of centrifugal forces an archipelagic state must constantly confront.

EAST TIMOR

The easternmost of the lesser Sunda Islands is Timor, the eastern part of which was a Portuguese colony, overrun by Indonesia in 1975 and annexed in 1976, which be-

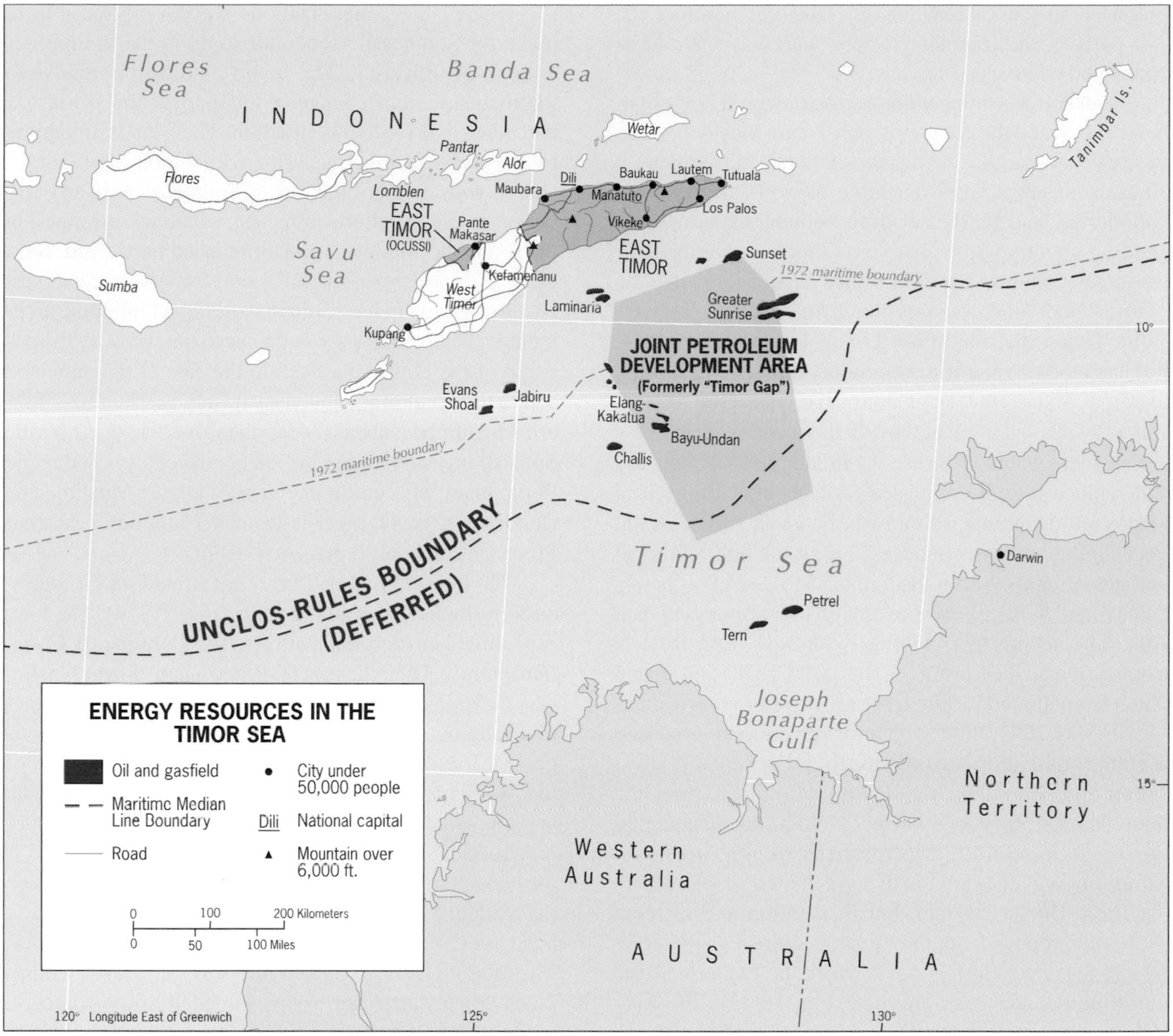

FIGURE 10-13 © H.J. de Blij, P.O. Muller, and John Wiley & Sons, Inc.

came the scene of a bitter struggle for independence. When the people of East Timor in 1999 were allowed to express their views on independence, this Connecticut-sized territory with only about 800,000 inhabitants took the first steps toward nationhood.

Indonesia's armed forces reacted violently, adding hundreds more to the tens of thousands who had already died in the struggle against Jakarta. The future country's modest infrastructure was devastated, and its people were dislocated. Foreign intervention, led by Australia (once Indonesia's sole supporter in its claim to East Timor), established a degree of order. Reconstruction and social reorganization, under UN auspices, have been in progress since.

In Indonesia, many viewed the "loss" of East Timor as the first step in the disintegration of the state, to be followed by secession in Aceh, the Malukus, and Papua. But in truth, East Timor was a special case, and if devolution were indeed to fragment Indonesia, East Timor was not the prototype. Unlike Papua, the Malukus, or Aceh, Indonesia's control over East Timor defied UN rules. This was not merely an internal matter.

East Timor became an independent state in 2002. As the map shows, its sovereignty creates several politico-geographical complications (Fig. 10-13). The political entity of East Timor consists of a main territory, where the capital named Dili is located, and a small exclave on the north coast of (Indonesian) West Timor called Ocussi (spellings vary, and it is sometimes mapped as Ocussi-Ambeno or Ambeno Province). Although there is a road from the exclave through the center of the island to the main territory, relations between East Timor and

Indonesia may not make this link feasible; therefore, the two parts of the fragmented new state may have to be connected by boat traffic.

East Timor's independence also creates a new alignment at sea, and the country's very future may depend on this realignment. As we discuss in detail in Chapter 12, all coastal states have maritime as well as terrestrial boundaries, and these maritime boundaries sometimes endow a country with valuable offshore oil, gas, or other resources. South of East Timor lies an area known as the "Timor Gap" that contains major oil and gas reserves. While Indonesia ruled East Timor, Indonesia and Australia agreed on a maritime boundary that greatly favored the Australians (the 1972 boundary on Figure 10-13). But after East Timor achieved independence, its new government wanted to erase the bilateral 1972 boundary and replace it with one based on UN regulations. That would put the boundary midway between Australia and East Timor, and assign to East Timor the bulk of the oil and gas reserves within the Timor Gap as well as newly discovered fields elsewhere under the Timor Sea. But Australia, seeing this coming, withdrew from the UN Convention on the Law of the Sea (UNCLOS) and argued that it is not bound by any UN regulations in this matter.

In May 2005, the East Timor government proposed a compromise the Australians found acceptable. East Timor agreed to defer the delimitation of the maritime boundary for 50 years, in effect yielding to Australia on this issue. In return, the Australians would award East Timor not only the previously agreed income share from the Bayu-Undan gasfield, but in addition a 50-percent share of revenues from the Greater Sunrise gas reserve, which lies under East Timor's waters no matter how the maritime boundary is delimited (Fig. 10-13). The Australians approved this proposal, and the "Timor Gap" was renamed the Joint Petroleum Development Area. To East Timor, this was the key to a brighter future—especially if oil and gas prices continue to rise.

FRAGMENTED PHILIPPINES

North of Indonesia, across the South China Sea from Vietnam, and south of Taiwan lies an archipelago of more than 7000 islands (only about 460 of them larger than one square mile in area) inhabited by 87.1 million people. The inhabited islands of the Philippines can be viewed as three groups: (1) Luzon, largest of all, and Mindoro in the north, (2) the Visayan group in the center, and (3) Mindanao, second largest, in the south (Fig. 10-14). Southwest of Mindanao lies a small group of islands, the Sulu Archipelago, nearest to Indonesia, where a Muslim-based insurgency has kept the area in turmoil.

Few of the generalizations we have been able to make for Southeast Asia could apply in the Philippines without qualification. The country's location relative to the mainstream of change in this part of the world has had much to do with this situation. The islands, inhabited by peoples of Malay ancestry with Indonesian strains, shared with much of the rest of Southeast Asia an early period of Hindu cultural influence, which was strongest in the south and southwest and diminished northward. Next came a Chinese invasion, felt more strongly on the largest island of Luzon in the northern part of the Philippine archipelago. Islam's arrival was delayed somewhat by the position of the Philippines well to the east of the mainland and to the north of the Indonesian islands. The few southern Muslim beachheads were soon overwhelmed by the Spanish invasion during the sixteenth century. Today the Philippines, adjacent to the world's largest Muslim state (Indonesia), is 81 percent Roman Catholic, 7 percent Protestant, and only 5 percent Muslim.

The Muslim population, concentrated on the southeastern flank of the archipelago (Fig. 10-14), has long proclaimed its marginalization in this dominantly Christian country. Over the past 30 years, a half-dozen Muslim organizations have promoted the Muslim cause with tactics ranging from political pressure to violent insurgency. The campaigns of violence have affected all of the Muslim areas, from southern Mindanao to Palawan, and the strategies of the rebels have ranged from bombings to kidnappings. Densely forested Basilan Island is the headquarters of a group known as Abu Sayyaf, which demanded a separate Muslim state and is believed to have received support through the al-Qaeda network. The War on Terror reached into this part of the Philippines when American troops joined Filipino forces to pursue Abu Sayyaf terrorists. Also active on behalf of Muslim causes in this area are the Moro National Liberation Front and the Moro Islamic Liberation Front; the Philippine government has entered into negotiations with both. Cease-fire agreements, however, have not held. The Muslim challenge is pirating a disproportionate share of Manila's operating budget.

Out of the Philippines melting pot, where Mongoloid-Malay, Arab, Chinese, Japanese, Spanish, and American elements have met and mixed, has emerged the distinctive Filipino culture. It is not a homogeneous or a unified culture, but in Southeast Asia it is in many ways unique. One example of its absorptive qualities is demonstrated by the way the Chinese infusion has been accommodated: although the "pure" Chinese minority numbers less than 2 percent of the population (far lower than in most Southeast Asian countries), a much larger portion of the Philippine population carries a decidedly Chinese ethnic imprint. What has happened is that the Chinese

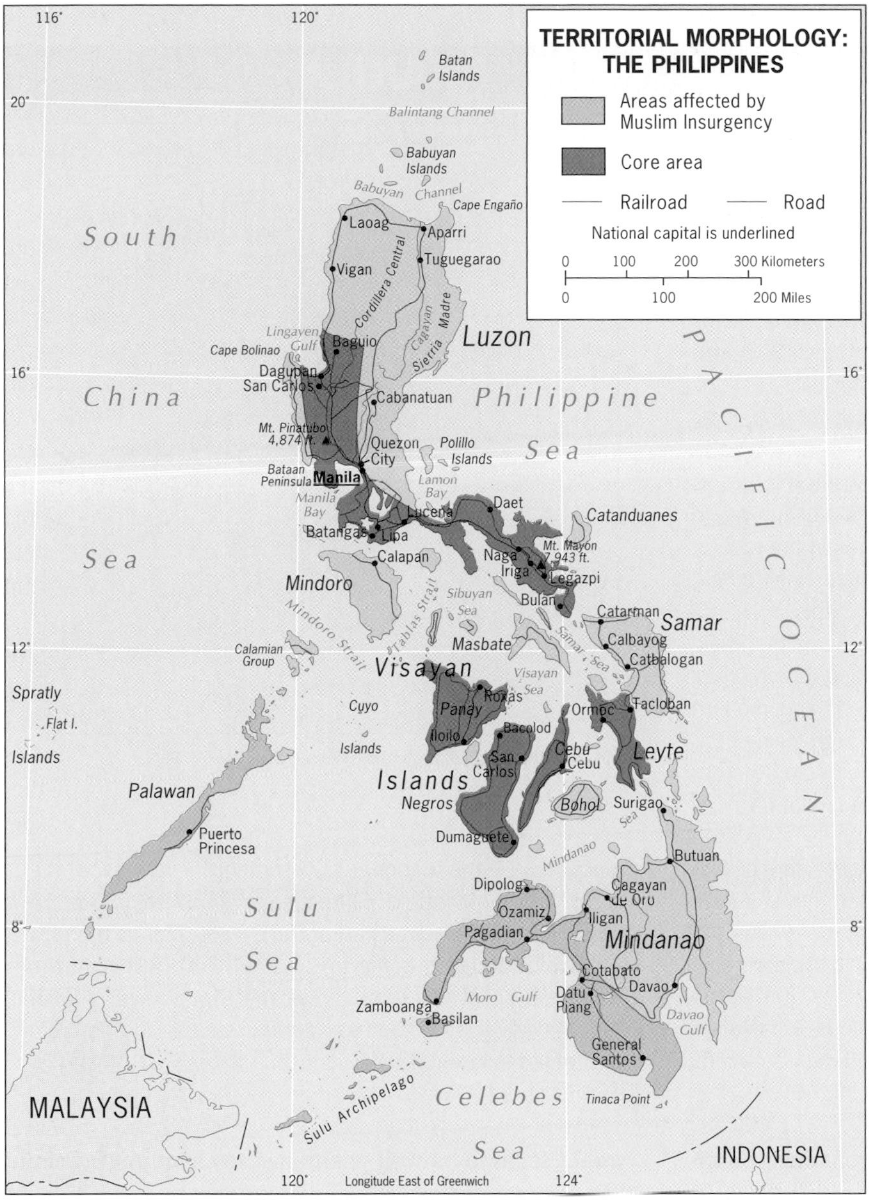

FIGURE 10-14 © H.J. de Blij, P.O. Muller, and John Wiley & Sons, Inc.

have intermarried, producing a sort of Chinese-mestizo element that constitutes more than 10 percent of the total population. In another cultural sphere, the country's ethnic mixture and variety are paralleled by its great linguistic diversity. Nearly 90 Malay languages, major and minor, are spoken by the 87 million people of the Philippines; only about 1 percent still use Spanish. At independence in 1946, the largest of the Malay languages, Tagalog, or Pilipino, became the country's official language, and the educational system promotes its general use. English is learned as a subsidiary language and remains the chief *lingua franca*; an English-Tagalog hybrid ("Taglish") is increasingly heard today, cutting across all levels of society.

The widespread use of English in the Philippines, of course, results from a half-century of American rule and influence, beginning in 1898 when the islands were ceded to the United States by Spain under the terms of the treaty that followed the Spanish-American War. The United States took over a country in open revolt against its former colonial master and proceeded to destroy the Filipino independence struggle, now directed against the new foreign rulers. It is a measure of the subsequent success of U.S. administration in the Philippines that this was the only dependency in Southeast Asia that sided against the Japanese during World War II in favor of the colonial power. The U.S. rule had its good and bad features, but the Americans did initiate reforms that were long overdue,

AMONG THE REALM'S GREAT CITIES . . . MANILA

Manila, capital of the Philippines, was founded by the Spanish invaders of Luzon more than four centuries ago. The colonists made a good choice in terms of site and situation. Manila sprawls at the mouth of the Pasig River where it enters one of Asia's finest natural harbors. To the north, east, and south a crescent of mountains encircles the city, which lies just 600 miles (975 km) across the South China Sea from Hong Kong.

Manila, named after a flowering shrub in the local marshlands, is bisected by the Pasig, which is bridged in numerous places. The old walled city, Intramuros, lies to the south. Despite heavy bombardment during World War II, some of the colonial heritage survives in the form of churches, monasteries, and convents. St. Augustine Church, completed in 1599, is one of the city's landmarks.

The CBD of Manila lies on the north side of the Pasig River. Although Manila has a well-defined commercial center with several avenues of luxury shops and modern buildings, the skyline does not reflect the high level of energy and activity common in Pacific Rim cities on the opposite side of the South China Sea. Neither is Manila a city of notable architectural achievements. Wide, long, and straight avenues flanked by palm, banyan, and acacia trees give it a look not unlike parts of San Juan, Puerto Rico.

In 1948, a newly built city immediately to the northeast of Manila was inaugurated as the *de jure* capital of the Philippines and called Quezon City. The new facilities were eventually to house all government offices, but many functions of the national government never made the move. In the meantime, Manila's growth overtook Quezon City's, so that it became part of the Greater Manila metropolis (which today is home to 10.9 million). Although the proclamation of Quezon City as the Philippines' official capital was never rescinded, Manila remains the *de facto* capital of the country today.

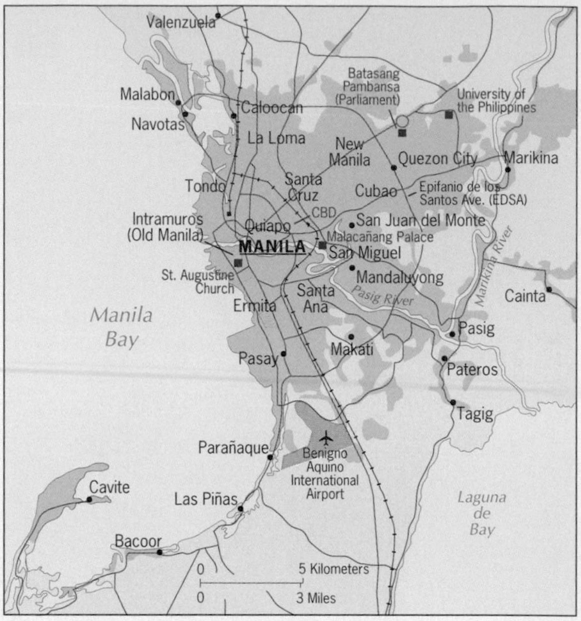

and they were already in the process of negotiating a future independence for the Philippines when the war intervened in 1941.

The Philippines' population, concentrated where the good farmlands lie in the plains, is densest in three general areas (Fig. G-9): (1) the northwestern and south-central part of Luzon, (2) the southeastern extension of Luzon, and (3) the islands of the Visayan Sea between Luzon and Mindanao. Luzon is the site of the capital, Manila-Quezon City (10.9 million, almost one-eighth of the entire national population), a major metropolis facing the South China Sea. Alluvial as well as volcanic soils, together with ample moisture in this tropical environment, produce self-sufficiency in rice and other staples and make the Philippines a net exporter of farm products despite a fairly high population growth rate of 2.0 percent.

The Philippines seems to get little mention in discussions of developments on the Pacific Rim, and yet it would seem to be well positioned to share in the Pacific Rim's economic growth. Governmental mismanagement and political instability have slowed the country's participation, but during the 1990s the situation improved. Despite a series of jarring events—the ouster of U.S. military bases, the damaging eruption of a volcano near the capital, the violence of Muslim insurgents, and a dispute over the nearby Spratly Islands in the South China Sea—the Philippines made substantial economic progress during the decade. Its electronics and textile industries (mostly in the Manila hinterland) expanded continuously, and more foreign investment arrived. The centrally positioned Visayan island of Cebu experienced particularly rapid growth, based on its central location, good port, expanded airport, large and literate labor force, and a cadre of managers experienced in the service industries. A free-trade zone attracted foreign companies from the United States and Japan, and now products from toys to semiconductors flow from Cebu's manufacturers to world markets. But agriculture continues to dominate the Philippines' economy, unemployment remains high, further land reform is badly needed, and social restructuring (reducing the controlling influence over national affairs by a comparatively small group of families) must occur. However, progress is being made. The country now is a lower-middle-income economy, and given a longer period of stability and success in reducing the population growth rate, it will rise to the next level and finally take its place among Pacific Rim growth poles.

In recent years, the population issue has divided this dominantly Roman Catholic society, with the government promoting family planning and the clergy opposing it. But behind this debate lies another of the Philippines' assets: in a realm of mostly undemocratic regimes, the Philippines has come out of its period of authoritarian rule a rejuvenated, if not yet robust, democracy.

WHAT YOU CAN DO

RECOMMENDATION: Participate in National Geography Awareness Week! By an Act of Congress, a week in November is designated National Geography Awareness Week, and all over the country organizations ranging from colleges to corporations mark the occasion by offering programs highlighting geography's importance and usefulness. Activities include competitions, exhibits, lectures, GIS demonstrations, and field trips, and faculty and students alike make a special effort to reach out to the general public to show what professional geographers do. This always requires assistance, from posting announcements to meeting speakers and from supervising exhibits to planning field trips. How about asking your local Geography Department what is being planned for next November, and lending a hand?

GEOGRAPHIC CONNECTIONS

1 In this chapter we divide the Southeast Asian geographic realm into a Mainland region and an Insular region. A major factor in this regionalization is religion: Buddhism dominates in Mainland Southeast Asia, and Islam in Insular Southeast Asia. As Figure 7-2 reveals, however, the religious geography of Southeast Asia is more complicated than that, and important vestiges of one major religion cannot even be shown at this small scale. Look very carefully at the political boundaries of Southeast Asia and the religious regions in Figure 7-2, and identify those countries that experience significant internal religious division of a regional nature. How is such division finding expression? Is religious regionalism absent in countries dominated by a single faith?

2 When we studied colonialism and its legacies in Subsaharan Africa, it was noted that, although most African states have been independent for nearly half a century, the imprint of colonial power still remains—and continues to divide the realm. Southeast Asia, too, was occupied by colonial powers (with one significant exception), but the cultural impress of colonialism appears to be less durable here. To what geographic factors do you attribute this contrast between two formerly colonized realms?

FIGURE 11-1 Reprinted with permission from *Goode's World Atlas*, 21st edition, pp. 220–221. © Rand McNally, 2005. License R.L. 05-S-64.

Scale 1:16 000 000; one inch to 250 miles. Lambert's Azimuthal, Equal Area Projection
Elevations and depressions are given in feet

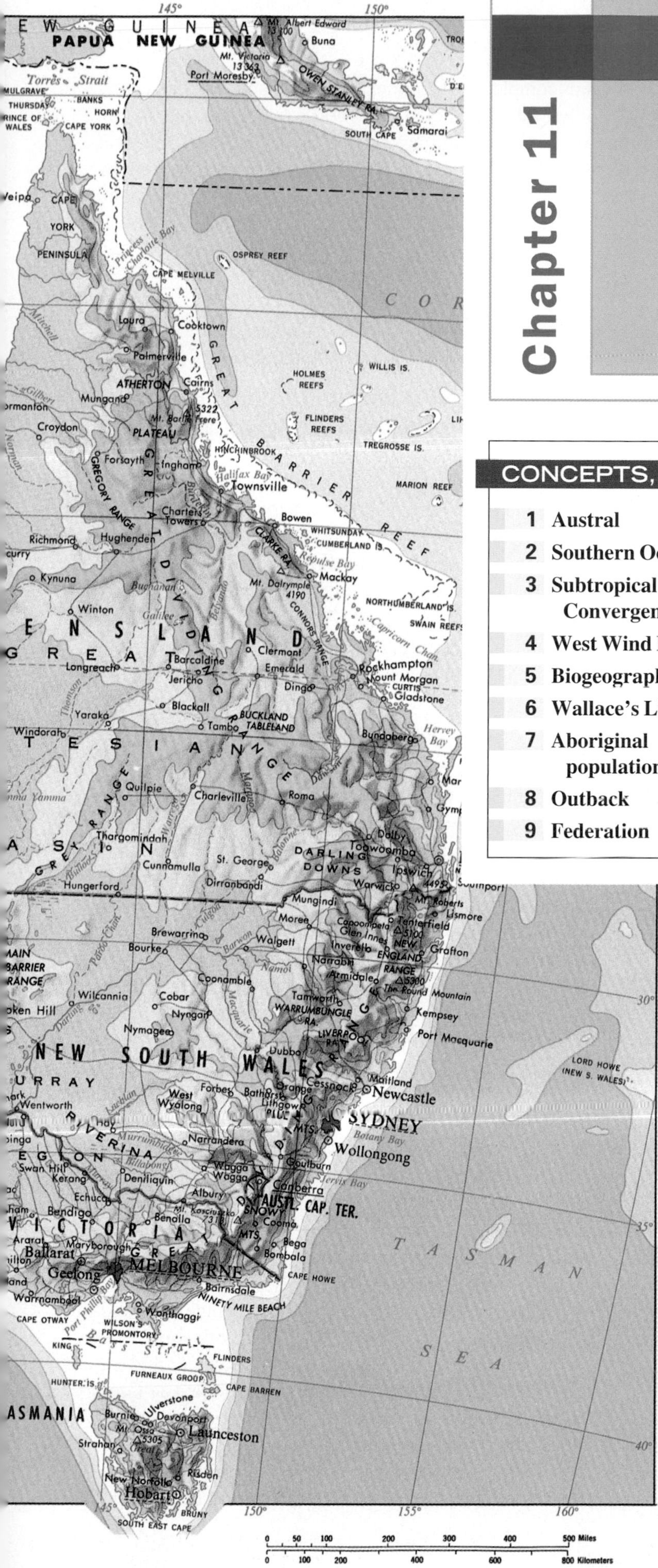

Chapter 11

The Austral Realm

CONCEPTS, IDEAS, AND TERMS

1. Austral
2. Southern Ocean
3. Subtropical Convergence
4. West Wind Drift
5. Biogeography
6. Wallace's Line
7. Aboriginal population
8. Outback
9. Federation
10. Unitary state
11. Import-substitution industries
12. Aboriginal land issue
13. Immigration policies
14. Environmental degradation
15. Peripheral development

REGIONS

- AUSTRALIA
  - CORE AREA
  - OUTBACK
- NEW ZEALAND

The Austral Realm is geographically unique. It is the only geographic realm that lies entirely in the Southern Hemisphere. It is also the only realm that has no land link of any kind to a neighboring realm and is thus completely surrounded by ocean and sea. It is second only to the Pacific as the world's least populous realm. Appropriately, its name refers to its location (**Austral** 1 means south)—a location far from the sources of its dominant cultural heritage but close to its newfound economic partners on the western Pacific Rim.

DEFINING THE REALM

Two countries constitute the Austral Realm: Australia, in every way the dominant one, and New Zealand, physiographically more varied than its giant partner. Between them lies the Tasman Sea. To the west lies the Indian Ocean, to the east the Pacific, and to the south the frigid Southern Ocean (see box titled "Is There a Southern Ocean?").

This southern realm is at a crossroads. On the doorstep of populous Asia, its Anglo-European legacies are now infused by other cultural strains. Polynesian Maori in New Zealand and Aboriginal communities in Australia are demanding better terms of life. Pacific Rim markets are buying huge quantities of raw materials. Japanese and other Asian tourists fill hotels and resorts. Queensland's tropical Gold Coast resembles Honolulu's Waikiki. The streets of Sydney and Melbourne display a multicultural panorama unimagined just two generations ago. All these changes have stirred political debate. Issues ranging from immigration quotas to indigenous land rights dominate, exposing social fault lines (city versus Outback in Australia, North and South in New Zealand). Aborigines and Maori were here first, and the Europeans came next. Now Asia looms in Australia's doorway.

MAJOR GEOGRAPHIC QUALITIES OF The Austral Realm

1. Australia and New Zealand constitute a geographic realm by virtue of territorial dimension, relative location, and dominant cultural landscape.
2. Despite their inclusion in a single geographic realm, Australia and New Zealand differ physiographically. Australia has a vast, dry, low-relief interior; New Zealand is mountainous.
3. Australia and New Zealand are marked by peripheral development—Australia because of its aridity, New Zealand because of its topography.
4. The populations of Australia and New Zealand are not only peripherally distributed but also highly clustered in urban centers.
5. The realm's human geography is changing—in Australia because of Aboriginal activism and Asian immigration, and in New Zealand because of Maori activism and Pacific-islander immigration.
6. The economic geography of Australia and New Zealand is dominated by the export of livestock products (and in Australia also by wheat production and mining).
7. Australia and New Zealand are being integrated into the economic framework of the western Pacific Rim, principally as suppliers of raw materials.

LAND AND ENVIRONMENT

Physiographic contrasts between massive, compact Australia and elongated, fragmented New Zealand are related to their locations with respect to the Earth's tectonic plates (consult Fig. G-4). Australia, with some of the geologically most ancient rocks on the planet, lies at the center of its own plate, the Australian Plate. New Zealand, younger and less stable, lies at the convulsive convergence of the Australian and Pacific plates. Earthquakes are rare in Australia and volcanic eruptions are unknown; New Zealand has plenty of both. This locational contrast is also reflected by differences in relief (Fig. 11-1). Australia's highest relief occurs in what Australians call the Great Dividing Range, the mountains that line the east coast from the Cape York Peninsula to southern Victoria, with an outlier in Tasmania. The highest point along these old, now eroding mountains is Mount Kosciusko, 7318 feet (2230 m) tall. In New Zealand, entire ranges are higher than this, and Mount Cook reaches 12,316 feet (3754 m).

West of Australia's Great Dividing Range, the physical landscape generally has low relief, with some local exceptions such as the Macdonnell Ranges near the center; plateaus and plains dominate (Fig. 11-2). The Great Artesian Basin is a key physiographic region, providing underground water sources in what would otherwise be desert country; to the south lies the continent's predominant river system, the Murray-Darling. The area mapped

Is There a Southern Ocean?

Here is an interesting geographic experiment: try to find the Southern Ocean on commonly used maps of the world—commonly used, that is, here in the Northern Hemisphere. You are unlikely to find it, even on maps and globes published by the National Geographic Society, Rand McNally, and other major mapmaking organizations.

Now look in the library for atlases and maps published in Southern Hemisphere countries, for example, New Zealand and Australia. There the Southern Ocean is a prominent geographic feature. And it is not just a matter of opinion: to people whose countries adjoin it, the Southern Ocean is a significant factor in daily life. Its persistent westerly winds influenced the course of history. It is a crucial weather maker. Beneath its vast and frigid surface lie resources yet unknown. Its waters contain biota that are crucial to the global web of marine life.

2 How is it that the primary issue involving the **Southern Ocean** is whether it exists at all? The all-too-obvious answer is that this ocean, unlike the others on our planet, is not neatly bounded by continental coasts. Of course, it borders Antarctica to the south, but its northern limits are not visible on a relief map. So mapmakers simply extend the Atlantic, Pacific, and Indian oceans all the way to Antarctic shores, ignoring the very existence of an ocean as large as the Indian.

For us geographers, it is a good exercise to turn the globe upside down now and then. After all, the usual orientation is quite arbitrary. Modern mapmaking started in the Northern Hemisphere, and the cartographers put their hemisphere on top and the other at the bottom. That is now the norm, and it can distort our view of the world. In bookstores in the Southern Hemisphere, you sometimes see tongue-in-cheek maps showing Australia and Argentina at the top, and Europe and Canada at the bottom. But this matter has a serious side. A reverse view of the globe shows us how vast the ocean encircling Antarctica is. The Southern Ocean may be remote, but its existence is real.

Where do the northward limits of the Southern Ocean lie? This ocean is bounded not by land but by a marine transition called the **Subtropical Convergence**. Here the 3
cold, extremely dense waters of the Southern Ocean meet the warmer waters of the Atlantic, Pacific, and Indian oceans. It is quite sharply defined by changes in temperature, chemistry, salinity, and marine fauna. Flying over it, you can actually observe it in the changing colors of the water: the Antarctic side is a deep gray, the northern side a greenish blue.

Although the Subtropical Convergence moves seasonally, its position does not vary far from latitude 40° South, which also is the approximate northern limit of Antarctic icebergs. Defined this way, the great Southern Ocean is a huge body of water that moves clockwise (from west to east) around Antarctica, which is why we also call it the **West Wind Drift**. 4

as *Western Plateau and Margins* in Figure 11-2 contains much of Australia's mineral wealth.

Figure G-8 reveals the effects of latitudinal location and interior isolation on Australia's climatology. In this respect, Australia is far more varied than New Zealand, its climates ranging from tropical in the far north, where rainforests flourish, to Mediterranean in parts of the south. The interior is dominated by desert and steppe conditions, the steppes providing the grasslands that sustain tens of millions of livestock. Only in the east does Australia have an area of humid temperate climate, and here lies most of the country's economic core area. New Zealand, by contrast, is totally under the influence of the Southern and Pacific oceans, creating moderate, moist conditions, temperate in the north and colder in the south.

Biogeography

One of this realm's defining characteristics is its wildlife. Australia is the land of kangaroos and koalas, wallabies and wombats, possums and platypuses. These and numerous other *marsupials* (animals whose young are born very early in their development and then carried in an abdominal pouch) owe their survival to Australia's early isolation during the breakup of Gondwana (see Fig. 6-3). Before more advanced mammals could enter Australia and replace the marsupials, as happened in other parts of the world, the landmass was separated from Antarctica and India, and today it contains the world's largest assemblage of marsupial fauna.

Australia's vegetation also has distinctive qualities, notably the hundreds of species of eucalyptus trees native to this geographic realm. Many other plants form part of Australia's unique flora, some with unusual adaptation to the high temperatures and low humidity that characterize much of the continent.

The study of fauna and flora in spatial perspective combines the disciplines of biology and geography in a field known as **biogeography**, and Australia is a giant 5
laboratory for biogeographers. In the Introduction (pp. 10–16), we noted that several of the world's climatic zones are named after the vegetation that marks them:

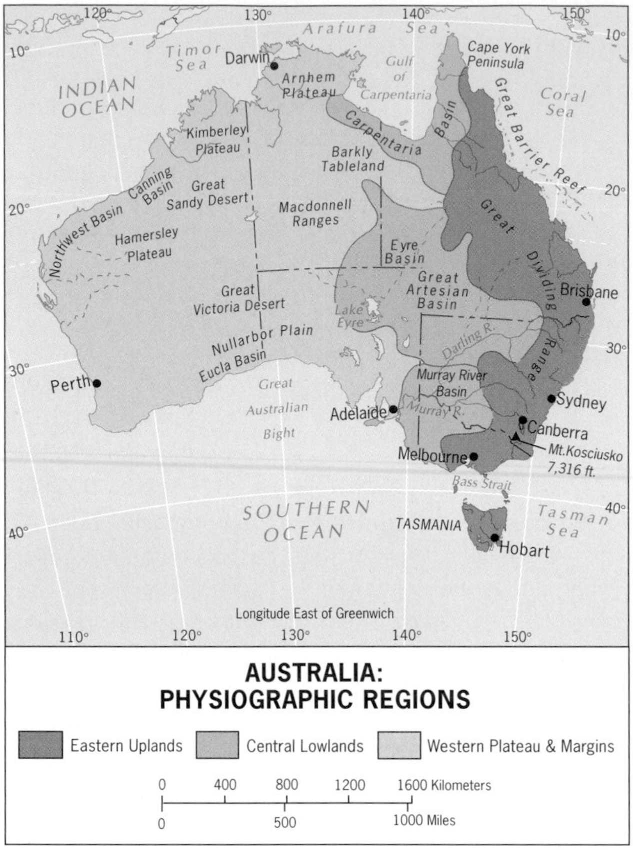

FIGURE 11-2 © H. J. de Blij, P. O. Muller, and John Wiley & Sons, Inc.

tropical savanna, steppe, tundra. When climate, soil, vegetation, and animal life reach a long-term, stable adjustment, vegetation forms the most visible element of this ecosystem.

Biogeographers are especially interested in the distribution of plant and animal species, and in the relationships between plant and animal communities and their natural environments. (The study of plant life is called *phytogeography*; the study of animal life is called *zoogeography*.) These scholars seek to explain the distributions the map reveals. In 1876 one of the founders of biogeography, Alfred Russel Wallace, published a book entitled *The Geographical Distribution of Animals* in which he fired the first shot in a long debate: where does the zoogeographic boundary of Australia's fauna lie? Wallace's fieldwork in the area revealed that Australian forms exist not only in Australia itself but also in New Guinea and in some islands to the west. So Wallace proposed that the faunal boundary should lie between Borneo and Sulawesi, and just east of Bali (Fig. 11-3).

Wallace's Line soon was challenged by other re- 6
searchers, who found species Wallace had missed and who visited islands Wallace had not. There was no question that Australia's zoogeographic realm ended somewhere in the Indonesian archipelago, but where? Western Indonesia was the habitat of nonmarsupial animals such as tigers, rhinoceroses, and elephants, as well as primates; New Guinea clearly was part of the realm of the marsupials. How far had the more advanced mammals progressed eastward along the island stepping stones toward New Guinea? The zoogeographer Max Weber found evidence that led him to postulate his own *Weber's Line*, which, as Figure 11-3 shows, lay very close to New Guinea.

Not all research in zoogeography or phytogeography deals with such large questions. Much of it focuses on the relationships between particular species and their habitats, that is, the environment they normally occupy and of which they form a part. Such environments change, and the changes can spell disaster for the species.
In Australia, the arrival of the **Aboriginal population** 7
(between 50,000 and 60,000 years ago) had limited effect on the habitats of the extant fauna. But the invasion of

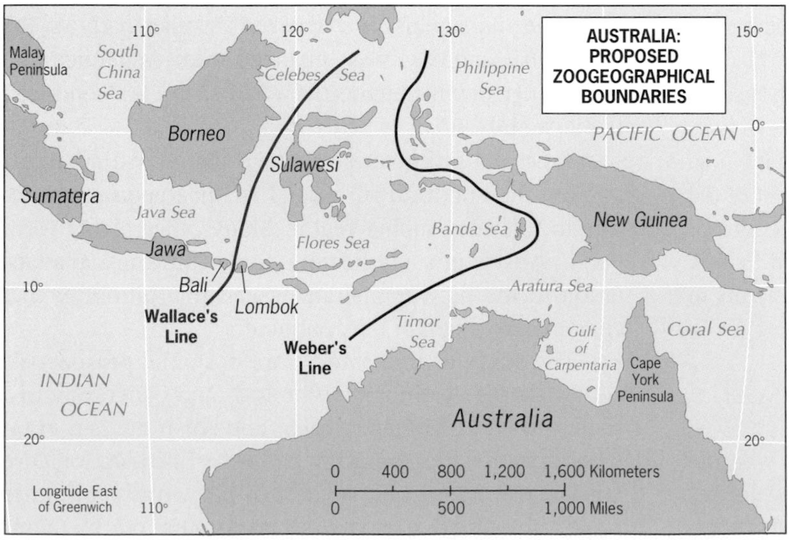

FIGURE 11-3 © H. J. de Blij, P. O. Muller, and John Wiley & Sons, Inc.

the European colonizers and the introduction of their livestock led to the destruction of habitats and the extinction of many native species. Whether in East Africa or in Western Australia, the key to the conservation of what remains of natural flora and fauna lies in the knowledge embodied by the field of biogeography.

REGIONS OF THE REALM

Australia is the dominant component of the Austral Realm, a continent-scale country in a size category that also includes China, Canada, the United States, and Brazil. For two reasons, however, Australia has fewer regional divisions than do the aforementioned countries: Australia's relatively uncomplicated physiography and its diminutive human numbers. Our discussion, therefore, uses the core-periphery concept as a basis for investigating Australia and focuses on New Zealand as a region by itself.

AUSTRALIA

On January 1, 2001, Australia celebrated its hundredth birthday as a state, the Commonwealth of Australia, recognizing (still) the British monarch as the head of state and entering its second century with a strong economy, stable political framework, high standard of living for most of its people, and favorable prospects ahead. Positioned on the Pacific Rim, ten times the size of Texas, well endowed with farmlands and vast pastures, major rivers, ample underground water, minerals, and energy resources, served by good natural harbors, and populated by 20.3 million mostly-well-educated people, Australia is one of the most fortunate countries on Earth.

Not everyone in Australia shares adequately in all this good fortune, however, and the less advantaged made their voices heard during the celebrations. The country's indigenous (Aboriginal) population, though a small minority today of about 400,000, remains disproportionately disadvantaged in almost every way, from lower life expectancies to higher unemployment than average, from lower high school graduation rates to higher imprisonment ratios. But the nation is now embarked on a campaign to address these ills, with actions ranging from public demonstrations supporting reconciliation to official expressions of regret for past mistreatment and from enhanced social services to favorable court decisions over Aboriginal land claims.

MAJOR CITIES OF THE REALM

City	Population* (in millions)
Adelaide, Australia	1.1
Auckland, New Zealand	1.2
Brisbane, Australia	1.8
Canberra, Australia	0.3
Melbourne, Australia	3.7
Perth, Australia	1.5
Sydney, Australia	4.5
Wellington, New Zealand	0.3

*Based on 2006 estimates.

When Australia was born as a federal state, its per-capita GNI, as reckoned by economic geographers, was the highest in the world. Australia's bounty fueled a huge flow of exports to Europe, and Australians prospered. That golden age could not last forever, and eventually the country's share of world trade declined. Still, Australia today ranks among the top 15 countries in the world in terms of GNI, and for the vast majority of Australians life is comfortable.

This is not to suggest that Australia has escaped every downturn in the global economy of which it is a part or that there are no concerns at all for the future. During the Pacific Rim boom of the 1980s and early 1990s, some locals wryly called Australia an NDC, a Newly Declining Country, a seller of raw materials, not finished ones; a purveyor of livestock, meat, and wheat on variable world markets; a society sinking deeply into debt. When the Pacific Rim economies stalled in the late 1990s, there were dire predictions about the impact on Australia. But Australia's economy proved to be remarkably adaptable. When its Asian markets dried up, Western markets soon took up the slack. The Australian economy weathered the Pacific Rim crisis better than expected. Today, a new economic partner is on the scene: China, whose voracious appetite for raw materials seems to have no limits.

The statistics bear witness to Australia's current good fortune. In terms of the indicators of development discussed in Chapter 9, Australia is far ahead of all its western Pacific Rim competitors except Japan (which has dropped in the rankings because of its own economic stagnation). As Australians celebrated their first century, they were, on average, earning far more than Thais, Malaysians, or Koreans. In terms of consumption of energy per person, number of automobiles and miles of roads, levels of health, and literacy, Australia had all the properties of a developed country. Australian cities and towns, where 91 percent of all Australians live, are not encircled by crowded shantytowns. Nor is the Australian countryside inhabited by a poverty-stricken peasantry.

Distance

Australians often talk about distance. One of their leading historians, Geoffrey Blainey, labeled it a "tyranny"—an imposed remoteness from without and a divisive part of life within. Even today, Australia is far from nearly everywhere on Earth. A jet flight from Los Angeles to Sydney takes more than 14 hours nonstop and is correspondingly expensive. Freighters carrying products to European markets take ten days to two weeks to get there. Inside Australia, distances also are of continental proportions, and Australians pay the price—literally. Until some upstart private airlines started a price war, Australians paid more per mile for their domestic flights than air passengers anywhere else in the world.

But distance also was an ally, permitting Australians to ignore the obvious. Australia was a British progeny, a European outpost. Once you had arrived as an immigrant from Britain or Ireland, there were a wide range of environments, magnificent scenery, vast open spaces, and seemingly limitless opportunities. When the Japanese Empire expanded, Australia's remoteness saved the day. When immigration became an issue, Australia in its comfortable isolation could adopt an all-white admission policy that was not officially terminated until 1976. When boat people by the hundreds of thousands fled Vietnam in the aftermath of the Indochina War, almost none reached Australian shores.

Today Australia is changing and rapidly so. Immigration policy now focuses on the would-be immigrants' qualifications, skills, financial status, age, and facility in the English language. With regard to skills, high-technology specialists, financial experts, and medical personnel are especially welcome. Relatives of earlier immigrants, as well as a quota of genuine asylum-seekers, also are readily admitted. In recent years, total immigration has been limited to about 80,000 annually, but the number may have to increase if only to keep the country's population growing. According to the latest data, Australia's declining natural rate of increase now stands at 0.6 percent.

Already, Australia's changed immigration policies have dramatically altered cultural landscapes, especially in the urban areas (see photo p. 554). The country is fast becoming a truly multicultural society; in Sydney, for example, one in seven residents is of Asian ancestry, a ratio that will rise to one in five by 2010. During the 1990s, when for a time Japan was Australia's leading trade partner, Australian schools began to teach Japanese to tens of thousands of children. Today, Chinese business representatives fill the hotel rooms of Canberra and Sydney. The Austral Realm is clearly in transition.

Core and Periphery

Australia is a large landmass, but its population is heavily concentrated in a core area that lies in the east and southeast, most of which faces the Pacific Ocean (here named the Tasman Sea between Australia and New Zealand). As

FROM THE FIELD NOTES

"My most vivid memory from my first visit to Alice Springs is spotting vineyards and a winery in this parched, desert environment as the plane approached the airport. I asked a taxi driver to take me there, and got a lesson in economic geography. Drip irrigation from an underground water supply made viticulture possible; the tourist industry made it profitable. None of this, however, is evident from the view seen here: a spur of the Macdonnell Ranges overlooks a town of bare essentials under the hot sun of the Australian desert. What Alice Springs has is centrality: it is the largest settlement in a vast area Australians often call 'the centre.' Not far from the midpoint on the nearly 2000-mile (3200-km) Stuart Highway from Darwin on the Northern Territory's north coast to Adelaide on the Southern Ocean, Alice Springs also was the northern terminus of the Central Australian Railway (before the line was extended north to Darwin in 2003), seen in the middle distance. The shipping of cattle and minerals is a major industry here. You need a sense of humor to live here, and the locals have it: the town actually lies on a river, the intermittent Todd River. An annual boat race is held, and in the absence of water the racers carry their boats along the dry river bed. No exploration of Alice Springs would be complete without a visit to the base of the Royal Flying Doctor Service, which brings medical help to outlying villages and homesteads."

Figure 11-4 shows, this crescent-like Australian heartland extends from north of the city of Brisbane to the vicinity of Adelaide and includes the largest city, Sydney, the capital, Canberra, and the second largest city, Melbourne. A secondary core area has developed in the far southwest, centered on Perth and its outport, Fremantle. Beyond lies the vast periphery, which the Australians call
8 the **Outback**.

To better understand the evolution of this spatial arrangement, it helps to refer again to the map of world climates (Fig. G-8). Environmentally, Australia's most favored strips face the Pacific and Southern oceans, and they are not large. We can describe the country as a coastal rimland with cities, towns, farms, and forested slopes giving way to the vast, arid, interior Outback. On the western flanks of the Great Dividing Range lie the

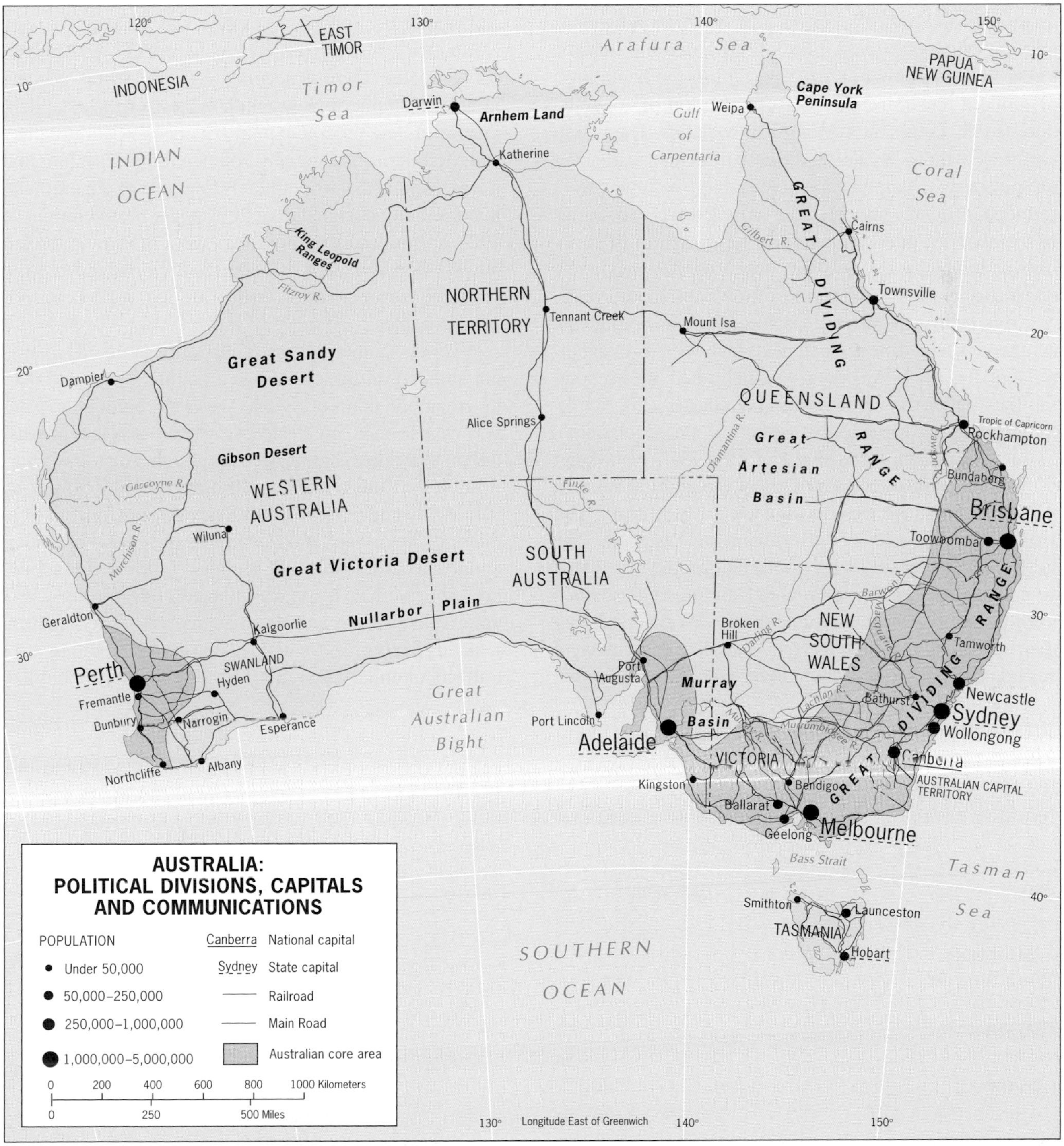

FIGURE 11-4 © H. J. de Blij, P. O. Muller, and John Wiley & Sons, Inc.

extensive grassland pastures that catapulted Australia into its first commercial age—and on which still graze one of the largest sheep herds on Earth (over 160 million sheep, producing more than one-fifth of all the wool sold in the world). Where it is moister, to the north and east, cattle by the millions graze on ranchlands. This is frontier Australia, over which livestock have ranged for almost two centuries.

Aboriginal Australians reached this landmass as long as 50,000 to 60,000 years ago, crossed the Bass Strait into Tasmania, and had developed a patchwork of indigenous cultures when Captain Arthur Phillip sailed into what is today Sydney Harbor (1788) to establish the beginnings of modern Australia. The Europeanization of Australia doomed the continent's Aboriginal societies. The first to suffer were those situated in the path of British settlement on the coasts, where penal colonies and free towns were founded. Distance protected the Aboriginal communities of the northern interior longer than elsewhere; in Tasmania, the indigenous Australians were exterminated in just decades after having lived there for perhaps 45,000 years.

Eventually, the major coastal settlements became the centers of seven different colonies, each with its own hinterland; by 1861, Australia was delimited by its now-familiar pattern of straight-line boundaries (Fig. 11-4). Sydney was the focus for New South Wales; Melbourne, Sydney's rival, anchored Victoria. Adelaide was the heart of South Australia, and Perth lay at the core of Western Australia. Brisbane was the nucleus of Queensland, and Hobart was the seat of government in Tasmania. The largest clusters of surviving Aboriginal people were in the so-called Northern Territory, with Darwin, on Australia's tropical north coast, its colonial city. Notwithstanding their shared cultural heritage, the Australian colonies were at odds not only with London over colonial policies but also with each other over economic and political issues. The building of an Australian nation during the late nineteenth century was a slow and difficult process.

A Federal State

On January 1, 1901, following years of difficult negotiations, the Australia we know today finally emerged: the Commonwealth of Australia, consisting of six States and two Federal Territories (Table 11-1; Fig. 11-4). The special status of Federal Territory was assigned to the Northern Territory to protect the interests of the Aboriginal population there. A second Federal Territory was delimited in southern New South Wales for the establishment of a new federal capital (Canberra). Both Sydney and Melbourne had fiercely competed for this honor, but it was decided that a completely new seat of government, in a specially designated area, was the best solution. In 1927, the government buildings were ready and the administration moved to Canberra, a city planned from scratch to serve a nation built from a group of quarrelsome colonies.

Today Canberra, Australia's only large city that is not situated on the coast, has a population of 345,000, less than one-tenth of Sydney's. As a growth pole, Canberra has been no great success. But as a symbol of Australian federalism, it has been a triumph. Australia's unification was made possible through the adoption of an idea with ancient Greek and Roman roots, one also familiar to Americans and Canadians: the notion of association. The term to describe it comes from the Latin *foederis*; in practice, it means alliance and coexistence, a union of consensus and common interest—a **federation**. 9
It stands in contrast to the idea that states should be centralized, or unitary. For this, too, the ancient Romans had

Table 11-1
STATES AND TERRITORIES OF FEDERAL AUSTRALIA, 2006

State	Area (1000 sq mi)	Estimated Population (millions)	Capital	Estimated Population (millions)
New South Wales	309.5	6.9	Sydney	4.5
Queensland	666.9	3.8	Brisbane	1.8
South Australia	379.9	1.5	Adelaide	1.1
Tasmania	26.2	0.5	Hobart	0.2
Victoria	87.9	5.1	Melbourne	3.7
Western Australia	975.1	2.0	Perth	1.5
Territory				
Australian Capital Territory	0.9	0.3	Canberra	0.3
Northern Territory	519.8	0.2	Darwin	0.1

a term: *unitas*, meaning unity. Most European countries
10 are **unitary states**, including the United Kingdom of Great Britain and Northern Ireland. Although the majority of Australians came from that tradition (a kingdom, no less), they managed to overcome their differences and establish a Commonwealth that was, in effect, a federation of States with different viewpoints, economies, and objectives, separated by vast distances along the rim of an island continent. And yet, the experiment succeeded.

An Urban Culture

During this century-plus of federal association, the Australians developed an urban culture. Despite those vast open spaces and romantic notions of frontier and Outback, no less than 91 percent of all Australians live in cities and towns. On the map, Australia's areal functional organization is similar to Japan's: large cities lie along the coast, the centers of manufacturing complexes as well as the foci of agricultural areas. Contributing to this situation in Japan was mountainous topography; in Australia, it was an arid interior. There, however, the similarity ends. Australia's territory is 20 times larger than Japan's, and Japan's population is more than six times that of Australia's. Japan's port cities are built to receive raw materials and to export finished products. Australia's cities forward minerals and farm products from the Outback to foreign markets and import manufactures from overseas. Distances in Australia are much greater, and spatial interaction (which tends to decrease with increasing distance) is less. In comparatively small, tightly organized Japan, you can travel from one end of the country to the other along highways, through tunnels, and over bridges with utmost speed and efficiency. In

AMONG THE REALM'S GREAT CITIES . . . SYDNEY

More than two centuries ago, Sydney was founded by Captain Arthur Phillip as a British outpost on one of the world's most magnificent natural harbors (photo p. 548). The free town and penal colony that struggled to survive evolved into Australia's largest city. Today metropolitan Sydney (4.5 million) is home to more than one-fifth of the country's entire population. An early start, the safe harbor, fertile nearby farmlands, and productive pastures in its hinterland combined to propel Sydney's growth. Later, as road and railroad links made Sydney the focus of Australia's growing core area, industrial development and political power augmented its primacy.

With its incomparable setting and mild, sunny climate, its many urban beaches, and its easy reach to the cool Blue Mountains of the Great Dividing Range, Sydney is one of the world's most livable cities. Good public transportation (including an extensive cross-harbor ferry system from the doorstep of the waterfront CBD), fine cultural facilities headed by the multitheatre Opera House complex, and many public parks and other recreational facilities make Sydney attractive to visitors as well. A healthy tourist trade, much of it from Japan and other Asian countries, bolsters the city's economy. Sydney's hosting of the 2000 Olympic Games was further testimony to its rising visibility.

Increasingly, Sydney also is a multicultural city. Its small Aboriginal sector is being overwhelmed by the arrival of large numbers of Asian immigrants. The Sydney suburb of Cabramatta symbolizes the impact: more than half of its nearly 100,000 residents were born elsewhere, mostly in Vietnam. Unemployment is high, drug use is a problem, and

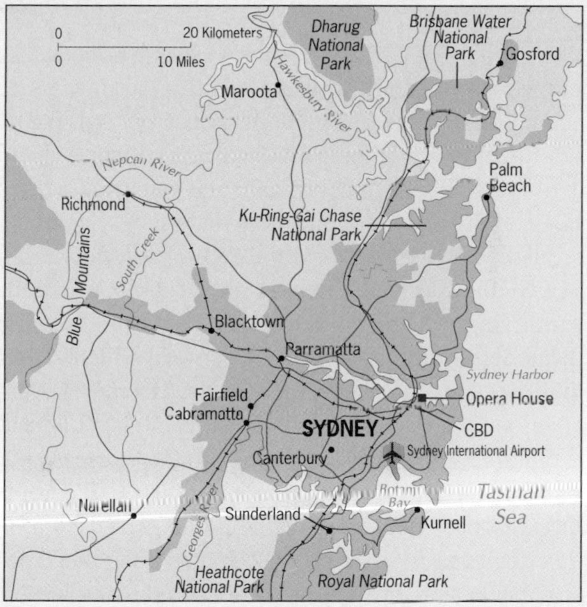

crime and gang violence persist. Yet, despite the deviant behavior of a small minority, tens of thousands of Asian immigrants have established themselves in some profession. As the photo and caption on p. 554 reveal, today things are definitely improving in this dynamic community.

These developments underscore Sydney's coming of age. The end of Australia's isolation has brought Asia across the country's threshold, and again the leading metropolis will show the way.

FROM THE FIELD NOTES

"See a scene like this, and you realize why Australians refer to their land as 'the lucky country.' Australia's periphery has much magnificent scenery ranging from spectacular cliffs to dune-lined beaches, but nothing matches Sydney Harbor on a sunny, breezy day when sailboats by the hundreds emerge from coves and inlets and, if you are fortunate enough to be on the water yourself, every turn around a headland presents still another memorable view. Ask the captains of ocean liners plying the world what port is their favorite, and most will point to this magnificent estuary with its narrow entrance and secluded bays as the grandest of all." © H. J. de Blij.

Australia, the overland trip from Sydney to Perth, or from Darwin to Adelaide, is time-consuming and slow. Nothing in Australia compares to Japan's high-speed bullet trains.

For all its vastness and youth, Australia nonetheless developed a remarkable cultural identity, a sameness of urban and rural landscapes that persists from one end of the continent to the other. Sydney, often called the New York of Australia, lies on a spectacular estuarine site, its compact, high-rise central business district overlooking a port bustling with ferry and freighter traffic (photo above). Sydney is a vast, sprawling metropolis with multiple outlying centers studding its far-flung suburbs; brash modernity and reserved British ways blend here. Melbourne, sometimes regarded as the Boston of Australia, prides itself on its more interesting architecture and more cultured ways. Brisbane, the capital of Queensland, which also anchors Australia's Gold Coast and adjoins the Great Barrier Reef, is the Miami of Australia; unlike Miami, however, its residents can find nearby relief from the summer heat in the mountains of its immediate hinterland (as well as at its beaches). Perth, Australia's San Diego, is one of the world's most isolated cities, separated from its nearest Australian neighbor by two-thirds of a continent and from Southeast Asia and Africa by thousands of miles of ocean.

And yet, each of these cities—as well as the capitals of South Australia (Adelaide), Tasmania (Hobart), and, to a lesser extent, the Northern Territory (Darwin)—exhibits an Australian character of unmistakable quality. Life is orderly and unhurried. Streets are clean, slums are few, graffiti rarely seen. By American and even European standards, violent crime (though rising) is uncommon. Standards of public transportation, city schools, and health-care provision are high. Spacious parks, pleasing waterfronts, and plentiful sunshine make Australia's urban life more acceptable than that almost anywhere else in the world. Critics of Australia's way of life say that this very pleasant state of affairs has persuaded Australians that hard work is not really necessary. Some experience in the commercial centers of the major cities contradicts that assertion: the pace of life is quickening. The country's cultural geography evolved as that of a European outpost, prosperous and secure in its isolation. Now Australia must reinvent itself as a major link in an Australo-Asian chain, a Pacific partner in a transformed regional economic geography.

Economic Geography

From the very beginning, however, goods imported from Britain (and later from the United States) were expensive, largely because of transport costs. This encouraged local entrepreneurs to set up their own industries in and near the developing cities—industries economic geographers call **import-substitution industries**. 11

When the prices of foreign goods became lower because transportation was more efficient and therefore cheaper, local businesses demanded protection from the colonial governments, and high tariffs were erected against imported goods. Local products now could continue to be made inefficiently because their market was guaranteed. If Japan could not afford this, how could Australia? We can see the answer on the map (Fig. 11-5). Even before federation in 1901, all the colonies could export valuable minerals whose earnings shored up those in-

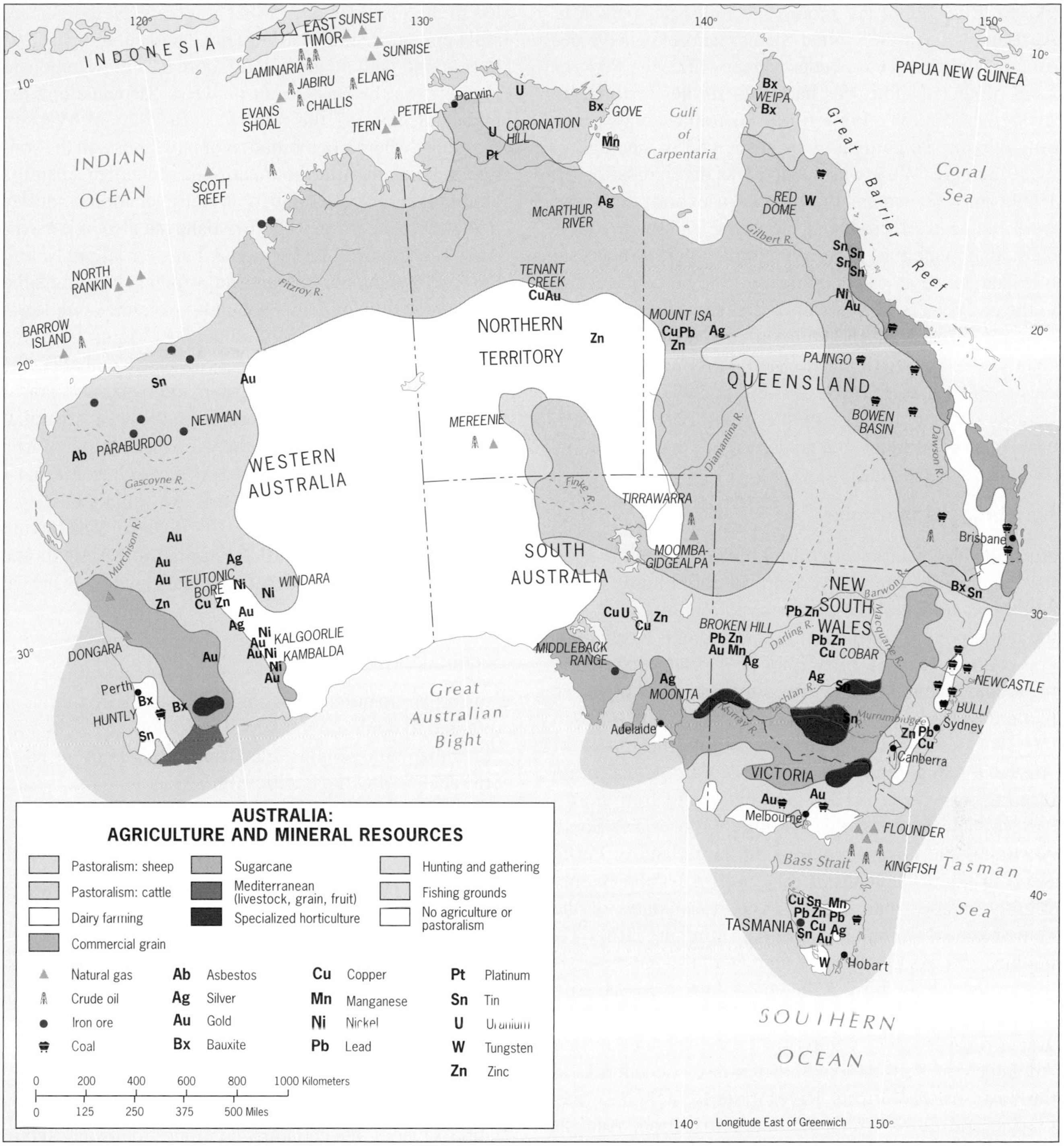

FIGURE 11-5 Adapted (in part) and updated with permission from *The Jacaranda Atlas*, 4 rev. ed., p. 29. © Jacaranda Press (Milton, Qld., Aust.), 1992.

efficient, uncompetitive local industries. By the time the colonies unified, pastoral industries were also contributing income. So the miners and the farmers paid for those imports Australians could not produce themselves, plus the products made in the cities. No wonder the cities grew: here were secure manufacturing jobs, jobs in state-run service enterprises, and jobs in the growing government bureaucracy. When we noted earlier that Australians once had the highest per-capita GNI in the world, this was achieved in the mines and on the farms, not in the cities.

But the good times had to come to an end. The prices of farm products fluctuated, and international market competition increased. The cost of mining, transporting, and shipping ores and minerals also rose. Australians like to drive (there are more road miles per person in Australia than in the United States), and expensive petroleum imports were needed. Meanwhile, the government-protected industries had been further fortified by strong labor unions. For a time, Australia saw its economy decline, its national debt grow, and its unemployment increase. What was needed was diversification: in 1999, only 22 percent of Australian exports by value were the kind of products, including high-tech equipment, that made East Asia's economic tigers so successful. That, however, was double the 1985 percentage, and in the early years of the twenty-first century Australia's economy prospered, the national debt relieved by substantially increased domestic energy production. Today, Australia's per-capita GNI of U.S. $28,290 compares favorably not only to those of successful East and Southeast Asian economies, but is also higher than the majority of European countries.

Agricultural Abundance

Australia has material assets of which other countries on the Pacific Rim can only dream. In agriculture, sheep-raising was the earliest commercial venture, but it was the technology of refrigeration that brought world markets within reach of Australian beef producers. Wool, meat, and wheat have long been the country's big three income earners; Figure 11-5 displays the vast pastures in the east, north, and west that constitute the ranges of Australia's huge herds. The zone of commercial grain farming forms a broad crescent extending from northeastern New South Wales through Victoria into South Australia, and covers much of the hinterland of Perth. Keep in mind the scale of this map: Australia is only slightly smaller than the 48 contiguous States of the United States! Commercial grain farming in Australia is big business. As the climatic map would suggest, sugarcane grows along most of the warm, humid coastal strip of Queensland, and Mediterranean crops (including grapes for Australia's wines) cluster in the hinterlands of Adelaide and Perth. Mixed horticulture concentrates in the basin of the Murray River system, including rice, grapes, and citrus fruits, all under irrigation. And, as elsewhere in the world, dairying has developed near the large urban areas. With its considerable range of environments, Australia yields a wide variety of crops.

Mineral Wealth

Australia's mineral resources, as Figure 11-5 shows, also are diverse. Major gold discoveries in Victoria and New South Wales produced a ten-year gold rush starting in 1851 and ushered in a new economic era. By the middle of that decade, Australia was producing 40 percent of the world's gold. Subsequently, the search for more gold led to the discoveries of other minerals. New finds are still being made today, and even oil and natural gas have been found both inland and offshore (see the symbols in Fig. 11-5 in the Bass Strait between Tasmania and the mainland, and off the northwestern coast of Western Australia, where a combination of new finds and the configuration of maritime boundaries has benefited Australia enormously). Coal is mined at many locations, notably in the east near Sydney and Brisbane but also in Western Australia and even in Tasmania; before coal prices fell, this was a valuable export. Major deposits of metallic and nonmetallic minerals abound—from the complex at Broken Hill and the mix of minerals at Mount Isa to the huge nickel deposits at Kalgoorlie and Kambalda, the copper of Tasmania, the tungsten and bauxite of northern Queensland, and the asbestos of Western Australia. A glance at the map reveals the wide distribution of iron ore (the red dots), and for this raw material as for many others, Japan for many years was Australia's best customer. As recently as the late 1990s, Japan was buying more than one-third of all Australian mineral exports, but during the current decade China's share has grown substantially.

Manufacturing's Limits

Australian manufacturing, as we noted earlier, remains oriented to domestic markets. One cannot expect to find Australian automobiles, electronic equipment, or cameras challenging the Pacific Rim's economic tigers for a place on world markets—not yet, at any rate. Australian manufacturing is diversified, producing some machinery and equipment made of locally produced steel as well as textiles, chemicals, paper, and many other items. These industries cluster in and near the major urban areas where the markets are. The domestic market in Australia is not large, but it is affluent by world standards; this makes it attractive to foreign producers, and Australian stores are full of high-priced goods from Japan, South Korea, Taiwan, and Hong Kong. Indeed, despite its long-term protectionist practices, Australia still does not produce many goods that could be manufactured at home. Overall, the economy continues to display symptoms of a still-developing rather than a fully developed country.

Australia's Future

A new age is dawning in Australia, a country that not long ago (1988) celebrated the bicentennial of its first European settlement—two centuries as a European outpost facing the Pacific Ocean. Now, with Australia well into its third century, its European bonds are weakening and its Asian ties are strengthening.

On the face of it, Australia and its northern neighbors on the Pacific Rim would seem to exhibit a geographic complementarity: Japan and the economic tigers need Australia's excess food, metals, and minerals, and Australia needs the cheap manufactures Asia produces. But it is not that simple. Australia still has tariff barriers against imported goods, and the Asian countries on the Pacific Rim maintain import barriers against processed foods and minerals, thus discouraging Australia from refining its exports and earning more from them. Furthermore, Asia's manufacturers are more interested in the potential of large markets (such as Indonesia, with 226 million people) than Australia, and its investors like the low wages of East and Southeast Asia. In any case, Australia would find it difficult to open its economy and lower its protective tariffs without reciprocation from its Asian trading partners.

Australia's allies in the global core also have not been invariably helpful. U.S. government subsidies to American wheat farmers have enabled U.S. farmers to sell their wheat at lower prices than Australian farmers can afford to sell theirs. The U.S. government's attempts to address this issue have failed because of vigorous bargaining by senators representing grain-producing States; this is only part of a giant farm-subsidy program that also hurts Mexican, Caribbean, and Brazilian farmers among others. But as Australia's overall economy grows, it also becomes less dependent on farm exports, so that an America that advocates free trade and then acts against this principle can still be regarded as a satisfactory trading partner and global ally.

FROM THE FIELD NOTES

"While in Darwin I had to visit a government office, and waiting in line with me was this Aboriginal girl, her brother, and her father. They lived about two hours from the city, she said (which I gathered was the length of their bus ride), and they had to come here from time to time for filing forms and visiting the doctor. When I asked whether they enjoyed coming to Darwin, all three shook their heads vigorously. 'It is not a friendly place,' she said. 'But we must come here.' In fact I had been surprised at the comparatively small number of Aboriginal people I had seen in Darwin, far fewer than in Alice Springs. Certainly laid-back Alice Springs is a very different place from larger, busier, and more modern Darwin."

Aboriginal Issues

In this first decade of the twenty-first century, Australia faces other, more serious challenges. The Aboriginal population of 400,000 (including many of mixed ancestry) has been gaining influence in the country's affairs. In the 1980s, Aboriginal leaders began a campaign to obstruct exploration and mining on ancestral and sacred lands. Until 1992, Australians had taken it for granted that Aborigines had no rights to land ownership; the continent had been open and undemarcated, and there was no evidence of prior title. In that year, however, the Australian High Court ruled in favor of an Aborigine, Eddie Mabo, and his co-inhabitants of the Murray Islands in the Torres Strait. The court ruled that Mabo and his community owned customary title to their land. The ruling implied that Aborigines elsewhere, too, could claim title to traditional land.

When the federal government passed the law that would codify this far-reaching and astounding ruling, it protected Australia's pastoral leases (the tracts of land leased to ranchers of cattle and sheep by the Commonwealth) by excluding them—but not the holdings of mining companies. This led to an immediate Aboriginal victory at Coronation Hill in the Northern Territory (Fig. 11-5) where the administration prohibited the development of platinum, gold, and palladium deposits because they were deemed to lie on an Aboriginal community's sacred land.

Under the Native Title Bill, passed by the Australian Senate in 1993, Aborigines who had long lived on "vacant" land could petition the courts for title to that land: those who could prove that their land was taken away were entitled to compensation. And when a mining company's lease expired, Aborigines could reclaim the land. The result was a crisis of confidence in the mining industry and a sharp drop in investment in exploration, resource development, and infrastructure.

Soon the Aborigines tested the government's move to exclude the pastoral leases, and again they won in the High Court. In 1996, ruling on a case brought before it by the Wik Aboriginal community in Queensland, the court decided that pastoral leases and native title could coexist, thereby voiding the government's exclusion of these leases from the Native Title Bill. This decision implied that vast areas (potentially as much as 78 percent of Australia) could be subject to Aboriginal claims (Fig. 11-6). Now not only miners and ranchers had reason to be concerned: activist Aboriginal groups announced that they would claim ownership to parcels of land in the heart of Sydney and other major cities.

Again the federal government tried to mitigate the impact of the High Court's ruling, this time by passing legislation that would abolish native title and thus preclude claims on pastoral leases where such title would "interfere" with the rights and work of the leaseholding pastoralist. Aborigines would be allowed to enter such land to hold ceremonies and gather food in their traditional ways, but they must not in any way obstruct pastoral activities by taking cultivated crops, moving livestock away, cutting fences, or opening gates. Moreover, all Aboriginal claims on offshore water and marine resources were voided. Aboriginal leaders responded by threatening to inundate the court system with numerous title claims that would cost those affected, and the government, hundreds of millions of dollars.

12 The **Aboriginal land issue** is essentially (though not exclusively) an Outback issue, notably in Western Australia, South Australia, and Queensland States. It is having a major impact on Australia's political geography; views on the matter in the large cities tend to differ sharply from those in the interior. It has had a polarizing effect on politics and society, strengthening political parties and movements opposed to making further "concessions" to Aborigines. In Queensland, the One Nation Party in 1998 got 25 percent of the vote on such a platform, forcing the ruling conservative party (the Liberals) that controlled the State government to make further concessions to its right wing, including a reduction in welfare awards to Aborigines. Polls indicate that most Australians nevertheless oppose the kind of extremism the anti-Aborigine (and anti-immigration) parties represent. But the national government must navigate a difficult course toward reconciliation (see the Issue Box entitled "Indigenous Rights and Wrongs").

It should be noted that only a few Aboriginal communities still pursue their original life of hunting, gathering, and fishing in the remotest corners of the Northern Territory, northern Queensland, and northernmost Western Australia. The remainder are scattered across the continent, subsisting on reserves the government set aside for them (much of the Northern Territory is so designated), working on cattle stations, or performing mostly menial jobs in cities and towns. In comparatively affluent Australia, the Aboriginal peoples suffer from poverty, disease, inadequate education, and even malnutrition. The Aboriginal rights question will be a major one in Australia's future.

Immigration Issues

The population question has preoccupied Australia as long as Australia has existed—and even longer. Fifty years ago, when Australia had less than half the population it has today, 95 percent of the people were of European ancestry, and more than three-quarters of them came from the British Isles. Eugenic (race-specific) **immigration policies** maintained this situation until **13** the 1970s. Today, the picture is dramatically different: of 20.3 million Australians, only about one-third have British-Irish origins, and Asian immigrants outnumber both European immigrants and the natural increase each year. During the early 1990s, nearly 150,000 legal immigrants arrived in Australia annually, most of them coming from Hong Kong, Vietnam, China, the Philippines, India, and Sri Lanka. Immigration quotas have since been reduced, most recently to 80,000, but Asian immigrants continue to outnumber those from Western sources. Immigration from Britain and Ireland varies from 20,000 to 25,000 annually, and between 5000 and 10,000 New Zealanders move to Australia each year.

LAND AND SEA AREAS SUBJECT TO ABORIGINAL CLAIMS

■ Aboriginal claims □ Other land

▨ Northern Territory already Aboriginal land

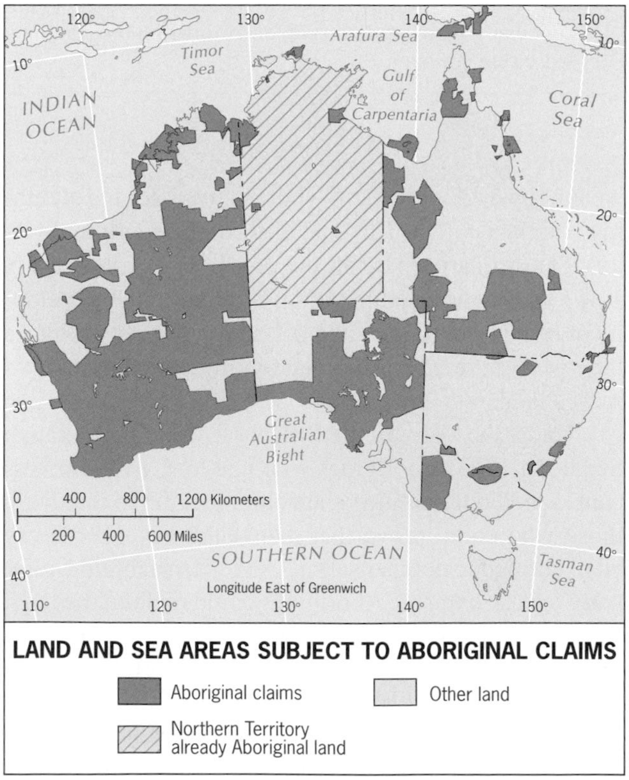

FIGURE 11-6 © H. J. de Blij, P. O. Muller, and John Wiley & Sons, Inc.

Indigenous Rights and Wrongs

THE LEAST WE SHOULD DO IS APOLOGIZE

"I can't believe this. Here we are more than a dozen years beyond our two-hundredth-year anniversary, the evidence of our mistreatment of the Aborigines is everywhere, and we can't, as a nation, say we're sorry? How could we get ourselves into a public debate over this? What does it take to offer sincere apologies to the original Australians and to tell them (and not only to tell them) that we'll do better?

"We should remind ourselves more often than we do of what happened here. When our ancestors arrived on these shores there were more than a million Aboriginal people in Australia, with numerous clans and cultures. They had many virtues, but one in particular: they didn't appropriate land. Land was assigned to them by the creator (what they call the Dreaming), and their relationship to it was spiritual, not commercial. They didn't build fences or walls. Neither had they adopted some bureaucratic religion. Just think, a world where land was open and free and religion was local and personal. Wouldn't that do something for the Middle East and Northern Ireland! Well, we showed up and started claiming and fencing off the land that, under British rules, was there for the taking. If the Aborigines got in the way, they were pushed out, and if they resisted, they got killed. You don't even want to think about what happened in Tasmania. A campaign of calculated extermination. Between 1800 and 1900 the Aboriginal population dropped from 1 million to around 50,000. And when we became a nation, they weren't even accorded citizenship.

"None of this stopped some white Australian men from getting Aboriginal women pregnant. But from 1910 on church and state managed to make things even worse. They took the young children away from their mothers and put them in institutions, where they would be 'Europeanized' and then married off to white partners, so that they would lose their Aboriginal inheritance. This, if you'll believe it, went on into the 1960s! Think of the scenes, these kids being kidnapped by armed officials from their mothers. And we have to argue about saying we're sorry?

"Well, at least we're discussing it in the open, and the misdeeds of the past are coming out. Aboriginal Australians finally got the vote in 1962, and they're able to get elected to parliament. I work in a State government office that assists Asian immigrants here in Sydney, and I only wish that we'd done for the Aborigines what we've done for the immigrants we admit."

Regional ISSUE

ENOUGH IS ENOUGH

"Prime Minister John Howard was right when he said that apologizing for history should be a private matter. I was born on this Outback sheep station 25 years ago, and we employ a dozen Aboriginal workers, most of whom were born in this area. I had nothing to do with what happened more than a century ago, or even the first 75 years of the past century. What I would have done is irrelevant, and what my great-great-grandparents may have done has nothing to do with me. All over the world people are born into situations not of their making, and they should all go around apologizing?

"And as a matter of fact I believe that this country has bent over backwards, in my time at least, to undo the alleged wrongs of the past. Look, the Aboriginal minority counts about 400,000 or 2 percent of the population. Compare this to the map: they've got the Northern Territory and other parts of the country too, and that's close to 15 percent of Australia. And now we're going to give them more? I could see disaster in the making when that guy Eddie Mabo got the High Court to agree to his so-called 'native title' claim to land with which he and his people hadn't done a thing. Since when do Aboriginal customs take precedence over the law of the land? But even worse was that decision supporting the Wiks up there in Cape York. That was unbelievable: native title on land already held in pastoral lease? That affects all of us. Pretty soon the whole of Australia will be targeted by these Aboriginal claimants and their lawyers. And by the way, in the old days those people moved around all the time. Who's to say what clans owned what land when it comes to claiming native title? These court cases are going to tie us up in legal knots for generations to come.

"And now I gather that some Aboriginal people are calling for a treaty between themselves and the government, for what purpose I'm not sure—I guess it'll be like the one the Kiwis made with the Maori there. Is this really what we need if we're going to put the past behind us? Look, I like the fellows working here, but you've got to realize that no laws or treaties are going to change all of the problems they have. They're getting preference for many kinds of government employment, remedial help in all kinds of areas, but still they wind up in jail, leaving school, giving up their jobs. Yes, I'm sure wrongs were done in the past. But apologizing isn't going to make any difference. They have their opportunity now. In a lot of countries they would have never gotten it: Aussies are pretty decent people. It's up to them to make the best of it."

Vote your opinion at www.wiley.com/college/deblij

The immigration issue roils Australian society. In the mid-1990s, when hundreds of "boat people" landed on Australian shores as illegal immigrants, their arrival and subsequent treatment led to a wave of soul-searching. In 1998, when immigrant workers were among strike breakers at a major dockyard, labor unions denounced their role. Restrictions on foreign ownership of Australian real estate reflect Australian fears of what extremist groups call the "Asianization" of the country. At the same time, the success of many Asian settlers in Australia evinces the opportunities still available in this open and free society.

The name Cabramatta conjures up varied reactions among Australians. During the 1950s and 1960s, many immigrants from southern Europe settled in this western suburb of Sydney attracted by affordable housing. During the 1970s and 1980s, Southeast Asians arrived in large numbers, and during this period Cabramatta, in the eyes of many, became synonymous with gang violence and drug dealing. More recently, however, Cabramatta's ethnic diversity has come to be viewed in a more favorable light, and is seen as the "multicultural capital of Australia," a tourist attraction and proof of Australia's capacity to accommodate non-Europeans. Meanwhile, Cabramatta has been spruced up with Oriental motifs of various kinds. This 'Freedom Gate' in the Vietnamese community is flanked by a Ming horse and a replica of a Forbidden City lion—all reflecting better times for an old transit point for immigrants and refugees. © Trip/Eric Smith.

The Asian immigration issue is essentially an urban one: most Asian immigrants have settled in the cities and towns. Sydney has by far the largest contingent, and many ethnic districts have become part of its urban-cultural mosaic. In the 1970s Sydney, always Australia's most multicultural city, had Italian, Greek, and Yugoslavian neighborhoods. Today it also has Vietnamese, Laotian, Indian, and Chinese districts (see photo above). Assimilation continues along a bumpy road, and the economy is the barometer: when the economy lags, social problems tend to worsen.

The issue of asylum-seekers suddenly returned to the forefront of Australian concerns in 2001 when a Norwegian freighter saved more than 400 Afghans from a sinking Indonesian fishing boat and took them toward Australian-administered Christmas Island in the Indian Ocean south of Jawa. Australian prime minister John Howard declared that none of the migrants would be allowed to set foot on Australian soil, and the government sent a military force to transfer them to a troop ship that took about half of them to Nauru in the mid-Pacific and the other half to New Zealand. An Australian federal court objected and international observers protested, but three-quarters of Australians supported this action. Since 2000, there have been riots, breakouts, and other violence at several remote refugee camps; the public's view of illegal immigration had hardened even before the Christmas Island incident, however.

Prosperous Australia's lengthy and open coastline lies at the end of a chain of land and water that starts in South and East Asia and converges on Indonesia. Australia today is open to legal immigration from all parts of the world, but it must constrain what would otherwise become a flood of asylum seekers whose invasion could disrupt the country's steady march toward multiculturalism.

Environmental Degradation

Another growing issue involves the environment and conservation. Australia's wealth was derived not only from its ores and minerals but from its soils and pastures as well. Not surprisingly, its ecology also paid a heavy price.

Great stands of magnificent forest were destroyed. In Western Australia, centuries-old trees were simply "ringed" and left to die, so that the sun could penetrate through their leafless crowns to nurture the grass below. Then the sheep could be driven into these new pastures. In Tasmania, where Australia's native eucalyptus tree reaches its greatest dimensions (comparable to North American redwood stands), tens of thousands of acres of this irreplaceable treasure have been lost to chain saws and pulp mills. Many of Australia's unique marsupial species have been destroyed, and many more are endangered or threatened. "Never have so few people wreaked so much havoc on the ecology of so large an area in so short a time," observed a geographer in Australia recently; today, awareness of this **environmental degradation** is growing. In Tasmania, 14
the "Green" environmentalist political party has become a force in State affairs, and its activism has slowed deforestation, dam-building, and other "development" projects. Still, many Australians fear the en-

vironmentalist movement as an obstacle to economic growth at a time when the economy needs stimulation. This, too, is an issue for the future.

Status and Role

Several issues involving Australia's status at home, relations with neighbors, and position in the world are stirring up national debate as well. A domestic question is whether Australia should become a republic, ending the status of the British monarch as the head of state, or continue its status quo in the British Commonwealth. A 1999 referendum proved that a majority of Australians were not prepared to abandon the monarchy—at least not in favor of "a president appointed by a two-thirds majority of the membership of the Commonwealth Parliament" as the ballot put it. Although polls had predicted a republican victory, the language on the ballot probably changed the outcome; most Australians seem to favor a republic, but more of them distrust their politicians. If the ballot had said "in favor of a popularly elected president," chances are that Australia would be on its way to becoming a federal republic. This issue will undoubtedly arise again.

Relations with neighboring Indonesia and East Timor have been complicated. For many years, Australia had what may be called a special relationship with Indonesia, whose help it needs in curbing illegal seaborne immigration. It was also profitable for Australia to counter international (UN) opinion and recognize Indonesia's 1976 annexation of Portuguese East Timor, for in doing so Australia could deal directly with Jakarta for the oil reserves under the Timor Sea. Thus Australia gave neither recognition nor support to the rebel movement that fought for independence in East Timor. But this story has a relatively happy ending. When the East Timorese campaign for independence succeeded in 1999 and Indonesian troops began an orgy of murder and destruction, Australia sent a peacekeeping force and spearheaded the UN effort to stabilize the situation. Today, as we saw in Chapter 10, Australia and East Timor are negotiating directly to resolve the seabed energy resources issue.

Australia also has a long-term relationship with Papua New Guinea (PNG), as we will note in Chapter 12. This association, too, has gone through difficult times. In recent years, the inhabitants of Papua New Guinea have strongly resisted privatization, World Bank involvement, and globalization generally. Australia assists PNG in several spheres, but its motives are sometimes questioned. In 2001 the construction of a projected gas pipeline from PNG to the Australian State of Queensland precipitated fighting among tribespeople over land rights, resulting in dozens of casualties, and Australian public opinion reflected doubts over the appropriateness of this venture.

Immediately after the September 2001 attack on U.S. targets in New York and Washington, Australian leaders expressed strong support for the American campaign against terrorism, but they made it a point also to assure the Indonesian government that the War on Terror was not a war on Islam. Although the great majority of Australians supported this stance, their will was severely tested in late 2002 when a terrorist attack on a nightclub in Kuta Beach on the island of Bali killed dozens of vacationing Australians (see photo p. 529).

Territorial dimensions, relative location, and raw-material wealth have helped determine Australia's place in the world and, more specifically, on the Pacific Rim. Australia's population, still just barely 20 million in the first decade of our new century, is smaller than Malaysia's and not much larger than that of the Caribbean island of Hispaniola. But Australia's importance in the international community far exceeds its human numbers.

NEW ZEALAND

Fifteen hundred miles east-southeast of Australia, in the Pacific Ocean across the Tasman Sea, lies New Zealand. In an earlier age, New Zealand would have been part of the Pacific geographic realm because its population was Maori, a people with Polynesian roots. But New Zealand, like Australia, was invaded and occupied by Europeans. Today, its population of 4.2 million is almost 75 percent European, and the Maori form a minority of less than 600,000, with many of mixed Euro-Polynesian ancestry.

New Zealand consists of two large mountainous islands and many scattered smaller islands (Fig. 11-7). The two large islands, with the South Island somewhat larger than the North Island, look diminutive in the great Pacific Ocean, but together they are larger than Britain. In contrast to Australia, the two main islands are mainly mountainous or hilly, with several peaks rising far higher than any on the Australian landmass. The South Island has a spectacular snowcapped range appropriately called the Southern Alps, with numerous peaks reaching beyond 10,000 feet (3300 m). The smaller North Island has proportionately more land under low relief, but it also has an area of central highlands along whose lower slopes lie the pastures of New Zealand's chief dairying district. Hence, while Australia's land lies relatively low in elevation and exhibits much low relief, New Zealand's is on the average high and is dominated by rugged relief (photo p. 557).

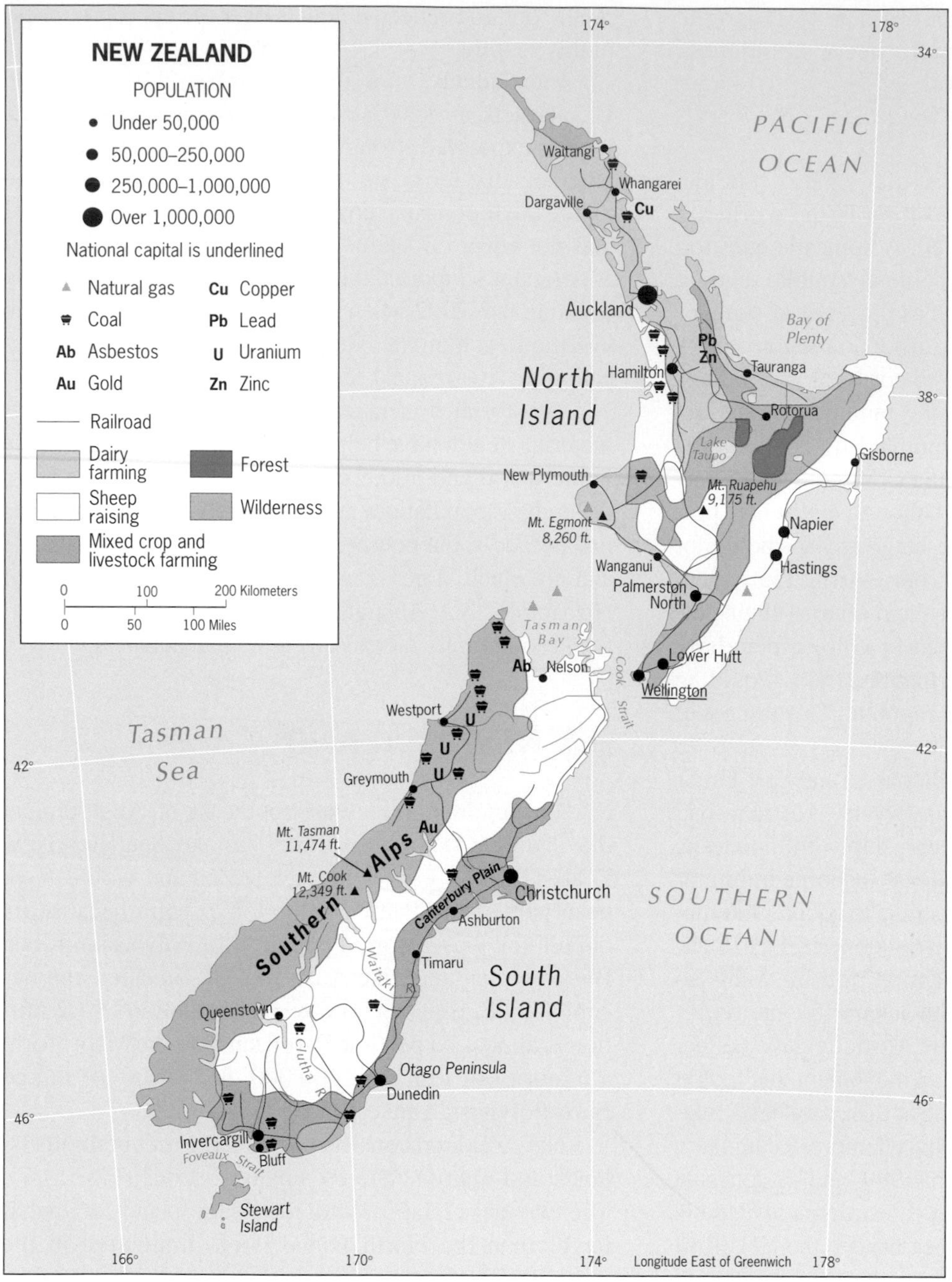

FIGURE 11-7 © H. J. de Blij, P. O. Muller, and John Wiley & Sons, Inc.

Human Spatial Organization

Thus the most promising areas for habitation are the lower-lying slopes and lowland fringes on both islands. On the North Island, the largest urban area, Auckland (1.2 million), occupies a comparatively low-lying peninsula. On the South Island, the largest lowland is the agricultural Canterbury Plain, centered on Christchurch. What makes these lower areas so attractive, apart from their availability as cropland, is their magnificent pastures. The range of soils and pasture plants allows both summer and winter grazing. Moreover, the Canterbury Plain, the chief farming region, also produces a wide variety of vegetables, cereals, and fruits. About half of all New Zealand is pasture land, and much of the farming provides fodder for the pastoral industry. Sixty million sheep and eight million cattle dominate these livestock-raising activities, with wool, meat, and dairy products providing nearly two-thirds of the islands' export revenues.

Despite their contrasts in size, shape, physiography, and history, New Zealand and Australia have much in common. Apart from their joint British heritage, they

FROM THE FIELD NOTES

"The drive from Christchurch to Arthur's Pass on the South Island of New Zealand was a lesson in physiography and biogeography. Here, on the east side of the Southern Alps, you leave the Canterbury Plain and its agriculture and climb into the rugged topography of the glacier-cut, snowcapped mountains. A last pasture lies on a patch of flatland in the foreground; in the background is the unmistakable wall of a U-shaped valley sculpted by ice. Natural vegetation ranges from pines to ferns, becoming even more luxuriant as you approach the moister western side of the island." © H. J. de Blij.

share a sizeable pastoral economy, a small local market, the problem of great distances to world markets, and a desire to stimulate (through protection) domestic manufacturing. The high degree of urbanization in New Zealand (78 percent of the total population) again resembles Australia: substantial employment in city-based industries, mostly the processing and packing of livestock and farm products, and government jobs.

More remote even than Australia, New Zealand also has been affected by Pacific Rim development, though less so than its giant neighbor. Still, by the mid-1990s, Japan and other Pacific Rim economies were buying more of New Zealand's exports (mainly agricultural products) than either Australia or the United States, a pattern that persists today. Britain currently takes only about 6 percent of New Zealand's exports. Australia and the United States still send New Zealand most of its imports (about 40 percent combined), but the Pacific Rim's contribution has been rising. As Figure G-11 and Table G-1 show, New Zealand is a high-income economy, but it is not in the upper tier of the world's richer economies.

Spatially, New Zealand shares with Australia its
15 pattern of **peripheral development** (Fig. G-9), imposed not by desert but by high rugged mountains. The country's major cities—Auckland and the capital of Wellington (together with its satellite of Hutt) on the North Island, and Christchurch and Dunedin on the South Island—are all located on the coast, and the entire railway and road system is peripheral in its configuration (Fig. 11-7). This is more pronounced on the South Island than in the north because the Southern Alps are New Zealand's most formidable barrier to surface communications.

The Maori Factor and New Zealand's Future

One of New Zealand's historic sites is the wooden Treaty House at Waitangi, in the northern peninsula of the North Island. There, following a series of brutal conflicts, the British and the Maori signed the Treaty of Waitangi in 1840. The treaty granted the British sovereignty over New Zealand, but it also guaranteed Maori rights over tribal lands. February 6, the date of the signing of the treaty, is a public holiday in New Zealand called Waitangi Day. A ceremony at the Treaty House, attended by the prime minister and by Maori leaders, is a highlight of the day's events. Visitors to the ceremony in 1995 were in for a shock, however. Maoris disrupted the proceedings; the prime minister was jostled by members of the crowd; and an attempt was made to burn the Treaty House. For the first time in more than half a century, the ceremony had to be abandoned.

What precipitated this show of anger? In fact, it was only the latest in a series of events signaling a rise in Maori ethnic consciousness. Maori leaders insist that the terms of the Waitangi Treaty (partly abrogated in 1862) be enforced, that large tracts of land in urban as well as rural areas, amounting to more than half of the national territory, be awarded to their rightful Maori owners, and that Maori fishing rights be paramount over offshore waters. Judicial rulings during the 1990s supported the Maori position, and Maori successes in one arena have led to new claims in others. Today, the Maori question has become the leading national issue in New Zealand, and Maori activism has risen sharply.

FROM THE FIELD NOTES

"It was Sunday morning in Christchurch on New Zealand's South Island, and the city center was quiet. As I walked along Linwood Street I heard a familiar sound, but on an unfamiliar instrument: the Bach sonata for unaccompanied violin in G minor—played magnificently on a guitar. I followed the sound to the artist, a Maori musician of such technical and interpretive capacity that there was something new in every phrase, every line, every tempo. I was his only listener; there were a few coins in his open guitar case. Shouldn't he be playing before thousands, in schools, maybe abroad? No, he said, he was happy here, he did all right. A world-class talent, a street musician playing Bach on a Christchurch side street, where tourists from around the world were his main source of income. Talk about globalization." © H. J. de Blij

In this context, the events of February 6, 1995 were no surprise.

The Maori appear to have reached New Zealand during the tenth century A.D., and by the time the European colonizers arrived they had had an enormous impact on the islands' ecosystems, especially on the North Island where most Maori lived. One of the Maori complaints is the slow pace of integration of the Maori minority of more than 550,000 (13.5 percent) into modern New Zealand society. In fact, the Maori have been joined by other Pacific islanders, so that Auckland today not only is New Zealand's largest city; it also may be the largest urban concentration of Polynesians anywhere, with over 200,000 Maori as well as Samoans, Cook Islanders, Tongans, and others making up one-sixth of the metropolitan population. Their neighborhoods give Auckland an ethnic patchwork that evinces the Maoris' complaint about nonintegration.

The Maori issue also has other geographic overtones. Nearly all Maori still live on the North Island (where three-quarters of all New Zealanders reside), but Maori leaders are making huge claims, based on historic primacy, over much of the South Island. Remote and comparatively conservative, South Islanders have reacted to these claims much as their counterparts in Western Australia have to the Mabo and Wik decisions on Aboriginal rights.

The Waitangi Treaty was intended as the framework for a partnership between the British and the Maoris. If the current disputes lead to a reconsideration and fulfillment of the treaty's terms in the context of modern New Zealand's human geography, then the country's future could be shaped by a harmonious, mainly bicultural society. If this effort fails, the specter of serious ethnic polarization threatens against a demographic background that projects the rapidly growing Polynesian population of New Zealand will double to more than 25 percent of the national total by 2012.

Dominant cultural heritage and prevailing cultural landscape form two criteria on which the delimitation of the Austral Realm is based. But in both Australia and New Zealand, the cultural mosaic is changing, and the convergence with neighboring realms is proceeding.

WHAT YOU CAN DO

RECOMMENDATION: Become a member! Just by taking this course, you are joining a still-small minority of geographically literate Americans (Canadians are far ahead), and you may wish to consider joining some organizations that support this cause. The National Geographic Society (1145 Seventeenth Street N.W., Washington, DC 20036) invites members to join and sends them its famous Magazine *with its remarkable foldout maps. Gamma Theta Upsilon is a students' geographical society (c/o Larry Handley, USGS/NWRC, 700 Cajundome Boulevard, Lafayette, LA 70506) that publishes* The Geographical Bulletin, *a journal that invites students to submit their work for publication. The American Geographical Society (120 Wall Street, Suite 100, New York, NY 10005-3904) publishes two journals that can be appreciated by the general public,* The Geographical Review *and* Focus. *And if you want to get really technical, there is the Association of American Geographers (1710 Sixteenth Street N.W., Washington, DC 20009-3198) which publishes the quarterly* Annals of the Association of American Geographers, The Professional Geographer, *and a monthly* Newsletter *that tells you what's going on in the discipline and where the jobs are. Student members, at appropriately reduced annual dues, are especially welcome!*

GEOGRAPHIC CONNECTIONS

1 Both Australia and New Zealand are plural societies, and there were indigenous peoples in both countries when the European settlers arrived. In virtually every respect, however, the historical geographies of contact and interaction in the two countries have progressed in very different ways. The differences relate to contrasts in indigenous as well as immigrant cultures, divergent population numbers, contrasting territorial dimensions and physiographies, and dissimilar governmental policies and practices. Comment on the differences between Australia and New Zealand in this context. Are there ways in which indigenous demands are similar? How are they expressed on the map?

2 Australia may have a modest population, but its territorial dimensions, resource base, economic wealth, military capacity, and relative location make it a regional power. Its relationships with its neighbors, however, have at times been problematic. To what extent does relative location affect Australia's interaction with Indonesia? How has Australia's relationship with East Timor, before and after the latter's independence in 2002, been affected by its links with Indonesia, and what was the key issue in this context? What interests does Australia maintain in Papua New Guinea? Why would Australia be involved in problems in the Solomon Islands?

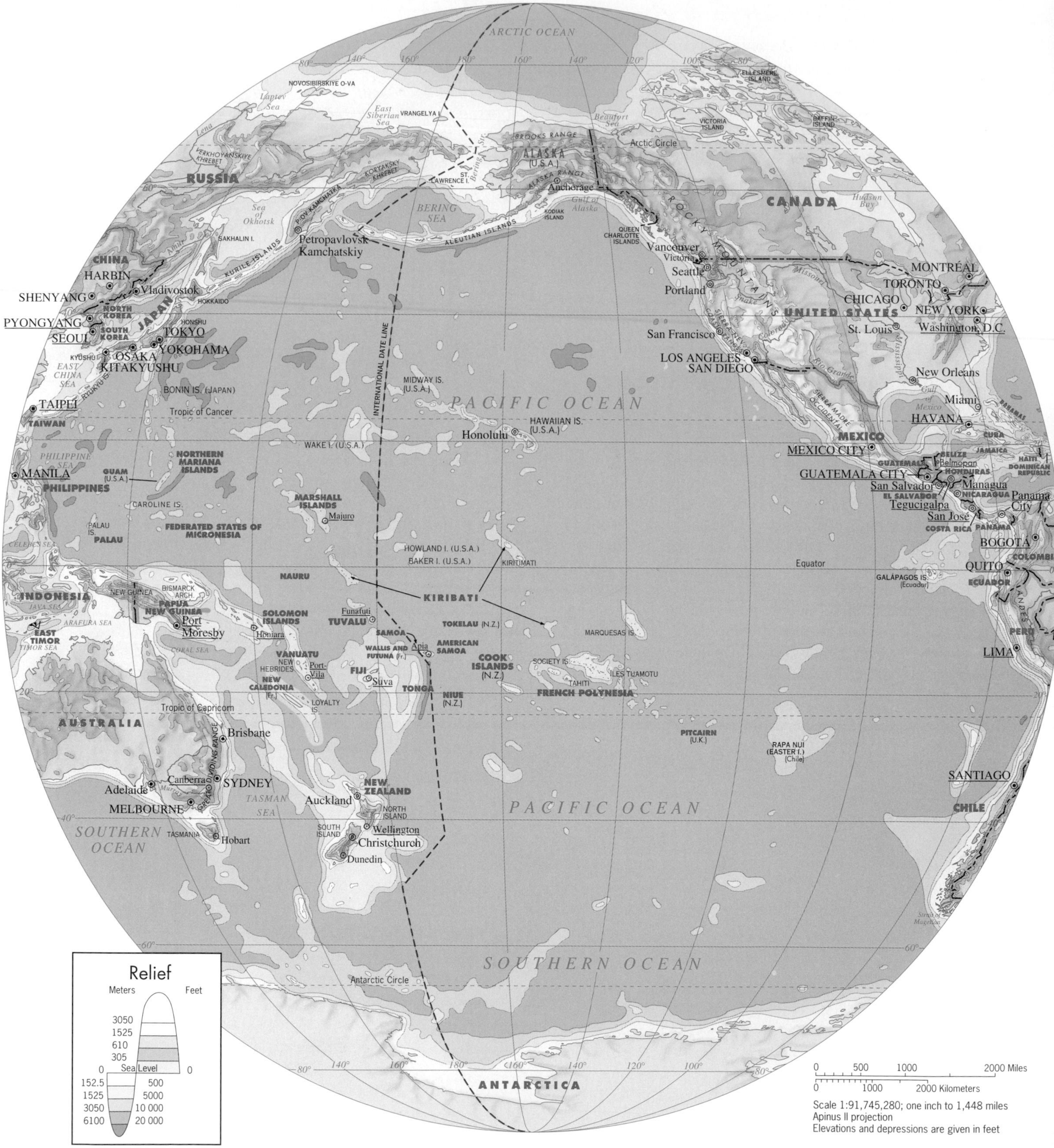

FIGURE 12-1 © H. J. de Blij, P. O. Muller, and John Wiley & Sons, Inc.

Chapter 12

The Pacific Realm

CONCEPTS, IDEAS, AND TERMS

1. **Marine geography**
2. **Territorial sea**
3. **High seas**
4. **Continental shelf**
5. **Exclusive Economic Zone (EEZ)**
6. **Maritime boundary**
7. **Median-line boundary**
8. **High-island cultures**
9. **Low-island cultures**
10. **Antarctic Treaty**

REGIONS

- **MELANESIA**
- **MICRONESIA**
- **POLYNESIA**

BETWEEN THE AMERICAS to the east and the western Pacific Rim to the west lies the vast Pacific Ocean, larger than all the world's land areas combined. In this greatest of all oceans lie tens of thousands of islands, some large (New Guinea is by far the largest), most small (many are uninhabited). Together, the land area of these islands is a mere 376,000 square miles (974,000 sq km), about the size of Texas plus New Mexico, and over 90 percent of this lies in New Guinea.*

DEFINING THE REALM

The Pacific geographic realm—land and water—covers nearly an entire hemisphere of this world, the one commonly called the Sea Hemisphere (Fig. 12-1). This Sea Hemisphere meets the Russian and North American realms in the far north and merges into the Southern Ocean in the south. Despite the preponderance of water, this fragmented, culturally complex realm does possess regional identities. It includes the Hawaiian Islands, Tahiti, Tonga, and Samoa—fabled names in a world apart.

In terms of modern cultural and political geography, Indonesia and the Philippines are not part of the Pacific Realm, although Indonesia's political system reaches into it; nor are Australia and New Zealand part of it. Before the European invasion and colonization, Australia would have been included because of its Aboriginal population and New Zealand because of its Maori population's Polynesian affinities. But the Europeanization of their countries has engulfed black Australians and Maori New Zealanders,

*The figures in Table G-1 do not match these totals because only the political entity of Papua New Guinea is listed, not the Indonesian province of Papua that occupies the western part of the island. Here, as Figure 10-2 showed, the political and the realm boundaries do not coincide.

MAJOR GEOGRAPHIC QUALITIES OF The Pacific Realm

1. The Pacific Realm's total area is the largest of all geographic realms. Its land area, however, is the smallest, as is its population.
2. The island of New Guinea, with 8.4 million people, alone contains over 80 percent of the Pacific Realm's population.
3. The Pacific Realm, with its wide expanses of water and numerous islands, has been strongly affected by United Nations Law of the Sea provisions regarding states' rights over economic assets in their adjacent waters.
4. The highly fragmented Pacific Realm consists of three regions: Melanesia (including New Guinea), Micronesia, and Polynesia.
5. Melanesia forms the link between Papuan and Melanesian cultures in the Pacific.
6. The Pacific Realm's islands and cultures may be divided into volcanic *high-island* cultures and coral-based *low-island* cultures.
7. In Micronesia, U.S. influence has been particularly strong and continues to affect local societies.
8. In Polynesia, local cultures nearly everywhere are severely strained by external influences. In Hawai'i, as in New Zealand, indigenous culture has been largely submerged by Westernization.
9. Indigenous Polynesian culture continues to exhibit a remarkable consistency and uniformity throughout the Polynesian region, its enormous dimensions and dispersal notwithstanding.

and the regional geography of Australia and New Zealand today is decidedly not Pacific. In New Guinea, on the other hand, Pacific peoples remain numerically and culturally the dominant element.

The Pacific islands were colonized by the French, British, and Americans; an indigenous Polynesian kingdom in the Hawaiian Islands was annexed by the United States and is now the fiftieth State. Still, today, the political map is an assemblage of independent and colonial territories. Paris controls New Caledonia and French Polynesia. The United States administers Guam and American Samoa, the Line Islands, Wake Island, Midway Islands, and several smaller islands; the United States also has special relationships with other territories, former dependencies that are now nominally independent. The British, through New Zealand, have responsibility for the Pitcairn group of islands, and New Zealand administers and supports the Cook, Tokelau, and Niue Islands. Easter Island, that storied speck of land in the southeastern Pacific, is part of Chile. Indonesia rules Papua in western New Guinea.

Other island groups have become independent states. The largest are Fiji, once a British dependency, the Solomon Islands (also formerly British), and Vanuatu (until 1980 ruled jointly by France and Britain). Also on the current map, however, are such microstates as Tuvalu, Kiribati, Nauru, and Palau. Foreign aid is crucial to the survival of most of these countries. Tuvalu, for example, has a total area of about 10 square miles, a population of some 10,000, and a per-capita GNI of about U.S. $1200, derived from fishing, copra sales, and some tourism. But what really keeps Tuvalu going is an international trust fund set up by Australia, New Zealand, the United Kingdom, Japan, and South Korea. Annual grants from that fund, as well as money sent back to families by workers who have left for New Zealand and elsewhere, allow Tuvalu to survive.

THE PACIFIC REALM AND ITS MARINE GEOGRAPHY

Certain land areas may not be part of the Pacific Realm (we cited the Philippines and New Zealand), but the Pacific Ocean extends from the shores of North and South America to mainland East and Southeast Asia and from the Bering Sea to the Subtropical Convergence (see p. 541). This means that several seas, including the Sea of Japan (East Sea), the East China Sea, and the South China Sea, are part of the Pacific Ocean. As we will see later, this relationship matters. Pacific coastal countries, large and small, mainland and island, compete for jurisdiction over the waters that bound them.

The Pacific Realm and its ocean, therefore, form an
ideal place to focus on **marine geography**. This field en- 1
compasses a variety of approaches to the study of oceans and seas; some marine geographers focus on the biogeography of coral reefs, others on the geomorphology of beaches, and still others on the movement of currents and drifts. A particularly interesting branch of marine geography has to do with the definition and delimitation of political boundaries at sea. Here geography meets political science and maritime law.

Littoral (coastal) states do not end where atlas maps suggest they do. States have claimed various forms of jurisdiction over coastal waters for centuries, closing off bays and estuaries and ordering foreign fishing fleets to stay away from nearby fishing grounds. Thus arose the
notion of the **territorial sea**, where all the rights of a 2
coastal state would prevail. Beyond lay the **high seas**, 3
free, open, and unfettered by national interests.

It was in the interest of colonizing, mercantile states to keep territorial seas narrow and high seas wide, thus interfering as little as possible with their commercial fleets. In the seventeenth and eighteenth centuries, the territorial sea was 3, 4, or at most 6 nautical miles wide, and the colonizing powers claimed the same widths for their colonies (1 nautical mile = 1.15 statute miles [1.85 km]).

In the twentieth century these constraints weakened. States without trading fleets saw no reason to limit their territorial seas. States with nearby fishing grounds traditionally exploited by their own fleets wanted to keep the increasing number of foreign trawlers away. States with
4 shallow **continental shelves**, offshore continuations of coastal plains, wished to control the resources on and below the seafloor, which were made more accessible by improving technology. States disagreed on the methods by which offshore boundaries, whatever their width, should be defined. Early efforts by the League of Nations in the 1920s to resolve these issues had only partial success, mainly in the technical area of boundary delimitation.

In 1945 the United States helped precipitate what has become known as the "scramble for the oceans." President Harry S Truman issued a proclamation that claimed U.S. jurisdiction and control over all the resources "in and on" the continental shelf down to its margin, around 100 fathoms (600 ft) deep. In some areas, the shallow continental shelf of the United States extends more than 200 miles offshore, and Washington did not want foreign countries drilling for oil just beyond the 3-mile territorial sea.

Few observers foresaw the impact the Truman Proclamation would have, not only on U.S. waters but on the oceans everywhere, including the Pacific. It set off a rush of other claims. In 1952 a group of South American countries, some with little continental shelf to claim, issued the Declaration of Santiago, claiming exclusive fishing rights up to a distance of 200 nautical miles (230 statute miles) off their coasts. Meanwhile, as part of the Cold War competition, the Soviet Union urged its allies to claim a 12-mile territorial sea.

The UNCLOS Convention

Now the United Nations intervened, and a series of UNCLOS (United Nations Conference on the Law of the Sea) meetings began. These meetings addressed issues ranging from the closure of bays to the width and delimitation of the territorial sea, and after three decades of negotiations they achieved a convention that changed the political and economic geography of the oceans forever. Among its key provisions were the authorization of a 12-mile territorial sea for all countries and the establishment
5 of a 200-mile (230 statute-mile) **Exclusive Economic Zone (EEZ)** over which a coastal state has certain economic rights. It has total control over the resources on and beneath the seabed, such as oil, natural gas, or manganese nodules. It also has substantial rights to the fish in the EEZ, but the UNCLOS Convention does stipulate that coastal states must cooperate through international organizations to ensure the conservation of highly migratory species such as tuna, marlin, and swordfish.

These provisions had a far-reaching impact on the world's oceans and seas (Fig. 12-2), especially on the Pacific. Unlike the Atlantic Ocean, the Pacific is studded with islands large and small, and a microstate consisting of one small island suddenly acquired an EEZ covering 166,000 square nautical miles. European colonial powers still holding minor Pacific possessions (notably France) saw their maritime jurisdictions vastly expanded. Small low-income archipelagos could now bargain with large fishing nations over fishing rights in their EEZs. Under UNCLOS rules, such bargaining should involve (in the

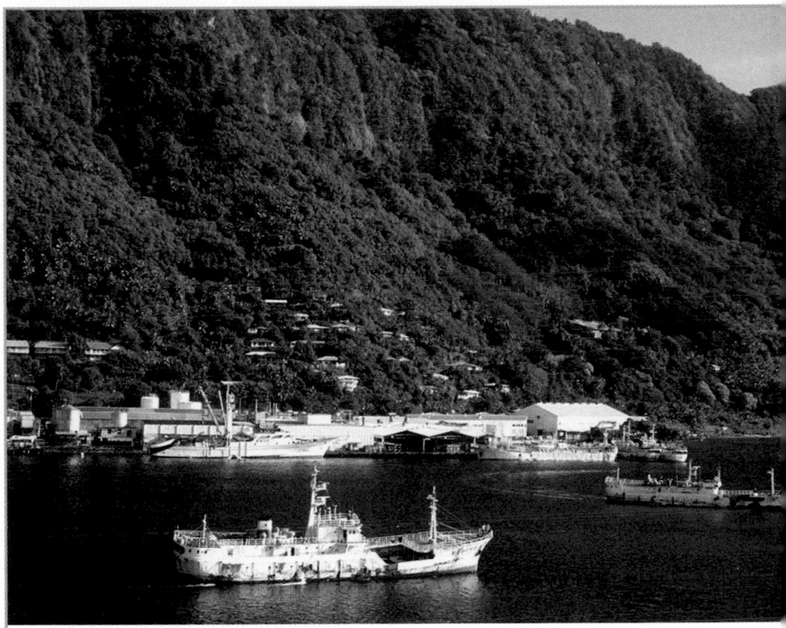

FROM THE FIELD NOTES

"Having arrived late at night, I did not have a perspective yet of the setting of Pago Pago—but even before I got up I knew what the leading industry here had to be. An intense, overpowering smell of fish filled the air. Sure enough, daylight revealed a huge tuna processing plant right across the water, with fishing boats arriving and departing continuously. American Samoa has a huge Pacific Ocean EEZ, one of the richest fishing grounds in the whole realm, and seafood exports rank ahead of tourism as the leading source of external income for the territory, an industry that is mostly American-owned. As the photograph shows, American Samoa is a group of 'high-islands' of volcanic origin, with considerable relief and environmental variety. Tuvalu, the country I had just left, was the complete opposite—a 'low-island' territory whose EEZ has been leased to Japanese fishing fleets, which do their processing on board and employ no locals." © H. J. de Blij.

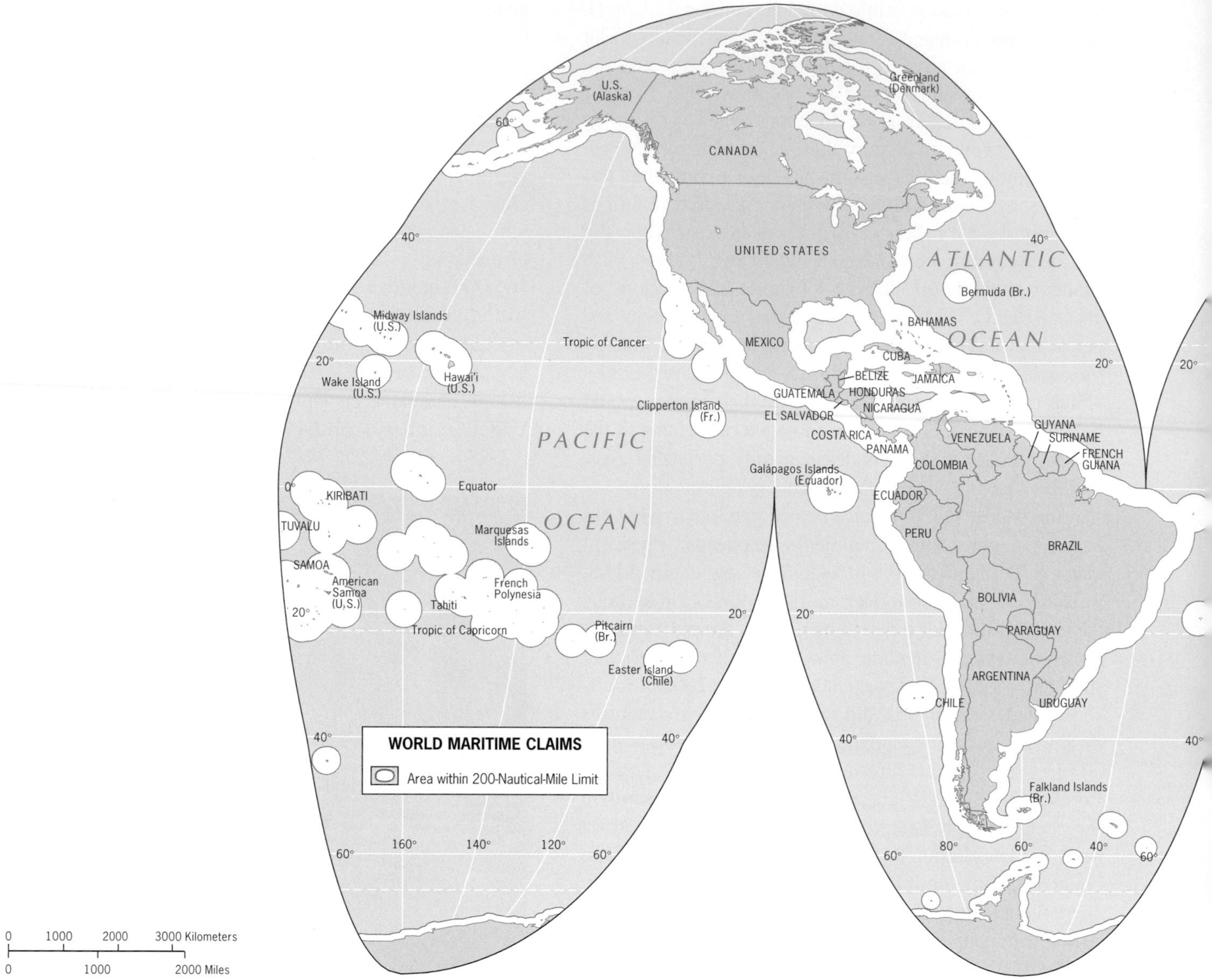

FIGURE 12-2 © H. J. de Blij, P. O. Muller, and John Wiley & Sons, Inc.

case of the Pacific) such organizations as the South Pacific Forum Fisheries Agency, and the United States insists on this procedure, which exists to protect highly migratory species. But some other countries, notably Japan and Taiwan, flout that rule and negotiate directly with the island-states over fishing rights in their EEZs. In 2003, the journal *Nature* published the results of a study that reported a 90 percent decline in large species over the past 50 years as fishing nations with high-technology equipment are scouring the oceans and destroying marine ecosystems even in remote reaches of the globe.

The extension of the territorial sea to 12 nautical miles and the EEZ to an additional 188 nautical miles
6 (both are measured from the coastline) created new **maritime boundary** problems. Waters less than 24 miles wide separate many countries all over the world, so that
median lines, equidistant from opposite shores, have 7
been delimited to establish their territorial seas. And even more countries lie closer than 400 nautical miles apart, requiring further maritime-boundary delimitation to determine their EEZs. In such maritime regions as the North Sea, the Caribbean Sea, and the Japan, East China, and South China seas, a maze of maritime boundaries emerged, some of them subject to dispute. Political changes on land can lead to significant modifications at sea and sometimes result in serious disputes.

A case in point involves East Timor and its neighbor across the Timor Sea, Australia. As we noted in Chapter 10, when East Timor became an independent state in 2002, this seemed to void the maritime bound-

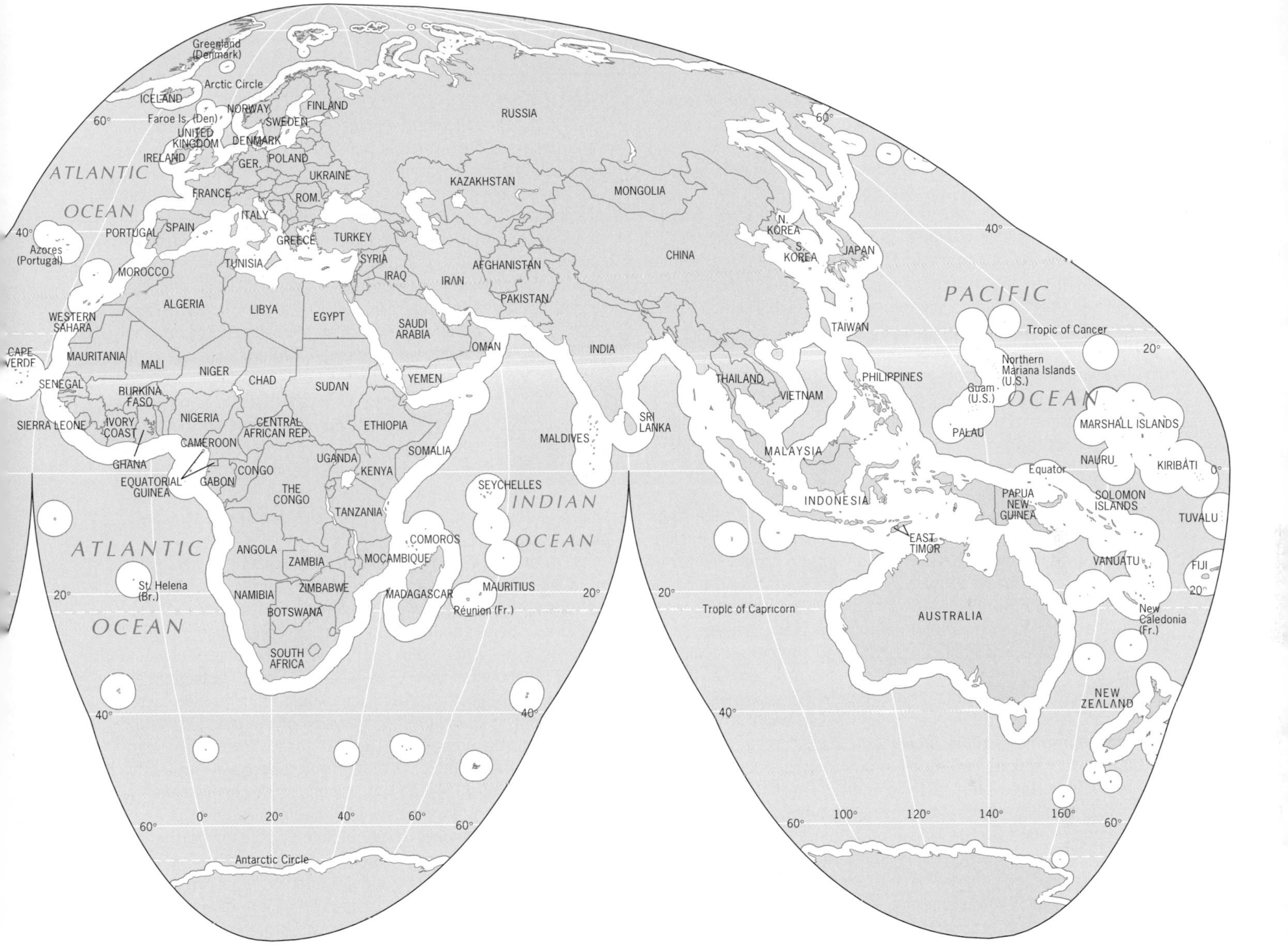

ary between Indonesia and Australia negotiated bilaterally and delimited in 1972 (Fig. 10-13). According to UNCLOS principles, the new boundary between East Timor and Australia would be the median line between the two countries. But Australia was unwilling to give up the wealth it gained from the 1972 boundary. It not only withdrew from UNCLOS, but also announced that it would not be subject to any maritime-boundary decisions made by the International Court of Justice. As far as Australia was concerned, the 1972 boundary would stay in place, and most of the reserves in the so-called Timor Gap and elsewhere in the Timor Sea would continue to be exploited by Australia. Although the Australians put the East Timorese at a disadvantage, this came with a price: Australia's reputation for fairness. Eventually the East Timorese, as noted in Chapter 10, proposed to defer the maritime boundary issue for 50 years in return for a greater share of revenues from the known fields. The Australians agreed, but they had won the key battle: where UNCLOS regulations would have put the border, no delimitation will take place until 2055. Small, poor countries do not have it easy when they confront large, rich ones.

The UNCLOS provisions also created opportunities for some states to expand their spheres of influence. Wider territorial-sea and EEZ allocations raised the stakes: claiming an island now entailed potential control over a huge maritime area. In Chapters 9 and 10, we referred to the large number of island disputes off mainland East Asia: between Japan and Russia, Japan and

Who Should Own the Oceans?

Regional ISSUE

WE WHO LIVE HERE SHOULD OWN THE WATERS

"Without exclusive fishing rights in our waters up to 200 nautical miles offshore, our economy would be dead in the water, no pun intended, sir. As it is, it's bad enough. I'm an official in the Fisheries Department of the Ministry of Natural Resources Development here in Tuvalu, and this modest building you're visiting is the most modern in all of the capital, Funafuti. That's because we're the only ones making some money for the state other than our Internet domain (*tv*), which has been the big story around here since we commercialized it. We don't have much in the way of natural resources and we export some products from our coconut and breadfruit trees, but otherwise we need donors to help us and major fishing nations to pay us for the use of our 750,000 square kilometers of ocean.

"Which brings me to one of our big concerns. Three-quarters of a million square kilometers sound like a lot, but look at the map. It's a mere speck in the vast Pacific Ocean. Enormous stretches of our ocean seem to belong to nobody. That is, they're open to fishing by nations that have the technology to exploit them, but we don't have those fast boats and factory ships they use nowadays. Still, the fish they take there might have come to our exclusive grounds, but we don't get paid for them. It seems to us that the Pacific Ocean should be divided among the countries that exist here, not just on the basis of nearby Exclusive Economic Zones but on some other grounds. For example, islands that are surrounded by others, like we are, ought to get additional maritime territory someplace else. Islands that aren't surrounded should get much more than just 200 nautical miles. To us the waters of the Pacific mean much more than they do to other countries, not only economically but historically too. I don't like Japanese and American and Norwegian fishing fleets on these nearby 'high seas.' They don't obey international regulations on fishing, and they overfish wherever they go. When there's nothing left, they'll simply end their agreement with us, and we'll be left without income from a natural resource over which we never had control.

"I don't expect anything to change, but we feel that the era of the 'high seas' should end and all waters and subsoil should be assigned to the nearest states. The world doesn't pay much attention to our small countries, I realize. But we've put up with colonialism, world wars, nuclear weapons testing, pollution. It's time we got a break."

THE OCEAN SHOULD BE OPEN AND UNRESTRICTED

"Ours is the Blue Planet, the planet of oceans and seas, and the world has fared best when its waters were free and open for the use of all. Encroachment from land in the form of 'territorial' seas (what a misnomer!) and other jurisdictions only interfered with international trade and commercial exploitation and caused far more problems than they solved. Already much of the open ocean is assigned to coastal countries that have no maritime tradition and don't know what to do with it other than to lease it. And when coastal countries lie closer than 400 nautical miles to each other, they divide the waters between them for EEZs and don't even leave a channel for international passage. Yes, I realize that there are guarantees of uninhibited passage through EEZs. But mark my words: the time will come when small coastal states will start charging vessels, for example cruise ships, for passage through 'their' waters. That's why, as the first officer of a passenger ship, I am against any further allocation of maritime territory to coastal states. We've gone too far already.

"In the 1960s, when those UNCLOS meetings were going on, a representative from Malta named Arvid Pardo came up with the notion that the whole ocean and seafloor beyond existing jurisdictions should be declared a 'common heritage of mankind' to be governed by an international agency that would control all activities there. The idea was that all economic activity on and beneath the high seas would be controlled by the UN, which would ensure that a portion of the proceeds would be given to 'Geographically Disadvantaged States' like landlocked countries, which can't claim any seas at all. Fortunately nothing came of this—can you imagine a UN agency administering half the world for the benefit of humanity?

"So you can expect us mariners to oppose any further extensions of land-based jurisdiction over the seas. The maritime world is complicated enough already, what with choke points, straits, narrows, and various kinds of prohibitions. To assign additional maritime territory to states on historic or economic grounds is to complicate the future unnecessarily, and to invite further claims. From time to time I see references to something called the 'World Lake Concept,' the preliminary assignment of all the waters of the world to the coastal states through the delimitation of median lines. Such an idea may be the natural progression from the EEZ concept, but it gives me nightmares."

Vote your opinion at www.wiley.com/college/deblij

South Korea, Japan and China, China and Vietnam, and China and the Philippines. Ownership of many islands there is uncertain, and small specks of island territory have become large stakes in the scramble for the oceans. In the case of the Spratly Islands (Fig. 10-2), six countries claim ownership, including both Taiwan and China. China's island claims in the South China Sea support Beijing's contention that this body of water is part and parcel of the Chinese state—a position that worries other states with coasts facing it.

Figure 12-2 reveals what EEZ regulations have meant to Pacific Realm countries such as Tuvalu, Kiribati, and Fiji, with nearly circular EEZs now surrounding the clusters of islands in this vast ocean space. Japan, Taiwan, and other fishing nations have purchased fishing rights in these EEZs from the island governments, although violations of EEZ rights do occur. Recently, Vanuatu and the Philippines were at odds over unauthorized Filipino fishing in Vanuatu's EEZ.

The process of boundary delimitation continues. In an earlier edition of this book, we included a map of the South China Sea and nearby waters off East and Southeast Asia that showed the median-line boundaries delimited according to UNCLOS specifications. But that map changed when coastal states engaged in bilateral and multilateral negotiations—and argued over island ownership with major boundary implications. The Pacific—and world—map of maritime boundaries remains a work in progress (see the Issue Box entitled "Who Should Own the Oceans?").

This point is illustrated by a development that occurred late in 1999, when the implications of provisions in Article 76 of the 1982 UNCLOS Convention were reexamined. Where a continental shelf extends beyond the 200-mile limit of the EEZ, the Convention apparently allows a coastal state to claim that extension as a "natural prolongation" of its landmass. Although there is as yet no regulation that would also allow the state to extend its EEZ to the edge of the continental shelf, it is not difficult to foresee such an amendment to the Convention. As it stands, states are now delimiting their proposed "natural prolongations" under a 2009 deadline, which again puts poorer states at a disadvantage since the required marine surveys are very expensive. In any case, an article in *Science* in 2002 reported that the big winners are likely to include the United States, Canada, Australia, New Zealand, Russia, and India—although a total of 60 coastal countries could ultimately benefit to some extent from the provision. Thus the "scramble for the oceans" continues, and with it the constriction of the world's high seas and open waters.

REGIONS OF THE REALM

In this realm as a whole, tourism may well be the largest overall revenue earner. Sail across the Pacific Ocean, and one spectacular vista follows another. Dormant and extinct volcanoes, sculpted by erosion into basalt spires draped by luxuriant tropical vegetation and encircled by reefs and lagoons, tower over azure waters. Low atolls with nearly snow-white beaches, crowned by stands of palm trees, seem to float in the water. Pacific islanders, where foreign influences have not overtaken them, appear to take life with enviable ease.

More serious investigations reveal that such Pacific sameness is more apparent than real. Even the Pacific Realm, with its long sailing traditions, its still-diffusing populations, and its historic migrations, has a persistent regional framework. Figure 12-3 outlines the three regions that constitute the Pacific Realm:

Melanesia Papua Province (Indonesia), Papua New Guinea, Solomon Islands, Vanuatu, New Caledonia (France), Fiji

Micronesia Palau, Federated States of Micronesia, Northern Mariana Islands, Republic of the Marshall Islands, Nauru, western Kiribati, Guam (U.S.)

Polynesia Hawaiian Islands (U.S.), Samoa, American Samoa, Tuvalu, Tonga, eastern Kiribati, Cook and other New Zealand-administered islands, French Polynesia

Ethnic, linguistic, and physiographic criteria are among the bases for this regionalization of the Pacific Realm, but we should not lose sight of the dimensions. Not only is the land area small, but also in 2006 the population of this entire realm (including the Indonesian province of Papua) was only about 11.5 million—9.2 million without Papua Province—about the same as one very large city. Fewer people live in this realm than in another vast area of far-flung settlements, the oases of North Africa's Sahara.

MAJOR CITIES OF THE REALM

City	Population* (in millions)
Honolulu, Hawai'i (U.S.)	0.9
Nouméa, New Caledonia	0.2
Port Moresby, Papua New Guinea	0.3
Suva, Fiji	0.2

*Based on 2006 estimates.

MELANESIA

The large island of New Guinea lies at the western end of a Pacific region that extends eastward to Fiji and includes the Solomon Islands, Vanuatu, and New Caledonia (Fig. 12-3). The human mosaic here is complex, both ethnically and culturally. Most of the 8.4 million people of New Guinea (Indonesia's Papua plus the independent state of Papua New Guinea) are Papuans, and a large minority is Melanesian. Altogether there are as many as 700 communities speaking different languages; the Papuans are most numerous in the densely forested highland

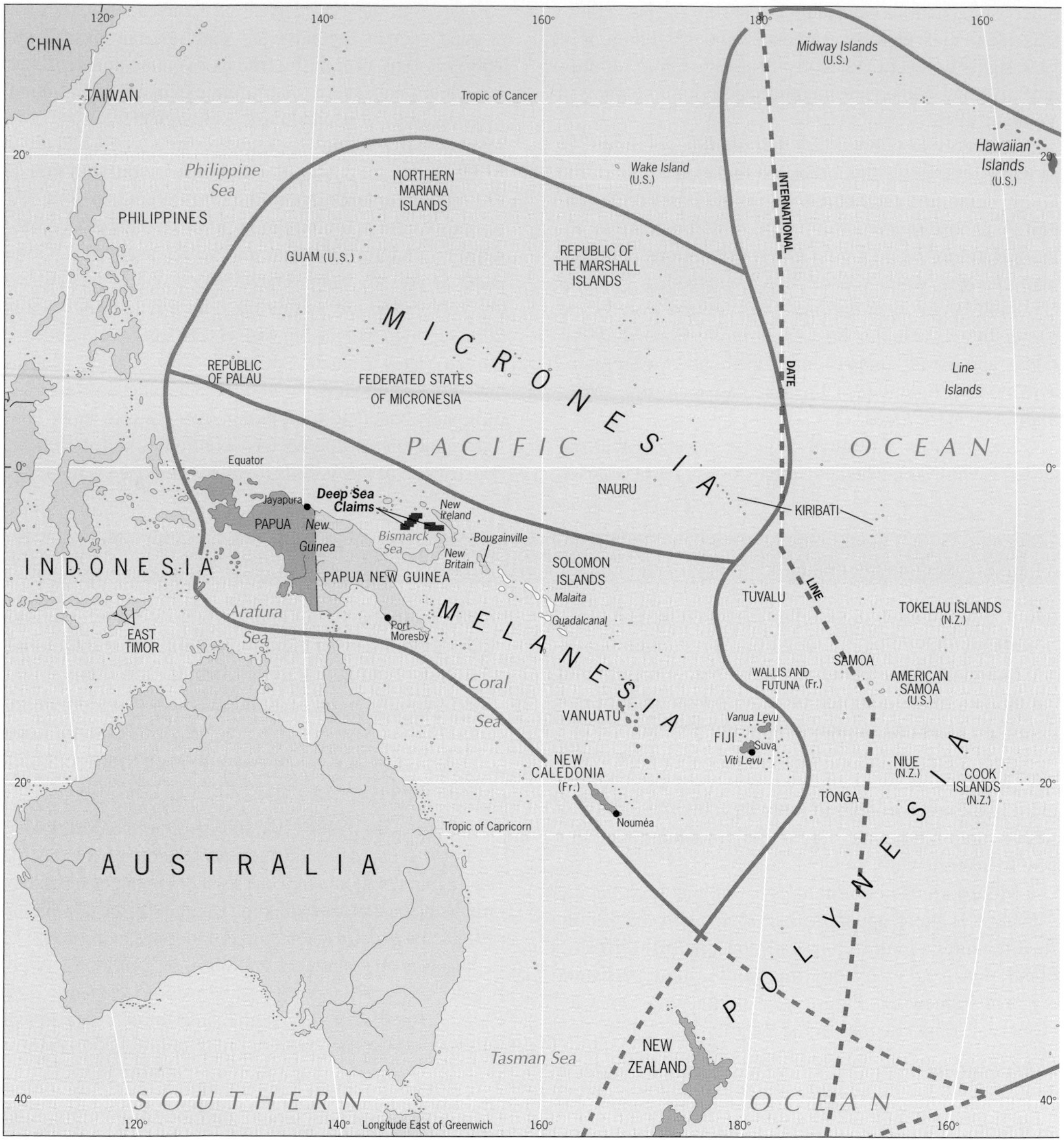

FIGURE 12-3 © H. J. de Blij, P. O. Muller, and John Wiley & Sons, Inc.

interior and in the south, while the Melanesians inhabit the north and east. The region as a whole has more than 7 million inhabitants, making this the most populous Pacific region by far.

With 6.0 million people today, **Papua New Guinea (PNG)** became a sovereign state in 1975 after nearly a century of British and Australian administration. Almost all of PNG's limited development is taking place along the coasts, while most of the interior remains hardly touched by the changes that transformed neighboring Australia. Perhaps four-fifths of the population lives in what we may describe as a self-sufficient subsistence economy, growing root crops and hunting wildlife, raising pigs, and gathering forest products. Old traditions of the kind lost in Australia persist here, protected by remoteness and the rugged terrain.

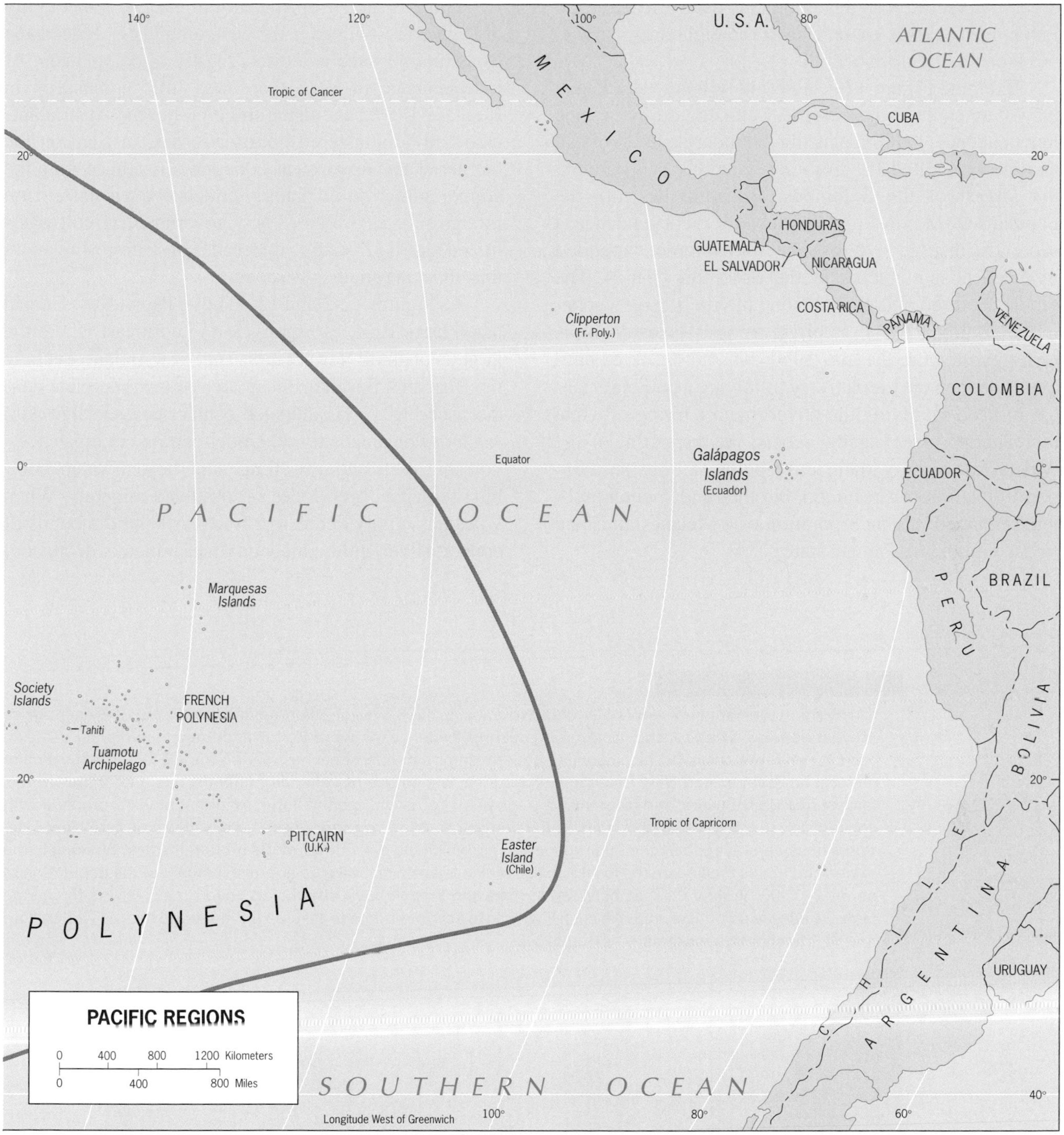

Welding this disparate population into a nation is a task hardly begun, and PNG faces numerous obstacles in addition to its cultural complexity. Not only are hundreds of languages in use, but more than half the population is illiterate. English, the official language, is used by the educated minority but is of little use beyond the coastal zone and its towns. The capital, Port Moresby, has about 330,000 inhabitants, reflecting the low level of urbanization (15 percent) in this low-income economy.

Yet Papua New Guinea is not without economic opportunities. Oil was discovered in the 1980s, and by the late 1990s crude oil was PNG's largest export by value. Gold now ranks second, followed by copper, silver, timber, and several agricultural products, including coffee and cocoa, reflecting the country's resource and environmental diversity. Pacific Rim developments have affected even PNG: most exports go to the nearest neighbor, Australia, but Japan ranks close behind.

Examine Figure 12-3, and you will see why Papua New Guinea has been beset by centrifugal forces since independence. PNG's maritime boundary cuts across an archipelago called the Solomon Islands, incorporating the largest of the Solomons, Bougainville, into the Papuan state. Almost immediately after independence, a separatist insurgency arose on Bougainville, supported by irredentist elements in the Solomon Islands. The rebels captured and closed Bougainville's large copper mine, cutting PNG's export revenues. A costly insurgency throughout the mid-1990s was not resolved until 1997 and then only tentatively following a political scandal in PNG involving the government's hiring of white mercenaries to defeat the rebels. At the outset of the twenty-first century, there was tenuous stability and copper production had resumed, but the fundamental problem remained: the incorporation of a Melanesian island into a dominantly Papuan state.

This destructive episode highlights the vulnerability of Papua New Guinea in the postcolonial era. Indeed, the Bougainville issue is a direct legacy of colonialism. At independence in 1975 there was little enthusiasm in Bougainville for inclusion in PNG, but the Australians, who had administered both territories, insisted on the merger. Their reason was economic: revenues from the copper mine would supply much-needed cash from foreign sources to PNG's new government. But it was a forced marriage, leading to decades of strife costing more than the mine could ever repay.

As Figures 12-2 and 12-3 show, Papua New Guinea has a large EEZ, especially to the northeast where the large islands of New Britain and New Ireland encircle the Bismarck Sea. This is an area of tectonic-plate contact (see Fig. G-4), and such contact is associated with seafloor volcanic activity. Under certain circumstances, volcanic vents and upwelling, superheated waters combine to deposit rich lodes of dissolved minerals. When volcanic activity decreases, it leaves behind mounds of copper, silver, gold, zinc, and other minerals in concen-

FROM THE FIELD NOTES

"Arriving in the capital of New Caledonia, Nouméa, in 1996, was an experience reminiscent of French Africa 40 years earlier. The French tricolor was much in evidence, as were uniformed French soldiers. European French residents occupied hillside villas overlooking palm-lined beaches, giving the place a Mediterranean cultural landscape. And New Caledonia, like Africa, is a source of valuable minerals. It is one of the world's largest nickel producers, and from this vantage point you could see the huge treatment plants, complete with concentrate ready to be shipped (left, under conveyor). What you cannot see here is how southern New Caledonia has been ravaged by the mining operations, which have denuded whole mountainsides. Working in the mines and in this facility are the local Kanaks, Melanesians who make up about 45 percent of the population of about 240,000. Violent clashes between Kanaks and French have obstructed government efforts to change New Caledonia's political status in such a way as to accommodate pressures for independence as well as continued French administration." © H. J. de Blij.

trations far greater than anything found on land. Now mining technology is making their exploitation possible, and in 1997 Papua New Guinea issued the first mining certification to an Australian company to exploit the *chimneys* (vent-related stacks of ore) deep below the surface of the sea (Fig. 12-3). Although the economic rewards will be high—the eastern claim has as much as 15 percent copper, 19 percent iron, 26 percent zinc, 7 percent silver—scientists fear that this kind of deep-sea mining will spread worldwide and damage the unique ecosystems associated with the still-hot vents. But under UNCLOS regulations, states have the rights to all economic assets in the EEZ, and for many the riches of the seafloor will prove irresistible.

Turning eastward, it is a measure of Melanesia's cultural fragmentation that as many as 120 languages are spoken in the approximately 1000 islands that make up the **Solomon Islands** (about 80 of these islands support almost all the people, numbering about 500,000). The islands' natural separation has helped limit ethnic conflict, but World War II bequeathed the Solomons a problem. When American forces took Guadalcanal from the Japanese, they brought in thousands of people from the neighboring island of Malaita (Fig. 12-3). After the war, most of these Malaitans stayed on and came to dominate life on Guadalcanal, controlling commerce and buying up land. Resentment among the local people grew, and in 1999 it exploded into violence. Attacks on Malaitans caused some 10,000 to flee the capital, Honiara; many sailed for Malaita. Then in 2000, a group of Malaitans still on Guadalcanal took the country's prime minister hostage in a bid to extract concessions. In 2003 the

FROM THE FIELD NOTES

"Back in Suva, Fiji more than 20 years after I had done a study of its CBD, I noted that comparatively little had changed, although somehow the city seemed more orderly and prosperous than it was in 1978. At the Central Market I watched a Fijian woman bargain with the seller over a batch of taro, the staple starch-provider in local diets. I asked her how she would serve it. 'Well, it's like your potato,' she said. 'I can make a porridge, I can cut it up and put it in a stew, and I can even fry pieces of it and make them look like the French fries they give you at the McDonald's down the street.' Next I asked the seller where his taro came from. 'It grows over all the islands,' he said. 'Sometimes, when there's too much rain, it may rot—but this year the harvest is very good.' . . . Next I walked into the crowded Indian part of the CBD. On a side street I got a reminder of the geographic concept of agglomeration. Here was a colonial-period building, once a hotel, that had been converted into a business center. I counted 15 enterprises, ranging from a shoestore to a photographer and including a shop where rubber stamps were made and another selling diverse tobacco products. Of course (this being the Indian sector of downtown Suva) tailors outnumbered all other establishments." © H. J. de Blij.

Australian government, concerned over instability in the region and possible terrorist activity in the failed state the Solomon Islands appeared to be becoming, sent a military force to reestablish order. But this was only the latest foreign intervention here. In the Solomon Islands as elsewhere in the Pacific Realm, outsiders came to rule, to make war, to relocate people, to establish an independent "nation" following decolonization—and then left the locals to deal with the consequences.

New Caledonia, still under French rule, is in a very different situation. Only about 45 percent of the population of about 240,000 are Melanesian; 34 percent are of French ancestry, many of them descended from the inhabitants of the penal colony France established here in the nineteenth century. Nickel mines, based on reserves that rank among the world's largest, dominate New Caledonia's export economy (see photo p. 570). The mining industry attracted additional French settlers, and social problems arose. Most of the French population lives in or near the capital city of Nouméa, steeped in French cultural landscapes, in the southeastern quadrant of the island. The miners, mainly Melanesians, live in dormitory-style buildings near the huge refinery in the capital; those not directly involved farm for subsistence in the island's middle and northern areas. Melanesian demands for an end to colonial rule have led to violence, and the two communities are still in the process of coming to terms.

On its eastern margins, Melanesia includes one of the Pacific Realm's most interesting countries, **Fiji**. On two larger and over 100 smaller islands live nearly 800,000 Fijians, of whom 52 percent are Melanesians and 42 percent South Asians. The South Asians were brought to Fiji from India during the British colonial occupation to work on the sugar plantations. When Fiji achieved independence in 1970, the native Fijians owned most of the land and held political control, while the Indians were concentrated in the towns (chiefly Suva, the capital; photo p. 571) and dominated commercial life. It was a recipe for trouble, and it was not long in coming when, in a later election, the politically active Indians outvoted the Fijians for seats in the parliament. A coup by the Fijian military was followed by a revision of the constitution, which awarded a majority of seats to ethnic Fijians. Before long, however, the Fijian majority splintered, but a coalition government for some time proved the constitution workable—until 2000. In 1999, Fiji's first prime minister of Indian ancestry had taken office, angering some ethnic Fijians to the point of staging another coup. The prime minister and members of the government were taken hostage at the parliament building in Suva, and the perpetrators demanded that Fijians would henceforth govern the country.

This action, and Fiji's inability to counter it, had a devastating impact on the country. Foreign trading partners stopped buying Fijian products. The tourist industry suffered severely. And in the end, although the coup leaders were ousted and arrested, they secured a deal that gave ethnic Fijians the control they had sought. Fiji's future thus remains clouded.

Melanesia, the most populous region in the Pacific Realm, also is bedeviled by centrifugal forces of many kinds. No two countries present the same form of multiculturalism; each has its own challenges to confront, and some of these challenges spill over into neighboring (or more distant foreign) islands.

MICRONESIA

North of Melanesia and east of the Philippines lie the islands that constitute the region known as Micronesia (Fig. 12-3). The name (*micro* means small) refers to the size of the islands: the 2000-plus islands of Micronesia are not only tiny (many of them no larger than 1 square mile), but they are also much lower-lying, on average, than those of Melanesia. Some are volcanic islands
(**high islands**, as the people call them), but they are out- 8
numbered by islands composed of coral, the **low islands** 9
that barely lie above sea level. Guam, with 210 square miles (550 sq km), is Micronesia's largest island, and no island elevation anywhere in Micronesia reaches 3300 feet (1000 m).

The high-island/low-island dichotomy is useful not only in Micronesia, but also throughout the realm. Both the physiographies of these islands and the economies they support differ in crucial ways. High islands wrest substantial moisture from the ocean air; they tend to be well watered and have good volcanic soils. As a result, agricultural products show some diversity, and life is reasonably secure. Populations tend to be larger on these high islands than on the low islands, where drought is the rule and fishing and the coconut palm are the mainstays of life. Small communities cluster on the low islands, and over time, many of these have died out. The major migrations, which sent fleets to populate islands from Hawai'i to New Zealand, tended to originate in the high islands.

Until the mid-1980s, Micronesia was largely a United States Trust Territory (the last of the post–World War II trusteeships supervised by the United Nations), but that status has now changed. As Figure 12-3 shows, today Micronesia is divided into countries bearing the names of independent states. The **Marshall Islands**, where the United States tested nuclear weapons (giving prominence to the name *Bikini*), now is a republic in "free association" with the United States, having the same status as the Federated States of Micronesia and (since 1994) Palau. The **Northern Mariana Islands** are a commonwealth "in political union" with the United States. In effect, the United States provides billions of dollars in assistance to these countries, in return for which they commit themselves to avoid foreign policy actions that are contrary to U.S. interests. There are other conditions as well: **Palau**, for example, granted the United States rights to existing military bases for 50 years following independence.

Also part of Micronesia are the U.S. territory of **Guam**, where independence is not in sight and where U.S. military installations and tourism provide the bulk of income, and the remarkable Republic of **Nauru**. With a population of barely 12,000 and 8 square miles of land, this microstate got rich by selling its phosphate deposits to Australia and New Zealand, where they are used as fertilizer. Per-capita incomes rose to U.S. $11,500, making Nauru one of the Pacific's high-income societies. But the phosphate deposits ran out, and an island scraped bare faced an economic crisis.

In this region of tiny islands, most people subsist on farming or fishing, and virtually all the countries need infusions of foreign aid to survive. The natural economic complementarity between the high-island farming cultures and the low-island fishing communities all too often is negated by distance, spatial as well as cultural. Life here may seem idyllic to the casual visitor, but for the Micronesians it often is a daily challenge.

POLYNESIA

To the east of Micronesia and Melanesia lies the heart of the Pacific, enclosed by a great triangle stretching from the Hawaiian Islands to Chile's Easter Island to New Zealand. This is Polynesia (Fig. 12-3), a region of numerous islands (*poly* means many), ranging from volcanic mountains rising above the Pacific's waters (Mauna Kea on Hawai'i reaches nearly 13,800 feet [over 4200 m]), clothed by luxuriant tropical forests and drenched by well over 100 inches of rainfall each year, to low coral atolls where a few palm trees form the only vegetation and where drought is a persistent problem. The Polynesians have somewhat lighter-colored skin and wavier hair than do the other peoples of the Pacific Realm; they are often also described as having an excellent physique. Anthropologists differentiate between these original Polynesians and a second group, the Neo-Hawaiians, who are a blend of Polynesian, European, and Asian ancestries. In the U.S. State of Hawai'i—actually an archipelago of more than 130 islands—Polynesian culture has been both Europeanized and Orientalized.

Its vastness and the diversity of its natural environments notwithstanding, Polynesia clearly constitutes a geographic region within the Pacific Realm. Polynesian culture, though spatially fragmented, exhibits a remarkable consistency and uniformity from one island to the next, from one end of this widely dispersed region to the other. This consistency is particularly expressed in vocabularies, technologies, housing, and art forms. The Polynesians are uniquely adapted to their maritime environment, and long before European sailing ships began to arrive in their waters, Polynesian seafarers had learned to navigate their wide expanses of ocean in huge double canoes as long as 150 feet (45 m). They traveled hundreds of miles to favorite fishing zones and engaged in inter-island barter trade, using maps constructed from bamboo sticks and cowrie shells and navigating by the stars. However, modern descriptions of a Pacific Polynesian paradise of emerald seas, lush landscapes, and gentle people distort harsh

FROM THE FIELD NOTES

"Space is at a premium in Tuvalu, a country of about 10 square miles (26 sq km) on 9 small islands forming a 360-mile (575-km) chain. I had the unusual opportunity to go ashore on the largest of these low islands, Funafuti Atoll, where about 4500 of the total population of about 10,000 live. The first building in the capital I saw was marked *Department of Fisheries*, and for good reason. Tuvalu may be territorially small, but it has a huge Exclusive Economic Zone, profitably leased to major fishing operations. Two other impressions: the obvious problems people here have with waste disposal (there is little room for dumps, and plastics are making this a crisis); and reminders of World War II. Rusting hulks of war materiel still lie where they were abandoned, a jarring sight in this isolated Pacific outpost."

realities. Polynesian society was forced to get used to much loss of life at sea when storms claimed their boats; families were ripped apart by accident as well as migration; hunger and starvation afflicted the inhabitants of smaller islands; and the island communities were often embroiled in violent conflicts and cruel retributions.

The political geography of Polynesia is complex. In 1959 the Hawaiian Islands became the fiftieth State to join the United States. The State's population is now 1.4 million, with over 80 percent living on the island of Oahu. There, the superimposition of cultures is symbolized by the panorama of Honolulu's skyscrapers against the famous extinct volcano at nearby Diamond Head. The Kingdom of **Tonga** became an independent country in 1970 after seven decades as a British protectorate; the British-administered Ellice Islands were renamed **Tuvalu**, and along with the Gilbert Islands to the north (now renamed **Kiribati**,) they received independence from Britain in 1978. Other islands continued under French control (including the Marquesas Islands and Tahiti), under New Zealand's administration (Rarotonga), and under British, U.S., and Chilean flags.

In the process of politico-geographical fragmentation, Polynesian culture has suffered severe blows. Land developers, hotel builders, and tourist dollars have set **Tahiti** on a course along which Hawai'i has already traveled far. The Americanization of eastern **Samoa** has created a new society different from the old. Polynesia has lost much of its ancient cultural consistency; today, the region is a patchwork of new and old—the new often bleak and barren, with the old under intensifying pressure.

The countries and cultures of the Pacific Realm lie in an ocean on whose rim a great drama of economic and political transformation will play itself out during the twenty-first century. Already, the realm's own former margins—in Hawai'i in the north and in New Zealand in the south—have been so recast by foreign intervention that little remains of the kingdoms and cultures that once prevailed. Now the Pacific world faces changes far greater even than those brought here by European colonizers. Once upon a time the shores and waters of the Mediterranean Sea formed an arena of regional transformation that changed the world. Then it was the Atlantic, avenue of the Industrial Revolution and stage of fateful war. Now it seems to be the turn of the Pacific as the world's largest country and next superpower (China) faces the richest and most powerful (the United States). Giants will jostle for advantage in the Pacific; how will the weak societies of the Pacific Realm fare?

A FINAL CAVEAT: PACIFIC AND ANTARCTIC

South of the Pacific Realm lies Antarctica and its encircling Southern Ocean. The combined area of these two

FROM THE FIELD NOTES

"The Antarctic Peninsula is geologically an extension of South America's Andes Mountains, and this vantage point in the Gerlach Strait leaves you in no doubt: the mountains rise straight out of the frigid waters of the Southern Ocean. We have been passing large, flat-topped icebergs, but these do not come from the peninsula's shores; rather, they form on the leading edges of ice shelves or on the margins of the mainland where continental glaciers slide into the sea. Here along the peninsula, the high-relief topography tends to produce jagged, irregular icebergs. The mountain range that forms the Antarctic Peninsula continues across Antarctica under thousands of feet of ice and is known as the Transantarctic Mountains. Looking at this place you are reminded that beneath all this ice lies an entire continent, still little-known, with fossils, minerals, even lakes yet to be discovered and studied." © H. J. de Blij.

geographic expanses constitutes 40 percent of the entire planet—two-fifths of the Earth's surface containing one one-thousandth of the world's population.

Is Antarctica a geographic realm? In physiographic terms, yes, but not on the basis of the criteria we use in this book. Antarctica is a continent, nearly twice as large as Australia, but virtually all of it is covered by a dome-shaped icesheet nearly 2 miles (3.2 km) thick near its center. The continent often is referred to as the "white desert" because, despite all of its ice and snow, annual precipitation is low (less than 6 inches [15 cm] per year). Temperatures are frigid, with winds so strong that Antarctica also is called the "home of the blizzard." For all its size, no functional regions have developed here, no towns, no transport networks except the supply lines of research stations. And Antarctica still is a frontier, even a scientific frontier still slowly giving up its secrets. Underneath all that ice lie some 70 lakes of which Lake Vostok is the largest with over 5400 square miles (14,000 sq km). It may be as much as 2000 feet (600 m) deep. No one has yet seen a sample of its water.

Like virtually all frontiers, Antarctica has always attracted pioneers and explorers. Whale and seal hunters destroyed huge populations of Southern Ocean fauna during the eighteenth and nineteenth centuries, and explorers planted the flags of their countries on Antarctic shores. Between 1895 and 1914, the quest for the South Pole became an international obsession; Roald Amundsen, the Norwegian, reached it first in 1911. All this led to national claims in Antarctica during the interwar period (1918–1939).

The geographic effect was the partitioning of Antarctica into pie-shaped sectors centered on the South Pole (Fig. 12-4). In the least frigid area of the continent, the Antarctic Peninsula (highlighted in the inset map; and see photo p. 574), British, Argentinean, and Chilean claims overlapped—and still do. One sector, Marie Byrd Land, was never formally claimed by any country.

Why should states be interested in territorial claims in so remote and difficult an area? Both land and sea contain raw materials that may some day become crucial: proteins in the waters, and fuels and minerals beneath the ice. Antarctica (5.5 million sq mi/14.2 million sq km) is almost twice as large as Australia, and the Southern Ocean (see box p. 541) is nearly as large as the North and South Atlantic. However distant actual exploitation may be, countries want to keep their stakes here.

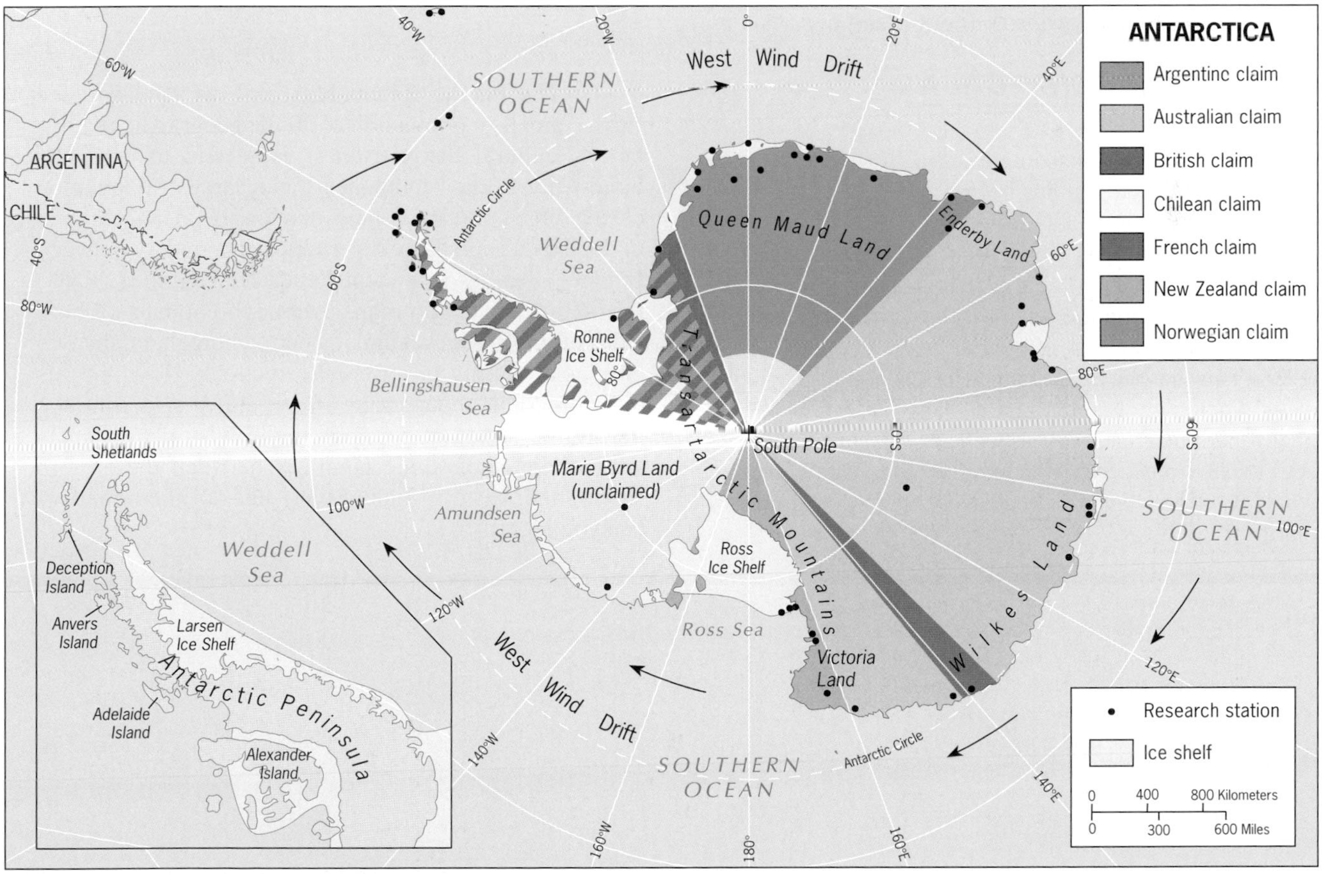

FIGURE 12-4 © H. J. de Blij, P. O. Muller, and John Wiley & Sons, Inc.

But the claimant states (those with territorial claims) recognize the need for cooperation. During the late 1950s, they joined in the International Geophysical Year (IGY) that launched major research programs and established a number of permanent research stations throughout the continent. This spirit of cooperation led to the 1961 sign-
10 ing of the **Antarctic Treaty**, which ensures continued scientific collaboration, prohibits military activities, protects the environment, and holds national claims in abeyance. In 1991, when the treaty was extended under the terms of the Wellington Agreement, concerns were raised that it does not do enough to control future resource exploitation.

In an age of growing national self-interest and increasing consumption of raw materials, the possibility exists that Antarctica and its offshore waters may yet become an arena for international rivalry. Until now, its remoteness and its forbidding environments have saved it from that fate. The entire world benefits from this because evidence is mounting that Antarctica plays a critical role in the global environmental system, so that human modifications may have global (and unpredictable) consequences.

WHAT YOU CAN DO

RECOMMENDATION: Attend an AAG Meeting! The Association of American Geographers is the largest organization for professional geographers and students in North America, and it holds an Annual Meeting every spring (its 100th meeting took place in Philadelphia, site of the first one, in March 2004). The meeting sites rotate around North America: Chicago in 2006 (March 7–11), San Francisco in 2007 (April 17–21), and Boston in 2008 (spring), so one of them is likely to convene relatively close to your college or home in the not too distant future. Several thousand geographers and students participate, and while these meetings have their professional side, they are also a lot of fun, and there are jobs available as well, as the AAG can tell you. If you're within a reasonable distance, why not join the crowd and enjoy the exhibits, plenary lectures (Kofi Annan spoke at a recent meeting), and parties, and meet some of the people who keep geography in the limelight! Details are available on the AAG website at www.aag.org.

GEOGRAPHIC CONNECTIONS

1 One of the most important achievements of the United Nations has been a series of conferences out of which grew the UN Convention on the Law of the Sea (UNCLOS). This Convention redrew the maritime map of the world, stabilizing what had been a scramble for the oceans, limiting the width of the Territorial Sea, and introducing the concept of the Exclusive Economic Zone (EEZ). How has the EEZ changed the economic geography of the Pacific geographic realm? To what extent has the EEZ functioned to expand as well as limit states' offshore activities? Considering the map of the Pacific Realm, what countries are favored over others in terms of maritime holdings?

2 The Pacific Realm has the least land and the smallest population of all the geographic realms, but its cultural distinctiveness is second to none. Its largest territorial component and largest population cluster constitute the island of New Guinea, which is divided by a geometric, superimposed boundary between Indonesia and independent Papua New Guinea (PNG) (Fig. 10-6, lower-left map). Although Papuans on both sides of this border dream of a united homeland, the two entities are going in different directions. How did this division come about? Why was it allowed to continue during the period of decolonization? What demographic and cultural-geographic contrasts between the eastern (PNG) and western (Indonesian) sides of the boundary can you identify?

Appendix A

Using the Maps

If, as is often said, a picture is worth a thousand words, then a map is worth a million. This book contains more than 200 maps, locational and thematic, to help you understand the text. Please make it a point to study each map when it is referred to and do so as carefully as you would a paragraph of text.

Maps display locations and patterns in geographic space. They are a kind of shorthand through which encoded spatial messages are transmitted to the reader. Such shorthand is necessary because the real world is so complex that a great deal of geographic information must be compressed into the small confines of maps that can fit onto the pages of this book. Cartographers (mapmakers) must carefully choose which information to include; these decisions force them to omit many other things in order to prevent cluttering a map with less relevant information. For instance, Figure A shows several city blocks in central London but avoids mapping individual buildings because they would interfere with the key information being presented—the spatial distribution of cholera deaths in that part of London during an epidemic.

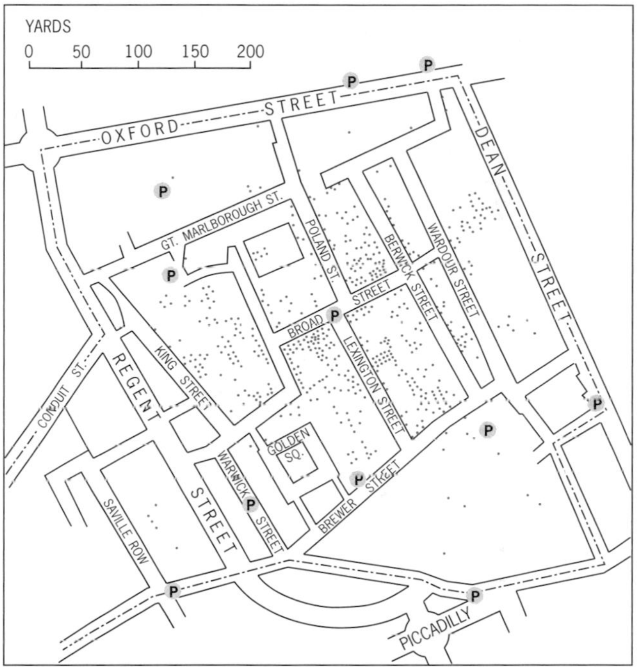

FIGURE A Adapted with permission from L. D. Stamp, *The Geography of Life and Death.* © Cornell University Press, 1964.

MAP READING

Deciphering the coded messages contained in our maps—*map reading*—is not difficult, but a few rules of the road do help. The need to miniaturize portions of the world on small maps is discussed in the section on map scale (pp. 7–8), and Figures A and B offer two additional and contrasting examples.

Orientation, or direction, on maps can usually be discerned by reference to the geographic grid of latitude and longitude. *Latitude* is measured from 0° to 90° north and south of the equator (parallels of latitude are always drawn in an east-west direction), with the equator being 0° and the North and South Poles being 90°N and 90°S, respectively. Meridians of *longitude* (always drawn north-south) are measured 180° east and west of the *prime meridian* (0°), which passes through the Greenwich Observatory next to London; the 180th meridian, for the most part, serves as the *international date line* that lies in the middle of the Pacific Ocean (the red dashed line in Fig. 12-1). Inspection of Figure B shows that north is not automatically at the top of a map; instead, the direction of north curves along every meridian, with all such lines of longitude converging at the North Pole. The many minor directional distortions in this map are unavoidable: it is geometrically impossible to transfer the grid of a three-dimensional sphere (globe) onto a two-dimensional flat map. Therefore, compromises in the form of *map projections* must be devised in which, for example, properties such as areal size and distance are preserved but directional constancy is sacrificed.

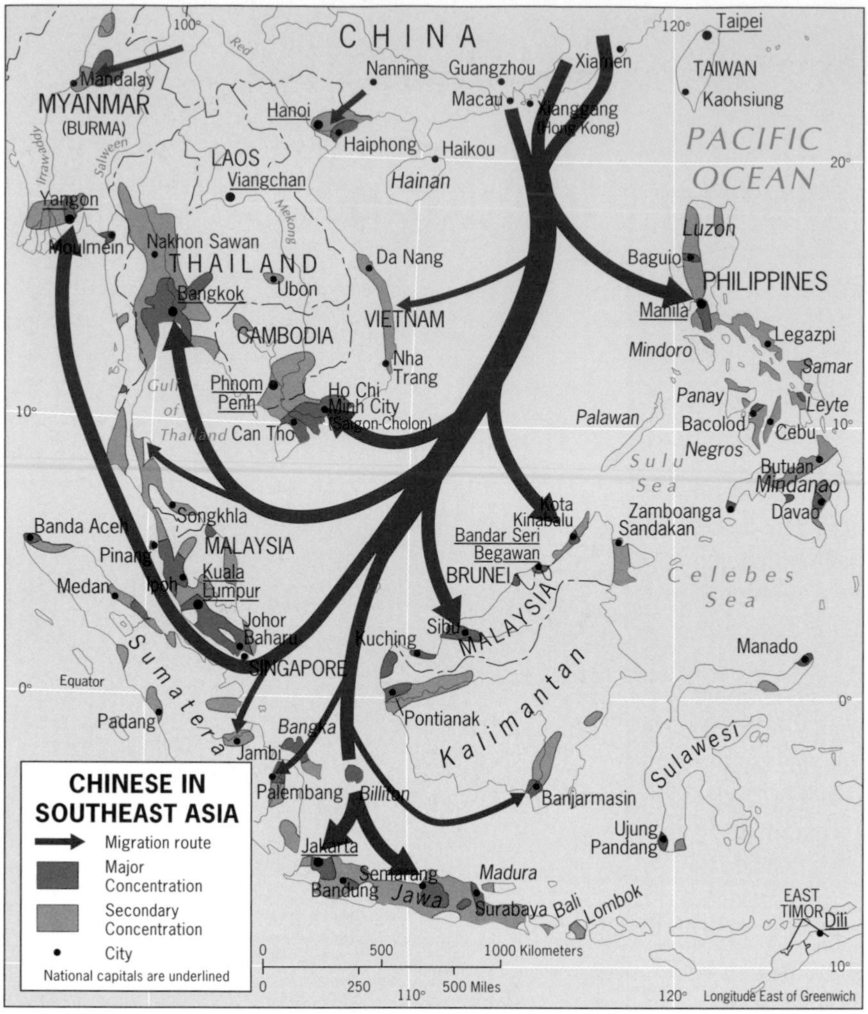

FIGURE B © H. J. de Blij, P. O. Muller, and John Wiley & Sons, Inc.

MAP SYMBOLS

Once the background mechanics of scale and orientation are understood, the main task of decoding the map's content can proceed. The content of most maps in this book is organized within the framework of point, line, and area symbols, which are made especially clear through the use of color. These symbols are usually identified in the map's *legend* (as in Figure B), but sometimes accompanying text must tell the map reader more. In Figure A, for example, we need to be informed that the dots reflect cholera deaths, and that each **P** symbol represents a municipal water pump at the time of this epidemic in 1854. *Point symbols*, shown as dots on maps, can tell us two things: the location of each phenomenon and, sometimes, its quantity. Look, for example, at Figure 1-14 (p. 68). Not only does this map show the locations of Western European cities but also their relative population size.

Line symbols connect places or areas between which some sort of movement or flow has occurred or is occurring. Figure B, taken from the chapter on Southeast Asia, shows the migration of Chinese into this realm, a mass movement that originated mainly in southeastern China and produced coastal Chinese communities on the islands and peninsulas of this multicultural part of the world. Note that we make no attempt to rank the sizes of Southeast Asian cities, so as not to obscure the key theme. *Area symbols* are used to classify two-dimensional spaces. Primary and secondary concentrations of Chinese residents in Southeast Asia are differentiated by color in Figure B. Area symbols can also be used to communicate quantitative information: for example, the light-tan-colored zones in Figure G-7 (p. 13) delimit semiarid areas within which annual precipitation averages from 12 to 20 inches (30 to 50 cm).

MAP INTERPRETATION

The explanation of cartographic patterns is one of the geographer's most important tasks. Although that task is performed for you throughout this book, readers should be aware that today's practitioners use many sophisticated techniques and cutting-edge computer technology to analyze vast quantities of areal data. These modern methods notwithstanding, geographic inquiry still focuses on the search for meaningful *spatial relationships*. This longstanding concern of the discipline is classically demonstrated in Figure A. By showing on his map that cholera fatalities clustered around municipal water pumps, Dr. John Snow was able to persuade city authorities to shut them off; almost immediately, the number of new disease victims dwindled to zero, thereby confirming Snow's theory that contaminated drinking water was crucial in the spread of cholera.

Sometimes maps can reveal information or insights beyond those anticipated or intended by the cartographer. Take a look at the ethnolinguistic map on page 372, Figure 7-20. That map was drawn before the events of September 11, 2001, and subsequent developments in and around Afghanistan. Its relevance and meaning now exceed its original, illustrative intent. So study maps such as this in detail, and you will vastly increase what you get out of this book and this course.

Appendix B

Opportunities in Geography

The chapters of this book give an idea of the wide range of topics and interests that geographers pursue, particularly in the many discussions of concepts (regional and otherwise) and frequent references to the content of geography's systematic fields. There are specializations within each of those topical fields as well as in regional studies—but an introductory book such as this lacks enough space to discuss all of them. This appendix, therefore, is designed to help you, should you decide to major or minor in geography and/or to consider it as a career option.

AREAS OF SPECIALIZATION

As in all disciplines, areas of concentration or specialization change over time. In North American geography early in the twentieth century, there was a period when most geographers were physical geographers, and the natural landscape was the main object of geographic analysis. Then the pendulum swung toward human (cultural) geography, and students everywhere focused on the imprints of human activity on the surface of the Earth. Still later the analysis of spatial organization became a major concern. In the meantime, geography's attraction for some students lay in technical areas: in cartography, in aerial and satellite remote sensing, in computer-assisted spatial data analysis, and, most recently, in geographic information systems (GIS) and GIScience.

All this meant that geography posed (and continues to pose) a challenge to its professionals. New developments require that we keep up to date, but we must also continue to build on established foundations.

Regional Geography

One of these established foundations, of course, is regional geography, which encompasses a large group of specializations. Some geographers specialize in the theory of regions: how they should be defined, how they are structured, and how their internal components work. This leads in the direction of *regional science*, and some geographers have preferred to call themselves regional scientists. But make no mistake: regional science is regional geography.

Another, older approach to regional geography involves specialization in an area of the world ranging in size from a geographic realm to a single region or even a State or part of a State. There was a time when regional geographers, because of the interdisciplinary nature of their knowledge, were sought after by government agencies. Courses in regional geography abounded in universities' geography departments; regional geographers played key roles in international studies and research programs. But then the drive to make geography a more rigorous science and to search for universal (rather than regional) truths contributed to a decline in regional geography. The results were not long in coming, and in recent years you have probably seen the issue of "geographic illiteracy" discussed in newspapers and magazines. Now the pendulum is swinging back again, and regional geography and regional specialization are reviving. This is a propitious time to consider regional geography as a professional field.

Your Personal Interests

Geography, as we have pointed out, is united by several bonds, of which regional geography is but one. Regional geography exemplifies the spatial view that all geographers hold; the spatial approach to study and research binds physical and human geographers, regionalists, and topical specialists. Another unifying theme is an abiding interest in the relationships between human societies and natural environments. We have referred to that topic frequently in this book; as an area of specialization it has gone through difficult times. Perhaps more than anything, geography remains a field of *synthesis*, of understanding interrelationships.

Geography also is a field science, using "field" in another context. In the past, almost all major geography departments required a student's participation in a "field camp" as part of a master's degree program; thus many undergraduate programs included field experience. It was one of those bonding practices in which students and faculty with diverse interests met, worked together, and learned from one another. Today, few field camps of this sort are offered, but that does not change what geography is all about. If you see an opportunity for field experience with professional geographers—even just a one-day reconnaissance—take it. But realize this: a few days in the field with geographic instruction may hook you for life.

Some geographers, in fact, are far better field-data gatherers than analysts or writers. In this respect they are not alone: this also happens in archeology, geology, and biology (among other field disciplines). This does not mean that these fieldworkers do not contribute significantly to knowledge. Often on a research team some of the members are better in the field, and others excel in subsequent analysis. From all points of view, however, fieldwork is important.

Geography, then, is practiced in the field and in the office, in physical and human contexts, in generality and detail. Small wonder that so many areas of specialization have developed! If you check the undergraduate catalogue of your college or university, you will see some of these specializations listed as semester-length courses. But no geography department, no matter how large, could offer them all.

How does an area of specialization develop, and how can one become a part of it? The way in which geographic specializations have developed tells us much about the entire discipline. Some major areas, now old and established, began as research and theory building by one scholar and his or her students. These graduate students dispersed to the faculties of other universities and began teaching what they had learned. Thus, for example, did Carl Sauer's cultural geography spread from the University of California, Berkeley during the middle decades of the twentieth century.

It is one of the joys of geography that the basics and methods, once learned, are applicable to so many features of the human and physical world. Geographers have specialized in areas as disparate as shopping-center location and glacier movement, tourism and coastal erosion, real estate and wildlife, retirement communities and sports. Many of these specializations began with the interests and energies of a single scholar. Some thrived and grew into major geographic pursuits; others remained one-person shows but with potential. When you discuss your own interests with a faculty advisor, you may refer to a university where you would like to do graduate work. "Oh yes," the answer may be, "Professor X is in their geography department, working on just that." Or perhaps your advisor will suggest another university where a member of the faculty is known to be working on the topic in which you are interested. That is the time to write a letter or e-mail of inquiry. What is the professor working on now? Are graduate students involved? Are research funds available? What are the career prospects after graduation?

The Association of American Geographers, or **AAG** (1710 16th Street, N.W., Washington, D.C. 20009-3198 [**www.aag.org**]), recognizes no less than 53 so-called Specialty Groups. In academic year 2005–2006, the Specialty Group roster included the following branches of geography:

Africa
Applied Geography
Asian Geography
Bible Geography
Biogeography
Canadian Studies
Cartography
China
Climate
Coastal and Marine Geography
Communication Geography
Cryosphere (Earth's Ice System)
Cultural and Political Ecology
Cultural Geography
Developing Areas
Disability
Economic Geography
Energy and Environment
Environmental Perception and Behavioral Geography
Ethics, Justice, and Human Rights
Ethnic Geography
European Geography
Geographic Information Science and Systems
Geographic Perspectives on Women
Geography Education
Geography of Religions and Belief Systems
Geomorphology (Landform Analysis)
Hazards
Historical Geography
History of Geography
Human Dimensions of Global Change
Indigenous Peoples
Latin American Geography
Medical Geography
Middle East Geography
Military Geography
Mountain Geography
Political Geography
Population
Qualitative Research
Recreation, Tourism, and Sport
Regional Development and Planning

Remote Sensing
Rural Geography
Russian, Central Eurasian, and East European Geography
Sexuality and Space
Socialist and Critical Geography
Spatial Analysis and Modeling
Transportation Geography
Urban Geography
Water Resources
Wine Geography
World Wide Web

If you contact the AAG at the Internet address given above, click on the first tab at the top of the home page ("About the AAG") and then click on the sixth item in the drop-down box that is entitled "Specialty Groups." Then scroll down the alphabetical listing to obtain the e-mail address of the current chairperson of any group. We encourage you to contact the professional geographer who chairs the group(s) you are interested in. These elected leaders are ready, enthusiastic, and willing to provide you with the information you are looking for.

All this may seem far in the future. Still, the time to start planning for graduate school is now. Applications for admission and financial assistance must be made shortly after the *beginning* of your senior year. That makes your junior year a year of decision.

AN UNDERGRADUATE PROGRAM

The most important concern for any geography major or minor is basic education and training in the field. An undergraduate curriculum contains all or several of the following courses (titles may vary):

1. Physical Geography of the Global Environment (weather and climate, natural landscapes, landforms, soils, moisture and surface water flows, elementary biogeography)
2. Introduction to Human Geography (basic principles of cultural and economic geography)
3. World Regional Geography (world geographic realms)

These beginning courses are followed by more specialized courses, including both methodological and substantive ones:

4. Geographic Information Technology
5. Cartographic Theory and Techniques
6. Introduction to Quantitative Methods of Analysis
7. Introduction to Geographic Information Systems (GIS)
8. Analysis of Remotely Sensed Data
9. Cultural Geography
10. Political Geography
11. Urban Geography
12. Economic Geography
13. Historical Geography
14. Geomorphology
15. Global Environmental Change
16. Geography of United States-Canada, Europe, and/or other world geographic realms

You can see how Physical Geography of the Global Environment would be followed by the more specialized analysis of landforms covered in Geomorphology, and how Human Geography now divides into such areas as intermediate and/or advanced cultural and economic geography. As you progress, the focus becomes even more specialized. Thus, Economic Geography may be followed by:

17. Industrial Geography
18. Transportation Geography
19. Geography of Development

At the same time, regional concentrations may come into sharper focus:

20. Geography of Western Europe (or other major regions)

In these more advanced courses, you will use the technical knowledge acquired from courses numbered **4** through **8** (and perhaps others). Now you can avail yourself of the opportunity to develop these skills further. Many departments offer such courses as these:

21. Advanced Quantitative Methods
22. Computer Cartography
23. Intermediate GIS
24. Advanced GIS
25. Advanced Satellite Imagery Interpretation

From this list (which represents only part of a comprehensive curriculum), it is evident that you cannot, even in four years of undergraduate study, register for all courses. The geography major in many universities requires a minimum of only 30 (semester-hour) credits—just 10 courses, fewer than half those listed here. This is another reason to begin thinking about specialization at an early stage.

Because of the number and variety of possible geography courses, most departments require that their majors complete a core program that includes courses in substantive areas as well as theory and methods. That core program is important, and you should not be tempted to put off these courses until your last semesters. What you learn in the core program will make what follows (or should follow) much more meaningful.

You should also be aware of the flexibility of many undergraduate programs, something that can be especially important to geographers. Imagine that you are majoring in geography and develop an interest in Southeast Asia. But the regional specialization of the geographers in your department may be focused somewhere else—say, Middle and South America. However, courses on Southeast Asia are indeed offered by other departments, such as anthropology, history, or political science. If you are going to be a regional specialist, those courses will be very useful and should be part of your curriculum—but you are not able to receive geography credit for them. After successfully completing those courses, however, you may be able to register for an independent study or reading course in the Geography of Southeast Asia if a faculty member is willing to guide you. Always discuss such matters with the undergraduate advisor or chairperson of your geography department.

LOOKING AHEAD

By now, as a geography major, you will be thinking of the future—either in terms of graduate school or a salaried job. In this connection, if there is one important lesson to keep in mind, it is to *plan ahead* (a redundancy for emphasis!). The choice of graduate school is one of the most important you will make in your life. The professional preparation you acquire as a graduate student will affect your competitiveness in the job market for years to come.

Choosing a Graduate School

Your choice of a graduate school hinges on several factors, and the geography program it offers is one of them. Possibly, you are constrained by residency factors, and your choice involves the schools of only one State. Your undergraduate record and grade point average affect the options. As a geographer, you may have strong feelings in favor of or against particular parts of the country. And although some schools may offer you financial support, others may not.

Certainly the programs and specializations of the prospective graduate department are extremely important. If you have settled on your own area of interest, it is wise to find a department that offers opportunities in that direction. If you have yet to decide, it is best to select a large department with several options. Some students are so impressed by the work and writings of a particular geographer that they go to his or her university solely to learn from and work with that scholar. In every case, information and preparation are crucial. Many a prospective graduate student has arrived on campus eager to begin work with a favorite professor, only to find that the professor is away on a sabbatical leave!

Fortunately, information can be acquired with little difficulty. One of the most useful publications of the AAG is its *Guide to Geography Programs in North America*, published yearly. A copy of this annual directory should be available in the office of your geography department, but if you plan to enter graduate school, a personal copy would be an asset. Not only does the *Guide* describe the programs, requirements, financial aid, and other aspects of geography departments in the United States, Canada, and Mexico, but it also lists all faculty members and their current research and teaching specializations. Moreover, it also contains a complete listing of the Association's 7000+ members and their specializations. Internet addresses are also provided for accessing the websites of geography departments, which they increasingly utilize to supplement their entries in the *Guide* and provide even more detailed information about their programs and faculty. Although you will find the *Guide* to be very useful in your decision-making, we also strongly recommend that you explore the websites of departments you wish to consider. In contacting departments, a particularly useful website to consult is *Geography Departments Worldwide* (http://univ.cc/geolinks). Just click on "Simple Search," which brings up a list of the world's countries. Under "United States" no fewer than 291 geography departments are listed in alphabetical order, each only an additional click away.

You may discover one particular department that stands out as the most interesting and most appropriate for you. But do not limit yourself to one school. Consult the *Guide*'s 18-page table (entitled *Geography Program Specialties List*) that links 35 geographic specializations to every department listed, and then explore each appropriate matchup. After careful investigation, it is best to rank a half-dozen schools (or more), write or e-mail all of them for admission application forms, and apply to several. Multiple applications are a bit costly, but the investment is worth it.

Assistantships and Scholarships

One reason to apply to several universities has to do with financial (and other) support for which you may be eligible. If you have a reasonably well-rounded undergraduate program behind you and a good record of achievement, you are eligible to become a teaching assistant (TA) in a graduate department. Such assistantships usually offer full or partial tuition plus a monthly stipend (during the nine-month academic year). Conditions vary, but this position may make it possible for you to attend a university that would otherwise be out of reach. A tuition waiver alone can be well over $10,000 annually. Application for an assistantship is made directly to the department; you should write or e-mail the department contact listed in the latest AAG *Guide* (or on the department's website), who will either respond directly to you or forward your inquiry to the departmental committee that evaluates applications.

What does a TA do? The responsibilities vary, but often teaching assistants are expected to lead discussion sections (of a larger class taught by a professor), laboratories, or other classes. They prepare and grade examinations and help undergraduates deal with problems arising from their courses. This is an excellent way to determine your own ability and interest in a teaching career.

In some instances, especially in larger geography departments, research assistantships are available. When a member of the faculty is awarded a large grant (e.g., by the National Science Foundation) for a research project, that grant may make possible the appointment of one or more research assistants (RAs). These individuals perform tasks generated by the project and are rewarded by a modest salary (usually comparable to that of a TA). Normally, RAs do not receive tuition waivers, but sometimes the department and the graduate school can arrange a waiver to make the research assistantship more attractive to better students. Usually, RAs are chosen from among graduate students already on campus who have proven their interest and ability. Sometimes, however, an incoming student is appointed. Always ask about opportunities.

Geography students also are among those eligible for many scholarships and fellowships offered by universities and off-campus organizations. When you write your introductory letter or e-mail, be sure to inquire about all forms of financial aid.

JOBS FOR GEOGRAPHERS

Upon completing your bachelor's degree, you may decide to take a job rather than go on to graduate school. Again, this is a decision best made early in your junior year, for two main reasons: (1) so that you can tailor your curriculum for a vocational objective, and (2) so that you can start searching for a job well before graduation.

Internships

A very good way to enter the job market—and to become familiar with the working environment—is by taking an internship in an agency, office, or firm. Many organizations find it useful to have interns. Internships help organizations train beginning professionals and give companies an opportunity to observe the performance of trainees. Many an intern has ultimately been employed by his or her organization. Some employers have even suggested what courses the intern should take in the next academic year to improve future performance. For example, an urban or regional planning agency that employs an intern might suggest that the intern add a relevant GIS course or urban-planning course to his or her program of study.

Some organizations will appoint interns on a continuing basis, say, two afternoons a week around the year; others make full-time, summer-only appointments available. Occasionally, an internship can be linked to a departmental curriculum, yielding academic credit as well as vocational experience. Your department undergraduate advisor or the chairperson will be the best source of assistance.

One of the most interesting internship programs is offered by the National Geographic Society in Washington, D.C. Every year, the Society invites three groups of about eight interns each to work with the permanent staff of its various departments. Application forms are available in every geography department in the United States, and competition is strong. The application itself is a useful exercise, as it tells you what the Society (and other organizations) look for in your qualifications.

Professional Opportunities

A term you will sometimes see in connection with jobs is *applied geography*. This, one supposes, distinguishes geography teaching from practical geography. In fact, however, all professional geography in education, business, government, and elsewhere is "applied." In the past a large majority of geography graduates became teachers—in elementary and high schools, colleges, and universities. More recently, geographers have entered other arenas in increasing numbers. In part, this is related to the decline in geographic education in schools, but it also reflects the growing recognition of geographic skills by employers in business and government.

Nevertheless, what you as a geographer can contribute to a business is not yet as clear to the managers of many companies as it should be. The anybody-can-do-geography attitude is a form of ignorance you will undoubtedly confront. (This is so even in precollegiate education, where geography was submerged in social studies—and often taught by teachers who had never taken a course in the discipline!)

So, once employed, you may have to prove not only yourself but also the usefulness of the skills and capacities you bring to your job. There is a positive side to this. Many employers, once dubious about hiring a geographer, learn how geography can contribute—and become enthusiastic users of geographic talent. Where one geographer is employed, whether in a travel firm, publishing house, or planning office, you will soon find more.

Geographers are employed in business, government, and education. Planning is a profession that employs many geographers. Employment in business has grown in recent years (see the following section). Governments at national, State, and local levels have always been major employers of geographers. And in education, where geography was long in decline, the demand for geography teachers will grow again.

For more detailed information about employment, you should contact the AAG (at the address given earlier) for the inexpensive booklet entitled *Careers in Geography*; or check with your geography department, whose office is very likely to have a copy.

Business and Geography

With their education in global and international affairs, their knowledge of specialized areas of interest to business, and their training in cartography, GIS, methods of quantitative analysis, and writing, geographers with a bachelor's degree are attractive business-employment prospects. Some undergraduate students have already chosen the type of business they will enter; for them, there are departments that offer concentrations in their areas of interest. Several geography departments, for instance, offer a curriculum that concentrates on tourism and travel—a business in which geographic skills are especially useful.

These days, many companies want graduates with a strong knowledge of international affairs and fluency in at least one foreign language in addition to their skills in other areas. The business world is quite different from academia, and the transition is not always easy. Your employer will want to use your abilities to enhance profits. You may at first be placed in a job where your geographic skills are not immediately applicable, and it will be up to you to look for opportunities to do so.

One of our students was in such a situation some years ago. She was one of more than a dozen new employees doing what was essentially clerical work. (Some companies will use this type of work to determine a new employee's punctuality, work habits, adaptability, and productivity.) One day she heard that the company was considering establishing a branch to sell its product in East Africa. The student had a regional interest in East Africa as an undergraduate and had even taken a year of Swahili-language training. On her own time she wrote a carefully documented memorandum to the company's president and vice presidents, describing factors that should be taken into consideration in the projected expansion. That report showcased her regional skills, locational insights, and knowledge of the local market and transport problems, along with the probable cultural reaction to the company's product, the country's political circumstances, and (last but significantly) this employee's ability to present such issues effectively. She supported her report with good maps and several illustrations. Soon she received a special assignment to participate in the planning process, and her rise in the company's ranks had begun. She had seized the opportunity and demonstrated the utility of her geographic skills.

Geography graduates have established themselves in businesses of all kinds: banking, international trade, manufacturing, retailing, and many more. Should you join a large firm, you may be pleasantly surprised by the number of other geographers who hold jobs there—not under the title of geographer but under countless other titles ranging from analyst and cartographer to market researcher and program manager. These are positions for which the appointees have competed with other graduates, including business graduates. As we noted earlier, once an employer sees the assets a geographer can bring, the role of geographers in the company is assured.

Government and Geography

Government has long been a major employer of geographers at the national and State as well as local level. *Careers in Geography* (the AAG booklet) estimates that at least 2500 geographers are working for governments, about half of them for the federal government. In the U.S. State Department, for instance, there is an Office of the Geographer staffed by professional geographers. Other agencies where geographers are employed include the Defense Mapping Agency, the Bureau of the Census, the U.S. Geological Survey, the Central Intelligence Agency, and the Army Corps of Engineers. Still other employers are the Library of Congress, the National Science Foundation, and the Smithsonian Institution. Many other branches of government also have positions for which geographers are eligible.

Opportunities also exist at State and local levels. Several States now have their own Office of the State Geographer; all States have agencies engaged in planning, resource analysis, environmental protection, and transportation policy making. All these agencies need geographers who have skills in cartography, remote sensing, spatial database analysis, and the operation of geographic information systems.

Securing a position in government requires early action. If you want a job with the federal government, start at the beginning of your senior year. Every State capital and many other large cities have a Federal Job Information Center (FJIC). (The Washington office is at 1900 E Street, N.W., Washington, D.C. 20415.) You may request information about a particular agency and its job opportunities; it is appropriate to write or e-mail directly to the personnel office of the agency or agencies in which you are interested. You may also write to the Office of Personnel Management (OPM), Washington, D.C. 20415, which also has offices in most large cities.

Planning and Geography

Planning has become one of geography's allied professions. The planning process is a complex one comprising people trained in many fields. Geographers, with their cartographic, locational, regional, and analytical skills, are sought after by planning agencies. Many an undergraduate student gets that first professional opportunity as an intern in a planning office.

Planning is done by many agencies and offices at levels ranging from the federal to the municipal. Cities have planning offices, as do regional authorities. Working in a planning office can be a very rewarding experience because it involves the solving of social and economic problems, the conservation and protection of the environment, the weighing of diverse and often conflicting arguments and viewpoints, and much interaction with workers trained in other fields. Planning is a superb learning experience.

A career in planning can be much enhanced by a background in geography, but you will have to adjust your undergraduate curriculum to include courses in such areas as public administration, public finance, and other related fields. Thus a career in planning itself requires early planning on your part. At many universities, the geography department is closely associated with the planning department, and your faculty advisor can inform you about course requirements. But if you have your eye on a particular office or agency, you should also request information from its director about desired and required skills.

Planning is by no means a monopoly of government. Government-related organizations such as the Agency for International Development (AID), the World Bank, and the International Monetary Fund (IMF) have planning offices, as do nongovernmental organizations (NGOs), banks, airline companies, industrial firms, multinational corporations, and research institutes. Private-sector opportunities for planners have been expanding, and you may wish to explore them. An important organization is the planners' equivalent of the geographers' AAG: the American Planning Association, 1776 Massachusetts Avenue, N.W., Washington, D.C. 20006. The AAG's *Careers in Geography* provides additional information on this expanding field and where you might pursue its study.

Teaching Geography

If you are presently a freshman or sophomore, your graduation may coincide with the end of a long decline in the geography-teaching profession. Just a few decades ago, teaching geography in elementary or high school was the goal of thousands of undergraduate geography majors. But then began the merging of geography into the hybrid field called "social studies," and prospective teachers no longer needed to have any training in geography in many States. In Florida, for instance, teachers were formerly required to take courses in regional geography and conservation (as taught in geography departments), but in the 1970s those requirements were dropped. Education planners fell victim to the myth that the teaching of geography does not require any training. What was left of geography often was taught by teachers whose own fields were history, civics, or even basketball coaching!

If you read the daily newspapers, you have seen reports of the predictable results. As mentioned earlier, "geographic illiteracy" has become a common complaint (often made by the same education planners who pushed geography into the social studies program and eliminated teacher-education requirements). Now States are returning to the education requirement so that teachers will learn some geography. And geography is returning to elementary and high school curricula—assisted in particular by the National Geographic Society, which supports State-level Geographic Alliances of educators and academic geographers. The need for geography teachers will soon be on the upswing again.

So this may be a good time to consider teaching geography as a career. You should do some research, however, because States vary in their progressiveness in this arena. You should also visit the School of Education in your college or university and ask questions about instructional opportunities in geography. In addition, contact not only the AAG but also the *National Council for Geographic Education (NCGE)* at 206-A Martin Hall, Jacksonville State University, Jacksonville, AL 36265-1602 (**www.ncge.org**). Your State geographic society or Geographic Alliance also may be helpful; ask your department advisor or chairperson for details.

SOME FINAL THOUGHTS

The opportunities in geography are many, but often they are not as obvious as those in other fields. You will find that good and timely preparation produces results and, frequently, unexpected rewards.

The discussion in this appendix has provided a comprehensive answer to the oft-asked question, "What can one do with geography?" If you wish to explore this question further, along with the AAG's *Careers in Geography* we recommend *On Becoming a Professional Geographer*, a book edited by Martin S. Kenzer (Merrill/Macmillan, 1989; reprinted in 2005 by Blackburn Press).

We wish you every success in all your future endeavors. And, speaking for the entire community of professional geographers, we would be delighted to have you join our ranks should you choose a geography-related career.

Appendix C

Pronunciation Guide

Abacha (ah-BAH-chah)
Abadan (ahba-DAHN)
Abakan (ahb-uh-KAHN)
Abdullah (ahb-DOO-lah)
Abidjan (abb-ih-JAHN)
Abkhazia (ahb-KAHZ-zee-uh)
Aboriginal (abb-uh-RIDGE-uh-null)
Abu Dhabi (ah-boo-DAH-bee)
Abu Sayyaf (AH-boo sye-YAHF)
Abu Dis (ah-boo-DEESE)
Abuja (uh-BOO-juh)
Abyssinia (abb-ih-SINNY-uh)
Acacia (uh-KAY-shuh)
Acadia (uh-KAY-dee-uh)
Acapulco (ah-kah-POOL-koh)
Accra (uh-KRAH)
Aceh (ah-CHEH)
Acehnese (ah-chay-NEEZ)
Acre (AH-kray)
Acropolis (uh-CROP-uh-liss)
Adamawa (add-uh-MAH-wuh)
Adan (AH-dahn/AYD'N)
Adelaide (ADDLE-ade)
Adirondack (add-ih-RONN-dak)
Adis Abeba (adda-SAB-uh-buh)
Adriatic (ay-dree ATTIK)
Adygeya (ah-duh-GAY-uh)
Aegean (uh-JEE-un)
Afars (uh-FAHRZ)
Afghanistan (aff-GHANN-uh-stan)
Afrikaans (uff-rih-KAHNZ)
Afrikaner (uff-rih-KAHN-nuh)
Agios Dimitrios (AH-jee-ohss deh-MEE-tree-ohss)
Agra (AHG-ruh)
Agri (AH-gree)
Agurchand (ah-ghoor-SHAHND)
Ahad (ah-HAHD)
Ahaggar (uh-HAH-gahr)
Ahmadabad (AH-muh-duh-bahd)
Ahvaz (ah-VAHZ)
Ainu (EYE-noo)
Ajaria (uh-JAR-ree-uh)
Akan (AH-kahn)
Akashi Kaikyo (AH-kah-shee KYE-kyoh)
Akhbari (ahk-BAH-ree)
Akihito (ah-kee-HEE-toh)
Al Andalus (ahl ahnda-LOOSE)
Al Basrah (ahl-BAHZ-ruh)
Al Hudaydah (ahl-hoh-DAY-duh)
Al Madinah (ahl-mah-DEENA)
Al Sistani, Ali (ahl-see-STAHN-nee, ah-Lee)
Alagoas (ah-LAH-goh-ahss)
Alawite (AH-lah-wyte)
Alcazar (AHL-kah-zahr)
Aleijadinho (ah-lay-jah-DEEN-yoh)
Aleppo (uh-LEP-poh)
Alevis (ah-LEE-veez)
Alexandria (ah-leg-ZAN-dree-uh)
Algae (AL-jee)
Algeria (al-JEERY-uh)
Algiers (al-JEERZ)
Ali (ah-LEE)
Allah (AHL-ah)
Allahabad (ALLA-huh-bahd)
Allemanni (alla-MAH-nee)
Alluvial (uh-LOO-vee-ull)
Alma-Ata (ahl-muh-uh-TAH)
Almaty (ahl-mah-TUH)
Alpaca (al-PAK-uh)
Al-Qaeda (ahl-KYE-duh)
Al-Quds (ahl-KOOTSS)
Al-Salam (ahl-sah-LAHM)
Altaic (al-TAY-ik)
Altay (AL-tye)
Altaya (al-TYE-uh)
Altiplano (ahl-tee-PLAH-noh)
Altun (ahl-TOON)
Al-Wahhab (ahl-wah-HAHB)
Amador (AMMA-dor)
Amazonas (ahma-ZOAN-ahss)
Ambarawa (am-bah-RAH-wah)
Ambeno (AHM-bay-noh)
Ambon (ahm-BON)
Amerind (AMMER-rind)
Amhara (am-HAH-ruh)
Amharic (am-HAH-rik)
Amin, Idi (uh-MEEN, iddy)
Amman (uh-MAHN)
Amoy (ah-MOY)
Amritsar (um-RIT-sahr)
Amu-Darya (uh-moo-DAHR-yuh)
Amundsen, Roald (AH-moon-sun, ROH-ahl)
Amur (uh-MOOR)
An Najaf (un-nuh-JAHF)
Anatolia (anna-TOH-lee-uh)
Ancona (ahng-KOH-nuh)
Ancud (ahn-KOOD)
Andalusia (ahn-duh-loo-SEE-uh)
Andaman (ANN-duh-mun)
Andes (ANN-deez)
Andhra Pradesh (ahn-druh pruh-DESH)
Angara (ahng-guh-RAH)
Angkor [Wat] (ANG-kor [WOT])
Angola (ang-GOH-luh)
Anhui (ahn-HWAY)
Ankara (ANG-kuh-ruh)
Annam (uh-NAHM)
Annam[ese]/[ite] (anna-[MEEZE]/[MITE])
Anshan (ahn-SHAHN)
Antalya (ahn-tahl-YAH)
Antananarivo (ahn-TAH-nah-nah-REEV)
Antecedent (an-tee-SEE-dunt)
Antigua (an-TEE-gwuh)
Antilles (an-TILL-eeze)
Antimony (ANN-tih-moh-nee)
Antioquia (ahn-tee-OH-kee-ah)
Antofagasta (untoh-fah-GAHSS-tah)
Antwerp (ANN-twerp)
Anuradhapura (UH-NOO-RAH-thah-poor-uh)
Anyang (ahn-YAHNG)
Aozou (OO-zoo)
Apapa (uh-PAHP-uh)
Apartheid (APART-hate)
Appennines (APP-uh-neinz)
Apure (ah-POOR-ray)
Apurimac (ah-poo-REE-mahk)
Aqaba (AH-kuh-buh)
Aqmola (ahk-moh-LAH)
Aquifer (ACK-kwuh-fer)
Arakan (ah-ruh-KAHN)
Aral (ARREL)
Aramco (uh-RAM-koh)
Arauca (ah-RAU-kah)

Araucania (ah-rau-KAH-nee-ah)
Arawak (ARRA-wak)
Arc de Triomphe (ark duh tree-AWMFF)
Archangel (ARK-ain-jull)
Archipelago (ark-uh-PELL-uh-goh)
Ardennes (ar-DEN)
Arequipa (ah-ray-KEE-pah)
Argentine (AR-jen-tyne)
Arguedas, Alcides (ahr-GWAY-dahss, ahl-SEE-duss)
Arica (ah-REE-kah)
Aridisol (uh-RIDDY-sol)
Aristide, Jean-Betrand (ah-ree-STEED, zhawn-bair-TRAWN)
Arkhangelsk (ahr-KAHN-gyil-sk)
Armenia (ar-MEENY-uh)
Armonk (AHR-munk)
Armuelles (ahr-MWAY-yace)
Artesian (ahr-TEE-zhun)
Aruba (uh-ROO-buh)
Arunachal Pradesh (AHRA-NAHTCH-ull pruh-DESH)
Aryan (AHR-yun)
Aryeetey-Attoh (ahr-YEE-tee ATTOH)
As Sulaymaniyah (ahss soo-lee-mah-NEE-yah)
Asante (ah-SAHN-tay)
Ashanti (ah-SHAHN-tee)
Ashgabat (AHSH-gah-baht)
Ashkelon (AHSH-kih-lonn)
Aśoka (uh-SHOH-kuh)
Assab (uh-SAHB)
Assad, Bashar (uh-SAHD, buh-SHAR)
Assad, Hafez al- (uh-SAHD, huh-FEZ ahl)
Assam (uh-SAHM)
Assyrian (uh-SEERY-un)
Astana (ah-STAH-nah)
Astrakhan (ASTRA-kahn)
Asturias (uh-STOOR-ree-uss)
Asunción (ah-soohn-see-OAN)
Aswan (as-SWAHN)
Atacama (ah-tah-KAH-mah)
Atatürk (ATTA-tyoork)
Atheism (AYTHEE-izm)
Athena (uh-THEENA)
Atoll (AY-toal)
Attenuated (uh-TEN-yoo-ay-ted)
Atyrau (ah-tah-RAU)
Auckland (AWK-lund)
Augelli (aw-JELLY)
Austral (AW-strull)
Australopithecene (aw-strallo-PITH-uh-seen)
Autocracy (aw-TOCK-ruh-see)
Avenida de 9 Julio (ah-veh-NEEDA day noo-AY-vay HOO-lee-oh)
Avenida Nicolas de Pierola (ah-veh-NEEDA NEEKO-lahss day pyair-ROH-lah)
Avignon (ah-veen-YAW)
Axum (AHKSS-oom)
Ayatollah (eye-uh-TOH-luh)

Aymara (ai-MAH-rah)
Ayodhya (uh-YODE-yuh)
Azerbaijan (ah-zer-bye-JAHN)
Azeri (ah-ZAREY)
Azov (uh-ZAWF)
Ba'ath (BATH)
Bab el Mandeb (bab ull MAN-dub)
Babylon (BABBA-lon)
Bac Bo (BAHK-BOH)
Badawi, Abdullah (buh-DAH-wee, ahb-DOOL-uh)
Baden-Württemberg (BAHDEN VEERT-um-bairk)
Baganda (bah-GAHN-duh)
Bahamas (buh-HAH-muzz)
Bahasa (bah-HAH-sah)
Bahia (bah-EE-yah)
Bahr el Ghazal (bar-ahl-gah-ZAHL)
Bahrain (bah-RAIN)
Baht (BAHT)
Baja (BAH-hah)
Bajau (bah-YAU)
Bakassi (bah-KAH-see)
Baki (bah-KEE)
Baku (bah-KOO)
Bali [nese] (BAH-lee [NEEZ])
Balikpapan (bah-LIK-pah-pahn)
Balkan (BAWL-kun)
Balkhan (bahl-KAHN)
Balsas (BAHL-suss)
Baluchistan (buh-loo-chih-STAHN)
Bamako (BAH-mah-koh)
Banda (BAHN-duh)
Banda Aceh (BAHN-dah uh-CHEH)
Bandar Lampung (BAHN-dahr lahm-PUNG)
Bandar Seri Begawan (BUN-dahr SERRY buh-GAH-wun)
Bandung (BAHN-doong)
Bangalore (BANG-guh-loar)
Bangladesh [i] (bang-gluh-DESH [ee])
Banja Luka (BUN-yuh LOO-kuh)
Bantu (ban-TOO)
Banyamulenge (bahn-yah-moo-LENG-geh)
Baotou (bao-TOH)
Barbados (bar-BAY-dohss)
Barcelona (bar-suh-LOH-nuh)
Barents (BARRENS)
Barim (buh-REEM)
Barquisimeto (bar-key-suh-MAY-toh)
Barranquilla (bah-rahn-KEE-yah)
Barrio [*s*] (BAHR-ree-oh[ss])
Basalt (buh-SAWLT)
Bashkortostan (bahsh-KORT-uh-stahn)
Basilan (bah-see-LAHN)
Basilicata (bah-ZEE-lee-kah-tah)
Basque (BASK)
Basra (BAHZ-ruh)
Batak (buh-TAHK)
Batavia (buh-TAY-vee-uh)
Batista, Fulgencio (bah-TEESTA, fool-HEN-see-oh)

Batticaloa (buh-tih-kuh-LOH-uh)
Batumi (bah-TOO-mee)
Bauxite (BAWKS-site)
Bavaria (buh-VAIRY-uh)
Baykal [iya] (bye-KAHL [ee-ah]))
Bayou (BYE-yoo)
Bayóvar (bye-YOH-vahr)
Bayu-Undan (bye-YOO oon-DAHN)
Beaumont (BOH-mont)
Bedouin (BEH-doo-in)
Beihai (bay-HYE)
Beijing (bay-ZHING)
Beira (BAY-ruh)
Beirut (bay-ROOT)
Bekaa (buh-KAH)
Bekasi (beh-KAH-see)
Belarus (bella-ROOSE)
Belarussia (bella-RUSH-uh)
Belém (bay-LEM)
Belgian (BEL-jun)
Belize (beh-LEEZE)
Belmopan (bell-moh-PAN)
Belo Horizonte (BAY-loh haw-ruh-ZONN-tee)
Bengal [i] (beng-GAHL [ee])
Benghazi (ben-GAH-zee)
Benguela (beng-GWELLA)
Beni (BAY-nee)
Benin (beh-NEEN)
Benue (BANE-way)
Berber (BERR-berr)
Berimbau (beh-RIM-bau)
Bering (BERRING)
Bern (BAIRN)
Beslan (bess-LAHN)
Betsimisaraka (bye-tsee-mee-SAH-rah-kah)
Beyoglu (bay-uh-GLOO)
Bharatiya Janata (bah-rah-TEE-uh jah-NAH-tah)
Bhutan (boo-TAHN)
Bhutia (BOOH-tee-uh)
Biafra (bee-AH-fruh)
Bihar (bih-HAHR)
Bihe (bee-HAY)
Bilbao (bil-BAU)
Bilharzia (bill-HARZEE-uh)
Bin Laden, Usama (bin-LAHD-nn, oo-SAH-mah)
Bío Bío (bee-oh-BEE-oh)
Bioko (bee-OH-koh)
Biota (bye-OH-tuh)
Bishkek (BISH-kek)
Bitumen (bye-TOO-men)
Bloemfontein (BLOOM-fon-tane)
Boeing (BOH-ing)
Boer (BOOR)
Bogor (BOH-goar)
Bogotá (boh-goh-TAH)
Bohai (bwoh-HYE)
Bohemia (boh-HEE-mee-uh)
Bolívar, Simón (boh-LEE-vahr, see-MOAN)

Bolshevik (BOAL-shuh-vick)
Bom Bahia (bom-bah-EE-yah)
Bombay (bom-BAY)
Bonaire (bun-AIR)
Bora Bora (boar-ruh-BOAR-ruh)
Boran (boar-RAN)
Bordeaux (boar-DOH)
Borneo (BOAR-nee-oh)
Bosnia-Herzegovina (BOZ-nee-uh hert-suh-goh-VEE-nuh)
Bosporus (BAHSS-puh-russ)
Botha (BAW-tuh)
Botswana (bah-TSWAHN-uh)
Bouchard, Lucien (boo-SHAR, looss-YAN)
Bougainville (BOO-gun-vil)
Boulangerie (boo-LAHN-zheh-ree)
Boulevard St. Laurent (boo-leh-VAHR san-law-RAH)
Bourgogne (boor-GOAN-yuh)
Bové, José (boh-VAY, zhoh-ZAY)
Brahmaputra (brahm-uh-POOH-truh)
Brandenburg (BRAHN-den-boorg)
Brasília (bruh-ZEAL-yuh)
Bratislava (BRUDDIS-lahva)
Bratsk (BRAHTSK)
Brazzaville (BRAHZ-uh-veel)
Brcko (BIRCH-koh)
Bremen (BRAY-mun)
Brisbane (BRIZZ-bun)
Brno (BURR-noh)
Brunei (BROO-nye)
Brunswick (BROON-svik)
Bubonic (byoo-BON-nik)
Bucharest (BOO-kuh-rest)
Budapest (BOODA-pest)
Buddha (BOOD-uh)
Buddhism (BOOD-izm)
Buddhist (BOOD-ist)
Buenos Aires (BWAY-nohss EYE-race)
Buganda (boo-GAHN-duh)
Bulawayo (boo-luh-WAY-oh)
Bulgaria (bahl-GHAIR-ree-uh)
Bund (BUND)
Bunia (BOO-nee-uh)
Bureya (buh-RAY-yuh)
Burkina Faso (ber-keena FAHSSO)
Burkinabe (ber-kee-NAH-bay)
Burmans (BURR-munz)
Burundi (buh-ROON-dee)
Buryat [iya] (boor-YAHT [ee-uh])
Busan (BOO-SAHN)
Bushehr (boo-SHAIR)
Buthelezi, Mangosuthu (boo-teh-LAY-zee, mungo-SOO-too)
Butte (BYOOT)
Byzantine (BIZZ-un-teen)
Cabimas (kah-BEE-mahss)
Cabinda (kuh-BIN-duh)
Cabora Bassa (kuh-boar-rah BAHSSA)
Cabramatta (kabbra-MADDA)
Cacao (kuh-KAY-oh/kuh-KAU)
Cadre (KAH-dray)
Caicos (KAY-kohss)
Cairo (KYE-roh)
Calabria (kuh-LAH-bree-uh)
Calais (kah-LAY)
Calbuco (kahl-BOO-koh)
Calcutta (kal-KUTT-uh)
Caldas (KAHL-dahss)
Calgary (KAL-guh-ree)
Cali (KAH-lee)
Caliph (KAY-liff)
Caliphate (KALLA-fate)
Callao (kah-YAH-oh)
Cambodia (kam-BOH-dee-uh)
Campeche (kahm-PAY-chee)
Canberra (KAN-burruh)
Candomblé (kan-DOM-blay)
Caño Limón (KAHN-yoh lee-MONE)
Cantabria (kahn-TAH-bree-uh)
Canterbury (KAN-ter-berry)
Canton (kan-TONN)
Capoeira (kah-poh-AY-ruh)
Caprivi (kuh-PREE-vee)
Caquetá (kah-kway-TAH)
Caracas (kah-RAH-kuss)
Carajás (kuh-ruh-ZHUSS)
Carara (kuh-RAHR-ruh)
Caribbean (kuh-RIB-ee-un/karra-BEE-un)
CARICOM (CARRY-komm)
Carioca (kah-ree-OH-kah)
Caritas (kuh-REE-tuss)
Carpathians (kar-PAY-thee-unz)
Carreño, Teresa (kah-RAIN-yoh, teh-RAY-sah)
Cartagena (karta-HAY-nuh)
Casamance (KAH-zah-mahnss)
Casanare (kah-sah-NAH-ray)
Caspian (KASS-spee-un)
Caste (CAST)
Castile (kuh-STEEL)
Castile-La Mancha (kuh-STEEL luh-MAHN-chuh)
Castile-Leon (kuh-STEEL lay-OAN)
Catalan (katta-LAHN)
Catalonia (katta-LOH-nee-uh)
Cauca (KOW-kah)
Caucasian (kaw-KAY-zhun)
Caucasus (KAW-kuh-zuss)
Cay (KEE)
Cayambe (kah-YAHM-bee)
Ceará (see-ah-RAH)
Ceausescu, Nicolae (chow-SHESS-koo, NICK-oh-lye)
Cebu (seh-BOO)
Celebes (SELL-uh-beeze)
Celt [ic] (KELT [ick])
Cenozoic (senno-ZOH-ik)
Centavo (sen-TAH-voh)
Central (sen-TRAHL)
Centrifugal (sen-TRIFFA-gull)
Centripetal (sen-TRIPPA-tull)
Centro-Oeste (SENTRO oh-ESS-tee)
Cerrado (seh-RAH-doh)
Cerro de Pasco (serro day PAH-skoh)
Ceuta (see-YOO-tah)
Ceyhan (jay-HAHN)
Ceylon (seh-LONN)
Chaco (CHAH-koh)
Chaebol (JAY-BOAL)
Champagne (shahm-PAHN-yuh)
Champs Elysées (SHAWZ elly-ZAY)
Chang Jiang (chang jee-AHNG)
Ch'angan (CHAHNG-GAHN)
Changchun (CHAHNG-CHOON)
Chao Phraya (CHOW pruh-yah)
Chaparé (chah-pah-RAY)
Charisma (kuh-RIZZ-muh)
Charleroi (SHARL-rwah)
Chateau Frontenac (shat-TOH FRAWN-teh-nahk)
Chavez (SHAH-vez)
Chechen (CHEH-chen)
Chechenya-Ingushetiya (cheh-CHEN-yuh in-goo-SHETTY-uh)
Chechnya (CHETCH-nee-uh)
Chelyabinsk (chel-YAH-bunsk)
Chengdu (chung-DOO)
Chennai (cheh-NYE)
Cherkessk (cher-KESK)
Chernobyl (see Chornobyl)
Chhattisgarh (CHADDISS-gahr)
Chiang Kai-shek (jee-AHNG kye-SHECK)
Chiang Mai (chee-AHNG mye)
Chiapas (chee-AHP-uss)
Chihuahua (chuh-WAH-wah)
Chile (CHILLI/CHEE-lay)
Chiloé (chee-luh-WAY)
Chilung [Jilong] (JEE-LOONG)
Chisinau (kee-shih-NAU)
Chittagong (CHITT-uh-gahng)
Cholla (JOH-luh)
Cholon (choh-LONN)
Cholula (choh-LOO-lah)
Choluteca (choh-loo-TAY-kah)
Chongqing (chong-CHING)
Chordata (kor-DATTA)
Chornobyl (CHAIR-noh-beel)
Chota Nagpur (choat-uh-NAHG-poor)
Chubu (CHOO-BOO)
Chukotka (chuh-KAHT-kuh)
Chukotskiy (chuh-KAHT-skee)
Chungyang (joong-YAHNG)
Chuquicamata (choo-kee-kah-MAH-tah)
Chuvash [iya] (choo-VAHSH [ee-uh])
Ciliwung (SILL-uh-wong)
Ciudad Alta (see-yoo-DAHD AHL-tuh)
Ciudad Baixa (see-yoo-DAHD bah-ZHEE-uh)
Ciudad del Este (see-yoo-DAHD del-ESS-stay)
Ciudad Guayana (see-yoo-DAHD gwuh-YAHNA)
Ciudad Juárez (see-you-DAHD WAH-rez)
Ciudades perdidas (see-you-DAH-dayss pair-DEE-duss)

Cliburn (KLYE-bern)
Clonshaugh (KLONZ-haw)
Coahuila (koh-uh-WEE-luh)
Coatzacoalcos (koh-aht-sah-koh-AHL-kohss)
Cochabamba (koh-chah-BUM-bah)
Cochin (KOH-chin)
Cocos (KOH-kuss)
Collor de Mello, Fernanado (KAW-lohr deh MELLOH, fair-NAHN-doh)
Colombo (kuh-LUM-boh)
Colón (kuh-LOAN)
Comodoro Rivadavia (comma-DORE-oh ree-vah-DAH-vee-ah)
CONAIE (koh-NYE)
Conakry (koh-NAK-ree)
Confucius (kun-FEW-shuss)
Coniferous (kuh-NIFF-uh-russ)
Conquistador (koan-KEE-stuh-doar)
Constantinople (kon-stant-uh-NOPLE)
Conurbation (konner-BAY-shun)
Copenhagen (koh-pen-HAHGEN)
Copra (KOH-pruh)
Corcovado (koar-koh-VAH-doh)
Cordillera (kor-dee-YERRA)
Cordillera del Litoral (kor-dee-YERRA dale lee-toh-RAHL)
Córdoba (KORR-doh-buh)
Coromandel (kor-uh-MANDLE)
Corsica (KORR-sih-kuh)
Cortés, Hernán (kor-TAYSS, air-NAHN)
Costa del Sol (koh-stuh-del-SOAL)
Costa Rica (koh-stuh-REE-kuh)
Côte d'Ivoire (KOAT deev-WAH)
Coup d'état (koo-day-TAH)
Coveñas (koh-VAIN-yahss)
Creole [s] (KREE-oal[z])
Crete (KREET)
Crimea (cry-MEE-uh)
Crimean (cry-MEE-un)
Croat (KROH-aht/KROH-at)
Croatia (kroh-AY-shuh)
Cruz (KROOZ)
Cuiabá (koo-yuh-BAH)
Curaçao (koor-uh-SAU)
Cusiana-Cupiagua (koo-see-AHNA, koopee-YAH-gwah)
Cuyo (KOO-yoh)
Cuzco (KOO-skoh)
Cyclades (SIK-luh-deeze)
Cyprus (SYE-pruss)
Cyrenaica (sear-uh-NAY-ih-kuh)
Cyrillic (suh-RILL-ick)
Czar (ZAHR)
Czarina (zah-REE-nuh)
Czech (CHECK)
Czechoslovakia (check-uh-sloh-VAH-kee-uh)
Daegu (DAH-goo)
Dagestan (dag-uh-STAHN)
Dahomey (dah-HOH-mee)
Dakar (duh-KAHR)
Dalai Lama (dah-lye LAHMA)
Dalian (dah-lee-ENN)
Dalit (DAH-lit)
Dalmatia (dal-MAY-shuh)
Damascus (duh-MASK-uss)
Damietta (dam-ee-ETTA)
Danube (DAN-yoob)
Daoism (DAU-ism)
Daqing (dah-CHING)
Dar es Salaam (dahr ess suh-LAHM)
Dardanelles (dahr-duh-NELZ)
Darfur (dahr-FOOR)
Darien (dar-YEN)
Daw Aung San Suu Kyi (daw-ong-SAHN soo-CHEE)
Dayak (DYE-ack)
De Gaulle, Charles (duh-GAWL, SHARL)
Deccan (DECKEN)
Delhi (DELLY)
Delphi (DELL-fye)
Deltaic (dell-TAY-ick)
Deng Xiaoping (DUNG shau-PING)
Département (day-part-MAW)
Devolution (dee-voh-LOOH-shun)
Dhahran (dah-RAHN)
Dhaka (DAHK-uh)
Dhow (DAU)
Diaspora (dee-ASP-uh-ruh)
Dichotomy (dye-KOTT-uh-mee)
Dien Bien Phu (d'yen-b'yen-FOOH)
Dijon (dee-ZHAW)
Dili (DIH-lee)
Dinaric (dih-NAHR-rick)
Diyarbakir (dih-yahr-buh-KEER)
Djezira-al-Maghreb (juh-ZEER-uh ahl-mahg-GRAHB)
Djibouti (juh-BOODY)
Djouf (JOOF)
Dnieper (duh-NYEPPER)
Dniester (duh-NYESS-truh)
Dnipropetrovsk (duh-nep-roh-puh-TRAWFSSK)
Dodoma (DOH-duh-mah)
Dominica (duh-MIN-ih-kuh)
Donbas (DAHN-bass)
Donets (duh-NETTS)
Donetsk (duh-NETTSK)
Dostoyevsky (doss-stoy-YEFF-skee)
Douala (doo-AHLA)
Drakensberg (DRAHK-unz-berg)
Dravidian (druh-VIDDY-un)
Dresden (DREZ-den)
Druzba (DROOZE-bah)
Druze (DROOZE)
Dubai (doo-BYE)
Duchy (DUTCH-ee)
Dunedin (duh-NEED-nn)
Durban (DER-bun)
Dushanbe (doo-SHAHM-buh)
Dvina (duh-vee-NAH)
Dzibilchaltun (zeeb-eel-chahl-TOON)
Dzungaria (joong-GAH-ree-uh)
Ebola (ee-BOH-luh)
Ecuador (ECK-wah-dor)
Ecumene (ECK-yoo-meen)
Edgecumbe (EDGE-kum)
Edinburgh (EDDIN-burruh)
Edo (EDD-oh)
Eelam (EE-lum)
Efik (EFFIK)
Egoli (ee-GOH-lee)
Eilat (AY-laht)
Eire (AIR)
Ejido [*s*] (eh-HEE-doh [ss])
El Aqsa (el AHK-suh)
El Cojo (ell-KOH-hoh)
El Duqqi (el-DOO-kee)
El Niño (ell-NEEN-yoh)
El Paso (ell-PASSO)
El Puente (el-PWEN-tay)
El Qahira (el-KAH-hee-ruh)
Elam (EE-lum)
Elbe (ELB)
Elburz (el-BOORZ)
Ellesmere (ELZ-mear)
Ellinikon (eh-LINNY-konn)
Emirate (EMMA-rate)
Endemism (en-DEM-izm)
Endoreic (en-doh-RAY-ick)
England (ING-glund)
Enkare (en-KAH-ray)
Entebbe (en-TEBBA)
Entre Rios (en-truh-REE-ohss)
Entrepôt (AHNTRA-poh)
Erdogan (er-DAU-un)
Eritrea (erra-TRAY-uh)
Erlitou (air-lee-TOH)
Esmeraldas (ezz-may-RAHL-dahss)
Estonia (eh-STOH-nee-uh)
Ethiopia (eeth-ee-OH-pea-uh)
Etorofu (etta-ROH-foo)
Eucalyptus (yoo-kuh-LIP-tuss)
Eunuch (YOO-nuck)
Euphrates (yoo-FRATE-eeze)
Eurasia (yoo-RAY-zhuh)
Evens (eh-VENSS)
Extremadura (ess-truh-muh-DOORA)
Façade (fuh-SAHD)
Faeroe (FAR-roh)
Faisalabad (fye-SAHL-ah-bahd)
Falkland [s] (FAWK-lund[z])
Farakka (fah-RAH-kuh)
Fare, Wahiba (FAH-ray, wah-HEE-buh)
Faro del Comercio (FAH-roh del koh-MAIR-see-oh)
Fauna (FAW-nuh)
Favela [*s*] (fah-VAY-lah[ss])
Fazenda (fah-ZENN-duh)
Fellaheen (fella-HEEN)
Fergana (fahr-guh-NAH)
Ferronorte (feh-roh-NOR-tay)
Feyzabad (FAY-zuh-bahd)
Fezzan (fuh-ZANN)
Ficus (FYE-kuss)
Fiji (FEE-jee)
Fijian (fuh-JEE-un)

Filipinos (filla-PEA-noze)
Finno-Ugric (finno-YOO-grick)
Fjord (FYORD)
Flores (FLAW-rihss)
Florianópolis (flaw-ree-ah-NOH-poh-leese)
Foederis (feh-DARE-iss)
Fonseca (fahn-SAY-kuh)
Formosa (for-MOH-suh)
Fortaleza (for-tuh-LAY-zuh)
Fox, Vicente (FOX, vee-SEN-tay)
Foz do Iguaçu (fahz-doo-ig-wah-SOO)
Francophone (FRANK-uh-foan)
Frankfurt (FRUNK-foort)
Fría (FREE-uh)
Fuji (FOODGY)
Fujian (foo-jee-ENN)
Fujimori, Alberto (foo-jee-MAW-ree, ahl-BAIR-toh)
Fukuoka (foo-kuh-WOH-kuh)
Fulani (foo-LAH-nee)
Funafuti Atoll (foo-nah-FOO-tee AY-toal)
Fungi (FUN-jye)
Fur (FOOR)
Fushun (foo-SHUN)
Futa Jallon (food-uh juh-LOAN)
Fuzhou (foo-ZHOH)
Gabon (gah-BAW)
Gaelic (GALE-ick)
Gaillard (gil-YARD)
Galicia (guh-LEE-see-uh)
Galla (GAH-lah)
Gambia (GAM-bee-uh)
Gandhi (GONDY)
Ganga (GUNG-guh)
Ganges/Gangetic (GAN-jeez/gan-JETTICK)
Gansu (gahn-SOO)
Gao (GAU)
Gaoliang (gow-lee-AHNG)
Gaoxiong (see Kaohsiung)
Garagum (gah-rah-GOOM)
Garonne (guh-RON)
Gasohol (GAS-uh-hoal)
Gatun (guh-TOON)
Gauteng (GAU-teng)
Gavan (guh-VAHN)
Gaza (GAH-zuh)
Gdansk (guh-DAHNSK)
Gdynia (guh-DINNY-uh)
Gedo (GEDDO)
Geiger (GHYE-gherr)
Geneva (jeh-NEE-vuh)
Genoa (JENNO-uh)
Georges (ZHOR-zh)
Georgia (GEORGE-uh)
Gerardi, Juan José (heh-RAHR-dee, HWAHN hoh-ZAY)
Gerlach (GHERR-lahk)
Gezira (juh-ZEER-uh)
Ghana (GAH-nuh)
Ghanaian (gah-NAY-un)
Ghats (GAHTSS)
Gibraltar (jih-BRAWL-tuh)
Gilgit (GILL-gutt)
Ginza (GHIN-zuh)
Giralda (hih-RAHL-duh)
Giza (GHEE-zuh)
Glasgow (GLASS-skau)
Glasnost (GLUZZ-nost)
Goa (GOH-uh)
Gobi (GOH-bee)
Godthab (GAWT-hawb)
Goiás (goy-AHSS)
Golan (goh-LAHN)
Gondwana (gond-WON-uh)
Gorbachev, Mikhail (GOR-buh-choff, meek-HYLE)
Gorkiy (GORE-kee)
Gorno-Altay (gore-noh-AL-tye)
Gorno-Badakhshan (gore-noh-bah-dahk-SHAHN)
Gorod (guh-RAHD)
Göteborg (GOAT-uh-borg)
Gouda (GOO-dah)
Grande Arche (GRAWND-ARSH)
Grande Carajás (GRUNN-dee kuh-ruh-ZHUSS)
Greenwich (GREN-itch)
Grenada (gruh-NAY-duh)
Groznyy (GRAWZ-nee)
Guadalajara (gwah-duh-luh-HAHR-uh)
Guadalcanal (gwaddle-kuh-NAL)
Guadalquivir (gwahddle-kee-VEER)
Guadeloupe (GWAH-duh-loop)
Guajira (gwah-HEAR-ah)
Guam (GWAHM)
Guanabara (gwah-nah-BAR-uh)
Guando (GWAHN-doh)
Guangdong (gwahng-DUNG)
Guangxi Zhuang (gwahng-shee JWAHNG)
Guangzhou (gwahng-JOH)
Guaraní (gwah-rah-NEE)
Guatemala (gwut-uh-MAH-lah)
Guayaquil (gwye-ah-KEEL)
Guayas (GWYE-ahss)
Guelph (GWELF)
Guerrero (geh-RARE-roh)
Guiana [s] (ghee-AH-nah[z])
Guilin (gway-LIN)
Guinea (GHINNY)
Guinea-Bissau (ghinny-bih-SAU)
Guizhou (gway-JOH)
Gujarat (goo-juh-RAHT)
Gulag (GHOO-lahg)
Guri (GOOR-ree)
Gurkha (GHOOR-kuh)
Guryev (GHOOR-yeff)
Guyana (guy-AHNA)
Habibie (huh-BEE-bee)
Habomai (HAH-boh-mye)
Hacienda (ah-see-EN-duh)
Hägerstrand, Torsten (HAYGER-strand, TOR-stun)
Hague (HAIG)
Haifa (HYE-fah)
Haikou (HYE-KOH)
Hainan (HYE-NAHN)
Haiphong (hye-FONG)
Haiti (HATE-ee)
Haitian (HAY-shun)
Halaib (huh-LABE)
Hamad (hah-MAHD)
Hamas (hah-MAHSS)
Hambali (hahm-BAH-lee)
Hamburg (HAHM-boorg)
Hamoud (hah-MOOD)
Han (HAHN)
Hang Bai (HAHNG-bye)
Hangzhou (hahng-JOH)
Hankou (hahn-KOH)
Hanoi (han-NOY)
Hanseatic (han-see-ATTIK)
Hanyang (hahn-YAHNG)
Harappa (huh-RAP-uh)
Harare (huh-RAH-ray)
Harbin (HAR-bin)
Harer (HAH-rahr)
Hargeysa (hahr-GAY-suh)
Harijan (hah-ree-JAHN)
Hartshorne (HARTSS-horn)
Haryana (hah-ree-AHNA)
Hashish (hah-SHEESH)
Hausa (HOW-sah)
Hawai'i (huh-WAH-ee)
Hazara (huh-ZAHR-ah)
Hebei (huh-BAY)
Hegemony (heh-JEH-muh-nee)
Heian (HAY-ahn)
Heilongjiang (hay-long-jee-AHNG)
Hejira (heh-JEER-ruh)
Helsinki (hel-SINKEE)
Hema (HAY-muh)
Henan (heh-NAHN)
Herat (heh-RAHT)
Herodotus (heh-RODDA-tuss)
Hesse (HESS)
Hexi (huh-SHEE)
Hidrovia (ee-DROH-vee-uh)
Hierarchical (hire-ARK-uh-kull)
Himachal Pradesh (huh-MAHTCH-ull pruh-DESH)
Himalayas (him-AHL-yuzz/himma-LAY-uzz)
Hindu (HIN-doo)
Hindu Kush (hin-doo KOOSH)
Hindustan (hin-doo-STAHN)
Hindustani (hin-doo-STAHN-nee)
Hindutva (hin-DOOT-vah)
Hippocrates (hih-POCK-ruh-teez)
Hiroshima (hirra-SHEE-muh/huh-ROH-shuh-muh)
Hispaniola (iss-pahn-YOH-luh)
Hizbollah (HIZZ-boh-luh)
Hohhot (huh-HOO-tuh)
Hokkaido (hoh-KYE-doh)
Holistic (hoh-LISS-tick)

Holocene (HOLLO-seen)
Hominid (HOM-ih-nid)
Homo sapiens (hoh-moh SAY-pea-enz)
Homogeneity (hoh-moh-juh-NAY-uh-tee)
Honduras (hon-DURE-russ)
Hongkou (hong-KOH)
Honiara (hoh-nee-AHR-ah)
Honolulu (honn-uh-LOO-loo)
Honshu (HONN-shoo)
Hooghly (HOO-glee)
Hormuz (hoar-MOOZE)
Hotan (HOH-TAN)
Houghton (HOH-t'n)
Houphouët-Boigny, Félix (oo-FWAY bwah-NYEE, fay-LEEKS)
Hsinchu [Xinzhu] (shin-JOO)
Hu Jintao (hoo-jin-DAU)
Huai (HWYE)
Huallaga (wah-YAH-gah)
Huancayo (wahn-KYE-oh)
Huang [He] (HWAHNG [huh])
Huangpu (hwahng-POO)
Hubei (hoo-BAY)
Hué (HWAY)
Huizhou (hway-JOH)
Humboldt (HUMM-bolt)
Humus (HYOO-muss)
Hunan (hoo-NAHN)
Husayn (hoo-SINE)
Hussein, Saddam (hoo-SAIN, suh-DAHM)
Hutu (HOO-too)
Hwange (WAHNG-ghee)
Hyderabad (HIDE-uh-ruh-bahd)
Hyundai (HUN-dye)
Ibadan (ee-BAHD'N)
Iberia (eye-BEERY-uh)
Ibn Saud (ib'n sah-OOD)
Ibo [land] (EE-boh [land])
Ife (EE-fay)
Iguaçu (EE-gwah-soo)
Iki (EE-kee)
Ikoyi (ee-KOY-yee)
Île de France (EEL duh-FRAWSS)
Île de la Cité (EEL duh-la-see-TAY)
Illyrian (ih-LEER-ree-un)
Ilmen (yill-MEN)
Ilyich (ILL-yitch)
Imam (ih-MAHM)
Inacio da Silva, Luiz (ee-NAH-see-oh dah-SIL-vah, loo-EESE)
Inchon (int-CHON)
Indonesia (indo-NEE-zhuh)
Ingush (in-GOOSH)
Ingushetiya (in-goo-SHETTY-uh)
Inkatha (in-KAH-tah)
Inle (IN-lay)
Interahamwe (in-terra-HAHM-way)
Interdigitate (intuh-DID-juh-tate)
Intramuros (in-trah-MOOR-rohss)
Inuit (IN-yoo-it)
Ipanema (ee-pah-NAY-muh)
Iquitos (ih-KEE-tohss)
Iran (ih-RAN/ih-RAHN)
Iranian (ih-RAIN-ee-un)
Iraq (ih-RAK/ih-RAHK)
Iraqi (ih-RAKKY)
Irbil (EAR-beel)
Irian Jaya (IH-ree-ahn JYE-uh)
Irkutsk (ear-KOOTSK)
Irrawaddy (ih-ruh-WODDY)
Irredentism (irruh-DEN-tism)
Irtysh (ear-TISH)
Isfahan (iz-fuh-HAHN)
Iskenderun (iz-ken-duh-ROON)
Islam (iss-LAHM)
Islamabad (iss-LAHM-uh-bahd)
Islington (IZZ-ling-tunn)
Ismaili (izz-MYE-lee)
Isohyet (EYE-so-hyatt)
Israel (IZ-rail)
Issas (ih-SAHZ)
Istanbul (iss-tum-BOOL)
Isthmus/isthmian (ISS-muss/ ISS-mee-un)
Itaipu (ee-TYE-pooh)
Ituri (ih-TOORY)
Ivanovo (ee-VAH-nuh-voh)
Ivoirian (ih-VWAH-ree-un)
Izhevsk (EE-zheffsk)
Jabal (JAB-ull)
Jabotabek (juh-BOH-tah-bek)
Jaffna (JAHF-nuh)
Jains (JYE-nz)
Jakarta (juh-KAHR-tuh)
Jakota (juh-KOH-tuh)
Jalisco (huh-LISS-koh)
Jamaica (juh-MAKE-uh)
Jammu (JUH-mooh)
Jamshedpur (JAHM-shed-poor)
Janjaweed (JAN-juh-weed)
Java (JAH-vuh)
Jawa (JAH-vuh)
Jawanese (jah-vuh-NEEZE)
Jayapura (jye-uh-POOR-ruh)
Jazirah (juh-ZEER-uh)
Jebel Ali (JEH-bel ah-LEE)
Jemaah Islamiah (jeh-MAH ee-slah-MEE-uh)
Jerusalem (juh-ROO-suh-lum)
Jesuit (JEH-zoo-it)
Jharkhand (JAHR-kahnd)
Jiang Zemin (jee-AHNG zuh-MIN)
Jiangsu (jee-ahng-SOO)
Jiangxi (jee-ahng-SHEE)
Jihad (jee-HAHD)
Jilin (jee-LIN)
Jilong (jee-LUNG)
Jinmen Dao (JIN-MEN-dau)
Joao, Dom (ZHWOW, dom)
Johannesburg (joh-HANNIS-berg)
Johor (juh-HOAR)
Jomon (JOH-mon)
Jubail (joo-BILE)
Judaic (joo-DAY-ick)
Judaism (JOODY-ism)
Junggar (JOONG-gahr)
Junta (HOON-tah)
Jurong (juh-RONG)
Jutland (JUT-lund)
Kabardino-Balkar [iya] (kabber-DEE-noh bawl-KAR [ree-uh])
Kabila, Laurent (kuh-BEE-luh law-RAH)
Kabol (KAH-bull)
Kabylia (kuh-BEE-lee-uh)
Kachins (kuh-CHINZ)
Kaduna (kah-DOO-nah)
Kai Tak (KYE TAK)
Kaifeng (kye-FUNG)
Kakadu (KAH-kuh-doo)
Kalaallit Nunaat (kuh-LAHT-lit noo-NAT)
Kalahari (kalla-HAH-ree)
Kalenjin (kuh-LEN-jin)
Kalgoorlie (kal-GHOOR-lee)
Kalimantan (kalla-MAN-tan)
Kaliningrad (kuh-LEEN-in-grahd)
Kalmyk (KAL-mik)
Kalmykiya (kal-MIK-ee-uh)
Kamba (KAHM-bah)
Kambalda (kahm-BAHL-duh)
Kamchatka (kum-CHAHT-kuh)
Kamehameha (kah-MAY-hah-may-hah)
Kampala (kahm-PAH-luh)
Kampuchea (kahm-pooh-CHEE-uh)
Kanaks (KAH-nahkss)
Kanarese (KAHN-uh-reece)
Kandahar (KAHN-duh-hahr)
Kannada (KAHN-uh-duh)
Kano (KAH-noh)
Kansai (KAHN-SYE)
Kanto (KAN-toh)
Kaohsiung [Gaoxiong] (GAU-see-OOHNG)
Kara (KAHR-ruh)
Kara Kum (kahr-ruh KOOM)
Karachay (kah-ruh-CHYE)
Karachayevo-Cherkessiya (kahra-CHAH-yeh-vuh cheer-KESS-ee-uh)
Karachi (kuh-RAH-chee)
Karafuto (kahra-FOO-toh)
Karaganda (karra-gun-DAH)
Karakalpak (karra-kal-PAK)
Karakoram (kahra-KOR-rum)
Karamay (kah-RAH-may)
Kareliya (kuh-REE-lee-uh)
Karen (kuh-REN)
Karens (kuh-RENZ)
Kariba (kuh-REE-buh)
Karnataka (kahr-NAHT-uh-kuh)
Karpas (KAHR-pahss)
Karzai, Hamid (KAR-zye, HAH-mid)
Kashagan (KAH-shah-gahn)
Kashgar (KAHSH-gahr)
Kashi (KAH-shee)
Kashmir (KASH-meer)
Katanga (kuh-TAHNG-guh)
Kathmandu (kat-man-DOOH)
Katowice (kah-toh-VEE-tsuh)
Kattegat (KAT-ih-gat)

Kauai (KAU-eye)
Kawasaki (kah-wah-SAH-kee)
Kazakh (KUZZ-uck)
Kazakhstan (KUZZ-uck-STAHN/ KUZZ-uck-stahn)
Kazan (kuh-ZAHN)
Kefamenanu (kuh-fahm-uh-NAH-noo)
Kelang (kuh-LAHNG)
Kelantan (keh-LAHN-tahn)
Kemal, Mustafa (keh-MAHL, moo-stah-FAH)
Kenya (KEN-yuh)
Kerala (KEH-ruh-luh)
Keratsinion (keh-rut-SINNY-onn)
Kerinci (kuh-REEN-chee)
Khabarovsk (kuh-BAHR-uffsk)
Khakassiya (kuh-KAHSS-ee-uh)
Khalistan (kahl-ee-STAHN)
Khan, Genghis (KAHN, JING-guss)
Khanate (KAHN-ate)
Khartoum (kar-TOOM)
Khatami (kah-TAH-mee)
Khiva (KEE-vuh)
Khmer [Rouge] (kuh-MAIR [ROOZH])
Khoi (KHOY)
Khoikhoi (KHOY-khoy)
Khoisan (khoy-SAHN)
Kholmsk (KAWLMSK)
Khomeini (hoh-MAY-nee)
Khorat (koh-RAHT)
Khulna (KOOL-nah)
Khurasan (koor-uh-SAHN)
Khuzestan (KOOH-ih-stahn)
Khyber (KYE-burr)
Kiev (Kyyiv) (KEE-yeff)
Kievan (kee-EVAN)
Kigali (kih-GAH-lee)
Kikuyu (kee-KOO-yoo)
Kikwit (KIK-wit)
Kilimanjaro (kil-uh-mun-JAH-roh)
Kilinochchi (kih-luh-NOCK-chee)
Kilwa (KEEL-wah)
Kinabalu (kin-ah-BAH-loo)
Kindu (KIN-doo)
Kinki (kin-KEE)
Kinmen (kin-MEN)
Kinneret (kin-uh-RETH)
Kinshasa (kin-SHAH-suh)
Kiosk (KEE-osk)
Kirghiz (keer-GEEZE)
Kiribati (KIH-ruh-bahss)
Kirkuk (keer-KOOK)
Kiruna (kih-ROONA)
Kisangani (kee-sahn-GAH-nee)
Kismaayo (keese-MYE-oh)
Kitakyushu (kee-TAH-KYOO-shoo)
Kivu (KEE-voo)
Kiwi (KEE-wee)
Klaipeda (KLYE-puh-duh)
Klyuchevskaya (klee-ooh-CHEFF-skuh-yuh)
Koala (kuh-WAH-luh)
Kobe (KOH-bay)
Kolkata (kol-KUTTA)
Kolkhoz (KOLL-koze)
Kolok (KOH-lok)
Kolyma (koh-LEE-mah)
Komi (KOH-mee)
Kompong Som (kahm-pong SAWM)
Kompong Thom (kahm-pong TAWM)
Komsomolsk (komm-suh-MAWLSK)
Kongfuzi (kung-FOODZEE)
Kongzi (KUNG-dzee)
Königsberg (KAY-nix-bairk)
Konkan (KAHNG-kun)
Köppen (KER-pun)
Koran (kaw-RAHN)
Kordofan (kor-duh-FAN)
Korea (kuh-REE-uh)
Korla (KOOR-LAH)
Koryakiya (kor-YAH-kee-uh)
Koryakskaya (kor-YAHK-skuh-yuh)
Kosciusko (kuh-SHOO-skoh)
Kosovar (KAW-suh-vahr)
Kosovo (KAW-suh-voh)
Kostunica, Vojislav (koss-too-NEE-chah, VOY-slahv))
Kota Kinabalu (KOH-tuh kin-ah-BAH-loo)
Kourou (koo-ROO)
Koutoubia (koo-TOO-bee-yuh)
Kowloon (kau-LOON)
Kra Isthmus (KRAH ISS-muss)
Krabi (KRAH-bee)
Krakow (KRAH-koov)
Krasnodar (KRASS-nuh-dahr)
Krasnovodsk (kruzz-noh-VAUGHTSK)
Krasnoyarsk (krass-nuh-YARSK)
Kribi (KREE-bee)
Krivoy Rog (krih-voy-ROAG)
Kronstadt (KROAN-shtaht)
Kruger (KROO-guh)
Krugersdorp (KROO-guzz-dorp)
Kryvyy Rih (kree-VEE-REE)
Kuala Lumpur (KWAHL-uh LOOM-poor)
Kublai Khan (koob-lye KAHN)
Kufra (KOO-fruh)
Kumartuli (koo-MAR-too-lee)
Kunashiri (koo-NAH-shuh-ree)
Kunlun (KOON-LOON)
Kunming (koon-MING)
Kura (KOOR-uh)
Kurd [istan] (KERD [uh-stahn])
Kure (KOOH-ray)
Kurile (KYOOR-reel)
Kuroshio (koo-roh-SHEE-oh)
Kush (KOOSH)
Kushites (KOO-sheits)
Kuta (KOO-tuh)
Kuwait (koo-WAIT)
Kuybyshev (KWEE-buh-sheff)
Kuzbas (kooz-BASS)
Kuznetsk (kooz-NETSK)
Kwangju (GWONG-JOO)
Kwazulu (kwah-ZOO-loo)
Kyongju (GYOONG-JOO)
Kyongsang (GYOONG-SAHN)
Kyoto (kee-YOH-toh)
Kyrgyz (KEER-geeze)
Kyrgyzstan (KEER-geeze-stahn)
Kyushu (kee-YOO-shoo)
Kyyiv [See Kiev]
La Coruña (lah-kor-ROON-yah)
La Défense (lah-day-FAWSS)
La Paz (lah-PAHZ)
La Violencia (lah vee-oh-LENN-see-uh)
Ladakh (luh-DAHK)
Ladino (luh-DEE-noh)
Laem Chabang (lay-EMM chuh-BAHNG)
Lafaiete (lah-fuh-YAY-tuh)
Lagos (LAY-gohss)
Lahore (luh-HOAR)
Lamanai (LAH-mah-nay)
Lancang (LAHN-ZAHNG)
Land (LAHNT)
Länder (LEN-derr)
Lantau (LAHN-DAU)
Lanzhou (lahn-JOH)
Lao (LAU)
Laos (LAUSS)
Laotian (lay-OH-shun)
Las Colinas (lahss-koh-LEE-nahss)
Latvia (LATT-vee-uh)
Lautoka (lau-TOH-kuh)
Laval (lah-VAHL)
Legume (LEG-gyoom)
Le Havre (luh-HAHV)
Leipzig (LYPE-sik)
Leith (LEETH)
Leizhou (lay-JOH)
Lemur (LEE-mer)
Lena (LAY-nuh)
Lendu (LEN-doo)
Lenin (LENNIN)
León (lay-OAN)
Lesotho (leh-SOO-too)
Levant (luh-VAHNT)
Lhasa (LAH-suh)
Li (LEE)
Lianyungang (lee-en-yoong-GAHNG)
Liao (lee-AU)
Liaodong (lee-au-DUNG)
Liaoning (lee-au-NING)
Liberia (lye-BEERY-uh)
Libreville (LEE-bruh-veel)
Lichens (LYE-kenz)
Liechtenstein (LIK-ten-shtine)
Liège (lee-EZH)
Lima (LEE-muh)
Limpopo (lim-POH-poh)
Lingala (ling-GAH-lah)
Lingua franca (LEENG-gwuh FRUNK-uh)
Linguae francae (LEENG-gwee FRUNK-kee)
Lisas (LEE-suss)
Litani (lih-TAH-nee)

Lithuania (lith-oo-AINY-uh)
Littoral (LIT-oh-rull)
Livingstonia (lih-ving-STOH-nee-uh)
Ljubljana (lee-oo-blee-AHNA)
Llama (LAH-muh)
Llano [*s*] (YAH-noh [ss])
Lobito (loh-BEE-toh)
Loess (LERSS)
Loihi (loh-EE-hee)
Loire (luh-WAHR)
Lombardy (LOM-bar-dee)
Lombok (LAHM-bahk)
Lourenço Marques (loh-REN-soh mahr-KESS)
Louvre (LOOV)
Lozi (LOH-zee)
Lualaba (loo-uh-LAH-buh)
Luanda (loo-AN-duh)
Lubumbashi (loo-boom-BAH-shee)
Lucknow (LUCK-nau)
Luhya (LOO-yuh)
Lujiazui (loo-jee-ah-ZWAY)
Lula da Silva, Luiz Inácio (LOO-lah dah SIL-vah, loo-EESE ee-NAH-see-oh)
Luleå (LOO-lee-oh)
Lund (LOOND)
Luo (LOO-oh)
Lusitania (loo-sih-TANE-yuh)
Luxembourg (LUX-em-borg)
Luxor (LUK-soar)
Luzon (loo-ZAHN)
Lyon (lee-AW)
Maale (MAH-lay)
Maas (MAHSS)
Maasai (muh-SYE)
Maastricht (mah-STRICT)
Mabo (MAY-boh)
Macau (muh-KAU)
Macedonia (massa-DOH-nee-uh)
Machu Picchu (MAH-choo PEEK-choo)
Mackinder, Halford (muh-KIN-der, HAL-ferd)
Mactan (mahk-TAHN)
Macumba (mah-KOOM-bah)
Madagascar (madda-GAS-kuh)
Madhya Pradesh (mahd-yuh-pruh-DESH)
Madras (muh-DRAHSS)
Madrid (muh-DRID)
Madura (muh-DOORA)
Madurai (mahd-uh-RYE)
Madurese (muh-dooh-REECE)
Magaz (muh-GAHZ)
Magdalena (mahg-dah-LAY-nah)
Magellan (muh-JELL-un)
Maghreb (mahg-GRAHB)
Magyar (MAG-yahr)
Mahakam (MAH-hah-kahm)
Maharajah (mah-hah-RAH-juh)
Maharashtra (mah-huh-RAH-shtra)
Mahathir Mohamad (MAH-hah-theer moh-HAH-mud)
Maize (MAYZ)
Makhachkala (muh-kahtch-kuh-LAH)
Makkah (MEK-ah)
Makung (mah-GOONG)
Mala Strana (mah-lah-STRAH-nah)
Malabar (MAL-uh-bahr)
Malabo (muh-LAH-boh)
Malacca (muh-LAH-kuh)
Málaga (MAHL-uh-guh)
Malagasy (malla-GASSY)
Malaita(ns) (mah-LAY-tah [nz])
Malawi (muh-LAH-wee)
Malay (muh-LAY)
Malaya (muh-LAY-uh)
Malayalam (mal-uh-YAH-lum)
Malaysia (muh-LAY-zhuh)
Maldives (MAWL-deevz)
Mali (MAH-lee)
Malmö (MAHL-meh)
Maluku(s) (mah-LOO-koo[z])
Malvinas (mahl-VEE-nahss)
Manado (muh-NAH-doh)
Managua (mah-NAH-gwuh)
Manaus (muh-NAUSS)
Manchu (man-CHOO)
Manchukuo (mahn-JOH-kwoh)
Manchuria (man-CHOORY-uh)
Mandalay (man-duh-LAY)
Mandarin (MAN-duh-rin)
Mandela (man-DELLA)
Manipur (man-uh-POOR)
Manitoba (manna-TOH-buh)
Manzanillo (mun-zuh-NEE-yoh)
Mao Zedong (MAU zee-DUNG)
Maori (MAH-aw-ree/MAU-ree)
Mapuche (mah-POO-chay)
Maputo (mah-POOH-toh)
Maquiladora (mah-kee-luh-DORR-uh)
Mara (MAHRA)
Maracaibo (mah-rah-KYE-boh)
Maracay (mah-rah-KYE)
Marajó (mah-rah-ZHOH)
Maranhão (mah-rahn-YAU)
Mariana (marry-ANNA)
Mariinsk (muh-ryee-YEENSK)
Marikana (mah-ree-KAH-nuh)
Maritsa (muh-REET-suh)
Mariy [-el] (MAH-ree [el])
Marmagao (marma-GAU)
Marmara (MAH-muh-ruh)
Marquesas (mahr-KAY-suzz)
Marrakech (mahr-uh-KESH)
Marseille (mar-SAY)
Marsupial (mar-SOOPY-ull)
Martinique (mahr-tih-NEEK)
Massif (mass-SEEF)
Matadi (muh-TAH-dee)
Matão (mah-TAU)
Mato Grosso (mutt-uh-GROH-soh)
Mato Grosso do Sul (mutt-uh-GROH-soh duh-SOOL)
Matsu (mah-TSOO)
Matsu Tao (mah-TSOO DAU)
Mauna Kea (mau-nuh-KAY-uh)
Mauritania (maw-ruh-TAY-nee-uh)
Mauritius (maw-REE-shuss)
Mauryan (MAW-ree-un)
Maya (MYE-uh)
Mayan (MYE-un)
Mazar-e-Sharif (mah-ZAHR-ee-shah-REEF)
Mbeki, Thabo (mm-BEH-kee, TAH-boh)
Mecca (MEK-ah)
Mecsek (MAH-chek)
Medan (MAY-dahn)
Medellín (meh-deh-YEEN)
Medina (muh-DEENA)
Megalopolis (meh-guh-LOPP-uh-liss)
Meghalaya (may-guh-LAY-uh)
Meghna (MAIG-nuh)
Meiji (may-EE-jee)
Mejía, Hipólito (meh-HEE-yah, ee-POH-lee-toh)
Mekong (MAY-kong)
Melaka (muh-LAH-kuh)
Melanesia (mella-NEE-zhuh)
Melbourne (MEL-bun)
Melilla (meh-LEE-yuh)
Mendoza (men-DOH-zah)
Menem, Carlos (MENNEM, KAHR-lohss)
Mengkabong (MENG-kah-bong)
Menshevik (MEN-shuh-vick)
Merauke (muh-RAU-kuh)
Mercosul (mair-koh-SOOL)
Mercosur (mair-koh-SOOR)
Mérida (MAY-ree-dah)
Meridian (meh-RIDDY-un)
Merina (meh-REE-nuh)
Meroe (MEH-roh-ay)
Meru (MAY-roo)
Mesa (MAY-suh)
Mesabi (meh-SAH-bee)
Meseta (meh-SAY-tuh)
Meshketian (mesh-KETTY-un)
Mesoamerica (MEZZOH-america)
Mesopotamia (messo-puh-TAY-mee-uh)
Mesquita (meh-SKEE-tuh)
Mestizo (meh-STEE-zoh)
Meuse (MERZZ)
Mezzogiorno (met-soh-JORR-noh)
Miao (m'YOW)
Michoacán (mee-chuh-wah-KAHN)
Micronesia (mye-kroh-NEE-zhuh)
Milan (mih-LAHN)
Millau (mee-LOH)
Milosevic, Slobodan (mih-LAW-suh-vitch, SLOH-boh-dahn)
Minahasa (MEE-nah-hah-sah)
Minaret (MINNA-ret)
Minas Gerais (MEE-nuss zhuh-RICE)
Mindanao (min-duh-NAU)
Minh, Ho Chi (MINN, hoh chee)
Minifundia (minny-FOON-dee-uh)
Mirador(es) (meera-DOAR [ayss])
Miraflores (meera-FLAW-rayss)
Miskito (mih-SKEE-toh)
Mitrovica (mee-troh-VEET-sah)

Mitú (mee-TOO)
Mius (mee-OOS)
Mizoram (mih-ZOR-rum)
Mobutu (moh-BOO-too)
Moçambique (moh-sum-BEEK)
Mogadishu (moo-gah-dee-SHOH)
Mohenjo Daro (moh-hen-joh-DAHRO)
Moi (MOY)
Mojave (moh-HAH-vee)
Moldavia (moal-DAY-vee-uh)
Moldova (moal-DOH-vuh)
Moluccans (muh-LUCK-unz)
Molybdenum (muh-LIB-dun-um)
Mombasa (mahm-BAHSSA)
Mongol (MUNG-goal)
Mongolia (mung-GOH-lee-uh)
Monogamy (muh-NOG-ah-mee)
Montagnards (MON-tun-yardz)
Montaña (mon-TAHN-yah)
Montego (mon-TEE-goh)
Montenegro (mon-teh-NAY-groh)
Monterrey (mon-teh-RAY)
Montevideo (moan-tay-vee-DAY-oh)
Montparnasse (mawn-pahr-NAHSS)
Montpellier (maw-pell-YAY)
Montreal (mun-tree-AWL)
Montserrat (mont-seh-RAHTT)
Moped (MOH-ped)
Moravia (more-RAY-vee-uh)
Mordoviya (mor-DOH-vee-uh)
Moscow (MAW-skau)
Moselle (moh-ZELL)
Mosque (MOSK)
Mount Isa (mount EYE-suh)
Mpumalanga (mm-pooma LAHNG-guh)
Muara Muntai (moo-WAH-rah MOON-tye)
Mugabe (moo-GAH-bay)
Muglad (moo-GLAHD)
Muhajir (MOO-hah-jeer)
Muhammad (moo-HAH-mid)
Mujahideen (moo-jah-heh-DEEN)
Mulatto (moo-LAH-toh)
Mullah (MOOL-ah)
Multan (mool-TAHN)
Mulukas (MAH-loo-kooz)
Mumbai (MOOM bye)
Munich (MYOO-nik)
Murmansk (moor-MAHNTSK)
Murray (MUH-ree)
Musandam (muh-SAND-um)
Muscat (MUH-skaht)
Muscovy (muh-SKOH-vee)
Museveni (moo-SAY-veh-nee)
Musharraf, Pervez (moo-SHAR-rahf, PER-vez)
Muslim (MUZZ-lim)
Mutrah (MUTT-truh)
Muuga (MOO-guh)
Muwahhidun (moo-wah-hih-DOON)
Myanmar (mee-ahn-MAH)
Mymensingh (mye-men-SING)
Naga [land] (NAHGA [-land])
Nagasaki (nah-guh-SAHKEE)
Nagorno-Karabakh (nuh-GORE-noh KAH-ruh-bahk)
Nagoya (nuh-GOYA)
Nagpur (NAHG-poor)
Nairobi (nye-ROH-bee)
Najaf (nuh-JAHF)
Nakhodka (nuh-KAUGHT-kuh)
Nakhon Sawan (NUH-KAWN suh-WOON)
Nam Bo (nahm-BOH)
Namib (nah-MEEB)
Namibia (nuh-MIBBY-uh)
Nanjing (nahn-ZHING)
Naoroji (nau-ROH-jee)
Nara (NAHRA)
Nassau (NASS-saw)
Nasser, Gamal Abdel (NASS-er, guh-MAHL AB-dul)
Natal (nuh-TAHL)
Nauru (nah-OO-roo)
Nautical (NAW-dih-kull)
Navi (NAH-vee)
Naxcivan (nah-kee-chuh-VAHN)
Nazi (NAH-tsee)
Nazran (nahz-RAHN)
Ndebele (en-duh-BEH-leh)
N'Djamena (en-jah-MAY-nah)
Ndola (en-DOH-luh)
Negev (NEH-ghev)
Negro (NAY-groh)
Nehru, Jawaharlal (NAY-roo, juh-WAH-hur-lahl)
Nei Mongol (nay-MUNG-goal)
Nepal (nuh-PAHL)
Neva (NAY-vuh)
Nevis (NEE-vuss)
Nevsky Prospekt (NEFF-skee PROSS-spekt)
New Caledonia (noo-kalla-DOAN-yuh)
New Guinea (noo-GHINNY)
Newfoundland (NYOO-fun-lund)
Ngodi (eng-GOH-dee)
Ngorongoro (eng-gore-ong-GORE-roh)
Nha Trang (nah-TRAHNG)
Niamey (NEE-ah-may)
Nicaragua (nick-uh-RAH-gwuh)
Niger [Country] (nee-ZHAIR)
Niger [River] (NYE-jer)
Nigeria (nye-JEERY-uh)
Niger-Kordofanian (NYE-jer kor-doh-FAN-ee-un)
Nilotic (nye-LODDIK)
Ningbo (ning-BWOH)
Ningxia Hui (NING-shee-AH HWAY)
Nistru (NEE-stroo)
Niue (nee-OOH-ay)
Nizhniy Novgorod (NIZH-nee NAHV-guh-rahd)
Nkrumah, Kwame (en-KROO-muh, KWAH-mee)
Nobi (NOH-bee)
Nord-Pas de Calais (NORD pah-duh-kah-LAY)
Noriega (noar-ree-AY-gah)
Norilsk (nuh-REELSK)
Norte (NOR-tay)
Notre Dame (NOH-truh DAHM)
Nouakchott (noo-AHK-shaht)
Nouméa (noo-MAY-uh)
Nouveau Quebec (noo-VOH kwih-BECK)
Nova Scotia (nova-SKOH-shuh)
Novaya Zemlya (NOH-vuh-yuh zem-lee-AH)
Novgorod (NAHV-guh-rahd)
Novi Arbat (NOH-vee AHR-baht)
Novokuznetsk (noh-voh-kooz-NETSK)
Novorossiysk (noh-voh-ruh-SEESK)
Novosibirsk (noh-voh-suh-BEERSK)
Nubia (NOO-bee-uh)
Nueva Mendez (noo-AY-vuh MEN-dez)
Nuevo León (noo-AY-voh lay-OAN)
Nunavut (NOON-uh-voot)
Nungambakkam (noon-GAHM-bah-KAHM)
Nuuk (NEWK)
Nyasaland (nye-ASSA-land)
Nyika (nye-YEE-kuh)
Oahu (uh-WAH-hoo)
Oaxaca (wuh-HAH-kuh)
Obasanjo (oh-BAH-sahn-joh)
Oblast (OB-blast)
Obsidian (ob-SIDDY-un)
Occidental (oak-see-den-TAHL)
Ocho Rios (oh-choh REE-ohss)
Ochre (OH-ker)
Ocussi-Ambeno (oh-KOOH-see AHM-bay-noh)
Oder-Neisse (OH-der NEISS)
Odesa (oh-DESSA)
Ogaden (oh-gah-DEN)
Ogoni (oh GOH-nee)
Oguz (uh-GOOZ)
Ohmae, Kenichi (OH-may, keh-NEE-chee)
Oirot (AW-ih-rut)
Oise (WAHZ)
Okavango (oh-kuh-VAHNG-goh)
Okhotsk (oh-KAHTSK)
Okrug (AW-krook)
Olduvai (OLE-duh-way)
Oman (oh-MAHN)
Omdurman (om-duhr-MAHN)
Omote Nippon (oh-MOH-tay NIP-on)
Oran (aw-RAHN)
Orangutan (aw-RANG-gyoo-tan)
Ordos (ORD-uss)
Ordzhonikidze (or-johnny-KIDD-zuh)
Øresund (ERR-uh-sun)
Organisasi Papua Merdeka (or-kah-nee-SAH-see pahp-OO-uh mair-DAY-kah)
Oriental (orry-en-TAHL)
Oriente (orry-EN-tay)
Orinoco (orry-NOH-koh)
Orissa (aw-RISSA)

Oromo (AW-ruh-moh)
Orontes (aw-RAHN-teeze)
Osaka (oh-SAH-kuh)
Osirak (OH-see-rahk)
Öskemen (ERSKEE-min)
Oslo (OZ-loh)
Oslofjord (OZ-loh-fyord)
Osman (oz-MAHN)
Osorno (oh-SOAR-noh)
Ossetia (oh-SEE-shuh)
Ossetians (oh-SEE-shunz)
Ossies (OH-seez)
Ostrava (AW-struh-vuh)
Otavalo (oh-tah-VAH-loh)
Ottawa (OTTA-wuh)
Oweinat (oh-WEE-naht)
Oxisol (OXY-soll)
Oyo (OH-joh)
Padania (pah-DAHN-yah)
Pago Pago (PAH-goh PAH-goh)
Pagoda (puh-GOH-duh)
Pahlavi, Shah Muhammad Reza (puh-LAH-vee, shah mooh-HAH-mid RAY-zuh)
Pakhtuns (puck-TOONZ)
Pakistan (PAH-kih-stahn)
Paktia (pahk-TEE-uh)
Palau (puh-LAU)
Palawan (pah-LAH-wahn)
Pale (PAH-lay)
Palembang (pah-LEM-bahng)
Palenque (puh-LENG-kay)
Palestine (PAL-uh-stine)
Palestinian (pal-uh-STINNY-un)
Palk (PAWK)
Palo Alto (PALLOH AL-toh)
Palu (pah-LOO)
Pamir (pah-MEER)
Pamirs (pah-MEERZ)
Pampa (PAHM-pah)
Panacea (pan-uh-SEE-uh)
Panache (pah-NAHSH)
Panaji (pah-NAH-jee)
Pangaea (pan-GAY-uh)
Papeete (pahp-ee-ATE-tee)
Papua New Guinea (pahp-OO-uh noo-GHINNY)
Papua [ns] (pahp-OO-uh [-unz])
Papuan (pahp-OO-un)
Pará (puh-RAH)
Paracas (pah-RAH-kuss)
Paraguay (PAHRA-gwye)
Paraíba (pah-rah-EE-buh)
Paramillos (pah-rah-MEE-yohss)
Paraná (pah-rah-NAH)
Paria (PAHR-yah)
Pariah (puh-RYE-uh)
Parliament (PAR-luh-ment)
Paseo del Norte (pah-SAY-oh del NOR-tay)
Pashtuns [see Pushtuns]
Pasig (PAH-sig)
Patagonia (patta-GOH-nee-uh)
Pathan (puh-TAHN)
Pattani (pah-TAH-nee)
Pattaya (puh-TYE-uh)
Paulista [*s*] (pow-LEASH-tah [ss])
Pechora (peh-CHORE-ruh)
Pedro Miguel (PAY-droh mee-GWELL)
Peking (pea-KING)
Peloponnesus (pelloh-puh-NEEZE-uss)
Pelourinho (peh-loo-REEN-yoh)
Penghu (pung-HOO)
Peón (pay-OAN)
Peones (pay-OH-nayss)
Perejil (peh-reh-HEEL)
Perestroika (perra-STROY-kuh)
Perm (PAIRM)
Permyakiya (pairm-YAH-kee-uh)
Pernambuco (pair-nahm-BOO-koh)
Perón, Juan (puh-ROAN, WAHN)
Peronista (peh-roh-NEE-stah)
Persepolis (per-SEPP-uh-luss)
Persia (PER-zhuh)
Peshawar (puh-SHAH-wahr)
Peso (PAY-soh)
Petén (peh-TEN)
Petrograd (PETTRO-grahd)
Petronas (peh-TROH-nuss)
Petropavlovsk-Kamchatskiy (pit-ruh-PAHV-lufsk kahm-CHAHT-skee)
Pettah (PET-uh)
Phaleron (fuh-LEER-un)
Pharaonic (fair-ray-ONNICK)
Philippines (FILL-uh-peenz)
Phnom Penh (puh-NOM PEN)
Phoenicians (fuh-NEE-shunz)
Phu Bia (POO-bee-uh)
Phuket (POO-KETT)
Physiography (fizzy-OGG-ruh-fee)
Phytogeography (FYE-toh-jee-OG-gruh-fee)
Piazza (pea-AH-tsuh)
Piedmont (PEED-mont)
Pietermaritzburg (peeh-tuh-MAHR-utss-berg)
Pilipino (pill-uh-PEA-noh)
Pinang (puh-NANG)
Pinatubo (pin-uh-TOO-boh)
Pinheiros (PEEN-yay-ross)
Pinochet (pee-noh-SHAY)
Pinyin (pin-YIN)
Piraeus (puh-RAY-uss)
Pisac (pea-SAHK)
Pitcairn (PITT-kairn)
Pizarro, Francisco (pea-SAHRO, frahn-SEECE-koh)
Place de la Concorde (PLAHSS duh lah kon-KORD)
Placer [ing] (PLASS-uh [ring])
Plata, Rio de la (PLAH-tah, REE-oh day-lah)
Platypus (PLATT-uh-pus)
Plaza Azoguejo (PLAH-sah ah-soh-GAY-hoh)
Plaza Bolívar (PLAH-sah boh-LEE-vahr)
Plaza de Armas (PLAH-sah day AR-mahss)
Plaza de Mayo (PLAH-sah day MYE-oh)
Plaza de San Martin (PLAH-sah day sahn mahr-TEEN)
Plaza Mayor (PLAH-sah mye-YOAR)
Pleistocene (PLY-stoh-seen)
Plymouth (PLIH-muth)
Pointe-Noire (pwahnt-nuh-WAHR)
Politburo (POLLIT-byoor-roh)
Polonoroeste (POLLOH-nuh-roh-ESS-tee)
Polygamy (puh-LIG-uh-mee)
Polynesia (polla-NEE-zhuh)
Polytheism (polly-THEE-izm)
Pontianak (pahn-tee-AH-nahk)
Popocatépetl (poh-poh-kah-TEH-peh-til)
Port Moresby (port MORZ-bee)
Port Said (port-sah-EED)
Port-au-Prince (por-toh-PRANSS)
Porteños (por-TAIN-yohss)
Pôrto Alegre (POR-too uh-LEG-ruh)
Pôrto Velho (POR-too VELL-yoo)
Posavina (poh-suh-VEENA)
Potgietersrus (POT-gheeters-roost)
Potosí (poh-toh-SEE)
Povolzhye (puh-VOLL-zhuh)
Poza Rica (POH-zah REE-kuh)
Prague (PRAHG)
Prayagraj (PRAY-ag-rahdge)
Pretoria (prih-TOR-ree-uh)
Primorskiy (pree-MOHR-skee)
Príncipe (PREEN-see-pea)
Progeny (PRAH-juh-nee)
Prokofiev (pruh-KOFFEY-ev)
Proselytism (PRAH-sell-eh-tizm)
Provence (pro-VAHNSS)
Prudhoe (PROO-doh)
Prut (PRROOT)
Pudong (poo-DUNG)
Puebla (poo-EBB-luh)
Puerto Barrios (pwair-toh bah-REE-uss)
Puerto Caldera (pwair-toh kahl-DERRA)
Puerto Rico (pwair-toh REE-koh)
Puerto Montt (pwair-toh-MAWNT)
Puget (PYOO-jet)
Punjab (pun-JAHB)
Punjabi (poon-JAH-bee)
Punta Arenas (POON-tah ah-RAY-nahss)
Punta del Este (POON-tah del ESS-tay)
Pushtun (PAH-shtoon)
Putin, Vladimir (POO-t'n, VLAH-duh-meer)
Putrajaya (poo-truh-JYE-UH)
Putumayo (poo-too-MYE-oh)
Pyatigorsk (pyih-tyih-GORSK)
Pygmy (PIG-mee)
Pyongyang (pea-AWHNG-yahng)
Pyrenees (PEER-uh-neez)
Qaddafi, Muammar (guh-DAHFI, MOO-uh-mar)
Qaidam (CHYE-DAHM)
Qanat (KAH-naht)
Qaraghandy (kah-rah-GONDY)
Qasim (kah-SEEM)

Qatar (KOTTER)
Qattarah (kuh-TAR-ruh)
Qiang (chee-AHNG)
Qin (CHIN)
Qing (CHING)
Qingdao (ching-DAU)
Qinghai (ching-HYE)
Qingyi (ching-YEE)
Qinhuangdao (chin-hwahng-DAU)
Qiqihar (chee-CHEE-har)
Qoraqalpog (kora-kal-PAHG)
Quaternary (kwuh-TER-nuh-ree)
Quebec (kwih-BECK)
Quebecer (kwih-BECK-er)
Québécois (kay-beh-KWAH)
Quebracho (kay-BROTCH-oh)
Quechua (KAYTCH-wah)
Quemoy (keh-MOY)
Quetzal (kay-DZAHL)
Quezon (KAY-soan)
Quinary (KWYE-nuh-ree)
Quito (KEE-toh)
Quran (kor-RAHN)
Raisina (rye-SEENA)
Raj (RAHDGE)
Rajang (rah-JAHNG)
Rajasthan (RAH-juh-stahn)
Rajin-Sonbong (RAH-jin SAWM-bong)
Rajiv (ruh-ZHEEV)
Rajshahi (rahdge-SHAH-hee)
Ramadan (rahma-DAHN)
Randstad (RUND-stud)
Rangoon (rang-GOON)
Rangpur (RHANG-poor)
Rarotonga (rarra-TAHNG-guh)
Rashid (ruh-SHEED)
Ratzel, Friedrich (RAHT-sull, FREED-rish)
Ravenstein (RAVVEN-steen)
Rawalpindi (rah-wull-PIN-dee)
Real (ray-AHL)
Recife (ruh-SEE-fuh)
Reg (REGG)
Réunion (ray-yoon-YAW)
Reykjavik (RAKE-yah-veek)
Rhaeto-Romansch (RAY-shoh roh-MAHN-ssh)
Rhine (RYNE)
Rhodesia (roh-DEE-zhuh)
Rhône-Alpes (ROAN AHLP)
Rhône-Saône (ROAN say-OAN)
Riau (REE-au)
Riga (REEGA)
Rio Branco (REE-oh BRUNG-koh)
Rio de Janeiro (REE-oh day zhah-NAIR-roh)
Rio Grande do Norte (REE-oh GRUN-dee duh NORTAH)
Rio Grande do Sul (REE-oh GRUN-dee duh SOOL)
Rioja (ree-OH-hah)
Riverine (RIVER-reen)
Riyadh (ree-AHD)
Rocinho (haw-SEEN-yah)
Roma (ROH-muh)
Romania (roh-MAIN-yuh)
Rondônia (roh-DOAN-yuh)
Roraima (raw-RYE-muh)
Rosario (roh-SAH-ree-oh)
Rostow (ROSS-stoff)
Rouen (roo-AW)
Rub al Khali (roob ahl KAH-lee)
Rue Notre Dame (ROO noh-truh DAHM)
Rue Saint-Louis (ROO sah-loo-WEE)
Rue Saint Louis (roo-saw-loo-EE)
Ruhr (ROOR)
Ruijin (rway-JEEN)
Rumailah (roo-MYE-luh)
Rus (ROOSE)
Ruwenzori (roo-when-ZOHR-ree)
Rwanda (roo-AHN-duh)
Ryukyu (ree-YOO-kyoo)
Saami (SAH-mee)
Saarland (ZAHR-lunt)
Saba (SAY-buh/SAH-buh)
Sabah (SAHB-ah)
Sabinas (sah-BEE-nuss)
Saddhu (SAH-dooh)
Sahara (suh-HARRA)
Sahel (suh-HELL)
Saigon (sye-GAHN)
Saint Dominique (san-dommi-NEEK)
Saint-Denis (san-deh-NEE)
Sakartvelos (sah-KART-vuh-lohss)
Sakha (SAH-kuh)
Sakhalin (SOCK-uh-leen)
Salalah (sah-LAH-luh)
Salina Cruz (sah-LEE-nuh CROOZ)
Salinas (sah-LEE-nuss)
Salisbury (SAWLZ-buh-ree)
Salta (SAHL-tah)
Salvador (SULL-vuh-dor)
Salween (SAHL-wain)
Samara (suh-MAH-ruh)
Samarinda (sah-mah-RIN-dah)
Samarqand (sah-mahr-KAHND)
Samarra (sah-MAH-rah)
Samoa (suh-MOH-uh)
Samoan (suh-MOH-un)
Samsung (SAHM-SOONG)
San (SAHN)
San Cristóbal (sahn kree-STOH-bahl)
San Joaquin (san-wah-KEEN)
San José (sahn hoh-ZAY)
San Juan (sahn HWAHN)
San Pedro Sula (sahn pay-droh-SOO-luh)
San Rafael (sahn rah-fye-ELL)
San Martin, José de (sahn-mahr-TEEN, hoh-ZAY day)
San'a (suh-NAH)
Sandinista (sahn-dee-NEE-stuh)
Santa Catarina (SUN-tuh kuh-tuh-REE-nuh)
Santa Cruz (SAHN-tah CROOZ)
Santa Tecla (SAHN-tah TEK-lah)
Santiago (sahn-tee-AH-goh)
Santos (SUNT-uss)
Sanxia (sahn-SHAH)
São Francisco (sau frahn-SEECE-koh)
São Luis (sau loo-EECE)
São Paulo (sau PAU-loh)
São Tomé (sau TOH-may)
Sapporo (suh-POAR-roh)
Sarajevo (sahra-YAY-voh)
Saratov (suh-RAHT-uff)
Sarawak (suh-RAH-wahk)
Sardinia (sahr-DINNY-uh)
Sarmatian (sahr-MAY-shee-un)
Saskatchewan (suss-KATCH-uh-wunn)
Saudi Arabia (SAU-dee uh-RAY-bee-uh)
Sauer (SAU er)
Savimbi (suh-VIM-bee)
Savoy (suh-VOY)
Saxony (SAX-uh-nee)
Scania (SKAIN-yuh)
Schengen (SHENG-gen)
Schistosomiasis (shistoh-soh-MYE-uh-siss)
Scythian (SITH-ee-un)
Sedentary (SEDDEN-terry)
Seikan (say-KAHN)
Seine (SENN)
Semarang (seh-MAHR-rahng)
Semeru (suh-MAY-roo)
Semitic (seh-MITTICK)
Sendero Luminoso (sen-DARE-oh loo-mee-NOH-soh)
Senegal (sen-ih-GAWL)
Sen-en (sen-NENN)
Seoul (SOAL)
Sepoy (SEE-poy)
Sepulchre (SEP-ul-kurr)
Serbia (SER-bee-uh)
Serer (seh-RAIR)
Sergipe (SAIR-zhee-pay)
Serov (SAIR-roff)
Serra dos Carajás (SERRA doo kuh-ruh-ZHUSS)
Sertão (sair-TOWNG)
Sesame (SESS-uh-mee)
Seto (SET-oh)
Sevastopol (seh-VASS-toh-pawl)
Seville (suh-VILL)
Seychelles (say-SHELLZ)
Sha Mian (shah mee-AHN)
Shaanxi (shahn-SHEE)
Shaba (SHAH-bah)
Shan (SHAHN)
Shandong (shahn-DUNG)
Shanghai (shang-HYE)
Shang-Yin (SHANG-YIN)
Shantou (SHAHN-TOH)
Shanxi (shahn-SHEE)
Shari (SHAH-ree)
Sharia (SHAH-ree-uh)
Shatt al Arab (shot ahl uh-RAHB)
Sheikdom (SHAKE-dum)
Sheikh (SHAKE)

Shenyang (shun-YAHNG)
Shenzhen (shun-ZHEN)
Shevardnadze, Eduard (sheh-vart-NAHD-zeh, ED-wahrd)
Shi (SHERR)
Shia (SHEE-uh)
Shi'a (SHEE-uh)
Shihezi (shee-HAY-zee)
Shi'ism (SHEE-izm)
Shi'ite (SHEE-ite)
Shikoku (shick-KOH-koo)
Shikotan (shee-koh-TAHN)
Shimonoseki (shim-uh-noh-SECKEE)
Shiraz (shih-RAHZ)
Shirioshi (shih-ree-OH-shee)
Shogun (SHOH-goon)
Shona (SHOH-nuh)
Shostakovich (shosta-KOH-vitch)
Shwedogon (SHWAY-duh-gonn)
Siam (sye-AMM)
Siberia (sye-BEERY-uh)
Sichuan (zeh-CHWAHN)
Sicily (SISS-uh-lee)
Siddhartha (sid-DAHR-tuh)
Sidra (SIDD-ruh)
Siem Reap (SERM-REEP)
Sierra Madre Occidental (see-ERRA MAH-dray oak-see-den-TAHL)
Sierra Madre Oriental (see-ERRA MAH-dray aw-ree-en-TAHL)
Sierra Maestra (see-ERRA mah-AY-strah)
Sikh (SEEK)
Sikhism (SEEK-izm)
Siking (see-KING)
Sikkim (SICK-um)
Silesia (sye-LEE-zhuh)
Sinai (SYE-nye)
Singapore (SING-uh-poar)
Sinhala (sin-HAHLA)
Sinhalese (sin-hah-LEEZE)
Sinicization (sine-ih-sye-ZAY-shun)
Sinicized (SYE-nuh-sized)
Sinuiji (seen-WEE-joo)
Sioux (SOO)
Sisal (SYE-sull)
Sitka (SIT-kuh)
Sivas (SEE-vahss)
Sjaelland (ZEE-lund)
Skagerrak (SKAG-uh-rak)
Skopje (SKAWP-yay)
Slamet (SLAH-met)
Slav (SLAHV)
Slavic (SLAH-vick)
Slovakia (sloh-VAH-kee-uh)
Slovenia (sloh-VEE-nee-uh)
Sofala (soh-FAH-luh)
Sofia (SOH-fee-uh)
Somali (suh-MAH-lee)
Somalia (suh-MAHL-yuh)
Somoza (suh-MOH-zah)
Song Qiang (SAWNG chee-AHNG)
Songdo (song-DOH)
Songhai (SAWNG-hye)
Songhua (SAWNG-hwah)
Sonora (suh-NORA)
Sotho (SOO-too)
Soufrière (soo-free-AIR)
Sovkhoz (SOV-koze)
Soweto (suh-WETTO)
Spatial (SPAY-shull)
Spratly (SPRAT-lee)
Spykman (SPIKE-mun)
Sri Lanka (sree-LAHNG-kuh)
Srinagar (srih-NUG-arr)
Srpska (SERP-skuh)
St. Eustatius (saint yoo-STAY-shuss)
St. Louis (sah-loo-EE)
St. Lucia (saint LOO-shuh)
St. Maarten (sint MAHRT-un)
Stalin (STAH-lin)
Stamboul (STAHM-bool)
Stanovoy (stuh-nah-VOY)
Steppe (STEP)
Stockholm (STOCK-hoam)
Strasbourg (STRAHSS-boorg)
Stupa (STOO-puh)
Stuttgart (SHTOOT-gart)
Sucre (SOO-kray)
Sudan (soo-DAN)
Sudanese (soo-duh-NEEZE)
Sudd (SOOD)
Sudeten (soo-DAYTEN)
Suez (SOO-ez)
Suharto (soo-HAHR-toh)
Sui (SWAY)
Sukarno (soo-KAR-noh)
Sukarnoputri, Megawati (soo-kar-noh-POO-tree, mega-WATTY)
Sulawesi (soo-luh-WAY-see)
Suleyman (SOO-lay-mahn)
Sultan (SULL-tun)
Sultanate (SULL-tuh-nut)
Sulu (SOO-loo)
Sumatera (suh-MAH-tuh-ruh)
Sumatra (suh-MAH-truh)
Sumbawa (soom-BAH-wuh)
Sumer (SOO-mer)
Sun Yat-sen (SOON yaht-SENN)
Sunda [nese] (SOON-duh [NEEZ])
Sung (SOONG)
Sunni (SOO-nee)
Supsa (SOOP-suh)
Suriname (soor-uh-NAHM-uh)
Suryavarman (soory-AHVA-mun)
Susten (SOOS-ten)
Sutlej (SUTT-ledge)
Suva (SOO-vuh)
Suzhou (SOO-JOH)
Swahili (swah-HEE-lee)
Swazi [land] (SWAH-zee [land])
Syktyvkar (sik-tiff-KAR)
Syr-Darya (seer-DAHR-yuh)
Syria (SEARY-uh)
Szczecin (SHCHE-tseen)
Tabasco (tuh-BAH-skoh)
Taejon (TAH-JON)
Tagalog (tuh-GAH-log)
Tahiti (tuh-HEET-tee)
Taiga (TYE-guh)
Taipa (TYE-pah)
Taipei [Taibei] (tye-BAY)
Taiwan (tye-WAHN)
Tajik (TUDGE-ick)
Tajikistan (tah-JEEK-ih-stahn)
Takbai (TAHK-bye)
Takla Makan (tahk-luh-muh-KAHN)
Taliban (TALLA-ban)
Tallinn (TALLEN)
Tamaulipas (tah-mau-LEEPUS)
Tamil [Nadu] (TAMMLE [NAH-doo])
Tampere (TAHM-puh-ray)
Tampico (tam-PEEK-oh)
Tana (TAHNA)
Tang (TAHNG)
Tanganyika (tan-gun-YEEKA)
Tangerang (tahn-guh-RAHNG)
Tangshan (tahng-SHAHN)
Tannu Tuva (tan-ooh-TOO-vuh)
Tanzania (tan-zuh-NEE-uh)
Tao (DAU)
Tarim (TAH-REEM)
Tashkent (Toshkent) (tahsh-KENT)
Tasman (TAZZ-mun)
Tasmania (tazz-MAY-nee-uh)
Tatar (TAHT-uh)
Tatarstan (TAHT-uh-STAHN)
Tatra (TAHT-truh)
Taxco (TAHSS-koh)
Tayshet (tye-SHET)
Tbilisi (tuh-BILL-uh-see)
Tchaikovsky (chye-KOFF-skee)
Teatro Colón (tay-AH-troh koh-LOAN)
Tecnópolis (tek-NOH-poh-leece)
Teda (TAY-duh)
Tegucigalpa (tuh-goose-ih-GAHL-puh)
Tehran (tay-uh-RAHN)
Tehuantepec (tuh-WHAHN-tuh-pek)
Tel Aviv-Jaffa (tella-VEEVE-JOFF-uh)
Telugu (TELLOO-goo)
Tema (TAY-muh)
Templada (tem-PLAH-dah)
Tengah (TENG-gah)
Tengiz (TEN-ghiz)
Tenochtitlán (tay-noh-chit-LAHN)
Teotihuacán (tay-uh-tee-wah-KAHN)
Terai (teh-RYE)
Terek (TEH-rek)
Terengganu (teh-reng-GAH-noo)
Tertiary (TER-shuh-ree)
Tete (TATE-uh)
Thailand (TYE-land)
Thames (TEMZ)
Thamesmead (TEMZ-meed)
Thebes (THEEBZ)
Theocracy (thee-OCK-ruh-see)
Thessaloniki (thess-uh-luh-NEE-kee)
Thimphu (thim-POOH)
Thünian (TOO-nee-un)

Tian Shan (TYAHN SHAHN)
Tiananmen (TYAHN-un-men)
Tianjin (tyahn-JEEN)
Tiber (TYE-ber)
Tibet (tuh-BETT)
Tierra caliente (tee-ERRA kahl-YEN-tay)
Tierra del Fuego (tee-ERRA dale FWAY-goh)
Tierra fría (tee-ERRA FREE-uh)
Tierra helada (tee-ERRA ay-LAH-dah)
Tierra nevada (tee-ERRA neh-VAH-dah)
Tierra templada (tee-ERRA tem-PLAH-dah)
Tigrean (tih-GRAY-un)
Tigris (TYE-gruss)
Tijuana (tee-WHAHN-uh)
Tikhvin (TIK-vun)
Tikrit (tih-KREET)
Tilbury (TILL-buh-ree)
Timbuktu (tim-buck-TOO)
Timor (TEE-moar)
Timur (TEE-moor)
Tirane (tih-RAHN-uh)
Titicaca (tiddy-KAH-kuh)
Tlingit (TLING-git)
Tobago (tuh-BAY-goh)
Tocantins (toke-un-TEENS)
Tokaido (toh-KYE-doh)
Tokelau (TOH-kuh-lau)
Tokyo (TOH-kee-oh)
Toledo (toh-LAY-doh)
Tolédo, Alejandro (toh-LAY-doh, ah-lay-HAHN-droh)
Tolstoy (TAWL-stoy)
Tolyatti (tawl-YAH-tee)
Tomolon (TOM-oh-loan)
Tondano (tonn-DAH-noh)
Tonga (TAHNG-guh)
Tongan (TONG-gun)
Tonkin (TAHN-KIN)
Tonle Sap (tahn-lay SAP)
Toponym (TOH-poh-nim)
Toponymy (toh-PONN-uh-mee)
Tordesillas (tor-day-SEE-yahss)
Torii (taw-REE)
Torres (TOAR-russ)
Torrijos (tor-REE-hohss)
Toshka (TOSH-kah)
Toshkent (tahsh-KENT)
Toyama (toh-YAH-muh)
Train à grande vitesse (TRAN ah-grawnd-vee-TESS)
Trajan (TRAY-junn)
Transcaucasia (tranz-kaw-KAY-zhuh)
Transdniestra (tranz-duh-NYESS-truh)
Transhumance (tranz-hyoo-MANSS)
Transvaal (TRUNZ-vahl)
Trias Monge, José (TREE-ahss MON-hay, hoh-ZAY)
Trichardt (TREE-shart)
Tripoli (TRIPPA-lee)
Tripolitania (trip-olla-TANEY-yuh)
Tripura (TRIP-uh-ruh)
Trondheim (TRAHN-hame)
Trung Bo (TROONG BOH)
Trypanosomiasis (try-pan-noh-soh-MYE-uh-sis)
Tselinograd (seh-LEENO-grahd)
Tsetse (TSETT-see)
Tsugaru (tsoo-GAH-roo)
Tsukuba (tsoo-KOOB-uh)
Tsumeb (SOO-meb)
Tsunami (tsoo-NAH-mee)
Tsushima (tsoo-SHEE-muh)
Tswana (SWAH-nuh)
Tuanku, Abdul Rahman (TWAHN-koo, ahb-DOOL RAH-mun)
Tuareg (TWAH-regg)
Tubarão (too-buh-RAWNG)
Tuchueh (too-CHOO-way)
Tucson (TOO-sonn)
Tucumán (too-koo-MAHN)
Tucuruí (too-koo-roo-EE)
Tufo (TOO-foh)
Tukiu (too-KYOO)
Tula (TOO-luh)
Tun Abdul Razak (toon ahb-DOOL rah-ZAHK)
Tunis (TOO-niss)
Tunisia (too-NEE-zhuh)
Turfan (TER-fan)
Turin (TOOR-rin)
Turkana (ter-KANNA)
Turkestan (TER-kuh-stahn)
Türkmenbashi (tyoork-men-BAH-shee)
Turkmenistan (terk-MEN-uh-stahn)
Turku (TOOR-koo)
Tuscany (TUSS-kuh-nee)
Tutsi (TOOTSIE)
Tuva (TOO-vuh)
Tuvalu (too-VAHL-oo)
Tuzla (TOOZ-lah)
Twa (TOO-wah)
Tyrol (tih-ROLL)
Tyumen (tyoo-MEN)
Tyva (TOO-vuh)
Tyvinian (too-VINNY-un)
Ubangi (oo-BANG-ghee)
Ubundu (oo-BOON-doo)
Udaipur (uh-DYE-poor)
Udmurt [iya] (ood-MOORT [ee-uh])
Ufa (oo-FAH)
Uganda (yoo-GAHN-duh/yoo-GANDA)
Uhuru (oo-HOO-roo)
Ujungpandang (oo-JUNG PAHN-dahng)
Ukraine (yoo-CRANE)
Ulaanbaatar (oo-lahn-BAH-tor)
Ulsan (OOL-SAHN)
Uluru (ooh-LOO-roo)
Ulyanov (ool-YAH-noff)
UNCLOS (UNN-klohss)
Ungava (ung-GAH-vuh)
Ura Nippon (OOH-ruh NIP-on)
Urabá (oo-rah-BAH)
Ural (YOOR-ull)
Ural-Altaic (YOOR-ull-al-TAY-ick)
Urdu (OOR-doo)
Uribe, Alvaro (oo-REE-bay, AHL-vah-roh)
Uruguay (OO-rah-gwye)
Ürümqi (oo-ROOM-chee)
Ushuaia (oo-SWAH-yah)
Usinsk (oo-SINSK)
Ussuri (ooh-SOOR-ree)
Ust-Kamenogorsk
Usuli (oo-SOO-lee)
Utrecht (YOO-trekt)
Uttar Pradesh (ootar-pruh-DESH)
Uttaranchal (OO-tahr-rahn-CHARL)
Uyghur (WEE-ghoor)
Uzbek (OOZE-beck)
Uzbekistan (ooze-BECK-ih-stahn)
Uzen (oo-ZEN)
Vaal (VAHL)
Vakhan (wah-KAHN)
Valencia (vuh-LENN-see-uh)
Valle Central (VAH-yay sen-TRAHL)
Valletta (vuh-LEH-tuh)
Valparaíso (vahl-pah-rah-EE-so)
Vanino (VAH-nih-noh)
Vanni (VAH-nee)
Vanuatu (vahn-uh-WAH-too)
Varanasi (vuh-RAHN-uh-see)
Varangian (vuh-RANGE-ee-un)
Veld (VELT)
Veneto (VENN-uh-toh)
Venezuela (veh-neh-SWAY-lah)
Venizelos, Eleftherios (venni-SAY-lohss, ellef-TEER-ree-ohss)
Ventspils (VENT-spilz)
Veracruz (verra-CROOZE)
Verde (VER-dee)
Vereeniging (fuh-REEN-ih-king)
Versailles (vair-SYE)
Viangchan (vyung-CHAHN)
Vienna (vee-ENNA)
Vietnam (vee-et-NAHM)
Vietnamese (vee-et-nuh-MEEZE)
Villahermosa (vee-yuh-air-MOH-suh)
Villa-Lobos, Heitor (vill-ah-LOH-bohss, AY-tohr)
Vilnius (VILL-nee-uss)
Vindhya (VIN-dyuh)
Visayan (vuh-SYE-un)
Vistula (VIST-yulluh)
Vladikavkaz (vlad-uh-kuff-KAHZ)
Vladivostok (vlad-uh-vuh-STAHK)
Vltava (VULL-tuh-vuh)
Vojvodina (VOY-vuh-deena)
Volgograd (VOLL-guh-grahd)
Vologda (VAW-lug-duh)
Volta Redonda (vahl-tuh rih-DONN-duh)
Von Thünen, Johann Heinrich (fon-TOO-nun, YOH-hahn HINE-rish)
Voortrekker (FOR-trecker)
Wabenzi (wah-BENZ-zee)
Wahhabism (wah-HAH-b'izm)
Wahid, Abdurrahman (wah-HEED, ahb-doo-RAH-mun)

Waikiki (wye-kuh-KEE)
Waitangi (WYE-tonggy)
Wakisihu (wah-kee-SEE-hoo)
Walachia (wuh-LAH-kee-yuh)
Wallaby (WALLA-bee)
Wallonia (wah-LOANY-uh)
Walloon (wah-LOON)
Walvis (WAHL-vuss)
Wasatch (WAW-satch)
Weber (VAY-buh)
Wegener (VAY-ghenner)
Wei (WAY)
Wenzhou (whunn-JOH)
Weser (VAY-zuh)
Westermann (VESS-tair-munn)
Westphalia (west-FAIL-yuh)
Wildebeest (WIL-duh-beest)
Willamette (wuh-LAMM-ut)
Windhoek (VINT-hook)
Winnipeg (WIN-uh-peg)
Witbank (WHIT-bank)
Witwatersrand (WITT-waw-terz-rand)
Wolof (WOH-loff)
Wroclaw (VROH-tswahf)
Wuchang (woo-CHAHNG)
Wuhan (woo-HAHN)
Wuzhou (woo-JOH)
Xenophobia (zee-nuh-FOH-bee-uh)
Xhosa (SHAW-suh)
Xi (SHEE)
Xi Jiang (SHEE jee-AHNG)
Xia (SHAH)
Xiamen (shah-MEN)
Xian (shee-AHN)
Xianggang (see-AHNG-gahng)
Xiangkhoang (SHEE-AHNG-kwahng)
Xiaolangdi (SH-YAU-lahng-dee)
Xinjiang (shin-jee-AHNG)
Xizang (sheedz-AHNG)
Yacyretá (yah-see-ray-TAH)
Yakut [sk] (yuh-KOOT [sk])
Yakutiya (yuh-KOOTY-uh)
Yamalo-Nenetskiy (yuh-MAH-luh-nuh-NET-skee)
Yamoussoukro (yahm-uh-SOO-kroh)
Yamuna (YAH-muh-nuh)
Yanan (yen-AHN)
Yanbu (YAN-boo)
Yangon (yahn-KOH)
Yangpu (YAHNG-poo)
Yangzi [Yangtze] (YANG-dzee)
Yanomami (yah-noh-MAH-mee)
Yantai (yahn-TYE)
Yaohan (yau-HAHN)
Yaoundé (yown-DAY)
Yaroslavl (yar-uh-SLAHV-ull)
Yawata (yuh-WAH-tuh)
Yayoi (yah-YOH-ee)
Yeates (YATES)
Yekaterinburg (yeh-KAHTA-rin-berg)
Yeltsin, Boris (YELT-sinn, BAW-reese)
Yemen (YEMMEN)
Yemeni (YEH-meh-nee)
Yenisey (yen-uh-SAY)
Yerba maté (YAIR-bah mah-TAY)
Yerevan (yair-uh-VAHN)
Yerushalayim (yeh-roo-shuh-LYE-im)
Yevreyskaya (yev-RAY-sky-yah)
Yibin (EE-BIN)
Yichang (yee-CHAHNG)
Yi-Luo (YEE-loo-oh)
Yogyakarta (yah-gyuh-KAR-tuh)
Yokohama (yoh-kuh-HAH-muh)
Yongbyon (yong-B'YON)
Yonne (YAHN)
Yoruba (YAH-rooba)
Yuan (YOO-ahn)
Yucatán (yoo-kuh-TAHN)
Yudhoyono, Susilo Bambang (YOOD-hoy-onno, SOO-seelo bahm-BAHNG)
Yugoslavia (yoo-goh-SLAH-vee-uh)
Yulara (yoo-LAH-rah)
Yumen (YOO-mun)
Yungas (YOONG-gahss)
Yunnan (yoon-NAHN)
Zagreb (ZAH-grebb)
Zagros (ZAH-gruss)
Zaïre (zah-EAR)
Zambezi (zam-BEEZY)
Zambia (ZAM-bee-uh)
Zamboanga (zahm-boh-AHNG-guh)
Zamfara (zahm-FAHR-rah)
Zanzibar (ZANN-zih-bar)
Zapata, Emiliano (zah-PAH-tah, eh-mee-lee-AH-noh)
Zapatista (zah-pah-TEESTA)
Zayed (zye-ED)
Zedillo, Ernesto (zeh-DEE-yoh, air-NES-toh)
Zeeland (ZAY-lund)
Zhanjiang (JAHN-jee-AHNG)
Zhejiang (JEJ-ee-AHNG)
Zhongshan (jong-SHAHN)
Zhou (JOH)
Zhou Enlai (JOH en-lye)
Zhuhai (joo-HYE)
Zimbabwe (zim-BAHB-way)
Zionism (ZYE-un-izm)
Zoogeography (ZOH-oh-jee-OG-gruh-fee)
Zoroaster (zorro-AST-tuh)
Zoroastrian (zorro-ASTREE-un)
Zuider Zee (ZYDER ZEE)
Zulu (ZOO-loo)
Zürich (ZOOR-ick)

References and Further Readings

(Bullets [•] denote basic introductory works)

INTRODUCTION

Brunn, Stanley D., Williams, Jack F., & Zeigler, Donald J. *Cities of the World: World Regional Urban Development* (Lanham, Md.: Rowman & Littlefield, 3rd rev. ed., 2003).

Clark, Gordon L., et al., eds. *The Oxford Handbook of Economic Geography* (New York: Oxford University Press, 2000).

Claval, Paul, ed. *Introduction to Regional Geography* (Malden, Mass.: Blackwell, 1998).

de Blij, H. J. *Why Geography Matters* (New York: Oxford University Press, 2005).

• de Blij, H. J. & Murphy, Alexander B. *Human Geography: Culture, Society, and Space* (Hoboken, N.J.: John Wiley & Sons, 7th rev. ed., 2002).

• de Blij, H. J., Muller, Peter O., & Williams, Richard S., Jr. *Physical Geography: The Global Environment* (New York: Oxford University Press, 3rd rev. ed., 2004).

Diamond, Jared. *Guns, Germs and Steel: The Fates of Human Societies* (New York: W. W. Norton, 1997).

Fenneman, Nevin M. "The Circumference of Geography," *Annals of the Association of American Geographers*, 9 (1919): 3–11.

Friedman, Thomas L. *The World Is Flat: A Brief History of the Twenty-First Century* (New York: Farrar, Straus & Giroux, 2005).

Gaile, Gary L., & Willmott, Cort J., eds. *Geography in America at the Dawn of the Twenty-First Century* (New York: Oxford University Press, 2nd rev. ed., 2004).

• *Goode's World Atlas* (Skokie, Ill.: Rand McNally, 21st rev. ed., 2005).

Grove, Jean M. *The Little Ice Age* (London & New York: Methuen, 1988). Quotation taken from pp. 1–2.

Holt-Jensen, Arild. *Geography: History and Concepts—A Student's Guide* (Thousand Oaks, Calif.: Sage, 3rd rev. ed., trans. Brian Fullerton, 1999).

Jackson, John Brinckerhoff. *Discovering the Vernacular Landscape* (New Haven, Conn.: Yale University Press, 1984).

Johnston, Ron J., et al., eds. *The Dictionary of Human Geography* (Malden, Mass.: Blackwell, 4th rev. ed., 2000).

Jordan, Terry G. "The Concept and Method," in Lich, Glen E., ed., *Regional Studies: The Interplay of Land and People* (College Station: Texas A & M Press, 1992), pp. 8–24.

• Kimerling, A. Jon, Muehrcke, Phillip C., & Muehrcke, Juliana O. *Map Use: Reading–Analysis–Interpretation* (Madison, Wis.: JP Publications, 5th rev. ed., 2005).

Livingstone, David. *The Geographical Tradition* (Cambridge, Mass.: Blackwell, 1992).

Martin, Geoffrey J. *All Possible Worlds: A History of Geographical Ideas* (New York: Oxford University Press, 4th rev. ed., 2005).

National Research Council. *Rediscovering Geography: New Relevance for Science and Society* (Washington, D.C.: National Academy Press, 1997).

• Pattison, William D. "The Four Traditions of Geography," *Journal of Geography*, 63 (1964): 211–216.

Rogers, Alisdair, & Viles, Heather A., eds. *The Student's Companion to Geography* (Malden, Mass.: Blackwell, 2nd rev. ed., 2003).

Sauer, Carl Ortwin. "Cultural Geography," *Encyclopedia of the Social Sciences*, Vol. 6 (New York: Macmillan, 1931), pp. 621–623.

Sheppard, Eric, & Barnes, Trevor, eds. *A Companion to Economic Geography* (Malden, Mass.: Blackwell, 2000).

Wegener, Alfred. *The Origin of Continents and Oceans* (New York: Dover, reprint of the 1915 original, trans. John Biram, 1966).

• Wheeler, James O., Muller, Peter O., Thrall, Grant I., & Fik, Timothy J. *Economic Geography* (New York: John Wiley & Sons, 3rd rev. ed., 1998).

Whittlesey, Derwent S., et al. "The Regional Concept and the Regional Method," in James, Preston E. & Jones, Clarence F., eds., *American Geography: Inventory and Prospect* (Syracuse, N.Y.: Syracuse University Press, 1954), pp. 19–68.

CHAPTER 1

"A Survey of the EU's Eastern Borders," *The Economist*, June 25, 2005, special insert, 16 pp.

Aslund, Anders. *Building Capitalism: The Transformation of the Former Soviet Bloc* (New York: Cambridge University Press, 2001).

Baldersheim, Harald, & Stahlberg, Krister, eds. *Nordic Region-Building in a European Perspective* (Brookfield, Vt.: Ashgate, 1999).

- Berentsen, William H., ed. *Contemporary Europe: A Geographic Analysis* (New York: John Wiley & Sons [7th rev. ed. of George W. Hoffman, *Europe: A Geographic Analysis*], 1997).

Bernstein, Richard. "An Aging Europe May Find Itself on the Sidelines," *New York Times*, June 29, 2003, 3.

Blacksell, Mark, & Williams, Allan M., eds. *The European Challenge: Geography and Development in the European Community* (New York: Oxford University Press, 1994).

Bruni, Frank. "Persistent Drop in Fertility Reshapes Europe's Future," *New York Times*, December 26, 2002, A1, A10.

Burtenshaw, David, et al. *The European City: A Western Perspective* (New York: John Wiley & Sons, 1991).

Butlin, Robin A., & Dodgson, R. A. *An Historical Geography of Europe* (New York: Oxford University Press, 1999).

Carter, Frank W., & Turnock, David. *Environmental Problems in Eastern Europe* (London & New York: Routledge, 2nd rev. ed., 1997).

Champion, Tony, et al. *The New Regional Map of Europe* (Amsterdam & New York: Elsevier, 1996).

Chisholm, Michael. *Rural Settlement and Land Use: An Essay in Location* (London: Hutchinson University Library, 3rd rev. ed., 1979).

- Clout, Hugh D., et al. *Western Europe: Geographical Perspectives* (New York: Wiley/Longman, 3rd rev. ed., 1994).

- Cole, John P., & Cole, Francis. *A Geography of the European Union* (London & New York: Routledge, 2nd rev. ed., 1997).

Delamaide, Darrell. *The New Superregions of Europe* (New York: Dutton, 1994).

- Diem, Aubrey. *Western Europe: A Geographical Analysis* (New York: John Wiley & Sons, 1979).

Embleton, Clifford, ed. *Geomorphology of Europe* (New York: Wiley-Interscience, 1984).

Emerson, Michael. *Redrawing the Map of Europe* (New York: St. Martin's Press, 1998).

Fernandez-Armesto, Felipe, ed. *The Times Guide to the Peoples of Europe* (Boulder, Colo.: Westview Press, 1995).

Glebe, Günther, & O'Loughlin, John, eds. *Foreign Minorities in Continental European Cities* (Wiesbaden, West Germany: Franz Steiner Verlag, 1987).

- Gottmann, Jean. *A Geography of Europe* (New York: Holt, Rinehart & Winston, 4th rev. ed., 1969).

Graham, Brian, ed. *Modern Europe: Place, Culture, and Identity* (New York: Oxford University Press, 1998).

Grove, Alfred T., & Rackham, Oliver. *The Nature of Mediterranean Europe: An Ecological History* (New Haven, Conn.: Yale University Press, 2001).

Guttman, Robert J., ed. *Europe in the New Century: Visions of an Emerging Superpower* (Boulder, Colo.: Lynne Rienner, 2001).

Haggett, Peter. *Locational Analysis in Human Geography* (London: Edward Arnold, 1965). Definition from p. 19.

Hall, Derek, & Danta, Darrick, eds. *Reconstructing the Balkans: A Geography of the New Southeast Europe* (New York: John Wiley & Sons, 1996).

Hall, Ray, & White, Paul, eds. *Europe's Population: Towards the Next Century* (London: UCL Press/Taylor & Francis, 1995).

"Happy Family? A Survey of the Nordic Countries," *The Economist*, January 23, 1999, special insert, 16 pp.

Heffernan, Michael. *The Meaning of Europe: Geography and Geopolitics* (New York: Oxford University Press, 1998).

Heffernan, Michael. *Twentieth-Century Europe: A Political Geography* (New York: John Wiley & Sons, 1997).

Hoggart, Keith, et al. *Rural Europe: Identity and Change* (New York: John Wiley & Sons, 1995).

Hudson, Ray, & Williams, Allan M., eds. *Divided Europe: Society and Territory* (Thousand Oaks, Calif.: Sage, 1998).

Hunter, Shireen T., ed. *Islam, Europe's Second Religion: The New Social, Cultural, and Political Landscape* (Westport, Conn.: Praeger, 2002).

Hupchick, Dennis P., & Cox, Harold E., eds. *A Concise Historical Atlas of Eastern Europe* (New York: St. Martin's Press, 1996).

Jefferson, Mark. "The Law of the Primate City," *Geographical Review*, 29 (1939): 226–232.

Johnson, Lonnie R. *Central Europe: Enemies, Neighbors, Friends* (New York: Oxford University Press, 1996).

Jones, Philip N., & Wild, Trevor. "Regional and Local Variations in the Emerging Economic Landscape of the New German *Länder*," *Applied Geography*, 17 (1997): 283–299.

Jönsson, Christer, Tägil, Sven, & Törnqvist, Gunnar. *Organizing European Space* (Thousand Oaks, Calif.: Sage, 2000).

- Jordan-Bychkov, Terry G., & Bychkova Jordan, Bella. *The European Culture Area: A Systematic Geography* (Lanham, Md.: Rowman & Littlefield, 4th rev. ed., 2001).

Keating, Michael. *Nations Against the State: The New Politics of Nationalism in Quebec, Catalonia, and Scotland* (New York: Macmillan, 1996).

Keating, Michael, ed. *Regions and Regionalism in Europe* (Northampton, Mass.: Edward Elgar, 2004).

Kucera, Tomas, et al., eds. *New Demographic Faces of Europe* (New York: Springer Verlag, 2000)

Kürti, Laszlo, & Langman, Juliet, eds. *Beyond Borders: Remaking Cultural Identities in the New East and Central Europe* (Boulder, Colo.: Westview Press, 1997).

Livi-Bacci, Massimo. *The Population of Europe* (Malden, Mass.: Blackwell, trans. Carl Ipsen, 2000).

Matvejevic, Predag. *The Mediterranean: A Cultural Landscape* (Berkeley: University of California Press, trans. Michael H. Heim, 1999).

McCarthy, Linda, & Danta, Darrick. "Cities of Europe," in Brunn, Stanley D., Williams, Jack F., & Zeigler, Donald J., *Cities of the World: World Regional Urban Development* (Lanham, Md.: Rowman & Littlefield, 3rd rev. ed., 2003), pp. 176–221.

• McDonald, James R. *The European Scene: A Geographic Perspective* (Upper Saddle River, N.J.: Prentice-Hall, 1997).

Mortimer, Edward, & Fine, Robert, eds. *People, Nation and State: The Meaning of Ethnicity and Nationalism* (London: I. B. Tauris, 1999).

Murphy, Alexander B. "Emerging Regional Linkages Within the European Community: Challenging the Dominance of the State," *Tijdschrift voor Economische en Sociale Geografie*, 84 (1993): 103–118.

Murphy, Alexander B. "Rethinking the Concept of European Identity," in Herb, Guntram H., & Kaplan, David H., eds., *Nested Identities: Nationalism, Territory, and Scale* (Lanham, Md.: Rowman & Littlefield, 1999), pp. 53–73.

O'Dowd, Liam & Wilson, Thomas M., eds. *Borders, Nations and States: Frontiers of Sovereignty in the New Europe* (Brookfield, Vt.: Ashgate, 1996).

Ohmae, Kenichi. *The End of the Nation-State: The Rise of Regional Economies* (New York: Free Press, 1996).

Ohmae, Kenichi. "The Rise of the Region State," *Foreign Affairs*, Spring 1993, 78–87.

• Ostergren, Robert C., & Rice, John G. *The Europeans: A Geography of People, Culture, and Environment* (New York: Guilford Press, 2004).

Petrakos, George, ed. *Integration and Transition in Europe: Economic Geography of Interaction* (London & New York: Routledge, 2000).

Pinder, David, ed. *The New Europe: Economy, Society and Environment* (Chichester, UK & New York: John Wiley & Sons, 1998).

Pond, Elizabeth. *The Rebirth of Europe* (Washington, D.C.: Brookings Institution Press, 1999).

"Poverty in Eastern Europe: The Land That Time Forgot," *The Economist*, September 23, 2000, 27–30.

Rhodes, Martin, ed. *Regions and the New Europe: Patterns in Core and Periphery Development* (Manchester, UK: Manchester University Press, 1996).

• Shaw, Denis J. B., ed. *The Post-Soviet Republics: A Systematic Geography* (New York: John Wiley & Sons/Longman, 1995).

"The Luck of the Irish: A Survey of Ireland," *The Economist*, October 16, 2004, special insert, 12 pp.

Townsend, Alan R. *Making a Living in Europe: Human Geographies of Economic Change* (London & New York: Routledge, 1997).

Turnock, David, ed. *East Central Europe and the Former Soviet Union: Environment and Society* (London: Arnold, 2001).

Turnock, David. *The Human Geography of East-Central Europe* (London & New York: Routledge, 2003).

Unwin, Tim, ed. *A European Geography* (Harlow, UK & New York: Longman, 1998).

Wheeler, James O., Muller, Peter O., Thrall, Grant I., & Fik, Timothy J. *Economic Geography* (New York: John Wiley & Sons, 3rd rev. ed., 1998), Chapter 13.

Williams, Allan M. *The West European Economy: A Geography of Post-War Development* (Savage, Md.: Rowman & Littlefield, 1988).

Wintle, Michael, ed. *Culture and Identity in Europe: Perceptions of Divergence and Unity in Past and Present* (Brookfield, Vt.: Ashgate, 1996).

CHAPTER 2

"A Caspian Gamble: A Survey of Central Asia," *The Economist*, February 7, 1998, special insert, 18 pp.

Aslund, Anders. *Building Capitalism: The Transformation of the Former Soviet Bloc* (New York: Cambridge University Press, 2001).

• Bater, James H. *Russia and the Post-Soviet Scene: A Geographical Perspective* (New York: John Wiley & Sons, 1996).

Brawer, Moshe. *Atlas of Russia and the Independent Republics* (New York: Simon & Schuster, 1994).

Brooke, James. "Russia's Latest Oil and Gas Oasis: Sakhalin Fields Have Japan, China, and Korea Lining Up," *New York Times*, May 13, 2003, W1, W7.

Brooke, James. "The Asian Battle for Russia's Oil and Gas," *New York Times*, January 3, 2004, B1, B3.

• Brown, Archie, et al., eds. *The Cambridge Encyclopedia of Russia and the Former Soviet Union* (New York: Cambridge University Press, 1994).

Bychkova Jordan, Bella, & Jordan-Bychkov, Terry G. *Siberian Village: Land and Life in the Sakha Republic* (Minneapolis: University of Minnesota Press, 2001).

Chew, Allen F. *Atlas of Russian History: Eleven Centuries of Changing Borders* (New Haven, Conn.: Yale University Press, 1967).

Chinn, Jeff, & Kaiser, Robert. *Russians as the New Minority: Ethnicity and Nationalism in the Soviet Successor States* (Boulder, Colo.: Westview Press, 1996).

Davis, Sue. *The Russian Far East: The Last Frontier* (London & New York: Routledge, 2002).

Demko, George J., et al., eds. *Population under Duress: The Geodemography of Post-Soviet Russia* (Boulder, Colo.: Westview Press, 1998).

Dmitrieva, O. *Regional Development: The U.S.S.R. and After* (London: University College London Press, 1996).

Dunlop, John B. *The Rise of Russia and the Fall of the Soviet Empire* (Princeton, N.J.: Princeton University Press, 1993).

Engelmann, Kurt E., & Pavlakovic, Vjeran, eds. *Rural Development in Eurasia and the Middle East: Land Reform, Demographic Change, and Environmental Constraints* (Seattle: University of Washington Press, 2001).

Forsyth, James. *A History of the Peoples of Siberia: Russia's North Asian Colony, 1581–1990* (New York: Cambridge University Press, 1992).

Gachechiladze, Revaz. *The New Georgia: Space, Society, Politics* (College Station: Texas A&M University Press, 1995).

Hanson, Philip, & Bradshaw, Michael, eds. *Regional Change in Russia* (Northampton, Mass.: Edward Elgar, 2000).

Harris, Chauncy D. "A Geographic Analysis of Non-Russian Minorities in Russia and Its Ethnic Homelands," *Post-Soviet Geography*, 34 (1993): 543–597.

Hill, Fiona, & Gaddy, Clifford. *The Siberian Curse: How Communist Planners Left Russia Out in the Cold* (Washington, D.C.: Brookings Institution Press, 2003).

Horensma, Pier. *The Soviet Arctic* (London & New York: Routledge, 1991).

Hosking, Geoffrey. *Russia and the Russians: A History* (Cambridge, Mass.: Belknap/Harvard University Press, 2001).

Hunter, Shireen T. *Transcaucasia in Transition: Nation-Building and Conflict* (Boulder, Colo.: Westview Press, 1994).

Huttenbach, Henry. *The Caucasus: A Region in Crisis* (Boulder, Colo.: Westview Press, 1996).

Ioffe, Gregory, & Nefedova, Tatyana. *Continuity and Change in Rural Russia* (Boulder, Colo.: Westview Press, 1997).

Kaiser, Robert J. *The Geography of Nationalism in Russia and the U.S.S.R.* (Princeton, N.J.: Princeton University Press, 1994).

Kraus, Michael, & Liebowitz, Ronald D., eds. *Russia and Eastern Europe After Communism: The Search for New Political, Economic, and Security Systems* (Boulder, Colo.: Westview Press, 1995).

Lincoln, W. Bruce. *The Conquest of a Continent: Siberia and the Russians* (New York: Random House, 1994).

Lloyd, John. "The Russian Devolution," *The New York Times Magazine*, August 15, 1999, 34–41, 52, 61, 64.

Lydolph, Paul E. *Climates of the Soviet Union* (Amsterdam: Elsevier, 1977).

• Lydolph, Paul E. *Geography of the U.S.S.R.* (Elkhart Lake, Wis.: Misty Valley Publishing, 1990).

Mackinder, Halford J. *Democratic Ideals and Reality* (New York: Holt, 1919).

Mastyugina, Tatiana, & Perepelkin, Lev. *An Ethnic History of Russia: Pre-Revolutionary Times to the Present* (Westport, Conn.: Greenwood Press, 1996).

Meier, Andrew. *Black Earth: A Journey Through Russia After the Fall* (New York: W. W. Norton, 2003).

Mote, Victor L. *Siberia: Worlds Apart* (Boulder, Colo.: Westview Press, 1998).

Myers, Steven Lee. "Siberians Tell Moscow: Like It or Not, It's Home," *New York Times*, January 28, 2004, A1, A9.

Newell, Josh. *The Russian Far East: A Reference Guide for Conservation and Development* (McKinleyville, Calif.: Daniel & Daniel, 2 rev. ed ., 2004).

Peterson, D. J. *Troubled Lands: The Legacy of Soviet Environmental Destruction* (Boulder, Colo.: Westview Press, 1993).

Pryde, Philip R., ed. *Environmental Resources and Constraints in the Former Soviet Republics* (Boulder, Colo.: Westview Press, 1995).

Remnick, David. *Lenin's Tomb: The Last Days of the Soviet Empire* (New York: Random House, 1993).

Remnick, David. *Resurrection: The Struggle to Build a New Russia* (New York: Random House, 1997).

• Shaw, Denis J. B. *Russia in the Modern World: A New Geography* (Malden, Mass.: Blackwell, 1999).

• Shaw, Denis J. B., ed. *The Post-Soviet Republics: A Systematic Geography* (New York: John Wiley & Sons/Longman, 1995).

Shlapentokh, Vladimir, et al. *From Submission to Rebellion: The Provinces Versus the Center in Russia* (Boulder, Colo.: Westview Press, 1997).

Smith, Graham. *The Post Soviet States: Mapping the Politics of Transition* (London: Arnold, 1999).

Stewart, John Massey, ed. *The Soviet Environment: Problems, Policies and Politics* (New York: Cambridge University Press, 1992).

• Symons, Leslie, ed. *The Soviet Union: A Systematic Geography* (London & New York: Routledge, 2nd rev. ed., 1990).

"The Caucasus: Where Worlds Collide," *The Economist*, August 19, 2000, 17–19.

"The New Russian Heartland (special issue)." *Eurasian Geography and Economics*, 46 (2005): 83–163.

Thompson, John M. *Russia and the Soviet Union: An Historical Introduction from the Kievan State to the Present* (Boulder, Colo.: Westview Press, 3rd rev. ed., 1994).

Thornton, Judith, & Ziegler, Charles E., eds. *Russia's Far East: A Region at Risk* (Seattle: University of Washington Press, 2002).

Thubron, Colin. *Among the Russians* (New York: HarperCollins, 2000).

Thubron, Colin. *In Siberia* (New York: HarperCollins, 2000).

Tomikel, John, & Henderson, Bonnie. *Russia and the Near Abroad* (Elgin, Penn.: Allegheny Press, 2nd rev. ed., 2003).

Trenin, Dmitri. *The End of Eurasia: Russia on the Border Between Geopolitics and Globalization* (Washington, D.C.: Carnegie Endowment for International Peace, 2002).

Turnock, David, ed. *East Central Europe and the Former Soviet Union: Environment and Society* (London: Arnold, 2001).

Wixman, Ronald. *The Peoples of the U.S.S.R.: An Ethnographic Handbook* (Armonk, N.Y.: M. E. Sharpe, 1984).

CHAPTER 3

"A Survey of Canada," *The Economist*, July 24, 1999, special insert, 18 pp.

Adams, John S. "Residential Structure of Midwestern Cities," *Annals of the Association of American Geographers*, 60 (1970): 37–62. Model diagram adapted from p. 56.

Agnew, John A., & Smith, Jonathan M., eds. *American Space/American Place: Geographies of the Contemporary United States* (London & New York: Routledge, 2002).

Allen, James P., & Turner, Eugene J. *We the People: An Atlas of America's Ethnic Diversity* (New York: Macmillan, 1987).

- Atwood, Wallace W. *The Physiographic Provinces of North America* (New York: Ginn, 1940).

Berry, Brian J. L. "The Decline of the Aging Metropolis: Cultural Bases and Social Process," in Sternlieb, George, & Hughes, James W., eds., *Post-Industrial America: Metropolitan Decline and Inter-Regional Job Shifts* (New Brunswick, N.J.: Center for Urban Policy Research, Rutgers University, 1975), pp. 175–185.

- Birdsall, Stephen S., Palka, Eugene, Malinowski, Jon, & Price, Margo L. *Regional Landscapes of the United States and Canada* (Hoboken, N.J.: John Wiley & Sons, 6th rev. ed., 2005).

- Boal, Frederick W., & Royle, Stephen A., eds. *North America: A Geographical Mosaic* (London: Arnold, 1999).

- Bone, Robert M. *The Regional Geography of Canada* (New York: Oxford University Press, 3rd rev. ed., 2005).

Bone, Robert M. *The Geography of the Canadian North: Issues and Challenges* (Toronto: Oxford University Press, 1992). Quotation taken from p. 12.

Borchert, John R. "American Metropolitan Evolution," *Geographical Review*, 57 (1967): 301–332.

Borchert, John R. "Futures of American Cities," in Hart, John Fraser, ed., *Our Changing Cities* (Baltimore, Md.: Johns Hopkins University Press, 1991), pp. 218–250.

Brewer, Cynthia A., & Suchan, Trudy A. *Mapping Census 2000: The Geography of U.S. Diversity* (Redlands, Calif.: ESRI Press, 2001).

Britton, John N. H., ed. *Canada and the Global Economy: The Geography of Structural and Technological Change* (Montreal: McGill-Queen's University Press, 1996).

Castells, Manuel, & Hall, Peter. *Technopoles of the World: The Making of Twenty-First-Century Industrial Complexes* (London & New York: Routledge, 1994).

Conrad, Margaret, & Hiller, James. *Atlantic Canada: A Region in the Making* (New York: Oxford University Press, 2001).

de Blij, H. J., ed. *Atlas of North America* (New York: Oxford University Press, 2005).

"Degrees of Separation: A Survey of America," *The Economist*, July 16, 2005, special insert, 20 pp.

DePalma, Anthony. *Here: A Biography of the North American Continent* (New York: Public Affairs, 2001).

Garreau, Joel. *Edge City: Life on the New Frontier* (New York: Doubleday, 1991).

Garreau, Joel. *The Nine Nations of North America* (Boston: Houghton Mifflin, 1981).

Gastil, Raymond D. *Cultural Regions of the United States* (Seattle: University of Washington Press, 1975).

Gaustad, Edwin Scott, & Barlow, Philip L. *New Historical Atlas of Religion in America* (New York: Oxford University Press, 2001).

- Getis, Arthur, & Getis, Judith, eds. *The United States and Canada: The Land and the People* (Dubuque, Iowa: Wm. C. Brown, 1995).

Gober, Patricia. "Americans on the Move," *Population Bulletin*, 48 (November 1993): 1–40.

Gottmann, Jean. *Megalopolis: The Urbanized Northeastern Seaboard of the United States* (New York: Twentieth Century Fund, 1961).

Harris, R. Cole. "Regionalism and the Canadian Archipelago," in McCann, Lawrence D., ed., *Heartland and Hinterland: A Geography of Canada* (Scarborough, Ont.: Prentice-Hall Canada, 1982), pp. 458–484.

Hart, John Fraser. *The Changing Scale of American Agriculture* (Charlottesville: University of Virginia Press, 2003).

Hartshorn, Truman A. "The Changed South, 1947–1997," *Southeastern Geographer*, 37 (November 1997): 122–139.

Harvey, Thomas. "The Changing Face of the Pacific Northwest," *Journal of the West*, 37 (July 1998): 22–32.

Historical Atlas of the United States (Washington, D.C.: National Geographic Society, centennial ed., 1988).

Hodge, Gerald, & Robinson, Ira M. *Planning Canadian Regions* (Vancouver: University of British Columbia Press, 2001).

- Hudson, John C. *Across This Land: A Regional Geography of the United States and Canada* (Baltimore: Johns Hopkins University Press, 2002).

- Hunt, Charles B. *Natural Regions of the United States and Canada* (San Francisco: W. H. Freeman, 2nd rev. ed., 1974).

Janelle, Donald G., ed. *Geographical Snapshots of North America* (New York: Guilford Press, 1992).

Kotkin, Joel. *The New Geography: How the Digital Revolution Is Reshaping the American Landscape* (New York: Random House, 2000).

Krauss, Clifford. "Quebec Seeking to End Its Old Cultural Divide," *New York Times*, April 13, 2003, A6.

Kritz, Mary M., & Gurak, Douglas T. "Immigration and a Changing America," *The American People: Census 2000* (New York & Washington, D.C.: Russell Sage Foundation/Population Reference Bureau, 2004), 43 pp.

Magocsi, Paul Robert, ed. *Encyclopedia of Canada's Peoples* (Toronto: University of Toronto Press, 1998).

Malecki, Edward J., & Gorman, Sean P. "Maybe the Death of Distance, But Not the End of Geography: The Internet as a Network," in Thomas R. Leinbach & Stanley D. Brunn, eds., *Worlds of E-Commerce: Economic, Geographical, and Social Dimensions* (New York: John Wiley & Sons, 2001), pp. 87–105.

Martin, Philip, & Midgley, Elizabeth. "Immigration: Shaping and Reshaping America," *Population Bulletin*, 58 (June 2003): 1–44.

Martin, Philip, & Widgren, Jonas. "International Migration: Facing the Challenge," *Population Bulletin*, 57 (March 2002): 1–40.

- McCann, Lawrence D., & Gunn, Angus, eds. *Heartland and Hinterland: A Regional Geography of Canada* (Scarborough, Ont.: Prentice Hall Canada, 3rd rev. ed., 1998).

McIlwraith, Thomas F., & Muller, Edward K., eds. *North America: The Historical Geography of a Changing Continent* (Lanham, Md.: Rowman & Littlefield, 2nd rev. ed., 2001).

McKee, Jesse O., ed. *Ethnicity in Contemporary America: A Geographical Appraisal* (Lanham, Md.: Rowman & Littlefield, 2nd rev. ed., 2000).

- McKnight, Tom L. *Regional Geography of the United States and Canada* (Upper Saddle River, N.J.: Prentice-Hall, 4th rev. ed., 2004).

Meinig, Donald W. *The Shaping of America: A Geographical Perspective on 500 Years of History; Vol. 1: Atlantic America, 1492–1800* (New Haven, Conn.: Yale University Press, 1986).

Meinig, Donald W. *The Shaping of America: A Geographical Perspective on 500 Years of History; Vol. 2: Continental America, 1800–1867* (New Haven, Conn.: Yale University Press, 1993).

Meinig, Donald W. *The Shaping of America: A Geographical Perspective on 500 Years of History. Vol. 3: Transcontinental America, 1850–1915* (New Haven, Conn.: Yale University Press, 1998).

Meinig, Donald W. *The Shaping of America: A Geographical Perspective on 500 Years of History. Volume 4: Global America, 1915–2000* (New Haven, Conn.: Yale University Press, 2004).

Muller, Peter O. *Contemporary Suburban America* (Englewood Cliffs, N.J.: Prentice-Hall, 1981).

Ohmae, Kenichi. *The End of the Nation-State: The Rise of Regional Economies* (New York: Free Press, 1996).

Orme, Antony R., ed. *The Physical Geography of North America* (New York: Oxford University Press, 2002).

Pacione, Michael. *Urban Geography: A Global Perspective* (London & New York: Routledge, 2 rev. ed., 2005).

Pollard, Kelvin M., & O'Hare, William P. "America's Racial and Ethnic Minorities," *Population Bulletin*, 54 (September 1999): 1–48.

Rooney, John F., Jr., et al., eds. *This Remarkable Continent: An Atlas of United States and Canadian Society and Cultures* (College Station: Texas A&M University Press, 1982).

Shelley, Fred M., et al. *Political Geography of the United States* (New York: Guilford Press, 1996).

"The New Geography of the Information Technology Industry," *The Economist*, July 19, 2003, 47–49.

"The New Map of High-Tech: From Billville to Silicon Alley, The 13 Hottest Regions in America," *Wall Street Journal*, November 23, 1999, B1, B12.

Vance, James E., Jr. *This Scene of Man: The Role and Structure of the City in the Geography of Western Civilization* (New York: Harper's College Press, 1977). Urban realms model discussed on pp. 411–416.

Wallace, Iain. *A Geography of the Canadian Economy* (New York: Oxford University Press, 2002).

- Warkentin, John. *Canada: A Regional Geography* (Scarborough, Ont.: Prentice Hall Canada, 1997).

"Welcome to Amexia: The Border Is Vanishing Before Our Eyes, Creating a New World for All of Us," *Time*, June 11, 2001, 36–79.

Wheeler, James O., Muller, Peter O., Thrall, Grant I., & Fik, Timothy J. *Economic Geography* (New York: John Wiley & Sons, 3rd rev. ed., 1998).

Yeates, Maurice H. *Main Street: Windsor to Quebec City* (Toronto: Macmillan of Canada, 1975).

Yeates, Maurice H. *The North American City* (New York: Harper & Row, 4th rev. ed., 1990). Canadian urban evolution model discussed on pp. 60–67.

Zelinsky, Wilbur. *Exploring the Beloved Country: Geographic Forays into American Society and Culture* (Iowa City, Iowa: University of Iowa Press, 1995).

• Zelinsky, Wilbur. *The Cultural Geography of the United States: A Revised Edition* (Englewood Cliffs, N.J.: Prentice-Hall, 2nd rev. ed, 1992).

Zook, Matthew A. *The Geography of the Internet Industry* (Malden, Mass.: Blackwell, 2005).

CHAPTER 4

"A Silicon Republic . . . Lessons from [Costa Rica's] High-Tech Frontier," *Newsweek*, August 28, 2000, 42–44.

"After the Revolution: A Survey of Mexico," *The Economist*, October 28, 2000, special supplement, 16 pp.

Alleyne, Mervyn C. *The Construction and Representation of Race and Ethnicity in the Caribbean* (Kingston, Jamaica: University of the West Indies Press. 2002).

Arreola, Daniel D. & Curtis, James R. *The Mexican Border Cities: Landscape Anatomy and Place Personality* (Tucson: University of Arizona Press, 1993).

Augelli, John P. "The Rimland-Mainland Concept of Culture Areas in Middle America," *Annals of the Association of American Geographers*, 52 (1962): 119–129.

Barker, David, et al., eds. *A Reader in Caribbean Geography* (Kingston, Jamaica: Ian Randle, 1998).

Barton, Jonathan R. *A Political Geography of Latin America* (London & New York: Routledge, 1997).

• Blakemore, Harold, & Smith, Clifford T., eds. *Latin America: Geographical Perspectives* (London & New York: Methuen, 2nd rev. ed., 1983).

• Blouet, Brian W., & Blouet, Olwyn M. *Latin America and the Caribbean: A Systematic and Regional Survey* (Hoboken, N.J.: John Wiley & Sons, 4th rev. ed., 2002).

Boswell, Thomas D., & Conway, Dennis. *The Caribbean Islands: Endless Geographical Diversity* (New Brunswick, N.J.: Rutgers University Press, 1992).

Brea, Jorge A. "Population Dynamics in Latin America," *Population Bulletin*, 58 (March 2003): 1–36.

Butler, Edgar W., et al. *Mexico and Mexico City in the World Economy* (Boulder, Colo.: Westview, 2001).

• Clawson, David L. *Latin America and the Caribbean: Lands and Peoples* (Dubuque, Iowa: WCB/McGraw-Hill, 3rd rev. ed., 2004).

Collier, Simon, et al., eds. *The Cambridge Encyclopedia of Latin America and the Caribbean* (New York: Cambridge University Press, 2nd rev. ed., 1992).

Davidson, William V., & Parsons, James J., eds. *Historical Geography of Latin America* (Baton Rouge, La.: Louisiana State University Press, 1980).

Elbow, Gary S. "Regional Cooperation in the Caribbean: The Association of Caribbean States," *Journal of Geography*, 96 (January/February 1997): 13–22.

Galloway, Jock H. "The Lesser Antilles," in Donald G. Janelle, ed., *Geographical Snapshots of North America* (New York: Guilford Press, 1992), pp. 16–19.

Gilbert, Alan. *Latin America* (London & New York: Routledge, 1990).

Gilbert, Alan, ed. *The Mega-City in Latin America* (New York & Tokyo: United Nations University Press, 1996).

Goodbody, Ivan, & Hope, Elizabeth T., eds. *Natural Resource Management for Sustainable Development in the Caribbean* (Kingston, Jamaica: University of the West Indies Press, 2001).

Greenfield, Gerald M., ed. *Latin American Urbanization: Historical Profiles of Major Cities* (Westport, Conn.: Greenwood Press, 1994).

Griffin, Ernst C., & Ford, Larry R. "Cities of Latin America," in Brunn, Stanley D., & Williams, Jack F., eds., *Cities of the World: World Regional Urban Development* (New York: HarperCollins, 2nd rev. ed., 1993), pp. 224–265.

Gwynne, Robert N., & Kay, Cristobal. *Latin America Transformed: Globalization and Modernity* (New York: Oxford University Press, 2nd rev. ed., 2004).

Hall, Carolyn, & Brignoli, Hector Perez. *Historical Atlas of Central America* (Norman, Okla.: University of Oklahoma Press, 2003).

Herzog, Lawrence A. *From Aztec to High-Tech: Architecture and Landscape Across the Mexico-United States Border* (Baltimore, Md.: Johns Hopkins University Press, 1999).

Hillman, Richard S., & D'Agostino, Thomas J., eds. *Understanding the Contemporary Caribbean* (Boulder, Colo.: Lynne Rienner, 2003).

• James, Preston E., & Minkel, Clarence W. *Latin America* (New York: John Wiley & Sons, 5th rev. ed., 1986), Chaps. 2–16.

• Kent, Robert B. *Latin America* (New York: Guilford Press, 2006).

Knapp, Gregory, ed. *Latin America in the Twenty-First Century: Challenges and Solutions* (Austin: University of Texas Press, 2002).

Kopinak, Kathryn. *Desert Capitalism: Maquiladoras in North America's Western Industrial Corridor* (Tucson: University of Arizona Press, 1996).

Lewis, Patsy. *Surviving Small Size: Regional Integration in Caribbean Ministates* (Kingston, Jamaica: University of the West Indies Press, 2002).

Lowenthal, David. *West Indian Societies* (New York: Oxford University Press, 1972).

MacLachlan, Ian, & Aguilar, Adrian G. "Maquiladora Myths: Locational and Structural Change in Mexico's Export Manufacturing Industry," *The Professional Geographer*, 50 (August 1998): 315–323.

Malkin, Elisabeth. "Manufacturing Jobs Are Exiting Mexico: Business Leaders Try to Stop the Exodus of Factories to China," *New York Times*, November 5, 2002, W1, W7.

McCullough, David G. *The Path Between the Seas: The Creation of the Panama Canal, 1870–1914* (New York: Simon & Schuster, 1977).

"Mexico's Indians: One Nation, or Many?," *The Economist*, January 20, 2001, 33–34.

Myers, Norman. *The Primary Source: Tropical Forests and Our Future* (New York: W. W. Norton, 1984).

Payne, Anthony, & Sutton, Paul. *Charting Caribbean Development* (Gainesville: University Press of Florida, 2001).

Pick, James B., & Butler, Edgar W. *Mexico Megacity* (Boulder, Colo.: Westview Press, 1997).

Portes, Alejandro, et al., eds. *The Urban Caribbean: Transition to a New Global Economy* (Baltimore, Md.: Johns Hopkins University Press, 1997).

Potter, Robert B. *The Urban Caribbean in an Era of Global Change* (Brookfield, Vt.: Ashgate, 2000).

- Preston, David A., ed. *Latin American Development: Geographical Perspectives* (London & New York: Longman, 2nd rev. ed., 1995).
- Richardson, Bonham C. *The Caribbean in the Wider World, 1492–1992: A Regional Geography* (New York: Cambridge University Press, 1992).

Rohter, Larry. "A Tiger in a Sea of Pussy Cats: Trinidad and Tobago Bid Goodbye to Oil, Hello to Gas," *New York Times*, September 4, 1998, C1–C2.

Rohter, Larry. "Asia Moves in on the Big Ditch," *New York Times*, December 19, 1999, WK 3.

Sargent, Charles S., Jr. "The Latin American City," in Blouet, Brian W., & Blouet, Olwyn M., eds., *Latin America and the Caribbean: A Systematic and Regional Survey* (New York: John Wiley & Sons, 3rd rev. ed., 1997), pp. 139–180. Diagram adapted from p. 173.

Sealey, Neil. *Caribbean World: A Complete Geography* (New York: Cambridge University Press, 1992).

Skelton, Tracey. *Introduction to the Pan-Caribbean* (New York: Oxford University Press, 2004).

"The Caribbean: Trouble in Paradise," *The Economist*, November 23, 2002, 36–37.

"The Mexicans Treated as Aliens in Their Own Country: The Right Not to be Hispanic," *The Economist*, March 7, 1998, 88–89.

Thompson, Ginger. "Chasing Mexico's Dream into Squalor: Misery on the Border," *New York Times*, February 11, 2001, 1, 6.

Trias Monge, José. *Puerto Rico: The Trials of the Oldest Colony in the World* (New Haven, Conn.: Yale University Press, 1997).

Warren, Kay B., & Jackson, Jean E. *Indigenous Movements, Self-Representation, and the State in Latin America* (Austin: University of Texas Press, 2003).

Watts, David. *The West Indies: Patterns of Development, Culture and Environmental Change Since 1492* (New York: Cambridge University Press, 1987).

West, Robert C. *Sonora: Its Geographical Personality* (Austin: University of Texas Press, 1993).

- West, Robert C., Augelli, John P., et al. *Middle America: Its Lands and Peoples* (Englewood Cliffs, N.J.: Prentice-Hall, 3rd rev. ed., 1989).

Winn, Peter. *Americas: The Changing Face of Latin America and the Caribbean* (Berkeley: University of California Press, updated ed., 1999).

CHAPTER 5

Allen, Christian M. *An Industrial Geography of Cocaine* (London & New York: Routledge, 2005).

Augelli, John P. "The Controversial Image of Latin America: A Geographer's View," *Journal of Geography*, 62 (1963): 103–112. Quotation from p. 111.

Barton, Jonathan R. *A Political Geography of Latin America* (London & New York: Routledge, 1997).

Becker, Bertha K., & Egler, Claudio A. G. *Brazil: A New Regional Power in the World-Economy: A Regional Geography* (New York: Cambridge University Press, 1992).

- Blouet, Brian W., & Blouet, Olwyn M. *Latin America and the Caribbean: A Systematic and Regional Survey* (Hoboken, N.J.: John Wiley & Sons, 4th rev. ed., 2002).

Box, Ben. "Latin America's New Transportation Links," *1999 Britannica Book of the Year* (Chicago: Encyclopedia Britannica, 1999), pp. 412–413.

Brawer, Moshe. *Atlas of South America* (New York: Simon & Schuster, 1991).

"Brazil: The Northeast—Politics, Water and Poverty," *The Economist*, August 29, 1998, 36–38.

Brea, Jorge A. "Population Dynamics in Latin America," *Population Bulletin*, 58 (March 2003): 1–36.

Brockerhoff, Martin P. "An Urbanizing World," *Population Bulletin*, 55 (September 2000): 1–44.

- Bromley, Rosemary D. F., & Bromley, Ray. *South American Development: A Geographical Introduction* (New York: Cambridge University Press, 2nd rev. ed., 1988).

Brooke, James. "The New South Americans: Friends and Partners," *New York Times*, April 8, 1994, A3.

Brookfield, Harold. *Exploring Agrodiversity* (New York: Columbia University Press, 2001).

Browder, John D., & Godfrey, Brian J. *Rainforest Cities: Urbanization, Globalization, and Development of the Amazon Basin* (New York: Columbia University Press, 1997).

Brunn, Stanley D., Williams, Jack F., & Zeigler, Donald J. *Cities of the World: World Regional Urban Development* (Lanham, Md.: Rowman & Littlefield, 3rd rev. ed., 2003).

- Caviedes, César N. & Knapp, Gregory. *South America* (Upper Saddle River, N.J.: Prentice Hall, 1995).

- Clawson, David L. *Latin America and the Caribbean: Lands and Peoples* (Dubuque, Iowa: WCB/McGraw-Hill, 3rd rev. ed., 2004).

Clawson, Patrick L., & Lee, Rensselaer W. *The Andean Cocaine Industry* (New York: St. Martin's Press, 1996).

Collier, Simon, et al., eds. *The Cambridge Encyclopedia of Latin America and the Caribbean* (New York: Cambridge University Press, 2nd rev. ed., 1992).

Denevan, William M. *Cultivated Landscapes of Native Amazonia and the Andes: Triumph Over the Soil* (New York: Oxford University Press, 2001).

"Drugs in Latin America: Battles Won, A War Still Lost," *The Economist*, February 12, 2005, 35–36.

Eakin, Marshall C. *Brazil: The Once and Future Country* (New York: St. Martin's Press, 1997).

Ford, Larry R. "A New and Improved Model of Latin American City Structure," *Geographical Review*, 86 (July 1996): 437–440.

Forero, Juan. "Colombia: The Next Oil Patch," *New York Times*, November 1, 2001, W1, W8.

"Free Trade on Trial," *The Economist*, January 3, 2004, 13.

"From Chaos, Order: What Can the World Do About State Failure?," *The Economist*, March 5, 2005, 45–47.

Gade, Daniel W. *Nature and Culture in the Andes* (Madison: University of Wisconsin Press, 1999).

- Gilbert, Alan. *Latin America* (London & New York: Routledge, 1990).

Gilbert, Alan, ed. *The Mega-City in Latin America* (New York & Tokyo: United Nations University Press, 1996).

Godfrey, Brian J. "Brazil," *Focus*, Summer 1999, 28 pp.

Godfrey, Brian J. "Revisiting Rio de Janeiro and São Paulo," *Geographical Review*, 89 (January 1999): 94–121.

Greenfield, Gerald M., ed. *Latin American Urbanization: Historical Profiles of Major Cities* (Westport, Conn.: Greenwood Press, 1994).

Griffin, Ernst. "Testing the von Thünen Theory in Uruguay," *Geographical Review*, 63 (October 1973): 500–516.

Griffin, Ernst C., & Ford, Larry R. "A Model of Latin American City Structure," *Geographical Review*, 70 (1980): 397–422. Model diagram adapted from p. 406.

Griffin, Ernst C., & Ford, Larry R. "Cities of Latin America," in Brunn, Stanley D., & Williams, Jack F., eds. *Cities of the World: World Regional Urban Development* (New York: HarperCollins, 2nd rev. ed., 1993), pp. 224–265.

Gugler, Josef, ed. *The Urban Transformation of the Developing World* (New York: Oxford University Press, 1996).

Gwynne, Robert N., & Kay, Cristobal. *Latin America Transformed: Globalization and Modernity* (New York: Oxford University Press, 2nd rev. ed., 2004).

"Indigenous Peoples in South America: A Political Awakening," *The Economist*, February 21, 2004, 35–37.

- James, Preston E., & Minkel, Clarence W. *Latin America* (New York: John Wiley & Sons, 5th rev. ed., 1986), Chaps. 17–37.

Kelly, Philip. *Checkerboards and Shatterbelts: The Geopolitics of South America* (Austin: University of Texas Press, 1997).

- Kent, Robert B. *Latin America* (New York: Guilford Press, 2006).

Knapp, Gregory, ed. *Latin America in the Twenty-First Century: Challenges and Solutions* (Austin: University of Texas Press, 2002).

McColl, Robert W. 'The Insurgent State: Territorial Bases of Revolution," *Annals of the Association of American Geographers*, 59 (1969): 613–631.

- Morris, Arthur S. *South America* (Totowa, N.J.: Barnes & Noble, 3rd rev. ed., 1987).

Myers, Norman. *The Primary Source: Tropical Forests and Our Future* (New York: W. W. Norton, 1984).

Pacione, Michael. *Urban Geography: A Global Perspective* (London & New York: Routledge, 2nd rev. ed., 2005).

Potter, Robert B. *Third World Urbanization: Contemporary Issues in Geography* (New York: Oxford University Press, 1991).

- Preston, David A., ed. *Latin American Development: Geographical Perspectives* (London & New York: Longman, 2nd rev. ed., 1995).

Ramos, Alcida Rita. "South America's Indigenous Peoples," *2000 Britannica Book of the Year* (Chicago: Encyclopedia Britannica, 2000), pp. 520–521.

Ribeiro, Darcy. *The Brazilian People: The Formation and Meaning of Brazil* (Gainesville: University Press of Florida, trans. Gregory Rabassa, 2000).

Rohter, Larry. "South America, A Rising Force in Agriculture, Seeks to Fill the World's Table," *New York Times*, December 12, 2004, 1, 22.

Schiffer, Sueli Ramos. "São Paulo: Articulating a Cross-Border Region," in Saskia Sassen, ed., *Global Networks, Linked Cities* (London & New York: Routledge, 2002), pp. 209–236.

Smith, Nigel J. H., et al. *Amazonia: Resiliency and Dynamism of the Land and Its People* (New York & Tokyo: United Nations University Press, 1995).

Sponsel, Leslie E., ed. *Indigenous Peoples and the Future of Amazonia* (Tucson: University of Arizona Press, 1995).

Tenenbaum, Barbara, ed. *Encyclopedia of Latin American History and Culture* (New York: Scribners, 5 vols., 1996).

"The Andean Coca Wars: A Crop That Refuses to Die," *The Economist*, March 4, 2000, 23–25.

"The Brazilian Amazon: Asphalt and the Jungle," *The Economist*, July 24, 2004, 33–35.

Warren, Kay B., & Jackson, Jean E. *Indigenous Movements, Self-Representation, and the State in Latin America* (Austin: University of Texas Press, 2003).

Wilkie, Richard W. *Latin American Population and Urbanization Analysis: Maps and Statistics, 1950–1982* (Westwood, Calif.: UCLA Latin American Center, 1984).

Winn, Peter. *Americas: The Changing Face of Latin America and the Caribbean* (Berkeley: University of California Press, updated ed., 1999).

Wood, Charles H., & Porro, Roberto, eds. *Deforestation and Land Use in the Amazon* (Gainesville: University Press of Florida, 2002).

World Bank. *World Resources, 1996–1997: The Urban Environment* (New York: Oxford University Press, 1996).

CHAPTER 6

Adams, William M., Goudie, Andrew S., & Orme, Antony R., eds. *The Physical Geography of Africa* (New York: Oxford University Press, 1996).

Adams, William M., & Mortimore, Michael J. *Working the Sahel: Environment and Society in Northern Nigeria* (London & New York: Routledge, 1999).

"Africa for Africans: A Survey of Sub-Saharan Africa," *The Economist*, September 7, 1996, special insert, 18 pp.

Agyei-Mensah, Samuel, & Casterline, John B., eds. *Reproduction and Social Context in Sub-Saharan Africa* (Westport, Conn.: Praeger, 2003).

- Aryeetey-Attoh, Samuel, ed. *Geography of Sub-Saharan Africa* (Upper Saddle River, N.J.: Prentice-Hall, 2nd rev. ed., 2003).

Azarya, Victor. *Nomads and the State in Africa: The Political Roots of Marginality* (Brookfield, Vt.: Ashgate, 1996).

Bassett, Thomas J., & Crummey, Donald, eds. *African Savannas: Global Narratives and Local Knowledge of Environmental Change* (Westport, Conn.: Praeger, 2003).

Berkeley, Bill. *The Graves Are Not Yet Full: Race, Tribe and Power in the Heart of Africa* (New York: Basic Books, 2001).

- Best, Alan C. G., & de Blij, H. J. *African Survey* (New York: John Wiley & Sons, 1977).

Binns, Tony, ed. *People and Environment in Africa* (New York: John Wiley & Sons, 1995).

- Bohannan, Paul, & Curtin, Philip D. *Africa and Africans* (Prospect Heights, Ill.: Waveland Press, 4th rev. ed., 1996).

Brookfield, Harold. *Exploring Agrodiversity* (New York: Columbia University Press, 2001).

Chapman, Graham P., & Baker, Kathleen M., eds. *The Changing Geography of Africa and the Middle East* (London & New York: Routledge, 1992).

Christopher, Anthony J. *Colonial Africa: An Historical Geography* (Totowa, N.J.: Barnes & Noble, 1984).

Christopher, Anthony J. *The Atlas of Changing South Africa* (London & New York: Routledge, 2nd rev. ed., 2001).

Cloudsley-Thompson, J. L., ed. *Sahara Desert* (Oxford: Pergamon Press, 1984).

Curtin, Philip D., et al. *African History: From Earliest Times to Independence* (London & New York: Longman, 2nd rev. ed., 1995).

Curtin, Philip D. *The Atlantic Slave Trade* (Madison: University of Wisconsin Press, 1969).

Curtis, Sarah, & Taket, Ann. *Health and Societies: Changing Perspectives* (London & New York: Edward Arnold, 1996).

- de Blij, H. J. "Africa's Geomosaic under Stress," *Journal of Geography*, 90 (January/February 1991): 2–9.

Djurfeldt, G., et al., eds. *The African Food Crisis: Lessons from the Asian Green Revolution* (Wallingford, UK: CABI Publishing, 2005).

Ezzell, Carol. "Care for a Dying Continent," *Scientific American*, May 2000, 96–105.

Fardon, Richard, & Furniss, Graham, eds. *African Languages, Development, and the State* (London & New York: Routledge, 1994).

Foster, Harold D. *Health, Disease and the Environment* (New York: John Wiley & Sons, 1992).

Fox, Roddy, & Rowntree, Kate, eds. *The Geography of South Africa in a Changing World* (New York: Oxford University Press, 2000).

Freeman-Grenville, G. S. P. *The New Atlas of African History* (New York: Simon & Schuster, 1991).

French, Howard W. *A Continent for the Taking: The Tragedy and Hope of Africa* (New York: Alfred A. Knopf, 2004).

Gates, Henry Louis, Jr., & Appiah, Kwame A., eds. *Africana: Encyclopedia of the African and African-American Experience* (New York: Basic Books, 1999).

Gesler, Wilbert M. *The Cultural Geography of Health Care* (Pittsburgh: University of Pittsburgh Press, 1991).

Gooneratne, Wilbert, & Obudho, Robert A., eds. *Contemporary Issues in Regional Development Policy: Perspectives from Eastern and Southern Africa* (Brookfield, Vt.: Ashgate, 1997).

• Gould, Peter R. *The Slow Plague: A Geography of the AIDS Pandemic* (Cambridge, Mass.: Blackwell, 1993).

Griffith, Daniel A., & Newman, James L., eds. *Eliminating Hunger in Africa: Technical and Human Perspectives* (Syracuse, N.Y.: Maxwell School, Syracuse University, Foreign and Comparative Studies/African Series 45, 1994).

Griffiths, Ieuan L. L. *The Atlas of African Affairs* (London & New York: Routledge, 2nd rev. ed., 1994).

• Grove, Alfred T. *The Changing Geography of Africa* (New York: Oxford University Press, 2nd rev. ed., 1994).

Gurdon, Charles G., ed. *The Horn of Africa* (New York: St. Martin's Press, 1994).

Haggett, Peter. *The Geographical Structure of Epidemics* (New York: Oxford University Press, 2000).

Harrison Church, Ronald J. *West Africa: A Study of the Environment and Man's Use of It* (London: Longman, 8th rev. ed., 1980).

Huke, Robert E. "The Green Revolution," *Journal of Geography*, 84 (1985): 248–254.

Kalipeni, Ezekiel, et al., eds. *HIV and AIDS in Africa: Beyond Epidemiology* (Malden, Mass.: Blackwell, 2003).

Kapuscinski, Ryszard. *The Shadow of the Sun* (New York: Alfred A. Knopf, trans. Klara Glowczewska, 2001).

Kearns, Robin A., & Gesler, Wilbert M., eds. *Putting Health into Place: Landscape, Identity, and Well-Being* (Syracuse, N.Y.: Syracuse University Press, 1998).

Knight, C. Gregory, & Newman, James L., eds. *Contemporary Africa: Geography and Change* (Englewood Cliffs, N.J.: Prentice-Hall, 1976).

Lamb, David. *The Africans* (New York: Random House, 1983).

Learmonth, Andrew T. A. *Disease Ecology: An Introduction* (New York: Blackwell, 1988).

Lemon, Anthony, ed. *Geography of Change in South Africa* (New York: John Wiley & Sons/Longman, 1995).

Lemon, Anthony, & Rogerson, C. M., eds. *Geography and Economy in South Africa and Its Neighbours* (Brookfield, Vt.: Ashgate, 2002).

Lewis, Laurence A., & Berry, Leonard. *African Environments and Resources* (Winchester, Mass.: Unwin Hyman, 1988).

Martin, Esmond Bradley, & de Blij, H. J., eds. *African Perspectives: An Exchange of Essays on the Economic Geography of Nine African States* (London & New York: Methuen, 1981).

McEvedy, Colin. *The Penguin Atlas of African History* (New York: Penguin Putnam, rev. ed., 2000).

Meade, Melinda S., & Earickson, Robert J. *Medical Geography* (New York: Guilford Press, 2nd rev. ed., 2000).

Mehretu, Assefa. *Regional Disparity in Sub-Saharan Africa: Structural Readjustment of Uneven Development* (Boulder, Colo.: Westview Press, 1989).

Middleton, John, ed. *Encyclopedia of Sub-Saharan Africa* (New York: Simon & Schuster, 4 vols., 1994).

Moon, Graham, & Jones, Kelvyn. *Health, Disease and Society: An Introduction to Medical Geography* (New York: Routledge & Kegan Paul, 1988).

• Mountjoy, Alan, & Hilling, David. *Africa: Geography and Development* (Totowa, N.J.: Barnes & Noble, 1987).

Murdock, George P. *Africa: Its Peoples and Their Culture History* (New York: McGraw-Hill, 1959).

Newman, James L. *The Peopling of Africa: A Geographic Interpretation* (New Haven, Conn.: Yale University Press, 1995).

O'Connor, Anthony M. *Poverty in Africa: A Geographical Approach* (New York: Columbia University Press, 1991).

Olson, James S. *The Peoples of Africa: An Ethnohistorical Dictionary* (Westport, Conn.: Greenwood Press, 1996).

Onishi, Norimitsu. "Rising Muslim Power in Africa Causes Unrest in Nigeria and Elsewhere," *New York Times*, November 1, 2001, A12.

Pakenham, Thomas. *The Scramble for Africa: The White Man's Conquest of the Dark Continent from 1876 to 1912* (New York: Random House, 1991).

Press, Robert M. *The New Africa: Dispatches from a Changing Continent* (Gainesville: University Press of Florida, 1999).

Pritchard, J. M. *Landform and Landscape in Africa* (London: Edward Arnold, 1979).

Rakodi, Carole, ed. *The Urban Challenge in Africa: Growth and Management of Its Large Cities* (New York & Tokyo: United Nations University Press, 1996).

Saff, Grant R. *Changing Cape Town: Urban Dynamics, Policy and Planning During the Political Transition in South Africa* (Lanham, Md.: University Press of America, 1998).

Senior, Michael, & Okunrotifa, P. *A Regional Geography of Africa* (London & New York: Longman, 1983).

Siddle, David, & Swindell, Ken. *Rural Change in Tropical Africa* (Cambridge, Mass.: Blackwell, 1994).

• Stock, Robert. *Africa South of the Sahara: A Geographical Interpretation* (New York: Guilford Press, 2nd rev. ed., 2004).

Stren, Richard, & White, Rodney, eds. *African Cities in Crisis: Managing Rapid Urban Growth* (Boulder, Colo.: Westview Press, 1989).

Theroux, Paul. *Dark Star Safari: Overland from Cairo to Cape Town* (Boston: Houghton Mifflin, 2003).

Turner, Bill L. II, Hyden, Goran, & Kates, Robert W., eds. *Population Growth and Agricultural Change in Africa* (Gainesville: University Press of Florida, 1993).

Waldmeir, Patti. *Anatomy of a Miracle: The End of Apartheid and the Birth of the New South Africa* (New York: W. W. Norton, 1997).

Wisner, Ben, Toulmin, Camilla, & Chitiga, Rutendo, eds. *Towards A New Map of Africa* (London: Earthscan, 2005).

CHAPTER 7

Ahmed, Akbar S. *Islam Today: A Short Introduction to the Muslim World* (London & New York: I. B. Tauris, 2nd rev. ed., 1999).

Amery, Hussein A., & Wolf, Aaron T. *Water in the Middle East: A Geography of Peace* (Austin: University of Texas Press, 2000).

• Anderson, Ewan W. *The Middle East: Geography and Geopolitics* (London & New York: Routledge, 2000).

"At the Crossroads: A Survey of Central Asia," *The Economist*, July 26, 2003, special insert, 16 pp.

Atlas of the Middle East (Washington, D.C.: National Geographic Society, 2003).

• Beaumont, Peter, et al. *The Middle East: A Geographical Study* (New York: John Wiley & Sons/Halsted, 2nd rev. ed., 1988).

Blake, Gerald H., et al. *The Cambridge Atlas of the Middle East and North Africa* (New York: Cambridge University Press, 1988).

Bonine, Michael E., ed. *Population, Poverty, and Politics in Middle East Cities* (Gainesville: University Press of Florida, 1997).

Chapman, Graham P., & Baker, Kathleen M., eds. *The Changing Geography of Africa and the Middle East* (London & New York: Routledge, 1992).

Chinn, Jeff, & Kaiser, Robert J. *Russians as the New Minority: Ethnicity and Nationalism in the Soviet Successor States* (Boulder, Colo.: Westview Press, 1996).

Cloudsley-Thompson, J. L., ed. *Sahara Desert* (Oxford: Pergamon Press, 1984).

Cohen, Saul B. "Middle East Geopolitical Transformation: The Disappearance of a Shatter Belt," *Journal of Geography*, 91 (January/February 1992): 2–10.

Collins, Robert O. *The Nile* (New Haven, Conn.: Yale University Press, 2002).

• Cressey, George B. *Crossroads: Land and Life in Southwest Asia* (Philadelphia: J. B. Lippincott, 1960).

Dekmejian, R. Hrair, & Simonian, Hovann H. *Troubled Waters: The Geopolitics of the Caspian Region* (London & New York: I. B. Tauris, 2001).

Dowry, Alan. *The Jewish State: A Century Later* (Berkeley: University of California Press, 1998).

• Drysdale, Alasdair, & Blake, Gerald H. *The Middle East and North Africa: A Political Geography* (New York: Oxford University Press, 1985).

"Dubai: Arabia's Field of Dreams," *The Economist*, May 29, 2004, 61–62.

Ebel, Robert, & Menon, Rajan, eds., *Energy and Conflict in Central Asia and the Caucasus* (Lanham, Md.: Rowman & Littlefield, 2000).

Engelmann, Kurt E., & Pavlakovic, Vjeran, eds. *Rural Development in Eurasia and the Middle East: Land Reform, Demographic Change, and Environmental Constraints* (Seattle: University of Washington Press, 2001).

Esposito, John L. *What Everyone Needs to Know About Islam* (New York: Oxford University Press, 2002).

• Fisher, William B. *The Middle East: A Physical, Social and Regional Geography* (London & New York: Methuen, 7th rev. ed., 1978).

Freeman-Grenville, G. S. P. *The Historical Atlas of the Middle East* (New York: Simon & Schuster, 1993).

Fuller, Graham E., & Francke, Rend Rahim. *The Arab Shi'a: The Forgotten Muslims* (New York: St. Martin's Press, 2000).

Gilbert, Martin. *Israel* (New York: William Morrow, 1998).

Gleason, Gregory. *The Central Asian States: Discovering Independence* (Boulder, Colo.: Westview Press, 1997).

Goldscheider, Calvin. *Israel's Changing Society: Population, Ethnicity, and Development* (Boulder, Colo.: Westview Press, 2nd rev. ed., 2002).

Goldschmidt, Arthur, Jr. *A Concise History of the Middle East* (Boulder, Colo.: Westview Press, 7th rev. ed., 2001).

• Gould, Peter R. *Spatial Diffusion* (Washington, D.C.: Association of American Geographers, Commission on College Geography, Resource Paper No. 4, 1969).

Gradus, Yehuda, & Lipshitz, Gabi, eds. *The Mosaic of Israeli Geography* (Beersheva, Israel: Ben Gurion University Press, 1996).

Heathcote, Ronald L. *The Arid Lands: Their Use and Abuse* (London & New York: Longman, 1983).

• Held, Colbert C. *Middle East Patterns: Places, Peoples, and Politics* (Boulder, Colo.: Westview Press, 3rd ed., 2000).

Hourani, Albert H. *A History of the Arab Peoples* (Cambridge, Mass.: Belknap/Harvard University Press, 1991).

Joffé, George, ed. *North Africa: Nation, State and Region* (London & New York: Routledge, 1993).

Khalidi, Rashid I. "The Middle East as an Area in an Era of Globalization," in Ali Mirsepassi et al., eds., *Localizing Knowledge in a Globalizing World: Recasting the Area Studies Debate* (Syracuse, N.Y.: Syracuse University Press, 2003), pp. 171–190.

Kleveman, Lutz. *The New Great Game: Blood and Oil in Central Asia* (Boston: Atlantic Monthly Press, 2003).

Lamb, David. *The Arabs: Journeys Beyond the Mirage* (New York: Random House, 1987).

Lemarchand, Philippe, ed. *The Arab World, the Gulf, and the Middle East: An Atlas* (Boulder, Colo.: Westview Press, 1992).

Lewis, Bernard. *What Went Wrong? Western Impact and Middle East Response* (New York: Oxford University Press, 2002).

Lewis, Robert A., ed. *Geographic Perspectives on Soviet Central Asia* (London & New York: Routledge, 1992).

• Longrigg, Stephen H. *The Middle East: A Social Geography* (Chicago: Aldine, 2nd rev. ed., 1970).

Matvejevic, Predag. *The Mediterranean: A Cultural Landscape* (Berkeley: University of California Press, trans. Michael H. Heim, 1999).

Meiselas, Susan. *Kurdistan: In the Shadow of History* (New York: Random House, 1998).

Mostyn, Trevor, ed. *The Cambridge Encyclopedia of the Middle East and North Africa* (New York: Cambridge University Press, 1988).

Newman, David. *The Dynamics of Territorial Change: A Political Geography of the Arab-Israeli Conflict* (Boulder, Colo.: Westview Press, 1998).

Polat, Necati. *Boundary Issues in Central Asia* (Ardsley, N.Y.: Transnational Publishers, 2002).

Prescott, J. R. V. *Political Frontiers and Boundaries* (Winchester, Mass.: Allen & Unwin, 1987).

Pryde, Philip R., ed. *Environmental Resources and Constraints in the Former Soviet Republics* (Boulder, Colo.: Westview Press, 1995).

Rashid, Ahmed. *Taliban: Militant Islam, Oil, and Fundamentalism in Central Asia* (New Haven, Conn.: Yale University Press, 2000).

Robinson, Francis, ed. *Atlas of the Islamic World Since 1500* (New York: Facts on File, 1982).

Rubin, Barnett R. *The Fragmentation of Afghanistan* (New Haven, Conn.: Yale University Press, 2nd rev. ed., 2002).

Sicker, Martin. *The Middle East in the Twentieth Century* (Westport, Conn.: Praeger, 2001).

Smart, Ninian. *The World's Religions* (New York: Cambridge University Press, 2nd rev. ed., 1998).

Soffer, Arnon. *Rivers of Fire: The Conflict over Water in the Middle East* (Lanham, Md.: Rowman & Littlefield, trans. Murray Rosovsky & Nina Copaken, 1999).

Swearingen, Will D., & Bencherifa, Abdellatif, eds. *The North African Environment at Risk* (Boulder, Colo.: Westview Press, 1996).

Waines, David. *An Introduction to Islam* (New York: Cambridge University Press, 1995).

Wines, Michael. "Grand Soviet Scheme for Sharing Water in Central Asia Is Foundering," *New York Times*, December 9, 2002, A14.

Wright, Robin. *Sacred Rage: The Wrath of Militant Islam* (New York: Simon & Schuster, 2nd rev. ed., 2002).

Wright, Robin. *The Last Great Revolution: Turmoil and Transformation in Iran* (New York: Alfred A. Knopf, 2000).

Zoubir, Yahia H., ed. *North Africa in Transition: State, Society, and Economic Transformation in the 1990s* (Gainesville: University Press of Florida, 1999).

CHAPTER 8

"A Survey of India's Economy: The Plot Thickens," *The Economist*, June 2, 2001, special insert, 22 pp.

Baxter, Craig. *Bangladesh: From a Nation to a State* (Boulder, Colo.: Westview Press, 1997).

Bennett Jones, Owen. *Pakistan: The Eye of the Storm* (New Haven, Conn.: Yale University Press, 2002).

Chapman, Graham P. *The Geopolitics of South Asia: From Early Empires to India, Pakistan and Bangladesh* (Brookfield, Vt.: Ashgate, 2nd rev. ed., 2003).

Chapman, Graham P., Dutt, Ashok K., & Bradnock, Robert W., eds. *Urban Growth and Development in Asia. Volume I: Making the Cities; Volume II: Living in the Cities* (Brookfield, Vt.: Ashgate, 1999).

Corbridge, Stuart E., guest ed. "India: 1947–1997," *Environment and Planning A*, 27, No. 12 (1997).

Corbridge, Stuart E., & Harriss, John. *Reinventing India: Liberalization, Hindu Nationalism and Popular Democracy* (Malden, Mass.: Blackwell/Polity, 2000).

Crossette, Barbara. *India: Facing the Twenty-First Century* (Bloomington: Indiana University Press, 1993).

Crossette, Barbara. *So Close to Heaven: The Vanishing Buddhist Kingdoms of the Himalayas* (New York: Alfred A. Knopf, 1995).

Deshpande, C. D. *India: A Regional Interpretation* (New Delhi: Northern Book Centre/Indian Council of Social Science Research, 1992).

Dumont, Louis. *Homo Hierarchus: The Caste System and Its Implications* (Chicago: University of Chicago Press, 1970).

Dutt, Ashok K., & Geib, Margaret. *An Atlas of South Asia* (Boulder, Colo.: Westview Press, 1987).

Dutt, Ashok K., & Pomeroy, George M. "Cities of South Asia," in Brunn, Stanley D., Williams, Jack F., & Ziegler, Donald J., eds., *Cities of the World: World Regional Urban Development* (Lanham, Md.: Rowman & Littlefield, 3 rev. ed., 2003), pp. 330–371.

Dyson, Tim, Cassen, Robert, & Visaria, Leela, eds. *Twenty-First Century India: Population, Economy, Human Development, and the Environment* (New York: Oxford University Press, 2004).

Er-Rashid, Haroun. *Geography of Bangladesh* (Boulder, Colo.: Westview Press, 1977).

• Farmer, B. H. *An Introduction to South Asia* (London & New York: Routledge, 2nd rev. ed, 1993).

Gupta, Akhil. *Postcolonial Developments: Agriculture in the Making of Modern India* (Durham, N.C.: Duke University Press, 1998).

Haque, C. Emdad. *Hazards in a Fickle Environment: Bangladesh* (Hingham, Mass.: Kluwer, 1998).

Hasan, Mushirul. *Legacy of a Divided Nation: India's Muslims from Independence to Ayodhya* (Boulder, Colo.: Westview Press, 1997).

Huke, Robert E. "The Green Revolution," *Journal of Geography*, 84 (1985): 248–254.

Ives, Jack D., & Messerli, Bruno. *Himalayan Dilemma: Reconciling Development and Conservation* (London & New York: Routledge, 1989).

Jaffrelot, Christopher. *The Hindu Nationalist Movement in India* (New York: Columbia University Press, 1997).

Javed, Shaid. *Pakistan: Fifty Years of Nationhood* (Boulder, Colo.: Westview, 3rd rev. ed., 1999).

Johnson, Basil L. C. *Bangladesh* (Totowa, N.J.: Barnes & Noble, 2nd rev. ed., 1982).

Johnson, Basil L. C. *India: Resources and Development* (Totowa, N.J.: Barnes & Noble, 1979).

Johnson, Gordon, ed. *Cultural Atlas of India* (New York: Facts on File, 1996).

Karan, Pradyumna P., & Ishii, Hiroshi, eds. *Nepal: A Himalayan Kingdom in Transition* (New York & Tokyo: United Nations University Press, 1996).

Karan, Pradyumna P., & Ishii, Hiroshi. *Nepal: Development and Change in a Landlocked Himalayan Kingdom* (Tokyo: Institute for the Studies of Languages and Cultures of Asia and Africa, Tokyo University of Foreign Studies, 1994).

Khilnani, Sunil. *The Idea of India* (New York: Farrar, Straus & Giroux, 1997).

Kosinski, Leszek A., & Elahi, K. Maudood, eds. *Population Redistribution and Development in South Asia* (Hingham, Mass.: D. Reidel, 1985).

Lall, Arthur S. *The Emergence of Modern India* (New York: Columbia University Press, 1981).

Lipner, Julius J. *Hinduism* (London & New York: Routledge, 1994).

Lukacs, John R., ed. *The People of South Asia: The Biological Anthropology of India, Pakistan, and Nepal* (New York: Plenum Press, 1984).

- McFalls, Joseph A., Jr. "Population: A Lively Introduction—Fourth Edition," *Population Bulletin*, 58 (December 2003): 1–40.

McGowan, William. *Only Man Is Vile: The Tragedy of Sri Lanka* (New York: Farrar, Straus & Giroux, 1992).

Mitra, Subrata K., ed. *Subnational Movements in South Asia* (Boulder, Colo.: Westview Press, 1994).

Mumtaz, Khavar, et al. *Pakistan* (Sterling, Va.: Stylus, 2nd rev. ed., 2003).

Murphey, Rhoads. *A History of Asia* (New York: HarperCollins, 1992).

- Muthiah, S., ed. *An Atlas of India* (New York: Oxford University Press, 1992).
- Newman, James L., & Matzke, Gordon E. *Population: Patterns, Dynamics, and Prospects* (Englewood Cliffs, N.J.: Prentice-Hall, 1984).

Noble, Allen G., & Dutt, Ashok K., eds. *India: Cultural Patterns and Processes* (Boulder, Colo.: Westview Press, 1982).

Roberts, Paul W. *Empire of the Soul: Some Journeys in India* (New York: Riverhead, 1999).

Robinson, Francis, ed. *The Cambridge Encyclopedia of India, Pakistan, Bangladesh, Sri Lanka, Nepal, Bhutan, and the Maldives* (New York: Cambridge University Press, 1989).

Samad, Yunas. *A Nation in Turmoil: Nationalism and Ethnicity in Pakistan* (Thousand Oaks, Calif.: Sage, 1995).

Sanderson, Warren C., & Tan, Jee-Peng. *Population in Asia* (Brookfield, Vt.: Ashgate, 1996).

Schwartzberg, Joseph E. *A Historical Atlas of South Asia: 2nd Impression, With Additional Material* (New York: Oxford University Press, 1992).

Shurmer-Smith, Pamela. *India: Globalization and Change* (London: Arnold, 2000).

Sopher, David E., ed. *An Exploration of India: Geographical Perspectives on Society and Culture* (Ithaca, N.Y.: Cornell University Press, 1980).

- Spate, Oskar H. K., & Learmonth, Andrew T. A. *India and Pakistan: A General and Regional Geography* (London: Methuen, 2 vols., 1971).
- Spencer, Joseph E., & Thomas, William L., Jr. *Asia, East by South: A Cultural Geography* (New York: John Wiley & Sons, 2nd rev. ed., 1971).

Srinivas, Smriti. *Landscapes of Urban Memory: The Sacred and the Civic in India's High-Tech City* (Minneapolis: University of Minnesota Press, 2001).

Thapar, Valmik. *Land of the Tiger: A Natural History of the Indian Subcontinent* (Berkeley: University of California Press, 1998).

"The Tiger in Front: A Survey of India and China," *The Economist*, March 5, 2005, special insert, 16 pp.

"Time to Let Go: A Survey of India," *The Economist*, February 22, 1997, special insert, 26 pp.

- Weightman, Barbara A. *Dragons and Tigers: A Geography of South, East, and Southeast Asia*, 2nd ed., (Hoboken, N.J.: John Wiley & Sons, 2006).

Wirsing, Robert G. *India, Pakistan, and the Kashmir Dispute: On Regional Conflict and Its Resolution* (New York: St. Martin's Press, 1994).

Wirsing, Robert G. *War or Peace on the Line of Control? The India-Pakistan Dispute Over Kashmir Turns Fifty* (Durham, UK: International Boundaries Research Unit,

University of Durham, Vol. 2, Boundary and Territory Briefing No. 5, 1998).

- Wolpert, Stanley. *India* (Berkeley: University of California Press, updated ed., 1999).

Zachariah, K. C., & Rajan, S. Irudaya, eds. *Kerala's Demographic Transition* (Thousand Oaks, Calif.: Sage, 1998).

Zurick, David, & Karan, Pradyumna P. *Himalaya: Life on the Edge of the World* (Baltimore, Md.: Johns Hopkins University Press, 1999).

CHAPTER 9

Atlas of Population, Environment, and Sustainable Development of China (Beijing & New York: Science Press, 2000).

Bleiker, Roland. *Divided Korea: Toward a Culture of Reconciliation* (Minneapolis: University of Minnesota Press, 2005).

Brooke, James. "Mongolia's Shifting Ties: More China, Less Russia," *New York Times*, July 9, 2004, W1, W7.

Brooke, James. "The Asian Battle for Russia's Oil and Gas," *New York Times*, January 3, 2004, B1, B3.

- Burks, Ardath W. *Japan: A Postindustrial Power* (Boulder, Colo.: Westview Press, 3rd rev. ed., 1991).

Cartier, Carolyn. *Globalizing South China* (Malden, Mass.: Blackwell, 2001).

Chapman, Graham P., Dutt, Ashok K., & Bradnock, Robert W., eds. *Urban Growth and Development in Asia. Volume I: Making the Cities; Volume II: Living in the Cities* (Brookfield, Vt.: Ashgate, 1999).

"China's New Revolution: Remaking Our World, One Deal at a Time," *Time*, June 27, 2005, 24–53.

Cressey, George B. *Asia's Lands and Peoples: A Geography of One-Third of the Earth and Two-Thirds of Its People* (New York: McGraw-Hill, 3rd rev. ed., 1963).

Cumings, Bruce. *Korea's Place in the Sun: A Modern History* (New York: W. W. Norton, 1997).

Cybriwsky, Roman A. *Tokyo: The Changing Profile of an Urban Giant* (Boston: G. K. Hall, 1991).

Dirlik, Arif, ed. *What Is in a Rim? Critical Perspectives on the Pacific Region Idea* (Lanham, Md.: Rowman & Littlefield, 2nd rev. ed., 1998).

Donald, Stephanie, H., & Benewick, Robert. *The State of China Atlas: Mapping the World's Fastest-Growing Economy* (Berkeley: University of California Press, 2005).

Dwyer, Denis J., ed. *China: The Next Decades* (New York: John Wiley & Sons/Longman, 1993).

Eccleston, Bernard, Dawson, Michael, & McNamara, Deborah, eds. *The Asia-Pacific Profile* (London & New York: Routledge, 1998).

Enright, Michael J., Scott, Edith E., & Chang, Ka-Mun. *Regional Powerhouse: The Greater Pearl River Delta and the Rise of China* (Hoboken, N.J.: John Wiley & Sons, 2005).

Fairbank, John King, & Goldman, Merle. *China: A New History* (Cambridge, Mass.: Harvard University Press, 2nd rev. ed., 2005).

Fishman, Ted C. "The China Bind: The Global Economy Is Not Likely to Be Led by the U.S. Forever. Is This the Beginning of the End?," *New York Times Magazine*, July 4, 2004, 24–31, 46, 50–51.

Ginsburg, Norton S., ed. *The Pattern of Asia* (Englewood Cliffs, N.J.: Prentice-Hall, 1958).

"Go West, Young Han: Plans to Develop China's Western Provinces Are About More than Economics," *The Economist*, December 23, 2000, 45–46.

Harrell, Stevan, ed. *Cultural Encounters on China's Ethnic Frontiers* (Seattle: University of Washington Press, 1995).

Harris, Chauncy D. "The Urban and Industrial Transformation of Japan," *Geographical Review*, 72 (January 1982): 50–89.

Hillel, Daniel. "Lash of the Dragon: China's Yellow River Remains Untamed," *Natural History*, August 1991, 28–37.

Hoare, James, & Pares, Susan. *Korea: An Introduction* (London & New York: Routledge, 1988).

- Hodder, Rupert. *The West Pacific Rim: An Introduction* (New York: John Wiley & Sons, 1992).

Hoh, Erling. "The Long River's Journey Ends: China's Three Gorges Dam Will Soon Transform the Yangtze," *Natural History*, July 1996, 28–38.

"Hong Kong and Shanghai: Rivals More Than Ever," *The Economist*, March 30, 2002, 19–21.

Hook, Brian, ed. *The Cambridge Encyclopedia of China* (New York: Cambridge University Press, 2nd rev. ed., 1994).

Hsieh, Chiao-min, & Hsieh, Jean Kan. *China: A Provincial Atlas* (New York: Macmillan, 1992).

- Hsieh, Chiao-Min, & Lu, Max. *Changing China: A Geographical Appraisal* (Boulder, Colo.: Westview Press, 2nd rev. ed., 2004).

Hsu, Francis L. K., & Serrie, Hendrick. *The Overseas Chinese: Ethnicity in National Context* (Lanham, Md.: University Press of America, 1998).

"Inside the New China," *Fortune*, special issue, October 4, 2004.

- Karan, Pradyumna P. *Japan in the 21st Century: Environment, Economy, and Society* (Lexington: University Press of Kentucky, 2005).

Karan, Pradyumna P., & Stapleton, Kristin, eds. *The Japanese City* (Lexington: University Press of Kentucky, 1997).

Knapp, Ronald G., ed. *Chinese Landscapes: The Village as Place* (Honolulu: University of Hawai'i Press, 1992).

- Kornhauser, David H. *Japan: Geographical Background to Urban-Industrial Development* (London & New York: Longman, 2nd rev. ed., 1982).

Kristof, Nicholas D., & WuDunn, Sheryl. *Thunder from the East: Portrait of a Rising Asia* (New York: Alfred A. Knopf, 2000).

• Leeming, Frank. *The Changing Geography of China* (Cambridge, Mass.: Blackwell, 1993).

Le Heron, Richard, & Park, Sam Ock, eds. *The Asian Pacific Rim and Globalization: Enterprise, Governance, and Territoriality* (Brookfield, Vt.: Ashgate, 1995).

Lin, George C. S. *Red Capitalism in South China: Growth and Development of the Pearl River Delta* (Vancouver: University of British Columbia Press, 1997).

Linge, Godfrey J. R., ed. *China's New Spatial Economy: Heading Towards 2020* (New York: Oxford University Press, 1997).

Linge, Godfrey J. R., & Forbes, D. K., eds. *China's Spatial Economy* (New York: Oxford University Press, 1990).

Lo, Chor-Pang. *Hong Kong* (New York: John Wiley & Sons/Belhaven, 1992).

Lo, Fu-chen, & Yeung, Yue-man, eds. *Emerging World Cities in Pacific Asia* (New York & Tokyo: United Nations University Press, 1996).

MacDonald, Donald. *A Geography of Modern Japan* (Ashford, UK: Paul Norbury, 1985).

Mackerras, Colin. *China's Minority Cultures: Identities and Integration Since 1912* (New York: St. Martin's Press, 1995).

Mackerras, Colin. *The New Cambridge Handbook of Contemporary China* (New York: Cambridge University Press, 2001).

Marton, Andrew M. *China's Spatial Economic Development: Restless Landscapes in the Lower Yangzi Delta* (London & New York: Routledge, 2000).

McColl, Robert W., guest ed. "Special Theme Issue: Mongolia," *Focus*, Fall 2002, 1–37.

Murphey, Rhoads. *A History of Asia* (New York: HarperCollins, 1992).

Noh, Toshio, & Kimura, John C., eds. *Japan: A Regional Geography of an Island Nation* (Tokyo: Teikoku-Shoin, 1985).

Nolan, Peter. *Transforming China: Globalization, Transition and Development* (Sterling, Va.: Stylus, 2003).

Ohmae, Kenichi. *The End of the Nation-State: The Rise of Regional Economies* (New York: Free Press, 1996).

Olds, Kris, et al., eds. *Globalization and the Asia Pacific: Contested Territories* (London & New York: Routledge, 2000).

Pannell, Clifton W. "China's Urban Transition," *Journal of Geography*, 94 (May/June 1995): 394–403.

• Pannell, Clifton W., & Ma, Laurence J. C. *China: The Geography of Development and Modernization* (New York: Halsted Press/V. H. Winston, 1983).

• Pannell, Clifton W., & Veeck, Gregory. *East Asia* (Lanham, Md.: Rowman & Littlefield, 1999).

Porter, Jonathan. *Macau: The Imaginary City* (Boulder, Colo.: Westview Press, 1996).

Potter, Robert B., et al. *Geographies of Development* (Harlow, UK & New York: Longman, 1999).

Preston, Peter W. *Pacific Asia in the Global System: An Introduction* (Malden, Mass.: Blackwell, 1998).

Riley, Nancy E. "China's Population: New Trends and Challenges," *Population Bulletin*, 59 (June 2004): 1–36.

Rozman, Gilbert, ed. *Japan and Russia: The Tortuous Path to Normalization, 1949–1999* (New York: St. Martin's Press, 2000).

Rumley, Dennis, et al., eds. *Global Geopolitical Change and the Asia-Pacific: A Regional Perspective* (Brookfield, Vt.: Ashgate, 1996).

"Shanghai: FT Survey." *Financial Times*, special supplement, October 16, 2001, 10 pp.

Shen, Jianfa. *Internal Migration and Regional Population Dynamics in China* (Amsterdam & New York: Elsevier, 1996).

Si-ming Li, & Wing-shing Tang, eds. *China's Regions, Polity, and Economy: A Study of Spatial Transformation in the Post-Reform Era* (Hong Kong: Chinese University Press, 2000).

Sit, Victor F. S. *Beijing: The Nature and Planning of a Chinese Capital City* (New York: John Wiley & Sons, 1995).

• Smith, Christopher J. *China in the Post-Utopian Age* (Boulder, Colo.: Westview Press, 2000).

Smith, Patrick. *Japan: A Reinterpretation* (New York: Pantheon, 1997).

• Spencer, Joseph E., & Thomas, William L., Jr. *Asia, East by South: A Cultural Geography* (New York: John Wiley & Sons, 2nd rev. ed., 1971).

Starr, John Bryan. *Understanding China: A Guide to China's Economy, History, and Political Structure* (New York: Hill & Wang, 1997).

Terrill, Ross. *The New Chinese Empire: Beijing's Political Dilemma and What It Means for the United States* (Boulder, Colo.: Westview Press, 2003).

"The Other [Northeast] China: Parts That the Bulldozers Have Not Yet Reached," *The Economist*, January 10, 2004, 59–61.

Theroux, Paul. "Going to See the Dragon: A Journey Through South China, Land of Capitalist Miracles, Where Yesterday's Rice Paddy Becomes Tomorrow's Metropolis, and a Thousand Factories Bloom," *Harper's Magazine*, October 1993, 33–56.

Trewartha, Glenn T. *Japan: A Geography* (Madison: University of Wisconsin Press, 2nd rev. ed., 1965).

van Kemenade, Willem. *China, Hong Kong, Taiwan Inc.: The Dynamics of a New Empire* (New York: Alfred A. Knopf, 1997). Quotation taken from p. 184.

Veeck, Gregory, ed. *The Uneven Landscape: Geographic Studies in Post-Reform China* (Baton Rouge, La.: Geoscience Publications, 1991).

Watters, R. F., & McGee, T. G. *Asia-Pacific: New Geographies of the Pacific Rim* (London: Hurst, 1997).

Wei, Yehua Dennis. *Regional Development in China: States, Globalization, and Inequality* (London & New York: Routledge, 2000).

- Weightman, Barbara A. *Dragons and Tigers: A Geography of South, East, and Southeast Asia*, 2nd ed., (Hoboken, N.J.: John Wiley & Sons, 2006).

Weiping, Wu. *Pioneering Economic Reform in China's Special Economic Zones: The Promotion of Foreign Investment and Technology Transfer in Shenzhen* (Brookfield, Vt.: Ashgate, 1999).

Witherick, M., & Carr, M. *The Changing Face of Japan* (London: Hodder & Stoughton, 1993).

Wittwer, Sylvan, et al. *Feeding a Billion: Frontiers of Chinese Agriculture* (East Lansing: Michigan State University Press, 1987).

Yatsko, Pamela. *New Shanghai: The Rocky Rebirth of China's Legendary City* (Singapore: John Wiley & Sons [Asia], 2001).

Yeung, Yue-man, & Chu, David K. Y., eds. *Guangdong: Survey of a Province Undergoing Rapid Change* (Hong Kong: Chinese University Press, 1994).

Yeung, Yue-man, & Hu, Xu-Wei, eds. *China's Coastal Cities: Catalysts for Modernization* (Honolulu: University of Hawai'i Press, 1992).

Yeung, Yue-man, & Jianfa, Shen, eds. *Developing China's West: A Critical Path to Balanced National Development* (Hong Kong: Chinese University Press, 2004).

Yeung, Yue-man, Jianfa, Shen, & Li, Zhang. *The Western Pearl River Delta: Growth and Opportunities for Cooperation Development with Hong Kong* (Hong Kong: Chinese University Press, 2005).

Yeung, Yue-man, & Yun-wing, Sung, eds. *Shanghai: Transformation and Modernization under China's Open Policy* (Hong Kong: Chinese University Press, 1996).

- Zhao Songqiao. *Geography of China: Environment, Resources, Population, and Development* (New York: John Wiley & Sons, 1994).

Zhao Songqiao. *Physical Geography of China* (New York & Beijing: John Wiley & Sons/Science Press, 1986).

CHAPTER 10

Barton, Greg. *Indonesia's Struggle: Jemaah Islamiyah and the Soul of Islam* (Sydney: University of New South Wales Press, 2005).

Broek, Jan O. M. "Diversity and Unity in Southeast Asia," *Geographical Review*, 34 (1944): 175–195.

Brookfield, Harold C., et al. *In Place of the Forest: Environmental and Socio-Economic Transformation in Borneo and the Eastern Malay Peninsula* (New York: United Nations University Press, 1995).

Bunnell, Tim. *Malaysia, Modernity and the Multimedia Super Corridor: The Construction of a High-Tech City-Region* (London & New York: Routledge, 2002).

Cleary, Mark, & Eaton, Peter. *Tradition and Reform: Land Tenure and Rural Development in South-East Asia* (New York: Oxford University Press, 1997).

Cohen, Saul B. *Geopolitics of the World System* (Lanham, Md.: Rowman & Littlefield, 2002).

Daws, Gavan, & Fujita, Marty. *Archipelago: The Islands of Indonesia* (Berkeley: University of California Press, 1999).

- Dixon, Chris. *South East Asia in the World Economy* (New York: Cambridge University Press, 1991).

Dixon, Chris, & Drakakis-Smith, David, eds. *Uneven Development in Southeast Asia* (Brookfield, Vt.: Ashgate, 1998).

Dwyer, Denis J., ed. *South East Asian Development* (New York: John Wiley & Sons/Longman, 1990).

- Fisher, Charles A. *Southeast Asia: A Social, Economic and Political Geography* (New York: E. P. Dutton, 2nd rev. ed., 1966).

Friend, Theodore. *Indonesian Destinies* (Cambridge, Mass.: Belknap/Harvard University Press, 2003).

- Fryer, Donald W. *Emerging South-East Asia: A Study in Growth and Stagnation* (New York: John Wiley & Sons, 2nd rev. ed., 1979).

Gargan, Edward A. *The River's Tale: A Year on the Mekong* (New York: Alfred A. Knopf, 2002).

Ginsburg, Norton S., et al., eds. *The Extended Metropolis: Settlement Transition in Asia* (Honolulu: University of Hawai'i Press, 1991).

- Glassner, Martin I., & de Blij, H. J. *Systematic Political Geography* (New York: John Wiley & Sons, 4th rev. ed., 1989).

"Good Fences: Borders Are Arbitrary Abstractions, Economic Impediments and Surprisingly Ineradicable," *The Economist*, December 19, 1998, 19–22.

Gupta, Avijit, ed. *The Physical Geography of Southeast Asia* (New York: Oxford University Press, 2005).

Hartshorne, Richard. "Suggestions on the Terminology of Political Boundaries," *Annals of the Association of American Geographers*, 26 (1936): 56–57.

Hill, Ronald D., ed. *South-East Asia: A Systematic Geography* (New York: Oxford University Press, 1979).

Hodder, Rupert. *The West Pacific Rim: An Introduction* (New York: John Wiley & Sons, 1992).

Hsu, Francis L. K., & Serrie, Hendrick. *The Overseas Chinese: Ethnicity in National Context* (Lanham, Md.: University Press of America, 1998).

Karnow, Stanley. *In Our Image: America's Empire in the Philippines* (New York: Random House, 1989).

Leinbach, Thomas R., ed. *The Indonesian Rural Economy: Mobility, Work and Enterprise* (Seattle: University of Washington Press, 2004).

Leinbach, Thomas R., & Sien, Chia Lin, eds. *South-East Asian Transport* (New York: Oxford University Press, 1989).

- Leinbach, Thomas R., & Ulack, Richard, eds. *Southeast Asia: Diversity and Development* (Upper Saddle River, N.J.: Prentice Hall, 2000).

Logan, William S., ed. *The Disappearing 'Asian' City: Protecting Asia's Urban Heritage in a Globalizing World* (New York: Oxford University Press, 2002).

McGee, Terence G. *The Mega-Urban Regions of Southeast Asia* (Vancouver, Canada: University of British Columbia Press, 1995).

McGee, Terence G. *The Southeast Asian City: A Social Geography* (New York: Praeger, 1967).

Mulder, Neils. *Inside Southeast Asia: Religion, Everyday Life, and Cultural Change* (Seattle: University of Washington Press, 2001).

Murray, Geoffrey, & Perera, Audrey. *Singapore: The Global City-State* (New York: St. Martin's Press, 1996).

Owen, Norman G., ed. *The Emergence of Modern Southeast Asia: A New History* (Honolulu: University of Hawai'i Press, 2005).

Parnwell, Michael J. G., & Bryant, Raymond L., eds. *Environmental Change in South-East Asia: People, Politics and Sustainable Development* (London & New York: Routledge, 1996).

- Prescott, J. R. V. *Political Frontiers and Boundaries* (Winchester, Mass.: Allen & Unwin, 1987).

Preston, Peter W. *Pacific Asia in the Global System: An Introduction* (Malden, Mass.: Blackwell, 1998).

Rigg, Jonathan. *Southeast Asia—A Region in Transition: A Thematic Human Geography of the ASEAN Region* (New York: HarperCollins Academic, 1991).

- Rigg, Jonathan. *Southeast Asia: The Human Landscape of Modernization and Development* (London & New York: Routledge, 2nd rev. ed., 2003).

Rodan, Gary, ed. *Singapore* (Brookfield, Vt.: Ashgate, 2001).

SarDesai, D. R. *Southeast Asia: Past and Present* (Boulder, Colo.: Westview Press, 5th rev. ed., 2003).

- Spencer, Joseph E., & Thomas, William L., Jr. *Asia, East by South: A Cultural Geography* (New York: John Wiley & Sons, 2nd rev. ed., 1971).

"Southeast Asia: The Geography of Terror," *The Economist*, August 23, 2003, 31–32.

Taylor, John G. *East Timor: The Price of Freedom* (London: Zed Books, 2000).

"The Sweet Serpent of South-East Asia: How Much Longer Will the Mekong Remain the World's Last Great Unspoilt River?," *The Economist*, January 3, 2004, 28–30.

"Time to Deliver: A Survey of Indonesia," *The Economist*, December 11, 2004, special insert, 16 pp.

Tyner, James. "Cities of Southeast Asia," in Brunn, Stanley D., Williams, Jack F., & Zeigler, Donald J., eds., *Cities of the World: World Regional Urban Development* (Lanham, Md.: Rowman & Littlefield, 3rd rev. ed., 2003), pp. 370–411.

Ulack, Richard & Pauer, Gyula. *Atlas of Southeast Asia* (New York: Macmillan, 1988).

Watters, R. F., & McGee, Terence G. *Asia-Pacific: New Geographies of the Pacific Rim* (London: Hurst, 1997).

- Weightman, Barbara A. *Dragons and Tigers: A Geography of South, East, and Southeast Asia*, 2nd ed., (Hoboken, N.J.: John Wiley & Sons, 2006).

Wurfel, David, & Burton, Bruce, eds. *Southeast Asia in the New World Order* (New York: St. Martin's Press, 1996).

CHAPTER 11

"A Survey of Australia," *The Economist*, September 9, 2000, special insert, 16 pp.

Bambrick, Susan, ed. *The Cambridge Encyclopedia of Australia* (New York: Cambridge University Press, 1994).

- Barrett, Rees D., & Ford, Roslyn A. *Patterns in the Human Geography of Australia* (South Melbourne: Macmillan of Australia, 1987).

Beadle, N. C. W. *The Vegetation of Australia* (New York: Cambridge University Press, 1981).

Britton, Steve, et al., eds. *Changing Places in New Zealand: A Geography of Restructuring* (Dunedin, N.Z.: Allied Press, 1992).

Burnley, Ian. *The Impact of Immigration in Australia: A Demographic Approach* (New York: Oxford University Press, 2001).

Camm, J. C. R., & McQuilton, J., eds. *Australia: An Historical Atlas* (Broadway, N.S.W.: Fairfax, Syme & Weldon Associates, 1987).

Connell, John, ed. *Sydney: The Emergence of a World City* (New York: Oxford University Press, 2000).

Courtenay, Percy P. *Northern Australia: Patterns and Problems of Tropical Development in an Advanced Country* (London & New York: Longman, 1983).

Cumberland, Kenneth B., & Whitelaw, James S. *New Zealand* (Chicago: Aldine, 1970).

"Environment and Development in Australia," *Australian Geographer*, 19 (May 1988): 3–220.

Fitzpatrick, Judith M., ed. *Endangered Peoples of Oceania: Struggles to Survive and Thrive* (Westport, Conn.: Greenwood Press, 2000).

Forster, Clive. *Australian Cities: Continuity and Change* (Melbourne: Oxford University Press, 2nd rev. ed., 1999).

Greiner, Alyson L., & Jordan-Bychkov, Terry G. *Anglo-Celtic Australia: Colonial Immigration and Cultural Regionalism* (Chicago: University of Chicago Press, 2003).

Head, Lesley. *Second Nature: The History and Implications of Australia as Aboriginal Landscape* (Syracuse, N.Y.: Syracuse University Press, 2000).

- • Heathcote, Ronald L. *Australia* (New York: John Wiley & Sons/Longman, 2nd rev. ed., 1994).

Heathcote, Ronald L., ed. *The Australian Experience: Essays in Australian Land Settlement and Resource Management* (Melbourne: Longman Cheshire, 1988).

Heathcote, Ronald L., & Mabbutt, J. A. *Land, Water and People: Essays in Australian Resource Management* (Winchester, Mass.: Allen & Unwin, 1988).

Hofmeister, Burkhard. *Australia and Its Urban Centers* (Berlin: Gebrüder Borntraeger, 1988).

Holland, P. G., & Johnston, W. B., eds. *Southern Approaches: Geography in New Zealand* (Christchurch, N.Z.: New Zealand Geographical Society, 1987).

Hughes, Robert. *The Fatal Shore* (New York: Alfred A. Knopf, 1987).

Hughes, Robert. "The Real Australia," *Time*, September 11, 2000, 98–111.

Inglis, C., et al., eds. *Asians in Australia: The Dynamics of Migration and Settlement* (Sydney: Allen & Unwin, 1992).

- • Jeans, Dennis N., ed. *Australia: A Geography. Volume 1: The Natural Environment* (Sydney: Sydney University Press, 1986); *Volume 2: Space and Society* (Sydney: Sydney University Press, 1987).
- • MacDonald, Glen M. *Biogeography: Introduction to Space, Time and Life* (New York: John Wiley & Sons, 2003).

McKnight, Tom L. *Australia's Corner of the World* (Englewood Cliffs, N.J.: Prentice-Hall, 1970).

- • McKnight, Tom L. *Oceania: The Geography of Australia, New Zealand, and the Pacific Islands* (Englewood Cliffs, N.J.: Prentice Hall, 1995).

Meinig, Donald W. *On the Margins of the Good Earth: The South Australian Wheat Frontier, 1869–1884* (Chicago: Rand McNally, 1962).

Moran, Anthony. *Australia: Nation, Belonging and Globalization* (London & New York: Routledge, 2005).

- • Morton, Harry, & Johnston, Carol M. *The Farthest Corner: New Zealand, a Twice Discovered Land* (Honolulu: University of Hawai'i Press, 1989).
- • O'Connor, Kevin, Stimson, Robert, & Daly, Maurice. *Australia's Changing Economic Geography: A Society Dividing* (South Melbourne: Oxford University Press, 2001).

Powell, Joseph M. *An Historical Geography of Modern Australia: The Restive Fringe* (New York: Cambridge University Press, 1988).

Powell, Joseph M. "Revisiting the Australian Experience: Transmillennial Conjurings," *Geographical Review*, 90 (January 2000): 1–17.

Rainnie, Al, & Grobbelaar, Mardelene, eds. *New Regionalism in Australia* (Burlington, Vt.: Ashgate, 2005).

Rich, David C. *The Industrial Geography of Australia* (Sydney: Methuen, 1987).

- • Robinson, Guy M., Loughran, Robert J., & Tranter, Paul J. *Australia and New Zealand: Economy, Society and Environment* (New York: Oxford University Press, 2000).

Serventy, J. *Landforms of Australia* (New York: American Elsevier, 1968).

- • Spate, Oskar H. K. *Australia* (New York: Praeger, 1968).

Stratford, Elaine, ed. *Australian Cultural Geographies* (New York: Oxford University Press, 1999).

Walmsley, D. J., & Sorenson, A. D. *Contemporary Australia: Explorations in Economy, Society and Geography* (Melbourne: Longman Cheshire, 2nd rev. ed., 1993).

Young, Ann. *Environment and Change in Australia Since 1788* (New York: Oxford University Press, 2nd rev. ed., 2000).

CHAPTER 12

Arnberger, Hertha, & Arnberger, Erik. *The Tropical Islands of the Indian and Pacific Oceans* (Seattle: University of Washington Press, 2001).

- • Bier, James A., cartographer. *Reference Map of Oceania: The Pacific Islands of Micronesia, Polynesia, Melanesia* (Honolulu: University of Hawai'i Press, 1995).

Blake, Gerald H., ed. *Maritime Boundaries: World Boundaries, Vol. 5* (London & New York: Routledge, 1994).

Blake, Gerald H., ed. *Maritime Boundaries and Ocean Resources* (Totowa, N.J.: Rowman & Littlefield, 1988).

Blank, Paul W., & Spier, Fred, eds. *Defining the Pacific: Opportunities and Constraints* (Brookfield, Vt.: Ashgate, 2002).

Brookfield, Harold C. *Exploring Agrodiversity* (New York: Columbia University Press, 2001).

Brookfield, Harold C., ed. *The Pacific in Transition: Geographical Perspectives on Adaptation and Change* (New York: St. Martin's Press, 1973).

Brookfield, Harold C., & Hart, Doreen. *Melanesia: A Geographical Interpretation of an Island World* (New York: Barnes & Noble, 1971).

Bunge, Frederica M., & Cooke, Melinda W., eds. *Oceania: A Regional Study* (Washington, D.C.: U.S. Government Printing Office, 1984).

Carter, John, ed. *Pacific Islands Yearbook* (Sydney: Pacific Publications, annual).

- • Chaturvedi, Sanjay. *The Polar Regions: A Political Geography* (New York: John Wiley & Sons, 1997).

Cole, George M. *Water Boundaries* (New York: John Wiley & Sons, 1997).

Connell, John. *Papua New Guinea: The Struggle for Development* (London & New York: Routledge, 1997).

Couper, Alastair D., ed. *Development and Social Change in the Pacific Islands* (London & New York: Routledge, 1989).

Crossley, L. *Explore Antarctica* (New York: Cambridge University Press, 1995).

Culliney, John, L. *Islands in a Fair Sea: The Fate of Nature in Hawai'i* (Honolulu: University of Hawai'i Press, 2nd rev. ed., 2006).

Damas, David. *Bountiful Island: A Study of Land Tenure on a Micronesian Atoll* (Waterloo, Ont., Canada: Wilfrid Laurier University Press, 1996).

de Blij, H. J. "A Regional Geography of Antarctica and the Southern Ocean," *University of Miami Law Review*, 33 (1978): 299–314.

Dodds, Klaus J. *Geopolitics of Antarctica: Views from the Southern Oceanic Rim* (New York: John Wiley & Sons, 1997).

Fitzpatrick, Judith M., ed. *Endangered Peoples of Oceania: Struggles to Survive and Thrive* (Westport, Conn.: Greenwood Press, 2000).

Freeman, Otis W., ed. *Geography of the Pacific* (New York: John Wiley & Sons, 1951).

Friis, Herman R., ed. *The Pacific Basin: A History of Its Geographical Exploration* (New York: American Geographical Society, Special Publication No. 38, 1967).

- Glassner, Martin I. *Neptune's Domain: A Political Geography of the Sea* (Winchester, Mass.: Unwin Hyman, 1990).

Grossman, Lawrence S. *Peasants, Subsistence Ecology, and Development in the Highlands of Papua New Guinea* (Princeton, N.J.: Princeton University Press, 1984).

Hanlon, David, & White, Geoffrey M., eds. *Voyaging Through the Contemporary Pacific* (Lanham, Md.: Rowman & Littlefield, 2000).

Howard, A., ed. *Polynesia: Readings on a Culture Area* (Scranton, Pa.: Chandler, 1971).

Howlett, Diana. *Papua New Guinea: Geography and Change* (Melbourne, Australia: Thomas Nelson, 1973).

Johnston, Douglas M. & Saunders, Philip M. *Ocean Boundary-Making: Regional Issues and Developments* (London & New York: Routledge, 1988).

Karolle, Bruce G. *Atlas of Micronesia* (Honolulu: Bess Press, 2nd rev. ed., 1995).

Kirch, Patrick V. *On the Road of Winds: An Archaeological History of the Pacific Islands Before European Contact* (Berkeley: University of California Press, 2000).

Kirch, Patrick V. *The Wet and the Dry: Irrigation and Agricultural Intensification in Polynesia* (Chicago: University of Chicago Press, 1994).

Kissling, Christopher C., ed. *Transport and Communications for Pacific Microstates: Issues in Organization and Management* (Suva, Fiji: University of the South Pacific, Institute of Pacific Studies, 1984).

Kluge, P. F. *The Edge of Paradise: America in Micronesia* (New York: Random House, 1991).

Lea, John, & Connell, John. *Urbanization in the Pacific* (London & New York: Routledge, 2001).

Leibowitz, Arnold H. *Embattled Island: Palau's Struggle for Independence* (Westport, Conn.: Praeger, 1996).

Lockhart, Douglas G., et al., eds. *The Development Process in Small Island States* (London & New York: Routledge, 1993).

- McEvedy, Colin. *The Penguin Historical Atlas of the Pacific* (New York: Penguin Putnam, 1999).

- McKnight, Tom L. *Oceania: The Geography of Australia, New Zealand, and the Pacific Islands* (Englewood Cliffs, N.J.: Prentice Hall, 1995).

Mitchell, Andrew. *The Fragile South Pacific: An Ecological Odyssey* (Austin: University of Texas Press, 1991).

"Mobility and Identity in the Island Pacific," special issue, *Pacific Viewpoint*, 26, No. 1 (1985).

Morgan, Joseph R. *Hawai'i: A Unique Geography* (Honolulu: Bess Press, 1996).

Nunn, Patrick. *Oceanic Islands* (Cambridge, Mass.: Blackwell, 1994).

Oliver, Douglas L. *Native Cultures of the Pacific Islands* (Honolulu: University of Hawai'i Press, 1989).

- Peake, Martin. *Pacific People and Society* (New York: Cambridge University Press, 1992).

Prescott, J. R. V. *The Maritime Political Boundaries of the World* (London & New York: Methuen, 1986).

Robillard, Albert B., ed. *Social Change in the Pacific Islands* (London & New York: Kegan Paul International, 1991).

Sager, Robert J. "The Pacific Islands: A New Geography," *Focus*, Summer 1988, pp. 10–14.

Spate, Oskar H. K. *The Pacific Since Magellan, Vol. II: Monopolists and Freebooters* (Minneapolis: University of Minnesota Press, 1983).

Spate, Oskar H. K. *The Pacific Since Magellan, Vol. III: Paradise Found and Lost* (Minneapolis: University of Minnesota Press, 1989).

Spate, Oskar H. K. *The Spanish Lake: A History of the Pacific Since Magellan, Vol. I* (Beckenham, UK: Croom Helm, 1979).

Sugden, David E. *Arctic and Antarctic: A Modern Geographical Synthesis* (Totowa, N.J.: Barnes & Noble, 1982).

Theroux, Paul. *The Happy Isles of Oceania: Paddling the Pacific* (New York: G. P. Putnam's Sons, 1992).

Ward, R. Gerard, ed. *Man in the Pacific Islands: Essays on Geographical Change in the Pacific* (New York: Oxford University Press, 1972).

Woodcock, Deborah W., ed. *Hawai'i: New Geographies* (Manoa, Hawai'i: University of Hawai'i, Department of Geography, 1999).

Zurick, David N. "Preserving Paradise," *Geographical Review*, 85 (April 1995): 157–172.

Glossary

Aboriginal land issue The legal campaign in which Australia's **indigenous peoples*** have claimed title to traditional land in several parts of that country. The courts have upheld certain claims, fueling Aboriginal activism that has raised broader issues of indigenous rights (see Regional Issue box on p. 553).

Aboriginal population See **indigenous peoples**.

Absolute location The position or place of a certain item on the surface of the Earth as expressed in degrees, minutes, and seconds of **latitude**, 0° to 90° north or south of the equator, and **longitude**, 0° to 180° east or west of the *prime meridian* passing through Greenwich, England (a suburb of London).

Accessibility The degree of ease with which it is possible to reach a certain location from other locations. *Inaccessibility* is the opposite of this concept.

Acculturation Cultural modification resulting from intercultural borrowing. In **cultural geography**, the term refers to the change that occurs in the culture of **indigenous peoples** when contact is made with a society that is technologically superior.

Advantage The most meaningful distinction that can now be made to classify a country's level of economic **development**. Takes into account geographic **location**, **natural resources**, government, political stability, productive skills, and much more.

Agglomeration Process involving the clustering or concentrating of people or activities.

Agrarian Relating to the use of land in rural communities, or to **agricultural** societies in general.

Agriculture The purposeful tending of crops and livestock in order to produce food and fiber.

Alluvial Refers to the mud, silt, and sand (collectively *alluvium*) deposited by rivers and streams. *Alluvial plains* adjoin many larger rivers; they consist of these renewable deposits that are laid down during floods, creating fertile and productive soils. Alluvial **deltas** mark the mouths of rivers such as the Nile and the Ganges.

Altiplano High-elevation plateau, basin, or valley between even higher mountain ranges, especially in the Andes of South America.

Altitudinal zonation Vertical regions defined by physical-environmental zones at various elevations (see Fig. 4-13), particularly in the highlands of South and Middle America. See ***tierra caliente***, ***tierra templada***, ***tierra fría***, ***tierra helada***, and ***tierra nevada***.

American Manufacturing Belt North America's near-rectangular Core Region, whose corners are Boston, Milwaukee, St. Louis, and Baltimore. Dominated the industrial geography of the U.S. and Canada during the industrial age; still a formidable economic powerhouse that remains the realm's geographic heart.

*Words in boldface type are defined elsewhere in this Glossary.

Antarctic Treaty International cooperative agreement on the use of Antarctic territory (see pp. 575-576).

Antecedent boundary A political boundary that existed before the **cultural landscape** emerged and stayed in place while people moved in to occupy the surrounding area.

Anthracite coal Hardest and highest carbon-content coal, and therefore of the highest quality.

Apartheid Literally, *apartness*. The Afrikaans term for South Africa's pre-1994 policies of racial separation, a system that produced highly segregated socio-geographical patterns.

Aquaculture The use of a river segment or an artificial pond for the raising and harvesting of food products, including fish, shellfish, and even seaweed (particularly in Japan).

Aquifer An underground reservoir of water contained within a porous, water-bearing rock layer.

Arable Land fit for cultivation by one farming method or another. See **physiologic density**.

Archipelago A set of islands grouped closely together, usually elongated into a *chain*.

Area A term that refers to a part of the Earth's surface with less specificity than **region**. For example, *urban area* alludes generally to a place where urban development has occurred, whereas *urban region* requires certain specific criteria upon which such a designation is based (e.g., the spatial extent of commuting or the built townscape).

Areal functional organization A geographic principle for understanding the evolution of regional organization, whose five interrelated tenets are applied to the spatial development of Japan on pp. 477-478.

Areal interdependence A term related to **functional specialization**. When one area produces certain goods or has certain raw materials and another area has a different set of raw materials and produces different goods, their needs may be *complementary*; by exchanging raw materials and products, they can satisfy each other's requirements.

Arithmetic density A country's population, expressed as an average per unit area, without regard for its **distribution** or the limits of **arable** land—see also **physiologic density**.

Aryan From the Sanskrit *Arya* ("noble"), a name applied to an ancient people who spoke an **Indo-European language** and who moved into northern India from the northwest.

Atmosphere The Earth's envelope of gases that rests on the oceans and land surface and penetrates open spaces within soils. This layer of nitrogen (78 percent), oxygen (21 percent), and traces of other gases is densest at the Earth's surface and thins with altitude.

Austral South.

Autocratic A government that holds absolute power, often ruled by one person or a small group of persons who control the country by despotic means.

Balkanization The fragmentation of a **region** into smaller, often hostile political units.
Barrio Term meaning "neighborhood" in Spanish. Usually refers to an urban community in a Middle or South American city; also applied to low-income, inner-city concentrations of Hispanics in such western U.S. cities as Los Angeles.
Bauxite Aluminum ore; usually deposited at shallow depths in the wet tropics.
Biogeography The study of *flora* (plant life) and *fauna* (animal life) in spatial perspective.
Birth rate The *crude birth rate* is expressed as the annual number of births per 1000 individuals within a given population.
Bituminous coal Softer coal of lesser quality than **anthracite**, but of higher grade than **lignite**. When heated and converted to coking coal or *coke*, it is used to make steel.
Break-of-bulk point A location along a transport route where goods must be transferred from one carrier to another. In a port, the cargoes of oceangoing ships are unloaded and put on trains, trucks, or perhaps smaller river boats for inland distribution.
Buffer state See **buffer zone**.
Buffer zone A country or set of countries separating ideological or political adversaries. In southern Asia, Afghanistan, Nepal, and Bhutan were parts of a buffer zone between British and Russian-Chinese imperial spheres. Thailand was a *buffer state* between British and French colonial domains in mainland Southeast Asia.
Caliente See ***tierra caliente***.
Cartogram A specially transformed map not based on traditional representations of **scale** or area.
Cartography The art and science of making maps, including data compilation, layout, and design. Also concerned with the interpretation of mapped patterns.
Caste system The strict **social stratification** and segregation of people—specifically in India's Hindu society—on the basis of ancestry and occupation.
Cay A low-lying small island usually composed of coral and sand. Pronounced *kee* and often spelled "key."
Central business district (CBD) The downtown heart of a central city, the CBD is marked by high land values, a concentration of business and commerce, and the clustering of the tallest buildings.
Centrality The strength of an urban center in its capacity to attract producers and consumers to its facilities; a city's "reach" into the surrounding region.
Centrifugal forces A term employed to designate forces that tend to divide a country—such as internal religious, linguistic, ethnic, or ideological differences.
Centripetal forces Forces that unite and bind a country together—such as a strong national culture, shared ideological objectives, and a common faith.
Chaebol Giant corporation controlling numerous companies and benefiting from government connections and favors, dominant in South Korea's **economic geography**; key to the country's **development** into an **economic tiger**, but more recently a barrier to free-market growth.
Charismatic Personal qualities of certain leaders that enable them to capture and hold the popular imagination, to secure the allegiance and even the devotion of the masses. Gandhi, Mao Zedong, and Franklin D. Roosevelt were good examples in the 20th century.
China Proper The eastern and northeastern portions of China that contain most of the country's huge population; mapped in Figure 9-4.
Choke point A narrowing of an international waterway causing marine traffic congestion, requiring reduced speeds and/or sharp turns, and increasing the risk of collision as well as vulnerability to attack. When the waterway narrows to a distance of less than 24 miles (38 km), this necessitates the drawing of a **median line (maritime) boundary**. Examples are the Hormuz Strait between Oman and Iran at the entrance to the Persian Gulf, and the Strait of Malacca between Malaysia and Indonesia.
City-state An independent political entity consisting of a single city with (and sometimes without) an immediate **hinterland**. The ancient city-states of Greece have their modern equivalent in Singapore.
Climate The long-term conditions (over at least 30 years) of aggregate **weather** over a region, summarized by averages and measures of variability; a synthesis of the succession of weather events we have learned to expect at any given location.
Climate change theory An alternative to the **hydraulic civilization theory**; holds that changing **climate** (rather than a monopoly over **irrigation** methods) could have provided certain cities in the ancient Fertile Crescent with advantages over others.
Climate region A **formal region** characterized by the uniformity of the **climate** type within it. Figure G-8 maps the global distribution of such regions.
Climatology The geographic study of **climates**. Includes not only the classification of climates and the analysis of their regional distribution, but also broader environmental questions that concern climate change, interrelationships with soil and vegetation, and human-climate interaction.
Coal See **anthracite coal**, **bituminous coal**, **fossil fuels**, and **lignite**.
Collectivization The reorganization of a country's **agriculture** under communism that involves the expropriation of private holdings and their incorporation into relatively large-scale units, which are farmed and administered cooperatively by those who live there.
Colonialism Rule by an autonomous power over a subordinate and alien people and place. Although often established and maintained through political structures, colonialism also creates unequal cultural and economic relations. Because of the magnitude and impact of the European colonial thrust of the last few centuries, the term is generally understood to refer to that particular colonial endeavor. Also see **imperialism**.
Command economy The tightly controlled economic system of the former Soviet Union, whereby central planners in Moscow assigned the production of particular goods to particular places, often guided more by socialist ideology than the principles of **economic geography**.
Commercial agriculture For-profit **agriculture**.
Common market A **free-trade area** that not only has created a **customs union** (a set of common tariffs on all imports from outside the area) but also has eliminated restrictions on the movement of capital, labor, and enterprise among its member countries.
Compact state A politico-geographical term to describe a **state** that possesses a roughly circular, oval, or rectangular territory in which the distance from the geometric center to any point on the boundary exhibits little variance. Poland and Cambodia are examples of this shape category.
Complementarity Exists when two regions, through an exchange of raw materials and/or finished products, can specifically satisfy each other's demands.
Confucianism A philosophy of ethics, education, and public service based on the writings of Confucius (*Kongfuzi*); traditionally regarded as one of the cornerstones of Chinese **culture**.

Congo See the footnote on p. 281, which makes the distinction between the two Equatorial African countries, Congo and The Congo.
Coniferous forest A forest of cone-bearing, needleleaf evergreen trees with straight trunks and short branches, including spruce, fir, and pine. Also see **taiga**.
Contagious diffusion The distance-controlled spreading of an idea, innovation, or some other item through a local population by contact from person to person—analogous to the communication of a contagious illness.
Conterminous United States The 48 **contiguous** or adjacent States that occupy the southern half of the North American realm. Alaska is not contiguous to these States because western Canada lies in between; neither is Hawai'i, separated from the mainland by over 2000 miles of ocean.
Contiguous Adjoining; adjacent.
Continental drift The slow movement of continents controlled by the **processes** associated with **plate tectonics**.
Continental shelf Beyond the coastlines of many landmasses, the ocean floor declines very gently until the depth of about 660 feet (200 m). Beyond the 660-foot line the sea bottom usually drops off sharply, along the *continental slope*, toward the much deeper mid-oceanic basin. The submerged continental margin is called the continental shelf, and it extends from the shoreline to the upper edge of the continental slope.
Continentality The variation of the continental effect on air temperatures in the interior portions of the world's landmasses. The greater the distance from the moderating influence of an ocean, the greater the extreme in summer and winter temperatures. Continental interiors also tend to be dry when the distance from oceanic moisture sources becomes considerable.
Conurbation General term used to identify a large multimetropolitan complex formed by the coalescence of two or more major **urban areas**. The Atlantic Seaboard **Megalopolis**, extending along the northeastern U.S. coast from southern Maine to Virginia, is a classic example.
Copra The dried-out, fleshy interior of a coconut that is used to produce coconut oil.
Cordillera Mountain chain consisting of sets of parallel ranges, especially the Andes in northwestern South America.
Core See **core area**; **core-periphery relationships**.
Core area In geography, a term with several connotations. *Core* refers to the center, heart, or focus. The core area of a **nation-state** is constituted by the national heartland, the largest population cluster, the most productive region, and the part of the country with the greatest **centrality** and **accessibility**—probably containing the capital city as well.
Core-periphery relationships The contrasting spatial characteristics of, and linkages between, the *have* (core) and *have-not* (periphery) components of a national or regional **system**.
Corridor In general, refers to a spatial entity in which human activity is organized in a linear manner, as along a major transport route or in a valley confined by highlands. More specifically, the politico-geographical term for a land extension that connects an otherwise **landlocked state** to the sea.
Cultural diffusion The **process** of spreading and adoption of a cultural element, from its place of origin across a wider area.
Cultural ecology The multiple interactions and relationships between a **culture** and its **natural environment**.
Cultural environment See **cultural ecology**.
Cultural geography The wide-ranging and comprehensive field of geography that studies spatial aspects of human **cultures**.
Cultural landscape The forms and artifacts sequentially placed on the **natural landscape** by the activities of various human occupants. By this progressive imprinting of the human presence, the physical (natural) landscape is modified into the cultural landscape, forming an interacting unity between the two.
Cultural pluralism See **plural(istic) society**.
Cultural revival The regeneration of a long-dormant **culture** through internal renewal and external infusion.
Culture The sum total of the knowledge, attitudes, and habitual behavior patterns shared and transmitted by the members of a society. This is anthropologist Ralph Linton's definition; hundreds of others exist.
Culture area See **culture region**.
Culture hearth Heartland, source area, innovation center; place of origin of a major **culture**.
Culture region A distinct, culturally discrete spatial unit; a **region** within which certain cultural norms prevail.
Customs union A **free-trade area** in which member countries set common tariff rates on imports from outside the area.
Death rate The *crude death rate* is expressed as the annual number of deaths per 1000 individuals within a given population.
Deciduous A deciduous tree loses its leaves at the beginning of winter or the onset of the dry season.
Definition In **political geography**, the written legal description (in a treaty-like document) of a boundary between two countries or territories—see **delimitation**.
Deforestation The clearing and destruction of forests (especially tropical rainforests) to make way for expanding settlement frontiers and the exploitation of new economic opportunities.
Deglomeration Deconcentration.
Delimitation In **political geography**, the translation of the written terms of a boundary treaty (the **definition**) into an official cartographic representation (map).
Delta Alluvial lowland at the mouth of a river, formed when the river deposits its alluvial load on reaching the sea. Often triangular in shape, hence the use of the Greek letter whose symbol is $\triangle$.
Demarcation In **political geography**, the actual placing of a political boundary on the **cultural landscape** by means of barriers, fences, walls, or other markers.
Demographic transition model Multi-stage model, based on Western Europe's experience, of changes in population growth exhibited by countries undergoing industrialization. High **birth rates** and **death rates** are followed by plunging death rates, producing a huge net population gain; this is followed by the convergence of birth and death rates at a low overall level. See Figure 8-7.
Demography The interdisciplinary study of population—especially **birth rates** and **death rates**, growth patterns, longevity, **migration**, and related characteristics.
Desert An arid area supporting sparse vegetation, receiving less than 10 inches (25 cm) of precipitation per year. Usually exhibits extremes of heat and cold because the moderating influence of moisture is absent.
Desertification The **process** of **desert** expansion into neighboring **steppelands** as a result of human degradation of fragile semiarid environments.
Development The economic, social, and institutional growth of national **states**.
Devolution The **process** whereby regions within a **state** demand and gain political strength and growing autonomy at the expense of the central government.
Dhows Wooden boats with characteristic triangular sails, plying the seas between Arabian and East African coasts.

Dialect Regional or local variation in the use of a major language, such as the distinctive accents of many residents of the U.S. South or New England.

Diffusion The spatial spreading or dissemination of a **culture** element (such as a technological innovation) or some other phenomenon (e.g., a disease outbreak). For the various channels of outward geographic spread from a source area, see **contagious**, **expansion**, **hierarchical**, and **relocation diffusion**.

Distance decay The various degenerative effects of distance on human spatial structures and interactions.

Diurnal Daily.

Divided capital In **political geography**, a country whose central administrative functions are carried on in more than one city is said to have divided capitals. The Netherlands and South Africa are examples.

Domestication The transformation of a wild animal or wild plant into a domesticated animal or a cultivated crop to gain control over food production. A necessary evolutionary step in the development of humankind: the invention of **agriculture**.

Domino theory The belief that political destabilization in one **state** can result in the collapse of order in a neighboring state, triggering a chain of events that, in turn, can affect a series of **contiguous** states.

Double cropping The planting, cultivation, and harvesting of two crops successively within a single year on the same plot of farmland.

Dry canal An overland rail and/or road **corridor** across an **isthmus** dedicated to performing the transit functions of a canalized waterway. Best adapted to the movement of containerized cargo, there must be a port at each end to handle the necessary **break-of-bulk** unloading and reloading.

Ecology The study of the many interrelationships between all forms of life and the natural environments in which they have evolved and continue to develop. The study of *ecosystems* focuses on the interactions between specific organisms and their environments. See also **cultural ecology**.

Economic geography The field of geography that focuses on the diverse ways in which people earn a living, and how the goods and services they produce are expressed and organized spatially.

Economic restructuring The transformation of China into a market-driven economy in the post-Mao era, beginning in the late 1970s.

Economic tiger One of the burgeoning beehive countries of the western **Pacific Rim**. Following Japan's route since 1945, these countries have experienced significant modernization, industrialization, and Western-style economic growth since 1980. Three leading economic tigers today are South Korea, Taiwan, and Singapore. Term is increasingly used more generally to describe any fast-developing economy.

Economies of scale The savings that accrue from large-scale production wherein the unit cost of manufacturing decreases as the level of operation enlarges. Supermarkets operate on this principle and are able to charge lower prices than small grocery stores.

Ecosystem See **ecology**.

Ecumene The habitable portions of the Earth's surface where permanent human settlements have arisen.

Elite A small but influential upper-echelon social class whose power and privilege give it control over a country's political, economic, and cultural life.

El Niño-Southern Oscillation (ENSO) A periodic, large-scale, abnormal warming of the sea surface in the low latitudes of the eastern Pacific Ocean that has global implications, disturbing normal **weather** patterns in many parts of the world, especially South America.

Elongated state A **state** whose territory is decidedly long and narrow in that its length is at least six times greater than its average width. Chile and Vietnam are two classic examples.

Emigrant A person **migrating** away from a country or area; an out-migrant.

Empirical Relating to the real world, as opposed to theoretical abstraction.

Enclave A piece of territory that is surrounded by another political unit of which it is not a part.

Endemism Refers to a disease in a host population that affects many people in a kind of equilibrium without causing rapid and widespread deaths.

Entrepôt A place, usually a port city, where goods are imported, stored, and transshipped; a **break-of-bulk point**.

Environmental degradation The accumulated human abuse of a region's **natural landscape** that, among other things, can involve air and water pollution, threats to plant and animal ecosystems, misuse of **natural resources**, and generally upsetting the balance between people and their habitat.

Epidemic A local or regional outbreak of a disease.

Epochs of metropolitan evolution The five-stage Borchert model that conceptualizes the growth **process** of the national U.S. urban system. Consists of the Sail-Wagon Epoch, Iron Horse Epoch, Steel-Rail Epoch, Auto-Air-Amenity Epoch, and Satellite-Electronic-Jet Propulsion Epoch.

Eras of intraurban structural evolution The four-stage Adams model that conceptualizes the growth **process** of the U.S. city. Consists of the Walking-Horsecar Era, Electric Streetcar Era, Recreational Automobile Era, and Freeway Era.

Escarpment A cliff or very steep slope; frequently marks the edge of a plateau.

Estuary The widening mouth of a river as it reaches the sea; land subsidence or a rise in sea level has overcome the tendency to form a **delta**.

Ethnic cleansing The slaughter and/or forced removal of one **ethnic** group from its homes and lands by another, more powerful ethnic group bent on taking that territory.

Ethnicity The combination of a people's **culture** (traditions, customs, language, and religion) and racial ancestry.

European state model A **state** consisting of a legally defined territory inhabited by a population governed from a capital city by a representative government.

European Union (EU) **Supranational** organization constituted by 25 European countries to further their common economic interests. In alphabetical order, these countries are: Austria, Belgium, Cyprus, the Czech Republic, Denmark, Estonia, Finland, France, Germany, Greece, Hungary, Ireland, Italy, Latvia, Lithuania, Luxembourg, Malta, the Netherlands, Poland, Portugal, Slovakia, Slovenia, Spain, Sweden, and the United Kingdom.

Exclave A bounded (non-island) piece of territory that is part of a particular **state** but lies separated from it by the territory of another state. Alaska is an exclave of the United Sates.

Exclusive Economic Zone (EEZ) An oceanic zone extending up to 200 **nautical miles** from a shoreline, within which the coastal **state** can control fishing, mineral exploration, and additional activities by all other countries.

Expansion diffusion The spreading of an innovation or an idea through a fixed population in such a way that the number of those adopting grows continuously larger, resulting in an expanding area of dissemination.

Extraterritoriality Politico-geographical concept suggesting that the property of one **state** lying within the boundaries of another actually forms an extension of the first state.
Failed state A country whose institutions have collapsed and in which anarchy prevails. Afghanistan during the late-1990s rule of the Taliban is a recent example.
Fatwa Literally, a legal opinion or proclamation issued by an Islamic cleric, based on the holy texts of Islam, long applicable only in the *Umma*, the realm ruled by the laws of Islam. In 1989 the Iranian ayatollah Khomeini extended the reach of the *fatwa* by condemning to death a British citizen and author living in the United Kingdom.
Favela Shantytown on the outskirts or even well within an urban area in Brazil.
Fazenda Coffee plantation in Brazil.
Federal state A political framework wherein a central government represents the various subnational entities within a **nation-state** where they have common interests—defense, foreign affairs, and the like—yet allows these various entities to retain their own identities and to have their own laws, policies, and customs in certain spheres.
Federation See **federal state**.
Fertile Crescent Crescent-shaped zone of productive lands extending from near the southeastern Mediterranean coast through Lebanon and Syria to the **alluvial** lowlands of Mesopotamia (in Iraq). Once more fertile than today, this is one of the world's great source areas of **agricultural** and other innovations.
First Nations Canada's indigenous peoples of American descent, whose U.S. counterparts are called Native Americans.
Fjord Narrow, steep-sided, elongated, and inundated coastal valley deepened by glacier ice that has since melted away, leaving the sea to penetrate.
Floodplain Low-lying area adjacent to a mature river, often covered by **alluvial** deposits and subject to the river's floods.
Forced migration Human **migration** flows in which the movers have no choice but to relocate.
Formal region A type of **region** marked by a certain degree of homogeneity in one or more phenomena; also called *uniform region* or *homogeneous region*.
Forward capital Capital city positioned in actually or potentially contested territory, usually near an international border; it confirms the **state's** determination to maintain its presence in the region in contention.
Fossil fuels The energy resources of **coal**, natural gas, and petroleum (oil), so named collectively because they were formed by the geologic compression and transformation of tiny plant and animal organisms.
Four Motors of Europe Rhône-Alpes (France), Baden-Württemberg (Germany), Catalonia (Spain), and Lombardy (Italy). Each is a high-technology-driven region marked by exceptional industrial vitality and economic success not only within Europe but on the global scene as well.
Fragmented state A **state** whose territory consists of several separated parts, not a **contiguous** whole. The individual parts may be isolated from each other by the land area of other states or by international waters. The United States and Indonesia are examples.
Francophone French-speaking. Quebec constitutes the heart of Francophone Canada.
Free-trade area A form of economic integration, usually consisting of two or more **states**, in which members agree to remove tariffs on trade among themselves. Usually accompanied by a **customs union** that establishes common tariffs on imports from outside the trade area, and sometimes by a **common market** that also removes internal restrictions on the movement of capital, labor, and enterprise.
Free Trade Area of the Americas (FTAA) The ultimate goal of **supranational** economic integration in North, Middle, and South America: the creation of a single-market trading bloc that would involve every country in the Western Hemisphere between the Arctic shore of Canada and Cape Horn at the southern tip of Chile.
Fría See ***tierra fría***.
Frontier Zone of advance penetration, usually of contention; an area not yet fully integrated into a national **state**.
FTAA See **Free Trade Area of the Americas**.
Functional region A **region** marked less by its sameness than its dynamic internal structure; because it usually focuses on a central node, also called *nodal region* or *focal region*.
Functional specialization The production of particular goods or services as a dominant activity in a particular location. See also **local functional specialization**.
Fundamentalism See **revivalism (religious)**.
Gentrification The upgrading of an older residential area through private reinvestment, usually in the downtown area of a central city. Frequently this involves the displacement of established lower-income residents, who cannot afford the heightened costs of living, and conflicts are not uncommon as such neighborhood change takes place.
Geographic change Evolution of **spatial** patterns over time.
Geographic realm The basic spatial unit in our world regionalization scheme. Each realm is defined in terms of a synthesis of its total human geography—a composite of its leading cultural, economic, historical, political, and appropriate environmental features.
Geography of development The subfield of economic geography concerned with spatial aspects and regional expressions of **development**.
Geometric boundaries Political boundaries **defined** and **delimited** (and occasionally **demarcated**) as straight lines or arcs.
Geomorphology The geographic study of the configuration of the Earth's solid surface—the world's landscapes and their constituent landforms.
Ghetto An intraurban region marked by a particular **ethnic** character. Often an inner-city poverty zone, such as the black ghetto in U.S. central cities. Ghetto residents are involuntarily segregated from other income and racial groups.
Glaciation See **Pleistocene Epoch**.
Globalization The gradual reduction of regional contrasts at the world scale, resulting from increasing international cultural, economic, and political exchanges.
Green Revolution The successful recent development of higher-yield, fast-growing varieties of rice and other cereals in certain developing countries.
Gross domestic product (GDP) The total value of all goods and services produced in a country by that state's economy during a given year.
Gross national product (GNP) The total value of all goods and services produced in a country by that state's economy during a given year, plus all citizens' income from foreign investment and other external sources.
Growth pole An urban center with certain attributes that, if augmented by a measure of investment support, will stimulate regional economic development in its **hinterland**.
Hacienda Literally, a large estate in a Spanish-speaking country. Sometimes equated with the **plantation**, but there are important differences between these two types of agricultural enterprise (see pp. 203-205).

Heartland theory The hypothesis, proposed by British geographer Halford Mackinder during the early twentieth century, that any political power based in the heart of Eurasia could gain sufficient strength to eventually dominate the world. Further, since Eastern Europe controlled access to the Eurasian interior, its ruler would command the vast "heartland" to the east.

Hegemony The political dominance of a country (or even a region) by another country. The former Soviet Union's postwar grip on Eastern Europe, which lasted from 1945 to 1990, was a classic example.

Helada See *tierra helada*.

Hierarchical diffusion A form of **diffusion** in which an idea or innovation spreads by trickling down from larger to smaller adoption units. An urban **hierarchy** is usually involved, encouraging the leapfrogging of innovations over wide areas, with geographic distance a less important influence.

Hierarchy An order or gradation of phenomena, with each level or rank subordinate to the one above it and superior to the one below. The levels in a national urban hierarchy are constituted by hamlets, villages, towns, cities, and (frequently) the **primate city**.

High island Volcanic islands of the Pacific Realm that are high enough in elevation to wrest substantial moisture from the tropical ocean air (see **orographic precipitation**). They tend to be well watered, their volcanic soils enable productive agriculture, and they support larger populations than **low islands**—which possess none of these advantages and must rely on fishing and the coconut palm for survival.

High sea Areas of the oceans away from land, beyond national jurisdiction, open and free for all to use.

Highveld A term used in Southern Africa to identify the high, grass-covered plateau that dominates much of the region. The lowest-lying areas in South Africa are called *lowveld*; areas that lie at intermediate elevations are the *middleveld*.

Hinterland Literally, "country behind," a term that applies to a surrounding area served by an urban center. That center is the focus of goods and services produced for its hinterland and is its dominant urban influence as well. In the case of a port city, the hinterland also includes the inland area whose trade flows through that port.

Historical inertia A term from manufacturing geography that refers to the need to continue using the factories, machinery, and equipment of heavy industries for their full, multiple-decade lifetimes to cover major initial investments—even though these facilities may be increasingly obsolete.

Holocene The current *interglaciation* epoch (the warm period of glacial contraction between the glacial expansions of an **ice age**); extends from 10,000 years ago to the present. Also known as the *Recent Epoch*.

Human evolution The long-term biological maturation of the human species. Geographically, all evidence points toward East Africa as the source of humankind. Our species, *Homo sapiens*, emigrated from this hearth to eventually populate the rest of the **ecumene**.

Humus Dark-colored upper layer of a soil that consists of decomposed and decaying organic matter such as leaves and branches, nutrient-rich and giving the soil a high fertility.

Hydraulic civilization theory The theory that cities able to control **irrigated** farming over large **hinterlands** held political power over other cities. Particularly applies to early Asian civilizations based in such river valleys as the Chang (Yangzi), the Indus, and those of Mesopotamia.

Hydrologic cycle The **system** of exchange involving water in its various forms as it continually circulates between the **atmosphere**, the oceans, and above and below the land surface.

Ice age A stretch of geologic time during which the Earth's average atmospheric temperature is lowered; causes the equatorward expansion of continental ice sheets in the higher latitudes and the growth of mountain glaciers in and around the highlands of the lower latitudes.

Iconography The identity of a region as expressed through its cherished symbols; its particular **cultural landscape** and personality.

Immigrant A person **migrating** into a particular country or area; an in-migrant.

Immigration policies Australia's policies to regulate in-migration. The issue continues to roil Australian society and is discussed on pp. 552-554.

Imperialism The drive toward the creation and expansion of a colonial empire and, once established, its perpetuation.

Import-substitution industries The industries local entrepreneurs establish to serve populations of remote areas when transport costs from distant sources make these goods too expensive to import.

Inaccessibility See **accessibility**.

Indentured workers Contract laborers who sell their services for a stipulated period of time.

Indigenous peoples Native or *Aboriginal* peoples; often used to designate the inhabitants of areas that were conquered and subsequently colonized by the **imperial** powers of Europe.

Indo-European languages The major world language family that dominates the European **geographic realm** (Fig. 1-8). This language family is also the most widely dispersed globally, and about half of humankind speaks one of its languages.

Industrial Revolution The term applied to the social and economic changes in agriculture, commerce, and especially manufacturing and urbanization that resulted from technological innovations and specialization in late-eighteenth-century Europe.

Informal sector Dominated by unlicensed sellers of homemade goods and services, the primitive form of capitalism found in many developing countries that takes place beyond the control of government.

Infrastructure The foundations of a society: urban centers, transport networks, communications, energy distribution systems, farms, factories, mines, and such facilities as schools, hospitals, postal services, and police and armed forces.

Insular Having the qualities and properties of an island. Real islands are not alone in possessing such properties of **isolation**: an **oasis** in the middle of a **desert** also has qualities of insularity.

Insurgent state Territorial embodiment of a successful guerrilla movement. The establishment by antigovernment insurgents of a territorial base in which they exercise full control; thus a state within a state.

Intercropping The planting of several types of crops in the same field; commonly used by **shifting cultivators**.

Interglaciation See **Pleistocene Epoch**.

Intermontane Literally, between mountains. The location can bestow certain qualities of natural protection or **isolation** to a community.

Internal migration **Migration** flow within a country, such as ongoing westward and southward movements toward the **Sunbelt** in the United States.

International migration **Migration** flow involving movement across an international boundary.

Intervening opportunity In trade or **migration** flows, the presence of a nearer opportunity that greatly diminishes the attractiveness of sites farther away.

Inuit **Indigenous** peoples of North America's Arctic zone, formerly known as Eskimos.

Irredentism A policy of cultural extension and potential political expansion by a **state** aimed at a community of its nationals living in a neighboring state.

Irrigation The artificial watering of croplands.

Islamic Front The southern border of the African Transition Zone that marks the religious **frontier** of the **Muslim** faith in its southward penetration of Subsharan Africa (see Fig. 6-18).

Islamization Introduction and establishment of the **Muslim** religion (see Fig. 7-5). A **process** still under way, most notably along the **Islamic Front**, that marks the southern border of the African Transition Zone.

Isohyet A line connecting points of equal rainfall total.

Isolated state See **von Thünen's Isolated State model**.

Isolation The condition of being geographically cut off or far removed from mainstreams of thought and action. It also denotes a lack of receptivity to outside influences, caused at least partially by poor **accessibility**.

Isotherm A line connecting points of equal temperature.

Isthmus A **land bridge**; a comparatively narrow link between larger bodies of land. Central America forms such a link between Mexico and South America.

Jakota Triangle The easternmost region of the East Asian realm, consisting of *Ja*pan, (South) *Ko*rea, and *Ta*iwan.

Juxtaposition Contrasting places in close proximity to one another.

Karst The distinctive natural landscape associated with the chemical erosion of soluble limestone rock.

Land alienation One society or culture group taking land from another. In Subsaharan Africa, for example, European **colonialists** took land from **indigenous** Africans and put it to new uses.

Land bridge A narrow **isthmian** link between two large landmasses. They are temporary features—at least in terms of geologic time—subject to appearance and disappearance as the land or sea level rises and falls.

Land Hemisphere The half of the globe that contains the greatest amount of land surface, centered on Western Europe (Fig. 1-3). In **geomorphology** can also refer to the position of the African continent, which lies central to the world's landmasses (Fig. 6-3).

Land reform The spatial reorganization of **agriculture** through the allocation of farmland (often expropriated from landlords) to **peasants** and tenants who never owned land.

Land tenure The way people own, occupy, and use land.

Landlocked An interior **state** surrounded by land. Without coasts, such a country is disadvantaged in terms of **accessibility** to international trade routes, and in the scramble for possession of areas of the **continental shelf** and control of the **exclusive economic zone** beyond.

"Latin" American city model The Griffin-Ford model of intraurban spatial structure in the Middle American and South American realms, diagrammed in Figure 5-7.

Latitude Lines of latitude are **parallels** that are aligned east-west across the globe, from 0° latitude at the equator to 90° North and South latitude at the poles.

Leached soil Infertile, reddish-appearing, tropical soil whose surface consists of oxides of iron and aluminum; all other soil nutrients have been dissolved and transported downward into the subsoil by percolating water associated with heavy rainfall.

Leeward The protected or downwind side of a **topographic** barrier with respect to the winds that flow across it.

Lignite Low-grade, brown-colored variety of **coal**.

Lingua franca A "common language" prevalent in a given area; a second language that can be spoken and understood by many peoples, although they speak other languages at home.

Littoral Coastal or coastland.

Llanos The interspersed **savanna** grasslands and scrub woodlands of the Orinoco River's wide basin that covers much of interior Colombia and Venezuela.

Local functional specialization A hallmark of Europe's **economic geography** that later spread to many other parts of the world, whereby particular people in particular places concentrate on the production of particular goods and services.

Location Position on the Earth's surface; see **absolute location** and **relative location**.

Location theory A logical attempt to explain the locational pattern of an economic activity and the manner in which its producing areas are interrelated. The agricultural location theory that underlies the **von Thünen model** is a leading example.

Loess Deposit of very fine silt or dust that is laid down after having been windborne for a considerable distance. Notable for its fertility under **irrigation** and its ability to stand in steep vertical walls.

Longitude Angular distance (0° to 180°) east or west as measured from the *prime meridian* (0°) that passes through the Greenwich Observatory in suburban London, England. For much of its length across the mid-Pacific Ocean, the 180th meridian functions as the *international date line*.

Low island Low-lying coral islands of the Pacific Realm that—unlike **high islands**—cannot wrest sufficient moisture from the tropical ocean air to avoid chronic drought. Thus productive agriculture is impossible, and their modest populations must rely on fishing and the coconut palm for survival.

Lusitanian The Portuguese sphere, which by extension includes Brazil.

Madrassa **Revivalist** (**fundamentalist**) religious school where the curriculum focuses on Islamic religion and law and requires rote memorization of the Qu'ran (Koran), Islam's holy book. Founded in former British India, these schools were most numerous in present-day Pakistan but have diffused to Turkey in the west and Indonesia in the east.

Maghreb The region occupying the northwestern corner of Africa, consisting of Morocco, Algeria, and Tunisia.

Main Street Canada's dominant **conurbation** that is home to nearly two-thirds of the country's inhabitants; extends southwestward from Quebec City in the mid-St. Lawrence Valley to Windsor on the Detroit River.

Mainland-Rimland framework Twofold regionalization of the Middle American realm based on its modern cultural history. The Euro-Amerindian *Mainland*, stretching from Mexico to Panama (minus the Caribbean coastal strip), was a self-sufficient zone dominated by **hacienda land tenure**. The Euro-African *Rimland*, consisting of that Caribbean coastal zone plus all of the Caribbean islands to the east, was the zone of the **plantation** that heavily relied on trade with Europe. See Figure 4-6.

Maquiladora The term given to modern industrial plants in Mexico's U.S. border zone. These foreign-owned factories assemble imported components and/or raw materials, and then export finished manufactures, mainly to the United States. Import duties are disappearing under **NAFTA**, bringing jobs to Mexico and the advantages of low wage rates to the foreign entrepreneurs.

Marchland An area or **frontier** of uncertain boundaries that is subject to various national claims and an unstable political history. Refers specifically to the movement of various armies across such zones.

Marine geography The geographic study of oceans and seas. Its practitioners investigate both the physical (e.g., coral-reef **biogeography**, ocean-**atmosphere** interactions, coastal **geomorphology**) and human (e.g., **maritime boundary-making**, fisheries, beachside development) aspects of oceanic environments.

Maritime boundary An international boundary that lies in the ocean. Like all boundaries, it is a vertical plane, extending from the seafloor to the upper limit of the air space in the atmosphere above the water.

Median line boundary An international **maritime boundary** drawn where the width of a sea is less than 400 **nautical miles**. Because the **states** on either side of that sea claim **exclusive economic zones** of 200 nautical miles, it is necessary to reduce those claims to a (median) distance equidistant from each shoreline. **Delimitation** on the map almost always appears as a set of straight-line segments that reflect the configurations of the coastlines involved.

Medical geography The study of health and disease within a geographic context and from a spatial perspective. Among other things, this field of geography examines the sources, **diffusion** routes, and distributions of diseases.

Megacity Informal term referring to the world's most heavily populated cities; in this book, the term refers to a **metropolis** containing a population of greater than 10 million.

Megalopolis When spelled with a lower-case *m*, a synonym for **conurbation**, one of the large coalescing supercities forming in diverse parts of the world. When capitalized, refers specifically to the multi-metropolitan corridor that extends along the northeastern U.S. seaboard from north of Boston to south of Washington, D.C. (Fig. 3-9).

Mercantilism Protectionist policy of European **states** during the sixteenth to the eighteenth centuries that promoted a state's economic position in the contest with rival powers. Acquiring gold and silver and maintaining a favorable trade balance (more exports than imports) were central to the policy.

Meridian Line of **longitude**, aligned north-south across the globe, that together with **parallels** of **latitude** forms the global grid system. All meridians converge at both poles and are at their maximum distances from each other at the equator.

Mestizo Derived from the Latin word for *mixed*, refers to a person of mixed white and Amerindian ancestry.

Metropolis Urban **agglomeration** consisting of a (central) city and its suburban ring. See **urban (metropolitan) area**.

Metropolitan area See **urban (metropolitan) area**.

Migration A change in residence intended to be permanent. See also **forced**, **internal**, **international**, and **voluntary migration**.

Migratory movement Human relocation movement from a source to a destination without a return journey, as opposed to cyclical movement (see **nomadism**).

Model An idealized representation of reality built to demonstrate its most important properties. A **spatial** model focuses on a geographical dimension of the real world, such as the **von Thünen model** that explains agricultural location patterns in a commercial economy.

Modernization In the eyes of the Western world, the Westernization **process** that involves the establishment of **urbanization**, a market (money) economy, improved circulation, formal schooling, adoption of foreign innovations, and the breakdown of traditional society. Non-Westerners mostly see 'modernization' as an outgrowth of **colonialism** and often argue that traditional societies can be modernized without being Westernized.

Monsoon Refers to the seasonal reversal of wind and moisture flows in certain parts of the subtropics and lower-middle latitudes. The *dry monsoon* occurs during the cool season when dry offshore winds prevail. The *wet monsoon* occurs in the hot summer months, which produce onshore winds that bring large amounts of rainfall. The air-pressure differential over land and sea is the triggering mechanism, with windflows always moving from areas of relatively higher pressure toward areas of relatively lower pressure. Monsoons make their greatest regional impact in the coastal and near-coastal zones of South Asia, Southeast Asia, and East Asia.

Mosaic culture The emerging cultural-geographic framework of the United States, dominated by the fragmentation of specialized social groups into homogeneous communities of interest marked not only by income, race, and **ethnicity** but also by age, occupational status, and lifestyle. The result is an increasingly heterogeneous socio-spatial complex, which resembles an intricate mosaic composed of myriad uniform—but separate—tiles.

Mulatto A person of mixed African (black) and European (white) ancestry.

Multilingualism A society marked by a mosaic of local languages. Constitutes a **centrifugal force** because it impedes communication within the larger population. Often a ***lingua franca*** is used as a "common language," as in many countries of Subsharan Africa.

Multinationals Internationally active corporations that can strongly influence the economic and political affairs of many countries they operate in.

Muslim An adherent of the Islamic faith.

Muslim Front See **Islamic Front**.

NAFTA (North American Free Trade Agreement) The **free-trade area** launched in 1994 involving the United States, Canada, and Mexico.

Nation Legally a term encompassing all the citizens of a **state**, it also has other connotations. Most definitions now tend to refer to a group of tightly-knit people possessing bonds of language, **ethnicity**, religion, and other shared **cultural** attributes. Such homogeneity actually prevails within very few states.

Nation-state A country whose population possesses a substantial degree of **cultural** homogeneity and unity. The ideal form to which most **nations** and **states** aspire—a political unit wherein the territorial state coincides with the area settled by a certain national group or people.

NATO (North Atlantic Treaty Organization) Established in 1950 at the height of the Cold War as a U.S.-led **supranational** defense pact to shield postwar Europe against the Soviet military threat. NATO is now in transition, expanding its membership while modifying its objectives in the post-Soviet era.

Natural hazard A natural event that endangers human life and/or the contents of a **cultural landscape**.

Natural increase rate Population growth measured as the excess of live births over deaths per 1000 individuals per year. Natural increase of a population does not reflect either **emigrant** or **immigrant** movements.

Natural landscape The array of landforms that constitutes the Earth's surface (mountains, hills, plains, and plateaus) and the physical features that mark them (such as water bodies, soils, and vegetation). Each **geographic realm** has its distinctive combination of natural landscapes.

Natural resource Any valued element of (or means to an end using) the environment; includes minerals, water, vegetation, and soil.

Nautical mile By international agreement, the nautical mile—the standard measure at sea—is 6076.12 feet in length, equivalent to approximately 1.15 statute miles (1.85 km).

Near Abroad The 14 former Soviet republics that, together with the dominant Russian Republic, constituted the U.S.S.R. Since the 1991 breakup of the Soviet Union, Russia has asserted a sphere of influence in these now-independent countries (listed in Table 2-1), based on its proclaimed right to protect the interests of ethnic Russians who were settled there in substantial numbers during Soviet times.

Neocolonialism The term used by developing countries to underscore that the entrenched **colonial** system of international exchange and capital flow has not changed in the postcolonial era—thereby perpetuating the huge economic advantages of the developed world.

Network (transport) The entire regional **system** of transportation connections and nodes through which movement can occur.

Nevada See ***tierra nevada***.

New World Order A description of the international system resulting from the collapse of the Soviet Union in which the balance of nuclear power theoretically no longer determines the destinies of **states**.

Nomadism Cyclical movement among a definite set of places. Nomadic peoples mostly are **pastoralists**.

North American Free Trade Agreement See **NAFTA**.

Nucleation Cluster; **agglomeration**.

Oasis An area, small or large, where the supply of water (from an **aquifer** or a major river such as the Nile) permits the transformation of the immediately surrounding **desert** into productive cropland.

Occidental Western. Also see ***Oriental***.

Offshore banking Term referring to financial havens for foreign companies and individuals, who channel their earnings to accounts in such a country (usually an "offshore" island-state) to avoid paying taxes in their home countries.

Oligarchs Opportunists in post-Soviet Russia who used their ties to government to enrich themselves.

OPEC (Organization of Petroleum Exporting Countries) The international oil *cartel* or syndicate formed by a number of producing countries to promote their common economic interests through the formulation of joint pricing policies and the limitation of market options for consumers. The 11 member-states (as of mid-2005) are: Algeria, Indonesia, Iran, Iraq, Kuwait, Libya, Nigeria, Qatar, Saudi Arabia, United Arab Emirates (UAE), and Venezuela.

Organic theory Friedrich Ratzel's theory of **state** development that conceptualized the state as a biological organism whose life—from birth through maturation to eventual senility and collapse—mirrors that of any living thing.

Oriental The root of the word *oriental* is from the Latin for *rise*. Thus it has to do with the direction in which one sees the sun "rise"—the east; *oriental* therefore means Eastern. ***Occidental*** originates from the Latin for *fall*, or the "setting" of the sun in the west; *occidental* therefore means Western.

Orographic precipitation Mountain-induced precipitation, especially where air masses are forced to cross **topographic** barriers. Downwind areas beyond such a mountain range experience the relative dryness known as the **rain shadow effect**.

Outback The name given by Australians to the vast, peripheral, sparsely-settled interior of their country.

Outer city The non-central-city portion of the American **metropolis**; no longer "sub" to the "urb," this outer ring was transformed into a full-fledged city during the late twentieth century.

Overseas Chinese The more than 50 million ethnic Chinese who live outside China. Over half live in Southeast Asia (see Fig. 10-5), and many have become quite successful. A large number maintain links to China, and as investors played a major economic role in stimulating the growth of **SEZs** and Open Cities in China's Pacific Rim.

Pacific Rim A far-flung group of countries and parts of countries (extending clockwise on the map from New Zealand to Chile) sharing the following criteria: they face the Pacific Ocean; they evince relatively high levels of economic development, industrialization, and urbanization; their imports and exports mainly move across Pacific waters.

Pacific Ring of Fire Zone of crustal instability along tectonic **plate** boundaries, marked by earthquakes and volcanic activity, that ring the Pacific Ocean basin.

Paddies (paddyfields) Ricefields.

Pandemic An outbreak of a disease that spreads worldwide.

Pangaea A vast, singular landmass consisting of most of the areas of the present-day continents. This supercontinent began to break up more than 200 million years ago when still-ongoing **plate** divergence and **continental drift** became dominant processes (see Fig. 6-3).

Parallel An east-west line of **latitude** that is intersected at right angles by **meridians** of **longitude**.

Pastoralism A form of **agricultural** activity that involves the raising of livestock.

Peasants In a **stratified** society, peasants are the lowest class of people who depend on **agriculture** for a living. But they often own no land at all and must survive as tenants or day workers.

Peninsula A comparatively narrow, finger-like stretch of land extending from the main landmass into the sea. Florida and Korea are examples.

Peon (*peone*) Term used in Middle and South America to identify people who often live in serfdom to a wealthy landowner; landless **peasants** in continuous indebtedness.

Per capita Capita means *individual*. Income, production, or some other measure is often given per individual.

Perforated state A **state** whose territory completely surrounds that of another state. South Africa, which encloses Lesotho and is perforated by it, is a classic example.

Periodic market Village market that opens every third day or at some other regular interval. Part of a regional network of similar markets in a preindustrial, rural setting where goods are brought to market on foot and barter remains a major mode of exchange.

Peripheral development Spatial pattern in which a country's or region's development (and population) is most heavily concentrated along its outer edges rather than in its interior. Australia, with its peripheral population distribution, is a classic example (note its **core area** in Fig. 11-4), and nearby New Zealand exhibits a similar configuration of people and activities.

Periphery See **core-periphery relationships**.

Permafrost Permanently frozen water in the near-surface soil and bedrock of cold environments, producing the effect of completely frozen ground. Surface can thaw during brief warm season.

Physical geography The study of the geography of the physical (natural) world. Its subfields include **climatology**, **geomorphology**,

biogeography, **soil geography**, **marine geography**, and water **resources**.

Physical landscape Synonym for **natural landscape**.

Physiographic political boundaries Political boundaries that coincide with prominent physical features in the **natural landscape**—such as rivers or the crest ridges of mountain ranges.

Physiographic region (province) A **region** within which there prevails substantial **natural-landscape** homogeneity, expressed by a certain degree of uniformity in surface **relief**, **climate**, vegetation, and soils.

Physiography Literally means *landscape description*, but commonly refers to the total **physical geography** of a place; includes all of the natural features on the Earth's surface, including landforms, **climate**, soils, vegetation, and water bodies.

Physiologic density The number of people per unit area of **arable** land.

Pilgrimage A journey to a place of great religious significance by an individual or by a group of people (such as a pilgrimage to Mecca for **Muslims**).

Plantation A large estate owned by an individual, family, or corporation and organized to produce a cash crop. Almost all plantations were established within the tropics; in recent decades, many have been divided into smaller holdings or reorganized as cooperatives.

Plate tectonics Plates are bonded portions of the Earth's mantle and crust, averaging 60 miles (100 km) in thickness. More than a dozen such plates exist (see Fig. G-4), most of continental proportions, and they are in motion. Where they meet one slides under the other, crumpling the surface crust and producing significant volcanic and earthquake activity; a major mountain-building force.

Pleistocene Epoch Recent period of geologic time that spans the rise of humankind, beginning about 2 million years ago. Marked by *glaciations* (repeated advances of continental ice sheets) and milder *interglaciations* (ice sheet contractions). Although the last 10,000 years are known as the **Holocene** Epoch, Pleistocene-like conditions seem to be continuing and we are most probably now living through another Pleistocene interglaciation; thus the glaciers likely will return.

Plural(istic) society A society in which two or more population groups, each practicing its own **culture**, live adjacent to one another without mixing inside a single **state**.

Polder Land reclaimed from the sea adjacent to the shore of the Netherlands by constructing dikes and then pumping out the water trapped behind them.

Political geography The study of the interaction of geographical area and political **process**; the spatial analysis of political phenomena and processes.

Pollution The release of a substance, through human activity, which chemically, physically, or biologically alters the air or water it is discharged into. Such a discharge negatively impacts the environment, with possible harmful effects on living organisms—including humans.

Population decline A decreasing national population. Russia, which now loses about 1 million people per year, is the best example. Also see **population implosion**.

Population density The number of people per unit area. Also see **arithmetic density** and **physiologic density** measures.

Population distribution The way people have arranged themselves in geographic space. One of human geography's most essential expressions because it represents the sum total of the adjustments that a population has made to its natural, cultural, and economic environments.

Population explosion The rapid growth of the world's human population during the past century, attended by accelerating *rates* of increase.

Population geography The field of geography that focuses on the spatial aspects of **demography** and the influences of demographic change on particular places.

Population implosion The opposite of the **population explosion**; refers to the declining populations of many European countries and Russia in which the **death rate** exceeds the **birth rate** and **immigration** rate.

Population movement See **migration**; **migratory movement**.

Population projection The future population total that demographers forecast for a particular country. For example, in Table G-1 such projections are given for all the world's countries for 2025.

Population (age-sex) structure Graphic representation (*profile*) of a population according to age and gender.

Postindustrial economy Emerging economy, in the United States and a handful of other highly advanced countries, as traditional industry is increasingly eclipsed by a higher-technology productive complex dominated by services, information-related, and managerial activities.

Primary economic activity Activities engaged in the direct extraction of **natural resources** from the environment such as mining, fishing, lumbering, and especially **agriculture**.

Primate city A country's largest city—ranking atop the urban **hierarchy**—most expressive of the national culture and usually (but not always) the capital city as well.

Process Causal force that shapes a spatial pattern as it unfolds over time.

Productive activities The major components of the spatial economy. For individual components see: **primary economic activity**, **secondary economic activity**, **tertiary economic activity**, **quaternary economic activity**, and **quinary economic activity**.

Protruded state Territorial shape of a **state** that exhibits a narrow, elongated land extension (or *protrusion*) leading away from the main body of territory. Thailand is a leading example.

Push-pull concept The idea that **migration** flows are simultaneously stimulated by conditions in the source area, which tend to drive people away, and by the perceived attractiveness of the destination.

Qanat In **desert** zones, particularly in Iran and western China, an underground tunnel built to carry **irrigation** water by gravity flow from nearby mountains (where **orographic precipitation** occurs) to the arid flatlands below.

Quaternary economic activity Activities engaged in the collection, processing, and manipulation of *information*.

Quinary economic activity Managerial or control-function activity associated with decision making in large organizations.

Rain shadow effect The relative dryness in areas downwind of mountain ranges caused by **orographic precipitation**, wherein moist air masses are forced to deposit most of their water content as they cross the highlands.

Rate of natural population increase See **natural increase rate**.

Realm See **geographic realm**.

Refugees People who have been dislocated involuntarily from their original place of settlement.

Region A commonly used term and a geographic concept of central importance. An **area** on the Earth's surface marked by specific criteria.

Regional boundary In theory, the line that circumscribes a **region**. But razor-sharp lines are seldom encountered, even in nature (e.g., a coastline constantly changes depending upon the tide). In the **cultural landscape**, not only are regional boundaries rarely self-evident, but when they are ascertained by geographers they most often turn out to be **transitional** borderlands.

Regional character The personality or "atmosphere" of a **region** that makes it distinct from all other regions.

Regional complementarity See **complementarity**.

Regional concept The geographic study of **regions** and regional distinctions, as discussed on pp. 5-7.

Regional disparity The spatial unevenness in standard of living that occurs within a country, whose "average," overall income statistics invariably mask the differences that exist between the extremes of the wealthy **core** and the poorer **periphery**.

Regional geography Approach to geographic study based on the spatial unit of the **region**. Allows for an all-encompassing view of the world, because it utilizes and integrates information from geography's topical (**systematic**) fields; diagrammed in Figure G-13.

Regional state A "natural economic zone" that defies political boundaries, and is shaped by the global economy of which it is a part; its leaders deal directly with foreign partners and negotiate the best terms they can with the national governments under which they operate.

Regionalism The consciousness and loyalty to a **region** considered distinct and different from the **state** as a whole by those who occupy it.

Relative location The regional position or **situation** of a place relative to the position of other places. Distance, **accessibility**, and connectivity affect relative location.

Relict boundary A political boundary that has ceased to function, but the imprint of which can still be detected on the **cultural landscape**.

Relief Vertical difference between the highest and lowest elevations within a particular area.

Religious revivalism See **revivalism (religious)**.

Relocation diffusion Sequential **diffusion process** in which the items being diffused are transmitted by their carrier agents as they relocate to new areas. The most common form of relocation diffusion involves the spreading of innovations by a **migrating** population.

Restrictive population policies Government policy design to reduce the **rate of natural population increase**. China's one-child policy, instituted in 1979 after Mao's death, is a classic example.

Revivalism (religious) Religious movement whose objectives are to return to the foundations of that faith and to influence **state** policy. Often called *religious fundamentalism*; but in the case of Islam, **Muslims** prefer the term revivalism.

Rift valley The trough or trench that forms when a strip of the Earth's crust sinks between two parallel faults (surface fractures).

Rural-to-urban migration The dominant **migration** flow from countryside to city that continues to transform the world's population, most notably in the less advantaged geographic realms.

Russification Demographic resettlement policies pursued by the central planners of the Soviet Empire, whereby ethnic Russians were encouraged to emigrate from the Russian Republic to the 14 non-Russian republics of the U.S.S.R. (listed in Table 2-1).

Sahel Semiarid **steppeland** zone extending across most of Africa between the southern margins of the arid Sahara and the moister tropical **savanna** and forest zone to the south. Chronic drought, **desertification**, and overgrazing have contributed to severe famines in this area since 1970.

Savanna Tropical grassland containing widely spaced trees; also the name given to the tropical wet-and-dry climate (*Aw*).

Scale Representation of a real-world phenomenon at a certain level of reduction or generalization. In **cartography**, the ratio of map distance to ground distance; indicated on a map as a bar graph, representative fraction, and/or verbal statement. *Macroscale* refers to a large area of national proportions; *microscale* refers to a local area no bigger than a county.

Scale economies See **economies of scale**.

Secondary economic activity Activities that process raw materials and transform them into finished industrial products. The *manufacturing* sector.

Sedentary Permanently attached to a particular area; a population fixed in its location. The opposite of **nomadic**.

Separate development The spatial expression of South Africa's "grand" **apartheid** scheme, whereby nonwhite groups were required to settle in segregated "homelands." The policy was dismantled when white-minority rule collapsed in the early 1990s.

Sequent occupance The notion that successive societies leave their cultural imprints on a place, each contributing to the cumulative **cultural landscape**.

Shantytown Unplanned slum development on the margins of cities in disadvantaged countries, dominated by crude dwellings and shelters mostly made of scrap wood, iron, and even pieces of cardboard.

Sharecropping Relationship between a large landowner and farmers on the land whereby the farmers pay rent for the land they farm by giving the landlord a share of the annual harvest.

Sharia The criminal code based in Islamic law that prescribes corporal punishment, amputations, stonings, and lashing for both major and minor offenses. Its occurrence today is associated with the spread of **religious revivalism** in **Muslim** societies.

Shatter belt **Region** caught between stronger, colliding external cultural-political forces, under persistent stress, and often fragmented by aggressive rivals. Eastern Europe and Southeast Asia are classic examples.

Shifting agriculture Cultivation of crops in recently cut and burned tropical-forest clearings, soon to be abandoned in favor of newly cleared nearby forest land. Also known as *slash-and-burn agriculture*.

Sinicization Giving a Chinese cultural imprint; Chinese **acculturation**.

Site The internal locational attributes of an urban center, including its local spatial organization and physical setting.

Situation The external locational attributes of an urban center; its **relative location** or regional position with reference to other non-local places.

Social stratification See **stratification (social)**.

South American culture spheres The fivefold Augelli scheme of cultural regionalization in the South American geographic realm. Consists of the tropical-plantation sphere, European-commercial sphere, Amerind-subsistence sphere, Mestizo-transitional sphere, and undifferentiated sphere.

Southern Ocean The ocean that surrounds Antarctica (see box on p. 541).

Spatial Pertaining to space on the Earth's surface. Synonym for *geographic(al)*.
Spatial diffusion See **diffusion**.
Spatial interaction See **complementarity**, **intervening opportunity**, and **transferability**.
Spatial model See **model**.
Spatial process See **process**.
Spatial system The components and interactions of a **functional region**, which is defined by the areal extent of those interactions. Also see **system**.
Special Administrative Region (SAR) Status accorded the former dependencies of Hong Kong and Macau that were taken over by China, respectively, from the United Kingdom in 1997 and Portugal in 1999. Both SARs received guarantees that their existing social and economic systems could continue unchanged for 50 years following their return to China.
Special Economic Zone (SEZ) Manufacturing and export center within China, created in the 1980s to attract foreign investment and technology transfers. Six SEZs—all located on southern China's Pacific coast—currently operate: Shenzhen, adjacent to Hong Kong; Zhuhai; Shantou; Xiamen; Hainan Island, in the far south; and still-building Pudong, across the river from Shanghai.
Squatter settlement See **shantytown**.
State A politically organized territory that is administered by a sovereign government and is recognized by a significant portion of the international community. A state must also contain a permanent resident population, an organized economy, and a functioning internal circulation system.
State boundaries The borders that surround **states** which, in effect, are derived through contracts with neighboring states negotiated by treaty. See **definition**, **delimitation**, and **demarcation**.
State capitalism Government-controlled corporations competing under free-market conditions, usually in a tightly regimented society. South Korea is a leading example. Also see ***chaebol***.
State formation The creation of a **state** based on traditions of human **territoriality** that go back thousands of years.
State planning Involves highly centralized control of the national planning process, a hallmark of communist economic systems. Soviet central planners mainly pursued a grand political design in assigning production to particular places; their frequent disregard of the principles of economic geography contributed to the eventual collapse of the U.S.S.R.
State territorial morphology A **state's** geographical shape, which can have a decisive impact on its spatial cohesion and political viability. A **compact** shape is most desirable; among the less efficient shapes are those exhibited by **elongated**, **fragmented**, **perforated**, and **protruded** states.
Stateless nation A national group that aspires to become a **nation-state** but lacks the territorial means to do so; the Palestinians and Kurds of Southwest Asia are classic examples.
Steppe Semiarid grassland; short-grass prairie. Also the name given to the semiarid climate type (*BS*).
Stratification (social) In a layered or stratified society, the population is divided into a **hierarchy** of social classes. In an industrialized society, the working class is at the lower end; **elites** that possess capital and control the means of production are at the upper level. In the traditional **caste system** of Hindu India, the "untouchables" form the lowest class or caste, whereas the still-wealthy remnants of the princely class are at the top.
Subduction In **plate tectonics**, the **process** that occurs when an oceanic plate converges head-on with a plate carrying a continental landmass at its leading edge. The lighter continental plate overrides the denser oceanic plate and pushes it downward.
Subsequent boundary A political boundary that developed contemporaneously with the evolution of the major elements of the **cultural landscape** through which it passes.
Subsistence Existing on the minimum necessities to sustain life; spending most of one's time in pursuit of survival.
Subtropical Convergence A narrow marine **transition zone**, girdling the globe at approximately latitude 40°S, that marks the equatorward limit of the frigid **Southern Ocean** and the poleward limits of the warmer Atlantic, Pacific, and Indian oceans to the north.
Suburban downtown In the United States (and increasingly in other advantaged countries), a significant concentration of major urban activities around a highly accessible suburban location, including retailing, light industry, and a variety of leading corporate and commercial operations. The largest are now coequal to the American central city's **central business district (CBD)**.
Sunbelt The popular name given to the southern tier of the United States, which is anchored by the mega-States of California, Texas, and Florida. Its warmer climate, superior recreational opportunities, and other amenities have been attracting large numbers of relocating people and activities since the 1960s; broader definitions of the Sunbelt also include much of the western U.S., particularly Colorado and the coastal Pacific Northwest.
Superimposed boundary A political boundary emplaced by powerful outsiders on a developed human landscape. Usually ignores preexisting cultural-spatial patterns, such as the border that still divides North and South Korea.
Supranational A venture involving three or more **states**—political, economic, and/or cultural cooperation to promote shared objectives. The **European Union** is one such organization.
System Any group of objects or institutions and their mutual interactions. Geography treats systems that are expressed spatially, such as in **functional regions**.
Systematic geography Topical geography: **cultural**, **political**, **economic geography**, and the like.
Taiga The subarctic, mostly **coniferous** snowforest that blankets northern Russia and Canada south of the **tundra** that lines the Arctic shore.
Takeoff Economic concept to identify a stage in a country's **development** when conditions are set for a domestic Industrial Revolution.
Taxonomy A **system** of scientific classification.
Technopole A planned techno-industrial complex (such as California's Silicon Valley) that innovates, promotes, and manufactures the products of the **postindustrial** informational economy.
Tectonics See **plate tectonics**.
Templada See ***tierra templada***.
Terracing The transformation of a hillside or mountain slope into a step-like sequence of horizontal fields for intensive cultivation.
Territoriality A country's or more local community's sense of property and attachment toward its territory, as expressed by its determination to keep it inviolable and strongly defended.
Territorial sea Zone of seawater adjacent to a country's coast, held to be part of the national territory and treated as a segment of the sovereign **state**.
Tertiary economic activity Activities that engage in *services*—such as transportation, banking, retailing, education, and routine office-based jobs.

Tierra caliente The lowest of the **altitudinal zones** into which the human settlement of Middle and South America is classified according to elevation. The ***caliente*** is the hot humid coastal plain and adjacent slopes up to 2500 feet (750 m) above sea level. The natural vegetation is the dense and luxuriant tropical rainforest; the crops include sugar and bananas in the lower areas, and coffee, tobacco, and corn along the higher slopes.

Tierra fría The cold, high-lying **altitudinal zone** of settlement in Andean South America, extending from about 6000 feet (1800 m) in elevation up to nearly 12,000 feet (3600 m). **Coniferous** trees stand here; upward they change into scrub and grassland. There are also important pastures within the *fría*, and wheat can be cultivated.

Tierra helada In Andean South America, the highest-lying habitable **altitudinal zone**—ca. 12,000 to 15,000 feet (3600 to 4500 m)—between the tree line (upper limit of the tierra fría) and the snow line (lower limit of the ***tierra nevada***). Too cold and barren to support anything but the grazing of sheep and other hardy livestock.

Tierra nevada The highest and coldest **altitudinal zone** in Andean South America (lying above 15,000 feet [4500 m]), an uninhabitable environment of permanent snow and ice that extends upward to the Andes' highest peaks of more than 20,000 feet (6000 m).

Tierra templada The intermediate **altitudinal zone** of settlement in Middle and South America, lying between 2500 feet (750 m) and 6000 feet (1800 m) in elevation. This is the "temperate" zone, with moderate temperatures compared to the ***tierra caliente*** below. Crops include coffee, tobacco, corn, and some wheat.

Topography The surface configuration of any segment of **natural landscape**.

Toponym Place name.

Transculturation Cultural borrowing and two-way exchanges that occur when different cultures of approximately equal complexity and technological level come into close contact.

Transferability The capacity to move a good from one place to another at a bearable cost; the ease with which a commodity may be transported.

Transhumance Seasonal movement of people and their livestock in search of pastures. Movement may be vertical (into highlands during the summer and back to lower elevations in winter) or horizontal, in pursuit of seasonal rainfall.

Transition zone An area of spatial change where the **peripheries** of two adjacent realms or regions join; marked by a gradual shift (rather than a sharp break) in the characteristics that distinguish these neighboring geographic entities from one another.

Transmigration The now-ended policy of the Indonesian government to induce residents of the overcrowded, **core-area** island of Jawa to move to the country's other islands.

Treaty ports **Extraterritorial enclaves** in China's coastal cities, established by European colonial invaders under unequal treaties enforced by gunboat diplomacy.

Tropical deforestation See **deforestation**.

Tropical savanna See **savanna**.

Tsunami A seismic (earthquake-generated) sea wave that can attain gigantic proportions and cause coastal devastation. The tsunami of December 26, 2004, centered in the Indian Ocean near the Indonesian island of Sumatera, produced the first great natural disaster of the twenty-first century.

Tundra The treeless plain that lies along the Arctic shore in northernmost Russia and Canada, whose vegetation consists of mosses, lichens, and certain hardy grasses.

Turkestan Northeasternmost region of the North Africa/Southwest Asia realm. Known as Soviet Central Asia before 1992, its five (dominantly Islamic) former S.S.R.'s have become the independent countries of Kazakhstan, Uzbekistan, Turkmenistan, Kyrgyzstan, and Tajikistan. Today Turkestan has expanded to include a sixth state, Afghanistan.

Unitary state A **nation-state** that has a centralized government and administration that exercises power equally over all parts of the state.

Urbanization A term with several connotations. The proportion of a country's population living in urban places is its level of urbanization. The process of urbanization involves the movement to, and the clustering of, people in towns and cities—a major force in every geographic realm today. Another kind of urbanization occurs when an expanding city absorbs rural countryside and transforms it into suburbs; in the case of cities in disadvantaged countries, this also generates peripheral **shantytowns**.

Urban (metropolitan) area The entire built-up, nonrural area and its population, including the most recently constructed suburban appendages. Provides a better picture of the dimensions and population of such an area than the delimited municipality (central city) that forms its heart.

Urban realms model A spatial generalization of the contemporary large American city. It is shown to be a widely dispersed, multicentered metropolis consisting of increasingly independent zones or *urban realms*, each focused on its own **suburban downtown**; the only exception is the shrunken central realm, which is focused on the central city's **central business district** (see Figs. 3-11 and 3-12).

Veld See **highveld**.

Voluntary migration Population movement in which people relocate in response to perceived opportunity, not because they are forced to migrate.

Von Thünen's Isolated State model Explains the location of **agricultural** activities in a commercial economy. A **process** of spatial competition allocates various farming activities into concentric rings around a central market city, with profit-earning capability the determining force in how far a crop locates from the market. The original (1826) Isolated State model (see Fig. 1-5) now applies to the continental scale (see Fig. 1-6).

Wahhabism A particularly virulent form of (Sunni) **Muslim revivalism** that was made the official faith when the modern **state** of Saudi Arabia was founded in 1932. Adherents call themselves "Unitarians" to signify the strict fundamentalist nature of their beliefs.

Wallace's Line As shown in Figure 11-3, the zoogeographical boundary proposed by Alfred Russel Wallace that separates the marsupial fauna of Australia and New Guinea from the nonmarsupial fauna of Indonesia.

Weather The immediate and short-term conditions of the **atmosphere** that impinge on daily human activities.

West Wind Drift The clockwise movement of water around Antarctica in the **Southern Ocean**.

Wet monsoon See **monsoon**.

Windward The exposed, upwind side of a **topographic** barrier that faces the winds that flow across it.

World geographic realm See **geographic realm**.

Index

A reference with an italicized number refers to figures, tables, boxes, and captions.

List of Maps and Figures

List of Maps and Figures (*continued*)